Catalogue of the Pseudoscorpionida

To Māra

Catalogue of the Pseudoscorpionida

MARK S. HARVEY

Western Australian Museum,
Perth, Western Australia

Edited by
V. MAHNERT

Muséum d'Histoire Naturelle,
Genève

Manchester University Press

Manchester and New York

Distributed exclusively in the USA and Canada by St. Martin's Press

Published by Manchester University Press
Oxford Road, Manchester M13 9PL, UK
and Room 400, 175 Fifth Avenue,
New York, NY 10010, USA

Distributed exclusively in the USA and Canada
by St. Martin's Press, Inc.,
175 Fifth Avenue, New York, NY 10010, USA

British Library cataloguing in publication data
Harvey, Mark S.
    Catalogue of the pseudoscorpionida.
    1. Pseudoscorpions
    I. Title
    595.47

Library of Congress cataloging in publication data
Harvey, Mark S.
    Catalogue of the pseudoscorpionida / Mark S. Harvey.
        p. cm.
    Includes index.
    ISBN 0-7190-2935-X
    1. Pseudoscorpions—Classification.   I. Title.
    QL 458.6.H37   1990
    595.4'7'012—dc20   90-43680

ISBN 0 7190 2935 X *hardback*

Typeset by Williams Graphics,
Llanddulas, North Wales, GB

Printed in Great Britain
by Hartnoll Ltd., Bodmin

Contents

Foreword *Peter Merrett, President, British Arachnological Society*   *page* vi

Introduction   1

Bibliography   10

Pseudoscorpionida   129
Chthonioidea   131
   Chthoniidae   131
   Tridenchthoniidae   217
Feaelloidea   230
   Feaellidae   230
   Pseudogarypidae   232
Garypoidea   235
   Garypidae   235
   Geogarypidae   249
   Menthidae   261
   Olpiidae   262
Neobisioidea   309
   Bochicidae   309
   Gymnobisiidae   312
   Hyidae   314
   Ideoroncidae   315
   Neobisiidae   323
   Syarinidae   417
   Vachoniidae   430
Cheiridioidea   433
   Cheiridiidae   433
   Pseudochiridiidae   444
   Sternophoridae   446
Cheliferoidea   451
   Atemnidae   451
   Cheliferidae   482
   Chernetidae   534
   Withiidae   640

Nomina dubia and nomina nuda   667

Summary of taxonomic changes   673

Index   677

Foreword

Following the successful publication of the valuable catalogues of spiders by Professor Brignoli in 1983 and by Dr Norman Platnick in 1989, we were delighted when Dr Mark Harvey offered us this *Catalogue of the Pseudoscorpionida*. While the spider catalogues of Brignoli and Platnick were important contributions as supplements to the earlier works of Roewer and Bonnet, this present volume, covering the whole period from 1758 to 1988, is perhaps an even more significant landmark in the study of pseudoscorpions, as there were no earlier detailed catalogues comparable to those of spiders.

This catalogue is primarily systematic, but other aspects of pseudoscorpion biology are also included. Unlike the recent spider catalogues, fossil taxa are included, and detailed information is provided on the distribution of each species.

The British Arachnological Society is glad, once again, to have been involved in the publication of this important work. Our thanks are due especially to Dr Volker Mahnert for undertaking the scientific editing and for translating the Introduction into French and German, and to Dr Peter Gabbutt for again handling the negotiations with Manchester University Press. We are also pleased to acknowledge the receipt of a grant from the Parliamentary Grant-in-aid (Scientific Publications) administered by the Royal Society, as a contribution towards the publication costs.

It is hoped that this catalogue of pseudoscorpions may be followed, in due course, by a series of catalogues dealing with the other smaller orders of arachnids.

Peter Merrett
President, British Arachnological Society

Introduction

'O! full of (pseudo)scorpions is my mind'

Macbeth, Act III, Scene 2
(with apologies to Shakespeare)

Introduction (English)

Although rarely attaining a length greater than 1 cm, members of the chelicerate order Pseudoscorpionida are a conspicuous component of many terrestrial ecosystems, and have received much attention from taxonomists, ecologists and, to a lesser extent behaviorists. Although several regional catalogues have appeared over the past fifty years, those compiled by Beier (1932a, 1932e) are the most thorough available for the entire world fauna. Roewer (1936, 1937, 1940) produced a useful list that added those species described after 1932 up to the late 1930s. However, there is no single source which lists the multitude of new names and new combinations that have occurred since these early catalogues.

This catalogue attempts to fill the void left by the great interest in pseudoscorpions since the 1930s and provides a complete list of all pseudoscorpion taxa described from 1758 to the end of 1988 (I have added a few papers available to me published during 1989). Although some of the data duplicates that presented by Beier (1932a, 1932e), I feel that so much progress has been achieved in the last sixty or so years that this complete catalogue will not seem superfluous. The scope of the catalogue is primarily systematic, but papers dealing with all other aspects of pseudoscorpions, such as ecology, histology, etc., have been included.

Bibliography

References have been obtained from many sources, primarily the *Zoological Record* and the CIDA lists. Many were taken from major works such as Beier (1932a, 1932e), J. C. Chamberlin (1931a), Ellingsen (1910a), Kew (1916a), and, in particular, Schawaller (1980a) (a few references listed by Schawaller have been found to lack citations to pseudoscorpions; consequently these have been omitted from the Catalogue).

Theses have been included in the reference list, but their contents have not been incorporated in the catalogue.

Although I have made a concerted effort to obtain as many of the more important references as possible, I have not been able to personally observe many papers, in particular, some of the older, general texts. This may be considered a shortcoming of this publication, but I am confident that these titles will add very little to the Catalogue.

1

Type localities and distributions

The type localities for each species (or subspecies) are taken (where possible) from the original publication. However, some type localities have been amended or refined in later publications, and I have adopted these changes. Naturally, new type localities have arisen in a few species from neotype designations. I have attempted to check each type locality in a variety of atlases at my disposal and amend the spelling to modern-day use. Where discrepancies arise, I have adopted the spelling advocated by the *Times Atlas of the World*. The type localities are presented with the most precise locality first, followed by the next nearest town, followed by the state or region, followed by the country.

Under each species-group taxon, the known distribution is presented based on literature records. Although few problems have arisen where the area has been politically stable and regional boundaries have remained static, considerable difficulties were faced when dealing with old publications, especially those dealing with European species. In several instances obscure place names could not be traced with certainty, and I trust that my distributions are fairly complete.

For those taxa which occur in one of the six largest countries (the USSR, Canada, China, the USA, Brazil and Australia), the states in which the species are known to occur have also been given.

Taxonomic arrangement

I have adopted the recommendation of Muchmore (1982a) and simply listed each super-family without placing them within suborders. The order of the superfamilies is that given by Muchmore (1982a), which essentially follows the order of Chamberlin (1931a) and Beier (1932a, 1932e). The only exception is the position of the Feaelloidea, which I have placed near the Chthonioidea due to their sister-group relationship (Harvey, 1986a). Within each superfamily, I have listed each family alphabetically. Similarly, the genera and species are listed alphabetically.

Few changes to the familial classification have occurred since the early 1930s. Two new families have been described (Gymnobisiidae and Vachoniidae), and several subfamilies have been elevated to familial level (Tridenchthoniidae, Miratemnidae, Bochicidae and Geogarypidae). The only change with which I disagree is that of the Miratemnidae which I treat as a synonym of the Atemnidae.

Genera are simply listed alphabetically under each family, and no attempt has been made to provide a subfamilial classification.

Taxonomic changes

The number of taxonomic changes, in particular generic recombinations, have been limited to a bare minimum, but several problems have arisen that deserve attention:

Blothrus and *Neobisium*: The publication of a short note by Harvey (1985a) high-lighting the priority of *Blothrus* Schiödte, 1847, over *Neobisium* J. C. Chamberlin, 1930, has been virtually ignored in the literature, with the exception of Legg & Jones (1988) who suggest that the current use of *Neobisium* be maintained over *Blothrus*. In the interests of stability, I have not transferred the 200 or so *Neobisium* species to *Blothrus*, but express the hope that this problem be tackled in the near future, either by applying to the International Commission of Zoological Nomenclature to conserve *Neobisium*, or by taking the tumultuous step of adopting the necessary changes.

Dichela and *Oligochelifer*: *Dichela* Menge, in C. L. Koch & Berendt (1854) and *Oligochelifer* Beier (1937d), both claim *Dichela berendtii* Menge, in C. L. Koch & Berendt (1854), as type species, and thus the latter is treated as a synonym of the former.

Lissochelifer and *Lophochelifer*: *Lissochelifer* was errected by J. C. Chamberlin (1932a) for *Chelifer mortensenii* With, 1906, and two other species, but was synonymized by Beier (1932e) with *Lophochernes*. Later, Beier (1940a) described *Lophochelifer* for *L. insularis* Beier, 1940a, and the three species that J. C. Chamberlin (1932a) included in *Lissochelifer*. Thus *Lophochelifer* is clearly a junior synonym of *Lissochelifer*.

Homonyms

Several cases of primary and secondary homonymy were detected that apparently have not been previously dealt with in the literature. They involve authors who are no longer alive. Replacement names are provided in the text, and a summary is provided on page 673.

New type species

Type species are lacking for several subgenera which are dealt with in the text. A summary is provided on page 673.

Fossil taxa

Species known only from fossils (mostly from amber) are denoted with '!'.

Numbers of taxa

Table 1 depicts the number of valid genera, species and subspecies (including the fossil taxa) treated in this catalogue.

Table 1

	Genera	Species	Subspecies
Chthoniidae	32	529	34
Tridenchthoniidae	17	72	0
Feaellidae	1	11	1
Pseudogarypidae	2	10	0
Garypidae	12	86	3
Geogarypidae	3	65	3
Menthidae	3	7	0
Olpiidae	51	311	16
Bochicidae	7	21	0
Gymnobisiidae	4	11	0
Hyidae	4	7	0
Ideoroncidae	9	52	0
Neobisiidae	36	461	65
Syarinidae	14	83	3
Vachoniidae	2	12	0
Cheiridiidae	6	69	0
Pseudochiridiidae	2	12	0
Sternophoridae	3	20	0
Atemnidae	20	183	5
Cheliferidae	63	284	13
Chernetidae	110	603	22
Withiidae	33	155	4
	434	3064	169

Acknowledgements

Many people have contributed to the production of this volume. To those colleagues who contributed reprints or copies of their papers, I am most grateful. Several people provided photocopies of rare publications (at my request) which have rendered the catalogue more complete than it otherwise might have been. In particular, Dr Andrew Austin (London and Adelaide), Dr Penny Gullan (Canberra), Dr Bruce Halliday (Canberra), Dr Jacqueline Heurtault (Paris), Mr Mark Judson (Leeds), Dr Peter Gabbutt (Manchester), Mr Paul Hillyard (London), Dr Volker Mahnert (Geneva), Dr William Muchmore (Rochester), Dr Norman Platnick (New York), Dr Robert Raven (Brisbane), Dr Robert Taylor (Canberra) and Dr Alan Yen (Melbourne) diligently found papers for me or allowed me access to their reprint collections. The staff of several libraries associated with institutions at which I have worked over the past decade have provided assistance above and (often) beyond the call of duty. Dr Norman Platnick graciously shared his cataloguing expertise with me. Dr Peter Gabbutt handled much of the initial work with Manchester University Press, and deserves much credit for seeing this project into print. I am grateful to the British Arachnological Society for their support. Finally, Māra Blosfelds assisted in numerous ways, and without her patience and understanding this catalogue could never have been completed.

Introduction (Française)

Bien qu'ils de dépassent guère 1 cm, les Pseudoscorpionida (ordre des Chélicérates) constituent un composant manifeste de plusieurs écosystèmes terrestres et de ce fait ont retenu l'attention des taxonomistes, écologistes et, dans une moindre mesure, des éthologistes. Bien que plusieurs catalogues régionaux aient paru durant les cinquantes dernières années, les deux ouvrages du *Tierreich* par Beier (1932a, 1932e) sont, aujourd'hui encore, indispensables à l'étude de la faune mondiale. Roewer (1936, 1937, 1940) y ajouta une liste des espèces décrites entre 1932 et 1938/1939. Toutefois, il n'existe pas actuellement une source unique donnant la liste de la multitude des nouveaux noms et des nouvelles combinaisons ayant paru depuis la publication de ces ouvrages.

Ce catalogue tente de combler le vide créé par les nombreuses nouvelles données récoltés sur les pseudoscorpions depuis 1930 et fournit une liste complète de tous les pseudoscorpions décrits de 1758 à fin 1988. Quelques publications de 1989 ont également été prises en considération. Bien qu'il existe un parallélisme certain avec le Catalogue de Beier (1932a, 1932e), je pense qu'il y a eu un tel progrès dans les connaissances qu'une liste principalement sur la systématique, mais des publications traitant d'autres aspectes des pseudoscorpions, tels que l'écologie ou l'histologie, y ont été incluses.

Bibliographie

Les références proviennent de plusieurs sources, principalement du Zoological Record et des listes du CIDA. Un certain nombre a été repris de travaux importants tels ceux de Beier (1932a, 1932e), J. C. Chamberlin (1931a), Ellingsen (1910a), Kew (1916a) et, en particulier, Schawaller (1980a) (quelques références inscrites par Schawaller manquant de citations sur les pseudoscorpions ont été omises du catalogue).

Des thèses ont été incluses dans la liste des références mais leur contenu n'a pas été incorporé dans le catalogue.

Bien que j'aie mis un soin tout particulier à obtenir le plus grand nombre possible des références les plus importantes, je n'ai pas pu consulter personnellement certaines publications, en particulier quelques-uns des textes généraux les plus anciens. Ceci peut être considéré comme une lacune, mais je suis persuadé que ces titres n'augmentaraient pas considérablement l'information fournie par ce catalogue.

Localités-types et distribution

Les localités-types de chaque espèce (ou sous-espèce) sont reprises autant que possible de la publication originale. Toutefois, certaines localités-types ont été rectifiées ou précisées dans des publications ultérieures et j'ai adopté ces modifications. Naturellement, de nouvelles localités-types ont apparu pour quelques espèces, suite à des désignations de néotypes. J'ai tenté de vérifier chaque localités-types dans plusieurs atlas à ma disposition et j'ai corrigé l'orthographe en usage actuellement. Lorsque des désaccords surgissaient, j'ai adopté l'orthographe préconisée par le *Times Atlas of the World*. Les localités-types sont présentées avec la localité la plus précise en premier, suivie de la ville la plus proche, puis de l'état ou de la région et ensuite du pays.

Sous chaque groupe d'espèces, la distribution connue est présentée, basée sur la littérature. Bien que peu de problèmes aient surgi lorsque les régions étaient politiquement stables et que les frontières régionales restaient statiques, des difficultés considérables apparaissant lors de l'utilisation de publications anciennes, spécialement de celles traitant d'espèces européennes. Dans quelques cas certains lieux n'ont pas pu être localisés avec certitude; toutefois je crois que la distribution indiquée est assez précise.

Arrangement taxonomique

J'ai adopte les recommandations de Muchmore (1982a) et listé chaque super-famille sans les placer dans des sous-ordres. L'arrangement des super-familles est celui donné par Muchmore (1982a), qui suit essentiellement celui de Chamberlin (1931a) et de Beier (1932a, 1932e). La seule exception est la position des Feaelloidea proche des Chthoniodea, en raison de leur parenté avec ce dernier groupe (Harvey, 1986a). A l'intérieur de chaque super-famille, j'ai listé chaque famille par ordre alphabétique, les genres et les espèces sont également arrangés par ordre alphabétique.

Peu de changements sont intervenus dans la classification des familles depuis 1930. Deux nouvelles familles ont été décrite (Gymnobisiidae et Vachoniidae), plusieurs sous-familles ont été élevées au rang de famille (Tridenchthoniidae, Miratemnidae, Bochicidae et Geogarypidae). Le seul changement que je désapprouve est celui de Miratemnidae que je traite comme synonyme de Atemnidae.

Les genres sont simplement listés par ordre alphabétique dans chaque famille et aucune tentative n'a été faite pour donner une classification selon des sous-familles.

Changements taxonomiques

Les nombre de changements taxonomiques, en particulier des recombinaisons de genres, a été limité au strict minimum, mais plusieurs problèmes ont surgi qui méritent l'attention:

Blothrus et *Neobisium*: La publication d'une courte note de Harvey (1985a) mettant en évidence la priorité de *Blothrus* Schiödte, 1847 sur *Neobisium* J. C. Chamberlin, 1930, a été virtuellement ignorée dans le littérature, à l'exception de Legg & Jones (1988) qui suggèrent que l'usage courant de *Neobisium* soit maintenu. Pour des raisons de stabilité, je n'ai pas transféré les quelques 200 espèces de *Neobisium* dans *Blothrus*, mais j'exprime l'espoir que ce problème soit abordé dans un futur proche, soit en se référant à la Commission Internationale de Nomenclature Zoologique pour conserver le nom *Neobisium*, soit en entreprenant les fastidieuses démarches pour adopter les changements nécessaires.

Dichela et *Oligochelifer*: *Dichela* Menge, in C. L. Koch & Berendt (1854) et *Oligochelifer* Beier (1937d) réclament tous deux *Dichela berendtii* Menge, in C. L. Koch & Berendt (1854) comme espéce-type, le second étant traité comme synonyme du premier.

Lissochelifer et *Lophochelifer*: *Lissochelifer* a été créé par J. C. Chamberlin (1932a) pour *Chelifer mortensenii* With, 1906, et deux autres espèces, mais ce genre a été mis en synonymie avec *Lophochernes* par Beier (1932a), Plus tard, Beier (1940) décrivit *Lophochelifer* pour *L. insularis* Beier, 1940a en incluant aussi les trois espèces que J. C. Chamberlin (1932a) avait placées dans *Lissochelifer*. De ce fait, *Lophochelifer* est clairement un synonyme récent de *Lissochelifer*.

Homonymes

Plusieurs cas d'homonymie primaire et secondaire ont été détectés qui, apparemment, n'avaient pas été résolus dans la littérature précédente. Ceci concerne des auteurs décédés. Les nouveaux noms sont données dans le texte, ainsi qu'in résumé un page 673.

Nouvelles espèces-types

Les espèces-types manquent pour plusieurs sous-genres qui sont traités dans le texte. Un résumé est donné en page 673.

Taxa fossiles

Les espèces, connues seulement de fossiles (la plupart de l'ambre), sont indiquées par un '!'.

Nombre de taxa

Le tableau 1 indique le nombre de genres, d'espèces et de sous-espèces valides (incluant les taxa fossiles) traités dans le catalogue.

Remerciements

De nombreuses personnes ont contribué à la réalisation de ce volume. Je suis très reconnaissant envers les collègues qui m'ont envoyé des tirés-à-part ou des copies de leurs publications. Plusieurs personnes, à ma demande, m'ont fourni des photocopies de publications rares, permettant ainsi à ce catalogue d'être plus complet. En particular le Dr Andrew Austin (Londres et Adelaide), le Dr Penny Gullan (Canberra), le Dr Bruce Halliday (Canberra), le Dr Jacqueline Heurtault (Paris), M. Mark Judson (Leeds), le Dr Peter Gabbutt (Manchester), M. Paul Hillyard (Londres), le Dr Volker Mahnert (Genève), le Dr William Muchmore (Rochester), le Dr Norman Platnick (New York), le Dr Robert Raven (Brisbane), le Dr Robert Taylor (Canberra) et le Dr Alan Yen (Melbourne) ont assidûment cherché pour moi des publications ou m'ont laissé accéder à leurs collections de tirés-à-part. Le personnel de plusieurs bibliothèques liées aux institutions où j'ai travaillé durant la dernière décade m'a prêté assistance dans le cadre de ses tâches (voire même en dehors). Le Dr Norman Platnick a gracieusement partagé ses connaissances en catalogue avec moi. Le Dr Peter Gabbutt a pris en main les démarches initiales auprès de Manchester University Press et mérite d'être grandement remercié pour avoir ainsi permis l'impression de cet ouvrage. Je remercie la British Arachnological Society pour leur support, et la Société Royale de Londres pour faire part des coûts de la publication. Beaucoup d'éloges sont dûs à mon éditeur, Volker Mahnert, qui patiemment à résoud beaucoup de problèmes qui souvent étaient tres obscures, mais quand même importants pour la publication, et à ma demandes, donnant la traduction franıaise et allemagnes de l'Introduction. Enfin, Māra Blosfelds m'a aidé de diverses manières et, sans sa patience et sa compréhension, ce catalogue n'aurait jamais pu être complété.

Einleitung (Deutsche)

Trotz ihrer geringen Körpergrösse (kaum eine Länge von 1 cm überschreitend), sind die Arten der Cheliceratenordnung Pseudoscorpionida wichtiger Bestand zahlreicher terrestrischer Ökosysteme und haben deshalb grosses Interesse bei Taxonomen, Ökologen und (in geringerem Masse) bei Verhaltensforschern gefunden. In den letzten 50 Jahren sind mehrere Landesfaunen und Kataloge erschienen, und trotzdem sind die *Tierreich*-Bearbeitungen Beiers (1932a, 1932e) heute noch unentbehrlich für das Studium der Weltfauna. Die von Roewer (1936, 1937, 1940) publizierten Listen der zwischen 1932 und dem Ende der '30er Jahre beschriebenen Arten stellen eine wertvolle Ergänzung dar. Seither erschien kein einziges Werk, das die zahlreichen neuen Namen und Kombinationen zusammenfasst, die seit der Veröffentlichung dieser frühen Kataloge erschienen sind.

Der vorliegende Katalog soll die grosse Lücke schliessen, die durch das grosse Interesse an Pseudoskorpionen seit 1930 entstanden ist. Er bringt eine vollständige Liste aller Pseudoskorpiontaxa, die zwischen 1758 und Ende 1988 beschrieben wurden, einige wenige Veröffentlichungen aus dem Jahr 1989 konnten noch mitberücksichtigt werden. Auch wenn eine gewisse Parallelität mit Beiers Weltkatalogen (1932a, 1932e) vorliegt, so ist der in den letzten 60 Jahren erreichte Kenntnisfortschritt so gross, dass er eine neue, komplette Liste durchaus rechtfertigt. Dieser Katalog ist vorwiegend auf Systematik ausgerichtet, doch wurden auch Veröffentlichungen über alle anderen Gebiete der Pseudoskorpione, wie Ökologie, Histologie usw. berücksichtigt.

Literatur

Zahlreiche Quellen wurden für Literaturhinweise ausgeschöpft, in erster Linie der Zoological Record und die CIDA Listen. Viele jedoch stammen auch aus umfangreichen Publikationen, wie z.B. Beier (1932a, 1932e), J. C. Chamberlin (1931a), Ellingsen (1910a), Kew (1916a) und besonders Schawaller (1980a) (einige der von Schawaller angeführten Arbeiten beziehen sich nicht auf Pseudoskorpione und werden daher in diesem Katalog nicht erwähnt).

Dissertationen wurden in die Literaturliste aufgenommen, ihr Inhalt jedoch nicht in den Katalog eingearbeitet.

Obwohl ich grossen Wert darauf legte, möglichst viele der wichtigeren Veröffentlichungen zu überprüfen, gelang mir dies, besonders bei älteren, allgemeinen Texten, nicht immer. Dies könnte als Nachteil dieser Liste ausgelegt werden, doch glaube ich, dass dies keinen allzu grossen Informationsverlust mit sich bringt.

Typuslokalitäten und Verbreitung

Der locus typicus jeder Art (oder Unterart) wurde (wenn möglich) der Originalpublikation entnommen. Einige der Typuslokalitäten wurden jedoch in späteren Arbeiten geändert und präzisiert, und ich habe diese Änderungen berücksichtigt. Neue Typuslokalitäten wurden für einige wenige Arten durch Neotypen-Designierung festgelegt. Jeder dieser Fundorte wurde in verschiedensten Atlanten kontrolliert, ihre Namen an den heutigen Gebrauch angepasst. Bei unterschiedlichen Schreibweisen habe ich mich auf den *Times Atlas of the World* berufen. Die Typuslokalitäten wird folgendermassen definiert: der genaue Fundort zuerst, gefolgt von der nächstgelegenen Stadt, der Provinz oder Region, der politische Staat wird zuletzt erwähnt.

Für jede Art wird die aus der Literatur bekannte Verbreitung angeführt. Wenige Probleme tauchten auf in politisch stabilen Gegenden, ohne bedeutende Grenzveränderungen; wesentlich schwieriger waren jedoch die Angaben der älteren Literatur zu deuten, vorwiegend europäische Arten betreffend. In einigen wenigen Fällen war die Ortsfestlegung mit Sicherheit nicht vorzunehmen, doch halte ich meine Verbreitungsangaben für ziemlich vollständig.

Für Arten, die in einem der sechs grössten Länder verbreitet sind (USSR, Kanada, China, USA, Brasilien und Australien) werden auch die Bundesstaaten angeführt, in denen sie nachgewiesen sind.

Systematische Anordnung

Der Empfehlung von Muchmore (1982a) folgend, führe ich einfach die Überfamilien an, ohne sie in die jeweiligen Unterordnungen zu stellen. Die Reihung der Überfamilien folgt ebenfalls Muchmore (1982a), die sich wiederum vorwiegend auf J. C. Chamberlin (1931a) und Beier (1932a, 1932e) stützt. Eine Ausnahme stellen die Feaelloidea dar, die ich als Schwestergruppe der Chthonioidea aufgefasst habe (Harvey, 1986a). Innerhalb jeder Überfamilie werden die Familien in alphabetischer Reihenfolge angeführt. Ebenso wurden die Gattungen und Arten alphabetisch geordnet.

Seit der frühen '30er Jahre sind nur wenige Änderungen in der Familienklassifikation aufgetreten. Zwei neue Familien wurde beschrieben (Gymnobisiidae und Vachoniidae), mehrere Unterfamilien wurden auf Familienrang erhoben (Tridenchthoniidae, Miratemnidae, Bochicidae und Geogarypidae). Die einzige Änderung, der ich nicht zustimme, betrifft die Miratemnidae, die ich als Synonym der Atemnidae betrachte.

Gattungen werden in jeder Familie in alphabetischer Reihenfolge angeführt, ohne auf Unterfamilien einzugehen.

Taxonomische Änderungen

Die Zahl der taxonomischen Änderungen, besonders Neukombinationen auf Gattungsniveau, wurden auf ein absolutes Minimum beschränkt, aber einige Probleme verdienen besonderer Beachtung:

Blothrus und *Neobisium*: Eine kurze Veröffentlichung von Harvey (1985a) wies auf die Priorität von *Blothrus* Schiödte, 1847 über *Neobisium* J. C. Chamberlin, 1930 hin. Dieser Hinweis wurde bislang in der Literatur überhaupt nicht berücksichtigt, nur Legg & Jones (1988) traten für eine Bei-behaltung des Namens *Neobisium* ein. Im Interesse der Stabilität habe ich die ungefähr 200 *Neobisium*-Arten nicht in die Gattung *Blothrus* versetzt, bringe jedoch meine Hoffnung zum Ausdruck, dass dieses Problem in naher Zukunft gelöst wird, sei es durch eine Anfrage an die Internationale Kommission für zoologische Nomenklatur, den Namen *Neobisium* beizubehalten, sei es durch den umwälzenden Schritt, diese Priorität anzuerkennen.

Dichela und *Oligochelifer*: *Dichela* Menge, in C. L. Koch & Berendt (1854) und *Oligochelifer* Beie (1937d) haben beide *Dichela berendtii* Menge, in C. L. Koch & Berendt, zur Typusart, weshalb die zweitgenannte Gattung als Synonym von *Dichela* angesehen wird.

Lissochelifer und *Lophochelifer*: *Lissochelifer* wurde von J. C. Chamberlin (1932a) für *Chelifer mortensenii* With, 1906, und zwei weitere Arten errichtet, von Beier (1932a) jedoch mit *Lophochernes* synonymisiert. Einige Jahre später beschrieb Beier (1940a) *Lophochelifer* für *insularis* Beier, 1940a, und die drei Arten, die J. C. Chamberlin (1932a) in die Gattung *Lissochelifer* gestellt hatte. *Lophochelifer* ist daher zweifellos ein jüngeres Synonym von *Lissochelifer*.

Homonyme

Mehrere Fälle von primärer oder sekundärer Homonymie wurden bislang in der Literatur übersehen. Sie betreffen die in der Zwischenzeit verstorbenen Autoren. Die neuen Namen werden im Text angeführt, eine Zusammenfassung wird auf Seite 673 gegeben.

Neue Typusarten

Für einige im Katalog behandelte Untergattungen fehlte die Festlegung der Typusarten, was nun bereinigt wurde. Eine Zusammenfassung wird auf Seite 673 gegeben.

Fossile Taxa

Arten, die nur als Fossile bekannt sind (meist aus Bernstein) sind mit '!' gekennzeichnet.

Zahl der Taxa

Tabelle 1 gibt die Zahl der gültigen Gattungen, Arten und Unterarten (fossile Taxa inbegriffen) an, die im Katalog behandelt sind.

Danksagung

Zahlreich sind die Personen, die zum Gelingen dieses Werkes beigetragen haben. Für die Übersendung von Sonderdrucken oder Kopien ihrer Publikationen bin ich allen Kollegen dankbar. Mehrere Personen haben auf meine Bitte hin Kopien seltener Werke verschafft, und nur auf diesem Weg war eine Vollständigkeit dieses Kataloges erreichbar. Dr Andrew Austin (London und Adelaide), Dr Penny Gullan (Canberra), Dr Bruce Halliday (Canberra), Dr Jacqueline Heurtault (Paris), Herr Mark Judson (Leeds), Dr Peter Gabbutt (Manchester), Herr Paul Hillyard (London), Dr Volker Mahnert (Geneva), Dr William Muchmore (Rochester), Dr Norman Platnick (New York), Dr Robert Raven (Brisbane), Dr Robert Taylor (Canberra) und Dr Alan Yen (Melbourne) suchten für mich die Veröffentlichungen oder öffneten mir ihre Sonderdrucksammlungen. Die Angestellten mehrerer Bibliotheken der Institute, an denen ich während des letzten Jahrzehnts gearbeitet habe, halfen mir im Rahmen (und oft weit darüber hinaus) ihrer Aufgaben. Dr Norman Platnick unterstützte mich mit seinen Katalog-Erfahrungen. Dr Peter Gabbutt führte die ersten Verhandlungen mit Manchester University Press, ihm gilt der Dank, dass dieses Werk in Druck gehen konnte. Aufrichtig dankbar bin ich der British Archnological Society für ihre Unterstützung, und der Royal Society of London, die die teilweise Rechnung der Druckwerke begleichte. Viel Lob gehört meinem Herausgeber, Volker Mahnert, der ohne Geduld zu verlieren, viele Probleme löste, einige unbekannte aber nichts destoweniger wichtige Drucksachen ausfindig machte und auf mein Ersuchen die französische und deutsche Übersetzung der Einleitung bereitstellte. Māra Blosfelds half in vielen Beziehungen, und ohne ihre Geduld und ihr Verständnis hätte dieser Katalog wohl nicht das Licht der Welt erblickt.

Bibliography

Abendroth, R. E. (1868). *Über Morphologie und Verwandtschaftsverhältnisse der Arachniden.* Inaugural Dissertation: Leipzig. (not seen)

Acloque, A. (1896–1900). *Faune de France*, vol. **3**. Paris. (not seen)

Adams, G. (1787). *Essays on the Microscope.* London. (not seen)

Adis, J. (1979). *Vergleichende Ökologische Studien an der Terrestrischen Arthropodenfauna Zentralamozonischer Überschwemmungswälder.* Dissertation University Ulm.

Adis, J. (1981). Comparative ecological studies of the terrestrial arthropod fauna in central Amazonian inundation-forests. *Amazoniana* **7**: 87–173.

Adis, J. & Funke, W. (1983). Jahresperiodische Vertikalwanderungen von Arthropoden in Überschwemmungswäldern Zentralamazoniens. *Verhandlungen der Gesellschaft für Ökologie* **10**: 351–359.

Adis, J., Junk, W. J. & Penny, N. D. (1987). Material zoológico depositado nas coleçoes sistemáticas de entomologia do INPA, resultante do 'Projeto INPA/Max-Planck'. *Acta Amazonica* **15**: 481–504.

Adis, J. & Mahnert, V. (1986). On the natural history and ecology of Pseudoscorpiones (Arachnida) from an Amazonian blackwater inundation forest. *Amazoniana* **9**: 297–314.

Adis, J., Mahnert, V., de Morais, J. W. & Rodrigues, J. M. G. (1988). Adaptation of an Amazonian pseudoscorpion (Arachnida) from dryland forests to inundation forests. *Ecology* **69**: 287–291.

Aellen, V. (1952). La faune de la grotte de Moron (Jura Suisse). *Bulletin de la Société Neuchâteloise des Sciences Naturelles* (3) **75**: 139–151.

Aellen, V. (1976). Biospéléologie. In: *Inventaire Spéléologique de la Suisse*: 22–24. Commission de Spéléologie de la Société Helvétique des Sciences Naturelles: Neuchâtel.

Ahern, L. D. & Yen, A. L. (1977). A comparison of the invertebrate fauna under *Eucalyptus* and *Pinus* forests in the Otway Ranges, Victoria. *Proceedings of the Royal Society of Victoria* **89**: 127–136.

Aitchison, C. W. (1979). Low temperature activity of pseudoscorpions and phalangids in southern Manitoba. *Journal of Arachnology* **7**: 85–86.

Alfonsus, A. (1891). Der Feind der Bienenlaus. *Deutsche Illustrierte Bienenzeitung* **8**: 503–506.

Alfonsus, A. (1922). An enemy of the mites in the bee hive. *Bee World* **4**: 2–3. (not seen)

Allred, D. M. & Beck, D. E. (1964). Arthropod associated of plants at the Nevada test site. *Brigham Young University, Science Bulletin, Biological Series* **5**(2): 1–16. (not seen)

Almquist, S. (1982). Spindeldjursfaunan i granplanteringar i södra Skåne. *Entomologisk Tidskrift* **103**: 97–105.

Alzona, C. (1903). Nota sulla fauna delle caverne italiane. *Rivista Italiana di Speleologia* **1**(3): 10–17.

Anciaux de Faveaux, M. (1980). Notes éco-éthologiques et parasitologiques sur les Chiroptères cavernicoles du Shaba (Zaïre). *International Journal of Speleology* **10**: 331–350.

Andersen, M. (1987). *Lamprochernes nodosus* (Schrank, 1761) found in Denmark (Pseudoscorpiones). *Entomologiske Meddelelser* **55**: 23–25. (not seen)

Andersson, J.S. (1961). The occurrence of some invertebrate animal groups in the south bluffs in Northern Sweden. *Oikos* **12**: 126–156.

André, E. (1908). Sur la piqûre des Chélifères. *Zoologischer Anzeiger* **33**: 289–290.

André, E. (1909a). Sur la piqûre des Chélifères. *Archives de Parasitologie* **12**: 478–479.

André, E. (1909b). Les faux-scorpions et leur morsure. *Bulletin de l'Institut National Genevois* **38**: 277–280.

Anonymous (1831a). A lobster-like insect attacking the leg of a house-fly. *Magazine of Natural History* **4**: 94.

Anonymous (1831b). A lobster-like insect attacking the leg of a house-fly. *Magazine of Natural History* **4**: 283–284.

Anonymous (1832). *Chélifer cancroïdes*, a lobster-like insect, parasitic on the common house-fly (*Músca doméstica* L.). *Magazine of Natural History* **5**: 754.

Anonymous (1834). *Chélifer cancroïdes*. *Magazine of Natural History* **7**: 162.

Anonymous (1835). Beobachtungen englischer Naturforscher über die Afterskorpione (*Chelifer*). *Archiv für Naturgeschichte* **1/2**: 186.

Anonymous (1879). House-flies and their parasites. *Science Gossip* **15**: 88.

Anonymous (1875). *Chelifer cancroides*. *Entomologist* **8**: 185.

Anonymous (1929). Some notes on the pseudoscorpion, *Chelifer sculp.*, in relation to the honey bee. *South African Journal of Natural History* **6**: 293–296. (not seen)

Anonymous (1983a). Quelques récoltes Ursus de pseudoscorpions. *Ursus Spelaeus* **8**: 27.

Anonymous (1983a). Pseudoscorpions cavernicoles du Vaucluse. *Ursus Spelaeus* **8**: 28.

Armas, L. F. de & Alayón García, G. (1984). Sinopsis de los arácnidos cavernicolas de Cuba (excepto ácaros). *Poeyana* **276**: 1–25.

Artault de Vevey, S. (1901). Pseudo-parasitisme du *Chelifer cancroïdes* chez l'homme. *Compte Rendu Hebdomadaire de Séances et Mémoires de la Société de Biologie, Paris* **53**: 105.

Aru, L. (1968). Ebaskorpionilistest. *Archiv für die Naturkunde Eestis* **1968**(10): 607–608.

Ass, M. J. (1973). Die Fangbien der Arthropoden, ihre Entstehung, Evolution und Funktion. *Deutsche Entomologische Zeitschrift* **20**: 127–152.

Astley, W. N. T. (1979). *A Numerical Classification of the British Family Chernetidae (Pseudoscorpionidea)*. M.Sc. Thesis: The City of London Polytech. (not seen)

Atkin, L. & Proctor, J. (1988). Invertebrates in the litter and soil on Volcán Barva, Costa Rica. *Journal of Tropical Ecology* **4**: 307–310.

Audouin, V. (1826). Explication sommaire des planches arachnides de l'Éogypte et de la Syrie. In: *Description de l'Égypte ou Recueil des Observations et des Recherches qui ont été Faites en Égypte Pendant l'Expédition de l'Armée Française*, 1st edition, vol. **1**(4): 99–186. C.L.F. Panckoucke: Paris.

Audouin, V. (1829). Explication sommaire des planches d'arachnides de l'Égypte et de la Syrie. In: *Description de l'Égypte ou Recueil des Observations et des Recherches qui ont été Faites en Égypte Pendant l'Expédition de l'Armée Française*, 2nd edition, vol. **22**: 291–430. C.L.F. Panckoucke: Paris.

Auer, A. & Gaisberger, K. (1978). Weitere Aufsammlungen von Kleintieren in Höhlen des Salzkammergutes. *Mitteilungen der Sektion Ausseerland des Landesvereines für Höhlenkunde in Steiermark* **16:** 11–15. (not seen)

Baccetti, B. & Lazzeroni, G. (1967). Primi reperti ultrastrutturali sul canale alimentare di uno pseudoscorpione. *Redia* **50:** 351–363.

Baccetti, B. & Lazzeroni, G. (1968). Primi reperti sugli ultrastrutturali sul canale alimentare di uno pseudoscorpione. *Atti della Reale Accademia Nazionale d'Italia Rendiconti della Entomologia* **15:** 119. (not seen)

Baccetti, B. & Lazzeroni, G. (1969). The envelopes of the nervous system of pseudo-scorpions and scorpions. *Tissue and Cell* **1:** 417–424.

Bacelar, A. (1928). Aracnidios Portuguêses. III. Catálogo sistemático dos Aracnidios de Portugal citados por diversos autores (1831-1926). *Bulletin de la Société Portugaise des Sciences Naturelles* **10:** 169–203.

Bachofen-Echt, A. (1934). Beobachtungen über im Bernstein vorkommende Spinnen-gewebe. *Biologia Generalis* **10:** 179–184.

Bachofen-Echt, A. (1949). *Der Bernstein und seine Einschlüsse.* Wien. (not seen)

Badian, Z. & Ogorzalek, A. (1982). Fine structure of the ovary in *Chelifer cancroides* (Linnaeus, 1761) (Arachnida, Pseudoscorpionidea). *Zoologica Poloniae* **29:** 137–146. (not seen)

Báez, M. (1984). Los Artrópodos. In: Bacallado, J. J. (ed.), *Fauna (marina y terrestre) del Archipiélago Canario:* 101–254. Edirca Ediciones: Las Palmas. (not seen)

Bagnall, R. S. (1907). *Anurida maritima,* Guér. and its enemies. *Entomologist's Record and Journal of Variations* **19:** ?–? (not seen)

Bagnall, R. S. & Turner, W. L. (1913). Preliminary list of spiders, harvestmen and pseudo-scorpions found in the Derwent Valley. *Transactions of the Vale of Derwent Naturalists' Field Club* **1:** 129–151. (not seen)

Bailey, G. (1865). Fly parasites. *Science Gossip* **1:** 227.

Balfour, F. (1885). *Treatise of Comparative Embryology.* (not seen)

Balazuc, J. (1962). Troglobies des cavités artificielles. *Spelunca Mémoires* **2:** 104–107.

Balazuc, J., Dresco, E., Henrot, H. & Nègre, J. (1951). Biologie des carrières souterraines de la région Parisienne. *Vie et Milieu* **2:** 301–334.

Balazuc, J., Miré, P., Sigwalt, J. & Théodoridès, J. (1951). Trois campagnes biospeleologiques dans la Bas-Vivarais (Avril 1949 – Décembre 1949, Juin–Juillet–Août 1950). *Bulletin Mensuel de la Société Linnéenne de Lyon* **20:** 187–192, 215–220, 238–242.

Balzan, L. (1887a). *Chernetidae Nonnullae Sud-Americanae, I.* Asuncion.

Balzan, L. (1887b). *Chernetidae Nonnullae Sud-Americanae, II.* Asuncion.

Balzan, L. (1888a). *Chernetidae Nonnullae Sud-Americanae, III.* Asuncion.

Balzan, L. (1888b). *Osservazioni Morfologiche e Biologiche Sui Pseudo-Scorpioni del Bacino dei Fiumi Paranà e Paraguay.* Ascension.

Balzan, L. (1889). *Sui Pseudo-Scorpioni del Bacino dei Fiumi Paraná e Paraguay.* Padova. (not seen)

Balzan, L. (1890). Revisione dei Pseudoscorpioni del Bacino dei Fiumi Paranà e Paraguay nell'America meridionale. *Annali del Museo Civico di Storia Naturale di Genova* (2a) **9:** 401–454.

Balzan, L. (1892). Voyage de M. E. Simon au Venezuela (Décembre 1887 – Avril 1888). Arachnides. Chernetes (Pseudoscorpiones). *Annales de la Société Entomologique de France* **60:** 497–552.

Banks, N. (1890). A new pseudoscorpion. *Canadian Entomologist* **22:** 152.

Banks, N. (1891). Notes on North American Chernetidae. *Canadian Entomologist* **23:** 161–166.

Banks, N. (1893). New Chernetidae from the United States. *Canadian Entomologist* **25:** 64–67.

Banks, N., in Hubbard, H.G. (1894). The insect guests of the Florida Land Tortoise. *Insect Life* **6:** 302–331.

Banks, N. (1895a). Notes on the Pseudoscorpionida. *Journal of the New York Entomological Society* **3:** 1–13.

Banks, N. (1895b). Chernetid attached to a fly. *Entomological News* **6:** 115.

Banks, N. (1895c). The Arachnida of Colorado. *Annals of the New York Academy of Sciences* **8:** 417–434.

Banks, N. (1898). Arachnida from Baja California and other parts of Mexico. *Proceedings of the Californian Academy of Sciences* (3) **1:** 205–308.

Banks, N. (1900). Papers from the Harriman Alaska Expedition. XI. Arachnida. *Proceedings of the Washington Academy of Sciences* **2:** 477–486.

Banks, N. (1901a). Some Arachnida from New Mexico. *Proceedings of the Academy of Natural Sciences of Philadelphia* **53:** 568–597.

Banks, N. (1901b). Some spiders and other Arachnida from southern Arizona. *Proceedings of the United States National Museum* **23:** 581–590.

Banks, N. (1902a). Papers from the Hopkins Stanford Galapagos Expedition, 1898–1899. VII. Entomological Results (6). Arachnida. *Proceedings of the Washington Academy of Sciences* **4:** 49–86.

Banks, N. (1902b). A list of spiders collected in Arizona by Messrs. Schwarz and Barber during the summer of 1901. *Proceedings of the United States National Museum* **25:** 211–221.

Banks, N. (1904a). The Arachnida of Florida. *Proceedings of the Academy of Natural Sciences of Philadelphia* **56:** 120–147.

Banks, N. (1904b). Some Arachnida from California. *Proceedings of the California Academy of Sciences* (3) **3:** 331–374.

Banks, N. (1908). The pseudoscorpions of Texas. *Bulletin of the Wisconsin Natural History Society* **6:** 39–42.

Banks, N. (1909a). New tropical pseudoscorpions. *Journal of the New York Entomological Society* **17:** 145–148.

Banks, N. (1909b). New Pseudoscorpionida. *Canadian Entomologist* **41:** 303–307.

Banks, N. (1909c). Arachnida of Cuba. *Report, Estación Central Agronómica de Cuba* **2:** 150–174.

Banks, N. (1911). The pseudoscorpions of California. *Pomona College Journal of Entomology* **3:** 633–640.

Banks, N. (1913). Notes on some Costa Rican Arachnida. *Proceedings of the Academy of Natural Sciences of Philadelphia* **65:** 676–688.

Banks, N. (1914). A new pseudoscorpion from California. *Pomona College Journal of Entomology* **6:** 203.

Bapsolle, P. (1967). Notes sur un chernète. *Bulletin de Mayenne-Sciences* **1967:** 63–66.

Bardenfleth, K. S. (1925). Gaest eller snylter? *Danske Biavls-Tidende* **59:** 89–90. (not seen)

Barnes, H. F. (1925). *Obisium maritimum* Leach found at Wembury, near Plymouth, together with its original description, and short notes on its geographical distribution. *Journal of the Marine Biological Association* **13:** 746–749.

Barrois, J. (1894). Sur le développement des *Chelifer*. *Comptes Rendus Hebdomadaires des Séances de l'Académie des Sciences, Paris* **99:** 1082–1083.

Barrois, J. (1896). Memoire sur le développement des *Chelifer*. *Revue Suisse de Zoologie* **3:** 461–498.

Barrows, W. M. (1925). Modification and development of the arachnid palpal claw, with especial reference to spiders. *Annals of the Entomological Society of America* **18:** 483–516.

Bartels, M. (1931). Beitrag zur Kenntnis der Schweizerischen Spinnenfauna. *Revue Suisse de Zoologie* **38**: 1–30.

Bartolomei, G. (1957). La Grotta della Poscola. *Rassegna Speleologica Italiana* **9**: 51–60.

Bassett, Y. (1988). A composite interception trap for sampling arthropods in tree canopies. *Journal of the Australian Entomological Society* **27**: 213–219.

Bawa, S. R., Sjöstrand, F. S., Kanwar, U. & Kanwar, K. C. (1971a). Pseudoscorpion spermatozoon – phase and electron microscopy. *Septième Congrès International de Microscopie Élctronique, Grenoble (1970) Résumé Comm. III, Société Française de Microscopie Électronique, Paris*: 649–650.

Bawa, S. R., Sjöstrand, F. S., Kanwar, U. & Kanwar, K. C. (1971b). Electron microscopic studies on the 'flagellate-nonflagellate' pseudoscorpion sperm. *Journal of Ultratructure Research* **37**: 251.

Beck, L. (1968). Aus den Regenwäldern am Amazonas I. *Natur und Museum* **98**: 24–32.

Beck, L. (1983). Zur Bodenbiologie des Laubwaldes. *Verhandlungen der Deutschen Zoologischen Gesellschaft* **1983**: 37–54.

Becker, L. (1879). Arachnides recoltées par Mm. de Borre et Donckier. *Annales de la Société Entomologique de Belgique* **22**: clxiii.

Becker, L. (1880a). Communications arachnologiques. Arachnides de Hongrie, recueillies par M. Dr. Horvath. *Annales de la Société Entomologique de Belgique* **23**: xii–xiv.

Becker, L. (1880b). Communications arachnologiques. *Annales de la Société Entomologique de Belgique* **23**: cxxxix–cxlii.

Becker, L. (1882). Les Arachnides de Belgique. *Annales de la Musée Royal d'Histoire Naturelle de Belgique* **10**: 1–246. (not seen)

Becker, L. (1884). Catalogue des Arachnides de Belgique, cinquième partie. Chernètes. *Comptes Rendus des Séances de la Société entomologique de Belgique* (3) no. 49 **3**: cclxiii–cclxiv.

Becker, L. (1896). Les Arachnides de Belgique. *Annales de la Musée Royal d'Histoire Naturelle de Belgique* **12**: 1–378. (not seen)

Beeby, W. M. (1879). House-flies and their parasites. *Science Gossip* **15**: 21.

Beier, M. (1928). Die Pseudoskorpione des Wiener Naturhistorischen Museums. I. Hemictenodactyli. *Annalen des Naturhistorischen Museums in Wien* **42**: 285–314.

Beier, M. (1929a). Die Pseudoskorpione des Wiener Naturhistorischen Museums. II. Panctenodactyli. *Annalen des Naturhistorischen Museums in Wien* **43**: 341–367.

Beier, M. (1929b). Sopra alcuni pseudoscorpioni della Cirenaica. *Bollettino del Laboratoria di Zoologia Generale e Agraria della R. Scuola sup. d'Agricoltura, Portici* **24**: 78–81.

Beier, M. (1929c). Bemerkungen über einige *Obisium*-Arten. *Zoologischer Anzeiger* **80**: 215–221.

Beier, M. (1929d). Alcuni pseudoscorpioni raccolti da C. Menozzi. *Bollettino della Società Entomologica Italiana* **61**: 154–156.

Beier, M. (1929e). Zoologische Forschungsreise nach den Jonischen Inseln und dem Peloponnes. I. und II. Teil. *Sitzungsberichte der Akademie der Wissenschaften in Wien* **128**: 425–456.

Beier, M. (1930a). Die Pseudoskorpione des Wiener Naturhistorischen Museums. III. *Annalen des Naturhistorischen Museums in Wien* **44**: 199–222.

Beier, M. (1930b). Neue Höhlenformen der Gattung *Chthonius* (Pseudoscorp.). *Annali del Museo Civico di Storia Naturale di Genova* **55**: 71–74.

Beier, M. (1930c). Zwei neue troglobionte *Parablothrus*-Arten aus Ligurien. *Annali del Museo Civico di Storia Naturale di Genova* **55**: 94–95.

Beier, M. (1930d). Alcuni Pseudoscorpioni esotici raccolti dal Prof. F. Silvestri. *Bollettino del Laboratoria di Zoologia Generale e Agraria della R. Scuola sup. d'Agricoltura, Portici* **23**: 197–209.

Beier, M. (1930e). Die Pseudoskorpione der Sammlung Roewer. *Zoologischer Anzeiger* **91**: 284–300.

Beier, M. (1930f). Pseudoscorpione aus Marocco nebst einer Art von der Insel Senafir. *Bulletin de la Société des Sciences Naturelles du Maroc* **10**: 70–77.

Beier, M. (1930g). Neue Höhlen-Pseudoscorpione der Gattung *Chthonius*. *Eos, Madrid* **6**: 323–327.

Beier, M. (1930h). Due nuovi pseudoscorpioni della Tunisia. *Bollettino del Laboratoria di Zoologia Generale e Agraria della R. Scuola sup. d'Agricoltura, Portici* **24**: 95–98.

Beier, M. (1930i). Nuovi pseudoscorpioni dell'Africa tropicale. *Bollettino del Laboratoria di Zoologia Generale e Agraria della R. Scuola sup. d'Agricoltura, Portici* **25**: 44–48.

Beier, M. (1931a). Zur Kenntnis der troglobionten Neobisien (Pseudoscorp.). *Eos, Madrid* **7**: 9–23.

Beier, M. (1931b). Zur Kenntnis der Chthoniiden (Pseudoskorpione). *Zoologischer Anzeiger* **93**: 49–56.

Beier, M. (1931c). Neue Pseudoscorpione der U. O. Neobisiinea. *Mitteilung aus dem Zoologischen Museum in Berlin* **17**: 299–318.

Beier, M. (1931d). Zoologische Streifzüge in Attika, Morea und besonders auf der Insel Kreta. III. Pseudoscorpionidea. *Abhandlungen hrsg. vom Naturwissenschaftlichen Verein zu Bremen* **28**: 91–100.

Beier, M. (1932a). Pseudoscorpionidea I. Subord. Chthoniinea et Neobisiinea. *Tierreich* **57**: i–xx, 1–258.

Beier, M. (1932b). Revision der Atemnidae (Pseudoscorpionidea). *Zoologische Jahrbücher, Systematik (Ökologie), Geographie und Biologie* **62**: 547–610.

Beier, M. (1932c). Zur Kenntnis der Lamprochernetinae (Pseudoscorp.). *Zoologischer Anzeiger* **97**: 258–267.

Beier, M. (1932d). Zur Kenntnis der Cheliferidae (Pseudoscorpionidea). *Zoologischer Anzeiger* **100**: 53–67.

Beier, M. (1932e). Pseudoscorpionidea I. Subord. C. Cheliferinea. *Tierreich* **58**: i–xxi, 1–294.

Beier, M. (1932f). Pseudoscorpionidea. Spedizione scientifica all'Oasi di Cufra (Marzo-Lugio 1931). *Annali del Museo Civico di Storia Naturale di Genova* **55**: 487–489.

Beier, M. (1932g). 5. Ordnung der Arachnida: Pseudoscorpionidea – Afterskorpione. In: Kükenthal, W. & Krumbach, T. (eds), *Handbuch der Zoologie*, vol. *3*(2): 117–192. Berlin.

Beier, M. (1933a). Revision der Chernetidae (Pseudoscorp.). *Zoologische Jahrbücher, Systematik (Ökologie), Geographie und Biologie* **64**: 509–548.

Beier, M. (1933b). Pseudoskorpione aus Mexiko. *Zoologischer Anzeiger* **104**: 91–101.

Beier, M. (1933c). Mission Robert Ph. Dollfuss en Égypte. Pseudoscorpionidea (Chelonethi). *Mémoires de l'Institut Égyptien* **21**: 85–87.

Beier, M. (1933e). Two new species of Cheliferinea (Pseudoscorpionidea). *Annals and Magazine of Natural History* (10) *11:* 644–647.

Beier, M. (1934a). Neue cavernicole und subterrane Pseudoscorpione. *Mitteilungen über Höhlen- und Karstforschung* **1934**: 53–59.

Beier, M. (1934b). Exploration biologique des cavernes de la Belgique et du Limbourg Hollandais. XVIIe contribution: Pseudoscorpionidea. *Bulletin et Annales de la Société Entomologique de Belgique* **74**: 283–285.

Beier, M. (1935a). Four new tropical Pseudoscorpionidea. *Annals and Magazine of Natural History* (10) *15:* 484–489.

Beier, M. (1935b). Neue Pseudoskorpione von Mauritius. *Zoologischer Anzeiger* **110**: 253–256.

Beier, M. (1935c). Arachnida I. Pseudoscorpionidea. In: *Mission Scientifique de l'Omo*, vol. **2**: 117–129. Paris.

Beier, M. (1935d). Chernetidae. In: Visser, C. & Visser-Hooft, J. (eds), *Wissenschaftliche Ergebnisse der Niederländischen Expeditionen in den Karakorum und die angrenzenden Gebiete in den Jahren 1922, 1925 und 1929/30*, vol. **1**: 176. Brockhaus: Leipzig.

Beier, M. (1935e). New Pseudoscorpionidea from the Solomon Islands. *Annals and Magazine of Natural History* (10) *16:* 637–641.

Beier, M. (1935f). Ein neuer Pseudoskorpion aus *Atta*-Nestern. *Zoologischer Anzeiger* **111**: 45–46.

Beier, M. (1935g). Drei neue Psudoscorpione aus Rumänien. *Bulletin de la Section Scientifique de l'Académie Roumaine* **17**: 31–34.

Beier, M. (1936a). Zoologische Ergebnisse einer Reise nach Bonaire, Curaçao und Aruba im Jahre 1930. No. 21. Einige neue neotropische Pseudoscorpione. *Zoologische Jahrbücher, Systematik (Ökologie), Geographie und Biologie* **67**: 443–447.

Beier, M. (1936b). Zwei neue Pseudoskorpione aus deutschen Kleinsäuger-Höhlen. *Zoologischer Anzeiger* **114**: 85–87.

Beier, M. (1937a). Zwei neue Neobisien (Pseudoscorp.) aus dem Kaukasus. *Zoologischer Anzeiger* **117**: 107–109.

Beier, M. (1937b). Neue ostasiatische Pseudoscorpione aus dem Zoologischen Museum Berlin. *Mitteilung aus dem Zoologischen Museum in Berlin* **22**: 268–279.

Beier, M. (1937c). Some remarks on *Ellingsenius indicus* Chamberlin (Pseudoscorpionidea). *Annals and Magazine of Natural History* (10) **20**: 633–634.

Beier, M. (1937d). Pseudoscorpione aus dem baltischen Bernstein. *Festschrift zum 60. Geburtstage von Professor Dr. Embrik Strand, Riga* **2**: 302–316.

Beier, M. (1938a). Vorläufige Mitteilung über neue Höhlenpseudoscorpione der Balkanhalbinsel. *Studien aus dem Gebiete der Allgemeinen Karstforschung, der Wissenschaftlichen Höhlenkunde, der Eiszeitforschung und den Nachbargebieten* **3** (8): 5–8.

Beier, M. (1938b). Zwei neue Neobisien (Pseudoscorp.) aus der Ostmark. *Zoologischer Anzeiger* **123**: 78–80.

Beier, M. (1939a). 5. Ordnung der Arachnida: Pseudoscorpionidea – Afterskorpione. In: Kükenthal, W. & Krumbach, T. (eds), *Handbuch der Zoologie*, vol. **3** (3): 169–185. Berlin.

Beier, M. (1939b). The Pseudoscorpionidea collected by the Percy Sladen Trust Expedition to Lake Titicaca. *Annals and Magazine of Natural History* (11) **3**: 288–290.

Beier, M. (1939c). Pseudoscorpionidea de Roumanie. *Bulletin du Musée Royal d'Histoire Naturelle de Belgique* **15** (39): 1–21.

Beier, M. (1939d). Die Höhlenpseudoscorpione der Balkanhalbinsel. *Studien aus dem Gebiete der Allgemeinen Karstforschung, der Wissenschaftlichen Höhlenkunde, der Eiszeitforschung und den Nachbargebieten* **4** (10): 1–83.

Beier, M. (1939e). Die Pseudoscorpione des oberösterreichischen Landesmuseums in Linz. *Jahrbuch des Vereines für Landeskunde und Heimatpflege im Gau Oberdonau, Linz* **88**: 303–312.

Beier, M. (1939f). Die Pseudoscorpioniden-Fauna der iberischen Halbinsel. *Zoologische Jahrbücher, Systematik (Ökologie), Geographie und Biologie* **72**: 157–202.

Beier, M. (1940a). Die Pseudoscorpionidenfauna der landfernen Inseln. *Zoologische Jahrbücher, Systematik (Ökologie), Geographie und Biologie* **74**: 161–192.

Beier, M. (1940b). Zur Phylogenie der troglobionten Pseudoscorpione. *6th International Congress of Entomology, 1935* **2**: 519–527.

Beier, M. (1942). Pseudoscorpione aus italischen Höhlen. *Bollettino del Laboratoria di Zoologia Generale e Agraria della R. Scuola sup. d'Agricoltura, Portici* **32**: 130–136.

Beier, M. (1943). Neue Pseudoskorpione aus West-, Zentral- und Ostasien. *Annalen des Naturhistorischen Museums in Wien* 53: 74–81.

Beier, M. (1944). Über Pseudoscorpioniden aus Ostafrika. *Eos, Madrid* 20: 173–212.

Beier, M. (1946). Some pseudoscorpions from the Upper Nile Territory. *Annals and Magazine of Natural History* (11) 13: 567–571.

Beier, M. (1947a). Die mit *praecipuum* Simon verwandten Arten der Gattung *Neobisium* (Pseudoscorp.). *Eos, Madrid* 23: 165–183.

Beier, M. (1947b). Zur Kenntnis der Pseudoscorpionidenfauna des südlichen Afrika, insbesondere der südwest- und südafrikanischen Trockengebiete. *Eos, Madrid* 23: 285–339.

Beier, M. (1947c). Results of the Armstrong College Expedition to Siwa Oasis (Libyan Desert), 1935. Pseudoscorpionidea (Arachnida). *Bulletin de la Société Fouad Ier d'Entomologie* 31: 127–128.

Beier, M. (1947d). Neue Pseudoscorpione aus der Steiermark. *Annalen des Naturhistorischen Museums in Wien* 55: 296–301.

Beier, M. (1947e). Pseudoskorpione im Baltischen Bernstein und die Untersuchung von Bernstein-Einschlüssen. *Mikroscopie, Wien* 1: 188–199.

Beier, M. (1948a). Zur Kenntnis der Pseudoscorpionidenfauna Sardiniens und Korsikas. *Annalen des Naturhistorischen Museums in Wien* 56: 188–191.

Beier, M. (1948b). Phoresie und Phagophilie bei Pseudoscorpionen. *Österreichische Zoologische Zeitschrift* 1: 441–497.

Beier, M. (1948c). Über Pseudoscorpione der australischen Region. *Eos, Madrid* 24: 525–562.

Beier, M. (1949). Türkiye Psevdoscorpion'lari hakkinda. Türkische Pseudoscorpione. *Revue de la Faculté des Sciences de l'Université d'Istanbul* (B) 14: 1–20.

Beier, M. (1950). Zur Phänologie einiger *Neobisium*-Arten (Pseudoscorp.). *Proceedings of the 8th Internattional Congress of Entomology, Stockholm 1950*: 1002–1007.

Beier, M. (1951a). Die Pseudoscorpione Indochinas. *Mémoires du Muséum National d'Histoire Naturelle, Paris (n.s.)* 1: 47–123.

Beier, M. (1951b). On some Pseudoscorpionidea from Kilimanjaro. *Annals and Magazine of Natural History* (12) 4: 606–609.

Beier, M. (1951c). Zur Kenntnis der ostalpinen Chthoniiden (Pseudoscorp). *Entomologisches Nachrichtenblatt, Wien* 3: 163–166.

Beier, M. (1951d). Ergebnisse der Österreichischen Iran-Expedition 1949/50. Pseudoscorpione und Mantiden. *Annalen des Naturhistorischen Museums in Wien* 58: 96–102.

Beier, M. (1951e). Der Bücherskorpion, ein willkommener Gast der Bienenvölker. *Österreichische Imker* 1: 209–211. (not seen)

Beier, M. (1952a). On some Pseudoscorpionidea from Malaya and Borneo. *Bulletin of the Raffles Museum* 24: 96–108.

Beier, M. (1952b). Eine neue *Garypus*-Art (Pseudoscorp.) aus Japan. *Zoologischer Anzeiger* 149: 235–239.

Beier, M. (1952c). Neue Pseudoscorpione von den Dolomiten. *Studi Trentini di Scienze Naturali* 29: 56–60.

Beier, M. (1952d). Über die von L. di Caporiacco aus Apulien beschriebenen Höhlen-Pseudoscorpione. *Memorie di Biogeografica Adriatica* 2: 103–108.

Beier, M. (1952e). Ordn.: Pseudoscorpionidea, Afterskorpione. In: Strouhal, H. (ed.), *Catalogus Faunae Austriae*, vol. 9a: 2–6. Springer-Verlag: Wien.

Beier, M. (1952f). The 3rd Danish Expedition to Central Asia. Zoological Results 7. Pseudoscorpionidea (Chelicerata) aus Afghanistan. *Videnskabelige Meddelelser fra Dansk Naturhistorisk Forening i Kjøbenhavn* 114: 245–250.

Beier, M. (1953a). Weiteres zur Kenntnis der iberischen Pseudoscorpioniden-Fauna. *Eos, Madrid* 28: 293–302.

Beier, M. (1953b). Pseudoscorpione aus El Salvador und Guatemala. *Senckenbergiana Biologica* **34**: 15–28.

Beier, M. (1953c). Ueber einige phoretische und phagophile afrikanische Pseudoscorpione. *Revue de Zoologie et de Botanique Africaines* **48**: 73–78.

Beier, M. (1953d). Neue und bemerkenswerte Pseudoscorpione aus oberitalienischen Hoehlen. *Bollettino della Società Entomologica Italiana* **83**: 35–38.

Beier, M. (1953e). Ueber eine Pseudoscorpioniden-Ausbeute aus ligurischen Höhlen. *Bollettino della Società Entomologica Italiana* **83**: 105–108.

Beier, M. (1953f). *Neobisium (Blothrus) patrizii*, ein neuer Hoehlen-Pseudoscorpion aus Mittelitalien. *Bollettino della Società Entomologica Italiana* **83**: 139–140.

Beier, M. (1953g). Pseudoscorpionidea von Sumba und Flores. *Verhandlungen der Naturforschenden Gesellschaft in Basel* **64**: 81–88.

Beier, M. (1954a). Eine neue *Neobisium*-Art (Pseudoscorp.) aus der Dauphiné. *Annalen des Naturhistorischen Museums in Wien* **59**: 155–156.

Beier, M. (1954b). Report from Prof. T. Gislén's expedition to Australia in 1951–1952. 7. Pseudoscorpionidea. *Acta Universitatis Lundensis, n.f.* (2) **50**: 1–26.

Beier, M. (1954c). Einige neue Pseudoscorpione aus dem Genueser Museum. *Annali del Museo Civico di Storia Naturale di Genova* **66**: 324–330.

Beier, M. (1954d). Pseudoscorpioniden aus dem Belgischen Congo. *Annales du Musée du Congo Belge, Sciences Zoologiques* **1**: 132–139.

Beier, M. (1954e). Zwei neue Pseudscorpione aus Ostafrika. *Zoologischer Anzeiger* **152**: 84–88.

Beier, M. (1954f). Eine Pseudoscorpioniden-Ausbeute aus Venezuela. *Memorie del Museo Civico di Storia Naturale di Verona* **4**: 131–142.

Beier, M. (1954g). Ein neuer Olpiide (Pseudoscorp.) aus dem Hochlande von Perú. *Senckenbergiana Biologica* **34**: 325–326.

Beier, M. (1955a). Pseudoscorpionidea. *Exploration de Parc National de l'Upemba. I. Mission G.F. de Witte* **32** (1): 3–19.

Beier, M. (1955b). Pseudoscorpione von den Juan-Fernandez-Inseln (Arachnida Pseudoscorpionida). *Revista Chilena de Entomología* **4**: 205–220.

Beier, M. (1955c). Pseudoscorpionidea, gesammelt während der schwedischen Expeditionen nach Ostafrika 1937–1938 und 1948. *Arkiv för Zoologi* (2) **7**: 527–558.

Beier, M. (1955d). Über Pseudoscorpione aus Syrien und Palästina. *Annalen des Naturhistorischen Museums in Wien* **60**: 212–219.

Beier, M. (1955e). A second collection of Pseudoscorpionidea from Malaya. *Bulletin of the Raffles Museum* **25**: 38–46.

Beier, M. (1955f). Pseudoscorpione von Tristan da Cunha. *Results of the Norwegian Scientific Expedition to Tristan da Cunha 1937–1938, no. 35*: 7–10.

Beier, M. (1955g). Pseudoscorpione im baltischen Bernstein aus dem Geologischen Staatsinstitut in Hamburg. *Mitteilungen aus dem Mineralogisch-Geologischen Staatsinstitut in Hamburg* **24**: 48–54.

Beier, M. (1955h). Ein neuer myrmecophiler Pseudoscorpion aus Ostafrika. *Bollettino della Società Entomologica Italiana* **85**: 7–9.

Beier, M. (1955i). Ein neuer *Incachernes* aus El Salvador (Pseudoscorp.). *Senckenbergiana Biologica* **36**: 369–370.

Beier, M. (1955j). *Neobisium (Blothrus) cerrutii*, ein weiterer neuer Höhlen-Pseudoscorpion aus Lazio. *Fragmenta Entomologica* **2**: 25–28.

Beier, M. (1955k). Höhlen-Pseudoscorpione aus Sardinien. *Fragmenta Entomologica* **2**: 41–46.

Beier, M. (1955l). Pseudoscorpionidea. In: Hanstrom, B., Brinck, P. & Rudebeck, G. (eds), *South African Animal Life. Results of the Lund Expedition in 1950–1951*, vol. **1**: 263–328. Almquist and Wiksell: Stockholm.

Beier, M. (1956a). Eine neue *Minniza* (Pseudoscorp.) aus Transvaal. *Entomologische Berichten, Amsterdam* **16**: 29–20.

Beier, M. (1956b). Ein neuer *Blothrus* (Pseudoscorp.) aus Sardinien, und ueber zwei Pseudoscorpione des westmediterranen litorals. *Fragmenta Entomologica* **2**: 55–63.

Beier, M. (1956c). A new phagophilus *Plesiochernes* (Pseudoscorpionidea) from Natal. *Annals of the Natal Museum* **13**: 437–439.

Beier, M. (1956d). Neue Beiträge zur Kenntnis der Iberischen Pseudoscorpioniden-Fauna. *Eos, Madrid* **31**: 87–122.

Beier, M. (1956e). Ueber Pseudoscorpione aus Spanische-Marocco. *Eos, Madrid* **31**: 303–310.

Beier, M. (1956f). Neue Troglobionte Pseudoscorpione aus Mexico. *Ciencia, México* **16**: 81–85.

Beier, M. (1956g). Über einige Pseudoscorpione von Kreta. *Entomologische Nachrichtsblatt, Wien* **8**: 8–9.

Beier, M. (1956h). Bemerkenswerte Pseudoscorpioniden-Funde aus Niederösterreich. *Entomologische Nachrichtsblatt, Wien* **8**: 24–25.

Beier, M. (1956i). Pseudoscorpionidea. 1. Nachtrag. In: Strouhal, H. (ed.), *Catalogus Faunae Austriae*, vol. **9a**: 8–9. Springer-Verlag: Wien.

Beier, M. (1956j). Weiteres zur Kenntnis der Hoehlenpseudoscorpione Sardiniens. *Fragmenta Entomologica* **2**: 131–135.

Beier, M. (1957a). Ueber Hoehlenbewohnende Pseudoscorpione aus Venezien. *Bollettino dei Musei Civico di Storia Naturale di Venezia* **10**: 161–163.

Beier, M. (1957b). Pseudoscorpione gesammelt von Dr. K. Lindberg 1956. *Förhandlingar vid Kungliga Fysiografiska Sällskapets i Lund* **27**: 145–151.

Beier, M. (1957c). Los Insectos de las Islas Juan Fernandez. 37. Die Pseudo-scorpioniden-Fauna der Juan-Fernandez-Inseln (Arachnida Pseudoscorpionida). *Revista Chilena de Entomología* **5**: 451–464.

Beier, M. (1957d). Pseudoscorpionidea. *Insects of Micronesia* **3**: 1–64.

Beier, M. (1957e). A new false scorpion of the genus *Olpium* from Tanganyika. *Annals and Magazine of Natural History (12)* **10**: 471–472.

Beier, M. (1958a). The Pseudoscorpionidea (false-scorpions) of Natal and Zululand. *Annals of the Natal Museum* **14**: 155–187.

Beier M. (1958b). Pseudoscorpione aus Gargano (Apulien). *Memorie di Biogeografica Adriatica* **4**: 27–31.

Beier, M. (1958c). Eine neue *Neobisium*-art aus den Picentinischen Bergen in Süd-Italien. *Memorie del Museo Civico di Storia Naturale di Verona* **6**: 135–137.

Beier, M. (1959a). Zur Kenntnis der Pseudoscorpioniden-Fauna Afghanistans. *Zoologische Jahrbücher, Systematik (Ökologie), Geographie und Biologie* **87**: 257–282.

Beier, M. (1959b). Neues über Sardinische Höhlenpseudoscorpione. *Annales de Spéléologie* **14**: 245–246.

Beier, M. (1959c). Ein neuer *Allochernes* (Pseudoscorp.) aus dem Karakorum-Gebirge. *Annalen des Naturhistorischen Museums in Wien* **63**: 407–408.

Beier, M. (1959d). Pseudoscorpione aus dem Belgischen Congo gesammelt von Herrn N. Leleup. *Annales du Musée du Congo Belge, Sciences Zoologiques* **72**: 5–69.

Beier, M. (1959e). Zur Kenntnis der Pseudoscorpioniden-Fauna des Andengebietes. *Beiträge zur Neotropischen Fauna* **1**: 185–228.

Beier, M. (1959f). Ergänzungen zur iberischen Pseudoscorpioniden-Fauna. *Eos, Madrid* **35**: 113–131.

Beier, M. (1960a). Pseudoscorpionidea. Contribution à l'étude de la fauna d'Afghanistan. 27. *Förhandlingar vid Kungliga Fysiografiska Sällskapets i Lund* **30**: 41–45.

Beier, M. (1960b). *Chernes cimicoides* (F.) und *Chernes hahni* (C. L. Koch), zwei gut unterschiedene Arten. *Zeitschrift der Arbeitsgemeinschaft Österreichischer Entomologen* **12**: 100–102.

Beier, M. (1961a). Pseudoscorpione von den Azoren und Madeira. *Boletim do Museu Municipal do Funchal* **14**: 67–74.

Beier, M. (1961b). Nochmals über iberische und marokkanische Pseudoscorpione. *Eos, Madrid* **37**: 21–39.

Beier, M. (1961c). Pseudoscorpione von der Insel Ascension. *Annals and Magazine of Natural History* (13) **3**: 593–598.

Beier, M. (1961d). Ueber Pseudoscorpione aus sizilianischen Höhlen. *Bolletino Accademia Gioenia di Scienze Naturali in Catania* (4) **6**: 89–96.

Beier, M. (1961e). Pseudoscorpionidea II. Contribution à l'étude de la faune d'Afghanistan. 56. *Förhandlingar vid Kungliga Fysiografiska Sällskapets i Lund* **31**: 1–4.

Beier, M. (1961f). Höhlenpseudoscorpione aus der Toscana. *Monitore Zoologico Italiano* **68**: 123–127.

Beier, M. (1961g). Biogeografica delle Isole Pelagie. Arachnida, Chernetes. *Rendiconti della Accademia Nazionale delle Scienze* (4) **11**: 411.

Beier, M. (1962a). Ergebnisse der von Dr. O. Paget und Dr. E. Kritscher auf Rhodos durchgeführten zoologischen Exkursionen. V. Pseudoscorpionidea. *Annalen des Naturhistorischen Museums in Wien* **64**: 139–142.

Beier, M. (1962b). Über kaukasische Pseudoskorpione. *Annalen des Naturhistorischen Museums in Wien* **64**: 146–153.

Beier, M. (1962c). Eine neue *Microcreagris* aus Portugal. Voyage au Portugal du Dr. K. Lindberg. *Boletim de Sociedades Portuguesa des Ciencias Naturais* (2) **9**: 25–26.

Beier, M. (1962d). Pseudoscorpioniden aus der Namib-Wüste. *Annals of the Transvaal Museum* **24**: 223–230.

Beier, M. (1962e). Ergebnisse der Zoologischen Nubien-Expedition 1962. Teil III. Pseudoscorpionidea. *Annalen des Naturhistorischen Museums in Wien* **65**: 297–303.

Beier, M. (1962f). Pseudoscorpionidea. Mission zoologique de l'I.R.S.A.C. en Afrique orientale. (P. Basilewsky et N. Leleup, 1957). *Annales du Musée de l'Afrique Centrale, Zoologie* (8) **107**: 9–37.

Beier, M. (1962g). On some Pseudoscorpionidea from New Zealand. *Records of the Canterbury Museum* **7**: 399–402.

Beier, M. (1962h). Pseudoscorpionidea. In: Delamare Debouteville, C. & Rapoport, E. (eds), *Biologie de l'Amérique Australe, Etudes sur la Faune du Sol*, vol. **1**: 131–137.

Beier, M. (1962i). Ein Höhlen-Pseudoskorpion aus den Nördlichen Kalkalpen. *Höhle* **13**: 1–3.

Beier, M. (1963a). Sizilianische Pseudoscorpione. *Bolletino Accademia Gioenia di Scienze Naturali in Catania* (4) **7**: 253–263.

Beier, M. (1963b). Ordnung Pseudoscorpionidea (Afterskorpione). In: *Bestimmungsbücher zur Bodenfauna Europas*, vol. **1**. Akademie-Verlag: Berlin.

Beier, M. (1963c). Pseudoscorpione von den Batu-Höhlen in Malaya. *Pacific Insects* **5**: 51–52.

Beier, M. (1963d). Pseudoscorpione aus Vogelnestern von Malaya. *Pacific Insects* **5**: 507–511.

Beier, M. (1963e). Eine neue Art der Pseudoscorpioniden-Gattung *Albiorix* aus Höhle Acuitlapan, Gro., Mexico. *Ciencia, México* **22**: 133–134.

Beier, M. (1963f). Die Pseudoscorpioniden-Fauna Israels und einiger angrenzender Gebiete. *Israel Journal of Zoology* **12**: 183–212.

Beier, M. (1963g). Pseudoskorpione aus Anatolien. *Annalen des Naturhistorischen Museums in Wien* **66**; 267–277.

Beier, M. (1963h). Pseudoscorpione aus dem Museum 'Enrico Caffi' in Bergamo. *Rendiconti dell'Instituto Lombardo di Scienze e Lettere* **97B:** 147–156.

Beier, M. (1963i). Appenninische Pseudoscorpione. *Memorie del Museo Civico di Storia Naturale di Verona* **10:** 283–286.

Beier, M. (1964a). Pseudoscorpione von der Insel San Ambrosio. *Annalen des Naturhistorischen Museums in Wien* **67:** 303–306.

Beier, M. (1964b). Die Pseudoscorpioniden-Fauna Chiles. *Annalen des Naturhistorischen Museums in Wien* **67:** 307–375.

Beier, M. (1964c). Pseudoskorpione aus dem Bucegi-Gebirge in Rumänien. *Zoologischer Anzeiger* **173:** 210–212.

Beier, M. (1964d). Some further nidicolous Chelonethi (Pseudoscorpionidea) from Malaya. *Pacific Insects* **6:** 312–313.

Beier, M. (1964e). Pseudoscorpione von Neu-Caledonien. *Pacific Insects* **6:** 403–411.

Beier, M. (1964f). Further records of Pseudoscorpionidea from the Solomon Islands. *Pacific Insects* **6:** 592–598.

Beier, M. (1964g). Insects of Campbell Island. Pseudoscorpionidea. *Pacific Insects Monograph* **7:** 116–120.

Beier, M. (1964h). False scorpions (Pseudoscorpionidea) from the Auckland Islands. *Pacific Insects Monograph* **7** (supplement): 628–629.

Beier, M. (1964i). Ein neuer Pseudoscorpion aus Termiten-Bauten. *Revue de Zoologie et de Botanique Africaines* **69:** 198–200.

Beier, M. (1964j). The zoological results of Gy. Tópal's collectings in South Argentina. 15. Pseudoscorpionidea. *Annales Historico-Naturales Musei Nationalis Hungarici* **56:** 487–500.

Beier, M. (1964k). Weiteres zur Kenntnis der Pseudoscorpioniden-Fauna des südlichen Afrika. *Annals of the Natal Museum* **16:** 30–90.

Beier, M. (1965a). Pseudoscorpione aus ostmediterranen Grotten. *Fragmenta Entomologica* **4:** 85–90.

Beier, M. (1965b). Pseudoskorpione aus dem Tschad-Gebiet. *Annalen des Naturhistorischen Museums in Wien* **68:** 365–374.

Beier, M. (1965c). Über Pseudoskorpione von den Kanaren. *Annalen des Naturhistorischen Museums in Wien* **68:** 375–381.

Beier, M. (1965d). Ergebnisse der von Dr. O. Paget und Dr. E. Kritscher auf Rhodos durchgeführten zoologischen Expedition. Teil XI. Pseudoscorpionidea (2. Teil). *Annalen des Naturhistorischen Museums in Wien* **68:** 631–633.

Beier, M. (1965e). Eine neue *Microcreagris* (Pseudoscorpionidea) aus Frankreich. *Zoölogische Mededeelingen* **40:** 301–303.

Beier, M. (1965f). Anadolu'nun Pseudoscorpion faunasi. Die Pseudoscorpioniden-Fauna Anatoliens. *Istanbul Üniversitesi Fen Fakültesi Mecmuasi* **29B:** 81–105.

Beier, M. (1965g). Die Pseudoscorpioniden Neu-Guineas und der benachbarten Inseln. *Pacific Insects* **7:** 749–796.

Beier, M. (1966a). On the Pseudoscorpionidea of Australia. *Australian Journal of Zoology* **14:** 275–303.

Beier, M. (1966b). Über Pseudoscorpione von den Philippinen. *Pacific Insects* **8:** 340–348.

Beier, M. (1966c). Zur Kenntnis der Pseudoscorpioniden-Fauna Neu-Seelands. *Pacific Insects* **8:** 363–379.

Beier, M. (1966d). Die Pseudoscorpioniden der Salomon-Inseln. *Annalen des Naturhistorischen Museums in Wien* **69:** 133–159.

Beier, M. (1966e). Über Pseudoskorpione der Insel Rhodos. *Annalen des Naturhistorischen Museums in Wien* **69:** 161–167.

Beier, M. (1966f). Zoologische Aufsammlungen auf Kreta. Pseudoscorpionidea. *Annalen des Naturhistorischen Museums in Wien* **69**: 343–346.

Beier, M. (1966g). Ergebnisse der österreichischen Neukaledonien-Expedition 1965. Pseudoscorpionidea. *Annalen des Naturhistorischen Museums in Wien* **69**: 363–371.

Beier, M. (1966h). Neues über Höhlen-Pseudoscorpione aus Veneto. *Atti della Società Italiana di Scienze Naturali, e del Museo Civile di Storia Naturale, Milano* **105**: 175–178.

Beier, M. (1966i). Ein neuer *Hysterochelifer* (Pseudoscorp.) aus Afghanistan. *Casopis Moravského Zemskeho Musea, Brné* **51**: 259–260.

Beier, M. (1966j). Ein neuer *Pselaphochernes* (Pseudoscorp.) aus Süditalien. *Fragmenta Entomologica* **4**: 109–111.

Beier, M. (1966k). Ergänzungen zur Pseudoscorpioniden-Fauna des südlichen Afrika. *Annals of the Natal Museum* **18**: 455–470.

Beier, M. (1966l). Ein neuer Höhlen-Pseudoscorpion aus den Abruzzen. *Bollettino della Società Entomologica Italiana* **96**: 35–36.

Beier, M. (1966m). Ein neuer nidikoler *Allochernes*. Ergebnisse der zoologischen Forschungen von Dr. Z. Kaszab in der Mongolei (Pseudoscorpionidea). *Reichenbachia* **7**: 225–227.

Beier, M. (1967a). Some Pseudoscorpionidea from Australia, chiefly from caves. *Australian Zoologist* **14**: 199-205.

Beier, M. (1967b). Contributions to the knowledge of the Pseudoscorpionidea from New Zealand. *Records of the Dominion Museum* **5**: 277–303.

Beier, M. (1967c). Die Pseudoscorpione der Noona Dan Expedition nach den Philippinen und Bismarck Inseln. *Entomologiske Meddelelser* **35**: 315–324.

Beier, M. (1967d). Pseudoskorpione aus dem tropischen Ostafrika (Kenya, Tansania, Uganda, etc.). *Annalen des Naturhistorischen Museums in Wien* **70**: 73–93.

Beier, M. (1967e). Zwei neue Chernetiden (Pseudoscorp.) von Argentinien. *Annalen des Naturhistorischen Museums in Wien* **70**: 95–98.

Beier, M. (1967f). Ergebnisse zoologischer Sammelreisen in der Türkei. *Annalen des Naturhistorischen Museums in Wien* **70**: 301–323.

Beier, M. (1967g). Pseudoscorpione vom kontinentalen Südost-Asien. *Pacific Insects* **9**: 341–369.

Beier, M. (1967h). Ein phoretischer *Allochernes* (Pseudoscorp.) aus Afghanistan. *Beiträge zur Naturkundlichen Forschung in Südwestdeutschland* **26**: 17–18.

Beier, M. (1968a). Some cave-dwelling Pseudoscorpionidea from Australia and New Caledonia. *Records of the South Australian Museum* **15**: 757–765.

Beier, M. (1968b). Ein neues Chernetiden-Genus (Pseudoscorp.) aus Nepal. *Khumbu Himal* **3**: 17–18.

Beier, M. (1969a). Neue Pseudoskorpione aus Australien. *Annalen des Naturhistorischen Museums in Wien* **73**: 171–187.

Beier, M. (1969b). Weitere Beiträge zur Kenntnis der Pseudoskorpione Anatoliens. *Annalen des Naturhistorischen Museums in Wien* **73**: 189–198.

Beier, M. (1969c). Pseudoscorpionidea. Ergebnisse der zoologischen Forschungen von Dr. K. Kaszab in der Mongolei. *Reichenbachia* **11**: 283–286.

Beier, M. (1969d). Ein neuer *Allochernes* (Pseudoscorp.) vom Elbursgebirge. *Beiträge zur Naturkundlichen Forschung in Südwestdeutschland* **28**: 121–122.

Beier, M. (1969e). Additional remarks to the New Zealand Pseudoscorpionidea. *Records of the Auckland Institute and Museum* **6**: 413–418.

Beier, M. (1969f). Ein wahrscheinlich troglobionter *Pseudochthonius* (Pseudoscorp.) aus Brasilien. *Revue Suisse de Zoologie* **76**: 1–2.

Beier, M. (1970a). Ergänzungen zur Pseudoskorpionidenfauna der Kanaren. *Annalen des Naturhistorischen Museums in Wien* **74**: 45–49.

Beier, M. (1970a). Ergänzungen zur Pseudoskorpionidenfauna der Kanaren. *Annalen des Naturhistorischen Museums in Wien* **74**: 45−49.

Beier, M. (1970b). Myrmecophile Pseudoskorpione aus Brasilien. *Annalen des Naturhistorischen Museums in Wien* **74**: 51−56.

Beier, M. (1970c). Zur Kenntnis der afrikanischen Arten der Gattung *Cheiridium* Menge. *Annalen des Naturhistorischen Museums in Wien* **74**: 57−61.

Beier, M. (1970d). Trogloxene Pseudoscorpione aus Südamerika. *Anales de la Escuela Nacional de Ciencias Biologicas, México* **17**: 51−54.

Beier, M. (1970e). Ergebnisse der zoologischen Forschungen von Dr. Z. Kaszab in der Mongolei (Pseudoscorpionidea). *Reichenbachia* **13**: 15−18.

Beier, M. (1970f). Reliktformen in der Pseudoscorpioniden-Fauna Europas. *Memorie della Società Entomologica Italiana* **48**: 317−323.

Beier, M. (1970g). Die Pseudoscorpione der Royal Society Expedition 1965 zu den Salomon-Inseln. *Journal of Natural History* **4**: 315−328.

Beier, M. (1971a). Pseudoskorpione aus dem Iran. *Annalen des Naturhistorischen Museums in Wien* **75**: 357−366.

Beier, M. (1971b). Pseudoskorpione unter Araucarien-Rinde in Neu-Guinea. *Annalen des Naturhistorischen Museums in Wien* **75**: 367−373.

Beier, M. (1971c). Ein neuer *Mundochthonius* (Arachnida, Pseudoscorpionidea) aus der Steiermark. *Mitteilungen des Naturwissenschaftlichen Vereins für Steiermark* **100**: 383−387.

Beier, M. (1971d). A new *Synsphyronus* Chamberlin (Pseudoscorpiones) from the Great Victoria Desert. *Journal of the Australian Entomological Society* **10**: 161−162.

Beier, M. (1971e). A new chthoniid pseudoscorpion from Western Australia. *Journal of the Australian Entomological Society* **10**: 233−234.

Beier, M. (1972a). Pseudoscorpionidea aus dem Parc National Garamba. *Exploration Parc National de la Garamba, Mission H. de Saeger*, vol. **56**: 2−19.

Beier, M. (1972b). Ein neuer troglobionter Pseudoscorpion aus Tarragona. *Eos, Madrid* **46**: 15−17.

Beier, M. (1972c). Neue Funde von *Roncus (Parablothrus) ghidinii* Beier 1942. *Natura, Bresciana* **8**: 3−5.

Beier, M. (1973a). Weiteres zur Kenntnis der Pseudoscorpioniden Südwestafrikas. *Cimbebasia, A* **2**: 97−101.

Beier, M. (1973b). Zwei neue höhlenbewohnende Chthoniiden aus Oberitalien. *Annalen des Naturhistorischen Museums in Wien* **77**: 159−161.

Beier, M. (1973c). Neue Funde von Höhlen-Pseudoskorpionen auf Sardinien. *Annalen des Naturhistorischen Museums in Wien* **77**: 163−166.

Beier, M. (1973d). Pseudoscorpionidea von Ceylon. *Entomologica Scandinavica, Supplement* **4**: 39−55.

Beier, M. (1973e). Beiträge zur Pseudoscorpionidenfauna Anatoliens. *Fragmenta Entomologica* **8**: 223−236.

Beier, M. (1973f). Pseudoscorpione aus der Mongolei. *Annalen des Naturhistorischen Museums in Wien* **77**: 167−172.

Beier, M. (1974a). Eine neue *Compsaditha* von den Seychellen (Arachnida, Pseudoscorpiones). *Entomologische Zeitschrift* **84**: 144−145.

Beier, M. (1974b). Pseudoscorpione aus Nepal. *Senckenbergiana Biologica* **55**: 261−280.

Beier, M. (1974c). Ein neuer *Paraliochthonius* aus Guatemala. *Revue Suisse de Zoologie* **81**: 101−102.

Beier, M. (1974d). Brasilianische Pseudoscorpione aus dem Museum in Genf. *Revue Suisse de Zoologie* **81**: 899−909.

Beier, M. (1974e). Pseudoscorpione aus Südindien des Naturhistorischen Museums in Genf. *Revue Suisse de Zoologie* **81**: 999−1017.

Beier, M. (1975a). Weitere bemerkenswerte Pseudoscorpione von Sizilien. *Animalia, Catania* **2**: 55–58.

Beier, M. (1975b). Neue Pseudoskorpione aus Australien und Neu-Guinea. *Annalen des Naturhistorischen Museums in Wien* **78**: 203–213.

Beier, M. (1976a). Die Pseudoscorpione der macaronesischen Inseln. *Vieraea, Tenerife* **5**: 23–32.

Beier, M. (1976b). A cavernicolous atemnid pseudoscorpion from New South Wales. *Journal of the Australian Entomological Society* **15**: 271–272.

Beier, M. (1976c). Neue und bemerkenswerte zentralamerikanische Pseudoskorpione aus dem Zoologischen Museum in Hamburg. *Entomologische Mitteilungen aus dem Staatsinstitut und Zoologischen Museum in Hamburg* **5** (91): 1–5.

Beier, M. (1976d). Pseudoscorpione von der Dominicanischen Republik (Insel Haiti). *Revue Suisse de Zoologie* **83**: 45–58.

Beier, M. (1976e). Ergebnisse der Bhutan-Expedition 1972 des Naturhistorischen Museums Basel. Pseudoscorpionidea. *Verhandlungen der Naturforschenden Gesellschaft in Basel* **85**: 95–100.

Beier, M. (1976f). The pseudoscorpions of New Zealand, Norfolk and Lord Howe. *New Zealand Journal of Zoology* **3**: 199–246.

Beier, M. (1977a). Pseudoscorpione aus einer Höhle der Philippinen-Insel Pagbilao. *Revue Suisse de Zoologie* **84**: 187–190.

Beier, M. (1977b). Pseudoscorpiones. In: La faune terrestre de l'île de Sainte-Hélène IV. *Annales du Musée Royal de l'Afrique Centrale, Zoologie* (8) **220**: 2–11.

Beier, M. (1977c). Pseudoscorpionidea. In: *Mission Zoologique Belge aux îles Galapagos et en Ecuador (N. et J. Leleup, 1964–1965)*, vol. **3**: 93–112. Bruxelles.

Beier, M. (1978a). A new *Cheiridium* (Pseudoscorpionidea) from South West Africa. *Annals of the Natal Museum* **23**: 429–430.

Beier, M. (1978b). Pseudoskorpione aus Kashmir und Ladakh (Arachnida). *Senckenbergiana Biologica* **58**: 415–417.

Beier, M. (1978c). Pseudoskorpione von den Galapagos-Inseln. *Annalen des Naturhistorischen Museums in Wien* **81**: 533–547.

Beier, M. (1978d). Zwei neue orientalische Pseudoscorpione aus dem Basler Museum. *Entomologica Basiliensia* **3**: 231–234.

Beier, M. (1979a). Ein neuer *Hebridochernes* von Neu-Kaledonien (Pseudoscorp.). *Annalen des Naturhistorischen Museums in Wien* **82**: 549–552.

Beier, M. (1979b). Pseudoskorpione aus der Küstenprovinz im Osten der USSR. *Annalen des Naturhistorischen Museums in Wien* **82**: 553–557.

Beier, M. (1979c). Neue afrikanische Pseudoskorpione aus dem Musée Royal de l'Afrique Central in Tervuren. *Revue de Zoologie Africaine* **93**: 101–113.

Beier, M. (1981). Eine Pseudoscorpioniden-Ausbeute von den Andaman-Inseln. *Bollettino del Museo Civico di Storia Naturale, Verona* **7**: 293–295.

Beier, M. (1982). Zoological results of the British Speleological Expedition to Papua New Guinea 1975. 9. Pseudoscorpionidea. *Acta Zoologica Bulgarica* **19**: 43–45.

Beier, M. & Franz, H. (1954). 16. Ordnung: Pseudoscorpionidea. In: H. Franz, *Die Nordost-Alpen im Spiegel ihrer Landtierwelt*, vol. **1**: 453–459. Universitätsverlag Wagner: Innsbruck.

Beier, M. & Turk, F. A. (1952). On two collections of Cyprian Pseudoscorpionidea. *Annals and Magazine of Natural History* (12) **5**: 766–771.

Beliawskij, ? (1927). *Wragi Ptschel (Bienenschädlinge)*. Leningrad. (not seen)

Bellés, X. (1978). Notas ecológicas sobre la Cova de la Torre (Sant Feliu de Pallarols, Gironá). Incidencia de los niveles de materia organica sobre la diversidad de la faune terrestre. *Speleon* **24**: 77–91.

Bellés, X. (1987). *Fauna Cavernícola i Intersticial de la Península Ibèrica i les Illes Balears*. C.S.I.C.: Mallorca.

Bellés, X. & Comas, J. (1980). Una campanya entomològica Tunisia (1979). Prospeccions biospeleològiques. *Boletin S.I.E.P., Barcelona* 12: 35−42.

Bellhouse Lemare, S. (1923). The false scorpion. *Bee World* 5: 232.

Beneden, P.J. van (1876). *Die Schmarotzer des Thierreichs*. Leipzig. (not seen)

Benedict, E.M. (1978a). A new pseudoscorpion genus *Malcolmochthonius* n.g., with three new species from the western United States. *Transactions of the American Microscopical Society* 97: 250−255.

Benedict, E.M. (1978b). False scorpions of the genus *Apocheiridium* Chamberlin from western North America (Pseudoscorpionida, Cheiridiidae). *Journal of Arachnology* 5: 231−241.

Benedict, E.M. (1978c). *A biogeographical study of currently identified Oregon pseudoscorpions with emphasis on western Oregon forms*. Ph.D. Thesis: Portland State University. (not seen)

Benedict, E.M. (1979). A new species of *Apochthonius* Chamberlin from Oregon (Pseudoscorpionida, Chthoniidae). *Journal of Arachnology* 7: 79−83.

Benedict, E. M. & Malcolm, D.R. (1970). Some pseudotyrannochthoniine false scorpions from western North America (Chelonethida: Chthoniidae). *Journal of the New York Entomological Society* 78: 38−51.

Benedict, E. M. & Malcolm, D.R. (1973). A new cavernicolous species of *Apochthonius* (Chelonethida: Chthoniidae) from the western United States with reference to troglobitic tendencies in the genus. *Transactions of the American Microscopical Society* 92: 620−628.

Benedict, E. M. & Malcolm, D. R. (1974). A new cavernicolous species of *Mundochthonius* from the eastern United States (Pseudoscorpionida, Chthoniidae). *Journal of Arachnology* 2: 1−4.

Benedict, E. M. & Malcolm, D.R. (1978a). Some garypoid false scorpions from western North America (Pseudoscorpionida: Garypidae and Olpiidae). *Journal of Arachnology* 5: 113−132.

Benedict, E. M. & Malcolm, D.R. (1978b). The family Pseudogarypidae (Pseudoscorpionida) in North America with comments on the genus *Neopseudogarypus* Morris from Tasmania. *Journal of Arachnology* 6: 81−104.

Benedict, E. M. & Malcolm, D.R. (1978c). Troglobitic tendencies in pseudoscorpions of the genus *Pseudogarypus* (Pseudogarypidae). *Bulletin of the National Speleological Society* 40: 91.

Benedict, E. M. & Malcolm, D. R. (1979). Pseudoscorpions of the family Cheliferidae from Oregon (Pseudoscorpionida, Cheliferoidea). *Journal of Arachnology* 7: 187−198.

Benedict, E. M. & Malcolm, D. R. (1982). Pseudoscorpions of the family Chernetidae newly identified from Oregon (Pseudoscorpionida, Cheliferoidea). *Journal of Arachnology* 10: 97−109.

Bennett, W.H. (1908). Occurrence of *Chernes cyrneus* L. Koch and *C. cimicoides* (Fabricius) in Richmond Park, Surrey. *Hastings and East Sussex Naturalist* 1: 114.

Berendt, G.C. (1830). *Die Insekten im Bernstein, ein Beitrag zur Thiergeschichte der Vorwelt*, vol. 1. Nicolai: Danzig.

Berg, C. (1893). Pseudoscorpionidenkniffe. *Zoologischer Anzeiger* 16: 446−448.

Berger, E. W. (1905). Habits and distribution of Pseudoscorpionidae, principally *Chelanops oblongus*, Say. *Ohio Naturalist* 6: 407−419.

Berger, E. W. (1906). A pseudoscorpion from Guatemala. *Ohio Naturalist* 6: 489−491.

Berland, L. (1925). Note sur un Pseudoscorpionide vivant dans les terriers de Taupe: *Chelifer (Chernes) falcomontanus* Haselhaus. *Bulletin de la Société Entomologique de France* 30: 212−216.

Berland, L. (1932). Les Arachnides (Scorpions, Araignées, etc.), biologie systématique. *Encyclopédie Entomologique* (A) **16**: 1–485.

Berland, L. (1955). Les Arachnides de l'Afrique noire français. *Initiations Africaines, Institut Français d'Afrique Noire* **12**: 1–127.

Bernard, H. M. (1893a). Additional notes on the origin of the tracheae from setiparous glands. *Annals and Magazine of Natural History* (6) **11**: 24–28.

Bernard, H.M. (1893b). Notes on some of the digestive processes in arachnids. *Journal of the Royal Microscopical Society* **4**: 427–443.

Bernard, H. M. (1893c). The stigmata of the Arachnida, as a clue to their ancestry. *Nature, London* **49**: 68–69.

Bernard, H. (1894). Notes on the Chernetidae, with special reference to the vestigial stigmata and to a new form of trachea. *Journal of the Linnean Society of London, Zoology* **24**: 410–430.

Bernard, H. (1896). The comparative morphology of the Galeodidae. *Transactions of the Linnean Society of London, Zoology* (2) **6**: 305–417.

Beron, P. (1968). Etudes sur les Pseudoscorpions. I. Sur les espèces du sous-ordre Cheliferinea en Bulgarie. *Bulletin de l'Institut de Zoologie et Musée, Sofia* **27**: 103–106.

Beron, P. (1972). Aperçu sur la faune cavernicole de la Corse. *Publication Laboratoire Souterrain du C.N.R.S., Moulis* **3**: 1–55.

Beron, P. (1985). On the cave fauna of the Greek islands of Santorin and Iraklia, with preliminary description of a new pseudoscorpion. *Grottes Bulgares* **3**: 64–71.

Beron, P. & Guéorguiev, V. (1967). Essai sur la faune cavernicole de Bulgarie. 2. Résultats des recherches biospéologiques de 1961 à 1965. *Izvestiya na Zoologicheskiya Institut, Sofia* **24**: 151–212. (not seen)

Bertkau, P. (1887a). Über die Chernetiden oder Bücherskorpione. *Verhandlungen des Naturhistorischen Vereins der Preussischen Rheinlande* **44**: 112. (not seen)

Bertkau, P. (1887b). Über den Bau der Chernetiden oder Pseudoskorpione. *Sitzungsberichte der Niederrheinischen Gesellschaft für Naturwissenschaft und Heilkunde zu Bonn* **44**: 112–117. (not seen)

Bertkau, P. (1891). Zur Entwicklungsgeschichte der Pseudoskorpione. *Verhandlungen des Naturhistorischen Vereins der Preussischen Rheinlande* **48**: 45–46. (not seen)

Bertrand, O. (1958). Le peuplement hivernal des écorces de platane. *Annales de la Société d'Horticulture et d'Histoire Naturelle de l'Hérault* **4**: 205–210. (not seen)

Besch, W. (1969). South American Arachnida. In: Fittkau, E. J., Illies, J., Klinge, H., Schwabe, G. H. & Sioli, H. (eds), *Biogeography and Ecology in South America*: 723–740. Junk: The Hague.

Betts, M. M. (1955). The food of titmice in oak woodland. *Journal of Animal Ecology* **24**: 282–323.

Bianchi, C., Caporiacco, L. di, Massera, M. G. & Valle, A. (1949). Raccolte faunistiche della Grotta della Spipola (Bologna). *Commentationes Pontificiae Academia Scientarum* **13**: 493–527.

Bielz, E. A. (1852). Naturhistorische Reiseskizzen. *Verhandlungen und Mitteilungen des Siebenburgischen Vereins für Naturwissenschaften Hermannstadt* **3**: 171–176, 187–192. (not seen)

Bignotti, G. (1909). Elenco dei Pseudoscorpioni trovati in Italia e loro distribuzione geografica. *Atti della Società dei Naturaliste e Matematici, Modena* (4) **11**: 56–76. (not seen)

Bishop, C.P. (1967). The cheliceral flagellum of pseudoscorpions. *Journal of Natural History* **1**: 393–397.

Biström, O. & Väisänen, R. (1988). Ancient-forest invertebrates of the Pyhän-Häkki National Park in central Finland. *Acta Zoologica Fennica* **185**: 1–69.

Bizzi, A. (1960). Richerche sulla fauna cavernicola. In: Allegranzi, A., *Cinque Anni di Attività del Gruppo Grotte G. Trevisiol. Le Alpi Venete* **14:** 58–60. (not seen)

Blanchard, E. (1852–1855). *L'Organisation du Règne Animal. Les Arachnides.* Paris. (not seen)

Blatchley, W.S. (1897). Indiana caves and their fauna. *Annual Report of the Indiana Department of Geology and Natural Resources* **21:** 121–212. (not seen)

Bliss, P. & Lippold, K. (1987). Pseudoskorpione (Arachnida, Pseudoscorpiones) aus dem Hakelwald im Nordharzvorland. *Hercynia, N.F.* **24:** 42–47.

Börner, C. (1902). Arachnologische Studien. (II und III). *Zoologischer Anzeiger* **25:** 433–466.

Börner, C. (1903). Über die Beingliederung der Arthropoden. *Sitzungsberichte der Gesellschaft Naturforschender der Freunde zu Berlin* **7:** 292–341. (not seen)

Börner, C. (1921). Die Gliedmassen der Arthropoden. In: Lang, A., *Handbuch der Morphologie der Wirbellosen Tiere*, vol. **4:** 649–694. Jena. (not seen)

Boissin, L. (1964). Remarques sur la morphologie des adultes et des nymphes d'*Hysterochelifer meridianus* (L. Koch) (Arachnides, Pseudoscorpions, Cheliferidae). *Bulletin de la Société Zoologique de France* **89:** 650–669.

Boissin, L. (1967). Cycle vital d'*Hysterochelifer meridianus* (L. Koch) (Arachnides, Pseudoscorpion, Cheliferidae): note preliminaire. *Revue d'Ecologie et de Biologie du Sol* **4:** 479–487.

Boissin, L. (1970). *Gamétogenèse au Cours du Développement Post-embryonnaire et Biologie de la Reproduction chez* Hysterochelifer meridianus *(L. Koch) (Arachnides, Pseudoscorpions)*. Thèse doctorat: Université Montpellier.

Boissin, L. (1971). Organogenese de la gonade et des annexes genitales males au cours du developpement post-embryonnaire chez le pseudoscorpion *Hysterochelifer meridianus* (L. Kock). *Bulletin Biologique* **105:** 113–123.

Boissin, L. (1973). Biologie sexuelle du pseudoscorpion *Hysterochelifer meridianus* (L. Koch); accouplement et description du spermatophore. *Bulletin de la Société Zoologique de France* **98:** 521–529.

Boissin, L. (1974). Etude ultrastructurale de la spermiogenèse de *Garypus beauvoisi* (Sav.) (Arachnides, Pseudoscorpions). *Archives de Zoologie Expérimentale et Générale* **115:** 169–184. (not seen)

Boissin, L., Bouix, G. & Maurand, J. (1970). Recherches histologiques et histochimiques sur le tractus génital mâle du pseudoscorpion *Hysterochelifer meridianus* (L. Koch). *Bulletin du Muséum National d'Histoire Naturelle, Paris* **42:** 491–501.

Boissin, L. & Cazal, M. (1969). Étude du système nerveux et des glandes endocrines céphaliques de l'adulte femelle d'*Hysterochelifer meridianus* (L. Koch) (Arachnide, Pseudoscorpion, Cheliferidae). *Bulletin de la Société Zoologique de France* **94:** 263–268.

Boissin, L. & Manier, J.-F. (1966a). Spermatogenèse d'*Hysterochelifer meridianus* (L. Koch) (Arachnide, Pseudoscorpion, Cheliferidae). I. Étude caryologique. *Bulletin de la Société Zoologique de France* **91:** 469–476.

Boissin, L. & Manier, J.-F. (1966b). Spermatogenèse d'*Hysterochelifer meridianus* (L. Koch) (Arachnide, Pseudoscorpion, Cheliferidae). II. Étude de l'évolution du chondriome. *Bulletin de la Société Zoologique de France* **91:** 697–706.

Boissin, L. & Manier, J.-F. (1967). Spermatogenèse d'*Hysterochelifer meridianus* (L. Koch) (Arachnide, Pseudoscorpion, Cheliferidae). III. Évolution de l'appareil de Golgi. *Bulletin de la Société Zoologique de France* **92:** 705–712.

Boissin, L. & Manier, J.-F. (1970). Ovogenèse et fécondation chez *Hysterochelifer meridianus* (L. Koch) (Arachnides, Pseudoscorpions, Cheliferidae). *Bulletin du Muséum National d'Histoire Naturelle, Paris* **41** (suppl. 1): 49–53.

Boldori, L. (1927). Contributo alla conoscenza della fauna cavernicola Lombarda. Quattro anni di ricerche nelle caverne Lombarde. *Memorie della Società Entomologica Italiana* **6:** 90–111.

Boldori, L. (1932). Altri quattro anni di ricerche nelle caverne italiane. *Grotte d'Italia* 6: 111–129.

Boldori, L. (1934). Ricerche in caverne italiane. III serie (1932–1933). *Bollettino della Società Entomologica Italiana* 66: 58–61.

Boldori, L. (1935). Animali cavernicoli in schiavitù. II. *Bollettino della Società Entomologica Italiana* 67: 26–30.

Boldori, L. (1946). Case speleologiche. *Natura, Milano* 36: 23–27. (not seen)

Bolivar y Pieltain, C. (1924). Estudios sobre *Obisium* (Pseudosc.) cavernicolas de la región Vasca. *Boletin de la Real Sociedad Española de Historia Natural* 24: 101–104.

Bologna, M. A. & Bonzano, C. (1976). Attivà biospeleologica nel 1976. *Boll. Gruppo Speleol. Imperiese C.A.I.* 6: 66–69. (not seen)

Bologna, M. A. & Taglianti, A. V. (1982). Il popolamento cavernicolo delle Alpi Occidentali. *Lavori della Società Italiana di Biogeografia* 7: 515–544. (not seen)

Bologna, M. A. & Taglianti, A. V. (1985). Fauna cavernicola delle Alpi Liguri. *Annali del Museo Civico di Storia Naturale di Genova* 84: 1–389.

Bolokan, V.I. (1984). Order False-scorpions/Pseudoscorpiones. In: *Animal World of Moldavia. Bryozoa, Mollusca, Arthropoda*: 203–204. Kishinev: 'Stiinsa'.

Bonnet, C. (1773). *Abhandlungen aus der Insektologie*, vols 1–2. Halle. (not seen)

Bonzano, C., (1983). Consideraizioni generali sulla fauna cavernicola delle Alpi Apuane. *Grotte d'Italia* (4) 9: 123–132. (not seen)

Bonzano, C., Grippa, C. & Ramella, L. (1977). La Tana Bertrand sul Monte Faudo (IM). *Riviera dei Fiori* 6: 1–12. (not seen)

Borchert, A. (1974). *Schädigungen der Bienenzucht durch Krankheiten, Vergiftungen und Schädlinge der Honigbiene.* Hirzel: Leipzig. (not seen)

Bornemissza, G. F. (1957). An analysis of arthropod succession in carrion and the effect of its decomposition on the soil fauna. *Australian Journal of Zoology* 5: 1–12.

Boscolo, L. F. (1968). Ricerche faunistiche nel Covolo della Guerra (n. 127 V–VI), Colli Berici. *Rassegna Speleologica Italiana* 20: 155–177. (not seen)

Boscolo, L.F. (1969). Esplorazioni faunistiche nel Grotta di San Gottardo (n. 186 V–VI), Colli Berici. *Rassegna Speleologica Italiana* 21: 18–24. (not seen)

Bourne, J. D. & Cherix, D. (1981). La vie cavernicole dans les grottes du Jura. *Bulletin Romand d'Entomologie* 1: 23–25.

Bouvier, E.A. (1896). Sur la ponte et le développement d'un Pseudoscorpionide, le *Garypus saxicola*, Waterhouse [Arachn.]. *Bulletin de la Société Entomologique de France* 1: 304–307, 342–343.

Bowman, C. E. (1979). Unusual feeding in a pseudoscorpion. *Newsletter of the British Arachnological Society* 23: 11–12.

Brach, V. (1978). Social behavior in the pseudoscorpion *Paratemnus elongatus* (Banks) (Pseudoscorpionida: Atemnidae). *Insectes Sociaux* 25: 3–11.

Brach, V. (1979). Species diversity and distributional relationships of pseudoscorpions from slash pine (*Pinus elliottii* Eng.) in Florida (Arachnida, Pseudoscorpionida). *Bulletin of the Southern California Academy of Sciences* 78: 32–39.

Braun, M. & Beck, L. (1986). Zur Biologie eines Buchenwaldbodens 9. Die Pseudo-skorpione. *Carolinea* 44: 139–148.

Brébisson, J. (1827). Catalogue des Arachnides, des Myriapodes et des Insectes-Aptères que l'on trouve dans le département du Calvados. *Mémoires de la Société Linnéenne de Normandie* 3: 254–274.

Brian, A. (1914). Elenco di animali cavernicoli delle grotte situate in vicinanza di Genova. *Monitore Zoologico Italiano* 25: 8–12. (not seen)

Brian, A. (1940). Le grotte di Toirano (Liguria). *Annali del Museo Civico di Storia Naturale di Genova* 60: 379–437.

Brimley, C. S. (1938). *The Insects of North Carolina*. Department of Agriculture: Raleigh. (not seen)

Bristowe, W. S. (1931a). Notes on the biology of spiders. IV. Further notes on aquatic spiders, with a description of a new species of pseudoscorpion from Singapore. *Annals and Magazine of Natural History* (10) **8**: 457–465.

Bristowe, W. S. (1931b). A preliminary note on the spiders of Krakatau. *Proceedings of the Zoological Society of London* **1931**: 1387–1400.

Bristowe, W. S. (1934). The spiders of Ramsey and Bardsey Islands, with notes on others from Montgomery and Merioneth. *Proceedings of the Zoological Society of London* **1934**: 1–9.

Bristowe, W. S. (1941). *The Comity of Spiders*, vol. 2. Ray Society: London. (not seen)

Bristowe, W.S. (1952). The arachnid fauna of the Batu Caves in Malaya. *Annals and Magazine of Natural History* (12) *5:* 697–707.

Britten, H. (1912). The arachnids of Cumberland. *Transactions of the Carlisle Natural History Society* **2**: 30–65. (not seen)

Browning, E. (1956). On a collection of Arachnida and Myriapoda from Jersey, Channel Islands, with a check list of the Araneae. *Bulletin Annuel de la Société Jersiaise* **16**: 377–394. (not seen)

Brünnich, ? (1914). Der Bücherskorpion. *Schweizerische Bienen-Zeitung* **37**: 483.

Bruntz, L. (1903). Contribution à l'étude de l'excrétion chez les Arthropodes. Arachnida. *Archives de Biologie* **20**: 359–395.

Burr, A. (1919). Epizoisme des Pseudoscorpions. *Bulletin de l'Association Philomatique d'Alsace et de Lorraine* **6**: 24–26. (not seen)

Butterfield, W. R. (1908). A preliminary list of the false-scorpions (Chernetidea) of the Hastings district. *Hastings and East Sussex Naturalist* **1**: 111–114.

Butterfield, W. R. (1909). Occurrence of the false-scorpion *Chernes cimicoides* (Fabr.) in Dallington Forest. *Hastings and East Sussex Naturalist* **1**: 195–196. (not seen)

Callaini, G. (1979a). Note preliminari sugli pseudoscorpioni della Sardegna: *Roncus dallaii*, nuova specie della Sardegna meridionale (Notulae Chernetologicae, I). *Redia* **62**: 111–119.

Callaini, G. (1979b). Osservazioni su alcuni pseudoscorpioni delle Isole Eolie (Notulae Chernetologicae II). *Redia* **62**: 129–145.

Callaini, G. (1979c). Notulae Chernetologicae. III. Gli pseudoscorpioni della Farma (Arachnida). *Redia* **62**: 339–354.

Callaini, G. (1980a). Considerazioni sugli pseudoscorpioni dell'altopiano del Cansiglio (Notulae Chernetologicae. IV). *Animalia, Catania* **6**: 219–241.

Callaini, G. (1980b). Notulae Chernetologicae VII. Un nuovo *Chthonius* dell'Italia settentrionale. *Redia* **63**: 203–214.

Callaini, G. (1981a). Notulae Chernetologicae V. Il sottogenere *Ephippiochthonius* in Corsica (Arachnida, Pseudoscorpionida, Chthoniidae). *Annali del Museo Civico di Storia Naturale di Genova* **83**: 307–323.

Callaini, G. (1981b). Notulae Chernetologicae VI. Una nuova specie di Neobisiidae delle Alpi Apuane (Arachnida, Pseudoscorpionida). *Fragmenta Entomologica* **16**: 9–17.

Callaini, G. (1981c). Notulae Chernetologicae VIII. *Neoccitanobisium ligusticum* n. gen. n. sp. della Liguria occidentale (Arachn. Pseudoscorp. Neobisiidae). *Annali del Museo Civico di Storia Naturale di Genova* **83**: 523–538.

Callaini, G. (1981d). L'ultrastruttura dell'occhio di *Neobisium muscorum* Leach (Aracnida, Pseudoscorpionida, Neobisiidae). *Redia* **64**: 217–228.

Callaini, G. (1982a). Etude comparative de la membrane pleurale des pseudoscorpions au microscope électronique à balayage. *Memorie, Atti della Società Toscana di Scienze Naturali residente in Pisa* B **88** supplemento: 16–26.

Callaini, G. (1982b). Due nuovi *Neobisium* della Corsica (Arachnida, Pseudoscorpionida). Notulae Chernetologicae X. *Bollettino del Museo Civico di Storia Naturale, Verona* **9**: 449–459.

Callaini, G. (1983a). Pseudoscorpioni dell'isola di Montecristo (Arachnida). Notulae Chernetologicae IX. *Redia* **66**: 147–165.

Callaini, G. (1983b). Notulae Chernetologicae XI. Il sottogenere *Ephippiochthonius* in Sardegna (Arachnida, Pseudoscorpionida, Chthoniidae). *Annali del Museo Civico di Storia Naturale di Genova* **84**: 401–423.

Callaini, G. (1983c). Notulae Chernetologicae XII. Nuovi reperti sugli pseudoscorpioni della Sardegna. *Lavori della Società Italiana di Biogeografia* **8**: 279–322.

Callaini, G. (1983d). Osservazioni sulla fauna Chernetologica di alcuni Rilievi Abruzzesi. Notulae Chernetologicae XIII. *Bollettino del Museo Civico di Storia Naturale, Verona* **10**: 221–239.

Callaini, G. (1983e). Contributo alla conoscenza degli Pseudoscorpioni d'Algeria (Arachnida). Notulae Chernetologicae XVI. *Animalia, Catania* **10**: 211–235.

Callaini, G. (1983f). L'ultrastruttara delle ghiandole genitali accessorie di *Neobisium muscorum* (Leach) (Arachnida, Pseudoscorpionida). *Redia* **66**: 375–388.

Callaini, G. (1984). Osservazioni su alcune specie di *Chthonius* del sottogenere *Ephippiochthonius* Beier (Arachnida, Pseudoscorpionida, Chthoniidae). Notulae Chernetologicae XVII. *Annali del Museo Civico di Storia Naturale di Genova* **85**: 125–159.

Callaini, G. (1985). Speleobiologica della Somalia. *Cryptocheiridium somalicum* n. sp. (Arachnida Pseudoscorpionida) delle grotte di Mugdile e Showli Berdi. *Monitore Zoologico Italiano, n.s. Supplemento* **20**: 181–189.

Callaini, G. (1986a). *Mesochelifer insignis* une nouvelle espèce de l'Algérie septentrionale (Arachnida, Pseudoscorpionida, Cheliferidae). Notulae Chernetologicae XX. *Revue Arachnologique* **7**: 1–8.

Callaini, G. (1986b). Osservazioni su alcune specie italiane del genere *Acanthocreagris* Mahnert. Notulae Chernetologicae XIV. *Bollettino del Museo Civico di Storia Naturale, Verona* **11**: 349–377.

Callaini, G. (1986c). Appunti su alcune specie italiane della famiglia Chernetidae Menge (Arachnida, Pseudoscorpionida). Notulae Chernetologicae XV. *Bollettino del Museo Civico di Storia Naturale, Verona* **11**: 379–401.

Callaini, G. (1986d). Note sugli pseudoscorpioni raccolti in alcune grotte della Toscana settentrionale (Arachnida) (Notulae Chernetologicae XXV). *Redia* **69**: 523–542.

Callaini, G. (1986e). Pseudoscorpioni dell'Italia settentrionale nel Museo Civico di Storia Naturale di Verona (Arachnida). Notulae Chernetologicae XIX. *Bollettino del Museo Civico di Storia Naturale, Verona* **12**: 229–255.

Callaini, G. (1987a). Pseudoscorpioni della Grotta di Trecchina (Italia meridionale) (Notulae Chernetologicae. XX). *Bollettino del Museo Civico di Storia Naturale, Verona* **13**: 69–79.

Callaini, G. (1987b). Su alcune specie di Cheliferidae della regione Mediterranea (Arachnida, Pseudoscorpionida). (Notulae Chernetologicae. XXII). *Bollettino del Museo Civico di Storia Naturale, Verona* **13**: 273–294.

Callaini, G. (1988a). Deux nouveaux *Neochthonius* de la région méditerranéenne occidentale (Arachnida, Pseudoscorpionida, Chthoniidae). (Notulae Chernetologicae XXIV). *Revue Arachnologique* **7**: 175–184.

Callaini, G. (1988b). Gli Pseudoscorpioni del Marocco. (Notulae Chernetologicae, XXVII). *Annali del Museo Civico di Storia Naturale di Genova* **87**: 31–66.

Callaini, G. (1989). Il popolamento delle isole Egadi. Un esempio dell'interesse biogeografico degli Pseudoscorpioni (Arachnida). (Notulae Chernetologicae, XXIX). *Annali del Museo Civico di Storia Naturale di Genova* **87**: 137–148.

Callaini, G. & Dallai, R. (1984). Spermatozoïdes et phylogenèse chez les Garypides (Arachnida, Pseudoscorpions). *Revue Arachnologique* **4**: 335–342.

Callaini, G. & Dallai, R. (1989). Les spermatozoïdes des Pseudoscorpions: étude comparative et considérations phylogénétiques. *Revue Arachnologique* **8**: 85–97.

Cambridge, F. O. Pickard- (1901). Arachnida. Araneida. In: Godman, F. D. & Salvin, O. (eds), *Biologia Centralia-Americana.* (not seen)

Cambridge, F. O. Pickard- (1902). In: Page, W. (ed.), *Victorian History of the County of Hertford.* (not seen)

Cambridge, O. Pickard- (1875). In: *Encyclopaedia Brittanica*, 11th edition, vol. 2. (not seen)

Cambridge, O. Pickard- (1885). Pseudoscorpions new to Britain. *Naturalist, London* **10**: 103. (not seen)

Cambridge, O. Pickard- (1886). A contribution towards the knowledge of the Arachnida of Epping Forest. *Transactions of the Essex Field Club* **4**: 41–45. (not seen)

Cambridge, O. Pickard- (1889). On new and rare British spiders. *Proceedings of the Dorset Natural History and Antiquarian Field Club and Archaeological Society* **10**: 107–138. (not seen)

Cambridge, O. Pickard- (1892). On the British species of false-scorpions. *Proceedings of the Dorset Natural History and Antiquarian Field Club and Archaeological Society* **13**: 199–231.

Cambridge, O. Pickard- (1905). On new and rare British Arachnida. *Proceedings of the Dorset Natural History and Antiquarian Field Club and Archaeological Society* **26**: 40–74. (not seen)

Cambridge, O. Pickard- (1906). In: Dyer, W. T. T., *The Wild Fauna and Flora of the Royal Botanic Gardens, Kew. Kew Bulletin, Add. Ser.* **4**: 53–65. (not seen)

Cambridge, O. Pickard- (1907). On new and rare British Arachnida. *Proceedings of the Dorset Natural History and Antiquarian Field Club and Archaeological Society* **28**: 121–148. (not seen)

Cambridge, O. Pickard- (1908). On new and rare British Arachnida noted and observed in 1907. *Proceedings of the Dorset Natural History and Antiquarian Field Club and Archaeological Society* **29**: 161–194. (not seen)

Canestrini, G. (1874). Osservazioni aracnologiche. *Atti della Accademia Scientifica Veneto-Trentino-Istriana, Padova* **3**: 206–232.

Canestrini, G. (1875a). Intorno ai Chernetidi ed Opilionidi della Calabria. *Atti della Accademia Scientifica Veneto-Trentino-Istriana, Padova* **4**: 1–12. (not seen)

Canestrini, G. (1875b). Intorno alla fauna del Trentino, notize bibliogr. *Atti della Accademia Scientifica Veneto-Trentino-Istriana, Padova* **4**: 13–35. (not seen)

Canestrini, J. (1883). Chernetidi Italiani. In: Berlese, A., *Acari, Myriapoda et Scorpiones Hucusque in Italia Reperta*, fascicolo 7. A. Berlese: Padova.

Canestrini, J. (1884). Chernetidi Italiani. In: Berlese, A., *Acari, Myriapoda et Scorpiones Hucusque in Italia Reperta*, fascicolo 10. A. Berlese: Padova.

Canestrini, J. (1885). Chernetidi Italiani. In: Berlese, A., *Acari, Myriapoda et Scorpiones Hucusque in Italia Reperta*, fascicolo 19. A. Berlese: Padova.

Cantoni, E. (1881). Aracnidi delle Madonie (Sicilie). *Bollettino della Società Entomologica Italiana* **13**: 278–289. (not seen)

Cantoni, E. (1882). Escursione in Calabria. Chernetidi ed Opilionidi. *Bollettino della Società Entomologica Italiana* **14**: 191–203. (not seen)

Caplin, G. (1974). *General Physiology of Some British Pseudoscorpions in Relation to Feeding and Environmental Factors.* Ph.D. Thesis: University of Manchester. (not seen)

Capolongo, D. (1969). Studio ecologico delle cantine del Napoletano (Primo contributo). *Bollettino della Società Entomologica Italiana* **99/101**: 193–205.

Caporiacco, L. di (1923). Una nuova specie di Chernetide italiano. *Bollettino della Società Entomologica Italiana* **55**: 131–134.

Caporiacco, L. di (1925). Una nuova specie di Chernetide dei dintorni di Firenze. *Bollettino della Società Entomologica Italiana* **57**: 123–124.

Caporiacco, L. di (1928a). Aracnidi di Giarabub e di Porto Bardia. *Annali del Museo Civico di Storia Naturale di Genova* **53**: 77–107.

Caporiacco, L. di (1928b). Aracnidi della Capraja. *Bollettino della Società Entomologica Italiana* **60**: 124–127.

Caporiacco, L. di (1929). Aracnidi delle Canarie. *Memorie della Società Entomologica Italiana* **6**: 240–241.

Caporiacco, L. di (1935). Aracnidi dell'Himalaia e de Karakoram. *Memorie della Società Entomologica Italiana* **13**: 113–160, 161–263.

Caporiacco, L. di (1936a). Aracnidi raccolti durante le primavera 1933 nelle oasi del deserto Libico. *Memorie della Società Entomologica Italiana* **15**: 93–122.

Caporiacco, L. di (1936b). Saggio sulla fauna aracnologica del Casentino, Val d'Arno Superiore e Alta Val Tiberina. *Festschrift zum 60. Geburtstage von Professor Dr. Embrik Strand, Riga*, vol. **1**: 326–369. (not seen)

Caporiacco, L. di (1936c). Aracnidi cavernicoli della provincia di Verona. *Grotte d'Italia* **2**: 85–92.

Caporiacco, L. di (1937a). Scorpioni, Pedipalpi, Solifugi e Chernetidi di Somalia e Dancalia. *Annali del Museo Civico di Storia Naturale di Genova* **58**: 135–149.

Caporiacco, L. di (1937b). Aracnidi cavernicoli e lucifugi di Postumia. *Grotte d'Italia* (2) **2**: 1–8.

Caporiacco, L. di (1939a). Aracnidi di Mogadiscio. *Memorie della Società Entomologica Italiana* **17**: 115–117.

Caporiacco, L. di (1939b). Arachnida. *Missione Biologica nel Paese dei Borana, Raccolte Zoologiche*, vol. **3**: 303–385. Reale Accademia d'Italia: Rome.

Caporiacco, L. di (1940). Arachniden aus der Provinz Verona (Norditalien). *Folia Zoologica et Hydrobiologica* **10**: 1–37.

Caporiacco, L. di (1941). Arachnida (esc. Acarina). *Missione Biologica Sagan-Omo, Roma, Zoologie* **6**: 21–175.

Caporiacco, L. di (1947a). Arachnida africae orientalis, a dominibus Kittenberger, Kovács et Bornemisza lecta, in Museo Nationali Hungarico servata. *Annales Historico-Naturales Musei Nationalis Hungarici* **40**: 97–257.

Caporiacco, L. di (1947b). Diagnosi preliminari di specie nuove di Aracnidi della Guiana Britannica raccolte dai Professori Beccari e Romiti. *Monitore Zoologico Italiano* **56**: 20–34.

Caporiacco, L. di (1947c). Alcuni Arachnidi cavernicoli di Toscana. *Commentationes Pontificiae Academia Scientarum* **11**: 251–258.

Caporiacco, L. di (1947d). Seconda nota su Aracnidi cavernicoli Veronesi. *Memorie del Museo Civico di Storia Naturale di Verona* **1**: 131–140.

Caporiacco, L. di (1948a). Arachnida of British Guiana collected in 1931 and 1936 Professors Beccari and Romiti. *Proceedings of the Zoological Society of London* **118**: 607–747.

Caporiacco, L. di (1948b). L'aracnofauna di Rodi. *Redia* **33**: 27–75.

Caporiacco, L. di (1948c). L'aracnofauna della Romagna in base alle raccolte Zangheri. *Redia* **34**: 237–288.

Caporiacco, L. di (1948d). *Troglohyphantes zorzii* nuova specie cavernicola veronese e notizie su altri ragni cavernicoli veronesi. *Memorie del Museo Civico di Storia Naturale di Verona* **1**: 237–239.

Caporiacco, L. di (1949a). Aracnidi della Colonia del Kenya raccolti da Toschi e Meneghetti negli anni 1944–1946. *Commentationes Pontificiae Academia Scientarum* **13**: 309–492.

Caporiacco, L. di (1949b). Alcuni Aracnidi Albanesi. *Atti del Museo Civico di Storia Naturel di Trieste* **17**: 122–125.

Caporiacco, L. di (1949c). Una piccola raccolta aracnologica dei monti di Calabria. *Atti del Museo Civico di Storia Naturel di Trieste* **17**: 132–136.

Caporiacco, L. di (1949d). Aracnidi delle Venezia Giulia. *Atti del Museo Civico di Storia Naturel di Trieste* **17**: 137–151.

Caporiacco, L. di (1950). Gli Aracnidi della laguna di Venezia. II Nota. *Bollettino della Società Veneziana di Storia Naturale e del Museo Civico di Storia Naturale* **5**: 114–140.

Caporiacco, L. di (1951a). Una raccolta di Aracnidi Umbri. *Annali del Museo Civico di Storia Naturale di Genova* **64**: 62–84.

Caporiacco, L. di (1951b). Aracnidi cavernicoli Liguri. *Annali del Museo Civico di Storia Naturale di Genova* **64**: 101–110.

Caporiacco, L. di (1951c). Seconda nota su Aracnidi cavernicoli Pugliesi. *Memorie del Museo Civico di Storia Naturale di Verona* **2**: 1–5.

Caporiacco, L. di (1951d). Studi sugli Aracnidi del Venezuela raccolti dalla Sezione di Biologia (Universitá Centrale del Venezuela). 1 Parte: Scorpiones, Opiliones, Solifuga y Chernetes. *Acta Biologica Venezuelica* **1**: 1–46.

Caporiacco, L. di (1951e). Aracnidi Pugliesi raccolti dai Signori Conci, Giordini-Soika, Gridelli, Ruffo e dall'autore. *Memorie di Biogeografica Adriatica* **2**: 63–94.

Caporiacco, L. di (1951f). Aracnidi cavernicoli Pugliesi. *Memorie di Biogeografica Adriatica* **2**: 95–101.

Caporiacco, L. di (1952a). Aracnidi cavernicola del Trentino. *Bollettino dei Musei e degli Istituti Biologici della Università di Genova* **24**: 55–62.

Caporiacco, L. di (1952b). Aracnidi della Grotta di Agnano (Ostuni, Puglie). *Bollettino dei Musei e degli Istituti Biologici dell'Università di Genova* **24**: 63–65.

Carrière, J. (1886). Kurze Mittheilungen aus fortgesetzten Untersuchungen über die Sehorgane. *Zoologischer Anzeiger* **9**: 496–500. (not seen)

Caron, D. M. (1978). Arachnids: Araneida and Pseudoscorpionida (spiders and pseudoscorpions). In: Morse, R.A. (ed.), *Honey Bee Pests, Predators, and Diseases*: 186–196. Cornell University Press: London.

Carpenter, G. H. (1896). ['*Chernes phaleratus* (Sim.)' in Ireland]. *Irish Naturalist* **5**: 215. (not seen)

Carpenter, G. H. (1902). Diptera, Neuroptera, Thysanura, Arachnida and Pantop. In: *Guide to Belfast and the Counties of Down and Arntrim*. Belfast. (not seen)

Carpenter, G. H. & Evans, W. (1897). A list of Phalangidea (harvestmen) and Chernetidea (false scorpions) collected in the neighbourhood of Edinburgh. *Proceedings of the Royal Physical Society of Edinburgh* **13**: 114–122. (not seen)

Carpenter, G. H. & Evans, W. (1899). Additional records of spiders and other arachnids from the Edinburgh district (second instalment). *Proceedings of the Royal Physical Society of Edinburgh* **14**: 168–181. (not seen)

Carr, J. W. (1906). New Nottinghamshire spiders and false-scorpions. *54th Report and Transactions Nottingham Naturalists' Society for 1905–1906*: 47–48.

Carrara, F.D. (1846). *La Dalmatia Descritta*. Zara. (not seen)

Carter, H. (1963). Pseudoscorpiones of the Henley area. *Reading Naturalist* **15**: 29. (not seen)

Carus, J. V. (1872). *Geschichte der Zoologie bis Müller und Darwin*. München. (not seen)

Caruso, D. & Costa, G. (1979). Ricerche faunistiche ed ecologiche sulle grotte di Sicilia. VI. Fauna cavernicola di Sicilia (Catalogo ragionato). *Animalia, Catania* **5**: 423–513. (not seen)

Casale, A. (1972). Vision d'insieme del complesso ecologico e faunistico della grotta del Bue Marino (Cala Gonone, Dorgali, NU). *Boll. Soc. Sarda Sci. Nat.* **10:** 111–136. (not seen)

Cecconi, G. (1908). Contributo alla fauna delle Isole Tremiti. *Bollettino dei Musei di Zoologia e di Anatomia Comparata della R. Università di Torino* **23**(583): 1–53. (not seen)

Cederhielm, J. (1798). *Faunae Ingricae Prodomus Exhibens Methodicam Descriptionem Insectorum Agri Petropolensis.* Lipsiae. (not seen)

Cekalovic K., T. (1976). Catalogo de los Arachnida: Scorpiones, Pseudoscorpiones, Opiliones, Acari, Araneae y Solifugae de la XII region de Chile, Magallenes incluyendo la antarctica chilena (Chile). *Gayana Zoologia* **37:** 3–108.

Cekalovic K., T. (1984). Catalogo de los Pseudoscorpiones y Palpigradi de Chile (Chelicerata). *Boletín de la Sociedad de Biología de Concepción* **55:** 7–35.

Cerruti, M. (1959). Aggiunta al i elenco della fauna cavernicola del Lazio e delle regioni limitrofe (Toscana esclusa). *Fragmenta Entomologica* **3:** 49–63.

Cerruti, M. (1968). Materiali per un primo elenco degli Artropodi speleobii della Sardegna. *Fragmenta Entomologica* **5:** 207–257.

Chamberlin, J. C. (1921). Notes on the genus *Garypus* in North America (Pseudoscorpionida - Cheliferidae). *Canadian Entomologist* **53:** 186–191.

Chamberlin, J. C. (1923a). The genus *Pseudogarypus* Ellingsen (Pseudoscorpionida – Feaellidae). *Entomological News* **34:** 146–149, 161–166.

Chamberlin, J. C. (1923b). On two species of pseudoscorpion from Chile with a note on one from Sumatra. *Revista Chilena de Historia Natural* **27:** 185–192.

Chamberlin, J. C. (1923c). New and little known pseudoscorpions, principally from the islands and adjacent shores of the Gulf of California. *Proceedings of the California Academy of Sciences* (4) **12:** 353–387.

Chamberlin, J. C. (1924a). The Cheiridiinae of North America (Arachnida – Pseudoscorpionida). *Pan-Pacific Entomologist* **1:** 32–40.

Chamberlin, J. C. (1924b). *Hesperochernes laurae*, a new species of false scorpion from California inhabiting the nest of *Vespa*. *Pan-Pacific Entomologist* **1:** 89–92.

Chamberlin, J. C. (1924c). Preliminary note upon the pseudoscorpions as a venomous order of the Arachnida. *Entomological News* **35:** 205–209.

Chamberlin, J. C. (1924d). Giant *Garypus* of the Gulf of California. *Nature Magazine* **2:** 171–172, 175.

Chamberlin, J. C. (1925a). On a collection of pseudoscorpions from the stomach contents of toads. *University of California Publications in Entomology* **3:** 327–332.

Chamberlin, J. C. (1925b). Notes on the status of genera in the Chelonethid family Chthoniidae together with a description of a new genus and species from New Zealand. *Videnskabelige Meddelelser fra Dansk Naturhistorisk Forening i Kjøbenhavn 81:* 333–338.

Chamberlin, J. C. (1929a). A synoptic classification of the false scorpions or chelaspinners, with a report on a cosmopolitan collection of the same. Part 1. The Heterosphyronida (Chthoniidae) (Arachnida-Chelonethida). *Annals and Magazine of Natural History* (10) **4:** 50–80.

Chamberlin, J. C. (1929b). *Dinocheirus tenoch*, an hitherto undescribed genus and species of false scorpion from Mexico. *Pan-Pacific Entomologist* **5:** 171–173.

Chamberlin, J. C. (1929c). *Dasychernes inquilinus* from the nest of meliponine bees in Columbia. *Entomological News* **40:** 49–51.

Chamberlin, J. C. (1929e). On some false scorpions of the suborder Heterosphyronida (Arachnida – Chelonethida). *Canadian Entomologist* **61:** 152–155.

Chamberlin, J. C. (1929f). The genus *Pseudochthonius* Balzan (Arachnida – Chelonethida). *Bulletin de la Société Zoologique de France* **54:** 173–179.

Chamberlin, J. C. (1930). A synoptic classification of the false scorpions or chela-spinners, with a report on a cosmopolitan collection of the same. Part II. The Diplosphyronida (Arachnida-Chelonethida). *Annals and Magazine of Natural History* (10) **5**: 1–48, 585–620.

Chamberlin, J. C. (1931a). The arachnid order Chelonethida. *Stanford University Publications, Biological Sciences* **7**(1): 1–284.

Chamberlin, J. C. (1931b). *Parachernes ronnaii*, a new genus and species of false scorpion from Brazil (Arachnida - Chelonethida). *Entomological News* **42**: 192–195.

Chamberlin, J. C. (1931c). A synoptic revision of the generic classification of the chelonethid family Cheliferidae Simon (Arachnida). *Canadian Entomologist* **63**: 289–294.

Chamberlin, J. C. (1932a). A synoptic revision of the generic classification of the chelonethid family Cheliferidae Simon (Arachnida) (continued). *Canadian Entomologist* **64**: 17–21, 35–39.

Chamberlin, J. C. (1932b). On some false scorpions of the superfamily Cheiridioidea (Arachnida – Chelonethida). *Pan-Pacific Entomologist* **8**: 137–144.

Chamberlin, J. C. (1933). Some false scorpions of the atemnid subfamily Miratemninae (Arachnida – Chelonethida). *Annals of the Entomological Society of America* **26**: 262–268.

Chamberlin, J. C. (1934a). On two species of false scorpions collected by birds in Montana, with notes on the genus *Dinocheirus*. *Pan-Pacific Entomologist* **10**: 125–132.

Chamberlin, J. C. (1934b). Check list of the false scorpions of Oceania. *Occasional Papers of the Bernice P. Bishop Museum* **10**(22): 1–14.

Chamberlin, J. C. (1935a). A new species of false scorpion (*Hesperochernes*) from a bird's nest in Montana. *Pan-Pacific Entomologist* **11**: 37–39.

Chamberlin, J. C. (1935b). Chelonethida. In: Pratt, H.S., *A Manual of the Common Invertebrate Animals (Exclusive of Insects)*: 477–481. Blakiston: Philadelphia.

Chamberlin, J. C. (1938a). New and little-known false-scorpions from the Pacific and elsewhere. *Annals and Magazine of Natural History* (11) **2**: 259–285.

Chamberlin, J. C. (1938b). A new genus and three new species of false scorpion from Yucatan Caves. *Publications of the Carnegie Institution of Washington* no. **491**: 109–121.

Chamberlin, J. C. (1939a). Tahitian and other records of *Haplochernes funafutensis* (With) (Arachnida: Chelonethida). *Bulletin of the Bernice P. Bishop Museum* **142**: 203–205.

Chamberlin, J. C. (1939b). New and little-known false scorpions from the Marquesas Islands. *Bulletin of the Bernice P. Bishop Museum* **142**: 207–215.

Chamberlin, J. C. (1943). The taxonomy of the false scorpion genus *Synsphyronus*, with remarks of the sporadic loss of stability in generally constant morphological characters. *Annals of the Entomological Society of America* **36**: 486–500.

Chamberlin, J. C. (1946). The genera and species of the Hyidae: a family of the arachnid order Chelonethida. *Bulletin of the University of Utah, Biological Series* **37**: 1–16.

Chamberlin, J. C. (1947a). The Vachoniidae – a new family of false scorpions represented by two new species from caves in Yucatan. *Bulletin of the University of Utah, Biological Series* **38**: 1–15.

Chamberlin, J. C. (1947b). Three new species of false scorpions from the islands of Guam. *Occasional Papers of the Bernice P. Bishop Museum* **18** (20): 305–316.

Chamberlin, J. C. (1949). New and little-known false scorpions from various parts of the world (Arachnida, Chelonethida), with notes on structural abnormalities in two species. *American Museum Novitates* **1430**: 1–57.

Chamberlin, J.C. (1952). New and little-known false scorpions (Arachnida, Chelonethida) from Monterey County, California. *Bulletin of the American Museum of Natural History* **99**: 259–312.

Chamberlin, J. C. (1962). New and little-known false scorpions, principally from caves, belonging to the families Chthoniidae and Neobisiidae (Arachnida, Chelonethida). *Bulletin of the American Museum of Natural History* **123**: 303–352.

Chamberlin, J. C. & Chamberlin, R. V. (1945). The genera and species of the Tridenchthoniidae (Dithidae). A family of the arachnid order Chelonethida. *Bulletin of the University of Utah, Biological Series* **35**: 1–67.

Chamberlin, J. C. & Malcolm, D. R. (1960). The occurrence of false scorpions in caves with special reference to cavernicolous adaptation and to cave species in the North American fauna. (Arachnida - Chelonethida). *American Midland Naturalist* **64**: 105–115.

Chamberlin, R. V. (1925). Diagnoses of new American Arachnida. *Bulletin of the Museum of Comparative Zoology* **67**: 211–248.

Chandrashekhar, S., Murthy, V. A. & Suryanarayanan, T. S. (1988). Behavioural changes in the pseudoscorpion *Oratemnus indicus* due to *Penicillium citrinum* infection. *Comparative Physiology and Ecology* **13**: 145–148. (not seen)

Chapman, P. (1984). The invertebrate fauna of the caves of Gunung Mulu National Park. *Sarawak Museum Journal* **30**(51): 1–18.

Chapman, P. (1984b). Invertebrates at Niah Great Cave, Butu, Niah National Park, Sarawak. *Cave Science* **11**: 89–91. (not seen)

Christian, E. (1986). Az ausztriai barlangok kisallatvilaga. *Karszt és Barlang, Budapest* **II (1984)**: 121–122. (not seen)

Christian, E. & Potočnik, F. (1985). Ein Beitrag zur Kenntnis der Höhlenfauna der Insel Krk. *Biološki Věstnik* **33**: 13–20.

Cîrdei, F. (1948). Beiträge zur Fauna der Pseudoscorpioniden aus dem Urwald 'Slatioara' Bezirk Cimpulung (Bucovina). *Revista Stiintifica 'V. Adamachi'* **4**: 293–295. (not seen)

Cîrdei, F., Bulimar, F. & Malcoci, E. (1967). Contribuţii la studiul pseudoscorpionidelor (ord. Pseudoscorpionidea) din Moldova (Masivul Repedea). *Anale Stiintifice, Universitatii 'Al I Cuza' (Series Noua) (2) Biol.* **13**: 237–242.

Cîrdei, F., Bulimar, F. & Malcoci, E. (1970). Contribuţii la studiul pseudoscorpionidelor (ord. Pseudoscorpionidea) din Carpaţii orientali (Rarău). *Comunicări de Zoologie, Bucaresti* **9**: 7–16.

Cîrdei, F. & Guţu, E. (1959). Contribuţii la cunoaşterea faunei pseudoscorpionidelor din Moldova şi Maramureş. *Studii şi Cercetari de Biologie, Academia Republicii Populare Romane Filiala (Cluj)* **10**(1): 1–11.

Clarke, L. L. (1858). *A Descriptive Catalogue of the Most Instructive and Beautiful Objects for the Microscope.* Routledge and Co.: London. (not seen)

Clements, D.K. (1987). A case of phoresy in the pseudoscorpion *Lamprochernes nodosus* (Schrank). *Entomologist's Monthly Magazine* **123**: 222. (not seen)

Clerck, C. (1757). *Svenska spindlar. Aranei Suecici descr. et figuris illustr.* Stockholm. (not seen)

Cloudsley-Thompson, J. L. (1956a). Notes on Arachnida, 25. – An unusual case of phoresy by false-scorpions. *Entomologist's Monthly Magazine* **92**: 71.

Cloudsley-Thompson, J. L. (1956b). Notes on Arachnida, 28. – Biological observations and records. *Entomologist's Monthly Magazine* **92**: 193. (not seen)

Cloudsley-Thompson, J. L. (1958). The behaviour of false-scorpions. *Naturalist, London* **866**: 96.

Cloudsley-Thompson, J. L. (1959). Notes on Arachnida, 33. - The fauna of Spaunton Moor, Yorkshire, and of the Galloway Hills. *Entomologist's Monthly Magazine* **95**: 179.

Cloudsley-Thompson, J. L. (1960). Some aspects of the fauna of the coastal dunes of the Bay of Biscay. *Entomologist's Monthly Magazine* **96**: 49–53.

Cloudsley-Thompson, J. L. & Chadwick, M. J. (1964). *Life in Deserts.* Foulis: London. (not seen)

Cockerell, T. D. A. (1907). Some Coleoptera and Arachnida from Florissant, Colorado. *Bulletin of the American Museum of Natural History* **23**: 617–621.

Cockerell, T. D. A. (1917). Arthropods in Burmese amber. *American Journal of Science* (4) **44**: 360–368.

Cockerell, T. D. A. (1920). Fossil arthropods in the British Museum. I. *Annals and Magazine of Natural History* (9) **5**: 273–279.

Cockerell, T. D. A. (1940). The insects of the Californian islands. *Proceedings of the 6th Pacific Science Congress* **4**: 283–295. (not seen)

Comotti, G. (1984). Nuovi dati per une fauna cavernicola bergamasca. *Rivista Mus. Sci. Nat. Bergamo* **6**: 75–94. (not seen)

Comotti, G. (1986). Appunti sulla fauna di alcune cavità Lombarde. *Rivista Mus. Civ. Sci. Nat. 'E. Caffi', Bergamo* **10**: 61–71. (not seen)

Comstock, J. H. (1913). *The Spider Book*. Doubleday: New York.

Conci, C. (1949). Ricerche speleologiche sulla catena dello Zugna. *Atti della I.R. Accademia Roveretana di Scienze, Lettere ed Arti degli Agiati* (4) **17**: 109–127. (not seen)

Conci, C. (1951). Contributo alla conoscenza della speleofauna della Venezia Tridentina. *Memorie della Società Entomologica Italiana* **30**: 5–76.

Conci, C. (1952). Le Arene Candide n. 34 Li. *Doriana, Genova* **1**(24): 1–12.

Conci, C. & Franceschi, T. (1953). Le grotte di Pignone e la loro fauna (La Spezia). *Rassegna Speleologica Italiana* **5**: 43–49. (not seen)

Conci, C. & Galvagni, A. (1956). La Grotta G.B. Trener n. 244 V.T. in Valsugana (o Gro ta del Calgeron). *Memorie del Museo di Storia Naturale della Venezia Tridentina* **11**: 4–23. (not seen)

Cooke, M. C. (1867). On pseudoscorpions. *Journal of the Quekett Microscopical Club* (1) **1**: 8–15. (not seen)

Coolidge, K. R. (1908). A list of the North American Pseudoscorpionida. *Psyche, Cambridge* **15**: 108–114.

Cooreman, J. (1946a). Les pseudoscorpions. *Naturalistes Belges* **27**: 106–113. (not seen)

Cooreman, J. (1946b). Note sur les pseudoscorpions de la faune Belge. *Bulletin du Musée Royal d'Histoire Naturelle de Belgique* **22**(2): 1–8.

Cooreman, J. (1947). *Chernes lasiophilus* n. sp., Chélonèthe myrmécophile de Belgique. *Bulletin du Musée Royal d'Histoire Naturelle de Belgique* **23**(28): 1–6.

Corda, A. J. C. (1839). Über eine fossile Gattung der Afterscorpione. *Verhandlungen des Vaterländischen Museums in Böhmen.* **17**: 14–19. (not seen)

Corey, D. T. & Taylor, W. K. (1987). Scorpion, pseudoscorpion and opilionid faunas in three central Florida plant communities. *Florida Science* **50**: 162–167. (not seen)

Costa, M. & Nevo, E. (1969). Nidicolous arthropods associated with different chromosomal types of *Spalax ehrenbergi* Nehring. *Journal of the Linnean Society of London, Zoology* **48**: 199–215.

Costantini, G. P. (1976). Gli scorpioni e pseudoscorpioni della provincia di Brescia. Note di Aracnologia. II. *Natura, Bresciana* **13**: 121–124.

Courtois de Langlade, F. de (1883). Untitled. *Nature, London* **11**: 71. (not seen)

Cowden, D. R. (1983). Further notes and records on spiders from Worcestershire. *British Arachnological Society, Secretary's Newsletter* **37**: 5–6.

Craig, J. L. (1977). Invertebrate faunas of caves to be inundated by the Merome Park Lake in eastern Missouri. *Bulletin of the National Speleological Society* **39**: 81–89. (not seen)

Crocker, J. (1976). False scorpions. *Heritage, Quarterly Bulletin of the Loughborough Naturalists' Club* **61**: 1–11.

Crocker, J. (1978). Introduction to false scorpions. *British Arachnological Society, Secretary's News Letter* **21**: 4–9.

Crocker, J. (1979). The Sherwood Forest arachnid survey. *British Arachnological Society, Secretary's News Letter* **25**: 4–7.

Crome, W. (1961). In: Stresemann, E., *Exkursionsfauna von Deutschland, Bd. 1 Wirbellose*: xix–xxxiv, 290–393. Berlin. (not seen)

Croneberg, A. (1880). Ueber die Mundtheile der Arachniden. *Archiv für Naturgeschichte* **46**: 285–300.

Croneberg, A. (1887). Vorläufige Mittheilung über den Bau der Pseudoscorpione. *Zoologischer Anzeiger* **10**: 147–151.

Croneberg, A. (1888). Beitrag zur Kenntniss des Baues der Pseudoscorpione. *Byulleten' Moskovskogo Obshchestva Ispytatelei Prirody, n.s.* **2**: 416–461.

Crowther, H. (1882a). *Chelifer Degeerii*, C. Koch, a species new to Britain. *Zoologist* (3) **6**: 465.

Crowther, H. (1882b). *Chelifer Degeerii* C.L. Koch, a species new to Britain. *Science Gossip* **18**: 277.

Cuní y Martornell, D.M. (1897). Fauna entomológica de la villa de Calella (Cataluña, Provincia de Barcelona). *Anales de la Sociedad Española de Historia Natural* **26**: 281–339.

Ćurčić, B. P. M. (1972a). *Pselaphochernes hadzii*, nouveau pseudoscorpion des montagnes du sud-est de la Bosnie. *Razprave Slovenska Akademija Znanosti in Umetnosti* **15**: 76–93.

Ćurčić, B. P. M. (1972b). Nouveaux pseudoscorpions cavernicoles de la Serbie et de la Macédoine. *Acta Musei Macedonici Scientiarum Natularium, Skopje* **12**: 141–161.

Ćurčić, B. P. M. (1972c). Un pseudoscorpion cavernicole nouveau pour la péninsule des Balkans, *Chthonius (C.) bogovinae* n. sp. (Chthoniidae, Pseudoscorpiones, Arachnida). *Annales de Spéléologie* **27**: 341–350.

Ćurčić, B. P. M. (1972d). Deux nouveaux pseudoscorpions habitant des localités souterraines de la péninsule balkanique: *Chthonius caecus iugoslavicus* n. ssp. et *Chthonius bogovinae latidentatus* n. ssp. *Glasnik Muzeja Srpske Zemlje, Beograd* (B) **27**: 125–142.

Ćurčić, B. P. M. (1972e). *Neobisium (Blothrus) stankovici*, nouvelle espèce de pseudoscorpions cavernicoles de la Serbie orientale. *Fragmenta Balcanica* **9**: 85–96.

Ćurčić, B. P. M. (1973a). Le sous-genre *Globochthonius* Beier 1931 dans la Méditerranée nord-occidentale: *Chthonius (G.) globifer* Simon 1879 (Chthoniidae, Pseudoscorpiones, Arachnida). *Wissenschaftliche Mitteilungen des Bosnisch-herzegovinischen Landesmuseums* **3C**: 77–84.

Ćurčić, B. P. M. (1973b). A new cavernicolous species of the pseudoscorpion genus *Roncus* L. Koch, 1873 (Neobisiidae, Pseudoscorpiones) from the Balkan Peninsula. *International Journal of Speleology* **5**: 127–134.

Ćurčić, B. P. M. (1974a). Arachnoidea. Pseudoscorpiones. In: *Catalogus Faunae Jugoslaviae*, vol. 3(4): 1–35. Academie Slovène: Ljubljana.

Ćurčić, B. P. M. (1974b). Pseudoscorpions cavernicoles de la Macédoine. *International Journal of Speleology* **6**: 193–215.

Ćurčić, B. P. M. (1974c). [The subgenus *Globochthonius* Beier, 1931 (Chthoniidae, Pseudoscorpiones): taxonomic considerations and biogeographic implications]. *Glasnik 289 de l'Académie Serbe des Sciences et des Arts, Classe de Sciences Mathématiques et Naturelles* **36**: 105–112.

Ćurčić, B. P. M. (1975a). Répartition de quelques Pseudoscorpions et les changements paléogéographiques dans la région méditerranéenne. *Glasnik Muzeja Srpske Zemlje, Beograd* (B) **30**: 135–142.

Ćurčić, B. P. M. (1975b). *Balkanoroncus* (Arachnida, Pseudoscorpiones, Neobisiidae), a new genus of pseudoscorpions based on *Roncus bureschi* Hadži, 1939. *Glasnik Muzeja Srpske Zemlje, Beograd* (B) **30**: 143–145.

Ćurčić, B. P. M. (1976a). *Acanthocreagris ludiviri* (Neobisiidae, Pseudoscorpiones, Arachnida), a new species of false scorpions from Serbia. *Glasnik Muzeja Srpske Zemlje, Beograd* (B) **31**: 159–168.

Ćurčić, B. P. M. (1976b). Une contribution a la connaissance de la faune des pseudoscorpions en Serbie. *Glasnik Muzeja Srpske Zemlje, Beograd* (B) **31:** 169–184.

Ćurčić, B. P. M. (1977a). *Uporedno-Morfoloska Obeležja-Njihov Značaj i Preimena u Klasifikaciji Kaksona Porodice Neobisiidae (Pseudoscorpiones, Arachnida).* Univerzitet Beograd. (not seen)

Ṕurčić, B. P. M. (1977b). Les voies de l'évolution morphologique des Pseudoscorpions méditerranéens. I. Le sous-genre *Globochthonius* Beier, 1931 (Chthoniidae, Pseudoscorpiones). *Proceedings of the 6th International Congress of Speleology* **5:** 47–49.

Ćurčić, B. P. M. (1978a). *Tuberocreagris*, a new genus of pseudoscorpions from the United States (Arachnida, Pseudoscorpiones, Neobisiidae). *Fragmenta Balcanica* **10:** 111–121.

Ćurčić, B. P. M. (1978b). Cavernicole pseudoscorpions from Bulgaria. *Glasnik Muzeja Srpske Zemlje, Beograd* (B) **33:** 119–142.

Ćurčić, B. P. M. (1978c). Criteria for determining affinity within groups of genera and gaps between groups of genera in the family Neobisiidae (Pseudoscorpiones). *Symposia of the Zoological Society of London* **42:** 503.

Ćurčić, B. P. M. (1978d). On the affinity within and between groups of genera of Neobisiidae (Pseudoscorpiones, Arachnida). *Proceedings of the First European Congress of Entomology, Reading:* 15.

Ćurčić, B. P. M. (1978e). Certains criteres d'identification des rapports de parente entre les genres de la famille des Neobisiidae (Pseudoscorpiones, Arachnida). *Proceedings of the 7th International Speleological Congress, Sheffield:* 134–136.

Ćurčić, B. P. M. (1979a). The genus *Pararoncus* Chamberlin 1938 (Pseudoscorpiones, Neobisiidae) in Japan. *Glasnik Muzeja Srpske Zemlje, Beograd* (B) **34:** 169–180.

Ćurčić, B. P. M. (1979b). On some changes in the late postembryogenesis of the pseudoscorpion *Ditha proxima* (Beier 1951). *Glasnik Muzeja Srpske Zemlje, Beograd* (B) **34:** 191–200.

Ćurčić, B. P. M. (1979c). Growth and pedal tactile setae in pseudoscorpions. *Glasnik Muzeja Srpske Zemlje, Beograd* (B) **34:** 223–229.

Ćurčić, B.P.M. (1980a). Accidental and teratological changes in the family Neobisiidae (Pseudoscorpiones, Arachnida). *Bulletin of the British Arachnological Society* **5:** 9–15.

Ćurčić, B. P. M. (1980b). Pseudoscorpions from Nepal. *Glasnik Muzeja Srpske Zemlje, Beograd* (B) **35:** 77–101.

Ćurčić, B. P. M. (1980c). A new species of cave-dwelling pseudoscorpion from Serbia (Arachnida: Pseudoscorpiones: Neobisiidae). *Senckenbergiana Biologica* **60:** 249–254.

Ćurčić, B. P. M. (1980d). The genus *Neobisium* Chamberlin 1930 (Pseudoscorpiones, Arachnida): post-embryonic development and taxonomy of subgenera. *Verhandlungen des 8. Internationalen Arachnologen-Kongresses, Wien:* 465.

Ćurčić, B. P. M. (1981a). New cave-dwelling pseudoscorpions from Serbia. *Bulletin de l'Académie Serbe des Sciences et des Arts, Classe des Sciences Naturelles et Mathématiques* **75(21):** 105–114.

Ćurčić, B. P. M. (1981b). A revision of some North American pseudoscorpions (Neobisiidae, Pseudoscorpiones). *Glasnik Muzeja Srpske Zemlje, Beograd* (B) **36:** 101–107.

Ćurčić, B. P. M. (1982a). New and little-known cave pseudoscorpions from Serbia. *Revue Arachnologique* **3:** 181–189.

Ćurčić, B. P. M. (1982b). A new cavernicole pseudoscorpion from Macedonia. *Fragmenta Balcanica* **11:** 145–150.

Ćurčić, B. P. M. (1982c). *Trisetobisium* (Pseudoscorpiones, Neobisiidae), a new genus of pseudoscorpions based on *Microcreagris fallax* Chamberlin. *Glasnik Muzeja Srpske Zemlje, Beograd* (B) **37:** 57–61.

Ćurčić, B. P. M. (1982d). *Americocreagris*, a new genus of pseudoscorpions from the United States. *Bulletin de l'Académie Serbe des Sciences et des Arts, Classe des Sciences Naturelles et Mathématiques* **80(22)**: 47–50.

Ćurčić, B. P. M. (1982e). Postembryonic development in the Neobisiidae (Pseudoscorpiones, Arachnida). *Serbian Academy of Sciences and Arts, Monographs* **56**: 1–90.

Ćurčić, B. P. M. (1983a). A revision of some Asian species of *Microcreagris* Balzan, 1892 (Neobisiidae, Pseudoscorpiones). *Bulletin of the British Arachnological Society* **6**: 23–36.

Ćurčić, B. P. M. (1983b). The biospeleological features of eastern Serbia. *Proceedings of the European Regional Conference on Speleology, Sofia, 1980*: 105–109.

Ćurčić, B. P. M. (1983d). Relic and endemic pseudoscorpions in Serbia. *Verh. X. S.I.E.E.C. Budapest 1983*: 280–282.

Ćurčić, B. P. M. (1983e). Nove pecinske pseudoskorpije iz Srbije. *Zbornik Radova Odb. za Krad i Speleol, SANU, Beograd* **1**: 135–145. (not seen)

Ćurčić, B. P. M. (1984a). On two new species of *Roncus* L. Koch 1873 from Macedonia (Arachnida: Pseudoscorpiones: Neobisiidae). *Senckenbergiana Biologica* **65**: 97–104.

Ćurčić, B. P. M. (1984b). A revision of some North American species of *Microcreagris* Balzan, 1892 (Arachnida: Pseudoscorpiones: Neobisiidae). *Bulletin of the British Arachnological Society* **6**: 149–166.

Ćurčić, B. P. M. (1984c). O poreklu i genezi nekih rodova Pseudoskorpija u evroaziji. *Zbornik Predavanja Deveti Jugoslavenski Speleoloski Kongres*: 521–527.

Ćurčić, B. P. M. (1984d). The genus *Neobisium* Chamberlin, 1930 (Neobisiidae, Pseudoscorpiones, Arachnida): on new species from the USSR and the taxonomy of its subgenera. *Glasnik Muzeja Srpske Zemlje, Beograd* (B) **39**: 124–153.

Ćurčić, B. P. M. (1985a). A revision of some species of *Microcreagris* Balzan, 1892 (Neobisiidae, Pseudoscorpiones) from the USSR and adjacent regions. *Bulletin of the British Arachnological Society* **6**: 331–352.

Ćurčić, B. P. M. (1985b). On the presence of *Neobisium korabense* Ćurčić, 1982 (Pseudoscorpiones, Neobisiidae) in Macedonia. *Fragmenta Balcanica* **12**: 169–177.

Ćurčić, B. P. M. (1986a). On the origin and biogeography of some pseudoscorpions of the Balkan Peninsula. *Biologia Gallo-Hellenica* **12**: 85–92.

Ćurčić, B. P. M. (1986b). [*Chthonius (C.) stevanovici* (Chthoniidae, Pseudoscorpiones), a new pseudoscorpion species from East Serbia.] *Recueil des Rapports du Comité pour la Karst et la Spéléologie* **DLXVIII**: 141–154.

Ćurčić, B. P. M. (1986c). Taxonomy and geographic distribution of some *Microcreagris*-related genera (Neobisiidae, Pseudoscorpiones). *Actas X Congreso Internacional de Aracnologia, Jaca* **1**: 321–326.

Ćurčić, B. P. M. (1986d). On the taxonomy and biogeography of *Microcreagris* related genera in Eurasia (Neobisiidae, Pseudoscorpiones). *Mémoires de la Société Royale Belge d'Entomologie* **33**: 75–79. (not seen)

Ćurčić, B. P. M. (1987). *Insulocreagris*, a new genus of pseudoscorpions from the Balkan Peninsula (Pseudoscorpiones, Neobisiidae). *Revue Arachnologique* **7**: 47–57.

Ćurčić, B. P. M. (1988a). Les Pseudoscorpions cavernicoles de la Yougoslavie: développement historique et implications biogéographiques. *Revue Arachnologique* **7**: 163–174.

Ćurčić, B. P. M. (1988b). *Cave-Dwelling Pseudoscorpions of the Dinaric Karst.* Slovenska Akademija Znanosti in Umetnosti: Ljubljana.

Ćurčić, B. P. M. (1988c). On the taxonomic status of *Chthonius caecus iugoslavicus* Ćurčić, 1972 (Chthoniidae, Pseudoscorpiones). *Fragmenta Balcanica* **14(1)**: 1–10.

Ćurčić, B. P. M. (1988d). On the origin and evolution of some cave pseudoscorpions of the Dinaric and Carpatho-Balkanic Karst. *Rec. Rapp Com. Karst Spéléol., III, Ed. Spec., Acad Serbe Sci. Arts, Belgrade* 3: 167–177. (not seen)

Ćurčić, B. P. M. (1988e). Edaphism and cave pseudoscorpions. *Rec. Rapp Com. Karst Spéléol., III, Ed. Spec., Acad Serbe Sci. Arts, Belgrade* 3: 179–185. (not seen)

Ćurčić, B. P. M. (1988f). Some remarks on the evolution of Dinaric cave pseudoscorpions. *Comptes Rendus du XIIème Colloque d'Arachnologie, Berlin*: 287–292. (not seen)

Ćurčić, B. P. M. (1989a). Segmental anomalies in some European Neobisiidae (Pseudoscorpiones, Arachnida) – Part I. *Acta Arachnologica* 37: 77–87.

Ćurčić, B. P. M. (1989b). Segmental anomalies in some European Neobisiidae (Pseudoscorpiones, Arachnida) – Part II. *Acta Arachnologica* 38: 1–10.

Ćurčić, B. P. M. (1989c). Further revision of some North American false scorpions originally assigned to *Microcreagris* Balzan (Pseudoscorpiones, Neobisiidae). *Journal of Arachnology* 17: 351–362.

Ćurčić, B. P. M. & Beron, P. (1981). [New and little-known cavernicole pseudoscorpions in Bulgaria (Neobisiidae, Pseudoscorpiones, Arachnida).] *Glasnik CCCXXIX de l'Académie Serbe des Sciences et des Arts, Classe des Sciences Naturelles et Mathématiques* 48: 63–85.

Ćurčić, B. P. M. & Dimitrijević, R.N. (1982). On abnormalities of abdominal segmentation in *Neobisium carpaticum* Beier (Neobisiidae, Pseudoscorpiones, Arachnida). *Revue Arachnologique* 4: 143–150.

Ćurčić, B. P. M. & Dimitrijević, R. N. (1983a). Three more examples of abnormal segmentation in *Neobisium carpaticum* Beier, 1934 (Arachnida: Pseudoscorpiones: Neobisiidae). *Proceedings of the Entomological Society of Washington* 85: 362–365.

Ćurčić, B. P. M. & Dimitrijević, R. N. (1983b). Two more examples of sternal anomaly in *Neobisium carpaticum* Beier and *Roncus lubricus* L. Koch (Pseudoscorpiones, Arachnida). *Verh. X. S.I.E.E.C. Budapest, 1983*: 283–285.

Ćurčić, B. P. M. & Dimitrijević, R. N. (1984a). An abnormal carapaco-abdominal junction in *Neobisium carpaticum* Beier, 1934 (Neobisiidae, Pseudoscorpiones). *Arhiv Bioloskih Nauka, Beograd* 36: 9–10.

Ćurčić, B. P. M. & Dimitrijević, R. N. (1984b). Endemicni i reliktni rodovi Pseudoskorpija u Jugosalviji. *Zbornik Predavanja Deveti Jugoslavenski Speleoloski Kongres, Karlovac*: 529–534.

Ćurčić, B. P. M. & Dimitrijević, R. N. (1985). Abdominal deficiencies in four species of the Neobisiidae (Pseudoscorpiones, Arachnida). *Revue Arachnologique* 6: 91–98.

Ćurčić, B. P. M. & Dimitrijević, R. N. (1986a). Biogeography of cave pseudoscorpions of the Balkan Peninsula. *Proceedings of the 3rd European Congress of Entomology* 3: 425–428.

Ćurčić, B. P. M. & Dimitrijević, R. N. (1986b). Abnormalities of carapacal and abdominal segmentation in *Neobisium* Chamberlin (Neobisiidae, Pseudoscorpiones). *Actas X Congreso Internacional de Aracnologia, Jaca* 1: 17–23.

Ćurčić, B. P. M. & Dimitrijević, R. N. (1986c). Teratology of abdominal tergites and sternites in *Neobisium carpaticum* Beier (Neobisiidae, Pseudoscorpiones). *Mémoires de la Société Royale Belge d'Entomologie* 33: 81–84. (not seen)

Ćurčić, B. P. M. & Dimitrijević, R. N. (1987). Two segmental anomolies in *Chthonius ischnocheles* (Hermann) (Chthoniidae, Pseudoscorpiones). *Arhiv Bioloskih Nauka, Beograd* 39: 1–2. (not seen)

Ćurčić, B. P. M., Dimitrijević, R. N. (1988a). Segmental anomolies in *Neobisium carpaticum* Beier (Neobisiidae, Pseudoscorpiones) from the Botanical Garden in Belgrade, Yugoslavia. *Fragmenta Balcanica* 15: 135–150. (not seen)

Ćurčić, B. P. M. & Dimitrijević, R. N. (1988b). Segmental deficiencies in some Neobisiidae (Pseudoscorpiones, Arachnida). *Comptes Rendus du XIème Colloque d'Arachnologie, Berlin*: 89–97. (not seen)

Ćurčić, B. P. M., Krunić, M. D. & Brajković, M. M. (1981). Further records of teratological changes in the Neobisiidae (Arachnida, Pseudoscorpiones). *Bulletin of the British Arachnological Society* 5: 280–284.

Ćurčić, B. P. M., Krunić, M. D. & Brajković, M. M. (1983). Tergal and sternal anomolies in *Neobisium* Chamberlin (Neobisiidae, Pseudoscorpiones, Arachnida). *Journal of Arachnology* 11: 243–250.

Curry, S. J., Humphreys, W. F., Koch, L. E. & Main, B. Y. (1985). Changes in arachnid communities resulting from forestry practices in Karri forest, south-west Western Australia. *Australian Forestry Research* 15: 469–480.

Curtis, J. (1849). Observations on the natural history and economy of various insects affecting the potato-crops, including plant-lice, plant-bugs, frog-flies, caterpillars, crane-flies, wireworms, millipedes, mites, beetles, flies, etc. *Journal of the Royal Agricultural Society* 10: 70–118.

Cuthbertson, D. R. (1982). Pseudoscorpions attached to flies: phoresy or predation? *Newsletter of the British Arachnological Society* 34: 4.

Cuthbertson, D. R. (1984). Catalepsy and phoresy in pseudoscorpions. *Newsletter of the British Arachnological Society* 39: 3.

Cuvier, G. (1797). *Tableau Élémentaire de l'Histoire Naturelle des Animaux*. Paris. (not seen)

Daday, E. (1880a). A magyarországi álskorpiók. *Értesítö az Erdélyi Múzeum-egyesület Orvos-Természet-Tudományi Szakosztályából* 5: 191–193.

Daday, E. (1880b). Az álskorpiók vérkeringési szervéröl. *Természetrajzi Füzetek* 4: 277–284, 331–339. (not seen)

Daday, E. (1882). Az álskorpiók boncztana. *Orvos-Termeszetudomanyi* 7: 1–76. (not seen)

Daday, E. (1888). A Magyar Nemzeti Muzeum álskorpióinak áttekintése. *Természetrajzi Füzetek* 11: 111–136, 165–192.

Daday, E. (1889a). Adatok a Kaukázus álskorpió-faunájának ismeretéhez. *Természetrajzi Füzetek* 12: 16–22.

Daday, E. (1889b). Egy Braziliai új álskorpio-faj a Magyar Nemzeti Muzeum állattárában. *Természetrajzi Füzetek* 12: 23–24.

Daday, E. (1889c). Ujabb adatok o magyar-fauna álskorpióinak ismeretéhez. *Természetrajzi Füzetek* 12: 25–28.

Daday, E. (1889d). Adatok a Balkán-félsziget álskorpió-faunájának ismeretéhez. *Természetrajzi Füzetek* 12: 80–84, 113.

Daday, E. (1896). Pseudoscorpiones and Opiliones. In: *Fauna Regni Hungariae*, vol. 3: 1–2. Budapest. (not seen)

Daday, E. (1897). Pseudoscorpiones e Nova-Guinea. *Természetrajzi Füzetek* 20: 475–480.

Daday, E. (1918). Ordo Pseudoscorpiones. In: *A Magyar Birodalom Allatvilága*: 1–2. Budapest.

Dahl, F. (1883). Über die Hörhaare bei den Arachnoiden. *Zoologischer Anzeiger* 6: 267–270. (not seen)

Daiber, M. (1921). Arachnoidea. In: Lang, A. (ed.), *Handbuch der Morphologie der Wirbellosen Tiere*, vol. 4: 269–350. Jena. (not seen)

Dale, C. W. (1878). *History of Glanville's Wootton*. London. (not seen)

Dalla Torre, K. v. (1882). Beiträge zur Arthropoden-Fauna Tirols. *Verhandlungen des Naturwissenschaftlich-Medizinischen Vereins in Innsbruck* 12: 32–73. (not seen)

Dalman, J. W. (1824). *Memoire on Copal Insects*. (not seen)

Dalman, J. W. (1825). Om Insekter inneslutne i Copal; jemte beskrifning på några deribland förekommande nya slägten och arter. *Kungliga Svenska Vetenskapsakademiens Handlingar* **46**: 375–410.

Dambach, K. W. (1914). Der Bücherskorpion. *Schweizerische Bienen-Zeitung* **37**: 483.

Dammerman, K. W. (1922). The fauna of Krakatau, Verlaten Island and Sebesy. *Treubia* **3**: 61–112.

Dammerman, K. W. (1929). Krakatau's new fauna. *4th Pacific Science Congress*: 83–118. (not seen)

Dammerman, K. W. (1948). The fauna of Krakatau 1883–1933. *Verhandelingen deer Koninklijke Nederlandsche Akademie van Wetenschappen, afd. Natuurkunde* **44**: 1–594.

Dartnall, A. J. (1969). Pseudoscorpions. *Tasmanian Naturalist* **17**: 2.

Dartnall, A. J. (1970). Some Tasmanian chthoniid pseudoscorpions. *Papers and Proceedings of the Royal Society of Tasmania* **104**: 65–68.

Dartnall, A. J. (1975). A note on the natural history of a pseudoscorpion. *Tasmanian Naturalist* **40**: 5–6.

Dashdamirov, S. (1988). A new pseudoscorpion species of the genus *Acanthocreagris* (Pseudoscorpiones, Neobisiidae). *Zoologiceskij Zhurnal* **67**: 1414–1416.

Davis, A. H. (1831). Lobster-like insect attacking the leg of a house-fly. *Magazine of Natural History* **4**: 283.

Dawydoff, C. (1940). Quelques observations sur le biologie d'un pseudoscorpion marin d'Indochine (*Apocheiridium pelegicum* Redikorzeff). *Comptes Rendus Hebdomadaires des Séances, Académie des Sciences, Paris* **210**: 447–449.

Decary, R. & Kiener, A. (1970). Les cavités souterraines de Madagascar. *Annales de Spéléologie* **25**: 409–440. (not seen)

Decou, A. & Decou, V. (1964). Recherches sur la synusie du guano des grottes d'Olténie et du Banat (Roumanie). (Note préliminaire). *Annales de Spéléologie* **19**: 781–797.

Decu, V. G. & Iliffe, T. M. (1983). A review of the terrestrial cavernicolous fauna of Romania. *Bulletin of the National Speleological Society* **45**: 86–97. (not seen)

de Geer, C. (1778). *Mémoires pour Servir á l'Histoire des Insectes*, vol. 7. Stockholm.

de Geer, C. (1783). *Abhandlungen zur Geschichte de Insecten, übers von Goeze*, vol. 7. Nürnberg. (not seen)

Delany, M. J. (1960). The food and feeding habits of some heath dwelling invertebrates. *Proceedings of the Zoological Society of London* **135**: 303–311.

Delhez, F. (1973). Pseudoscorpion nouveau pour la faune belge. *Bulletin et Annales de la Société Royale d'Entomologie de Belgique* **109**: 220–221. (not seen)

Dell'oca, S. & Pozzi, A. (1959). Note speleologiche di una escurcione attraverso la Sardegna (1956). *Rassegna Speleologica Italiana* **11**: 130–147. (not seen)

Demoor, J. (1891). Recherches sur la marche des Insectes et des Arachnides. *Archiv Néerlandaises de Biologie* **10**: 567–608. (not seen)

di Castri, F. & Mooney, H. A. (1973). Ecological studies: analysis and synthesis. In: Castri, F. di & Mooney, H. (eds), *Mediterranean Type Ecosystems, Origin and Structure*: ?–?. Springer-Verlag: Berlin. (not seen)

Distant, W. L. (1908). Curious habit of a Chelifer. *Zoologist* **12**: 77–78.

Dobson, R. M. (1986). The natural history of the Muck Islands, North Ebudes, Scotland, United Kingdom. 2. Spiders, harvestmen and pseudoscorpions. *Glasgow Naturalist* **21**: 173–181. (not seen)

Dohrn, A. (1870). Untersuchungen über Bau und Entwickelung der Arthropoden. *Jenaische Zeitschrift für Naturwissenschaft* **5**: 54–81, 138–157, 277–292, 471–491. (not seen)

Doleschal, L. (1852). Systematisches Verzeichniss der im Kaiserthum Oesterreich vorkommenden Spinnen. *Sitzungsberichte der Mathematischen-Naturwissenschaftlichen Classe der Königlichen Akademie der Wissenschaften, Wien* **9**: 622–651. (not seen)

Dondale, C. D. (1979). Canada and its insect fauna. Part 3. Status of taxa in Canada. 11. Opiliones, Pseudoscorpionida, Scorpionida, Solifugae. *Memoirs of the Entomological Society of Canada* **108**: 250–251.

Donisthorpe, H. (1902). Notes on the British myrmecophilous fauna (excluding Coleoptera). *Entomologist's Record and Journal of Variations* **14**: 14–18, 37–40, 67–70.

Donisthorpe, H. (1907). Myrmecophilous notes for 1907. *Entomologist's Record and Journal of Variations* **19**: 254–256.

Donisthorpe, H. (1910). On the founding of nests by ants; and a few notes on myrmecophiles. *Entomologist's Record and Journal of Variations* **22**: 82–85.

Donisthorpe, H. (1915). *British Ants, their Life-History and Classification*. W. Brendon and Son: Plymouth.

Donisthorpe, H. (1926). A further study of the habits of *Acanthomyops (Donisthorpea) brunneus*, Latr., and the myrmecophiles inhabiting its nests. *Entomologist's Record and Journal of Variations* **38**: 52–55.

Donisthorpe, H. (1927). *The Guests of British Ants, their Habits and Life-Histories.* Routledge and Sons: London.

Donovan, E. (1797). *The Natural History of British Insects*, vol. 6. Donovan: London.

Dresco, E. & Derouet, L. (1960). Araignées et Opiliones des cavités souterraines de Varzo (Piémont, Italie). *Annales de Spéléologie* **15**: 107–115. (not seen)

Drift, J. (1951). Analysis of the animal community in a beech forest floor. *Mededeelingen van het Instituut voor Toegepast Biologisch Onderzoek in der Natuur* **9**: 1–168. (not seen)

Drift, J. (1963). A comparative study of the soil fauna in forests and cultivated land on sandy soils in Suriname. *Natuurwetenschappelijke Studiekring voor Suriname et Curaçao* **32**: 1–42. (not seen)

Drogla, R. (1977). Zur Pseudoskorpion-Fauna des Naturschutzgebietes 'Tiefental'. *Veröffentlichungen aus dem Museum Westlausitz* **1**: 87–90.

Drogla, R. (1984). Erstnachweis von drei Pseudoskorpion-Arten für die DDR (Arachnida, Pseudoscorpiones). *Faunistische Abhandlungen, Staatliches Museum für Tierkunde Dresden* **11**: 191.

Drogla, R. (1988). Pseudoskorpione des Deutsch Paulsdorfer Waldes (Oberlausitz) mit Beschreibung einer Pedipalpenanomalie (Arachnida, Pseudoscorpiones). *Abhandlungen und Berichte der Naturkunde-Museums Görlitz* **62**: 17–20. (not seen)

Duchac, V. (1988). Prispevek k poznani stirku ceskeho stredohori. *Fauna Bohemiae Septentrionalis* **13**: 103–108. (not seen)

Dugès, A. & Edwards, M. (1836). Arachnides. In: Cuvier, G., *Le Règne Animal Distribue d'Après son Organisation*, 3rd edition, vols **15–16**. Fortin, Masson: Paris.

Duméril, A. M. C. (1826). Pince, porte-pince ou Chélifère. In: Levrault, F. G. (ed.), *Dictionnaire des Sciences Naturelles*, vol. **41**: 49–50. Normant: Paris.

Dumitresco, M. (19??). Ordinul Pseudoscorpionida. In: *Fauna Illustratà a României*, vol. **1**. Ed Stiint. Enciclop. Bucuresti. (not seen)

Dumitresco, M. & Orghidan, T. (1964). Contribution a la connaissance des Pseudoscorpions de la Dobroudja. 1re note. *Annales de Spéléologie* **19**: 599–630.

Dumitresco, M. & Orghidan, T. (1966). Quelques observations sur l'écologie des Pseudescorpions vivant dans les lithoclases (Arach., Chelonethi). *Senckenbergiana Biologica* **47**: 81–83.

Dumitresco, M. & Orghidan, T. (1969). Sur deux espèces nouvelles de Pseudoscorpions (Arachnides) lithoclasicoles de Roumanie: *Diplotemnus vachoni* (Atemnidae) et *Dactylochelifer marlausicolus*. *Bulletin du Muséum National d'Histoire Naturelle, Paris* **41**: 675–687.

Dumitresco, M. & Orghidan, T. (1970a). Cycle du développement de *Diplotemnus vachoni* Dumitresco et Orghidan, 1969, appartenant à la nouvelle famille des Miratemnidae (Arachnides, Pseudoscorpions). *Bulletin du Muséum National d'Histoire Naturelle, Paris* (2) **41** (supplément 1): 128–134.

Dumitresco, M. & Orghidan, T. (1970b). Contribution à la connaissance des Pseudoscorpions souterrains de Roumanie. *Travaux de l'Institute de Spéologie 'Émile Racovitza'* **9**: 97–111.

Dumitresco, M. & Orghidan, T. (1972). Sur l'espèce *Neobisium (Blothrus) beieri* Dumitresco et Orghidan. *Travaux de l'Institute de Spéologie 'Émile Racovitza'* **11**: 247.

Dumitresco, M. & Orghidan, T. (1977). Pseudoscorpions de Cuba. In: Orghidan, T., Núñez Jiménez, A., Decou, V., St. Negrea & Bayés, N. V. (eds), *Résultats des Expéditions Biospéologiques Cubano-Roumaines à Cuba*, vol. **2**: 99–124. Bucaresti: Editura Academiei Republicii Socialiste România.

Dumitresco, M. & Orghidan, T. (1981). Représentants de la fam. Cheiridiidae Chamberlin (Pseudoscorpionidea) de Cuba. In: Orghidan, T., Núñez Jiménez, A., Decou, V., St. Negrea & Bayés, N. V. (eds), *Résultats des Expéditions Biospéologiques Cubano-Roumaines à Cuba*, vol. **3**: 77–87. Bucaresti: Editura Academiei Republicii Socialiste România.

Dumitresco (as Dumitrescu), M. & Orghidan, T. (1986). *Acanthocreagris mahnerti* sp. n. (Pseudoscorpions, Neobisiidae). *Revue Suisse de Zoologie* **93**: 51–58.

Dumitrescu, D. (1976). Opilionida, Pseudoscorpionida et Acari. In: Contributions a la connaissance de la faune du département Vrancea. *Travaux du Muséum d'Histoire Naturelle Grigore Antipa* **17**: 273–276.

Dumitrescu, M. (1979). Bibliographica Arachnologica Romanica (1). *Travaux du Muséum d'Histoire Naturelle Grigore Antipa* **20**: 43–84. (not seen)

Dumitrescu, M. – see Dumitresco, M.

Dunkle, S. W. (1984). First record of pseudoscorpions phoretic on dragonflies. *Notulae Odonatologicae* **2**: 48.

Durden, L. A. (1986). Ectoparasites and other arthropod associates of tropical rain forest mammals in Sulawesi Utara, Indonesia. *National Geographic Research* **2**: 320–331.

Durden, L. A. (1987). Predator-prey interactions between ectoparasites. *Parasitology Today* **3**: 306–308.

Edwards, J. S. (1955). An instance of phoresy. *New Zealand Entomologist* **1**: 9–10.

Eisenbeis, G. & Wichard, W. (1987). *Atlas on the Biology of Soil Arthropods*. Springer-Verlag: Berlin.

El-Hennawy, H. K. (1988a). Key to pseudoscorpionid families (Arachnida: Pseudo scorpionida). *Serket* **1**(3): 1–8. (not seen)

El-Hennawy, H. K. (1988b). Pseudoscorpions of Egypt, key and list of species. *Serket* **1**(3): 9–18. (not seen)

El-Hennawy, H. K. (1988c). *Hysterochelifer tuberculatus* (Lucas, 1846) (Pseudo-scorpionida: Cheliferidae) in Jordan. *Serket* **1**(3): 20. (not seen)

El-Kifl, A. H. (1969). The soil Arachnoidea of a farm at Giza, U.A.R. *Bulletin de la Société Entomologique d'Égypte* **52**: 413–428.

Ellingsen, E. (1895). Description d'une espèce nouvelle de l'ordre des Chernètes. *Bulletin de la Société Zoologique de France* **20**: 137–138.

Ellingsen, E. (1897). Norske Pseudoscorpioner. *Forhandlinger i Videnskabsselskabet i Kristiania* **1896**(5): 1–21.

Ellingsen, E. (1901a). Sur une espèce nouvelle d'*Ideobisium* genre des pseudoscorpions de l'Europe. *Bulletin de la Société Zoologique de France* **26**: 86–89.

Ellingsen, E. (1901b). Sur deux espèces de pseudoscorpions de l'Asie. *Bulletin de la Société Zoologique de France* **26**: 205–209.

Ellingsen, E. (1902). Sur la faune de pseudoscorpions de l'Équateur. *Mémoires de la Société Zoologique de France* **15**: 146–168.

Ellingsen, E. (1903). Norske Pseudoscorpioner. II. *Forhandlinger i Videnskabsselskabet i Kristiania* **1903**(5): 1–18.

Ellingsen, E. (1904). On some pseudoscorpions from Patagonia collected by Dr. Filippo Silvestri. *Bollettino dei Musei di Zoologia e di Anatomia Comparata della R. Università di Torino* **19**(480): 1–7.

Ellingsen, E. (1905a). On a pseudoscorpion from Congo. *Bollettino dei Musei di Zoologia e di Anatomia Comparata della R. Università di Torino* **20**(496): 1–3.

Ellingsen, E. (1905b). Viaggio del Dr. Enrico Festa nell'Ecuador e regioni vicine. XXIX. Pseudoscorpiones. *Bollettino dei Musei di Zoologia e di Anatomia Comparata della R. Università di Torino* **20**(497): 1–3.

Ellingsen, E. (1905c). Pseudoscorpions from South America collected by Dr. A. Borelli, A. Bertoni de Winkelried, and Prof. Goeldi. *Bollettino dei Musei di Zoologia e di Anatomia Comparata della R. Università di Torino* **20**(500): 1–17.

Ellingsen, E. (1905d). Pseudoscorpions from Italy and southern France conserved in the R. Museo Zoologico in Torino. *Bollettino dei Musei di Zoologia e di Anatomia Comparata della R. Università di Torino* **20**(503): 1–13.

Ellingsen, E. (1905e). On some pseudoscorpions from South America in the collection of Prof. F. Silvestri. *Zoologischer Anzeiger* **29**: 323–328.

Ellingsen, E. (1906). Report on the pseudoscorpions of the Guinea Coast (Africa) collected by Leonardo Fea. *Annali del Museo Civico di Storia Naturale di Genova* (3) **2**: 243–265.

Ellingsen, E. (1907a). Notes on pseudoscorpions, British and foreign. *Journal of the Quekett Microscopical Club* (2) **10**: 155–172.

Ellingsen, E. (1907b). Über einige Pseudoskorpione aus Deutsch-Ostafrika. *Zoologischer Anzeiger* **32**: 28–30.

Ellingsen, E. (1907c). On some pseudoscorpions from Japan collected by Hans Sauter. *Nytt Magasin for Naturvidenskapene* **45**: 1–17.

Ellingsen, E. (1908a). Biospéologica. VII. Pseudoscorpiones (seconde série). *Archives de Zoologie Expérimentale et Générale* (4) **8**: 415–420.

Ellingsen, E. (1908b). Arachniden aus Madagaskar. Pseudoscorpiones. *Zoologische Jahrbücher, Systematik (Ökologie), Geographie und Biologie* **26**: 487–488.

Ellingsen, E. (1908c). Materiali per una fauna dell'Arcipelago Toscano. VIII. Isola del Giglio. Notes on pseudoscorpions. *Annali del Museo Civico di Storia Naturale di Genova* (3) **3**: 668–670.

Ellingsen, E. (1908d). Über Pseudoskorpione aus West-Deutschland. *Sitzungsberichte hrsg. vom Naturhistorischen Verein Preussischen Rheinlände und Westfalens* **1908**: 69–70.

Ellingsen, E. (1908e). Two Canadian species of pseudoscorpions. *Canadian Entomologist* **40**: 163.

Ellingsen, E. (1909a). Contributions to the knowledge of the pseudoscorpions from material belonging to the Museo Civico in Genova. *Annali del Museo Civico di Storia Naturale di Genova* (3) **4**: 205–220.

Ellingsen, E. (1909b). On some North American pseudoscorpions collected by Dr. F. Silvestri. *Bollettino del Laboratoria di Zoologia Generale e Agraria della R. Scuola sup. d'Agricoltura, Portici* **3**: 216–221.

Ellingsen, E. (1910a). Die Pseudoskorpione des Berliner Museums. *Mitteilung aus dem Zoologischen Museum in Berlin* **4**: 357–423.

Ellingsen, E. (1910b). Pseudoskorpione und Myriopoden des Naturhistorischen Museums der Stadt Wiesbaden. *Jahrbuch des Nassauischen Vereins für Naturkunde* **63**: 62–65.

Ellingsen, E. (1910c). Myriapoda und Pseudoscorpiones. In: Strand, E., Neue Beiträge zur Arthropoden-Fauna Norwegens nebst gelegentlichen Bemerkungen über deutsche Arten. *Nytt Magasin for Naturvidenskapene* **48**: 344–348.

Ellingsen, E. (1910d). Pseudoscorpions from Uganda collected by Dr. E. Bayon. *Annali del Museo Civico di Storia Naturale di Genova* (3a) **4**: 536–538.

Ellingsen, E. (1911a). Pseudoscorpions from Sumatra. *Annali del Museo Civico di Storia Naturale di Genova* (3a) **5**: 34–40.

Ellingsen, E. (1911b). Pseudoscorpions collected by Leonardo Fea in Birma. *Annali del Museo Civico di Storia Naturale di Genova* (3a) **5**: 141–144.

Ellingsen, E. (1911c). Pseudoscorpionina. In: Le Roi, O. (ed.), Zur Fauna des Vereinsgebietes. *Sitzungsberichte hrsg. vom Naturhistorischen Verein Preussischen Rheinlände und Westfalens* **1911**: 173–174.

Ellingsen, E. (1912a). H. Sauter's Formosa-Ausbeute. Pseudoscorpions from Formosa. I. *Nytt Magasin for Naturvidenskapene* **50**: 121–128.

Ellingsen, E. (1912b). The pseudoscorpions of South Africa, based on the collections of the South African Museum, Cape Town. *Annals of the South African Museum* **10**: 75–128.

Ellingsen, E. (1912c). Pseudoscorpiones (troisième série). *Archives de Zoologie Expérimentale et Générale* (5) **10**: 163–175.

Ellingsen, E. (1913a). Note on some pseudoscorpions in the British Museum. *Annals and Magazine of Natural History* (8) **11**: 451–455.

Ellingsen, E. (1913b). Chelifera. Zoological results of the Abor expedition. *Records of the Indian Museum* **8**: 127.

Ellingsen, E. (1914). On the pseudoscorpions of the Indian Museum, Calcutta. *Records of the Indian Museum* **10**: 1–14.

Elliott, W. R. (1973). Damming up caves. *Caving International Magazine* **10**: 38–41. (not seen)

Elliott, W. R. & Reddell, J. R. (1973). A checklist of the cave fauna of Mexico. VI. Valle de los Fantasmas region, San Luis Potosi. *Bulletin of the Association for Mexican Cave Studies* **5**: 191–201.

Enghoff, H. (1983). Mosskorpioner (Pseudoscorpiones). *Entomologiske Meddelelser* **51**: 7.

Erichson, W. F. (1844). Bericht über die wissenschaftlichen Leistungen in der Naturgeschichte der Insecten, Arachniden, Crustaceen und Entomostraceen während des Jahres 1843. *Archiv für Naturgeschichte* **10**: 249–346. (not seen)

Esaki, T. (1922). [The book-scorpion being a parasite on the beetle.] *Zoological Magazine, Tokyo* **34**: 974–975. (not seen)

Esaki, T. (1930). ['Pseudoscorpii', in Myriapoda and Araneia.] *Iwanami, Tokyo*: 34–49. (not seen)

Essig, E. O. (1926). *Insects of Western North America*. Macmillan: New York. (not seen)

Essig, E. O. (1958). *Insects and mites of Western North America*, revised edition. Macmillan: New York.

Estany, J. (1977a). Sobre una poblacion trogloxena de *Allochernes masi* Navás, (Arachnida, Pseudoscorpionida). *Comunicacions del 6è. Simposium d'Espeleologica, Bioespeleologia, Terrassa*: 149–152.

Estany, J. (1977b). Sobre algunos Pseudoscorpiones de las islas Baleares. *Publicaciones del Departmento de Zoologia* **2**: 29–33.

Estany, J. (1978). Sobre algunos Neobisiidae cavernólócolas del País Valenciano. *Speleon* **24**: 33–37.

Estany, J. (1979a). A propos de quelques Pseudoscorpions des îles Canaries. *Revue Arachnologique* **2**: 221–223.

Estany, J. (1979b). Sobre la presencia de *Rhacochelifer maculatus* L. Koch (Arachnida, Pseudoscorpionida) en Cataluña. *Publicaciones del Deptartamento de Zoologia, Universidad de Barcelona* **4**: 47–50.

Estany, J. (1980a). Quelques remarques à propos de *Larca hispanica* Beier et *Larca spelaea* Beier (Pseudoscorpionida, Garypidae). *Vème Colloque d'Arachnologie d'Expression Française, Barcelone*: 65–70.

Estany, J. (1980b). Contribución al conocimiento de la fauna cavernícola del País Vasco. Arachnida Pseudoscorpionida. *Kobie, Bilbao* **10**: 526–528.

Evans, G. O. & Browning, E. (1954). *Synopses of the British Fauna. No. 10. Pseudoscorpiones.* Linnean Society: London.

Evans, W. (1901a). *Roncus cambridgii*, L. K., and other chernetids in Scotland. *Annals of Scottish Natural History* **10**: 53–54.

Evans, W. (1901b). *Chthonius tetrachelatus*, Preyss., and other chernetids in Scotland. *Annals of Scottish Natural History* **10**: 241.

Evans, W. (1903a). *Chelifer (Chernes) tullgreni*, Strand, in Scotland. *Annals of Scottish Natural History* **12**: 120–121.

Evans, W. (1903b). *Chernes dubius*, Cambr. (= *C. tullgreni*, Strand) in Scotland. *Annals of Scottish Natural History* **12**: 249–250.

Evans, W. (1905). *Chelifer latraillii*, Leach, in Fife. *Annals of Scottish Natural History* **14**: 247.

Ewing, H. E. (1911). Notes on pseudoscorpions; a study on the variations of our common species, *Chelifer cancroides* Linn., with systematic notes on other species. *Journal of the New York Entomological Society* **19**: 65–81.

Ewing, H. E. (1928). The legs and leg-bearing segments of some primitive arthropod groups, with notes on leg-segmentation in the Arachnida. *Smithsonian Miscellaneous Collections* **80** (11): 1–41.

Fabiani, R. (1902). Le grotte dei Colli Berici nel Vicentino. *Antologia Veneta* **3**(2): 3–16. (not seen)

Fabiani, R. (1904). Contributo alla conoscenza della fauna delle grotte di Malo, Priabona e Cereda nel Vicentino. *Rivista Italiana Speleol.* **2**: 8–13. (not seen)

Fabricius, J. C. (1775). *Systema Entomologiae Sistens Insectorum Classes, Ordines, Genera, Species, Adjectis Synonymis, Locis, Descriptionibus, Observationibus.* Flensburgi et Lipsiae. (not seen)

Fabricius, J.C. (1781). *Species Insectorum*, vol. **1**. Hamburg. (not seen)

Fabricius, J. C. (1787). *Mantissa Insectorum Sistens Eorum Species Nuper Detectas Adjectis Characteribus Genericis, Differentiis Specifieis, Emendationibus, Observationibus*, vol. **1**. Hafniae. (not seen)

Fabricius, J. C. (1793). *Entomologia Systematica Emendata et Aucta. Secundum Classes, Ordines, Genera, Species Adjectis Synonimis, Locis, Obervationibus, Descriptionibus*, vol. **2**. Hafniae: C.G. Proft.

Fage, L. (1921). Travaux scientifiques de l'Armée d'Orient (1916–1918). Arachnides. *Bulletin du Muséum National d'Histoire Naturelle, Paris* **27**: 96–102, 173–177, 227–232.

Fage, L. (1933). Les Arachnides cavernicoles de Belgique. *Bulletin de la Société Entomologique de France* **38**: 53–56.

Fahringer, J. (1925). Beobachtungen über einige Bewohner von Bienenstöcken. I. Bücherskorpione. *Bienenvater, Wien* **57**: 83–84. (not seen)

Falconer, W. (1907a). A pseudo-scorpion new to Northumberland. *Naturalist, London* **32**: 388. (not seen)

Falconer, W. (1907b). A pseudo-scorpion new to Yorkshire. *Naturalist, London* **32**: 432. (not seen)

Falconer, W. (1908). *Chiridium museorum* Leach at Huddersfield. *Naturalist, London* **33**: 110. (not seen)

Falconer, W. (1910). Keys to the families and genera of British spiders, and to the families, genera and species of British harvestmen and pseudoscorpions. *Naturalist, London* **35**: 233–242, 323–332, 438–447.

Falconer, W. (1916). The harvestmen and pseudoscorpions of Yorkshire. *Naturalist, London* **41**: 103–106, 135–140, 155–158, 191–193.

Falconer, W. (1931). Isle of Wight Arachnida. New and additional records. *Proceedings of the Isle of Wight Natural History Society* **2**: 118–124. (not seen)

Falcoz, L. (1912a). Contribution à la faune des terriers de Mammifères. *Comptes Rendus Hebdomadaires des Séances de l'Académie des Sciences, Paris* **154**: 1380–1383.

Falcoz, L. (1912b). La recherche des arthropodes dans les terriers. *Feuille des Naturalistes* (5) **42**: 178–180. (not seen)

Falcoz, L. (1913). La recherche des arthropodes dans les terriers. *Feuille des Naturalistes* (5) **43**: 1–5. (not seen)

Falcoz, L. (1914). *Contribution à l'étude de la faune des microcavernes*. Thèse: Université de Lyon. (not seen)

Fanfani, A. & Groppali, R. (1979). La fauna di Montecristo. Arcipelago Toscano. *Pubblicazioni dell'Istituto Ent. Universita Pavia* **9**: 1–52. (not seen)

Fava, C. & Regalin, R. (1984). Ricerche biospeleologiche. V. Artropodi raccolti presso due sorgenti calde del Mte Reit (Alta Valtellina). *Rivista del Museo Civico di Scienze Naturali 'E. Caffi'* **8**: 109–117. (not seen)

Feio, J. L. de Araújo (1941). Sôbre um curiosa pseudoscorpião. *Geogarypus (Geogarypus) itapemirinensis* sp. n. (Garypidae: Neobisiinea). *Papéis Avulsos do Departamento de Zoologia, Secretaria de Agricultura, Sao Paulo* **1**: 241–244.

Feio, J. L. de Araújo (1942). Sôbre o apresamento e sucçao em algumas espécies de *Pachyolpium* e *Lustrochernes* (Pseudoscorpiones). *Boletim do Museu Nacional Rió de Janeiro, n.s. Zoologia* **3**: 113–120.

Feio, J. L. de Araújo (1944). *Victorwithius monoplacophorus* n. gen., n. sp. da subfamilia Withiinae Chamberlin, 1931 (Pseudoscorpiones: Cheliferidae). *Boletim do Museu Nacional Rió de Janeiro, n.s. Zoologia* **28**: 1–7.

Feio, J. L. de Araújo (1945). Novos pseudoscorpiões de região neotropical (com a descriçao de uma subfamilia, dois géneros e sete espécies). *Boletim do Museu Nacional Rió de Janeiro, n.s. Zoologia* **44**: 1–47.

Feio, J. L. de Araújo (1946). Sôbre o género *Pycnochernes* Beier, 1932, com a descrição de *P. guarany* n. sp., do Paraguai (Chernetidae: Pseudoscorpiones). *Livro de Homenagem a R. F. d'Almeida S. Paulo* no. **15**: 167–172.

Feio, J. L. de Araújo (1950). Informe fornecido a C. Hoff sobre pseudoscorpiones. *Arthropoda* **12**: 4. (not seen)

Feio, J. L. de Araújo (1960). Consideraciones sobre Chernetinae con la descripcion de *Maxchernes birabeni* genero y especies nuevos (Arachnida, Pseudoscorpiones). *Neotropica* **5**: 71–82.

Feio, J. L. de Araújo (1974). A quetotaxia do campo genital i a taxinomia em Pseudoscorpiones. *Arquivos do Museu de História Natural* **1**: 127–134.

Fenizia, C. (1902). Un caso di simbiosi utilitara reciproca. *Bollettino del Naturalista, Siena* **22**: 55–58. (not seen)

Fenstermacher, D. J. (1959). *Survey of the Pseudoscorpions of Michigan*. M.Sc. Thesis: Michigan State University. (not seen)

Fernandez, G. E. (1910). Datos para el conocimiento de la distribución geográfica de los Arácnidos de España. *Memorias de la Real Sociedad Española* de Historia Natural **6**: 343–424. (not seen)

Ferrer, F. (1904). Alguns articulats dels voltants de Barcelona. *Butlletí de la Institució Catalana d'Historia Natural* (2) **1**: 14–16. (not seen)

Ferronière, G. (1899). IIe contribution à l'étude de la faune de la Loire-Inférieure (Pseudoscorpion, Myriopodes, Annélides). *Bulletin de la Société des Sciences Naturelles de l'Ouest de la France* **9**: 137–146.

Filleul, E. (1922). The false scorpion. *Bee World* **4**: 140. (not seen)

Firstman, B. (1973). The relationship of the chelicerate arterial system to the evolution of the endosternite. *Journal of Arachnology* **1**: 1–54.

Flower, S. S. (1901). Notes on the millipedes, centipedes, scorpions, etc., of the Malay Peninsula and Siam. *Journal of the Straits Branch of the Asiatic Society* **36**: 1–48.

Földi, J. (1801). *Természeti Historia*. Pozsoni és Komàrom. (not seen)

Forcart, L. (1961). Katalog der Typusexemplare in der Arachnida-Sammlung des Naturhistorischen Museums zu Basel: Scorpionidea, Pseudoscorpionidea, Solifuga, Opilionidea und Araneida. *Verhandlungen der Naturforschenden Gesellschaft in Basel* **72**: 47–87.

Foster, N. H. (1912). *Obisium lubricum*, a false-scorpion new to the Irish fauna. *Irish Naturalist* **21**: 245.

Fourcroy, A. F. de (1785). *Entomologica Parisiensis: Sive Catalogus Insectorum, Quae in Agro Parisiensi Reperiunter; Secundum Methodum Geoffraeanam in Sectiones, Genera et Species Distributus: Cui Addita Sunt Nomina Trivialia et Fere Trecentae Novae Species*, vols **1, 2**. Academiae: Parisiis.

Franciscolo, M. (1949). Su alcune grotte dei dintorni di Bardineto (Provincia di Savona). *Rassegna Speleologica Italiana* **1**: 43–52. (not seen)

Franciscolo, M. (1951). La fauna della Arma Pollera n. 24 Li, presso Finale Ligure. *Rassegna Speleologica Italiana* **3**: 40–53. (not seen)

Franciscolo, M. (1952). Su alcune grotte nuove o poco note della provincia di Savona (Liguria occidentale). *Rassegna Speleologica Italiana* **4**: 57–70. (not seen)

Franciscolo, M. (1955). Fauna cavernicola del Savonese. *Annali del Museo Civico di Storia Naturale di Genova* **67**: 1–223.

Franganillo, P. (1913). Arácnidos de Asturias y Galicia. *Broteria* **11**: 119–133. (not seen)

Franke, U., Friebe, B. & Beck, L. (1988). Methodisches zur Ermittlung der Siedlungsdichte von Bodentieren aus Quadratproben und Barberfallen. *Pedobiologia* **32**: 253–264. (not seen)

Franz, H. & Beier, M. (1948). Zur Kenntnis der Bodenfauna in pannonischen Klimagebiet Österreichs. II. Die Arthropoden. *Annalen des Naturhistorischen Museums in Wien* **56**: 440–549.

Franz, H., Grunhold, P. & Pschorn-Walcher, H. (1959). Die Kleintiergemeinschaft der Auwaldböden der Umgebung von Linz und benachbarter Flussgebiete. *Naturkundliches Jahrbuch der Stadt Linz* **1959**: 7–64. (not seen)

Frauenfeld, G.R. von (1867). Zoologische miscellen. XI. Das Insektenleben zur See. *Verhandlungen der K. K. Zoologischen-Botanischen Gesellschaft in Wien* **17**: 425–502.

Frickhinger, H. W. (1919). *Chelifer cancroides* als Feind der Bettwanze (*Cimex lectularius* L.). *Zeitschrift für Angewandte Entomologie* **6**: 170–171.

Fritsch, A. (1904). *Palaeozoische Arachniden*. Prague. (not seen)

Fritsch, P. (1980). Tanzfliege und Anhalter-Moosskorpion. *Bericht der Naturforschenden Gesellschaft Augsberg* **35**: 44–46. (not seen)

Frivaldsky, I. (1876). Jellemzö adatok Magyarország faunájához. *Magyar Tudomànyos Akademia Evnönyvei* **11**: ?–? (not seen)

Frivaldsky, J. (1865a). Adatok a magyarhoni barlangok faunájához. *Mathematikai és Természettudományi Közlemények, Vonatkozólag a Hazai Viszonyokra* **3**: 17–53.

Frivaldsky, J. (1865b). Title? *Mathematikai és Természettudományi Közlemények, Vonatkozólag a Hazai Viszonyokra* **5**: ?–? (not seen)

Frøiland, O. (1976). Pseudoscorpions. *Fauna of the Hardangervidda* **8**: 11–12.

Frost, S.W. (1966). Additions to Florida insects taken in light traps. *Florida Entomologist* **49**: 243–251.

Fuller, C. (1901). Entomology. *Report of the Natal Government Entomologist* **1**: ?–? (not seen)

Fussey, G. D. (1982). A new pseudoscorpion for the north of England. *Naturalist, London* **107**: 111–112.

Gabbutt, P. D. (1962). 'Nests' of the marine false-scorpion. *Nature, London* **196**: 87–88.
Gabbutt, P. D. (1965). The external morphology of two pseudoscorpions *Neobisium carpenteri* and *Neobisium maritimum*. *Proceedings of the Zoological Society of London* **145**: 359–386.
Gabbutt, P. D. (1966a). An investigation of the silken chambers of the marine pseudoscorpion *Neobisium maritimum*. *Journal of Zoology, London* **149**: 337–343.
Gabbutt, P. D. (1966b). A new species of pseudoscorpion from Britain. *Journal of Zoology, London* **150**: 165–181.
Gabbutt, P. D. (1967). Quantitative sampling of the pseudoscorpion *Chthonius ischnocheles* from beech litter. *Journal of Zoology, London* **151**: 469–478.
Gabbutt, P. D. (1969a). A key to all stages of the British species of the family Neobisiidae (Pseudoscorpiones: Diplosphyronida). *Journal of Natural History* **3**: 183–195.
Gabbutt, P. D. (1969b). Chelal growth in pseudoscorpions. *Journal of Zoology, London* **157**: 413–427.
Gabbutt, P. D. (1969c). Life histories of some British pseudoscorpions inhabiting leaf litter. In: Sheals, J.G. (ed.), *The Soil Ecosystem*: 229–235. Systematics Association Publication no. 8.
Gabbutt, P. D. (1970a). Pseudoscorpions: growth and trichobothria. *Bulletin du Muséum National d'Histoire Naturelle, Paris* (2) **41**, supplement 1: 135–140.
Gabbutt, P. D. (1970b). The external morphology of the pseudoscorpion *Dactylochelifer latreillei*. *Journal of Zoology, London* **160**: 313–335.
Gabbutt, P. D. (1970c). Sampling problems and the validity of life history analyses of pseudoscorpions. *Journal of Natural History* **4**: 1–15.
Gabbutt, P. D. (1972a). The disposition of trichobothria in the Chernetidae (Pseudoscorpiones). *Arachnologorum Congressus Internationalis V, Brno*: 37–42.
Gabbutt, P. D. (1972b). Differences in the disposition of trichobothria in the Chernetidae (Pseudoscorpiones). *Journal of Zoology, London* **167**: 1–13.
Gabbutt, P. D. (1972c). Some observations of taxonomic importance on the family Chernetidae (Pseudoscorpiones). *Bulletin of the British Arachnological Society* **2**: 83–86.
Gabbutt, P. D. & Aitchison, C. W. (1980). The effect of temperature and season on the number of hibernation chambers built by adult pseudoscorpions. *Verhandlungen des 8. Internationalen Arachnologen-Kongresses, Wien*: 57–60.
Gabbutt, P. D. & Vachon, M. (1963). The external morphology and life history of the pseudoscorpion *Chthonius ischnocheles* (Hermann). *Proceedings of the Zoological Society of London* **140**: 75–98.
Gabbutt, P. D. & Vachon, M. (1965). The external morphology and life history of the pseudoscorpion *Neobisium muscorum*. *Proceedings of the Zoological Society of London* **145**: 335–358.
Gabbutt, P. D. & Vachon, M. (1967). The external morphology and life history of the pseudoscorpion *Roncus lubricus*. *Journal of Zoology, London* **153**: 475–498.
Gabbutt, P. D. & Vachon, M. (1968). The external morphology and life history of the pseudoscorpion *Microcreagris cambridgei*. *Journal of Zoology, London* **154**: 421–441.
Gaisberger, K. (1973). Über Beobachtungen des Pseudoskorpiones im Toten Gebirge. *Mitteilungen der Sektion Ausseerland des Landesvereines für Höhlenkunde in Steiermark* **11**: 33–34. (not seen)
Gaisberger, K. (1977). Beobachtungen am Verhalten des Pseudoskorpions *Neobisium (Blothrus) aueri* Beier. *Mitteilungen der Sektion Ausseerland des Landesvereines für Höhlenkunde in Steiermark* **15**: 66. (not seen)

Gaisberger, K. (1979). Schon 15 Fundstellen des Höhlen-Pseudoskorpions *Neobisium aueri* Beier im Toten Gebirge. *Mitteilungen der Sektion Ausseerland des Landesvereines für Höhlenkunde in Steiermark* **17**: 44–45. (not seen)

Gaisberger, K. (1980a). Die Arbeit am Katalog der rezenten Höhlenfauna. *Mitteilungen der Sektion Ausseerland des Landesvereines für Höhlenkunde in Steiermark* **18**: 33. (not seen)

Gaisberger, K. (1980b). Weitere Höhlen-Pseudoskorpionfunde im Toten Gebirge. *Mitteilungen der Sektion Ausseerland des Landesvereines für Höhlenkunde in Steiermark* **18**: 91–92. (not seen)

Gaisberger, K. (1981). Eine neue Fundstelle des Höhlenlaufkäfers *Arctaphaenops nihilumalbi* im Toten Gebirge. *Mitteilungen der Sektion Ausseerland des Landesvereines für Höhlenkunde in Steiermark* **19**: 33. (not seen)

Gaisberger, K. (1984a). Katalog der rezeneten Höhlentiere (Wirbellose) des Toten Gebirge. *Schriftenreihe des Heimatmuseums 'Ausseerland'* **6**: 1–30. (not seen)

Gaisberger, K. (1984b). Bemerkungen zum Vorkommen von Pseudoskorpionen im Toten Gebirge (Österreich). *Höhle* **35**: 57–58. (not seen)

Gardini, G. (1975). Pseudoscorpioni dell'Isola di Capraia (Arcipelago Toscano) (Arachnida). *Lavori della Societa Italiana di Biogeografia, nuova serie* **5**: 1–12.

Gardini, G. (1976a). Note su alcuni Pseudoscorpioni raccolti in Romagna (Arachnida). *Bollettino del Museo Civico di Storia Naturale, Verona* **2**: 251–257.

Gardini, G. (1976b). Descrizione di *Chthonius (Ephippiochthonius) bartolii*, nuova specie di Pseudoscorpionida Chthoniidae (Arachnida) della Liguria orientale. (Note sugli Pseudoscorpioni d'Italia. V). *Bollettino dei Musei e degli Istituti Biologici della Università di Genova* **44**: 93–101.

Gardini, G. (1977a). Note sugli Pseudoscorpioni d'Italia. III. Su un ♂ di *Spelyngochthonius* di Sardegna: *S. sardous* Beier? (Pseudoscorpionida, Chthoniidae). *Bollettino della Società Sarda di Scienze Naturali* **16**: 39–49.

Gardini, G. (1977b). *Chthonius (Neochthonius) caprai*, n. sp., della Liguria orientale. (Note sugli Pseudoscorpioni d'Italia. IV). *Memorie della Società Entomologica Italiana* **55**: 216–222.

Gardini, G. (1979). Ridescrizione di *Chthonius (s. str.) lanzai* di Cap., 1948 e *C. (s. str.) elongatus* Lazzeroni, 1969 (Pseudoscorpioni d'Italia VI). *Bollettino della Società Entomologica Italiana* **111**: 126–133.

Gardini, G. (1980a). Ridescrizione di *Chthonius (C.) irregularis* Beier, 1961 e *C. (E.) concii* Beier, 1953 (Pseudoscorpioni d'Italia IX). *Atti della Società Italiana di Scienze Naturali, e del Museo Civile di Storia Naturale, Milano* **121**: 193–200.

Gardini, G. (1980b). Identita' di *Chthonius tetrachelatus fuscimanus* Simon, 1900 e redescrizione di *C. (E.) nanus* Beier, 1953. (Pseudoscorpionida Chthoniidae) (Pseudoscorpioni d'Italia IX). *Annali del Museo Civico di Storia Naturale di Genova* **88**: 261–270.

Gardini, G. (1980c). Catalogo degli Pseudoscorpioni cavernicoli italiani (Pseudoscorpioni d'Italia VIII). *Memorie della Società Entomologica Italiana* **58**: 95–140.

Gardini, G. (1981a). Pseudoscorpioni cavernicole sardi. I. Chthoniidae (Pseudoscorpioni d'Italia, X). *Revue Arachnologique* **3**: 101–114.

Gardini, G. (1981b). *Roncus caralitanus* n. sp. della Sardegna meridionale (Pseudoscorpionida Neobisiidae) (Pseudoscorpioni d'Italia XIII). *Bollettino della Società Entomologica Italiana* **113**: 129–135.

Gardini, G. (1981c). Raccolta, conservazione, allevamento e studio degli Pseudoscorpioni. *Bollettino della Società Entomologica Italiana, Supplemento* **22**: 13–16.

Gardini, G. (1982a). Raccolta, conservazione, allevamento studio degli Pseudoscorpioni (continued). *Bollettino della Società Entomologica Italiana, Supplemento* **23**: 1–7.

Gardini, G. (1982b). Compléments à la description de *Roncus euchirus* (Simon, 1879) (Pseudoscorpionida, Neobisiidae). *Revue Arachnologique* **4**: 151–155.

Gardini, G. (1982c). Pseudoscorpioni cavernicoli Sardi. II. Neobisiidae e Chernetidae, con considerazioni sui Neobisiinae cavernicoli. (Pseudoscorpioni d'Italia XII). *Fragmenta Entomologica* **16:** 89–115.

Gardini, G. (1982d). *Balkanoroncus baldensis* n. sp. delle Prealpi Venete (Pseudoscorpionida, Neobisiidae) (Pseudoscorpioni d'Italia XIV). *Bollettino del Museo Civico di Storia Naturale, Verona* **9:** 161–173.

Gardini, G. (1982e). Pseudoscorpioni cavernicoli italiani. *Lavori della Società Italiana di Biogeografia* **7:** 15–32.

Gardini, G. (1983a). Redescription of *Roncus lubricus* L. Koch, 1873, type-species of the genus *Roncus* L. Koch, 1873 (Pseudoscorpionida, Neobisiidae). *Bulletin of the British Arachnological Society* **6:** 78–82.

Gardini, G. (1983b). *Larca italica* n. sp. cavernicola dell'Appennino Abruzzese (Pseudoscorpionida, Garypidae) (Pseudoscorpioni d'Italia XV). *Bollettino della Società Entomologica Italiana* **115:** 63–69.

Gardini, G. (1985a). Su alcuni Pseudoscorpioni cavernicoli di Grecia (Pseudoscorpionida, Neobisiidae). *Bollettino del Museo Regionale di Scienze Naturali, Torino* **3:** 53–64.

Gardini, G. (1985b). Segnalazioni faunistiche Italiane. *Bollettino della Società Entomologica Italiana* **117:** 60–61.

Gardini, G. (1987). Segnalazioni faunistiche Italiane no. 103: *Mesochelifer ressli* Mahnert. *Bollettino della Società Entomologica Italiana* **119:** 123. (not seen)

Gardini, G. (1989). Pseudoscorpioni cavernicoli greci, con descrizione di *Chthonius (E.) gasparoi* n. sp. della Macedonia (Arachnida, Pseudoscorpionida) (Pseudoscorpioni di Grecia II). *Atti e Memorie della Commissione Grotte 'E. Boegan'* **27:** 57–62.

Gardini, G., Lattes, A. & Rizzerio, R. (1981). Variabilità morfometrica nel genere *Roncus* (Pseudoscorpionida, Neobisiidae). *Bollettino di Zoologia* 48, supplement **1:** 59.

Gardini, G. & Rizzerio, R. (1985). Materiali per una revisione del genere *Roncus* L. Koch, 1873. I. Ridescrizione dei tipi di alcune specie italiane non cavernicole (Pseudoscorpionida, Neobisiidae). *Fragmenta Entomologica* **18:** 47–79.

Gardini, G. & Rizzerio, R. (1986a). Materiali per una revisione del genere *Roncus* L. Koch, 1873. II. Ridescrizione dei tipi delle specie parablothroidi alpine e appenniniche. *Fragmenta Entomologica* **19:** 1–56.

Gardini, G. & Rizzerio, R. (1986b). *Neobisium (O.) zoiai* n. sp. delle Alpi Liguri e note su *Roncus ligusticus* Beier, 1930 (Pseudoscorpionida Neobisiidae). *Bollettino della Società Entomologica Italiana* **118:** 5–16.

Gardini, G. & Rizzerio, R. (1987). *Roncus zoiai*, nuova specie cavernicola del Monte Albo, Sardegna nord-orientale (Pseudoscorpionida, Neobisiidae). *Fragmenta Entomologica* **19:** 283–292.

Gardini, G. & Rizzerio, R. (1987b). I *Roncus* eucavernicoli del gruppo *siculus*. *Bollettino della Società Entomologica Italiana* **119:** 67–80.

Garin, H. (1937). Über den Bienen-Skorpion *Chelifer sculpturatus*. *Archiv für Bienenkunde* **18:** 33–48.

Garneri, G.A. (1902). Contribuzione alla fauna Sarda, Aracnidi. *Bollettino della Società Zoologica Italiana* **11:** 57–103. (not seen)

Gasperini, R. (1892). Prilog k Dalmatinskoj fauna (Isopoda, Myriapoda, Arachnida). *Godišnji Izveštaj Gimn.* **1892:** 1–75. (not seen)

Gaubert, P. (1890). Sur les fentes qui se trouvent sur le céphalothorax des Araneides et de *Chelifer*. *Compte Rendu de la Société Philomatique* (7) **2:** 47–53. (not seen)

Gaubert, P. (1892a). Recherches sur les organes des sens et sur les systèmes tégumentaire, glandulaire et musculaire des appendices des Arachnides. *Annales des Sciences Naturelle, Zoologie* (7) **13:** 31–185.

Gaubert, P. (1892b). Sur les pièces buccales des Arachnides. *Compte Rendu de la Société Philomatique* (7) **4**: 2. (not seen)

Gebhardt, A. (1932a). Die spelaeobiologische Erforschung der Abaligeter Höhle (Södungarn). *Sitzungsberichte der Gesellschaft Naturforschender der Freunde zu Berlin* **1931**: 304–317. (not seen)

Gebhardt, A. (1932b). Ökologische und faunistische Untersuchungen im Zenoga-Becken. *Allattini Közlemények* **29**: 42–58. (not seen)

Geoffroy, E. L. (1762). *Histoire Abregée des Insectes Qui se Trouvent aux Environs de Paris; dans Laquelle ces Animaux Sont Rangés Suivant un Ordre Méthodique*, vol. **2**. Durand: Paris.

Geoffroy, J. J., Christophe, T., Molfetas, S. & Blandin, P. (1981). Etude d'un écosystème mixte. III. Traits généraux du peuplement de macroarthropodes édaphiques. *Revue d'Ecologie et de Biologie du Sol* **18**: 39–59. (not seen)

George, R. S. (1955). Some additional records of Arachnida from Gloucestershire, including four species of mites new to Great Britain. *Entomologist's Monthly Magazine* **91**: 121–124.

George, R. S. (1957). A brief list of the harvestmen (Opiliones) and falsescorpions (Pseudoscorpiones) of Gloucestershire. *Proceedings of the Cotteswood Naturalists Field Club* **32**: 79–81.

George, R. S. (1961). More records of Gloucestershire falsescorpions. *Report of the North Gloucestershire Naturalists' Society* **1959–1960**: 38.

Gerdes, G. & Krumbein, W. E. (1984). Animal communities in recent potential stromatolites of hypersaline origin. In.: Cohen, Y., Castenholz, R. W. & Halvorson, D. (eds), *MBL Lectures in Biology* **3**: 59–83.

Gering, R. L. (1948). *A Comparative Morphological Study of Three Species of the Chelonethid Genus* Parachelifer. M.A. Thesis: University of Utah. (not seen)

Gering, R. L. (1956). Arachnids: spiders, pseudoscorpions, scorpions, solpugids. In: Woodbury, A.M. (ed.), *Ecological Check Lists. The Great Salt Lake Desert Series*. University of Utah: Dugway. (not seen)

Gerstäcker, C. E. A. (1859). *Bericht über die Wissenschaftlichen Leistungen im Gebiete der Entomologie*. (not seen)

Gerstäcker, C. E. A. (1863). Arthropoden. In: Peters, W. C. H., Carus, J. V. & Gerstäcker, C. E. A., *Handbuch der Zoologie*, vol. **2**. Engelmann: Leipzig.

Gerstaecker, A. (1873). Arachnoidea. In: C. C. von der Decken, *Reisen in Ost-afrika in den Jahren 1859–1865*, vol. **3**: 461–503. C. F. Winter: Leipzig & Heidelberg.

Gervais, P. (1842). Untitled. *Annales de la Société Entomologique de France* **11**: xlv–xlviii.

Gervais, P. (1844). In: Walckenaer, C.A., *Histoire Naturelle des Insectes. Aptères*, vol. **3**. Librairie Encyclopédique de Roret: Paris.

Gervais, P. (1849). Aracnidos. II. Quelifereos. In: Gay, C., *Historia Fisica y Politica de Chile, Zoologia*, vol. **4**: 10–13. Paris.

Gestro, R. (1887). Gli *Anophthalmus* trovati finora in Liguria. *Annali del Museo Civico di Storia Naturale di Genova* (2) **5**: 487–508.

Gestro, R. (1904). Una gita in Sardegna. Divagazioni biogeografiche. *Bollettino della Societá Geografica Italiana* (4) **5**(4): 1–39. (not seen)

Gestro, R. (1907). Una gita in Garfagnana. *Annali del Museo Civico di Storia Naturale di Genova* (3) **3**: 168–177.

Ghabbour, S. I., Mikhaïl, W., & Rizk, M. A. (1977). Ecology of soil fauna of Mediterranean desert ecosystems in Egypt. I. – Summer populations of soil mesofauna associated with major shrubs in the littoral sand dunes. *Revue d'Ecologie et de Biologie du Sol* **14**: 429–459.

Gigon, R., Strinati, P. & Aellen, V. (1980). Contribution suisse à la spéléologie de la région de Taza (Moyen Atlas marocain). *Cavernes, Neuchâtel* **24**: 9–26. (not seen)

Gilbert, O. (1951). Observations on the feeding of some British false scorpions. *Proceedings of the Zoological Society of London* **121**: 547–555.

Gilbert, O. (1952a). Studies on the histology of the mid-gut of the Chelonethi or Pseudoscorpiones. *Quarterly Journal of Microscopical Science* **93**: 31–45.

Gilbert, O. (1952b). Three examples of abnormal segmentation of the abdomen in *Dactylochelifer latreilli* (Leach), (Chelonethi). *Annals and Magazine of Natural History* (12) **5**: 47–49.

Ginet, R. (1952). La grotte de la Balme (Isère): topographie et faune. *Bulletin Mensuel de la Société Linnéenne de Lyon* **21**: 4–17. (not seen)

Giovannoli, L. (1933). Invertebrate life of Mammoth and other neighboring caves. *American Midland Naturalist* **14**: 600–634.

Gisin, H. (1949). Les pseudoscorpions. *Musées de Genève* **6(9)**: 4.

Goddard, S. J. (1976a). Feeding in *Neobisium muscorum* (Leach) (Arachnida: Pseudoscorpiones). *Bulletin of the British Arachnological Society* **3**: 232–234.

Goddard, S. J. (1976b). Population dynamics, distribution patterns and life cycles of *Neobisium muscorum* and *Chthonius orthodactylus* (Pseudoscorpiones: Arachnida). *Journal of Zoology, London* **178**: 295–304.

Goddard, S. J. (1979). The population metabolism and life history tactics of *Neobisium muscorum* (Leach) (Arachnida: Pseudoscorpiones). *Oecologia* **42**: 91–105.

Godfrey, R. (1901a). *Obisium muscorum* on Edinburgh Castle Rock. *Annals of Scottish Natural History* **10**: 118.

Godfrey, R. (1901b). Chernetidae or false scorpions of West Lothian. *Annals of Scottish Natural History* **10**: 214–217.

Godfrey, R. (1904). Chernetidae in Ayrshire. *Annals of Scottish Natural History* **13**: 195.

Godfrey, R. (1907). The false-scorpions of the west of Scotland. *Annals of Scottish Natural History* **16**: 162–163.

Godfrey, R. (1908). The false scorpions of Scotland. *Annals of Scottish Natural History* **17**: 90–100, 155–161.

Godfrey, R. (1909). The false scorpions of Scotland (continued). *Annals of Scottish Natural History* **18**: 22–26, 153–163.

Godfrey, R. (1910). The false scorpions of Scotland (continued). *Annals of Scottish Natural History* **19**: 23–33.

Godfrey, R. (1921). False scorpions. *South African Journal of Natural History* **2**: 118–120.

Godfrey, R. (1923). The book-scorpion and its South African allies. *South African Journal of Natural History* **4**: 93–99.

Godfrey, R. (1927). The false-scorpions of Lovedale. *South African Outlook* **57**: 17–18.

Goede, A. & Goede, T. (1974). Tasmanian cave fauna, part 4 - pseudoscorpions. *Speleo Spiel* **93**: 2–4.

Gomez, P. & Diego, L. (1975). La biota bromelicola except anfibios y reptiles. *Historia Naturale de Costa Rica* **1**: 45–62. (not seen)

Goodier, R. (1970a). Notes on mountain spiders from Wales. I. Spiders caught in pitfall traps on the Snowdon National Nature Reserve. *Bulletin of the British Arachnological Society* **1**: 85–87.

Goodier, R. (1970b). Notes on mountain spiders from Wales. II. Spiders caught in pitfall traps on the Cwn Idwal National Nature Reserve, Caernarvonshire. *Bulletin of the British Arachnological Society* **1**: 97–98.

Gossel, P. (1936). Beitrage zur Kenntnis der Hautsinnesorgane und Hautdrüsen der Cheliceraten und der Aagen der Ixodiden. *Zeitschrift für Morphologie und Ökologie der Tiere* **30**: 177--205. (not seen)

Goulding, R. W. (1901). Lincolnshire Naturalists' Union at Mablethorpe. *Naturalist, London* **26**: ?-? (not seen)

Gozo, A. (1908). Gli aracnidi di caverne italiane. *Bollettino della Società Entomologica Italiana* **38**: 109–139. (not seen)

Graham-Smith, G. S. (1916). Observations on the habits and parasites of common flies. *Parasitology* **8**: 440–544.

Grasshoff, M. (1978). A model of the evolution of the main Chelicerate groups. *Symposia of the Zoological Society of London* **42**: 273–284.

Gravenhorst, J. L. C. (1835). Bericht über die in Bernstein erhaltenen Insecten der phys.-ökon. Gesellscaft zu Königsberg. *Uebersicht der Arbeiten und Veräderungen Schlesische Gesellschaft für Vaterländische Kultur* **1834**: 92–93. (not seen)

Graves, R. C. (1969). A pseudoscorpion found with *Cicindela hirticollis*. *Cicindela* **1**: 6. (not seen)

Graves, R. C. & Graves, A. C. F. (1969). Pseudoscorpions and spiders from moss, fungi, Rhododendron leaf litter, and other microcommunities in the highlands area of western North Carolina. *Annals of the American Entomological Society* **62**: 267–269.

Green, E. E. (1908). Curious habits of Chelifers. *Zoologist* (4) **12**: 159–160.

Grimpe, G. (1921). Chelifer als Schmarotzer. *Naturwissenschaftliche Wochenschrift* (NF) **20**: 628–631. (not seen)

Gross, G. F., Lee, D. C. & Zeidler, W. (1979). Invertebrates. In: Tyler, M. J., Twidale, C. R. & Ling, J. K. (eds), *Natural History of Kangaroo Island*: 129–137. Royal Society of South Australia: Adelaide.

Grossinger, J. B. (1794). *Universa Historica Physica Regni Hungariae*, vol. 4. (not seen)

Grube, A. E. (1859). Verzeichniss der Arachniden Liv-, Kur- und Ehstlands. *Archiv für die Naturkunde Liv-, Ehst- und Kurlands, Dorpat* (2) **1**: 415–486. (not seen)

Grube, A. E. (1872). Mittheilungen über die Meeresfauna von St-Waast-la-Hougue, St-Malo und Roscoff. *Veränderungen, Schlesische Gesellschaft für Vaterländische Kultur, Breslau* **1869–1872**: ?–? (not seen)

Gtowacinski, Z. & Witkowski, Z. (1969). [The fauna of the western Bieszczady Mountains, and problems of its conservation.] *Ochrona Przyrody* **34**: 127–160. (not seen)

Guéorguiev, V. (1966). Aperçu sur la faune cavernicole de la Bulgarie. *Izvestiya na Zoologicheskiya Institut, Sofia* **21**: 157–184. (not seen)

Guéorguiev, V. & Beron, P. (1962). Essai sur la faune cavernicole de Bulgarie. *Annales de Spéléologie* **17**: 285–441. (not seen)

Guérin-Méneville, F. E. (1838). *Iconographie du Régne Animal de Cuvier*, vols **2, 3**. Paris. (not seen)

Guilmette, J. E., Jr, Holzapfel, E. P. & Tsula, D. M. (1970). Trapping of air-borne insects on ships in the Pacific, part 8. *Pacific Insects* **12**: 303–325.

Gulicka, J. (1977). *Neobisium (Blothrus) slovacum* sp. n., eine neue Art des blinden Höhlenafterskorpions aus der Slowakei (Pseudoscorpionida). *Annotationes Zoologicae et Botanicae* **117**: 1–9.

Haack, R. A. & Wilkinson, R. C. (1987). Phoresy by *Dendrochernes* pseudoscorpions on Cerambycidae (Coleoptera) and Aulacidae (Hymenoptera) in Florida. *American Midland Naturalist* **117**: 369–373.

Hadži, J. (1929). [*Obisium (Blothrus) karamani*, pseudoscorpion nouveaux de la Serbie du Sud.] *Acta Societatis Entomologicae Jugoslavensis* **3/4**: 61–71.

Hadži, J. (1930a). Prirodoslovna istrazivanja sjevernodalmatinskog otocja. I. Dugi i kornati. Pseudoscorpiones. *Prirodoslovna Istraživanja Kraljevine Jugoslavije* **16**: 65–79.

Hadži, J. (1930b). [Contribution à la connaissance des Pseudoscorpions cavernicoles]. *Glasnik Académie Royale Serbe, Belgrade* **140**: 1–36. (in Russian)

Hadži, J. (1930c). Pseudoscorpiones. *Bulletin International de l'Académie Yugoslave des Sciences et des Beaux-arts* **1930**: 19–24.

Hadži, J. (1932). [Prilog poznavanju pecinske faune Vjetrenice.] *Glasnik Srpske Akademije Nauka* **151**: 101–157.

Hadži, J. (1933a). Beitrag zur Kenntnis der Fauna der Höhle Vjetrenica. *Bulletin de l'Académie des Sciences Mathématiques et Naturelles, Belgrade* **1**: 49–79.

Hadži, J. (1933b). Prinos poznavanju pseudoskorpijske faune Primorja. *Prirodoslovna Istraživanja Kraljevine Jugoslavije* **18**: 125–192.

Hadži, J. (1933c). Beitrag zur Kenntnis der Pseudoskorpionen-Fauna des Küstenlandes. *Bulletin International de l'Académie Yugoslave des Sciences et des Beaux-arts* **27**: 173–199.

Hadži, J. (1937). Pseudoskorpioniden aus Südserbien. *Glasnik Skopskog Naucnog Drustva* **17**: 151–187.

Hadži, J. (1938). Pseudoskorpioniden aus Südserbien (cont.). *Glasnik Skopskog Naucnog Drustva* **18**: 13–38.

Hadži, J. (1939a). Pseudoskorpione aus Karpathenrussland. *Věstník Československé Zoologické Spolecnosti v Praze* **6–7**: 183–208.

Hadži, J. (1939b). Pseudoskorpioniden aus Bulgarien. *Mitteilungen aus dem Königl. Naturwissenschaftlichen Institut in Sofia* **13**: 18–48.

Hadži, J. (1940). Eine neue Art von Höhlen-Pseudoskorpioniden aus Südserbien. *Neobisium (Blothrus) ohridanum* sp. n. *Glasnik Skopskog Naucnog Drustva* **22**: 129–135.

Hadži, J. (1941). Biospeološki prisperek. *Zbornikna Prirodne Društvo* **2**: 83–91. (not seen)

Hagen, H. (1859). *Chelifer* als Schmarotzer auf Insekten. *Stettiner Entomologische Zeitung* **20**: 202.

Hagen, H. (1867). Untitled. *Proceedings of the Boston Society of Natural History* **11**: 323–325.

Hagen, H. (1868a). Untitled. *Proceedings of the Boston Society of Natural History* **11**: 435.

Hagen, H. (1868b). The American pseudo-scorpions. *Record of American Entomology for the Year 1868*: 48–52.

Hagen, H. (1870). Synopsis pseudoscorpionidum synonymica. *Proceedings of the Boston Society of Natural History* **13**: 263–272.

Hagen, H. (1879). Hoehlen-Chelifer in Nord-America. *Zoologischer Anzeiger* **2**: 399–400.

Hagen, H. (1880). Untitled. *American Entomologist* **3**: 83–84.

Hahn, C. W. (1834). *Die Arachniden. Getreu nach der Natur abgebildet und beschrieben*, vol. **2**. C.H. Zeh'schen: Nürnberg. (not seen)

Hahn, N. S. (1983). *Estudos Biológicos e Morfológicos Sobre* Paratemnus minor *(Balzan, 1891) (Pseudoscorpiones, Atemnidae)*. Thesis: Universidade Estadual Paulista.

Hahn, N. S. & Matthiesen, F. A. (1980). Estudos biológicos sobre pseudo-escorpioes brasileiros. *Ciência e Cultura* **32**: 844.

Hahn, N. S. & Matthiesen, F. A. (1982). Aspectos do comportamento de *Paratemnus minor* (Balzan, 1891) (Pseudoscorpiones, Atemnidae). *IX Congresso Brasileiro de Zoologia, Porto Aligre*: 98–99.

Haldeman, S. S. (1848a). *Cremastochilus* in ant nests. *American Journal of Science* (2) **6**: 148.

Haldeman, S. S. (1848b). *Cremastochilus* in ant nests. *Annals and Magazine of Natural History* (2) **2**: 221–222.

Halperin, J. & Mahnert V. (1987). On some bark-inhabiting Pseudoscorpiones (Arachnida) from Israel. *Israel Journal of Entomology* **21**: 127–128.

Hamann, O. (1896). *Europäische Höhlenfauna. Eine Dartellung der in den Höhlen Europas lebenden Tierwelt mit besonderer Berücksichtigung der Höhlenfauna Krains.* H. Costenoble: Jena. (not seen)

Hamilton-Smith, E. (1967). The Arthropoda of Australian caves. *Journal of the Australian Entomological Society* **6**: 103–118.

Hamlyn-Harris, (1900). Title? *British Bee Journal and Bee-Keepers' Adviser* **27**: ?–? (not seen)

Hammen, L. van der (1949). On Arachnida collected in Dutch greenhouses. *Tijdschrift voor Entomologie* **91**: 72–82.

Hammen, L. van der (1969). Bijdrage tot de kennis van de Nederlandse bastaard-schorpioenen (Arachnida, Pseudoscorpionida). *Zoologische Bijdragen* **11**: 15–24.

Hammen, L. van der (1977). A new classification of Chelicerata. *Zoologische Mededeelingen* **51**: 307–319.

Hammen, L. van der (1983). De Spinachtigen (Arachnida) van de ondergrondse kalksteengroeven in Zuid-Limburg. *Zoologische Bijdragen* **29**: 5–51.

Hammen, L. van der (1985a). Functional morphology and affinities of extant Chelicerata in evolutionary perspective. *Transactions of the Royal Society of Edinburgh* **76**: 137–146.

Hammen, L. van der (1985b). Comparative studies in Chelicerata III. Opilionida. *Zoologische Verhandelingen* **220**: 1–60.

Hammen, L. van der (1986a). Comparative studies in Chelicerata IV. Apatellata, Arachnida, Scorpionida, Xiphosura. *Zoologische Verhandelingen* **226**: 1–52.

Hammen, L. van der (1986b). Acarological and arachnological notes. *Zoologische Mededelingen* **60**: 217–230.

Hansen, H. (1988). Über die Arachniden-fauna von urbanen lebensräumen in Venedig (Arachnida: Pseudoscorpiones, Araneae). *Bollettino del Museo Civico di Storia Naturale di Venezia* **38**: 183–219.

Hansen, H. J. (1884). Arthrogastra Danica: en monographisk fremstillning af de i Danmark levende Meiere og Mosskorpioner med bidrag til sidstaevnte underorders systematik. *Natuurwetenschappelijk Tijdschrift* (3) **14**: 491–554.

Hansen, H. J. (1885). Chernetidae. In: Schiödte, J. C. & Hansen, H. J., *Zoologia Danica*, vol. **4**: 101–117. (not seen)

Hansen, H. J. (1893). Organs and characters in different orders of arachnids. *Entomologiske Meddelelser* **4**: 135–251.

Hansen, H. J. & Sørensen, W. (1904). *On Two Orders of Arachnida*. Cambridge University Press: Cambridge.

Hanström, B. (1919). *Zur Kenntnis des centralen Nervensystems der Arachnoiden und Pantopoden, nebst Schussfolgerungen betreffs der Phylogenie der genannten Gruppen*. Inaugural Diss.: Lund. (not seen)

Hanström, B. (1928). *Vergleichende Anatomie des Gehirns Kenntnis Nervensystems der Wirbellosen Tiere*. Berlin. (not seen)

Harms, K. H. (1978). Zur Verbreitung und Gefährdung der Spinnentiere Baden-Württembergs (Arachnida: Araneae, Pseudoscorpiones, Opiliones). *Beihefte Veröffentlichungen zum Naturschutz und Landschaftspflege Baden-Württemberg* **11**: 313–322. (not seen)

Harnisch, O. (1926). Studien zur Ökologie und Tiergeographie der Moore. *Zoologische Jahrbücher, Systematik (Ökologie), Geographie und Biologie* **61**: 1–166. (not seen)

Harvey, M. S. (1981a). *Geogarypus rhantus*, sp. nov. (Pseudoscorpionida: Garypidae: Geogarypinae): a generic addition to the Australian fauna. *Memoirs of the Queensland Museum* **20**: 279–283.

Harvey, M. S. (1981b). A checklist of the Australian Pseudoscorpionida. *Bulletin of the British Arachnological Society* **5**: 237–252.

Harvey, M. S. (1982). A parasitic nematode (Mermithidae) from the pseudoscorpion 'Sternophorus' hirsti Chamberlin (Sternophoridae). *Journal of Arachnology* **10**: 192.

Harvey, M. S. (1983). *Contributions to the Systematics of the Pseudoscorpionida: the Genus* Synsphyronus *Chamberlin (Garypidae) and the Family Sternophoridae (Arachnida)*. Ph.D. Thesis: Monash University.

Harvey, M. S. (1984). The genus *Nannochelifer* Beier, with a new species from the Coral Sea (Pseudoscorpionida, Cheliferidae). *Journal of Arachnology* 12: 291–296.

Harvey, M. S. (1985a). The priority of *Blothrus* Schiödte, 1847, over *Neobisium* Chamberlin, 1930 (Neobisiidae: Pseudoscorpionida). *Bulletin of the British Arachnological Society* 6: 367–368.

Harvey, M. S. (1985b). Pseudoscorpionida. In: Walton, D.W. (ed.), *Zoological Catalogue of Australia*, vol. 3: 126–155. Australian Government Publishing Service: Canberra.

Harvey, M. S. (1985c). The systematics of the family Sternophoridae (Pseudo-scorpionida). *Journal of Arachnology* 13: 141–209.

Harvey, M. S. (1986). The Australian Geogarypidae, new status, with a review of the generic classification (Arachnida: Pseudoscorpionida). *Australian Journal of Zoology* 34: 753–778.

Harvey, M. S. (1987a). A revision of the genus *Synsphyronus* Chamberlin (Garypidae: Pseudoscorpionida: Arachnida). *Australian Journal of Zoology, Supplementary Series* 126: 1–99.

Harvey, M. S. (1987b). The occurrence in Australia of *Chthonius tetrachelatus* (Preyssler) (Pseudoscorpionida:Chthoniidae). *Australian Entomological Magazine* 13: 68–70.

Harvey, M.S. (1987c). Redescription and new synonyms of the cosmopolitan species *Lamprochernes savignyi* (Simon) (Chernetidae: Pseudoscorpionida). *Bulletin of the British Arachnological Society* 7: 111–116.

Harvey, M. S. (1987d). Redescriptions of *Geogarypus bucculentus* Beier and *G. pustulatus* Beier (Geogarypidae: Pseudoscorpionida). *Bulletin of the British Arachnological Society* 7: 137–141.

Harvey, M. S. (1987e). *Chelifer* Geoffroy, 1762 (Arachnida, Pseudoscorpionida): proposed conservation. *Bulletin of Zoological Nomenclature* 44: 188–189.

Harvey, M. S. (1988a). The systematics and biology of pseudoscorpions. In: Austin, A. D. & Heather, N. W. (eds), *Australian Arachnology*: 75–85. Australian Entomological Society: Brisbane.

Harvey, M. S. (1988b). Pseudoscorpions from the Krakatau Islands and adjacent regions, Indonesia (Chelicerata: Pseudoscorpionida). *Memoirs of the Museum of Victoria* 49: 309–353.

Harvey, M. S. (1989a). Two new cavernicolous chthoniids from Australia, with notes on the generic placement of the south-western Pacific species attributed to the genera *Paraliochthonius* Beier and *Morikawia* Chamberlin (Pseudoscorpionida: Chthoniidae). *Bulletin of the British Arachnological Society* 8: 21–29.

Harvey, M. S. (1989b). A new species of *Feaella* Ellingsen from north-western Australia (Pseudoscorpionida: Feaellidae). *Bulletin of the British Arachnological Society* 8: 41–44.

Harvey, M. S. & Mahnert, V. (1985). *Olpium* L. Koch, 1873 (Arachnida, Pseudo-scorpionida, Olpiidae): proposed designation of type species and related problems. *Bulletin of Zoological Nomenclature* 42: 85–88.

Harvey, M. S. & Yen, A. L. (1989). *Worms to Wasps: an Illustrated Guide to Australia's Terrestrial Invertebrates.* Oxford University Press: Melbourne.

Haupt, ? (1909). Beobachtungen am Bücherskorpion. *Zeitschrift für Naturwissen-schaften* 81: 181–182. (not seen)

Hayward, C. L. (1948). Biotic communities of the Wasatch Chaparral, Utah. *Ecological Monographs* 18: 473–506.

Heatwole, H. (1987). Major components and distributions of the terrestrial fauna. In: Dyne, G. R. & Walton, D. W. (eds), *Fauna of Australia*: vol. 1a: 101–135. Australian Government Printing Service: Canberra.

Heerdt, P. F. van & Mörzer Bruyns, M. F. (1960). A biocenological investigation in the yellow dune region of Terschelling. *Tijdschrift voor Entomologie* 103: 225–275. (not seen)

Helversen, O. von (1965). Scientific expedition to the Salvage Islands, July 1963. VI. Einige Pseudoskorpione von den Ilhas Selvagens. *Boletim do Museu Municipal do Funchal* **19**: 95–103.

Helversen, O. von (1966a). Pseudoskorpione aus dem Rhein-Main-Gebiet. *Senckenbergiana Biologica* **47**: 131–150.

Helversen, O. von (1966b). Über die Homologie der Tasthaare bei Pseudoskorpionen (Arach.). *Senckenbergiana Biologica* **47**: 185–195.

Helversen, O. von (1968). *Troglochthonius doratodactylus* n. sp., ein troglobionter Chthoniide (Arachnida: Pseudoscorpiones: Chthoniidae). *Senckenbergiana Biologica* **49**: 59–65.

Helversen, O. von (1969). *Roncus (Parablothrus) peramae* n. sp., ein troglobionter Neobisiide aus einer griechischen Tropfsteinhöhle (Arachnida: Pseudoscorpiones: Neobisiidae). *Senckenbergiana Biologica* **50**: 225–233.

Helversen, O. von & Martens, J. (1971). Pseudoskorpione und Weberknechte. *Die Wutach* **1**: 377–385.

Helversen, O. von & Martens, J. (1972). Unrichtige Fundort-Angaben in der Arachniden-Sammlung Roewer. *Senckenbergiana Biologica* **53**: 109–123.

Henriksen, K. L. (1929). Opiliones and Chernetes. In: Spärck R. & Tuxen, S. L. (eds), *The Zoology of the Faroes*, vol. **2**(46): 1–4. Andr. Fred. Host. and Son: Copenhagen. (not seen)

Henrikson, K. L., Lindroth, C. H. & Braendegaard, J. (1932). Isländische Spinnentiere. I. Opiliones, Chernetes, Araneae. *Göteborgs Kungl. Vetenskaps- och Vitterhetssamhälles Handlingar* (5B) **2**(7): 1–36. (not seen)

Henslow, J. S. (1831). A lobster-like insect, &c. *Magazine of Natural History* **4**: 284.

Hentschel, E. (1979). *Biología del Pseudoscorpion* Dinocheirus *sp. Asociado a* Neotomodon alstoni *(Mammalia Rodentia).* Thesis: Universidad Nacional Autónoma de México.

Hentschel Ariza, E. (1981). La evolución de la foresia en pseudoscorpiones (Arachnida, Pseudoscorpionida). *Folia Entomologica Mexicana* **48**: 44–45.

Hentschel, E. & Muchmore, W. B. (1989). *Cocinachernes foliosus*, a new genus and species of pseudoscorpion (Chernetidae) from Mexico. *Journal of Arachnology* **17**: 345–349.

Hermann, J. F. (1804). *Mémoire Aptérologique.* F. G. Levrault: Strasbourg.

Hernandez, J. J., Martin, J. L. & Medina, A. L. (1986). Fauna subterrania de Canarias. La Fauna de las Cuevas Volcanicas en Tenerife (Islas Canarias). *Congr. Espeleol., Barcelona* **2**: 139–142. (not seen)

Heselhaus, F. (1913). Über Arthropoden in Maulwurfnestern. *Tijdschrift voor Entomologie* **56**: 195–237.

Heselhaus, E. (1914). Über Arthropoden in Nestern. *Tijdschrift voor Entomologie* **57**: 62–88.

Hess, W. (1894). Über die Pseudoscorpioniden als Räuber. *Zoologischer Anzeiger* **17**: 119–121.

Heurtault-Rossi, J. (1963). Description de *Chthonius (E.) vachoni*, nouvelle espèce de Pseudoscorpions (Heterosphyronida, Chthoniidae) découverte en France, dans le Département de la Gironde. *Bulletin du Muséum National d'Histoire Naturelle, Paris* (2) **35**: 419–428.

Heurtault-Rossi, J. (1966a). *Roncus (R.) lucifugus* Simon, 1879, pseudoscorpion cavernicole de la faune française n'appartient pas au genre *Roncus* L. Koch, mais au genre *Microcreagris* Balzan. *Bulletin du Muséum National d'Histoire Naturelle, Paris* (2) **37**: 659–666.

Heurtault-Rossi, J. (1966b). Description d'une nouvelle espèce: *Neobisium (N.) caporiaccoi* (Arachnides, Pseudoscorpions, Neobisiidae) de la Province de Belluno, en Italie. *Bulletin du Muséum National d'Histoire Naturelle, Paris* (2) **38**: 606–628.

Heurtault-Rossi, J. (1968a). Quelques remarques sur deux espèces cavernicoles de *Chthonius* des départements des Bouches-du Rhône et du Gard: *Chthonius (C.) cephalotes* (Simon, 1875) et *Chthonius mayi* sp. nov. (Pseudoscorpions, Chthoniidae). *Bulletin du Muséum National d'Histoire Naturelle, Paris* (2) **39**: 912–922.

Heurtault, J. (1968b). Contribution a l'étude de *Neobisium (N.) praecipuum* Simon, 1879 (Pseudoscorpion, Neobisiidae). *Bulletin du Muséum National d'Histoire Naturelle, Paris* (2) **39**: 1077–1083.

Heurtault, J. (1968c). Une nouvelle espèce de pseudoscorpion du Gard: *Neobisium (N.) vachoni* (Neobisiidae). *Bulletin du Muséum National d'Histoire Naturelle, Paris* (2) **40**: 315–319.

Heurtault, J. (1969a). Neurosécrétion et glandes endocrines chez *Neobisium caporiaccoi* (Arachnides, Pseudoscorpions). *Comptes Rendus des Séances Hebdomadaires de l'Académie des Sciences de Paris* (D) **268**: 1105–1108.

Heurtault, J. (1969b). Une nouvelle espèce de l'Ardèche: *Neobisium (N.) balazuci* (Arachnides, Pseudoscorpions, Neobisiidae). *Bulletin du Muséum National d'Histoire Naturelle, Paris* (2) **40**: 955–961.

Heurtault, J. (1969c). Une nouvelle espèce de pseudoscorpion de l'Hérault, *Neobisium (N.) boui* (Neobisiidae). *Bulletin du Muséum National d'Histoire Naturelle, Paris* (2) **40**: 1171–1174.

Heurtault, J. (1970a). Pseudoscorpions du Tibesti (Tchad). I. Olpiidae. *Bulletin du Muséum National d'Histoire Naturelle, Paris* (2) **41**: 1164–1174.

Heurtault, J. (1970b). Pseudoscorpions du Tibesti (Tchad). II. Garypidae. *Bulletin du Muséum National d'Histoire Naturelle, Paris* (2) **41**: 1361–1366.

Heurtault, J. (1970c). Pseudoscorpions du Tibesti (Tchad). III. Miratemnidae et Chernetidae. *Bulletin du Muséum National d'Histoire Naturelle, Paris* (2) **42**: 192–200.

Heurtault, J. (1970d). Recherches préliminaires sur la neurosécrétion et les glandes endocrines chez un pseudoscorpion, *Neobisium caporiaccoi* Heurtault. *Bulletin du Muséum National d'Histoire Naturelle, Paris* (2) **42**, supplement 1: 59.

Heurtault, J. (1971a). Pseudoscorpions de la région du Tibesti (Sahara méridionale). IV. Cheliferidae. *Bulletin du Muséum National d'Histoire Naturelle, Paris* (2) **42**: 685–707.

Heurtault, J. (1971b). Chambre génitale, armature génitale et caractères sexuels secondaires chez quelques espèces de Pseudoscorpions (Arachnida) du genre *Withius*. *Bulletin du Muséum National d'Histoire Naturelle, Paris* (2) **42**: 1037–1053.

Heurtault, J. (1971c). Une nouvelle espèce cavernicole de Suisse *Neobisium (N.) helveticum* (Arachnide, Pseudoscorpion, Neobisiidae). *Revue Suisse de Zoologie* **78**: 903–907.

Heurtault, J. (1971d). Données complémentaires sur le complexe neuro-endocrine rétrocérébral des pseudoscorpions. *Comptes Rendus des Séances Hebdomadaires de l'Académie des Sciences de Paris* (D) **272**: 1981–1983.

Heurtault, J. (1972a). *Chthonius (C.) petrochilosi* (Arachnide, Pseudoscorpion, Chthoniidae), nouvelle espèce cavernicole de Grèce. *Biologia Gallo-Hellenica* **4**: 19–25.

Heurtault, J. (1972b). Etude histologique de quelques caractères sexuels mâles des Cheliferidae (Pseudoscorpions). *Arachnologorum Congressus Internationalis V, Brno*: 43–52.

Heurtault, J. (1972c). *Contribution à la Connaissance Biologique et Anatomo-Physiologique des Pseudoscorpions*. Thèse: no. C.N.R.S. AO 7261. (not seen)

Heurtault, J. (1973). Contribution à la connaissance biologique et anatomo-physiologique des pseudoscorpions. *Bulletin du Muséum National d'Histoire Naturelle, Paris* (3) **124**: 561–670.

Heurtault, J. (1974a). *Simonobisium* genre nouveau pour l'espèce *Neobisium myops* Simon, 1881 (Arachnides, Pseudoscorpions, Neobisiidae). *Bulletin du Muséum National d'Histoire Naturelle, Paris* (3) **164**: 1085–1093.

Heurtault, J. (1974b). *Roncus (R.) longidigitatus* (Ellingsen, 1908) (Arachnide, Pseudoscorpion cavernicole de la faune française) appartient au genre *Neobisium* J. C. Chamberlin (sous-genre *Blothrus*). *Annales de Spéléologie* **29**: 631–635.

Heurtault, J. (1975a). Deux nouvelles espèces de pseudoscorpions Chthoniidae (Arachnides) cavernicoles de Corse: *Chthonius (E.) remyi* et *Chthonius (E.) siscoensis*. *Annales de Spéléologie* **30**: 313–318.

Heurtault, J. (1975b). La digestion extra-intestinale: les dispositifs de filtrage et de triage chez les Pseudoscorpions (Arachnides). *Proceedings of the 6th International Arachnological Congress, Amsterdam*: 133–135.

Heurtault, J. (1976). Nouveaux caractères taxonomiques pour la sous-famille des Olpiinae (Arachnides, Pseudoscorpions). Note préliminaire. *Comptes Rendus 3ème Réunion des Arachnologistes d'Expression française, Les Eyzies, 1976*: 62–73.

Heurtault, J. (1977). *Occitanobisium coiffaiti* n. gen. n. sp. de Pseudoscorpions (Arachnides, Neobisiidae, Neobisiinae) du département de l'Hérault, France. *Bulletin du Muséum National d'Histoire Naturelle, Paris* (3) **497** (Zool. 346): 1121–1134.

Heurtault, J. (1979a). *R. leclerci*, deuxième espèce connue en France du genre *Roncobisium* (Arachnides, Pseudoscorpions, Neobisiidae). *Revue Arachnologique* **2**: 225–230.

Heurtault, J. (1979b). Le sous-genre *Ommatoblothrus* en France (Pseudoscorpions, Neobisiidae). *Revue Arachnologique* **2**: 231–238.

Heurtault, J. (1979c). Complément à la description de *Olpium pallipes* Lucas, 1845, type de la famille Olpiidae (Arachnides, Pseudoscorpions). *Revue Suisse de Zoologie* **86**: 925–931.

Heurtault, J. (1980a). Quelques remarques sur les espèces françaises du genre *Rhacochelifer* Beier (Arachnides, Pseudoscorpions, Cheliferidae). *Bulletin du Muséum National d'Histoire Naturelle, Paris* (4) **2**: 161–173.

Heurtault, J. (1980b). Complément à la description de *Minniza vermis* Simon, 1881, espèce-type du genre (Arachnides, Pseudoscorpions, Olpiidae). *Bulletin du Muséum National d'Histoire Naturelle, Paris* (4) **2**: 175–184.

Heurtault, J. (1980c). Données nouvelles sur les genres *Xenolpium, Antiolpium, Indolpium* et *Euryolpium* (Arachnides, Pseudoscorpions). *Revue Suisse de Zoologie* **87**: 143–154.

Heurtault, J. (1980d). La néochétotaxie majorante prosomatique chez les Pseudoscorpions Nesbisiidae: *Neobisium pyrenaicum* et *N. mahnerti* sp. n. Veme *Colloque d'Arachnologie d'Expression Française, Barcelone*: 87–97.

Heurtault, J. (1981). Présence et signification dans la France méditerranéenne des espèces des genres *Beierochelifer, Cheirochelifer* et *Calocheiridius* (Arachnides, Pseudoscorpions). *Atti della Società Toscana di Scienze Naturali, Memorie B* **88** supplement: 209–222.

Heurtault, J. (1982). Le développement postembryonnaire chez duex espèces nouvelles de Pseudoscorpions Olpiinae du Venezuela. *Rèvue de Nordest Biologie* **3**: 57–85.

Heurtault, J. (1983). Pseudoscorpions de Côte d'Ivoire. *Revue Arachnologique* **5**: 1–27.

Heurtault, J. (1986a). Les Pseudoscorpions de Madagascar: réflexions sur la répartition géographique. In: Eberhard, W. G., Lubin, Y. D. & Robinson, B. C. (eds), *Proceedings of the Ninth International Congress of Arachnology, Panama 1983*: 127–129. Smithsonian Institution Press: Washington D.C.

Heurtault, J. (1986b). *Petterchernes brasiliensis*, genre et espèce nouveaux de Pseudoscorpions du Brésil (Arachnides, Pseudoscorpionida, Chernetidae). *Bulletin du Muséum National d'Histoire Naturelle, Paris* (4) **8**: 351–355.

Heurtault, J. (1986c). Pseudoscorpions cavernicoles de France: revue synoptique. *Mémoires de Biospéologie* **12**: 19–32.

Heurtault, J. (1990a). *Neobisium (N.) maxvachoni*, new name for *Neobisium (N.) vachoni* Heurtault, 1968 (Arachnida, Pseudoscorpionida, Neobisiidae). *Bulletin of the British Arachnological Society* **8**: 128.

Heurtault, J. (1990b). *Chamberlinarius*, new name for *Chamberlinius* Heurtault, 1983 (Arachnida, Pseudoscorpionida, Cheliferidae). *Bulletin of the British Arachnological Society* **8**: 128.

Heurtault-Rossi, J. & Jézéquel, J. F. (1965). Observations sur *Feaella mirabilis* Ell. (Arachnide, Pseudoscorpion). Les chélicères et les pattes-mâchoires des nymphes et des adultes. Description de l'appareil reproducteur. *Bulletin du Muséum National d'Histoire Naturelle, Paris* (2) **37**: 450–461.

Heurtault, J. & Jézéquel, J. F. (1970). Les organes propriorécepteurs des pseudo-scorpions. *Bulletin du Muséum National d'Histoire Naturelle, Paris* (2) **41**, supplement 1: 54–58.

Heurtault, J. & Kovoor, J. (1980). Ultrastructure du complexe mécanorécepteur des chélicères de pseudoscorpions. *Verhandlungen des 8. Internationalen Arachnologen-Kongresses, Wien*: 325–330.

Heurtault, J. & Rebière, J. (1983). Pseudoscorpions des Petites Antilles. I. Chernetidae, Olpiidae, Neobisiidae, Syarinidae. *Bulletin du Muséum National d'Histoire Naturelle, Paris* (4) **5**: 591–609.

Hewitt, J. & Godfrey, R. (1929). South African pseudoscorpions of the genus *Chelifer* Geoffroy. *Annals of the Natal Museum* **6**: 305–336.

Heyden, L. von (1869). Ueber neue, von Herrn v. Frivaldszky in den Schriften der ungarischen Academie 1865 beschriebenen Insekten-Arten. *Berliner Entomologische Zeitschrift* **13**: 53–64.

Hickman, J. L. & Hill, L. (1978). *Terrestrial Invertebrates. Lower Gordon River Scientific Survey.* Hydro-electric Commission: Tasmania.

Hickson, S. J. (1889). *A Naturalist in North Celebes.* (not seen)

Hickson, S. J. (1893). Note on the parasitism of chelifers on beetles. *Zoologischer Anzeiger* **16**: 93.

Hickson, S. J. (1905a). A parasite of the house-fly. *Nature, London* **72**: 429.

Hickson, S. J. (1905b). Chelifers and house-flies. *Nature, London* **72**: 629–630.

Hill, M. D. (1905). A parasite of the house-fly. *Nature, London* **72**: 397.

Hilton, W. A. (1913). The nervous system of *Chelifer. Journal of Entomology and Zoology* **5**: 189–201.

Hilton, W. A. (1931). Nervous system and sense organs. Pseudoscorpionida. *Journal of Entomology and Zoology* **23**: 67–75.

Hippa, H., Koponen S. & Mannila, R. (1984). Invertebrates of Scandinavian caves I. Araneae, Opiliones, and Pseudoscorpionida (Arachnida). *Annales Entomologici Fennici* **50**: 23–29.

Hirst, S. (1911). The Araneae, Opiliones and Pseudoscorpiones. *Transactions of the Linnean Society of London, Zoology* (2) **14**: 379–395.

Hirst, S. (1913). Second report on the Arachnida - the scorpions, pedipalpi, and supplementary notes on the opiliones and pseudoscorpions. *Transactions of the Linnean Society of London, Zoology* (2) **16**: 31–37.

Hirst, S. (1922). Pseudoscorpions and bees. *Bee World* **4**: 36–37. (not seen)

Hölzel, E. (1963). Tierleben im Eiskeller des Matzen in der Karawankennordkette. *Carinthia, II* **73**: 161–187.

Hölzel, E. (1967). Die Fauna des Hochmoores von St. Lorenzen in den Gurker Alpen. *Carinthia, II* **77**: 195–211.

Hoff, C. C. (1944a). New pseudoscorpions of the subfamily Lamprochernetinae. *American Museum Novitates* **1271**: 1–12.

Hoff, C. C. (1944b). Notes on three pseudoscorpions from Illinois. *Transactions of the Illinois Academy of Science* **37**: 123–128.

Hoff, C. C. (1945a). Additional notes on pseudoscorpions from Illinois. *Transactions of the Illinois Academy of Science* **38**: 103–110.

Hoff, C. C. (1945b). *Hesperochernes canadensis*, a new chernetid pseudoscorpion from Canada. *American Museum Novitates* **1273**: 1–4.

Hoff, C. C. (1945c). New species and records of pseudoscorpions from Arkansas. *Transactions of the American Microscopical Society* **64**: 34–57.

Hoff, C. C. (1945d). The pseudoscorpion genus *Albiorix* Chamberlin. *American Museum Novitates* **1277**: 1–12.

Hoff, C. C. (1945e). New neotropical Diplosphyronida (Chelonethida). *American Museum Novitates* **1288**: 1–17.

Hoff, C. C. (1945f). The pseudoscorpion subfamily Olpiinae. *American Museum Novitates* **1291**: 1–30.

Hoff, C. C. (1945g). New species and records of cheliferid pseudoscorpions. *American Midland Naturalist* **34**: 511–522.

Hoff, C. C. (1945h). Two new pseudoscorpions of the genus *Dolichowithius*. *American Museum Novitates* **1300**: 1–7.

Hoff, C. C. (1945i). Pseudoscorpions from North Carolina. *Transactions of the American Microscopical Society* **64**: 311–327.

Hoff, C. C. (1946a). A redescription of two of Hagen's pseudoscorpion species. *Proceedings of the New England Zoölogical Club* **23**: 99–107.

Hoff, C. C. (1946b). A redescription of *Atemnus elongatus* Banks, 1895. *Proceedings of the New England Zoölogical Club* **23**: 109–113.

Hoff, C. C. (1946c). New pseudoscorpions, chiefly neotropical, of the suborder Monosphyronida. *American Museum Novitates* **1318**: 1–32.

Hoff, C. C. (1946d). A study of the type collections of some pseudoscorpions originally described by Nathan Banks. *Journal of the Washington Academy of Sciences* **36**: 195–205.

Hoff, C. C. (1946e). Three new species of heterosphyronid pseudoscorpions from Trinidad. *American Museum Novitates* **1322**: 1–13.

Hoff, C. C. (1946f). The pseudoscorpion tribe Cheliferini. *Bulletin of the Chicago Academy of Sciences* **7**: 485–490.

Hoff, C. C. (1946g). American species of the pseudoscorpion genus *Microbisium* Chamberlin, 1930. *Bulletin of the Chicago Academy of Sciences* **7**: 493–497.

Hoff, C. C. (1946h). Descripcion de una especie nueva del genero *Pachychernes* Beier, 1932 (Pseudoscorpionida). *Ciencia, México* **7**: 13–14.

Hoff, C. C. (1947a). New species of diplosphyronid pseudoscorpions from Australia. *Psyche, Cambridge* **54**: 36–56.

Hoff, C. C. (1947b). The species of the pseudoscorpion genus *Chelanops* described by Banks. *Bulletin of the Museum of Comparative Zoology* **98**: 471–550.

Hoff, C. C. (1947c). Two new pseudoscorpions of the subfamily Lamprochernetinae from Venezuela. *Zoologica, New York* **32**: 61–64.

Hoff, C. C. (1948). *Hesperochernes thomomysi*, a new species of chernetid pseudoscorpion from California. *Journal of the Washington Academy of Sciences* **38**: 340–345.

Hoff, C. C. (1949a). *Wyochernes hutsoni*, a new genus and species of chernetid pseudoscorpion. *Transactions of the American Microscopical Society* **68**: 40–48.

Hoff, C. C. (1949b). The pseudoscorpions of Illinois. *Bulletin of the Illinois Natural History Survey* **24**: 407–498.

Hoff, C. C. (1950a). Some North American cheliferid pseudoscorpions. *American Museum Novitates* **1448**: 1–18.

Hoff, C. C. (1950b). Pseudoescorpionidos nuevos o poco conocidos de la Argentina (Arachnida, Pseudoscorpionida). *Arthropoda, Buenos Aires* **1**: 225–237.

Hoff, C. C. (1950c). A new species and some records of tridenchthoniid pseudo-scorpions. *Annals of the Entomological Society of America* **43**: 534–536.

Hoff, C. C. (1951). New species and records of chthoniid pseudoscorpions. *American Museum Novitates* **1483**: 1–13.

Hoff, C. C. (1952a). Some heterosphyronid pseudoscorpions from New Mexico. *Great Basin Naturalist* **12**: 39–45.

Hoff, C. C. (1952b). Two new species of pseudoscorpions from Illinois. *Transactions of the Illinois Academy of Sciences* **45**: 188–195.

Hoff, C. C. (1956a). The heterosphyronid pseudoscorpions of New Mexico. *American Museum Novitates* **1772**: 1–13.

Hoff, C. C. (1956b). Diplosphyronid pseudoscorpions from New Mexico. *American Museum Novitates* **1780**: 1–49.

Hoff, C. C. (1956c). Pseudoscorpions of the family Chernetidae from New Mexico. *American Museum Novitates* **1800**: 1–66.

Hoff, C. C. (1956d). Pseudoscorpions of the family Cheliferidae from New Mexico. *American Museum Novitates* **1804**: 1–36.

Hoff, C. C. (1957). *Tejachernes* (Arachnida – Chelonethida – Chernetidae – Chernetinae), a new genus of pseudoscorpion based on *Dinocheirus stercoreus*. *Southwestern Naturalist* **2**: 83–88.

Hoff, C. C. (1958). List of the pseudoscorpions of North America north of Mexico. *American Museum Novitates* **1875**: 1–50.

Hoff, C. C. (1959a). The pseudoscorpions of Jamaica. Part 1. The genus *Tyrannoch-thonius*. *Bulletin of the Institute of Jamaica, Science Series* **10**(1): 1–39.

Hoff, C. C. (1959b). The ecology and distribution of the pseudoscorpions of north-central New Mexico. *University of New Mexico Publications in Biology* **8**: 1–68.

Hoff, C. C. (1961). Pseudoscorpions from Colorado. *Bulletin of the American Museum of Natural History* **122**: 409–464.

Hoff, C. C. (1963a). Pseudoscorpions from the Black Hills of South Dakota. *American Museum Novitates* **2134**: 1–10.

Hoff, C. C. (1963b). Sternophorid pseudoscorpions, chiefly from Florida. *American Museum Novitates* **2150**: 1–14.

Hoff, C. C. (1963c). The pseudoscorpions of Jamaica. Part 2. The genera *Pseudoch-thonius, Paraliochthonius, Lechytia,* and *Tridenchthonius. Bulletin of the Institute of Jamaica, Science Series* **10**(2): 1–35.

Hoff, C. C. (1964a). The pseudoscorpions of Jamaica. Part 3. The suborder Diplosphyronida. *Bulletin of the Institute of Jamaica, Science Series* **10**(3): 1–47.

Hoff, C. C. (1964b). Atemnid and cheliferid pseudoscorpions, chiefly from Florida. *American Museum Novitates* **2198**: 1–43.

Hoff, C. C. (1964c). A new pseudochiridiid pseudoscorpion from Florida. *Transactions of the American Microscopical Society* **83**: 89–92.

Hoff, C. C. & Bolsterli, J. E. (1956). Pseudoscorpions of the Mississippi River drainage basin area. *Transactions of the American Microscopical Society* **75**: 155–179.

Hoff, C. C. & Clawson, D. L. (1952). Pseudoscorpions from rodent nests. *American Museum Novitates* **1585**: 1–38.

Hoff, C. C. & Jennings, D. T. (1974). Pseudoscorpions phoretic on a spider. *Entomological News* **85**: 21–22.

Hoff, C. C. & Parrack, D. (1958). Results of the Archbold Expeditions. No. 77. Two species of *Megachernes* (Pseudoscorpionida, Chernetidae). *American Museum Novitates* **1881**: 1–9.

Holdhaus, K. (1932). Die europäische Höhlenfauna in ihren Beziehungen zur Eiszeit. *Zoogeografica* **1**: 1–53. (not seen)

Holmberg, E. L. (1874). Descriptions et notices d'Arachnides de la République Argentine. *Periódico Zoológico* **1**: 283–302.

Holmberg, E. L. (1876). Aracnidos Argentinos. *Anales de Agricultura de la República Argentina* 4: 1–28.

Holsinger, J. R. (1963). Annotated checklist of the macroscopic troglobites of Virgina with notes on their geographic distribution. *Bulletin of the National Speleological Society* 25: 23–36.

Holsinger, J. R. & Culver, D. C. (1988). The invertebrate cave fauna of Virgina and a part of eastern Tennessee: zoogeography and ecology. *Brimleyana* 14: 1–162.

Hong, Y. (1983). [Discovery of new fossil pseudoscorpionods in amber]. *Bulletin of the Tianjin Institute of Geology and Mineral Resources* 8: 24–29. (in Chinese)

Höregott, H. (1963). Zur Ökologie und Phänologie einiger Chelonethi und Opiliones (Arach.) des Gonsenheimer Waldes und Sandes bei Mainz. *Senckenbergiana Biologica* 44: 545–551.

Horvath, G. (1885). *Alskorpiók növenyeken*. Rovartani lapok. (not seen)

Hounsome, M. V. (1980). The terrestrial fauna (excluding birds and insects) of Little Cayman. In: Stoddart, D. R. & Giglioli, M. E. C. (eds), *Geography and Ecology of Little Cayman. Atoll Research Bulletin* 241: 81–90.

Howarth, F. G. (1987a). The evolution of non-relictual tropical troglobites. *International Journal of Speleology* 16: 1–16.

Howarth, F. G. (1987b). Evolutionary ecology of aeolian and subterranean habitats in Hawaii. *Trends in Ecology and Evolution* 2: 220–223.

Howarth, F. G. (1988). Environmental ecology of north Queensland caves: or why are there so many troglobites in Australia. In: Pearson, L. (ed.), *Australian Speleological Federation Tropicon Conference, Lake Tinaroo, Far North Queensland*: 76–84. Australian Speleological Federation: Cairns.

Howes, C. A. (1972a). A review of Yorkshire pseudoscorpions. *Naturalist, Hull* 918: 107–110.

Howes, C. A. (1972b). The pseudoscorpion *Lamprochernes godfreyi* Kew in Britain. *Naturalist, Hull* 919: 122.

Howes, C. A. (1972c). The pseudoscorpion *Chthonius kewi* Gabbutt in North Nottinghamshire and notes on its breeding. *Naturalist, Hull* 923: 142.

Hulley, P. E. (1983). A survey of the flies breeding in poultry manure, and their potential natural enemies. *Journal of the Entomological Society of Southern Africa* 46: 37–47.

Humphreys, W. F. (1987). The accoutrements of spiders' eggs (Araneae) with an exploration of their functional importance. *Zoological Journal of the Linnean Society* 89: 171–201.

Hunt, S. (1970). Amino acid composition of silk from the pseudoscorpion *Neobisium maritimum* (Leach): a possible link between the silk fibroins and the keratins. *Comparative Biochemistry and Physiology* (B) 34: 773–776.

Hunt, S. and Legg, G. (1971). Characterization of the structural protein component in the spermatophore of the pseudoscorpion *Chthonius ischnocheles* (Hermann). *Comparative Biochemistry and Physiology* (B) 40: 475–479.

Ihering, H. (1893). Zum Commensalismus der Pseudoscorpioniden. *Zoologischer Anzeiger* 16: 346–347.

Illiger, J. K. W. (1798). In: Kugelaan, J. G., *Verzeichniss der Käfer Preussens. Entworfen von Johann Gottlieb Kugelaan Apotheker in Ossterode. Ausgearbeit von Johann Karl Wilhelm Illiger.* J. J. Gebauer: Halle.

Illiger, J. K. W. (1807). In: Rossius, P., *Fauna Etrusca*, vol. 2. Masi: Liburni.

Imms, A. D. (1905). On a marine pseudoscorpion from the Isle of Man. *Annals and Magazine of Natural History* (7) 15: 231–232.

International Commission on Zoological Nomenclature (1987). Opinion 1423. *Olpium* Koch, 1873 (Arachnida): *Obisium pallipes* Lucas, [1846] designated as type species;

interpretation of the nominal species *Olpium kochi* Simon, 1881. *Bulletin of Zoological Nomenclature* **44**: 53–54.

International Commission on Zoological Nomenclature (1989). Opinion 1542. *Chelifer* Geoffroy, 1762 (Arachnida, Pseudoscorpionida): conserved. *Bulletin of Zoological Nomenclature* **46**: 143–144.

Inzaghi, S. (1981). Pseudoscorpioni raccolti dal Sig. M. Valle in nidi di *Talpa europaea* L. nella provincia di Bergamo con descrizione di una nuova specie del gen. *Chthonius* C. L. Koch. *Bollettino della Società Entomologica Italiana* **113**: 67–73.

Inzaghi, S. (1983). *Pseudoblothrus regalini* n. sp., da grotte della Provincia di Bergamo (Italia sett.) (Pseudoscorpiones Syarinidae). *Atti della Società Italiana di Scienze Naturali, e del Museo Civile di Storia Naturale, Milano* **124**: 38–48.

Inzaghi, S. (1987). Una nuova specie del genere *Chthonius* s. str. delle preapli Lombarde. *Natura Bresciana* **23**: 165–182.

Ionescu, M. A. (1936). Contribuţiuni la studiul Pseudoscorpionilor din România. *Buletinul Societaţii Naturaliştilor din România* **8**: 1–5.

Ishikawa, J. (1948). [Study of the fauna in the lime-stone caves of Kochi Prefecture.] *Shizen to Jinbun, Shizen Kagaku Hen* **1**: 9–17. (not seen)

Jackson, A. R. (1906). The spiders of the Tyne Valley. *Transactions of the Natural History Society of Northumberland, Durham, and Newcastle-upon-Tyne* **1**: 337–405. (not seen)

Jackson, A. R. (1907). On some rare arachnids captured during 1906. *Proceedings of the Chester Society of Natural Science 6:* 1–7. (not seen)

Jackson, A. R. (1908). On some rare arachnids captured during 1907. *Transactions of the Natural History Society of Northumberland, Durham, and Newcastle-upon-Tyne* **3**: 49–78. (not seen)

Jacot, A. P. (1938). A pseudogarypin pseudoscorpion in the White Mountains. *Occasional Papers of the Boston Society for Natural History* **8**: 301–303.

Jäger, H. (1932). Untersuchungen über die geotaktischen Reaktionen verschiedner Evertebraten auf schiefer Ebene. *Zoologische Jahrbücher, Allgemeine Zoologie und Physiologie* **51**: 289–320. (not seen)

Janetschek, H. (1948). Zur Brutbiologie von *Neobisium jugorum* (L. Koch) (Arachnoidea, Pseudoscorpiones). *Annalen des Naturhistorischen Museums in Wien* **56**: 309–316.

Jaquet, M. (1898). Faune de la Roumanie. Arachnides recueillies en 1897 par M. Jaquet et déterminées par M. le Prof. Pavesi de l'Université de Pavie. *Buletinul Societatii de Sciinte din Bucaresti* **7**: 274–282. (not seen)

Jaquet, M. (1905). Faune de la Roumanie. Arachnides recueillies par M. Jaquet et déterminées par M. le Dr E. Corti. *Buletinul Societatii de Sciinte din Bucaresti* **14**: 204–226. (not seen)

Jędryczkowski, W. B. (1985). Zaleszczotki (Pseudoscorpiones) Mazowsza. *Fragmenta Faunistica Musei Zoologici Polonici* **29**: 77–83.

Jędryczkowski, W. B. (1987a). Zaleszczotki (Pseudoscorpiones) Bieszczadów. *Fragmenta Faunistica Musei Zoologici Polonici* **30**: 341–349.

Jędryczkowski, W. B. (1987b). Zaleszczotki (Pseudoscorpiones) Gór Świętokryskich. *Fragmenta Faunistica Musei Zoologici Polonici* **31**: 135–157.

Jennings, T. (1968). False scorpions. *Wildlife and the Countryside* **1968**: 3–5.

Jenyns, L. (1846). *Observations in Natural History*. John van Voorst: London.

Jéquier, J.-P. (1964). Étude écologique et statistique de la faune terrestre d'une caverne du Jura Suisse au cours d'une année d'observation. *Revue Suisse de Zoologie* **71**: 313–370. (not seen)

Johnson, D. L. & Wellington, W. G. (1980a). Predation of *Apochthonius minimus* (Pseudoscorpionida: Chthoniidae) on *Folsomia candida* (Collembola: Isotomidae) I. Predation rate and size-selection. *Researches on Population Ecology, Kyoto University* **22**: 339–352.

Johnson, D. L. & Wellington, W. G. (1980b). Predation of *Apochthonius minimus* (Pseudoscorpionida: Chthoniidae) on *Folsomia candida* (Collembola: Isotomidae) II. Effects of predation on prey populations. *Researches on Population Ecology, Kyoto University* **22**: 353–365.

Johnson, H. E. (1901). East Riding pseudoscorpions. *Transactions of the Hull Scientific and Field Naturalists' Club* **1**: 228. (not seen)

Jones, P. E. (1970a). The occurrence of *Chthonius ischnocheles* (Hermann) (Chelonethi: Chthoniidae) in two types of hazel coppice leaf litter. *Bulletin of the British Arachnological Society* **1**: 77–80.

Jones, P. E. (1970b). *Lamprochernes nodosus* (Schrank) – an example of phoresy in pseudoscorpions. *Bulletin of the British Arachnological Society* **1**: 118–119.

Jones, P. E. (1975a). The occurrence of pseudoscorpions in the nests of British birds. *Proceedings and Transactions of the British Entomological and Natural History Society* **8**: 87–89.

Jones, P. E. (1975b). Notes on the predators and prey of British pseudoscorpions. *Bulletin of the British Arachnological Society* **3**: 104–105.

Jones, P. E. (1975c). The false scorpions of Norfolk. *Transactions of the Norfolk Norwich Naturalists' Society* **27**: 67–71. (not seen)

Jones, P. E. (1978). Phoresy and commensalism in British pseudoscorpions. *Proceedings and Transactions of the British Entomological and Natural History Society* **1978**: 90–96.

Jones, P. E. (1979a). Pseudoscorpions from Little Wood, Eye, with some additional records from Northamptonshire. *Journal of the Northamptonshire Natural History Society and Field Club* **37**: 198–201.

Jones, P. E. (1979b). The ecology and distribution pseudoscorpions. *Annual Report, Institute for Terrestrial Ecology* **1978**: 40–41. (not seen)

Jones, P. E. (1980a). The ecology and distribution of the pseudoscorpion *Dendrochernes cyrneus* (L. Koch) in Great Britain. *Proceedings and Transactions of the British Entomological and Natural History Society* **1980**: 33–37.

Jones, P. E. (1980b). Recent pseudoscorpion records from Huntingdonshire. *Annual Report, Huntingdon Fauna and Flora Society* **32**: 14–18. (not seen)

Jones, P. E. (1980c). *Provisional Atlas of the Arachnida of the British Isles. Part 1. Pseudoscorpiones.* Institute of Terrestrial Ecology: Monks Wood.

Joseph, G. (1871). Giebt es augenlose Arthropoden in Schlesien? *Jahrbücher der Schlesische Gesellschaft für Vaterländische Kultur, Breslau* **48**: 160–162. (not seen)

Joseph, G. (1881). Erfahrungen im wissenschaftlichen Sammeln und Beobachten der den Krainer Tropfsteingrotten eigenen Arthropoden. *Berliner Entomologische Zeitschrift* **25**: 233–282.

Joseph, G. (1882). Systematiches Verzeichniss der in den Tropfstein-Grotten von Krain einheimischen Arthropoden nebst Diagnosen der vom Verfasser entdeckten und bisher noch nicht beschriebenen Arten. *Berliner Entomologische Zeitschrift* **26**: 1–50.

Joseph, H. C. (1927). Observaciones sobre el *Chelonops coecus* Gerv. *Revista Chilena de Historia Natural* **31**: 53–56.

Juberthie, C., Delay, D. Decon, V. & Racovitza, G. (1981). Premières données sur la faune des microespeces du milieu souterrain superficiel de Roumanie. *Travaux de l'Institut de Spéologie E. Racovitza* **20**: 103–111. (not seen)

Juberthie, C. & Heurtault, J. (1975). Ultrastructure des plaques paraganglionnaires d'un Pseudoscorpion souterrain, *Neobisium cavernarum* (L. K.). *Annales de Spéléologie* **30**: 433–439.

Judson, M. L. I. (1979a). Pseudoscorpions in Hertfordshire. *Transactions of the Hertfordshire Natural History Society and Field Club* **28**: 58–64.

Judson, M. L. I. (1979b). On the status of *Microbisium dumicola* (C. L. Koch) 1835 (Neobisiidae: Chelonethi) in the British fauna. *Newsletter of the British Arachnological Society* **25**: 7–8.

Judson, M. L. I. (1980). On some changes in the names of British Chelonethi (Pseudoscorpionida) with a note on the status of *Chthonius (C.) dacnodes* Navás, 1918, in Britain. *Newsletter of the British Arachnological Society* **28**: 7–9.

Judson, M. L. I. (1985). Redescription of *Myrmochernes* Tullgren (Chelonethida: Chernetidae). *Bulletin of the British Arachnological Society* **6**: 321–327.

Judson, M. L. I. (1987). Further records of pseudoscorpions (Arachnida) from Hertfordshire. *Transactions of the Hertfordshire Natural History Society and Field Club* **29**: 368–370.

Judson, M. L. I. (1988). Sur la présence en France de *Chthonius (C.) halberti* Kew et de *Chthonius (C.) ressli* Beier avec remarques sur le rang de *Kewochchthonius* Chamberlin et de *Neochthonius* Chamberlin (Arachnida, Chelonethida, Chthoniidae). *Comptes Rendu Xème Colloque Europ. Arachnol., Rennes. Bulletin de la Société de Scientifique de Bretagne* **59**: 131.

Judson, M. L. I. (1990). On the presence of *Chthonius (C.) halberti* Kew and *Chthonius (C.) ressli* Beier in France with remarks on the status of *Kewochthonius* Chamberlin and *Neochthonius* Chamberlin (Arachnida, Chelonethida, Chthoniidae). *Bulletin du Muséum National d'Histoire Naturelle, Paris* (4) **11**: 593–603.

Kästner, A. (1923). Beiträge zur Kenntnis der Locomotion der Arachniden. II. *Obisium muscorum* C. Koch. *Zoologischer Anzeiger* **57**: 247–253.

Kästner, A. (1927a). Pseudoscorpiones. In: *Insects of Samoa and other Samoan Terrestrial Arthropoda*, vol. **8**(1): 15–24. British Museum (Natural History): London.

Kästner, A. (1927b). Pseudoscorpiones. In: Schulze, P., *Biologie der Tiere Deutschlands*, vol. **18**: 1–68. Berlin. (not seen)

Kästner, A. (1928). 2. Ordnung: Moos- oder Afterskorpione, Pseudoscorpiónes Latr. (Chernétes Simon; Chelonéti Thorell; Chernetídea Camb.). In: Brohmer, P., Ehrmann, P. & Ulmer, G. (eds), *Die Tierwelt Mitteleuropas*, vol. 3(1, IV): 1–13.

Kästner, A. (1931a). Studien zur Ernährung der Arachniden. 2. Der Fressakt von *Chelifer cancroides* L. *Zoologischer Anzeiger* **96**: 73–77.

Kästner, A. (1931b). Die Hüfte und ihre Umformung zu Mundwerkzeugen bei den Arachniden. Versuch einer Organgeschichte. *Zeitschrift für Morphologie und Ökologie der Tiere* **22**: 721–758. (not seen)

Kaimal, P. P. (1986). *Some aspects of ecology and biology of pseudoscorpions of Kerala*. Dissertation, University of Kerala. (not seen)

Kaisila, J. (1947). Turun Yliopistan kokoelmien valeskorpioniaineisto. *Annales Entomologici Fennici* **13**: 86–88.

Kaisila, J. (1949a). Muutamia Suomen faunalle uusia valeskorpionilajeja. *Archivum Societatis Zoologico-Botanicae Fennicae* **2**: 74–76.

Kaisila, J. (1949b). A revision of the pseudoscorpion fauna of eastern Fennoscandia. *Annales Entomologici Fennici* **15**: 72–92.

Kaisila, J. (1964). Some pseudoscorpionids from Newfoundland. *Annales Zoologici Fennici* **1**: 52–54.

Kaisila, J. (1966). A new species of the genus *Mesochelifer* Vachon (Pseudosc., Cheliferidae). *Annales Entomologici Fennici* **32**: 260–263.

Kanmacher, F. (1798). In: Adams, G., *Essays on the Microscope*, 2nd edition. London. (not seen)

Kanwar, U. (1966). 'Pseudochromosomes' in pseudoscorpions. *Microscope* **15**: 203–205.

Kanwar, U. (1968). Origin of the spiral thread in the pseudoscorpion sperm. *Microscope* **16**: 369–371.

Kanwar, K. C. & Kanwar, U. (1968). 'Flagellate non-flagellate' pseudoscorpion sperm. *Microscope* **16**: 373–375.

Kappler, ?. (1887). *Suriname: Seine Natur, Bevölkerung und seine Kulturverhältnisse.* Stuttgart. (not seen)

Karsch, F. (1879). Zwei neue Arachniden des Berliner Museums. *Mitteilungen des Münchener Entomologischen Verein* **3**: 95–96.

Karsch, F. (1881). Diagnoses Arachnoidarum Japoniae. *Berliner Entomologische Zeitschrift* **25**: 35–40.

Karsch, F. (1882). Ueber ein neues Spinnenthier aus der schlesischen Steinkohle und die Arachnoiden der Steinkohlen-formation überhaupt. *Zeitschrift der Deutschen Geologischen Gesellschaft* **34**: 556–561.

Kathariner, L. (1898). Zur Lebensweise der After-Skorpione. *Illustrierte Zeitschrift für Entomologie* **3**: 250.

Kauri, H. (1971). Vevkjerringene og mosskorpionene. In: *Norges Dyr*, vol. **4**: 142–147. J. W. Cappelen: Oslo. (not seen)

Kawasawa, T. (1969). [The arthropodous fauna of Rakan-ana Cave in western Shikoku]. *Journal Ent. Soc. Kochi* **20**: 1–8. (in Japanese) (not seen)

Keisl, G. (1933). Der Bücherskorpion als Vernichter der Wachsmottenraupen. *Méhészet* **30**: 140. (not seen)

Kennedy, C. M. A. (1989). *Pycnodithella harveyi*, a new Australian species of the Tridenchthoniidae (Pseudoscorpionida: Arachnida). *Proceedings of the Linnean Society of New South Wales* **110**: 289–296.

Kennel, J. von (1891). Über die Abstammung der Arthropoden und deren Verwandschaftsbeziehungen. *Sitzungsberichte der Naturforscher Gesellschaft zu Dorpat* **9**: 48 pp. (not seen)

Kensler, C. B. (1964). A crevice habitat on Anglesey, N. Wales. *Journal of Animal Ecology* **33**: 200–202. (not seen)

Kensler, C. B. (1967). Dessication resistance of intertidal crevice species as a factor in their zonation. *Journal of Animal Ecology* **36**: 391–406. (not seen)

Kensler, C. B. & Crisp, D. J. (1965). The colonization of artificial crevices by marine invertebrates. *Journal of Animal Ecology* **34**: 507–516.

Kerzhner, I. M. (1988). Comment on the proposed conservation of *Chelifer* Geoffroy, 1762 (Arachnida, Pseudoscorpionida). *Bulletin of Zoological Nomenclature* **45**: 49.

Ketterer, C. E. (1955). A propos d'un cas de phoresie chez *Chelifer cancroides* (Pseudoscorpionidea). *Bulletin de la Murithienne* **72**: 93–97.

Kew, H. W. (1886). *Chelifer DeGeerii* Koch near the Lincolnshire coast. *Naturalist, London* **11**: 339.

Kew, H. W. (1901a). Notes on spinning animals. VI. Pseudoscorpions. *Science Gossip, (new series)* **7**: 228–229. (not seen)

Kew, H. W. (1901b). Lincolnshire pseudoscorpions: with an account of the association of such animals with other arthropods. *Naturalist, London* **26**: 193–215.

Kew, H. W. (1903). North of England pseudoscorpions. *Naturalist, London* **28**: 293–300.

Kew, H. W. (1904). *Chernes nodosus* at Louth, Lincs. *Naturalist, London* **29**: 292.

Kew, H. W. (1906). *Chernes cyrneus* in Nottinghamshire: a recent addition to the known false-scorpions of Britain. *Report and Transactions of the Nottingham Naturalists' Society* **1905–1906**: 41–46.

Kew, H. W. (1909a). Notes on the Irish false-scorpions in the National Museum of Ireland. *Irish Naturalist* **18**: 249–250.

Kew, H. W. (1909b). Pseudoscorpiones (False-scorpions). In: Grinling, C. *et al.* (eds), *Survey and Record of Woolwich and West Kent*: 258–259. South Eastern Union of Scientific Societies: Woolwich.

Kew, H. W. (1910a). A holiday in south-western Ireland. Notes on some false-scorpions and other animals observed in the counties of Kerry and Cork. *Irish Naturalist* **19**: 64–73.

Kew, H. W. (1910b). On the Irish species of *Obisium*; with special reference to one from Glengariff new to the Britannic fauna. *Irish Naturalist* **19**: 108–112.

Kew, H. W. (1911a). A synopsis of the false scorpions of Britain and Ireland. *Proceedings of the Royal Irish Academy* (B) **29**: 38–64.

Kew, H. W. (1911b). Clare Island Survey. Pseudoscorpiones. *Proceedings of the Royal Irish Academy* **31**(38): 1–2.

Kew, H. W. (1912). On the pairing of Pseudoscorpiones. *Proceedings of the Zoological Society of London* **25**: 376–390.

Kew, H. W. (1914). On the nests of Pseudoscorpiones: with historical notes on the spinning-organs and observations on the building and spinning of the nests. *Proceedings of the Zoological Society of London* **27**: 93–111.

Kew, H. W. (1916a). An historical account of the pseudoscorpion-fauna of the British islands. *Journal of the Quekett Microscopical Club* (2) **13**: 117–136.

Kew, H. W. (1916b). A synopsis of the false-scorpions of Britain and Ireland; supplement. *Proceedings of the Royal Irish Academy* (B) **33**: 71–85.

Kew, H. W. (1929a). On the external features of the development of the Pseudoscorpiones: with observation on the ecdyses and notes on the immature forms. *Proceedings of the Zoological Society of London* **42**: 33–38.

Kew, H. W. (1929b). Notes on some Coleoptera and a *Chelifer* observed on a Richmond Park oak after nightfall. *Entomologist's Monthly Magazine* **65**: 83–86.

Kew, H. W. (1929c). Observations on Mr. Donisthorpe's 'Guests of British ants, chapter XII. Pseudoscorpiones'. *Entomologist's Record and Journal of Variations* **41**: 21–22.

Kew, H. W. (1930). On the spermatophores of the Pseudoscorpiones *Chthonius* and *Obisium*. *Proceedings of the Zoological Society of London* **43**: 253–256.

Kishida, K. (1915). Notes on the pseudoscorpions of Japan. *Science, Kyoto* **5**: 362–369. (not seen)

Kishida, K. (1920). On a pseudoscorpion, *Microcreagris formosana* from Formosa: 1–6. Tokyo. (not seen)

Kishida, K. (1927). [Cheliferidae.] *Illus. Ency. Fauna Jap.*: 1001. (in Japanese) (not seen)

Kishida, K. (1928). Pseudoscorpions (*Microcreagris*) of Japan. *Annotationes Zoologicae Japonenses* **11**: 407–413.

Kishida, K. (1929a). [On the localities of a Japanese *Chelifer*.] *Lansania* **1**: 39. (in Japanese) (not seen)

Kishida, K. (1929b). [On the criteria to classify Chelifers.] *Lansania* **1**: 124. (in Japanese) (not seen)

Kishida, K. (1930). 26 families of the order Chelonethida: 1–16. Tokyo. (not seen)

Kishida, K. (1942). [Biotic systems of Chelonethida.] *Lansania* **20**: 1–5. (in Japanese) (not seen)

Kishida, K. (1947). [Cheliferidae.] *Illus. Ency. Fauna Jap.*: 1001. (in Japanese) (not seen)

Kishida, K. (1966). On the altitudinal distribution of the Chelonethida in Japan. *Acta Arachnologica* **20**: 6–8.

Klausen, F. E. (1975). Notes on the Pseudoscorpiones of Norway. *Norwegian Journal of Entomology* **22**: 63–65.

Klausen, F. E. (1977). *Lamprochernes nodosus* (Schrank 1761) (Pseudoscorpionida, Chernetidae) new to Norway. *Norwegian Journal of Entomology* **24**: 83–84.

Klausen, F. E. & Totland, G. K. (1977). A scanning electron microscopic study of the setae of some chernetid pseudoscorpions. *Bulletin of the British Arachnological Society* **4**: 101–108.

Klein, K. (1934). Über die Helligkeitsreaktionen einiger Arthropoden. *Zeitschrift für Wissenschaftliche Zoologie* **145**: 1–38. (not seen)

Kleisl, G. (1933). Der Bücherskorpion als Vernichter der Wachsmottenraupen. *Méhészet* **30:** 140. (not seen)

Knab, F. (1897). Untitled. *Entomological News* **8:** 13.

Knowlton, G. F. (1972). Some terrestrial arthropods of Curlew Valley. *Utah State University Ecology Center, Terrestrial Arthropod Series* **4:** 1–7.

Knowlton, G. F. (1974). Some pseudoscorpions of Curlew Valley. *Utah State University Ecology Center, Terrestrial Arthropods Series* **11:** 1–3.

Knudsen, J. W. (1954). *Pseudoscorpions, a Natural Control of Siphonaptera in* Neotoma *Nests*. Masters Thesis: University of Southern California. (not seen)

Knudsen, J. W. (1956). Pseudoscorpions, a natural control of Siphonaptera in *Neotoma* nests. *Bulletin of the Southern California Academy of Sciences* **55:** 1–6.

Kobakhidze (as Kobachidze), D. (1960a). Materialien zur Hohenstufenverbreitung der Pseudoscorpionidea in der Georgischen SSR. *Zeitschrift der Arbeitsgemeinschaft Österreichischer Entomologen* **12:** 103–106.

Kobakhidze, D. N. (1960b). [A new species of pseudoscorpion from Baniskhevi.] *Trudy Instituta Zoologii Akademii Nauk Gruzinskoi S.S.R.* **17:** 239–240. (in Russian)

Kobakhidze, D. N. (1960c). [New species of pseudoscorpionid from Bathum botanical garden.] *Soobscenija Akademiji Nauk Gruzinskoj S.S.R.* **24:** 465–466. (in Russian)

Kobakhidze, D. N. (1960d). [New species of pseudoscorpions from Kelasuri.] *Soobscenija Akademiji Nauk Gruzinskoj S.S.R.* **25:** 457–459. (in Russian)

Kobakhidze, D. (1961a). Die Standorte des *Chthonius tetrachelatus* (Preissler) in den verschiedenen Landschaftstypen der Georgischen SSR. *Zoologischer Anzeiger* **167:** 166–169.

Kobakhidze, D. N. (1961b). [The distribution of *Chelifer cancroides* (L.) in territory of Georgian SSR.] *Soobscenija Akademiji Nauk Gruzinskoj S.S.R.* **27:** 471–472. (in Russian)

Kobakhidze, D. N. (1961c). [The distribution of pseudoscorpions – *Dendrochernes cyrneus* (L. Koch) in Georgian SSR.] *Trudy Instituta Zoologii Akademii Nauk Gruzinskoi S.S.R.* **18:** 209–211. (in Russian)

Kobakhidze, D. N. (1963). [*Toxochernes panzeri caucasicus* Kobakhidze, new subspecies from Caucasus.] *Soobscenija Akademiji Nauk Gruzinskoj S.S.R.* **30:** 645–649. (in Russian)

Kobakhidze, D. N. (1964a). [A new sub-species *Allochernes wideri transcaucasius* from the Transcaucasus.] *Soobscenija Akademiji Nauk Gruzinskoj S.S.R.* **33:** 449–452. (in Russian)

Kobakhidze, D. N. (1964b). [Geographic distribution of the pseudoscorpion *Dactylochelifer latreillei* (Leach) in Georgia.] *Soobscenija Akademiji Nauk Gruzinskoj S.S.R.* **34:** 445–448. (in Russian)

Kobakhidze, D. N. (1965a). [A new sub-species of pseudoscorpion, *Chernes cimicoides caucasicus* Kobakhidze ssp. n. from Caucasus.] *Soobscenija Akademiji Nauk Gruzinskoj S.S.R.* **37:** 441–443. (in Russian)

Kobakhidze, D. (1965b). Ecological and zoogeographical characteristics of Pseudoscorpionidea from Georgian SSR. *Revue d'Ecologie et de Biologie du Sol* **2:** 541–543.

Kobakhidze, D. (1965c). [A new species of pseudoscorpions, *Withius lohmanderi* Kobakhidze from Sotschi. *Soobscenija Akademiji Nauk Gruzinskoj S.S.R.* **38:** 417–419. (in Russian)

Kobakhidze, D. N. (1966). [Material for the faunistic records of Pseudoscorpionidea in the Gruzian Republic.] *Soobscenija Akademiji Nauk Gruzinskoj S.S.R.* **41:** 701–708. (in Russian)

Kobari, H. (1983). A seasonal change of the age composition in a population of the pseudoscorpion, *Neobisium (Parobisium) pygmaeum* (Ellingsen), in a temperate deciduous forest. *Acta Arachnologica* **31:** 65–71.

Kobari, H. (1984a). Redescription of the male and redesignation of *Neobisium (Parobisium) pygmaeum* (Ellingsen) (Arachnida: Pseudoscorpionida). *Acta Arachnologica* 32: 55–64.

Kobari, H. (1984b). Seasonal fluctuations of some soil pseudoscorpions at Mt Tsukuba, central Japan. *Edaphologia* 30: 1–10. (not seen)

Koch, C. (1873). Beiträge zur Kenntnis der Arachniden Nord-Afrikas, insbesondere einiger in dieser Richtung bisher noch unbekannt gebliebener Gebiete des Atlas und der Küstenländer von Marocco. *Berichte der Senckenbergischen naturforschenden Gesellschaft* 1873: 104–118. (not seen)

Koch, C. L. (1835). *Deutschlands Crustaceen, Myriapoden und Arachniden*, vol. 2. Regensberg.

Koch, C. L. (1837). *Deutschlands Crustaceen, Myriapoden und Arachniden*, vol. 7. Regensberg.

Koch, C. L. (1839). *Uebersicht des Arachnidensystems*, vol. 2. C.H. Zeh'schen: Nürnberg.

Koch, C. L. (1843). *Die Arachniden. Getreu nach der Natur Abgebildet und Beschrieben*, vol. 10. C.H. Zeh'schen: Nürnberg.

Koch, C. L. (1850). *Übersicht des Arachnidensystems*, vol. 5. J. L. Lotzbeck: Nürnberg.

Koch, C. L. & Berendt, G. C. (1854). Die im Bernstein befindlichen Myriapoden, Arachniden und Apteren der Vorwelt. In: Berendt, G. C., *Die im Bernstein Befindlichen Organischen Reste der Vorwelt Gesammelt in Verbindung mit Mehreren Bearbeitet und Herausgegeben*, vol. 1(2): 1–124. Nicolai: Berlin.

Koch, L. (1856). Arachnoidea. In: Rosenhauer, W. G., *Die Thiere Andalusiens*: 407–410. Erlangen. (not seen)

Koch, L. (1870). Die Arachniden Galiziens. *Jahrbuch der K. K. Gelehrten Gesellschaft in Krakau* 41: 1–56. (not seen)

Koch, L. (1873). *Uebersichtliche Dartstellung der Europäischen Chernetiden (Pseudoscorpione)*. Bauer & Raspe: Nürnberg.

Koch, L. (1876). Verzeichniss der in Tirol bis jetzt beobachteten Arachniden nebst Beschreibung einiger neuen oder weniger bekannten Arten. *Zeitschrift des Ferdinandeums für Tirol und Vorarlberg* 20: 219–354. (not seen)

Koch, L. (1877). Verzeichniss der bei Nürnberg bis jetzt beobachteten Arachniden (mit Ausschluss der Ixodiden und Acariden) und Beschreibungen von neuen, hier vorkommenden Arten. *Abhandlungen der Naturhistorischen Gesellschaft zu Nürnberg* 6: 113–198.

Koch, L. (1878). Kaukasische Arachnoideen. In: Schneider, O., *Naturwissenschaftliche Beiträge zur Kenntnis der Kaukasusländer*: 36–71. Burdach'schen: Dresden.

Koch, L. (1881). Zoologische Ergebnisse von Excursionen auf den Balearen. II. Arachniden und Myriapoden. *Verhandlungen der K. K. Zoologischen-Botanischen Gesellschaft in Wien* 31: 625–678.

Koch, L. (1885). Ordo Chelonethi. In: Koch, L. & Keyserling, E., *Die Arachniden Australiens nach der Natur beschrieben und abgebildet*, vol. 2: 44–51. Bauer and Raspe: Nürnberg.

Koch, L. E. & Majer, J. D. (1980). A phenological investigation of various invertebrates in forest and woodland areas in the south-west of Western Australia. *Journal of the Royal Society of Western Australia* 63: 21–28.

Kofler, A. (1968). Zur Begleitfauna von *Quedius (Microsaurus) ventralis* (Arag.) (Col., Staphylinidae). *Berichte des Naturwissenschaftlich-Medizinischen Vereins in Innsbruck* 56: 355–360.

Kofler, A. (1972). Die Pseudoskorpione Osttirols. *Mitteilungen der Zoologischen Gesellschaft Braunau* 1: 286–289.

Kolenati, F. A (1857). Meletemata entomologica. VII. *Byulleten' Moskovskogo Obshchestva Ispytatelei Prirody* 30: 399–444.

Kolenati, A. F. (1859). Fauna des Altvaters (hohen Gesenkes der Sudeten). *Jahresheft des Naturwissenschaftlichen Section d. K. K. Mähr. Schles. Gesellschaft z. Beförd. d. Ackerbaues, d. Natur und Landeskunde f. d. Jahr 1858*. Brünn. (not seen)

Komatsu, T. (1952). [The spiders from Saisho-do.] *Acta Arachnologica* **7**: 56. (in Japanese) (not seen)

Koponen, S. & Sharkey, M. J. (1989). Northern records of *Microbisium brunneum* (Pseudoscorpionida, Neobisiidae) from eastern Canada. *Journal of Arachnology* **16**: 388–390.

Korschelt, E. & Heider, K. (1899). *Text-book of the Embryology of Invertebrates*, vol. **3**. Swan Sonnenschein: London. (English translation).

Korunić, Z. (1975). [Fauna of insects, Acarina and Pseudoscorpiones found in stores and mills in Croatia in the course of the years 1971–1974.] *Zastita Bilja* **26**: 219–227. (in Croatian) (not seen)

Kovoor, J. (1977). La soie et les glandes séricigènes des Arachnides. *Annales de Biologie* **16**: 97–171. (not seen)

Kowalski, K. (1955). Fauna jaskín Tatr Polskich. *Ochrona Przyrody* **23**: 283–332. (not seen)

Kratochvíl, J. (ed.) (1971). *Klíczvíreny CSSR*. Akademie Ved: Praha. (not seen)

Krauss, H. (1896). Einiges über Chernetiden nebst einem Auszug der Sammelergebnisse hierüber durch den Entomologischen Verein, Sektion Nürnberg. *Illustrierte Wochenschrift für Entomologie* **1**: 627–628.

Krauss, H. (1901). Über Chernetiden, eine interessante Gruppe der Arthropoden. *Entomologisches Jahrbuch* **9**: 237–248. (not seen)

Krausse-Heldrungen, A. H. (1912). Sardische Chernetiden. *Archiv für Naturgeschichte* **21**: 65–66.

Kreissl, E. (1969). Ein weiterer steirischer Fund des Höhlen-Pseudoskorpions *Neobisium hermanni* Beier (Arachnoidea – Pseudoscorp.). *Mitteilungen der Abteilung für Zoologie und Botanik am Landesmuseum 'Joanneum', Graz* **31**: 43–44.

Krumpál, M. (1979a). *Neobisium polonicum* Rafalski, 1937 (Pseudoscorpionidea) novy druh pre faunu CSSR. *Biológia* **34**: 429–435.

Krumpál, M. (1979b). *Gobichelifer dashdorzhi* (Pseudoscorpionidea, Cheliferidae) eine neue Gattung und species aus der Mongolei. *Biológia* **34**: 667–672.

Krumpál, M. (1980). Šťúriky (Pseudoscorpionidea) gaderskej doliny (Vel'ká Fatra). *Entomologicke Problemy* **16**: 23–29.

Krumpál, M. (1983a). Ein neuer *Calocheiridius* (Pseudoscorpionidea, Olpiidae) aus der USSR. *Acta Biologica* **13**: 58–61.

Krumpál, M. (1983b). Zwei neue *Diplotemnus*-Arten der UdSSR (Pseudoscorpiones, Miratemnidae). Über Pseudoscorpioniden-Fauna der UdSSR. II. *Biológia* **38**: 173–179.

Krumpál, M. (1983c). *Neobisium (N.) vilcekii* sp. n., ein neuer Pseudoscorpion aus der UdSSR (Neobisiidae, Pseudoscorpiones). Über Pseudoscorpioniden-Fauna der UdSSR IV. *Biológia* **38**: 607–612.

Krumpál, M. (1984a). Zwei neue Höhlen-Pseudoskorpionen aus der UdSSR (Pseudoscorpiones). Über Pseudoscorpioniden-Fauna der UdSSR VI. *Biológia* **38**: 637–646.

Krumpál, M. (1984b). Einige bemerkenswerte Pseudoscorpione aus der UdSSR. *Acta Entomologica Bohemoslovaca* **81**: 63–69.

Krumpál, M. (1986). Pseudoscorpione (Arachnida) aus Höhlen der UdSSR. Über Pseudoscorpioniden-Fauna der UdSSR V. *Biológia* **41**: 163–172.

Krumpál, M. (1987). Ein neuer *Dactylochelifer* aus Nepal Himalaya (Arachnida, Pseudoscorpiones). *Acta Entomologica Bohemoslovaca* **81**: 221–226.

Krumpál, M. & Cyprich, D. (1987). Pseudoscorpiones, Chernetidae. *Biológia* **42**: 196. (not seen)

Krumpál, M. & Cyprich, D. (1988). O výskyte šťúrikov (Pseudoscorpiones) v hniezdach vtákov (Aves) v podmienkach Slovenska. *Zbornik Slov. Nár. Múz., Prir. Vedy* **34:** 41–48.

Krumpál, M. & Kiefer, M. (1981). Príspevok k Poznaniu Šťúrikov Čel'ade Chthoniidae v ČSSR (Pseudoscorpionidea). *Zprávy Československé Společnosti Entom. Pri. CSAV, Praha* **17:** 127–130.

Krumpál, M. & Kiefer, M. (1982). Pseudoskorpione aus der Mongolei (Arachnida, Pseudoskorpiones). Ergebnisse der gemeinsamen Mongolisch-Slowakischen biologischen expedition. *Annotationes Zoologicae et Botanicae* **146:** 1–27.

Krunić, M. & Ćurčić, B. P. M. (1981). Correlation between the amount of super-cooling and hibernation sites in insects and arachnids. *Acta Entomologica Jugoslavica* **17:** 131–135.

Kúhnelt, W. (1961). *Soil Biology, with Special Reference to the Animal Kingdom.* Faber: London. (not seen)

Kulczynski, V. (1876). Dodatek do Fauny Pajeczaków Galicyi. *Sprawazdanie Komisyi Fizyograficznej, Kraków* **10:** 41–67. (not seen)

Kulczynski, V. (1899). Arachnoidea opera Rev. E. Schmitz collecta insulis Maderianis et in insulis Selvages dictis. *Rozpravy Akademia Umiejetnosei* **36:** 319–461.

Kurir, A. (1954). Der Pseudoskorpion *Dactylochelifer latreillei* Leach als Säftesauger der Larven von *Helicomyia saliciperda* Duf. *Anzeiger für Schädlingskunde* **27:** 137–138.

Kusch, H. (1982). Ergebnisse speläologischer Forschungen in Thailand (Stand 1978). *Höhle* **33:** 59–69. (not seen)

Kuschel, G. (1963). Composition and relationship of the terrestrial faunas of Easter, Juan Fernandez, Desventuradas, and Galapágos islands. *Occasional Papers of the California Academy of Sciences* **44:** 79–95.

Lagar, A. (1972a). Contribución al conocimiento de los Pseudoescorpiones de España. I. *Miscelanea Zoologica* **3:** 17–21.

Lagar, A. (1972b). Contribución al conocimiento de los Pseudoescorpiones de España. II. *Speleon* **19:** 45–52.

Lagerspetz, K. (1953). Biocoenological notes on the *Parmelia saxatilis* – *Dactylochelifer latreillei* community of seashore rocks. *Archivum Societatis Zoologico-Botanicae Fennicae* **7:** 131–142. (not seen)

Laloy, L. (1904). Insectes, Arachnides et Myriapodes marins. *Nature, Paris* **32:** 154–155. (not seen)

Lamarck, J. B. P. A. de (1818). *Histoire Naturelle des Animaux sans Vertèbres*, vol. **5.** Lanoe: Paris.

Lamarck, J. B. P. A. de (1838). *Histoire Naturelle des Animaux sans Vertèbres*, vol. **5,** 2nd edition. Paris.

Lamarck, J. B. P. A. de (1839). *Histoire Naturelle des Animaux sans Vertèbres*, 3rd edition. Meline: Bruxelles.

Lameere, A. (1895). *Manuel de la Faune de Belgique*, vol. **1.** Lamertin: Bruxelles.

Lampio, T. (1945). Suurikokoinen valeskorpioni Taivalsoskella. *Luonnon Ystävä* **49:** 187. (not seen)

Lancelevée, T. (1885). *Arachnides Recueillis aux environs d'Elbeuf et sur quelques points des Départements de la Seine Inférieure et de l'Eure.* Rouen. (not seen)

Lankester, E. R. (1904). The structure and classification of the Arachnida. *Quarterly Journal of Microscopical Science (new series)* **48:** 165–269.

Lanza, B. (1947). Nota preliminare sulla fauna di alcune grotte dei monti della Calvana. *Atti della Società Italiana di Scienze Naturali, e del Museo Civile di Storia Naturale, Milano* **86:** 180–184. (not seen)

Lanza, B. (1949). Speleofauna toscana. I. Cenni stotici ed elenco ragionato dei Protozoi, dei Molluschi, dei Crostacei, dei Miriapodi e degli Aracnidi (Acari esclusi) cavernicoli della Toscana. *Archivio Zoologico Italiano* **6:** 161–223. (not seen)

Lapschoff, I. I. (1940). Biospelogica Sovietica. V. [Die Höhlen-Pseudoscorpiones Transkaukasiens.] *Byulleten' Moskovskogo Obshchestva Ispytatelei Prirody, Biologii, n.s.* **49**: 61–74.

Larsson, S. G. (1978). Baltic amber – a palaeobiological study. *Entomonograph* **1**: 1–192.

Lasebikan, B. A. (1977). The arthropod fauna of a decaying log of an oil palm tree (*Elais guineensis* Jacq.) in Nigeria. *Ecological Bulletin* **25**: 530–533. (not seen)

Lasebikan, B. A. (1981). Comparative studies on the arthropod fauna of the soil and the decaying log of an oil palm tree in a tropical forest. *Pedobiologia* **21**: 110–116. (not seen)

Latreille, P. A. (1795). Observations sur la variété des organes de la bouche des Tiques, et distribution méthodique des insectes de cette famille d'après des caractères établis sur la conformation de ces organes. *Revue Encyclopédique* **4**: 15–20.

Latreille, [P. A.] (1796). *Précis des Caractères Génériques des Insectes, Disposés dans un Ordre Naturel*. F. Bordeaux: Brive.

Latreille, P. A. (1804). *Histoire Naturelle, Generale et Particulière, de Crustacés et des Insectes*, vol. **7**. F. Dufart: Paris.

Latreille, P. A. (1806). *Genera Crustacearum et Insectorum Secundum Ordinem Naturalem in Familias Disposita, Iconibus Exemplisque Plurimis Explicata*, vol. **1**. Paris. (not seen)

Latreille, P. A. (1810). *Considérations Générales sur l'Ordre Naturel des Animaux Composant les Classes des Crustacés, des Arachnides et des Insectes*. Paris.

Latreille, P. A. (1817). Arachnides. In: Cuvier, G., *Le Règne Animal Distribué d'Après son Organisation*, 1st edition, vol. **3**. Deterville: Paris.

Latreille, P. A. (1825a). Pince. In: *Encyclopédie Méthodique. Histoire Naturelle. Entomologie, ou Histoire Naturelle de Crustacés, des Arachnides et des Insectes*, vol. **10**: 131–133. Agasse: Paris.

Latreille, P. A. (1825b). *Familles Naturelles du Règne Animal, Exposées Succintement et dans un Ordre Analytique avec Indication de leurs Genres*. Paris. (not seen)

Latreille, P. A. (1827). *Natürliche Familien des Thierreichs. Übersetzt von Dr. Arn. Ad. Berthold*. Weimar. (not seen)

Latreille, P. A. (1829). Arachnides. In: Cuvier, G., *Le Règne Animal Distribué d'Après son Organisation*, 2nd edition, vol. **4**. Paris. (not seen)

Latreille, P. A. (1837). In: Cuvier, G., *The Animal Kingdom*, vol. **3**. G. Henderson: London.

Lawrence, R. F. (1935). A cavernicolous false scorpion from Table Mountain, Cape Town. *Annals and Magazine of Natural History* (10) **15**: 549–555.

Lawrence, R. F. (1937). A collection of Arachnida from Zululand. *Annals of the Natal Museum* **8**: 211–273.

Lawrence, R. F. (1964). New cavernicolous spiders from South Africa. *Annals of the South African Museum* **48**: 57–75.

Lawrence, R. F. (1967). The pseudoscorpions (false-scorpions) of the Kruger National Park. *Koedoe* **10**: 87–91.

Lawson, J. E. (1968). *Systematic Studies of Some Pseudoscorpions (Arachnida: Pseudoscorpionida) from the Southern United States*. Dissertation Abstracts 28B: 4351. (not seen)

Lawson, J. E. (1969). Description of a male belonging to the genus *Microbisium* (Arachnida: Pseudoscorpionida). *Research Division Bulletin of the Virginia Polytechnic Institute* **35**: 1–7.

Lazzeroni, G. (1966). Una nuova specie di *Chthonius* dell'Italia centrale. (Ricerche sugli Pseudoscorpioni. I). *Memorie del Museo Civico di Storia Naturale di Verona* **14**: 497–501.

Lazzeroni, G. (1967). Primi reperti sugli Pseudoscorpioni di Sardegna. *Bollettino di Zoologia* **34**: 129–130.

Lazzeroni, G. (1969a). Sur la faune de pseudoscorpions de la région apenninique méridionale. (Recherches sur les Pseudoscorpions. III.) *Memorie del Museo Civico di Storia Naturale di Verona* **16**: 321–344.

Lazzeroni, G. (1969b). Contributo alla conoscenza degli pseudoscorpioni della regione Veronese. (Ricerche sugli Pseudoscorpioni. IV.) *Memorie del Museo Civico di Storia Naturale di Verona* **16**: 379–418.

Lazzeroni, G. (1969c). Ricerche sugli Pseudoscorpioni. VI. Il popolamento della Sardegna. *Fragmenta Entomologica* **6**: 223–251.

Lazzeroni, G. (1970a). Ricerche sugli Pseudoscorpioni. III. Considerazioni biogeografiche sulla fauna della regione appenninica meridionale. *Bulletin du Muséum National d'Histoire Naturelle, Paris* (2) **41**, supplément 1: 205–208.

Lazzeroni, G. (1970b). *Chthonius* (s. str.) *elongatus*, nuova specie cavernicola della Toscana. (Ricerche sugli Pseudoscorpioni. VII). *Memorie del Museo Civico di Storia Naturale di Verona* **17**: 141–146.

Lazzeroni, G. (1970c). Ricerche sugli Pseudoscorpioni. V. L'isola di Giannutri. *Memorie, Atti della Societa Toscana di Scienze Naturali* (B) **76**: 101–112.

Lazzeroni, G. (1970d). Ricerche sugli Pseudoscorpioni. VIII. Su alcune interessanti specie raccolte allo Scoglio d'Affrica Archipelago Toscano. *Memorie, Atti della Societa Toscana di Scienze Naturali* (B) **77**: 37–50.

Leach, W. E. (1814). Crustaceology. In: Brewster, D. (ed.), *The Edinburgh Encyclopaedia*, vol. **7**: 383–437. Blackwood: Edinburgh.

Leach, W. E. (1815). A tabular view of the external characters of four classes of animals, which Linné arranged under Insecta; with the distribution of the genera composing three of these classes into orders, &c. and descriptions of several new genera and species. *Transactions of the Linnean Society of London* **11**: 306–400.

Leach, W. E. (1816). Annulosa. In: *Encyclopaedia Britannica; Supplement* **1**: 401–453. Edinburgh. (not seen)

Leach, M. D. (1817). On the characters of the genera of the family Scorpionidea, with descriptions of the British species of *Chelifer* and *Obisium*. In: *The Zoological Miscellany; Being Descriptions of New or Interesting Animals*: 48–53. R.P. Nodder: London.

Leakey, R. J. G. & Proctor, J. (1987). Invertebrates in the litter and soil at a range of altitudes on Gunung Silam, a small ultrabasic mountain in Sabah. *Journal of Tropical Ecology* **3**: 119–129. (not seen)

Lebedinsky, J. (1904). [Zur Höhlenfauna der Krym]. *Zapiski Novorossiiskago Obshchestva Estestvoispytatelei* **25**: 75–88. (in Russian)

Lebert, H. (1874). Über den Werth und die Bereitung des Chitinskelettes der Arachniden für mikroskopische Studien. *Sitzungsberichte der Kais. Akademie der Wissenschaften in Wien* **59**: 605–659. (not seen)

Lebert, H. (1877). Die Spinnen der Schweiz. *Neue Denkschriften der Schweizerischen Naturforschenden Gesellschaft* **27**: 1–321. (not seen)

Leclerc, P. (1979). *Les phénomènes de spéciation chez les Pseudoscorpions cavernicoles des karsts de la bordure orientale des Cévennes (France)*. Rapport de DEA de Biologie evolutive des populations et des speces animales, Université Paris, vol. 7.

Leclerc, P. (1981). Nouveaux Chthoniidae cavernicoles de la bordure orientale des Cévennes (France) (Arachnides, Pseudoscorpions). *Revue Arachnologique* **3**: 115–131.

Leclerc, P. (1982a). Une nouvelle espèce de Pseudscorpion cavernicole de la Drôme: *Neobisium (Blothrus) auberti* (Pseudoscorpions, Arachnides). *Revue Arachnologique* **4**: 39–45.

Leclerc, P. (1982b). Les pseudoscorpions des grottes des Sadoux. *Ursus Spelaeus* **7**: 43–46.

Leclerc, P. (1983a). *Neochthonius chamberlini* espèce nouvelle du sud de la France (Arachnides, Pseudoscorpions). *Revue Arachnologique* **5**: 45–53.

Leclerc, P. (1983b). A propos d'une collecte de Pseudoscorpion. *Ursus Spelaeus* **8**: 25.

Leclerc, P. (1983c). Les Chthoniidae de la grotte du Rendez-Vous de Chasse (Hérault) (Arachnides, Pseudoscorpions). *Bulletin de la Société d'Etude des Sciences Naturelles Béziers, n.s.* **9**: 11–19. (not seen)

Leclerc, P. (1984a). Notes chernetologiques. *Ursus Spelaeus* **9**: 53–56.

Leclerc, P. (1984b). *Etude biométrique de populations de faible effectif: le cas des Pseudoscoprions cavernicoles du sous-genre* Chthonius *(Arachnida: Chthoniidae).* Thèse, no. 1448: Université Claude Bernard, Lyon. (not seen)

Leclerc, P. (1986). Arachnides. In: *Expédition Thai-Maors 85*: 181–185. A.P.S.: Toulouse. (not seen)

Leclerc, P. (1989). *Neobisium (N.) atlasense* nouvelle espèce de Neobisiidae cavernicole du Maroc (Pseudoscorpions, Arachnides). *Revue Arachnologique* **8**: 45–51.

Leclerc, P. & Ginet, R. (1981). Les pseudo-scorpions cavernicoles. *Spelunca* **1**: 27–29. (not seen)

Leclerc, P. & Heurtault, J. (1979). Pseudoscorpions de l'Ardèche. *Revue Arachnologique* **2**: 239–247.

Leclerc, P. & Mahnert, V. (1988). A new species of the genus *Levigatocreagris* (Pseudoscorpiones: Neobisiidae) from Thailand, with remarkable sexual dimorphism. *Bulletin of the British Arachnological Society* **7**: 273–277.

Lee, D. C. (1983). Spiders, scorpions and other arachnids. In: *Natural History of the South East*: 183–185. Royal Society of South Australia: Adelaide.

Lee, D. C. & Southcott, R. V. (1983). Spiders and other arachnids of South Australia. In: *South Australian Year Book*: 29–43. Australian Bureau of Statistics: Adelaide.

Lee, D. J. (1961). *Cause and Effect Relating to Arthropod Bites and Stings*. School of Public Health and Tropical Medicine, University of Sydney: Sydney. (not seen)

Lee, V. F. (1972). *Systematic Studies of the Litoral Chelonethida of Baja California, México*. M.Sc. Thesis: California State University. (not seen)

Lee, V. F. (1979a). The maritime pseudoscorpions of Baja California, México (Arachnida: Pseudoscorpionida). *Occasional Papers of the California Academy of Sciences* **131**: i–iv, 1–38.

Lee, V. F. (1979b). Unusual habitats of some *Garypus* pseudoscorpions of Baja California, México. *American Arachnology* **20**: 13.

Lee, W. K. (1981). [A taxonomic study on the pseudoscorpions in Korea.] *Basic Science Review* **4**: 129–132. (in Korean)

Lee, W. K. (1982). Pseudoscorpions (Arachnida) from Korea II. A new species of the genus *Allochthonius*. *Basic Science Review* **5**: 75–80.

Legendre, R. (1966). Mission scientifique à l'Ile Europa. *Mémoires du Muséum Naturelle d'Histoire Naturelle, Paris* (A) **41**: 1–219. (not seen)

Legendre, R. (1968). Morphologie et développement des Chélicerates. Embryologie, développement et anatomie des Xiphosures, Scorpions, Pseudoscorpions, Opilions, Palpigrades, Uropyges, Amblypyges, Solifuges et Pycnogonides. *Fortschritte der Zoologie* **19**: 1–50.

Legendre, R. (1972). Les Arachnides de Madagascar. In: Battistini, R. & Richard-Vindard, G. (eds), *Biogeography and Ecology in Madagascar*: 427–457. Junk: The Hague.

Legg, G. (1970a). False-scorpions. *Countryside* **21**: 262–266.

Legg, G. (1970b). False-scorpions: their capture and care, and identification of families. *Countryside* **21**: 367–372.

Legg, G. (1971a). False-scorpions: the families Chthoniidae, Neobisiidae and Cheiridiidae. *Countryside* **21**: 472–477.

Legg, G. (1971b). *The Comparative and Functional Morphology of the Genitalia of the British Pseudoscorpiones*. Ph.D. Thesis: University of Manchester. (not seen)

Legg, G. (1972). False-scorpions: the families Chernetidae and Cheliferidae. *Countryside* **21**: 576–583.

Legg, G. (1973a). Spermatophore formation in the pseudoscorpion *Chthonius ischnocheles* (Chthoniidae). *Journal of Zoology, London* **170**: 367–394.

Legg, G. (1973b). The structure of encysted sperm of some British Pseudoscorpiones (Arachnida). *Journal of Zoology, London* **170**: 429–440.

Legg, G. (1974a). The genitalia and accessory glands of the pseudoscorpion *Cheiridium museorum* (Cheiridiidae). *Journal of Zoology, London* **173**: 323–339.

Legg, G. (1974b). An account of the genital musculature of pseudoscorpions (Arachnida). *Bulletin of the British Arachnological Society* **3**: 38–41.

Legg, G. (1974c). A generalised account of the female genitalia and associated glands of pseudoscorpions (Arachnida). *Bulletin of the British Arachnological Society* **3**: 42–48.

Legg, G. (1975a). A generalised account of the male genitalia and associated glands of pseudoscorpions (Arachnida). *Bulletin of the British Arachnological Society* **3**: 66–74.

Legg, G. (1975b). The genitalia and associated glands of five British species belonging to the family Chthoniidae (Pseudoscorpiones: Arachnida). *Journal of Zoology, London* **177**: 99–121.

Legg, G. (1975c). The genitalia and associated glands of five British species belonging to the family Neobisiidae (Pseudoscorpiones: Arachnida). *Journal of Zoology, London* **177**: 123–151.

Legg, G. (1975d). The possible significance of spermathecae in pseudoscorpions (Arachnida). *Bulletin of the British Arachnological Society* **3**: 91–95.

Legg, G. (1975e). Spermatophore formation in pseudoscorpions. *Proceedings of the 6th International Arachnological Congress, Amsterdam*: 141–144.

Legg, G. (1978). British pseudoscorpions. *Newsletter of the British Arachnological Society* **21**: 8–9. (not seen)

Legg, G. (1987). Proposed taxonomic changes to the British pseudoscorpion fauna (Arachnida). *Bulletin of the British Arachnological Society* **7**: 179–182.

Legg, G. & Jones, R. E. (1988). *Synopses of the British Fauna (new series). 40. Pseudoscorpions (Arthropoda; Arachnida)*. Brill/Backhuys: Leiden.

Lehnert, W. (1933). Beobachtungen über die Biocönose der Vogelnester. *Ornithologische Monatsberichte* **41**: 161–166. (not seen)

Lehtinen, P. T. (1964). The phalangids and pseudoscorpionids of Finnish Lapland. *Annales Universitatis Turkuensis* (2A) **32**: 279–287.

Leidy, [J.] (1877). Remarks on the Seventeen-year Locust, the Hessian Fly, and a Chelifer. *Proceedings of the Academy of Natural Scieneces of Philadelphia* **1877**: 260–261.

Leleup, N. (1947). Contribution a l'étude des Arthropodes nidicoles et microcavernicoles de Belgique. *Bulletin et Annales de la Société Entomologique de Belgique* **83**: 304–343.

Leleup, N. (1948). Contribution à l'étude des Arthropodes nidicoles et microcavernicoles de Belgique. *Mémoires de la Société Entomologique de Belgique* **25**: 1–55.

Lerma, B. (1932). Opilionidi e pseudoscorpionidi del Trentino. *Studi Trentini di Scienze Naturali* **13**: 219–222. (not seen)

Leruth, R. (1935). Exploration biologique des cavernes de la Belgique et du Limbourg Hollandais. XXVIIᵉ contribution: Arachnida. *Bulletin du Musée Royal d'Histoire Naturelle de Belgique* **11**(39): 1–34.

Lesne, P. (1896). Moers du *Limosina sacra*, Meig. (famille Muscidae, tribu Borborinae). Phénomènes de transport mutuel chez les animaux articulés. Origines du parasitisme chez les Insectes diptères. *Bulletin de la Société Entomologique de France* **65**: 162–165.

Lessert, R. de (1911). Pseudoscorpions. *Catalogue des Invertébrés de la Suisse* **5**: 1–50.

Levi, H. W. (1948). Notes on the life history of the pseudoscorpion *Chelifer cancroides* (Linn.) (Chelonethida). *Transactions of the American Microscopical Society* **67**: 290–298.

Levi, H. W. (1949). *Studies on the Life History of Three Species of Wisconsin Pseudoscorpions.* Thesis: University of Wisconsin. (not seen)

Levi, H. W. (1953). Observations on two species of pseudoscorpions. *Canadian Entomologist* **85**: 55–62.

Levi, H. W. & Levi, L. R. (1968). *Spiders and their Kin.* Golden Press: New York.

Lewis, R. J. (1903). On an undescribed species of *Chelifer. Journal of Quekett Microscopical Club* (2) **8**: 497–498.

Leydig, F. (1855). Zum feineren Bau der Arthropoden. *Archiv für Anatomie und Physiologie* **1855**: 376–480. (not seen)

Leydig, F. (1867). *Skizze zu einer Fauna Tübingensis.* Stuttgart. (not seen)

Leydig, F. (1881a). *Chelifer* als Schmarotzer. *Verhandlungen des Naturhistorischen Vereins der Preussischen Rheinlande, Westfalens und des Reg.-Bez. Osnabrück* **38**: 180. (not seen)

Leydig, F. (1881b). Über Verbreitung der Thiere im Rhöngebirge und Mainthal mit Hinblick auf Eifel und Rheinthal. *Verhandlungen des Naturhistorischen Vereins in Bonn* **?**: ?–? (not seen)

Leydig, F. (1893). Zum Parasitismus der Pseudoscorpioniden. *Zoologischer Anzeiger* **16**: 36–37.

Liebherr, J. K. (1988). General patterns in West Indian insects, and graphical biogeographic analysis of some circum-Caribbean *Platynus* beetles (Carabidae). *Systematic Zoology* **37**: 385–409.

Lindberg, K. (1955). Notes sur les grottes de l'île de Crète. *Fragmenta Balcanica* **1**: 165–174.

Lindberg, K. (1961). Recherches biospéléologiques en Afghanistan. *Acta Universitatis Lundensis, n.f.* **57**: 1–39.

Linnaeus, C. (1758). *Systema Naturae*, 10th edition, vol. **1**. Holmiae: Salvii.

Linnaeus, C. (1761). *Fauna Suecica Sistens Animalia Sveciae Regni: Mammalia, Aves, Amphibia, Pisces, Insecta, Vermes*, 2nd edition. Holmiae: Stockholm.

Linnaeus, C. (1767). *Systema Naturae*, 12th edition, vol. **1**. Holmiae: Salvii.

Linnaeus, C. (1788). *Systema Naturae*, 13th edition, cura J.F. Gmelin, vol. **1** (5). Lipsiae. (not seen)

Lippold, K. (1985). Pseudoscorpione aus dem NSG 'Ostufer der Müritz'. *Zool. Rdbrf. Bez. Neubrandenburg* **4**: 40.

Lloyd, J. E., Correale, S. & Muchmore, W. B. (1975). Pseudoscorpions phoretic on fireflies II. *Florida Entomologist* **58**: 241–242.

Lloyd, J. E. & Muchmore, W. B. (1974). Pseudoscorpions phoretic on fireflies. *Florida Entomologist* **57**: 381.

Loew, H. (1845). *Dipterologische Beiträge.* Posen.

Löw, F. (1866). Zoologische Notizen. Erste Serie. Arachnoidea. *Verhandlungen der K. K. Zoologischen-Botanischen Gesellschaft in Wien* **16**: 944–945.

Löw, F. (1867). Zoologische Notizen. Zweite Serie. *Verhandlungen der K. K. Zoologischen-Botanischen Gesellschaft in Wien* **17**: 746–752.

Löw, F. (1871). Zoologische Notizen. Dritte Serie. I. Beobachtungen über das Eierlegen und Spinnen der After- oder Bücherskorpione (Pseudoscorpiones v. Obisida). *Verhandlungen der K. K. Zoologischen-Botanischen Gesellschaft in Wien* **21**: 841–843.

Lohmander, H. (1939a). Zwei neue Chernetiden der nordwesteuropäischen Fauna. *Göteborgs Kungl. Vetenskaps- och Vitterhetssamhälles Handlingar* (5B) **6**(11): 1–11.

Lohmander, H. (1939b). Zur Kenntnis der Pseudoskorpionfauna Schwedens. *Entomologisk Tidskrift* **60**: 279–323.

Lohmander, H. (1945). Arachnologische Fragmente. I. Über eine für die schwedische Fauna neue Pseudoskorpionart. *Göteborgs Kungl. Vetenskaps- och Vitterhetssamhälles Handlingar* (6B) **3**(9): 3–14.

Loksa, I. (1960). Faunistisch-systematische und ökologische Untersuchungen in der Lóczy-Höhle bei Balatonfüred. *Annales Universitatis Scientarum Budapest (Biologie)* **3**: 253–266. (not seen)

Loksa, I. (1961). Quantitative Untersuchungen streuschichtbewohnender Arthropoden-Bevölkerungen in eigen Ungarischen Waldbeständen. *Annales Universitatis Scientarum Budapest (Biologie)* **4**: 99–112. (not seen)

Loksa, I. (1962). Über die Landarthropoden der István-, Forrás-, und Szeleta-Höhle bei Lillafüred. *Karzt Barlangk* **3**: 59–81. (not seen)

Loksa, I. (1979). Quantitative Untersuchungen über die Makrofauna der Laubstreu in Zerreichen- und Hainsimsen-Eichen-Beständen des Bükk-Gebirges. *Opuscula Zoologica, Budapest* **16**: 87–95.

Lomnicki, A. (1963). The distribution and abundance of ground-surface-inhabiting arthropods above the timber line in the region of Zolta Turnia in the Tatra Mts. *Acta Zoologica, Cracoviensia* **8**: 183–249. (not seen)

Lubbock, J. (1861). Notes on the generative organs and of the formation of eggs in the Annulosa. *Philosophical Transactions of the Royal Society of London* **151**: 595–627. (not seen)

Lucas, H. (1839). Arachnides, Myriapodes et Thysanoures. In: Webb, P. B. Berthelot, S. (eds), *Histoire Naturelle des Iles Canaries*, vol. **2**(2). Bethune. (not seen)

Lucas, H. (1849). Histoire naturelle des animaux articulés. Crustacés, Arachnides, Myriapodes et Hexapodes. In: *Exploration Scientifique de l'Algérie Pendant les Années 1840, 1841, 1842. Sciences Physiques. Zoologie*, vol. **2**(1). Imprimerie Nationale: Paris.

Lukis, F. C. (1834). *Chélifer cancröides. Magazine of Natural History* **7**: 162–163.

Lukjanov, N. (1895). Spisok paukow ... [Liste des araignées (Araneina, Pseudoscorpiona et Phalangina) du sud-ouest de la Russie et des gouvernements voissins.] *Zapiski Kievskavo Obcestva Jestestvouspytatelej* **14**: 559–577. (in Russian) (not seen)

Ma, M., Burkholder, W. E. & Carlson, S. D. (1978). Supra-anal organ: a defensive mechanism of the furniture carpet beetle, *Anthrenus flavipes* (Coleoptera: Dermestidae). *Annals of the Entomological Society of America* **71**: 718–723. (not seen)

MacGillavry, D. (1914). De entomologische fauna van het eiland Terschelling voor zover zij tot nu toe bekend is. *Tijdschrift voor Entomologie* **57**: 89–106. (not seen)

Machado, A. (1987). *Bibliografía Entomológica Canaria*. Instituto de Estudios Canarios: La Laguna.

Mackie, D. W. (1960). *Neobisium carpenteri* (Kew) (Pseudoscorpiones): the first record for Great Britain. *Naturalist, London* **874**: 109.

MacLeod, J. (1884). La structure de l'intestin antérieur des Arachnides. Communication préliminaire. *Bulletin de l'Académie Royal de Belgique* (3) **8**: 377–391. (not seen)

Macrae, G. (1869). Untitled. *Science Gossip* **1869**: 223. (not seen)

Mahnert, V. (1972a). *Neobisium (Blothrus) kwartirnikovi* nov. spec. (Pseudoscorpionidea) aus Bulgarien. *Archives des Sciences, Genève* **24**: 383–389.

Mahnert, V. (1972b). Über griechische Pseudoskorpione I: *Microcreagris leucadia* nov. spec. (Arachnida: Pseudoscorpiones, Neobisiidae). *Berichte des Naturwissenschaftlich-Medizinischen Vereins in Innsbruck* **59**: 51–56.

Mahnert, V. (1973a). Über griechische Pseudoskorpione II: Höhlenpseudoskorpione (Pseudoscorpionides, Neobisiidae) von Korfu. *Revue Suisse de Zoologie* **80**: 207–220.

Mahnert, V. (1973b). Drei neue Neobisiidae (Arachnida: Pseudoscorpiones) von den Ionischen Inseln (Über griechische Pseudoskorpione III). *Berichte des Naturwissenschaftlich-Medizinischen Vereins in Innsbruck* **60**: 27–39.

Mahnert, V. (1974a). *Roncus viti* n. sp. (Arachnida, Pseudoscorpiones) aus dem Iran. *Berichte des Naturwissenschaftlich-Medizinischen Vereins in Innsbruck* **61**: 87–91.

Mahnert, V. (1974b). Über höhlenbewohnende Pseudoskorpione (Neobisiidae, Pseudoscorpiones) aus Süd- und Osteuropa. *Revue Suisse de Zoologie* **81**: 205–218.

Mahnert, V. (1974c). Einige Pseudoskorpione aus Israel. *Revue Suisse de Zoologie* **81**: 377–386.

Mahnert, V. (1974d). *Acanthocreagris* nov. gen. mit Bemerkungen zur Gattung *Microcreagris* (Pseudoscorpiones, Neobisiidae) (Über griechische Pseudoskorpione IV). *Revue Suisse de Zoologie* **81**: 845–885.

Mahnert, V. (1975a). Griechische Höhlenpseudoskorpione. *Revue Suisse de Zoologie* **82**: 169–184.

Mahnert, V. (1975b). Pseudoskorpione der Insel Réunion und von T.F.A.I. (Djibouti). *Revue Suisse de Zoologie* **82**: 539–561.

Mahnert, V. (1975c). Pseudoscorpione von den maltesischen Inseln. *Fragmenta Entomologica* **11**: 185–197.

Mahnert, V. (1976a). Zwei neue Pseudoskorpion-Arten (Arachnida) aus griechischen Höhlen (Über griechische Pseudoskorpione VII). *Berichte des Naturwissenschaftlich-Medizinischen Vereins in Innsbruck* **63**: 177–183.

Mahnert, V. (1976b). Zwei neue Pseudoskorpion-Arten (Arachnida, Pseudoscorpiones) aus marokkanischen Höhlen. *International Journal of Speleology* **8**: 375–381.

Mahnert, V. (1976c). Zur Kenntnis der Gattungen *Acanthocreagris* und *Roncocreagris* (Arachnida, Pseudoscorpiones, Neobisiidae). *Revue Suisse de Zoologie* **83**: 193–214.

Mahnert, V. (1976d). Pseudoscorpions des grottes de la Sardaigne. *Fragmenta Entomologica* **12**: 309–316.

Mahnert, V. (1976e). Quelques remarques sur les problèmes taxonomiques chez les pseudoscorpions d'Europe et sur la structure des trichobothries des pseudoscorpions. *Comptes Rendus 3ème Réunion des Arachnologistes d'Expression française, Les Eyzies, 1976*: 97–99.

Mahnert, V. (1977a). Pseudoskorpione (Arachnida) aus dem Tien-Shan. *Berichte des Naturwissenschaftlich-Medizinischen Vereins in Innsbruck* **64**: 89–95.

Mahnert, V. (1977b). Über einige Atemnidae und Cheliferidae Griechenlands (Pseudoscorpiones). *Mitteilungen der Schweizerischen Entomologischen Gesellschaft* **50**: 67–74.

Mahnert, V. (1977c). Spanische Höhlenpseudoskorpione. *Miscelanea Zoologica* **4**: 61–104.

Mahnert, V. (1977d). Etude comparative des trichobothries de pseudoscorpions au microscope électronique à balayage. *Comptes Rendus des Séances de la Société de Physique et d'Histoire Naturelle de Genève, Nouvelle Série* **11**: 96–99.

Mahnert, V. (1978a). Die Pseudskorpiongattung *Toxochernes* Beier, 1932. *Symposia of the Zoological Society of London* **42**: 309–315.

Mahnert, V. (1978b). Pseudoskorpione (Arachnida) aus der Höhle Sisco (Korsika). *Revue Suisse de Zoologie* **85**: 381–384.

Mahnert, V. (1978c). Zwei neue *Dactylochelifer*-Arten aus Spanien und von Mallorca (Pseudoscorpiones). *Eos, Madrid* **52**: 149–157.

Mahnert, V. (1978d). Contributions à l'étude de la faune terrestre des îles granitiques de l'archipel des Séchelles. Pseudoscorpiones. *Revue de Zoologie Africaine* **92**: 867–888.

Mahnert, V. (1978e). Weitere Pseudoskorpione (Arachnida Pseudoscorpiones) aus griechischen Höhlen. *Annales Musei Goulandris* **4**: 273–298.

Mahnert, V. (1978f). Pseudoskorpione (ausgenommen Olpiidae, Garypidae) aus Congo-Brazzaville (Arachnida, Pseudoscorpiones). *Folia Entomologica Hungarica* **31**: 69–133.

Mahnert, V. (1978g). Zwei neue *Rhacochelifer*-Arten aus dem westlichen Mediterrangebiet und Wiederbeschreibung von *Chelifer heterometrus* L. Koch. *Comptes Rendus des Séances de la Société de Physique et d'Histoire Naturelle de Genève, nouvelle série* **12**: 14–24.

Mahnert, V. (1978h). Zur Verbreitung höhlenbewohnender Pseudoskorpione der iberischen Halbinsel. *Comunicacions del 6è. Simposium d'Espeleologia, Terrassa*: 21–23.

Mahnert, V. (1979a). The identity of *Microcreagris gigas* Balzan (Pseudoscorpiones, Neobisiidae). *Bulletin of the British Arachnological Society* **4**: 339–341.

Mahnert, V. (1979b). Pseudoskorpione (Arachnida) aus Höhlen der Türkei und des Kaukasus. *Revue Suisse de Zoologie* **86**: 259–266.

Mahnert, V. (1979c). Zwei neue Chthoniiden-Arten aus der Schweiz (Pseudoscorpiones). *Revue Suisse de Zoologie* **86**: 501–507.

Mahnert, V. (1979d). Pseudoskorpione (Arachnida) aus dem Amazonas-Gebiet (Brasilien). *Revue Suisse de Zoologie* **86**: 719–810.

Mahnert, V. (1979e). L'identité de *Olpium chironomum* L. Koch (Pseudoscorpions, Neobisiidae). *Revue Arachnologique* **2**: 249–252.

Mahnert, V. (1980a). Pseudoscorpions from the Canary Islands. *Entomologica Scandinavica* **11**: 259–264.

Mahnert, V. (1980b). Zwei neue *Chthonius*-Arten (Pseudoscorpiones) aus Höhlen Marokkos. *Mitteilungen der Schweizerischen Entomologischen Gesellschaft* **53**: 215–219.

Mahnert, V. (1980c). Arachnids of Saudi Arabia. Pseudoscorpiones. In: *Fauna of Saudi Arabia*, vol. **2**: 32–48.

Mahnert, V. (1980d). Pseudoskorpione (Arachnida) aus Höhlen Italiens, mit Bemerkungen zur Gattung *Pseudoblothrus*. *Grotte d'Italia* (4a) **8**: 21–38.

Mahnert, V. (1980e). Pseudoskorpione (Arachnida) aus Höhlen Griechenlands, insbesondere Kretas. *Archives des Sciences, Genève* **32**: 213–233.

Mahnert, V. (1980f). Verbreitung der Pseudoskorpione (Arachnida) in Kenya (Ostafrika). *Verhandlungen des 8. Internationalen Arachnologen-Kongresses, Wien*: 470.

Mahnert, V. (1980g). Höhlenpseudoskorpione aus Norditalien und der dalmatinischen Insel Krk. *Atti e Memorie della Commissione Grotte 'E. Boegan'* **20**: 95–100.

Mahnert, V. (1980h). In the tracks of Balzan in Paraguay. *Newsletter of the British Arachnological Society* **27**: 11–12.

Mahnert, V. (1981a). Die Pseudoskorpione (Arachnida) Kenyas. I. Neobisiidae und Ideoroncidae. *Revue Suisse de Zoologie* **88**: 535–559.

Mahnert, V. (1981b). Taxonomische Irrwege: *Olpium savignyi* Simon, *O. kochi* Simon, *O. bicolor* Simon (Pseudoscorpiones). *Folia Entomologica Hungaricae* **42**: 95–99.

Mahnert, V. (1981c). Sigles trichobothriaux chez les pseudoscorpions (Arachnida). *Memorie, Atti della Società Toscana di Scienze Naturali residente in Pisa* B **88** supplemento: 185–192.

Mahnert, V. (1981d). *Mesochelifer ressli* n. sp., eine mit *Chelifer cancroides* (L.) verwechselte Art aus Mitteleuropa (Pseudoscorpiones, Cheliferidae). *Veröffentlichungen des Tiroler Landesmuseums Ferdinandeum* **61**: 47–53.

Mahnert, V. (1981e). *Chthonius (C.) hungaricus* sp. n., eine neue Afterskorpion-Art aus Ungarn (Arachnida). *Folia Entomologica Hungarica* **41**: 279–282.

Mahnert, V. (1982a). The pseudoscorpion genus *Corosoma* Karsch, 1879, with remarks on *Dasychernes* Chamberlin, 1929 (Pseudoscorpiones, Chernetidae). *Journal of Arachnology* **10**: 11–14.

Mahnert, V. (1982b). Die Pseudoskorpione (Arachnida) Kenyas II. Feaellidae; Cheiridiidae. *Revue Suisse de Zoologie* **89**: 115–134.

Mahnert, V. (1982c). Die Pseudoskorpione (Arachnida) Kenyas. 3. Olpiidae. *Monitore Zoologico Italiano, n.s. Supplemento* **16**: 263–304.

Mahnert, V. (1982d). Die Pseudoskorpione (Arachnida) Kenyas, IV. Garypidae. *Annales Historico-Naturales Musei Nationalis Hungarici* **74**: 307–329.

Mahnert, V. (1982e). Die Pseudoskorpione (Arachnida) Kenyas V. Chernetidae. *Revue Suisse de Zoologie* **89**: 691–712.

Mahnert, V. (1982f). Neue höhlenbewohnende Pseudoskorpione aus Spanien, Malta und Griechenland (Arachnida, Pseudoscorpiones). *Mitteilungen der Schweizerischen Entomologischen Gesellschaft* **55**: 297–304.

Mahnert, V. (1983a). Die Pseudoskorpione Kenyas VI. Dithidae (Arachnida). *Revue de Zoologie Africaine* **97**: 141–157.

Mahnert, V. (1983b). Die Pseudoskorpione (Arachnida) Kenyas VII. Miratemnidae und Atemnidae. *Revue Suisse de Zoologie* **90**: 357–398.

Mahnert, V. (1983c). The genus *Caffrowithius* Beier, 1932, with the description of a new species from South Africa (Arachnida: Pseudoscorpiones). *Annals of the Natal Museum* **25**: 501–510.

Mahnert, V. (1983d). Pseudoscorpions from the Hortobágy National Park (Arachnida). In: *The Fauna of the Hortobágy National Park*: 361–363. Akadémiai Kiadó: Budapest.

Mahnert, V. (1984a). Beitrag zu einer besseren Kenntnis der Ideoroncidae (Arachnida: Pseudoscorpiones), mit Beschreibung von sechs neuen Arten. *Revue Suisse de Zoologie* **91**: 651–686.

Mahnert, V. (1984b). Forschungen an der Somalilandküste. Am Strand und auf den Dünen bei Sar Uanle. 36. Pseudoscorpiones (Arachnida). *Monitore Zoologico Italiano, n.s. Supplemento* **19**: 43–66.

Mahnert, V. (1984c). Pseudoscorpions (Arachnida) récoltés durant la mission spéologique espagnole au Pérou en 1977. *Revue Arachnologique* **6**: 17–28.

Mahnert, V. (1985a). *Roncus (Parablothrus) comasi*, espèce nouvelle d'une grotte de la Tunisie (Pseudoscorpiones, Neobisiidae). *Speleon* **26–27**: 17–20.

Mahnert, V. (1985b). Pseudoscorpions (Arachnida) from the lower Amazon region. *Revista Brasileira de Entomologia* **29**: 75–80.

Mahnert, V. (1985c). Weitere Pseudoskorpione (Arachnida) aus dem zentralen Amazonasgebiet (Brasilien). *Amazoniana* **9**: 215–241.

Mahnert, V. (1986a). Die Pseudoskorpione (Arachnida) Kenyas. VIII. Chthoniidae. *Revue Suisse de Zoologie* **92**: 823–843.

Mahnert, V. (1986b). *Parachernes gracilimanus* n. sp., espèce nouvelle de Pseudo-scorpion (Arachnida, Chernetidae) de l'Equateur. *Revue Suisse de Zoologie* **93**: 813–816.

Mahnert, V. (1986c). Une nouvelle espèce du genre *Tyrannochthonius* Chamb. des îles Canaries, avec remarques sur les genres *Apolpiolum* Beier et *Calocheirus* Chamberlin (Arachnida, Pseudoscorpiones). *Mémoires de la Société Royale Entomologique de Belgique* **33**: 143–153.

Mahnert, V. (1986d). Arthropodes epigés du Massif de 'San Juan de la Peña' (Jaca, Huesca). *Pirineos* **124**: 73–86.

Mahnert, V. (1987). Neue oder wenig bekannte, vorwiegend mit Insekten vergesell-schaftete Pseudoskorpione (Arachnida) aus Südamerika. *Mitteilungen der Schweizerischen Entomologischen Gesellschaft* **60**: 403–416.

Mahnert, V. (1988a). Die Pseudoskorpione (Arachnida) Kenyas. Familien Withiidae und Cheliferidae. *Tropical Zoology* **1**: 39–89.

Mahnert, V. (1988b). Zwei neue Garypininae-Arten (Pseudoscorpiones: Olpiidae) aus Afrika mit Bemerkungen zu den Gattungen *Serianus* Chamberlin und *Paraserianus* Beier. *Stuttgarter Beiträge zur Naturkunde, Serie A (Biologie)* **420**: 1–11.

Mahnert, V. (1988c). *Neobisium carcinoides* (Hermann, 1804) (Pseudoscorpionida, Neobisiidae) – une espece polymorphe? *Comptes Rendus Xème Colloque Europeen Arachnologie. Bulletin de la Société de Sciences de Bretagne* **59**: 161–174.

Mahnert, V. (1988d). Une nouvelle espèce du genre *Tyrannochthonius (Lagynochthonius)* (Pseudoscorpiones: Chthoniidae) des grottes de Sarawak (Malaysia). *Archives des Sciences, Genève* **41**: 383–386.

Mahnert, V. (1989a). Les pseudoscorpions (Pseudoscorpiones, Arachnida) récoltés pendant la campagne biospéologique 1987 à Minorque. *Endins* **14–15**: 85–87.

Mahnert, V. (1989b). Les pseudoscorpions (Arachnida) des grottes des iles Canaries, avec description de deux especes nouvelles du genre *Paraliochthonius* Beier. *Mémoires de Biospéologie* **16**: 41–46.

Mahnert, V. & Adis, J. (1986). On the occurrence and habitat of Pseudoscorpiones (Arachnida) from Amazonian forest of Brazil. *Studies on Neotropical Fauna and Environment* **20**: 211–215.

Mahnert, V., Adis, J. & Bührnheim, P. F. (1986). Key to the families of Amazonian Pseudoscorpiones (Arachnida). *Amazoniana* **10**: 21–40.

Mahnert, V. & Aguiar, N. O. (1986). Wiederbeschreibung von *Neocheiridium corticum* (Balzan, 1890) und Beschreibung von zwei neuen Arten der Gattung aus Südamerika (Pseudoscorpiones, Cheiridiidae). *Mitteilungen der Schweizerischen Entomologischen Gesellschaft* **59**: 499–509.

Mahnert, V. & Leclerc, P. (1988). A new species of the genus *Levigatocreagris* Ćurčić (Pseudoscorpiones: Neobisiidae) from Thailand, with remarkable sexual dimorphism. *Bulletin of the British Arachnological Society* **7**: 273–277.

Mahnert, V. & Schuster, R. (1981). *Pachyolpium atlanticum* n. sp., ein Pseudoskorpion aus der Gezeitenzone der Bermudas – Morphologie und Ökologie (Pseudoscorpiones: Olpiidae). *Revue Suisse de Zoologie* **88**: 265–273.

Main, A. R. (1954). *A Guide for Naturalists*, 1st edition. Western Australian Naturalists' Club: Perth.

Main, A. R. & Edward, D. H. (1968). *A Guide for Naturalists*, 2nd edition. Western Australian Naturalists' Club: Perth.

Main, B. Y. (1981). A comparative account of the biogeography of terrestrial invertebrates in Australia: some generalizations. In: Keast, A. (ed.), *Ecological Biogeography of Australia*, vol. **2**: 1055–1077. Junk: The Hague.

Makioka, T. (1968). Morphological and histochemical studies on embryos and ovaries during the embryo-breeding of the pseudoscorpion, *Garypus japonicus*. *Science Report of the Tokyo Kyoiku Daigaku (B)* **13**: 207–227.

Makioka, T. (1969). A temporary gonopodium in a pseudoscorpion, *Garypus japonicus*. *Science Report of the Tokyo Kyoiku Daigaku (B)* **14**: 113–120.

Makioka, T. (1976). Alternative occurrence of two ovarian functions in the adult pseudoscorpion, *Garypus japonicus* Beier. *Acta Arachnologica* **27**: 8–15.

Makioka, T. (1977). [The mode of breeding for the embryos and larvae and the breeding stages in the pseudoscorpion, *Garypus japonicus* Beier.] *Acta Arachnologica* **27** (special number): 185–197. (in Japanese)

Makioka, T. (1979). Structures of the adult ovaries in different functional phases of the pseudoscorpion, *Garypus japonicus* Beier. I. The ovary in the resting phase. *Acta Arachnologica* **28**: 71–81.

Malcolm, D. R. & Chamberlin, J. C. (1960). The pseudoscorpion genus *Chitrella* (Chelonethida, Syarinidae). *American Museum Novitates* **1989**: 1–19.

Malcolm, D. R. & Chamberlin, J. C. (1961). The pseudoscorpion genus *Kleptochthonius* (Chelonethida, Chthoniidae). *American Museum Novitates* **2063**: 1–35.

Malcolm, D. R. & Muchmore, W. B. (1985). An unusual species of *Tyrannochthonius* from Florida (Pseudoscorpionida, Chthoniidae). *Journal of Arachnology* **13**: 403–405.

Mani, M. S. (1959). On a collection of high altitude scorpions and pseudo-scorpions (Arachnida) from the North-West Himalaya. *Agra University Journal of Reseacrh Science* **8**: 11–16.

Mani, M. S. (1962). *Introduction to High Altitude Entomology*. Methuen: London.

Mani, M. S. (1968). *Ecology and Biogeography of High Altitude Insects.* Junk: The Hague.

Manley, (1969). A pictorial key and annotated list of Michigan pseudoscorpions (Arachnida: Pseudoscorpionida). *Michigan Entomologist* **2**: 2–13.

Mann, B. P. (1868). Untitled. *Proceedings of the Boston Society of Natural History* **11**: 325.

Marchesetti, C. (1890). La caverna di Gabrovizza presso Trieste. *Atti del Museo Civico di Storia Naturel di Trieste* **8**: 143–184. (not seen)

Marcuzzi, G. (1956). Fauna della Dolomiti. *Memorie dell'Istituto Veneto Classe di Scienze Matematica e Naturali* **31**: 1–595. (not seen)

Marcuzzi, G. (1961). Supplemento allo Fauna delle Dolomiti (Aggiunte e commenti). *Memorie dell'Istituto Veneto Cl. Sci. Mat. Nat.* **32**: 1–136. (not seen)

Marcuzzi, G., Lorenzoni, A. M. & di Castri, F. (1970). La fauna del suole di une regione delle Prealpe Venete (M. Spitz, Recoara). Aspetti autoecologici. *Atti del Istituto Veneto di Scienze, Lettere ed Arti, Scienze Matematica e Naturali* **128**: 411–567. (not seen)

Marquand, E. D. (1908). The spiders of Guernsey. *Report and Transactions of the Guernsey Society of Natural Science* **5**: 367–383. (not seen)

Martens, J. (1975). Phoretische Pseudoskorpione auf Kleinsäugern des Nepal-Himalaya. *Zoologischer Anzeiger* **194**: 84–90.

Martin, B. (1782). *The Young Gentleman and Lady's Philosophy*, vol. 3. London. (not seen)

Martin, J. L., Oromi, P. & Hernandez, J. J. (1986). Ei tubo volcanico de la Cueva de San Marcos (Tenerife, Islas Canarias): origin geologico de la cavidad y estudio de su biocenosis. *Vieraea* **16**: 295–308. (not seen)

Matthiesen, F. A. & Hahn, N. S. (1981). Forésia em pseudo-escorpioes brasileiros. *Ciência e Cultura* **33**: 689–690.

Maury, A. (1979). Faune de la résurgence d'Hannetot (près Norville, Seine-Maritime). *Bulletin Trimestrial de la Société Géologique de Normandie et des Amis du Muséum du Havre* **66**: 24. (not seen)

May, A. F. (1969). *Beekeeping.* Haum: Capetown. (not seen)

McClain, L. (1988). A Namib desert pseudoscorpion: between a rock and a hard place. *Newsletter, Research Group for the Study of African Arachnids* **3**: 15. (not seen)

McClure, H. E. (1943). Aspection in the biotic communities of the Churchill area, Manitoba. *Ecological Monographs* **13**: 1–35.

McDaniel, V. R., Paige, K. N. & Tumlinson, C. R. (1980). Cave fauna of Arkansas: additional invertebrate and vertebrate records. *Proceedings of the Arkansas Academy of Science* **33**: 84–85. (not seen)

McIntire, S. J. (1868). On pseudo-scorpiones. *Journal of the Quekett Microscopical Club* **1**: 8–15.

McIntire, S. J. (1869). Pseudoscorpions. *Science Gossip* **5**: 243–247.

McIntire, S. J. (1871). An incident in the life of a Chelifer. *Monthly Microscopical Journal* **6**: 209–210.

Mégnin, J. P. (1886). Sur les prétendus parasites des Mouches. *Nature, Paris* **14**: 241–243. (not seen)

Meinertz, T. (1962a). Beiträge zur Verbreitung der Pseudoskorpioniden in Dänemark. *Videnskabelige Meddelelser fra Dansk Naturhistorisk Forening i Kjøbenhavn* **126**: 387–402.

Meinertz, T. (1962b). Mosskorpioner og mejere (Pseudoscorpionidae og Opiliones). *Danmarks Fauna* **67**: 193. (not seen)

Mello-Leitão, C. (1925). Dois interessantes arachnideos myrmecophiles. *Physis, Buenos Aires* **8**: 228–237.

Mello-Leitão, C. (1937). Novo pseudoscorpião do Brasil. *Annaes da Academia Brasileira de Sciencias* **9**: 269–270.

Mello-Leitão, C. (1939a). Pseudoscorpionidos de Argentina. *Notas del Museo de La Plata* **4:** 115–122.

Mello-Leitão, C. (1939b). Les Arachnides et la zoogéographie de l'Argentine. *Physis, Buenos Aires* **17:** 601–630.

Mello-Leitão, C. & Feio, J. de Araújo (1949). Notas sôbre pequena colecção de arachnídios do Perú. *Boletim do Museu Goeldi de Historia Naturel e Ethnographia* **10:** 313–324.

Menge, A. (1855). Ueber die Scheerenspinnen, Chernetidae. *Neueste Schriften der Naturforschenden Gesellschaft* **5**(2): 1–43.

Menozzi (as Minozzi), C. (1917). Contributo allo studio della speleologia italiana. La grotta di S. Maria sul Monte Vallestra (Reggio E.). *Bollettino della Società Entomologica Italiana* **48:** 164–174.

Menozzi (as Minozzi), C. (1920). Nota complementare alla topografia e alla fauna della grotta di S. Maria M. sul monte Vallestra. *Atti Soc. Nat. Mat. Modena* (5) *5:* 70–74.

Menozzi, C. (1924). Nuova specie di Pseudoscorpione alofilo. *Annuario R. Museo Zoologico della R. Università di Napoli* **5:** 1–3.

Menozzi, C. (1933). Alcuni aspetti della vita in relazione all'ambiente nella Grotta di S. Maria Maddalena sul Monte Vallestra (Reggio Emilia). *Atti i Congresso di Speleologia Nazionale, Trieste*: 194–198. (not seen)

Menozzi, C. (1939). La fauna della grotta della Suja sul Monte Fascie (Genova) ed osservazioni biologiche sulla *Parabathyscia doderoi* Fairm. (Coleopt. Catopidae), con descrizione della larva e delle caratteristiche morfologiche del suo intestino e di quello dell'adulto. *Memorie della Società Entomologica Italiana* **18:** 129–154.

Merwe, J. H. (1908). Pseudoscorpion in beehive. *Agricultural Journal of the Cape of Good Hope* **33:** 517–518. (not seen)

Messana, G., Chelazzi, L. & Baccetti, N. (1985). Biospeleology of Somalia, Mugdile and Showli Berdi Caves. *Monitore Zoologico Italiana, n.s., supplemento* **20:** 325–340. (not seen)

Metschnikoff, E. (1871). Entwicklungsgeschichte des *Chelifer*. *Zeitschrift für Wissenschaftliche Zoologie* **21:** 513–525.

Meyer, E., Schwarzenberger, I., Stark, G. & Wechselberger, G. (1984). Bestand und jahreszeitliche Dynamik der Bodenmakrofauna in einem inneralpinen Eichen-mischwald (Tirol, Österreich). *Pedobiologica* **27:** 115–132. (not seen)

Meyer, E., Wäger, H. & Thaler, K. (1985). Struktur und jahreszeitliche Dynamik von *Neobisium*-Populationen in zwei Höhenstufen in Nordtirol (Österreich) (Arachnida: Pseudoscorpiones). *Revue d'Ecologie et Biologie du Sol* **22:** 221–232. (not seen)

Michael, A. D. (1879). House-flies and their parasites. *Science Gossip* **15:** 39.

Millot, J. (1948). Revue générale des Arachnides de Madagascar. *Mémoires de l'Institut Scientifique de Madagascar* (A) **1:** 137–155. (not seen)

Millot, J. (1949). Morphologie générale et anatomie interne. In: Grassé, P.-P. (ed.), *Traité de Zoologie*, vol. **6:** 263–385. Masson: Paris.

Milstead, W. M. (1958). A list of the arthropods found in the stomach of whiptail lizards from four stations in southwestern Texas. *Texas Journal of Science* **10:** 443–446. (not seen)

Minelli, A. (1977). Corologia ed ecologia di alcuni artropodi geofili di Fusine in Valromana. *Lavori de la Società Veneziana di Scienze Naturali* **2:** 43–49. (not seen)

Minozzi, C. – see Menozzi, C.

Mirande, M. (1905). Sur la présence d'un 'corps réducteur' dans le tégument chitineux des Arthropodes. *Archives d'Anatomie Microscopique* **7:** 208–231. (not seen)

Mitchell, R. W. (1969). A comparison of temperate and tropical cave communities. *Southwestern Naturalist* **14:** 73–88.

Mitchell, R. W. (1970). Total number and density estimates of some species of cavernicoles inhabiting Fern Cave, Texas. *Annales de Spéléologie* **25:** 73–90.

Möllendorf, O. (1873). *Beiträge zur Fauna Bosniens.* Görlitz. (not seen)

Moles, M. (1914a). A new species of pseudoscorpion from Laguna Beach, California. *Journal of Entomology and Zoology* **6**: 42–44.

Moles, M. (1914b). A pseudoscorpion from Poplar trees. *Journal of Entomology and Zoology* **6**: 81–83.

Moles, M. (1914c). Pseudoscorpions in the Claremont-Laguna region. *Journal of Entomology and Zoology* **6**: 187–197.

Moles, M. & Moore, W. (1921). A list of California Arachnida. I. Pseudoscorpionida. *Journal of Entomology and Zoology* **13**: 6–9.

Moniez, R. (1889). Sur un Pseudo-scorpion marin (*Obisium littorale* nov. sp.). *Revue Biologique du Nord de la France* **2**: 102–109.

Moniez, R. (1893). A propos des publications récentes sur le faux parasitisme des Chernétides sur différents Arthropodes. *Revue Biologique du Nord de la France* **6**: 47–54.

Montagu, G. (1815). Descriptions of several new or rare animals, principally marine, discovered on the south coast of Devonshire. *Transactions of the Linnean Society of London* **11**: 1–26.

Montandon, I. A. (1909). Les Pseudoscorpions de Roumanie. *Buletinul Societatii de Sciinte din Bucaresti* **18**: 147–148. (not seen)

Moore, G. (1835). *Chelifer cancroides. Entomological Magazine* **2**: 321–322.

Morais, J. W. (1985). *Abundancia e distribuiçao vertical de artropodos de solo numa floresta primaria nao inundada.* Dissertation: INPA, Manaus. (not seen)

Morais, J. W., Adis, J. & Mahnert, V. (1986). Abundancia e distribuiçao de pseudoscorpiones do solo numa floresta neotropical nao inundada. *XII. Congr. Brasil. Zool. Cuiaba-Mato-Grosso*: 259. (not seen)

Morikawa, K. (1952a). Three new species of false-scorpions from the island of Marcus in the West Pacific Ocean. *Memoirs of Ehime University* (2B) **1**: 241–248.

Morikawa, K. (1952b). Notes on the Japanese Pseudoscorpiones I. *Memoirs of Ehime University* (2B) **1**: 249–258.

Morikawa, K. (1953a). Notes on Japanese Pseudoscorpiones. II. Family Cheiridiidae, Atemnidae and Chernetidae. *Memoirs of Ehime University* (2B) *1*: 345–354.

Morikawa, K. (1953b). [On a new pseudoscorpion of family Olpiidae.] *Zoological Magazine, Tokyo* **62**: 327–328. (in Japanese)

Morikawa, K. (1954a). Notes on Japanese Pseudoscorpions III. Family Cheliferidae. *Memoirs of Ehime University* (2B) **2**: 71–77.

Morikawa, K. (1954b). On some pseudoscorpions in Japanese lime-grottoes. *Memoirs of Ehime University* (2B) **2**: 79–87.

Morikawa, K. (1954c). Two new species of Chthoniinea from Japan. *Japanese Journal of Zoology* **11**: 329–331.

Morikawa, K. (1954d). [Environment and biology of pseudoscorpions.] *Atypus* **6**: 27–35. (not seen)

Morikawa, K. (1955a). Pseudoscorpions of forest soil in Shikoku. *Memoirs of Ehime University* (2B) **2**: 215–222.

Morikawa, K. (1955b). [On a new Garypidae (Pseudoscorp.) from Japan.] *Zoological Magazine, Tokyo* **64**: 225–228. (in Japanese)

Morikawa, K. (1956). Cave pseudoscorpions of Japan (I). *Memoirs of Ehime University* (2B) **2**: 271–282.

Morikawa, K. (1957a). Cave pseudoscorpions of Japan (II). *Memoirs of Ehime University* (2B) **2**: 357–365.

Morikawa, K. (1957b). [*Kashimachelifer cinnamoneus*, a new genus and species of cheliferid pseudoscorpions from Japan.] *Zoological Magazine, Tokyo* **66**: 399–402. (in Japanese)

Morikawa, K. (1958). Maritime pseudoscorpions from Japan. *Memoirs of Ehime University* (2B) **3**: 5–11.

Morikawa, K. (1960). Systematic studies of Japanese pseudoscorpions. *Memoirs of Ehime University* (2B) **4**: 85–172.

Morikawa, K. (1962). Ecological and some biological notes on Japanese pseudo-scorpions. *Memoirs of Ehime University* (2B) **4**: 417–435.

Morikawa, K. (1963). Pseudoscorpions from Solomon and New Britain. *Bulletin of the Osaka Museum of Natural History* **16**: 1–8.

Morikawa, K. (1968). On some pseudoscorpions from Rolwaling Himal. *Journal of the College of Arts and Sciences, Chiba University* **2**: 259–263.

Morikawa, K. (1970). Results of the speleological survey in South Korea 1966. XX. New pseudoscorpions from South Korea. *Bulletin of the National Science Museum of Tokyo* **13**: 141–148.

Morikawa, K. (1972). Pseudoscorpions from Mt. Poroshiri-daké of the Hidaka Mountain Range, northern Japan. *Memoirs of the National Science Museum of Tokyo* **5**: 33–35.

Morris, J. C. H. (1948a). The taxonomic position of *Idiogarypus hansenii* (With). *Papers and Proceedings of the Royal Society of Tasmania* **1947**: 37–41.

Morris, J. C. H. (1948b). A new genus of pseudogarypin pseudoscorpions possessing pleural plates. *Papers and Proceedings of the Royal Society of Tasmania* **1947**: 43–47.

Morris, P. (1973). An unusual partnership. *Animals, London* **14**: 554–555. (not seen)

Morton, J. E. (1954). The crevice faunas of the upper intertidal zone at Wembury. *Journal of the Marine Biological Association of the United Kingdom* **33**: 187–224. (not seen)

Moyle, B. (1989a). A description of *Tyrannochthonius noaensis* (Arachnida: Pseudoscorpionida: Chthoniidae). *New Zealand Entomologist* **12**: 58–59.

Moyle, B. (1989b). A description of *Tyrannochthonius tekauriensis* (Pseudoscorpionida: Chthoniidae). *New Zealand Entomologist* **12**: 60–62.

Muchmore, W. B. (1962). A new cavernicolous pseudoscorpion belonging to the genus *Microcreagris*. *Postilla* **70**: 1–6.

Muchmore, W. B. (1963a). Redescription of some cavernicolous pseudoscorpions (Arachnida, Chelonethida) in the collection of the Museum of Comparative Zoology. *Breviora* **188**: 1–16.

Muchmore, W. B. (1963b). Two European arachnids new to the United States. *Entomological News* **74**: 208–210.

Muchmore, W. B. (1965). North American cave pseudoscorpions of the genus *Kleptochthonius*, subgenus *Chamberlinochthonius* (Chelonethida, Chthoniidae). *American Museum Novitates* **2234**: 1–27.

Muchmore, W.B. (1966a). A cavernicolous pseudoscorpion of the genus *Microcreagris* from southern Tennessee. *Entomological News* **77**: 97–100.

Muchmore, W. B. (1966b). Two new species of *Kleptochthonius* (Arachnida, Chelonethida) from a cave in Tennessee. *Journal of the Tennessee Academy of Science* **41**: 68–69.

Muchmore, W. B. (1967a). Two new species of the pseudoscorpion genus *Paralioch-thonius*. *Entomological News* **78**: 155–162.

Muchmore, W. B. (1967b). *Novobisium* (Arachnida, Chelonethida, Neobisiidae, Neobisiinae), a new genus of pseudoscorpions based on *Obisium carolinensis* Banks. *Entomological News* **78**: 211–215.

Muchmore, W. B. (1967c). Pseudotyrannochthoniine pseudoscorpions from the western United States. *Transactions of the American Microscopical Society* **86**: 132–139.

Muchmore, W. B. (1967d). New cave pseudoscorpions of the genus genus *Apochthonius* (Arachnida: Chelonethida). *Ohio Journal of Science* **67**: 89–95.

Muchmore, W. B. (1967e). Phoresy in American pseudoscorpions. *American Zoologist* **7**: 773.

Muchmore, W. B. (1968a). Recent progress in our understanding of cavernicolous pseudoscorpions in North America. *Bulletin of the National Speleological Society* **29**: 101. (not seen)

Muchmore, W. B. (1968a). A new species of the pseudoscorpion genus *Aphrastochthonius* (Arachnida, Chelonethida), from a cave in Alabama. *Bulletin of the National Speleological Society* **30**: 17–18.

Muchmore, W. B. (1968b). Two new species of chthoniid pseudoscorpions from the western United States (Arachnida: Chelonethida: Chthoniidae). *Pan-Pacific Entomologist* **44**: 51–57.

Muchmore, W. B. (1968c). Redescription of the type species of the pseudoscorpion genus *Kewochthonius* Chamberlin. *Entomological News* **79**: 71–76.

Muchmore, W. B. (1968d). A new species of the pseudoscorpion genus *Serianus* (Arachnida, Chelonethida, Olpiidae) from North Carolina. *Entomological News* **79**: 145–150.

Muchmore, W. B. (1968e). A new species of the pseudoscorpion genus *Syarinus* (Arachnida, Chelonethida, Syarinidae) from the northeastern United States. *Journal of the New York Entomological Society* **76**: 112–116.

Muchmore, W. B. (1968f). A cavernicolous species of the pseudoscorpion genus *Mundochthonius* (Arachnida, Chelonethida, Chthoniidae). *Transactions of the American Microscopical Society* **87**: 110–112.

Muchmore, W. B. (1968g). A new species of the pseudoscorpion genus *Parobisium* from Utah (Arachnida, Chelonethida, Neobisiidae). *American Midland Naturalist* **79**: 531–534.

Muchmore, W. B. (1969a). New species and records of cavernicolous pseudoscorpions of the genus *Microcreagris* (Arachnida, Chelonethida, Neobisiidae, Ideobisiinae). *American Museum Novitates* **2392**: 1–21.

Muchmore, W. B. (1969b). A cavernicolous *Tyrannochthonius* from Mexico (Arach., Chelon., Chthon.). *Ciencia, México* **27**: 31–32.

Muchmore, W. B. (1969c). The pseudoscorpion genus *Macrochernes*, with the description of a new species from Puerto Rico (Arachnida, Chelonethida, Chernetidae). *Caribbean Journal of Science and Mathematics* **1**(2): 9–14.

Muchmore, W. B. (1969d). A population of a European pseudoscorpion established in New York. *Entomological News* **80**: 66.

Muchmore, W. B. (1969e). The pseudoscorpion genus *Neochthonius* Chamberlin (Arachnida, Chelonethida, Chthoniidae) with description of a cavernicolous species. *American Midland Naturalist* **81**: 387–394.

Muchmore, W. B. (1970a). New cavernicolous *Kleptochthonius* spp. from Virginia (Arachnida, Pseudoscorpionida, Chthoniidae). *Entomological News* **81**: 210–212.

Muchmore, W. B. (1970b). An unusual new *Pseudochthonius* from Brazil (Arachnida, Pseudoscorpionida, Chthoniidae). *Entomological News* **81**: 221–223.

Muchmore, W. B. (1971a). Phoresy by North and Central American pseudoscorpions. *Proceedings of the Rochester Academy of Sciences* **12**: 79–97.

Muchmore, W. B. (1971b). The identity of *Olpium obscurum* (Pseudoscorpionida, Olpiidae). *Florida Entomologist* **54**: 241–243.

Muchmore, W. B. (1971c). On phoresy in pseudoscorpions. *Bulletin of the British Arachnological Society* **2**: 38.

Muchmore, W. B. (1972a). A phoretic *Metatemnus* (Pseudoscorpionida, Atemnidae) from Malaysia. *Entomological News* **83**: 11–14.

Muchmore, W. B. (1972b). The pseudoscorpion genus *Paraliochthonius* (Arachnida, Pseudoscorpionida, Chthoniidae). *Entomological News* **83**: 248–256.

Muchmore, W. B. (1972c). A new *Lamprochernes* from Utah (Pseudoscorpionida, Chernetidae). *Entomological News* **82**: 327–329.

Muchmore, W. B. (1972d). European pseudoscorpions from New England. *Journal of the New York Entomological Society* **80**: 109–110.

Muchmore, W. B. (1972e). A remarkable pseudoscorpion from the hair of a rat (Pseudoscorpionida, Chernetidae). *Proceedings of the Biological Society of Washington* **85**: 427–432.

Muchmore, W. B. (1972f). The unique, cave-restricted genus *Aphrastochthonius* (Pseudoscorpionida, Chthoniidae). *Proceedings of the Biological Society of Washington* **85**: 433–444.

Muchmore, W. B. (1972g). New diplosphyronid pseudoscorpions, mainly cavernicolous, from Mexico (Arachnida, Pseudoscorpionida). *Transactions of the American Microscopical Society* **91**: 261–276.

Muchmore, W. B. (1972h). Observations on the classification of some European chernetid pseudoscorpions. *Bulletin of the British Arachnological Society* **2**: 112–115.

Muchmore, W. B. (1973a). A new genus of pseudoscorpions based upon *Atemnus hirsutus* (Pseudoscorpionida: Chernetidae). *Pan Pacific Entomologist* **49**: 43–48.

Muchmore, W. B. (1973b). New and little known pseudoscorpions, mainly from caves in Mexico (Arachnida, Pseudoscorpionida). *Bulletin of the Association for Mexican Cave Studies* **5**: 47–62.

Muchmore, W. B. (1973c). The pseudoscorpion genus *Mexobisium* in middle America (Arachnida, Pseudoscorpionida). *Bulletin of the Association for Mexican Cave Studies* **5**: 63–72.

Muchmore, W. B. (1973d). A second troglobiotic *Tyrannochthonius* from Mexico (Arachnida, Pseudoscorpionida, Chthoniidae). *Bulletin of the Association for Mexican Cave Studies* **5**: 81–82.

Muchmore, W. B. (1973e). The genus *Chitrella* in America (Pseudoscorpionida, Syarinidae). *Journal of the New York Entomological Society* **81**: 183–192.

Muchmore, W. B. (1973f). Cavernicolous pseudoscorpions in the eastern United States. *Bulletin of the National Speleological Society* **35**: 18.

Muchmore, W. B. (1973g). Ecology of pseudoscorpions – a review. In: Dindal, D. L. (ed.), *Proceedings of the First Soil Microcommunities Conference, Syracuse, New York*: 121–127. U.S. Atomic Energy Commission.

Muchmore, W. B. (1974a). Pseudoscorpions from Florida. 1. The genus *Aldabrinus* (Pseudoscorpionida: Olpiidae). *Florida Entomologist* **57**: 1–7.

Muchmore, W. B. (1974b). Pseudoscorpions from Florida. 2. A new genus and species *Bituberochernes mumae* (Chernetidae). *Florida Entomologist* **57**: 77–80.

Muchmore, W. B. (1974c). Pseudoscorpions from Florida. 3. *Epactiochernes*, a new genus based upon *Chelanops tumidus* Banks (Chernetidae). *Florida Entomologist* **57**: 397–407.

Muchmore, W. B. (1974d). Clarification of the genera *Hesperochernes* and *Dinocheirus* (Pseudoscorpionida, Chernetidae). *Journal of Arachnology* **2**: 25–36.

Muchmore, W. B. (1974e). New cavernicolous species of *Kleptochthonius* from Virginia and West Virginia (Pseudoscorpionida, Chthoniidae). *Entomological News* **85**: 81–84.

Muchmore, W. B. (1975a). A new genus and species of chthoniid pseudoscorpion from Mexico (Pseudoscorpionida, Chthoniidae). *Journal of Arachnology* **3**: 1–4.

Muchmore, W. B. (1975b). The genus *Lechytia* in the United States (Pseudoscorpionida, Chthoniidae). *Southwestern Naturalist* **20**: 13–27.

Muchmore, W. B. (1975c). Two miratemnid pseudoscorpions from the western Hemisphere (Pseudoscorpionida, Miratemnidae). *Southwestern Naturalist* **20**: 231–239.

Muchmore, W. B. (1975d). Use of the spermathecae in the taxonomy of chernetid Pseudoscorpions. *Proceedings of the 6th International Arachnological Congress, Amsterdam*: 17–20.

Muchmore, W. B. (1975e). Pseudoscorpions from Florida. 4. The genus *Dinochernes* (Chernetidae). *Florida Entomologist* **58**: 275–279.

Muchmore, W. B. (1976a). *Aphrastochthonius pachysetus*, a new cavernicolous species from New Mexico (Pseudoscorpionida, Chthoniidae). *Proceedings of the Biological Society of Washington* **89**: 361–364.

Muchmore, W. B. (1976b). Pseudoscorpions from Florida and the Caribbean area. 5. *Americhernes*, a new genus based upon *Chelifer oblongus* Say (Chernetidae). *Florida Entomologist* **59**: 151–163.

Muchmore, W. B. (1976c). Pseudoscorpions from Florida and the Caribbean area. 6. *Caribchthonius*, a new genus with species from St. John and Belize (Chthoniidae). *Florida Entomologist* **59**: 361–367.

Muchmore, W. B. (1976d). New cavernicolous species of *Kleptochthonius*, and recognition of a new species group within the genus (Pseudoscorpionida: Chthoniidae). *Entomological News* **87**: 211–217.

Muchmore, W. B. (1976e). New species of *Apochthonius*, mainly from caves in central and eastern United States (Pseudoscorpionida, Chthoniidae). *Proceedings of the Biological Society of Washington* **89**: 67–80.

Muchmore, W. B. (1977). Preliminary list of the pseudoscorpions of the Yucatan Peninsula and adjacent regions, with descriptions of some new species (Arachnida: Pseudoscorpionida). *Bulletin of the Association for Mexican Cave Studies* **6**: 63–78.

Muchmore, W. B. (1978). A second species of the genus *Mexichthonius* (Pseudoscorpionida, Chthoniidae). *Journal of Arachnology* **6**: 155–156.

Muchmore, W. B. (1979a). Pseudoscorpions from Florida and the Caribbean area. 7. Floridian diplosphyronids. *Florida Entomologist* **62**: 193–213.

Muchmore, W. B. (1979b). Pseudoscorpions from Florida and the Caribbean area. 8. A new species of *Bituberochernes* from the Virgin Islands (Chernetidae). *Florida Entomologist* **62**: 313–316.

Muchmore, W. B. (1979c). Pseudoscorpions from Florida and the Caribbean area. 9. *Typhloroncus*, a new genus from the Virgin Islands (Ideoroncidae). *Florida Entomologist* **62**: 317–320.

Muchmore, W. B. (1979d). The cavernicolous fauna of Hawaiian lava tubes. 11. A troglobitic pseudoscorpion (Pseudoscorpionida: Chthoniidae). *Pacific Insects* **20**: 187–190.

Muchmore, W. B. (1980a). A new species of *Apochthonius* with paedomorphic tendencies (Pseudoscorpionida: Chthoniidae). *Journal of Arachnology* **8**: 87–90.

Muchmore, W. B. (1980b). A new cavernicolous *Apochthonius* from California (Pseudoscorpionida: Chthoniidae). *Journal of Arachnology* **8**: 93–95.

Muchmore, W. B. (1980c). *Interchernes*, a new genus of pseudoscorpion from Baja California (Pseudoscorpionida: Chernetidae). *Southwestern Naturalist* **25**: 89–94.

Muchmore, W. B. (1980d). Pseudoscorpions from Florida and the Caribbean area. 10. New *Mexobisium* species from Cuba. *Florida Entomologist* **63**: 125–127.

Muchmore, W. B. (1980e). Three new olpiid pseudoscorpions from California (Pseudoscorpionida, Olpiidae). *Pan-Pacific Entomologist* **56**: 161–169.

Muchmore, W. B. (1980f). An unusual new *Parachernes* from El Salvador (Pseudoscorpionida: Chernetidae). *Transactions of the American Microscopical Society* **99**: 227–229.

Muchmore, W. B. (1981a). Cavernicolous species of *Larca, Archeolarca* and *Pseudogarypus* with notes on the genera, (Pseudoscorpionida, Garypidae and Pseudogarypidae). *Journal of Arachnology* **9**: 47–60.

Muchmore, W. B. (1981b). Redescription of *Chthonius virginicus* Chamberlin (Pseudoscorpionida, Chthoniidae). *Journal of Arachnology* **9**: 110–112.

Muchmore, W. B. (1981c). New pseudoscorpion synonymies (Pseudoscorpionida, Chernetidae and Cheliferidae). *Journal of Arachnology* **9**: 335–336.

Muchmore, W. B. (1981d). The identity of *Olpium minutum* Banks (Pseudoscorpionida, Olpiidae). *Journal of Arachnology* **9**: 337–338.

Muchmore, W. B. (1981e). Pseudoscorpions from Florida and the Caribbean area. 11. A new *Parachelifer* from the Virgin Islands (Cheliferidae). *Florida Entomologist* **64**: 189–191.

Muchmore, W. B. (1981f). Cavernicolous pseudoscorpions in North and Middle America. *Proceedings of the 8th International Congress of Speleology* **1**: 381–384. (not seen)

Muchmore, W. B. (1982a). Pseudoscorpionida. In: Parker, S.P. (ed.), *Synopsis and Classification of Living Organisms*, vol. **2**: 96–102. McGraw-Hill: New York.

Muchmore, W. B. (1982b). A new *Rhinochernes* from Ecuador (Pseudoscorpionida, Chernetidae). *Journal of Arachnology* **10**: 87–88.

Muchmore, W. B. (1982c). The genera *Ideobisium* and *Ideoblothrus*, with remarks on the family Syarinidae (Pseudoscorpionida). *Journal of Arachnology* **10**: 193–221.

Muchmore, W. B. (1982d). A new cavernicolous *Sathrochthonius* from Australia (Pseudoscorpionida: Chthoniidae). *Pacific Insects* **24**: 156–158.

Muchmore, W. B. (1982e). The genus *Anagarypus* (Pseudoscorpionida: Garypidae). *Pacific Insects* **24**: 159–163.

Muchmore, W. B. (1982f). Some new species of pseudoscorpions from caves in Mexico (Arachnida, Pseudoscorpionida). *Bulletin for the Association of Mexican Cave Studies* **8**: 63–78.

Muchmore, W. B. (1982g). Survey of terrestrial invertebrates of St. John. *Abstracts of the Colloquium on Long-term Ecological Research in the Virgin Islands, Maho Bay*: 19. (not seen)

Muchmore, W. B. (1983). The cavernicolous fauna of Hawaiian lava tubes. 14. A second troglobitic *Tyrannochthonius* (Pseudoscorpionida: Chthoniidae). *International Journal of Entomology* **25**: 84–86.

Muchmore, W. B. (1984a). Further data on *Mucrochernes hirsutus* (Banks) (Pseudoscorpionida, Chernetidae). *Pan-Pacific Entomologist* **60**: 20–22.

Muchmore, W. B. (1984b). Pseudoscorpions from Florida and the Caribbean area. 12. *Antillochernes*, a new genus with setae on the pleural membranes (Chernetidae). *Florida Entomologist* **67**: 106–118.

Muchmore, W. B. (1984c). Pseudoscorpions from Florida and the Caribbean area. 13. New species of *Tyrannochthonius* and *Paraliochthonius* from the Bahamas, with discussion of the genera (Chthoniidae). *Florida Entomologist* **67**: 119–126.

Muchmore, W. B. (1984d). The pseudoscorpions described by R. V. Chamberlin (Pseudoscorpionida, Olpiidae and Chernetidae). *Journal of Arachnology* **11**: 353–362.

Muchmore, W. B. (1984e). *Troglobochica*, a new genus from caves in Jamaica, and redescription of the genus *Bochica* Chamberlin (Pseudoscorpionida, Bochicidae). *Journal of Arachnology* **12**: 61–68.

Muchmore, W. B. (1984f). New cavernicolous pseudoscorpions from California (Pseudoscorpionida, Chthoniidae and Garypidae). *Journal of Arachnology* **12**: 171–175.

Muchmore, W. B. (1986a). Redefinition of the genus *Olpiolum* and description of a new genus *Banksolpium* (Pseudoscorpionida, Olpiidae). *Journal of Arachnology* **14**: 83–92.

Muchmore, W. B. (1986b). Additional pseudoscorpions, mostly from caves, in Mexico and Texas (Arachnida: Pseudoscorpionida). *Texas Memorial Museum, Speleological Monographs* **1**: 17–30.

Muchmore, W. B. (1989). A *Sathrochthonius* north of the equator (Pseudoscorpionida, Chthoniidae). *Journal of Arachnology* **17**: 251–253.

Muchmore, W. B. & Alteri, C. H. (1969). *Parachernes* (Arachnida, Chelonethida, Chernetidae) from the coast of North Carolina. *Entomological News* **80**: 131–137.

Muchmore, W. B. & Alteri, C. H. (1974). The genus *Parachernes* (Pseudoscorpionida, Chernetidae) in the United States, with descriptions of new species. *Transactions of the American Entomological Society* **99**: 477–506.

Muchmore, W. B. & Benedict, E. M. (1976). Redescription of *Apochthonius moestus* (Banks), type of the genus *Apochthonius* Chamberlin (Pseudoscorpionida, Chthoniidae). *Journal of the New York Entomological Society* **84**: 67–74.

Muchmore, W. B. & Hentschel, E. (1982). *Epichernes aztecus*, a new genus and species of pseudoscorpion from Mexico (Pseudoscorpionida, Chernetidae). *Journal of Arachnology* **10**: 41–45.

Müller, F. & Schenkel, E. (1895). Verzeichnis der Spinnen von Basel und Umgegend. *Verhandlungen der Naturforschenden Gesellschaft in Basel* **10**: 691–824. (not seen)

Müller, G. (1931). Nuovi pseudoscorpioni cavernicoli appartenenti al sottogenere *Blothrus* Schioedte (diagnosi preliminari). *Bollettino della Società Entomologica Italiana* **63**: 125–127.

Müller, O. F. (1764). *Fauna Insectorum Fridrichsdalina.* Hafniae et Lipsiae. (not seen)

Müller, O. F. (1776). *Zoologiae Danicae Prodromus, seu Animalium Daniae et Norvegiae Indigenarum Characteres, Nomina, et Synonyma Imprimis Popularium.* Hallager: Havniae.

Murthy, V. A. (1960). On two new species of pseudoscorpions from Madras. *Bulletin of Entomology, Madras* **1**: 28–31.

Murthy, V. A. (1961). Two new species of pseudoscorpions from south India. *Annals and Magazine of Natural History* (13) *4*: 221–224.

Murthy, V. A. (1962). On the genus *Tullgrenius* Chamberlin (Chelonethi) with the description of a new species. *Bulletin of Entomology, Madras* **3**: 62–65.

Murthy, V. A. & Ananthakrishnan, T. N. (1977). Indian Chelonethi. *Oriental Insects Monograph* **4**: 1–210.

Myers, J. G. (1927). Ethological observations on some Pyrrhocoridae of Cuba. *Annals of the Entomological Society of America* **20**: 279–300.

Myers, J. G. (1934). The arthropod fauna of a rice ship, trading from Burma to the West Indies. *Journal of Animal Ecology* **3**: 146–149.

Navás, L. (1909). Una visita a Valdealgorfa (Ternel). *Boletín de la Sociedad Aragonese de Ciencias Naturales* **8**: 195–197. (not seen)

Navás, L. (1918). Algunos Quernetos (Arácnidos) de la provincia de Zaragoza. *Boletín de la Sociedad Entomológica de España* **1**: 83–90, 106–119, 131–136.

Navás, L. (1919). Excursiones entomológicas por Cataluña durante el verano de 1918. *Memorias de la Real Academia de Ciencias y Artes de Barcelona* (3) *15*: 181–214.

Navás, L. (1921). Mis excursiones científicas del verano de 1919. *Memoria de la Real Academia de Ciencias y Artes de Barcelona* (3) *17*: 143–169.

Navás, L. (1923). Excursions entomològiques de l'Istiu de 1922. *Arxius de l'Institut de Ciencas, Barcelona* **8**: 1–34.

Navás, L. (1924). Quernets de la val d'Aran (Leyda) rec. pel Gmà. *Butlletí de la Institució Catalana d'Historio Naturale* **41**: 43. (not seen)

Navás, L. (1925). Sinopsis de los Quernetos (Arácnidos) de la Península Ibérica. *Broteria, Zoologica* **22**: 99–130.

Nelson, S. O., Jr. (1971). Phoresy by pseudoscorpions. *Michigan Entomologist* **4**: 95–96.

Nelson, S. O., Jr. (1972). *A Systematic Study of Michigan Chelonethida (Arachnida), and the Population Structure of* Microbisium confusum *Hoff in a Beech Maple Woodlot.* Diss. Abstr. Int. 32: 3707B. (not seen)

Nelson, S. O., Jr. (1973). Population structure of *Microbisium confusum* Hoff in a beech-maple woodlot. *Revue d'Ecologie et de Biologie du Sol* **10**: 231–236.

Nelson, S. O., Jr. (1975). A systematic study of Michigan Pseudoscorpionida (Arachnida). *American Midland Naturalist* **93**: 257–301.

Nelson, S. O., Jr. (1982). The external morphology and life history of the pseudo-scorpion *Microbisium confusum* Hoff. *Journal of Arachnology* **10**: 261–274.

Nelson, S. O., Jr. (1984). The genus *Microbisium* in North and Central America (Pseudoscorpionida, Neobisiidae). *Journal of Arachnology* **12**: 341–350.

Nelson, S. O., Jr. & Manley, G. V. (1972). *Dinocheirus horricus* n. sp., a new species of pseudoscorpion (Arachnida, Pseudoscorpionida, Chernetidae) from Michigan. *Transactions of the American Microscopical Society* **91**: 217–221.

Nester, H. G. (1932). Some cytoplasmic structures in the male reproductive cells of a pseudoscorpion (*Chelanops corticis*). *Journal of Morphology* **53**: 97–131.

Newlands, G. (1978). Arachnida (except Acari). In: Werger, M. J. A. (ed.), *Biogeography and Ecology of Southern Africa*. Junk: The Hague. *Monographiae Biologicae* **31**: 685–702.

Newman, E. (1875). *Chelifer cancroides*. *Entomologist* **8**: 186.

Niculescu, F. & Fuhn, I. E. (1963). Étude de la nourriture chez la grenouille verte (*Rana r. rillibunda* Pall.) dans la réservation naturelle 'I. Mai'. *Studii si Cercetari Stiintifice, Filiala, Iaşi Biologie si Stiinte Agricole 14:* 193–211. (not seen)

Nobre, A. (1899). *Catalogo do Gabinete de Zoologia 1898–1899*. Academia Polytechnica do Porta, Coimbra. (not seen)

Nonidez, J. F. (1917). Pseudoscorpiones de España. *Trabajos del Museo Nacional de Ciencas Naturales, Madrid* **32**: 1–46.

Nonidez, J. F. (1925). Los *Obisium* españoles del subgénero *Blothrus* (Pseudosc. Obisidae) con descripción de nuevas especies. *Eos, Madrid* **1**: 43–83.

Nordberg, S. (1936). Biologisch-ökologische Untersuchungen über die Vogelnidicolen. *Acta Zoologica Fennica* **21**: 1–168.

Nosek, A. (1901). *Přehled Štírkuv a Jich Rozšíření Zeměpisné. Conspectus Chelonethium (Pseudoscorpionum) et Eorum Distributio Geographica*. Starcha: Čáslavi.

Nosek, A. (1902). Seznam štírků. Catalogus chelonethium seu pseudoscorpionum. *Zvláštni Otisk z Věstníku Klubu Přírodeovedeckého v. Prostějově* **3**: 35–75. (not seen)

Nosek, A. (1903). Proní doplněk katalogu štírků. Primum supplementum catalogi chelonethium seu pseudoscorpionum. *Zvláštni Otisk z Věstníku Klubu Přírodeovedeckého v. Prostějově* **5**: 55–64. (not seen)

Nosek, A. (1905). Araneiden, Opilionen und Chernetiden. In: Penther, A., Zederbauer, E. Ergebnisse einer naturwissenschaftliche Reise zum Erdschais-Dagh (Kleinasien). *Annalen des Naturhistorischen Museums in Wien* **20**: 114–154.

Nowicki, M. (1874). Dodatek do fauny pajeczakow Galicyi. *Sprawozdanie Komisyi fizyograficznej, Krakow* **8**: 1–11. (not seen)

Oken, L. (1815). *Lehrbuch der Naturgeschichte, 3. Teil Zoologie, 1. Abt. Fleischlose Tiere*. A. Schmid & Co.: Jena. (not seen)

Oldham, C. (1912). On the occurrence of *Chernes godfreyi* Kew and other false-scorpions in Hertfordshire. *Transactions of the Hertfordshire Natural History Society* **14**: 299–300. (not seen)

Oldham, C. (1938). *Chernes nodosus* at Berkhansted and Letchworth. *Transactions of the Hertfordshire Natural History Society 20:* 354. (not seen)

Orghidan, T. (1975). Développement des recherches spéologiques en Roumanie. *Annales de Spéléologie* **30**: 553–560. (not seen)

Orghidan, T. & Dumitresco (as Dumitrescu), M. (1964a). Das lithoklasische Lebensreich. *Zoologischer Anzeiger* **173**: 325–332.

Orghidan, T. & Dumitresco, M. (1964b). Données préliminaires concernant la faune des espaces lithoclasiques des schistes verts de Dobrogea. *Spelunca Mémoires* **4**: 188–196.

Orghidan, T., Dumitresco, M. & Georgesco, M. (1975). Mission biospéologique 'Constantin Dragan' à Majorque (1970–1971). Première note: Arachnides (Araneae et Pseudoscorpionidea). *Travaux de l'Institute de Spéologie 'Émile Racovitza'* **14**: 9–33.

Örösi-Pal, Z. (1938). Afterskorpione (Chelonethi) in der Wohnung der Honigbiene. Eine Zusammenfassung und eigene Untersuchungen. *Zeitschrift für Angewandte Entomologie* **25**: 142–150.

Örösi-Pal, Z. (1939). *Bienenfunde und Tierwelt des Bienenstockes.* Budapest. (not seen)

Oudemans, A. C. (1886). Die gegenseitige Verwandschaft, Abstammung und Classification der sogenannten Arthropoden. *Tijdschrift der Nederlandsche Dierkundige Vereeniging* (2) **1**: 37–56. (not seen)

Oudemans, A. C. (1906). Über Genitaltracheen bei Chernetiden und Acari. *Zoologischer Anzeiger* **30**: 135–140.

Packard, A. S. (1869). The false scorpion. *American Naturalist* **2**: 216.

Packard, A. S. (1884). New cave arachnids. *American Naturalist* **18**: 202–204.

Packard, A. S. (1886). The cave fauna of North America, with remarks on the anatomy of the brain and origin of the blind species. *Memoirs of the National Academy of Sciences* **4**: 1–156.

Palacios-Vargas, J. G. (1981). Sistematica y morfologia. Los artropodes de la gruta de Acuitlapan. *Folia Entomologica Mexicana, XVI Congr. Nacionale de Entomologia* **48**: 64–65. (not seen)

Palacios-Vargas, J. G. (1983). Microartropodos de la Gruta de Aguacachil, Guerrero, Mexico. *Anales de Escuela Nacional de Ciencias Biologicas, Mexico* **27**: 55–60. (not seen)

Palacios-Vargas, J. G. & Morales-Malacara, J. B. (1983). Biocenosis de algunas cuevas de Morelos. *Mémoires de Biospéologie* **10**: 163–169. (not seen)

Pallas, P. S. (1767–1780). *Spicilegia Zoologica, Quibus Novae et Obsc. Anim. Spec. Illustr.* Berol. (not seen)

Palmgren, P. (1973). Über die Biotopverteilung waldbodenlebender Pseudoscorpionidea (Arachnoidea) in Finnland und Österreich. *Commentationes Biologicae* **61**: 1–11.

Paoletti, M. G. (1978). Cenni sulla fauna ipogea delle Prealpi Bellunesi e colli subalpini. *Grotte d'Italia* (4) **7**: 45–198. (not seen)

Paoletti, M. G. (1979). Micorarthropodi ipogei delle Alpi orientali. *Mondo Sotterraneo, n.s.* (2) **3**: 23–32. (not seen)

Paoletti, M. G. (1980). La diffusion des troglobies dans les cavernes et le sol des Préalpes vénitiennes (Italie nord-orientale). *Mémoires de Biospéologie* **7**: 63–75. (not seen)

Papp, R. P. (1978). New records for pseudoscorpions from the Sierra Nevada. *Pan-Pacific Entomologist* **54**: 325.

Parenzan, P. (1953). Fauna del sottosuolo di Napoli (Primo contributo). *Bollettino della Società di Naturalisti in Napoli* **62**: 89–93. (not seen)

Parenzan, P. (1954). Ricerche biologiche nell'Italia meridionale della sezione speleologica dell'I.R.B. *Bollettino della Società di Naturalisti in Napoli* **63**: 96–101. (not seen)

Parenzan, P. (1955). Speleobiologia dell Murge. *Bollettino di Zoologia* **22**: 293–307. (not seen)

Parenzan, P. (1958). Attività speleologica meridionale. *Studia Speleol.* **3**: 111–118. (not seen)

Park, O. & Auerbach, S. (1954). Further study of the tree-hole complex with emphasis on quantitative aspects of the fauna. *Ecology* **35**: 208–222.

Parker, J. R. (1982). What's in a name? - scorpions and pseudoscorpions. *Newsletter of the British Arachnological Society* **34**: 1–2.

Patrizi, S. (1954). Materiali per un primo elenco della fauna cavernicola del Lazio e delle regioni limitrofe (esclusa la Toscana). *Notizario Circ. Speleol. Romano* **7**: 22–34. (not seen)

Paulian, R. (1947). Observations écologiques en forêt de Basse Côte d'Ivoire. *Encyclopédie Biogéographie et Écologique* **2**: 1–147. (not seen)

Pavan, M. (1939). Le caverne della regione M. Palosso – M. Doppo (Brescia) e la loro fauna. *Commenatri dell'Ateneo di Brescia, Suppl.*: 5–95. (not seen)

Pavan, M., Pavan, M. & Scossiroli, R. (1953). Il Buco del Corno n. 1004 Lo (Lombardia – Italia). *Rassegna Speleologica Italiana* **5**: 4–27. (not seen)

Pavesi, P. (1876). Gli aracnidi turchi. *Atti della Società Italiana di Scienze Naturali, e del Museo Civile di Storia Naturale, Milano* **19**: 50–74. (not seen)

Pavesi, P. (1878). Aracnidi aggiunto un catalogo sistematico della specie di Grecia. *Annali del Museo Civico di Storia Naturale di Genova* **11**: 335–396. (not seen)

Pavesi, P. (1879). Saggio di una fauna arachnologica del Varesotto. *Atti della Società Italiana di Scienze Naturali, e del Museo Civile di Storia Naturale, Milano* **21**: 789–817. (not seen)

Pavesi, P. (1880). Aracnidi di Tunisia. *Annali del Museo Civico di Storia Naturale di Genova* **15**: 310–388.

Pavesi, P. (1884). Materiali per lo studio della fauna Tunisina raccolti da G. e L. Doria. *Annali del Museo Civico di Storia Naturale di Genova* **20**: 446–486.

Pavesi, P. (1897). Aracnidi di Somali e Galla. *Annali del Museo Civico di Storia Naturale di Genova* **38**: 151–188.

Pax, F. (1936). Die Reyersdorfer Tropfsteinhöhle und ihre Tierbevölkerung. *Mitteilungen über Höhlen- und Karstforschung* **1936**: 97–122. (not seen)

Pax, F. (1937). Die Moorfauna des Glatzer Schneeberges. 2. Allgemaine Charakteristik der Hochmoore. *Beitr. Biol. Glatzer Schneeberges* **3**: 237–266. (not seen)

Pax, F. & Paul, H. (1961). Die Stollenfauna des Siebengebirges. *Dechiana, Beihefte* **9**: 69–76.

Peacock, E. A. W. (1902a). Lincolnshire naturalists at Torksey. *Naturalist, London* **27**: 133–138.

Peacock, E. A. W. (1902b). Lincolnshire naturalists at Scunthorpe. *Naturalist, London* **27**: ?–? (not seen)

Pearse, A. S. (1943). Chelonethida from the Duke Forest. *Anatomical Record* **87**: 464. (not seen)

Pearse, A. S. (1946). Notes on Chelonethida from the Duke Forest. *Ecology* **22**: 257–258.

Peck, S. B. (1973). A review of the invertebrate fauna of volcanic caves in western North America. *Bulletin of the National Speleological Society* **35**: 99–107. (not seen)

Peck, S. B. (1974). The invertebrate fauna of tropical American caves. II. Puerto Rico, an ecological and zoogeographic analysis. *Biotropica* **6**: 14–31. (not seen)

Peck, S. B. & Lewis, J. L. (1977). Zoogeography and evolution of the subterranean invertebrate fauna of Illinois and southwestern Missouri. *Bulletin of the National Speleological Society* **40**: 39–63. (not seen)

Pedder, I. J. (1965). Abnormal segmentation of the abdomen in six species of British pseudoscorpions. *Entomologist* **98**: 108–112.

Pellegrini, B. (1968). Cattura di un carabide cavernicolo effettuata alla Spluga Carpenè a S. Mauro di Saline. *Rassegna Speleologica Italiana* **20**: 266–267. (not seen)

Penny, N. D. & Arias, J. R. (1982). *Insects of an Amazon Forest*. Columbia University Press: New York. (not seen)

Pereira, C. & Castro, M. P. de (1944). Morfologia externa e análise dos caracteres taxonômicos de *Pycnochernes eidmanni* Beier, 1935 (Chelonethida: Chernetidae), das panelas de lixo dos formigueiros de *Atta sexdens rubropilosa* Forel, 1908. *Archivos do Instituto Biológico* **15**: 239–261.

Perry, R. H., Surdick, R. W. & Anderson, D. M. (1974). Observations on the biology, ecology, behavior, and larvae of *Dryobius sexnotatus* Linsley (Coleoptera: Cerambycidae). *Coleopterists' Bulletin* **28**: 169–176. (not seen)

Petersen, B. (1968). The Pseudoscorpionidea of Rennell and Bellona Islands. In: Wolff, T. (ed.), *The Natural History of Rennell Island, British Solomon Islands*, vol. **5**: 119–120. University of Copenhagen: Copenhagen.

Petrunkevitch, A. (1913). A monograph of terrestrial Palaeozoic Arachnida of North America. *Transactions of the Connecticut Academy of Arts and Sciences* **18**: 1–137. (not seen)

Petrunkevitch, A. (1953). Paleozoic and Mesozoic Arachnida of Europe. *Memoirs of the Geological Society of America* **53**: 1–128. (not seen)

Petrunkevitch, A. (1955). Arachnida. In: Moore, R.C. (ed.), *Treatise on Invertebrate Paleontology. P, Arthropoda*, vol. **2**: 42–162. University of Kansas Press: Lawrence.

Peus, F. (1928). Beiträge zur Kenntnis der Tierwelt nordwestdeutscher Hochmoore. Eine ökologische Studie. Insekten, Spinnentiere (teilw.), Wirbeltiere. *Zeitschrift für Morphologie und Ökologie der Tiere* **12**: 533–683. (not seen)

Pfeiffer, A. P. (1901). Verzeichnis der Arachniden von Oberösterreich. *Programm des k. k. Benedictiner Gymnasiums in Kremmsmünster*. (not seen)

Pieper, H. (1980). Neue Pseudoskorpion-Funde auf den Ilhas Selvagens und Bemerkungen zur Zoogeographie dieser Inselgruppe. *Vieraea, Tenerife* **8**: 261–270.

Pieper, H. (1981). Die Pseudoskorpione von Madeira und nachbarinseln. *Bocagiana, Funchal* **60**: 1–7.

Pierce, W. D. & Gibron, J., Sr. (1962). Fossil arthropods of California. 24. Some unusual fossil arthropods from the Calico Mountains nodules. *Bulletin of the Southern California Academy of Sciences* **61**: 143–151.

Pinnerup, S. P. (1982). Bogskorpion (*Chelifer cancroides*) beuytter stueflue (*Musca domestica*) som transportmiddel. *Flora og Fauna* **88**: 21. (not seen)

Plachter, H. & Plachter, J. (1988). Ökologische Studien zur terrestrischen Höhlenfauna Süddeutschlands. *Zoologica, Stuttgart* **139**: 1–67.

Planet, L. (1905). Araignées, Chernètes, Scorpions, Opilions. In: *Histoire Naturelle de la France*, vol. *14*. Paris. (not seen)

Platnick, N. I. (1984). On the pseudoscorpion-mimicking spider *Cheliferoides* (Araneae: Salticidae). *Journal of the New York Entomological Society* **92**: 169–173.

Pliginsky, V. G. (1927). [Contributions to the cave fauna of the Crimea. III.] *Revue Russe d'Entomologie* **21**: 171–180. (in Russian) (not seen)

Pocock, R. I. (1893). On some points in the morphology of the Arachnida (s.s.), with notes on the classification of the group. *Annals and Magazine of Natural History* (6) **11**: 1–19.

Pocock, R. I. (1898). List of the Arachnida and 'Myriapoda' obtained in Funafuti by Prof. W. J. Sollas and Mr. Stanley Gardiner, and in Rotuma by Mr. Stanley Gardiner. *Annals and Magazine of Natural History* (7) **1**: 321–329.

Pocock, R. I. (1900). Chilopoda, Diplopoda, and Arachnida. In: Andrews, C. W. (ed.), *A Monograph of Christmas Island (Indian Ocean)*. British Museum (Natural History): London.

Pocock, R. I. (1902a). Studies of the arachnid endosternite. *Quarterly Journal of Microscopical Science (new series)* **46**: 225–262. (not seen)

Pocock, R. I. (1902b). On some points in the anatomy of the alimentary and nervous systems of the arachnidan suborder Pedipalpi. *Proceedings of the Zoological Society of London* **1902**(2): 169–188.

Pocock, R. I. (1904). Arachnida. In: Gardiner, J.S. (ed.), *The Fauna and Geography of the Maldive and Laccadive Archipelagoes: Being an Account of the Work Carried on and of the Collections Made by an Expedition During the Years 1899 and 1900*, vol. **2**: 797–805. Cambridge University Press: Cambridge.

Pocock, R. I. (1905). A parasite of the house-fly. *Nature, London* **72**: 604.

Poda, N. (1761). *Insecta Musei Graecensis, Quae in Ordines, Ggenera et Species Juxta Systema Naturae Caroli Linnaei*. Graecii. (not seen)

Poinar, G. O., Jr., Thomas, G. M. & Lee, V. F. (1985). Laboratory infection of *Garypus californicus* (Pseudoscorpionida, Garypidae) with neoaplectanid and heterorhabditid nematodes (Rhabditoidea). *Journal of Arachnology* **13**: 400–401.

Pontin, A. J. (1961). The prey of *Lasius niger* (L.) and *L. flavus* (F.) (Hym., Formicidae). *Entomologist's Monthly Magazine* **97**: 135–137.

Popple, E. (1914). Hertfordshire false-scorpions. *Transactions of the Hertfordshire Natural History Society* **15**: 124.

Porter, C. E. (1898). Contribucion a la faune de la provincia de Valparaiso. *Revista Chilena de Historia Natural* **2**: 31–33. (not seen)

Pozzi, A. (1952). Grotte del circondario di Brunate (Como). *Rassegna Speleologica Italiana* **4**: 92–101. (not seen)

Pratt, H. S. (1916). *A Manual of the Common Invertebrate Animals Exclusive of Insects*. A.C. McClurg & Co.: Chicago.

Preston-Mafham, R. & Preston-Mafham, K. (1984). *Spiders of the World*. Blandford Press: Poole.

Pretner, E. & Strasser, K. (1931). Die Fauna der Nordfriauler Höhlen. *Mitteilungen über Höhlen- und Karstforschung* **9**: 84–90. (not seen)

Preudhomme de Borre, A. (1873). Note biologique relative aux moeurs des Arachnides du genre *Chelifer*. *Verhandlungen der K. K. Zoologischen-Botanischen Gesellschaft in Wien* **23**: 36. (not seen)

Preyssler, J. D. (1790). *Verzeichniss Böhmischer Insekten*. Prague.

Protescu, O. (1937). Étude géologique et paléobiologique de l'ambre Roumain. *Bulletin de la Société Romaine de Géologie* **3**: 65–110. (not seen)

Pschorn-Walcher, H. & Gunhold, P. (1957). Zur Kenntnis der Tiergemeinschaft in Moos- und Flechtenrasen an Park- und Waldbäumen. *Zeitschrift für Morphologie und Ökologie der Tiere* **46**: 342–354.

Puddu, S. & Pirodda, G. (1973). Catalogo sistematico ragionato della fauna cavernicola della Sardegna. *Rendiconti del Seminario della Facoltà di Scienze dell'Università di Cagliari* **43**: 151–205.

Purushothama Kaimal, P. (1988). *Some Aspects of Ecology and Biology of Pseudoscorpions of Kerala*. Ph.D. Thesis: University of Kerala.

Rabeler, W. (1951). Biozönotische Untersuchungen im hannoverschen Kiefernhorst. *Zeitschrift für Angewandte Entomologie* **32**: 591–598.

Rack, G. (1960). Über einen in Aschaffenburg gefundenen Pseudoscorpion. *Nachrichten, Naturwissenschaftliches Museum der Stadt, Aschaffenburg* **65**: 99–107. (not seen)

Rack, G. (1971). Die Entomologischen Sammlungen des Zoologischen Instituts und Zoologischen Museums Hamburg. I. und II. Teil (Nachtrag). *Mitteilungen aus dem Hamburgischen Zoologischen Museum und Institute* **67**: 109–133.

Rafalski, J. (1936). *Neobisium (Neobisium) polonicum* nov. spec. Nowy gatunek zaleszczotka (Pseudoscorpionidea). *Sprawozdania Poznanskiego Towarzystwa Przyjaciol Nauk* **9**: 119. (not seen)

Rafalski, J. (1937). *Neobisium (Neobisium) polonicum* nov. spec. Nowy gatunek zaleszczotka (Pseudoscorpionidea). *Prace Komisji Matematyczano-Pryzyrodniczej, Poznanskie Towarzystwo Przyjaciol Nauk* B **8**: 159–172.

Rafalski, J. (1948). *Mundochthonius carpaticus* sp. nov., nowy gatunek zaleszczotka (Pseudoscorpionidea). *Annales Musei Zoologici Polonici* **14**: 13–20.

Rafalski, J. (1949). Pseudoscorpionidea z Kaukazu w zbiorach Państwowego Muzeum Zoologicznego. *Annales Musei Zoologici Polonici* **14**: 75–120.

Rafalski, J. (1953). Fauna pajęczaków Parku Narodowego na wyspie Wolinie świetle dotychczasowych badań. *Ochrona Przyrody* **21**: 217–248. (not seen)

Rafalski, J. (1967). Zaleszczotki. Pseudoscorpionidea. In: *Katalog Fauny Polski*, vol. **32**(1): 1–34. Polska Akademia Nauk: Warszawa.

Rafalski, J. (1981). Pajęczaki (Arachnida). *Przeglad Zoologiczny* **25**: 222–228.

Rainbow, W. J. (1897). The arachnidan fauna of Funafuti. *Memoirs of the Australian Museum* **3**: 105–124.

Rainbow, W. J. (1916). Arachnida from northern Queensland. *Records of the Australian Museum* **11**: 31–64, 79–119.

Ralph, C. P., Nagata, S. E. & Ralph, C. J. (1985). Analysis of droppings to describe diets of small birds. *Journal of Field Ornithology* **56**: 165–174. (not seen)

Rapp, J. L. C. (1946). Notes on pseudoscorpions. *Entomological News* **57**: 197.

Rapp, W. F. (1978a). Preliminary studies on pseudoscorpion populations in the soil-grass interface as observed in the Nebraska prairies of the U.S.A. *Newsletter of the British Arachnological Society* **23**: 5–7.

Rapp, W. F. (1978b). Preliminary studies on pseudoscorpion populations in the soil-grass interface as observed in the Nebraska prairies. *American Arachnology* **18**: 12. (not seen)

Rapp, W. F. (1986). Pseudoscorpion population in oak-hickory woodlands. In: Eberhard, W. G., Lubin, Y. D. & Robinson, B. C. (eds), *Proceedings of the Ninth International Congress of Arachnology, Panama 1983*: 219–221. Smithsonian Institution Press: Washington D.C.

Reddell, J. R. (1965a). A checklist of the cave fauna of Texas. I. The Invertebrata (exclusive of Insecta). *Texas Journal of Science* **17**: 143–187. (not seen)

Reddell, J. R. (1965b). The caves of Medina County. *Texas Speleological Survey* **3**(11): 1–58. (not seen)

Reddell, J. R. (1970). A checklist of the cave fauna of Texas. IV. Additional records of Invertebrata (exclusive of Insecta). *Texas Journal of Science* **21**: 389–415. (not seen)

Reddell, J. R. (1971). A checklist of the cave fauna of Mexico. III. New records from southern Mexico. *Bulletin of the Association for Mexican Cave Studies* **4**: 217–230.

Reddell, J. R. (1981). A review of the cavernicole fauna of Mexico, Guatemala and Belize. *Bulletin of the Texas Memorial Museum* **27**: 1–327. (not seen)

Reddell, J. R. & Cokendolpher, J. C. (1984). A new species of troglobitic *Schizomus* (Arachnida: Schizomida) from Ecuador. *Bulletin of the British Arachnological Society* **6**: 172–177.

Reddell, J. R. & Elliott, W. R. (1973a). A checklist of the cave fauna of Mexico. IV. Additional records from the Sierra de El Abra, Tamaulipas and San Luis Potosi. *Bulletin of the Association for Mexican Cave Studies* **5**: 171–180.

Reddell, J. R. & Elliott, W. R. (1973b). A checklist of the cave fauna of Mexico. V. Additional records from the Sierra de Guatemala, Tamaulipas. *Bulletin of the Association for Mexican Cave Studies* **5**: 181–190.

Reddell, J. R. & Mitchell, R. W. (1971a). A checklist of the cave fauna of Mexico. I. Sierra de El Albra, Tamaulipas and San Luis Potosi. *Bulletin of the Association for Mexican Cave Studies* **4**: 137–180.

Reddell, J. R. & Mitchell, R. W. (1971b). A checklist of the cave fauna of Mexico. II. Sierra de Guatemala, Tamaulipas. *Bulletin of the Association for Mexican Cave Studies* **4**: 181–215.

Redikorzev, V. (1918). Pseudoscorpions nouveaux. I. *Ezhegodnik Zoologicheskago Muzeya* **22**: 91−101.

Redikorzev, V. (1922a). Pseudoscorpions nouveaux. II. *Ezhegodnik Zoologicheskago Muzeya* **23**: 257−272.

Redikorzev, V. (1922b). Two new species of pseudoscorpion from Sumatra. *Ezhegodnik Zoologicheskago Muzeya* **23**: 545−554.

Redikorzev, V. (1924a). Pseudoscorpions nouveaux de l'Afrique Orientale tropicale. *Revue Russe d'Entomologie* **18**: 187−200.

Redikorzev, V. (1924b). [Les pseudoscorpions de l'Oural.] *Bulletin de la Société Ouralienne d'Sciences Naturelles* **39**: 11−27 (in Russian).

Redikorzev, V. (1926). Pseudoscorpion nouveaux du Caucase. *Revue Russe d'Entomologie* **20**: 1−4.

Redikorzev, V. (1928). Beiträge zur Kenntnis der Pseudoscorpionenfauna Bulgariens. *Mitteilungen aus dem Königl. Naturwissenschaftlichen Institut in Sofia* **1**: 118−141.

Redikorzev, V. (1930). Contribution à l'étude de la faune des pseudoscorpions du Caucase. *Bulletin du Muséum de Géorgie* **6**: 97−106.

Redikorzev, V. (1934a). Neue paläarktische Pseudoscorpione. *Zoologische Jahrbücher, Systematik (Ökologie), Geographie und Biologie* **65**: 423−440.

Redikorzev, V. (1934b). Verbesserungen zu dem Aufsatz von 'Neue paläarktische Pseudoskorpione'. *Zoologische Jahrbücher, Systematik (Ökologie), Geographie und Biologie* **66**: 152.

Redikorzev, V. (1934c). Schwedisch-Chinesische wissenschaftliche Expedition nach den nord-westlichen Provinzen Chinas, unter Leitung von Dr. Sven Hedin und Prof. Sü Ping-Chang. Pseudoscorpiones. *Arkiv för Zoologi* **27A**(20): 1−4.

Redikorzev, V. (1935a). *Apocheiridium rossicum* sp. n. *Compte Rendu de l'Académie des Sciences de l'U.R.S.S., n.s.* **1**: 184−186.

Redikorzev, V. (1935b). Pseudoscorpiones. In: Sjöstedt, Y., Entomologische Ergebnisse der schwedischen Kamtchatka-Expedition 1920−1922. 37. Abschluss und Zusammenfassung. *Arkiv för Zoologi* **28A**(7): 1−19.

Redikorzev, V. (1937). Die erste neotropische *Roncus*-Art. *Entomologisk Tidskrift* **58**: 146−147.

Redikorzev, V. (1938). Les pseudoscorpions des l'Indochine française recueillis par M. C. Dawydoff. *Mémoires du Muséum National d'Histoire Naturelle, Paris* **10**: 69−116.

Redikorzev, V. (1949). [Pseudoscorpionidea of Central Asia.] *Travaux de l'Institute de Zoologique de l'Académie Sciences de l'U.R.S.S.* **8**: 638−668 (in Russian).

Reeker, H. (1894). Zur Lebensweise der Afterskorpione. *Jahresbericht des Westfälischen Provinzialvereins für Wissenschaft und Kunst, Münster* **22**: 103−108. (not seen)

Ressl, F. (1963). Können Vögel als passive Verbreiter von Pseudoscorpioniden betrachtet werden? *Vogelwarte* **84**: 114−119. (not seén)

Ressl, F. (1965). Über Verbreitung, Variabilität und Lebensweise einiger österreichischer Afterskorpione. *Deutsche Entomologische Zeitschrift* **12**: 289−295.

Ressl, F. (1970). Weitere Pseudoskorpion-Funde aus dem Bezirk Scheibbs (Niederösterreich). *Berichte des Naturwissenschaftlich-Medizinischen Vereins in Innsbruck* **58**: 249−254.

Ressl, F. (1974). Myrmecophile Pseudoscorpione aus dem Bezirk Scheibbs (Niederösterreich). *Entomologische Nachrichten* **18**: 26−31.

Ressl, F. (1983). Die Pseudoskorpione Niederösterreichs mit besonderer Berücksichtigung des Bezirkes Scheibbs. *Naturkunde des Bezirkes Scheibbs, Tierwelt* **2**: 174−202. (not seen)

Ressl, F. & Beier, M. (1958). Zur Ökologie, Biologie und Phänologie der heimischen Pseudoskorpione. *Zoologische Jahrbücher, Systematik (Ökologie), Geographie und Biologie* **86**: 1−26.

Rey, J. R. & McCoy, E. D. (1983). Terrestrial arthropods of northwest Florida salt marshes: Araneae and Pseudoscorpiones (Arachnida). *Florida Entomologist* **66**: 497–503.

Reyes-Castillo, P. & Hendrichs, J. (1975). Pseudoscorpiones asociados con pasálidos. *Acta Politécnica Mexicana* **26**: 129–133.

Richard, J. & Neuville, H. (1897). Sur l'histoire naturelle de l'île d'Alborán. *Memoires de la Société Zoologique de France* **10**: 75–87.

Richards, A. M. (1971). An ecological study of the cavernicolous fauna of the Nullarbor Plain, southern Australia. *Journal of Zoology* **164**: 1–60.

Richters, F. (1902a). Beiträge zur Kenntnis der Fauna der Umgebung von Frankfurt a. M. *Bericht der Senckenbergischen Naturforschenden Gesellschaft in Frankfurt a. Main* **2**(2): 3–21. (not seen)

Richters, F. (1902b). Wandern die Chernetiden freiwillig? *Prometheus* **13**: 349–350. (not seen)

Ripper, W. (1931). Versuch einer Kritik der Homologiefrage der Arthropodentracheen. *Zeitschrift für Wissenschaftliche Zoologie* **138**: 303–369. (not seen)

Risso, A. (1826). Animaux articulés: description de quelques Myriapodes, Scorpionides, Arachnides et Acarides, habitant les Alpes Maritimes. In: Risso, A., *Histoire Naturelle des Principales Productions de l'Europe Méridionale et Principalement de Celles des Environs de Nice et des Alpes Maritimes.* Levrault: Paris.

Ritsema, C. (1874). Naamlijst der tot Heden in Nederland waargenomen bastaard-schorpioenen (Chernetiden). *Tijdschrift voor Entomologie* **18**: xxxiii–xxxiv.

Robson, M. (1879). On the development of the house fly and its parasite. *Science Gossip* **1879**: ?–? (not seen)

Rodrigues, J. M. G. (1986). *Abundancia e distribuçao vertical de arthropoda do solo, em capoeira de terra firme.* M. Sci. Dissertation: INPA, Manaus. (not seen)

Roemer, J. S. (1789). *Genera Insectorum Linnaei et Fabricii Iconibus Illustrata.* Vitoduri Helvetorum. (not seen)

Roesel von Rosenhof, A. J. (1755). *Monatlichherausgegebe Insectenbelustigungen*, vol. **4**. Nürnberg. (not seen)

Roewer, C. F. (1929). Süd-indische Skorpione, Chelonethi und Opilioniden. *Revue Suisse de Zoologie* **36**: 609–639.

Roewer, C. F. (1931). Arachnoiden aus südostalpinen Höhlen. *Mitteilungen über Höhlen- und Karstforschung* **9**: 40–46. (not seen)

Roewer, C. F. (1936, 1937, 1940). Chelonethi oder Pseudoskorpione. In: Bronns, H. G. (ed.), *Klassen und Ordnungen des Tierreichs. 5: Arthropoda. IV: Arachnoidea*, vol. **6**(1). Akademische Verlagsgesellschaft M.B.H.: Leipzig.

Ross, H. H. (1944). How to collect and preserve insects. *Illinois Natural History Survey, Divisional Circular* **39**: 1–55.

Rossi, A. (1909). Materiali per una fauna aracnologica della Provincia di Roma. Chernetes (Pseudoscorpioni). *Bollettino della Società Entomologica Italiana* **11**: 182–194. (not seen)

Roth, V. D. & Brown, W. L. (1976). Other intertidal air-breathing arthropods. In: Cheng, L. (ed.), *Marine Insects.* (not seen)

Rowland, J. M. & Reddell, J. R. (1976). Annotated checklist of the arachnid fauna of Texas (excluding Acarida and Araneida). *Occasional Papers of the Museum of the Texas Technical University* **38**: 1–25.

Rowland, K. (1983). The Roman River Valley - arachnids (spiders, harvestmen and pseudoscorpions). *Nature in North Essex* **1983**: 122–126. (not seen)

Rüeger, H. (1914). Der Bücherskorpion. *Schweizerische Bienen-Zeitung* **37**: 483.

Ruffo, S. (1950). Sull'interesse biogeografico del popolomento cavernicolo pugliese. *Rassegna Speleologica Italiana* **2**: 57–62. (not seen)

Ruffo, S. (1953). Le attuali conoscenze sulla fauna cavernicole della regione Pugliese. *Memorie di Biogeografica Adriatica* **3**: 1–143. (not seen)

Ruffo, S. (1960). La fauna. In: Allegranzi, A., Bartolomei, G., Broglio, A., Pasa, A., Rigobello, A., Ruffo, S., Il Buso della Rana (40 V−VI). *Rassegna Speleologica Italiana* **12**: 99−164. (not seen)

Rundle, A. J. (1979). Pseudoscorpions and harvestmen in Bedfordshire. *Bedfordshire Naturalist* **33**: 47−50.

Ryckman, R. E. (1956). Parasitic and some nonparasitic arthropods from bat caves in Texas and Mexico. *American Midland Naturalist* **56**: 186−190. (not seen)

Sacher, P. (1987). *Neobisium crassifemoratum* (Beier, 1928) in der polnischen Tatra (Arachnida, Pseudoscorpiones). *Polskie Pismo Entomologiczne* **57**: 417.

Sacher, P. & Breinl, K. (1986). Über Nachweise von Pseudoskorpionen in Ostthüringen (Arachnida, Pseudoscorpiones). *Abhandlungen und Berichte, Museums der Natur Gotha* **13**: 47−49.

Saint-Remy, G. (1889). Structure du cerveau chez les Myriapodes et les Arachnides. *Revue Biologique du Nord de la France* **1**: 281−298. (not seen)

Salmon, S. J. (1972). Recent records of British pseudoscorpions. *Bulletin of the British Arachnological Society* **2**: 66−68.

Salt, G. (1929). A contribution to the ethology of the Meliponinae. *Transactions of the Entomological Society of London* **77**: 431−470.

Salt, G. (1954). A contribution to the ecology of upper Kilimanjaro. *Journal of Ecology* **42**: 375−423.

Salt, G., Hollick, F. S. J., Raw, F. & Brian, M. V. (1948). The arthropod population of pasture soil. *Journal of Animal Ecology* **17**: 139−150. (not seen)

Sanfilipo, N., Timossi, G. & Conci, C. (1943). La grotta del Brigidun e la grotta Dragonara (Esplorazioni speleologiche nella Provincia di Genova − I). *Annali del Museo Civico di Storia Naturale di Genova* **61**: 307−319.

Sankey, J. H. P. (1949). Observations on food, enemies and parasites of British harvest-spiders (Arachnida, Opiliones). *Entomologist's Monthly Magazine* **85**: 246−247.

Sanocka-Woloszynowa, E. (1981). Badania pajeczaków (Aranei, Opiliones, Pseudoscorpionida) jaskin Wyzyny Krakowsko-Czestochowskiej. *Prace Zoologiczne* **11**: 1−90.

Sareen, M. L. (1965). Histochemical studies on the female germ cells of the pseudoscorpion, *Diplotemnus insolitus* Chamberlin (Chelonethida, Arachnida). *Research Bulletin of Panjab University (n.s.)* **16**: 9−18.

Sato, H. (1976). [Observation on the nest of Japanese pseudoscorpions.] *Atypus* **67**: 51−52. (in Japanese) (not seen)

Sato, H. (1977a). [Variations of carapacal spots on the Japanese pseudoscorpion (*Garypus japonicus*) from Enoshima Island.] *Atypus* **68**: 42−44. (in Japanese)

Sato, H. (1977b). [Notes on the colony building of *Garypus japonicus*.] *Atypus* **70**: 36. (in Japanese)

Sato, H. (1978a). [Life history of a Japanese arboreal pseudoscorpion, *Haplochernes boncicus* Karsch.] *Acta Arachnologica* **28**: 31−38. (in Japanese)

Sato, H. (1978b). [Variations of galeae on the pseudoscorpion *Microcreagris japonica* (Ellingsen).] *Atypus* **71**: 44. (in Japanese) (not seen)

Sato, H. (1978c). [Faunistic data on Japanese pseudoscorpions.] *Atypus* **72**: 39−42. (in Japanese) (not seen)

Sato, H. (1978d). [Notes on the Japanese maritime pseudoscorpions, *Garypus japonicus* and *Nipponogarypus enoshimaensis*.] *Atypus* **71**: 45−48. (in Japanese)

Sato, H. (1978e). [Extracting experiment of soil pseudoscorpions by dry funnel method.] *Edaphologia* **18**: 15−20. (in Japanese)

Sato, H. (1979a). [Pseudoscorpions from Mt. Takao, Tokyo. (An introduction to morphology of pseudoscorpion.)] *Memoirs of the Education Institute for Private Schools in Japan* **64**: 79−105. (in Japanese)

Sato, H. (1979b). [Quantitative survey of soil pseudoscorpions at Mt Daisen by sifting method.] *Edaphologia* **19**: 13–24. (in Japanese)

Sato, H. (1979c). [Altitudinal distribution of soil pseudoscorpions on Yakushima Island.] *Edaphologia* **20**: 13–18. (in Japanese)

Sato, H. (1979d). [Faunistic data on Japanese pseudoscorpions. II.] *Atypus* **74**: 42–44. (in Japanese)

Sato, H. (1980a). [Influence of humidity on three pseudoscorpions *Microcreagris japonica*, *Garypus japonicus* and *Haplochernes boncicus*.] *Memoirs of the Education Institute for Private Schools in Japan* **72**: 57–63. (in Japanese)

Sato, H. (1980b). [Life history of arboreal pseudoscorpion *Apocheiridium pinium* based on the investigation of the nests and exuviae.] *Memoirs of the Education Institute for Private Schools in Japan* **72**: 65–71. (in Japanese)

Sato, H. (1980c). [Altitudinal distribution of soil pseudoscorpions on Mt Chokai.] *Edaphologia* **22**: 9–14. (in Japanese)

Sato, H. (1980d). [Title?] *Heredity* **34**: 85–91. (in Japanese)

Sato, H. (1981). A new species of the genus *Ditha* from Japan (Pseudoscorpionidea: Dithidae). *Edaphologia* **24**: 11–14.

Sato, H. (1982a). A new species of the genus *Dactylochelifer* (Pseudoscorpionidea, Cheliferidae) from Japan. *Acta Arachnologica* **30**: 105–110.

Sato, H. (1982b). [Seasonal fluctuations of some soil pseudoscorpions in Karuizawa, central Japan.]*Edaphologia* **25**: 57–64.

Sato, H. (1982c). [Faunistic data on Japanese pseudoscorpions. III.] *Atypus* **81**: 31–34. (in Japanese)

Sato, H. (1983a). *Hesperochernes shinjoensis*, a new pseudoscorpion (Chernetidae) from Japan. *Bulletin of the Biogeographical Society of Japan* **38**: 31–34.

Sato, H. (1983b). Altitudinal distribution of soil pseudoscorpions at Mt. Fuji. *Edaphologia* **28**: 13–22.

Sato, H. (1983c). *Tyrannochthonius (Lagynochthonius) nagaminei*, a new pseudoscorpion (Chthoniidae) from Mt. Kirishima, Japan. *Edaphologia* **29**: 7–11.

Sato, H. (1984a). *Allochthonius borealis*, a new pseudoscorpion (Chthoniidae) from Tohoku District, Japan. *Bulletin of the Biogeographical Society of Japan* **39**: 17–20.

Sato, H. (1984b). Pseudoscorpions from the Ogasawara Islands. *Proceedings of the Japanese Society of Systematic Zoology* **28**: 49–56.

Sato, H. (1984c). Population dynamics of the soil pseudoscorpions at Mt. Takao. *Edaphologia* **31**: 13–19.

Sato, H. (1984d). [Study on life history of pseudoscorpions – with special reference to the brooding and the number of moulting.] *Atypus* **85**: 75–77.

Sato, H. (1985). [Altitudinal distribution of soil pseudoscorpions at Mt. Funagata, Yamagata Prefecture.] *Bulletin of the Biogeographical Society of Japan* **40**: 21–24.

Sato, H. (1986). Soil pseudoscorpions in the cool temperate forests of Japan. In: Hanxi, Y., Zhan, W., Jeffers, J. N. R., and Ward, P. A. (eds), *The Temperate Forest Ecosystem, ITE Symposium no. 20, Proceedings of International Symposium on Temperate Forest Ecosystem Management and Environmental Protection, China*: 94–96. Institute of Terrestrial Ecology: Grange-over-Sands.

Sato, H. (1988). [Seasonal fluctuations of some pseudoscorpions in Yokohama, central Japan.] *Edaphologia* **38**: 11–16. (in Japanese) (not seen)

Savory, T. H. (1934). List of the Arachnida of Worcestershire. *Transactions of the Worstershire Naturalists' Club* **9**: 85–92. (not seen)

Savory, T. H. (1943). On a collection of Arachnida from Christmas Island. *Annals and Magazine of Natural History* (11) **10**: 355–360.

Savory, T. H. (1949). Notes on the biology of Arachnida. *Journal of the Quekett Microscopical Club* (4) **3**: 18–24. (not seen)

Savory, T. H. (1964a). *Arachnida*. Academic Press: London.

Savory, T. H. (1964b). *Spiders and Other Arachnids.* English Universities Press: London. (not seen)

Savory, T. H. (1966). False scorpions. *Scientific American* **214**(3): 95–100.

Savory, T. H. (1977). *Arachnida,* 2nd edition. Academic Press: London.

Savory, T. H. & Gros, A. E. le (1957). The Arachnida of London. *Naturalist, London* **36:** 41–50. (not seen)

Say, T. (1821). An account of the Arachnides of the United States. *Journal of the Academy of Natural Sciences of Philadelphia* **2:** 59–82.

Sbordoni, V. & Cobolli, M. (1969). Note sull'allevamento sperimentale degli animali cavernicoli in laboratorio. *Archivo Zoologico Italiano* **54:** 33–57. (not seen)

Schaefer, M. (1971). Pseudoscorpiones - Araneae - Opiliones. In: Brohmer, P. & Tischler, W. (eds), *Fauna von Deutschland, Heidelberg*: 355–381. (not seen)

Schaeffer, J. C. (1766). *Elementa entomologica.* Regensburg. (not seen)

Schaller, F. (1965). Mating behaviour of lower terrestrial arthropods from the phylogenetic point of view. *XII International Congress of Entomology*: 297–298.

Schawaller, W. (1978). Neue Pseudoskorpione aus dem Baltischen Bernstein der Stuttgarter Bernsteinsammlung (Arachnida: Pseudoscorpionidea). *Stuttgarter Beiträge zur Naturkunde* (B) **42:** 1–21.

Schawaller, W. (1979). Einige Pseudoskorpione und Weberknechte aus dem landschaftsschutzgebiet Poppenweiler bei Ludwigsburg. *Mitteilungen Entomologischer Verein Stuttgart 14:* 12-13.

Schawaller, W. (1980a). Bibliographie der rezenten und fossilen Pseudoscorpionidea 1890–1979 (Arachnida). *Stuttgarter Beiträge zur Naturkunde* (A) **338:** 1–61.

Schawaller, W. (1980b). Erstnachweis tertiärer Pseudoskorpione (Chernetidae) in Dominikanischem Bernstein (Stuttgarter Bernsteinsammlung: Arachnida, Pseudoscorpionidea). *Stuttgarter Beiträge zur Naturkunde* (B) **57:** 1–20.

Schawaller, W. (1980c). Fossile Chthoniidae in Dominikanischem Bernstein, mit phylogenetischen Anmerkungen (Stuttgarter Bernsteinsammlung: Arachnida, Pseudoscorpionidea). *Stuttgarter Beiträge zur Naturkunde* (B) **63:** 1–19.

Schawaller, W. (1980d). Eine Pseudoskorpion-Art, *Neobisium erythrodactylum* L. Koch 1873, in Süddeutschland aktiv auf Schnee (Arachnida: Pseudoscorpiones: Neobisiidae). *Entomologische Zeitschrift* **90:** 54–56.

Schawaller, W. (1981a). Pseudoskorpione (Cheliferidae) phoretisch auf Käfern (Platypodidae) in Dominikanischem Bernstein (Stuttgarter Bernsteinsammlung: Arachnida, Pseudoscorpionidea und Coleoptera). *Stuttgarter Beiträge zur Naturkunde* (B) **71:** 1–17.

Schawaller, W. (1981b). Pseudoskorpione von Korsika (Arachnida, Pseudoscorpionidea). *Entomologica Basiliensia* **6:** 42–51.

Schawaller, W. (1981c). Eine neue troglobionte *Roncus*-Art und weitere Pseudoskorpione von den Nördlichen Sporaden (ägäis) (Arachnida: Pseudoscorpionidea). *Stuttgarter Beiträge zur Naturkunde* (A) **344:** 1–9.

Schawaller, W. (1981d). Cheiridiidae in Dominikanischem Bernstein, mit Anmerkungen zur morphologischen Variabilität (Stuttgarter Bernsteinsammlung: Arachnida, Pseudoscorpionidea). *Stuttgarter Beiträge zur Naturkunde* (B) **75:** 1–14.

Schawaller, W. (1982a). Der erste Pseudoskorpion (Chernetidae) aus Mexicanischem Bernstein (Stuttgarter Bernsteinsammlung: Arachnida, Pseudoscorpionidea). *Stuttgarter Beiträge zur Naturkunde* (B) **85:** 1–9.

Schawaller, W. (1982b). Eine neue höhlenbewohnende *Chthonius*-Art aus den italienischen Südalpen. *Bollettino della Società Entomologica Italiana* **114:** 49–55.

Schawaller, W. (1982c). Eine für Deutschland neue Pseudoskorpion-Art aus dem Allgäu (Arachnida). *Jahreshefte, Gesellschaft für Naturkunde in Württemberg* **137:** 159–160.

Schawaller, W. (1983a). Neue Pseudoskorpion-Funde aus dem Nepal-Himalaya. *Senckenbergiana Biologica* **63:** 105–111.

Schawaller, W. (1983b). Pseudoskorpione aus dem Kaukasus (Arachnida). *Stuttgarter Beiträge zur Naturkunde* (A) *362:* 1–24.

Schawaller, W. (1983c). Pseudoskorpione aus dem Norden des Iran. *Senckenbergiana Biologica* **63**: 367–371.

Schawaller, W. (1985a). Pseudoskorpione aus der Sowjetunion (Arachnida: Pseudoscorpiones). *Stuttgarter Beiträge zur Naturkunde* (A) **385**: 1–12.

Schawaller, W. (1985b). Liste griechischer Neobisiidae mit neuen Höhlenfunden in Epirus, auf Samos und Kreta (Arachnida: Pseudoscorpiones). *Stuttgarter Beiträge zur Naturkunde* (A) **386**: 1–8.

Schawaller, W. (1986). Pseudoskorpione aus der Sowjetunion, Teil 2 (Arachnida: Pseudoscorpiones). *Stuttgarter Beiträge zur Naturkunde* (A) **396**: 1–15.

Schawaller, W. (1987a). Eine neue *Dactylochelifer*-Art aus Spanien (Prov. Tarragona). *Eos, Madrid* **63**: 277–280.

Schawaller, W. (1987b). Erstnachweis der Familie Syarinidae in Deutschland: ein Reliktvorkommen von *Syarinus strandi* im Oberen Donautal (Arachnida: Pseudoscorpiones). *Jahreshefte, Gesellschaft für Naturkunde in Württemberg* **142**: 287–292.

Schawaller, W. (1987c). Neue Pseudoskorpion-Funde aus dem Nepal-Himalaya, II (Arachnida: Pseudoscorpiones). *Senckenbergiana Biologica* **68**: 199–221.

Schawaller, W. (1988). Neue Pseudoskorpion-Funde aus dem Kashmir-Himalaya (Arachnida: Pseudoscorpionida). *Annalen des Naturhistorischen Museums in Wien* **90**: 157–162.

Schawaller, W. (1989). Pseudoskorpione aus der Sowjetunion, Teil 3 (Arachnida: Pseudoscorpiones). *Stuttgarter Beiträge zur Naturkunde* (A) **440**: 1–30.

Schawaller, W. & Dashdamirov, S. (1988). Pseudoskorpione aus dem Kaukasus, Teil 2 (Arachnida). *Stuttgarter Beiträge zur Naturkunde* (A) **415**: 1–51.

Schenkel, E. (1926). Beitrag zur Kenntnis der Schweizerischen Spinnenfauna. II. Teil. *Revue Suisse de Zoologie* **33**: 301–316.

Schenkel, E. (1928). Pseudoscorpionida (Afterskorpione). In: Dahl, F., *Die Tierwelt Deutschlands* **8**: 52–72. Jena.

Schenkel, E. (1929a). Pseudoskorpione des Zehlaubruches. *Schriften der Physikalischen-Ökonomischen Gesellschaft zu Königsberg* **66**: 320–323.

Schenkel, E. (1929b). Beitrag zur Kenntnis der Schweizerischen Spinnenfauna. IV. Teil. Spinnen von Bedretto. *Revue Suisse de Zoologie* **36**: 1–24.

Schenkel, E. (1933). Beitrag zur Kenntnis der Schweizerischen Spinnenfauna. V. Teil. Spinnen aus dem Saas-Tal (Wallis) und von der Gegend zwischen Trins und Flims (Graubünden). *Revue Suisse de Zoologie* **40**: 11–29.

Schenkel, E. (1937). Schwedisch-chinesische wissenschaftliche Expedition nach den nordwestlichen Provinzen Chinas. Araneae. *Arkiv för Zoologi* **29**(1): 1–314.

Schenkel, E. (1938). Die Arthropodenfauna von Madeira nach den Ergebnissen der Reise von Prof. Dr. O. Lundblad Juli-August 1935. IV. Araneae, Opiliones und Pseudoscorpiones. *Arkiv för Zoologi* **30**(7): 1–42.

Schenkel, E. (1953a). Bericht über einige Spinnentiere aus Venezuela. *Verhandlungen der Naturforschenden Gesellschaft in Basel* **64**: 1–57.

Schenkel, E. (1953b). Chinesische Arachnoidea aus dem Museum Hoangho-Peiho in Tientsin. *Boletim do Museu Nacional Rió de Janeiro, nova série* **119**: 1–108.

Scheuchzer, J. J. (1777). *Dissertatio Physicae Sacrae Specimen de Locustis.* (not seen)

Scheuring, L. (1913). Die Augen der Arachnoideen. *Zoologische Jahrbücher, Anatomie und Ontogenie* **33**: 553–636. (not seen)

Schimkewitsch, W. (1894). Sur la structure et la signification de l'endosternite des Arachnides. *Zoologischer Anzeiger* **17**: 127–128. (not seen)

Schimkewitsch, W. (1895). Über Bau und Entwicklung des endosternits der Arachniden. *Zoologische Jahrbücher, Anatomie und Ontogenie* **8**: 191–216. (not seen)

Schiner, T. (1872). Vorkommen von *Chelifer* an Fliegen. *Verhandlungen der K. K. Zoologischen-Botanischen Gesellschaft in Wien* **22**: 75–76. (not seen)

Schiödte, J. C. (1847). Undersögelser over den underjordiske Fauna i Hulerne i Krain og Istrien. *Oversigt over det Konigl. Danske Videnskabernes Selskabs Forhandlingar* **6**: 75–81.

Schiödte, J. C. (1851a). Bidrag til den underjordiske Fauna. *Videnskabelige Meddelelser fra Danske Naturhistorisk Forening i Kjøbenhavn* (5) **2**: 1–39.

Schiödte, J. C. (1851b). Specimen faunae subterraneae; being a contribution towards the subterranean fauna. *Transactions of the Royal Entomological Society of London* **1**: 134–157.

Schlee, D. & Glöckner, W. (1978). Bernstein. *Stuttgarter Beiträge zur Naturkunde* (C) **8**: 1–72.

Schloss, I. P. (1942). Pseudoscorpions. *Bulletin of the Natural History Society of Maryland* **13**: 12–14. (not seen)

Schlottke, E. (1933). Der Fressakt des Bücherskorpions (*Chelifer cancroides* L.). *Zoologischer Anzeiger* **104**: 109–112.

Schlottke, E. (1940). Zur Biologie des Bücherskorpions (*Chelifer cancroides* L.). *Bericht des Westpreussischen Botanisch-Zoologischens Vereins* **62**: 57–69.

Schmalfuss, H. & Schawaller, W. (1984). Die Fauna der Ägäis-Insel Santorin. Teil 5. Arachnida und Crustacea. *Stuttgarter Beiträge zur Naturkunde* (A) **371**: 1–16.

Schmidt, F. (1848). Naturhistorisches aus Krain. *Illyrisches Blatt* **2**: 10.

Schmiedeknecht, O. (1931–1933). *Opuscula Ichneumonologica*, vol. **1b**, supplement. Thuringen: Blandkenburg. (not seen)

Schrank, F. P. (1781). *Enumeratio Insectorum Austriae Indigenorum*. Augustae Vindelicorum.

Schrank, F. P. (1803). *Fauna Boica. Durchgedachte, Geschichte der in Baiern einheimischen und Zahmen Theire*, vol. **3**. Krüll: Landshut.

Schubert, K. (1933). Beiträge zur Kenntnis der Tierwelt des Moosebruches im Altvatergebirge (Ostsudeten). (Spinnentiere teilweise, Insekten, Wirbeltiere). *Zeitschrift für Morphologie und Ökologie der Tiere* **27**: 325–372. (not seen)

Schüller, L. (1951). Ein Beitrag zur Kenntnis der Pseudoskorpione im Lande Salzburg (mit einer Kartenskizze). *Mitteilungen der Naturwissenschaftlichen Arbeitsgemeinschaft von Haus der Natur in Salzburg* **2**: 1–9. (not seen)

Schulte, G. (1976). Litoralzonierung von Pseudoskorpionen an der nordamerikanischen Pazifikküste (Arachnida: Pseudoscorpiones: Neobisiidae, Garypidae). *Entomologica Germanica* **3**: 119–124.

Schulte, G. (1978). Die Küstenbindung terrestricher Arthropoden und ihre Bedeutung für den Wandel des Ökosystems 'marines Felslitoral' in unterschiedlichen geographischen Breiten. *Mitteilungen der Deutschen Gesellschaft für allgemeine und angewandte Entomologie* **1**: 211–219. (not seen)

Schuster, R. (1956). Das Kalkalgen-Trottoir an der Côte des Albères als Lebensraum terricoler Kleintiere. *Vie et Milieu* **7**: 242–257.

Schuster, R. (1959). Ökologisch-faunistische Untersuchungen an bodenbewohnenden Kleinarthropoden (speziell Oribatiden) des Salzlachergebietes im Seewinkel. *Sitzungsberichte der Österreichischen Akademie der Wissenschaften, Math.-Naturw. Kl. Abt. 1* **168**: 27–? (not seen)

Schuster, R. (1962). Das marine Litoral als Lebensraum terrestrischer Kleinarthropoden. *Internatione Revue der Gesamten Hydrobiologie* **47**: 359–412. (not seen)

Schuster, R. (1964). Die Ökologie der terrestrischen Kleinfauna des Meeresstrandes. *Verhandlungen der Deutschen Zoologischen Gesellschaft* **1964**: 492–521. (not seen)

Schuster, R. (1965). Faunistische Studien am Roten Meer (im Winter 1961/62). Teil I. Litoralbewohnende Arthropoden terrestrischer Herkunft. *Zoologische Jahrbücher, Systematik (Ökologie), Geographie und Biologie* **92**: 327–343.

Schuster, R. (1972). Faunistische Nachrichten aus der Steiermark (XVII/12): neue Spinnentier-Funde (Arachnida div.). *Mitteilungen des Naturwissenschaftlichen Vereins für Steiermark* **102**: 239–241.

Schuster, R. (1986a). Order Pseudoscorpiones. In: Sterrer, W. (ed.), *Marine Fauna and Flora of Bermuda*. John Wiley & Sons.

Schuster, R. (1986b). Comment on the proposed designation of type species of *Olpium* Koch, 1873 (Arachnida, Pseudoscorpionida). *Bulletin of Zoological Nomenclature* **43**: 118.

Schuster, R. O. (1962). New species of *Kewochthonius* Chamberlin from California (Arachnida: Chelonethida). *Proceedings of the Biological Society of Washington* **75**: 223–226.

Schuster, R. O. (1966a). A new species of *Allochthonius* from the Pacific northwest of North America (Arachnida: Chelonethida). *Pan-Pacific Entomologist* **42**: 172–175.

Schuster, R. O. (1966b). New species of *Apochthonius* from western North America (Arachnida: Chelonethida). *Pan-Pacific Entomologist* **42**: 178–183.

Schuster, R. O. (1966c). New species of *Parobisium* Chamberlin (Arachnida: Chelonethida). *Pan-Pacific Entomologist* **42**: 223–228.

Schuster, R. O. (1968). The identity of *Roncus pacificus* Banks (Arachnida: Chelonethida). *Pan-Pacific Entomologist* **44**: 137–139.

Scopoli, J. A. (1763). *Entomologia Carniolica Exhibens Insecta Carniolae Indigena et Distributa in Ordines, Genera, Species, Varietates. Methodo Linnaeano.* Vindobonae.

Scudder, S. H. (1885). Systematische Übersicht der fossilen Arthropoden, Arachniden und Insecten. In: Zittel, K.A. von, *Handbuch der Palaeontologie*, vol. **1**. München and Leipzig. (not seen)

Scudder, S. H. (1891). Index to the known fossil insects of the world including myriapods and arachnids. *Bulletin of the United States Geological Survey* **71**: 1–744. (not seen)

Sewell, A. C. (1899). Bee parasites in South Africa. *British Bee Journal and Bee-Keepers' Adviser* **27**: 211–212. (not seen)

Sgardelis, S., Stamon, G. & Margaris, N. S. (1981). Structure and spatial distribution of soil arthropods in a phryganic (East Mediterranean) ecosystems. *Revue d'Ecologie et de Biologie du Sol* **18**: 221–230. (not seen)

Sharkey, M. J. (1987). Subclass Chelonethida, order Pseudoscorpionida (pseudo-scorpions): 17. In: *The Insects, Spiders and Mites of Cape Breton Highlands National Park*. Agriculture Canada, Biosystematics Research Centre, Report 1. (not seen)

Sharma, B. D. & Sharma, T. (1975). Note on the pseudoscorpion *Chelifer cancroides* related to *Chelifer orientalis* from Jammu and Kashmir State (India). *Indian Journal of Animal Research* **9**: 56. (not seen)

Shaw, G. (1791). *The Naturalist's Miscellany*, vol. **3**. Nodder: London.

Shaw, G. (1806). *General Zoology or Systematic History*, vol. **6**(1): 474–475. London. (not seen)

Shear, W. A. (1986). A fossil fauna of early terrestrial arthropods from the Givetian (Upper Middle Devonian) of Gilboa, New York, USA. *Actas X Congreso International de Aracnologia, Jaca* **1**: 387–392.

Shear, W. A., Schawaller, W. & Bonamo, P. M. (1989). Record of Palaeozoic pseudoscorpions. *Nature* **341**: 527–529.

Shipley, A. E. (1909). Introduction to Arachnida and Xiphosura. *The Cambridge Natural History*, vol. **4**: 255–279. (not seen)

Siebold, C. T. E. von & Stannius, F. H. (1846). *Lehrbuch der Vergleichenden Anatomie*, vol. **1**. Berlin. (not seen)

Sill, V. (1861). Beiträge zur Kenntniss der Crustaceen, Arachniden und Myriapoden Siebenbürgens. *Verhandlungen und Mitteilungen des Siebenbürgischen Vereins für Naturwissenschaften in Hermannstadt* **12**: 6–7, 201–206. (not seen)

Sill, V. (1865). Systematisches Verzeichniss bisher bekannter Arachniden Siebenbürgens. *Ugyanitt* **16**: 50–74. (not seen)

Silvestri, F. (1918). Contribuzione alla conoscenza die termitidi e termitofili dell'-
Africa occidentale. II. Termitofili. Parte prima. *Bollettino del Laboratoria di
Zoologia Generale e Agraria della R. Scuola sup. d'Agricoltura, Portici* (2) **12**:
287–346.

Silvestri, F. (1919). Contribuzione alla conoscenza di Termitidi e Termitofili dell'Africa
occidentale. II. *Annali della Scuola Superiore di Agricoltura in Portici* (2) **15**:
10–20. (not seen)

Silvestri, F. (1922). Contribuzione allo studio della fauna delle caverne in Liguria.
Bollettino della Società Entomologica Italiana **54**: 18–20.

Simberloff, D. S. & Wilson, E. O. (1969). Experimental zoogeography of islands:
the colonization of empty islands. *Ecology* **50**: 278–296.

Simon, A. (1878). Das Hautskelett der Arthrogastrischen Arachniden. *28. Programm
des Kaiserlich-Koeniglicher Staats Gymnasiums in Salzburg.* (not seen)

Simon, E. (1872). Notice sur les Arachnides cavernicoles et hypogés. *Annales de la
Société Entomologique de France* (5) **2**: 215–244.

Simon, E. (1875). Untitled. *Annales de la Société Entomologique de France* (5) **5**:
ccvi–ccvii.

Simon, E. (1878a). Liste des espèces de la famille des Cheliferidae qui habitant l'Algérie
et le Maroc. *Annales de la Société Entomologique de France* (5) **8**: 144–153.

Simon, E. (1878b). Descriptions de quelques Cheliferidae de Californie. *Annales
de la Société Entomologique de France* (5) **8**: 154–158.

Simon, E. (1878c). Description d'un genre nouveau de la famille des Cheliferidae.
Bulletin de la Société Zoologique de France **3**: 66.

Simon, E. (1879a). *Les Arachnides de France*, vol. 7, *les Ordres des Chernetes,
Scorpiones et Opiliones.* Paris: Librairie Encyclopédique de Roret.

Simon, E. (1879b). Arachnides nouveaux de France, d'Espagne et d'Algérie. *Bulletin
de la Société Entomologique de France* **4**: 251–263.

Simon, E. (1880a). Matériaux pour servir à une faune arachnologique de la Nouvelle-
Calédonie. *Annales de la Société Entomologique de Belgique* **23**: 164–175.

Simon, E. (1880b). Sur les Arachnides recueillis à Sebenico, en Dalmatie, par
M. Munier-Chalmas. *Annales de la Société Entomologique de France* (5) **10**:
xxxv–xxxvi.

Simon, E. (1880c). Arachnides rec. aux environs immédiats d'Alexandrie (Égypte)
par A. Letourneux. *Annales de la Société Entomologique de France* (5) **10**: xlvii–
xlvii. (not seen)

Simon, E. (1881a). Descriptions de deux nouvelles espèces d'*Obisium* anophthalmes
du sous-genre *Blothrus. Annali del Museo Civico di Storia Naturale di Genova*
16: 299–302.

Simon, E. (1881b). Descriptions d'Arachnides nouveaux d'Afrique. (Chernetes de la
Basse Égypte rec. par M. Letourneux). *Bulletin de la Société Zoologique de France*
6: 1–15.

Simon, E. (1881c). Arachnides nouveaux ou rares de la faune française. *Bulletin de
la Société Zoologique de France* **6**: 82–91.

Simon, E. (1882a). Viaggio ad Assab nel Mar Rosso dei signori G. Doria ed O. Beccari
con il R. Avviso 'Esploratore' dal 16 Novembre 1879 al 26 Febbraio 1880. II. Étude
sur les Arachnides de l'Yemen méridional. *Annali del Museo Civico di Storia
Naturale di Genova* **18**: 207–260.

Simon, E. (1882b). Arachnidae. In: Cavanna, G., Artropodi raccolti a Lavaiano
(Provincia di Pisa). *Bollettino della Società Entomologica Italiana* **14**: 356–366.

Simon, E. (1883). Matériaux pour servir à la faune arachnologique des îles de l'Océan
Atlantique (Açores, Madère, Salvages, Canaries, Cap Vert, Sainte-Hélène et
Bermudes). *Annales de la Société Entomologique de France* (6) **3**: 259–314.

Simon, E. (1884). Note sur les Arachnides recueillis par M. Weyers a Aguilas, Province de Murcie. *Annales de la Société Entomologique de Belgique* **28**: ccxxxi–ccxxxii.

Simon, E. (1885a). Matériaux pour servir a la faune des Arachnides de la Grèce. *Annales de la Société Entomologique de France* (6) **4**: 303–356.

Simon, E. (1885b). Arachnides recueillies dans la vallée de Tempé et sur le mont Ossa (Thessalie) par M. de Dr J. Stussiner (de Laibach). *Annales de la Société Entomologique de France* (6) **5**: 209–217.

Simon, E. (1885c). Étude sur les Arachnides recueillis en Tunisie en 1883 et 1884 par Mm. A. Letourneux, M. Sédillot et Valery Mayet. In: *Exploration Scientifique de la Tunisie*: 1–55. Imprimerie Nationale: Paris.

Simon, E. (1887). Arachnides. In: *Mission Scientifique du Cap Horn. 1882–1883*, vol. 4 (Zoologie): 1–42. Gauthier-Villars: Paris.

Simon, E. (1890). Études sur les Arachnides de l'Yemen. *Annales de la Société Entomologique de France* (6) **10**: 77–124.

Simon, E. (1894). Note sur les Arthropodes cavernicoles du Transvaal. *Annales de la Société Entomologique de France* (6) **14**: 63–67.

Simon, E. (1895). Arachnides recueillis à la Terre-de-Feu par M. Carlos Backhausen. *Anales del Museo Nacional de Historia Natural de Buenos Aires* **4**: 167–172.

Simon, E. (1896). Note sur quelques Chernetes de Ligurie. *Annali del Museo Civico di Storia Naturale di Genova* (2) **16**: 372–376.

Simon, E. (1898a). Sur quelques Arachnides du Portugal appartenant au Musée de Zoologie de l'Académie Polytechnique de Porto. *Annaes de Sciencias Naturaes, Porto* **5**: 92–102.

Simon, E. (1898b). Aracnidos de Pozuelo de Calatrava. *Anales de la Sociedad Española de Historia Natural* (2) **7**: 89–99. (not seen)

Simon, E. (1898c). Étude sur les Arachnides de la région des Maures (Var). *Feuille des Jeunes Naturalistes* **29**: 2–4.

Simon, E. (1898d). Studio sui Chernetes Italiani conservati nel Museo Civico di Genova con descrizione di nuova specie. *Annali del Museo Civico di Storia Naturale di Genova* (2) **19**: 20–24.

Simon, E. (1899a). Contribution a la faune de Sumatra. Arachnides recueillis par M. J.-L. Weyers, a Sumatra. *Annales de la Société Entomologique de Belgique* **43**: 78–125.

Simon, E. (1899b). Arachnides recueillis par M. C.-J. Dewitz en 1898, à Bir-Hooker (Wadi Natron), en Égypte. *Bulletin de la Société Entomologique de France* **68**: 244–247.

Simon, E. (1899c). Ergebnisse einer Reise nach dem Pacific (Schauinsland 1896–1897). Arachnoideen. *Zoologische Jahrbücher, Systematik (Ökologie), Geographie und Biologie* **12**: 411–437.

Simon, E. (1899d). Liste des Arachnides recueillis en Algérie par M. P. Lesne et description d'une espèce nouvelle. *Bulletin du Muséum National d'Histoire Naturelle, Paris* **5**: 82–87.

Simon, E. (1900a). Arachnida. In: Sharp, D. (ed.), *Fauna Hawaiiensis*, vol. **2**: 443–519. Cambridge University Press: Cambridge.

Simon, E. (1900b). Studio sui Chernetes Italiani conservati nel Museo Civico di Genova. II. *Annali del Museo Civico di Storia Naturale di Genova* (2) **20**: 593–595.

Simon, E. (1900c). Chernetes recueillis en Erythrée par le Lieutenant F. Derchi en 1896. *Annali del Museo Civico di Storia Naturale di Genova* (2) **20**: 596.

Simon, E. (1901). On the Arachnida collected during the 'Skeat Expedition' to the Malay Peninsula, 1899–1900. *Proceedings of the Zoological Society of London* **71**: 45–84.

Simon, E. (1902). Arachnoideen, excl. Acariden und Gonyleptiden. *Ergebnisse der Hamburger Magalhaensischen Sammelreise* **2**: 1–47. Friedrichsen: Hamburg.

Simon, E. (1903a). Arachnides de la Guinée Espagnole. *Memorias de la Real Sociedad Española de Historia Natural* **1**: 65–124.

Simon, E. (1903b). Liste des Arachnides recueillis par M. Schmitt dans l'île d'Anticosti. *Bulletin du Muséum National d'Histoire Naturelle, Paris* **9**: 386–387.

Simon, E. (1904). Etude sur les Arachnides recueillis au cours de la mission Du Bourg de Bozas en Afrique. *Bulletin du Muséum National d'Histoire Naturelle, Paris* **10**: 442–448.

Simon, E. (1905). Description d'un *Blothrus* nouveau (Arachn.) des grottes des Basses-Alpes. *Bulletin de la Société Entomologique de France* **74**: 282–283.

Simon, E. (1907). Araneae, Chernetes et Opiliones (première série). *Archives de Zoologie Expérimentale et Générale* (4) **6**: 537–553.

Simon, E. (1908). Etude sur les Arachnides, recueillis par Mr. le Dr. Klaptocz en Tripolitaine. *Zoologische Jahrbücher, Systematik (Ökologie), Geographie und Biologie* **26**: 419–438.

Simon, E. (1909). Étude sur les Arachnides recueillis au Maroc par M. Martínez de la Escalera en 1907. *Memorias de la Real Sociedad Española de Historia Natural* **6**: 1–43.

Simon, E. (1912). Arachnides recueillis par M. L. Garreta à l'île Grande-Salvage. *Bulletin de la Société Entomologique de France* **81**: 59–61.

Simon, H. R. (1966a). Der Pseudoskorpionide *Neobisium muscorum* (Leach) als Collembolenfeind. *Institut für Naturschutz Darmstadt, Schriftenreihe* **8**: 77–85.

Simon, H.R. (1966b). Wie gelangen Pseudoskorpion in Vogelnester? *Die Vogelwelt* **87**: 80–83.

Simon, H. R. (1969). Art und ökologische Umwelt: der Moosskorpion – *Neobisium muscorum* Leach – und seine Stellung im Ökosystem. *Mitteilungen der Pollichia* (3) **16**: 149–159.

Singh, J. N. & Venkataraman, T. V. (1947). Pseudoscorpions in bee hives in India. *Current Science, Bangalore* **16**: 122. (not seen)

Singh, J. N. & Venkataraman, T. V. (1948). Pseudoscorpions in bee hives in India. *Indian Bee Journal* **10**: 6. (not seen)

Sivaraman, S. (1979). *Studies on Some Indian Pseudoscorpions*. Ph.D. Thesis: Madras University. (not seen)

Sivaraman, S. (1980a). Pseudoscorpions from South India - four new species of the family Chernetidae Menge and Cheliferidae Hagen (Pseudoscorpionida, Monosphyronida). *Journal of the Bombay Natural History Society* **77**: 106–116.

Sivaraman, S. (1980b). Pseudoscorpions from South India: a new genus and some new species of the super-family Garypoidea Chamberlin (Pseudoscorpionida: Diplosphyronida). *Oriental Insects* **14**: 325–343.

Sivaraman, S. (1980c). Pseudoscorpions from South India: some new species of the family Atemnidae Chamberlin (Pseudoscorpionida: Monosphyronida). *Oriental Insects* **14**: 345–362.

Sivaraman, S. (1980d). Two new species of pseudoscorpions from South India (Pseudoscorpionida, Heterosphyronida). *Entomon* **5**: 237–242.

Sivaraman, S. (1981a). Systematics of some South Indian sternophorid pseudoscorpions (Pseudoscorpionida, Monosphyronida). *Revue Suisse de Zoologie* **88**: 313–325.

Sivaraman, S. (1981b). Replacement name for *Paratemnus robustus* Sivaraman (Pseudoscorpionida: Atemnidae). *Oriental Insects* **15**: 96.

Sivaraman, S. (1981c). Changes in the functional response and prey-predator interaction of pseudoscorpions. *American Arachnology* **24**: 22–23.

Sivaraman, S. (1982). Chelal growth in three species of South Indian pseudoscorpions and their taxonomic relevance. *Entomon* **7**: 187–196.

Sivaraman, S. & Murthy, V. A. (1980a). Observations on the silk chamber construction and brooding behaviour of pseudoscorpions (Cl. Arachnida). *Journal of the Bombay Natural History Society* **77**: 163–167.

Sivaraman, S. & Murthy, V. A. (1980b). Feeding behaviour of some pseudoscorpions. *Bulletin of the Ethological Society of India* **2**: 19–30. (not seen)

Skaife, S. H. (1922a). Title? *South African Bee Journal* **1**: 148. (not seen)

Skaife, S. H. (1922b). Title? *Bee World* **4**: 37. (not seen)

Skaife, S.H. (1929). Some notes on the pseudo-scorpion, *Chelifer sculp.*, in relation to the honey bee. *South African Journal of Natural History* **6**: 293–296. (not seen)

Skeel, M. E. (1974). An ecological analysis of the Arachnida of Open Bay Islands, New Zealand. *Journal of the Royal Society of New Zealand* **4**: 39–46.

Skinner, A. (1977). Protection of Tasmanian cave fauna. *Helictite* **15(1)**: 38.

Skuratowicz, W. & Urbanski, J. (1953). Rezerwat na Bukowej Gorze kolo Zwierzynca w woj. Lubelskim i jego fauna. *Ochrona Przyrody* **21**: 193–216. (not seen)

Sladen, F. W. L. (1899). Bee parasites. *British Bee Journal and Bee-Keepers' Adviser* **27**: 126. (not seen)

Smith, M. (1961). The cheliceral flagellum of some British Chernetidae (Pseudo-scorpiones). *Journal of Natural History* **3**: 295–300.

Smith, W. W. (1924). *Chelifer* in ants' nests. *New Zealand Journal of Science and Technology* **6**: 341–342.

Snell, N. (1933). A study of humus fauna. *Journal of Entomology and Zoology* **25**: 33–40.

Snodgrass, R. E. (1948). The feeding organs of Arachnida, including mites and ticks. *Smithsonian Miscellaneous Collections* **110** (10): 1–93.

Soerensen, W. (1906). Un animal fabuleux, des temps modernes; analyse critique. *Oversigt over det Konigelige Danske Videnskabernes Selskabs Forhandlinger* **4**: 197–232. (not seen)

Sokolow, I. (1926). Untersuchungen über die Spermatogenese bei den Arachniden. II. Über die Spermatogenese der Pseudoskorpione. *Zeitschrift für Zellforschung und Mikroskopische Anatomie* **3**: 615–681. (not seen)

Southcott, R. V. (1978). *Australian Harmful Arachnids and their Allies.* R. V. Southcott: Adelaide.

Southwood, T. R. E. (1962). Migration of terrestrial arthropods in relation to habitat. *Biological Reviews* **37**: 171–214.

Spaull, V. W. (1979). Distribution of soil and litter arthropods on Aldabra Atoll. *Philosophical Transactions of the Royal Society of London* (B) **285**: 109–117.

Spicer, W. W. (1867). Helps to distribution. *Science Gossip* **3**: 244–245.

Stadler, H. & Schenkel, E. (1940). Die Spinnentiere (Arachniden) Mainfrankens. *Mitteilungen des Naturwissenschaftlichen Museums der Stadt Aschaffenburg* (NF) **2**: 1–58. (not seen)

Stainton, H. T. (1865a). Untitled. *Entomologist's Monthly Magazine* **2**: 119–120.

Stainton, H. T. (1865b). Note on *Chelifer*. *Transactions of the Royal Entomological Society of London* (3) **2**: cxii. (not seen)

Standen, R. (1912). The false-scorpions of Lancashire and some adjoining counties, with a preliminary list of records. *Lancashire and Cheshire Naturalist* **5**: 7–16.

Standen, R. (1914). *Chelifer (Chernes) panzeri* C.L. Koch in Cheshire. *Lancashire and Cheshire Naturalist* **7**: 461–462.

Standen, R. (1916a). Report on the false-scorpions and woodlice for 1915–16. *Lancashire and Cheshire Naturalist* **9**: 17–20.

Standen, R. (1916b). *Chelifer (Chernes) powelli* Kew and *Chelifer (Withius) subruber* Simon in Lancashire and Cheshire. *Lancashire and Cheshire Naturalist* **9**: 124–125.

Standen, R. (1917). Report on the false-scorpions (Chelifers). *Lancashire and Cheshire Naturalist* **10**: 27–31.

Standen, R. (1918). Report on the false-scorpions (Chelifers) for 1917. *Lancashire and Cheshire Naturalist* **10**: 335.

Standen, R. (1922). Occurrence in Lancashire of the false-scorpion *Chernes scorpioides*. *Lancashire and Cheshire Naturalist* **14**: 23–24.

Stark, N. (1969). Microecosystems in Lehman Cave, Nevada. *Bulletin of the National Speleological Society* **31**: 73–82. (not seen)

Stecker, A. (1874a). Zur Kenntnis der Chernetidenfauna Böhmens. *Sitzungsberichte der Königliche Böhmischen Gesellschaft der Wissenschaften* **8**: 1–16. (not seen)

Stecker, A. (1874b). Zur Kenntnis der Chernetidenfauna Böhmens. *Sitzungsberichte der Königliche Böhmischen Gesellschaft der Wissenschaften* **8**: 227–241.

Stecker, A. (1874c). Ueber zweifelhafte Chernetiden-Arten, welche von A. Menge beschrieben wurden. *Berliner Entomologische Zeitschrift* **19**: 305–314.

Stecker, A. (1875a). Ueber die geographische Verbreitung der europäischen Chernetiden (Pseudoscorpione). *Archiv für Naturgeschichte* **41**: 157–182.

Stecker, A. (1875b). Über neue indische Chernetiden. *Sitzungsberichte der Akademie der Wissenschaften in Wien, Math.-Nat. Cl.* **72**: 512–526.

Stecker, A. (1875c). Über eine neue Arachnidgattung aus der Abtheilung der Arthrogastren. *Sitzungsberichte der Königlichen Böhmischen Gesellschaft der Wissenschaften* **6**: 239–255. (not seen)

Stecker, A. (1875d). Die Chernetidenfauna Böhmens. *Zeitschrift für die Gesamten Naturwissenschaften* **11**: 87–89.

Stecker, A. (1876a). Die Entwickelung der *Chthonius*-Eier im Mutterleibe, und die Bildung des Blastoderms. *Sitzungsberichte der Königlichen Böhmischen Gesellschaft der Wissenschaften* **3**: 1–13. (not seen)

Stecker, A. (1876b). Anatomisches und Histologisches ueber *Gibocellum*, eine neue Arachnide. *Archiv für Naturgeschichte* **42**: 293–346.

Stecker, A. (1876c). On a new genus of Arachnida of the Section Arthrogastra. *Annals and Magazine of Natural History* (4) **17**: 230–243.

Stecker, A. (1876d). The development of the ova of *Chthonius* in the body of the mother, and the formation of the blastoderm. *Annals and Magazine of Natural History* (4) **18**: 197–207.

Stecker, A. (1878). Über die Rückbildung von Sehorganen bei den Arachniden. *Morphologisches Jahrbuch* **4**: 279–287. (not seen)

Steinböck, O. (1939). Die Nunatak-Fauna der Venter Berge. In: *Das Venter-Tal, Festgabe des Zweiges Mark Brandenburg des Deutschen Alpenvereins*: 64–73. München. (not seen)

Stevens, S. (1866). Note on *Chelifer*. *Proceedings of the Entomological Society of London* (3) **5**: xxvii. (not seen)

Strand, E. (1900a). Arachnologisches. *Nytt Magasin for Naturvidenskapene* **38**: 95–102.

Strand, E. (1900b). Zur Kenntniss der Arachniden Norwegens. *Kongelige Norske Videnskabernes Selskabs Skrifter* **2**: 1–46. (not seen)

Strand, E. (1906). Die arktischen Araneae, Opiliones und Chernetes. In: *Fauna Arctica, Jena*, vol. **4**: 431–478. (not seen)

Strand, E. (1907). Verzeichnis der bis jetzt bei Marburg von Prof. Dr. H. Zimmermann aufgefundenen Spinnenarten. *Zoologischer Anzeiger* **32**: 216–243.

Strand, E. (1909). Spinnentiere von Süd-Afrika und einigen Inseln gesammelt bei der Deutschen Südpolar-Expedition 1901–1903. *Deutsche Südpolar-Expedition*, vol. **10** (Zoologie II): 541–596.

Strand, E. (1932). Miscellanea nomenclatorica zoologica et palaeontologica. IV. *Folia Zoologica et Hydrobiologica* **4**: 193–196.

Strebel, O. (1937). Beobachtungen am einheimischen Bücherskorpion *Chelifer cancroides* L. (Pseudoscorpiones). *Beiträge zur Naturkundlichen Forschung in Südwestdeutschland* **2**: 143–155.

Strebel, O. (1961). Pseudoscorpiones aus dem Siebengebirge. *Decheniana, Beihefte* **9**: 107–108.

Striganova, B. R., Pokarzhevskij, A. D. & Druk, A. Y. (1978). [The formation of soil invertebrate complexes in pistacchios in Central Asia.] *Doklady Moskovskogo Obshchestra Ispytatelei Prirody Obshchaya Biologiya* **1975**: 78–81. (not seen)

Strinati, P. (1953). Campagne d'explorations spéléologiques au Maroc (été 1950). *Annales de Spéléologie* **7**: 99–107. (not seen)

Strinati, P. (1955). La faune de la Grotte de Pertius (Jura neuchatelois). *Bulletin de la Société Neuchâteloise des Sciences Naturelles* (3) **78**: 5–16. (not seen)

Strinati, P. (1957). La faune de la grotte de Lajoux (Jura Bernois). *Rassegna Speleologica Italiana* **9**: 61–64.

Strinati, P. (1960). La faune actuelle de trois grottes d'Afrique Equatoriale Française. *Annales de Spéléologie* **15**: 533–538. (not seen)

Strinati, P. (1966). *Faune cavernicole de la Suisse.* Editions C.N.R.S.: Paris. (not seen)

Strinati, P. & Aellen, V. (1959). Faune cavernicole de la région de Taza (Maroc). *Revue Suisse de Zoologie* **66**: 765–777.

Strinati, P. & Aellen, V. (1983). *Voyage biospéologique autour du monde.* Memoires du Spéléo-Club: Paris. (not seen)

Strouhal, H. (1961). Die rezente Höhlenfauna Österreichs. *Österreichische Hochschulzeitung* **13**: 8–9. (not seen)

Stryker, S. L. & Taylor, R. W. (1977). The pseudoscorpions of West Virginia, U.S.A. (Arachnida, Pseudoscorpionida). *Proceedings of the West Virginia Academy of Sciences* **49**: 26–27. (not seen)

Stschelkanovzeff, J. P. (1897). Zur Entwicklungsgeschichte der Pseudoscorpioniden. *Mémoires de la Société Impériale des Naturalistes, Moscou* **86**: ?–? (not seen)

Stschelkanovzeff, J. P. (1898). Bau der weiblichen Genitalorgane der Pseudoscorpione. *Nachrichten aus der Gesellschaft Fr. Naturw. Moskau* **86**: ?–? (not seen)

Stschelkanovzeff, J. P. (1902a). Über den Bau der Respirationsorgane bei den Pseudoscorpionen. *Zoologischer Anzeiger* **25**: 126–135.

Stschelkanovzeff, J. P. (1902b). *Chernes multidentatus* n. sp. nebst einem Beitrage zur Systematik der *Chernes*-Arten. *Zoologischer Anzeiger* **25**: 350–355.

Stschelkanovtzeff, J. P. (1903a). Beiträge zur Kenntnis der Segmentierung und des Körperbaues der Pseudoscorpione. *Zoologischer Anzeiger* **26**: 318–334.

Stschelkanovzeff, J. P. (1903b). Matérial pour l'anatomie des pseudoscorpions. *Gelehrte Schriften Univ., Naturwissenschaftliche Abteilung* **26**: 1–202. (not seen)

Stschelkanovzeff, J. P. (1910). Der Bau der männlichen Geschlechtsorgane von *Chelifer* und *Chernes.* Zur Kenntnis der Stellung der Chelonethi im System. *Festschrift 60. Geburtstag R. Hertwig* **2**: 1–38. (not seen)

Sturany, R. (1891). Die Coxaldrüsen der Arachnoideen. *Arbeiten aus den Zoologischen Instituten der Univ. Wien und der Zoologischen Station in Triest* **9**: 129–150. (not seen)

Subbiah, M. S., Mahadevan, V. & Janakiraman, R. (1957). A note on the occurrence of an arachnid – *Ellingsenius indicus* Chamberlin – infesting bee hives in South India. *Indian Journal of Veterinary Science and Animal Husbandary* **27**: 155–156. (not seen)

Sundevall, C. J. (1833). *Conspectus Arachnidum.* C. F. Berling: Londini Gothorum.

Sundevall, J. C. (1863). *Die Thierarten des Aristoteles von den Klassen der Säugethiere, Vögel, Reptilien und Insecten.* Stockholm. (not seen)

Supino, F. (1899). Osservazione sopra l'anatomia degli Pseudoscorpioni. *Atti della Reale Accademia Nazionale dei Lincei* (5) **8**: 604–608. (not seen)

Sweeney, M. P., Taylor, R. W. & Counts, C. L., III (1977). Non-cavernicolous pseudoscorpions from West Virginia. *Entomological News* **88**: 55–56.

Sweeney, S. J., Stryker, S. L., Sweeney, M. P., Taylor, R. W. & Counts, C. L. (1977). New record for a non-cavernicolous pseudoscorpion in West Virginia. *Entomological News* **88**: 98.

Syms, E. E. (1950). Notes on pseudoscorpions. *Proceedings of the South London Entomological and Natural History Society* **1948–1949**: 142–145.

Szalay, L. (1931a). Beiträge zur Kenntnis der Arachnoideen-fauna der Aggteleker Höhle. *Annales Historico-Naturales Musei Nationalis Hungarici* **27**: 351–370.

Szalay, L. (1931b). Beiträge zur Kenntnis der Afterskorpion-und Milbenfauna des Retyezát-Gebirges. *Annales Historico-Naturales Musei Nationalis Hungarici* **27**: 371.

Szalay, L. (1932). Beiträge zur Kenntnis der Arachnoidenfauna der Aggteleker Höhle. *Allattani Közlemények* **29**: 15–31. (not seen)

Szalay, L. (1968). Pókszabásúak I. Arachnoidea I. In: *Magyarország Allat-világa (Faunae Hungariae)*, vol. *18*(1): 1–122. Akadémiai Kiadó: Budapest. (not seen)

Szent-Ivány, J. (1941). Neue Angaben zur Verbreitung der Pseudoscorpione im Karpaten-becken. *Fragmenta Faunistica Hungarica* **4**: 85–90. (not seen)

Sztarek, K. (1937). [Über den Afterskorpion.] *Méhészet, Budapest* **34**: 23. (not seen)

Takashima, H. (1947). [Studies on Japanese pseudoscorpions. I.] *Acta Arachnologica* **10**: 9–31. (in Japanese) (not seen)

Takashima, H. (1955). [Notes on *Garypus japonicus* (Garypidae, Pseudoscorpiones) and *Typopeltis stimpsonii* (Thelyphonidae, Pedipalpi.] *Bulletin of the Biogeographical Society of Japan* **16–19**: 175–181. (in Japanese)

Taylor, R. W., Sweeney, M. P. & Counts, C. L. (1977). Use of empty gastropod shells (Polygyridae) by pseudoscorpions. *Nautilus* **91**: 115.

Templeton, R. (1836). Catalogue of Irish Crustàcea, Myriápoda, and Arachnöìda, selected from the papers of the late John Templeton, Esq. *Magazine of Natural History* **9**: 9–14.

Tenorio, J. M. & Muchmore, W. B. (1982). Catalog of entomological types in the Bishop Museum. Pseudoscorpionida. *Pacific Insects* **24**: 377–385.

Thaler, K. (1966). Fragmenta faunistica Tirolensia (Diplopoda, Arachnida). *Berichte des Naturwissenschaftlich-Medizinischen Vereins in Innsbruck* **54**: 151–157. (not seen)

Thaler, K. (1979). Fragmenta faunistica Tirolensia, IV (Arachnida: Acari: Caeculidae; Pseudoscorpiones; Scorpiones; Opiliones; Aranei; Insecta: Dermaptera; Thysanoptera; Diptera Nematocera: Mycetophilidae, Psychodidae, Limoniidae und Tipulidae). *Veröffentlichungen des Museums Ferdinandeum 59*: 49–83.

Thaler, K. (1980). Die Spinnenfauna der Alpen: ein zoogeographischer Versuch. *Verhandlungen des 8. Internationalen Arachnologen-Kongresses, Wien*: 389–404.

Thaler, K. (1981). Neue Arachniden-Funde in der nivalen Stufe der Zentralalpen Nordtirols (Österreich) (Aranei, Opiliones, Pseudoscorpiones). *Berichte des Naturwissenschaftlich-Medizinischen Vereins in Innsbruck* **68**: 99–105.

Théis, C. de (1832). Lettre adressée à M. Audouin, sur quelques Arachnides des genres *Hydrachna* et *Chelifer*. *Annales des Sciences Naturelles* **27**: 57–78.

Théodoridès, J. (1950). Observations écologiques dans l'état de New Jersey (États-Unis). *Revue Canadienne de Biologie* **9**: 9–27.

Théodoridès, J. (1955). Contribution à l'étude des parasites et phorétiques de coléoptères terrestres. *Vie et Milieu, Supplément* **4**: 1–310.

Thorell, T. (1876). On the classification of scorpions. *Annals and Magazine of Natural History* (4) **17**: 1–15.

Thorell, T. (1877a). Sobre algunos Aracnidos de la República Argentina. 1. Scorpiones, Opiliones y Pseudoscorpiones. *Periódico Zoológico* **2**: 201–218.

Thorell, T. (1877b). Sobre algunos Aracnidos de la República Argentina. 1. Scorpiones, Opiliones y Pseudoscorpiones. *Boletín de la Academia Nacional de Ciencias de Córdoba* **2**(3): 255–272. (not seen)

Thorell, T. (1883). Descrizione di alcuni Aracnidi inferiori dell'Arcipelago Malese. *Annali del Museo Civico di Storia Naturale di Genova* **18**: 21–69.

Thorell, T. (1889). Aracnidi Artrogastri Birmani raccolti da L. Fea nel 1885–1887. *Annali del Museo Civico di Storia Naturale di Genova* (2) *7*: 521–729.

Thorell, T. (1890). Aracnidi di Pinang raccolti nel 1889 dia Sig.[ri] L. Loria e L. Fea. *Annali del Museo Civico di Storia Naturale di Genova* (2) **10**: 269–383.

Thornton, I. W. B. & New, T. R. (1988). Krakatau invertebrates: the 1980s fauna in the context of a century of recolonization. *Philosophical Transactions of the Royal Society of London* B **322**: 493–522.

Tichomirowa, O. (1895). Zur Biologie der Pseudoscorpione der Umgegend von Moskau. *Tagebl. Ges. Naturw. Moskau* **2**: 12–14. (not seen)

Tirini Pavan, M. (1958). Contributo alla conoscenza speleologica della regione fra il Lago d'Iseo e la Valle Trompia in provincia di Brescia. *Rassegna Speleologica Italiana* **10**: 3–54. (not seen)

Tömösváry, O. (1882a). Egy új alak hazánk Arachnoida faunájában Zemplén megyéböl. *Természetrajzi Füzetek* **6**: 226–228, 296–298.

Tömösváry, O. (1882b). A Magyar fauna álskorpiói. *Magyar Tudományos Akadémia Matematikai és Természettudományi Közlemények* **18**: 135–256.

Tömösváry, O. (1884). Adatok az álskorpiók ismeretéhez. (Data ad cognitionem Pseudoscorpionum). *Természetrajzi Füzetek* **8**: 16–27.

Torii, H. (1953). Fauna der Ryugado Sinterhöhle in Kochi Präfektur. (Die Berichte der Speläobiologischen Expeditionen VI). *Annotationes Zoologicae Japonenses* **26**: 246–255.

Traegardh, L. (1928). Studies in the fauna of the soil in Swedish forests. *Transactions of the 4th International Congress of Entomology, Ithaca*: 781–792. (not seen)

Trappen, A. von der (1906). Sonderbare Jäger. *Societas Entomologica* **21**: 52. (not seen)

Treat, A. E. (1956). A pseudoscorpion on moths. *Lepidopterists' News* **10**: 87–89.

Treviranus, G. R. (1816). *Chelifer*, der Bastard Scorpion. *Vermischte Schriften* **1**: 15. (not seen)

Tubb, J. A. (1937). Reports of the expedition of the McCoy Society for field investigation and research (Lady Julia Percy Island). 19. Arachnida. *Proceedings of the Royal Society of Victoria (new series)* **49**: 412–421.

Tulk, A. (1844). Note upon *Obisium orthodactylum* (Leach). *Annals and Magazine of Natural History* (1) **13**: 55–57.

Tullgren, A. (1899a). Bidrag till Kännedomen om Sveriges Pseudoscorpioner. *Entomologisk Tidskrift* **20**: 161–182.

Tullgren, A. (1899b). Tillägg till 'Bidrag till Kännedomen om Sveriges Pseudoscorpioner'. *Entomologisk Tidskrift* **20**: 297–298.

Tullgren, A. (1900a). Two new species of Chelonethi (Pseudoscorpions) from America. *Entomologisk Tidskrift* **21**: 153–157.

Tullgren, A. (1900b). Chelonethi (Pseudoscorpiones) from the Canary and the Balearic Islands. *Entomologisk Tidskrift* **21**: 157–160.

Tullgren, A. (1901). Chelonethi from Camerun in Westafrika collected by Dr. Yngve Sjöstedt. *Entomologisk Tidskrift* **22**: 97–101.

Tullgren, A. (1905). Einige Chelonethiden aus Java. *Mitteilungen aus dem Naturhistorischen Museum in Hamburg* **22**: 37–47.

Tullgren, A. (1906a). Svensk spindelfauna. Första Ordningen. Klokrypare Chelonethi. *Entomologisk Tidskrift* **27**: 195–205.

Tullgren, A. (1906b). Notiser rörande arter af Arachnidgrupperna Chelonethi och Phalangidea. *Entomologisk Tidskrift* **27**: 214–218.

Tullgren, A. (1907a). Zur Kenntnis aussereuropäischer Chelonethiden des Naturhistorischen Museums in Hamburg. *Mitteilungen aus dem Naturhistorischen Museum in Hamburg* **24**: 21–75.

Tullgren, A. (1907b). Chelonethiden aus Natal und Zululand. In: *Zoologiska Studier Tillägnade Professor T. Tullberg*: 216–236. Almquist and Wiksells: Uppsala.

Tullgren, A. (1907c). Solifugae, Scorpiones und Chelonethi aus ägypten und dem Sudan. In: *Results of the Swedish Zoological Expedition to Egypt and the White Nile 1901 Under the Direction of L.A. Jägerskiöld* (A), vol. **21**: 1–12.

Tullgren, A. (1907d). Arachnoidea. 1. Pedipalpi, Scorpiones, Solifugae, Chelonethi. *Wissenschaftliche Ergebnisse der Schwedischen Zoologischen Expedition nach dem Kilimandjaro, dem Meru und dem umgebenden Massaisteppen Deutsch-Ostafrikas 1905–1906*, vol. **3**(20): 1–15.

Tullgren, A. (1907e). Über einige Chelonethiden des Naturhistorischen Museums zu Wiesbaden. *Jahrbuch des Nassauischen Vereins für Naturkunde* **60**: 246–248.

Tullgren, A. (1908a). Eine neue *Olpium*-Art aus Java. *Notes from the Leyden Museum* **29**: 148–150.

Tullgren, A. (1908b). Pseudoscorpionina (Chelonethi). In: Schulze, L., Zoologische und Anthropologische ergebnisse einer Forschungsreise im westlichen und zentralen Südafrika ausgeführt in den Jahren 1903–1905. *Denskschriften der Medizinisch-Naturwissenschaftlichen Gesellschaft zu Jena* **13**: 283–288.

Tullgren, A. (1908c). Über einige exotische Chelonethiden. *Entomologisk Tidskrift* **29**: 57–64.

Tullgren, A. (1908d). Über *Chelifer patagonicus* Tullgr. *Entomologisk Tidskrift* **29**: 116.

Tullgren, A. (1909a). Eine neue *Chelifer*-Art aus Schweden. *Entomologisk Tidskrift* **30**: 92–94.

Tullgren, A. (1909b). Chelonethi. In: Michaelsen, W. & Hartmeyer, R. (eds), *Fauna Südwest-Australiens*, vol. **2**: 411–415. Gustav Fischer: Jena.

Tullgren, A. (1911). En för Sverige ny klokrypare (Pseudoscorpion). *Entomologisk Tidskrift* **32**: 125.

Tullgren, A. (1912a). Einige Chelonethiden aus Java und Krakatau. *Notes from the Leyden Museum* **34**: 259–267.

Tullgren, A. (1912b). Vier Chelonethiden-Arten auf einem Javanischen Käfer gefunden. *Notes from the Leyden Museum* **34**: 268–270.

Tumšs, V. (1934). Beitrag zur Kenntnis der Pseudoscorpionen-Fauna Lettlands. *Folia Zoologica et Hydrobiologica* **7**: 12–19.

Turk, F. A. (1946). On two new false scorpions of the genera *Tridenchthonius* and *Microcreagris*. *Annals and Magazine of Natural History* (11) **13**: 64–70.

Turk, F. A. (1949). *Dinocheirus stercoreus*, a new pseudoscorpion from the Bracken Cave, Texas, U.S.A. *Annals and Magazine of Natural History* (12) **2**: 120–126.

Turk, F. A. (1951). On the swarming of pseudoscorpions and their association with ants. *Entomologist's Monthly Magazine* **87**: 169.

Turk, F. A. (1953). A new genus and species of pseudoscorpion with some notes on its biology. *Proceedings of the Zoological Society of London* **122**: 951–954.

Turk, F. A. (1967). The non-aranean arachnid orders and the myriapods of British caves and mines. *Transactions of the Cave Research Group of Great Britain* **9**: 142–161.

Tuzet, O., Manier, J.-F. & Boissin, L. (1966). Étude du spermatozoïde mûr enkysté d'*Hysterochelifer meridianus* (L. Koch) (Arachnide, Pseudoscorpion, Cheliferidae). *Comptes Rendus Hebdomadaires des Séances de l'Académie des Sciences, Paris* (D) **262**: 376–378.

Vachon, M. (1932). Recherche sur la biologie des Pseudoscorpionides. 1re note. La nutrition chez *Chelifer cancroides* L. *Bulletin Scientifique de Bourgogne* **2**: 21–26.

Vachon, M. (1933a). Acte de nutrition d'un pseudoscorpionide: *Chelifer cancroides* L. *Comptes Rendus Hebdomadaires des Séances de l'Académie des Sciences, Paris* **198**: 1874–1875.

Vachon, M. (1933b). Recherches sur la biologie des Pseudoscorpionides. Deuxième note. *Bulletin Scientifique de Bourgogne* **3**: 59–64.

Vachon, M. (1934a). Sur la ponte et le sac ovigère d'un Pseudoscorpionide (*Chelifer cancroides* L.). *Revue Français d'Entomologie* **1**: 174–178.

Vachon, M. (1934b). Sur le développement post-embryonnaire des Pseudoscorpionides. Première note. Les formes immatures de *Chelifer cancroides* L. *Bulletin de la Société Zoologique de France* **59**: 154–160.

Vachon, M. (1934c). Sur la nutrition des Pseudoscorpionides (*Chelifer cancroides, Allochernes italicus*). *Bulletin Scientifique de Bourgogne* **4**: 38–55.

Vachon, M. (1934d). Pseudoscorpionides de la Côte-d'Or: 1re liste. *Bulletin Scientifique de Bourgogne* **4**: 133.

Vachon, M. (1934e). Sur la biologie d'un Pseudoscorpionide commun: *Chelifer cancroides* (L.). *Bulletin Scientifique de Bourgogne* **4**: 154–155.

Vachon, M. (1934f). Le phénomène de phorésie chez les Pseudoscorpionides. *Bulletin Scientifique de Bourgogne* **4**: 158–159.

Vachon, M. (1934g). Pseudoscorpionides de la Côte-d'Or (2e liste). *Bulletin Scientifique de Bourgogne* **4**: 162–163.

Vachon, M. (1934h). Une espèce géante de Pseudoscorpionides: *Titanatemnus monardi* n. sp. *Bulletin Scientifique de Bourgogne* **4**: 167–168.

Vachon, M. (1935a). Une particularité dans le développement d'un pseudoscorpion: *Cheiridium museorum* Leach. *Bulletin de la Société Zoologique de France* **60**: 330–333.

Vachon, M. (1935b). Deux nouvelles espèces de Pseudoscorpions Africains *Titanatemnus monardi* et *Titanatemnus alluaudi*. *Bulletin de la Société Zoologique de France* **60**: 510–521.

Vachon, M. (1935c). Sur le développement postembryonnaire des Pseudoscorpions. 3e note: la mue. *Bulletin Scientifique de Bourgogne* **5**: 21–29.

Vachon, M. (1935d). Une sous-espèce de Pseudoscorpions nouvelle pour la France. *Bulletin Scientifique de Bourgogne* **5**: 81.

Vachon, M. (1935e). Développement des Pseudoscorpionides: origine du lait maternel. *Bulletin Scientifique de Bourgogne* **5**: 149–150.

Vachon, M. (1935f). Sur la sac dit 'ovigère' des Pseudoscorpions. *Bulletin Scientifique de Bourgogne* **5**: 154–155.

Vachon, M. (1935g). Sur les stades des Pseudoscorpions. *Bulletin Scientifique de Bourgogne* **5**: 160–161.

Vachon, M. (1935h). Le rôle nourricier de l'ovaire chez les Pseudoscorpions. *Bulletin Scientifique de Bourgogne* **5**: 169.

Vachon, M. (1935i). Développement des Pseudoscorpions. *Bulletin Scientifique de Bourgogne* **5**: 180–181.

Vachon, M. (1935j). Sur les Pseudoscorpions qui habitent les maisons et leurs dépendances. *Bulletin Scientifique de Bourgogne* **5**: 187–188.

Vachon, M. (1936a). Sur le développement postembryonnaire des Pseudoscorpions (quatrième note). Les formules chaetotaxiques des pattes-mâchoires. *Bulletin du Muséum National d'Histoire Naturelle, Paris* (2) **8**: 77–83.

Vachon, M. (1936b). Description d'une nouvelle espèce de Pseudoscorpions *Epaphochernes bouvieri* suivie de quelques remarques sur les genres *Dendrochernes* Beier et *Epaphochernes* Beier. *Bulletin de la Société Zoologique de France* **61**: 140–145.

Vachon, M. (1936c). Sur l'anatomie des Pseudoscorpions. Première note préliminaire. Les organes génitaux externes de *Chelifer cancroides* L. ♀. *Bulletin de la Société Zoologique de France* **61**: 294–298.

Vachon, M. (1936d). Les stades du développement chez les Pseudoscorpions. *Livre Jubilaire, E. L. Bouvier*: 89–92.

Vachon, M. (1937a). Pseudoscorpions nouveaux des collections du Muséum National d'Histoire Naturelle de Paris (Première note). *Bulletin du Muséum National d'Histoire Naturelle, Paris* (2) **9**: 129–133.

Vachon, M. (1937b). Trois nouveaux Pseudoscorpions de la région Pyrénéenne Française. *Bulletin de la Société Zoologique de France* **62**: 39–44.

Vachon, M. (1937c). Pseudoscorpions nouveaux des collections du Muséum National d'Histoire Naturelle de Paris. Deuxième note. *Bulletin de la Société Zoologique de France* **62**: 307–310.

Vachon, M. (1937d). Pseudoscorpions nouveaux des collections du Muséum National d'Histoire Naturelle de Paris. (3ᵉ note). *Bulletin de la Société Entomologique de France* **42**: 188−190.

Vachon, M. (1937e). Deux espèces nouvelles de Pseudoscorpions Algériens. *Bulletin Scientifique de Bourgogne* **6**: 107−112.

Vachon, M. (1937f). Le contenu du sac ovigère chez les Pseudoscorpions. *Bulletin Scientifique de Bourgogne* **6**: 128−129.

Vachon, M. (1938a). Recherches anatomiques et biologiques sur la réproduction et le développement des Pseudoscorpions. *Annales des Sciences Naturelles, Zoologie* (11) **1**: 1−207.

Vachon, M. (1938b). Pseudoparasitisme de *Chelifer cancroides* L. *Bulletin Scientifique de Bourgogne* **7**: 200.

Vachon, M. (1938c). Voyage en A.O.F. de L. Berland et J. Millot. IV. Pseudoscorpions. Première note. Atemnidae. *Bulletin de la Société Zoologique de France* **63**: 304−315.

Vachon, M. (1938d). Remarques sur la famille des Cheiridiidae, Chamberlin, à propos d'un nouveau genre et d'une nouvelle espèce: *Paracheiridium decaryi* (Arachnida Pseudoscorpionidae). *Bulletin de la Société Entomologique de France* **43**: 235−241.

Vachon, M. (1938e). Récoltes de R. Paulian et A. Villiers dans le Haut Atlas Marocain, 1938 (deuxième note). *Bulletin de la Société des Sciences Naturelles du Maroc* **18**: 205−212.

Vachon, M. (1939a). Remarques sur la sous-famille des Goniochernetinae Beier a propos de la description d'un nouveau genre et d'une nouvelle espèce de Pseudo-scorpions (Arachnides): *Metagoniochernes picardi*. *Bulletin du Muséum National d'Histoire Naturelle, Paris* (2) **11**: 123−128.

Vachon, M. (1939b). Remarques sur le genre *Dactylochelifer* Beier, à propos d'une espèce nouvelle de Pseudoscorpions: *Dactylochelifer legrandi*. *Bulletin Scientifique de Bourgogne* **8**: 155−159.

Vachon, M. (1940a). Sur la présence du Mozambique de *Cheiridium museorum* Leach (Pseudoscorpions) dans les galeries de Colèoptères Bostrichides. *Bulletin du Muséum National d'Histoire Naturelle, Paris* (2) **12**: 53−54.

Vachon, M. (1940b). Sur une espèce mal connue de Pseudoscorpions des Açores: *Microcreagris caeca* E. Simon. *Bulletin du Muséum National d'Histoire Naturelle, Paris* (2) *12:* 108−110.

Vachon, M. (1940c). Remarques sur quelques Pseudoscorpions du Sahara Central a propos des récoltes du Professeur L.G. Seurat, au Hoggar (Mars−Avril 1928). *Bulletin du Muséum National d'Histoire Naturelle, Paris* (2) **12**: 157−160.

Vachon, M. (1940d). Remarques sur *Macrochelifer*, nouveau genre de Pseudoscorpions. *Bulletin du Muséum National d'Histoire Naturelle, Paris* (2) **12**: 412−414.

Vachon, M. (1940e). Voyage en A.O.F. de L. Berland et J. Millot. IV. Pseudoscorpions. Deuxième note. *Bulletin de la Société Zoologique de France* **65**: 62−72.

Vachon, M. (1940f). Remarques sur la phóresie des Pseudoscorpions. *Annales de la Société Entomologique de France* **109**: 1−18.

Vachon, M. (1940g). Éléments de la faune portugaise des Pseudoscorpions (Arachnides) avec description de quatre espèces nouvelles. *Anais da Faculdade de Ciencias do Porto Academia Polytechnica do Porto* **25**: 141−164.

Vachon, M. (1940h). Sur la présence au Congo Belge d'un pseudoscorpion appartement au genre *Horus* J.C. Chamberlin. *Revue de Zoologie et de Botanique Africaines* **33**: 225−226.

Vachon, M. (1941a). Remarques sur le genre sud-Africain *Beierus* (sic) Chamberlin (Pseudoscorpions). *Bulletin du Muséum National d'Histoire Naturelle, Paris* (2) *13:* 80−81.

Vachon, M. (1941b). *Chthonius tetrachelatus* P. (Pseudoscorpions) et ses formes immatures (1ʳᵉ note). *Bulletin du Muséum National d'Histoire Naturelle, Paris* (2) **13**: 442−449.

Vachon, M. (1941c). *Chthonius tetrachelatus* P. (Pseudoscorpions) et ses formes immatures (2ᵉ note). *Bulletin du Muséum National d'Histoire Naturelle, Paris* (2) *13*: 540–547.

Vachon, M. (1941d). Remarques biogeographiques sur quelques Scorpions et Pseudoscorpions pré-désertiques. *Compte Rendu de Séances, Société de Biogeographie, Paris* **18**: 50–53.

Vachon, M. (1941e). Pseudoscorpions récoltés en Afrique occidentale tropicale par P. Lepesme, R. Paulian et A. Villiers (Note préliminaire). *Bulletin Scientifique de Bourgogne* **9**: 29–35.

Vachon, M. (1942). A propos du *Cordylochernes octentoctus* Balzan (Pseudoscorpions). *Bulletin du Muséum National d'Histoire Naturelle, Paris* (2) **14**: 181–184.

Vachon, M. (1943). L'allongement des doigts des pinces au cours du développement post-embryonnaire chez *Chelifer cancroides* L. (Pseudoscorpions). *Bulletin du Muséum National d'Histoire Naturelle, Paris* (2) *15*: 299–302.

Vachon, M. (1944). Pseudoscorpions nouveaux des collections du Muséum National d'Histoire Naturelle de Paris. 4ᵉ note. *Bulletin du Muséum National d'Histoire Naturelle, Paris* (2) **16**: 439–441.

Vachon, M. (1945a). Remarques sur un pseudoscorpion des cavernes de France: *Pseudoblothrus peyerimhoffi* (E. S.) = *Blothrus peyerimhoffi* E. S. 1905. *Bulletin du Muséum National d'Histoire Naturelle, Paris* (2) **17**: 230–233.

Vachon, M. (1945b). Mission scientifique l'Omo, vol. 6, part 10. Chernètes. *Mémoires de Muséum National d'Histoire Naturelle, Paris, n.s.* **19**: 187–197.

Vachon, M. (1946). Description d'une nouvelle espèce de Pseudoscorpion (Arachnide) habitant les grottes portugaises: *Microcreagris cavernicola. Bulletin du Muséum National d'Histoire Naturelle, Paris* (2) **18**: 333–336.

Vachon, M. (1947a). Nouvelles remarques a propos de la phorésie des Pseudoscorpions. *Bulletin du Muséum National d'Histoire Naturelle, Paris* (2) **19**: 84–87.

Vachon, M. (1947b). Comment reconnaitre l'age chez les Pseudoscorpions (Arachnides). *Bulletin du Muséum National d'Histoire Naturelle, Paris* (2) **19**: 271–274.

Vachon, M. (1947c). A propos de quelques Pseudoscorpions (Arachnides) des cavernes de France, avec description d'une espèce nouvelle: *Neobisium (Blothrus) tuzeti. Bulletin du Muséum National d'Histoire Naturelle, Paris* (2) **19**: 318–321.

Vachon, M. (1947d). Remarques sur l'arthrogenèse des appendices. A propos d'un cas de symmélie partielle chez un pseudoscorpion *Chelifer cancroides* L. (Arachnide). *Bulletin Biologie de la France et de la Belgique* **81**: 177–194.

Vachon, M. (1948). Quelques remarques sur le 'nettoyage des pattes machoires' et les glandes salivaires, chez les Pseudoscorpions (Arachnides). *Bulletin du Muséum National d'Histoire Naturelle, Paris* (2) **20**: 162–164.

Vachon, M. (1949). Ordre des Pseudoscorpions. In: Grassé, P.-P. (ed.), *Traité de Zoologie*, vol. **6**: 437–481. Masson: Paris.

Vachon, M. (1950). Contribution a l'étude de l'Air (Mission L. Chopard et A. Villiers). Scorpions, Pseudoscorpions et Solifuges. *Mémoires de l'Institut Français d'Afrique Noire* **10**: 93–107.

Vachon, M. (1951a). A propos d'une 'association' phorétique: Coléoptère – Acariens – Pseudoscorpions. *Bulletin du Muséum National d'Histoire Naturelle, Paris* (2) **22**: 728–733.

Vachon, M. (1951b). Sur les nids et spécialement les nids de ponte chez les Pseudoscorpions (Arachnides). *Bulletin du Muséum National d'Histoire Naturelle, Paris* (2) **23**: 196–199.

Vachon, M. (1951c). Les Pseudoscorpions de Madagascar. I. Remarques sur la famille des Chernetidae J. C. Chamberlin, 1931, a propos de la description d'une nouvelle espece: *Metagoniochernes milloti. Mémoires de l'Institut Scientifique de Madagascar* **5**: 159–172.

Vachon, M. (1951d). Les glandes des chélicères de Pseudoscorpions (Arachnides). *Comptes Rendus Hebdomadaires des Séances de l'Académie des Sciences, Paris* **233**: 205–206.

Vachon, M. (1952a). La reserve naturelle intégrale du Mt. Nimba. II. Pseudoscorpions. *Mémoires de l'Institut Français d'Afrique Noire* **19**: 17–43.

Vachon, M. (1952b). A propos d'un pseudoscorpion cavernicole découvert par M. le Dr H. Henrot, dans une grotte de la Virginie occidentale, en Amérique du Nord. *Notes Biospéologiques* **7**: 105–112.

Vachon, M. (1952c). Remarques préliminaires sur l'anatomie et la biologie de deux pseudoscorpions très rare de la faune française: *Pseudoblothrus peyerimhoffi* (E. S.) et *Apocheiridium ferum* (E. S.). *Bulletin du Muséum National d'Histoire Naturelle, Paris* (2) **24**: 536–539.

Vachon, M. (1953). Nouveaux cas de phorésie chez les Pseudoscorpions. *Bulletin du Muséum National d'Histoire Naturelle, Paris* (2) **25**: 572–575.

Vachon, M. (1954a). Contribution à l'étude du peuplement de la Mauritanie. Pseudoscorpions. *Bulletin de l'Institut Français d'Afrique Noire* **16**: 1022–1030.

Vachon, M. (1954b). Remarques morphologiques et anatomiques sur les Pseudoscorpions (Arachnides) appartenant au genre *Pseudoblothrus* (Beier) (Fam. Syarinidae J.C.C.) (à propos de la description de *P. strinatii* n. sp., des cavernes de Suisse). *Bulletin du Muséum National d'Histoire Naturelle, Paris* (2) **26**: 212–219.

Vachon, M. (1954c). Remarques sur un pseudoscorpion vivant dans les ruches d'Abeilles au Congo Belge, *Ellingsenius hendrickxi* n. sp. *Annales du Musée du Congo Belge, Sciences Zoologiques* **1**: 284–287.

Vachon, M. (1954d). Nouvelles captures de Pseudoscorpions (Arachnides) transportés par des insectes. *Bulletin du Muséum National d'Histoire Naturelle, Paris* (2) **26**: 590–592.

Vachon, M. (1956). Quelques remarques préliminaires sur les Pseudoscorpions des îles du Cap-Vert. *Commentationes Biologicae* **15**(20): 1–9.

Vachon, M. (1957). Remarques sur les Chernetidae (Pseudoscorpions) de la fauna Britannique. *Annals and Magazine of Natural History* (12) **10**: 389–394.

Vachon, M. (1958). Sur deux Pseudoscorpions nouveaux des cavernes de l'Afrique équatoriale [Ideoroncidae]. *Notes Biospéologiques* **13**: 57–66.

Vachon, M. (1960a). Sur la présence a Madagascar d'un représentant de la famille des Faellidae Ellingsen (Pseudoscorpions). *Bulletin du Muséum National d'Histoire Naturelle, Paris* (2) **32**: 165–166.

Vachon, M. (1960b). Sur une nouvelle espèce halophile de Pseudoscorpions de l'Archipel de Madère: *Paraliochthonius hoestlandti* (Fam. des Chthoniidae). *Bulletin du Muséum National d'Histoire Naturelle, Paris* (2) **32**: 331–337.

Vachon, M. (1961). Remarques sur les Pseudoscorpions de Madère, des Açores et les Canaries (première note). *Bulletin du Muséum National d'Histoire Naturelle, Paris* (2) **33**: 98–104.

Vachon, M. (1963). *Chthonius (C.) balazuci*, nouvelle espèce de Pseudoscorpion cavernicole du département français de l'Ardèche (Heterosphyronida, Chthoniidae). *Bulletin du Muséum National d'Histoire Naturelle, Paris* (2) **35**: 394–399.

Vachon, M. (1964a). *Roncus (R.) barbei* nouvelle espèce de Pseudoscorpion Neobisiidae des cavernes du Lot-et-Garonne, France. *Bulletin du Muséum National d'Histoire Naturelle, Paris* (2) **36**: 72–79.

Vachon, M. (1964b). Sur l'établissement de formules précisant l'ordre d'apparition des trichobothries au cours du développement post-embryonnaire chez les Pseudoscorpions (Arachnides). *Comptes Rendus Hebdomadaires des Séances de l'Académie des Sciences, Paris* **258**: 4839–4842.

Vachon, M. (1966a). Quelques remarques sur le genre *Neobisium* J. C. Chamberlin (Arachnides, Pseudoscorpions, Neobisiidae) a propos d'une espèce nouvelle *Neobisium (N.) gineti*, habitant les cavernes de l'est de la France. *Bulletin du Muséum National d'Histoire Naturelle, Paris* (2) **37**: 645–658.

Vachon, M. (1966b). *Olpium minnizioides* nouvelle espèce de Pseudoscorpions Olpiidae habitant l'île Hasikaya (sud de l'Arabie). *Annals and Magazine of Natural History* (13) **9**: 183–188.

Vachon, M. (1966c). Les conduits évacuateurs des glandes chélicériennes chez les Pseudoscorpions (Arach.). *Senckenbergiana Biologica* **47**: 29–33.

Vachon, M. (1967a). *Neobisium (Roncobisium) allodentatum* n. sg., n. sp. de Pseudoscorpion Neobisiidae (Arachnides) habitant une caverne du département de Saône-et-Loire, France. *International Journal of Speleology* **2**: 363–367.

Vachon, M. (1967b). *Spelyngochthonius heurtaultae*, nouvelle espèce de Pseudo-scorpions cavernicoles, habitant l'Espagne (Famille des Chthoniidae). *Bulletin du Muséum National d'Histoire Naturelle, Paris* (2) **39**: 522–527.

Vachon, M. (1969). Remarques sur la famille des Syarinidae J.C. Chamberlin (Arachnides, Pseudoscorpions) à propos de la description d'une nouvelle espèce: *Pseudoblothrus thiebaudi* habitant des cavernes de Suisse. *Revue Suisse de Zoologie* **76**: 387–396.

Vachon, M. (1970). Remarques sur *Withius piger* (Simon, 1878) nov. comb. (Pseudo-scorpion Cheliferidae) et sur le genre *Diplotemnus* J. C. Chamberlin, 1933, à propos de *Diplotemnus beieri* nov. nom. (Pseudoscorpion Miratemnidae). *Bulletin du Muséum National d'Histoire Naturelle, Paris* (2) **42**: 185–191.

Vachon, M. (1973). Étude des caractères utilisés pour classer des familles et les genres de Scorpions (Arachnides). 1. La trichobothriotaxie en Arachnologie. Sigles trichobothriaux et types de trichobothriotaxie chez les Scorpions. *Bulletin du Muséum National d'Histoire Naturelle, Paris* (3) **140**: 857–958.

Vachon, M. (1976). Quelques remarques sur la Pseudoscorpions (Arachnides) cavernicoles de la Suisse à propos de la description de deux espèces nouvelles: *Neobisium (N.) aelleni* et *Neobisium (N.) strausaki*. *Revue Suisse de Zoologie* **83**: 243–253.

Vachon, M. & Gabbutt, P. D. (1964). Sur l'utilisation des soies flagellaires chélicériennes dans la distinction des genres *Neobisium* J. C. Chamberlin et *Roncus* L. Koch (Arachnides, Pseudoscorpions, Neobisiidae). Importance d'une connaissance précise de la genèse d'un caractère pour en déceler la valeur taxonomique. *Bulletin de la Société Zoologique de France* **89**: 174–188.

Vachon, M. & Heurtault-Rossi, J. (1964). Une nouvelle espèce française de Pseudo-scorpion cavernicole: *Spelyngochthonius provincialis* (Chthoniidae) du départment de l'Hérault. *Bulletin du Muséum National d'Histoire Naturelle, Paris* **36**: 80–85.

Väänänen, H. (1928a). Suomen valeskorppioonilajit. *Luonnon Ystävä* **32**: 9–19.

Väänänen, H. (1928b). Piirteitä valeskorppioonien elintavoista. *Luonnon Ystävä* **32**: 139–146.

Valle, A. (1911). Note sulla fauna e flora della grotta di Trebiciano presso Trieste. *Alpi Giulie* **16**: 22–26. (not seen)

Vance, T. C. (1971). Pseudoscorpion predation on thrips. *Entomological News* **82**: 322.

Veeresh, G. K. (1983). Microarthropods. In: Veeresh, G.K. & Rajagopal, D. (eds), *Applied Soil Biology and Ecology*: 141–164. Sharada Publications: Gangenahalli. (not seen)

Vejdowsky, J. F. (1892a). Sur la question de la segmentation de l'oeuf et la formation du blastoderme des Pseudoscorpionides. *IV International Congress of Zoology* **1**: 120–125.

Vejdowsky, J. F. (1892b). Sur un organe embryonnaire des Pseudoscorpionides. *IV International Congress of Zoology* **1**: 126–131.

Verdcourt, B. (1949). Miscellaneous records of Arachnida and insects from N. W. Hertfordshire. *Entomologist's Monthly Magazine* **85**: 249–353. (not seen)

Verner, P. H. (1958a). Eine neue Pseudoskorpionart aus der Tschechoslowakei. *Zoologischer Anzeiger* **161**: 198–199.

Verner, P. H. (1958b). *Neobisium (Neobisium) crassifemoratum* (Beier), ein neuer Pseudoskorpion für die Fauna der Tschechoslowakei. *Acta Faunistica Entomologica Musei Nationalis Pragae* 3: 109–110.

Verner, P. H. (1959). Ein interessanter Fund eines Pseudoscorpions in der Tschechoslowakei (Pseudoscorpionidea). *Acta Faunistica Entomologica Musei Nationalis Pragae* 5: 61–63.

Verner, P. H. (1960). Příspevek k poznání štírku Československa. *Vestnik Československé Zoologické Spolecnosti v Praze* 24: 167–169.

Verner, P. H. (1971). Pseudoscorpionidea. In: Daniel, M. & Černý, V. (eds), Klič Zvířeny ČSSR, vol. 4: 19–31. Československenka Akademie Ved. Academia: Praha. (not seen)

Viggiani, G. (1973). Le specie descritte da Filippo Silvestri (1873–1949). *Bollettino di Laboratorio di Entomologia Agraria Portici* 30: 351–418. (not seen)

Vigna Taglianti, A. (1969). Un nuovo *Doderotrechus* cavernicolo delle Alpi occidentali (Coleoptera, Carabidae). *Fragmenta Entomologica* 6: 253–269.

Villers, C. de (1789). *Caroli Linnaei Entomologia, Faunae Suecicae Descriptionibus Aucta*. C. de Villers: Lugduni. (not seen)

Viré, A. (1896). La faune des catacombes de Paris. *Bulletin du Muséum National d'Histoire Naturelle, Paris* 2: 226–234.

Vitali-di Castri, V. (1962). La familia Cheiridiidae (Pseudoscorpionida) en Chile. *Investigaciones Zoológicas Chilenas* 8: 119–142.

Vitali-di Castri, V. (1963). La familia Vachoniidae (= Gymnobisiidae) en Chile (Arachnidea, Pseudoscorpionida). *Investigaciones Zoológicas Chilenas* 10: 27–82.

Vitali-di Castri, V. (1965a). *Cheiridium danconai* n. sp. (Pseudoscorpionida) con consideraciones sobre su desarrollo postembrionario. *Investigaciones Zoológicas Chilenas* 12: 67–92.

Vitali-di Castri, V. (1965b). Consideraciones sobre la trichobothriotaxia de los Pseudoscorpiones (Arach.). *Investigaciones Zoológicas Chilenas* 12: 85–96.

Vitali-di Castri, V. (1966). Observaciones biogeograficas y filogeniticas sobre la familia Cheiridiidae (Pseudoscorpionida). *Progresos en Biologia del Suelo* 1966: 379–386.

Vitali-di Castri, V. (1968). *Austrochthonius insularis*, nouvelle espèce de pseudoscorpions de l'Archipel de Crozet (Heterosphyronida, Chthoniidae). *Bulletin du Muséum National d'Histoire Naturelle, Paris* (2) 40: 141–148.

Vitali-di Castri, V. (1969a). Remarques sur la famille des Menthidae (Arachnida Pseudoscorpionida) a propos de la présence au Chili d'une nouvelle espèce, *Oligomenthus chilensis*. *Bulletin du Muséum National d'Histoire Naturelle, Paris* (2) 41: 498–506.

Vitali-di Castri, V. (1969b). Tercera nota sobre los Cheiridiidae de Chile (Pseudoscorpionida) con descripcion de *Apocheiridium (Chiliocheiridium) serenense* n. subgen., n. sp. *Boletín de la Sociedad de Biología de Concepción* 41: 265–280.

Vitali-di Castri, V. (1970a). Un nuevo genero de Gymnobisiinae (Pseudoscorpionida) de las Islas Malvinas. Revision taxonomica de la subfamilia. *Physis, Buenos Aires* 30: 1–9.

Vitali-di Castri, V. (1970b). Pseudochiridiinae (Pseudoscorpionida) du Muséum National d'Histoire Naturelle. Remarques sur la sous-famille et description de deux nouvelles espèces de Madagascar et d'Angola. *Bulletin du Muséum National d'Histoire Naturelle, Paris* (2) 41: 1175–1199.

Vitali-di Castri, V. (1970c). Revision de la sistematica y distribucion de los Gymnobisiinae (Pseudoscorpionida, Vachoniidae). *Boletín de la Sociedad de Biología de Concepción* 42: 123–135.

Vitali-di Castri, V. (1972). El genero sudamericano *Gigantochernes* (Pseudoscorpionida, Chernetidae) con descripcion de dos nuevas especies. *Physis, Buenos Aires* 31: 23–38.

Vitali-di Castri, V. (1973). Biogeography of pseudoscorpions in the mediterranean regions of the world. In: Castri, F. di & Mooney, H. (eds), *Mediterranean Type Ecosystems, Origin and Structure*: 295–305. Springer-Verlag: Berlin.

Vitali-di Castri, V. (1974). Presencia en America del sur del genero *Sathrochthonius* (Pseudoscorpionida) con descripcion de una nueva especie. *Physis, Buenos Aires* 33: 193–201.

Vitali-di Castri, V. (1975a). Nuevos *Austrochthonius* sudamericanos (Pseudoscorpionida, Chthoniidae). *Physis, Buenos Aires* 34: 117–127.

Vitali-di Castri, V. (1975b). Deux nouveaux genres de Chthoniidae du Chili: *Chiliochthonius* et *Francochthonius* (Arachnida, Pseudoscorpionida). *Bulletin du Muséum National d'Histoire Naturelle, Paris* (3) 334: 1277–1291.

Vitali-di Castri, V. (1984). Chthoniidae et Cheiridiidae (Pseudoscorpionida, Arachnida) des Petites Antilles. *Bulletin du Muséum National d'Histoire Naturelle, Paris* (4) 5: 1059–1078.

Vitali-di Castri, V. & Castri, F. di (1970). L'évolution du dimorphisme sexuel dans une lignée de pseudoscorpions. *Bulletin du Muséum National d'Histoire Naturelle, Paris* (2) 42: 382–391.

Vitali-di Castri, V. & Castri, F. di (1976). Estudio preliminar de los Pseudoscorpiones del Parque Nacional 'Vicente Perez Rosales' (Llanquihue, Chile). *Anales Museo de Historia Natural de Valparaiso* 9: 61–64.

Vogel, M. (1983). Die Pseudoskorpione des Roten Moores. In: Nentwig, W. & Droste, M. (eds), *Die Fauna des Roten Moores in der Rhön*: 123–125. Marburg. (not seen)

Volz, P. (1965). Von der Fauna der Kleinen Kalmit bei Landau/Pf. *Mitteilungen der Pollichia* (3) 12: 132–150. (not seen)

Vornatscher, J. (1950). *Neobisium hermanni*, ein Höhlen-Pseudoskorpione vom Alpenostrand. *Höhle* 1: 50–51. (not seen)

Wäger, H. (1982). *Populationsdynamik und Entwicklungszyklus der Pseudoscorpiones im Stamser Eichenwald (Tirol)*. Examensarbeit: Univ. Innsbruck. (not seen)

Wagenaar-Hummelinck, P. (1948). Studies on the fauna of Curaçao, Aruba, Bonaire and the Venezuelan Islands: no. 13. Pseudoscorpions of the genera *Garypus, Pseudochthonius, Tyrannochthonius* and *Pachychitra*. *Natuurwetenschappelijke Studiekring voor Suriname en Curaçao* 5: 29–77.

Wagner, F. (1892). Biologische Notiz. *Zoologischer Anzeiger* 15: 434–436.

Wagner, F. (1894). Beiträge zur Phylogenie der Arachniden. *Zeitschrift für Naturwissenschaften* 29: 123–153. (not seen)

Walckenaer, C. A. (1802). *Faune Parisienne. Insectes*. Imprimeur-Libraire: Paris.

Wallwork, J. A. (1970). *Ecology of Soil Animals*. McGraw-Hill: London. (not seen)

Wallwork, J. A. (1982). *Desert Soil Fauna*. Praegar: New York.

Walsh, G. B. (1924). A new Yorkshire pseudo-scorpion. *Naturalist, London* 49: 140.

Warburton, C. (1909). Arachnida Embolobranchiata. *The Cambridge Natural History*, vol. 4: 297–473. (not seen)

Ward, J.R. (1887). Bees in Natal: their companions and parasites. *British Bee Journal and Bee-Keepers' Adviser* 15: 563. (not seen)

Ward, R. D. (1969). A pseudoscorpion found with *Cicindela hirticollis* Say. *Cicindela* 1(4): 6. (not seen)

Wasmann, E. (1894). *Kritisches Verzeichnis der Myrmekophilen und Termitophilen Arthropoden*. F. L. Dames: Berlin. (not seen)

Wasmann, E. (1899). Ein neuer *Melipona*-Gast (*Scotocryptus Goeldii*) aus Parà. *Deutsche Entomologische Zeitschrift* 1899: 411.

Waterhouse, C. O. (1875). Untitled. *Entomologist's Monthly Magazine* 12: 20.

Waterhouse, C. O. (1878). Description of a new species of Chernetidae (Pseudoscorpionidae) from Spain. *Transactions of the Entomological Society of London* 26: 181–182.

Waterlot, G. (1953). Classe des Arachnides (Arachnida Cuvier 1812). In: Piveteau, J. (ed.), *Traité de Paléontologie*, vol. **3**: 555–584. Masson: Paris.

Waterston, J. (1903). *Roncus cambridgii* L. K. in Argyllshire. *Annals of Scottish Natural History* **12**: 187. (not seen)

Weber, D. (1988). Die Höhlenfuana und -flora des Höhlenkatastergebietes Rheinland-Pfalz/Saarland. *Abhandlungen der Karst- und Höhlenkunde* **22**: 1–157. (not seen)

Webster, F. M. (1897). Pseudoscorpions attached to flies. *Entomological News* **8**: 59.

Weidner, H. (1954a). Ein Beitrag zur Spinnenfauna von Unterfranken. *Nachrichten Naturwissenschaftliches Museum der Stadt, Aschaffenburg* **42**: 45–48. (not seen)

Weidner, H. (1954b). Die Pseudoskorpione, Wederknechte und Milben der Umgebung von Hamburg mit besonderer Berücksichtigung der für den Menschen wichtigen Arten. *Entomologische Mitteilungen aus dem Zoologischen Staatsinstitut und Zoologischen Museum in Hamburg* **1**: 103–156.

Weidner, H. (1959). Die Entomologischen Sammlungen des Zoologischen Staatsinstituts und Zoologischen Museums Hamburg. I. Teil. Pararthropoda und Chelicerata I. *Mitteilungen aus dem Hamburgischen Zoologischen Museum und Institute* **57**: 89–142.

Weidner, H. (1969). Über den Stand der Erforschung der Landarthropodenfauna von Hamburg und seiner weiteren Umgebung. *Entomologische Mitteilungen aus dem Staatsinstitut und Zoologischen Museum in Hamburg* **4**: 26–34.

Weissenborn, B. (1887). Beiträge zur Phylogenie der Arachniden. *Zeitschrift für Naturwissenschaften* **20**: 33–119. (not seen)

Welbourne, W. C. (1978). Biology of Ogle Cave with list of the cave fauna of Slaughter Canyon. *Bulletin of the National Speleological Society* **40**: 27–34. (not seen)

Wells, ? (1899). Title? *British Bee Journal and Bee-Keepers' Adviser* **27**: 126. (not seen)

Werner, G. & Bawa, S. R. (1988a). Acrosome formation in the pseudoscorpion *Diplotemnus* sp. *Journal of Ultrastructure Research* **98**: 105–118. (not seen)

Werner, G. & Bawa, S. R. (1988b). Spermatogenisis in the pseudoscorpion *Diplotemnus* sp. with special reference to nuclear changes. *Journal of Ultrastructure Research* **98**: 119–136. (not seen)

West, L. S. (1951). *The Housefly, its Natural History, Medical Importance and Control*. Ithaca. (not seen)

Westwood, J. O. (1836). Cheliferidae. In: *British Cyclopaedia of Natural History*, vol. **2**: 10. London. (not seen)

Westwood, J. O. (1838). *The Entomologist's Text Book; an Introduction to the Natural History, Structure, Physiology and Classification of Insects, Including the Crustacea and Arachnida*. London. (not seen)

Wettstein, L. (1955). Un pseudoscorpion de notre région: *Obisium lubricum* L. Koch. *Revue Verviétoise d'Histoire Naturelle* **12**: 26–27. (not seen)

Weygoldt, P. (1961). Zucht und Beobachtung von Pseudoskorpionen. *Mikrokosmos* **12**: 361–364. (not seen)

Weygoldt, P. (1962). Beobachtungen am Pumporgan der Embryonen der Pseudoscorpione (Chelonethi). *Zoologische Beiträge* **7**: 293–309. (not seen)

Weygoldt, P. (1963). Entwicklung und Funktion der Embryonalhülle der Pseudoscorpione. *Verhandlungen der Deutschen Zoologischen Gesellschaft* **26**: 447–452.

Weygoldt, P. (1964a). Vergleichend-embryologische Untersuchungen an Pseudoscorpionen II. Das zweite Embryonalstadium von *Lasiochernes pilosus* Ellingsen und *Cheiridium museorum* Leach. *Zoologische Beiträge* **10**: 353–368.

Weygoldt, P. (1964b). Vergleichend-embryologische Untersuchungen an Pseudoscorpionen (Chelonethi). *Zeitschrift für Morphologie und Ökologie der Tiere* **54**: 1–106. (not seen)

Weygoldt, P. (1965a). Vergleichend-embryologische Untersuchungen an Pseudo-scorpionen. III. Die Entwicklung von *Neobisium muscorum* Leach (Neobisiinae, Neobisiidae). Mit dem Versuch einer Deutung der Evolution des embryonalen Pumporgans. *Zeitschrift für Morphologie und Ökologie der Tiere* **55**: 321–383. (not seen)

Weygoldt, P. (1965b). Mechanismus der Spermienübertragung bei einum Pseudo-scorpion. *Naturwissenschaften* **52**: 218.

Weygoldt, P. (1965c). Das Fortpflanzungsverhalten der Pseudoskorpione. *Natur-wissenschaften* **52**: 436.

Weygoldt, P. (1966a). Vergleichende Untersuchungen zur Fortpflanzungsbiologie der Pseudoscorpione: Beobachtungen über das Verhalten, die Samenubertragungs-weisen und die Spermatophoren einiger einheimischer Arten. *Zeitschrift für Morphologie und Ökologie der Tiere* **56**: 39–92. (not seen)

Weygoldt, P. (1966b). Die Ausbildung transitorischer Pharynxapparate bei Embryonen. *Zoologischer Anzeiger* **176**: 147–160.

Weygoldt, P. (1966c). Spermatophore web formation in a pseudoscorpion. *Science* **153**: 1647–1649.

Weygoldt, P. (1966d). *Moos- und Bücherskorpione.* Die Neue Brehm-Bücherei 365, A. Ziemsen Verlag: Wittenberg-Lutherstadt. (not seen)

Weygoldt, P. (1966e). Mating behaviour and spermatophore morphology in the pseudoscorpion *Dinocheirus tumidus* Banks (Cheliferinea, Chernetidae). *Biological Bulletin of the Marine Biological Laboratory, Wood's Hole* **130**: 462–467.

Weygoldt, P. (1968). Vergleichend-embryologische Untersuchungen an Pseudo-scorpionen. IV. Die Entwicklung von *Chthonius tetrachelatus* Preyssl., *Chthonius ischnocheles* (Hermann) (Chthoniinea, Chthoniidae) und *Verrucaditha spinosa* Banks (Chthoniinea, Tridenchthoniidae). *Zeitschrift für Morphologie und Ökologie der Tiere* **63**: 111–154. (not seen)

Weygoldt, P. (1969a). *The Biology of Pseudoscorpions.* Harvard University Press: Cambridge, Massachusetts.

Weygoldt, P. (1969b). Paarungsverhalten und Samenübertragung beim Pseudoskorpion *Withius subruber* Simon (Cheliferidae). *Zeitschrift für Tierpsychologie* **26**: 230–235.

Weygoldt, P. (1970a). Vergleichende Untersuchungen zur Fortpflanzungsbiologie der Pseudoscorpione II. *Zeitschrift für die Zoologische Systematik und Evolution-forschung* **8**: 241–259. (not seen)

Weygoldt, P. (1970b). Vergleichend-embryologische Untersuchungen an Pseudo-scorpionen. V. Das 2. Embryonalstadium mit seinem Pumporgan bei verschiedenen Arten und sein Wert als taxonomisches Merkmal. *Zeitschrift für die Zoologische Systematik und Evolutionforschung* **9**: 3–29. (not seen)

Weygoldt, P. (1970c). Evolution des Paarungsverhaltens bei Pseudoscorpionen. *Bulletin du Muséum National d'Histoire Naturelle, Paris* (2) **41**, supplément 1: 141.

Weygoldt, P. (1980). Towards a cladistic classification of the Chelicerata. *Verhand-lungen des 8. Internationalen Arachnologen-Kongresses, Wien*: 331–334.

Weygoldt, P. & Paulus, H. F. (1979a). Untersuchungen zur Morphologie, Taxonomie und Phylogenie der Chelicerata. I. Morphologische Untersuchungen. *Zeitschrift für die Zoologische Systematik und Evolutionforschung* **17**: 85–116.

Weygoldt, P. & Paulus, H. F. (1979b). Untersuchungen zur Morphologie, Taxonomie und Phylogenie der Chelicerata. II. Cladogramme und die Entfaltung der Chelicerata. *Zeitschrift für die Zoologische Systematik und Evolutionforschung* **17**: 177–200.

Wheeler, W. M. (1911). Pseudoscorpions in ant nests. *Psyche, Cambridge* **18**: 166–168.

White, A. (1849). Descriptions of apparently new species of Aptera from New Zealand. *Proceedings of the Zoological Society of London* **17**: 3–6.

Whyte, G. A. (1907). *Chelifer cancroides* (Linn.) in Manchester. *Zoologist* (4) **11**: 388–389. (not seen)

Whyte, G. A. & Whyte, R. B. (1907). The false-scorpions of Cumberland. *Naturalist, London* **1907**: 203–204.

Winckler, W. (1886). Das Herz der Acarinen nebst vergleichenden Bemerkungen über das Herz der Phalangiden und Chernetiden. *Arbeiten aus dem Zoologischen Institut der Universität Wien* **7**: 111–118. (not seen)

Wise, K. A. J. (1970). On the terrestrial invertebrate fauna of White Island, New Zealand. *Records of the Auckland Institute and Museum* **7**: 217–252.

With, C. J. (1905). On Chelonethi, chiefly from the Australian region, in the collection of the British Museum, with observations on the 'coxal sac' and on some cases of abnormal segmentation. *Annals and Magazine of Natural History* (7) **15**: 94–143, 328.

With, C. J. (1906). The Danish expedition to Siam 1899-1900. III. Chelonethi. An account of the Indian false-scorpions together with studies on the anatomy and classification of the order. *Oversigt over det Konigelige Danske Videnskabernes Selskabs Forhandlinger* (7) **3**: 1–214.

With, C. J. (1907). On some new species of the Cheliferidae, Hans., and Garypidae, Hans., in the British Museum. *Journal of the Linnean Society of London* **30**: 49–85.

With, C. J. (1908a). An account of the South-American Cheliferinae in the collections of the British and Copenhagen Museums. *Transactions of the Zoological Society of London* **18**: 217–340.

With, C. J. (1908b). Remarks on the Chelonethi. *Videnskabelige Meddelelser fra Dansk Naturhistorisk Forening i Kjøbenhavn* (6) **10**: 1–25.

Wolf, B. (1937). *Animalium Cavernarum Catalogus*, vol. **2**. Junk: Wien.

Wood, J. G. (1861–1863). *The Illustrated Natural History*. London. (not seen)

Wood, P. A. (1971). *Studies on Laboratory and Field Populations of Three British Pseudoscorpions with Particular Reference to their Gonadial Cycles*. Ph.D. Thesis: University of Manchester. (not seen)

Wood, P. A. (1975). Cyclical gonadial development of *Chthonius ischnocheles* (Hermann). *Proceedings of the 6th International Arachnological Congress, Amsterdam*: 145–149.

Wood, P. A. & Gabbutt, P. D. (1978). Seasonal vertical distribution of pseudoscorpions in beech litter. *Bulletin of the British Arachnological Society* **4**: 176–183.

Wood, P. A. & Gabbutt, P. D. (1979a). Silken chambers built by adult pseudo-scorpions in laboratory culture. *Bulletin of the British Arachnological Society* **4**: 285–293.

Wood, P. A. & Gabbutt, P. D. (1979b). Silken chambers built by nymphal pseudo-scorpions in laboratory culture. *Bulletin of the British Arachnological Society* **4**: 329–336.

Woodroffe, G. E. (1953). An ecological study of the insects and mites in the nests of certain birds in Britain. *Bulletin of Entomological Research* **44**: 739–772.

Worth, C. B. (1975). Pseudoscorpions on a Dark-eyed Junco, *Junco hyemalis*. *Bird-banding* **46**: 76.

Yatsu, N. (1908). On Japanese pseudoscorpions. *Zoological Magazine, Tokyo* **20**: 327. (not seen)

Yoshikura, M. (1975). Comparative embryology and phylogeny of Arachnida. *Kumamoto Journal Science, Biology* **12**: 71–142. (not seen)

Zabriskie, J. L (1884). Nest of the pseudoscorpion. *American Naturalist* **18**: 427.

Zamudio Villanueva, M. de la Luz (1963). *Pseudoscorpiones de Mexico de la subfamilia Lamprochernetinae (Arachnida Pseudoscorpionidea)*. Tesis: Univ. Nacional autonoma de Mexico. (not seen)

Zangheri, P. (1966). Repertorio sistematico e topografico della flora e fauna vivente e fossile della Romagna. *Museo Civico di Storia Naturale di Verona, Memorie Fuori Serie 1* **2**: 485–854.

Zaragoza, J. A. (1982). *Roncus (Parablothrus) setosus* n. sp., otro caso de 'néochéto-taxie majorante prosomatique' (Heurtault) en los Pseudoscorpiones Neobisiidae. *Mediterranea, Serie de Estudios Biologicos* 6: 101–108.

Zaragoza, J. A. (1984). Un nuevo *Chthonius* cavernícola de la Provincia de Alicante (Arachnida: Pseudoscorpionidea, Chthoniidae). *Mediterranea, Serie de Estudios Biologicos* 7: 49–54.

Zaragoza, J. A. (1985). Nuevos o interesantes Chthoniidae cavernícolas del País Valenciano (Arachnida, Pseudoscorpiones). *Miscelanea Zoologica* 9: 145–158.

Zaragoza, J. A. (1986). Distribucion de los Pseudoscorpiones cavernicolas de la peninsula Iberica e islas Baleares (Arachnida). *Actas X Congreso International de Aracnologia, Jaca* 1: 405–411.

Zaragoza, J. A. (1987). *Chthonius (Ephippiochthonius) verai* nueva especie cavernicola del sureste Español (Arachnida, Pseudoscorpiones, Chthoniidae). *Mediterranea, Serie de Estudios Biologicos* 8: 5–15.

Zeh, D. W. (1983). Ecological factors and the evolution of sexual dimorphism in chernetid pseudoscorpions. *American Zoologist* 23: 878. (not seen)

Zeh, D. W. (1986). *Ecological Factors, Pleiotropy, and the Evolution of Sexual Dimorphism in Chernetid Pseudoscorpions.* Dissertation, University of Arizona: Tuscon. (not seen)

Zeh, D. W. (1987a). Life history consequences of sexual dimorphism in a chernetid pseudoscorpion. *Ecology* 68: 1495–1501.

Zeh, D. W. (1987b). Aggression, density and sexual dimorphism in chernetid pseudoscorpion (Arachnida: Pseudoscorpionida). *Evolution* 41: 1072–1087.

Order PSEUDOSCORPIONIDA de Geer

Orders:

Faux-scorpions de Geer, 1778: 349−353; Latreille, 1817: 107.

Pseudoscorpiones de Geer: Dugès & Edwards, 1836: 81−82; Gerstäcker, 1863: 330−331; Tömösváry, 1882b: 181; Daday, 1888: 112−113; Balzan, 1892: 509; Lankester, 1904: 255; Kew, 1911a: 38−39; Navás, 1918: 85−86; Kästner, 1928: 1−3; G.O. Evans & Browning, 1954: 1; Savory, 1964a: 97; Levi & Levi, 1968: 120; Savory, 1977: 222; Weygoldt & Paulus, 1979b: 189; R. Schuster, 1986a: 269−270; Eisenbeis & Wichard, 1987: 48.

Pseudo-scorpiones de Geer: McIntire, 1868: 8.

Chernetes E. Simon, 1879a: 1−18; Lameere, 1895: 473.

Pseudoscorpions de Geer: Gaubert, 1892a: 84−85; Vachon, 1949: 437.

Chelonethi Thorell, 1883: 36; With, 1906: 51−56; Roewer, 1936: 1−4.

Chernetidea E. Simon: O. P.-Cambridge, 1892: 201−207.

Chelonethida Thorell: J. C. Chamberlin, 1929a: 56; J. C. Chamberlin, 1931a: 206−208; J. C. Chamberlin, 1935b: 477−479; Hoff, 1959a: 7; Main, 1954: 40; Main & Edward, 1968: 49.

Pseudoscorpionidea de Geer: Beier, 1932a: 1−22; Beier, 1932g: 118; Beier, 1963b: 1−15.

Pseudoscorpionida de Geer: Comstock, 1913: 39−44; Pratt, 1927: 409; Schenkel, 1928: 52−54; Ross, 1944: 48; Hoff, 1949b: 413−427; Petrunkevitch, 1955: 79−80; Nelson, 1975: 258−262; Muchmore, 1982a: 96; Harvey & Yen, 1989: 84.

Pseudoscorpion de Geer: Waterlot, 1953: 565.

Cheloneti Thorell: Grasshoff, 1978: 282.

Suborders:

Haplochelonethi Thorell, 1883: 36.

Diplochelonethi Thorell, 1883: 36.

Panctenodactyli Balzan, 1892: 509; H. J. Hansen, 1893: 230−231; With, 1906: 88; Kew, 1911a: 39; Lessert, 1911: 9; Kästner, 1928: 3; Schenkel, 1928: 55; Väänänen, 1928a: 17.

Emictenodactyli (sic) Balzan, 1892: 539; H. J. Hansen, 1893: 231.

Hemictenodactyli Balzan: With, 1906: 63; Kew, 1911a: 51−52; Lessert, 1911: 26; Kästner, 1928: 3; Schenkel, 1928: 62−63; Väänänen, 1928a: 16.

Heterosphyronida J. C. Chamberlin, 1929a: 57; J. C. Chamberlin, 1931a: 208−209; J. C. Chamberlin, 1935b: 479; Hoff, 1949b: 429; Hoff, 1956a: 1−2; Hoff, 1959a: 7; Murthy & Ananthakrishnan, 1977: 9.

Homosphyronida J. C. Chamberlin, 1929a: 78; J. C. Chamberlin, 1931a: 213 (Group).

Diplosphyronida J. C. Chamberlin, 1929a: 78; J. C. Chamberlin, 1930: 6; J. C. Chamberlin, 1931a: 213−214; J. C. Chamberlin, 1935b: 479; Hoff, 1949b: 443; Hoff, 1956b: 2; Hoff, 1964a: 6; Murthy & Ananthakrishnan, 1977: 23.

Monosphyronida J. C. Chamberlin, 1929a: 78; J. C. Chamberlin, 1931a: 228; J. C. Chamberlin, 1935b: 480; Hoff, 1949b: 449; Hoff, 1956c: 2; Murthy & Ananthakrishnan, 1977: 114.

Chthoniinea Beier, 1932a: 22−23; Beier, 1932g: 181; Roewer, 1937: 232; Morikawa, 1960: 92; Legg & Jones, 1988: 55.

Neobisiinea Beier, 1931e: 299; Beier, 1932a: 74−75; Beier, 1932g: 182; Roewer, 1937: 242; Morikawa, 1960: 111; Legg & Jones, 1988: 74.

Cheliferinea Beier, 1932e: 1−2; Beier, 1932g: 185; Roewer, 1937: 272−273; Morikawa, 1960: 133; Legg & Jones, 1988: 90.
Chthoniina Beier: Petrunkevitch, 1955: 80.
Neobisiina Beier: Petrunkevitch, 1955: 81.
Cheliferina Beier: Petrunkevitch, 1955: 82.

Superfamily CHTHONIOIDEA Daday

Chthonioidea Daday: J. C. Chamberlin, 1931a: 209; Muchmore, 1982a: 96; Legg & Jones, 1988: 55.

Family CHTHONIIDAE Daday

Chthoniinae Daday, 1888: 133; Balzan, 1892: 545; H. J. Hansen, 1893: 232; With, 1906: 64–66; Lessert, 1911: 37; Redikorzev, 1924b: 26; J. C. Chamberlin, 1929a: 62–63; J. C. Chamberlin, 1931a: 212; Beier, 1932a: 35; Morikawa, 1960: 93.
Chthoniidae Daday: H. J. Hansen, 1893: 232; Tullgren, 1906a: 205; With, 1906: 64; Kew, 1911a: 55; Lessert, 1911: 37; J. C. Chamberlin, 1929a: 56; J. C. Chamberlin, 1931a: 211–212; Beier, 1932a: 23; Beier, 1932g: 181–182; Roewer, 1937: 235–236; Hoff, 1949b: 431; Hoff, 1951: 1; G. O. Evans & Browning, 1954: 7; Petrunkevitch, 1955: 80; Hoff, 1956a: 2; Hoff, 1959a: 7; Morikawa, 1960: 93; Beier, 1963b: 17–18; Murthy & Ananthakrishnan, 1977: 11; Muchmore, 1982a: 97; Harvey, 1985b: 137; Legg, 1987: 179; Legg & Jones, 1988: 55.
Kewochthonini (sic) J. C. Chamberlin, 1929a: 63.
Tyrannochthoniini J. C. Chamberlin, 1962: 310; Muchmore, 1972b: 249.
Chthonini (sic) Daday: J. C. Chamberlin, 1929a: 68.
Chthoniini Daday: J. C. Chamberlin, 1931a: 212; Beier, 1932a: 35; Roewer, 1937: 236; Hoff, 1951: 2; Hoff, 1956a: 2; Hoff, 1959a: 7.
Lechytini (sic) J. C. Chamberlin, 1929a: 76; J. C. Chamberlin, 1931a: 212.
Lechytiini J. C. Chamberlin: Beier, 1932a: 73; Roewer, 1937: 241–242; Hoff, 1956a: 11; Hoff, 1963c: 25; Muchmore, 1975b: 13–14.
Pseudotyrannochthoniini Beier, 1932a: 69; Roewer, 1937: 240–241; Hoff, 1951: 9.
Chthonidae (sic) Daday: J. C. Chamberlin, 1935b: 479.
Lechytiinae J. C. Chamberlin: Morikawa, 1960: 111; Murthy & Ananthakrishnan, 1977: 20–21.

Genus Afrochthonius Beier

Afrochthonius Beier, 1930e: 286; Beier, 1932a: 71.

Type species: *Afrochthonius similis* Beier, 1930e, by original designation.

Afrochthonius brincki Beier

Afrochthonius brincki Beier, 1955l: 275–276, fig. 2; Beier, 1964k: 32.

Type locality: Makheka (as Makheke) Mountains, 10 miles ENE. of Mokhotlong, Lesotho.
Distribution: Lesotho, South Africa.

Afrochthonius ceylonicus Beier

Afrochthonius ceylonicus Beier, 1973b: 40–42, fig. 2.

Type locality: Horten Plains, 11 miles SSE. of Nuwara-Eliya, Central Province, Sri Lanka.
Distribution: Sri Lanka.

Family Chthoniidae

Afrochthonius godfreyi (Ellingsen)

Chthonius godfreyi Ellingsen, 1912b: 120–122.
Afrochthonius godfreyi (Ellingsen): Beier, 1930e: 287–288; Beier, 1932a: 72; Roewer, 1937: 241; Beier, 1958a: 159–160; Beier, 1964k: 32; Beier, 1966k: 456.

Type locality: Pirie, near King William's Town, Cape Province, South Africa.
Distribution: Lesotho, South Africa.

Afrochthonius inaequalis Beier

Afrochthonius inaequalis Beier, 1958a: 160–162, fig. 1; Beier, 1964k: 31.

Type locality: Champagne Castle Hotel, Drakensberg Mountains, Lesotho.
Distribution: Lesotho, South Africa.

Afrochthonius natalensis Beier

Afrochthonius natalensis Beier, 1931b: 56; Beier, 1932a: 72–73, figs 87–88; Roewer, 1937: 241, fig. 206; Beier, 1958a: 160; Beier, 1964k: 31.

Type locality: Van Reenen, Natal, South Africa.
Distribution: South Africa.

Afrochthonius reductus Beier

Afrochthonius reductus Beier, 1973b: 42, fig. 3.

Type locality: Hatton, Central Province, Sri Lanka.
Distribution: Sri Lanka.

Afrochthonius similis Beier

Afrochthonius similis Beier, 1930e: 286–288, figs 2a–c; Beier, 1932a: 71–72, fig. 86; Roewer, 1937: 241.

Type locality: Windhoek (as Windhuk), Namibia.
Distribution: Namibia.

Genus Allochthonius J. C. Chamberlin

Allochthonius J. C. Chamberlin, 1929e: 154–155; Beier, 1932a: 61; Morikawa, 1960: 97; J. C. Chamberlin, 1962: 319–320.
Allochthonius (Allochthonius): Morikawa, 1960: 98.

Type species: *Chthonius opticus* Ellingsen, 1907c, by original designation.

Subgenus Allochthonius (Allochthonius) J. C. Chamberlin

Allochthonius (Allochthonius) borealis Sato

Allochthonius borealis Sato, 1984a: 17–18, figs 1–9; Sato, 1985: 22, figs 1–2.

Type locality: Mt Iwaki, Hirosaki, Aomori, Japan.
Distribution: Japan.

Allochthonius (Allochthonius) buanensis W. K. Lee

Allochthonius buanensis W. K. Lee, 1982: 76–79, figs 1a–c, 2a–k; Sato, 1984a: fig. 9.

Type locality: Buan, Jeonbug, Korea.
Distribution: Korea.

Allochthonius (Allochthonius) opticus (Ellingsen)

Chthonius opticus Ellingsen, 1907c: 16–17.
Allochthonius opticus (Ellingsen): J. C. Chamberlin, 1929e: 155; Beier, 1932a: 62; Roewer, 1937: 240; Morikawa, 1955a: 215–216; Sato, 1978e: fig. 7; Sato, 1979a:

89–91, figs 3a–i, 4a–e, plate 1; Sato, 1979b: 13, figs 1–3, 5; Sato, 1979c: 13, fig. 2; Sato, 1979d: 42–43; Sato, 1980c: 10, figs 1–2; Sato, 1982b: 58, figs 1–3; Sato, 1982c: 31; Sato, 1984a: fig. 9; Sato, 1984c: 14, figs 1, 3; Sato, 1983b: 16, figs 2–4; Sato, 1985: 22, figs 1–2; Sato, 1986: 95.

Allochthonius anthracinus Morikawa, 1952b: 250, fig. 1 (synonymized by Morikawa, 1955a: 215).

Allochthonius yanoi Morikawa, 1952b: 251–252, figs 2–3, 4a–c (synonymized by Morikawa, 1955a: 215).

Allochthonius (Allochthonius) opticus (Ellingsen): Morikawa, 1960: 98; Morikawa, 1962: 418, figs 1, 3–5.

Allochthonius (Allochthonius) opticus opticus (Ellingsen): Morikawa, 1960: 99–100, plate 1 figs 3–4, plate 5 fig. 1, plate 9 fig. 1, plate 10 figs 10–12; Morikawa, 1962: fig. 5.

Type localities: of *Chthonius opticus*: Okayama, Japan.
of *Allochthonius anthracinus*: Dogo, Matsuyama, Shikoku, Japan.
of *Allochthonius yanoi*: Shiro-yama, Matsuyama, Shikoku, Japan.
Distribution: Japan.

Allochthonius (Allochthonius) opticus coreanus Morikawa

Allochthonius (Allochthonius) opticus coreanus Morikawa, 1970: 141–143, figs 1, 2a.

Type locality: Sinryeong-gul Cave, Baegsan-ri, Changseong-eub, Samcheog-gun, Kangweon-do, South Korea.
Distribution: South Korea.

Allochthonius (Allochthonius) opticus troglophilus Morikawa

Allochthonius (Allochthonius) troglophilus Morikawa, 1956: 280–281, fig. 4f.
Allochthonius (Allochthonius) opticus takasawadoensis Morikawa, 1956: 281, fig. 4g (synonymized by Morikawa, 1960: 100).
Allochthonius (Allochthonius) opticus troglophilus Morikawa: Morikawa, 1960: 100–101, plate 5 fig. 2; Morikawa, 1962: 418, fig. 4.

Type localities: of *Allochthonius (Allochthonius) troglophilus*: Kugô Cave, Taniai-mura, Gifu, Japan.
of *Allochthonius (Allochthonius) opticus takasawadoensis*: Takasawa-dô Cave, Kônosé, Kuma-mura, Kumamoto, Japan.
Distribution: Japan.

Allochthonius (Allochthonius) shintoisticus J. C. Chamberlin

Allochthonius shintoisticus J. C. Chamberlin, 1929e: 155; J.C. Chamberlin, 1931a: figs 21a, 21l; Beier, 1932a: 62; Roewer, 1937: 240; J.C. Chamberlin, 1962: 320–322, figs 6a–e; R. O. Schuster, 1966a: figs 6–8.

Type locality: Unzen, Kyushu, Japan.
Distribution: Japan.

Subgenus Allochthonius (Urochthonius) Morikawa

Allochthonius (Urochthonius) Morikawa, 1954b: 80.
Type species: *Allochthonius (Urochthonius) ishikawai* Morikawa, 1954b, by original designation.

Allochthonius (Urochthonius) biocularis Morikawa

Allochthonius (Urochthonius) biocularis Morikawa, 1956: 279–280, fig. 4e; Morikawa, 1960: 101, plate 5 fig. 7; Morikawa, 1962: fig. 4.

Type locality: Tsuruoka Mine, Saeki, Oita, Japan.
Distribution: Japan.

Allochthonius (Urochthonius) ishikawai Morikawa

Allochthonius (Urochthonius) ishikawai Morikawa, 1954b: 80–82, figs 1a–c, 2; Morikawa, 1960: 101–102; Morikawa, 1962: fig. 4.
Allochthonius (Urochthonius) ishikawai ishikawai Morikawa: Morikawa, 1960: 102, plate 1 fig. 6, plate 7 fig. 3, plate 5 fig. 5, plate 8 fig. 12.

Type locality: Ryuga-do, Shikoku, Japan.
Distribution: Japan.

Allochthonius (Urochthonius) ishikawai deciclavatus Morikawa

Allochthonius (Urochthonius) deciclavatus Morikawa, 1956: 277–279, fig. 4b.
Allochthonius (Urochthonius) ishikawai deciclavatus Morikawa: Morikawa, 1960: 104, plate 5 fig. 6; Morikawa, 1962: figs 4–5.

Type locality: Matsubarano-ana Cave, Mitô-chô, Yamaguchi, Japan.
Distribution: Japan.

Allochthonius (Urochthonius) ishikawai kyushuensis Morikawa

Allochthonius (Urochthonius) ishikawai kyushuensis Morikawa, 1960: 103–104, plate 1 fig. 5, plate 4 fig. 12, plate 5 fig. 9; Morikawa, 1962: figs 4–5.

Type locality: Goya daiichi-do Cave, Ita, Tagawa, Fukuoka, Japan.
Distribution: Japan.

Allochthonius (Urochthonius) ishikawai shiragatakiensis Morikawa

Allochthonius (Urochthonius) ishikawai shiragatakiensis Morikawa, 1954b: 82–83, fig. 1d; Morikawa, 1960: 102–103, plate 7 fig. 5.

Type locality: Shiragataki-do, Shikoku, Japan.
Distribution: Japan.

Allochthonius (Urochthonius) ishikawai uenoi Morikawa

Allochthonius (Urochthonius) uénoi Morikawa, 1956: 277, figs 2e, 4a.
Allochthonius (Urochthonius) ishikawai uenoi Morikawa: Morikawa, 1960: 103, plate 5 fig. 3, plate 10 fig. 5; Morikawa, 1962: fig. 4.

Type locality: Komakado-kaza-ana Cave, Fujioka, Shizuoka, Japan.
Distribution: Japan.

Allochthonius (Urochthonius) ishikawai uyamadensis Morikawa

Allochthonius (Urochthonius) uyamadensis Morikawa, 1954b: 83, fig. 1e.
Allochthonius (Urochthonius) ishikawai uyamadensis Morikawa: Morikawa, 1960: 103, plate 7 fig. 4, plate 5 figs 8, 11; Morikawa, 1962: fig. 4.

Type locality: Uyama-do, Honshu, Japan.
Distribution: Japan.

Genus Aphrastochthonius J. C. Chamberlin

Aphrastochthonius J. C. Chamberlin, 1962: 307–308; Muchmore, 1972f: 433–435; Muchmore, 1973b: 47; Muchmore, 1986b: 17.

Type species: *Aphrastochthonius tenax* J. C. Chamberlin, 1962, by original designation.

Aphrastochthonius alteriae Muchmore

Aphrastochthonius alteriae Muchmore, 1977: 64, figs 1–3; Muchmore, 1982f: 63, fig. 1.

Type locality: Group IV ruins at Palenque, Chiapas, Mexico.
Distribution: Mexico.

Aphrastochthonius cubanus Dumitresco & Orghidan

Aphrastochthonius cubanus Dumitresco & Orghidan, 1977: 99–101, figs 1a–g.

Type locality: Isla de Pinos, Cuba.
Distribution: Cuba.

Aphrastochthonius grubbsi Muchmore

Aphrastochthonius grubbsi Muchmore, 1984f: 171–173, figs 1–2.

Type locality: Lost Piton Cave, 6.5 km SE. of Angels Camp, Calaveras County, California, U.S.A.
Distribution: U.S.A. (California).

Aphrastochthonius major Muchmore

Aphrastochthonius major Muchmore, 1973b: 47–48, fig. 1; Reddell & Elliott, 1973b: 183.

Type locality: Cueva de la Capilla, 13.5 km NW. of Gómez Farías, Tamaulipas, Mexico.
Distribution: Mexico.

Aphrastochthonius pachysetus Muchmore

Aphrastochthonius pachysetus Muchmore, 1976a: 361–363, figs 1–3.

Type locality: Doc Brito Cave, 4.5 miles NE. of Whites City, Eddy County, New Mexico, U.S.A.
Distribution: U.S.A. (New Mexico).

Aphrastochthonius palmitensis Muchmore

Aphrastochthonius palmitensis Muchmore, 1986b: 17–19, fig. 1.

Type locality: Grutas del Palmito, 7 km SSW. of Bustamante, Nuevo Léon, Mexico.
Distribution: Mexico.

Aphrastochthonius parvus Muchmore

Aphrastochthonius parvus Muchmore, 1972f: 437–438, figs 4–7; Reddell & Elliott, 1973a: 173.

Type locality: La Cueva de la Florida, 15 km SSW. of Mante, Tamaulipas, Mexico.
Distribution: Mexico.

Aphrastochthonius patei Muchmore

Aphrastochthonius patei Muchmore, 1982f: 63–64, fig. 2.

Type locality: Cueva de Oyamel, SW. of El Barretal, Tamaulipas, Mexico.
Distribution: Mexico.

Aphrastochthonius pecki Muchmore

Aphrastochthonius pecki Muchmore, 1968a: 17–18, figs 1–2.

Type locality: Crystal Caverns, Clay, Jefferson County, Alabama, U.S.A.
Distribution: U.S.A. (Alabama).

Aphrastochthonius russelli Muchmore

Aphrastochthonius russelli Muchmore, 1972f: 440–441, figs 10–12; Reddell & Elliott, 1973a: 173.

Type locality: La Cueva Pinta, about 12 km NE. of Valles, San Luis Potosi, Mexico.
Distribution: Mexico.

Aphrastochthonius similis Muchmore

Aphrastochthonius similis Muchmore, 1984f: 173–174, fig. 3.

Type locality: Carlow's Cave, 7.5 km SE. of Angels Camp, Calaveras County, California, U.S.A.
Distribution: U.S.A. (California).

Aphrastochthonius tenax J. C. Chamberlin

Aphrastochthonius tenax J. C. Chamberlin, 1962: 308–310, figs 2a–j; Muchmore, 1972f: 435, figs 1–3.

Type locality: Bangor Cave, Blount County, Alabama, U.S.A.
Distribution: U.S.A. (Alabama).

Aphrastochthonius verapazanus Muchmore

Aphrastochthonius verapazanus Muchmore, 1972f: 438–440, figs 8–9; Muchmore, 1977: 64.

Type locality: La Cueva Sepacuite #2, Senahu Finca Sepacuite, Alta Verapaz, Guatemala.
Distribution: Guatemala.

Genus Apochthonius J. C. Chamberlin

Apochthonius J. C. Chamberlin, 1929a: 66–67; J. C. Chamberlin, 1923e: 152; Beier, 1932a: 41; Hoff, 1949b: 434; Hoff, 1956a: 2; Benedict & Malcolm, 1973: 621; Muchmore & Benedict, 1976: 68–69.
Apochthonius (Apochthonius): J. C. Chamberlin: Beier, 1932a: 41.

Type species: *Chthonius moestus* Banks, 1891, by original designation.

Apochthonius coecus (Packard)

Chthonius coecus Packard, 1884: 203; Packard, 1886: 46–48, plate 11 figs 4, 4a–c; Banks, 1895a: 13; Coolidge, 1908: 114; Roewer, 1937: 239.
Chthonius (?) *coecus* Packard: Beier, 1932a: 61; Hoff, 1958: 4.
Apochthonius coecus (Packard): Muchmore, 1963a: 9–11, figs 5–6; Holsinger, 1963: 34; Holsinger & Culver, 1988: 42, fig. 15.

Type locality: Weyer's Cave, Augusta County, Virginia, U.S.A.
Distribution: U.S.A. (Virginia).

Apochthonius colecampi Muchmore

Apochthonius colecampi Muchmore, 1967d: 89–90, fig. 1; Muchmore, 1976e: 76.

Type locality: Cole Camp Cave, SE. of Lincoln, Benton County, Missouri, U.S.A.
Distribution: U.S.A. (Missouri).

Apochthonius diabolus Muchmore

Apochthonius diabolus Muchmore, 1967d: 92, fig. 3.

Type locality: Devil's Den, Devil's Den State Park, Washington County, Arkansas.
Distribution: U.S.A. (Arkansas).

Apochthonius forbesi Benedict

Apochthonius forbesi Benedict, 1979: 79–82, figs 1–3.

Type locality: 14 km S., 11 km E. of Bend, Deschutes County, Oregon, U.S.A.
Distribution: U.S.A. (Oregon).

Apochthonius grubbsi Muchmore

Apochthonius grubbsi Muchmore, 1980b: 93–94, figs 1–3.

Type locality: Music Hall Cave, 4.5 miles SE. of Angels Camp, Calavera County, California, U.S.A.
Distribution: U.S.A. (California).

Apochthonius holsingeri Muchmore

Apochthonius holsingeri Muchmore, 1967d: 93–94, fig. 5; Muchmore, 1976e: 77; Holsinger & Culver, 1988: 42, fig. 15.

Type locality: Cave Run Pit Cave, about 8 miles N. of Warm Springs, Bath County, Virginia, U.S.A.
Distribution: U.S.A. (Virginia).

Apochthonius hypogeus Muchmore

Apochthonius hypogeus Muchmore, 1976e: 75–76, fig. 8.

Type locality: Great North Mountain, Augusta County, Virginia, U.S.A.
Distribution: U.S.A. (Virginia).

Apochthonius indianensis Muchmore

Apochthonius indianensis Muchmore, 1967d: 92–93, fig. 4; Muchmore, 1976e: 76.

Type locality: Donaldson's Cave, Spring Mill State Cave, Lawrence County, Indiana, U.S.A.
Distribution: U.S.A. (Indiana).

Apochthonius intermedius J. C. Chamberlin

Apochthonius (Apochthonius) intermedius J. C. Chamberlin, 1929e: 153; Beier, 1932a: 42.
Apochthonius intermedius J. C. Chamberlin: J. C. Chamberlin, 1931a: figs 22a–b, 46a; Beier, 1932g: fig. 223; Roewer, 1937: 238; Snodgrass, 1948: fig. 12a; Vachon, 1949: fig. 239; Hoff, 1958: 6; R. O. Schuster, 1966b: figs 4, 8–9.

Type locality: Friday Island, San Juan Island, Washington, U.S.A.
Distribution: U.S.A. (Washington).

Apochthonius irwini R. O. Schuster

Apochthonius irwini R. O. Schuster, 1966b: 179–182, figs 11–12.

Type locality: 5 miles MW. of Spanish Flat, Napa County, California, U.S.A.
Distribution: U.S.A. (California).

Apochthonius knowltoni Muchmore

Apochthonius knowltoni Muchmore, 1980a: 87–89, figs 1–5.

Type locality: 10 miles SE. of Smoot, Lincoln County, Wyoming, U.S.A.
Distribution: U.S.A. (Wyoming).

Apochthonius magnanimus Hoff

Apochthonius magnanimus Hoff, 1956a: 4–10, figs 2–4; Hoff, 1958: 6; Hoff, 1959b: 4, etc.; Zeh, 1987b: 1086.

Type locality: W. of Cowles, San Miguel County, New Mexico, U.S.A.
Distribution: U.S.A. (New Mexico).

Apochthonius malheuri Benedict & Malcolm

Apochthonius malheuri Benedict & Malcolm, 1973: 621–626, figs 1–6.

Type locality: Malheur Cave, 13 miles SE. of Princeton, Harney County, Oregon, U.S.A.

Distribution: U.S.A. (Oregon).

Apochthonius maximus R. O. Schuster

Apochthonius maximus R. O. Schuster, 1966b: 182, figs 3, 10, 13.

Type locality: Caspar Loop Road, Russian Gulch State Park, Mendocino County, California, U.S.A.

Distribution: U.S.A. (California).

Apochthonius minimus R. O. Schuster

Apochthonius minimus R.O. Schuster, 1966b: 178–179, figs 1, 2, 5, 7; D.L. Johnson & Wellington, 1980a: 339–349; D.L. Johnson & Wellington, 1980b: 353–362.

Type locality: 2 miles N. of Brinnon, Olympic Peninsula, Washington, U.S.A.

Distribution: Canada (British Columbia), U.S.A. (Oregon, Washington).

Apochthonius minor Muchmore

Apochthonius minor Muchmore, 1976e: 72–75, figs 6–7.

Type locality: Parker Cave, 2 miles NE. of Subligna, Chatooga County, Georgia, U.S.A.

Distribution: U.S.A. (Georgia).

Apochthonius moestus (Banks)

Chthonius moestus Banks, 1891: 165; Banks, 1895a: 13; Coolidge, 1908: 114; Ewing, 1911: 80.

Apochthonius moestus (Banks): J. C. Chamberlin, 1929a: 67, figs 2c, 3g (in part); J. C. Chamberlin, 1931a: figs 6a, 21f; Beier, 1932a: 41–42, figs 47–48 (in part); Roewer, 1937: 238; Hoff, 1944b: 125, figs 1c–d; Hoff, 1945a: 105–109, figs 4–7; Pearse, 1946: 257; Hoff, 1958: 6 (in part); Muchmore & Benedict, 1976: 69–73, figs 1–9; Rowland & Reddell, 1976: 14; Sweeney, Taylor & Counts, 1977: 55; Taylor, Sweeney & Counts, 1977: 115; Zeh, 1987b: 1086.

Apochthonius sp. prob. *moestus* (Banks): Graves & Graves, 1969: 267.

Not *Apochthonius moestus* (Banks): J. C. Chamberlin, 1929a: 67 (in part); Beier, 1930a: 202; Hoff, 1945i: 311–313; Hoff, 1949b, figs 11, 15a–d: 434; Hoff, 1951: 4–5; Hoff, 1952a: 42–43; Hoff, 1956a: 2–4, fig. 1; Hoff & Bolsterli, 1956: 158–159; Hoff, 1958: 6 (in part); Hoff, 1959b: 4, etc.; Nelson, 1975: 279–280, figs 14–17 (misidentifications; true identifications unknown).

Not *Apochthonius (Apochthonius) moestus* (Banks): J. C. Chamberlin, 1929e: 153; Beier, 1932a: 41–42 (in part) (misidentifications; true identifications unknown).

Type locality: Ithaca, New York, U.S.A.

Distribution: U.S.A. (Indiana, Illinois, Missouri, New Mexico, New York, North Carolina, Pennsylvania, West Virginia).

Apochthonius mysterius Muchmore

Apochthonius mysterius Muchmore, 1976e: 70, fig. 4.

Type locality: Red Fork Passage, Mystery Cave, about 5 miles SE. of Perryville, Perry County, Missouri, U.S.A.

Distribution: U.S.A. (Missouri).

Apochthonius occidentalis J. C. Chamberlin

Apochthonius occidentalis J. C. Chamberlin, 1929a: 67–68; Beier, 1932a: 42; Roewer, 1937: 238; Hoff, 1958: 6.

Type locality: Portland, Oregon, U.S.A.
Distribution: U.S.A. (Oregon).

Apochthonius paucispinosus Muchmore

Apochthonius paucispinosus Muchmore, 1967d: 90–92, fig. 2.

Type locality: Bennett Cave, about 2 miles N. of Dryfork, Tucker County, West Virginia, U.S.A.
Distribution: U.S.A. (West Virginia).

Apochthonius russelli Muchmore

Apochthonius russelli Muchmore, 1976e: 70–72, fig. 5.

Type locality: Pig Entrance, Russell Cave, Russell Cave National Monument, Jackson County, Alabama, U.S.A.
Distribution: U.S.A. (Alabama).

Apochthonius titanicus Muchmore

Apochthonius titanicus Muchmore, 1976e: 67–69, figs 1–3.

Type locality: near 'The Titans', Blanchard Springs Caverns, 3 miles E. of Fifty Six, Stone County, Arkansas, U.S.A.
Distribution: U.S.A. (Arkansas).

Apochthonius typhlus Muchmore

Apochthonius typhlus Muchmore, 1967d: 94–95, fig. 6.

Type locality: Old Spanish Cave, Stone County, Missouri, U.S.A.
Distribution: U.S.A. (Missouri).

Genus **Austrochthonius** J. C. Chamberlin

Austrochthonius J. C. Chamberlin, 1929a: 68; Beier, 1932a: 38; Beier, 1976f: 203.
Paraustrochthonius Beier, 1931b: 52; Beier, 1932a: 40 (synonymized by Beier, 1976f: 203).

Type species: of *Austrochthonius*: *Chthonius chilensis* J. C. Chamberlin, 1923b, by original designation.
of *Paraustrochthonius*: *Paraustrochthonius tullgreni* Beier, 1931b, by original designation.

Austrochthonius argentinae Hoff

Austrochthonius argentinae Hoff, 1950b: 225–228, figs 1–6; Beier, 1974d: 899.

Type locality: Pampa de Olaen, Córdoba, Argentina.
Distribution: Argentina, Brazil (Santa Catarina).

Austrochthonius australis Hoff

Austrochthonius australis Hoff, 1951: 5–9, figs 1–3; Beier, 1966a: 281–283, fig. 2; Beier, 1969a: 187; Dartnall, 1970: 65–66; Beier, 1975b: 203; Hickman & Hill, 1978: appendix a; Koch & Majer, 1980: 25; Harvey, 1981b: 239; Main, 1981: 1065; Harvey, 1985b: 137.

Type locality: Mt Slide, Victoria, Australia.
Distribution: Australia (Australian Capital Territory, New South Wales, Queensland, Tasmania, Victoria, Western Australia).

Family Chthoniidae

Austrochthonius bolivianus Beier

Austrochthonius bolivianus Beier, 1930a: 201–202; Beier, 1932a: 38–39, fig. 42; Roewer, 1937: 238; Beier, 1974d: 899.

Type locality: Bolivia.
Distribution: Bolivia, Brazil (Santa Catarina).

Austrochthonius cavicola Beier

Austrochthonius cavicola Beier, 1968a: 757–758, fig. 1; Harvey, 1981b: 239; D. C. Lee, 1983: 183; Harvey, 1985b: 137.

Type locality: Cathedral Cave, Naracoorte, South Australia, Australia.
Distribution: Australia (South Australia).

Austrochthonius chilensis (J. C. Chamberlin)

Chthonius chilensis J. C. Chamberlin, 1923b: 187–189, figs 20a–j.
Austrochthonius chilensis (J. C. Chamberlin): J. C. Chamberlin, 1929a: 68; J. C. Chamberlin, 1931a: fig. 21e; Beier, 1932a: 38; Roewer, 1937: 238; Beier, 1962h: 131; Beier, 1964b: 312–313; Cekalovic, 1984: 8.

Type locality: Butalcura, Ciloe Island, Chile.
Distribution: Argentina, Chile.

Austrochthonius chilensis magalhanicus Beier

Austrochthonius chilensis magalhanicus Beier, 1964b: 313; Cekalovic, 1976: 17; Cekalovic, 1984: 9.

Type locality: Magallanes (as Magellanes), Chile.
Distribution: Chile.

Austrochthonius chilensis transversus Beier

Austrochthonius chilensis transversus Beier, 1964j: 488, fig. 1.

Type locality: Mt Piltriquitron, El Bolsón, Rio Negro, Argentina.
Distribution: Argentina.

Austrochthonius iguazuensis Vitali-di Castri

Austrochthonius iguazuensis Vitali-di Castri, 1975a: 117–119, figs 1–2, 5–6, 9–10.

Type locality: Cataratas del Iguazú, Paraná, Brazil.
Distribution: Brazil (Paraná).

Austrochthonius insularis Vitali-di Castri

Austrochthonius insularis Vitali-di Castri, 1968: 141–144, figs 1–7.

Type locality: Possession Island, Crozet Islands.
Distribution: Crozet Islands.

Austrochthonius mordax Beier

Austrochthonius inversus mordax Beier, 1967b: 282–283, fig. 4.
Austrochthonius mordax Beier: Beier, 1976f: 203–204.

Type locality: Lake Wairarapa, North Island, New Zealand.
Distribution: New Zealand.

Austrochthonius paraguayensis Vitali-di Castri

Austrochthonius paraguayensis Vitali-di Castri, 1975a: 121–124, figs 3–4, 7–8.

Type locality: Puerto del Este (as Puerto Presidente Stroessner), Paraguay.
Distribution: Paraguay.

Austrochthonius persimilis Beier

Austrochtonius (sic) *persimilis* Beier, 1930d: 205–206, figs 10a–d.
Austrochthonius persimilis Beier: Beier, 1932a: 39, fig. 43; Roewer, 1937: 238; Beier, 1964b: 313–314, fig. 1; Cekalovic, 1984: 9.

Type locality: Pitrufquén (as Pitrufquen), Chile.
Distribution: Chile.

Austrochthonius rapax Beier

Austrochthonius rapax Beier, 1976f: 204, fig. 4.

Type locality: Tuna Saddle, Taumarunui, North Island, New Zealand.
Distribution: New Zealand.

Austrochthonius semiserratus Beier

Austrochthonius semiserratus Beier, 1930d: 206–207, figs 11a–b, 12–13; Beier, 1932a: 39, figs 44–45; Roewer, 1937: 238; Beier, 1974d: 899; Cekalovic, 1984: 9.
Austrochthonius semisseratus (sic) Beier: Vitali-di Castri & Castri, 1976: 62, 63.
Not *Austrochthonius semiserratus* Beier: Beier, 1964b: 314 (misidentification; see *Chiliochthonius montanus* Vitali-di Castri).

Type locality: Temuco, Chile.
Distribution: Brazil (Santa Catarina), Chile.

Austrochthonius tullgreni (Beier), new combination

Paraustrochthonius tullgreni Beier, 1931b: 53; Beier, 1932a: 40, fig. 46; Roewer, 1937: 238; Beier, 1958a: 159; Beier, 1964k: 31; Beier, 1966k: 455–456; Lawrence, 1967: 87.

Type locality: Pietermaritzburg, Natal, South Africa.
Distribution: South Africa.

Austrochthonius zealandicus Beier

Austrochthonius zealandicus Beier, 1966c: 363–364, fig. 1; Tenorio & Muchmore, 1982: 384.
Austrochthonius luxtoni Beier, 1967b: 279–280, fig. 2 (synonymized by Beier, 1976f: 203).
Austrochthonius inversus Beier, 1967b: 280–282, fig. 3; Beier, 1969e: 413, 414; Wise, 1970: 234 (synonymized by Beier, 1976f: 203).
Austrochthonius zealandicus zealandicus Beier: Beier, 1976f: 203.

Type localities: of *Austrochthonius zealandicus*: Birdling Flat, South Island, New Zealand.
of *Austrochthonius luxtoni*: Mackenzie's Bush, near Oparau, New Zealand.
of *Austrochthonius inversus*: Palmerston North, Pohangina Valley, Silverstream, South Island, New Zealand.
Distribution: New Zealand.

Austrochthonius zealandicus obscurus Beier

Austrochthonius obscurus Beier, 1966c: 364–365, fig. 2; Beier, 1967b: 279; Tenorio & Muchmore, 1982: 382.
Austrochthonius zealandicus obscurus Beier: Beier, 1976f: 203.

Type locality: Scenic Reserve, Kaituna Valley, Banks Peninsula, South Island, New Zealand.
Distribution: New Zealand.

Family Chthoniidae

Genus **Caribchthonius** Muchmore

Caribchthonius Muchmore, 1976c: 361−363.

Type species: *Caribchthonius butleri* Muchmore, 1976c, by original designation.

Caribchthonius butleri Muchmore

Caribchthonius butleri Muchmore, 1976c: 363−365, figs 1−7.

Type locality: Reef Bay Trail, Virgin Islands National Park, St John, U.S. Virgin Islands.
Distribution: U.S. Virgin Islands.

Caribchthonius orthodentatus Muchmore

Caribchthonius orthodentatus Muchmore, 1976c: 365−367, figs 8−12.

Type locality: Corozal, Belize.
Distribution: Belize.

Genus **Centrochthonius** Beier

Centrochthonius Beier, 1931b: 56; Beier, 1932a: 69.

Type species: *Chthonius kozlovi* Redikorzev, 1918, by original designation.

Centrochthonius kozlovi (Redikorzev)

Chthonius kozlovi Redikorzev, 1918: 100−101, figs 9, 10, 10a−c.
Centrochthonius kozlovi (Redikorzev): Beier, 1931b: 56; Beier, 1932a: 70, fig. 84; Roewer, 1937: 241; Mani, 1962: 235.

Type locality: Dulon-kit, Tibet, China.
Distribution: China (Tibet), India, Nepal.

Centrochthonius schnitnikovi (Redikorzev)

Chthonius (Centrochthonius) schnitnikovi Redikorzev, 1934a: 439−440, fig. 19.
Centrochthonius schnitnikovi (Redikorzev): Roewer, 1937: 241; Redikorzev, 1949: 641.

Type locality: Tujuk River, Semiretshje, R.S.F.S.R., U.S.S.R.
Distribution: U.S.S.R. (Kirghizia, R.S.F.S.R.).

Centrochthonius ussuriensis Beier

Centrochthonius ussuriensis Beier, 1979b: 554−556, fig. 2; Schawaller, 1985a: 2; Schawaller, 1989: 8, figs 18−19.

Type locality: Ussuri Station (as Ussurijsky-Station), Primorskiy Kray, R.S.F.S.R., U.S.S.R.
Distribution: U.S.S.R. (R.S.F.S.R.).

Genus **Chiliochthonius** Vitali-di Castri

Chiliochthonius Vitali-di Castri, 1975b: 1277−1278.

Type species: *Chiliochthonius centralis* Vitali-di Castri, 1975b, by original designation.

Chiliochthonius centralis Vitali-di Castri

Chiliochthonius centralis Vitali-di Castri, 1975b: 1278−1280, figs 1−6; Cekalovic, 1984: 9.

Type locality: Quebrada La Plata, Santiago, Chile.
Distribution: Chile.

Chiliochthonius montanus Vitali-di Castri

Austrochthonius semiserratus Beier: Beier, 1964b: 314 (misidentification).
Chiliochthonius montanus Vitali-di Castri, 1975b: 1280–1283, figs 7–12, plate 2; Cekalovic, 1984: 10.

Type locality: Farellones, Santiago, Chile.
Distribution: Chile.

Genus Chthonius C. L. Koch

Chthonius C. L. Koch, 1843: 76; Menge, 1855: 22–23; Stecker, 1874b: 237; E. Simon, 1879a: 69; Tömösváry, 1882b: 237; Canestrini, 1884: no. 6; H. J. Hansen, 1884: 553; Balzan, 1887a: no pagination; Daday, 1888: 133–134; Balzan, 1890: 446–447; Balzan, 1892: 545–546; O. P. Cambridge, 1892: 208; Tullgren, 1899a: 181; Kew, 1911a: 56; Lessert, 1911: 37; J. C. Chamberlin, 1925b: 334; Pratt, 1927: 410; Schenkel, 1928: 70; J. C. Chamberlin, 1929a: 69; Beier, 1932a: 43; J. C. Chamberlin, 1935b: 479; Beier, 1939f: 160; Hoff, 1949b: 432; Petrunkevitch, 1955: 80; Beier, 1963b: 19–20; Legg, 1987: 179; Legg & Jones, 1988: 59.
Kewochthonius J. C. Chamberlin, 1929a: 65; Hoff, 1951: 3–4; Legg & Jones, 1988: 56 (synonymized by Judson, 1988: 131).

Type species of *Chthonius*: *Obisium orthodactylum* Leach, 1817, by subsequent designation of E. Simon, 1879a: 69.
Type species of *Kewochthonius*: *Chthonius halberti* Kew, 1916b, by original designation.

Subgenus Chthonius (Chthonius) C. L. Koch

Chthonius (Chthonius) C. L. Koch: Beier, 1932a: 47–48; Hoff, 1949b: 433.

Chthonius (Chthonius) absoloni Beier

Chthonius (Chthonius) absoloni Beier, 1938a: 5; Beier, 1939d: 10, fig. 3; Beier, 1963b: 28, fig. 19; Ćurčić, 1974a: 11; Ćurčić, 1988b: 26–28, figs 26–30.

Type locality: Duzica Cave, Trebinja, Yugoslavia.
Distribution: Yugoslavia.

Chthonius (Chthonius) agazzii Beier

Chthonius (Chthonius) agazzii Beier, 1966h: 175–177, fig. 1; Callaini, 1980a: 226–227; Gardini, 1980c: 97, 125.

Type locality: Grotta la Bisongola, M. Vallina, Segusino, Treviso, Veneto, Italy.
Distribution: Italy.

Chthonius (Chthonius) alpicola Beier

Chthonius (Neochthonius) alpicola Beier, 1951c: 163–164; Beier, 1952e: 2; Beier & Franz, 1954: 453; Beier, 1956i: 8; Beier, 1963b: 23; Beier, 1963h: 147–148; Ressl, 1965: 289.
Neochthonius cfr. *alpicola* (Beier): Callaini, 1986e: 229–230.
Chthonius (Chthonius) alpicola Beier: Judson, 1990: 597.

Type locality: Krizersberg, Salzburg, Austria.
Distribution: Austria, Italy.

Chthonius (Chthonius) apollinis Mahnert

Chthonius (Chthonius) apollinis Mahnert, 1978e: 287–291, figs 36–46.
Chthonius (Chthonius) aff. *apollinis* Mahnert: Mahnert, 1980e: 214.

Type locality: Koryghion Antron Cave, 10 km from Arachova, Voiotia (as Beotien), Greece.
Distribution: Greece.

Chthonius (Chthonius) azerbaidzhanus Schawaller & Dashdamirov

Chthonius (Chthonius) azerbaidzhanus Schawaller & Dashdamirov, 1988: 4–6, figs 6–11, 27.

Type locality: Shemakha District, Azerbaijan, U.S.S.R.
Distribution: U.S.S.R. (Azerbaijan).

Chthonius (Chthonius) baccettii Callaini

Chthonius (Chthonius) baccettii Callaini, 1980b: 203–213, figs 1a–f, 2, 3a–g, 4a–g; Callaini, 1986e: 236.

Type locality: Andriano, Balzano, Alto Adige, Italy.
Distribution: Italy.

Chthonius (Chthonius) balazuci Vachon

Chthonius (Chthonius) balazuci Vachon, 1963: 394–398, figs 1–7; Leclerc & Heurtault, 1979: 240, map 1; Heurtault, 1986c: 19.

Type locality: Grotte des Baumas, Larnas, Ardèche, France.
Distribution: France.

Chthonius (Chthonius) bogovinae Ćurčić

Chthonius (Chthonius) bogovinae Ćurčić, 1972c: 342–347, figs 1, 2a–d, 3a–b, 4a–c, 5a–b; Ćurčić, 1976b: 171, fig. 1.
Chthonius (Chthonius) bogovinae bogovinae Ćurčić: Ćurčić, 1974a: 11.
Chthonius bogovinae Ćurčić: Ćurčić, 1983b: fig. 2.

Type locality: Bogovina Cave, 11 km N. of Boljevac, Serbia, Yugoslavia.
Distribution: Yugoslavia.

Chthonius (Chthonius) bogovinae latidentatus Ćurčić

Chthonius (Chthonius) bogovinae latidentatus Ćurčić, 1972d: 130–139, figs 7–14; Ćurčić, 1974a: 11; Ćurčić, 1976b: 171, fig. 1.

Type locality: Mitrova Pecina, near Cuprija, Serbia, Yugoslavia.
Distribution: Yugoslavia.

Chthonius (Chthonius) brandmayri Callaini

Chthonius (Chthonius) brandmayri Callaini, 1986e: 237–241, figs 2a–e.

Type locality: Monte Taiano, Istria, Yugoslavia.
Distribution: Yugoslavia.

Chthonius (Chthonius) caprai Gardini

Chthonius (Neochthonius) caprai Gardini, 1977b: 216–220, figs 1–7; Callaini, 1979b: 129–130.
Chthonius (Chthonius) caprai Gardini: Judson, 1990: 597.

Type locality: Monte di Portofino, Liguria, Italy.
Distribution: Italy.

Chthonius (Chthonius) caoduroi Callaini

Chthonius (Chthonius) caoduroi Callaini, 1987a: 69–73, figs 1a–f.

Type locality: Grotta di Trecchina, Potenza, Lucania, Italy.
Distribution: Italy.

Chthonius (Chthonius) cavernarum Ellingsen

Chthonius cavernarum Ellingsen, 1909a: 217–218; Ellingsen, 1910a: 401; Hadži, 1930b: 16–22, figs 7a–f, 8–9; Ionescu, 1936: 2–3; Roewer, 1937: 239.
Chthonius (Chthonius) cavernarum Ellingsen: Beier, 1932a: 53, fig. 61 (in part, see *Chthonius (Chthonius) ellingseni* Beier); Beier, 1939d: 16, fig. 13; Beier, 1963b: 49–50, fig. 46b; Beier, 1970f: 318; Ćurčić, 1974a: 11.
Not *Chthonius cavernarum* Ellingsen: Beier, 1928: 313; Beier, 1931d: 91–92 (misidentifications; see *Chthonius (Chthonius) ellingseni* Beier).

Type locality: Sinaia, Romania.
Distribution: Romania, Yugoslavia.

Chthonius (Chthonius) cavophilus Hadži

Chthonius (Chthonius) cavophilus Hadži, 1939b: 21–22, figs 2a–e; Beier, 1963b: 45–46.

Type locality: Glava Panega Cave, Bulgaria.
Distribution: Bulgaria.

Chthonius (Chthonius) cebenicus Leclerc

Chthonius (Chthonius) cebenicus Leclerc, 1981: 116–118, plate 1 figs 1–7; Heurtault, 1986c: 20.

Type locality: Grotte de Bedousse, Aujac, Gard, France.
Distribution: France.

Chthonius (Chthonius) cephalotes (E. Simon)

Blothrus cephalotes E. Simon, 1875: ccvi–ccvii.
Chthonius cephalotes (E. Simon): E. Simon, 1879a: 76, plate 19 fig. 20; Ellingsen, 1912c: 173; Roewer, 1937: 239.
Chthonius (Chthonius) cephalotes (E. Simon): Heurtault-Rossi, 1968a: 912–916, figs 1–6; Heurtault, 1986c: 20.
Not *Chthonius cephalotes* (E. Simon): Ellingsen, 1909a: 212 (misidentification; see *Chthonius (Chthonius) italicus* Beier); Boldori, 1932: 127; Costantini, 1976: 122 (misidentifications; see *Chthonius (Chthonius) doderoi* Beier).
Not *Chthonius (Chthonius) cephalotes* (E. Simon): Beier, 1932a: 49, fig. 53 (misidentification; see *Chthonius (Chthonius) mayi* Heurtault).
Not *Chthonius (Chthonius) cephalotes cephalotes* (E. Simon): Beier, 1932a: 49; Beier, 1963b: 28–29, fig. 20 (misidentification; see *Chthonius (Chthonius) mayi* Heurtault).

Type locality: Grotte des Baux, Bouches-du-Rhône, France.
Distribution: France.

Chthonius (Chthonius) chamberlini (Leclerc)

Neochthonius chamberlini Leclerc, 1983a: 45–48, figs 1–6.
Neochthonius cfr. *chamberlini* Leclerc: Callaini, 1983e: 212–215, figs 1a–d.
Chthonius (Chthonius) chamberlini (Leclerc): Judson, 1990: 597.

Type locality: Marières à Bruniquel, Tran et Garonne, France.
Distribution: Algeria, France.

Chthonius (Chthonius) comottii Inzaghi

Chthonius (Chthonius) comottii Inzaghi, 1987: 166–173, figs 1–17.

Type locality: Oneta, Bergamo, Lombardia, Italy.
Distribution: Italy.

Chthonius (Chthonius) cryptus J. C. Chamberlin

Chthonius caecus E. Simon, 1885b: 214; Roewer, 1937: 239 (junior primary homonym of *Chthonius coecus* Packard, 1884).
Chthonius (Chthonius) caecus E. Simon: Beier, 1932a: 54, fig. 63; Beier, 1939d: 14; Beier, 1963b: 41, fig. 34.
Chthonius (Chthonius) cryptus J. C. Chamberlin, 1962: 307 (replacement name for *Chthonius caecus* E. Simon, 1885b); Mahnert, 1975a: 169; Mahnert, 1978e: 283–286, figs 26–31; Mahnert, 1980e: 214.

Type locality: Kokkino-Vrachio, Mt Ossa, Greece.
Distribution: Greece.

Chthonius (Chthonius) dacnodes Navás

Chthonius dacnodes Navás, 1918: 133–134, figs 6a–b; Navás, 1925: 126, figs 19a–b; Vachon, 1961: 102, 103; Lazzeroni, 1970a: 207; Pieper, 1981: 2.
Chthonius (?) dacnodes Navás: Beier, 1932a: 61; Roewer, 1937: 240.
Chthonius (Chthonius) dacnodes Navás: Beier, 1939f: 161–162, fig. 1; Beier, 1955n: 88; Beier, 1959f: 113–114; Beier, 1961b: 21; Beier, 1963b: 35, fig. 27; Lazzeroni, 1969a: 323–324, fig. 1; Lagar, 1972a: 17; Orghidan, Dumitresco & Georgesco, 1975: 29; Mahnert, 1986d: 74–76, fig. 2.
Chthonius (Chthonius) cfr. *dacnodes* Navás: Callaini, 1979c: 342–343, figs 2a–f; Bologna & Taglianti, 1985: 226, fig. 8.
Not *Chthonius (Chthonius) dacnodes* Navás: G. O. Evans & Browning, 1954: 10, figs 10–11; Legg, 1971a: 475, fig. 1e (misidentifications; see *Chthonius (Chthonius) ischnocheles* (Hermann)).
Not *Chthonius* sp. cf. *dacnodes* Navás: Turk, 1967: 153 (misidentification; true identification unknown).

Type locality: Veruela, Zaragoza, Spain.
Distribution: Azores, Balearic Islands, Italy, Madeira Islands, Spain.

Chthonius (Chthonius) dalmatinus Hadži

Chthonius rayi dalmatinus Hadži, 1930a: 72–79, figs 10a–e, 11a–e, 12a–b, 13a–d, 14, 15a–b, 16a–c; Hadži, 1930c: 22–24.
Chthonius (Chthonius) ischnocheles dalmatinus Hadži: Beier, 1932a: 48.
Chthonius (Chthonius) dalmatinus Hadži: Hadži, 1933b: 126–130, figs 1a–f; Hadži, 1933c: 173–176; Beier, 1963b: 24; Ćurčić, 1974a: 11; Ćurčić, 1988b: 20–22, figs 14–17.

Type locality: Sali, Dugi Otok, Yugoslavia.
Distribution: Yugoslavia.

Chthonius (Chthonius) densedentatus Beier

Chthonius (Chthonius) densedentatus Beier, 1938a: 5; Beier, 1939d: 8–9, fig. 1; Beier, 1963b: 24–25; Ćurčić, 1974a: 11–12.
Chthonius cfr. *densedentatus* Beier: Callaini, 1986e: 234–236, figs 1g–n.

Type locality: Pjecaljina Cave, Popova Polja, Yugoslavia.
Distribution: Italy, Yugoslavia.

Chthonius (Chthonius) diophthalmus Daday

Chthonius diophthalmus Daday, 1888: 134–135, 191, plate 4 figs 21, 27; Daday, 1918: 2.
Chthonius (?) diophthalmus Daday: Beier, 1932a: 61; Roewer, 1937: 240.
Chthonius (Chthonius) orthodactylus gracilis Beier, 1935g: 31, fig. 1 (synonymized by Beier, 1939c: 2).

Chthonius (Chthonius) diophthalmus Daday: Beier, 1939c: 2−4, fig. 1; Beier, 1963b: 39, fig. 32; Mahnert, 1980e: 214−215; Krumpál & Kiefer, 1981: 127−128, figs 1−4; Schawaller, 1989: 6, figs 15−16.
Chthonius (Chthonius) ksenemani Hadži, 1939a: 184−188, figs 1a−e, 2, 3a−d; Krumpál & Kiefer, 1981: 128 (synonymized by Schawaller, 1989: 6).
Chthonius orthodactylus gracilis Beier: Roewer, 1940: 343.
Chthonius (Chthonius) knesemanni (sic) Hadži: Beier, 1963b: 40−41.
Chthonius (Chthonius) aff. *diophthalmus* Daday: Mahnert, 1978e: 286−287, figs 32−35.

Type localities: of *Chthonius diophthalmus*: Mehadia, Romania.
of *Chthonius (Chthonius) orthodactylus gracilis*: Sinaia, Romania.
of *Chthonius (Chthonius) ksenemani*: near Brno, Czechoslovakia.
Distribution: Czechoslovakia, Greece, Romania, U.S.S.R. (Ukraine).

Chthonius (Chthonius) doderoi Beier

Chthonius (Chthonius) cephalotes doderoi Beier, 1930b: 73; Beier, 1932a: 49; Beier, 1963b: 29.
Chthonius cephalotes E. Simon: Boldori, 1932: 127; Costantini, 1976: 122 (misidentifications).
Chthonius (Chthonius) doderoi doderoi Beier: Heurtault, 1968a: 921; Gardini, 1980c: 97−98, 122.

Type locality: Meailles Cave, Basses Alpes, France.
Distribution: France, Italy.

Chthonius (Chthonius) doderoi horridus Beier

Chthonius (Chthonius) cephalotes horridus Beier, 1934a: 54−55, fig. 2; Beier, 1963b: 30, fig. 21.
Chthonius (Chthonius) doderoi horridus Beier: Heurtault, 1968a: 921; Gardini, 1980c: 98.

Type locality: Oropa, Italy.
Distribution: Italy.

Chthonius (Chthonius) ellingseni Beier

Chthonius cavernarum Beier: Beier, 1928: 313; Beier, 1931d: 91−92 (misidentifications).
Chthonius (Chthonius) cavernarum Ellingsen: Beier, 1932a: 53 (misidentification, in part).
Chthonius (Chthonius) ellingseni Beier, 1939d: 16; Beier, 1952e: 2; Beier, 1956i: 8; Beier, 1963b: 49, fig. 45a; Ćurčić, 1974a: 12; Paoletti, 1979: 25 (not seen).
Chthonius (Chthonius) aff. *ellingseni* Beier: Ćurčić, 1988b: 22−24, figs 18−22.

Type locality: Zavrska Jama, Yugoslavia.
Distribution: Austria, Crete?, Italy, Yugoslavia.

Chthonius (Chthonius) elongatus Lazzeroni

Chthonius (Chthonius) elongatus Lazzeroni, 1970b: 141−145, figs 1−9; Gardini, 1979: 130−133, figs 13−19; Gardini, 1980c: 98, 128.
Chthonius elongatus Lazzeroni: Callaini, 1986d: 526−529, figs 2a−b.
Chthonius cfr. *elongatus* Lazzeroni: Callaini, 1986d: 529−531, figs 2c−f.

Type locality: Buca Tana, Maggiano, Lucca, Toscana, Italy.
Distribution: Italy.

Chthonius (Chthonius) exarmatus Beier

Chthonius (Chthonius) exarmatus Beier, 1939d: 15−16, fig. 12; Beier, 1963b: 48−49, fig. 44; Ćurčić, 1974a: 12.

Family Chthoniidae

Type locality: Izeta Cave, Orjena, Yugoslavia.
Distribution: Yugoslavia.

Chthonius (Chthonius) graecus Beier

Chthonius (Neochthonius) graecus Beier, 1963b: 21−22; Beier, 1965a: 85−86, fig. 1; Mahnert, 1978e: 274, figs 1−3.
Chthonius (Neochthonius?) graecus Beier: Mahnert, 1975a: 169.
Chthonius (Chthonius) graecus Beier: Judson, 1990: 597.

Type locality: Petalás (as Petalas), Akarnanía, Greece.
Distribution: Greece.

Chthonius (Chthonius) guglielmii Callaini

Chthonius (Chthonius) guglielmii Callaini, 1986e: 231−234, figs 1a−f.

Type locality: Monti Lessini, Trezzolano, Verona, Italy.
Distribution: Italy.

Chthonius (Chthonius) halberti Kew

Chthonius halberti Kew, 1916b: 77, fig. 3; Jones, 1980c: map 2; Roewer, 1937: 239.
Kewochthonius halberti (Kew): J. C. Chamberlin, 1929a: 65−66; Muchmore, 1968c: 71−75, figs 1−3.
Chthonius (Neochthonius) halberti Kew: Beier, 1932a: 46; G. O. Evans & Browning, 1954: 8; Beier, 1963b: 20−21, fig. 13; Crocker, 1978: 8; Judson, 1988: 131; Legg & Jones, 1988: 56−58, figs 9a, 9b(a−e).
Chthonius (Chthonius) halberti Kew: Legg, 1971a: 473; Judson, 1990: 598−599, fig. 11.

Type locality: Malahide, Dublin County, Ireland.
Distribution: France, Great Britain, Ireland.

Chthonius (Chthonius) herbarii Mahnert

Chthonius (Chthonius) herbarii Mahnert, 1980e: 220−222, figs 11−13.

Type locality: Aghios Ioannis Cave, Akrotiri, Crete.
Distribution: Crete.

Chthonius (Chthonius) heterodactylus Tömösváry

Chthonius heterodactylus Tömösváry, 1882b: 241−242, plate 5 figs 1−2; Jędryczkowski, 1987a: 342−343, fig. 1.
Chthonius (?) heterodactylus Tömösváry: Beier, 1932a: 61; Roewer, 1937: 240.
Chthonius (Chthonius) heterodactylus Tömösváry: Beier, 1939c: 4−5, fig. 2; Beier, 1963b: 27, fig. 18; Rafalski, 1967: 6; Dumitrescu, 1976: 275; Krumpál & Kiefer, 1981: 128.

Type localities: several localities in Hungary.
Distribution: Czechoslovakia, Hungary, Poland, Romania.

Chthonius (Chthonius) heurtaultae Leclerc

Chthonius (Chthonius) heurtaultae Leclerc, 1981: 123−126, plate 3 figs 1−7; Heurtault, 1986c: 20.

Type locality: Grotte de Trabuc à Mialet, Gard, France.
Distribution: France.

Chthonius (Chthonius) hungaricus Mahnert

Chthonius (Chthonius) hungaricus Mahnert, 1981e: 279−282, figs 1−5; Mahnert, 1983d: 361.

Type locality: Ujszentmargita, Hortobágy National Park, Hungary.
Distribution: Hungary.

Chthonius (Chthonius) ilvensis Beier

Chthonius (Neochthonius) ilvensis Beier, 1963h: 148–149, fig. 1; Costantini, 1976:
122; Callaini, 1979c: 339–341, figs 1a–b; Gardini, 1980c: 97, 123; Callaini, 1982a:
17, 18.
Chthonius (Chthonius) ilvensis Beier: Judson, 1990: 597.

Type locality: Canaloni, Mt Calamita, Elba, Italy.
Distribution: Italy.

Chthonius (Chthonius) imperator Mahnert

Chthonius (Chthonius) imperator Mahnert, 1978e: 279–283, figs 12–25.

Type locality: 'Spilia tou Garzeniko' Cave, Arkadhia (as Arkadien), Peloponnes,
Greece.
Distribution: Greece.

Chthonius (Chthonius) irregularis Beier

Chthonius (Chthonius) lanzai Caporiacco: Beier, 1953d: 35–36, fig. 1; Conci &
Franceschi, 1953: 47 (not seen) (misidentifications).
Chthonius (Chthonius) irregularis Beier, 1961f: 125; Beier, 1963b: 34–35, fig. 26;
Gardini, 1980a: 193–196, figs 1–9; Gardini, 1980c: 98, 120.

Type locality: Grotta di Pignone, near La Spezia, Liguria, Italy.
Distribution: Italy.

Chthonius (Chthonius) irregularis brevis Cîrdei, Bulimar & Malcoci

Chtonius (sic) *irregularis brevis* Cîrdei, Bulimar & Malcoci, 1967: 237–238.

Type locality: Masivul Repedea, Iasi, Romania.
Distribution: Romania.

Chthonius (Chthonius) ischnocheles (Hermann)

Chelifer ischnochelus Hermann, 1804: 118–119, plate 5 fig. p, plate 6 fig. 14.
Chelifer ischnocheles Hermann: Gervais, 1844: 81.
Chelifer trombidioides Latreille, 1804: 142–143; Latreille, 1806: 133 (not seen);
Latreille, 1825a: 133; Brébisson, 1827: 263 (synonymized by C. L. Koch, 1843: 77).
Chelifer trombidioidos (sic) Latreille: Leach, 1814: 429.
Obisium trombidioides (Latreille): Leach, 1815: 391; Dugès & Edwards, 1836: plate
20 bis figs 5, 5a–b.
Chthonius pensylvanicus (sic) Hagen, 1868b: 52.
Chthonius pennsylvanicus Hagen: Hagen, 1870: 268; Banks, 1890: 152; Banks, 1891:
164; Banks, 1895a: 13; Banks, 1904a: 141; Coolidge, 1908: 114; Ewing, 1911: 80;
Pratt, 1927: 410 (synonymized by J. C. Chamberlin, 1929a: 72).
Obisium ischnocheles (Hermann): Hagen, 1870: 270.
Chthonius rayi L. Koch, 1873: 48–49; Ritsema, 1874: xxxiv; Stecker, 1874b: 238;
Stecker, 1875d: 88; Canestrini, 1874: 228; E. Simon, 1879a: 74–75, plate 17 figs
2, 8, plate 19 fig. 19; Tömösváry: 1882b: 240–241, plate 5 figs 13–14; E. Simon,
1883: 279; Becker, 1884: cclxiv; Canestrini, 1885: no. 5, figs 1–3; O. P. Cambridge,
1892: 209–210, plate A figs 10–14; H. J. Hansen, 1893: plate 5 figs 12–13; Lameere,
1895: 474; Krauss, 1896: 628; Ellingsen, 1897: 18–20; E. Simon, 1898d: 22–23;
Kulczynski, 1899: 458; Tullgren, 1899a: 181; E. Simon, 1900b: 595; Evans, 1901a:
54; Evans, 1901b: 242; Godfrey, 1901b: 214–215; Kew, 1901b: 194; Donisthorpe,
1902: 67; Peacock, 1902a: 137; Kew, 1903: 298, 299, fig. 2; With, 1906: plate 1
fig. 2a; Ellingsen, 1907a: 172; Godfrey, 1907: 163; Tullgren, 1907e: 246; Ellingsen,

1908d: 70; Godfrey, 1908: 160; Gozo, 1908: 132 (not seen); Ellingsen, 1909a: 211; Kew, 1909a: 250; Kew, 1909b: 259; Donisthorpe, 1910: 84; Ellingsen, 1910a: 401; Ellingsen, 1910b: 62; Godfrey, 1910: 31–33; Kew, 1910a: 64, 73; Ellingsen, 1911c: 174; Kew, 1911a: 56–57, fig. 19; Lessert, 1911: 37–38, fig. 29; Ellingsen, 1912c: 174; Krausse-Heldrungen, 1912: 65; Ellingsen, 1913a: 455; Standen, 1912: 16, fig. 2; Falconer, 1916: 157–158; Kew, 1916a: 124; Kew, 1916b: 81; Standen, 1916a: 17; Standen, 1916b: 124; Nonidez, 1917: 43; Standen, 1917: 30; Standen, 1918: 335; Navás, 1919: 214; Navás, 1921: 166; Navás, 1923: 34; Navás, 1925: 126, fig. 18b; Donisthorpe, 1927: 183; Bacelar, 1928: 190; Beier, 1928: 313; Kästner, 1928: 9, fig. 31; Schenkel, 1928: 70, fig. 31; Beier, 1929d: 156; Kew, 1930: 253–256, figs 1b, 2; Weidner, 1959: 115 (synonymized by J. C. Chamberlin, 1929a: 72).

Chthonius (Chthonius) rayi L. Koch: Ellingsen, 1905d: 12.

Chthonius ischnocheles (Hermann): J. C. Chamberlin, 1929a: 71–72, figs 2a, 2g, 2k, 2m–n, 3a, 3d; J.C. Chamberlin, 1931a: figs 4a, 11a, 11t, 15a, 28d, 33o, 34d, 40a–c, 42a, 43a–b, 46f, 52a; Beier, 1932g: fig. 167; Vachon, 1934g: 162; J. C. Chamberlin, 1935b: 479; Vachon, 1935j: 187–188; Roewer, 1936: figs 16, 70a; Vachon, 1936a: 78; Roewer, 1937: 239; Vachon, 1937f: 128; Schenkel, 1938: 37–38; Vachon, 1938a: 73–79, figs 41–45, 60a–b, 61a; Beier, 1939a: fig. 166a; Lohmander, 1939b: 285–286, fig. 2; Caporiacco, 1940: 6; Leleup, 1947: 322; Beier, 1948a: 188; Beier, 1948b: 456; Caporiacco, 1949c: 133; Vachon, 1949: figs 206c, 207c; Balazuc, Miré, Sigwalt & Théodoridès, 1951: 187, 215, 216, 219; Parenzan, 1953: 90 (not seen); Parenzan, 1954: 99 (not seen); Betts, 1955: 298; George, 1955: 123; George, 1957: 80; Cloudsley–Thompson, 1958: 96; George, 1961: 38; Pontin, 1961: 135; Vachon, 1961: 102; Meinertz, 1962a: 388–389, map 1; Gabbutt & Vachon, 1963: 76–97, figs 1–44; Savory, 1964a: fig. 33b; Helversen, 1966a: 133–134; Helversen, 1966b: figs 1c, 4h, 6a; Zangheri, 1966: 529–530; Bishop, 1967: 393, figs 1–2; Gabbutt, 1967: 470–478, figs 1–2; Turk, 1967: 153; Helversen, 1968: fig. 5; Gabbutt, 1969c: 230; Weygoldt, 1969a: 70, 122, 123, figs 112a–d; Legg, 1970b: figs 3(2), 3(7); Gabbutt, 1970c: 7–8, fig. 2; Jones, 1970a: 77–79, figs 1–5; Hunt & Legg, 1971: 476–478; Salmon, 1972: 66, 67; Legg, 1973a: 368–392, figs 1–11, plates 1–3; Legg, 1973b: 430–440, figs 1, 2a, plates 1–3; Jones, 1975a: 87–88; Legg, 1975a: 66; Legg, 1975b: 100, figs 1a, 2, 7a, 9a–c, 10a–b, 11a–b; P. A. Wood, 1975: 145–149, figs 1–2; Beier, 1977b: 2; Savory, 1977: fig. 93b; Bellés, 1978: 87; Jones, 1978: 93, 94, 95; Paoletti, 1978: 90; P. A. Wood & Gabbutt, 1978: 176–182, figs 6–10, 21–25; Jones, 1979a: 200–201; Judson, 1979a: 60; Rundle, 1979: 47; Schawaller, 1979: 13; Thaler, 1979: 51; P. A. Wood & Gabbutt, 1979a: 285–292; P. A. Wood & Gabbutt, 1979b: 329–336; Gabbutt & Aitchison, 1980: 57–60, figs 1–4; Jones, 1980c: map 3; Judson, 1980: 8; Mahnert, 1981c: fig. 1; Schawaller, 1981b: 42–44; Pieper, 1981: 2; Callaini & Dallai, 1989: 87, figs 1–3.

Chthonius (Chthonius) ischnocheles (Hermann): Beier, 1932a: 48, fig. 52; Beier, 1934b: 283–285; Leruth, 1935: 26; Beier, 1939f: 161; Vachon, 1940g: 141; Cooreman, 1946b: 2; Leleup, 1948: 8; Hoff, 1949b: 433, fig. 13; Hoff, 1951: 2; Beier, 1952e: 2; Beier, 1953a: 293; G.O. Evans & Browning, 1954: 8, figs 3, 6–7; Beier, 1955n: 88; Beier, 1958b: 27; Beier, 1958c: 137; Hoff, 1958: 3; Beier, 1959f: 113; Beier, 1961a: 67–68; Beier, 1961b: 21; Beier, 1963a: 253–254; Beier, 1963b: 25, fig. 16; Beier, 1963h: 149; Beier, 1963i: 283; Beier, 1965c: 375; Beier, 1966h: 175; Capolongo, 1969: 202; Hammen, 1969: 16–17; Beier, 1970a: 45; Legg, 1971a: 475, fig. 1b; Lagar, 1972b: 46; Beier, 1976a: 23; Crocker, 1976: 7–8; Mahnert, 1977c: 62; Crocker, 1978: 8; Leclerc & Heurtault, 1979: 239; Krumpál, 1980: 23; Callaini, 1982a: 17, 18, fig. 1a; Gardini, 1981a: 102; Krumpál & Kiefer, 1981: 128; Leclerc, 1981: 127–128, plate 2; Anonymous, 1983a: 27; Hammen, 1983: 10; Heurtault, 1986c: 20; Mahnert, 1986d: 74, fig. 1; Legg & Jones, 1988: 65–66, figs 1–2, 3a–c, 4a–b, 5c, 12a, 12b(a–e); Mahnert, 1989a: 85; Mahnert, 1989b: 41; Judson, 1990: fig. 1.

Chthonius (Chthonius) ischnocheles ischnocheles (Hermann): Beier, 1932a: 48;
   Weidner, 1954b: 108; Rafalski, 1967: 5–6; Lazzeroni, 1969a: 321–322; Lazzeroni,
   1969b: 390–392; Lazzeroni, 1969c: 224; Lazzeroni, 1970c: 102; Ćurčić, 1974a: 12;
   Mahnert, 1975c: 185–186; Callaini, 1979b: 131–134, figs 1a–c; Callaini, 1979c:
   341; Gardini, 1980c: 98, 121, 125, 126, 129, 130, 131, 132; Callaini, 1983a: 148–149;
   Callaini, 1983c: 279–280; Callaini, 1983d: 221; Callaini, 1986e: 236.
Chthoniuss (sic) *ischnochele* (sic) (Hermann): Fage, 1933: 54.
Chthonius (Chthonius) rhodochelatus Hadži, 1933b: 133–139, figs 3a–c, 4a–b, 5a–c,
   6a–e, 7; Hadži, 1933c: 177–179; Ćurčić, 1974a: 13 (synonymized by Beier,
   1963b: 25).
Chthonius rhodochelatus Hadži: Roewer, 1937: 239.
Chthonius reyi (sic) L. Koch: Sanfilipo, Timossi & Conci, 1943: 312.
Chtonius (sic) *(Chthonius) ischnocheles* (Hermann): Beier, 1949: 2, fig. 1.
Chthonius ischnocheles ischnocheles (Hermann): Caporiacco, 1948c: 240; Lazzeroni,
   1970a: 207; Callaini, 1986d: 523–524, figs 1a–d.
Chthonius ischnochelus (sic) (Hermann): Gilbert, 1951: 548–551, figs 1–2; Betts,
   1955: 298; Cloudsley-Thompson, 1956a: 71.
Chthonius (Chthonius) dacnodes Navás: G. O. Evans & Browning, 1954: 10, figs
   10–11; Legg, 1971a: 475, fig. 1e (misidentifications).
Cthonius (sic) *ischnocheles* (Hermann): Cloudsley-Thompson, 1959: 179.
Chthonius ischonocheles (Hermann): Howes, 1972a: 108.
Chthonius iscnocheles (sic) (Hermann): Klausen, 1975: 63.
Chthonius cfr. *ischnocheles ischnocheles* (Hermann): Callaini, 1980a: 219–220.
Not *Obisium ischnocheles* Hermann: Théis, 1832: 63–66 (misidentification; see
   Chthonius (Chthonius) orthodactylus (Leach)).
Not *Chthonius trombidioides* (Latreille): C. L. Koch, 1843: 76–79, figs 806–807;
   L. Koch, 1873: 49–50; L. Koch, 1877: 180; Stecker, 1874b: 238; Stecker, 1875d:
   88; Daday, 1880a: 193; Tömösváry, 1882b: 238–239, plate 5 figs 9–12; Krauss,
   1896: 628 (misidentifications; see *Chthonius (Ephippiochthonius) tetrachelatus*
   (Preyssler)).
Not *Chthonius pennsylvanicus* Hagen: Berger, 1905: 407 (misidentification; see
   Tridenchthonius cubanus (J. C. Chamberlin)).
Not *Chthonius ischnocheles* (Hermann): Pearse, 1943: 464 (not seen) (misidentificatiʻɔn;
   see *Chthonius (Chthonius) paludis* (J. C. Chamberlin)).

Type localities: of *Chelifer ischnochelus*: not stated, but presumably near Strasbourg,
   France.
of *Chelifer trombidioides*: Paris, France.
of *Chthonius pennsylvanicus*: Philadelphia, Pennsylvania, U.S.A.
of *Chthonius rayi*: several localities in France; Italy; and Great Britain.
of *Chthonius (Chthonius) rhodochelatus* Hadži: Omis (as Omislja), Yugoslavia.
Distribution: Austria, Azores, Balearic Islands, Belgium, Canary Islands, Corsica,
   Czechoslovakia, Denmark, France, Great Britain, Ireland, Italy, Madeira Islands,
   Malta, Netherlands, Norway, Poland, Portugal, Sardinia, Sicily, St Helena, Spain,
   Sweden, Switzerland, Turkey, U.S.A. (Connecticut, Florida, Illinois, Minnesota,
   New Jersey, New York, North Carolina, Pennsylvania), West Germany, Yugoslavia.

Chthonius (Chthonius) ischnocheles reductus Beier

Chthonius (Chthonius) rhodochelatus reductus Beier, 1939d: 9–10, fig. 2.
Chthonius (Chthonius) ischnocheles reductus Beier: Beier, 1963b: 25–26; Beier, 1963h:
   149; Ćurčić, 1974a: 12; Costantini, 1976: 122; Gardini, 1980c: 98–99, 123; Christian
   & Potocnik, 1985: 15; Inzaghi, 1987: 180–181, figs 18–20.

Type locality: Giuppana Island, Yugoslavia.
Distribution: Greece, ɪtaly, Yugoslavia.

Family Chthoniidae

Chthonius (Chthonius) ischnocheles ruffoi Caporiacco

Chthonius (Chthonius) tenuis stammeri Beier, 1942: 130–131 (misidentification, in part; see *Chthonius (Chthonius) ischnocheles stammeri* Beier).
Chthonius (Globochthonius) ruffoi Caporiacco, 1951c: 4, fig. 2; Caporiacco, 1951e: 69; Caporiacco, 1951f: 95–97, fig. 2.
Chthonius (Chthonius) stammeri ruffoi Caporiacco: Beier, 1952d: 104–105, fig. 1; Beier, 1963b: 33–34.
Chthonius (Chthonius) stammeri Beier: Beier, 1961b: 90 (misidentification).
Chthonius (Chthonius) ischnocheles ruffoi Caporiacco: Beier, 1963a: 254; Lazzeroni, 1969a: 322; Gardini, 1980c: 99, 130.
Chthonius ischnocheles ruffoi Caporiacco: Lazzeroni, 1970a: 207.

Type locality: Grotta Zinzulusa, Apulia, Italy.
Distribution: Sicily, Italy.

Chthonius (Chthonius) ischnocheles stammeri Beier

Chthonius (Chthonius) tenuis stammeri Beier, 1942: 130–131 (in part; see *Chthonius (Chthonius) ischnocheles ruffoi* Caporiacco).
Chthonius (Chthonius) stammeri Beier: Beier, 1961d: 90; Beier, 1970f: 318.
Chthonius (Chthonius) ischnocheles stammeri Beier: Beier, 1963a: 254; Lazzeroni, 1969a: 322; Gardini, 1980c: 99, 130.
Chthonius (Chthonius) stammeri stammeri Beier: Beier, 1963b: 33; Mahnert, 1980d: 22.
Chthonius ischnocheles stammeri Beier: Lazzeroni, 1970a: 207.
Not *Chthonius (Chthonius) stammeri* Beier: Beier, 1961b: 90 (misidentification; see *Chthonius (Chthonius) ischnocheles ruffoi* Caporiacco).

Type locality: Grotta di Castelcivita, Salerno, Italy.
Distribution: Italy, Sicily.

Chthonius (Chthonius) ischnocheloides Beier

Chthonius (Chthonius) ischnocheloides Beier, 1974f: 160–161, fig. 2; Gardini, 1980c: 99, 127.

Type locality: Bus del pal, Cansiglio, Belluno, Italy.
Distribution: Italy.

Chthonius (Chthonius) italicus Beier

Chthonius cephalotes (E. Simon): Ellingsen, 1909a: 212 (misidentification).
Chthonius (Chthonius) italicus Beier, 1930b: 72; Beier, 1932a: 50, fig. 55; Beier, 1963b: 33, fig. 25; Beier, 1970f: 318; Gardini, 1980c: 99, 119; Bologna & Taglianti, 1985: 224, fig. 8.
Chthonius italicus Beier: Roewer, 1937: 239.

Type locality: near Certosa di Pesio, Italy.
Distribution: Italy.

Chthonius (Chthonius) iugoslavicus Ćurčić

Chthonius (Chthonius) caecus iugoslavicus Ćurčić, 1972d: 126–130, figs 1–6; Ćurčić, 1974a: 11.
Chthonius (Chthonius) cryptus iugoslavicus Ćurčić: Ćurčić, 1976b: 171, fig. 1.
Chthonius cryptus iugoslavicus Ćurčić: Ćurčić, 1983b: fig. 2.
Chthonius iugoslavicus Ćurčić: Ćurčić, 1988c: 2–6, figs 1–7.

Type locality: Sveta Dupka Cave, 2 km NE. of Gulenovci, Serbia, Yugoslavia.
Distribution: Yugoslavia.

Chthonius (Chthonius) jalzici Ćurčić

Chthonius (Chthonius) jalzici Ćurčić, 1988b: 17–19, figs 2–8.

Type locality: Veternica Cave, Gornji Stenjevac, Mt Medvednica, near Zagreb, Yugoslavia.
Distribution: Yugoslavia.

Chthonius (Chthonius) jonicus Beier

Chthonius (Neochthonius) jonicus Beier, 1931b: 53; Beier, 1932a: 47, fig. 50; Roewer, 1937: 239; Beier, 1955d: 212; Beier, 1963a: 253; Beier, 1963b: 21, fig. 14; Beier, 1963i: 283; Callaini, 1979b: 130; Gardini, 1980c: 97, 131; Mahnert, 1982f: 297.
Neochthonius jonicus (Beier): Dumitresco & Orghidan, 1964: 600–602, figs 1–6; Orghidan & Dumitresco, 1964a: 328; Orghidan & Dumitresco, 1964b: 192; Callaini, 1989: 137.
Chthonius (Kewochthonius) jonicus Beier: Mahnert, 1974c: 378–379.
Chthonius (Chthonius) jonicus Beier: Judson, 1990: 597.

Type locality: Kaligoni, Levkás, Greece.
Distribution: Greece, Israel, Italy, Lebanon, Malta, Romania, Sicily.

Chthonius (Chthonius) jugorum Beier

Chthonius (Chthonius) jugorum Beier, 1952c: 56–57, fig. 1; Beier, 1963b: 37, fig. 30; Kofler, 1972: 286; Gardini, 1980c: 99, 125.
Chthonius cf. *jugorum* Beier: Paoletti, 1978: 90 (not seen); Paoletti, 1980: 67 (not seen).

Type locality: Popera, Italy.
Distribution: Austria, Italy.

Chthonius (Chthonius) karamanianus Hadži

Chthonius (Neochthonius) karamanianus Hadži, 1937: 152–156, figs 1a–d, 2a–g; Roewer, 1940: 343; Beier, 1963b: 22; Ćurčić, 1974a: 15.
Chthonius (Chthonius) karamanius (sic) Hadži: Judson, 1990: 597.

Type locality: Treska-Tel, Skopje (as Skoplje), Yugoslavia.
Distribution: Yugoslavia.

Chthonius (Chthonius) lanzai Caporiacco

? *Chthonius* sp.: Gestro, 1907: 169.
Chthonius tenuis L. Koch: Ellingsen, 1909a: 211, 215 (possible misidentification).
Chthonius lanzai Caporiacco, 1947c: 255–256, fig. 1; Callaini, 1986d: 526.
Chthonius (Chthonius) lanzai Caporiacco: Beier, 1961f: 123–125, fig. 1; Beier, 1963b: 42, fig. 35; Beier, 1970f: 318; Gardini, 1979: 126–130, figs 1–12; Gardini, 1980c: 99–100, 128, 129.
Not *Chthonius (Chthonius) lanzai* Caporiacco: Beier, 1953d: 35–36, fig. 1; Conci & Franceshi, 1953: 47 (not seen) (misidentifications; see *Chthonius (Chthonius) irregularis* Beier).

Type locality: Tana dei Pipistrelli (as Tana del Frate), Sassorosso-Corfino, Villa Collemandina, Toscana, Italy.
Distribution: Italy.

Chthonius (Chthonius) lanzai vannii Callaini

Chthonius lanzai vannii Callaini, 1986d: 534–537, figs 4a–f.

Type locality: Buca di Nadia Cave, Piteccio, Italy.
Distribution: Italy.

Chthonius (Chthonius) leoi (Callaini)

Neochthonius leoi Callaini, 1988a: 179–181, figs 2a–e.
Chthonius (Chthonius) leoi (Callaini): Judson, 1990: 597.

Type locality: Bauladu, Sardinia.
Distribution: Sardinia.

Chthonius (Chthonius) leruthi Beier

Chthonius (Chthonius) leruthi Beier, 1939c: 5–7, fig. 3.
Chthonius (Chthonius) leruhti (sic) Beier: Beier, 1963b: 46–47, fig. 40.

Type locality: Iara (as Jara), Turda, Romania.
Distribution: Romania.

Chthonius (Chthonius) lessiniensis Schawaller

Chthonius (Chthonius) lessiniensis Schawaller, 1982b: 49–54, figs 1–7; Gardini, 1985b: 60.

Type locality: Grotta Rovere Mille, near Rovere Veronese, Monte Lessini, Verona, Italy.
Distribution: Italy.

Chthonius (Chthonius) lindbergi Beier

Chthonius (Chthonius) lindbergi Beier, 1956g: 8; Beier, 1963b: 27–28; Mahnert, 1980e: 219–220, fig. 10.

Type locality: Neraido Spila Cave, 10 km from Iráklion (as Candia), Crete.
Distribution: Crete.

Chthonius (Chthonius) litoralis Hadži

Chthonius (Chthonius) litoralis Hadži, 1933b: 130–133, figs 2a–c; Hadži, 1933c: 176–177; Beier, 1963b: 26; Lazzeroni, 1969b: 393, figs 1–2; Ćurčić, 1974a: 12; Ćurčić, 1988b: 19–20, figs 9–13.
Chthonius litoralis Hadži: Roewer, 1937: 239.

Type locality: Dugi Otok, Yugoslavia.
Distribution: Italy, Yugoslavia.

Chthonius (Chthonius) lucifugus Mahnert

Chthonius (Chthonius) lucifugus Mahnert, 1977c: 62–63, figs 59–63; Heurtault, 1986c: 20; Zaragoza, 1986: fig. 1.
Chthonius (Chthonius) cf. *lucifugus* Mahnert: Leclerc, 1983c: 15–18, figs 2–6 (not seen).

Type locality: Cova Espluga, Sta. Maria de Meià, Lérida, Spain.
Distribution: France, Spain.

Chthonius (Chthonius) macedonicus Ćurčić

Chthonius (Chthonius) macedonicus Ćurčić, 1972b: 142–147, figs 1–7; Ćurčić, 1974a: 12.

Type locality: Bela Voda Cave, 1 km SE. of Demir-Kapija, Yugoslavia.
Distribution: Yugoslavia.

Chthonius (Chthonius) magnificus Beier

Chthonius (Chthonius) magnificus Beier, 1938a: 5; Beier, 1939d: 11–12, figs 6–7; Beier, 1963b: 44–45, fig. 38; Ćurčić, 1974a: 12; Ćurčić, 1988b: 24–26, figs 23–25.

Type locality: Sipun Cave, Cavtata, Yugoslavia.
Distribution: Yugoslavia.

Chthonius (Chthonius) malatestai Callaini

Chthonius (Chthonius) malatestai Callaini, 1980a: 220–225, figs 1a–c, 2a–d.

Type locality: Mt Pizzoc, Italy.
Distribution: Italy.

Chthonius (Chthonius) mauritanicus (Callaini)

Neochthonius mauritanicus Callaini, 1988a: 176–179, figs 1a–e; Callaini, 1988b: 31.
Chthonius (Chthonius) mauritanicus (Callaini): Judson, 1990: 597.

Type locality: Ouezzane, Morocco.
Distribution: Morocco.

Chthonius (Chthonius) mayi Heurtault-Rossi

Chthonius (Chthonius) cephalotes (E. Simon): Beier, 1932a: 49, fig. 53 (misidentification).
Chthonius (Chthonius) cephalotes cephalotes (E. Simon): Beier, 1932a: 49; Beier, 1963b: 28–29, fig. 20 (misidentifications).
Chthonius (Chthonius) mayi Heurtault-Rossi, 1968a: 916–920, figs 7–12; Leclerc & Heurtault, 1979: 241, map 2; Heurtault, 1986c: 20.

Type locality: Tharaux, Gard, France.
Distribution: France.

Chthonius (Chthonius) mazaurici Leclerc

Chthonius (Chthonius) mazaurici Leclerc, 1981: 119–121, plate 2 figs 1–7; Heurtault, 1986c: 21.

Type locality: Bruge à Monclus, Gard, France.
Distribution: France.

Chthonius (Chthonius) mazaurici coironi Leclerc

Chthonius (Chthonius) mazaurici coironi Leclerc, 1981: 121–123.

Type locality: Grotte du Loup, Saint-Laurent-sous-Coiron, Ardèche, France.
Distribution: France.

! Chthonius (Chthonius) mengei Beier

Chthonius mengei Beier, 1937d: 303–304, fig. 2; Roewer, 1940: 330; Schawaller, 1978: 3.

Type locality: Baltic Amber.
Distribution: Baltic Amber.

Chthonius (Chthonius) microphthalmus E. Simon

Chthonius microphthalmus E. Simon, 1879a: 75–76, 312, plate 19 fig. 17; Daday, 1889a: 16; E. Simon, 1896: 375; Alzona, 1903: 14; Nonidez, 1917: 43; Navás, 1919: 214; Navás, 1921: 166; Navás, 1925: 126; Roewer, 1937: 239; Sanfilipo, Timossi & Conci, 1943: 312.
Chthonius (Chthonius) microphthalmus E. Simon: Beier, 1932a: 51–52; Lazzeroni, 1969c: 225–226; Heurtault, 1986c: 21.

Family Chthoniidae

Chthonius (Chthonius) microphthalmus microphthalmus E. Simon: Beier, 1932a: 52, fig. 58; Beier, 1963b: 36, fig. 29.
Chtonius (sic) *(Chtonius)* (sic) *microphthalmus microphthalmus* E. Simon: Puddu & Pirodda, 1973: 170.
Chthonius (Chthonius) microphtalmus (sic) E. Simon: Leclerc, 1981: 127–128, plate 1.
Not *Chthonius microphthalmus* E. Simon: E. Simon, 1898d: 23; E. Simon, 1900b: 595; Gozo, 1908: 132 (not seen) (misidentifications; see *Chthonius (Chthonius) microphthalmus ligusticus* Beier).
Not *Chthonius microphthalmus* E. Simon: Beier, 1929a: 365–366 (misidentification; see *Chthonius (Globochthonius) spelaeophilus histricus* Beier)

Type localities: Grotte du Capucin, Tarn-et-Garonne, France; Grotte de Bruniquel, Tarn-et-Garonne, France; and Penne, Tarn, France.
Distribution: France, Sardinia, Spain.

Chthonius (Chthonius) microphthalmus ligusticus Beier

Chthonius microphthalmus E. Simon: E. Simon, 1898d: 23; E. Simon, 1900b: 595; Gozo, 1908: 132 (not seen) (misidentifications).
Chthonius (Chthonius) microphthalmus ligusticus Beier, 1930b: 72–73; Beier, 1932a: 52, fig. 59; Beier, 1953e: 105; Beier, 1963b: 36, fig. 28; Gardini, 1980c: 100, 120, 121; Bologna & Taglianti, 1985: 224, fig. 8.
Chthonius microphthalmus ligusticus Beier: Caporiacco, 1951b: 103.

Type locality: Grotta di Cassana, near Borghetto di Vara, Italy.
Distribution: Italy.

Chthonius (Chthonius) microphthalmus vallei Caporiacco

Chthonius microphthalmus vallei Caporiacco, in Bianchi, Caporiacco, Massera & Valle, 1949: 498–499, fig. 1.
Chthonius (Chthonius) microphthalmus vallei Caporiacco: Beier, 1963b: 35–36; Gardini, 1980c: 100, 128.

Type locality: Grotte Spipolae, Bologna, Italy.
Distribution: Italy.

Chthonius (Chthonius) motasi Dumitresco & Orghidan

Chthonius motasi Dumitresco & Orghidan, 1964: 602–606, figs 7–13, plate 1 fig. 2.

Type localities: Lake Razelm, Romania; Grotte de Casian, Casian, Romania; Mt Consul, Filimon Sirbu, Romania; Grotte de Pestera dintre Gorni, Cheia, Romania; and Grotte de Pestera Diaclaza, Cheia, Romania.
Distribution: Romania.

Chthonius (Chthonius) multidentatus Beier

Chthonius (Chthonius) multidentatus Beier, 1963a: 254–256, fig. 1; Gardini, 1980c: 100, 131.

Type locality: Grotta Monella, Siracusa, Sicily.
Distribution: Sicily.

Chthonius (Chthonius) occultus Beier

Chthonius (Chthonius) occultus Beier, 1939d: 10–11, fig. 4; Beier, 1963b: 43, fig. 36; Ćurčić, 1974a: 12.

Type locality: Baba Cave, Popovu Polju, Zavale (as Zavala), Yugoslavia.
Distribution: Yugoslavia.

156

Chthonius (Chthonius) orthodactyloides Beier

Chthonius (Chthonius) orthodactyloides Beier, 1967f: 302–303, fig. 1.

Type locality: Marmaris, Turkey.
Distribution: Turkey.

Chthonius (Chthonius) orthodactylus (Leach)

Obisium orthodactylum Leach, 1817: 51, plate 141 fig. 2; Tulk, 1844: 55–57.
Obisium ischnocheles Hermann: Théis, 1832: 63–66, plate 1 figs 3, 3a–c (misidentification).
Chthonius orthodactylus (Leach): C.L. Koch, 1843: 79–80, figs 808; Hagen, 1870: 268; L. Koch, 1873: 50–51; L. Koch, 1877: 181; E. Simon, 1879a: 73–74; Tömösváry, 1882b: 239–240, plate 5 figs 3–8; Becker, 1884: cclxiv; Pavesi, 1884: 457–458; Canestrini, 1885: no. 8, figs 1–4; E. Simon, 1885c: 50; Daday, 1888: 135, 191–192; Daday, 1889c: 25; Daday, 1889d: 81; O. P. Cambridge, 1892: 209, plate A figs 7–9; E. Simon, 1898d: 22; E. Simon, 1900b: 595; Evans, 1901a: 54; Godfrey, 1908: 160; Ellingsen, 1909a: 211, 216; Ellingsen, 1910a: 401; Godfrey, 1910: 30; Kew, 1911a: 57, fig. 20; Kew, 1916a: 124; Daday, 1918: 2; Beier, 1928: 312–313; Kästner, 1928: 9; Schenkel, 1928: 71; Beier, 1929a: 365; Beier, 1929e: 454; J. C. Chamberlin, 1929a: 71; Kew, 1930: 253–256, figs 1c–d; Vachon, 1934g: 163; Ionescu, 1936: 2; Roewer, 1937: 239; Caporiacco, 1950: 116; Caporiacco, 1951b: 103; Franciscolo, 1955: 125; Bishop, 1967: 393; Lazzeroni, 1970a: 207; Howes, 1972a: 108; Salmon, 1972: 67; Legg, 1973b: 430; Palmgren, 1973: 8; Legg, 1975a: 66; Legg, 1975b: 101, figs 1c, 4, 7c, 13a–b; Goddard, 1976b: 298–302, figs 2, 4; Goddard, 1979: 100–104; Judson, 1979a: 61; Rundle, 1979: 47; Jones, 1980c: map 4; Schawaller, 1981b: 44, 50; Judson, 1987: 368.
Chthonius (Chthonius) orthodactylus (Leach): Beier, 1932a: 50–51, figs 51, 56; Beier, 1939e: 305; Cooreman, 1946b: 3; Beier, 1952e: 2; Beier, 1953e: 105; Beier & Franz, 1954: 453–454; G. O. Evans & Browning, 1954: 10, figs 9, 12; Beier, 1958c: 137; Beier, 1961f: 123; Beier, 1963b: 30, fig. 22; Beier, 1963a: 256; Beier, 1963h: 150; Lazzeroni, 1969a: 323; Lazzeroni, 1969b: 393–394; Lazzeroni, 1969c: 224; Lazzeroni, 1970c: 102; Legg, 1971a: 475, figs 1d, 2e; Ćurčić, 1974a: 12; Gardini, 1975: 2; Costantini, 1976: 122; Ćurčić, 1976b: 172; Crocker, 1978: 8; Leclerc & Heurtault, 1979: 239; Callaini, 1980a: 226; Gardini, 1980c: 100, 120, 121; Krumpál, 1980: 23; Inzaghi, 1981: 72; Krumpál & Kiefer, 1981: 128–129; Callaini, 1983d: 221–222; Bologna & Taglianti, 1985: 224, fig. 8; Callaini, 1986e: 236–237; Legg & Jones, 1988: 71–72, figs 14a, 14b(a–f).
Chtonius (sic) orthodactylus (Leach): Cerruti, 1959: 51.
Not *Obisium orthodactylum* Leach: McIntire, 1868: 8; McIntire, 1869: 247, fig. 219 (misidentifications; see *Chthonius (Ephippiochthonius) tetrachelatus* (Preyssler)).

Type localities: Devon (as Danmonia), England, Great Britain; and Kent (as Cantia), England, Great Britain.
Distribution: Austria, Belgium, Corsica, Czechoslovakia, France, Great Britain, Greece, Hungary, Ireland, Italy, Romania, Sardinia, Sicily, Tunisia, West Germany, Yugoslavia.

Chthonius (Chthonius) pacificus Muchmore

Chthonius (Chthonius) pacificus Muchmore, 1968b: 55–57, figs 3–4.

Type locality: E. of San Juan Hot Springs, Orange County, California, U.S.A.
Distribution: U.S.A. (California).

Chthonius (Chthonius) paganus (Hoff), new combination

Kewochthonius paganus Hoff, 1961: 417–420, figs 1–2; Zeh, 1987b: 1086.

Type locality: Mesa Verde National Park, Montezuma County, Colorado, U.S.A.
Distribution: U.S.A. (Colorado).

Chthonius (Chthonius) paludis (J. C. Chamberlin)

Neochthonius paludis J. C. Chamberlin, 1929a: 66, figs 1d, 2i.
Chthonius (Neochthonius) paludis (J. C. Chamberlin): Beier, 1932a: 46.
Chthonius paludis (J. C. Chamberlin): Roewer, 1937: 238.
Chthonius (Chthonius) pearsei Hoff, 1945i: 319–323, figs 5–9 (synonymized by
    Hoff & Bolsterli, 1956: 157).
Chthonius pearsi (sic): Gering, 1956: 50 (not seen).
Chthonius ischnocheles (Hermann): Pearse, 1943: 464 (not seen) (misidentification).
Kewochthonius paludis (J. C. Chamberlin): Pearse, 1946: 257; Hoff, 1951: 4;
    Hoff & Bolsterli, 1956: 157; Hoff, 1958: 5; Muchmore, 1969e: 393; Zeh, 1987b:
    1086.
Chthonius (Chthonius) paludis (Chamberlin): Judson, 1990: 597.

Type localities: of *Neochthonius paludis*: Billy's Island, Okefinokee Swamp, Georgia,
    U.S.A.
of *Chthonius (Chthonius) pearsei*: Duke Forest, Durham, North Carolina, U.S.A.
Distribution: U.S.A. (Georgia, Mississippi, North Carolina, Tennessee, Utah).

Chthonius (Chthonius) paolettii Beier

Chthonius (Chthonius) paolettii Beier, 1974f: 159–160, fig. 1; Gardini, 1980c:
    101, 127.

Type locality: Grotta Eccelini (as Escalini), near Valstagna, Vicenza, Italy.
Distribution: Italy.

Chthonius (Chthonius) parvioculatus Beier

Chthonius (Chthonius) parvioculatus Beier, 1930b: 71; Beier, 1932a: 49–50, fig. 54;
    Roewer, 1937: 239; Beier, 1942: 131; Sanfilipo, Timossi & Conci, 1943: 312; Beier,
    1953e: 105; Beier, 1963b: 26, fig. 17; Gardini, 1980c: 101, 120, 122, 123; Bologna
    & Taglianti, 1985: 225, fig. 8.
Chthonius parvioculatus Beier: Franciscolo, 1955: 125.

Type locality: Grotta Dragonara, near Genoa (as Genua), Italy.
Distribution: Italy.

Chthonius (Chthonius) persimilis Beier

Chthonius (Chthonius) persimilis Beier, 1939d: 13–14, fig. 9; Beier, 1963b: 39–40,
    fig. 33; Ćurčić, 1974a: 12; Ćurčić, 1976b: 172, fig. 1.

Type locality: Bari, Sjenice, Yugoslavia.
Distribution: Yugoslavia.

Chthonius (Chthonius) petrochilosi Heurtault

Chthonius (Chthonius) petrochilosi Heurtault, 1972a: 19–21, figs 1–6; Mahnert,
    1975a: 169; Mahnert, 1978e: 274–279, figs 4–11; Mahnert, 1980e: 214.

Type locality: Kerateas Cave, Attica, Greece.
Distribution: Greece.

Chthonius (Chthonius) ponticoides Mahnert

Chthonius (Chthonius) ponticoides Mahnert, 1975a: 173–175, figs 10–16.

Type locality: 'Ton Limnon' Cave, near Kato Klitoria, 20 km S. of Kalavrita, Greece.
Distribution: Greece.

Chthonius (Chthonius) ponticus Beier

Chthonius (Chthonius) ponticus Beier, 1965f: 82–83, fig. 1; Beier, 1969b: 189; Beier, 1973c: 223–224.
Chthonius ponticus Beier: Schawaller, 1983b: 5, figs 3–4.

Type locality: Abant-Berge near Bolu, Turkey.
Distribution: Turkey, U.S.S.R. (Georgia).

! Chthonius (Chthonius) pristinus Schawaller

Chthonius (Chthonius) pristinus Schawaller, 1978: 8–11, figs 1–4.

Type locality: Baltic Amber.
Distribution: Baltic Amber.

Chthonius (Chthonius) pusillus Beier

Chthonius (Chthonius) pusillus Beier, 1947d: 298–300, fig. 2; Beier, 1952e: 2; Beier & Franz, 1954: 454; Beier, 1956h: 25; Beier, 1956i: 8; Beier, 1963b: 38–39, fig. 31; Ressl, 1965: 289.
Chthonius pusillus Beier: Ressl & Beier, 1958: 2; Palmgren, 1973: 8.

Type locality: Mitterndorf (as Mitterdorf), Austria.
Distribution: Austria.

Chthonius (Chthonius) pygmaeus Beier

Chthonius (Neochthonius) pygmaeus Beier, 1934a: 53–54, fig. 1; Beier, 1951c: 164–165, fig. 1b; Krumpál & Kiefer, 1981: 127.
Chthonius pygmaeus Beier: Roewer, 1937: 239.
Chthonius (Neochthonius) pygmaeus pygmaeus Beier: Beier, 1952e: 2; Beier, 1963b: 23, fig. 15; Ćurčić, 1974a: 15.
Chthonius (Chthonius) pygmaeus Beier: Mahnert, 1980e: 217.

Type locality: Vasberzence völgy, Hungary.
Distribution: Austria, Hungary, Yugoslavia.

Chthonius (Chthonius) pygmaeus carinthiacus Beier

Chthonius (Neochthonius) pygmaeus carinthiacus Beier, 1951c: 165–166, fig. 1c; Beier, 1952e: 2; Beier, 1956i: 8; Beier, 1963b: 23; Lazzeroni, 1969b: 389–390; Ćurčić, 1974a: 15; Costantini, 1976: 122; Gardini, 1980c: 97, 124.
Chthonius (Chthonius) pygmaeus carinthiacus Beier: Mahnert, 1980e: 217.

Type locality: Burgerboden, near Warmbad Villach, Austria.
Distribution: Austria, Czechoslovakia, Italy, Yugoslavia.

Chthonius (Chthonius) radjai Ćurčić

Chthonius (Chthonius) radjai Ćurčić, 1988b: 28–30, figs 31–35.

Type locality: Manita Pec Cave, Paklenica, Mt Velebit, Yugoslavia.
Distribution: Yugoslavia.

Chthonius (Chthonius) raridentatus Hadži

Chthonius raridentatus Hadži, 1930b: 22–29, figs 10–11, 12a–d, 13a–e; Roewer, 1937: 239.
Chthonius (Chthonius) raridentatus Hadži: Beier, 1932a: 53; Beier, 1939d: 15, fig. 11; Beier, 1957a: 161; Beier, 1963b: 48, fig. 43; Beier, 1974f: 159; Ćurčić, 1974a: 13; Minelli, 1977: 44 (not seen); Gardini, 1980c: 101, 128.

Type locality: Jama Borovca, Kocevju, Yugoslavia.
Distribution: Italy, Yugoslavia.

Family Chthoniidae

Chthonius (Chthonius) ressli Beier

Chthonius (Chthonius) ressli Beier, 1956h: 24–25, fig. 1; Beier, 1956i: 8; Beier, 1963b: 32; Ressl, 1965: 289; Judson, 1990: 599–600, figs 4–9.
Chthonius ressli Beier: Ressl & Beier, 1958: 2.
Chthonius (Chthonius) parvulus Inzaghi, 1981: 67–72, figs 1–11 (synonymized by Judson, 1988: 131).

Type localities: of *Chthonius (Chthonius) ressli*: Purgstall, Austria.
of *Chthonius (Chthonius) parvulus*: Astino, Bergamo, Italy.
Distribution: Austria, France, Italy.

Chthonius (Chthonius) satapliaensis Schawaller & Dashdamirov

Chthonius (Chthonius) satapliaensis Schawaller & Dashdamirov, 1988: 6–8, figs 12–17, 27.

Type locality: Sataplia Reserve, Kutaissi District, Georgia, U.S.S.R.
Distribution: U.S.S.R. (Georgia).

Chthonius (Chthonius) sestasi Mahnert

Chthonius (Chthonius) sestasi Mahnert, 1980e: 218–219, figs 5–9.

Type locality: Liparo Tripa Cave, Berg Kokkino Vrachos, near Spilia, Thessalia, Greece.
Distribution: Greece.

Chthonius (Chthonius) shelkovnikovi Redikorzev

Chthonius (Mundochthonius) shelkovnikovi Redikorzev, 1930: 103–106, figs 3–4, 5a–c.
Chthonius (Neochthonius) shelkovnikovi Redikorzev: Beier, 1932a: 47.
Chthonius shelkovnikovi Redikorzev: Roewer, 1937: 239.
Chthonius shelkovnikovi redikorzevi Kobakhidze, 1961a: 168–169 (synonymized by Schawaller & Dashdamirov, 1988: 3).
Chthonius (Neochthonius) tauricus Beier, 1963g: 267–268, fig. 1; Beier, 1965f: 82; Beier, 1966e: 161; Beier, 1967f: 301–302; Beier, 1969b: 189 (synonymized by Schawaller & Dashdamirov, 1988: 3).
Chthonius schelkovnikovi (sic) Redikorzev: Kobakhidze, 1965b: 541; Kobakhidze, 1966: 701.
Mundochthonius shelkovnikovi (Redikorzev): Beier, 1970f: 317, fig. 1.
Chthonius (Kewochthonius) tauricus Beier: Mahnert, 1974a: 87.
Chthonius (Chthonius) shelkovnikovi Redikorzev: Schawaller, 1983b: 5, figs 1–2.
Chthonius (Kewochthonius) shelkovnikovi Redikorzev: Schawaller & Dashdamirov, 1988: 3, figs 1–5, 27.
Chthonius (Chthonius) tauricus Beier: Judson, 1990: 597.

Type localities: of *Chthonius (Mundochthonius) shelkovnikovi*: Stepanavan (as Djelal-Ogly), Armenia, U.S.S.R.
of *Chthonius shelkovnikovi redikorzevi*: Kloster Grem, Kachetia, Georgia, U.S.S.R.
of *Chthonius (Neochthonius) tauricus*: Namrun, Turkey.
Distribution: Greece, Iran, Turkey, U.S.S.R (Armenia, Azerbaijan, Georgia, R.S.F.S.R.).

Chthonius (Chthonius) shulovi Beier

Chthonius (Neochthonius) shulovi Beier, 1963f: 184.
Chthonius (Chthonius) shulovi Beier: Judson, 1990: 597.

Type locality: Wadi Kelt, Israel.
Distribution: Israel.

Chthonius (Chthonius) stevanovici Ćurčić

Chthonius (Chthonius) stevanovici Ćurčić, 1986b: 142–145, 149–153, figs 1–6.

Type locality: near Kalna, Yugoslavia.
Distribution: Yugoslavia.

Chthonius (Chthonius) strinatii Mahnert

Chthonius (Chthonius?) strinatii Mahnert, 1975a: 177–179, figs 17–22.

Type locality: Glifada Cave, near Dirou, Greece.
Distribution: Greece.

Chthonius (Chthonius) submontanus Beier

Chthonius (Chthonius) submontanus Beier, 1963b: 31–32, fig. 24; Beier, 1963h: 150; Ressl, 1965: 289.
Chtonius (sic) *submontanus* Beier: Cîrdei, Bulimar & Malcoci, 1970: 7–8, fig. 1.
Chthonius submontanus Beier: Ressl, 1970: 250; Palmgren, 1973: 8; Drogla, 1984: 191.

Type locality: Gaming, Austria.
Distribution: Austria, East Germany, Italy, Romania.

Chthonius (Chthonius) subterraneus Beier

Chthonius (Chthonius) subterraneus Beier, 1931b: 53–54; Beier, 1932a: 52–53, fig. 60.
Chthonius (Chthonius) subterraneus subterraneus Beier: Beier, 1939d: 14; Beier, 1963b: 47, fig. 41; Ćurčić, 1974a: 13.
Chthonius subterraneus Beier: Roewer, 1937: 239.

Type locality: Castelnuovo (now Herceg Novi), Montenegro, Yugoslavia.
Distribution: Hungary, Yugoslavia.

Chthonius (Chthonius) subterraneus meuseli Beier

Chthonius (Chthonius) subterraneus meuseli Beier, 1939d: 14, fig. 10; Beier, 1963b: 47–48, fig. 42; Ćurčić, 1974a: 13.

Type locality: Mrkviste Cave, Croatia, Yugoslavia.
Distribution: Yugoslavia.

Chthonius (Chthonius) tamaninii Caporiacco

Chthonius (Chthonius) tamaninii Caporiacco, 1952b: 63–64, fig. 1; Beier, 1963b: 26–27; Lazzeroni, 1969a: 322; Lazzeroni, 1970a: 207; Gardini, 1980c: 101, 130.

Type locality: Grotta di Agnano, near Ostuni, Italy.
Distribution: Italy.

Chthonius (Chthonius) tenuis L. Koch

Chthonius tenuis L. Koch, 1873: 51–52; Stecker, 1874b: 238–239; Stecker, 1875d: 88–89; E. Simon, 1879a: 72–73, 312; O. P. Cambridge, 1885: 103 (not seen); O. P. Cambridge, 1892: 211, plate A figs 17–19; F. Müller & Schenkel, 1895: 822 (not seen); E. Simon, 1898c: 3; E. Simon, 1898d: 22; E. Simon 1900b: 595; Ellingsen, 1907a: 172; E. Simon, 1907: 551; Gozo, 1908: 132 (not seen); Kew, 1909b: 259; Ellingsen, 1910a: 402; Kew, 1911a: 57–58, fig. 21; Lessert, 1911: 39–40, fig. 30; Ellingsen, 1912c: 174; Krausse-Heldrungen, 1912: 65; Caporiacco, 1923: 132;

Schenkel, 1926: 311; Beier, 1928: 312; Kästner, 1928: 9; Redikorzev, 1928: 136; Schenkel, 1928: 71, fig. 33; Beier, 1929d: 156; Beier, 1929e: 454; Kew, 1930: 253–256, fig. 1a; Beier, 1930h: 95; Boldori, 1932: 127; Schenkel, 1933: 17; Vachon, 1934g: 162; Caporiacco, 1936b: 328; Roewer, 1937: 239; Caporiacco, 1948c: 240; Caporiacco, 1951a: 62; Caporiacco, 1951b: 103; Pedder, 1965: 108, 109, fig. 1; Zangheri, 1966: 530; Bishop, 1967: 393; Gabbutt, 1969c: 231; Lazzeroni, 1970a: 207; Gabbutt, 1970c: 11; Helversen & Martens, 1971: 382, fig. 4; Legg, 1973b: 430; Legg, 1975a: 66; Legg, 1975b: 100, figs 1b, 3, 7e, 12a–b; Judson, 1979a: 60; Rundle, 1979: 47; Jones, 1980c: map 5; Callaini, 1986d: 524–526, figs 1e–f; Judson, 1987: 369; Callaini, 1989: 137.

Chthonius (Chthonius) tenuis L. Koch: Ellingsen, 1905d: 11–12; Beier, 1932a: 51, fig. 57; Beier, 1939e: 305; Cooreman, 1946b: 3; Beier, 1952e: 2; Beier & Franz, 1954: 454; G. O. Evans & Browning, 1954: 10, fig. 8; Beier, 1963a: 256; Beier, 1963b: 31, fig. 23; Beier, 1963h: 150; Beier, 1963i: 283; Ressl, 1965: 289; Rafalski, 1967: 6; Lazzeroni, 1969a: 323; Lazzeroni, 1969c: 224–225; Legg, 1971a: 475, fig. 1c; Lagar, 1972a: 17; Lagar, 1972b: 46; Ćurčić, 1974a: 13; Mahnert, 1975c: 186; Costantini, 1976: 122; Vachon, 1976: 243; Hammen, 1977: 311; Callaini, 1979b: 134; Callaini, 1979c: 342; Crocker, 1978: 8; Leclerc & Heurtault, 1979: 240; Gardini, 1980c: 101, 120, 121, 123, 124; Inzaghi, 1981: 72; Callaini, 1983c: 280–281; Callaini, 1983d: 222; Callaini, 1983e: 215; Bologna & Taglianti, 1985: 225, fig. 8; Callaini, 1986e: 237; Heurtault, 1986c: 21; Hammen, 1986a: 4, figs 1b, 4b–c, 9a–c, 16a–b; Legg & Jones, 1988: 68–70, figs 13a, 13b(a–f).

Chthonius (Chthonius) tenuichelatus Hadži, 1937: 162–14, figs 6a–g; Roewer, 1940: 343; Ćurčić, 1974a: 13 (synonymized by Beier, 1963b: 31).

Chthonius (Chthonius) aff. *tenuis* L. Koch: Mahnert, 1980d: 22.

Not *Chthonius tenuis* L. Koch: Ellingsen, 1909a: 211, 215 (possible misidentification; see *Chthonius (Chthonius) lanzai* Caporiacco).

Type localities: of *Chthonius tenuis*: Corsica.
of *Chthonius (Chthonius) tenuichelatus*: Vodno, near Skopje (as Skoplje), Yugoslavia.
Distribution: Algeria, Austria, Belgium, Bulgaria, Corsica, Czechoslovakia, France, Great Britain, Greece, Italy, Malta, Poland, Tunisia, Romania, Sardinia, Sicily, Switzerland, Yugoslavia.

Chthonius (Chthonius) thessalus Mahnert

Chthonius (Chthonius) thessalus Mahnert, 1980e: 215–218, figs 2–4.

Type locality: Kirche Profitis Elais, near Ampelakia, Thessalia, Greece.
Distribution: Greece.

Chthonius (Chthonius) trebinjensis Beier

Chthonius (Chthonius) magnificus trebinjensis Beier, 1938a: 5.
Chthonius (Chthonius) trebinjensis Beier: Beier, 1939d: 10, fig. 8; Beier, 1963b: 45, fig. 39; Ćurčić, 1974a: 13.

Type locality: Hrupjela Cave, Popovu Polju, Trebinja, Yugoslavia.
Distribution: Yugoslavia.

Chthonius (Chthonius) troglobius Hadži

Chthonius (Chthonius) troglobius Hadži, 1937: 156–162, figs 3a–h, 4a–d, 5a–d; Beier, 1939d: 14; Roewer, 1940: 343; Beier, 1963b: 41; Ćurčić, 1974a: 13.

Type locality: Pecina Rasce, near Skopje (as Skoplje), Yugoslavia.
Distribution: Yugoslavia.

Family Chthoniidae

Chthonius (Chthonius) troglodites Redikorzev

Chthonius troglodites Redikorzev, 1928: 133–136, figs 11, 12a–c; Roewer, 1936: fig. 66a; Roewer, 1937: 239, fig. 205.
Chthonius (Chthonius) troglodites Redikorzev: Beier, 1932a: 53–54, fig. 62; Beier, 1939d: 14; Beier, 1963b: 43–44, fig. 37.
Chthonius (Chthonius) troglodytes (sic) Redikorzev: Hadži, 1939b: 20–21, figs 1a–f.

Type locality: Kassapnitzite Cave, near Karlukowo, Bulgaria.
Distribution: Bulgaria.

Chthonius (Chthonius) tzanoudakisi Mahnert

Chthonius (Chthonius) tzanoudakisi Mahnert, 1975a: 170–173, figs 1–9.

Type locality: 'tou Chajoti' Cave, Zakynthos, Greece.
Distribution: Greece.

Subgenus Chthonius (Ephippiochthonius) Beier

Chthonius (Ephippiochthonius) Beier, 1930g: 323; Beier, 1932a: 56; Beier, 1939d: 23; Hadži, 1937: 164–165; Hoff, 1949b: 433.

Type species: *Scorpio tetrachelatus* Preyssler, 1790, by subsequent designation of Hoff, 1949b: 433.

Chthonius (Ephippiochthonius) aegatensis Callaini

Chthonius (Chthonius) aegatensis Callaini, 1989: 138–142, figs 1a–f.

Type locality: Contrada Cisterna, Egadi Island, Sicily.
Distribution: Sicily.

Chthonius (Ephippiochthonius) anatolicus Beier

Chthonius (Ephippiochthonius) anatolicus Beier, 1969b: 190–191, fig. 1; Beier, 1971a: 359; Mahnert, 1974a: 87.

Type locality: 5 km NE. of Ulubey, Ordu, Turkey.
Distribution: Iran, Turkey.

Chthonius (Ephippiochthonius) anophthalmus Ellingsen

Chthonius tetrachelatus anophthalmus Ellingsen, 1908a: 419–420.
Chthonius (Ephippiochthonius) anophthalmus Ellingsen: Beier, 1932a: 61.
Chthonius anophthalmus Ellingsen: Roewer, 1937: 239.

Type localities: Grotte de la Madeleine, Algeria; and Bougie, Algeria.
Distribution: Algeria.

Chthonius (Ephippiochthonius) apulicus Beier

Chthonius (Ephippiochthonius) apulicus Beier, 1958b: 27–29, fig. 1; Beier, 1963b: 63–64, fig. 60; Lazzeroni, 1969a: 325; Beier, 1970f: 318; Beier, 1975a: 55; Gardini, 1980c: 102, 130.
Chthonius apulicus Beier: Lazzeroni, 1970a: 207.

Type locality: Grotta Paglicci, Rignano Garganico, Gargano, Italy.
Distribution: Italy, Sicily.

Chthonius (Ephippiochthonius) asturiensis Beier

Chthonius (Ephippiochthonius) asturiensis Beier, 1955n: 90–91, fig. 1; Beier, 1959f: 115–116; Beier, 1963b: 67–68.

Type locality: Monte Montera near Puente de los Fierros, Asturias, Spain.
Distribution: Spain.

Chthonius (Ephippiochthonius) atlantis Mahnert

Chthonius (Ephippiochthonius) atlantis Mahnert, 1980b: 217–218, figs 6–10; Callaini, 1988b: 32.

Type locality: Ifri Tselet, near Ain-Tesli, Morocco.
Distribution: Morocco.

Chthonius (Ephippiochthonius) balearicus Mahnert

Chthonius (Ephippiochthonius) balearicus Mahnert, 1977c: 67–69, figs 5–9; Zaragoza, 1986: fig. 1.

Type locality: Avenc des Travessets, Cap Ferrutz, Artá, Mallorca, Balearic Islands.
Distribution: Balearic Islands.

Chthonius (Ephippiochthonius) bartolii Gardini

Chthonius (Ephippiochthonius) bartolii Gardini, 1976b: 93–100, figs 8–9; Gardini, 1981a: 109.

Type locality: Semaforo di Punta Mesco, Monterosso al Mare, La Spezia, Italy.
Distribution: Italy.

Chthonius (Ephippiochthonius) bauneensis Callaini

Chthonius (Ephippiochthonius) bauneensis Callaini, 1983b: 415–420, figs 3a–g, 4a–b.

Type locality: Baunei, Sardinia.
Distribution: Sardinia.

Chthonius (Ephippiochthonius) beieri Lazzeroni

Chthonius (Ephippiochthonius) beieri Lazzeroni, 1966: 497–500, figs 1a–g, 2a–b.
Chthonius (Ephippiochthonius) tetrachelatus (Preyssler): Lazzeroni, 1969b: 395–396 (in part).

Type locality: Sirolo, Ancona, Italy.
Distribution: Italy.

Chthonius (Ephippiochthonius) bellesi Mahnert

Chthonius (Ephippiochthonius) bellesi Mahnert, 1989a: 86–87, figs 1–4.

Type locality: Cova Polida, Es Mercadal, Menorca, Balearic Islands.
Distribution: Balearic Islands.

Chthonius (Ephippiochthonius) berninii Callaini

Chthonius (Ephippiochthonius) berninii Callaini, 1983b: 406–412, figs 1a–g, 2a–b.

Type locality: 15 km from Narcao, Sardinia.
Distribution: Sardinia.

Chthonius (Ephippiochthonius) bidentatus Beier

Chthonius (Ephippiochthonius) bidentatus Beier, 1938a: 6; Beier, 1939d: 23, fig. 25; Beier, 1963b: 66, fig. 64; Ćurčić, 1974a: 13; Ćurčić, 1976b: 172, fig. 1.

Type locality: Spilja Petnica, Srbija (as Serbia), Yugoslavia.
Distribution: Yugoslavia.

Chthonius (Ephippiochthonius) boldorii Beier

Chthonius (Ephippiochthonius) austriacus boldorii Beier, 1934a: 56, fig. 4; Beier, 1963b: 69, fig. 66; Beier, 1963h: 150; Costantini, 1976: 122; Gardini, 1980c: 102, 122, 123, 124.
Chthonius (Ephippiochthonius) boldorii Beier: Mahnert, 1980g: 95–96, figs 2–3; Callaini, 1986e: 242; Christian & Potocnik, 1985: 15; H. Hansen, 1988: 187.

Type locality: II Pozzo del Budellone, Nuvolento, near Brescia, Italy.
Distribution: Italy, Yugoslavia.

Chthonius (Ephippiochthonius) bolivari Beier

Chthonius (Ephippiochthonius) bolivari Beier, 1930g: 325–326, fig. 3; Beier, 1932a:
   60, fig. 75; Beier, 1939f: 164; Beier, 1963b: 65–66, fig. 62; Zaragoza, 1986: fig. 1.
Chthonius bolivari Beier: Roewer, 1937: 239.

Type locality: Cueva del Espinoso, La Franca, Asturias, Spain.
Distribution: Spain.

Chthonius (Ephippiochthonius) cassolai Beier

Chtonius (sic) *cassolai* Beier: Puddu & Pirodda, 1973: 170 (nomen nudum).
Chthonius (Ephippiochthonius) cassolai Beier, 1974g: 164–165, fig. 1; Gardini,
   1980c: 102, 132.
Chthonius (Ephippiochthonius) cfr. *cassolai* Beier: Callaini, 1983b: 414, fig. 2c.

Type locality: Grotte e'Scusi, Villasalto, Sardinia.
Distribution: Sardinia.

Chthonius (Ephippiochthonius) catalonicus Beier

Chthonius (Ephippiochthonius) hispanus catalonicus Beier, 1939f: 163–164, figs 2–3.
Chthonius (Ephippiochthonius) catalonicus Beier: Beier, 1963b: 62–63; Lagar, 1972b:
   46; Zaragoza, 1986: fig. 1.

Type locality: Avenc de S. Roc near Begues, Barcelona, Spain.
Distribution: Spain.

Chthonius (Ephippiochthonius) concii Beier

Chthonius (Ephippiochthonius) tetrachelatus concii Beier, 1953d: 36, fig. 2.
Ephippiochthonius tetrachelatus concii (Beier): Franciscolo, 1955: 125.
Chthonius (Ephippiochthonius) concii Beier: Beier, 1963b: 58–59, fig. 55; Gardini,
   1980a: 196–200, figs 10–18; Gardini, 1980c: 102–103, 121; Bologna & Taglianti,
   1985: 225, fig. 8.
Chthonius (Ephippiochthonius) siculus Beier: Callaini, 1983b: 405 (misidentification).
Chthonius (Ephippiochthonius) cf. *concii* Beier: Callaini, 1984: 139–141, figs 4a–e.

Type locality: Grotta delle Arene Candide, near Finale, Liguria, Italy.
Distribution: Italy, Sardinia.

Chthonius (Ephippiochthonius) corcyraeus Mahnert

Chthonius (Ephippiochthonius) corcyraeus Mahnert, 1976a: 177–179, figs 1–4;
   Gardini, 1989: 61.

Type locality: Peristero Grava Cave, Lutses, Kérkira (as Corfu), Greece.
Distribution: Greece.

Chthonius (Ephippiochthonius) corsicus Callaini

Chthonius (Ephippiochthonius) corsicus Callaini, 1981a: 315–319, figs 3a–f, 4.

Type locality: San Martino di Lota, Corsica.
Distribution: Corsica.

Chthonius (Ephippiochthonius) creticus Mahnert

Chthonius (Ephippiochthonius) creticus Mahnert, 1980e: 228–229, figs 23–24.

Type locality: Milatos spilia Cave, near Milatos, Crete.
Distribution: Crete.

Family Chthoniidae

Chthonius (Ephippiochthonius) daedaleus Mahnert

Chthonius (Ephippiochthonius) daedaleus Mahnert, 1980e: 223–225, figs 18–22.

Type locality: Thergiospilia Cave, Kavoussi, Crete.
Distribution: Crete.

Chthonius (Ephippiochthonius) distinguendus Beier

Chthonius (Ephippiochthonius) distinguendus Beier, 1930g: 326, fig. 4; Beier, 1932a: 60, fig. 76; Beier, 1939f: 165; Beier, 1963b: 66, fig. 63; Estany, 1980b: 526, map 1; Zaragoza, 1986: fig. 1.
Chthonius distinguendus Beier: Roewer, 1937: 239.

Type locality: Cueva de Menicute near Albistur, Tolosa, Guipuzcoa, Spain.
Distribution: Spain.

Chthonius (Ephippiochthonius) elbanus Beier

Chthonius (Ephippiochthonius) elbanus Beier, 1963h: 151–152, fig. 2; Lazzeroni, 1969c: 226; Lazzeroni, 1970c: 103; Gardini, 1975: 3–4, fig. 1; Callaini, 1979c: 345–346, figs 1c–d; Callaini, 1980a: 227–228; Callaini, 1981a: 308–309; Callaini, 1983a: 149–150; Callaini, 1983b: 412–413; Callaini, 1984: 130–135, figs 2a–d; Callaini, 1986e: 242; Callaini, 1988b: 36–37.

Type locality: Chiessi, Elba, Italy.
Distribution: Corsica, Italy, Morocco, Sardinia.

Chthonius (Ephippiochthonius) fuscimanus E. Simon

Chthonius tetrachelatus fuscimanus E. Simon, 1900b: 595.
Chthonius (Ephippiochthonius) austriacus Beier, 1931b: 55; Beier, 1932a: 57, figs 67, 69; Beier, 1952e: 2; Beier & Franz, 1954: 453; Beier, 1956h: 24; Beier, 1956i: 8; Beier, 1962b: 146; Strebel, 1961: 107; Kofler, 1972: 286–287; Beier, 1973c: 224; Mahnert, 1980g: fig. 1; Callaini, 1986e: 242 (synonymized by Gardini, 1980a: 261).
Chthonius austriacus Beier: Beier, 1932a: fig. 16; Beier, 1932g: fig. 175; Roewer, 1937: 239; Caporiacco, 1940: 6; Ressl & Beier, 1958: 2; Kobakhidze, 1965b: 541; Kobakhidze, 1966: 701; Gardini, 1980b: 261–264, figs 1–7; Drogla, 1984: 191.
Chthonius (Ephippiochthonius) austriacus austriacus Beier: Beier, 1963b: 68–69, fig. 65; Ressl, 1965: 289; Rafalski, 1967: 7; Lazzeroni, 1969b: 396–397; Callaini, 1980a: 228–229; Gardini, 1980c: 102, 125; Krumpál & Kiefer, 1981: 127.
Chthonius (Chthonius) austriacus Beier: Beier, 1966h: 175.
Chthonius austriacus austriacus Beier: Palmgren, 1973: 8.

Type localities: of *Chthonius tetrachelatus fuscimanus*: Lavarone, Trentino, Italy.
of *Chthonius (Ephippiochthonius) austriacus*: Helenental near Baden, Austria.
Distribution: Austria, Czechoslovakia, East Germany, Italy, Turkey, U.S.S.R. (Georgia), West Germany.

Chthonius (Ephippiochthonius) gasparoi Gardini

Chthonius (Ephippiochthonius) gasparoi Gardini, 1989: 57–58, figs 1–4.

Type locality: Alistrati Cave n. 6389, Alistrati, Macedonia, Greece.
Distribution: Greece.

Chthonius (Ephippiochthonius) gestroi Simon

Chthonius gestroi E. Simon, 1896: 375–376; E. Simon, 1898d: 23; E. Simon, 1900b: 595 (in part; see *Chthonius (Ephippiochthonius) pieltaini* Beier); Alzona, 1903: 14; E. Simon, 1907: 551; Roewer, 1937: 239; Wolf, 1937: 222; Menozzi, 1939: 132.

Chthonius (Ephippiochthonius) gestroi E. Simon: Beier, 1932a: 59, fig. 73; Conci, 1952: 10; Beier, 1953e: 106–107; Beier, 1963b: 73, fig. 72; Beier, 1970f: 318; Callaini, 1980c: 103, 119, 121, 122; Bologna & Taglianti, 1985: 226, fig. 8.
Chthonius vaccai Wolf, 1937: 222 (nomen nudum).
Ephippiochthonius gestroi (Simon): Caporiacco, 1951b: 103; Franciscolo, 1955: 124–125.

Type locality: Grotta Pollera, Finalborgo, Italy.
Distribution: Italy.

Chthonius (Ephippiochthonius) gibbus Beier

Chthonius (Ephippiochthonius) gibbus Beier, 1953a: 293–295, fig. 1; Beier, 1955n: 89; Beier, 1961b: 22; Beier, 1963b: 60, fig. 56; Beier, 1975a: 55; Mahnert, 1975c: 186, figs 1a–c; Callaini, 1979b: 134–137, figs 2a–b; Callaini, 1981a: 308; Callaini, 1983b: 403–405; Callaini, 1983e: 215–216; Mahnert, 1986d: 76; Callaini, 1989: 138.
Chthonius (Ephippiochthonius) austriacus nanus Beier, 1953e: 105–106 (in part; see *Chthonius (Ephippiochthonius) nanus* Beier).
Chthonius (Ephippiochthonius) cfr. *gibbus* Beier: Gardini, 1980b: 268–269, figs 16–22.
Chthonius gibbus Beier: Callaini, 1988b: 36.

Type locality: El Escorial (as Escorial), Spain.
Distribution: Algeria, Balearic Islands, Corsica, Italy, Malta, Morocco, Sardinia, Sicily, Spain.

Chthonius (Ephippiochthonius) girgentiensis Mahnert

Chthonius (Ephippiochthonius) girgentiensis Mahnert, 1982f: 300–301, figs 6–10.

Type locality: Girgenti Cave, Malta.
Distribution: Malta.

Chthonius (Ephippiochthonius) giustii Callaini

Chthonius (Ephippiochthonius) giustii Callaini, 1981a: 309–315, figs 1a–g, 2a–b.

Type locality: Forêt de Cervallo, Corsica.
Distribution: Corsica.

Chthonius (Ephippiochthonius) grafittii Gardini

Chthonius (Ephippiochthonius) grafittii Gardini, 1981a: 102–109, figs 1–21.

Type locality: Grotta del Diavalo o dell'Inferno, Monte Tudurighe, Muros, Sardinia.
Distribution: Sardinia.

Chthonius (Ephippiochthonius) hiberus Beier

Chthonius (Ephippiochthonius) hiberus Beier, 1930g: 324–325, fig. 2; Beier, 1932a: 59, fig. 74; Beier, 1939f: 164; Beier, 1963b: 64–65, fig. 61; Zaragoza, 1986: fig. 1.
Chthonius hiberus Beier: Roewer, 1937: 239.
Chthonius (Ephippiochthonius) aff. *hiberus* Beier: Mahnert, 1977c: 65–66, figs 1–4; Zaragoza, 1986: fig. 1.

Type locality: Sima de la Raya, Tamajón, Spain.
Distribution: Spain.

Chthonius (Ephippiochthonius) hispanus Beier

Chthonius (Ephippiochthonius) hispanus Beier, 1930g: 323–324, fig. 1; Beier, 1932a: 58, fig. 70; Beier, 1939f: 163; Vachon, 1940g: 142; Beier, 1959f: 115; Beier, 1963b: 62, fig. 59.
Chthonius hispanus Beier: Roewer, 1937: 239.

Type locality: Cueva de la Loja, El Mazo, Spain.
Distribution: Balearic Islands, Portugal, Spain.

Chthonius (Ephippiochthonius) insularis Beier

Chthonius (Ephippiochthonius) insularis Beier, 1938a: 6; Beier, 1939d: 23–24, fig. 26; Beier, 1963b: 60–61, fig. 57; Ćurčić, 1974a: 13.

Type locality: Spilja Movrica, Mljet (as Meleda) Island, Yugoslavia.
Distribution: Yugoslavia.

Chthonius (Ephippiochthonius) iranicus Beier

Chthonius (Ephippiochthonius) iranicus Beier, 1971a: 357–358, fig. 1; Mahnert, 1974b: 88.

Type locality: Chalus, Elburz (as Elburs) Mountains, Iran.
Distribution: Iran.

Chthonius (Ephippiochthonius) kabylicus Callaini

Chthonius (Ephippiochthonius) kabylicus Callaini, 1983e: 216–220, figs 2a–f.

Type locality: Point de Vue de Djurdjura, Algeria.
Distribution: Algeria.

Chthonius (Ephippiochthonius) kewi Gabbutt

Chthonius (Ephippiochthonius) kewi Gabbutt, 1966b: 169–178, figs 1a–h, 2a–h, 3a–f, 4a–h, 5a–d; Legg, 1971a: 475, fig. 2d; Crocker, 1978: 8; Legg, 1987: 180; Legg & Jones, 1988: 62–64, figs 11a, 11b(a–f).
Chthonius kewi Gabbutt: Bishop, 1967: 393; Howes, 1972c: 142; Legg, 1973b: 430; Legg, 1975a: 66; Legg, 1975b: 101, figs 1e, 6, 7b, 15a–b; Jones, 1980c: map 7.
Chthonius (Ephippiochthonius) tetrachelatus (Preyssler): Lazzeroni, 1969b: 395–396 (in part).

Type locality: Colne Point, near St Osyth, Essex, Great Britain.
Distribution: Great Britain.

Chthonius (Ephippiochthonius) longesetosus Mahnert

Chthonius (Ephippiochthonius) longesetosus Mahnert, 1976b: 375–377, figs 1–4; Mahnert, 1980b: 218; Callaini, 1988b: 32.

Type locality: Sidi Mejbeur, Taza, Morocco.
Distribution: Morocco.

Chthonius (Ephippiochthonius) lucanus Callaini

Chthonius (Ephippiochthonius) lucanus Callaini, 1984: 144–148, figs 5a–e.

Type locality: Mormanno, Pollino Mountains, Cosenza/Potenza, Italy.
Distribution: Italy.

Chthonius (Ephippiochthonius) machadoi Vachon

Chthonius (Ephippiochthonius) machadoi Vachon, 1940g: 145–147, figs 1–6; Beier, 1955n: 88–89; Beier, 1959f: 115; Beier, 1961b: 22; Beier, 1962c: 25; Beier, 1963b: 67; Beier, 1965c: 375; Beier, 1970a: 45; Beier, 1976a: 23.

Type locality: Algarao da Ribeira de Alte, Paderne, Portugal.
Distribution: Canary Islands, Morocco, Portugal, Spain.

Chthonius (Ephippiochthonius) machadoi canariensis Beier

Chthonius (Ephippiochthonius) machadoi canariensis Beier, 1965c: 375–377, fig. 1; Beier, 1970a: 45; Beier, 1976a: 23; Mahnert, 1989b: 41.

Type locality: near Erjos, Tenerife, Canary Islands.
Distribution: Canary Islands.

Chthonius (Ephippiochthonius) mahnerti Zaragoza

Chthonius (Ephippiochthonius) mahnerti Zaragoza, 1984: 49–53, figs 1–5; Zaragoza, 1986: fig. 1.

Type locality: Cova del Bolumini, Beniarberg, Alicante, Spain.
Distribution: Spain.

Chthonius (Ephippiochthonius) maltensis Mahnert

Chthonius (Ephippiochthonius) maltensis Mahnert, 1975c: 186–192, figs 2–11.

Type locality: Il Karraba, Ghajn Tuffieha Bay, Malta.
Distribution: Malta.

Chthonius (Ephippiochthonius) maroccanus Mahnert

Chthonius (Ephippiochthonius) maroccanus Mahnert, 1980b: 215–217, figs 1–5.
Chthonius maroccanus Mahnert: Callaini, 1988b: 32.
Chthonius cfr. *maroccanus* Mahnert: Callaini, 1988b: 32–36, figs 1a–d.

Type locality: Ras el Oued, S. of Taza, Morocco.
Distribution: Morocco.

Chthonius (Ephippiochthonius) microtuberculatus Hadži

Chthonius (Ephippiochthonius) microtuberculatus Hadži, 1937: 172–175, figs 11a–f; Hadži, 1939b: 23–24, figs 3a–d; Ćurčić, 1974a: 13; Ćurčić, 1976b: 172.
Chthonius microtuberculatus Hadži: Roewer, 1940: 343.
Chthonius (Ephippiochthonius) romanicus microtuberculatus Hadži: Beier, 1963b: 59–60.

Type locality: Kloster Sveti Nikola, Treska, Yugoslavia.
Distribution: Bulgaria, Yugoslavia.

Chthonius (Ephippiochthonius) minous Mahnert

Chthonius (Ephippiochthonius) minous Mahnert, 1980e: 225–226, figs 15–16.

Type locality: Milatos spilia Cave, near Milatos, Crete.
Distribution: Crete.

Chthonius (Ephippiochthonius) minous peramae Mahnert

Chthonius (Ephippiochthonius) minous peramae Mahnert, 1980e: 226–228, fig 17; Gardini, 1989: 61.

Type locality: Melidoni spilia Cave, near Perama, Crete.
Distribution: Crete.

Chthonius (Ephippiochthonius) minutus Vachon

Chthonius (Ephippiochthonius) minutus Vachon, 1940g: 147–149, figs 7–11; Beier, 1963b: 70–71; Zaragoza, 1986: fig. 1.

Type locality: Mina dos Mouros, Mexilhoeira Grande, Portugal.
Distribution: Portugal.

Chthonius (Ephippiochthonius) nanus Beier

Chthonius (Ephippiochthonius) austriacus nanus Beier, 1953e: 105–106, fig. 1 (in part; see *Chthonius (Ephippiochthonius) gibbus* Beier); Beier, 1963b: 69–70; Gardini, 1980b: 264–268, figs 8–15; Gardini, 1980c: 103, 120.

Type locality: Tann-a da Reixe, Carsi, Genova, Italy.
Distribution: Italy.

Chthonius (Ephippiochthonius) nidicola Mahnert

Chthonius (Ephippiochthonius) nidicola Mahnert, 1979c: 502–504, figs 6–10.

Type locality: Plan-les-Ouates, Bois de Milly, Genève, Switzerland.
Distribution: Switzerland.

Chthonius (Ephippiochthonius) nudipes Mahnert

Chthonius (Ephippiochthonius) nudipes Mahnert, 1982f: 301–303, figs 11–16; Zaragoza, 1986: fig. 1.

Type locality: Cueva de las Canpanas, Granada, Spain.
Distribution: Spain.

Chthonius (Ephippiochthonius) parmensis Beier

Chthonius (Ephippiochthonius) parmensis Beier, 1963h: 152–153, fig. 3; Lazzeroni, 1969b: 396; Inzaghi, 1981: 72.
Chthonius (Ephippiochthonius) cfr. *parmensis* Beier: Callaini, 1984: 142–143, fig. 4f.

Type locality: Sasso di Neviano, Parma, Italy.
Distribution: Italy.

Chthonius (Ephippiochthonius) pieltaini Beier

Chthonius gestroi E. Simon: E. Simon, 1900b: 595 (misidentification, in part).
Chthonius sp.: Fabiani, 1902: 7 (not seen).
Chthonius (Ephippiochthonius) pieltaini Beier, 1930g: 327, fig. 5; Beier, 1932a: 58, fig. 71; Beier, 1963b: 72, fig. 70; Gardini, 1980c: 103, 127; Gardini, 1986e: 60.
Chthonius pieltaini Beier: Roewer, 1937: 239.

Type locality: Covolo di Costozza, Longare, Vicenza, Italy.
Distribution: Italy.

Chthonius (Ephippiochthonius) pinai Zaragoza

Chthonius (Ephippiochthonius) pinai Zaragoza, 1985: 150–153, figs 20–25, 30–31; Zaragoza, 1986: fig. 1.

Type locality: Masalatava, Bolulla, Alicante, Spain.
Distribution: Spain.

Chthonius (Ephippiochthonius) platakisi Mahnert

Chthonius (Ephippiochthonius) platakisi Mahnert, 1980e: 229–231, figs 25–27.
Chthonius (Ephippiochthonius) aff. *platakisi* Mahnert: Mahnert, 1980e: 28.

Type locality: Thergiospilia Cave, Kavoussi, Crete.
Distribution: Crete.

Chthonius (Ephippiochthonius) poeninus Mahnert

Chthonius (Ephippiochthonius?) poeninus Mahnert, 1979c: 505–507, figs 1–5.

Type locality: Susten, Wallis, Switzerland.
Distribution: Switzerland.

Chthonius (Ephippiochthonius) pyrenaicus Beier

Chthonius (Ephippiochthonius) pyrenaicus Beier, 1934a: 55–56, fig. 3; Beier, 1955n: 89; Beier, 1959f: 115; Beier, 1963b: 70, fig. 68; Mahnert, 1977c: 66–67; Orghidan, Dumitresco & Georgesco, 1975: 29; Heurtault, 1986c: 21.
Chthonius pyrenaeus (sic) Beier: Roewer, 1937: 239.

Type locality: Forêt de Sorède, Pyrénées Orientales, France.
Distribution: Balearic Islands, France, Spain.

Chthonius (Ephippiochthonius) remyi Heurtault

Chthonius (Ephippiochthonius) remyi Heurtault, 1975a: 315–317, figs 7–12; Gardini, 1980c: 103, 133; Heurtault, 1986c: 21.

Type locality: Conca, Corsica.
Distribution: Corsica.

Chthonius (Ephippiochthonius) romanicus Beier

Chthonius (Ephippiochthonius) romanicus Beier, 1935g: 32–33, fig. 2; Beier, 1939c: 7; Beier, 1963g: 268–269; Beier, 1965f: 83–84, fig. 2; Beier, 1966e: 161; Beier, 1967f: 303; Beier, 1969b: 189–190; Beier, 1971a: 357; Beier, 1973c: 224; Mahnert, 1974a: 87.
Chthonius romanicus Beier: Roewer, 1940: 343.
Chthonius (Ephippiochthonius) romanicus romanicus Beier: Beier, 1963b: 57–58.

Type locality: Comana, Romania.
Distribution: Greece, Iran, Romania, Turkey.

Chthonius (Ephippiochthonius) sacer Beier

Chthonius (Ephippiochthonius) sacer Beier, 1963f: 184–185, fig. 1.

Type locality: Jerusalem, Israel.
Distribution: Israel.

Chthonius (Ephippiochthonius) samius Mahnert

Chthonius (Ephippiochthonius) samius Mahnert, 1982f: 297–300, figs 1–5.

Type locality: Spilia tis Panaghias Cave, Drakei, Samos, Greece.
Distribution: Greece.

Chthonius (Ephippiochthonius) sendrai Zaragoza

Chthonius (Ephippiochthonius) sendrai Zaragoza, 1985: 149–150, figs 14–19; Zaragoza, 1986: fig. 1.

Type locality: Cueva de las Palomas, Millares, Valencia, Spain.
Distribution: Spain.

Chthonius (Ephippiochthonius) serbicus Hadži

Chthonius (Ephippiochthonius) tetrachelatus serbicus Hadži, 1937: 167–172, figs 8a–c, 9a–e, 10a–d; Hadži, 1939b: 22–23.
Chthonius trachelatus (sic) *serbicus* Hadži: Roewer, 1940: 343.
Chthonius (Ephippiochthonius) serbicus Hadži: Beier, 1963b: 58; Ćurčić, 1974a: 14.

Type locality: Rasce, Skopje (as Skoplje), Yugoslavia.
Distribution: Bulgaria, Yugoslavia.

Chthonius (Ephippiochthonius) siculus Beier

Chthonius (Ephippiochthonius) siculus Beier, 1961d: 90–92, fig. 1; Beier, 1963a: 256; Beier, 1963b: 61–62, fig. 58; Beier, 1963h: 150; Beier, 1974g: 163; Mahnert, 1980e: 223; Gardini, 1980c: 103–104, 131; Zaragoza, 1985: 154–155, figs 26–27.
Chtonius (sic) *siculus* Beier: Puddu & Pirodda, 1973: 170.
Not *Chthonius (Ephippiochthonius) siculus* Beier: Callaini, 1983b: 405 (misidentification; see *Chthonius (Ephippiochthonius) concii* Beier).

Type locality: Grotta Calafarina, Pachino-Siracusa, Sicily.
Distribution: Crete, Italy, Sardinia, Sicily, Spain.

Family Chthoniidae

Chthonius (Ephippiochthonius) siscoensis Heurtault

Chthonius (Ephippiochthonius) siscoensis Heurtault, 1975a: 313–315, figs 1–6; Mahnert, 1978b: 381–382, fig. 1; Gardini, 1980c: 104, 133; Callaini, 1981a: 309; Heurtault, 1986c: 21.

Chthonius siscoensis Heurtault: Schawaller, 1981b: 44–45.

Type locality: Sisco, Corsica.
Distribution: Corsica.

Chthonius (Ephippiochthonius) tetrachelatus (Preyssler)

Scorpio tetrachelatus Preyssler, 1790: 56, plate 3, fig. 3.

Chthonius trombidioides (Latreille): C.L. Koch, 1843: 76–79, figs 806–807; L. Koch, 1873: 49–50; Stecker, 1874b: 238; Stecker, 1875d: 88; L. Koch, 1877: 180; Daday, 1880a: 193; Tömösváry, 1882b: 238–239, plate 5 figs 9–12; Krauss, 1896: 628 (misidentifications).

Chthonius maculatus Menge, 1855: 23–25, plate 1 figs 10–14, plate 3 fig. 7, plate 4 fig. 1; Hagen, 1870: 268 (synonymized by E. Simon, 1879a: 70).

Obisium orthodactylum Leach: McIntire, 1868: 8; McIntire, 1869: 247, fig. 219 (misidentifications).

Chthonius trombidioides maculatus Menge: Stecker, 1874c: 308–310.

Chthonius longipalpis Banks, 1891: 164–165 (synonymized by J. C. Chamberlin, 1929a: 72).

Chthonius tetrachelatus (Preyssler): E. Simon, 1879a: 70–71, 312, plate 19 fig. 18; E. Simon, 1881b: 15; H.J. Hansen, 1884: 553–554; Canestrini, 1885: no. 9, figs 1–3; E. Simon, 1885b: 213; Daday, 1888: 134, 190–191; O. P. Cambridge, 1892: 210–211, plate A figs 15–16; Ellingsen, 1897: 22–21; E. Simon, 1898d: 22; Tullgren, 1899a: 181–182, plate 2 figs 10–14; E. Simon, 1900b: 594–595; Evans, 1901b: 242; Godfrey, 1901b: 214–215; Kew, 1903: 299; Godfrey, 1904: 195; Tullgren, 1906a: 205; Ellingsen, 1907a: 171; Godfrey, 1907: 163; Strand, 1907: 243; Tullgren, 1907e: 246; Whyte & Whyte, 1907: 204; Ellingsen, 1908a: 417; Ellingsen, 1908c: 670; Ellingsen, 1908d: 70; Godfrey, 1908: 159–160; Ellingsen, 1909a: 207, 211, 216; Ellingsen, 1909b: 220–221; Kew, 1909a: 250; Kew, 1909b: 259; Ellingsen, 1910a: 402; Ellingsen, 1910b: 62, 63; Ellingsen, 1910c: 347; Godfrey, 1910: 28–30; Kew, 1910a: 67, 68, 71, 73; Ellingsen, 1911c: 174; Kew, 1911a: 58, fig. 22; Kew, 1911b: 2; Lessert, 1911: 40–41, figs 31–32; Ellingsen, 1912b: 89; Krausse-Heldrungen, 1912: 65; Ellingsen, 1913a: 455; Standen, 1912: 16; Falconer, 1916: 158; Kew, 1916b: 82; Standen, 1916a: 18; Nonidez, 1917: 42–43; Standen, 1917: 30; Daday, 1918: 2; Navás, 1921: 166; Redikorzev, 1924b: 26, fig. 12; Navás, 1925: 125–126, fig. 18a; Schenkel, 1926: 313; Beier, 1928: 312; Caporiacco, 1928b: 127; Kästner, 1928: 9, fig. 32; Schenkel, 1928: 71, figs 32, 32a; Beier, 1929a: 365; Beier, 1929e: 454; J.C. Chamberlin, 1929a: 72–73, figs 1a, 1e–f, 3e; J. C. Chamberlin, 1929e: 154; J.C. Chamberlin, 1931a: 128, figs 1o, 19b, 21b, 21j, 22c, 33k, 47a; Szalay, 1931a: 353; Beier, 1932a: fig. 11; Nester, 1932: 98; Schenkel, 1933: 29; Caporiacco, 1936b: 328; Roewer, 1936: fig. 72; Roewer, 1937: 239; Vachon, 1938a: fig. 61b; Lohmander, 1939b: 286–288, fig. 1; Vachon, 1941b: 442–449, figs 1–23; Vachon, 1941c: 540–547, figs 24–29; Caporiacco, 1948c: 240; Hoff, 1950b: 228; Park & Auerbach, 1954: 210; Ressl & Beier, 1958: 2; Cirdei & Gutu, 1959: 10, fig. 9; Kobakhidze, 1961a: 166–169; Meinertz, 1962a: 390, map 1; Höregott, 1963: 546; Vachon, 1964b: 4841; Kobakhidze, 1965b: 541; Vitali–di Castri, 1965b: fig. 2; Helversen, 1966a: 134; Kobakhidze, 1966: 701; Zangheri, 1966: 530; Manley, 1969: 6, fig. 3; Weygoldt, 1969a: 26, 41, 66–70, 108, figs 21–22, 37a–b, 52, 59–61, 105; Lazzeroni, 1970a: 207; Ressl, 1970: 250; Helversen & Martens, 1971: 382, fig. 4; Muchmore, 1971a: 89, 91; Howes, 1972a: 108; Muchmore, 1973g: 122; Palmgren, 1973: 7; Sweeney, Taylor & Counts, 1977: 55;

Taylor, Sweeney & Counts, 1977: 115; Paoletti, 1978: 90 (not seen); Judson, 1979a: 61; Rundle, 1979: 47; Schawaller, 1979: 13; Thaler, 1979: 51–52; Jones, 1980c: map 6; Pieper, 1981: 2; Schawaller, 1983b: 5; Schawaller, 1983c: 368; Hippa, Koponen & Mannila, 1984: 26, 27; Schawaller, in Schmalfuss & Schawaller, 1984: 3; Jędryczkowski, 1985: 78; Lippold, 1985: 40; Braun & Beck, 1986: 140–148; Bliss & Lippold, 1987: 43, 45; Harvey, 1987b: 68–69; Judson, 1987: 369; Krumpál & Cyprich, 1988: 42.

Chthonius longipalpus (sic) Banks: Banks, 1895a: 13; Ewing, 1911: 77, 80, fig. 15.

Chthonius (Chthonius) tetrachelatus (Preyssler): Ellingsen, 1905d: 12; Vachon, 1976: 244.

Chthonius (Ephippiochthonius) tetrachelatus (Preyssler): Beier, 1932a: 56–57, fig. 68; Fage, 1933: 54; Hadži, 1933b: 139–142, figs 8a–d; Hadži, 1933c: 179–180; Beier, 1934b: 284–285; Leruth, 1935: 26; Ionescu, 1936: 3; Beier, 1939c: 7; Beier, 1939e: 305; Beier, 1939f: 163; Vachon, 1940g: 141–142; Hoff, 1945a: 109, fig. 8; Cooreman, 1946b: 3; Caporiacco, 1947c: 252; Franz & Beier, 1948: 473, 495; Leleup, 1948: 8; Hoff, 1949b: 433–434, figs 14a–f; Kaisila, 1949a: 74; Kaisila, 1949b: 76–77, map 1; Rafalski, 1949: 79–85, figs 1–2; Vachon, 1949: fig. 222b; Caporiacco, 1951e: 69; Hoff, 1951: 2–3; Beier, 1952e: 3; Beier & Turk, 1952: 766; Beier & Franz, 1954: 454; G. O. Evans & Browning, 1954: 8; Beier, 1955d: 212; Beier, 1955n: 88; Beier, 1956h: 24; Hoff & Bolsterli, 1956: 157; Hoff, 1958: 3–4; Beier, 1959f: 114–115; Beier, 1961a: 68; Vachon, 1961: 102–103; Beier, 1963b: 57, fig. 54; Beier, 1963f: 184; Beier, 1963h: 150; Beier, 1963i: 283–284; Dumitresco & Orghidan, 1964: 610–615, figs 21–26, 27a–b, plate 1 fig. 1 (these figs erroneously labelled *Chthonius (Ephippiochthonius) labilus* Dumitresco & Orghidan, nomen nudum); Orghidan & Dumitresco, 1964a: 328; Beier, 1965c: 375; Ressl, 1965: 289; Beier, 1966f: 343; Rafalski, 1967: 6–7; Hammen, 1969: 17; Lazzeroni, 1969a: 324–325; Lazzeroni, 1969b: 395–396 (in part; see *Chthonius (Ephippiochthonius) beieri* Lazzeroni and *Chthonius (Ephippiochthonius) kewi* Gabbutt); Lazzeroni, 1969c: 226; Beier, 1970a: 45; Lazzeroni, 1970c: 103; Legg, 1971a: 473–475, fig. 1a; Muchmore, 1971a: 89, 91; Beron, 1972: 13; Ćurčić, 1972b 153–159, figs 15–21; Lagar, 1972a: 17; Anonymous, 1983a: 27; Legg, 1973b: 430; Ćurčić, 1974a: 14; Mahnert, 1974c: 378; Ressl, 1974: 30; Beier, 1976a: 23; Gardini, 1975: 3; Legg, 1975a: 66; Legg, 1975b: 101, figs 1d, 5, 7d, 14a–b; Nelson, 1975: 278–279, figs 4–9; Crocker, 1976: 8; Ćurčić, 1976b: 172–173; Drogla, 1977: 89; Mahnert, 1977d: 97; Crocker, 1978: 8; Mahnert, 1978e: 291; Callaini, 1979c: 343–345; Leclerc & Heurtault, 1979: 241; Gardini, 1980c: 104, 119, 130; Mahnert, 1980a: 259; Callaini, 1981a: 307; Krumpál & Kiefer, 1981: 127; Sanocka–Woloszynowa, 1981: 36; Callaini, 1983a: 149; Callaini, 1983b: 403; Mahnert, 1983d: 361; Callaini, 1984: 126–130, figs 1a–c; Heurtault, 1986c: 21; Legg, 1987: 180; Legg & Jones, 1988: 59–60, figs 10a, 10b(a–f); Schawaller & Dashdamirov, 1988: 8–11, figs 18–27; Callaini, 1989: 138; Schawaller, 1989: 6–7, fig. 17.

Ephippiochthonius tetrachelatus (Preyssler): Vachon, 1934g: 163; Vachon, 1935j: 187–188; Vachon, 1936a: 78.

Chtonius (sic) *tetrachelatus* (Preyssler): Hammen, 1949: 73, 74.

Cothonius (sic) *(Ephippiochthonius) tetrachelatus* (Preyssler): Orghidan & Dumitresco, 1964b: 192.

Chthonius (Ephippiochthonius) cf. *tetrachelatus* (Preyssler): Mahnert, 1989a: 85–86.

Not *Chthonius longipalpis* Banks: Banks, 1909c: 174 (misidentification; see *Tridenchthonius cubanus* (J. C. Chamberlin)).

Not *Chthonius tetrachelatus* (Preyssler): Ellingsen, 1912c: 174–175 (misidentification; see *Spelyngochthonius heurtaultae* Vachon).

Type localities: of *Scorpio tetrachelatus*: Bohemia, Czechoslovakia.
of *Chthonius maculatus*: Bischofsberge, West Germany.

of *Chthonius longipalpis*: Long Island, New York, U.S.A.; Ithaca, New York, U.S.A.; and Washington, District of Columbia, U.S.A.

Distribution: Algeria, Argentina, Australia (Victoria), Austria, Azores, Balearic Islands, Belgium, Canary Islands, Corsica, Crete, Cuba, Cyprus, Czechoslovakia, Denmark, East Germany, Egypt, Finland, France, Great Britain, Greece, Hungary, Iran, Ireland, Israel, Italy, Lebanon, Madeira Islands, Netherlands, Norway, I oland, Portugal, Romania, Sardinia, Seychelles, Sicily, Spain, Sweden, Switzerland, Syria, U.S.A. (Ponnecticut, District of Columbia, Georgia, Illinois, Indiana, Kentucky, Maine, Maryland, Massachusetts, Michigan, New Jersey, New York, North Carolina, Ohio, Pennsylvania, Virginia, West Virginia), U.S.S.R. (Armenia, Azerbaijan, Georgia, Moldavia, R.S.F.S.R., Turkmenistan, Ukraine), West Germany, Yugoslavia.

Chthonius (Ephippiochthonius) troglophilus Beier

Chthonius (Ephippiochthonius) troglophilus Beier, 1920b: 73–74; Beier, 1932a: 58, fig. 72; Beier, 1953e: 106; Beier, 1963b: 72, fig. 71; Gardini, 1980c: 104, 118; Callaini, 1984: 143–144; Bologna & Taglianti, 1985: 226, fig. 8.
Chthonius troglophilus Beier: Roewer, 1937: 239.
Ephippiochthonius troglophilus (Beier): Franciscolo, 1955: 124.

Type locality: Tana dello Scopeto, near Albenga, Italy.
Distribution: Italy.

Chthonius (Ephippiochthonius) tuberculatus Hadži

Chthonius (Ephippiochthonius) tuberculatus Hadži, 1937: 175–179, figs 12a–e, 13a–d; Beier, 1963b: 63; Ćurčić, 1974a: 14; Mahnert, 1980e: 219; Mahnert, 1983d: 361.
Chthonius tuberculatus Hadži: Roewer, 1940: 343.
Chthonius aff. *tuberculatus* Hadži: Helversen, 1966a: 134–136, fig. 1.
Chtonius (sic) *tuberculatus* Hadži: Cîrdei, Bulimar & Malcoci, 1967: 238.
Chthonius (Ephippiochthonius) cf. *tuberculatus* Hadži: Mahnert, 1978e: 291–293, figs 47–48.

Type locality: Rasce, Skopje (as Skoplje), Yugoslavia.
Distribution: Greece, Hungary, Romania, West Germany, Yugoslavia.

Chthonius (Ephippiochthonius) vachoni Heurtault-Rossi

Chthonius (Ephippiochthonius) vachoni Heurtault-Rossi, 1963: 419–426, figs 1–14.

Type locality: Taussat-les-Bains, Gironde, France.
Distribution: France.

Chthonius (Ephippiochthonius) ventalloi Beier

Chthonius (Ephippiochthonius) ventalloi Beier, 1939f: 164–165, fig. 4; Beier, 1963b: 71, fig. 69; Lagar, 1972a: 17–18; Lagar, 1972b: 46; Zaragoza, 1985: 145–149, figs 1–13, 28–29; Zaragoza, 1986: fig. 1.
Chthonius ventalloi Beier: Roewer, 1940: 343.

Type locality: Cova del Candil, Barranco del Castellet near Tous, Valencia, Spain.
Distribution: Spain.

Chthonius (Ephippiochthonius) verai Zaragoza

Chthonius (Ephippiochthonius) verai Zaragoza, 1986: fig. 1 (nomen nudum).
Chthonius (Ephippiochthonius) verai Zaragoza, 1987: 5–13, figs 1–6.

Type locality: Cova del Tío Melxor, Castalla, Alicante, Spain.
Distribution: Spain.

Chthonius (Ephippiochthonius) virginicus J. C. Chamberlin

Chthonius virginicus J. C. Chamberlin, 1929a: 73–74; Pearse, 1946: 257; Roewer, 1937: 239.
Chthonius (Globochthonius) virginicus J. C. Chamberlin: Beier, 1931b: 54; Beier, 1932a: 55.
Chthonius (Globochthonius?) virginicus J. C. Chamberlin: Hoff, 1958: 4.
Chthonius (Ephippiochthonius) virginicus J. C. Chamberlin: Muchmore, 1981b: 110–111, figs 1–2.

Type locality: Great Falls, Virginia, U.S.A.
Distribution: U.S.A. (District of Columbia, Maryland, North Carolina, Virginia).

Subgenus Chthonius (Globochthonius) Beier

Chthonius (Globochthonius) Beier, 1931b: 54; Beier, 1932a: 54; Beier, 1939d: 18.
Type species: *Chthonius globifer* E. Simon, 1879a, by present designation.

Chthonius (Globochthonius) abnormis Beier

Chthonius (Globochthonius) abnormis Beier, 1939d: 18, fig. 15; Beier, 1963b: 51, fig. 46; Ćurčić, 1974a: 14; Ćurčić, 1974c: figs 1, 10.

Type locality: Sitnice, Yugoslavia.
Distribution: Yugoslavia.

Chthonius (Globochthonius) caligatus Beier

Chthonius (Globochthonius) caligatus Beier, 1938a: 5; Beier, 1939d: 20–21, figs 18–21; Beier, 1963b: 53–54, fig. 49; Beier, 1970f: 318; Ćurčić, 1974a: 14; Ćurčić, 1974c: figs 6, 10.

Type locality: Ilijina Cave, Trebinje, Yugoslavia.
Distribution: Yugoslavia.

Chthonius (Globochthonius) cavernicola Beier

Chthonius (Globochthonius) cavernicola Beier, 1938a: 5; Beier, 1939d: 22, fig. 23; Beier, 1963b: 55–56, fig. 52; Ćurčić, 1974a: 14; Ćurčić, 1974c: figs 7, 10.

Type locality: Kjecina Stijena, Sarajevo, Yugoslavia.
Distribution: Yugoslavia.

Chthonius (Globochthonius) cerberus Beier

Chthonius (Globochthonius) cerberus Beier, 1938a: 5–6; Beier, 1939d: 22–23; Beier, 1963b: 56, fig. 53; Ćurčić, 1974a: 14; Ćurčić, 1974c: figs 8, 10.

Type locality: Semec Planina, Bosnia, Yugoslavia.
Distribution: Yugoslavia.

Chthonius (Globochthonius) globifer E. Simon

Chthonius globifer E. Simon, 1879a: 72, plate 19 fig. 16; E. Simon, 1898d: 22; Roewer, 1937: 239.
Chthonius (Globochthonius) globifer E. Simon: Beier, 1931b: 54; Beier, 1932a: 55, fig. 65; Beier, 1963b: 51, fig. 47; Ćurčić, 1973a: 77–82, figs 1–14; Ćurčić, 1974c: figs 3, 10; Mahnert, 1979c: 502.

Type locality: Bourg-d'Oisans, Isère, France.
Distribution: France, Switzerland.

Chthonius (Globochthonius) polychaetus Hadži

Chthonius (Ephippiochthonius) polychaetus Hadži, 1937: 165–167, figs 7a–e.
Chthonius polychaetus Hadži: Roewer, 1940: 343.

Family Chthoniidae

Chthonius (Globochthonius) polychaetus Hadži: Beier, 1963b: 52–53; Ćurčić, 1974c: figs 9–10.
Chthonius (Chthonius) polychaetus polychaetus Hadži: Ćurčić, 1974a: 14; Ćurčić, 1976b: 174, fig. 1.

Type locality: Kovacica, Kacanik, Yugoslavia.
Distribution: Yugoslavia.

Chthonius (Globochthonius) polychaetus pancici Ćurčić

Chthonius (Globochthonius) polychaetus pancici Ćurčić, 1972b: 148–153, figs 8–14; Ćurčić, 1974a: 14; Ćurčić, 1976b: 174, fig. 1.

Type locality: Perucac, 11 km E. of Bajina Basta, Yugoslavia.
Distribution: Yugoslavia.

Chthonius (Globochthonius) simplex Beier

Chthonius (Globochthonius) simplex Beier, 1939d: 18–19, fig. 16; Beier, 1963b: 52, fig. 48; Ćurčić, 1974a: 14–15; Ćurčić, 1974c: figs 2, 10.

Type locality: Bazdovaca Jama, Brac (as Brazza) Island, Yugoslavia.
Distribution: Yugoslavia.

Chthonius (Globochthonius) spelaeophilus Hadži

Chthonius spelaeophilus Hadži, 1930b: 6–16, figs 1–2, 3a–e, 4–5, 6a–d; Roewer, 1937: 239.
Chthonius (Globochthonius) spelaeophilus Hadži: Beier, 1931b: 54; Beier, 1932a: 55; Ćurčić, 1974c: figs 5, 10.
Chthonius (Globochthonius) spelaeophilus spelaeophilus Hadži: Beier, 1939d: 19, fig. 17; Beier, 1963b: 54–55, fig. 51; Ćurčić, 1974a: 15.

Type locality: Jama Dacarica, Dobrepolju, Yugoslavia.
Distribution: Yugoslavia.

Chthonius (Globochthonius) spelaeophilus histricus Beier

Chthonius microphthalmus E. Simon: Beier, 1929a: 365–366 (misidentification).
Chthonius (Globochthonius) histricus Beier, 1931b: 54; Beier, 1932a: 55–56, fig. 66.
Chthonius histricus Beier: Roewer, 1937: 239.
Chthonius (Globochthonius) spelaeophilus histricus Beier: Beier, 1939d: 19; Beier, 1963b: 55, fig. 51; Beier, 1974f: 159; Ćurčić, 1974a: 15; Gardini, 1980c: 102, 128.

Type locality: Dante Cave, near Tolmin (as Tolmein), Yugoslavia.
Distribution: Italy, Yugoslavia.

Chthonius (Globochthonius) vandeli Dumitresco & Orghidan

Chthonius (Globochthonius) vandeli Dumitresco & Orghidan, 1964: 606–610, figs 14–20, plate 1 fig. 3; Ćurčić, 1974c: figs 4, 10.

Type locality: Grotte Pestera Liliecilor, Gura Dobrogei, Romania.
Distribution: Romania.

Subgenus Chthonius (Hesperochthonius) Muchmore

Chthonius (Hesperochthonius) Muchmore, 1968b: 51–52.

Type species: *Chthonius californicus* J. C. Chamberlin, 1929a, by present designation.

Chthonius (Hesperochthonius) californicus J. C. Chamberlin

Chthonius californicus J. C. Chamberlin, 1929a: 74; J. C. Chamberlin, 1929e: 154; J. C. Chamberlin, 1931a: fig. 9b; Roewer, 1937: 239.

Chthonius (Ephippiochthonius) californicus J. C. Chamberlin: Beier, 1932a: 57; J. C.
    Chamberlin, 1949: 2–4, figs 1a–c; Hoff, 1958: 4.
Chthonius (Hesperochthonius) californicus J. C. Chamberlin: Muchmore, 1968b: 51.

Type locality: Berkeley, California, U.S.A.
Distribution: U.S.A. (California).

Chthonius (Hesperochthonius) oregonicus Muchmore

Chthonius (Hesperochthonius) oregonicus Muchmore, 1968b: 52–54, figs 1–2.

Type locality: Coos Head, Coos County, Oregon, U.S.A.
Distribution: U.S.A. (Oregon).

Chthonius (Hesperochthonius) spingolus (R. O. Schuster)

Kewochthonius spingolus R. O. Schuster, 1962: 223–225, figs 1–2, 7, 7a.
Chthonius (Hesperochthonius) spingolus (R. O. Schuster): Judson, 1990: 597.

Type locality: between Ramona and Julian, California, U.S.A.
Distribution: U.S.A. (California).

Genus Congochthonius Beier

Congochthonius Beier, 1959d: 13–14.

Type species: *Congochthonius nanus* Beier, 1959d, by original designation.

Congochthonius nanus Beier

Congochthonius nanus Beier, 1959d: 14–15, fig. 4.

Type locality: Kundelungu, Zaire.
Distribution: Zaire.

Genus Drepanochthonius Beier

Drepanochthonius Beier, 1964b: 314–315.

Type species: *Drepanochthonius horridus* Beier, 1964b, by original designation.

Drepanochthonius horridus Beier

Drepanochthonius horridus Beier, 1964b: 14–15, fig. 2; Cekalovic, 1984: 10.

Type locality: Zapallar, Quebrada de Aguas Claras, Chile.
Distribution: Chile.

Genus Francochthonius Vitali-di Castri

Francochthonius Vitali-di Castri, 1975b: 1283–1285.

Type species: *Francochthonius hirsutus* Vitali-di Castri, 1975b, by original designation.

Francochthonius hirsutus Vitali-di Castri

Francochthonius hirsutus Vitali-di Castri, 1975b: 1285–1290, figs 13–26, plate 1;
    Cekalovic, 1984: 10.

Type locality: Cerro El Roble, Valparaiso, Chile.
Distribution: Chile.

Genus Kleptochthonius J. C. Chamberlin

Apochthonius (Heterochthonius) J. C. Chamberlin, 1929e: 153; Beier, 1932a: 42
    (junior homonym of *Heterochthonius* Berlese, 1910).
Heterochthonius J. C. Chamberlin: Hoff, 1945i: 313–314; Hoff, 1949b: 434.
Kleptochthonius J. C. Chamberlin, 1949: 4 (replacement name for *Apochthonius*
    (Heterochthonius) J. C. Chamberlin, 1929e); Malcolm & J. C. Chamberlin,
    1961: 2–3.

Type species: *Apochthonius (Heterochthonius) crosbyi* J. C. Chamberlin, 1929e, by original designation.

Subgenus **Kleptochthonius (Kleptochthonius)** J. C. Chamberlin

Kleptochthonius (Kleptochthonius) J. C. Chamberlin: Malcolm & J. C. Chamberlin, 1961: 5.

Kleptochthonius (Kleptochthonius) crosbyi J. C. Chamberlin)

Apochthonius (Heterochthonius) crosbyi J. C. Chamberlin, 1929e: 153–154; J. C. Chamberlin, 1931a: figs 16d, 34g–h, 45a, 50e; Beier, 1932a: 42.
Apochthonius crosbyi J. C. Chamberlin: Beier, 1932g: figs 169, 227; Roewer, 1936: fig. 140; Roewer, 1937: 238.
Heterochthonius crosbyi (J. C. Chamberlin): Hoff, 1945i: 313; Hoff, 1949b: 436.
Kleptochthonius crosbyi (J. C. Chamberlin): J. C. Chamberlin, 1949: 4; Hoff, 1958: 7.
Kleptochthonius (Kleptochthonius) crosbyi (J. C. Chamberlin): Malcolm & J. C. Chamberlin, 1961: 11–14, figs 3a–c.
Kleptochthonius (Kleptochthonius) crosbyi (J. C. Chamberlin) (?): Muchmore, 1976d: fig. 6.
Not *Apochthonius crosbyi* J. C. Chamberlin: Pearse, 1943: 464 (not seen) (misidentification; true identification unknown).

Type locality: Mt Mitchell, North Carolina, U.S.A.
Distribution: U.S.A. (Kentucky, North Carolina).

Kleptochthonius (Kleptochthonius) geophilus Malcolm & J. C. Chamberlin

Kleptochthonius (Kleptochthonius) geophilus Malcolm & J. C. Chamberlin, 1961: 7–11, figs 2a–f.

Type locality: 2 miles N. of Denmark, Curry County, Oregon, U.S.A.
Distribution: U.S.A. (Oregon).

Kleptochthonius (Kleptochthonius) magnus Muchmore

Kleptochthonius (Kleptochthonius) magnus Muchmore, 1966b: 68.

Type locality: Dry Cave, 1.8 miles N. of Sewanee, Franklin County, Tennessee, U.S.A.
Distribution: U.S.A. (Tennessee).

Kleptochthonius (Kleptochthonius) multispinosus (Hoff)

Heterochthonius multispinosus Hoff, 1945i: 314–318, figs 1–4; Pearse, 1946: 257; Hoff, 1949b: 434–436, figs 16a–b.
Kleptochthonius multispinosus (Hoff): Hoff, 1951: 5; Hoff & Bolsterli, 1956: 159; Hoff, 1958: 7; Zeh, 1987b: 1086.
Kleptochthonius (Kleptochthonius) multispinosus (Hoff): Malcolm & J. C. Chamberlin, 1961: 14–16.

Type locality: Duke Forest, Durham, North Carolina, U.S.A.
Distribution: U.S.A. (Alabama, Illinois, Kentucky, Mississippi, Missouri, North Carolina, Tennessee, Virginia).

Kleptochthonius (Kleptochthonius) oregonus Malcolm & J. C. Chamberlin

Kleptochthonius (Kleptochthonius) oregonus Malcolm & J. C. Chamberlin, 1961: 5–7, figs 1a–c.

Type locality: 1 mile N. of Selma, Josephine County, Oregon, U.S.A.
Distribution: U.S.A. (Oregon).

Subgenus **Kleptochthonius (Chamberlinochthonius)** Vachon

Chamberlinochthonius Vachon, 1952b: 105–106.
Kleptochthonius (Chamberlinochthonius) Vachon: Malcolm & J. C. Chamberlin, 1961: 16–17.

Type species: *Chamberlinochthonius henroti* Vachon, 1952b, by original designation.

Kleptochthonius (Chamberlinochthonius) affinis Muchmore

Kleptochthonius (Chamberlinochthonius) affinis Muchmore, 1976d: 215–216, fig. 7; Holsinger & Culver, 1988: 42, fig. 15.

Type locality: Chadwell's Cave, 3.5 miles NE. of Tazewell, Claiborne County, Tennessee, U.S.A.
Distribution: U.S.A. (Tennessee).

Kleptochthonius (Chamberlinochthonius) anophthalmus Muchmore

Kleptochthonius (Chamberlinochthonius) anophthalmus Muchmore, 1970a: 211; Holsinger & Culver, 1988: 42, fig. 15.

Type locality: Porter's Cave, about 7.5 miles WSW. of Millboro, Bath County, Virginia, U.S.A..
Distribution: U.S.A. (Virginia).

Kleptochthonius (Chamberlinochthonius) attenuatus Malcolm & J. C. Chamberlin

Kleptochthonius (Chamberlinochthonius) attenuatus Malcolm & J. C. Chamberlin, 1961: 31–33, figs 10a–d.

Type locality: Blind Snail Cave, 7 miles ENE. of Munfordville, Hart County, Kentucky, U.S.A.
Distribution: U.S.A. (Kentucky).

Kleptochthonius (Chamberlinochthonius) barri Muchmore

Kleptochthonius (Chamberlinochthonius) barri Muchmore, 1965: 13–15, figs 11–12.

Type locality: Tom Campbell Cave, 5 miles SE. of Viola, Grundy County, Tennessee, U.S.A.
Distribution: U.S.A. (Tennessee).

Kleptochthonius (Chamberlinochthonius) binoculatus Muchmore

Kleptochthonius (Chamberlinochthonius) binoculatus Muchmore, 1974e: 82, figs 1–3; Holsinger & Culver, 1988: 42, fig. 15.

Type locality: Hill Cave, 4 miles NNE. of Natural Tunnel, Scott County, Virginia, U.S.A.
Distribution: U.S.A. (Virginia).

Kleptochthonius (Chamberlinochthonius) cerberus Malcolm & J. C. Chamberlin

Chthonius packardii Hagen: Giovannoli, 1933: 604 (possible misidentification).
Kleptochthonius (Chamberlinochthonius) cerberus Malcolm & J. C. Chamberlin, 1961: 28–31, figs 9a–e.

Type locality: White Cave, Mammoth Cave National Park, Kentucky, U.S.A.
Distribution: U.S.A. (Kentucky).

Kleptochthonius (Chamberlinochthonius) charon Muchmore

Kleptochthonius (Chamberlinochthonius) charon Muchmore, 1965: 16–19, fig. 6.

Type locality: Lowe Gap Cave, 2 miles E. of Litton, Bledsoe County, Tennessee, U.S.A.
Distribution: U.S.A. (Tennessee).

Kleptochthonius (Chamberlinochthonius) daemonius Muchmore

Kleptochthonius (Chamberlinochthonius) daemonius Muchmore, 1965: 22–24, fig. 9.

Type locality: McElroy Cave, 2 miles S. of Quebeck, Van Buren County, Tennessee, U.S.A.

Distribution: U.S.A. (Tennessee).

Kleptochthonius (Chamberlinochthonius) erebicus Muchmore

Kleptochthonius (Chamberlinochthonius) erebicus Muchmore, 1965: 15–16, fig. 5.

Type locality: Hog Cave, about 2 miles NE. of Mill Springs, Wayne County, Kentucky, U.S.A.

Distribution: U.S.A. (Kentucky).

Kleptochthonius (Chamberlinochthonius) gertschi Malcolm & J. C. Chamberlin

Kleptochthonius (Chamberlinochthonius) gertschi Malcolm & J. C. Chamberlin, 1961: 17–19, figs 4a–d; Holsinger, 1963: 34; Holsinger & Culver, 1988: 42, fig. 15.

Type locality: Gilly's Cave, Pennington Gap, Virginia, U.S.A.
Distribution: U.S.A. (Virginia).

Kleptochthonius (Chamberlinochthonius) hageni Muchmore

Kleptochthonius (Chamberlinochthonius) hageni Muchmore, 1963a: 5–9, figs 3–4.

Type locality: Mammoth Dome, Edmonson County, Kentucky, U.S.A.
Distribution: U.S.A. (Kentucky).

Kleptochthonius (Chamberlinochthonius) henroti (Vachon)

Chamberlinochthonius henroti Vachon, 1952b: 107–111, figs 1–13; Hoff, 1958: 7.
Kleptochthonius (Chamberlinochthonius) henroti (Vachon): Malcolm & J. C. Chamberlin, 1961: 35; Muchmore, 1965: 3–6; Muchmore, 1970a: 210.

Type locality: McClung's Cave, West Virginia, U.S.A.
Distribution: U.S.A. (West Virginia).

Kleptochthonius (Chamberlinochthonius) hetricki Muchmore

Kleptochthonius (Chamberlinochthonius) hetricki Muchmore, 1974e: 83–84, figs 4–5.

Type locality: Greenbrier Caverns System, Masters Section, about 2 miles SE. of Ronceverte, Greenbrier County, West Virginia, U.S.A.
Distribution: U.S.A. (West Virginia).

Kleptochthonius (Chamberlinochthonius) hubrichti Muchmore

Kleptochthonius (Chamberlinochthonius) hubrichti Muchmore, 1965: 19–20, fig. 7.

Type locality: Duval Saltpeter Cave, 0.7 mile NW. of Becton, Barren County, Kentucky, U.S.A.
Distribution: U.S.A. (Kentucky).

Kleptochthonius (Chamberlinochthonius) infernalis Malcolm & J. C. Chamberlin

Kleptochthonius (Chamberlinochthonius) infernalis Malcolm & J. C. Chamberlin, 1961: 21–23, figs 6a–d.

Type locality: Saltpeter Cave, Grassy Cove, Tennessee, U.S.A.
Distribution: U.S.A. (Tennessee).

Kleptochthonius (Chamberlinochthonius) krekeleri Muchmore

Kleptochthonius (Chamberlinochthonius) krekeleri Muchmore, 1965: 8–11, figs 1–2.

Type locality: Yarberry's Cave 1, 1.5 miles NNE. of Columbia, Adair County, Kentucky, U.S.A.
Distribution: U.S.A. (Kentucky).

Kleptochthonius (Chamberlinochthonius) lutzi Malcolm & J. C. Chamberlin

Kleptochthonius (Chamberlinochthonius) lutzi Malcolm & J. C. Chamberlin, 1961: 19–21; Holsinger, 1963: 34; Holsinger & Culver, 1988: 42, fig. 15.

Type locality: Cudjo's Cave, Cumberland Gap, Tennessee, U.S.A.
Distribution: U.S.A. (Tennessee).

Kleptochthonius (Chamberlinochthonius) microphthalmus Malcolm & J. C. Chamberlin

Kleptochthonius (Chamberlinochthonius) microphthalmus Malcolm & J. C. Chamberlin, 1961: 26–28, figs 8a–d.

Type locality: Thomas Cave, 2 miles NE. of Hadley, Warren County, Kentucky, U.S.A.
Distribution: U.S.A. (Kentucky).

Kleptochthonius (Chamberlinochthonius) myopius Malcolm & J. C. Chamberlin

Kleptochthonius (Chamberlinochthonius) myopius Malcolm & J. C. Chamberlin, 1961: 23–26, figs 7a–d.

Type locality: Cathcart Cave, De Kalb County, Tennessee, U.S.A.
Distribution: U.S.A. (Tennessee).

Kleptochthonius (Chamberlinochthonius) orpheus Muchmore

Kleptochthonius (Chamberlinochthonius) orpheus Muchmore, 1965: 11–13, fig. 4.

Type locality: Patton's Cave, Monroe County, West Virginia, U.S.A.
Distribution: U.S.A. (West Virginia).

Kleptochthonius (Chamberlinochthonius) packardi (Hagen)

Blothrus packardi Hagen, 1879: 399.
Chthonius packardi (Hagen): Hagen, 1880: 83–84, figs a–g; Banks, 1895a: 13; Coolidge, 1908: 114.
Chthonius packardii (Hagen): Packard, 1886: 43–46, figs 12a–g, plate 11 figs 3, 3a–j.
Chthonius (?) *packardi* (Hagen): Beier, 1932a: 61; Roewer, 1937: 240; Hoff, 1958: 4.
Genus ? *packardi* Hagen: Hoff, 1949b: 443.
Kleptochthonius (Chamberlinochthonius) packardi (Hagen): Muchmore, 1963a: 2–5, figs 1–2.
Not *Chthonius packardii* Hagen: Giovannoli, 1933: 604 (possible misidentification; see *Kleptochthonius (Chamberlinochthonius) cerberus* Malcolm & J.C. Chamberlin).

Type locality: Mammoth Cave, Kentucky, U.S.A.
Distribution: U.S.A. (Indiana, Kentucky).

Kleptochthonius (Chamberlinochthonius) pluto Muchmore

Kleptochthonius (Chamberlinochthonius) pluto Muchmore, 1965: 25–27, fig. 10.

Type locality: Raven Bluff Cave, 3.5 miles NNW. of Allons, Overton County, Tennessee, U.S.A.
Distribution: U.S.A. (Tennessee).

Kleptochthonius (Chamberlinochthonius) proserpinae Muchmore

Kleptochthonius (Chamberlinochthonius) proserpinae Muchmore, 1965: 6–7, fig. 3.

Type locality: Pollock Cave, Greenbriar County, West Virginia, U.S.A.
Distribution: U.S.A. (West Virginia).

Family Chthoniidae

Kleptochthonius (Chamberlinochthonius) proximosetus Muchmore

Kleptochthonius (Chamberlinochthonius) proximosetus Muchmore, 1976d: 213–215, figs 1–3; Holsinger & Culver, 1988: 42, fig. 15.

Type locality: Gallohan Cave No. 1, 7 miles SE. of Rose Hill, Lee County, Virginia, U.S.A.
Distribution: U.S.A. (Virginia).

Kleptochthonius (Chamberlinochthonius) regulus Muchmore

Kleptochthonius (Chamberlinochthonius) regulus Muchmore, 1970a: 211; Holsinger & Culver, 1988: 42, fig. 15.

Type locality: Fallen Rock Cave, about 2 miles SW. of Pounding Mill, Tazewell County, Virginia, U.S.A.
Distribution: U.S.A. (Virginia).

Kleptochthonius (Chamberlinochthonius) rex Malcolm & J. C. Chamberlin

Kleptochthonius (Chamberlinochthonius) rex Malcolm & J. C. Chamberlin, 1961: 33–35, figs 11a–d.

Type locality: Bunkum Cave, Byrdstown, Tennessee, U.S.A.
Distribution: U.S.A. (Tennessee).

Kleptochthonius (Chamberlinochthonius) similis Muchmore

Kleptochthonius (Chamberlinochthonius) similis Muchmore, 1976d: 216–217, fig. 8; Holsinger & Culver, 1988: 42, fig. 15.

Type locality: Sweet Potato Cave, 6.5 miles ESE. of Rose Hill, Lee County, Virginia, U.S.A.
Distribution: U.S.A. (Virginia).

Kleptochthonius (Chamberlinochthonius) stygius Muchmore

Kleptochthonius (Chamberlinochthonius) stygius Muchmore, 1965: 20–22, fig. 8; Muchmore, 1976d: fig. 4.

Type locality: Sadler Cave, 0.85 mile ENE. of Sadler, Putnam County, Tennessee, U.S.A.
Distribution: U.S.A. (Tennessee).

Kleptochthonius (Chamberlinochthonius) tantalus Muchmore

Kleptochthonius (Chamberlinochthonius) tantalus Muchmore, 1966b: 68–69; Muchmore, 1976d: fig. 5.

Type locality: Dry Cave, 1.8 miles N. of Sewanee, Franklin County, Tennessee, U.S.A.
Distribution: U.S.A. (Tennessee).

Genus Lagynochthonius Beier

Tyrannochthonius (Lagynochthonius) Beier, 1951a: 61.
Lagynochthonius Beier: J. C. Chamberlin, 1962: 314; Murthy & Ananthakrishnan, 1977: 16–17; Harvey, 1989a: 21.

Type species: *Chthonius johni* Redikorzev, 1922b, by original designation.

Lagynochthonius annamensis (Beier)

Tyrannochthonius (Lagynochthonius) annamensis Beier, 1951a: 62–63, fig. 10.
Lagynochthonius annamensis (Beier): J. C. Chamberlin, 1962: 315.

Type locality: Pic de Langbian, Vietnam.
Distribution: Vietnam.

Lagynochthonius arctus (Beier), new combination

Tyrannochthonius (Lagynochthonius) arctus Beier, 1967c: 319–321, fig. 3.

Type locality: Lorengau, Manus, Bismarck Archipelago, Papua New Guinea.
Distribution: Papua New Guinea.

Lagynochthonius australicus (Beier)

Tyrannochthonius (Lagynochthonius) australicus Beier, 1966a: 284–285, fig. 4.
Tyrannochthonius australicus Beier: Beier, 1975b: 203.
Lagynochthonius australicus (Beier): Harvey, 1981b: 240; Harvey, 1985b: 138.

Type locality: Denmark, Western Australia, Australia.
Distribution: Australia (Northern Territory, Western Australia).

Lagynochthonius bakeri (J. C. Chamberlin)

Tyrannochthonius bakeri J. C. Chamberlin, 1929a: 75–76; J. C. Chamberlin, 1931a: 144, figs 34b–c; Beier, 1932a: 64; Roewer, 1937: 240.
Tyrannochthonius (Tyrannochthonius) bakeri J. C. Chamberlin: Beier, 1966b: 340.
Lagynochthonius bakeri (J. C. Chamberlin): J. C. Chamberlin, 1962: 315–316, figs 4a–c.

Type locality: Mt Makiling, Luzon, Philippines.
Distribution: Philippines.

Lagynochthonius brincki (Beier), new combination

Tyrannochthonius (Lagynochthonius) brincki Beier, 1973b: 39–40, fig. 1.

Type locality: Bopathella forest, 9 miles NNW. of Ratnapura, Sabaragamuwa, Sri Lanka.
Distribution: Sri Lanka.

Lagynochthonius dybasi (Beier)

Tyrannochthonius (Lagynochthonius) dybasi Beier, 1957d: 11–12, figs 2a–c.
Lagynochthonius dybasi (Beier): J. C. Chamberlin, 1962: 315.

Type locality: Peleliu, Palau Islands, Caroline Islands.
Distribution: Caroline Islands.

Lagynochthonius exiguus (Beier)

Tyrannochthonius (Lagynochthonius) exiguus Beier, 1952a: 97–98, fig. 2.
Lagynochthonius exiguus (Beier): J. C. Chamberlin, 1962: 315.

Type locality: Sungei Buloh Leper Settlement, Selangor, Malaysia.
Distribution: Malaysia.

Lagynochthonius flavus (Mahnert), new combination

Tyrannochthonius (Lagynochthonius?) flavus Mahnert, 1986a: 832–833, figs 27–31.

Type locality: Makadara Forest, Shimba Hills, Kenya.
Distribution: Kenya.

Lagynochthonius guasirih (Mahnert), new combination

Tyrannochthonius (Lagynochthonius) guasirih Mahnert, 1988d: 384–386, figs 1–4.

Type locality: Grotte Gua Sirah (= Sireh Cave), Serian-Bau District, Sarawak, Malaysia.
Distribution: Malaysia (Sarawak).

Lagynochthonius hamatus Harvey

Lagynochthonius hamatus Harvey, 1988b: 323–324, figs 34–39.

Type locality: 7.5 km W. of Liwa, Sumatra, Indonesia.
Distribution: Indonesia (Sumatra).

Lagynochthonius himalayensis (Morikawa), new combination

Tyrannochthonius (Lagynochthonius) himalayensis Morikawa, 1968: 62, figs 1b, 2c; Beier, 1974b: 262.
Tyrannochthonius (Lagynochthonius) himalaiensis Morikawa: Beier, 1976e: 96–97, fig. 2.
Tyrannochthonius himalayensis (Morikawa): Ćurčić, 1980b: 78–79, figs 1–9; Schawaller, 1983a: 107; Schawaller, 1987a: 202–203.

Type locality: Maedane Karka, near Mt Numbur, Nepal.
Distribution: Bhutan, Nepal.

Lagynochthonius hygricus Murthy & Ananthakrishnan

Lagynochthonius hygricus Murthy & Ananthakrishnan, 1977: 17–19, figs 4a–d.
Type locality: Shimoga, Karnataka, India.
Distribution: India.

Lagynochthonius indicus Murthy & Ananthakrishnan

Lagynochthonius indicus Murthy & Ananthakrishnan, 1977: 19–20, figs 5a–d.
Type locality: Rourkela, Orissa, India.
Distribution: India.

Lagynochthonius johni (Redikorzev)

Chthonius johni Redikorzev, 1922b: 550–554, figs 5–9.
Tyrannochthonius johni (Redikorzev): J. C. Chamberlin, 1929a: 75.
Tyrannochthonius (Lagynochthonius) johni (Redikorzev): Beier, 1951a: 62; Beier, 1966b: 340.
Lagynochthonius johni (Redikorzev): J. C. Chamberlin, 1962: 314; Harvey, 1988b: 320–322, figs 23–29.
Not *Tyrannochthonius johni* (Redikorzev): Beier, 1930e: 294; Beier, 1932a: 65, fig. 80; Roewer, 1936: fig. 68e; Roewer, 1937: 240 (misidentifications; see *Lagynochthonius roeweri* J. C. Chamberlin).

Type locality: Siak, Sumatra, Indonesia.
Distribution: Indonesia (Java, Sumatra), Philippines.

Lagynochthonius kapi Harvey

Lagynochthonius kapi Harvey, 1988b: 317–320, figs 1–6.

Type locality: Zwarte Hoek, Rakata, Krakatau Islands, Indonesia.
Distribution: Indonesia (Krakatau Islands).

Lagynochthonius kenyensis (Mahnert), new combination

Tyrannochthonius (Lagynochthonius?) kenyensis Mahnert, 1986a: 830–832, figs 21–26.
Type locality: Nairobi, Kenya.
Distribution: Kenya.

Lagynochthonius mordor Harvey

Lagynochthonius mordor Harvey, 1989a: 21–23, figs 1–9.

Type locality: Tier Cave, North Mordor Tower, Mt Mulgrave Station, Cape York, Queensland, Australia.
Distribution: Australia (Queensland).

Lagynochthonius nagaminei (Sato), new combination

Tyrannochthonius (Lagynochthonius) nagaminei Sato, 1983c: 7–10, figs 1–11.

Type locality: Makizono-cho, Aira-gun, Kagoshima, Japan.
Distribution: Japan.

Lagynochthonius novaeguineae (Beier), new combination

Tyrannochthonius (Lagynochthonius) novaeguineae Beier, 1965g: 758–760, fig. 3; Beier, 1982: 43.

Type locality: Ajamaroe, Irian Jaya, Indonesia.
Distribution: Indonesia (Irian Jaya), Papua New Guinea.

Lagynochthonius paucedentatus (Beier)

Tyrannochthonius (Lagynochthonius) paucedentatus Beier, 1955e: 39–40, fig. 2.
Lagynochthonius paucedentatus (Beier): J. C. Chamberlin, 1962: 315.

Type locality: Kuala Terla, Telom Valley, Cameron Highlands, Pahang, Malaysia.
Distribution: Malaysia.

Lagynochthonius ponapensis (Beier)

Tyrannochthonius (Lagynochthonius) ponapensis Beier, 1957d: 12–13, figs 3a–b.
Lagynochthonius ponapensis (Beier): J. C. Chamberlin, 1962: 315.

Type locality: Mt Pairot, Ponape, Caroline Islands.
Distribution: Caroline Islands.

Lagynochthonius roeweri J. C. Chamberlin

Tyrannochthonius johni (Redikorzev): Beier, 1930e: 294; Beier, 1932a: 65, fig. 80; Roewer, 1936: fig. 68e; Roewer, 1937: 240 (misidentifications).
Lagynochthonius roeweri J. C. Chamberlin, 1962: 316–317.

Type locality: Batavia (now Jakarta), Java, Indonesia.
Distribution: Indonesia (Java).

Lagynochthonius salomonensis (Beier), new combination

Tyrannochthonius (Lagynochthonius) salomonensis Beier, 1966d: 134–135, fig. 1; Beier, 1970g: 316.

Type locality: Mt Popomanaseu (as Mt Popamanusiu), Guadalcanal, Solomon Islands.
Distribution: Solomon Islands.

Lagynochthonius sinensis (Beier), new combination

Tyrannochthonius (Lagynochthonius) sinensis Beier, 1967g: 341–342, fig. 1.

Type locality: Suisapa, Li-ch'uan (as Lichuen), Hupeh, China.
Distribution: China (Hupeh).

Lagynochthonius thorntoni Harvey

Lagynochthonius thorntoni Harvey, 1988b: 322–323, figs 30–34.

Type locality: Pulau Peucang, Ujung Kulon, Java, Indonesia.
Distribution: Indonesia (Java).

Lagynochthonius tonkinensis (Beier)

Tyrannochthonius (Lagynochthonius) tonkinensis Beier, 1951a: 61–62, fig. 9.
Lagynochthonius tonkinensis (Beier): J. C. Chamberlin, 1962: 315.

Type locality: Fan Si Pan (as Fan-Si-Pan), Tonkin, Vietnam.
Distribution: Vietnam.

Genus **Lechytia** Balzan

Lechytia Balzan, 1892: 499; J. C. Chamberlin, 1925b: 337; J. C. Chamberlin, 1929a: 77; Beier, 1932a: 73; Morikawa, 1960: 111; Muchmore, 1973b: 50; Muchmore, 1975b: 14–15; Murthy & Ananthakrishnan, 1977: 21.

Type species: *Roncus chthoniiformis* Balzan, 1887a, by original designation.

Lechytia anatolica Beier

Lechytia anatolica Beier, 1965f: 84–85, fig. 3; Beier, 1967f: 303; Beier, 1970f: 319, fig. 1.

Type locality: Namrun, Toros Daglari (as Taurus), Turkey.
Distribution: Turkey.

Lechytia arborea Muchmore

Lechytia arborea Muchmore, 1975b: 19–21, figs 11–13; Rowland & Reddell, 1976: 14.

Type locality: Myakka State Park, Sarosota County, Florida, U.S.A.
Distribution: U.S.A. (Florida, Texas).

Lechytia asiatica Redikorzev

Lechytia asiatica Redikorzev, 1938: 74–76, fig. 3; Roewer, 1940: 343; Beier, 1951a: 63.

Type locality: Da Lat (as Dalat), Vietnam.
Distribution: Vietnam.

Lechytia cavicola Muchmore

Lechytia cavicola Muchmore, 1973b: 50–51, figs 6–14.

Type locality: Grutas de Cacahuamilpa, 17 km NE. of Taxco, Guerrero, Mexico.
Distribution: Mexico.

Lechytia chilensis Beier

Lechytia chilensis Beier, 1964b: 319, fig. 5; Cekalovic, 1984: 11.

Type locality: Los Maitenes, Santiago, Chile.
Distribution: Chile.

Lechytia chthoniiformis (Balzan)

Roncus chthoniiformis Balzan, 1887a: no pagination, figs; Balzan, 1890: 445–446, figs 24, 24a; Cekalovic, 1984: 14.
Lechytia chthoniiformis (Balzan): Balzan, 1892: 499, 549; J.C. Chamberlin, 1929a: 77; Beier, 1932a: 74, fig. 89; Roewer, 1936: fig. 112b; Roewer, 1937: 242; Mello-Leitão, 1939a: 117; Mello-Leitão, 1939b: 615; Hoff, 1950b: 228–229; Beier, 1959e: 186; Hoff, 1963c: 26–27; Beier, 1964b: 318; Beier, 1977c: 100.

Type localities: Resistencia, Argentina; and Rio Apa, Paraguay.
Distribution: Argentina, Chile, Galapagos Islands, Jamaica, Paraguay, Peru.

Lechytia delamarei Vitali-di Castri

Lechytia delamarei Vitali-di Castri, 1984: 1071–1073, figs 32–37.

Type locality: Trace des Contrebandiers, Guadeloupe.
Distribution: Guadeloupe.

Lechytia dentata Mahnert

Lechytia dentata Mahnert, 1978f: 88, figs 43–47.

Type locality: Loudima, Congo.
Distribution: Congo.

Lechytia garambica Beier

Lechytia garambica Beier, 1972a: 6–7, fig. 3.

Type locality: Parc National Garamba, Zaire.
Distribution: Zaire.

Lechytia himalayana Beier

Lechytia himalayana Beier, 1974b: 262–263, fig. 1; Schawaller, 1987a: 203–204, figs 1–5.

Type locality: Dhorpatan, Nepal.
Distribution: Nepal.

Lechytia hoffi Muchmore

Lechytia pacifica (Banks): J. C. Chamberlin, 1925b: 337; J. C. Chamberlin, 1929a: 77, figs 1g, 3c, 3f; J. C. Chamberlin, 1931a: 69, figs 13d, 15c, 17c, 25d, 28c, 34i–j, 43e–f, 47c; Beier, 1932a: 73–74; Roewer, 1937: 242; Hoff, 1952a: 43–45; Hoff & Clawson, 1952: 2; Hoff, 1956a: 11–12; Hoff, 1958: 7; Hoff, 1959b: 4, etc.; Hoff, 1961: 425–427; Hoff, 1963a: 2; Muchmore, 1971a: 90; Zeh, 1987b: 1086 (misidentifications).
Lechytia hoffi Muchmore, 1975b: 21–25, figs 14–17.

Type locality: North Cheyenne Canyon, about 5 miles SW. of Colorado Springs, El Paso County, Colorado, U.S.A.
Distribution: U.S.A. (Arizona, California, Colorado, Nevada, New Mexico, Oregon, South Dekota, Utah, Washington).

Lechytia indica Murthy & Ananthakrishnan

Lechytia indica Murthy & Ananthakrishnan, 1977: 21–23, figs 6a–d; Sivaraman & Murthy, 1980a: 163–166, figs 1c, 4.

Type locality: Madras, India.
Distribution: India.

Lechytia kuscheli Beier

Lechytia kuscheli Beier, 1957c: 453–454, fig. 1; Cekalovic, 1984: 11.

Type locality: Guindal de la Pólvora, Masatierra, Juan Fernandez Islands.
Distribution: Juan Fernandez Islands.

Lechytia leleupi Beier

Lechytia leleupi Beier, 1959d: 21–22, fig. 9; Mahnert, 1986a: 824, figs 1–2.

Type locality: Kundelungu, Zaire.
Distribution: Kenya, Zaire.

Lechytia madrasica Sivaraman

Lechytia madrasica Sivaraman, 1980d: 239–241, figs 3–4; Sivaraman, 1981: 22.

Type locality: Nungambakkam, Madras, Tamil Nadu, India.
Distribution: India.

Lechytia martiniquensis Vitali-di Castri

Lechytia martiniquensis Vitali-di Castri, 1984: 1068–1071, figs 26–31.

Type locality: Agsalon, Martinique.
Distribution: Martinique.

Lechytia maxima Beier

Lechytia maxima Beier, 1955c: 531–532, fig. 3; Beier, 1962f: 14; Mahnert, 1986a: 825, figs 3–7.

Type locality: Mt Elgon, Kenya.
Distribution: Kenya, Tanzania.

Lechytia natalensis (Tullgren)

Chthonius natalensis Tullgren, 1907b: 231–232, figs 9a–b; Ellingsen, 1912b: 89.
Lechytia natalensis (Tullgren): Beier, 1932a: 74, fig. 90; Roewer, 1937: 242, fig. 207; Beier, 1958a: 162; Beier, 1966k: 456.

Type locality: Stamford Hill, Natal, South Africa.
Distribution: South Africa.

Lechytia sakagamii Morikawa

Lechytia sakagamii Morikawa, 1952a: 244, figs 3, 5b; Beier, 1957d: 13, fig. 3c; Morikawa, 1960: 111, plate 2 fig. 1, plate 7 fig. 10; Morikawa, 1962: 418.

Type locality: Marcus Island.
Distribution: Caroline Islands, Marcus Island, Marshall Islands.

Lechytia serrulata Beier

Lechytia serrulata Beier, 1955a: 7–8, fig. 3.

Type locality: Mukana, Zaire.
Distribution: Zaire.

Lechytia sini Muchmore

Lechytia sini Muchmore, 1975b: 15–19, figs 1–10; Rowland & Reddell, 1976: 14.

Type locality: Watson Hammock, Big Pine Key, Monroe County, Florida, U.S.A.
Distribution: U.S.A. (Florida, Texas).

! Lechytia tertiaria Schawaller

Lechytia tertiaria Schawaller, 1980c: 9–12, figs 10–19.

Type locality: Dominican Amber.
Distribution: Dominican Amber.

Lechytia trinitatis Beier

Lechytia trinitatis Beier, 1970d: 51–52, fig. 1; Beier, 1976d: 46.

Type locality: Trinidad, West Indies.
Distribution: Dominican Republic, Trinidad.

Genus Malcolmochthonius Benedict

Malcolmochthonius Benedict, 1978a: 251.

Type species: *Malcolmochthonius malcolmi* Benedict, 1978a, by original designation.

Malcolmochthonius malcolmi Benedict

Malcolmochthonius malcolmi Benedict, 1978a: 252, figs 1–4.

Type locality: Arch Rock Point, Curry County, Oregon, U.S.A.
Distribution: U.S.A. (California, Oregon).

Malcolmochthonius oregonus Benedict

Malcolmochthonius oregonus Benedict, 1978a: 252–255, figs 5–8.

Type locality: SE. of Port Orford, Curry County, Oregon, U.S.A.
Distribution: U.S.A. (Oregon).

Malcolmochthonius perplexus Benedict

Malcolmochthonius perplexus Benedict, 1978a: 255, figs 9–12.

Type locality: 1.8 miles N. of Jenks Lake, San Bernadino County, California, U.S.A.
Distribution: U.S.A. (California).

Genus **Maorichthonius** J. C. Chamberlin

Maorichthonius J. C. Chamberlin, 1925b: 335; J. C. Chamberlin, 1929a: 64; Beier, 1932a: 36; Beier, 1976f: 210.

Type species: *Maorichthonius mortenseni* J. C. Chamberlin, 1925b, by original designation.

Maorichthonius mortenseni J. C. Chamberlin

Maorichthonius mortenseni J. C. Chamberlin, 1925b: 335–337; J. C. Chamberlin, 1931a: 65; Beier, 1932a: 36; J. C. Chamberlin, 1934b: 3; Beier, 1969e: 414–415, fig. 1; Beier, 1976f: 210.

Type locality: Onehunga, Auckland, North Island, New Zealand.
Distribution: New Zealand.

Genus **Mexichthonius** Muchmore

Mexichthonius Muchmore, 1975a: 1–2.

Type species: *Mexichthonius unicus* Muchmore, 1975a, by original designation.

Mexichthonius pacal Muchmore

Mexichthonius pacal Muchmore, 1978: 155–156, figs 1–3.

Type locality: near Chacomax River, Palenque, Chiapas, Mexico.
Distribution: Mexico.

Mexichthonius unicus Muchmore

Mexichthonius unicus Muchmore, 1975a: 2–4, figs 1–7; Muchmore, 1977: 70.

Type locality: 5 km SSW. of Ich-Ek, Campeche, Mexico.
Distribution: Mexico.

Genus **Microchthonius** Hadži

Chthonius (Microchthonius) Hadži, 1933b: 142; Hadži, 1933c: 181; Beier, 1939d: 17.
Microchthonius Hadži: Beier, 1963b: 73.

Type species: *Chthonius (Microchthonius) karamani* Hadži, 1933b, by monotypy.

Microchthonius karamani (Hadži)

Chthonius (Microchthonius) karamani Hadži, 1933b: 142–146, figs 9a–f; Hadži, 1933c: 181–182; Beier, 1939d: 17.
Chthonius karamani Hadži: Roewer, 1937: 239.
Microchthonius karamani (Hadži): Beier, 1963b: 74; Beier, 1970f: 318, fig. 1; Ćurčić, 1974a: 15.

Type locality: Trogir (as Trogira), Yugoslavia.
Distribution: Yugoslavia.

Microchthonius rogatus (Beier)

Chthonius (Microchthonius) rogatus Beier, 1938a: 5; Beier, 1939d: 17, fig. 14.
Microchthonius rogatus (Beier): Beier, 1963b: 73–74, fig. 73; Beier, 1970f: 318, fig. 1; Ćurčić, 1974a: 15.

Type locality: Jeskalovica Cave, Brac (as Brazza) Island, Yugoslavia.
Distribution: Yugoslavia.

Genus **Mundochthonius** J. C. Chamberlin

Mundochthonius J. C. Chamberlin, 1929a: 64; Beier, 1932a: 36–37; Hoff, 1949b: 436; Hoff, 1956a: 10; Morikawa, 1960: 94; Beier, 1963b: 18; Muchmore, 1973b: 48.

Type species: *Mundochthonius erosidens* J. C. Chamberlin, 1929a, by original designation.

Mundochthonius alpinus Beier

Mundochthonius alpinus Beier, 1947d: 296–298, figs 1a–c; Beier, 1952e: 2; Beier & Franz, 1954: 453; Beier, 1956i: 8; Beier, 1963b: 19, fig. 12a–c; Beier, 1970f: 318, fig. 1.

Type locality: Kraubath, Austria.
Distribution: Austria.

Mundochthonius basarukini Schawaller

Mundochthonius basarukini Schawaller, 1989: 2–6, figs 1–14.

Type locality: Golovnino Caldera (as Caldera von Golovnin), Kunashir, Kurilskiye Ostrova, R.S.F.S.R., U.S.S.R.
Distribution: U.S.S.R. (R.S.F.S.R.).

Mundochthonius carpaticus Rafalski

Mundochthonius carpaticus Rafalski, 1948: 13–20, figs 1–5; Beier, 1963b: 19; Rafalski, 1967: 5; Beier, 1970f: 317, fig. 1; Jędryczkowski, 1985: 78; Jędryczkowski, 1987a: 342; Jędryczkowski, 1987b: 137, map 1; Schawaller, 1989: 6.

Type locality: Pieniny Mountains, Poland.
Distribution: Poland, U.S.S.R. (Ukraine).

Mundochthonius cavernicola Muchmore

Mundochthonius cavernicolus (sic) Muchmore, 1968f: 110–112, figs 1–4.

Type locality: Saltpeter Cave, Fults, Monroe County, Illinois, U.S.A.
Distribution: U.S.A. (Illinois).

Mundochthonius decoui Dumitresco & Orghidan

Mundochthonius decoui Dumitresco & Orghidan, 1970b: 98–101, figs 1–8.

Type locality: Vîrtoape, Gorj-Olténie, Romania.
Distribution: Romania.

Mundochthonius erosidens J. C. Chamberlin

Mundochthonius erosidens J. C. Chamberlin, 1929a: 64; Beier, 1932a: 37; Roewer, 1937: 238; Hoff, 1958: 5.

Type locality: Jasper Ridge, San Mateo County, California, U.S.A.
Distribution: U.S.A. (California).

Mundochthonius holsingeri Benedict & Malcolm

Mundochthonius holsingeri Benedict & Malcolm, 1974: 1–3, figs 1–3; Holsinger & Culver, 1988: 43, fig. 15.

Type locality: Helsley Cave, Shenandoah County, Virginia, U.S.A.
Distribution: U.S.A. (Virginia).

Mundochthonius japonicus J. C. Chamberlin

Mundochthonius japonicus J. C. Chamberlin, 1929e: 152; Beier, 1932a: 37; Roewer, 1937: 238; Morikawa, 1955a: 216; Morikawa, 1960: 94–95; Morikawa, 1962: 418;

Sato, 1978e: 17, figs 3, 7; Sato, 1979a: 86–88, figs 1a–j, 2a–e, plate 13; Sato, 1979b: 13, figs 1–3, 5; Sato, 1979c: 1, fig. 2; Sato, 1979d: 42; Sato, 1980c: 10, figs 1–2; Sato, 1980d: fig. 1, plate 3; Sato, 1982b: 58, figs 1–3; Sato, 1982c: 31; Sato, 1983b: 16, fig. 2; Sato, 1984c: 14, figs 1, 3; Sato, 1984d: figs 2–3; Sato, 1985: 22, figs 1–2; Sato, 1986: 95, fig. 1.

Mundochthonius japonicus japonicus J. C. Chamberlin: Morikawa, 1960: 95, plate 7 fig. 2; Morikawa, 1962: fig. 2.

Type locality: Mt Kirishima, Japan.
Distribution: Japan.

Mundochthonius japonicus imadatei Morikawa

Mundochthonius imadatei Morikawa, 1956: 282, fig. 4i.
Mundochthonius japonicus imadatei Morikawa: Morikawa, 1960: 96–97.

Type locality: Gongen-ana Cave, Kagoshima, Japan.
Distribution: Japan.

Mundochthonius japonicus scolytidis Morikawa

Mundochthonius scolytidis Morikawa, 1954c: 329–331, figs 1–4.
Mundochthonius japonicus scolyditis Morikawa: Morikawa, 1960: 95–96, plate 1 fig. 2, plate 10 fig. 9; Morikawa, 1972: 33.

Type locality: Mt Saraga-miné, near Matsuyama, Shikoku, Japan.
Distribution: Japan.

Mundochthonius japonicus tripartitus Morikawa

Mundochthonius tripartitus Morikawa, 1956: 281–282, figs 2f, 4h.
Mundochthonius japonicus tripartitus Morikawa: Morikawa, 1960: 96, plate 10 fig. 6.

Type locality: Karyu-dô Cave, Saéki, Oita, Japan.
Distribution: Japan.

Mundochthonius magnus J. C. Chamberlin

Mundochthonius magnus J. C. Chamberlin, 1929a: 65, fig. 2b; J. C. Chamberlin, 1931a: fig. 21m; Beier, 1932a: 37; Roewer, 1937: 238; Hoff, 1958: 5.

Type locality: San Mateo County, California, U.S.A.
Distribution: U.S.A. (California).

Mundochthonius mexicanus Muchmore

Mundochthonius mexicanus Muchmore, 1973b: 48–50, figs 2–5; Reddell & Elliott, 1973b: 183; Muchmore, 1977: 63.

Type locality: Chipinque Mesa, Monterrey, Nuevo León, Mexico.
Distribution: Mexico.

Mundochthonius montanus J. C. Chamberlin

Mundochthonius montanus J. C. Chamberlin, 1929a: 65; J. C. Chamberlin, 1931a: fig. 21i; Beier, 1932a: 37; Roewer, 1937: 238; Hoff, 1952a: 40–42, figs 1–4; Hoff, 1956a: 10–11; Hoff, 1958: 5; Hoff, 1959b: 4, etc.; Hoff, 1961: 420–425; Zeh, 1987b: 1086.

Type locality: Manitou, Colorado, U.S.A.
Distribution: U.S.A. (Colorado, New Mexico).

Mundochthonius pacificus (Banks)

Roncus pacificus Banks, 1893: 66.
Lechytia pacifica (Banks): Banks, 1895a: 13; Coolidge, 1908: 114.

Mundochthonius pacificus (Banks): R. O. Schuster, 1968: 137–139, figs 1–6.
Not *Lechytia pacifica* (Banks): J. C. Chamberlin, 1925b: 337; J. C. Chamberlin, 1929a: 77, figs 1g, 3c, 3f; J. C. Chamberlin, 1931a: 69, figs 13d, 15c, 17c, 25d, 28c, 34i–j, 43e–f, 47c; Beier, 1932a: 73–74; Roewer, 1937: 242; Hoff, 1952a: 43–45; Hoff & Clawson, 1952: 2; Hoff, 1956a: 11–12; Hoff, 1958: 7; Hoff, 1959b: 4, etc.; Hoff, 1961: 425–427; Hoff, 1963a: 2; Zeh, 1987b: 1086 (misidentifications; see *Lechytia hoffi* Muchmore).

Type locality: Washington, U.S.A.
Distribution: U.S.A. (California, Idaho, Washington).

Mundochthonius rossi Hoff

Mundochthonius rossi Hoff, 1949a: 437–440, figs 17a–e; Hoff & Bolsterli, 1956: 158; Hoff, 1958: 5; Hoff, 1963a: 1–2; Nelson, 1975: 279, figs 10–13; Zeh, 1987b: 1086.

Type locality: Starved Rock State Park, Illinois, U.S.A.
Distribution: U.S.A. (Illinois, Iowa, Michigan, North Dakota, South Dakota, Wisconsin).

Mundochthonius sandersoni Hoff

Mundochthonius sandersoni Hoff, 1949a: 440–443, figs 18a–g; Hoff & Bolsterli, 1956: 158; Hoff, 1958: 6; Zeh, 1987b: 1086.

Type locality: Herod, Illinois, U.S.A.
Distribution: U.S.A. (Illinois).

Mundochthonius styriacus Beier

Mundochthonius styriacus Beier, 1971c: 386–387, fig. 1; Mahnert, 1979c: 501–502.

Type locality: Pöls near Zwaring, 20 km S. of Graz, Austria.
Distribution: Austria, Switzerland.

Mundochthonius ussuricus Beier

Mundochthonius ussuricus Beier, 1979b: 553–554, fig. 1.

Type locality: Ussuri Reservation, Primorskiy Kray (as Primorsky kraj), R.S.F.S.R., U.S.S.R.
Distribution: U.S.S.R. (R.S.F.S.R.).

Genus Neochthonius J. C. Chamberlin

Neochthonius J.C. Chamberlin, 1929a: 66; Muchmore, 1969e: 388; Judson, 1990: 594.
Chthonius (Neochthonius) J. C. Chamberlin: Beier, 1932a: 46.

Type species: *Neochthonius stanfordianus* J. C. Chamberlin, 1929a, by original designation.

Neochthonius amplus (R. O. Schuster)

Kewochthonius amplus R. O. Schuster, 1962: 225–226, figs 3–6, 6a.
Neochthonius amplus (R. O. Schuster): Muchmore, 1969e: 391.

Type locality: near Winters, Yolo County, California, U.S.A.
Distribution: U.S.A. (California).

Neochthonius stanfordianus J. C. Chamberlin

Neochthonius stanfordianus J. C. Chamberlin, 1929a: 66; Muchmore, 1969e: 388–391, figs 1–5; Judson, 1990: fig. 2.
Chthonius (Neochthonius) stanfordianus (J. C. Chamberlin): Beier, 1932a: 46.
Chthonius stanfordianus (J. C. Chamberlin): Roewer, 1937: 238.

Kewochthonius stanfordianus (J. C. Chamberlin): Hoff, 1958: 5; R. O. Schuster, 1962: fig. 5.

Type locality: San Francisquito Creek, near Palo Alto, San Mateo County, California, U.S.A.

Distribution: U.S.A. (California).

Neochthonius troglodytes Muchmore

Neochthonius troglodytes Muchmore, 1969e: 391–393, figs 6–9.

Type locality: Wool Hollow Cave, 3 miles SE. of Murphys, Calaveras County, California, U.S.A.

Distribution: U.S.A. (California).

Genus Paraliochthonius Beier

Paraliochthonius Beier, 1956b: 58–59; Beier, 1963b: 76; Hoff, 1963c: 23; Muchmore, 1972b: 250–252; Murthy & Ananthakrishnan, 1977: 12–13; Muchmore, 1984c: 121.
Morikawia J. C. Chamberlin, 1962: 311–312 (synonymized by Muchmore, 1972b: 250).

Type species: of *Paraliochthonius*: *Chthonius singularis* Menozzi, 1924, by original designation.
of *Morikawia*: *Chthonius johnstoni* J. C. Chamberlin, 1923c, by original designation.

Paraliochthonius azanius Mahnert

Paraliochthonius azanius Mahnert, 1986a: 827–830, figs 12–20.

Type locality: Mombasa Island, Kenya.
Distribution: Kenya.

Paraliochthonius canariensis Vachon

Paraliochthonius hoestlandti canariensis Vachon, 1961: 98–101, figs 1–6; Beier, 1963b: 78, fig. 78; Helversen, 1968: fig. 5; Beier, 1970f: 318; Beier, 1976a: 24.
Paraliochthonius canariensis Vachon: Muchmore, 1972b: 252.

Type locality: Arrieta, Lanzarote, Canary Islands.
Distribution: Canary Islands.

Paraliochthonius carpenteri Muchmore

Paraliochthonius carpenteri Muchmore, 1984c: 121–123, figs 5–8.

Type locality: Lighthouse Cave, San Salvador, Bahamas.
Distribution: Bahamas.

Paraliochthonius hoestlandti Vachon

Paraliochthonius hoestlandti Vachon, 1960b: 331–337, figs 1–10; Vachon, 1961: 102; Beier, 1970f: 318; Beier, 1976a: 24; Pieper, 1981: 2.
Paraliochthonius hoestlandi (sic) Vachon: Beier, 1961a: 69–70, fig. 1.
Paraliochthonius hoestlandti hoestlandti Vachon: Beier, 1963b: 77–78, fig. 75; Helversen, 1968: fig. 5.

Type locality: 2 km E. of Funchal, Madeira Islands.
Distribution: Madeira Islands.

Paraliochthonius hoestlandti giustii Lazzeroni

Paraliochthonius hoestlandti giustii Lazzeroni, 1970b: 38–42, figs 1a–h, 2a–e.

Type locality: Scoglio d'Africa, Toscana, Italy.
Distribution: Italy.

Paraliochthonius insulae Hoff

Paraliochthonius insulae Hoff, 1963c: 23–25, figs 7–8.

Type locality: Drunkenmans Cay, 2 miles S. of Port Royal, Jamaica.
Distribution: Jamaica.

Paraliochthonius johnstoni (J. C. Chamberlin)

Chthonius johnstoni J. C. Chamberlin, 1923c: 357–358, plate 2, fig. 17, plate 3, figs 11–13.
Tyrannochthonius johnstoni (J. C. Chamberlin): J. C. Chamberlin, 1929a: 75, figs 2d, 2f; J. C. Chamberlin, 1931a: 56, figs 21k; Beier, 1932a: 63–64, fig. 78; Roewer, 1936: fig. 68c; Roewer, 1937: 240.
Morikawia johnstoni J. C. Chamberlin: J. C. Chamberlin, 1962: 312–313, figs 3a–f.
Paraliochthonius johnstoni J. C. Chamberlin: Muchmore, 1972b: 252; V. F. Lee, 1979a: 19–21, figs 24–25, 31.
Paraliochthonius mexicanus Muchmore, 1972b: 253–254, figs 1–5 (synonymized by V. F. Lee, 1979a: 19).

Type locality: of *Chthonius johnstoni*: Escondido Gorge, Puerto Escondido, Baja California, Mexico.
of *Paraliochthonius mexicanus*: Manzanilla, Tenacatita Bay, Jalisco, Mexico.
Distribution: Mexico.

Paraliochthonius martini Mahnert

Paraliochthonius martini Mahnert, 1989b: 43–44, figs 5–11.

Type locality: Cueva de Don Justo, Hierro, Canary Islands.
Distribution: Canary Islands.

Paraliochthonius puertoricensis Muchmore

Paraliochthonius puertoricensis Muchmore, 1967a: 158–162, figs 5–7.

Type locality: Ramosito Key, Puerto Rico.
Distribution: Puerto Rico.

Paraliochthonius singularis (Menozzi)

Chthonius singularis Menozzi, 1924: 1–3, figs 1–4; Roewer, 1937: 239.
Chthonius (Chthonius) singularis Menozzi: Beier, 1932a: 49.
Paraliochthonius singularis Menozzi: Beier, 1956b: 59–61, fig. 3; R. Schuster, 1956: 246; Beier, 1963b: 76–77; Beier, 1965a: 86–88, fig. 2; Helversen, 1968: fig. 5; Lazzeroni, 1969a: 325; Lazzeroni, 1969c: 228; Beier, 1970f: 318, fig. 1; Lazzeroni, 1970a: 207; Ćurčić, 1974: 15; Callaini, 1983c: 281–282; Callaini, 1989: 142–144.

Type locality: Portici, Italy.
Distribution: France, Italy, Sardinia, Sicily, Turkey, Yugoslavia.

Paraliochthonius takashimai (Morikawa)

Tyrannochthonius johnstoni takashimai Morikawa, 1958: 6–7, figs 1a–b.
Tyrannochthonius takashimai Morikawa: Morikawa, 1960: 110, plate 5 fig. 4, plate 10 fig. 8.
Morikawia takashimai (Morikawa): J. C. Chamberlin, 1962: 313–314; Morikawa, 1962: 419.
Paraliochthonius takashimai J. C. Chamberlin: Muchmore, 1972b: 252.

Type locality: Manazuru Cape, Kanagawa, Japan.
Distribution: Japan.

Paraliochthonius tenebrarum Mahnert

Paraliochthonius tenebrarum Mahnert, 1989b: 42–43, figs 1–4.

Type locality: Cuevas Negras, Tenerife, Canary Islands.
Distribution: Canary Islands.

Paraliochthonius weygoldti Muchmore

Paraliochthonius weygoldti Muchmore, 1967a: 155–158, figs 1–4.

Type locality: Pig Pine Key, Dade County, Florida, U.S.A.
Distribution: U.S.A. (Florida).

Genus Pseudochthonius Balzan

Chthonius (Pseudochthonius) Balzan, 1892: 546; J. C. Chamberlin, 1925b: 338.
Chthonius (Typhlochthonius) Ellingsen, 1902: 164 (synonymized by J. C. Chamberlin, 1929f: 173).
Chthonius (Typlochthonius) (sic): J. C. Chamberlin, 1925b: 338.
Pseudochthonius Balzan: J. C. Chamberlin, 1929a: 69; J. C. Chamberlin, 1929e: 154; J. C. Chamberlin, 1929f: 173–175; Beier, 1932a: 66–67; Beier, 1959d: 18; Hoff, 1963c: 6; Muchmore, 1977: 64; Muchmore, 1986b: 19.
Typlochthonius (sic): J. C. Chamberlin, 1929a: 76.

Type species: of *Chthonius (Pseudochthonius)*: *Chthonius (Pseudochthonius) simoni* Balzan, 1892, by monotypy.
of *Typhlochthonius*: *Chthonius (Typhlochthonius) pulchellus* Ellingsen, 1902, by monotypy.

Pseudochthonius arubensis Wagenaar-Hummelinck

Pseudochthonius arubensis Wagenaar-Hummelinck, 1948: 47–52, figs 14a–g, 15a–f.

Type locality: Quadirikiri Cave, Aruba.
Distribution: Aruba.

Pseudochthonius beieri Mahnert

Pseudochthonius beieri Mahnert, 1978f: 84–87, figs 32–42.

Type locality: Sibiti, Congo.
Distribution: Congo.

Pseudochthonius billae Vachon

Pseudochthonius billae Vachon, 1941e: 33; Beier, 1959d: 21, fig. 8.

Type localities: Sassandra, Ivory Coast; and Port Bouet, Ivory Coast.
Distribution: Ivory Coast.

Pseudochthonius brasiliensis Beier

Pseudochthonius brasiliensis Beier, 1970b: 54–55, fig. 2; Beier, 1974d: 899.

Type locality: Barueri, Sao Paulo, Brazil.
Distribution: Brazil (Santa Catarina, Sao Paulo).

Pseudochthonius clarus Hoff

Pseudochthonius clarus Hoff, 1963c: 8–14, figs 1–2; Zeh, 1987b: 1086.

Type locality: Long Mountain, Parish of St Andrew, Jamaica.
Distribution: Jamaica.

Family Chthoniidae

Pseudochthonius congicus Beier

Pseudochthonius congicus Beier, 1959d: 18−19, fig. 6.

Type locality: Ile de M'Boko, Lake Tanganyika, Kivu, Zaire.
Distribution: Zaire.

Pseudochthonius doctus Hoff

Pseudochthonius doctus Hoff, 1963c: 14−18, figs 3−4; Zeh, 1987b: 1086.

Type locality: near Morant Point Lighthouse, Parish of St Thomas, Jamaica.
Distribution: Jamaica.

Pseudochthonius falcatus Muchmore

Pseudochthonius falcatus Muchmore, 1977: 68, figs 12−15.

Type locality: Belmopan, Belize.
Distribution: Belize.

Pseudochthonius galapagensis Beier

Pseudochthonius galapagensis Beier, 1977c: 98−99, fig. 4.

Type locality: Turtle Bay, Santa Cruz, Galapagos Islands.
Distribution: Galapagos Islands.

Pseudochthonius heterodentatus Hoff

Pseudochthonius heterodentatus Hoff, 1946e: 4−8, figs 4−8.

Type locality: St Augustine, Trinidad.
Distribution: Trinidad.

Pseudochthonius homodentatus J. C. Chamberlin

Pseudochthonius homodentatus J. C. Chamberlin, 1929f: 179; J. C. Chamberlin,
    1931a: fig. 11s; Beier, 1932a: 67; Roewer, 1937: 240; Mahnert, 1979d: 726−728;
    Mahnert & Adis, 1986: 213.

Type locality: Venezuela.
Distribution: Brazil (Amazonas), Venezuela.

Pseudochthonius insularis J. C. Chamberlin

Pseudochthonius insularis J. C. Chamberlin, 1929f: 178, figs c, g, i, l−n, t; J. C.
    Chamberlin, 1931a: figs 13c, 28e, 33i, 34e−f; Beier, 1932a: 68; Roewer, 1937: 240.

Type locality: St Vincent.
Distribution: St Vincent.

Pseudochthonius leleupi Beier

Pseudochthonius leleupi Beier, 1959d: 20−21, fig. 7; Mahnert, 1978f: 83−84.

Type locality: Thysville, Zaire.
Distribution: Congo, Zaire.

Pseudochthonius moralesi Muchmore

Pseudochthonius moralesi Muchmore, 1977: 64−66, figs 4−8.

Type locality: La Cañada, Palenque, Chiapas, Mexico.
Distribution: Mexico.

Pseudochthonius mundanus Hoff

Pseudochthonius mundanus Hoff, 1963c: 18−22, figs 5−6; Zeh, 1987b: 1086.

Type locality: Port Henderson Hill, Parish of St Catherine, Jamaica.
Distribution: Jamaica.

Pseudochthonius naranjitensis (Ellingsen)

Chthonius (Pseudochthonius) naranjitensis Ellingsen, 1902: 162–163.
Pseudochthonius (?) *naranjitensis* (Ellingsen): Beier, 1932a: 68; Roewer, 1937: 240.

Type locality: Naranjito, Ecuador.
Distribution: Ecuador.

Pseudochthonius orthodactylus Muchmore

Pseudochthonius orthodactylus Muchmore, 1970b: 221–223, figs 1–3; Mahnert, 1979d: 728–729; Mahnert, 1985b: 76; Mahnert & Adis, 1986: 213.

Type locality: Belém, Pará, Brazil.
Distribution: Brazil (Pará).

Pseudochthonius perreti Mahnert

Pseudochthonius perreti Mahnert, 1986a: 827, figs 8–11.

Type locality: near Kogari, Embu, Kenya.
Distribution: Kenya.

Pseudochthonius pulchellus (Ellingsen)

Chthonius (Typhlochthonius) pulchellus Ellingsen, 1902: 164–165.
Chthonius pulchellus Ellingsen: J. C. Chamberlin, 1923b: 186.
Pseudochthonius pulchellus (Ellingsen): J. C. Chamberlin, 1929f: 176; Beier, 1932a: 68; Roewer, 1937: 240.

Type locality: Naranjito, Ecuador.
Distribution: Ecuador.

Pseudochthonius simoni (Balzan)

Chthonius (Pseudochthonius) simoni Balzan, 1892: 546–547, figs 35, 35a; Ellingsen, 1905e: 327.
Chthonius simoni Balzan: J. C. Chamberlin, 1923b: 186.
Pseudochthonius simoni (Balzan): J. C. Chamberlin, 1929a: 69; J. C. Chamberlin, 1929e: 154; J. C. Chamberlin, 1929f: 176–178, figs a–b, h, j–k, r; Beier, 1932a: 67–68, fig. 83; Roewer, 1937: 240.

Type locality: Caracas, Venezuela.
Distribution: Uruguay, Venezuela.

! Pseudochthonius squamosus Schawaller

Pseudochthonius squamosus Schawaller, 1980c: 3–5, figs 1–9; Liebherr, 1988: 387.

Type locality: Dominican Amber.
Distribution: Dominican Amber.

Pseudochthonius strinatii Beier

Pseudochthonius strinatii Beier, 1969f: 1–2, fig. 1.

Type locality: Gruta das Areias, Sao Paulo, Brazil.
Distribution: Brazil (Sao Paulo).

Pseudochthonius thibaudi Vitali-di Castri

Pseudochthonius thibaudi Vitali-di Castri, 1984: 1059–1064, figs 1–17.

Type locality: Pointe le Gouffre, Guadeloupe.
Distribution: Guadeloupe.

Family Chthoniidae

Pseudochthonius troglobius Muchmore

Pseudochthonius troglobius Muchmore, 1986b: 19–20, figs 2–5.

Type locality: Cueva del Cenote Xtolok, Chichén Itzá, Yucatan, Mexico.
Distribution: Mexico.

Pseudochthonius tuxeni Mahnert

Pseudochthonius tuxeni Mahnert, 1979d: 729–730, figs 6–10; Mahnert, 1985c: 219, fig. 1; Mahnert & Adis, 1986: 213.

Type locality: Santarém, Pará, Brazil.
Distribution: Brazil (Amazonas, Pará).

Pseudochthonius yucatanus Muchmore

Pseudochthonius yucatanus Muchmore, 1977: 66–68, figs 9–11.

Type locality: 7 km SW. of Oxkutzcab, Yucatán, Mexico.
Distribution: Mexico.

Genus Pseudotyrannochthonius Beier

Pseudotyrannochthonius Beier, 1930d: 207–208; Beier, 1932a: 70; Beier, 1966a: 285; Muchmore, 1967c: 134.
Tubbichthonius Hoff, 1951: 10–11 (synonymized by Beier, 1966a: 285).
Spelaeochthonius Morikawa, 1954b: 83–84 (synonymized by Muchmore, 1967c: 134).
Allochthonius (Spelaeochthonius) Morikawa: Morikawa, 1960: 104–105.

Type species: of *Pseudotyrannochthonius*: *Chthonius (Chthonius) silvestrii* Ellingsen, 1905e, by original designation.
of *Tubbichthonius*: *Tubbichthonius solitarius* Hoff, 1951 by original designation.
of *Spelaeochthonius*: *Spelaeochthonius kubotai* Morikawa, 1954b, by original designation.

Pseudotyrannochthonius australiensis Beier

Pseudotyrannochthonius australiensis Beier, 1966a: 287–288, fig. 6; Harvey, 1981b: 240; Harvey, 1985b: 139.

Type locality: south-east Edith, New South Wales, Australia.
Distribution: Australia (New South Wales).

Pseudotyrannochthonius bornemisszai Beier

Pseudotyrannochthonius bornemisszai Beier, 1966a: 286–287, fig. 5; Harvey, 1981b: 240; Harvey, 1985b: 139.

Type locality: 'Laughing Waters', Boolarra, Victoria, Australia.
Distribution: Australia (Victoria).

Pseudotyrannochthonius dentifer (Morikawa), new combination

Allochthonius (Spelaeochthonius) dentifer Morikawa, 1970: 144–146, figs 2c–e, 3.

Type locality: Yong'yeon-gul Cave, Yong'yeon-gog, Changseong-eub, Samcheog-gun, Kangweon-do, South Korea.
Distribution: South Korea.

Pseudotyrannochthonius giganteus Beier

Pseudotyrannochthonius giganteus Beier, 1971e: 233–234, fig. 1; Harvey, 1981b: 240; Harvey, 1985b: 139.

Type locality: Calgadup Cave, near Augusta, Western Australia, Australia.
Distribution: Australia (Western Australia).

Pseudotyrannochthonius gigas Beier

Pseudotyrannochthonius gigas Beier, 1969a: 178−179, fig. 6; Harvey, 1981b: 240; Harvey, 1985b: 139.

Type locality: Harmon One Cave, Byaduk, Victoria, Australia.
Distribution: Australia (Victoria).

Pseudotyrannochthonius gracilis Benedict & Malcolm

Pseudotyrannochthonius gracilis Benedict & Malcolm, 1970: 47−51, figs 3a−g.

Type locality: Puget, Thurston County, Washington, U.S.A.
Distribution: U.S.A. (California, Washington).

Pseudotyrannochthonius hamiltonsmithi Beier

Pseudotyrannochthonius hamiltonsmithi Beier, 1968a: 759−760, fig. 3; Harvey, 1981b: 240; Harvey, 1985b: 139.

Type locality: Mt Widderin Cave, Skipton, Victoria, Australia.
Distribution: Australia (Victoria).

Pseudotyrannochthonius incognitus (R. O. Schuster)

Allochthonius incognitus R. O. Schuster, 1966a: 174−175, figs 1−5.
Pseudotyrannochthonius incognitus (R. O. Schuster): Muchmore, 1967c: 134; Benedict & Malcolm, 1970: 39−47, figs 1a−g, 2a−d.
Pseudotyrannochthonius newelli Muchmore, 1967c: 134−136, figs 1−5 (synonymized by Benedict & Malcolm, 1970: 39).

Type localities: of *Allochthonius incognitus*: Loon Lake, Douglas County, Oregon, U.S.A.
of *Pseudotyrannochthonius newelli*: 2 miles S. of Timber, Washington County, Oregon, U.S.A.
Distribution: U.S.A. (California, Idaho, Oregon, Washington).

Pseudotyrannochthonius jonesi (J. C. Chamberlin)

Tubbichthonius jonesi J. C. Chamberlin, 1962: 317−319, figs 5a−j.
Pseudotyrannochthonius jonesi (J. C. Chamberlin): Beier, 1966a: 276; Hamilton-Smith, 1967: 108; Harvey, 1981b: 240; Harvey, 1985b: 139.

Type locality: probably in Blue Mountains near Sydney, New South Wales, Australia.
Distribution: Australia (New South Wales).

Pseudotyrannochthonius kobayashii (Morikawa)

Spelaeochthonius kobayashii Morikawa, 1956: 274−275, figs 2b, 3b.
Allochthonius (Spelaeochthonius) kobayashii Morikawa: Morikawa, 1960: 107.
Allochthonius (Spelaeochthonius) kobayashii kobayashii (Morikawa): Morikawa, 1960: 107−108, plate 5 fig. 13, plate 10 fig. 1; Morikawa, 1962: fig. 4.
Pseudotyrannochthonius kobayashii (Morikawa): Muchmore, 1967c: 134.

Type locality: Samé-no-kômori-ana Cave, Taga-chô, Shiga, Japan.
Distribution: Japan.

Pseudotyrannochthonius kobayashii akiyoshiensis (Morikawa), new combination

Spelaeochthonius kobayashii akiyoshiensis Morikawa, 1956: 276−277, fig. 3d.
Allochthonius (Spelaeochthonius) kobayashii akiyoshiensis (Morikawa): Morikawa, 1960: 108, plate 6 fig. 6, plate 5 fig. 15; Morikawa, 1962: figs 4−5.

Type locality: Tanuki-ana Cave, Akiyoshi-dai, Yamaguchi, Japan.
Distribution: Japan.

Pseudotyrannochthonius kobayashii dorogawaensis (Morikawa), new combination

Spelaeochthonius kobayashii dorogawaensis Morikawa, 1956: 275−276, figs 2c, 3c.
Allochthonius (Spelaeochthonius) kobayashii dorogawaensis (Morikawa): Morikawa, 1960: 108, plate 1 fig. 7, plate 10 fig. 2; Morikawa, 1962: fig. 4.

Type locality: Menfudô-no-iwaya Cave, Dorogawa, Nara, Japan.
Distribution: Japan.

Pseudotyrannochthonius kubotai (Morikawa)

Spelaeochthonius kubotai Morikawa, 1954b: 84, figs 1f−i.
Allochthonius (Spelaeochthonius) kubotai (Morikawa): Morikawa, 1960: 108, plate 1 fig. 8, plate 5 fig. 12, plate 9 fig. 2, plate 10 fig. 3; Morikawa, 1962: fig. 4.
Pseudotyrannochthonius kubotai (Morikawa): Muchmore, 1967c: 134.

Type locality: Shimizu-do, Kyushu, Japan.
Distribution: Japan.

Pseudotyrannochthonius octospinosus Beier

Pseudotyrannochthonius octospinosus Beier, 1930d: 208−209, figs 15a−b; Beier, 1932a: 71; Roewer, 1937: 241; Beier, 1964b: 316−317, fig. 3; Cekalovic, 1984: 10.

Type locality: Temuco, Chile.
Distribution: Chile.

Pseudotyrannochthonius queenslandicus Beier

Pseudotyrannochthonius queenslandicus Beier, 1969a: 177−178, fig. 5; Harvey, 1981b: 240; Harvey, 1985b: 139.

Type locality: Mt Tamborine (as Jamborine), Joalah (as Toalah) National Park, Queensland, Australia.
Distribution: Australia (Queensland).

Pseudotyrannochthonius rossi Beier

Pseudotyrannochthonius rossi Beier, 1964b: 317−318, fig. 4; Cekalovic, 1984: 11.

Type locality: 30 km S. of Valdivia, Chile.
Distribution: Chile.

Pseudotyrannochthonius silvestrii (Ellingsen)

Chthonius (Chthonius) silvestrii Ellingsen, 1905e: 327−328.
Pseudotyrannochthonius silvestrii (Ellingsen): Beier, 1930d: 208, fig. 14; Beier, 1932a: 70−71, fig. 85; Roewer, 1937: 241; Beier, 1964b: 316; Cekalovic, 1984: 11.

Type locality: Santiago, Chile.
Distribution: Chile.

Pseudotyrannochthonius solitarius (Hoff)

Tubbichthonius solitarius Hoff, 1951: 9−13, figs 4−6.
Pseudotyrannochthonius solitarius (Hoff): Beier, 1966a: 286; Dartnall, 1970: 67; Beier, 1975b: 203; Hickman & Hill, 1978: appendix a; Harvey, 1981b: 240; Harvey, 1985b: 139.

Type locality: Mt Slide, Victoria, Australia.
Distribution: Australia (Australian Capital Territory, Tasmania, Victoria, Western Australia).

Pseudotyrannochthonius tasmanicus Dartnall

Pseudotyrannochthonius tasmanicus Dartnall, 1970: 66, figs 1a–e; Goede, 1974: 3, fig. 1; Skinner, 1977: 38; Hickman & Hill, 1978: appendix a; Harvey, 1981b: 241; Harvey, 1985b: 139–140.

Type locality: King George V Cave, Tasmania, Australia.
Distribution: Australia (Tasmania).

Pseudotyrannochthonius typhlus Dartnall

Pseudotyrannochthonius typhlus Dartnall, 1970: 67, figs 2a–e; Goede, 1974: 3; Harvey, 1981b: 241; Harvey, 1985b: 140.
Pseudotyrannochthonius typhus (sic) Dartnall: Skinner, 1977: 38.

Type locality: Sennacheribs Passage, Georgies Hall Cave, Mole Creek, Tasmania, Australia.
Distribution: Australia (Tasmania).

Pseudotyrannochthonius undecimclavatus (Morikawa)

Spelaeochthonius undecimclavatus Morikawa, 1956: 272–274, figs 2a, 3a, 3e.
Allochthonius (Spelaeochthonius) undecimclavatus Morikawa: Morikawa, 1960: 105.
Allochthonius (Spelaeochthonius) undecimclavatus undecimclavatus (Morikawa): Morikawa, 1960: 106, plate 5 figs 10, 14, plate 10 fig. 4; Morikawa, 1962: fig. 4.
Pseudotyrannochthonius undecimclavatus (Morikawa): Muchmore, 1967c: 134.

Type locality: Kwannoniwa-no-ana Cave, near Sennintogé, Iwaté, Japan.
Distribution: Japan.

Pseudotyrannochthonius undecimclavatus kishidai (Morikawa), new combination

Allochthonius (Spelaeochthonius) undecimclavatus kishidai Morikawa, 1960: 106–107, plate 5 fig. 16; Morikawa, 1962: fig. 4.

Type locality: Kurasawa-do Cave, Okutama, Tokyo, Japan.
Distribution: Japan.

Pseudotyrannochthonius utahensis Muchmore

Pseudotyrannochthonius utahensis Muchmore, 1967c: 136–138, figs 6–7.

Type locality: Logan Canyon, Cache County, Utah, U.S.A.
Distribution: U.S.A. (Utah).

Genus Sathrochthoniella Beier

Sathrochthoniella Beier, 1967b: 277–278; Beier, 1976f: 203.

Type species: *Sathrochthoniella zealandica* Beier, 1967b, by original designation.

Sathrochthoniella zealandica Beier

Sathrochthoniella zealandica Beier, 1967b: 277–278, fig. 1; Beier, 1976f: 203.

Type locality: Taurewa National Park, New Zealand.
Distribution: New Zealand.

Genus Sathrochthonius J. C. Chamberlin

Sathrochthonius J.C. Chamberlin, 1962: 303–304; Beier, 1976f: 201.

Type species: *Sathrochthonius tuena* J. C. Chamberlin, 1962, by original designation.

Sathrochthonius crassidens Beier

Sathrochthonius crassidens Beier, 1966a: 280–281, fig. 1; Harvey, 1981b: 241; Harvey, 1985b: 140.

Type locality: south-east Edith, New South Wales, Australia.
Distribution: Australia (New South Wales).

Sathrochthonius insulanus Beier

Sathrochthonius insulanus Beier, 1976f: 202–203, fig. 3; Harvey, 1985b: 140.

Type locality: Lord Howe Island.
Distribution: Lord Howe Island.

Sathrochthonius kaltenbachi Beier

Sathrochthonius kaltenbachi Beier, 1966g: 366–367, fig. 3.

Type locality: Forêt de la Rivière Tendea, New Caledonia.
Distribution: New Caledonia.

Sathrochthonius maoricus Beier

Sathrochthonius maoricus Beier, 1976f: 201–202, fig. 2.

Type locality: Pelorus Bridge, Marlborough, South Island, New Zealand.
Distribution: New Zealand.

Sathrochthonius pefauri Vitali-di Castri

Sathrochthonius pefauri Vitali-di Castri, 1974: 194–201, figs 1–14; Cekalovic, 1984: 11.

Type locality: Cordillera de Nahuelbuta, Malleco, Chile.
Distribution: Chile.

Sathrochthonius tuena J. C. Chamberlin

Sathrochthonius tuena J. C. Chamberlin, 1962: 304–306, figs 1a–g; Beier, 1967a: 199; Hamilton-Smith, 1967: 108; Beier, 1968a: 757; Harvey, 1981b: 241; Harvey, 1985b: 140.

Type locality: probably in the Blue Mountains, near Sydney, New South Wales, Australia.
Distribution: Australia (New South Wales).

Sathrochthonius tullgreni J. C. Chamberlin

Chthonius caecus Tullgren, 1909b: 414–415, fig. 3 (junior primary homonym of *Chthonius coecus* Packard, 1884, and *Chthonius caecus* E. Simon, 1885b).
Mundochthonius (?) *caecus* (Tullgren): Beier, 1932a: 38; Roewer, 1937: 238.
Mundochthonius caecus (Tullgren): J. C. Chamberlin, 1934b: 3; Weidner, 1959: 115.
Sathrochthonius tullgreni J. C. Chamberlin, 1962: 306–307 (replacement name for *Chthonius caecus* Tullgren); Harvey, 1981b: 241; Harvey, 1985b: 140–141.
Sathrochthonius (?) *tullgreni* J. C. Chamberlin: Beier, 1966a: 276; Muchmore, 1982d: 158.

Type locality: Brunswick Junction (as Brunswick), Western Australia, Australia.
Distribution: Australia (Western Australia).

Sathrochthonius venezuelanus Muchmore

Sathrochthonius venezuelanus Muchmore, 1989: 251–253, figs 1-3.

Type locality: 9 km from Chivaton Hotel toward Kavanayén, La Gran Sabana, Bolivar, Venezuela.
Distribution: Venezuela.

Sathrochthonius webbi Muchmore

Sathrochthonius webbi Muchmore, 1982d: 156–157, figs 1–2; Tenorio & Muchmore, 1982: 382; Harvey, 1985b: 140.

Type locality: Holy Jump Lava Cave, 25 km E. of Warwick, Queensland, Australia.
Distribution: Australia (Queensland).

Genus **Selachochthonius** J.C. Chamberlin

Selachochthonius J. C. Chamberlin, 1929a: 68; Beier, 1932a: 40.
Chthoniella Lawrence, 1935: 551 (synonymized by Beier, 1964k: 33).
Kafirchthonius J. C. Chamberlin, 1962: 319 (synonymized by Beier, 1964k: 33).

Type species: of *Selachochthonius*: *Chthonius serratidentatus* Ellingsen, 1912b, by original designation.
of *Chthoniella*: *Chthoniella cavernicola* Lawrence, 1935, by monotypy.
of *Kafirchthonius*: *Chthoniella heterodentata* Beier, 1955l, by original designation.

Selachochthonius cavernicola (Lawrence)

Chthoniella cavernicola Lawrence, 1935: 551−555, figs 1, 2a−i; Roewer, 1937: 241; Lawrence, 1964: 74.
Selachochthonius cavernicola (Lawrence): Beier, 1964k: 33.

Type locality: Wynberg Cave, Table Mountain, Cape Town, Cape Province, South Africa.
Distribution: South Africa.

Selachochthonius heterodentata (Beier)

Chthoniella heterodentata Beier, 1955l: 276−277, fig. 3.
Kafirchthonius heterodentata (Beier): J. C. Chamberlin, 1962: 319.
Selachochthonius heterodentatus (Beier): Beier, 1964k: 33.

Type locality: Wynberg Cave, Table Mountain, Cape Town, Cape Province, South Africa.
Distribution: South Africa.

Selachochthonius serratidentatus (Ellingsen)

Chthonius serratidentatus Ellingsen, 1912b: 123−124.
Selachochthonius serratidentatus (Ellingsen): J. C. Chamberlin, 1929a: 68; Beier, 1932a: 40−41; Roewer, 1937: 238; Beier, 1964k: 32−33, fig. 1; Beier, 1966k: 456.

Type locality: Pirie, near King William's Town, Cape Province, South Africa.
Distribution: Lesotho, South Africa.

Genus **Spelyngochthonius** Beier

Spelyngochthonius Beier, 1955k: 41−42; Beier, 1963b: 75; Vachon & Heurtault-Rossi, 1964: 84.

Type species: *Spelyngochthonius sardous* Beier, 1955k, by original designation.

Spelyngochthonius heurtaultae Vachon

Chthonius tetrachelatus (Preyssler): Ellingsen, 1912c: 174−175 (misidentifiation).
Spelyngochthonius heurtaultae Vachon, 1967b: 522−555, figs 1−7; Lazzeroni, 1969c: fig. 1; Beier, 1970f: 118, fig. 1; Zaragoza, 1986: fig. 1.
Spelyngochthonius cf. *heurtaultae* Vachon: Gardini, 1981a: 109−112, figs 23−32.

Type locality: Cova d'En Merla, Roda de Bara, Partido de Vendrell, Tarragona, Spain.
Distribution: Sardinia, Spain.

Spelyngochthonius provincialis Vachon & Heurtault-Rossi

Spelyngochthonius provincialis Vachon & Heurtault-Rossi, 1964: 80−84, figs 1−6; Lazzeroni, 1969c: fig. 1; Beier, 1970f: 118, fig. 1; Heurtault, 1986c: 21−22.

Type locality: near Matelles, Hérault, France.
Distribution: France.

Spelyngochthonius sardous Beier

Spelyngochthonius sardous Beier, 1955k: 42–43, fig. 1; Beier, 1963b: 75–76, fig. 74; Lazzeroni, 1969c: 226–227, fig. 1; Beier, 1970f: 118, fig. 1; Puddu & Pirodda, 1973: 171; Gardini, 1980c: 104, 132, 133.
Spelyngochtonius (sic) *sardous* Beier: Cerruti, 1968: 219.
Spelyngochthonius sardous Beier?: Gardini, 1977a: 39–47, figs 1–15.
Spelyngochthonius cf. *sardous* Beier: Gardini, 1981a: fig. 33.

Type locality: Grotta del Bue Marino, Cala Gonone, Dorgali, Sardinia.
Distribution: Sardinia.

Genus **Troglochthonius** Beier

Troglochthonius Beier, 1939d: 24–25; J.C. Chamberlin, 1962: 311; Beier, 1963b: 78.

Type species: *Troglochthonius mirabilis* Beier, 1939d, by original designation.

Troglochthonius doratodactylus Helversen

Troglochthonius doratodactylus Helversen, 1968: 59–63, figs 1, 2a–c, 3, 4b, 5; Lazzeroni, 1969c: 228, fig. 1; Weygoldt, 1969a: fig. 103; Beier, 1970f: 319, fig. 1; Puddu & Pirodda, 1973: 171; Gardini, 1980c: 105, 133; Mahnert, 1980g: 96–97.

Type locality: stated as Sardinia, but the type specimen is from the notoriously mislabelled Roewer collection.
Distribution: Italy.

Troglochthonius mirabilis Beier

Troglochthonius mirabilis Beier, 1939d: 25, figs 27–29; Beier, 1963b: 78–79, fig. 76; Helversen, 1968: fig. 5; Lazzeroni, 1969c: fig. 1; Beier, 1970f: 319, fig. 1; Ćurčić, 1974a: 15.

Type locality: Grabovica Pecina, Grepci, Yugoslavia.
Distribution: Yugoslavia.

Genus **Tyrannochthoniella** Beier

Tyrannochthoniella Beier, 1966c: 366–367; Beier, 1976f: 204.

Type species: *Tyrannochthoniella zealandica* Beier, 1966c, by original designation.

Tyrannochthoniella zealandica Beier

Tyrannochthoniella zealandica Beier, 1966c: 367, fig. 4; Beier, 1967b: 283–284; Tenorio & Muchmore, 1982: 384.
Tyrannochthoniella ligulifera Beier, 1967b: 284–285, fig. 5 (synonymized by Beier, 1976f: 205).
Tyrannochthoniella zealandica zealandica Beier: Beier, 1976f: 204–205.

Type localities: of *Tyrannochthoniella zealandica*: Lewis Pass, South Island, New Zealand.
of *Tyrannochthoniella ligulifera*: McKenzie's Bush, near Oparau, New Zealand.
Distribution: New Zealand.

Tyrannochthoniella zealandica foveauxana Beier

Tyrannochthoniella foveauxana Beier, 1967b: 285–287, fig. 6.
Tyrannochthoniella zealandica foveauxana Beier, 1976f: 205.

Type locality: Ruapuke Island, Foveaux Strait, South Island, New Zealand.
Distribution: New Zealand.

Genus **Tyrannochthonius** J. C. Chamberlin

Tyrannochthonius J.C. Chamberlin, 1929a: 74; Beier, 1932a: 62; Hoff, 1959a: 7–8; Morikawa, 1960: 108; Murthy & Ananthakrishnan, 1977: 13; Muchmore, 1984c: 119; Muchmore, 1986b: 20; Harvey, 1989a: 23.
Parachthonius Caporiacco, 1949a: 317 (synonymized by Mahnert, 1986a: 838).
Paraliochthonius (Pholeochthonius) Beier, 1976f: 209 (synonymized by Harvey, 1989a: 23).
Paraliochthonius Beier: Beier, 1976f: 205.

Type species: of *Tyrannochthonius*: *Chthonius terribilis* With, 1906, by original designation.
of *Parachthonius*: *Parachthonius meneghettii* Caporiacco, 1949a, by monotypy.
of *Paraliochthonius (Pholeochthonius)*: *Paraliochthonius (Pholeochthonius) cavernicola* Beier, 1976f, by original designation.

Tyrannochthonius albidus (Beier)

Morikawia albida Beier, 1977c: 96–98, fig. 3.
Tyrannochthonius albidus (Beier): Beier, 1978c: 533–534, fig. 1.

Type locality: Santa Cruz, Galapagos Islands.
Distribution: Galapagos Islands.

Tyrannochthonius amazonicus Mahnert

Tyrannochthonius (Tyrannochthonius) amazonicus Mahnert, 1979d: 734–736, figs 19–26; Mahnert & Adis, 1986: 213.
Thyrannochthonius (sic) *amazonicus* Mahnert: Adis, 1981: 117–118, etc.
Tyrannochthonius amazonicus Mahnert: Adis & Funke, 1983: 357; Adis & Mahnert, 1986: 301–304; Adis, Junk & Penny, 1987: 488; Mahnert, Adis, & Bührnheim, 1987: figs 3a, 7a.

Type locality: Taruma Mirim, Manaus, Amazonas, Brazil.
Distribution: Brazil (Amazonas).

Tyrannochthonius bagus Harvey

Tyrannochthonius bagus Harvey, 1988b: 315–317, figs 1–6.

Type locality: 7.5 km W. of Liwa, Sumatra, Indonesia.
Distribution: Indonesia (Sumatra).

Tyrannochthonius bahamensis Muchmore

Tyrannochthonius bahamensis Muchmore, 1984c: 119–121, figs 1–4.

Type locality: South Bimini Island, Bahamas.
Distribution: Bahamas.

Tyrannochthonius beieri Morikawa

Tyrannochthonius (Tyrannochthonius) beieri Morikawa, 1963: 2, figs 1a–c; Beier, 1964f: 592; Beier, 1965g: 758; Beier, 1966d: 133–134; Beier, 1967c: 319; Petersen, 1968: 119; Beier, 1970g: 315–316; Beier, 1982: 43.

Type locality: Honiala, Guadalcanal, Solomon Islands.
Distribution: Papua New Guinea, Solomon Islands.

Tyrannochthonius bispinosus (Beier)

Paraliochthonius bispinosus Beier, 1974e: 1001–1002, fig. 2; Murthy & Ananthakrishnan, 1977: 13.
Tyrannochthonius bispinosus (Beier): Mahnert, 1986a: 830.

Type locality: Alagarkovil, 21 km N. of Madurai, Madras, India.
Distribution: India.

Tyrannochthonius brasiliensis Mahnert

Tyrannochthonius (Tyrannochthonius) brasiliensis Mahnert, 1979d: 736–737, figs
27–32; Mahnert & Adis, 1986: 213.

Type locality: Santarém, Pará, Brazil.
Distribution: Brazil (Pará).

Tyrannochthonius brevimanus Beier

Tyrannochthonius brevimanus Beier, 1935c: 118–119, fig. 2; Roewer, 1940: 343; Beier,
1959d: 15; Beier, 1972a: 6.
Tyrannochthonius brevifemoratus (sic) Beier: Vachon, 1941e: 34.
Tyrannochthonius (Tyrannochthonius) brevimanus Beier: Beier, 1955c: 529; Mahnert,
1986a: 837–838.

Type locality: Mt Elgon, Uganda.
Distribution: Kenya, Uganda, Zaire.

Tyrannochthonius caecatus (Beier)

Paraliochthonius caecatus Beier, 1976f: 208–209, fig. 9.
Tyrannochthonius caecatus (Beier): Harvey, 1989a: 25.

Type locality: Brightwater near Nelson, South Island, New Zealand.
Distribution: New Zealand.

Tyrannochthonius callidus Hoff

Tyrannochthonius callidus Hoff, 1959a: 13–16, figs 7–9.

Type locality: 2 miles N. of Papine, Parish of St Andrew, Jamaica.
Distribution: Jamaica.

Tyrannochthonius cavernicola (Beier)

Paraliochthonius (Pholeochthonius) cavernicola Beier, 1976f: 209, fig. 10; Harvey,
1985b: 138.
Tyrannochthonius cavernicola (Beier): Harvey, 1989a: 25.

Type locality: North Bay (as station 3), Lord Howe Island.
Distribution: Lord Howe Island.

Tyrannochthonius cavicola (Beier)

Morikawia cavicola Beier, 1967a: 199–200, fig. 1.
Paraliochthonius cavicolus (sic) (Beier): Harvey, 1981b: 240.
Paraliochthonius (Paraliochthonius) cavicola (Beier): Harvey, 1985b: 138.
Tyrannochthonius cavicola Beier: Harvey, 1989a: 25–26.

Type locality: Grill Cave, Bungonia, New South Wales, Australia.
Distribution: Australia (New South Wales).

Tyrannochthonius centralis Beier

Tyrannochthonius centralis Beier, 1931b: 55–56; Beier, 1932a: 63, fig. 77; Roewer,
1937: 240; Mahnert, 1979d: 742.
Morikawia centralis (Beier): Beier, 1977c: 93–95, fig. 1.

Type locality: Faldas Vulcan Irazu, Costa Rica.
Distribution: Costa Rica, Ecuador.

Tyrannochthonius chamarro J. C. Chamberlin

Tyrannochthonius chamarro J. C. Chamberlin, 1947b: 305–308, figs 1a–f.
Tyrannochthonius (Tyrannochthonius) chamarro J. C. Chamberlin: Beier, 1957d:
9–10.

Type locality: Oca Point, Guam, Mariana Islands.
Distribution: Caroline Islands, Mariana Islands.

Tyrannochthonius chelatus Murthy & Ananthakrishnan

Tyrannochthonius chelatus Murthy & Ananthakrishnan, 1977: 14–16, figs 3a–b.

Type locality: Trivandrum, Kerala, India.
Distribution: India.

Tyrannochthonius confusus Mahnert

Tyrannochthonius (Tyrannochthonius) sokolovi (Redikorzev): Beier, 1955c: 529, fig. 1 (misidentification).
Tyrannochthonius confusus Mahnert, 1986a: 833–834, figs 32–34.

Type locality: Shimoni Cave A, Shimoni, Kenya.
Distribution: Kenya.

Tyrannochthonius contractus (Tullgren)

Chthonius contractus Tullgren, 1907b: 232–233, figs 10a–b; Ellingsen, 1909a: 219; Ellingsen, 1912b: 89, 120.
Tyrannochthonius contractus (Tullgren): Beier, 1932a: 65–66, fig. 82; Roewer, 1937: 240; Vachon, 1952a: 22–24, figs 8–9; Beier, 1958a: 159; Beier, 1964k: 31; Beier, 1966k: 455; Lawrence, 1967: 87; Spaull, 1979: 117.

Type localities: Van Reenen, Orange Free State, South Africa; Amanzimtoti, Natal, South Africa; Stamford Hill, Natal, South Africa; Lake Sibayi, Natal, South Africa; and junction of the black and white Umfolozi, Natal, South Africa.
Distribution: Aldabra Islands, Ethiopia?, Guinea?, Lesotho, South Africa, Zimbabwe.

Tyrannochthonius convivus Beier

Tyrannochthonius convivus Beier, 1974e: 1000–1001, fig. 1.

Type locality: Palni Hills, 16 km E. of Kodaikanal, Madras, India.
Distribution: India.

Tyrannochthonius curazavius Wagenaar-Hummelinck

Tyrannochthonius curazavius Wagenaar-Hummelinck, 1948: 53–59, figs 16a–b, 17g, 18a–c, 19a–c, 21d–e.

Type locality: Seroe Christoffel, Curaçao.
Distribution: Curaçao.

Tyrannochthonius densedentatus (Beier)

Morikawia densedentata Beier, 1967b: 287–288, fig. 7.
Paraliochthonius densedentatus Beier, 1976f: 205, fig. 5.
Tyrannochthonius densedentatus (Beier): Harvey, 1989a: 26.

Type locality: Omahuta Forest, North Auckland, North Island, New Zealand.
Distribution: New Zealand.

Tyrannochthonius ecuadoricus (Beier)

Morikawia ecuadorica Beier, 1977c: 95–96, fig. 2.
Tyrannochthonius ecuadoricus (Beier): Mahnert, 1979d: 742.
Tyrannochthonius (Tyrannochthonius) cf. *ecuadoricus* (Beier): Mahnert, 1984c: 18–19, figs 1–2.

Type locality: Santa Domingo, Ecuador.
Distribution: Ecuador, Peru.

Tyrannochthonius elegans Beier

Tyrannochthonius elegans Beier, 1944: 176–177, fig. 3; Beier, 1955h: 9; Beier, 1959d: 16; Weidner, 1959: 117.

Type locality: Amani, Tanzania.
Distribution: Tanzania, Zaire.

Tyrannochthonius ferox Mahnert

Tyrannochthonius ferox Mahnert, 1978f: 78, figs 18–23.

Type locality: Louolo River, Meya, Kindamba, Congo.
Distribution: Congo.

Tyrannochthonius floridensis Malcolm & Muchmore

Tyrannochthonius floridensis Malcolm & Muchmore, 1985: 403–405, figs 1–4.

Type locality: 3 miles NW. of Marianna, Jackson County, Florida, U.S.A.
Distribution: U.S.A. (Florida).

Tyrannochthonius gezei Vachon

Tyrannochthonius gezei Vachon, 1941e: 33.

Type locality: Mt Cameroun, Cameroun.
Distribution: Cameroun.

Tyrannochthonius gigas Beier

Tyrannochthonius gigas Beier, 1954f: 131–132, fig. 1.

Type locality: El Junquito, Venezuela.
Distribution: Venezuela.

Tyrannochthonius gomyi Mahnert

Tyrannochthonius (Tyrannochthonius) gomyi Mahnert, 1975b: 539–540, figs 1a–e.

Type locality: Piton Marmite, Cirque de Salazie, Réunion.
Distribution: Réunion.

Tyrannochthonius grimmeti J. C. Chamberlin

Tyrannochthonius grimmeti J. C. Chamberlin, 1929a: 76; Beier, 1932a: 64; J. C. Chamberlin, 1934b: 4; Roewer, 1937: 240; Harvey, 1989a: 26.
Morikawia grimmeti (J. C. Chamberlin): Beier, 1966c: 365–366, fig. 3; Beier, 1967b: 288.
Paraliochthonius grimmeti (J. C. Chamberlin): Beier, 1976f: 206.

Type locality: Day's Bay, Wellington, North Island, New Zealand.
Distribution: New Zealand.

Tyrannochthonius guadeloupensis Vitali-di Castri

Tyrannochthonius guadeloupensis Vitali-di Castri, 1984: 1064–1067, figs 18–24.

Type locality: Douville, 6 km NE. of Goyave, Guadeloupe.
Distribution: Guadeloupe.

Tyrannochthonius helenae (Beier)

Paraliochthonius helenae Beier, 1977b: 2–4, fig. 1.
Tyrannochthonius helenae (Beier): Mahnert, 1986a: 830.

Type locality: Mt Eternity, St Helena.
Distribution: St Helena.

Tyrannochthonius heterodentatus Beier

Tyrannochthonius heterodentatus Beier, 1930e: 294–295, fig. 1; Beier, 1932a: 65, fig. 81; Roewer, 1937: 240; Beier, 1974e: 999–1000; Murthy & Ananthakrishnan, 1977: 14.
Tyrannochthonius (Tyrannochthonius) madrasensis Murthy, 1961: 223–224, figs 2a–b; Beier, 1973b: 39 (synonymized by Beier, 1974e: 999).

Type localities: of *Tyrannochthonius heterodentatus*: Travancore, India.
of *Tyrannochthonius (Tyrannochthonius) madrasensis*: Tambaram, Madras, India.
Distribution: India, Sri Lanka.

Tyrannochthonius horridus (Beier)

Paraliochthonius luxtoni horridus Beier, 1976f: 207, fig. 7.
Tyrannochthonius horridus (Beier): Harvey, 1989a: 26.

Type locality: Waipoua State Forest, North Island, New Zealand.
Distribution: New Zealand.

Tyrannochthonius howarthi Muchmore

Tyrannochthonius howarthi Muchmore, 1979d: 187–189, figs 1–5; Tenorio & Muchmore, 1982: 382; Howarth, 1987a: 3.

Type locality: Petroglyph Cave, Hawaii Volcanoes National Park, Hawaii Island, Hawaii.
Distribution: Hawaii.

Tyrannochthonius imitatus Hoff

Tyrannochthonius imitatus Hoff, 1959a: 21–27, figs 13–17; Beier, 1976d: 45.
Tyrannochthonius fastuosus Hoff, 1959a: 28–31, figs 18–20 (synonymized by Beier, 1976d: 45).
Tyrannochthonius lautus Hoff, 1959a: 32–35, figs 21–23 (synonymized by Beier, 1976d: 45).

Type localities: *Tyrannochthonius imitatus*: 2 miles S. of Moneague, Parish of St Ann, Jamaica.
of *Tyrannochthonius fastuosus*: near Oxford Cave, 2 miles NE. of Balaclava, Manchester Parish, Jamaica.
of *Tyrannochthonius lautus*: Windsor Estate, 10–12 miles S. of Falmouth, Trelawny Parish, Jamaica.
Distribution: Dominican Republic, Jamaica.

Tyrannochthonius innoxius Hoff

Tyrannochthonius innoxius Hoff, 1959a: 9–12, figs 1–6.

Type locality: Portland Ridge, Clarendon Parish, Jamaica.
Distribution: Jamaica.

Tyrannochthonius insulae Hoff

Tyrannochthonius insulae Hoff, 1946e: 8–12, figs 9–11; Wagenaar-Hummelinck, 1948: 59–61, figs 17a–f, 17h–i, 18d–f, 20a–c, 21a–c, 21f.

Type locality: St Augustine, Trinidad.
Distribution: Trinidad.

Tyrannochthonius intermedius Muchmore

Tyrannochthonius intermedius Muchmore, 1986b: 21, fig. 8.

Type locality: Sótano de San Rafael de los Castros, 13 km WNW. of Ciudad Mante, Tamaulipas, Mexico.
Distribution: Mexico.

Tyrannochthonius irmleri Mahnert

Tyrannochthonius (Tyrannochthonius) irmleri Mahnert, 1979d: 739–741, figs 39–47; Mahnert & Adis, 1986: 213.
Tyrannochthonius irmleri Mahnert: Adis, Junk & Penny, 1987: 488.

Type locality: Taruma Mirim, Manaus, Amazonas, Brazil.
Distribution: Brazil (Amazonas).

Tyrannochthonius japonicus (Ellingsen)

Chthonius japonicus Ellingsen, 1907c: 14–15; Ellingsen, 1912a: 128.
Tyrannochthonius (?) *japonicus* (Ellingsen): Beier, 1932a: 66; Roewer, 1937: 240.
Tyrannochthonius dogoensis Morikawa: Morikawa, 1955a: 216 (misidentification).
Tyrannochthonius japonicus (Ellingsen): Morikawa, 1960: 109; Morikawa, 1962: 418; Sato, 1978e: fig. 7; Sato, 1979a: 91–92, figs 5a–i, 6a–e, plate 2; Sato, 1979c: 13, fig. 2; Sato, 1979d: 43; Sato, 1980c: 10, figs 1–2; Sato, 1982c: 32; Sato, 1983b: 16, figs 2, 4; Sato, 1984c: 14, figs 1, 3.
Tyrannochthonius japonicus japonicus (Ellingsen): Morikawa, 1960: 109–110.

Type localities: Ooyama, Sagami, Japan; and Negishi, S. of Yokohama, Japan.
Distribution: Japan, Taiwan.

Tyrannochthonius japonicus dogoensis Morikawa

Tyrannochthonius dogoensis Morikawa, 1954c: 331, figs 5–7.
Tyrannochthonius japonicus dogoensis Morikawa: Morikawa, 1960: 110, plate 1 fig. 9, plate 7 fig. 9, plate 10 fig. 7; Morikawa, 1962: fig. 1.
Not *Tyrannochthonius dogoensis* Morikawa: Morikawa, 1955a: 216 (misidentification; see *Tyrannochthonius japonicus* (Ellingsen)).

Type locality: Dogo, Matsuyama, Shikoku, Japan.
Distribution: Japan.

Tyrannochthonius kermadecensis (Beier)

Paraliochthonius kermadecensis Beier, 1976f: 205–206, fig. 6.
Paraliochthonius (Paraliochthonius) kermadecensis Beier: Harvey, 1985b: 138.
Tyrannochthonius kermadecensis (Beier): Harvey, 1989a: 26.

Type locality: Meyers Island, Kermadec Island, New Zealand.
Distribution: Lord Howe Island, New Zealand.

Tyrannochthonius krakatau Harvey

Tyrannochthonius krakatau Harvey, 1988b: 314–315, figs 1–6.

Type locality: Rakata, Krakatau Islands, Indonesia.
Distribution: Indonesia (Krakatau Islands).

Tyrannochthonius laevis Beier

Tyrannochthonius (Tyrannochthonius) laevis Beier, 1966a: 283–284, fig. 3.
Tyrannochthonius laevis Beier: Harvey, 1981b: 241; Harvey, 1985b: 140.

Type locality: Kimberley Research Station, 100 km S. of Wyndham, Western Australia, Australia.
Distribution: Australia (Western Australia).

Tyrannochthonius luxtoni (Beier)

Morikawia luxtoni Beier, 1967b: 289–290, fig. 8.
Paraliochthonius luxtoni luxtoni (Beier): Beier, 1976f: 206–207.
Tyrannochthonius luxtoni (Beier): Harvey, 1989a: 26.

Type locality: Whakatane, North Island, New Zealand.
Distribution: New Zealand.

Tyrannochthonius mahunkai Mahnert

Tyrannochthonius mahunkai Mahnert, 1978f: 78–80, figs 24–29.

Type locality: Sibiti, Congo.
Distribution: Congo.

Tyrannochthonius meneghettii (Caporiacco)

Parachthonius meneghettii Caporiacco, 1949a: 317.
Tyrannochthonius (Tyrannochthonius) holmi Beier, 1955c: 530–531, fig. 2 (synonymized by Mahnert, 1986a: 838).
Tyrannochthonius holmi Beier: Beier, 1959d: 16; Beier, 1962f: 11–12; Beier, 1972a: 6.
Tyrannochthonius (Tyrannochthonius) menegehettii (Caporiacco): Mahnert, 1986a: 838–839, figs 40–41.

Type localities: of *Parachthonius meneghettii*: Mau, Kenya.
of *Tyrannochthonius (Tyrannochthonius) holmi*: Mt Elgon, E. side, Kenya.
Distribution: Burundi, Kenya, Ruanda, Tanzania, Uganda, Zaire.

Tyrannochthonius meruensis Beier

Tyrannochthonius meruensis Beier, 1962f: 13–14, fig. 2.
Tyrannochthonius (Tyrannochthonius) meruensis Beier: Mahnert, 1986a: 836–837.

Type locality: Olkokola, Mt Meru, Tanzania.
Distribution: Kenya, Tanzania.

Tyrannochthonius migrans Mahnert

Tyrannochthonius (Tyrannochthonius) migrans Mahnert, 1979d: 731–733, figs 11–18; Adis, 1981: 117–118, etc.; Mahnert & Adis, 1986: 213.
Tyrannochthonius migrans Mahnert: Adis & Funke, 1983: 357; Adis & Mahnert, 1986: 304; Adis, Junk & Penny, 1987: 488.

Type locality: Taruma Mirim, Manaus, Amazonas, Brazil.
Distribution: Brazil (Amazonas).

Tyrannochthonius minor Mahnert

Tyrannochthonius (Tyrannochthonius) minor Mahnert, 1979d: 737–739, figs 33–38; Mahnert & Adis, 1986: 213.

Type locality: Reserva Ducke, Manaus, Amazonas, Brazil.
Distribution: Brazil (Amazonas, Pará).

Tyrannochthonius monodi Vachon

Tyrannochthonius monodi Vachon, 1941e: 34.

Type locality: Mt Cameroun, Cameroun.
Distribution: Cameroun.

Tyrannochthonius nanus (Beier)

Morikawia nana Beier, 1966d: 135–137, fig. 2; Beier, 1967c: 321; Beier, 1970g: 316.
Tyrannochthonius nanus (Beier): Harvey, 1989a: 26–27.

Type locality: Dala, Malaita, Solomon Islands.
Distribution: Papua New Guinea, Solomon Islands.

Tyrannochthonius noaensis Moyle

Tyrannochthonius noaensis Moyle, 1989a: 58–59, figs 1–3.

Type locality: not known.
Distribution: New Zealand.

Family Chthoniidae

Tyrannochthonius norfolkensis (Beier)

Paraliochthonius norfolkensis Beier, 1976f: 207–208, fig. 8.
Paraliochthonius (Paraliochthonius) norfolkensis Beier: Harvey, 1985b: 138.
Tyrannochthonius norfolkensis (Beier): Harvey, 1989a: 27.

Type locality: Mt Pitt, Norfolk Island.
Distribution: Norfolk Island.

Tyrannochthonius ovatus Vitali-di Castri

Tyrannochthonius ovatus Vitali-di Castri, 1984: 1067–1068, fig. 25.

Type locality: Montagne Pelée, Martinique.
Distribution: Martinique.

Tyrannochthonius pachythorax Redikorzev

Tyrannochthonius pachythorax Redikorzev, 1938: 73–74, figs 1–2; Roewer, 1940: 343.
Tyrannochthonius (Tyrannochthonius) pachythorax Redikorzev: Beier, 1951a: 58.

Type localities: Ream (as Réam), Cambodia; Nha Trang (as Nhatrang), Vietnam; and Ba Ngoi (as Bangoï), Vietnam.
Distribution: Cambodia, Vietnam.

Tyrannochthonius palauanus Beier

Tyrannochthonius (Tyrannochthonius) palauanus Beier, 1957d: 10–11, figs 1d–e.

Type locality: Ulebsehel (Aurapushekaru), Palau Islands, Caroline Islands.
Distribution: Caroline Islands.

Tyrannochthonius pallidus Muchmore

Tyrannochthonius pallidus Muchmore, 1973d: 81–82, figs 1–2.

Type locality:Cueva de El Jobo, 5 km NE. of Xilitla, San Luis Potosí, Mexico.
Distribution: Mexico.

Tyrannochthonius perpusillus Beier

Tyrannochthonius (Tyrannochthonius) perpusillus Beier, 1951a: 60–61, fig. 8.

Type localities: Da Lat (as Dalat), Vietnam; Cao Nguyên Lâm Viên (as Plateau von Langbian), Vietnam.
Distribution: Vietnam.

Tyrannochthonius philippinus (Beier)

Morikawia philippina Beier, 1966b: 340–342, fig. 1.
Genus ? *philippina* Beier: Tenorio & Muchmore, 1982: 382.
Tyrannochthonius philippinus (Beier): Harvey, 1989a: 27.

Type locality: Mt Katanglad, Bukidnon, Mindanao, Philippines.
Distribution: Philippines.

Tyrannochthonius procerus Mahnert

Tyrannochthonius procerus Mahnert, 1978f: 75–77, figs 6–11.

Type locality: Meya, Kindamba, Congo.
Distribution: Congo.

Tyrannochthonius proximus Hoff

Tyrannochthonius proximus Hoff, 1959a: 16–21, figs 10–12; Hoff, 1963c: 6; Beier, 1976d: 45–46.

Type locality: 5 miles S. of Hardwar Gap, Parish of St Andrew, Jamaica.
Distribution: Dominican Republic, Jamaica.

Tyrannochthonius pugnax Mahnert

Tyrannochthonius pugnax Mahnert, 1978f: 81−82, figs 12−17.

Type locality: Lefini reservation, Mbéokala forest, Congo.
Distribution: Congo.

Tyrannochthonius pupukeanus Muchmore

Tyrannochthonius pupukeanus Muchmore, 1983: 84−86, figs 1−4; Howarth, 1987a: 3.

Type locality: Pupukea Lava Tube, Pupukea, Oahu Island, Hawaii.
Distribution: Hawaii.

Tyrannochthonius pusillimus Beier

Tyrannochthonius (Tyrannochthonius) pusillimus Beier, 1951a: 58−59, fig. 6.

Type locality: Cao Nguyên Lâm Viên (as Plateau von Langbian), Vietnam.
Distribution: Vietnam.

Tyrannochthonius pusillus Beier

Tyrannochthonius pusillus Beier, 1955m: 2−3, fig. 2; Weidner, 1959: 117.

Type locality: Sivia, Peru.
Distribution: Peru.

Tyrannochthonius queenslandicus (Beier)

Morikawia queenslandica Beier, 1969a: 174−175, fig. 3.
Paraliochthonius queenslandicus (Beier): Beier, 1976f: 206; Harvey, 1981b: 240.
Paraliochthonius (Paraliochthonius) queenslandicus (Beier): Harvey, 1985b: 138.
Tyrannochthonius queenslandicus (Beier): Harvey, 1989a: 27.

Type locality: Mt Tamborine (as Jamborine), Joalah (as Toalah) National Park,
    Queensland, Australia.
Distribution: Australia (Queensland), Norfolk Island.

Tyrannochthonius rahmi Beier

Tyrannochthonius (Tyrannochthonius) rahmi Beier, 1976e: 96, fig. 1.
Tyrannochthonius rahmi Beier: Beier, 1974b: 262; Schawaller, 1983a: 108; Schawaller,
    1987a: 203.

Type locality: Phuntsholing, Bhutan.
Distribution: Bhutan, Nepal.

Tyrannochthonius rex Harvey

Tyrannochthonius rex Harvey, 1989a: 23−25, figs 10−19.

Type locality: Royal Arch Cave, Chillagoe, Queensland, Australia.
Distribution: Australia (Queensland).

Tyrannochthonius riberai Mahnert

Tyrannochthonius riberai Mahnert, 1984c: 20, figs 3−9.
Tyrannochthonius aff. *riberai* Mahnert: Mahnert, 1984c: 20−22.

Type locality: Cueva de la Ascuncion, El Pajonal, Peru.
Distribution: Peru.

Tyrannochthonius robustus Beier

Tyrannochthonius (Tyrannochthonius) robustus Beier, 1951a: 59−60, fig. 7.

Type locality: Cao Nguyên Läm Viên (as Plateau von Langbian), Vietnam.
Distribution: Vietnam.

Tyrannochthonius rotundimanus Mahnert

Tyrannochthonius (Tyrannochthonius) rotundimanus Mahnert, 1985c: 218–219, figs 2–4; Mahnert & Adis, 1986: 213.

Type locality: Reserve Florestal Ducke, 26 km on Manaus-Itacoatiara Highway, Amazonas, Brazil.
Distribution: Brazil (Amazonas).

Tyrannochthonius semidentatus (Redikorzev)

Chthonius semidentatus Redikorzev, 1924a: 197–199, figs 12–14.
Tyrannochthonius (?) *semidentatus* (Redikorzev): Beier, 1932a: 66.
Tyrannochthonius semidentatus (Redikorzev): Roewer, 1937: 240, fig. 208.

Type locality: Molo, Kenya.
Distribution: Kenya.

Tyrannochthonius semihorridus (Beier)

Morikawia semihorrida Beier, 1969a: 176–177, fig. 4.
Paraliochthonius semihorridus (Beier): Harvey, 1981b: 240.
Paraliochthonius (Paraliochthonius) semihorridus (Beier): Harvey, 1985b: 138.
Tyrannochthonius semihorridus (Beier): Harvey, 1989a: 27.

Type locality: Mt Nebo, Queensland, Australia.
Distribution: Australia (Queensland).

Tyrannochthonius similidentatus Sato

Tyrannochthonius similidentatus Sato, 1984b: 50–52, figs 1–13.

Type locality: Chibusa-yama, Haha-jima, Ogasawara, Japan.
Distribution: Japan.

Tyrannochthonius simillimus Beier

Tyrannochthonius (Tyrannochthonius) simillimus Beier, 1951a: 56–58, fig. 5.

Type locality: Sre Ambel (as Sre Umbell), Cambodia.
Distribution: Cambodia.

Tyrannochthonius simulans Mahnert

Tyrannochthonius (Tyrannochthonius) simulans Mahnert, 1986a: 834–836, figs 35–38.

Type locality: between Limuru and Naivasha/Narok Road, Kikuyu Escarpment, Kiambu, Kenya.
Distribution: Kenya.

Tyrannochthonius sokolovi (Redikorzev)

Chthonius sokolovi Redikorzev, 1924a: 195–197, figs 10–11, 14.
Tyrannochthonius sokolovi (Redikorzev): Beier, 1932a: 66; Roewer, 1937: 240; Beier, 1955h: 9; Beier, 1959d: 15–16; Beier, 1962f: 12; Beier, 1967d: 73.
Tyrannochthonius (Tyrannochthonius) sokolovi (Redikorzev): Mahnert, 1986a: 839–841, figs 42–48.
Tyrannochthonius (Tyrannochthonius) aff. *sokolovi* (Redikorzev): Mahnert, 1986a: 841.
Not *Tyrannochthonius (Tyrannochthonius) sokolovi* (Redikorzev): Beier, 1955c: 529, fig. 1 (misidentification; see *Tyrannochthonius confusus* Mahnert).

Type locality: Mabira, Uganda.
Distribution: Kenya, Tanzania, Uganda, Zaire.

Tyrannochthonius sparsedentatus Beier

Tyrannochthonius sparsedentatus Beier, 1959d: 17–18, fig. 5.

Type locality: Nyakagera River, Kabare, Kivu, Zaire.
Distribution: Zaire.

Tyrannochthonius strinatii (Beier)

Paraliochthonius strinatii Beier, 1974c: 101–102, fig. 1.
Tyrannochthonius strinatii Beier: Muchmore, 1984c: 125.

Type locality: Cueva Chirrepeck, Alta Verapaz, Guatemala.
Distribution: Guatemala.

Tyrannochthonius superstes Mahnert

Tyrannochthonius (Tyrannochthonius) superstes Mahnert, 1986c: 144–145, figs 1–5;
Mahnert, 1989b: 41.

Type locality: Cueva Felipe Revento'n, Tenerife, Canary Islands.
Distribution: Canary Islands.

Tyrannochthonius tekauriensis Moyle

Tyrannochthonius tekauriensis Moyle, 1989b: 60–62, figs 1–5.

Type locality: Te Kauri Park Scenic Reserve, South Island, New Zealand.
Distribution: New Zealand.

Tyrannochthonius terribilis (With)

Chthonius terribilis With, 1906: 69–73, figs 10–11, plate 1 figs 1a–m; Ellingsen,
1910a: 402.
Tyrannochthonius terribilis (With): J. C. Chamberlin, 1929a: 75; Beier, 1930e: 294;
Beier, 1932a: 64, figs 5–7, 79; Beier, 1932g: figs 206–207; J. C. Chamberlin, 1934b:
4; Roewer, 1936: figs 15a, 19a–b, 36a–b, 68d, 70b, 83f, 106a–b, 141; Roewer,
1937: 240.

Type localities: Ko Chang (as Koh Chang); Thailand; and Laem Ngop (as Lem Ngob),
Thailand.
Distribution: Papua New Guinea, Sumatra, Thailand.

Tyrannochthonius terribilis malaccensis Beier

Tyrannochthonius terribilis malaccensis Beier, 1952a: 96–97, fig. 1.

Type locality: Sungei Buloh Leper Settlement, Selangor, Malaysia.
Distribution: Malaysia.

Tyrannochthonius tlilapanensis Muchmore

Tyrannochthonius tlilapanensis Muchmore, 1986b: 23, fig. 10.

Type locality: Cueva Macinga, 2 km E. of Tlilipan, Veracruz, Mexico.
Distribution: Mexico.

Tyrannochthonius troglobius Muchmore

Tyrannochthonius troglobius Muchmore, 1969b: 31–32, figs 1–3; Reddell & Mitchell,
1971b: 184; Muchmore, 1986b: 20.

Type locality: Mine Cave, Rancho del Cielo, Tamaulipas, Mexico.
Distribution: Mexico.

Tyrannochthonius troglodytes Muchmore

Tyrannochthonius troglodytes Muchmore, 1986b: 23–24, fig. 11.

Family Chthoniidae

Type locality: Rock Slab Cave on Enchanted Rock, 19 miles SSE. of Llano, Llano County, Texas, U.S.A.
Distribution: U.S.A. (Texas).

Tyrannochthonius troglophilus (Beier)

Morikawia troglophila Beier, 1968a: 758–759, fig. 2.
Tyrannochthonius troglophilus (Beier): Harvey, 1989a: 27.

Type locality: Grotte de Ninrin-Reu, near Poya, New Caledonia.
Distribution: New Caledonia.

Tyrannochthonius vampirorum Muchmore

Tyrannochthonius vampirorum Muchmore, 1986b: 20–21, figs 6–7.

Type locality: Cueva de los Vampiros, 9.5 km NNE. of Chamal, Tamaulipas, Mexico.
Distribution: Mexico.

Tyrannochthonius volcancillo Muchmore

Tyrannochthonius volcancillo Muchmore, 1986b: 21–23, fig. 9.

Type locality: Cueva del Volcancillo, 5 km SE. of Las Vigas, Veracruz, Mexico.
Distribution: Mexico.

Tyrannochthonius volcanus Muchmore

Tyrannochthonius volcanus Muchmore, 1977: 68–70, figs 16–17.

Type locality: Volcán Tzontehuitz, 8 miles NE. of San Cristóbal de las Casas, Chiapas, Mexico.
Distribution: Mexico.

Tyrannochthonius wittei Beier

Tyrannochthonius wittei Beier, 1955a: 5–7, fig. 2; Beier, 1959d: 16–17.

Type locality: Kafwe, Zaire.
Distribution: Zaire.

Tyrannochthonius wlassicsi (Daday)

Chthonius wlassicsi Daday, 1897: 479–480, figs 1–4, 8–9; With, 1906: 73.
Tyrannochthonius (?) *wlassicsi* (Daday): Beier, 1932a: 66; J. C. Chamberlin, 1934b: 4; Roewer, 1937: 240.
Morikawia wlassicsi (Daday): Beier, 1965g: 760–761, fig. 4.
Paraliochthonius wlassicsi (Daday): Beier, 1982: 43.
Tyrannochthonius wlassicsi (Daday): Harvey, 1989a: 27–28.

Type localities: Madang (as Friedrich-Wilhelmshafen), Papua New Guinea; and Lemien, Papua New Guinea.
Distribution: Indonesia (Irian Jaya), Papua New Guinea.

Tyrannochthonius zicsii Mahnert

Tyrannochthonius zicsii Mahnert, 1978f: 80–81, figs 30–31.

Type locality: Sibiti, Congo.
Distribution: Congo.

Tyrannochthonius zonatus (Beier)

Morikawia zonata Beier, 1964e: 403–405, fig. 1.
Paraliochthonius zonatus (Beier); Beier, 1977b: 4.
Genus ? *zonata* Beier: Tenorio & Muchmore, 1982: 382.
Tyrannochthonius zonatus (Beier): Harvey, 1989a: 27.

Type locality: Mt Koghi, New Caledonia.
Distribution: New Caledonia.

216

Family TRIDENCHTHONIIDAE Balzan

Tridenchthoniidae Balzan, 1892: 505; J. C. Chamberlin & R. V. Chamberlin, 1945: 6–15; Hoff, 1949b: 429; Hoff, 1963c: 27; Murthy & Ananthakrishnan, 1977: 9–10; Muchmore, 1982a: 96–97; Harvey, 1985b: 152.
Tridenchthoniinae Balzan: H. J. Hansen, 1893: 232; With, 1906: 64; J. C. Chamberlin, 1931a: 212; J. C. Chamberlin & R. V. Chamberlin, 1945: 17; Hoff, 1963c: 27.
Dithinae J. C. Chamberlin, 1929a: 58; J. C. Chamberlin, 1931a: 209–210; Beier, 1932a: 24.
Dithini J. C. Chamberlin: J. C. Chamberlin, 1929a: 60.
Verrucadithini J. C. Chamberlin, 1929a: 59.
Verrucadithiini (sic) J. C. Chamberlin: J. C. Chamberlin & R. V. Chamberlin, 1945: 17.
Tridenchthoniini Balzan: J. C. Chamberlin & R. V. Chamberlin, 1945: 30; Hoff, 1963c: 27.
Cecodithiinae J. C. Chamberlin & R. V. Chamberlin, 1945: 65.
Dithidae J. C. Chamberlin: Beier, 1932g: 181; Roewer, 1937: 233; Petrunkevitch, 1955: 81; Morikawa, 1960: 92–93.

Genus **Anaulacodithella** Beier

Verrucadithella (Anaulacodithella) Beier, 1944: 175.
Xenoditha J. C. Chamberlin & R. V. Chamberlin, 1945: 17–18 (synonymized by Beier, 1947b: 287).
Anaulacodithella Beier: Beier, 1947b: 287; Beier, 1976f: 200.

Type species: of *Verrucadithella (Anaulacodithella)*: *Chthonius mordax* Tullgren, 1907b, by original designation.
of *Xenoditha*: *Chthonius mordax* Tullgren, 1907b, by original designation.

Anaulacodithella angustimana Beier

Anaulacodithella angustimana Beier, 1955l: 273–274, fig. 1.

Type locality: Cape Point Nature Reserve, Cape Province, South Africa.
Distribution: South Africa.

Anaulacodithella australica Beier

Anaulacodithella australica Beier, 1969a: 172–174, fig. 2; Beier, 1975b: 203; Harvey, 1981b: 241; Harvey 1985b: 152.

Type locality: Lamington National Park, Queensland, Australia.
Distribution: Australia (New South Wales, Queensland).

Anaulacodithella deserticola (Beier)

Verrucadithella (Anaulacodithella) deserticola Beier, 1944: 175 (nomen nudum).
Anaulacodithella deserticola (Beier): Beier, 1947b: 286–287, fig. 1; Beier, 1955l: 273; Beier, 1964k: 31.

Type locality: Viljoenspos (as Viljoens-Pass), Cape Province, South Africa.
Distribution: South Africa.

Anaulacodithella mordax (Tullgren)

Chthonius mordax Tullgren, 1907b: 234–235, figs 10a–d; Ellingsen, 1912b: 89, 122–123; Godfrey, 1927: 18.
Verrucadithella mordax (Tullgren): Beier, 1932a: 33, fig. 39; Roewer, 1937: 235.
Verrucadithella (Anaulacodithella) mordax (Tullgren): Beier, 1944: 175.
Anaulacodithella mordax (Tullgren): Beier, 1958a: 159; Beier, 1964k: 31; Beier, 1966k: 455.

Xenoditha mordax (Tullgren): J. C. Chamberlin & R. V. Chamberlin, 1945: 18–19, fig. 3.

Type localities: Caversham, South Africa; Stamford Hill, Natal, South Africa; and Lake Sibayi, Natal, South Africa.
Distribution: South Africa.

Anaulacodithella novacaledonica Beier

Anaulacodithella novacaledonica Beier, 1966g: 363–365, fig. 1.

Type locality: Forêt de Mt Pouédihi near Rivière Blanche, New Caledonia.
Distribution: New Caledonia.

Anaulacodithella plurisetosa Beier

Anaulacodithella plurisetosa Beier, 1976f: 200–201, fig. 1; Harvey, 1985b: 152.

Type locality: near Boat Harbour, Lord Howe Island.
Distribution: Lord Howe Island.

Anaulacodithella reticulata Beier

Anaulacodithella reticulata Beier, 1966g: 365–366, fig. 2.

Type locality: Point Gouli, Col d'Amieu, New Caledonia.
Distribution: New Caledonia.

Genus **Anisoditha** J. C. Chamberlin & R. V. Chamberlin

Anisoditha J. C. Chamberlin & R. V. Chamberlin, 1945: 26.

Type species: *Chthonius curvidigitatus* Balzan, 1887b, by original designation.

Anisoditha curvidigitata (Balzan)

Chthonius curvidigitatus Balzan, 1887b: no pagination, figs; Balzan, 1890: 449–450, figs 27, 27a; Balzan, 1892: 550.
Verrucaditha (?) *curvidigitata* (Balzan): Beier, 1932a: 32; Roewer, 1937: 235.
Anisoditha curvidigitata (Balzan): J. C. Chamberlin & R. V. Chamberlin, 1945: 26–27.

Type locality: Rio Apa, Paraguay.
Distribution: Paraguay.

Genus **Cecoditha** Mello-Leitão

Cecoditha Mello-Leitão, 1939a: 115–116; J. C. Chamberlin & R. V. Chamberlin, 1945: 65–66.

Type species: *Cecoditha parva* Mello-Leitão, 1939a, by original designation.

Cecoditha parva Mello-Leitão

Cecoditha parva Mello-Leitão, 1939a: 116–117, figs 1a–b; J. C. Chamberlin & R. V. Chamberlin, 1945: 66–67, figs 17a–b.

Type locality: Madryn, Chubut, Argentina.
Distribution: Argentina.

Genus **Chelignathus** Menge

Chelignathus Menge, in C. L. Koch & Berendt, 1854: 97; J. C. Chamberlin & R. V. Chamberlin, 1945: 30–31.

Type species: *Chelignathus kochii* Menge, in C. L. Koch & Berendt, 1854, by monotypy.

! Chelignathus kochii Menge

Chelignathus kochii Menge, in C. L. Koch & Berendt, 1854: 97; J. C. Chamberlin &
R. V. Chamberlin, 1945: 32, fig. 7; Beier, 1955g: 48; Schawaller, 1978: 3.
Chthonius kochii (Menge): Menge, 1855: 25–26, plate 4 fig. 2; Hagen, 1870: 268;
Beier, 1932a: 278.
Heterolophus (?) *kochii* (Menge): Beier, 1937d: 303, fig. 1; Roewer, 1940: 330.
Heterolophus kochii (Menge): Petrunkevitch, 1955: fig. 49(2).

Type locality: Baltic Amber.
Distribution: Baltic Amber.

Genus Compsaditha J. C. Chamberlin

Compsaditha J. C. Chamberlin, 1929a: 62; Beier, 1932a: 29; J. C. Chamberlin &
R. V. Chamberlin, 1945: 61–63; Murthy & Ananthakrishnan, 1977: 10.

Type species: *Compsaditha pygmaea* J. C. Chamberlin, 1929a, by original designation.

Compsaditha aburi J. C. Chamberlin & R. V. Chamberlin

Chtonius (sic) *serrulata* Silvestri, 1918: 294–296 (misidentification, in part).
Afroditha serrulata (Silvestri): Beier, 1930i: 48, figs 6a–b (misidentification).
Ditha serrulata (Silvestri): Beier, 1932a: 27 (misidentification).
Compsaditha aburi J. C. Chamberlin & R. V. Chamberlin, 1945: 63–64; Vachon,
1952a: 19–20, figs 1–2; Beier, 1959d: 10; Mahnert, 1978f: 70–72, figs 1–5;
Mahnert, 1983a: 148; Mahnert, Adis, & Bührnheim, 1987: fig. 4b.

Type locality: Aburi, Ghana.
Distribution: Ghana, Congo, Guinea, Zaire.

Compsaditha angustula Beier

Compsaditha angustula Beier, 1972a: 5–6, fig. 2.

Type locality: Parc National Garamba, Zaire.
Distribution: Zaire.

Compsaditha basilewskyi Beier

Compsaditha basilewskyi Beier, 1962f: 9–10, fig. 1; Mahnert, 1983a: 148–149,
fig. 17.

Type locality: Mt Hanang, Tanzania.
Distribution: Kenya, Tanzania.

Compsaditha camponota Sivaraman

Comsaditha (sic) *camponota* Sivaraman, 1980d: 237–239, figs 1–2.

Type locality: Madras, Tamil Nadu, India.
Distribution: India.

Compsaditha congica Beier

Compsaditha congica Beier, 1959d: 10–11, fig. 2.

Type locality: Kisanga, Lubumbashi (as Elisabethville), Zaire.
Distribution: Zaire.

Compsaditha elegantula Beier

Compsaditha elegantula Beier, 1972a: 3–4, fig. 1.

Type locality: Parc National Garamba, Zaire.
Distribution: Zaire.

Compsaditha fiebrigi (Beier)

Chthonius parvidentatus Balzan: Ellingsen, 1910a: 401 (misidentification).
Ditha fiebrigi Beier, 1931b: 50–51; Beier, 1932a: 27–28, fig. 33; Roewer, 1937: 234, fig. 204; Feio, 1945: 3.
Compsaditha fiebrigi (Beier): J. C. Chamberlin & R. V. Chamberlin, 1945: 63.

Type locality: San Bernardino, Paraguay.
Distribution: Brazil (Rio de Janeiro), Paraguay.

Compsaditha gressitti Beier

Compsaditha gressitti Beier, 1957d: 9, fig. 1c.

Type locality: Colonia, Ponape, Caroline Islands.
Distribution: Caroline Islands.

Compsaditha indica Murthy

Compsaditha indica Murthy, 1960: 28–30, figs 1a–b; Murthy & Ananthakrishnan, 1977: 10–11, figs 2a–b; Sivaraman, 1981: 22.

Type locality: Government Museum gardens, Egmore, Madras, India.
Distribution: India.

Compsaditha parva Beier

Compsaditha parva Beier, 1951a: 54–55, fig. 3.

Type locality: Lo-Ku-Ho, Cha Pa (as Chapa), Tonkin, Vietnam.
Distribution: Vietnam.

Compsaditha pygmaea J. C. Chamberlin

Compsaditha pygmaea J. C. Chamberlin, 1929a: 62; Beier, 1932a: 30; J. C. Chamberlin, 1931a: figs 19a, 21c, 21g; Roewer, 1937: 235; J. C. Chamberlin & R. V. Chamberlin, 1945: 64–65, figs 2a, 2f–g, 16a–c; Vachon, 1949: fig. 199a.

Type locality: Mt Makiling, Luzon, Philippines.
Distribution: Philippines.

Compsaditha seychellensis Beier

Compsaditha seychellensis Beier, 1974a: 144–145.

Type locality: Mahé Island, Seychelles.
Distribution: Seychelles.

Genus Cryptoditha J. C. Chamberlin & R. V. Chamberlin

Cryptoditha J. C. Chamberlin & R. V. Chamberlin, 1945: 29–30.

Type species: *Tridenchthonius elegans* Beier, 1931b, by original designation.

Cryptoditha elegans (Beier)

Tridenchthonius elegans Beier, 1931b: 52; Beier, 1932a: 34–35, figs 40–41; Roewer, 1937: 235; Helversen, 1966b: fig. 8e.
Cryptoditha elegans (Beier): J. C. Chamberlin & R. V. Chamberlin, 1945: 30, fig. 6; Beier, 1974d: 899.

Type locality: Passo Quatro, Brazil.
Distribution: Brazil (Santa Catarina).

Genus Ditha J. C. Chamberlin

Ditha J. C. Chamberlin, 1929a: 61; Beier, 1932a: 25; J. C. Chamberlin & R. V. Chamberlin, 1945: 37–39; Morikawa, 1960: 93.

Type species: *Ditha elegans* J. C. Chamberlin, 1929a, by original designation.

Subgenus **Ditha (Ditha)** J. C. Chamberlin

Ditha (Ditha) elegans J. C. Chamberlin

Ditha elegans J. C. Chamberlin, 1929a: 61; J. C. Chamberlin, 1931a: figs 9a, 11g–h, 34a; Beier, 1932a: 27; J. C. Chamberlin, 1934b: 3; Roewer, 1937: 234; J. C. Chamberlin & R. V. Chamberlin, 1945: 39–40.
Ditha (Ditha) elegans Beier: Beier, 1955e: 39.

Type locality: Banda, Moluccas, Indonesia.
Distribution: Indonesia.

Ditha (Ditha) loricata Beier

Ditha loricata Beier, 1965g: 757–758, fig. 2; Beier, 1982: 43; Tenorio & Muchmore, 1982: 383.

Type locality: Oransbari, Irian Jaya, Indonesia.
Distribution: Indonesia (Irian Jaya), Papua New Guinea.

Ditha (Ditha) novaeguineae Beier

Ditha novaeguineae Beier, 1965g: 756–757, fig. 1.

Type locality: Finschhafen, Papua New Guinea.
Distribution: Papua New Guinea.

Ditha (Ditha) ogasawarensis Sato

Ditha (Ditha) ogasawarensis Sato, 1981: 11–14, figs 1–14.

Type locality: Shindo-iriguchi, Chibusa-yama, Hahajima, Ogasawara, Japan.
Distribution: Japan.

Ditha (Ditha) palauensis Beier

Ditha (Ditha) palauensis Beier, 1957d: 7–8, figs 1a–b.

Type locality: Peleliu, Palau Islands.
Distribution: Palau Islands.

Ditha (Ditha) philippinensis J. C. Chamberlin

Ditha philippinensis J. C. Chamberlin, 1929a: 61, fig. 1b; J. C. Chamberlin, 1931a: figs 15b, 25a, 45g; Beier, 1932a: 27; Beier, 1932g: fig. 222; Roewer, 1937: 234; J. C. Chamberlin & R. V. Chamberlin, 1945: 40–41, figs 2e, 10a–b; Beier, 1966b: 340; Beier, 1967c: 317; Zeh, 1987b: 1086.
Ditha (Ditha) philippinensis Beier: Beier, 1955e: 39.

Type locality: Mt Makiling, Luzon, Philippines.
Distribution: Philippines.

Ditha (Ditha) proxima (Beier)

Compsaditha proxima Beier, 1951a: 56, fig. 4.
Ditha proxima (Beier): Beier, 1974b: 261; Beier, 1976e: 95–96; Ćurčić, 1979b: 192–198, figs 7–8; Ćurčić, 1980b: 78; Schawaller, 1983a: 107–108, figs 1–3; Schawaller, 1987a: 202.
Ditha numburensis Morikawa, 1968: 259–262, figs 1a, 2b (synonymized by Schawaller, 1987a: 202).

Type localities: of *Compsaditha proxima*: Da Lat (as Dalat), Vietnam; and Cao Nguyên Lâm Viên (as Plateau von Langbian), Vietnam.
of *Ditha numburensis*: Maedane Karka, near Mt Numbur, Nepal.
Distribution: Bhutan, Nepal, Vietnam.

Subgenus **Ditha (Paraditha)** Beier

Paraditha Beier, 1931b: 51; Beier, 1932a: 30; J. C. Chamberlin & R. V. Chamberlin, 1945: 35–36.
Ditha (Paraditha) Beier: Beier, 1955e: 39.

Type species: *Chthonius sinuatus* Tullgren, 1901, by original designation.

Ditha (Paraditha) laosana Beier

Ditha laosana Beier, 1951a: 53–54, fig. 2.
Ditha (Paraditha) laosana Beier: Beier, 1955e: 39.

Type locality: Pak Lay (as Paclay), Laos.
Distribution: Laos.

Ditha (Paraditha) latimana (Beier)

Paraditha latimana Beier, 1931b: 51; Beier, 1932a: 31, fig. 37; Roewer, 1937: 235; J. C. Chamberlin & R. V. Chamberlin, 1945: 37, fig. 9; Helversen, 1966b: fig. 8d.
Ditha (Paraditha) latimana (Beier): Beier, 1955e: 39.

Type locality: Mabira, Tanzania.
Distribution: Tanzania.

Ditha (Paraditha) marcusensis (Morikawa)

Verrucaditha marcusensis Morikawa, 1952a: 243, figs 2, 5a.
Ditha (Paraditha) marcusensis (Morikawa): Beier, 1957d: 8–9.
Ditha marcusensis (Morikawa): Morikawa, 1960: 93, plate 1 fig. 1, plate 7 fig. 1; Morikawa, 1962: 418.

Type locality: Marcus Island.
Distribution: Marcus Island.

Ditha (Paraditha) pahangica Beier

Ditha (Paraditha) pahangica Beier, 1955e: 38–39, fig. 1.

Type locality: Telom Valley, near Gunong Siku, Cameron Highlands, Pahang, Malaysia.
Distribution: Malaysia.

Ditha (Paraditha) sinuata (Tullgren)

Chthonius sinuatus Tullgren, 1901: 100–101; Ellingsen, 1906: 264–265; Ellingsen, 1912b: 89, 124–125.
Paraditha sinuata (Tullgren): Beier, 1931b: 50; Beier, 1932a: 30–31; Roewer, 1937: 235; J. C. Chamberlin & R. V. Chamberlin, 1945: 37.
Ditha (Paraditha) sinuata (Tullgren): Beier, 1955e: 39; Beier, 1959d: 9, fig. 1; Mahnert, 1978f: 70.

Type locality: Cameroun.
Distribution: Burundi, Cameroun, Congo, Guinea-Bissau, Sao Tomé, South Africa, Zaire.

Ditha (Paraditha) sumatraensis (J. C. Chamberlin)

Chthonius curvidigitatus E. Simon, 1899a: 122–123; With, 1906: 74; Ellingsen, 1911a: 35–36 (junior primary homonym of *Chthonius curvidigitatus* Balzan, 1890).
Chthonius sumatraensis J. C. Chamberlin, 1923b: 186–187 (replacement name for *Chthonius curvidigitatus* E. Simon, 1899a).
Heterolophus sumatraensis (J. C. Chamberlin): Beier, 1932a: 29, fig. 35; Roewer, 1937: 235; J. C. Chamberlin & R. V. Chamberlin, 1945: 35.
Ditha (Paraditha) sumatraensis (J. C. Chamberlin): Beier, 1955e: 39.

Type locality: Sumatra, Indonesia.
Distribution: Indonesia (Sumatra).

Ditha (Paraditha) tonkinensis Beier

Ditha tonkinensis Beier, 1951a: 52–53, fig. 1.
Ditha (Paraditha) tonkinensis Beier: Beier, 1955e: 39.
Type locality: Cha Pa (as Chapa), Tonkin, Vietnam.
Distribution: Vietnam.

Genus Dithella J. C. Chamberlin & R. V. Chamberlin

Dithella J. C. Chamberlin & R. V. Chamberlin, 1945: 41–42.
Type species: *Chthonius javanus* Tullgren, 1912b, by original designation.

Dithella javana (Tullgren)

Chthonius javanus Tullgren, 1912b: 269–270, fig. 1.
Compsaditha javana J. C. Chamberlin, 1929e: 152 (junior secondary homonym of
   Compsaditha javana (Tullgren)) (synonymized by J. C. Chamberlin & R. V.
   Chamberlin, 1945: 42).
Compsaditha javana (Tullgren): J. C. Chamberlin, 1931a: figs 13j, 17a–b, 28a–b;
   Beier, 1932a: 30; Roewer, 1937: 235.
Dithella javana (Tullgren): J. C. Chamberlin & R. V. Chamberlin, 1945: 42–43, figs
   11a–e; Beier, 1948b: 452.

Type localities: of *Chthonius javanus*: Banjuwangi, Java, Indonesia.
of *Compsaditha javana*: Banjuwangi (as Banjoewangi), Java, Indonesia.
Distribution: Indonesia (Java).

Dithella philippinica Beier

Dithella philippinica Beier, 1967c: 315–317, fig. 1.
Type locality: Mantalingajan, Pinigisan, Palawan, Philippines.
Distribution: Philippines.

Genus Haploditha Caporiacco

Haploditha Caporiacco, 1951d: 37.
Type species: *Haploditha chamberlinorum* Caporiacco, 1951d, by original designation.

Haploditha chamberlinorum Caporiacco

Haploditha chamberlinorum Caporiacco, 1951d: 37–38, figs 17a–b.
Type locality: Rancho Grande, Aragua, Venezuela.
Distribution: Venezuela.

Genus Heterolophus Tömösváry

Heterolophus Tömösváry, 1884: 24; J. C. Chamberlin, 1925b: 337–338; J. C.
   Chamberlin, 1929a: 60; Beier, 1932a: 28; J. C. Chamberlin & R. V. Chamberlin,
   1945: 32–33; Petrunkevitch, 1955: 81.
Type species: *Heterolophus guttiger* Tömösváry, 1884, by subsequent designation of
   J. C. Chamberlin, 1925b: 338.

Heterolophus australicus Beier

Heterolophus australicus Beier, 1969a: 171–172, fig. 1; Harvey, 1981b: 241; Harvey,
   1985b: 153.
Type locality: Mt Tamborine (as Jamborine), Joalah (as Toalah) National Park,
   Queensland, Australia.
Distribution: Australia (Queensland).

Family Tridenchthoniidae

Heterolophus clathratus (Tullgren)

Chthonius clathratus Tullgren, 1907b: 233–234, figs 11a–b; Ellingsen, 1912b: 88, 120.
Heterolophus clathratus (Tullgren): Beier, 1932a: 29, fig. 36; Roewer, 1937: 235; J. C. Chamberlin & R. V. Chamberlin, 1945: 35; Beier, 1958a: 159; Beier, 1966k: 455.
Not *Chthonius clathratus* Tullgren: Redikorzev, 1924a: 199 (misidentification; see *Tridenchthonius africanus* Beier).

Type localities: Pietermaritzburg (as Maritzburg), Natal, South Africa; and Lake Sibayi, Natal, South Africa.
Distribution: South Africa.

Heterolophus guttiger Tömösváry

Heterolophus guttiger Tömösváry, 1884: 24–25, figs 3–4; Beier, 1932a: 28, fig. 34; Roewer, 1937: 234; J. C. Chamberlin & R. V. Chamberlin, 1945: 34, fig. 8; Beier, 1962h: 131.
Chthonius guttiger (Tömösváry): Daday, 1888: 135, 191.

Type locality: Sao Paulo, Brazil.
Distribution: Argentina, Brazil (Sao Paulo).

Heterolophus nitens Tömösváry

Heterolophus nitens Tömösváry, 1884: 25–26, fig. 5; Roewer, 1937: 235; J. C. Chamberlin & R. V. Chamberlin, 1945: 34–35.
Heterolophus (?) *nitens* Tömösváry: Beier, 1932a: 29.
Chthonius nitens (Tömösváry): Daday, 1888: 135, 191.

Type locality: Sao Paulo, Brazil.
Distribution: Brazil (Sao Paulo).

Genus Neoditha Feio

Neoditha Feio, 1945: 11.

Type species: *Neoditha irusanga* Feio, 1945, by original designation.

Neoditha irusanga Feio

Neoditha irusanga Feio, 1945: 11–14, figs 1–5.

Type locality: Represa dos Ciganos, Jacarepaguá, Rio de Janeiro, Brazil.
Distribution: Brazil (Rio de Janeiro).

Genus Pycnodithella Beier

Pycnodithella Beier, 1947b: 287; Kennedy, 1989: 289.

Type species: *Verrucadithella abyssinica* Beier, 1944, by original designation.

Pycnodithella abyssinica (Beier)

Verrucadithella abyssinica Beier, 1944: 174–175, fig. 2.
Pycnodithella abyssinica (Beier): Beier, 1947b: 287.

Type locality: Mt Chillalo, Ethiopia.
Distribution: Ethiopia.

Pycnodithella harveyi Kennedy

Pycnodithella harveyi Kennedy, 1989: 291–295, figs a–w.

Type locality: Macquarie University, North Ryde, 13.5 km NW. of Sydney, New South Wales, Australia.
Distribution: Australia (New South Wales).

Genus **Sororoditha** J. C. Chamberlin & R. V. Chamberlin

Sororoditha J. C. Chamberlin & R. V. Chamberlin, 1945: 27–28.

Type species: *Chthonius hirsutus* Balzan, 1887a, by original designation.

Sororoditha hirsuta (Balzan)

Chthonius hirsutus Balzan, 1887a: no pagination, figs; Balzan, 1890: 448–449, figs 26, 26a; Balzan, 1892: 550.
Verrucaditha (?) *hirsuta* (Balzan): Beier, 1932a: 32; Roewer, 1937: 235.
Sororoditha hirsuta (Balzan): J. C. Chamberlin & R. V. Chamberlin, 1945: 28.

Type locality: Mato Grosso (as Mato-groso), Brazil.
Distribution: Brazil (Mato Grosso).

Genus **Tridenchthonius** Balzan

Tridenchthonius Balzan, 1887a: no pagination; Balzan, 1890: 450; Balzan, 1892: 505, 509; J. C. Chamberlin, 1925b: 334; Beier, 1932a: 34; J. C. Chamberlin & R. V. Chamberlin, 1945: 44–46; Hoff, 1963c: 28.
Afroditha Beier, 1930i: 47–48 (synonymized with *Ditha* by Beier, 1932a: 25; synonymized with *Tridenchthonius* by J. C. Chamberlin & R. V. Chamberlin, 1945: 44).

Type species: of *Tridenchthonius*: *Tridenchthonius parvulus* Balzan, 1887a, by monotypy.
of *Afroditha*: *Chthonius serrulatus* Silvestri, 1918, by original designation (based upon misidentified type species; an application is currently before the International Commission on Zoological Nomenclature).

Tridenchthonius addititius Hoff

Tridenchthonius addititius Hoff, 1950c: 535–536; Zeh, 1987b: 1086.

Type locality: Mount Coffee, Sao Tomé (as St Thomas Island).
Distribution: Ghana, Sao Tomé.

Tridenchthonius africanus (Beier)

Chthonius clathratus Tullgren: Redikorzev, 1924a: 199 (misidentification).
Ditha africana Beier, 1931b: 50; Beier, 1932a: 27, fig. 32; Roewer, 1937: 234; Helversen, 1966b: fig. 8a.
Tridenchthonius africanus (Beier): J. C. Chamberlin & R. V. Chamberlin, 1945: 50.

Type locality: Mabira, Tanzania.
Distribution: Tanzania.

Tridenchthonius beieri Mahnert

Tridenchthonius beieri Mahnert, 1983a: 149–153, figs 18–29.

Type locality: Wema, NE. of Garsen, Tana River, Kenya.
Distribution: Kenya.

Tridenchthonius brasiliensis Mahnert

Tridenchthonius brasiliensis Mahnert, 1979d: 725–726, figs 1–5; Adis, 1981: 119, etc.; Mahnert, 1985c: 218; Adis & Mahnert, 1986: 312; Mahnert & Adis, 1986: 214; Adis, Junk & Penny, 1987: 488; Mahnert, Adis, & Bührnheim, 1987: figs 3, 7c.

Type locality: Taruma Mirim, Manaus, Amazonas, Brazil.
Distribution: Brazil (Amazonas).

Family Tridenchthoniidae

Tridenchthonius buchwaldi (Tullgren)

Chthonius (Pseudochthonius) buchwaldi Tullgren, 1907a: 69−70, figs 23a−b.
Ditha buchwaldi (Tullgren): Beier, 1932a: 26, figs 20, 28; Beier, 1932g: fig. 176; Roewer, 1936: fig. 46; Roewer, 1937: 234.
Tridenchthonius buchwaldi (Tullgren): J. C. Chamberlin & R. V. Chamberlin, 1945: 49.
Ditha buchwaldti (sic) (Tullgren): Weidner, 1959: 115.

Type locality: Guayaquil, Ecuador.
Distribution: Ecuador.

Tridenchthonius cubanus (J. C. Chamberlin)

Chthonius pennsylvanicus Hagen: Berger, 1905: 407 (misidentification).
Chthonius longipalpis Banks: Banks, 1909c: 174 (misidentification).
Ditha cubana J. C. Chamberlin, 1929a: 61−62, figs 1h, 2e, 2j, 2l, 3b; J. C. Chamberlin, 1931a: figs 13e, 43c−d, 46i, 52b, 54; Beier, 1932a: 26; Beier, 1932g: fig. 251; Roewer, 1937: 234.
Tridenchthonius cubanus (J. C. Chamberlin): J. C. Chamberlin & R. V. Chamberlin, 1945: 52, figs 1, 2j; Hoff, 1963c: 29.

Type locality: Cayamas, Cuba.
Distribution: Cuba, Jamaica.

Tridenchthonius donaldi Turk

Tridenchthonius donaldi Turk, 1946: 64−67, figs 1−5.

Type locality: San Rafael, Trinidad.
Distribution: Trinidad.

Tridenchthonius gratus Hoff

Tridenchthonius gratus Hoff, 1963c: 29−33, figs 9−10; Zeh, 1987b: 1086.

Type locality: Johnson Mountain, 5 miles ENE. of Bath, Parish of St Thomas, Jamaica.
Distribution: Jamaica.

Tridenchthonius juxtlahuaca J. C. Chamberlin & R. V. Chamberlin

Tridenchthonius juxtlahuaca J. C. Chamberlin & R. V. Chamberlin, 1945: 53−57, figs 13a−k, 14a−f, 15o; Zeh, 1987b: 1086.

Type locality: Juxtlahuaca Cave, near Colotlipa, Mexico.
Distribution: Mexico.

Tridenchthonius mexicanus J. C. Chamberlin & R. V. Chamberlin

Tridenchthonius mexicanus J. C. Chamberlin & R. V. Chamberlin, 1945: 57−61, figs 15a−n; Hoff, 1946e: 4; Hoff, 1950c: 534; Reyes-Castillo & Hendrichs, 1975: 131; Beier, 1976c: 1; Muchmore, 1977: 63; Mahnert, 1985b: 75; Mahnert, 1985c: 218; Adis & Mahnert, 1986: 312; Mahnert & Adis, 1986: 214; Zeh, 1987b: 1086.

Type locality: El Potrero, Vera Cruz, Mexico.
Distribution: Brazil (Amazonas, Pará), Costa Rica, Mexico, Trinidad.

Tridenchthonius parvidentatus (Balzan)

Chthonius parvidentatus Balzan, 1887b: no pagination, figs; Balzan, 1890: 447−448, figs 25, 25a.
Chthonius (Chthonius) parvidentatus Balzan: Balzan, 1892: 547, fig. 36.
Ditha parvidentata (Balzan): Beier, 1932a: 25−26, fig. 30; Roewer, 1937: 234; Feio, 1945: 3; Helversen, 1966b: fig. 8c.
Tridenchthonius parvidentatus (Balzan): J. C. Chamberlin & R. V. Chamberlin, 1945: 49.

Not *Chthonius parvidentatus* Balzan: Ellingsen, 1910a: 401 (misidentification; see *Compsaditha fiebrigi* (Beier)).

Type locality: Rio Apa, Paraguay.
Distribution: Argentina, Paraguay, Venezuela.

Tridenchthonius parvulus Balzan

Tridenchthonius parvulus Balzan, 1887a: no pagination, figs; Balzan, 1890: 450–451, figs 28, 28a; Balzan, 1892: 550; J. C. Chamberlin, 1931a: 82, fig. 17d; Beier, 1932a: 34; Roewer, 1936: fig. 112d; Roewer, 1937: 235; J. C. Chamberlin & R. V. Chamberlin, 1945: 47–48.

Type locality: Rio Apa, Paraguay.
Distribution: Paraguay.

Tridenchthonius peruanus Beier

Tridenchthonius peruanus Beier, 1955m: 1–2, fig. 1; Weidner, 1959: 117.

Type locality: Sivia, Peru.
Distribution: Peru.

Tridenchthonius serrulatus (Silvestri)

Chtonius (sic) *serrulata* Silvestri, 1918: 294–296, plate 5, plate 6 figs 1–13 (in part; see *Compsaditha aburi* J. C. Chamberlin & R. V. Chamberlin).
Tridenchthonius serrulatus (Silvestri): J. C. Chamberlin & R. V. Chamberlin, 1945: 50–51, figs 12a–b; Beier, 1948b: 451, 456; Hoff, 1950c: 534; Vachon, 1952a: 20–21, figs 3–7; Zeh, 1987b: 1086.
Ditha serrulata (Silvestri): Roewer, 1937: 234; Vachon, 1940e: 68.
Not *Afroditha serrulata* (Silvestri): Beier, 1930i: 48, figs 6a–b (misidentification; see *Compsaditha aburi* J. C. Chamberlin & R. V. Chamberlin).
Not *Ditha serrulata* (Silvestri): Beier, 1932a: 27 (misidentification; see *Compsaditha aburi* J. C. Chamberlin & R. V. Chamberlin).

Type locality: Olokemeji, Nigeria.
Distribution: Guinea, Ivory Coast, Nigeria.

Tridenchthonius surinamus (Beier)

Ditha surinama Beier, 1931b: 49–50; Beier, 1932a: 26, fig. 31; Roewer, 1937: 234; Helversen, 1966b: fig. 8b.
Tridenchthonius surinamus (Beier): J. C. Chamberlin & R. V. Chamberlin, 1945: 49.

Type locality: Paramaribo, Surinam.
Distribution: Surinam.

Tridenchthonius trinidadensis Hoff

Tridenchthonius trinidadensis Hoff, 1946e: 1–4, figs 1–3.

Type locality: Navy Base, Trinidad.
Distribution: Trinidad.

Genus Typhloditha Beier

Typhloditha Beier, 1955a: 3–4.

Type species: *Typhloditha anophthalma* Beier, 1955a, by original designation.

Typhloditha anophthalma Beier

Typhloditha anophthalma Beier, 1955a: 4–5, fig. 1.

Type locality: Kenia River, Parc National Upemba, Zaire.
Distribution: Zaire.

Family Tridenchthoniidae

Typhloditha minima Beier

Typhloditha minima Beier, 1959d: 11–13, fig. 3; Mahnert, 1978f: 74–75.

Type locality: Bunia, Zaire.
Distribution: Congo, Zaire.

Typhloditha termitophila Beier

Typhloditha termitophila Beier, 1964k: 30–31.

Type locality: Wakfontein, E. of Pretoria, Transvaal, South Africa.
Distribution: South Africa.

Genus **Verrucaditha** J. C. Chamberlin

Alura J. C. Chamberlin, 1925b: 334 (junior homonym of *Alura* Möschler, 1884).
Verrucaditha J. C. Chamberlin, 1929a: 59; Beier, 1932a: 31; J. C. Chamberlin &
    R. V. Chamberlin, 1945: 22–24 (replacement name for *Alura* J. C. Chamberlin).

Type species: *Chthonius spinosus* Banks, 1893, by original designation.

Verrucaditha spinosa (Banks)

Chthonius spinosus Banks, 1893: 67; Banks, 1895a: 13; Banks, 1904a: 141; Coolidge,
    1908: 114.
Alura spinosa (Banks): J. C. Chamberlin, 1925b: 334.
Verrucaditha spinosa (Banks): J. C. Chamberlin, 1929a: 59, figs 1c, 2h; J. C.
    Chamberlin, 1931a: fig. 16a; Beier, 1932a: 31–32; Roewer, 1937: 235; J. C.
    Chamberlin & R. V. Chamberlin, 1945: 24–25, figs 2b–d, 2h–i, 5a–h; Hoff,
    1945a: 103–105, figs 1–3; Hoff, 1949b: 429–431, figs 12a–e; Hoff, 1950c: 534;
    Hoff & Bolsterli, 1956: 160; Hoff, 1958: 3; Weygoldt, 1969a: 64, 109, fig. 52;
    Muchmore, 1973g: 122; Brach, 1979: 34; Zeh, 1987b: 1086.
Verrucaditha megaloptera J. C. Chamberlin, 1929a: 59–60; J. C. Chamberlin, 1931a:
    figs 11e, 11m, 22d–f, 23g, 25b–c, 25h, 47y; Beier, 1932a: 32; Nester, 1932: 98;
    Roewer, 1937: 235 (synonymized by J. C. Chamberlin & R. V. Chamberlin,
    1945: 24).

Type localities: of *Chthonius spinosus*: Citrus County, Florida, U.S.A.
of *Verrucaditha megaloptera*: Bloomington, Indiana, U.S.A.
Distribution: U.S.A. (Florida, Illinois, Indiana, Louisiana, Mississippi, Missouri, North
    Carolina, Ohio).

Genus **Verrucadithella** Beier

Verrucadithella Beier, 1931b: 51–52; Beier, 1932a: 32; J. C. Chamberlin & R. V.
    Chamberlin, 1945: 19–21.

Type species: *Chthonius dilatimanus* Redikorzev, 1924a, by original designation.

Verrucadithella dilatimana (Redikorzev)

Chthonius dilatimanus Redikorzev, 1924a: 193–195, figs 7–9; J. C. Chamberlin,
    1931a: 128, fig. 35c.
Verrucadithella dilatimana (Redikorzev): Beier, 1931b: 52; Beier, 1932a: 33, fig. 38;
    Roewer, 1937: 235; J. C. Chamberlin & R. V. Chamberlin, 1945: 21, figs 4a–e;
    Vachon, 1949: fig. 203c; Beier, 1955c: 528; Beier, 1959d: 13; Beier, 1962f: 11;
    Mahnert, 1983a: 142–145, figs 1–11.
Verrucadithella (Verrucadithella) dilatimana (Redikorzev): Beier, 1944: 175.
Verrucadithella dilatimanus (sic) (Redikorzev): Zeh, 1987b: 1086.

Type locality: Molo, Kenya.
Distribution: Kenya, Ruanda, Zaire.

Verrucadithella francisi Feio

Verrucadithella francisi Feio, 1945: 14–16, figs 6a–c, 7.

Type locality: Represa da Covanca, Rio de Janeiro, Brazil.
Distribution: Brazil (Rio de Janeiro).

Verrucadithella jeanneli Beier

Verrucadithella jeanneli Beier, 1935c: 117–118, fig. 1; Roewer, 1940: 343; J. C. Chamberlin & R. V. Chamberlin, 1945: 22; Beier, 1955c: 528–529; Mahnert, 1983a: 142–145, figs 12–16.
Verrucadithella cannelli (sic) Beier: Vachon, 1937f: 128.
Verrucadithella (Verrucadithella) jeanneli Beier: Beier, 1944: 175.

Type locality: Koptawelil Valley, Mt Elgon, Uganda.
Distribution: Kenya, Uganda.

Verrucadithella sulcatimana Beier

Verrucadithella sulcatimana Beier, 1944: 173–174, fig. 1; Weidner, 1959: 117.

Type locality: Amani, Tanzania.
Distribution: Tanzania.

Superfamily FEAELLOIDEA Ellingsen

Feaelloidea Ellingsen: J. C. Chamberlin, 1931a: 230; Petrunkevitch, 1955: 82; Hoff, 1956c: 2–3; Muchmore, 1982a: 100.
Feaellides (sic) Beier, 1932a: 237; Beier, 1932g: 185; Roewer, 1937: 270.

Family FEAELLIDAE Ellingsen

Feaellidae Ellingsen, 1906: 259–260; J. C. Chamberlin, 1923a: 147–148; J. C. Chamberlin, 1931a: 232–233; Beier, 1932a: 240–241; Beier, 1932g: 185; Roewer, 1937: 272; Muchmore, 1982a: 100.
Feaellinae Ellingsen: J. C. Chamberlin, 1923a: 148–149.

Genus **Feaella** Ellingsen

Feaella Ellingsen, 1906: 260; J. C. Chamberlin, 1931a: 233; Beier, 1932a: 241.

Type species: *Feaella mirabilis* Ellingsen, 1906, by monotypy.

Subgenus **Feaella (Feaella)** Ellingsen

Feaella (Feaella) mirabilis Ellingsen

Feaella mirabilis Ellingsen, 1906: 260–263, figs 1–8; With, 1908b: 8–12, fig. 1, plate 1 figs 1–10; Ellingsen, 1912b: 86; J. C. Chamberlin, 1931a: 233; Beier, 1932a: 242, figs 268–270; Roewer, 1936: figs 10–11, 17, 41a–b, 63c, 78; Roewer, 1937: 272; Berland, 1955: 30, fig. 16; Heurtault-Rossi & Jezequel, 1965: 450–461, figs 1–15; Weygoldt, 1969a: 129; Mahnert, 1978f: 94; Mahnert, 1981c: fig. 1; Heurtault, 1983: 21–22, figs 48–49.
Feaella (Feaella) mirabilis Ellingsen: Beier, 1955c: 546.

Type locality: Bolama, Guinea-Bissau.
Distribution: Congo, Guinea-Bissau, Ivory Coast.

Feaella (Feaella) mombasica Beier

Feaella (Feaella) mombasica Beier, 1955c: 544–546, fig. 12; Mahnert, 1982b: 115.

Type locality: Bamburi Beach, Mombasa, Kenya.
Distribution: Kenya.

Subgenus **Feaella (Difeaella)** Beier

Feaella (Difeaella) Beier, 1966k: 463.

Type species: *Feaella (Difeaella) krugeri* Beier, 1966k, by original designation.

Feaella (Difeaella) krugeri Beier

Feaella (Difeaella) krugeri Beier, 1966k: 463–464, fig. 5; Lawrence, 1967: 89.

Type locality: Olifantspoort area, Kruger National Park, Transvaal, South Africa.
Distribution: South Africa.

Subgenus **Feaella (Tetrafeaella)** Beier

Feaella (Tetrafeaella) Beier, 1955c: 546.

Type species: *Feaella mucronata* Tullgren, 1907b, by present designation.

Feaella (Tetrafeaella) affinis Hirst

Feaella affinis Hirst, 1911: 394—395, fig. 11; Hirst, 1913: 37; J. C. Chamberlin, 1931a: 234; Beier, 1932a: 243; Roewer, 1937: 272; Mahnert, 1978d: 879—880, figs 21—23.
Feaella (Tetrafeaella) affinis Hirst: Beier, 1955c: 546.
Not *Feaella affinis* Hirst: Ellingsen, 1914: 13—14 (misidentification; see *Feaella (Tetrafeaella) indica* J. C. Chamberlin).

Type localities: Silhouette, Seychelles; and Praslin, Seychelles.
Distribution: Seychelles.

Feaella (Tetrafeaella) anderseni Harvey

Feaella (Tetrafeaella) anderseni Harvey, 1989b: 41—43, figs 1—13.

Type locality: Cape Bougainville, Western Australia, Australia.
Distribution: Australia (Western Australia).

Feaella (Tetrafeaella) capensis Beier

Feaella capensis Beier, 1955l: 303—305, figs 23—24.
Feaella (Tetrafeaella) capensis Beier: Beier, 1955c: 546.
Not *Feaella capensis* Beier: Beier, 1964k: 63 (misidentification; see *Feaella (Tetrafeaella) capensis nana* Beier).

Type locality: Cape Point, Cape Peninsula, Cape Province, South Africa.
Distribution: South Africa.

Feaella (Tetrafeaella) capensis nana Beier

Feaella capensis Beier: Beier, 1964k: 63 (misidentification).
Feaella (Tetrafeaella) capensis nana Beier, 1966k: 461—463, fig. 4; Lawrence, 1967: 89.

Type locality: Hlanganine Spruit, near Letaba, Kruger National Park, Transvaal, South Africa.
Distribution: South Africa, Zimbabwe.

Feaella (Tetrafeaella) indica J. C. Chamberlin

Feaella affinis Hirst: Ellingsen, 1914: 13—14 (misidentification).
Feaella indica J. C. Chamberlin, 1931a: 234, figs 1p, 6f, 7c—g, 9k, 11d, 14f, 16m, 18h, 20a, 29g, 38c—d, 64; Beier, 1932a: 243; Beier, 1932g: figs 165, 260; Roewer, 1936: figs 6, 13; Roewer, 1937: 272; Vachon, 1949: figs 200c, 201, 203e, 204d; Weygoldt, 1969a: figs 4a—b; Beier, 1973b: 46—47; Beier, 1974e: 1008.
Feaella (Tetrafeaella) indica J. C. Chamberlin: Beier, 1955c: 546; Murthy & Ananthakrishnan, 1977: 112—114.

Type locality: Jhenidah, Jessore District, India.
Distribution: India, Sri Lanka.

Feaella (Tetrafeaella) leleupi Beier

Feaella (Tetrafeaella) leleupi Beier, 1959d: 34—36, fig. 16.

Type locality: Ile de M'Boko, Kivu, Lake Tanganika, Zaire.
Distribution: Zaire.

Feaella (Tetrafeaella) mucronata Tullgren

Feaella mucronata Tullgren, 1907b: 226–228, figs 6a–g; Ellingsen, 1912b: 86, 104–105; J. C. Chamberlin, 1931a: 234; Beier, 1932a: 242–243, fig. 271; Beier, 1932g: fig. 171; Roewer, 1936: fig. 29; Roewer, 1937: 272; Beier, 1964k: 63.
Feaella (Tetrafeaella) mucronata Tullgren: Beier, 1955c: 546; Beier, 1958a: 171.

Type locality: Amanzimtoti, Natal, South Africa.
Distribution: South Africa.

Feaella (Tetrafeaella) parva Beier

Feaella parva Beier, 1947b: 321–322, figs 27–28.
Feaella (Tetrafeaella) parva Beier: Beier, 1955c: 546.

Type locality: Vioolsdrift (as Viodsdrift), Orange River, Namaqualand, Cape Province, South Africa.
Distribution: South Africa.

Feaella (Tetrafeaella) perreti Mahnert

Feaella (Tetrafeaella) perreti Mahnert, 1982b: 115–118, figs 1–8.

Type locality: Makadara Forest, Shimba Hills, Kenya.
Distribution: Kenya.

Family PSEUDOGARYPIDAE J. C. Chamberlin

Pseudogarypinae J. C. Chamberlin, 1923a: 161.
Pseudogarypidae J. C. Chamberlin: J. C. Chamberlin, 1931a: 230; Beier, 1932a: 239; Beier, 1932g: 185; Roewer, 1937: 271; Petrunkevitch, 1955: 82; Benedict & Malcolm, 1978b: 82–85; Muchmore, 1982a: 99; Harvey, 1985b: 150.

Genus Neopseudogarypus J. C. H. Morris

Neopseudogarypus J. C. H. Morris, 1948b: 43.

Type species: *Neopseudogarypus scutellatus* J. C. H. Morris, 1948b, by original designation.

Neopseudogarypus scutellatus J. C. H. Morris

Neopseudogarypus scutellatus J. C. H. Morris, 1948b: 44–46, figs 1–20; Benedict & Malcolm, 1978b: 83, 85, figs 1–3; Harvey, 1981a: 244; Harvey, 1985b: 150.

Type locality: near Launceston, Tasmania, Australia.
Distribution: Australia (Tasmania).

Genus Pseudogarypus Ellingsen

Pseudogarypus Ellingsen, 1909b: 217–218; J. C. Chamberlin, 1931a: 231; Beier, 1932a: 239; Hoff, 1961: 442–443; Benedict & Malcolm, 1978b: 85–89.
Cerogarypus Jacot, 1938: 301 (synonymized by Benedict & Malcolm, 1978b: 85).

Type species: of *Pseudogarypus*: *Garypus bicornis* Banks, 1895a, by original designation.
of *Cerogarypus*: *Cerogarypus agassizi* Jacot, 1938 (junior synonym of *Pseudogarypus bicornis* (Banks, 1895a)), by original designation.

Pseudogarypus banksi Jacot

Pseudogarypus banksi Jacot, 1938: 302–303, figs c, e; Hoff, 1958: 18; Weygoldt,
   1969a: 119; Muchmore, 1973g: 123; Benedict & Malcolm, 1978b: 92–95, figs 4,
   8–10, 14–20; Zeh, 1987b: 1086.
Pseudogarypus hesperus J. C. Chamberlin: Manley, 1969: 7, fig. 6 (misidentification).
Pseudogarypus sp.: Nelson, 1975: 283, figs 32–34.

Type locality: Gale River Experimental Forest, Pierce Bridge, New Hampshire,
   U.S.A.
Distribution: Canada (Quebec), U.S.A. (Michigan, New Hampshire, New York).

Pseudogarypus bicornis (Banks)

Garypus bicornis Banks, 1895a: 8–9; Coolidge, 1908: 112.
Pseudogarypus bicornis (Banks): Ellingsen, 1909b: 218; Banks, 1911: 637; Moles &
   Moore, 1921: 8; J. C. Chamberlin, 1923a: 162–166, figs 1–22; J. C. Chamberlin,
   1931a: 231, 232, figs 6e, 9l, 9n, 11c, 14e, 15q, 18b, 24d, 29j, 38k, 41a–d, 42k–l,
   51a–b; Beier, 1932a: 239–240, figs 266–267; Roewer, 1936: figs 81, 87, 138;
   Roewer, 1937: 271; Hoff, 1946d: 198–200, figs 4–5; Hoff, 1958: 17–18; Hoff,
   1961: 443–444; Benedict & Malcolm, 1978b: 95–100, figs 6, 11, 21–23; Papp,
   1978: 325; Zeh, 1987b: 1086.
Cerogarypus agassizi Jacot, 1938: 301; Hoff, 1958: 18 (synonymized by Benedict &
   Malcolm, 1978b: 95).

Type locality: of *Garypus bicornis*: Specimen Ridge, Yellowstone National Park,
   Wyoming, U.S.A.
of *Cerogarypus agassizi*: Bear Lake, Utah.
Distribution: U.S.A. (Arizona, California, Colorado, Idaho, Oregon, Utah,
   Washington, Wyoming).

! Pseudogarypus extensus Beier

Pseudogarypus extensus Beier, 1937d: 306–307, fig. 7; Roewer, 1940: 331; Schawaller,
   1978: 3.

Type locality: Baltic Amber.
Distribution: Baltic Amber.

! Pseudogarypus hemprichii (C. L. Koch & Berendt)

Chelifer hemprichii C. L. Koch & Berendt, 1854: 94–95, fig. 94; Menge, 1855: 34,
   plate 4 fig. 8; Hagen, 1870: 266; Beier, 1932e: 278.
Pseudogarypus hemprichii (C. L. Koch & Berendt): Beier, 1937d: 306, fig. 6; Roewer,
   1940: 331; Petrunkevitch, 1955: fig. 49(3); Schawaller, 1978: 3.

Type locality: Baltic Amber.
Distribution: Baltic Amber.

Pseudogarypus hesperus J. C. Chamberlin

Pseudogarypus hesperus J. C. Chamberlin, 1931a: 232, figs 20b, 21h, 26d, 38a–b,
   63; Beier, 1932a: 240; Beier, 1932g: fig. 259; Roewer, 1936: fig. 9; Roewer, 1937:
   271; Vachon, 1949: figs 199d, 240; Hoff, 1958: 18; Benedict & Malcolm, 1978b:
   100–103, figs 5, 12–13, 24–29; Zeh, 1987b: 1086.
Not *Pseudogarypus hesperus* J. C. Chamberlin: Manley, 1969: 7, fig. 6 (misidentification; see *Pseudogarypus banksi* Jacot).

Type locality: Puyallup, Washington, U.S.A.
Distribution: U.S.A. (Oregon, Washington).

Pseudogarypus hypogeus Muchmore

Pseudogarypus hypogeus Muchmore, 1981a: 59–60, figs 15–16.

Type locality: Doney Fissure, Wupatki National Monument, Coconino County, Arizona, U.S.A.
Distribution: U.S.A. (Arizona).

! Pseudogarypus minor Beier

Pseudogarypus minor Beier, 1947e: 193–194, figs 1–2; Schawaller, 1978: 3.

Type locality: Baltic Amber.
Distribution: Baltic Amber.

Pseudogarypus orpheus Muchmore

Pseudogarypus orpheus Muchmore, 1981a: 56–59, figs 13–14.

Type locality: Music Hall Cave, 5 miles E. of Parrots Ferry, Calaveras County, California, U.S.A.
Distribution: U.S.A. (California).

Pseudogarypus spelaeus Benedict & Malcolm

Pseudogarypus spelaeus Benedict & Malcolm, 1978b: 103, figs 7, 30–32; Benedict & Malcolm, 1978c: 91; Zeh, 1987b: 1086.

Type locality: Samwell Cave, approx. 20 miles NE. of Redding, California, U.S.A.
Distribution: U.S.A. (California).

Superfamily GARYPOIDEA E. Simon

Garypoidea E. Simon: J. C. Chamberlin, 1930: 585; J. C. Chamberlin, 1931a:222; Hoff, 1949b: 446; Petrunkevitch, 1955: 82; Hoff, 1956b: 26; Morikawa, 1960: 124–125; Hoff, 1964a: 18; Murthy & Ananthakrishnan, 1977: 29; Muchmore, 1982a: 98–99.
Garypides (sic) E. Simon: Beier, 1932a: 176; Beier, 1932g: 183; Roewer, 1937: 257–258.

Family GARYPIDAE E. Simon

Garypinae E. Simon, 1879a: 42; Balzan, 1892: 534; Redikorzev, 1924b: 24; J. C. Chamberlin, 1930: 612; J. C. Chamberlin, 1931a: 228; Beier, 1932a: 215; Roewer, 1937: 267; Hoff, 1956b: 44; Morikawa, 1960: 131; Hoff, 1964a: 40; Murthy & Ananthakrishnan, 1977: 101.
Garypina E. Simon: Tömösváry, 1882b: 208–209.
Garypidae E. Simon: H.J. Hansen, 1893: 231–232; Ellingsen, 1904: 2; With, 1906: 89–95; J. C. Chamberlin, 1930: 608; J. C. Chamberlin, 1931a: 226–227; Beier, 1932a: 215; Beier, 1932g: 184; J. C. Chamberlin, 1935b: 480; Roewer, 1937: 266; Hoff, 1949b: 447; Petrunkevitch, 1955: 82; Hoff, 1956b: 44; Morikawa, 1960: 130–131; Beier, 1963b: 237; Hoff, 1964a: 39–40; Murthy & Ananthakrishnan, 1977: 101; Benedict & Malcolm, 1978a: 113–114; Muchmore, 1979a: 211; Muchmore, 1982a: 99; Harvey, 1985b: 142.
Synsphyronidae Beier, 1932a: 238; Beier, 1932g: 185; Roewer, 1937: 271.

Genus **Ammogarypus** Beier
Ammogarypus Beier, 1962d: 226.
Type species *Ammogarypus lawrencei* Beier, 1962d, by original designation.

Ammogarypus kalaharicus Beier
Ammogarypus kalaharicus Beier, 1964k: 57–59, fig. 19; Beier, 1973a: 99.
Type locality: Auob (as Ouob) River, Gemsbok Plain, Cape Province, South Africa.
Distribution: Namibia, South Africa.

Ammogarypus lawrencei Beier
Ammogarypus lawrencei Beier, 1962d: 227–228, figs 4–5; Beier, 1973a: 99.
Type locality: Natab, Namibia.
Distribution: Namibia.

Ammogarypus minor Beier
Ammogarypus minor Beier, 1973a: 99, fig. 1.
Type locality: 75 km SE. of Gobabeb, Namibia.
Distribution: Namibia.

Genus **Anagarypus** J. C. Chamberlin
Anagarypus J. C. Chamberlin, 1930: 616; Beier, 1932a: 225; Muchmore, 1982e: 159.
Type species: *Anagarypus oceanusindicus* J. C. Chamberlin, 1930, by original designation.

Anagarypus australianus Muchmore

Anagarypus australianus Muchmore, 1982e: 160–162, figs 1–6; Tenorio & Muchmore, 1982: 382; Harvey, 1985b: 142.

Type locality: Nymph Island, Great Barrier Reef, Queensland, Australia.
Distribution: Australia (Northern Territory, Queensland).

Anagarypus heatwolei Muchmore

Anagarypus heatwolei Muchmore, 1982e: 162–163, fig. 7; Tenorio & Muchmore, 1982: 383; Harvey, 1985b: 142.

Type locality: Barrow Island, Western Australia, Australia.
Distribution: Australia (Western Australia).

Anagarypus oceanusindicus J. C. Chamberlin

Anagarypus oceanus-indicus J. C. Chamberlin, 1930: 615–616, figs 1a, 1g, 1l, 1q, 1r, 1y, 1ff, 2e, 2i, 2p, 2mm, 2nn, 3k, 3l; J.C. Chamberlin, 1931a: figs 15p, 37t–v, 47v, 52d; Beier, 1932a: 225; Roewer, 1937: 268; Spaull, 1979: 114, 117; Beier, 1981: 293–294, fig. 1.

Anagarypus oceanusindicus J. C. Chamberlin: Muchmore, 1982e: 160.

Type locality: Takamaka Island, Chagos Archipelago.
Distribution: Aldabra Islands, Andaman Islands, Chagos Archipelago.

Genus **Archeolarca** Hoff & Clawson

Archeolarca Hoff & Clawson, 1952: 2–3; Hoff, 1956b: 44; Benedict & Malcolm, 1978a: 118.

Type species: *Archeolarca rotunda* Hoff & Clawson, 1952, by original designation.

Archeolarca aalbui Muchmore

Archeolarca aalbui Muchmore, 1984f: 174–175, figs 4–5.

Type locality: Mitchell Caverns, Mitchell Caverns State Park, San Bernardino County, California, U.S.A.
Distribution: U.S.A. (California).

Archeolarca cavicola Muchmore

Archeolarca cavicola Muchmore, 1981a: 55–56, figs 11–12.

Type locality: Cave of the Domes, Grand Canyon National Park, Coconino County, Arizona, U.S.A.
Distribution: U.S.A. (Arizona).

Archeolarca guadalupensis Muchmore

Archeolarca guadalupensis Muchmore, 1981a: 54–55, figs 9–10.

Type locality: Lower Sloth Cave, Guadalupe Mountains National Park, Culberson County, Texas, U.S.A.
Distribution: U.S.A. (Texas).

Archeolarca rotunda Hoff & Clawson

Archeolarca rotunda Hoff & Clawson, 1952: 3–8, figs 1–4; Hoff, 1956b: 44–46; Hoff, 1958: 15; Hoff, 1959b: 4, etc.; Muchmore, 1971a: 90, 92; Benedict & Malcolm, 1978a: 118–119; Muchmore, 1981a: 52; Zeh, 1987b: 1086.

Type locality: American Fork Canyon, Utah County, Utah, U.S.A.
Distribution: U.S.A. (New Mexico, Oregon, Utah, Wyoming).

Archeolarca welbourni Muchmore

Archeolarca welbourni Muchmore, 1981a: 52–54, figs 7–8.

Type locality: Malmquist Fissure, Wupatki National Monument, Coconino County, Arizona, U.S.A.

Distribution: U.S.A. (Arizona).

Genus **Elattogarypus** Beier

Elattogarypus Beier, 1964k: 54–55.

Type species: *Elattogarypus cruciatus* Beier, 1964k, by original designation.

Elattogarypus cruciatus Beier

Elattogarypus cruciatus Beier, 1964k: 55–56, fig. 17.

Type locality: South Africa.

Distribution: South Africa.

Elattogarypus somalicus Mahnert

Elattogarypus somalicus Mahnert, 1984b: 51–55, figs 15–29.

Type locality: Sar Uanle, Somalia.

Distribution: Somalia.

Genus **Eremogarypus** Beier

Eremogarypus Beier, 1955l: 294–295.

Type species: *Eremogarypus gigas* Beier, 1955l, by original designation.

Eremogarypus eximius Beier

Eremogarypus eximius Beier, 1973: 100, fig. 3.

Type locality: Swartzbank, Namibia.

Distribution: Namibia.

Eremogarypus gigas Beier

Eremogarypus gigas Beier, 1955l: 295–296, figs 13–15; Beier, 1973a: 100.

Type locality: Khumib-Tal, 110 miles SW. of Ohopoho, Namibia.

Distribution: Namibia.

Eremogarypus perfectus Beier

Eremogarypus perfectus Beier, 1962d: 228–229, figs 6–7; Beier, 1973a: 99.

Type locality: Swartbankbergen, Namibia.

Distribution: Namibia.

Eremogarypus trichoideus Beier

Eremogarypus trichoideus Beier, 1973a: 99–100, fig. 2.

Type locality: Bullsport, Namibia.

Distribution: Namibia.

Genus **Garypus** L. Koch

Garypus L. Koch, 1873: 38; E. Simon, 1879a: 45–46; Tömösváry, 1882b: 211–212; H. J. Hansen, 1884: 549–550; Canestrini, 1885: no. 6; Balzan, 1887a: no pagination; Balzan, 1890: 441; Tullgren, 1899a: 179; Banks, 1891: 163; Balzan, 1892: 534–535; With, 1906: 101–103; J. C. Chamberlin, 1930: 612; Beier, 1932a: 216; J. C. Chamberlin, 1935b: 480; Beier, 1939f: 193; Morikawa, 1960: 131; Beier, 1963b: 238; Murthy & Ananthakrishnan, 1977: 102; V. F. Lee, 1979a: 4–5; Muchmore, 1979a: 211.

Type species: Garypus litoralis L. Koch, 1873 (junior synonym of *Chelifer beauvoisii* Audouin, 1826), by subsequent designation of E. Simon, 1879a: 45.

Garypus armeniacus Redikorzev

Garypus armeniacus Redikorzev, 1926: 1–3, figs 1–2; Beier, 1932a: 221–222, fig. 249; Roewer, 1937: 268.

Type locality: Dzhul'fa (as Dzhulfa), Armenia, U.S.S.R.
Distribution: U.S.S.R. (Armenia).

Garypus beauvoisii (Audouin)

Chelifer beauvoisii Audouin, 1826: 175, plate 8 fig. 6; Audouin, 1827: 414, plate 8 fig. 6; Gervais, 1844: 83, plate 25 fig. 3.
Chelifer bravaisii Gervais: Dugès & Edwards, 1836: plate 20 bis fig. 3.
Chelifer bravaisii Gervais, 1842: xlvi; Gervais, 1844: 84; Hagen, 1870: 264 (synonymized by E. Simon, 1885c: 49).
Obisium bravaisii (Gervais): Lucas, 1849: 277.
Obisium beauvoisii (Audouin): Hagen, 1870: 269.
Garypus litoralis L. Koch, 1873, 40–41; E. Simon, 1879a: 48, plate 17 figs 5, 7, plate 19 fig. 1; E. Simon, 1884: ccxxxii; Pocock, 1893: 6; E. Simon, 1898d: 21; Lankester, 1904: figs 70–72 (synonymized by E. Simon, 1885c: 49).
Garypus bravaisi (Gervais): E. Simon, 1878a: 152.
Garypus beauvoisi (Savigny): E. Simon, 1879a: 48; E. Simon, 1881b: 13; E. Simon, 1885c: 49; Ellingsen, 1909a: 207, 209; Ellingsen, 1910a: 387; Krausse-Heldrungen, 1912: 65; Nonidez, 1917: 37; Navás, 1925: 112; Beier, 1929a: 358; Beier, 1929b: 79; J. C. Chamberlin, 1930: 613; Beier, 1930h: 95; Beier, 1932a: 218–219, fig. 245 (in part; see *Garypus levantinus* Navás); Beier, 1932g: figs 203–204; Roewer, 1936: figs 26, 34b, 35; Vachon, 1936a: 78; Roewer, 1937: 268; Vachon, 1938a: 68–72, figs 39–40, 59a; Beier, 1939a: fig. 166c; Beier, 1939f: 193; Beier, 1948b: 457; Vachon, 1949: figs 217, 222a, 226; Beier, 1963b: 239, fig. 243; Beier, 1963f: 192; Beier, 1965d: 633; Beier, 1966f: 345; Helversen, 1966b: figs 3c, 4k, 6c; Lazzeroni, 1969c: 239; Weygoldt, 1969a: 27, fig. 104; Ćurčić, 1974a: 25; Mahnert, 1975c: 195; Estany, 1977b: 30; Estany, 1979a: 221–222, fig. 1; Mahnert, 1982f: 297; Schwaller, in Schmalfuss & Schwaller, 1984: 4.
Garypus littoralis L. Koch: Becker, 1880b: cxli; Daday, 1889d: 80; Berland, 1932: figs 51–52, 63, 69; Pieper, 1980: 263.
Garypus beauvoisii (Audouin): J. C. Chamberlin, 1931a: fig. 1k.
Garypus beauvoisi (Savigny)?: Hadži, 1933b: 183–192, figs 32a–c, 33a–e, 34a–b; Hadži, 1933c: 196–199.

Type localities: of *Chelifer beauvoisii*: Egypt.
of *Chelifer bravaisii*: Algeria.
of *Garypus litoralis*: Bonifacio, Corsica.
Distribution: Algeria, Balearic Islands, Crete, Egypt, Canary Islands, Corsica, France, Greece, Ilhas Selvagens, Israel, Italy, Malta, Libya, Sardinia, Spain, Tunisia, Yugoslavia.

Garypus bonairensis Beier

Garypus bonairensis Beier, 1936a: 444–446, fig. 2; Roewer, 1940: 345.
Garypus bonairensis bonairensis Beier: Wagenaar-Hummelinck, 1948: 31–38, figs 4a–c, 5a–c, 6a–d, 7a–g, 8a–b, 8d–g, 9a–k, 10a–e, 11a–b, 12a–d, plate 2.

Type locality: Gotomeer (as Lagoen di Goto), Bonaire.
Distribution: Bonaire.

Garypus bonairensis realini Wagenaar-Hummelinck

Garypus bonairensis realini Wagenaar-Hummelinck, 1948: 39, figs 8c, 9g, 10f–i, plates 1–2.

Type locality: near Boca Grandi, Seroe Plat, Aruba.
Distribution: Aruba.

Garypus bonairensis withi Hoff

Garypus floridensis Banks: With, 1906: figs 9a–b; With, 1907: 70–72, figs 26–28 (misidentifications).
Garypus withi Hoff, 1946d: 198.
Garypus bonairensis withi Hoff: Wagenaar-Hummelinck, 1948: 40–45, figs 9h–i, 13a–c, plate 2.

Type locality: Mustique Island.
Distribution: Mustique Island.

! Garypus burmiticus Cockerell

Garypus burmiticus Cockerell, 1920: 274, fig. 1; Schawaller, 1978: 3.

Type locality: Burmese Amber.
Distribution: Burmese Amber.

Garypus californicus Banks

Garypus californicus Banks, 1909b: 305; Banks, 1911: 635, fig. 210b; J. C. Chamberlin, 1921: 190, figs a–d; Moles & Moore, 1921: 8; J. C. Chamberlin, 1925a: 330; J. C. Chamberlin, 1930: 613, figs 2j, 2o, 2s, 2z, 3p, 3s; J. C. Chamberlin, 1931a: figs 3, 4c, 6c, 8b, 11p, 14c–d, 16n, 17w, 19j, 21d, 24a, 26a, 29f, 40u, 41e–f, 45o, 46l, 47f, 47t, 50f; Beier, 1932a: 219–220, fig. 246; Beier, 1932g: figs 159, 185; J. C. Chamberlin, 1935b: 480; Roewer, 1937: 268; Hoff, 1958: 15; Firstman, 1973: 16; Schulte, 1976: 119–123, figs 1–2; V. F. Lee, 1979a: 4–7, figs 2, 4, 6–8, 30; V. F. Lee, 1979b: 13; Poinar, Thomas & V. F. Lee, 1985: 400–401, fig. 1; Zeh, 1987b: 1086.

Type localities: Palo Alto, California, U.S.A.; and San Nicolas Island, California, U.S.A.
Distribution: Mexico, U.S.A. (California).

Garypus floridensis Banks

Garypus floridensis Banks, 1895a: 9; Banks, 1904a: 141; Coolidge, 1908: 112; J. C. Chamberlin, 1921: 191; J. C. Chamberlin, 1930: 612; Beier, 1932a: 220; J. C. Chamberlin, 1935b: 480; Roewer, 1936: figs 22e, 70l, 112h; Roewer, 1937: 268; Hoff, 1946d: 195–198, figs 1–3; Wagenaar-Hummelinck, 1948: 45–46; Hoff, 1958: 15; Muchmore, 1979a: 211; Zeh, 1987b, 1086.
Not *Garypus floridensis* Banks: With, 1906: figs 9a–b; With, 1907: 70–72, figs 26–28 (misidentifications; see *Garypus bonairensis withi* Hoff).

Type locality: near St Lucie River, Florida, U.S.A.
Distribution: U.S.A. (Florida).

Garypus giganteus J. C. Chamberlin

Garypus giganteus J. C. Chamberlin, 1921: 186–190, figs a–g; J. C. Chamberlin, 1923c: 360–361; J. C. Chamberlin, 1930: 613–614; Beier, 1932a: 218; Roewer, 1937: 268; V. F. Lee, 1979a: 8–10, figs 11–12, 30.

Type locality: Turtle Bay, Baja California, Mexico.
Distribution: Mexico.

Garypus gracilis V. F. Lee

Garypus gracilis V.F. Lee, 1979a: 15–18, figs 5, 17–21, 30.

Type locality: NW. side of Isla Danzante, Baja California, Mexico.
Distribution: Mexico.

Family Garypidae

Garypus guadalupensis J. C. Chamberlin

Garypus guadalupensis J. C. Chamberlin, 1930: 614–615; J. C. Chamberlin, 1931a:
figs 8a, 33d, 37p–q; Beier, 1932a: 221; Roewer, 1937: 268; V. F. Lee, 1979a: 7–8,
figs 9, 10, 30.

Type locality: Jack's Bay, Guadalupe Island, Baja California, Mexico.
Distribution: Mexico.

Garypus insularis Tullgren

Garypus insularis Tullgren 1907b: 62–63, figs 19a–b; Ellingsen, 1912b: 87; Ellingsen,
1914: 13; Beier, 1932a: 221, fig. 248; Roewer, 1937: 268; Weidner, 1959: 115.
Garypus sp. vic. *insularis* Tullgren: Beier, 1958a: 167.

Type locality: Seychelles.
Distribution: India, Seychelles, South Africa.

Garypus japonicus Beier

Garypus sp.: Morikawa, 1952b: 256–257, figs 13, 14a–b, 15.
Garypus japonicus Beier, 1952b: 235–238, fig. 1; Takashima, 1955: 176–178, figs
1–6; Morikawa, 1958: 8; Morikawa, 1960: 131–132, plate 3 fig. 1, plate 4 figs 7–8,
plate 6 fig. 12, plate 10 fig. 19; Morikawa, 1962: 419; Makioka, 1968: 208–222,
figs 1–6; Makioka, 1969: 113–118, figs 1a–c, plate I figs 1–6, plate II figs 1–8;
Makioka, 1976: 9–11, figs 1–9; Makioka, 1977: 186–196, figs 1–5; Sato, 1977a:
42–44, figs 1–2; Sato, 1977b: 36, fig. 1; Sato, 1978d: 45–48; Makioka, 1979:
72–80, figs 1–3; Sato, 1979a: plates 10, 19; Sato, 1979d: 44; Sato, 1980a: 60–61,
fig. 2; Sato, 1980d: plate 5; Sato, 1982c: 33.

Type locality: Enoshima Island, Japan.
Distribution: Japan.

Garypus krusadiensis Murthy & Ananthakrishnan

Garypus krusadiensis Murthy & Ananthakrishnan, 1977: 102–104, figs 34a–b.

Type locality: Krusadi Islands, Tamil Nadu, India.
Distribution: India.

Garypus levantinus Navás

Garypus levantinus Navás, 1925: 112–113, figs 10a–b; Beier, 1939f: 194, fig. 21;
Roewer, 1940: 345; Beier, 1953d: 38; Beier, 1961g: 411; Schuster, 1962: 379 (not
seen); Beier, 1963b: 238–239, fig. 242; Beier, 1963f: 192; Beier, 1965d: 633;
Weygoldt, 1969a: 23, 27; Gardini, 1975: 6–8; Callaini, 1979b: 137–138; Gardini,
1980c: 116, 121; Callaini, 1982a: 17, 21–22, fig. 3c; Pieper, 1981: 3–4; Callaini,
1983a: 151; Callaini, 1983c: 297–298; Callaini & Dallai, 1984: 336–341, fig. 4;
Callaini & Dallai, 1989: 90, fig. 12.
Garypus beauvoisi (Savigny): Beier, 1932a: 218–219 (in part).
Garypus baronii Lazzeroni, 1970d: 43–49, figs 3a–g, 4a–b, 5a–c (synonymized by
Gardini, 1975: 6).

Type localities: of *Garypus levantinus*: Cabo de Creus, Cadaqués, Gerona, Spain.
of *Garypus baronii*: Scoglio d'Affrica, Toscana, Italy.
Distribution: Balearic Islands, Greece, Israel, Italy, Sardinia, Spain.

Garypus longidigitus Hoff

Garypus longidigitus Hoff, 1947a: 41–44, figs 7–8; Harvey, 1981b: 241; Harvey,
1985b: 142.

Type locality: Weier, Murray Islands, Torres Strait, Queensland, Australia.
Distribution: Australia (Queensland).

Garypus maldivensis Pocock

Garypus maldivensis Pocock, 1904: 798–799, figs 1a–e; J. C. Chamberlin, 1930: 613, figs 1s, 1z; Beier, 1932a: 222; Roewer, 1937: 268; Beier, 1973b: 45; Mahnert, 1982d: 310, fig. 9; Harvey, 1988b: 324–327, figs 40–49.
Goryphus (sic) *maldivensis* Pocock: Thornton & New, 1988: 499–500.

Type locality: Midu Atoll, Maldive Islands.
Distribution: Indonesia (Krakatau Islands), Maldive Islands, Sri Lanka.

Garypus marmoratus Mahnert

Garypus marmoratus Mahnert, 1982d: 307–310, figs 1–8; Mahnert, 1984b: 50.

Type locality: Diani Beach, S. of Mombasa, Ukanda, Kenya.
Distribution: Kenya, Somalia.

Garypus nicobarensis Beier

Garypus nicobarensis Beier, 1930a: 204–205, figs 2–3; Beier, 1932a: 220, fig. 247; Roewer, 1937: 268.

Type locality: Camorta (as Kamorta), Nicobar Islands.
Distribution: Nicobar Islands.

Garypus occultus Mahnert

Garypus occultus Mahnert, 1982d: 310–313, figs 10–17; Mahnert, 1984b: 50–51.

Type locality: S. of Mpekatoni, Lamu, Kenya.
Distribution: Kenya, Somalia.

Garypus ornatus Beier

Garypus ornatus Beier, 1957d: 20–21, figs 8a–c.

Type locality: Bokonfuaaku Island, Bikini, Marshall Islands.
Distribution: Marshall Islands.

Garypus pallidus J. C. Chamberlin

Garypus pallidus J. C. Chamberlin, 1923c: 362–363, plate 3, fig. 7; J. C. Chamberlin, 1930: 613; Beier, 1932a: 218; Roewer, 1937: 268; V. F. Lee, 1979a: 10–12, figs 13–14, 30.

Type locality: Gordas Point, Ceralbo Island, Gulf of California, Mexico.
Distribution: Mexico.

Garypus saxicola Waterhouse

Garypus saxicola Waterhouse, 1878: 182, fig. 1; Bouvier, 1896: 304–307, 342–343; Richard & Neuville, 1897: 82; With, 1906: fig. 15; Nonidez, 1917: 27; Navás, 1921: 166; Navás, 1925: 113; J. C. Chamberlin, 1930: 612; Beier, 1932a: 219; Roewer, 1936: fig. 70k: Roewer, 1937: 268; Vachon, 1937f: 128; Beier, 1939f: 193; Vachon, 1940f: 2; Vachon, 1940g: 144; Beier, 1963b: 239–240.
Garypus sp.: Nonidez, 1917: 27–31, figs 6, 7a–b.
Garypus dubius Navás, 1925: 113 (synonymized by Beier, 1932a: 219).

Type localities: of *Garypus saxicola*: north Spain.
of *Garypus dubius*: San Vicente de la Barquera, Santander, Spain.
Distribution: Spain, Portugal.

Garypus saxicola salvagensis Helversen

Garypus saxicola salvagensis Helversen, 1965: 96–98, fig. 1; Beier, 1976a: 24.

Type locality: Ilheu de Fora, Ilhas Selvagens.
Distribution: Ilhas Selvagens.

Garypus sini J. C. Chamberlin

Garypus sini J. C. Chamberlin, 1923c: 361–362, plate 2, fig. 20; J. C. Chamberlin, 1924d: 171, figs 1–6; J. C. Chamberlin, 1930: 614; J. C. Chamberlin, 1931a: figs 23a–f, 24b, 24e, 27e–i; Beier, 1932a: 217–218, fig. 244; Beier, 1932g: fig. 210; Snodgrass, 1948: fig. 12c–e; V. F. Lee, 1979a: 12–15, figs 3, 15–16, 30; V. F. Lee, 1979b: 13.

Garypus sine (sic) J. C. Chamberlin: Roewer, 1937: 268.

Type locality: Puerto Ballandra, Carmen Island, Gulf of California, Mexico.
Distribution: Mexico.

Garypus titanius Beier

Garypus titanius Beier, 1961c: 594–595, fig. 1.

Type locality: Boatswain-bird (Bos'nbird) Island, Ascension Island.
Distribution: Ascension Island.

Genus Larca J. C. Chamberlin

Larca J. C. Chamberlin, 1930: 616; Beier, 1932a: 224; Beier, 1939f: 194; Hoff, 1949b: 447; Hoff, 1961: 435; Beier, 1963b: 240; Benedict & Malcolm, 1978a: 114–115.

Type species: *Garypus latus* H. J. Hansen, 1884, by original designation.

Larca chamberlini Benedict & Malcolm

Larca chamberlini Benedict & Malcolm, 1978a: 115–118, figs 2–8; Muchmore, 1981a: 47–48; Zeh, 1987b: 1086.

Type locality: 4 miles NE. of Corvallis, Oregon, U.S.A.
Distribution: U.S.A. (California, Oregon).

Larca granulata (Banks)

Grypus (sic) *granulatus* Banks, 1891: 163–164.
Garypus granulatus Banks: Banks, 1895a: 9; Coolidge, 1908: 112; J. C. Chamberlin, 1921: 191.
Larca granulata (Banks): J. C. Chamberlin, 1930: 616, figs 3m–n; Beier, 1932a: 224–225; Roewer, 1937: 268; Hoff, 1945a: 109–110, figs 9–10; Hoff, 1949b: 447–448, figs 26, 27a–d; Hoff & Bolsterli, 1956: 163–164; Hoff, 1958: 15; Weygoldt, 1969a: fig. 51; Muchmore, 1971a: 90, 92; Nelson, 1975: 282, figs 29–31; Benedict & Malcolm, 1978a: 117; Muchmore, 1981a: 50–51; Zeh, 1987b: 1086.

Chernes dentatus (Banks): Ross, 1944: fig. 56 (misidentification).

Type locality: Ithaca, New York, U.S.A.
Distribution: U.S.A. (Illinois, Kansas, Michigan, New Hampshire, New York, North Carolina, Pennsylvania, Tennessee, Virginia, West Virginia).

Larca hispanica Beier

Larca hispanica Beier, 1939f: 194–195, fig. 22; Roewer, 1940: 346; Beier, 1963b: 240–241, fig. 244; Beier, 1970f: 322; Estany, 1980a: 66–69; Zaragoza, 1986: fig. 1.
Larca spelaea Beier, 1939f: 195–196, fig. 23; Roewer, 1940: 346; Beier, 1963b: 241, fig. 245; Beier, 1970f: 322; Mahnert, 1977c: 102; Zaragoza, 1986, fig. 1 (synonymized by Estany, 1980a: 69).

Type localities: of *Larca hispanica*: Cova del Bisbe, Cataluña, Spain.
of *Larca spelaea*: Baumes de Aros, near Rielles, Barcelona, Spain.
Distribution: Spain.

Larca italica Gardini

Larca italica Gardini, 1983b: 63–67, figs 1–11.

Type locality: Grotta San Angelo, Civitella del Tronto, Teramo, Abruzzo, Italy.
Distribution: Italy.

Larca laceyi Muchmore

Larca laceyi Muchmore, 1981a: 48–50, figs 1–6.

Type locality: Music Hall Cave, Calaveras County, California, U.S.A.
Distribution: U.S.A. (California).

Larca lata (H. J. Hansen)

Garypus latus H. J. Hansen, 1884: 550–551; H. J. Hansen, 1893: plate 4 fig. 14.
Garypus lata (sic) H. J. Hansen: Tullgren, 1899a: 179.
Larca lata (H. J. Hansen): J. C. Chamberlin, 1930: 616, fig. 2k; J. C. Chamberlin, 1931a: figs 37r, 42i, 47w; Beier, 1932a: 225; Roewer, 1937: 268; Lohmander, 1939b: 291–292; Beier, 1956h: 25, fig. 2; Beier, 1956i: 8; Ressl & Beier, 1958: 2; Meinertz, 1962a: 392, map 3; Beier, 1963b: 240; Dumitresco & Orghidan, 1964: 620–626, figs 36–46, plate 1 fig. 8; Ressl, 1965: 289; Rafalski, 1967: 13–14; Beier, 1970f: 322.

Type locality: Denmark.
Distribution: Austria, Denmark, Poland, Romania, Sweden.

Larca notha Hoff

Larca notha Hoff, 1961: 435–438, figs 12–13; Muchmore, 1981a: 51–52.

Type locality: 32 miles W. of Fort Collins, Larimer County, Colorado, U.S.A.
Distribution: U.S.A. (Colorado, Oregon), Canada (Saskatchewan).

Genus Meiogarypus Beier

Meiogarypus Beier, 1955l: 296–298.

Type species: *Meiogarypus mirus* Beier, 1955l, by original designation.

Meiogarypus mirus Beier

Meiogarypus mirus Beier, 1955l: 298–299, figs 16–18; Beier, 1973a: 100.

Type locality: 35 miles W. of Orupembe, 10 miles SSW. of Ogams, Namibia.
Distribution: Namibia.

Genus Neogarypus Vachon

Neogarypus Vachon, 1937a: 129–130; Roewer, 1940: 346.

Type species: *Neogarypus gravieri* Vachon, 1937a, by original designation.

Neogarypus gravieri Vachon

Neogarypus gravieri Vachon, 1937a: 129–130, figs 1–2; Vachon, 1938a: fig. 59c; Roewer, 1940: 346.

Type locality: Makapau, Victoria, Zimbabwe.
Distribution: Zimbabwe.

Genus Paragarypus Vachon

Paragarypus Vachon, 1937d: 188–190; Roewer, 1940: 345–346.

Type species: *Paragarypus fagei* Vachon, 1937d, by original designation.

Paragarypus fagei Vachon

Paragarypus fagei Vachon, 1937d: 188–190, figs 1–3; Vachon, 1938a: fig. 59b; Roewer, 1940: 346; Legendre, 1972: 446.

Type locality: Ambohibato, Madagascar.
Distribution: Madagascar.

Genus **Synsphyronus** J. C. Chamberlin

Synsphyronus J. C. Chamberlin, 1930: 616; Beier, 1932a: 238; J. C. Chamberlin, 1943: 488–489; Hoff, 1947a: 44–45; J. C. H. Morris, 1948a: 37; Beier, 1954b: 6–7; Beier, 1976f: 211; Harvey, 1987a: 5.
Maorigarypus J. C. Chamberlin, 1930: 617; Beier, 1932a: 238 (synonymized by J. C. Chamberlin, 1943: 488).
Synsphyronus (*Synsphyronus*) J. C. Chamberlin: J. C. Chamberlin, 1943: 488–489.
Synsphyronus (*Maorigarypus*) J. C. Chamberlin: J. C. Chamberlin, 1943: 488–489.
Idiogarypus J. C. Chamberlin, 1943: 499 (synonymized by J. C. H. Morris, 1948a: 37).

Type species: of *Synsphyronus*: *Synsphyronus paradoxus* J. C. Chamberlin, 1930, by original designation.
of *Maorigarypus*: *Maorigarypus melanochelatus* J. C. Chamberlin, 1930, by original designation.
of *Idiogarypus*: *Garypus hansenii* With, 1908b, by original designation.

Synsphyronus absitus Harvey

Synsphyronus absitus Harvey, 1987a: 28–30, figs 87–94.

Type locality: Mundibarcooloo Waterhole, South Australia, Australia.
Distribution: Australia (Northern Territory, Queensland, South Australia, Western Australia).

Synsphyronus amplissimus Harvey

Synsphyronus amplissimus Harvey, 1987a: 33–34, figs 103–105, 111–112.

Type locality: Mt Conner, Northern Territory, Australia.
Distribution: Australia (Northern Territory).

Synsphyronus apimelus Harvey

Synsphyronus apimelus Harvey, 1987a: 37–38, figs 111–112, 116–119.

Type locality: Toolbrunup Peak, Western Australia, Australia.
Distribution: Australia (Western Australia).

Synsphyronus attiguus Harvey

Synsphyronus attiguus Harvey, 1987a: 31–32, figs 87, 93–98.

Type locality: Mernmerna, South Australia, Australia.
Distribution: Australia (South Australia).

Synsphyronus bounites Harvey

Synsphyronus fallaciosus Beier: Beier, 1975b: 203 (misidentification); Harvey, 1981b: 242 (in part); Harvey, 1985b: 144 (in part).
Synsphyronus bounites Harvey, 1987a: 27, figs 66, 79–83.

Type locality: Wilsons Valley, New South Wales, Australia.
Distribution: Australia (New South Wales, Victoria).

Synsphyronus callus Hoff

Synsphyronus (*Synsphyronus*) *callus* Hoff, 1947a: 50–53, figs 13–15; Harvey, 1981b: 242; Harvey, 1985b: 143.

Synsphyronus callus Hoff: Harvey, 1987a: 19–21, figs 26–27, 35, 50–54.

Type locality: Rottnest Island, Western Australia, Australia.
Distribution: Australia (Western Australia).

Synsphyronus dewae Beier

Synsphyronus (Maorigarypus) dewae Beier, 1969a: 181–182, fig. 8; Harvey, 1981b: 242; Harvey, 1985b: 143.
Synsphyronus dewae Beier: Harvey, 1987a: 36–37, figs 111–115.

Type locality: Truro, South Australia, Australia.
Distribution: Australia (South Australia).

Synsphyronus dorothyae Harvey

Synsphyronus dorothyae Harvey, 1987a: 27–28, figs 6, 84–87, 93.

Type locality: 67 km SE. of Coolgardie, Western Australia, Australia.
Distribution: Australia (South Australia, Western Australia).

Synsphyronus ejuncidus Harvey

Synsphyronus ejuncidus Harvey, 1987a: 10–12, figs 20–27, 35.

Type locality: 8 km S. of Knob Peak, Western Australia, Australia.
Distribution: Australia (South Australia, Western Australia).

Synsphyronus elegans Beier

Synsphyronus (Maorigarypus) elegans Beier, 1954b: 13–15, figs 4–5; Harvey, 1981b: 242; Harvey, 1985b: 144.
Synsphyronus elegans Beier: Harvey, 1987a: 40–42, figs 132–137.

Type locality: Yorkrakine Hill, Western Australia, Australia.
Distribution: Australia (Western Australia).

Synsphyronus gigas Beier

Synsphyronus (Maorigarypus) gigas Beier, 1971d: 161–162, fig. 1; Harvey, 1981b: 242; Harvey, 1985b: 144.
Synsphyronus gigas Beier: Harvey, 1987a: 34–36, figs 106–112.

Type locality: 18 km E. of Milne Rock (as Lennis 553623), Western Australia, Australia.
Distribution: Australia (Western Australia).

Synsphyronus gracilis Harvey

Synsphyronus gracilis Harvey, 1987a: 38, figs 111–112, 120–121.

Type locality: Marillana Station, Western Australia, Australia.
Distribution: Australia (Western Australia).

Synsphyronus greensladeae Harvey

Synsphyronus greensladeae Harvey, 1987a: 46–47, figs 137, 150–154.

Type locality: Marble Range, South Australia, Australia.
Distribution: Australia (South Australia).

Synsphyronus hadronennus Harvey

Synsphyronus hadronennus Harvey, 1987a: 13–14, figs 26–30, 35.

Type locality: 6 km W. of South Alligator, Northern Territory, Australia.
Distribution: Australia (Northern Territory).

Synsphyronus hansenii (With)

Garypus hansenii With, 1908b: 12–15, fig. 2, plate 1 figs 11–15, plate 2 figs 1–3.
Garypus hanseni With: J. C. Chamberlin, 1930: 612.
Maorigarypus hansenii (With): Beier, 1932a: 226–227; J. C. Chamberlin, 1934b: 6; Roewer, 1937: 268.
Idiogarypus hansenii (With): J. C. Chamberlin, 1943: 500.
Synsphyronus (Maorigarypus) hansenii (With): J. C. H. Morris, 1948a: 38–40, figs 1–5, 5a; Harvey, 1981b: 242; Harvey, 1985b: 144.
Synsphyronus (Maorigarypus) gisléni Beier, 1954b: 9–10, fig. 2 (synonymized by Harvey, 1987a: 23).
Synsphyronus (Maorigarypus) hanseni (With): Beier, 1954b: 11–12, fig. 3.
Synsphyronus (Maorigarypus) fallaciosus Beier, 1966a: 294–295, fig. 11. Harvey, 1981b: 242 (in part); Harvey, 1985b: 144 (in part) (synonymized by Harvey, 1987a: 23).
Synsphyronus hanseni (With): Dartnall, 1969: 2; Dartnall, 1975: 5–6.
Synsphyronus (Maorigarypus) gisleni Beier: Harvey, 1981b: 242; Harvey, 1985b: 144.
Synsphyronus hansenii (With): Harvey, 1987a: 23–26, figs 12–13, 66–80.
Not *Synsphyronus fallaciosus* Beier: Beier, 1975b: 202 (misidentification; see *Synsphyronus bounites* Harvey).
Not *Synsphyronus hanseni* (With): Hickman & Hill, 1978, appendix a (misidentification; true identification unknown).

Type localities: of *Garypus hansenii*: Tasmania, Australia.
of *Synsphyronus (Maorigarypus) gisleni*: Denmark, near mouth of Denmark River, Western Australia, Australia.
of *Synsphyronus (Maorigarypus) fallaciosus*: Gunyah Gunyah (as Gunyah), Victoria, Australia.
Distribution: Australia (Tasmania, Victoria, Western Australia).

Synsphyronus heptatrichus Harvey

Synsphyronus heptatrichus Harvey, 1987a: 17–18, figs 26–27, 35, 44–45.

Type locality: 13.5 km SE. of Roper Bar, Northern Territory, Australia.
Distribution: Australia (Northern Territory).

Synsphyronus lathrius Harvey

Synsphyronus lathrius Harvey, 1987a: 50–51, figs 137, 164–170.

Type locality: 98 km ENE. of Norseman, Western Australia, Australia.
Distribution: Australia (Western Australia).

Synsphyronus leo Harvey

Synsphyronus leo Harvey, 1987a: 47–48, figs 137, 154–156.

Type locality: Lion Island, Recherche Archipelago, Western Australia, Australia.
Distribution: Australia (Western Australia).

Synsphyronus lineatus Beier

Synsphyronus (Maorigarypus) lineatus Beier, 1966c: 368–369, fig. 5; Tenorio & Muchmore, 1982: 384.
Synsphyronus melanochelatus (J. C. Chamberlin): Beier, 1976f: 211 (misidentification, in part).
Synsphyronus lineatus Beier: Beier, 1976f: 211; Harvey, 1987a: 42–44, figs 138–146.

Type locality: Alexandra, Otago, South Island, New Zealand.
Distribution: New Zealand.

Synsphyronus magnus Hoff

Synsphyronus (Maorigarypus) magnus Hoff, 1947a: 47–50, figs 11–12; Harvey, 1981b: 242; Harvey, 1985b: 144.
Synsphyronus magnus Hoff: Harvey, 1987a: 48–49, figs 137, 158–164.
Synsphronus (sic) *magnus* Hoff: Humphreys, 1987: 199.

Type locality: Margaret River, Western Australia, Australia.
Distribution: Australia (Western Australia).

Synsphyronus meganennus Harvey

Synsphyronus meganennus Harvey, 1987a: 14–15, figs 26–27, 31–35.

Type locality: Great Sugarloaf Mt, New South Wales, Australia.
Distribution: Australia (New South Wales).

Synsphyronus melanochelatus (J. C. Chamberlin)

Maorigarypus melanochelatus J. C. Chamberlin, 1930: 617–618; J. C. Chamberlin, 1931a: figs 37s, 42h, 44e; Beier, 1932a: 226; J. C. Chamberlin, 1934b: 6; Roewer, 1937: 268.
Synsphyronus (Maorigarypus) melanochelatus (J. C. Chamberlin): J. C. Chamberlin, 1943: 498, fig. 21; Beier, 1969e: 413.
Synsphyronus melanochelatus (J. C. Chamberlin): Beier, 1976f: 211 (in part; see *Synsphyronus lineatus* Beier); Harvey, 1987a: 44–46, figs 144–149, 154.

Type locality: Okahune, North Island, New Zealand.
Distribution: New Zealand.

Synsphyronus mimetus J. C. Chamberlin

Synsphyronus paradoxus J. C. Chamberlin: Tubb, 1937: 412 (misidentification).
Synsphyronus (Synsphyronus) mimetus J. C. Chamberlin, 1943: 492–496, figs 3, 6, 8, 10–12, 15, 18, 20, 22, 25, 27; J. C. Chamberlin, 1949: 4–5, fig. 2 (in part; see *Synsphyronus mimulus* J. C. Chamberlin); Harvey, 1981b: 242; Harvey, 1985b: 143.
Synsphyronus mimetus J. C. Chamberlin: Harvey, 1987a: 18–19, figs 26–27, 35, 45–49.

Type locality: Corny (as Corney) Point, South Australia, Australia.
Distribution: Australia (South Australia, Victoria).

Synsphyronus mimulus J. C. Chamberlin

Synsphyronus (Maorigarypus) mimulus J. C. Chamberlin, 1943: 496–498, figs 1, 4, 5, 7, 13–14, 16–17, 23, 26, 28; Harvey, 1981b: 242–243; Harvey, 1985b: 144.
Synsphyronus (Synsphyronus) mimetus J. C. Chamberlin, 1943: 492 (in part).
Synsphyronus (Maorigarypus) grayi Beier, 1975b: 205–206, fig. 2; Harvey, 1981b: 242; Harvey, 1985b: 144 (synonymized by Harvey, 1987a: 21).
Synsphyronus mimulus J. C. Chamberlin: Harvey, 1987a: 21–23, figs 16–19, 26–27, 55–66, 93.

Type localities: of *Synsphyronus (Maorigarypus) mimulus*: Corny (as Corney) Point, South Australia, Australia.
of *Synsphyronus (Maorigarypus) grayi*: Mullamullang Cave, Western Australia, Australia.
Distribution: Australia (Northern Territory, Queensland, South Australia, Victoria, Western Australia).

Synsphyronus niger Hoff

Synsphyronus (Maorigarypus) niger Hoff, 1947a: 45–47, figs 9–10; Harvey, 1981b: 243; Harvey, 1985b: 144.

Synsphyronus (Maorigarypus) nullarborensis Beier: Beier, 1975b: 203 (misidentification); Harvey, 1981b: 243 (in part); Harvey, 1985b: 144−5 (in part).
Synsphyronus niger Hoff: Harvey, 1987a: 39−40, figs 122−132, 137.

Type locality: Adelaide, South Australia, Australia.
Distribution: Australia (New South Wales, South Australia, Victoria).

Synsphyronus nullarborensis Beier

Synsphyronus (Maorigarypus) nullarborensis Beier, 1969a: 179−181, fig. 7; Harvey, 1981b: 243 (in part); Harvey, 1985b: 144−145 (in part).
Synsphyronus nullarborensis Beier: Harvey, 1987a: 51−53, figs 137, 164, 171−176.
Not *Synsphyronus (Maorigarypus) nullarborensis* Beier: Beier, 1975b: 203 (misidentification; see *Synsphyronus niger* Hoff).

Type locality: Toolinna Rock Hole, Western Australia, Australia.
Distribution: Australia (South Australia, Western Australia).

Synsphyronus paradoxus J. C. Chamberlin

Synsphyronus paradoxus J. C. Chamberlin, 1930: 617, figs 1bb, 1cc, 3v; J. C. Chamberlin, 1931a: figs 11i, 29e, 44c−d, 47m, 47p; Beier, 1932a: 238; J. C. Chamberlin, 1934b: 6; Roewer, 1937: 271; Harvey, 1987a: 15−17, figs 7−11, 14−15, 26−27, 35−43.
Synsphyronus (Synsphyronus) paradoxus J. C. Chamberlin: J. C. Chamberlin, 1943: 490−492, figs 2, 24, 29−30; Harvey, 1981b: 242; Harvey, 1985b: 143.
Not *Synsphyronus paradoxus* J. C. Chamberlin: Tubb, 1937: 412 (misidentification; see *Synsphyronus mimetus* J. C. Chamberlin).

Type locality: Menindee (as Menindie), New South Wales, Australia.
Distribution: Australia (New South Wales, South Australia, Victoria).

Synsphyronus silveirai Harvey

Synsphyronus silveirai Harvey, 1987a: 32−33, figs 87, 93−94, 99−102.

Type locality: 16 km S. of Texas, New South Wales, Australia.
Distribution: Australia (New South Wales).

Genus Thaumastogarypus Beier

Thaumastogarypus Beier, 1947b: 312−313.

Type species: *Thaumastogarypus grandis* Beier, 1947b, by original designation.

Thaumastogarypus capensis (Ellingsen)

Garypus capensis Ellingsen, 1912b: 105−107.
Anagarypus capensis (Ellingsen): Beier, 1932a: 226; Roewer, 1937: 268.
Thaumastogarypus capensis (Ellingsen): Beier, 1947b: 314−315.

Type locality: Stompneus Bay in St Helena Bay, Cape Province, South Africa.
Distribution: South Africa.

Thaumastogarypus grandis Beier

Thaumastogarypus grandis Beier, 1947b: 315−316, figs 20, 22.

Type locality: Steinkopf, Namaqualand, Cape Province, South Africa.
Distribution: South Africa.

Thaumastogarypus longimanus Beier

Thaumastogarypus longimanus Beier, 1947b: 316−317, fig. 23.

Type locality: Doorn River, Montagu, Cape Province, South Africa.
Distribution: South Africa.

Thaumastogarypus mancus Mahnert

Thaumastogarypus mancus Mahnert, 1982d: 313, figs 18–23.

Type locality: ca. 20 km W. of Sala Gate, Tsavo East National Park, Kenya.
Distribution: Kenya.

Thaumastogarypus okahandjanus Beier

Thaumastogarypus okahandjanus Beier, 1964k: 56–57, fig. 18.

Type locality: Okahandja (as Okahandju), Namibia.
Distribution: Namibia.

Thaumastogarypus robustus Beier

Thaumastogarypus robustus Beier, 1947b: 314–315, fig. 21; Beier, 1964k: 56; Beier, 1973a: 98–99.

Type locality: Kleinsee (as Kleinzee), Namaqualand, Cape Province, South Africa.
Distribution: Namibia, South Africa.

Thaumastogarypus transvaalensis Beier

Thaumastogarypus transvaalensis Beier, 1955l: 292–294, figs 11–12; Beier, 1964k: 57; Beier, 1966k: 460–461, fig. 3; Lawrence, 1967: 89.

Type locality: Letaba Camp, Kruger National Park, Transvaal, South Africa.
Distribution: South Africa.

Thaumastogarypus zuluensis Beier

Thaumastogarypus zuluensis Beier, 1958a: 167–169, figs 5–6.

Type locality: Ingwavuma, Zululand, Natal, South Africa.
Distribution: South Africa.

Family GEOGARYPIDAE J. C. Chamberlin

Geogarypinae J. C. Chamberlin, 1930: 609; J. C. Chamberlin, 1931a: 228; Beier, 1932a: 227; Roewer, 1937: 268–269; Morikawa, 1960: 132; Murthy & Ananthakrishnan, 1977: 104.
Geogarypidae J. C. Chamberlin: Harvey, 1986: 754.

Genus Afrogarypus Beier

Afrogarypus Beier, 1931c: 317; Harvey, 1986: 758.
Geogarypus (Afrogarypus) Beier: Beier, 1932a: 236; Heurtault, 1970b: 1365.

Type species: *Garypus senegalensis* Balzan, 1892, by original designation.

Afrogarypus basilewskyi (Beier)

Geogarypus (Afrogarypus) basilewskyi Beier, 1962f: 16–18, figs 5–6; Mahnert, 1982d: 318–319, figs 27–28, 54, 57.
Afrogarypus basilewskyi (Beier): Harvey, 1986: 758.

Type locality: Ulugu-Ndogo Valley, Uluguru Mountains, Tanzania.
Distribution: Kenya, Tanzania.

Afrogarypus curtus (Beier)

Geogarypus (Afrogarypus) curtus Beier, 1955c: 544, fig. 11; Mahnert, 1982d: 325–326, figs 43, 52, 56.
Afrogarypus curtus (Beier): Harvey, 1986: 758.

Type locality: 7 miles N. of Mombasa, Kenya.
Distribution: Kenya.

Afrogarypus excelsus (Beier)

Geogarypus (Afrogarypus) excelsus Beier, 1964k: 61–62, fig. 22.
Afrogarypus excelsus (Beier): Harvey, 1986: 758.

Type locality: Paradise Kloof, Grahamstown, Cape Province, South Africa.
Distribution: South Africa.

Afrogarypus excelsus excellens (Beier)

Geogarypus (Afrogarypus) excelsus excellens Beier, 1964k: 62–63, fig. 23; Beier, 1966k: 461.
Afrogarypus excelsus excellens (Beier): Harvey, 1986: 758.

Type locality: no data.
Distribution: South Africa.

Afrogarypus impressus (Tullgren)

Garypus impressus Tullgren, 1907b: 228–229, fig. 7; Ellingsen, 1912b: 86, 107–108.
Geogarypus impressus (Tullgren): J. C. Chamberlin, 1930: 609; Roewer, 1936: fig. 77; Vachon, 1949: fig. 204b; Beier, 1964k: 60; Lawrence, 1967: 89; Spaull, 1979: 117.
Geogarypus (Afrogarypus) impressus (Tullgren): Beier, 1932a: 236–237, fig. 265; Roewer, 1937: 270, fig. 222b; Beier, 1947b: 320; Beier, 1955l: 301, fig. 21a; Beier, 1958a: 171; Beier, 1966k: 461.
Afrogarypus impressus (Tullgren): Harvey, 1986: 758.

Type localities: Van Reenen, Orange Free State, South Africa; Amanzimtoti, Natal, South Africa; and junction of Black and White Umfolozi, Natal, South Africa.
Distribution: Aldabra Islands, Mozambique, South Africa.

Afrogarypus intermedius (Beier)

Geogarypus (Afrogarypus) intermedius Beier, 1955c: 540–541, fig. 9; Beier, 1955h: 9; Beier, 1959d: 31.
Geogarypus (Afrogarypus) intermedius intermedius Beier: Mahnert, 1982d: 318.
Afrogarypus intermedius (Beier): Harvey, 1986: 758.

Type locality: Bundibugyo, Ruwenzori, Uganda.
Distribution: Kenya, Rwanda, Uganda, Zaire.

Afrogarypus intermedius nanus (Beier)

Geogarypus (Afrogarypus) intermedius nanus Beier, 1959d: 32–33, fig. 14.
Afrogarypus intermedius nanus (Beier): Harvey, 1986: 758.

Type locality: Kundelungu, Zaire.
Distribution: Zaire.

Afrogarypus monticola (Beier)

Geogarypus (Afrogarypus) monticola Beier, 1955c: 541–544, fig. 10; Beier, 1962f: 15–16, figs 3–4; Mahnert, 1982d: 314–318, figs 24–26, 53, 58.
Afrogarypus monticola (Beier): Harvey, 1986: 758.

Type locality: Mt Elgon, Kenya.
Distribution: Kenya, Uganda, Tanzania.

Afrogarypus plumatus (Mahnert)

Geogarypus (Afrogarypus) plumatus Mahnert, 1982d: 320–322, figs 29–35, 59.
Afrogarypus plumatus (Mahnert): Harvey, 1986: 758.

Type locality: Makadara Forest, Shimba Hills, Kenya.
Distribution: Kenya.

Afrogarypus pseudocurtus (Mahnert)

Geogarypus (Afrogarypus) pseudocurtus Mahnert, 1982d: 322–325, figs 44–48, 51.
Afrogarypus pseudocurtus (Mahnert): Harvey, 1986: 758.

Type locality: Ngao, Lac Shakababo, Tana River, Kenya.
Distribution: Kenya.

Afrogarypus senegalensis (Balzan)

Garypus senegalensis Balzan, 1892: 535–536, figs 27, 27a–b; Ellingsen, 1913a: 454.
Geogarypus senegalensis (Balzan): J. C. Chamberlin, 1930: 609.
Afrogarypus senegalensis (Balzan): Beier, 1931c: 317; Harvey, 1986: 758, fig. 2.
Geogarypus (Afrogarypus) senegalensis (Balzan): Beier, 1932a: 236, fig. 264; Roewer, 1937: 270; Vachon, 1952a: 27; Beier, 1965b: 371.
Not *Garypus senegalensis* Balzan: Ellingsen, 1912b: 87, 112–114 (misidentification; see *Geogarypus olivaceus* (Tullgren)).

Type locality: Senegal.
Distribution: Cameroun, Ghana, Guinea, Liberia, Senegal.

Afrogarypus seychellesensis (Beier)

Geogarypus seychellesensis Beier, 1940a: 173–174, fig. 173.
Geogarypus (Afrogarypus) seychellesensis Beier: Mahnert, 1978d: 878.
Afrogarypus seychellesensis (Beier): Harvey, 1986: 758.

Type locality: Praslin, Seychelles.
Distribution: Seychelles.

Afrogarypus stellifer (Mahnert)

Geogarypus (Afrogarypus) stellifer Mahnert, 1982d: 322, figs 36–42, 49, 55.
Afrogarypus stellifer (Beier): Harvey, 1986: 758.

Type locality: Mombasa, Kenya.
Distribution: Kenya.

Afrogarypus subimpressus (Beier)

Geogarypus (Afrogarypus) subimpressus Beier, 1955l: 302–303, figs 21b, 22.
Geogarypus subimpressus Beier: Beier, 1964k: 60.
Afrogarypus subimpressus (Beier): Harvey, 1986: 758.

Type locality: Cape Point, Cape Peninsula, Cape Province, South Africa.
Distribution: South Africa.

Afrogarypus sulcatus (Beier)

Geogarypus (Afrogarypus) sulcatus Beier, 1955a: 10–12, fig. 5.
Afrogarypus sulcatus (Beier): Harvey, 1986: 758.

Type locality: Mabwe, Zaire.
Distribution: Zaire.

Afrogarypus sulcatus rhodesiacus (Beier)

Geogarypus (Afrogarypus) sulcatus rhodesiacus Beier, 1964k: 60–61, fig. 21.
Afrogarypus sulcatus rhodesiacus (Beier): Harvey, 1986: 758.

Type locality: Vumba, Zimbabwe.
Distribution: Zimbabwe.

Afrogarypus zonatus (Beier)

Geogarypus (Afrogarypus) zonatus Beier, 1959d: 33–34, fig. 15.
Afrogarypus zonatus (Beier): Harvey, 1986: 758.

Type locality: Ht. Ngovi, Itombwe, Uvira, Zaire.
Distribution: Zaire.

Genus Geogarypus J. C. Chamberlin

Geogarypus J. C. Chamberlin, 1930: 609; Beier, 1932a: 227; Morikawa, 1960: 132; Beier, 1963b: 241–242; Murthy & Ananthakrishnan, 1977: 104–105; Sivaraman, 1980b: 340; Harvey, 1986: 760.
Geogarypus (Geogarypus) J. C. Chamberlin: Beier, 1932a: 229; Morikawa, 1960: 132; Heurtault, 1970b: 1365; Murthy & Ananthakrishnan, 1977: 105.

Type species: *Garypus minor* L. Koch, 1873, by original designation.

Geogarypus albus Beier

Geogarypus albus Beier, 1963d: 508–510, fig. 1; Tenorio & Muchmore, 1982: 382; Harvey, 1986: 760; Harvey, 1988b: 329–330, figs 55–58.

Type locality: 8 km N. of Klang, Rantau Panjang, Selangor, Malaysia.
Distribution: Indonesia (Java), Malaysia.

Geogarypus amazonicus Mahnert

Geogarypus (Geogarypus) amazonicus Mahnert, 1979d: 759–761, figs 85–90; Mahnert, 1985c: 226; Mahnert & Adis, 1986: 213.
Geogarypus amazonicus Mahnert: Adis, 1981: 119, etc.; Mahnert, 1985b: 78; Adis & Mahnert, 1986: 312; Harvey, 1986: 760; Adis, Junk & Penny, 1987: 488; Mahnert, Adis, & Bührnheim, 1985: fig. 6.

Type locality: Manaus, Amazonas, Brazil.
Distribution: Brazil (Amazonas, Pará).

Geogarypus angulatus J. C. Chamberlin

Geogarypus angulatus J. C. Chamberlin, 1930: 611, figs 2u–v, 3j, 3q; J. C. Chamberlin, 1931a: fig. 4b; Harvey, 1986: 760.
Geogarypus (Geogarypus) angulatus J. C. Chamberlin: Beier, 1932a: 234–235; Roewer, 1937: 270; Murthy & Ananthakrishnan, 1977: 108.
Geogarypus (Indogarypus) angulatus J. C. Chamberlin: Beier, 1957d: 25.

Type locality: Ootacamund, Nilgiris, India.
Distribution: India.

Geogarypus asiaticus Murthy & Ananthakrishnan

Geogarypus (Indogarypus) asiaticus Murthy & Ananthakrishnan, 1977: 108–110, figs 36a–b.
Geogarypus asiaticus Murthy & Ananthakrishnan: Harvey, 1986: 760.

Type locality: Madras, Tamil Nadu, India.
Distribution: India.

Geogarypus bucculentus Beier

Geogarypus (Geogarypus) bucculentus Beier, 1955b: 206–207, figs 1–2; Cekalovic, 1984: 16–17.
Geogarypus bucculentus Beier: Beier, 1957c: 454; Harvey, 1986: 760; Harvey, 1987d: 137–140, figs 1–8, 15.

Type locality: Plateau del Yunque, Masatierra, Juan Fernandez Islands.
Distribution: Juan Fernandez Islands.

Geogarypus canariensis (Tullgren)

Garypus canariensis Tullgren, 1900b: 157–159, figs 1–2.
Geogarypus canariensis (Tullgren): J. C. Chamberlin, 1930: 609; Beier, 1956e: 306, fig. 3; Beier, 1963b: 243; Beier, 1965c: 379; Beier, 1970a: 45; Beier, 1976a: 24; Estany, 1979a: 222, fig. 1; Pieper, 1981: 3; Harvey, 1986: 760; Callaini, 1988b: 50–53, figs 5a–g.
Geogarypus (Geogarypus) canariensis (Tullgren): Beier, 1932a: 230; Roewer, 1937: 269; Mahnert, 1980a: 259; Mahnert, 1989b: 42.

Type locality: Barranco de Ruiz, Tenerife, Canary Islands.
Distribution: Canary Islands, Madeira Islands, Morocco.

Geogarypus ceylonicus Beier

Geogarypus ceylonicus Beier, 1973b: 46, fig. 10; Beier, 1974e: 1008; Harvey, 1986: 760.
Geogarypus (Indogarypus) ceylonicus Beier: Murthy & Ananthakrishnan, 1977: 110.

Type locality: Yakkala, 18 miles NE. of Colombo, Western Province, Sri Lanka.
Distribution: India, Sri Lanka.

Geogarypus connatus Harvey

Geogarypus connatus Harvey, 1986: 772–775, figs 5, 41–53.

Type locality: Horseshoe Bend, Little Desert National Park, Victoria, Australia.
Distribution: Australia (New South Wales, Queensland, Victoria).

Geogarypus continentalis (Redikorzev)

Garypus (Geogarypus) continentalis Redikorzev, 1934a: 432–434, figs 11–12; Roewer, 1936: fig. 66c; Roewer, 1937: 270, fig. 219.
Geogarypus continentalis (Redikorzev): Roewer, 1936: fig. 70n: Redikorzev, 1949: 645; Schawaller, 1985a: 9–10, figs 19–21; Harvey, 1986: 760; Schawaller, 1989: 16.

Type locality: Tujuk River, R.S.F.S.R., U.S.S.R.
Distribution: U.S.S.R. (Kazakhstan, Kirghizia, R.S.F.S.R.).

Geogarypus cuyabanus (Balzan)

Garypus cuyabanus Balzan, 1887a: no pagination, figs; Balzan, 1890: 441–443, figs 22, 22a; Balzan, 1892: 549.
Geogarypus cuyabanus (Balzan): J. C. Chamberlin, 1930: 609; Beier, 1974d: 899; Harvey, 1986: 760.
Geogarypus (Geogarypus) cuyabanus (Balzan): Beier, 1932a: 231–232; Roewer, 1937: 270.
Not *Garypus cuyabanus* Balzan: Ellingsen, 1910a: 316 (misidentification; see *Geogarypus fiebrigi* Beier).

Type locality: Mato Grosso (as Mato-Groso), Brazil.
Distribution: Brazil (Mato Grosso, Santa Catarina).

Geogarypus elegans (With)

Garypus personatus E. Simon, 1901: 79 (junior primary homonym of *Garypus personatus* E. Simon, 1900a).
Garypus elegans With, 1906: 107–109, fig. 16, plate 2 figs 2a–b (replacement name for *Garypus personatus* E. Simon, 1901).
Geogarypus (Geogarypus) elegans (With): Beier, 1932a: 234; Roewer, 1937: 270.
Geogarypus elegans (With): Harvey, 1986: 760.

Type locality: Kuala Aring, Kelantan, Malaysia.
Distribution: Malaysia.

Geogarypus exochus Harvey

Geogarypus exochus Harvey, 1986: 769–771, figs 5, 35–37.

Type locality: 90 miles S. of Mackay, Queensland, Australia.
Distribution: Australia (Northern Territory, Queensland).

Geogarypus fiebrigi Beier

Garypus cuyabanus Balzan: Ellingsen, 1910a: 316 (misidentification).
Geogarypus fiebrigi Beier, 1931c: 316; Harvey, 1986: 760.
Geogarypus (Geogarypus) fiebrigi Beier: Beier, 1932a: 232, fig. 259; Roewer, 1937: 270.

Type locality: South Bernardino, Paraguay.
Distribution: Paraguay.

Geogarypus formosus (Mello-Leitão)

Ideobisium formosum Mello-Leitão, 1937: 269–270, fig. 1.
Geogarypus formosus (Mello-Leitão): Mahnert, 1979d: 761; Harvey, 1986: 760.

Type locality: Morro de Caxambú, Petrópolis (as Petropolis), Brazil.
Distribution: Brazil (Rio de Janeiro).

Geogarypus globulus Sivaraman

Geogarypus (Indogarypus) globulus Sivaraman, 1980b: 340–342, figs 7a–b.
Geogarypus globulus Sivaraman: Harvey, 1986: 760.

Type locality: Pondicherry, Pondicherry, India.
Distribution: India.

Geogarypus granulatus Murthy & Ananthakrishnan

Geogarypus (Geogarypus) granulatus Murthy & Ananthakrisnan, 1977: 105–108, figs 35a–b.
Geogarypus granulatus Murthy & Ananthakrishnan: Harvey, 1986: 760.

Type locality: Courtallam, Tamil Nadu, India.
Distribution: India.

Geogarypus heterodentatus Murthy & Ananthakrishnan

Geogarypus (Indogarypus) heterodentatus Murthy & Ananthakrishnan, 1977: 111–112, figs 37a–b.
Geogarypus heterodentatus Murthy & Ananthakrishnan: Sivaraman, 1981: 22; Harvey, 1986: 760.

Type locality: Madras, Tamil Nadu, India.
Distribution: India.

Geogarypus hungaricus (Tömösváry)

Garypus hungaricus Tömösváry, 1882b: 212–213, plate 2 figs 19–21; Daday, 1918: 2.
Geogarypus (Geogarypus) (?) hungaricus (Tömösváry): Beier, 1932a: 235; Roewer, 1937: 270.
Geogarypus (?) hungaricus (Tömösváry): Harvey, 1986: 760.

Type locality: Sátoraljaújhely (as Sátoralja-Ujhely), Hungary.
Distribution: Hungary.

Geogarypus incertus Caporiacco

Geogarypus incertus Caporiacco, 1947b: 21; Caporiacco, 1948a: 617, fig. 10; Harvey, 1986: 760.

Type locality: Guyana.
Distribution: Guyana.

Geogarypus irrugatus (E. Simon)

Garypus irrugatus E. Simon, 1899a: 122; With, 1906: 28–29, 104–107, figs 8a–b, plate 1 figs 6a–d, plate 2 figs 1a–d.
Geogarypus irrugatus (E. Simon): J. C. Chamberlin, 1930: 611; J. C. Chamberlin, 1931a: 114, figs 9m, 15o, 16g, 17z, 32k–l, 37w; Roewer, 1936: figs 40a–b, 42c–d; Beier, 1976e: 99–100; Mahnert, 1977a: 94; Harvey, 1986: 760.
Geogarypus (Geogarypus) irrugatus (E. Simon): Beier, 1932a: 233; Roewer, 1937: 270, fig. 221.

Type locality: Sumatra, Indonesia.
Distribution: Bhutan, China, Indonesia (Sumatra), Philippines, Thailand, U.S.S.R. (Kirghizia).

Geogarypus itapemirinensis Feio

Geogarypus (Geogarypus) itapemirinensis Feio, 1941: 241–244, fig. 1; Feio, 1945: 5.
Geogarypus itapemirinensis Feio: Harvey, 1986: 760.

Type locality: Cachoeiro de Itapemirim, Espirito Santo, Brazil.
Distribution: Brazil (Distrito Federal, Espirito Santo).

Geogarypus javanus (Tullgren)

Garypus javanus Tullgren, 1905: 43–44; Tullgren, 1907a: 66; Ellingsen, 1910a: 388; Ellingsen, 1912a: 122.
Geogarypus javanus (Tullgren): J. C. Chamberlin, 1930: 609; Beier, 1953g: 82; Weidner, 1959: 115; Beier, 1982: 43; Harvey, 1986: 760; Harvey, 1988b: 327–329, figs 50–54.
Geogarypus formosanus Beier, 1931c: 315–316, fig. 10 (synonymized by Harvey, 1988b: 327).
Geogarypus (Geogarypus) formosanus Beier: Beier, 1932a: 232–233, fig. 260; Roewer, 1937: 270.
Geogarypus (Geogarypus) javanus (Tullgren): Beier, 1932a: 233–234, fig. 261; Roewer, 1937: 270; Morikawa, 1963: 6–7, figs c–e.
Geogarypus audyi Beier, 1952a: 103–105, fig. 6 (synonymized by Harvey, 1988b: 327).
Geogarypus (Geogarypus) javanus formosanus Beier: Beier, 1957d: 21–24, figs 9a–b, 10a–d, 11a–d.
Geogarypus (Geogarypus) javanus javanus (Tullgren): Beier, 1957d: 25; Beier, 1965g: 764.
Geogarypus (Geogarypus) javanus audyi Beier: Beier, 1957d: 25; Harvey, 1986: 760.
Geogarypus (Geogarypus) elegans audyi Beier: Beier, 1963d: 507–508.
Geogarypus javanus javanus (Tullgren): Beier, 1966d: 140; Beier, 1970g: 318.
Geogarypus (Geogarypus) javanus takensis Beier, 1967g: 352, fig. 12 (synonymized by Harvey, 1988b: 328).
Geogarypus javanus formosanus Beier: Harvey, 1986: 760.
Geogarypus javanus takensis Beier: Harvey, 1986: 760.

Type localities: of *Garypus javanus*: Bogor (as Buitenzorg), Java, Indonesia.
of *Geogarypus formosanus*: Takao (now Kao-hsiung), Taiwan.
of *Geogarypus audyi*: Kuala Lumpur, Malaysia.
of *Geogarypus (Geogarypus) javanus takensis*: Langs weg var Tak noar Thoen, 65 km var. Tak, Thailand.
Distribution: Caroline Islands, Indonesia (Irian Jaya, Java, Krakatau Islands, Sumba), Malaysia, Mariana Islands, Marshall Islands, Papua New Guinea, Solomon Islands, Taiwan, Thailand.

Geogarypus longidigitatus (Rainbow)

Chelifer longidigitatus Rainbow, 1897: 108–109, fig. 2.
Garypus longidigitatus (Rainbow): Pocock, 1898: 323; With, 1906: fig. 3; With, 1907: 66–68, fig. 20.
Geogarypus longidigitatus (Rainbow): J. C. Chamberlin, 1930: 609; Roewer, 1936: figs 22d, 70m; Harvey, 1986: 760.
Geogarypus (Geogarypus) longidigitatus (Rainbow): Beier, 1932a: 233; J. C. Chamberlin, 1934b: 6; Roewer, 1937: 270.

Type locality: Funafuti, Tuvalu.
Distribution: Tuvalu.

! Geogarypus macrodactylus Beier

Geogarypus macrodactylus Beier, 1937d: 305, fig. 4; Roewer, 1940: 331; Beier, 1955g: 51; Schawaller, 1978: 3; Harvey, 1986: 760.

Type locality: Baltic Amber.
Distribution: Baltic Amber.

Geogarypus maculatus (With)

Garypus maculatus With, 1907: 168–170, figs 21–25.
Geogarypus maculatus (With): J. C. Chamberlin, 1930: 609; Harvey, 1986: 760.
Geogarypus (Geogarypus) maculatus (With): Beier, 1932a: 232; Roewer, 1937: 270.

Type locality: Balthazar, Grenada.
Distribution: Grenada.

! Geogarypus major Beier

Geogarypus (?) major Beier, 1937d: 306, fig. 5; Roewer, 1940: 331; Schawaller, 1978: 3.
Geogarypus major Beier: Harvey, 1986: 760.

Type locality: Baltic Amber.
Distribution: Baltic Amber.

Geogarypus maroccanus Beier

Geogarypus (Geogarypus) maroccanus Beier, 1961b: 31–32, fig. 3.
Geogarypus maroccanus Beier: Harvey, 1986: 760; Callaini, 1988b: 47–50, figs 4a–g.

Type locality: S. of Chechaouen (as Xauen), Morocco.
Distribution: Morocco.

Geogarypus marquesianus J. C. Chamberlin

Geogarypus (Geogarypus) marquesianus J. C. Chamberlin, 1939b: 208–210, figs 1a–g; Tenorio & Muchmore, 1982: 383.
Geogarypus marquesianus J. C. Chamberlin: Harvey, 1986: 760.

Type locality: Vaipee Valley, Putatauua, Uahuka, Marquesas Islands.
Distribution: Marquesas Islands.

Geogarypus micronesiensis Morikawa

Geogarypus (Geogarypus) micronesiensis Morikawa, 1952a: 245, figs 4, 5c; Beier, 1957d: 24; Morikawa, 1960: 132–133, plate 3 fig. 2, plate 7 fig. 15; Morikawa, 1962: 419.
Geogarypus micronesiensis Morikawa: Harvey, 1986: 760.

Type locality: Marcus Island.
Distribution: Marcus Island.

Geogarypus minor (L. Koch)

Garypus minor L. Koch, 1873: 38–40; E. Simon, 1878a: 152; E. Simon, 1879a: 46–47, plate 17 fig. 3; Becker, 1880b: cxli; Canestrini, 1884: no. 8, figs 1–4; Daday, 1888: 126, 182; E. Simon, 1898c: 3; E. Simon, 1898d: 21; E. Simon, 1900b: 593; Tullgren, 1907c: 4–7, fig. 2; Ellingsen, 1908c: 670; Ellingsen, 1909a: 207, 209; Ellingsen, 1910a: 388; Krausse-Heldrungen, 1912: 65; Nonidez, 1917: 27; Navás, 1921: 166; Navás, 1925: 112, fig. 9; Bacelar, 1928: 190; Caporiacco, 1928b: 126; Beier, 1929a: 358; Beier, 1929e: 450; Beier, 1930f: 72.

Garypus meridionalis Canestrini, 1885: no. 7, figs 1–5 (synonymized by Beier, 1963b: 242).

Cheiridium tetrophthalmum Daday, 1889c: 27–28, figs 8–10, 14; Daday, 1918: 2 (synonymized by Beier, 1932a: 230).

Garypus lusitanus Navás, 1923: 32–33, fig. 12 (synonymized by Navás, 1925: 112).

Geogarypus minor (L. Koch): J. C. Chamberlin, 1930: 610; J. C. Chamberlin, 1931a: 181; Beier, 1932g: figs 232, 238–245; Roewer, 1937: figs 162–163, 166–176, 178; Beier, 1939f: 196–197; Vachon, 1940g: 144; Beier, 1948a: 190; Caporiacco, 1948b: 34; Beier, 1949: 9; Beier, 1953a: 301; Beier, 1955n: 107; Beier, 1957b: 147; Beier, 1959f: 128; Beier, 1961b: 30; Beier, 1963b: 242; Beier, 1963i: 285; Beier, 1965f: 92; Beier, 1966f: 345; Beier, 1967f: 308; Lazzeroni, 1969a: 335; Lazzeroni, 1969c: 239; Lazzeroni, 1970a: 207; Lagar, 1972a: 20; Gardini, 1975: 8–9; Mahnert, 1975c: 195; Mahnert, 1977d: 97, fig. 4; Callaini, 1979b: 138; Callaini, 1983c: 298–299; Harvey, 1986: 760, fig. 4; Callaini, 1988b: 46.

Geogarypus (Geogarypus) minor (L. Koch): Beier, 1932a: 230, figs 253, 255; Roewer, 1937: 269, fig. 222a; Beier, 1952e: 4.

Type localities: of *Garypus minor*: Corsica.
of *Garypus meridionalis*: Rome, Italy.
of *Cheiridium tetrophthalmum*: Vadé, Portugal.
of *Garypus lusitanus*: San Martinho d'Anta, Portugal.
Distribution: Albania, Algeria, Austria, Corsica, France, Greece, Italy, Malta, Morocco, Portugal, Sardinia, Spain, Sudan, Turkey, Yugoslavia.

Geogarypus minutus (Tullgren)

Garypus minutus Tullgren, 1907a: 65–66, fig. 21; Tullgren, 1907b: 229; Ellingsen, 1912b: 87, 108–110; Godfrey, 1927: 18.

Geogarypus minutus (Tullgren): J. C. Chamberlin, 1930: 609; Weidner, 1959: 115; Beier, 1964k: 60; Harvey, 1986: 760.

Geogarypus (Geogarypus) minutus (Tullgren): Beier, 1932a: 231, fig. 257; Roewer, 1937: 269; Beier, 1958a: 171.

Type locality: Port Elizabeth, Cape Province, South Africa.
Distribution: South Africa.

Geogarypus mirei Heurtault

Geogarypus (?) mirei Heurtault, 1970b: 1361–1365, figs 1–7.
Geogarypus mirei Heurtault: Mahnert, 1978d: 879; Harvey, 1986: 760.

Type locality: Kaortchi, Borkou, Chad.
Distribution: Chad.

Geogarypus nepalensis Beier

Geogarypus nepalensis Beier, 1974b: 268–269, fig. 5; Harvey, 1986: 760; Schawaller, 1987a: 210.

Type locality: between Pina and Rarasee, Nepal.
Distribution: Nepal.

Geogarypus nigrimanus (E. Simon)

Garypus nigrimanus E. Simon, 1879a: 47, 312; E. Simon, 1898c: 3; E. Simon, 1898d: 21; E. Simon, 1899d: 86; E. Simon, 1900b: 592.

Geogarypus nigrimanus (E. Simon): J. C. Chamberlin, 1930: 610, figs 1m–n; J. C. Chamberlin, 1931a: 227, figs 17x–y, 19k, 44a–b, 62; Beier, 1932g: fig. 258; Vachon, 1936a: 78; Vachon, 1937f: 128; Beier, 1955n: 107; Beier, 1959f: 129; Beier, 1961b: 31; Beier, 1963a: 259; Beier, 1963b: 243, fig. 247; Beier, 1963h: 155; Beier, 1963i: 285; Beier, 1966e: 164; Beier, 1966f: 345; Beier, 1967f: 308; Lazzeroni, 1969a: 335; Lazzeroni, 1969c: 239; Lazzeroni, 1970a: 207; Mahnert, 1975c: 195; Estany, 1977b: 30–31; Callaini, 1979b: 138–139; Leclerc & Heurtault, 1979: 245; Callaini, 1982a: 17, 20–21, fig. 3b; Pieper, 1981: 3; Callaini, 1982b: 449; Callaini, 1983a: 151–152; Callaini, 1983c: 299–300; Callaini, 1983e: 224; Callaini & Dallai, 1984: 336–341, fig. 5; Harvey, 1986: 760; Callaini, 1988b: 47; Callaini, 1989: 145; Callaini & Dallai, 1989: 90, fig. 13.

Geogarypus (Geogarypus) nigrimanus (E. Simon): Beier, 1932a: 229, fig. 254; Roewer, 1937: 269.

Geogaripus (sic) *nigrimanus* (E. Simon): Orghidan, Dumitresco & Orghidan, 1975: 30–31.

Type locality: Corsica.
Distribution: Algeria, Balearic Islands, Corsica, Crete, France, Greece, Italy, Madeira Islands, Malta, Morocco, Sardinia, Sicily, Spain, Turkey.

Geogarypus ocellatus Mahnert

Geogarypus (Geogarypus) ocellatus Mahnert, 1978d: 874–879, 887, figs 14–20.
Geogarypus ocellatus Mahnert: Harvey, 1986: 760.

Type locality: Vallée de Mai, Praslin, Seychelles.
Distribution: Seychelles.

Geogarypus olivaceus (Tullgren)

Garypus olivaceus Tullgren, 1907a: 63–65, fig. 20; Tullgren, 1907b: 229.
Garypus senegalensis Balzan: Ellingsen, 1912b: 87, 112–114 (misidentification).
Geogarypus (Geogarypus) olivaceus (Tullgren): Beier, 1932a: 230–231, fig. 256; Roewer, 1937: 269; Beier, 1955l: 301, fig. 20; Beier, 1958a: 170–171.
Geogarypus (Geogarypus) flavus Beier, 1947b: 318–319, fig. 24; Beier, 1958a: 170 (synonymized by Beier, 1964k: 59).
Geogarypus olivaceus (Tullgren): Weidner, 1959: 115; Beier, 1964k: 59; Beier, 1966k: 461, 470; Lawrence, 1967: 89; Harvey, 1986: 760.

Type localities: of *Garypus olivaceus*: Bothaville, Orange Free State, South Africa.
of *Geogarypus (Geogarypus) flavus*: Keurboomis River, Knysna, Cape Province, South Africa.
Distribution: South Africa.

Geogarypus palauanus Beier

Geogarypus (Indogarypus) palauanus Beier, 1957d: 25–27, figs 12a–d.
Geogarypus palauanus Beier: Harvey, 1986: 760.

Type locality: Koror, Palau Islands, Caroline Islands.
Distribution: Caroline Islands.

Geogarypus paraguayanus Beier

Geogarypus paraguayanus Beier, 1931c: 316; Harvey, 1986: 760.
Geogarypus (Geogarypus) paraguayanus Beier: Beier, 1932a: 232, fig. 258; Roewer, 1937: 270; Feio, 1945: 4.

Type locality: Paraguay.
Distribution: Brazil (Distrito Federal), Paraguay.

Geogarypus personatus (E. Simon)

Garypus personatus E. Simon, 1900a: 518–519.
Geogarypus personatus (E. Simon): J. C. Chamberlin, 1930: 609; Harvey, 1986: 760.
Geogarypus (Geogarypus) (?) *personatus* (E. Simon): Beier, 1932a: 236.
Geogarypus (Geogarypus) personatus (E. Simon): J. C. Chamberlin, 1934b: 6.

Type locality: Kaala Mountains, Oahu, Hawaii.
Distribution: Hawaii.

Geogarypus pisinnus Harvey

Geogarypus pisinnus Harvey, 1986: 771–772, figs 5, 38–40.

Type locality: Arnhem Highway at South Alligator River, Northern Territory, Australia.
Distribution: Australia (Northern Territory).

Geogarypus pulcher Beier

Geogarypus (Geogarypus) pulcher Beier, 1963f: 192–193, fig. 6; Mahnert, 1974c: 379.
Geogarypus pulcher Beier: Harvey, 1986: 760.

Type locality: Kfar Mallal, Israel.
Distribution: Israel.

Geogarypus purcelli (Ellingsen)

Garypus purcelli Ellingsen, 1912b: 110–112.
Geogarypus (Geogarypus) purcelli (Ellingsen): Beier, 1932a: 235; Roewer, 1937: 270; Beier, 1955l: 300, fig. 19.
Geogarypus purcelli (Ellingsen): Beier, 1964k: 59; Beier, 1966k: 461; Harvey, 1986: 760.

Type locality: Beaufort West, Cape Province, South Africa.
Distribution: South Africa.

Geogarypus pustulatus Beier

Geogarypus (Geogarypus) pustulatus Beier, 1959e: 200–201, fig. 13.
Geogarypus pustulatus (Beier): Beier, 1962h: 133; Beier, 1964j: 491; Harvey, 1986: 760; Harvey, 1987d: 140–141, figs 9–15.

Type locality: Lago Moreno, Bariloche, Rio Negro, Argentina.
Distribution: Argentina, Chile.

Geogarypus rhantus Harvey

Geogarypus (Geogarypus) rhantus Harvey, 1981a: 279–282, figs 1a–f; Harvey, 1985b: 143.
Geogarypus rhantus Harvey, 1986: 761–764, figs 5–10, 18.

Type locality: Lamond Hill, Iron Range, Queensland, Australia.
Distribution: Australia (Queensland).

Geogarypus robustus Beier

Geogarypus (Geogarypus) robustus Beier, 1947b: 319–320, fig. 25; Beier, 1955l: 301.
Geogarypus robustus (Beier): Harvey, 1986: 760.

Type locality: Fernwood, Cape Province, South Africa.
Distribution: South Africa.

Geogarypus sagittatus Beier

Geogarypus (Indogarypus) sagittatus Beier, 1965g: 764–765, fig. 7; Tenorio & Muchmore, 1982: 383.
Geogarypus sagittatus Beier: Beier, 1982: 43; Harvey, 1986: 760.

Type locality: Ifar, Irian Jaya, Indonesia.
Distribution: Indonesia (Irian Jaya), Papua New Guinea.

Geogarypus shulovi Beier

Geogarypus (Geogarypus) shulovi Beier, 1963f: 193–195, fig. 7; Mahnert, 1974c: 379.
Geogarypus shulovi Beier: Beier, 1965f: 93; Beier, 1967f: 308; Beier, 1971a: 363; Beier, 1973c: 226; Schawaller, 1985a: 10, fig. 22; Harvey, 1986: 760.

Type locality: Daliyya (as Dalia), Israel.
Distribution: Iran, Israel, Turkey, U.S.S.R. (Turkmenistan).

Geogarypus taylori Harvey

Geogarypus taylori Harvey, 1986: 764–769, figs 5, 11–15, 17, 19–34.

Type locality: Lerderderg Gorge, 9 km NNW. of Bacchus Marsh, Victoria, Australia.
Distribution: Australia (New South Wales, Northern Territory, South Australia, Victoria, Western Australia).

Geogarypus tenuis J. C. Chamberlin

Geogarypus tenuis J. C. Chamberlin, 1930: 611–612; J. C. Chamberlin, 1931a: fig. 29i; Beier, 1973b: 45; Harvey, 1986: 760.
Geogarypus (Geogarypus) tenuis J. C. Chamberlin: Beier, 1932a: 234; Roewer, 1937: 270.

Type locality: 'presumably from Ceylon'.
Distribution: Sri Lanka.

Geogarypus triangularis (Ellingsen)

Garypus minutus triangularis Ellingsen, 1912b: 110.
Geogarypus (Geogarypus) triangularis (Ellingsen): Beier, 1932a: 231; Roewer, 1937: 269.
Geogarypus triangularis (Ellingsen): Beier, 1964k: 59–60, fig. 20; Harvey, 1986: 760.

Type localities: Izeli, near King William's Town, Cape Province, South Africa; Bushman's Rock, Blythswood, Butterworth, Cape Province, South Africa; and Naval Hill, Bloemfontein, Orange Free State, South Africa.
Distribution: South Africa.

Genus Indogarypus Beier

Geogarypus (Indogarypus) Beier, 1957d: 25; Murthy & Ananthakrishnan, 1977: 108.
Indogarypus Beier: Harvey, 1986: 759.

Type species: *Garypus indicus* Beier, 1930e, by original designation.

Indogarypus indicus (Beier)

Garypus indicus Beier, 1930e: 290–291, figs 3, 4a–b.
Geogarypus (Geogarypus) indicus (Beier): Beier, 1932a: 235, fig. 262; Roewer, 1937: 270.
Geogarypus (Indogarypus) indicus (Beier): Beier, 1957d: 25; Murthy & Ananthakrishnan, 1977: 108.
Geogarypus indicus (Beier): Beier, 1973b: 46; Beier, 1974e: 1007–1008.
Indogarypus indicus (Beier): Harvey, 1986: 759, fig. 3.

Type locality: Travancore, India.
Distribution: India, Sri Lanka.

Family MENTHIDAE J. C. Chamberlin

Menthidae J. C. Chamberlin, 1930: 585; J. C. Chamberlin, 1931a: 222; Beier, 1932a: 177; Beier, 1932g: fig. 184; Roewer, 1937: 258−259; Vitali-di Castri, 1969a: 501; Muchmore, 1982a: 99.

Genus **Menthus** J. C. Chamberlin

Menthus J. C. Chamberlin, 1930: 585−586; Beier, 1932a: 177.

Type species: *Minniza rossi* J. C. Chamberlin, 1923c, by original designation.

Menthus californicus J. C. Chamberlin

Menthus californicus J. C. Chamberlin, 1930: 587; J. C. Chamberlin, 1931a, figs 29a, 37a; Beier, 1932a: 178; Roewer, 1937: 259; Hoff, 1958: 17.

Type locality: Palm Canyon, near Palm Springs, Riverside County, California, U.S.A.
Distribution: U.S.A. (California).

Menthus lindahli (J. C. Chamberlin)

Minniza lindahli J. C. Chamberlin, 1923c: 365−366, plate 2 fig. 12.
Menthus lindahli (J. C. Chamberlin): J. C. Chamberlin, 1930: 587; Beier, 1932a: 178; Roewer, 1937: 259; V. F. Lee, 1979a: 18−19, figs 22−23, 31.

Type locality: Tepoca Bay, Sonora, Mexico.
Distribution: Mexico.

Menthus mexicanus Hoff

Menthus mexicanus Hoff, 1945e: 4−7, figs 6−7.

Type locality: Mexcala, Guerrero, Mexico.
Distribution: Mexico.

Menthus rossi (J. C. Chamberlin)

Minniza rossi J. C. Chamberlin, 1923c: 365, plate 1 fig. 5, plate 2 fig. 11, plate 3 figs 9, 17.
Menthus rossi (J. C. Chamberlin): J. C. Chamberlin, 1930: 586, figs 1o, 2b, 3g; J. C. Chamberlin, 1931a: figs 7b, 7h, 9g, 11y, 13r−s, 15k, 17r−s, 19h, 26e, 33n, 60; Beier, 1932a: 178, fig. 205; Beier, 1932g: figs 164a−b, 256; Roewer, 1936: fig. 14; Roewer, 1937: 259, fig. 216; Vachon, 1949: fig. 197; Weygoldt, 1969a: 6.

Type locality: San Esteban Island, Gulf of California, Mexico.
Distribution: Mexico.

Genus **Oligomenthus** Beier

Oligomenthus Beier, 1962h: 131−132.

Type species: *Oligomenthus argentinus* Beier, 1962h, by original designation.

Oligomenthus argentinus Beier

Oligomenthus argentinus Beier, 1962h: 132−133, fig. 1.

Type locality: Aconcagua, Mendoza, Argentina.
Distribution: Argentina.

Oligomenthus chilensis Vitali-di Castri

Oligomenthus chilensis Vitali-di Castri, 1969a: 502−506, figs 3−10; Cekalovic, 1984: 14.

Type locality: Paposo, Antofagasta, Chile.
Distribution: Chile.

Genus **Paramenthus** Beier

Paramenthus Beier, 1963f: 186.

Type species: *Paramenthus shulovi* Beier, 1963f, by original designation.

Paramenthus shulovi Beier

Paramenthus shulovi Beier, 1963f: 186–187, fig. 2; Mahnert, 1974c: 378.

Type locality: Yeroham (as 24 km Beer Sheva-Yeroham Road), Israel.
Distribution: Israel.

Family OLPIIDAE Banks

Olpiinae Banks, 1895a: 2; J. C. Chamberlin, 1930: 598; J. C. Chamberlin, 1931a: 225;
  Beier, 1932a: 179; Roewer, 1937: 259–260; Hoff, 1945f: 1; Hoff, 1956b: 26–27;
  Morikawa, 1960: 125; Hoff, 1964a: 19; Murthy & Ananthakrishnan, 1977: 30;
  Benedict & Malcolm, 1978a: 119–120; Muchmore, 1980e: 161.
Olpiidae Banks: J. C. Chamberlin, 1930: 588; J. C. Chamberlin, 1931a: 223–224;
  Beier, 1932a: 179; Beier, 1932g: 184; Roewer, 1937: 259; Petrunkevitch, 1955: 81;
  Morikawa, 1960: 125; Beier, 1963b: 229–230; Hoff, 1964a: 18; Murthy &
  Ananthakrishnan, 1977: 30–31; Benedict & Malcolm, 1978a: 119; Muchmore,
  1979a: 197; Muchmore, 1980e: 161; Muchmore, 1982a: 99; Harvey, 1985b: 146;
  R. Schuster, 1986a: 270.
Garypininae Daday, 1888: 123; J. C. Chamberlin, 1930: 590–591; J. C. Chamberlin,
  1931a: 225; Beier, 1932a: 203; Roewer, 1937: 263; Hoff, 1956b: 27; Morikawa,
  1960: 129; Hoff, 1964a: 35; Benedict & Malcolm, 1978a: 124–125; Muchmore,
  1980: 165.
Olpini (sic) Banks: Hoff, 1945f: 1.
Olpiini Banks: Hoff, 1964a: 19.
Xenolpini (sic): Hoff, 1945f: 1.
Xenolpiini Hoff: Hoff, 1964a: 28; Murthy & Ananthakrishnan, 1977: 80.
Hesperolpiini Hoff, 1964a: 30–31: Muchmore, 1980e: 161.

Genus **Aldabrinus** J. C. Chamberlin

Aldabrinus J. C. Chamberlin, 1930: 597; Beier, 1932a: 214; Muchmore, 1974a: 1–2;
  Muchmore, 1979a: 200.

Type species: *Aldabrinus aldabrinus* J. C. Chamberlin, 1930, by original designation.

Aldabrinus aldabrinus J. C. Chamberlin

Aldabrinus aldabrinus J. C. Chamberlin, 1930: 597–598; J. C. Chamberlin, 1931a:
  figs 16o, 37g; Beier, 1932a: 214; Roewer, 1937: 266; Muchmore, 1974a: 2–4,
  figs 1–5.

Type locality: El Esprit, Aldabra Islands.
Distribution: Aldabra Islands.

Aldabrinus floridanus Muchmore

Aldabrinus floridanus Muchmore, 1974a: 4–6, figs 6–12; Muchmore, 1979a: 208.

Type locality: north Key Largo, Monroe County, Florida, U.S.A.
Distribution: U.S.A. (Florida).

Genus **Amblyolpium** E. Simon

Amblyolpium E. Simon, 1898c: 3; J. C. Chamberlin, 1930: 593; Beier, 1932a: 204;
  Morikawa, 1960: 129; Murthy & Ananthakrishnan, 1977: 96–97.

Type species: *Amblyolpium dollfusi* E. Simon, 1898c, by monotypy.

Amblyolpium anatolicum Beier

Amblyolpium anatolicum Beier, 1967f: 306–307, fig. 2; Beier, 1970f: 322.

Type locality: Egridir, Turkey.
Distribution: Turkey.

Amblyolpium bellum J. C. Chamberlin

Amblyolpium bellum J. C. Chamberlin, 1930: 593–594; Beier, 1932a: 204–205; Roewer, 1937: 265; Harvey, 1988b: 330–333, figs 59–67.

Type locality: Kepulauan Banda (as Banda), Maluku, Indonesia.
Distribution: Indonesia (Java, Kepulauan Banda, Krakatau Islands).

Amblyolpium biaroliatum (Tömösváry)

Olpium biaroliatum Tömösváry, 1884: 23, figs 15–18.
Garypus biaroliatus (Tömösváry): Daday, 1888: 126, 182.
Olpium biareolatum (sic) Tömösváry: With, 1906: 109.
Amblyolpium biaroliatum (Tömösváry): Beier, 1932a: 205, fig. 233; Roewer, 1937: 265; Murthy & Ananthakrishnan, 1977: 97.

Type locality: east India.
Distribution: India.

Amblyolpium birmanicum (With)

Olpium birmanicum With, 1906: 109–111, plate 2 figs 7a–h; Ellingsen, 1914: 11–12; Beier, 1930e: 289.
Garypinus birmanicus (With): J. C. Chamberlin, 1930: 592–593.
Amblyolpium birmanicum (With): Beier, 1932a: 205, fig. 234; Roewer, 1936: figs 38a–b, 70i; Roewer, 1937: 265.

Type locality: Tharrawaddy (as Tharrawaddi), Burma.
Distribution: Burma, India, Indonesia (Sumatra).

Amblyolpium dollfusi E. Simon

Amblyolpium dollfusi E. Simon, 1898c: 3–4; J. C. Chamberlin, 1930: 593; Beier, 1932a: 204; Roewer, 1937: 265; Lazzeroni, 1969c: 239; Beier, 1970f: 322; Lazzeroni, 1970c: 106–110, figs 1–7; Callaini, 1982a: 17, 20, figs 2d, 3a; Schawaller, 1981b: 46–47, figs 5–6; Callaini, 1983a: 152; Callaini, 1983c: 297; Leclerc, 1984a: 53, fig. 1; Callaini & Dallai, 1989: 90, fig. 10.

Type locality: Collobrières, Var, France.
Distribution: Corsica, France, Italy, Sardinia.

Amblyolpium franzi Beier

Amblyolpium franzi Beier, 1970a: 47–49, fig. 2; Beier, 1970f: 322; Beier, 1976a: 24; Pieper, 1981: 3.

Type locality: Pico Branco, Porto Santo, Madeira Islands.
Distribution: Madeira Islands.

Amblyolpium graecum Mahnert

Amblyolpium graecum Mahnert, 1976a: 180–182, figs 5–9.

Type locality: Thea Dimitra spilija Cave, Peloponnes, Greece.
Distribution: Greece.

Amblyolpium japonicum Morikawa

Amblyolpium japonicum Morikawa, 1960: 129–130, plate 2 fig. 9, plate 7 fig. 12, plate 9 fig. 15; Morikawa, 1962: 418.

Type locality: Matsuyama, Shikoku, Japan.
Distribution: Japan.

Amblyolpium novaeguineae Beier

Amblyolpium novaeguineae Beier, 1971b: 367–368, fig. 1.

Type locality: Heads Hump, Bulolo, Morobe District, Papua New Guinea.
Distribution: Papua New Guinea.

Amblyolpium ortonedae (Ellingsen)

Olpium ortonedae Ellingsen, 1902: 159–162; Tullgren, 1907a: 67.
Garypinus ortonedae (Ellingsen): J. C. Chamberlin, 1930: 593.
Amblyolpium ortonedae (Ellingsen): Beier, 1932a: 206, fig. 235; Roewer, 1937: 265; Feio, 1945: 4; Beier, 1959e: 196, fig. 9.

Type locality: Naranjito, Ecuador.
Distribution: Brazil (Distrito Federal), Colombia, Ecuador.

Amblyolpium ruficeps Beier

Amblyolpium ruficeps Beier, 1966g: 370–371, fig. 5.

Type locality: Niaouli forest near Col Boa, Poya, New Caledonia.
Distribution: New Caledonia.

Amblyolpium salomonense Beier

Amblyolpium salomonense Beier, 1970g: 317–318, fig. 1.

Type locality: Kukum, Guadalcanal, Solomon Islands.
Distribution: Solomon Islands.

Amblyolpium simoni Heurtault

Amblyolpium simoni Heurtault, 1970a: 1165–1172, figs 1–17.

Type locality: Emi Koussi, Chad.
Distribution: Chad.

Genus **Aphelolpium** Hoff

Aphelolpium Hoff, 1964a: 31–32; Muchmore, 1979a: 200.

Type species: *Aphelolpium scitulum* Hoff, 1964a, by original designation.

Aphelolpium cayanum Muchmore

Aphelolpium cayanum Muchmore, 1979a: 200–202, figs 12–20.

Type locality: Marathon, Vaca Key, Monroe County, Florida, U.S.A.
Distribution: U.S.A. (Florida).

Aphelolpium scitulum Hoff

Aphelolpium scitulum Hoff, 1964a: 32–35, figs 11–12; Zeh, 1987b: 1086.

Type locality: near Port Henderson, Parish of St Catherine, Jamaica.
Distribution: Jamaica.

Aphelolpium thibaudi Heurtault & Rebière

Aphelolpium thibaudi Heurtault & Rebière, 1983: 594–596, figs 5–10.

Type locality: Régale, Marie-Galante.
Distribution: Marie-Galante, Martinique.

Genus **Apolpium** J. C. Chamberlin

Apolpium J. C. Chamberlin, 1930: 606; Beier, 1932a: 191.

Type species: *Olpium cordimanum* Balzan, 1892, by original designation.

Apolpium cordimanum (Balzan)

Olpium cordimanum Balzan, 1892: 536–537, fig. 28; With, 1906: fig. 1; With, 1907: 72–73.

Apolpium cordimanum (Balzan): J. C. Chamberlin, 1930: 606, fig. 2d; J. C. Chamberlin, 1931a: figs 32m, 37j; Beier, 1932a: 191; Roewer, 1936: figs 22b, 70h; Roewer, 1937: 262.

Type localities: Tovar, Venezuela; and San Esteban, Venezuela.
Distribution: Colombia, Venezuela.

Apolpium ecuadorense Hoff

Apolpium ecuadorensis Hoff, 1945e: 7–10, figs 8–10.

Apolpium ecuadorense Hoff: Mahnert, 1985c: 225–226, fig. 32; Mahnert & Adis, 1986: 214.

Type locality: Banos, Tungurahua, Ecuador.
Distribution: Brazil (Amazonas), Ecuador.

Apolpium leleupi Beier

Apolpium leleupi Beier, 1977c: 101–103, fig. 5.

Type locality: Baños, Ecuador.
Distribution: Ecuador.

Apolpium longidigitatum (Ellingsen)

Olpium longidigitatum Ellingsen, 1910a: 391–392.

Apolpium longidigitatum (Ellingsen): Beier, 1932a: 192, fig. 220; Roewer, 1937: 262; Beier, 1954f: 132; Beier, 1977c: 101; Beier, 1978c: 534–535; Hounsome, 1980: 85.

Type locality: St Thomas, Virgin Islands.
Distribution: Cayman Islands, Galapagos Islands, Venezuela, Virgin Islands.

Apolpium minutum Beier

Apolpium minutum Beier, 1931c: 309–310, fig. 6; Beier, 1932a: 192, fig. 221; Roewer, 1937: 262; Mahnert, 1985c: 226; Mahnert & Adis, 1986: 214.

Type locality: Jimenez, Costa Rica.
Distribution: Brazil (Amazonas), Costa Rica.

Apolpium parvum Hoff

Apolpium parvum Hoff, 1945e: 10–12, figs 11–14.

Type locality: Non Pareil Estate, Sangre Grande, Trinidad.
Distribution: Trinidad.

Apolpium rufeolum (Balzan)

Olpium rufeolum Balzan, 1892: 537.

Apolpium (?) *rufeolum* (Balzan): Beier, 1932a: 192; Roewer, 1937: 262.

Type locality: Caracas, Venezuela.
Distribution: Venezuela.

Apolpium vastum Beier

Apolpium vastum Beier, 1959e: 191–192, fig. 5.

Type locality: Colombia.
Distribution: Colombia.

Genus **Austrohorus** Beier

Austrohorus Beier, 1966a: 288–289.

Type species: *Austrohorus exsul* Beier, 1966a, by original designation.

Austrohorus exsul Beier

Austrohorus exsul Beier, 1966a: 289–290, fig. 7; Harvey, 1981b: 243; Harvey, 1985b: 146.

Type locality: Morawa, Western Australia, Australia.
Distribution: Australia (Western Australia).

Genus **Banksolpium** Muchmore

Banksolpium Muchmore, 1986a: 87.

Type species: *Olpium modestum* Banks, 1909a, by original designation.

Banksolpium magnum Muchmore

Banksolpium magnum Muchmore, 1986a: 89–91, figs 11–12.

Type locality: Viçosa, Minas Gerais, Brazil.
Distribution: Brazil (Minas Gerais).

Banksolpium modestum (Banks)

Olpium modestum Banks, 1909a: 148.
Olpiolum (?) *modestum* (Banks): Beier, 1932a: 197; Roewer, 1937: 262.
Banksolpium modestum (Banks): Muchmore, 1986a: 88–89, figs 6–10.
Not *Olpiolum modestum* (Banks): Beier, 1954f: 134–135, fig. 3 (misidentification; see *Olpiolum machadoi* Heurtault).

Type locality: Pernambuco, Brazil.
Distribution: Brazil (Pernambuco).

Genus **Beierolpium** Heurtault

Beierolpium Heurtault, 1976: 67.

Type species: *Garypinus oceanicus* With, 1907, by original designation.

Beierolpium benoiti Mahnert

Beierolpium benoiti Mahnert, 1978d: 868–873, figs 1–6.

Type locality: Baie Laraie, Curieuse, Seychelles.
Distribution: Seychelles.

Beierolpium bornemisszai (Beier)

Xenolpium bornemisszai Beier, 1966a: 292–294, fig. 10; Harvey, 1981b: 244; Harvey, 1985b: 148.
Beierolpium bornemisszai (Beier): Mahnert, 1978d: 873.

Type locality: Gnangara, 30 km N. of Perth, Western Australia, Australia.
Distribution: Australia (Western Australia).

Beierolpium clarum (Beier)

Indolpium clarum Beier, 1952a: 100–103, figs 4–5.
Xenolpium clarum (Beier): Beier, 1967g: 349.

Beierolpium clarum (Beier); Heurtault, 1980c: 149.

Type locality: Sungei Buloh Leper Settlement, Selangor, Malaysia.
Distribution: Malaysia.

Beierolpium deserticola (Beier)

Calocheiridius deserticola Beier, 1964k: 43–44, fig. 10.
Xenolpium deserticola (Beier): Beier, 1965b: 369; Beier, 1966k: 457.
Beierolpium deserticola (Beier): Mahnert, 1982c: 298.

Type locality: Tsotsoroga, Ngamiland, Botswana.
Distribution: Botswana, Zimbabwe.

Beierolpium flavum Mahnert

Beierolpium flavum Mahnert, 1984b: 47–50, figs 5–14.

Type locality: Sar Uanle, Somalia.
Distribution: Somalia.

Beierolpium holmi Mahnert

Beierolpium holmi Mahnert, 1982c: 290–293, figs 77–83.

Type locality: Warges, Kenya.
Distribution: Kenya.

Beierolpium kerioense Mahnert

Beierolpium kerioense Mahnert, 1982c: 295–298, figs 90–96.

Type locality: near Kabarnet, Val de Kerio, Kenya.
Distribution: Kenya.

Beierolpium lawrencei (Beier)

Calocheiridius lawrencei Beier, 1964k: 42–43, fig. 9.
Xenolpium lawrencei (Beier): Beier, 1965b: 369; Beier, 1966k: 457; Lawrence, 1967: 88.
Beierolpium lawrencei (Beier): Heurtault, 1976: 67.
Beierolpium cf. *lawrencei* (Beier): Mahnert, 1982c: 293–295, figs 84–89.

Type locality: Tshokwane, Kruger National Park, Transvaal, South Africa.
Distribution: Kenya, South Africa.

Beierolpium oceanicum (With)

Olpium longiventer L. Koch: Pocock, 1898: 323 (misidentification).
Garypinus oceanicus With, 1907: 77–79, figs 40–47; Kästner, 1927a: 15.
Horus oceanicus (With): J. C. Chamberlin, 1930: 600.
Xenolpium oceanicum (With): Beier, 1932a: 202; J. C. Chamberlin, 1934b: 5; Roewer, 1937: 263; Beier, 1957d: 16–17, fig. 6a.
Xenolpium oceanicum palauense Beier, 1957d: 18, fig. 6b; Morikawa, 1960: 126; Morikawa, 1962: 419 (synonymized by Harvey, 1988b: 333).
Xenolpium oceanicum reductum Beier, 1957d: 19, figs 6c, 7a–b (synonymized by Harvey, 1988b: 333).
Xenolpium oceanicum latum Beier, 1957d: 19–20, fig. 6d (synonymized by Harvey, 1988b: 333).
Beierolpium oceanicum (With): Heurtault, 1976: 67, fig. 4; Harvey, 1988b: 333–336, figs 68–78.

Type localities: of *Garypinus oceanicus*: Funafuti, Tuvalu.
of *Xenolpium oceanicum palauense*: Auluptagel (Aurapushekaru), Palau Islands, Caroline Islands.
Xenolpium oceanicum reductum: Hill 541, Kusaie, Caroline Islands.
Xenolpium oceanicum latum: Lwejap Island, Marshall Islands.

Distribution: Caroline Islands, Japan, Indonesia (Java, Krakatau Islands), Mariana Islands, Marshall Islands, Samoa, Tuvalu.

Beierolpium rossi (Beier)

Xenolpium rossi Beier, 1967d: 76–77, fig. 3.
Beierolpium rossi (Beier): Mahnert, 1982c: 284–288, figs 65–59.

Type locality: 11 miles E. of Moshi, Tanzania.
Distribution: Kenya, Tanzania.

Beierolpium soudanense (Vachon)

Horus soudanensis Vachon, 1940e: 65–67, figs e–f.
Xenolpium soudanense (Vachon): Beier, 1965b: 369.
Beierolpium soudanense (Vachon): Heurtault, 1976: 68, fig. 6.

Type locality: Bamako, Mali.
Distribution: Mali.

Beierolpium soudanense franzi (Beier)

Xenolpium soudanense franzi Beier, 1965b: 367–370, fig. 3.
Beierolium soudanense franzi (Beier): Mahnert, 1982c: 298.

Type locality: Mt Kelinguen, Chad.
Distribution: Chad.

Beierolpium squalidum (Beier)

Xenolpium squalidum Beier, 1966a: 291–292, fig. 9; Harvey, 1981b: 244; Harvey, 1985b: 148.
Beierolpium squalidum (Beier): Mahnert, 1978d: 873.

Type locality: Kimberley Research Station, 100 km S. of Wyndham, Western Australia, Australia.
Distribution: Australia (Western Australia).

Beierolpium tanense Mahnert

Beierolpium tanense Mahnert, 1982c: 288–290, figs 70–76.

Type locality: 10 km S. of Garsen, Tana River, Kenya.
Distribution: Kenya.

Beierolpium venezuelense Heurtault

Beierolpium venezuelensis (sic) Heurtault, 1976: 62 (nomen nudum).
Beierolpium venezuelensis (sic) Heurtault, 1982: 58–64, figs 1–25.

Type locality: Caracas, Venezuela.
Distribution: Venezuela.

Genus Calocheiridius Beier & Turk

Calocheiridius Beier & Turk, 1952: 767–768; Beier, 1963b: 234; Murthy & Ananthakrishnan, 1977: 32; Sivaraman, 1980b: 325–326.

Type species: *Calocheiridius mavromoustakisi* Beier & Turk, 1952, by original designation.

Calocheiridius africanus Beier

Calocheiridius africanus Beier, 1955c: 536–539, fig. 7; Mahnert, 1982c: 272–274, figs 34–36.
Xenolpium africanum (Beier): Beier, 1967d: 76.

Type locality: Bamburi Beach, Kenya.
Distribution: Kenya.

Calocheiridius amrithiensis Sivaraman

Calocheiridius amrithiensis Sivaraman, 1980b: 332–335, figs 4a–b.

Type locality: Amrithi forest, North Arcot district, Tamil Nadu, India.
Distribution: India.

Calocheiridius antushi Krumpál

Calocheiridius antushi Krumpál, 1983a: 58–60, figs 1–7; Schawaller, 1985a: 8, figs
17–18; Schawaller, 1989: 14.

Type locality: Firyuza (as Firjuza), Ashkhabad (as Aschabad), Turkmenistan, U.S.S.R.
Distribution: U.S.S.R. (Tadzhikistan, Turkemistan).

Calocheiridius badonneli Heurtault

Calocheiridius badonneli Heurtault, 1983: 22–24, figs 50–52.

Type locality: Bandama, Ivory Coast.
Distribution: Ivory Coast.

Calocheiridius beieri (Murthy)

Minniza beieri Murthy, 1960: 30–31, fig. 1A.
Calocheiridius beieri (Murthy): Beier, 1967g: 349–351, fig. 10; Beier, 1974e: 1006;
Murthy & Ananthakrishnan, 1977: 33–34.

Type locality: Nungambakkan, Madras, India.
Distribution: India.

Calocheiridius braccatus Beier

Calocheiridius braccatus Beier, 1959d: 27–29, fig. 12.
Calocheiridius cf. *braccatus* Beier: Mahnert, 1982c: 274–275, figs 37–44.

Type locality: Ile de M'Boko, Lake Tanganyika, Kivu, Zaire.
Distribution: Burundi, Kenya, Zaire.

Calocheiridius centralis (Beier)

Minniza centralis Beier, 1952f: 247–248, fig. 2.
Calocheiridius centralis (Beier): Beier, 1959a: 267; Beier, 1960a: 42; Beier, 1971a: 361;
Murthy & Ananthakrishnan, 1977: 33.

Type locality: Paghman near Kabul, Afghanistan.
Distribution: Afghanistan, India, Iran.

Calocheiridius congicus (Beier)

Pseudohorus congicus Beier, 1954d: 133–134, fig. 2; Beier, 1955a: 10.
Olpium congicum (Beier): Beier, 1957a: 471.
Calocheiridius congicus (Beier): Beier, 1959d: 29–30.

Type locality: Kambove, Zaire.
Distribution: Zaire.

Calocheiridius crassifemoratus Beier

Calocheiridius crassifemoratus Beier, 1955c: 539, fig. 8; Beier, 1959d: 27; Beier,
1962f: 14.

Type locality: Entebbe, Uganda.
Distribution: Uganda, Zaire.

Calocheiridius crassifemoratus moderatus Beier

Calocheiridius crassifemoratus moderatus Beier, 1965b: 367, fig. 2.

Type locality: Deli near Moundou, Chad.
Distribution: Chad.

Calocheiridius cyclopium (Beier)

Xenolpium cyclopium Beier, 1965g: 763–764, fig. 6; Tenorio & Muchmore, 1982: 383.
Calocheiridius sp. vic. *cyclopium* (Beier): Beier, 1982: 43.

Type locality: near Kota Nica, Cyclops Mountains, Irian Jaya, Indonesia.
Distribution: Indonesia (Irian Jaya), Papua New Guinea.

Calocheiridius elegans Murthy & Ananthakrishnan

Calocheiridius elegans Murthy & Ananthakrishnan, 1977: 34–36, figs 8a–b; Sivaraman
& Murthy, 1980a: 163–166, fig. 3; Sivaraman, 1981: 22.

Type locality: Bangalore, Karnataka, India.
Distribution: India.

Calocheiridius elegans pallens Murthy & Ananthakrishnan

Calocheiridius elegans pallens Murthy & Ananthakrishnan, 1977: 36–38, figs 9a–c.

Type locality: Bhubaneswar, Orissa, India.
Distribution: India.

Calocheiridius gabbutti Murthy & Ananthakrishnan

Calocheiridius gabbutti Murthy & Ananthakrishnan, 1977: 40–42, fig. 10a.

Type locality: Nungambakkam, Tamil Nadu, India.
Distribution: India.

Calocheiridius gracilipalpus Mahnert

Calocheiridius gracilipalpus Mahnert, 1982c: 276–278, figs 45–52.

Type locality: Mwingi, Kitui, Kenya.
Distribution: Kenya.

Calocheiridius granulatus Sivaraman

Calocheiridius granulatus Sivaraman, 1980b: 326–328, figs 1a–b.

Type locality: Avadi, Madras, Tamil Nadu, India.
Distribution: India.

Calocheiridius hygricus Murthy & Ananthakrishnan

Calocheiridius hygricus Murthy & Ananthakrishnan, 1977: 42–44, figs 11a–b.

Type locality: Madurai, Tamil Nadu, India.
Distribution: India.

Calocheiridius indicus Beier

Calocheiridius indicus Beier, 1967g: 351–352, fig. 11; Murthy & Ananthakrishnan,
1977: 34.

Type locality: 8 km SE. of Indapur, India.
Distribution: India.

Calocheiridius intermedius Sivaraman

Calocheiridius intermedius Sivaraman, 1980b: 331–332, figs 3a–b.

Type locality: Tirupathi, Andhra Pradesh, India.
Distribution: India.

Calocheiridius libanoticus Beier

Calocheiridius libanoticus Beier, 1955d: 213–215, fig. 2; Beier, 1957b: 146–147; Beier,
1962a: 141; Beier, 1963b: 234–235, fig. 239; Beier, 1963f: 189; Beier, 1963g: 272;

Beier, 1969b: 192; Lazzeroni, 1969a: 334; Lazzeroni, 1969c: 238; Lazzeroni, 1970a: 207; Beier, 1973c: 226; Mahnert, 1975c: 195; Mahnert, 1977d: 97, fig. 3; Callaini, 1982a: 17, 20; Schawaller & Dashdamirov, 1988: 32−34, figs 54−55.
Calocheiridius libanoiicus (sic) Beier: Beier, 1965f: 91.
Chalocheiridius (sic) *libanoticus* Beier: Beier, 1967f: 305.

Type locality: Vi Baht Atahe, Lebanon.
Distribution: Greece, Israel, Italy, Lebanon, Malta, Sardinia, Turkey, U.S.S.R. (Azerbaijan).

Calocheiridius loebli Beier

Calocheiridius loebli Beier, 1974e: 1006−1007, fig. 5; Murthy & Ananthakrishnan, 1977: 34.

Type locality: Pothundy Dam, Nelliampathi Hills, Kerala, India.
Distribution: India.

Calocheiridius mavromoustakisi Beier & Turk

Calocheiridius mavromoustakisi Beier & Turk, 1952: 768−769, fig. 2; Beier, 1963b: 235; Beier, 1963h: 155; Beier, 1963i: 285; Beier, 1966f: 343; Lazzeroni, 1969a: 334; Lazzeroni, 1970a: 207; Heurtault, 1976: fig. 5; Callaini, 1989: 144.
Calocheiridius cf. *mavromoustakisi* Beier & Turk: Gardini, 1975: 5; Gardini, 1985b: 61.

Type locality: Yermasoyia River, Cyprus.
Distribution: Crete, Cyprus, Italy, Sicily.

Calocheiridius murthii Sivaraman

Calocheiridius murthii Sivaraman, 1980b: 328−330, figs 2a−b.

Type locality: Redhills, Madras, India.
Distribution: India.

Calocheiridius mussardi Beier

Calocheiridius mussardi Beier, 1973b: 44−45, fig. 8.

Type locality: Palatupana, Yala National Park, Southern Province, Sri Lanka.
Distribution: Sri Lanka.

Calocheiridius nepalensis Beier

Calocheiridius nepalensis Beier, 1974b: 266, fig. 3.

Type locality: Amlekhganj, Nepal.
Distribution: Nepal.

Calocheiridius novaguineense (Beier)

Xenolpium nova-guineense Beier, 1935e: 484−485, fig. 1; Roewer, 1937: 263.
Xenolpium novaguineense Beier: Beier, 1965g: 762−763, fig. 5; Beier, 1966d: 138; Beier, 1967c: 322; Beier, 1970g: 317.
Xenolpium (Xenolpium) bougainvillense Morikawa, 1963: 5−6, figs 3a−b (synonymized by Beier, 1965g: 762).
Calocheiridius novaguineensis (Beier): Beier, 1982: 43.

Type localities: of *Xenolpium novaguineense*: Irian Jaya (as N.W. New Guinea), Indonesia.
of *Xenolpium (Xenolpium) bougainvillense*: Bouin, Bougainville Island, Papua New Guinea.
Distribution: Indonesia (Irian Jaya), Papua New Guinea, Solomon Islands.

Calocheiridius olivieri (E. Simon)

Olpium olivieri E. Simon, 1879b: 262; Beier, 1932a: 182; Roewer, 1937: 261; Beier, 1963b: 231.
Olpium ollivieri (sic) E. Simon: Becker, 1880b: cxli.
Calocheiridius olivieri (E. Simon): Heurtault, 1981: 218–221, figs 13–18; Callaini, 1983c: 296–297; Callaini & Dallai, 1984: 336–341, fig. 1; Callaini & Dallai, 1989: 89, fig. 8.

Type locality: Ile de Porquerolles, France.
Distribution: France, Sardinia.

Calocheiridius orientalis Murthy & Ananthakrishnan

Calocheiridius orientalis Murthy & Ananthakrishnan, 1977: 38–40, figs 10a–b.

Type locality: Tambaram, Tamil Nadu, India.
Distribution: India.

Calocheiridius rhodesiacus Beier

Calocheiridius rhodesiacus Beier, 1964k: 39–41, fig. 7.

Type locality: Birchenough Bridge, Zimbabwe.
Distribution: Zimbabwe.

Calocheiridius rhodesiacus fuliginosus Beier

Calocheiridius rhodesiacus fuliginosus Beier, 1966k: 456–457, fig. 1.

Type locality: 20 miles from Bochem toward Tonash via Liepzig Store, Transvaal, South Africa.
Distribution: South Africa.

Calocheiridius somalicus (Caporiacco)

Minniza somalica Caporiacco, 1941: 44, fig. 3.
Calocheiridius somalicus (Caporiacco): Mahnert, 1982c: 279, figs 55–57.

Type locality: El Banno, Tertale, Ethiopia.
Distribution: Ethiopia.

Calocheiridius sulcatus Beier

Calocheiridius sulcatus Beier, 1974b: 267, fig. 4.

Type locality: Mt Phulchoki, Nepal.
Distribution: Nepal.

Calocheiridius termitophilus Beier

Calocheiridius termitophilus Beier, 1964i: 198–200, fig. 1.
Calocheiriridius (sic) *termitophilus* Beier: Newlands, 1978: 700.

Type locality: near Kinshasa (as Leopoldville), Zaire.
Distribution: Zaire.

Calocheiridius viridis Murthy & Ananthakrishnan

Calocheiridius viridis Murthy & Ananthakrishnan, 1977: 44–46, fig. 12.

Type locality: Tinnevelly, Tamil Nadu, India.
Distribution: India.

Genus Calocheirus J. C. Chamberlin

Calocheirus J. C. Chamberlin, 1930: 607; Beier, 1932a: 197; Mahnert, 1986c: 145–146.

Apolpiolum Beier, 1963f: 189 (synonymized by Mahnert, 1986c: 146).

Type species: of *Calocheirus*: *Calocheirus atopos* J. C. Chamberlin, 1930, by original designation.

of *Apolpiolum*: *Apolpiolum peregrinum* Beier, 1963f, by original designation.

Calocheirus atopos J. C. Chamberlin

Calocheirus atopos J. C. Chamberlin, 1930: 607−608, fig. 2ll; J. C. Chamberlin, 1931a: 144, figs 31a−b, 43o−p, 46g, 50b; Beier, 1932a: 197−198; Roewer, 1937: 263; Mahnert, 1986c: 145−146, figs 11−12.

Apolpiolum peregrinum Beier, 1963f: 189−190, fig. 4; Mahnert, 1974c: 378; Mahnert, 1980c: 33−35, fig. 1 (synonymized by Mahnert, 1986c: 148).

Type localities: of *Calocheirus atopos*: near Port Sudan, Sudan.

of *Apolpiolum peregrinum*: Horvot Mezada Plateau, Israel.

Distribution: Israel, Saudi Arabia, Sudan.

Calocheirus canariensis (Beier)

Apolpiolum canariense Beier, 1970a: 46−47, fig. 1; Beier, 1976a: 24; Mahnert, 1980a: 259.

Calocheirus canariensis (Beier): Mahnert, 1986c: 151−152.

Type locality: Restinga, Hierro, Canary Islands.

Distribution: Canary Islands.

Calocheirus gigas (Mahnert)

Apolpiolum gigas Mahnert, 1980a: 259−261, figs 1−6.

Calocheirus gigas (Mahnert): Mahnert, 1986c: 148.

Type locality: Fataga, Gran Canaria, Canary Islands.

Distribution: Canary Islands.

Calocheirus mirus Mahnert

Calocheirus mirus Mahnert, 1986c: 149−151, figs 6−10.

Type locality: Valle Gran Rey, Gomera, Canary Islands.

Distribution: Canary Islands.

Genus **Cardiolpium** Mahnert

Cardiolpium Mahnert, 1986c: 148−149.

Type species: *Apolpiolum stupidum* Beier, 1963f, by original designation.

Cardiolpium aeginense (Beier)

Apolpiolum aeginense Beier, 1966f: 343−344, fig. 1.

Cardiolpium aeginense (Beier): Mahnert, 1986c: 148.

Type locality: Aíyina (as Ägina), Greece.

Distribution: Greece.

Cardiolpium stupidum (Beier)

Apolpiolum stupidum Beier, 1963f: 191, fig. 5; Beier, 1965f: 91−92, fig. 8; Beier, 1967f: 305; Beier, 1969b: 193; Schawaller, 1985a: 8, fig. 16.

Apolpiolum rhodium Beier, 1965d: 632−633, fig. 1; Beier, 1966e: 162 (synonymized by Mahnert, 1986c: 149).

Cardiolpium stupidum (Beier): Mahnert, 1986c: 148.

Type localities: of *Apolpium stupidum*: Aqua Bella, Israel.

of *Apolpium rhodium*: Apolakkia (as Apolakia), Ródhos (as Rhodes), Greece.

Distribution: Greece, Israel, Turkey, U.S.S.R. (Turkmenistan, Uzbekistan).

Genus **Ectactolpium** Beier

Ectactolpium Beier, 1947b: 295–296.

Type species: *Ectactolpium simile* Beier, 1947b, by original designation.

Ectactolpium astatum Beier

Ectactolpium astatum Beier, 1947b: 297–298, fig. 8; Beier, 1966k: 470; Beier, 1973a: 98.

Type locality: Steinkopf, Namaqualand, Cape Province, South Africa.
Distribution: Namibia, South Africa.

Ectactolpium brevifemoratum Beier

Ectactolpium brevifemoratum Beier, 1947b: 302–304, fig. 11; Beier, 1955l: 282; Beier, 1964k: 51; Beier, 1966k: 470; Beier, 1973a: 98.

Type locality: Koeboes, Namaqualand, Cape Province, South Africa.
Distribution: Namibia, South Africa.

Ectactolpium eximium Beier

Ectactolpium eximium Beier, 1962d: 225–226, fig. 3; Beier, 1973a: 98.

Type locality: 13 miles NE. of Good Hope Mine, Namibia.
Distribution: Namibia.

Ectactolpium flavum Beier

Ectactolpium flavum Beier, 1955l: 280–281, fig. 5; Beier, 1964k: 52; Beier, 1966k: 470; Beier, 1973a: 98.

Type locality: Sanitatas, 85 miles WSW. of Ohopoho, Namibia.
Distribution: Namibia.

Ectactolpium garypoides Beier

Ectactolpium garypoides Beier, 1947b: 298–299; Weidner, 1959: 115; Beier, 1966k: 470; Beier, 1973a: 98.

Type locality: Osona (as Osoni), Namibia.
Distribution: Namibia.

Ectactolpium kalaharicum Beier

Ectactolpium kalaharicum Beier, 1964k: 53–54, fig. 16.

Type locality: Auob (as Ouob) River, Gemsbok Plain, Cape Province, South Africa.
Distribution: South Africa.

Ectactolpium namaquense Beier

Ectactolpium namaquense Beier, 1947b: 301–302, fig. 10; Beier, 1955l: 282; Beier, 1973a: 98.

Type locality: Kamieskroon, Namaqualand, Cape Province, South Africa.
Distribution: Namibia, South Africa.

Ectactolpium namaquense obscurum Beier

Ectactolpium namaquense obscurum Beier, 1964k: 51–52, fig. 15; Beier, 1966k: 459.

Type locality: Pearston, Cape Province, South Africa.
Distribution: South Africa.

Ectactolpium schultzei (Tullgren)

Olpium schultzei Tullgren, 1908b: 286–287, figs 4a–b; Ellingsen, 1912b: 88; Beier, 1932a: 185–186; Roewer, 1937: 261.
Ectactolpium schultzei (Tullgren): Beier, 1947b: 296; Beier, 1962d: 225; 1964k: 52–53; Beier, 1973a: 98.

Type locality: Prince of Wales Bay, near Pomona, Namibia.
Distribution: Namibia, South Africa.

Ectactolpium simile Beier

Ectactolpium simile Beier, 1947b: 299–301, figs 7, 9; Beier, 1955l: 281–282; Beier, 1964k: 53; Beier, 1973a: 98.

Type locality: Lekkersing, Namaqualand, Cape Province, South Africa.
Distribution: Namibia, South Africa.

Ectactolpium zuluanum Beier

Ectactolpium zuluanum Beier, 1958a: 164–167, fig. 4; Beier, 1964k: 54; Beier, 1966k: 459–460; Lawrence, 1967: 88.

Type locality: Ingwavuma, Zululand, Cape Province, South Africa.
Distribution: South Africa.

Genus Euryolpium Redikorzev

Euryolpium Redikorzev, 1938: 82; Murthy & Ananthakrishnan, 1977: 80–81.
Euryolipium (sic) Redikorzev: Roewer, 1940: 345.

Type species: *Euryolpium agniae* Redikorzev, 1938, by monotypy.

Euryolpium agniae Redikorzev

Euryolpium agniae Redikorzev, 1938: 82–84, figs 10–13; Roewer, 1940: 345; Beier, 1967g: 352; Heurtault, 1980c: figs 10–13.
Xenolpium agniae (Redikorzev): Beier, 1951a: 67–69, fig. 13.

Type localities: Loc Ninh (as Lochninch), Vietnam; and Poulo-Condore, Vietnam.
Distribution: China (Fukien), Vietnam.

Euryolpium amboinense (J. C. Chamberlin)

Xenolpium amboinense J. C. Chamberlin, 1930: 601–602; Beier, 1932a: 202; J. C. Chamberlin, 1934b: 5; Roewer, 1937: 263.
Euryolpium amboinense (J. C. Chamberlin): Heurtault, 1980c: 152, figs 16–22.

Type locality: Ambon (as Amboin), Moluccas, Indonesia.
Distribution: Indonesia (Moluccas).

Euryolpium aureum Murthy & Ananthakrishnan

Euryolpium (Euryolpium) aureum Murthy & Ananthakrishnan, 1977: 90–92, figs 30a–b.

Type locality: Trivandrum, Kerala, India.
Distribution: India.

Euryolpium granulatum Murthy & Ananthakrishnan

Euryolpium (Euryolpium) granulatum Murthy & Ananthakrishnan, 1977: 92–94, figs 31a–c.

Type locality: Shingle Islands, Tamil Nadu, India.
Distribution: India.

Euryolpium granulosum (Hoff)

Xenolpium granulosum Hoff, 1947a: 39–41, figs 4–6.
Xenolpium robustum Beier, 1948c: 525–527, fig. 1 (synonymized by Beier, 1966a: 291).
Euryolpium granulosum (Hoff): Beier, 1966a: 291; Harvey, 1981b: 243; Harvey, 1985b: 146.

Type localities: of *Xenolpium granulosum*: Mullewa, Western Australia, Australia.
of *Xenolpium robustum*: Pender Bay, Kimberley, Western Australia, Australia.
Distribution: Australia (Western Australia).

Euryolpium indicum Murthy & Ananthakrishnan

Euryolpium (Euryolpium) indicum Murthy & Ananthakrishnan, 1977: 86–88, figs 28a–b.

Type locality: Nungambakkam, Tamil Nadu, India.
Distribution: India.

Euryolpium intermedium Murthy & Ananthakrishnan

Euryolpium (Euryolpium) intermedium Murthy & Ananthakrishnan, 1977: 88–90, fig. 29.

Type locality: Madras, Tamil Nadu, India.
Distribution: India.

Euryolpium michaelseni (Tullgren)

Olpium michaelseni Tullgren, 1909b: 412–414, fig. 2; J. C. Chamberlin, 1931a: figs 90, 17v; Beier, 1932a: 184, fig. 210; Roewer, 1937: 261.
Olpium michaelsoni (sic) Tullgren: J. C. Chamberlin, 1930: 604; J. C. Chamberlin, 1934b: 5.
Xenolpium michaelseni (Tullgren): Beier, 1948c: 527; Weidner, 1959: 116.
Euryolpium michaelseni (Tullgren): Beier, 1966a: 290–291, fig. 8; Harvey, 1981b: 243; Harvey, 1985b: 146.

Type locality: Yalgoo, Western Australia, Australia.
Distribution: Australia (New South Wales, Western Australia).

Euryolpium oceanicum Murthy & Ananthakrishnan

Euryolpium (Euryolpium) oceanicum Murthy & Ananthakrishnan, 1977: 84–86, figs 27a–c.

Type locality: Krusadi Islands, Tamil Nadu, India.
Distribution: India.

Euryolpium robustum Murthy & Ananthakrishnan

Euryolpium (Euryolpium) robustum Murthy & Ananthakrishnan, 1977: 82–84, figs 26a–b.

Type locality: Krusadi Islands, Tamil Nadu, India.
Distribution: India.

Euryolpium salomonis (Beier)

Xenolpium salomonis Beier, 1935e: 637–638, fig. 1; Roewer, 1937: 263; Beier, 1982: 43.
Xenolpium (Euryolpium) tokiokai Morikawa, 1963: 4–5, figs 2d–g, 3c, 4a (synonymized by Beier, 1964f: 594).
Euryolpium salomonis (Beier): Beier, 1964f: 594; Beier, 1965g: 762; Beier, 1966d: 138; Beier, 1967c: 321; Petersen, 1968: 119; Beier, 1970g: 317.

Type localities: of *Xenolpium salomonis*: Rendova Island, Solomon Islands.
of *Xenolpium (Euryolpium) tokiokai*: Honiala, Guadalcanal, Solomon Islands.
Distribution: Indonesia (Irian Jaya), Papua New Guinea, Solomon Islands.

Euryolpium striatum Murthy & Ananthakrishnan

Euryolpium (Euryolpium) striatum Murthy & Ananthakrishnan, 1977: 94–96, figs 32a–c.

Type locality: Srivaikundam, Tamil Nadu, India.
Distribution: India.

Genus **Galapagodinus** Beier

Galapagodinus Beier, 1978c: 538.

Type species: *Galapagodinus franzi* Beier, 1978c, by original designation.

Galapagodinus franzi Beier

Galapagodinus franzi Beier, 1978c: 538–540, fig. 4.

Type locality: Jabosillo, Santiago Island, Galapagos Islands.
Distribution: Galapagos Islands.

Genus **Garypinidius** Beier

Garypinidius Beier, 1955l: 290.

Type species: *Garypinidius mollis* Beier, 1955l, by original designation.

Garypinidius capensis (Ellingsen)

Garypinus capensis Ellingsen, 1912b: 114–115; Godfrey, 1927: 18.
Garypinus (?) *capensis* Ellingsen: Beier, 1932a: 211; Roewer, 1937: 265.
Garypinidius capensis (Ellingsen): Beier, 1955l: 292; Beier, 1958a: 164, fig. 3; Beier, 1964k: 51.

Type localities: Alice, Woodstock, Cape Province, South Africa; Lovedale, Cape Province, South Africa; Cwencwe, King William's Town, Cape Province, South Africa; and Xukwane, Cape Province, South Africa.
Distribution: South Africa.

Garypinidius mollis Beier

Garypinidius mollis Beier, 1955l: 290–292, fig. 10.

Type locality: Malagas Island, Saldanhabaai, Cape Province, South Africa.
Distribution: South Africa.

Genus **Garypinus** Daday

Garypinus Daday, 1888: 124, 179–180; Ellingsen, 1904: 5; With, 1906: 111–112; J. C. Chamberlin, 1930: 585; Beier, 1932a: 208; Beier, 1963b: 235–236.

Type species: *Olpium dimidiatum* L. Koch, 1873, by original designation.

Garypinus afghanicus Beier

Garypinus afghanicus Beier, 1959a: 269–270, fig. 11; Beier, 1961e: 2; Lindberg, 1961: 31; Beier, 1971a: 362.

Type locality: Tchehel Dokhteran Cave, Kouh-Ghoramban, Afghanistan.
Distribution: Afghanistan, Iran.

Garypinus afghanicus minor Beier

Garypinus afghanicus minor Beier, 1959a: 270, fig. 12; Beier, 1960a: 42; Beier, 1961e: 2–3.

Type locality: Soumi, 12 km W. of Khvadjéh Tchicht, Afghanistan.
Distribution: Afghanistan.

Garypinus asper Beier

Garypinus asper Beier, 1955d: 215–216, fig. 3; Beier, 1967f: 306; Mahnert, 1974b: 378.

Type locality: Anteljas, Syria.
Distribution: Israel, Syria, Turkey.

Garypinus biimpressus (E. Simon)

Olpium biimpressum E. Simon, 1890: 121.
Garypinus biimpressus (E. Simon): Beier, 1932a: 210, fig. 241; Roewer, 1937: 265;
  Beier, 1956e: 303–306, figs 1–2; Callaini, 1988b: 54.

Type locality: Aden, South Yemen.
Distribution: Morocco, South Yemen.

Garypinus dimidiatus (L. Koch)

Olpium dimidiatum L. Koch, 1873: 34–35; Tömösváry, 1884: 21–22, fig. 6.
Olpium semivittatum Tömösváry, 1884: 22–23, figs 1–2 (synonymized by Daday,
  1888: 124).
Garypinus dimidiatus (L. Koch): Daday, 1888: 124–125, 180–181, plate 4 figs 14,
  17, 19, 23; Ellingsen, 1910a: 388–389; Beier, 1929a: 359; Beier, 1929e: 450; J. C.
  Chamberlin, 1930: 592; Beier, 1931d: 98; Beier, 1932a: 209, figs 238–239; Roewer,
  1937: 265, fig. 218; Beier, 1949: 9; Beier & Turk, 1952: 769; Beier, 1956g: 8; Beier,
  1957b: 147; Beier, 1962a: 141; Beier, 1963b: 236, fig. 240; Beier, 1963f: 192; Beier,
  1963h: 155; Beier, 1965a: 85; Beier, 1965d: 633; Beier, 1965f: 92; Beier, 1966e:
  162–163; Beier, 1966f: 344; Helversen, 1966b: figs 3d, 4i; Beier, 1967f: 306; Beier,
  1969b: 193; Mahnert, 1974c: 378; Callaini, 1982a: 17, 20; Schawaller, 1981c: 9;
  Callaini & Dallai, 1984: 336–341, figs 2–3; Callaini, 1989: 144; Callaini & Dallai,
  1989: 89–90, figs 9, 11.
Garypinus dimidiatus kusceri Hadži, 1933b: 175–183, figs 28a–c, 29a–c, 30a–c,
  31a–c; Hadži, 1933c: 193–196; Ćurčić, 1974a: 25 (synonymized by Beier, 1963b:
  236).
Garypinus dimidiatus dimidiatus (L. Koch): Ćurčić, 1974a: 25.

Type localities: of *Olpium dimidiatum*: Greece.
of *Olpium semivittatum*: Kérkira (as Corfu), Greece.
of *Garypinus dimidiatus kusceri*: Dugi Otok, Yugoslavia.
Distribution: Crete, Cyprus, Greece, Israel, Italy, Sicily, Syria, Turkey, Yugoslavia.

! Garypinus electri Beier

Garypinus electri Beier, 1937d: 305, fig. 3; Roewer, 1940: 331; Schawaller, 1978: 3.

Type locality: Baltic Amber.
Distribution: Baltic Amber.

Garypinus mirabilis With

Amblyolpium longiventer (L. Koch): E. Simon, 1900a: 519 (misidentification).
Garypinus mirabilis With, 1907: 79–80, figs 48–53.
Garypinus (?) *mirabilis* With: Beier, 1932a: 211; J. C. Chamberlin, 1934b: 5; Roewer,
  1937: 265.

Type locality: Kauai, Hawaii.
Distribution: Hawaii.

Garypinus nicolaii Mahnert

Garypinus nicolaii Mahnert, 1988b: 2–4, figs 1–5.

Type locality: Station Nylsvley, near Nylstroom, Transvaal, South Africa.
Distribution: South Africa.

Garypinus nobilis With

Garypinus nobilis With, 1906: 29, 112−115, figs 7a−b, plate 1 figs 7a−c, plate 2 figs 8a−g; Tullgren, 1907a: 67; Ellingsen, 1910a: 389; Beier, 1930e: 289−290; J. C. Chamberlin, 1930: 594; Beier, 1932a: 210; Roewer, 1936: figs 15b, 39a−b, 42a−b, 51; Roewer, 1937: 265.

Type locality: Ko Chang (as Koh Chang), Thailand.
Distribution: Indonesia (Java), Papua New Guinea, Thailand.

Garypinus vachoni Redikorzev

Garypinus vachoni Redikorzev, 1938: 86−87, fig. 15; Beier, 1951a: 69−70, fig. 14.
Garypus (sic) *vachoni* Redikorzev: Roewer, 1940: 345.

Type locality: Ba Ngoi (as Bangoï), Vietnam.
Distribution: Cambodia, Vietnam.

Garypinus validus Beier

Garypinus validus Beier, 1971a: 362−363, fig. 3.

Type locality: 30 km NW. of Bandar'Abbas (as Bandar abbas), Iran.
Distribution: Iran.

Genus **Halominniza** Mahnert

Halominniza Mahnert, 1975b: 542.

Type species: *Halominniza parentorum* Mahnert, 1975b, by original designation.

Halominniza aegyptiacum (Ellingsen)

Olpium aegyptiacum Ellingsen, 1910a: 389−390; Beier, 1932a: 186; Roewer, 1937: 261.
Halominniza aegyptiacum (Ellingsen): Mahnert, 1975b: 543.

Type locality: Egypt.
Distribution: Egypt.

Halominniza aegyptiacum litorale (Beier)

Olpium aegyptiacum litorale Beier, 1963f: 187−188, fig. 3.
Halominniza aegyptiacum litoralis (Beier): Gerdes & Krumbein, 1984: ? (not seen).

Type locality: Aqaba shore, Israel.
Distribution: Israel, Jordan.

Halominniza parentorum Mahnert

Halominniza parentorum Mahnert, 1975b: 543−545, figs 2a−f.

Type locality: Moucha Island, F.T.A.I.
Distribution: F.T.A.I.

Genus **Haplogarypinus** Beier

Haplogarypinus Beier, 1959d: 30.

Type species: *Haplogarypinus pauperatus* Beier, 1959d, by original designation.

Haplogarypinus pauperatus Beier

Haplogarypinus pauperatus Beier, 1959d: 30−31, fig. 13.

Type locality: Kundelungu Mountains, Zaire.
Distribution: Zaire.

Genus **Hemisolinus** Beier

Hemisolinus Beier, 1977b: 4−5.

Type species: *Hemisolinus helenae* Beier, 1977b, by original designation.

Hemisolinus helenae Beier

Hemisolinus helenae Beier, 1977b: 5–6, fig. 2.

Type locality: East Prosperous Bay Plain, St Helena.
Distribution: St Helena.

Genus **Hesperolpium** J. C. Chamberlin

Hesperolpium J. C. Chamberlin, 1930: 606; Beier, 1932a: 198.

Type species: *Olpium slevini* J. C. Chamberlin, 1923c, by original designation.

Hesperolpium andrewsi Muchmore

Hesperolpium andrewsi Muchmore, 1980e: 164–165, fig. 4.

Type locality: Eureka Valley, Inyo County, California, U.S.A.
Distribution: U.S.A. (California).

Hesperolpium slevini (J. C. Chamberlin)

Olpium slevini J. C. Chamberlin, 1923c: 363–364, plate 2, figs 15–16, plate 3, figs 10, 16.
Hesperolpium slevini (J. C. Chamberlin): J. C. Chamberlin, 1930: 607; J. C. Chamberlin, 1931a: 144, figs 2, 4g, 16f, 32e, 37n–o, 42f–g; Beier, 1932a: 198; Beier, 1932g: fig. 169; Roewer, 1936: fig. 89; Roewer, 1937: 263; Vachon, 1949: figs 198, 207d; Hoff, 1958: 16; Muchmore, 1980e: 162–164, figs 1–3.

Type locality: Cuesta Blanca, 8 miles N. of Loreto, Baja California, Mexico.
Distribution: Mexico, U.S.A. (Arizona, California).

Genus **Heterolpium** Sivaraman

Heterolpium Sivaraman, 1980b: 337.

Type species: *Heterolpium indicum* Sivaraman, 1980b, by original designation.

Heterolpium indicum Sivaraman

Heterolpium indicum Sivaraman, 1980b: 337–340, figs 6a–b.

Type locality: Guindy, Madras, Tamil Nadu, India.
Distribution: India.

Genus **Hoffhorus** Heurtault

Hoffhorus Heurtault, 1976: 69.

Type species: *Novohorus cinereus* Hoff, 1945f, by original designation.

Hoffhorus cinereus (Hoff)

Novohorus cinereus Hoff, 1945f: 23–26, figs 27–29.
Hoffhorus cinereus (Hoff): Heurtault, 1976: 67, figs 8–9.

Type locality: St Augustine, Trinidad.
Distribution: Trinidad.

Genus **Horus** J. C. Chamberlin

Horus J. C. Chamberlin, 1930: 598; Beier, 1932a: 199; Vachon, 1940e: 67–68.

Type species: *Garypinus obscurus granulatus* Ellingsen, 1912b, by original designation.

Horus asper Beier

Horus asper Beier, 1947b: 309–311, fig. 17; Beier, 1973a: 98.

Type locality: Kleinsee (as Kleinzee), Namaqualand, Cape Province, South Africa.
Distribution: Namibia, South Africa.

Horus brevipes Beier

Horus brevipes Beier, 1964k: 47–48, fig. 12.

Type locality: Mkuzi Game Reserve, Natal, South Africa.
Distribution: South Africa.

Horus difficilis Vachon

Horus difficilis Vachon, 1940h: 225–226 (nomen nudum).
Horus difficilis Vachon, 1941e: 31.
Horus sp. ? *difficilis* Vachon: Vachon, 1952a: 27.

Type locality: Port Bouet, Ivory Coast.
Distribution: Guinea, Ivory Coast, Zaire.

Horus gracilis Beier

Horus gracilis Beier, 1958a: 163–164, fig. 2; Beier, 1964k: 46; Beier, 1966k: 469;
    Lawrence, 1967: 88.

Type locality: Estcourt, Natal, South Africa.
Distribution: South Africa.

Horus granulatus (Ellingsen)

Garypinus obscurus granulatus Ellingsen, 1912b: 116.
Horus granulatus (Ellingsen): J. C. Chamberlin, 1930: 599, figs 1e–f; J. C. Chamberlin,
    1931a: figs 4e, 15m, 29c; Beier, 1932a: 200; Roewer, 1937: 263; Beier, 1947b:
    307–309, figs 15–16; Beier, 1955l: 287–288, fig. 8; Beier, 1964k: 50–51, fig. 14;
    Beier, 1966k: 459; Beier, 1973a: 98; Heurtault, 1976: 64, fig. 1.

Type locality: unknown.
Distribution: Namibia, South Africa, Zimbabwe.

Horus modestus J. C. Chamberlin

Horus modestus J. C. Chamberlin, 1930: 599–600, fig. 2r; J. C. Chamberlin, 1931a:
    144, figs 37e–f, 43q–r, 52f; Beier, 1932a: 202; Roewer, 1937: 263; Vachon, 1937a:
    129; Heurtault, 1976: 64, fig. 2.

Type locality: Alicedale, Cape Province, South Africa.
Distribution: South Africa, Zimbabwe.

Horus montanus Beier

Horus montanus Beier, 1955l: 288–290, fig. 9; Beier, 1964k: 45–46.

Type locality: Mt Morosi, 15 miles NE. of Quthing, Lesotho.
Distribution: Lesotho, South Africa.

Horus obscurus (Tullgren)

Garypinus obscurus Tullgren, 1907a: 68–69, figs 22a–c; Tullgren, 1907b: 229;
    Ellingsen, 1912b: 87, 115–116.
Horus obscurus (Tullgren): J. C. Chamberlin, 1930: 599; Beier, 1932a: 200–201,
    fig. 231; Roewer, 1937: 263; Vachon, 1937a: 129; Beier, 1947b: 307; Beier, 1955l:
    288; Beier, 1958a: 162; Weidner, 1959: 115; Beier, 1964k: 45; Beier, 1966k: 458–459;
    Lawrence, 1967: 88.

Type locality: Bothaville, Orange Free State, South Africa.
Distribution: South Africa, Zimbabwe.

Horus transvaalensis Beier

Horus transvaalensis Beier, 1964k: 48–50, fig. 13; Beier, 1966k: 459, 470; Lawrence,
    1967: 88.

Type locality: Witpoort, Transvaal, South Africa.
Distribution: South Africa.

Horus zonatus Beier

Horus zonatus Beier, 1964k: 46–47, fig. 11; Beier, 1966k: 459; Lawrence, 1967: 88.

Type locality: Birchenough Bridge, Zimbabwe.
Distribution: South Africa, Zimbabwe.

Genus Indogarypinus Murthy & Ananthakrishnan

Indogarypinus Murthy & Anathakrishnan, 1977: 98.

Type species: *Indogarypinus minutus* Murthy & Ananthakrishnan, 1977, by original designation.

Indogarypinus minutus Murthy & Ananthakrishnan

Indogarypinus minutus Murthy & Ananthakrishnan, 1977: 98–101, figs 33a–e.

Type locality: Mahabalipuram, Tamil Nadu, India.
Distribution: India.

Genus Indolpium Hoff

Indolpium Hoff, 1945f: 19; Murthy & Ananthakrishnan, 1977: 63–64; Heurtault, 1980c: 149.

Type species: *Xenolpium funebrum* Redikorzev, 1938, by original designation.

Indolpium afghanicum Beier

Indolpium afghanicum Beier, 1961e: 1–2, fig. 1.

Type locality: Kouh-Qorough near Tang-Saïdan, 20 km W. of Kabul (as Kaboul), Afghanistan.
Distribution: Afghanistan.

Indolpium asiaticum Murthy & Ananthakrishnan

Indolpium asiaticum Murthy & Ananthakrishnan, 1977: 68–69.

Type locality: Egmore, Tamil Nadu, India.
Distribution: India.

Indolpium centrale Beier

Indolpium centrale Beier, 1967g: 347, fig. 7.

Type locality: 13 km W. of Edalabad, Maharashtra, India.
Distribution: India.

Indolpium decolor Beier

Indolpium decolor Beier, 1953g: 81–82, fig. 1.

Type locality: Waimangura, Sumba, Indonesia.
Distribution: Indonesia (Sumba).

Indolpium funebrum (Redikorzev)

Xenolpium funebrum Redikorzev, 1938: 84–85, fig. 14; Roewer, 1940: 345; Beier, 1951a: 69.
Indolpium funebrum (Redikorzev): Hoff, 1945e: 19; Beier, 1967g: 347; Heurtault, 1980c: fig. 14.

Type localities: Quan Phu Quoc (as Ile Phu Quoc), Vietnam; and Caûda, near Nha Trang (as Nhatrang), Vietnam.
Distribution: Vietnam.

Indolpium intermedium Murthy & Ananthakrishnan

Indolpium intermedium Murthy & Ananthakrishnan, 1977: 69–71, figs 21a–b.

Type locality: Nungambakkam, Tamil Nadu, India.
Distribution: India.

Indolpium loyolae (Murthy)

Minniza loyolae Murthy, 1961: 221–222, figs 1a–b.
Indolpium loyalae (sic) (Murthy): Beier, 1967g: 347.
Indolpium loyolae (Murthy): Beier, 1973b: 45; Murthy & Ananthakrishnan, 1977: 67.

Type locality: Nungambakkam, Madras, India.
Distribution: India, Sri Lanka.

Indolpium majusculum Beier

Indolpium majusculum Beier, 1967g: 348, fig. 8.

Type locality: 13 km NE. of Tuni, Andhra Pradesh, India.
Distribution: India.

Indolpium modestum Murthy & Ananthakrishnan

Indolpium modestum Murthy & Ananthakrishnan, 1977: 75–77, figs 24a–c.

Type locality: Guindy, Tamil Nadu, India.
Distribution: India.

Indolpium politum Murthy & Ananthakrishnan

Indolpium politum Murthy & Ananthakrishnan, 1977: 65–67, figs 20a–b.

Type locality: Tinnevelly, Tamil Nadu, India.
Distribution: India.

Indolpium robustum Murthy & Ananthakrishnan

Indolpium robustum Murthy & Ananthakrishnan, 1977: 73–75, figs 23a–b.

Type locality: Tambaram, Tamil Nadu, India.
Distribution: India.

Indolpium squalidum Murthy & Ananthakrishnan

Indolpium squalidum Murthy & Ananthakrishnan, 1977: 77–79, figs 25a–b.

Type locality: Tambaram, Tamil Nadu, India.
Distribution: India.

Indolpium thevetium Murthy & Ananthakrishnan

Indolpium thevetium Murthy & Ananthakrishnan, 1977: 73–75, figs 22a–b.

Type locality: Nungambakkam (as Nungambakam), Tamil Nadu, India.
Distribution: India.

Indolpium transiens Beier

Indolpium transiens Beier, 1967g: 348–349, fig. 9; Murthy & Ananthakrishnan, 1977: 67.

Type locality: 6 km W. of Khammam, Andhra Pradesh, India.
Distribution: India.

Genus **Minniza** E. Simon

Minniza E. Simon, 1881b: 14; J. C. Chamberlin, 1923c: 364; J. C. Chamberlin, 1930: 604; Beier, 1932a: 186; Beier, 1963b: 232.

Type species: *Minniza vermis* E. Simon, 1881b, by monotypy.

Minniza aequatorialis Beier

Minniza aequatorialis Beier, 1944: 179–180, fig. 5; Mahnert, 1982c: 284.

Type locality: Lodwar, Kenya.
Distribution: Kenya.

Minniza algerica Beier

Minniza algerica Beier, 1931c: 306–307, fig. 5; Beier, 1932a: 187–188, fig. 214; Roewer, 1937: 262; Beier, 1956e: 303; Beier, 1961g: 411; Beier, 1963b: 233–234, fig. 238; Mahnert, 1975c: 194–195; Callaini, 1988b: 53.

Type locality: Biskra, Algeria.
Distribution: Algeria, Italy, Malta, Morocco.

Minniza babylonica Beier

Olpium graecum L. Koch: Beier, 1929a: 358–359 (misidentification).
Minniza babylonica Beier, 1931c: 307–308; Beier, 1932a: 188, fig. 215; Roewer, 1937: 262; Beier, 1951d: 98; Beier, 1971a: 361.
Minniza babylonica babylonica Beier: Mahnert, 1980c: 35–36, figs 2–9.

Type locality: Iraq (as Babylon).
Distribution: Iran, Iraq, Saudi Arabia.

Minniza babylonica afghanica Beier

Minniza babylonica afghanica Beier, 1959a: 267–269, fig. 10; Beier, 1960a: 42; Beier, 1961e: 1; Lindberg, 1961: 31.

Type locality: Kandahar, Afghanistan.
Distribution: Afghanistan.

Minniza babylonica lindbergi Beier

Minniza babylonica lindbergi Beier, 1957b: 145–146, fig. 1; Beier, 1963f: 188–189; Beier, 1965f: 91; Beier, 1967f: 305; Beier, 1973c: 226; Mahnert, 1974c: 378.

Type locality: Viransehir, Turkey.
Distribution: Israel, Turkey.

Minniza ceylonica Beier

Minniza ceylonica Beier, 1973b: 44, fig. 7.

Type locality: Nanthi Kadal, 3 miles S. of Mullaittivu, Northern Province, Sri Lanka.
Distribution: Sri Lanka.

Minniza deserticola E. Simon

Minniza deserticola E. Simon, 1885c: 50; Beier, 1932a: 189, fig. 216; Beier, 1932f: 487; Roewer, 1937: 262; Beier, 1963b: 232, fig. 237; Beier, 1963i: 285; Lazzeroni, 1969a: 334; Heurtault, 1970a: 1164–1165; Lazzeroni, 1970a: 207; Heurtault, 1980b: 182, figs 15, 20–24; Mahnert, 1985a: 17.
Olpium deserticola (E. Simon): Ellingsen, 1909a: 207.
Not *Olpium deserticola* (E. Simon): Ellingsen, 1906: 263–264; Ellingsen, 1912b: 88 (misidentifications; see *Olpium pallipes* (Lucas)).

Type locality: Gabès, Tunisia.
Distribution: Chad, Italy, Libya, Sicily, Tunisia.

Minniza exorbitans Beier

Minniza exorbitans Beier, 1965b: 365–367, fig. 1.

Type locality: between Mardengai and Largeau, Chad.
Distribution: Chad.

Minniza graeca (L. Koch)

Olpium graecum L. Koch, 1873: 36–37.
Olpium (?) *graecum* L. Koch: Beier, 1932a: 186; Roewer, 1937: 262.
Minniza hirsti cypria Beier & Turk, 1952: 766–767, fig. 1 (synonymized by Schawaller, in Schmalfuss & Schawaller, 1984: 4).
Minniza cretica Beier, 1956g: 8–9 (synonymized by Beier, 1963b: 233).
Minniza cypria Beier & Turk: Beier, 1962a: 141; Beier, 1963b: 233; Beier, 1965d: 632; Beier, 1966e: 162.
Minniza graeca L. Koch): Beier, 1963b: 232–233; Beier, 1966f: 343; Schawaller, in Schmalfuss & Schawaller, 1984: 4, figs 4–10.
Not *Olpium graecum* L. Koch: Beier, 1929a: 358–359 (misidentification; see *Minniza babylonica* Beier).

Type localities: of *Olpium graecum*: Greece.
of *Minniza hirsti cypria*: Erimi, Cyprus.
of *Minniza cretica*: Sitia, Crete.
Distribution: Crete, Cyprus, Greece.

Minniza hirsti J. C. Chamberlin

Minniza hirsti J. C. Chamberlin, 1930: 605, fig. 2c; J. C. Chamberlin, 1931a: fig. 9i; Beier, 1932a: 188; Roewer, 1937: 262; Vachon, 1940c: 157; Vachon, 1940e: 65.

Type locality: Luxor, Egypt.
Distribution: Algeria, Egypt, Mali.

Minniza occidentalis Vachon

Minniza occidentalis Vachon, 1954a: 1022–1024, figs 1–5.

Type locality: Atar, Mauritania.
Distribution: Mauritania.

Minniza persica Beier

Minniza persica persica Beier, 1951d: 98–99, fig. 1c; Beier, 1971a: 361.

Type locality: Sabzawaran (now Giroft), Iran.
Distribution: Iran.

Minniza persica deminuta Beier

Minniza persica deminuta Beier, 1951d: 99.

Type locality: Lahidschan, Mazandaran, Iran.
Distribution: Iran.

Minniza rollei Caporiacco

Minniza rollei Caporiacco, 1936a: 119, figs 10a–b; Roewer, 1940: 345.

Type locality: Oasis Rolla, Al Kufra (as Cufra), Libya.
Distribution: Libya.

Minniza rubida (E. Simon)

Olpium rubidum E. Simon, 1882a: 211; Simon, 1890: 121.
Minniza rubida (E. Simon): Beier, 1932a: 189, fig. 217; Roewer, 1937: 262; Heurtault, 1980b: 182–183, figs 13, 16–19.

Type locality: Aden, South Yemen.
Distribution: South Yemen.

Minniza sola J. C. Chamberlin

Minniza sola J. C. Chamberlin, 1930: 605–606; Beier, 1932a: 188; Roewer, 1937: 262; Beier, 1965b: 365.
Minniza solus J. C. Chamberlin: J. C. Chamberlin, 1931a: fig. 29d.

Type locality: Khartoum, Sudan.
Distribution: Chad, Sudan.

Minniza sola distincta Vachon

Minniza sola distincta Vachon, 1940e: 63–64, figs a–d.

Type locality: Adrar des Iforhas, Algeria/Mali.
Distribution: Algeria/Mali.

Minniza syriaca Beier

Minniza syriaca Beier, 1951d: 97–98, fig. 1b; Beier, 1963f: 188; Beier, 1967f: 305; Beier, 1971a: 361; Schawaller, 1985a: 8; Schawaller, 1989: 14.

Type locality: near Ar Rutbah (as Rutbah), Iraq.
Distribution: Iran, Iraq, Israel, Turkey, U.S.S.R. (Turkmenistan).

Minniza vermis E. Simon

Minniza vermis E. Simon, 1881b: 14–15; E. Simon, 1885c: 50; E. Simon, 1899b: 244; E. Simon, 1908: 438; Beier, 1929b: 80; J. C. Chamberlin, 1930: 604–605, figs 2q, 2hh; Beier, 1931d: 98; J. C. Chamberlin, 1931a: figs 14b, 37k–m; Beier, 1932a: 187, fig. 213; Roewer, 1936: fig. 30f; Vachon, 1936a: 78; Roewer, 1937: 262; Caporiacco, 1948b: 34; Heurtault, 1980b: 175–181, figs 1–12, 14–15, 20–24; Callaini, 1988b: 54.
Olpium vermis (E. Simon): Ellingsen, 1906: 263; Tullgren, 1907c: 8–10, figs 3a–c; Ellingsen, 1912b: 88.

Type locality: El Mex (as Le Mex), Egypt.
Distribution: Crete, Egypt, Equatorial Guinea, Greece, Libya, Morocco, Sudan, Tunisia.

Genus Nanolpium Beier

Nanolpium Beier, 1947b: 293.

Type species: *Olpium pusillum* Ellingsen, in Strand, 1909, by original designation.

Nanolpium congicum Beier

Nanolpium congicum Beier, 1954d: 132–133, fig. 1.

Type locality: Kambove, Zaire.
Distribution: Zaire.

Nanolpium milanganum Beier

Nanolpium milanganum Beier, 1964k: 36–37, fig. 4.

Type locality: Mt Machinjiri, Milange, Mozambique.
Distribution: Mozambique.

Nanolpium nitens (Tullgren)

Olpium nitens Tullgren, 1908b: 287, figs 5a–b: Ellingsen, 1912b: 88, 116; Beier, 1932a: 185; Roewer, 1937: 261.
Nanolpium nitens (Tullgren): Beier, 1947b: 293; Beier, 1955l: 267.

Type locality: Lüderitz Bay, Namibia.
Distribution: Namibia.

Nanolpium pusillum (Ellingsen)

Olpium pusillum Ellingsen, in Strand, 1909: 596; Ellingsen, 1912b: 88; Beier, 1932a: 185; Roewer, 1937: 261.
Nanolpium pusillum (Ellingsen): Beier, 1947b: 293; Beier, 1964k: 36.
Nanolpium falsum Beier, 1947b: 294–295, figs 5–6; Beier, 1955l: 278 (synonymized by Beier, 1964k: 36).

Type localities: of *Olpium pusillum*: Fish Hoek (as Fishhoek), Cape Province, South Africa.
of *Nanolpium falsum*: Fish Hoek, Cape Peninsula, Cape Province, South Africa.
Distribution: South Africa.

Nanolpium rhodesiacum Beier

Olpium subgrande Tullgren: Ellingsen, 1912b: 88, 116–117 (misidentification).
Nanolpium rhodesiacum Beier, 1955l: 278–279.

Type locality: Maramba (as Livingstone), Zambia.
Distribution: Zambia, Zimbabwe.

Nanolpium smithersi Beier

Nanolpium smithersi Beier, 1964k: 38–39, fig. 6.

Type locality: Umtali Road, Ruwa River, Zimbabwe.
Distribution: Zimbabwe.

Nanolpium transvaalense Beier

Nanolpium transvaalense Beier, 1964k: 37–38, fig. 5.

Type locality: N.E. Pretoria, Transvaal, South Africa.
Distribution: South Africa.

Genus Nelsoninus Beier

Nelsoninus Beier, 1967b: 291–292; Beier, 1976f: 211.

Type species: *Nelsoninus maoricus* Beier, 1967b, by original designation.

Nelsoninus maoricus Beier

Nelsoninus maoricus Beier, 1967b: 291–292, fig. 9; Beier, 1976f: 211.

Type locality: Canaan Track, Abel Tasman National Park, Nelson, South Island, New Zealand.
Distribution: New Zealand.

Genus Neoamblyolpium Hoff

Neoamblyolpium Hoff, 1956b: 27–28.

Type species: *Neoamblyolpium alienum* Hoff, 1956b, by original designation.

Neoamblyolpium alienum Hoff

Neoamblyolpium alienum Hoff, 1956b: 28–33, figs 13–15; Hoff, 1958: 16; Hoff, 1959b: 4, etc.; Hoff, 1961: 438–439; Zeh, 1987b: 1086.

Type locality: Mt Taylor, near Grants, Valencia County, New Mexico, U.S.A.
Distribution: U.S.A. (Colorado, New Mexico).

Neoamblyolpium giulianii Muchmore

Neoamblyolpium giulianii Muchmore, 1980e: 165–166, figs 5–6.

Type locality: 10 miles S. of Independence, Inyo County, California, U.S.A.
Distribution: U.S.A. (California).

Genus **Neominniza** Beier

Neominniza Beier, 1930d: 202; Beier, 1932a: 207.

Type species: *Neominniza divisa* Beier, 1930d, by original designation.

Neominniza divisa Beier

Neominniza divisa Beier, 1930d: 203–204, figs 7a–b, 8a–b; Beier, 1932a: 207–208, fig. 237; Roewer, 1936: fig. 112g; Roewer, 1937: 265; Beier, 1964b: 326–327, fig. 10; Cekalovic, 1984: 15.

Type locality: Viña del Mar, Chile.
Distribution: Chile.

Neominniza halophila Beier

Neominniza halophila Beier, 1964b: 327–329, fig. 11; Cekalovic, 1984: 15.

Type locality: Punta Teatinos, 12 km N. of La Serena, Coquimbo, Chile.
Distribution: Chile.

Genus **Neopachyolpium** Hoff

Neopachyolpium Hoff, 1945f: 15.

Type species: *Neopachyolpium longum* Hoff, 1945f, by original designation.

Neopachyolpium longum Hoff

Neopachyolpium longum Hoff, 1945f: 15–19, figs 16–18.

Type locality: St Augustine, Trinidad.
Distribution: Trinidad.

Genus **Nipponogarypus** Morikawa

Nipponogarypus Morikawa, 1955b: 225, 227; Morikawa, 1960: 126–127.

Type species: *Nipponogarypus enoshimaensis* Morikawa, 1955b, by original designation.

Nipponogarypus enoshimaensis Morikawa

Nipponogarypus enoshimaensis Morikawa, 1955b: 225–228, fig. 1; Morikawa, 1958: 8; Morikawa, 1960: 127; Morikawa, 1962: 419; Sato, 1978d: 45–48; Sato, 1982c: 33; Sato, 1984b: 52.
Nipponogarypus enoshimaensis enoshimaensis Morikawa: Morikawa, 1960: 127–128, plate 2 fig. 8, plate 6 fig. 8, plate 7 fig. 14, plate 10 figs 32–33.

Type locality: Enoshima Island, Kanagawa, Japan.
Distribution: Japan.

Nipponogarypus enoshimaensis okinoerabensis Morikawa

Nipponogarypus enoshimaensis okinoerabensis Morikawa, 1960: 128.

Type locality: Okino-erabu Island, Amami gunto (Ryukyu Islands), Japan.
Distribution: Japan.

Genus **Novohorus** Hoff

Novohorus Hoff, 1945f: 19; Hoff, 1964a: 29; Muchmore, 1979a: 197–199.

Type species: *Novohorus suffuscus* Hoff, 1945f, by original designation.

Novohorus obscurus (Banks)

Olpium obscurum Banks, 1893: 65; Banks, 1895a: 10; Banks, 1904a: 141; Coolidge, 1908: 112.
Pachyolpium (?) *obscurum* (Banks): Beier, 1932a: 196; Hoff, 1958: 16.

Pachyolpium (?) *obscutum* (sic) (Banks): Roewer, 1937: 262.
Novohorus obscurus (Banks): Muchmore, 1971b: 241–243, figs 1–2; Muchmore, 1979a: 199–200, figs 8–11.

Type locality: Runnymede, Osceola County, Florida, U.S.A.
Distribution: U.S.A. (Florida).

Novohorus suffuscus Hoff

Novohorus suffuscus Hoff, 1945f: 19–23, figs 19–26; Hoff, 1964a: 29–30; Heurtault, 1976: fig. 7; Zeh, 1987b: 1087.

Type locality: Mona Island, Puerto Rico.
Distribution: Jamaica, Puerto Rico.

Genus **Olpiolum** Beier

Olpiolum Beier, 1931c: 312; Beier, 1932a: 196; Hoff, 1964a: 19–20; Muchmore, 1979a: 197; Muchmore, 1986a: 84–85.

Type species: *Olpiolum medium* Beier, 1931c, by original designation.

Olpiolum amplum Hoff

Olpiolum amplum Hoff, 1964a: 24–25, figs 9–10.

Type locality: Long Mountain, Parish of St Andrew, Jamaica.
Distribution: Jamaica.

Olpiolum aureum (Hoff)

Pachyolpium aureum Hoff, 1945f: 4–6, figs 1–4.
Olpiolum aureum (Hoff): Hoff, 1964a: 20.

Type locality: Mona Island, Puerto Rico.
Distribution: Puerto Rico.

Olpiolum confundens (Hoff)

Pachyolpium confundens Hoff, 1945f: 9–11, figs 9–10.
Olpiolum confundens (Hoff): Hoff, 1964a: 21.

Type locality: San Juan, Puerto Rico.
Distribution: Puerto Rico.

Olpiolum crassum Beier

Olpiolum crassum Beier, 1959e: 194–195, fig. 7.

Type locality: Río Pampas, Peru.
Distribution: Peru.

Olpiolum elegans (Balzan)

Olpium elegans Balzan, 1887a: no pagination, figs; Balzan, 1887b: no pagination; Balzan, 1890: 437–438, figs 19, 19a–b; Balzan, 1892: 549.
Olpiolum elegans (Balzan): Beier, 1932a: 196, fig. 227; Roewer, 1937: 262; Feio, 1945: 4; Mello-Leitão & Feio, 1949: 320; Muchmore, 1986a: 87.
Not *Olpium elegans* Balzan: Ellingsen, 1910a: 390–391; Beier, 1930a: 207 (misidentifications; see *Olpiolum medium* Beier).

Type locality: Rio Apa, Paraguay.
Distribution: Argentina, Brazil (Bahia), Paraguay, Peru.

Olpiolum fuscipalpum Muchmore

Olpiolum fuscipalpum Muchmore, 1977: 74–76, figs 24–25.

Type locality: 2 miles S. of Belmopan, Belize.
Distribution: Belize.

Olpiolum machadoi Heurtault

Olpiolum modestum (Banks): Beier, 1954f: 134–135, fig. 3 (misidentification).
Olpiolum machadoi Heurtault, 1976: 62 (nomen nudum).
Olpiolum machadoi Heurtault, 1982: 64–72, figs 26–49.

Type locality: Caracas, Venezuela.
Distribution: Venezuela.

Olpiolum medium Beier

Olpium elegans Balzan: Ellingsen, 1910a: 390–391; Beier, 1930a: 207 (misidentifications).
Olpiolum medium Beier, 1931c: 312–313, fig. 8; Beier, 1932a: 197, fig. 228; Roewer, 1937: 262; Muchmore, 1986a: 85–86, figs 1–5.

Type locality: Paraguay.
Distribution: Paraguay.

Olpiolum monae Hoff

Pachyolpium medium Hoff, 1945f: 11–15, figs 11–15 (junior secondary homonym of *Olpiolum medium* Beier).
Olpiolum monae Hoff, 1964a: 21–24; Beier, 1976d: 46; Muchmore, 1979a: 197, fig. 7; Zeh, 1987b: 1086 (replacement name for *Pachyolpium medium* Hoff).

Type locality: Mona Island, Puerto Rico.
Distribution: Dominican Republic, Jamaica, Puerto Rico, U.S.A. (Florida).

Olpiolum paucisetosum Muchmore

Olpiolum paucisetosum Muchmore, 1977: 74, figs 22–23.

Type locality: 1 km S. of Muna, Yucatán, Mexico.
Distribution: Mexico.

Olpiolum peruanum Beier

Olpiolum peruanum Beier, 1959e: 193–194, fig. 6.

Type locality: Huanuco, Peru.
Distribution: Peru.

Olpiolum puertoricensis (Hoff)

Pachyolpium puertoricensis Hoff, 1945f: 6–9, figs 5–8.
Olpiolum puertoricensis (Hoff): Hoff, 1964a: 21.

Type locality: Coamo Springs, Puerto Rico.
Distribution: Puerto Rico.

Genus Olpium L. Koch

Olpium L. Koch, 1873: 33; E. Simon, 1879a: 49; Tömösváry, 1882b: 245–246; Canestrini, 1885: no. 10; Balzan, 1887a: no pagination; Balzan, 1890: 434–435; Balzan, 1892: 536; With, 1906: 109; J. C. Chamberlin, 1930: 602; Beier, 1932a: 180–181; Beier, 1963b: 230–231; Murthy & Ananthakrishnan, 1977: 46; Sivaraman, 1980b: 335.

Type species: *Obisium pallipes* Lucas, 1849, by subsequent designation of International Commission of Zoological Nomenclature, 1987: 53.

Olpium afghanicum Beier

Olpium afghanicum Beier, 1952f: 245–247, fig. 1.

Type locality: Kajkai, Helmand, Afghanistan.
Distribution: Afghanistan.

Olpium angolense Beier

Olpium angolensis Beier, 1931c: 306, fig. 4.
Olpium angolense Beier: Beier, 1932a: 184, fig. 211; Roewer, 1937: 261.

Type locality: Angola.
Distribution: Angola.

Olpium asiaticum Murthy & Ananthakrishnan

Olpium asiaticum Murthy & Ananthakrishnan, 1977: 51−53, figs 14Aa−b.

Type locality: Tamil Nadu, India.
Distribution: India.

Olpium australicum Beier

Olpium australicum Beier, 1969a: 182−183, fig. 9; Beier, 1975b: 203; Harvey, 1981b: 243; Harvey, 1985b: 147.

Type locality: Trig Station NM/F/228 near Madura, Western Australia, Australia.
Distribution: Australia (South Australia, Western Australia).

Olpium canariense Beier

Olpium canariense Beier, 1965c: 377−378, fig. 2; Lagar, 1972b: 51; Beier, 1976a: 24; Estany, 1979a: 221, fig. 1; Mahnert, 1980a: 259.

Type locality: N. of Los Cristianos, Tenerife, Canary Islands.
Distribution: Canary Islands.

Olpium ceylonicum Beier

Olpium ceylonicum Beier, 1973b: 43−44, fig. 6.

Type locality: Kuchchaveli, 20 miles NW. of Trincomalee, Eastern Province, Sri Lanka.
Distribution: Sri Lanka.

Olpium crypticum Murthy & Ananthakrishnan

Olpium crypticum Murthy & Ananthakrishnan, 1977: 55−57, figs 16a−c.

Type locality: Rameswaram, Tamil Nadu, India.
Distribution: India.

Olpium digitum Murthy & Ananthakrishnan

Olpium digitum Murthy & Ananthakrishnan, 1977: 47−49, figs 13a−b.

Type locality: Madras, Tamil Nadu, India.
Distribution: India.

Olpium fuscimanum Beier

Olpium fuscimanum Beier, 1957a: 471−472, fig. 1; Mahnert, 1982c: 279−281, figs 53−54; Mahnert, 1984b: 44.

Type locality: Old Shinyanga, Tanzania.
Distribution: Kenya, Somalia, Tanzania.

Olpium gladiatum Murthy & Ananthakrishnan

Olpium gladiatum Murthy & Ananthakrishnan, 1977: 57−59, fig. 17.

Type locality: Goa, India.
Distribution: India.

Olpium graminum Murthy & Ananthakrishnan

Olpium graminum Murthy & Ananthakrishnan, 1977: 49−51, figs 14a−b.

Type locality: Madras, Tamil Nadu, India.
Distribution: India.

Olpium granulatum Murthy & Ananthakrishnan

Olpium granulatum Murthy & Ananthakrishnan, 1977: 61–63, figs 19a–c.

Type locality: Krusadi Islands, Tamil Nadu, India.
Distribution: India.

Olpium halophilum Mahnert

Olpium kochi E. Simon: Beier, 1967d: 74 (misidentification).
Olpium halophilum Mahnert, 1982c: 281–284, figs 58–64; Mahnert, 1984b: 45–47, figs 1–4.

Type locality: Mombasa, Kenya.
Distribution: Kenya, Somalia.

Olpium indicum Beier

Olpium indicum Beier, 1967g: 345–346, fig. 5.

Type locality: 8 km SE. of Indapur, India.
Distribution: India.

Olpium intermedium Beier

Olpium intermedium Beier, 1959a: 264–265, fig. 7; Beier, 1961e: 1; Lindberg, 1961: 31.

Type locality: N. of Nourgal, Konar, Afghanistan.
Distribution: Afghanistan.

Olpium jacobsoni Tullgren

Olpium jacobsoni Tullgren, 1908a: 148–150, fig. 1; Ellingsen, 1913a: 455; Ellingsen, 1914: 12–13; Beier, 1932a: 183–184; Roewer, 1937: 261; Weidner, 1959: 116.
Olpium longiventer L. Koch: Ellingsen, 1912a: 125 (misidentification).

Type locality: Semarang, Java, Indonesia.
Distribution: India, Indonesia (Java), Sri Lanka, Taiwan, Thailand.

Olpium kochi E. Simon

Olpium kochi E. Simon, 1881b: 13–14; E. Simon, 1908: 438; Beier, 1932a: 183, fig. 208; Roewer, 1937: 261; Ghabbour, Mikhaïl, & Rizk, 1977: 439; Mahnert, 1981b: 95–96, figs 1–10; Harvey & Mahnert, 1985: 86–87; International Commission on Zoological Nomenclature, 1987: 53–54.
Olpium turcicum Beier, 1949: 9, fig. 8; Beier, 1965d: 631; Beier, 1965f: 91; Helversen, 1966b: fig. 4a; Rack, 1971: 114 (synonymized by Mahnert, 1981b: 97).
Not *Olpium kochi* E. Simon: Beier, 1967d: 74 (misidentification; see *Olpium halophilum* Mahnert).

Type localities: of *Olpium kochi*: north of the Great Pyramid, Egypt.
of *Olpium turcicum*: Mersin, Anatolia, Turkey.
Distribution: Egypt, Greece, Libya, Turkey.

Olpium lindbergi Beier

Olpium lindbergi Beier, 1959a: 265–266, fig. 8; Beier, 1959c: 407; Beier, 1960a: 41.

Type locality: Angout near Bhougavi, Afghanistan.
Distribution: Afghanistan, India.

Olpium microstethum Pavesi

Olpium microstethum Pavesi, 1880: 314−315; Caporiacco, 1928a: 79.
Olpium (?) *microstethum* Pavesi: Beier, 1932a: 186; Roewer, 1937: 262.

Type locality: Tameghza, Tunisia.
Distribution: Libya, Tunisia.

Olpium minnizioides Vachon

Olpium minnizioides Vachon, 1966b: 183−186, figs 1−10.

Type locality: Hasikiyah (as Hasikaya) Island, Oman.
Distribution: Oman.

Olpium pallipes (Lucas)

Obisium pallipes Lucas, 1849: 277−278, plate 17 fig. 3; Hagen, 1870: 270.
Olpium pallipes (Lucas): E. Simon, 1878a: 152−153; E. Simon, 1879a: 49−50, 312,
    plate 19 fig. 2; E. Simon, 1881b: 13; Tömösváry, 1882b: 246−247; Canestrini,
    1884: no. 9, figs 1−4; E. Simon, 1885a: 349; E. Simon, 1885c: 49−50; Daday,
    1888: 126, 183; E. Simon, 1898d: 21; E. Simon, 1899d: 86; Tullgren, 1907c:
    10−12; Ellingsen, 1908c: 670; E. Simon, 1909: 42; Ellingsen, 1910a: 392; E. Simon,
    1912: 61; Ellingsen, 1913a: 455; Nonidez, 1917: 31; Navás, 1923: 33; Navás, 1925:
    114, fig. 11; Bacelar, 1928: 190; Caporiacco, 1928a: 79; Beier, 1929a: 358; Beier,
    1929d: 155; Beier, 1929e: 450; Beier, 1930f: 72; J. C. Chamberlin, 1930: 603,
    fig. 1d; Beier, 1932a: 181−182; Roewer, 1937: 261; Beier, 1939f: 191; Vachon,
    1940g: 144; Redikorzev, 1949: 644−645; Beier, 1953a: 300−301; Beier, 1955d:
    213; Beier, 1956e: 303; Beier, 1958b: 27; Beier, 1959f: 128; Beier, 1961b: 30;
    Vachon, 1961: 103; Beier, 1963a: 259; Beier, 1963i: 284; Beier, 1965c: 377;
    Helversen, 1965: 95−96; Weygoldt, 1969a: 42; Ćurčić, 1974a: 25; Beier, 1976a:
    24; Mahnert, 1975c: 194; Heurtault, 1979c: 925−930, figs 1−6; Heurtault, 1980c:
    fig. 15; Pieper, 1980: 263; Mahnert, 1982f: 297; Schawaller, 1983b: 19; Mahnert,
    1985a: 17; International Commission on Zoological Nomenclature, 1987: 53;
    Callaini, 1988b: 54; Callaini, 1989: 144.
Olpium pallidipes (sic) (Lucas): Pavesi, 1884: 457.
Olpium deserticola (E. Simon): Ellingsen, 1906: 263−264; Ellingsen, 1912b: 88
    (misidentifications).
Olpium pallipes pallipes (Lucas): Beier, 1932a: 182, fig. 206; Beier, 1955n: 106; Beier,
    1963b: 231, fig. 235; Lazzeroni, 1969a: 334; Lazzeroni, 1969c: 238; Lazzeroni,
    1970a: 207; Orghidan, Dumitresco & Georgesco, 1975: 30; Callaini, 1983c:
    292−294.
Not *Olpium pallipes* (Lucas): Tullgren, 1907d: 10, fig. 4 (misidentification; see *Olpium
    savignyi* E. Simon, nomen dubium).

Type locality: Algeria.
Distribution: Algeria, Balearic Islands, Canary Islands, Cape Verde Islands, Corsica,
    France, Egypt, Greece, Ilhas Selvagens, Israel, Italy, Libya, Madeira Islands,
    Malta, Morocco, Portugal, Sardinia, Sicily, Spain, Syria, Tunisia, U.S.S.R.
    (R.S.F.S.R., Turkmenistan, Uzbekistan), Yugoslavia.

Olpium pallipes balcanicum Beier

Olpium pallipes balcanicum Beier, 1931c: 305−306; Beier, 1932a: 182, fig. 207;
    Caporiacco, 1951e: 69; Beier, 1963b: 232, fig. 236; Beier, 1963f: 187; Lazzeroni,
    1970a: 207.

Type locality: Levkás, Greece.
Distribution: Greece, Israel.

Olpium philippinum Beier

Olpium philippinum Beier, 1967c: 317–318, fig. 2.

Type locality: Brooke's Point, Uring Uring, Palawan, Philippines.
Distribution: Philippines.

Olpium pusillulum Beier

Olpium pusillulum Beier, 1959a: 266–267, fig. 9; Beier, 1960a: 41.

Type locality: Kouh-Siah Pochtéh, Farah, Afghanistan.
Distribution: Afghanistan.

Olpium robustum Murthy & Ananthakrishnan

Olpium robustum Murthy & Ananthakrishnan, 1977: 59–61, figs 18a–b.

Type locality: Kodaikkanal, Tamil Nadu, India.
Distribution: India.

Olpium subgrande Tullgren

Olpium subgrande Tullgren, 1908b: 287–288, figs 6a–b; Beier, 1932a: 185; Roewer, 1937: 261.
Not *Olpium subgrande* Tullgren: Ellingsen, 1912b: 88, 116–117 (misidentification; see *Nanolpium rhodesiacum* Beier).

Type locality: between Kang and Khakea, Botswana.
Distribution: Botswana.

Olpium tenue J. C. Chamberlin

Olpium tenuis (sic) J. C. Chamberlin, 1930: 602–603, figs 1j, 1u–v, 3h–i, 3w.
Olpium tenue J. C. Chamberlin: Beier, 1932a: 185; Beier, 1946: 567–568, fig. 1; Beier, 1962e: 297–298, fig. 1; Roewer, 1937: 261; Heurtault, 1970a: 1164.

Type locality: Wadi Halfa, Sudan.
Distribution: Chad, Sudan.

Olpium tibium Sivaraman

Olpium tibium Sivaraman, 1980b: 335–336, figs 5a–b.

Type locality: Avadi, Madras, Tamil Nadu, India.
Distribution: India.

Olpium tropicum Murthy & Ananthakrishnan

Olpium tropicum Murthy & Ananthakrishnan, 1977: 53–55, figs 15a–b.

Type locality: Puttur, Andhra Pradesh, India.
Distribution: India.

Genus **Oreolpium** Benedict & Malcolm

Oreolpium Benedict & Malcolm, 1978a: 120.

Type species: *Oreolpium nymphum* Benedict & Malcolm, 1978a, by original designation.

Oreolpium nymphum Benedict & Malcolm

Oreolpium nymphum Benedict & Malcolm, 1978a: 120–124, figs 9–14; Zeh, 1987b: 1086.

Type locality: 4 miles N., 13 miles E. of Lowell, Lane County, Oregon, U.S.A.
Distribution: U.S.A. (Oregon).

Genus **Pachyolpium** Beier

Pachyolpium Beier, 1931c: 310; Beier, 1932a: 193; Hoff, 1945f: 2–3; Hoff, 1964a: 26.

Type species: *Pachyolpium reimoseri* Beier, 1931c, by original designation.

Pachyolpium arubense Beier

Pachyolpium arubense Beier, 1936a: 443–444, fig. 1; Roewer, 1940: 345.

Type locality: between Seroe Macuarina and Seroe Wara Wara, Aruba.
Distribution: Aruba.

Pachyolpium atlanticum Mahnert & R. Schuster

Pachyolpium atlanticum Mahnert & R. Schuster, 1981: 266–271, figs 1–13; Mahnert, 1981c: fig. 1; R. Schuster, 1986a: 270.

Type locality: Whalebone Bay, St George's Island, Bermuda.
Distribution: Bermuda.

Pachyolpium brevifemoratum (Balzan)

Olpium brevifemoratum Balzan, 1887b: no pagination, figs; Balzan, 1890: 440–441, figs 21, 21a–b; Balzan, 1892: 549; Ellingsen, 1910a: 390.
Pachyolpium brevifemoratum (Balzan): Beier, 1932a: 194; Roewer, 1937: 262.
Not *Olpium brevifemoratum* Balzan: Ellingsen, 1910a: 390 (misidentification; see *Pachyolpium incertum* Beier).

Type locality: Rio Apa, Paraguay.
Distribution: Paraguay, Virgin Islands.

Pachyolpium brevipes (With)

Olpium brevipes With, 1907: 73–75, figs 29–33.
Pachyolpium brevipes (With): Beier, 1932a: 194; Roewer, 1937: 262; Heurtault & Rebière, 1983: 591–593, figs 1–4.

Type locality: St Vincent.
Distribution: Martinique, St Vincent.

Pachyolpium crassichelatum (Balzan)

Olpium crassichelatum Balzan, 1887a: no pagination, figs; Balzan, 1890: 439–440, figs 20, 20a–b; Balzan, 1892: 537, fig. 29; With, 1906: fig. 13; Tullgren, 1907a: 67.
Pachyolpium crassichelatum (Balzan): Beier, 1932a: 195, figs 17, 21, 224; Roewer, 1937: 262; Feio, 1942: 115; Feio, 1945: 4; Beier, 1970b: 51.
Not *Olpium crassichelatum* Balzan: Ellingsen, 1910a: 390 (misidentification; see *Pachyolpium erratum* Beier).

Type locality: Mato Grosso (as Mato-groso), Brazil.
Distribution: Argentina, Brazil (Distrito Federal, Espírito Santo, Mato Grosso, Rio de Janeiro, Sao Paulo).

Pachyolpium erratum Beier

Olpium crassichelatum Balzan: Ellingsen, 1910a: 390 (misidentification).
Pachyolpium erratum Beier, 1931c: 311–312; Beier, 1932a: 195, fig. 225; Roewer, 1937: 262; Feio, 1942: 117; Feio, 1945: 4; Beier, 1954f: 132.

Type locality: Rio de Janeiro, Brazil.
Distribution: Brazil (Distrito Federal, Rio de Janeiro), Venezuela.

Pachyolpium furculiferum (Balzan)

Olpium furculiferum Balzan, 1892: 537–538, figs 30, 30a–b; Ellingsen, 1910a: 391.
Olpium furciliferum (sic) Balzan: H. J. Hansen, 1893: plate 5 figs 2–4; With, 1907: 73.

Pachyolpium furculiferum (Balzan): Beier, 1932a: 195, fig. 226; Roewer, 1937: 262; Feio, 1945: 4; Beier, 1954f: 132; Beier, 1970d: 52; Hounsome, 1980: 85.

Type localities: Caracas, Venezuela; Corozal, Venezuela; and La Guaira, Venezuela.
Distribution: Brazil (Distrito Federal), Cayman Islands, St Thomas, St Vincent, Venezuela.

Pachyolpium granulatum Beier

Pachyolpium granulatum Beier, 1954f: 132–134, fig. 2; Beier, 1959e: 192.

Type locality: Krashki, Los Roques, Venezuela.
Distribution: Colombia, Peru, Venezuela.

Pachyolpium incertum Beier

Olpium brevifemoratum Balzan: Ellingsen, 1910a: 390 (misidentification).
Pachyolpium incertum Beier, 1931c: 311; Beier, 1932a: 194, fig. 223; Roewer, 1937: 262; Beier, 1954f: 132.

Type locality: St Thomas.
Distribution: St Thomas, Venezuela.

Pachyolpium irmgardae Mahnert

Pachyolpium irmgardae Mahnert, 1979d: 755–759, figs 75–84; Adis, 1981: 118, etc.; Adis & Funke, 1983: 357; Adis & Mahnert, 1986: 310, fig. 13; Mahnert & Adis, 1986: 214; Adis, Junk & Penny, 1987: 489.

Type locality: Taruma Mirim, Manaus, Amazonas, Brazil.
Distribution: Brazil (Amazonas).

Pachyolpium isolatum (R. V. Chamberlin)

Olpium isolatum R. V. Chamberlin, 1925: 239.
Pachyolpium adiposum Hoff, 1945e: 12–15, figs 15–16; Hoff, 1964a: 26–28; Zeh, 1987b: 1086 (synonymized by Muchmore, 1984d: 353).
Pachyolpium isolatum (R. V. Chamberlin): Muchmore, 1984d: 353–355, fig. 1.

Type localities: of *Olpium isolatum*: Largo Remo Island, Panama.
of *Pachyolpium adiposum*: Barro Colorado Island, Panama.
Distribution: Jamaica, Panama.

Pachyolpium reimoseri Beier

Pachyolpium reimoseri Beier, 1931c: 310–311, fig. 7; Beier, 1932a: 193–194, fig. 222; Roewer, 1937: 262.

Type locality: Jimenez, Costa Rica.
Distribution: Costa Rica.

Genus **Paraldabrinus** Beier

Paraldabrinus Beier, 1966g: 367–368.

Type species: *Paraldabrinus novaecaledoniae* Beier, 1966g, by original designation.

Paraldabrinus novaecaledoniae Beier

Paraladabrinus novaecaledoniae Beier, 1966g: 368–370, fig. 4.

Type locality: Ndokoa Gorge between Pic Adio and Dent de poya, New Caledonia.
Distribution: New Caledonia.

Genus **Parolpium** Beier

Parolpium Beier, 1931c: 308; Beier, 1932a: 189; Murthy & Ananthakrishnan, 1977: 79–80.

Type species: *Olpium gracile* Beier, 1930f, by original designation.

Parolpium arabicum (E. Simon)

Olpium arabicum E. Simon, 1890: 121.

Parolpium arabicum (E. Simon): Beier, 1931c: 308; Beier, 1932a: 190; Roewer, 1937: 262.

Not *Olpium arabicum* E. Simon: Ellingsen, 1906: 263 (misidentification; see *Parolpium minor* (Ellingsen)); Ellingsen, 1910a: 390 (misidentification; see *Xenolpium madagascariense* (Beier)); Ellingsen, 1912b: 87, 116 (misidentification; see *Parolpium minor* (Ellingsen) and *Xenolpium madagascariense* (Beier)); Godfrey, 1927: 18 (misidentification; see *Parolpium minor* (Ellingsen)).

Type locality: Aden, South Yemen.
Distribution: South Yemen.

Parolpium gracile (Beier)

Olpium gracile Beier, 1930f: 75–77, figs 4, 5a–b; Beier, 1933c: 85–87.

Parolpium gracile (Beier): Beier, 1931c: 308; Beier, 1932a: 190, fig. 218; Roewer, 1937: 262.

Type locality: Sinafir (as Senafir) Island, Saudi Arabia.
Distribution: Saudi Arabia.

Parolpium minor (Ellingsen)

Olpium arabicum E. Simon: Ellingsen, 1906: 263; Ellingsen, 1912b: 87, 116; Godfrey, 1927: 18 (misidentifications).

Olpium arabicum minor Ellingsen, 1910d: 538.

Parolpium minor (Ellingsen): Beier, 1931c: 308.

Parolpium minus (Ellingsen): Beier, 1932a: 191; Roewer, 1937: 262.

Parolpium (?) *minus* (Ellingsen): Vachon, 1956: 3–4.

Type locality: Bugala, Uganda.
Distribution: Cape Verde Islands, South Africa, Uganda.

Parolpium pallidum Beier

Parolpium pallidum Beier, 1967g: 346–347, fig. 6; Murthy & Ananthakrishnan, 1977: 80.

Type locality: 65 km NE. of Dumka, India.
Distribution: India.

Genus **Planctolpium** Hoff

Planctolpium Hoff, 1964a: 40–41; Muchmore, 1979a: 203.

Type species: *Planctolpium arboreum* Hoff, 1964a, by original designation.

Planctolpium arboreum Hoff

Planctolpium arboreum Hoff, 1964a: 41–44, figs 15–16; Beier, 1976d: 46; Muchmore, 1977: 76; Zeh, 1987b: 1086.

Type locality: Rio Cobre Gorge, 4 miles N. of Spanish Town, Parish of St Catherine, Jamaica.
Distribution: Dominican Republic, Jamaica, Mexico.

Planctolpium peninsulae Muchmore

Planctolpium peninsulae Muchmore, 1979a: 203–205, figs 21–27.

Type locality: Winter Haven, Polk County, Florida, U.S.A.
Distribution: U.S.A. (Florida).

Planctolpium suteri Muchmore

Planctolpium suteri Muchmore, 1979a: 206–207, figs 28–30.

Type locality: Dauphin Island, Mobile County, Alabama, U.S.A.
Distribution: U.S.A. (Alabama, Florida).

Genus **Progarypus** Beier

Progarypus Beier, 1931c: 317; Beier, 1932a: 222–223.

Type species: *Olpium ramicola* Balzan, 1887a, by original designation.

Progarypus longipes Beier

Progarypus longipes Beier, 1964b: 340–341, fig. 18; Cekalovic, 1984: 17.

Type locality: El Portillo, Atacama, Chile.
Distribution: Chile.

Progarypus marginatus Beier

Progarypus marginatus Beier, 1964b: 338–339, fig. 17; Cekalovic, 1984: 17.

Type locality: Algarrobo, Chile.
Distribution: Chile.

Progarypus novus Beier

Olpium ramicola Balzan: Beier, 1930a: 205–206, figs 4a–b (misidentification).
Progarypus novus Beier, 1931c: 318; Beier, 1932a: 223–224, fig. 252; Roewer, 1937: 268; Feio, 1945: 4.

Type locality: Brazil.
Distribution: Brazil (Bahia).

Progarypus oxydactylus (Balzan)

Olpium oxydactylum Balzan, 1887b: no pagination, figs; Balzan, 1890: 435–436, figs 17, 17a–b; Balzan, 1892: 549.
Progarypus oxydactylus (Balzan): Beier, 1931c: 317; Beier, 1932a: 223; Roewer, 1937: 268.

Type locality: Rio Apa, Paraguay.
Distribution: Paraguay.

Progarypus peruanus Beier

Progarypus peruanus Beier, 1959e: 199–200, fig. 12.

Type locality: 60 miles N. of Puno, Peru.
Distribution: Peru.

Progarypus ramicola (Balzan)

Olpium ramicola Balzan, 1887a: no pagination, figs; Balzan, 1890: 436–437, figs 18, 18a–c; Balzan, 1892: 549.
Progarypus ramicola (Balzan): Beier, 1931c: 317–318; Beier, 1932a: 223, figs 250–251; Roewer, 1937: 268, fig. 220.
Not *Olpium ramicola* Balzan: Beier, 1930a: 205–206, figs 4a–b (misidentification; see *Progarypus novus* Beier).

Type localities: Asuncion, Paraguay; and Rio Apa, Paraguay.
Distribution: Paraguay.

Progarypus viridans (Banks)

Garypus viridans Banks, 1909a: 145–146.
Progarypus (?) *viridans* (Banks): Beier, 1932a: 224; Roewer, 1937: 268.

Type locality: Santa Marta, Colombia.
Distribution: Colombia.

Genus Protogarypinus Beier

Protogarypinus Beier, 1954b: 3–4.

Type species: *Protogarypinus giganteus* Beier, 1954b, by original designation.

Protogarypinus dissimilis Beier

Protogarypinus dissimilis Beier, 1975b: 204–205, fig. 1; Harvey, 1981b: 243; Harvey, 1985b: 147.

Type locality: Alligator Gorge, Flinders Ranges, South Australia, Australia.
Distribution: Australia (South Australia).

Protogarypinus giganteus Beier

Protogarypinus giganteus Beier, 1954b: 4–6, fig. 1; Harvey, 1981b: 243; Harvey, 1985b: 147.

Type locality: Denmark, near mouth of Denmark River, Western Australia, Australia.
Distribution: Australia (Western Australia).

Genus Pseudogarypinus Beier

Pseudogarypinus Beier, 1931c: 313–314; Beier, 1932a: 206; Hoff, 1956b: 33; Benedict & Malcolm, 1978a: 125.

Type species: *Pseudogarypinus costaricensis* Beier, 1931c, by original designation.

Pseudogarypinus cooperi Muchmore

Pseudogarypinus cooperi Muchmore, 1980e: 167–169, figs 7–14.

Type locality: Gavilan Hills, S. of Riverside, Riverside County, California, U.S.A.
Distribution: U.S.A. (California).

Pseudogarypinus costaricensis Beier

Pseudogarypinus costaricensis Beier, 1931c: 314–315, fig. 9; Beier, 1932a: 207, figs 232, 236; Beier, 1932g: fig. 183; Roewer, 1937: 265; Beier, 1976c: 1.

Type locality: Irazu, Costa Rica.
Distribution: Costa Rica.

Pseudogarypinus frontalis (Banks)

Olpium frontalis Banks, 1909b: 307.
Garypinus marianae J. C. Chamberlin, 1930: 591–592, figs 1t, 1aa; J. C. Chamberlin, 1931a: figs 11w, 15l, 19i, 26b, 37b, 47j (synonymized by Benedict & Malcolm, 1978a: 125).
Pseudogarypinus marinae (sic) (J. C. Chamberlin): Beier, 1931c: 314.
Pseudogarypinus marianae (J. C. Chamberlin): Beier, 1932a: 206–207; Roewer, 1937: 265; Hoff, 1958: 16; Hoff, 1961: 439–440; Weygoldt, 1969a: 43, 124; Muchmore, 1971a: 91; Knowlton, 1974: 3; Rowland & Reddell, 1976: 16; Zeh, 1987b: 1086.
Serianus (?) *frontalis* (Banks): Beier, 1932a: 213; Roewer, 1937: 266.

Family Olpiidae

Pseudogarypinus frontalis (Banks): Hoff, 1956b: 33–34; Benedict & Malcolm, 1978a: 125–132, figs 15–18.
Pseudogarypinus (?) *frontalis* (Banks): Hoff, 1958: 16–17; Hoff, 1959b: 4, etc.

Type localities: of *Olpium frontalis*: Las Vegas, New Mexico, U.S.A.
of *Garypinus marianae*: Madrone, Atherton, San Mateo County, California, U.S.A.
Distribution: U.S.A. (California, Colorado, New Mexico, Oregon, Texas, Utah, Washington).

Pseudogarypinus giganteus Hoff

Pseudogarypinus giganteus Hoff, 1961: 440–442, figs 14–15.

Type locality: Stoneham, Weld County, Colorado, U.S.A.
Distribution: U.S.A. (Colorado).

Genus Pseudohorus Beier

Pseudohorus Beier, 1947b: 304–305.

Type species: *Pseudohorus vermiformis* Beier, 1947b, by original designation.

Pseudohorus caecus Beier

Pseudohorus caecus Beier, 1967d: 75–76, fig. 2.

Type locality: 15 miles NW. of Morogoro, Tanzania.
Distribution: Tanzania.

Pseudohorus embuensis Mahnert

Pseudohorus embuensis Mahnert, 1982c: 268–270, figs 18–24.

Type locality: Kogari, Embu, Kenya.
Distribution: Kenya.

Pseudohorus excavatus Beier

Pseudohorus excavatus Beier, 1955l: 282–285, fig. 6; Beier, 1964k: 45; Beier, 1973a: 98.

Type locality: 35 miles SW. of Ananbib (Orupembe), Namibia.
Distribution: Namibia.

Pseudohorus gracilis Beier

Pseudohorus gracilis Beier, 1962d: 224–225, fig. 2; Beier, 1973a: 98.

Type locality: Gobabeb, Namibia.
Distribution: Namibia.

Pseudohorus incrassatus Beier

Pseudohorus incrassatus Beier, 1955l: 286–287, fig. 7; Beier, 1964k: 45; Beier, 1973a: 98.

Type locality: Kamanjab, 100 miles SW. of Outjo, Namibia.
Distribution: Namibia.

Pseudohorus luscus Mahnert

Pseudohorus luscus Mahnert, 1982c: 266–268, figs 10–17.

Type locality: 53 km W. of Garissa, Tana River, Kenya.
Distribution: Kenya.

Pseudohorus molliventer Beier

Pseudohorus molliventer Beier, 1947b: 306–307, fig. 14; Weidner, 1959: 116; Beier, 1962d: 223; Beier, 1973a: 98.

Type locality: Otavifontain, 5 km E. of Otavi, Namibia.
Distribution: Namibia.

Pseudohorus pilosus Mahnert

Pseudohorus pilosus Mahnert, 1982c: 264–266, figs 1–9.

Type locality: 10 km N. of Garsen, Tana River, Kenya.
Distribution: Kenya.

Pseudohorus strumosus Beier

Pseudohorus strumosus Beier, 1962d: 223–224, fig. 1; Beier, 1973a: 98.

Type locality: Natab, Kuiseb River, Namibia.
Distribution: Namibia.

Pseudohorus transvaalensis (Beier)

Minniza transvaalensis Beier, 1956a: 29–30, fig. 1.
Pseudohorus transvaalensis Beier, 1964k: 45; Beier, 1966k: 457; Lawrence, 1967: 88.

Type locality: Chipisé, between Louis Trichardt and Messina, Transvaal, South Africa.
Distribution: South Africa, Zimbabwe.

Pseudohorus transvaalensis fenestratus Beier

Pseudohorus transvaalensis fenestratus Beier, 1966k: 457–458, fig. 2; Lawrence, 1967: 88.

Type locality: N. of Mahlakuza, Kruger National Park, Transvaal, South Africa.
Distribution: South Africa.

Pseudohorus vermiformis Beier

Pseudohorus vermiformis Beier, 1947b: 305–306, figs 12–13; Beier, 1955l: 285; Beier, 1964k: 45; Beier, 1966k: 469.

Type locality: Viooolsdrift (as Viodsdrift), Orange River, Namaqualand, Cape Province, South Africa.
Distribution: Namibia, South Africa.

Pseudohorus vermis Mahnert

Pseudohorus vermis Mahnert, 1982c: 270–272, figs 25–33.

Type locality: Mwingi, Kitui, Kenya.
Distribution: Kenya.

Genus Serianus J. C. Chamberlin

Serianus J. C. Chamberlin, 1930: 594; Beier, 1932a: 211; Hoff, 1956b: 34–35; Hoff, 1964a: 35; Muchmore, 1979a: 207.
Paraserianus Beier, 1939b: 288; Roewer, 1940: 345 (synonymized by Mahnert, 1988b: 7–8).

Type species: of *Serianus*: *Garypinus serianus* J. C. Chamberlin, 1923c, by original designation.
of *Paraserianus*: *Paraserianus bolivianus* Beier, 1939b, by original designation.

Serianus arboricola (J. C. Chamberlin)

Garypinus arboricolus (sic) J. C. Chamberlin, 1923c: 369, plate 2, fig. 5.
Serianus arboricolus (sic) (J. C. Chamberlin): J. C. Chamberlin, 1930: 595; J. C. Chamberlin, 1931a: figs 17t–u, 37d; Beier, 1932a: 212; Roewer, 1937: 265.

Type locality: San Esteban Island, Gulf of California, Mexico.
Distribution: Mexico.

Serianus argentinae Muchmore

Serianus minutus Hoff, 1950b: 233–237, figs 10–14 (junior secondary homonym of *Serianus minutus* (Banks)).
Serianus argentinae Muchmore, 1981d: 238 (replacement name for *Serianus minutus* Hoff).

Type locality: Monte Veloz, Buenos Aires, Argentina.
Distribution: Argentina.

Serianus birabeni Feio

Serianus birabeni Feio, 1945: 17–20, figs 8–11, 12a–c.

Type locality: Penas, Pinchas, La Rioja, Argentina.
Distribution: Argentina.

Serianus bolivianus (Beier)

Paraserianus bolivianus Beier, 1939b: 288–289, fig. 1; Roewer, 1940: 345.
Serianus bolivianus (Beier): Mahnert, 1988b: 8, figs 13–18.

Type locality: Pazña, Bolivia.
Distribution: Bolivia.

Serianus carolinensis Muchmore

Serianus sp.: Weygoldt, 1966c: 1648–1649, figs 1–7.
Serianus carolinensis Muchmore, 1968d: 145–149, figs 1–3; Weygoldt, 1969a: 14, 15, 30, 42–44, 119, figs 39a–c, 40, 41a–c, 105, 109; Muchmore, 1979a: 207, fig. 31.

Type locality: Beaufort, Carteret County, North Carolina, U.S.A.
Distribution: U.S.A. (Florida, North Carolina).

Serianus dolosus Hoff

Serianus dolosus Hoff, 1956b: 35–43, figs 16–18; Hoff, 1958: 17; Hoff, 1959b: 4, etc.; Zeh, 1987b: 1086.

Type locality: 8 miles N. of Golden, Ortiz Mountains, Santa Fe County, New Mexico, U.S.A.
Distribution: U.S.A. (New Mexico).

Serianus galapagoensis Beier

Serianus galapagoensis Beier, 1978c: 536–538, fig. 3.

Type locality: Santa Fé Island, Galapagos Islands.
Distribution: Galapagos Islands.

Serianus gratus Hoff

Serianus gratus Hoff, 1964a: 36–39, figs 13–14; Muchmore, 1977: 76; Muchmore, 1979a: 207; Zeh, 1987b: 1086.

Type locality: near Morant Point Lighthouse, Parish of St Thomas, Jamaica.
Distribution: Belize, Jamaica, U.S.A. (Florida).

Serianus litoralis (J. C. Chamberlin)

Garypinus litoralis J. C. Chamberlin, 1923c: 368, plate 1 fig. 4, plate 2 fig. 4.
Serianus litoralis (J. C. Chamberlin): J. C. Chamberlin, 1930: 595; Beier, 1932a: 212; Roewer, 1937: 265; V.F. Lee, 1979a: 24–25, figs 28–29, 31.

Type locality: Monserrate Island, Gulf of California, Mexico.
Distribution: Mexico.

Serianus minutus (Banks)

Olpium minutum Banks, 1908: 42.
Pachyolpium (?) *minutum* (Banks): Beier, 1932a: 196; Roewer, 1937: 262; Beier, 1948b: 457; Hoff, 1958: 16.
Serianus minutus (Banks): Muchmore, 1971a: 87; Muchmore, 1981d: 337–338, figs 1–3.
Pachyolpium minutum (Banks): Rowland & Reddell, 1976: 16.

Type locality: Austin, Texas, U.S.A.
Distribution: U.S.A. (Texas).

Serianus patagonicus (Ellingsen)

Garypinus patagonicus Ellingsen, 1904: 5–7.
Neominniza (?) *patagonicus* (Ellingsen): Beier, 1932a: 208; Roewer, 1937: 265.
Neominniza patagonicus (Ellingsen): Mello-Leitão, 1939a: 117; Mello-Leitão, 1939b: 615.
Serianus patagonicus (Ellingsen): Beier, 1959e: 198–199, fig. 11; Beier, 1962h: 133; Beier, 1964b: 336–338, fig. 16; Beier, 1964j: 490–491; Cekalovic, 1976: 18; Cekalovic, 1984: 16.

Type locality: near Rio Santa Cruz, Patagonia, Argentina.
Distribution: Argentina, Chile.

Serianus pusillimus Beier

Serianus pusillimus Beier, 1959e: 197–198, fig. 10; Beier, 1978c: 536.

Type locality: Isla Puná, Ecuador.
Distribution: Ecuador, Galapagos Islands.

Serianus sahariensis Mahnert

Serianus sahariensis Mahnert, 1988b: 4–7, figs 6–12.

Type locality: 30 km S. of Djanet, Algeria.
Distribution: Algeria.

Serianus salomonensis Beier

Serianus salomonensis Beier, 1966d: 138–139, fig. 4; Beier, 1970g: 318.

Type locality: Kukum, Guadalcanal, Solomon Islands.
Distribution: Solomon Islands.

Serianus serianus (J. C. Chamberlin)

Garypinus serianus J. C. Chamberlin, 1923c: 367, fig. 1, plate 1 fig. 1, plate 2 fig. 6.
Serianus serianus (J. C. Chamberlin): J. C. Chamberlin, 1930: 595–596, figs 21, 30; J. C. Chamberlin, 1931a: figs 40m–n, 42j, 42m; Beier, 1932a: 212; Roewer, 1937: 265; Hoff, 1958: 17.

Type locality: Pelican Island, Kino Bay, Gulf of California, Mexico.
Distribution: Mexico, U.S.A. (Utah).

Serianus solus (J. C. Chamberlin)

Garypinus solus J. C. Chamberlin, 1923c: 367–368, plate 1 fig. 3, plate 2 fig. 3.
Serianus solus (J. C. Chamberlin): J. C. Chamberlin, 1930: 596; J. C. Chamberlin, 1931a: fig. 29b; Beier, 1932a: 212; Roewer, 1937: 266.

Type locality: South Santa Inez, Gulf of California, Mexico.
Distribution: Mexico.

Genus **Solinellus** Muchmore

Solinellus Muchmore, 1979a: 208–209.

Type species: *Solinellus simberloffi* Muchmore, 1979a, by original designation.

Solinellus simberloffi Muchmore

Solinellus simberloffi Muchmore, 1979a: 209–211, figs 33–42.

Type locality: Mud Keys, Monroe County, Florida, U.S.A.
Distribution: U.S.A. (Florida).

Genus **Solinus** J. C. Chamberlin

Solinus J.C. Chamberlin, 1930: 596; Beier, 1932a: 213; Morikawa, 1960: 130.

Type species: *Garypinus corticolus* J. C. Chamberlin, 1923c, by original designation.

Solinus africanus Beier

Solinus africanus Beier, 1967d: 77–79, fig. 4; Mahnert, 1982c: 298.

Type locality: Simu Beach, Kwale, Kenya.
Distribution: Kenya.

Solinus australiensis J. C. Chamberlin

Solinus australiensis J. C. Chamberlin, 1930: 597; Beier, 1932a: 214; J. C. Chamberlin, 1934b: 5; Roewer, 1937: 266; Tubb, 1937: 412; Harvey, 1981b: 243; Harvey, 1985b: 147.

Type locality: Barringun, New South Wales, Australia.
Distribution: Australia (New South Wales, Victoria).

Solinus corticolus (J. C. Chamberlin)

Garypinus corticolus J. C. Chamberlin, 1923c: 366, plate 1 fig. 2, plate 2 fig. 1.
Solinus corticolus (J. C. Chamberlin): J. C. Chamberlin, 1930: 596–597; J. C. Chamberlin, 1931a: figs 9h, 37c, 42e, 61; Beier, 1932a: 213–214, fig. 243; Beier, 1932g: fig. 257; Roewer, 1937: 266, fig. 217.

Type locality: La Paz, Baja California, Mexico.
Distribution: Mexico.

Solinus cyrenaicus (Beier)

Garypinus cyrenaicus Beier, 1929b: 80–81, figs 2a–b; Beier, 1932a: 209–210, fig. 240; Roewer, 1937: 265.
Solinus cyrenaicus (Beier): Beier, 1966e: 164.

Type locality: Colfia bengasi, Cyrenaica, Libya.
Distribution: Libya.

Solinus hispanus Beier

Solinus hispanus Beier, 1939f: 192–193, fig. 20; Roewer, 1940: 345; Beier, 1963b: 237, fig. 241; Beier, 1970f: 322.

Type locality: Tortosa, Cataluña, Spain.
Distribution: Spain.

Solinus japonicus Morikawa

Solinus japonicus Morikawa, 1953b: 327–328, figs 1–2; Morikawa, 1960: 130, plate 3 fig. 3, plate 7 fig. 13.

Type locality: Fukaura, Shikoku, Japan.
Distribution: Japan.

Solinus pusillus Beier

Solinus pusillus Beier, 1971b: 368–369.

Type locality: Mt Dayman, Milne Bay District, Papua New Guinea.
Distribution: Papua New Guinea.

Solinus rhodius Beier

Solinus rhodius Beier, 1966e: 163–164, fig. 1.

Type localities: Ródhos (as Rhodos), Greece; and Kritica region, Ródhos, Greece.
Distribution: Greece.

Genus Stenolpiodes Beier

Stenolpiodes Beier, 1959e: 190.

Type species: *Stenolpiodes gracillimus* Beier, 1959e, by original designation.

Stenolpiodes gracillimus Beier

Stenolpiodes gracillimus Beier, 1959e: 190–191, fig. 4.

Type locality: 4 miles W. of Otusco, La Libertad, Peru.
Distribution: Peru.

Genus Stenolpium Beier

Stenolpium Beier, 1955m: 5.

Type species: *Stenolpium peruanum* Beier, 1955m, by original designation.

Stenolpium asperum Beier

Stenolpium asperum Beier, 1954g: 325–326, fig. 1.

Type locality: near Arequipa, Peru.
Distribution: Peru.

Stenolpium asperum nitrophilum Beier

Stenolpium asperum nitrophilum Beier, 1964b: 325, fig. 9; Cekalovic, 1984: 14.

Type locality: Pampa del Tamarugal, Tarapacá, Chile.
Distribution: Chile.

Stenolpium fasciculatum Mahnert

Stenolpium (?) *fasciculatum* Mahnert, 1984c: 22–24, figs 10–15.

Type locality: Cherrepe, Chiclayo, Lambayeque, Peru.
Distribution: Peru.

Stenolpium insulanum Beier

Stenolpium insulanum Beier, 1978c: 535–536, fig. 2.

Type locality: Playa Ocohava, San Cristobál Island, Galapagos Islands.
Distribution: Galapagos Islands.

Stenolpium mediocre Beier

Stenolpium mediocre Beier, 1959e: 187–188, fig. 2.

Type locality: 38 miles N. of Olmos, Peru.
Distribution: Peru.

Stenolpium peruanum Beier

Stenolpium peruanum Beier, 1955m: 5.

Type locality: Paita (as Payte), Peru.
Distribution: Peru.

Stenolpium robustum Beier

Stenolpium robustum Beier, 1959e: 186–187, fig. 1.

Type locality: 8 km NE. of Pucusana, Lima, Peru.
Distribution: Peru.

Stenolpium rossi Beier

Stenolpium rossi Beier, 1959e: 188–189, fig. 3.

Type locality: 10 miles N. of Barranca, Peru.
Distribution: Peru.

Genus **Teratolpium** Beier

Teratolpium Beier, 1959e: 195.

Type species: *Teratolpium andinum* Beier, 1959e, by original designation.

Teratolpium andinum Beier

Teratolpium andinum Beier, 1959e: 195–196, fig. 8.

Type locality: Jahua-Kocha, Huayhuash Cordillero, Peru.
Distribution: Peru.

Genus **Thaumatolpium** Beier

Thaumatolpium Beier, 1931c: 313; Beier, 1932a: 198–199; Beier, 1964b: 329.

Type species: *Ideoroncus silvestrii* Beier, 1930d, by original designation.

Thaumatolpium caecum Beier

Thaumatolpium caecum Beier, 1964b: 335–336, fig. 15; Cekalovic, 1984: 15.

Type locality: 10 miles W. of Vicuña, Chile.
Distribution: Chile.

Thaumatolpium kuscheli Beier

Thaumatolpium kuscheli Beier, 1964b: 330–331, fig. 12; Cekalovic, 1984: 15.

Type locality: Corrizal Bajo, Chile.
Distribution: Chile.

Thaumatolpium longesetosum Beier

Thaumatolpium longesetosum Beier, 1964b: 333–335, fig. 14; Cekalovic, 1984: 15.

Type locality: Talinay, Coquimbo, Chile.
Distribution: Chile.

Thaumatolpium robustius Beier

Thaumatolpium robustius Beier, 1964b: 331–333, fig. 13; Cekalovic, 1984: 16.

Type locality: Huasco, Atacama, Chile.
Distribution: Chile.

Thaumatolpium silvestrii (Beier)

Ideoroncus silvestrii Beier, 1930d: 204–205, fig. 9.
Thaumatolpium silvestrii (Beier): Beier, 1931c: 313; Beier, 1932a: 199, figs 229–230;
    Roewer, 1936: fig. 76; Roewer, 1937: 263; Beier, 1964b: 329–330; Cekalovic,
    1984: 16.

Type locality: Viña del Mar, Chile.
Distribution: Chile.

Genus **Xenolpium** J. C. Chamberlin

Xenolpium J. C. Chamberlin, 1930: 600; Beier, 1932a: 201; Morikawa, 1960: 126; Beier, 1976f: 210.
Antiolpium Beier, 1962g: 399 (synonymized by Beier, 1976f: 210).

Type species: of *Xenolpium*: *Olpium pacificum* With, 1907, by original designation.
of *Antiolpium*: of *Olpium zealandiensis* Hoff, 1947a, by original designation.

Xenolpium graniferum Beier

Xenolpium graniferum Beier, 1965b: 370−371, fig. 4.

Type locality: Fort Lamy, Chad.
Distribution: Chad.

Xenolpium incrassatus (Beier)

Calocheiridius incrassatus Beier, 1964k: 41−42, fig. 8.
Xenolpium incrassatum (Beier): Beier, 1965b: 369.

Type locality: Bushman's River, Natal, South Africa.
Distribution: South Africa.

Xenolpium insulare Beier

Xenolpium insulare Beier, 1940a: 171−173, fig. 1; Mahnert, 1978d: 873, figs 12−13.

Type locality: Silhouette, Seychelles.
Distribution: Seychelles.

Xenolpium longiventer (L. Koch)

Olpium longiventer L. Koch, 1885: 50−51, plate 6, figs 9, 9a−c.
Amblyolpium longiventer (L. Koch): E. Simon, 1899a: 121.
Xenolpium longiventer (L. Koch): Beier, 1932a: 202; J. C. Chamberlin, 1934b: 5; Roewer, 1937: 263; Beier, 1966a: 291; Beier, 1969a: 187; Hickman & Hill, 1978: appendix a; Harvey, 1981b: 244; Harvey, 1985b: 148.
Not *Olpium longiventer* L. Koch: Pocock, 1898: 323 (misidentification; see *Beierolpium oceanicum* (With)); Ellingsen, 1912a: 125 (misidentification; see *Olpium jacobsoni* Tullgren); Ellingsen, 1914: 12 (misidentification; true identification unknown).
Not *Amblyolpium longiventer* (L. Koch): E. Simon, 1900a: 519 (misidentification; see *Garypinus mirabilis* With)).

Type locality: Peak Downs (as Peack Downs), Queensland, Australia.
Distribution: Australia (Queensland, Tasmania, Victoria).

Xenolpium madagascariense (Beier)

Olpium arabicum E. Simon: Ellingsen, 1910a: 390 (misidentification); Ellingsen, 1912b: 87, 116 (misidentification, in part).
Parolpium madagascariense Beier, 1931c: 308−309; Beier, 1932a: 190−191, fig. 219; Roewer, 1937: 262.
Parolpius (sic) *madagascariensis* Beier: Legendre, 1972: 446.*Xenolpium madagascarienses* (sic) (Beier): Spaull, 1979: 117.
Xenolpium madagascariense (Beier): Beier, 1981: 293.

Type locality: St Marie, Madagascar.
Distribution: Aldabra Islands, Andaman Islands, Madagascar.

Xenolpium pacificum (With)

Olpium pacificum With, 1907: 75−77, figs 34−38.
Xenolpium pacificum (With): J. C. Chamberlin, 1930: 600−601; J. C. Chamberlin, 1931a: figs 14a, 15n, 16k, 37h−i, 52e; Beier, 1932a: 201−202; J. C. Chamberlin,

1934b: 5; Roewer, 1937: 263; Heurtault, 1976: fig. 3; Heurtault, 1980c: 144–147, figs 1–9.

Olpium zealandiensis Hoff, 1947a: 36–39, figs 1–3 (synonymized by Beier, 1976f: 210).

Antiolpium zealandiense (Hoff): Beier, 1962g: 400, fig. 1; Beier, 1966c: 367–368.

Euryolpium (Antiolpium) zealandiense (Hoff): Beier, 1967b: 290; Beier, 1969e: 413, 414; Wise, 1970: 234.

Xenolpium pacificum pacificum (With): Beier, 1976f: 210.

Type localities: of *Olpium pacificum*: Stewart Island, New Zealand.
of *Olpium zealandiensis*: Tera Kihi Island, Hauraki Gulf, North Island, New Zealand.
Distribution: Aldabra Islands?, New Zealand.

Xenolpium pacificum norfolkense Beier

Xenolpium pacificum norfolkense Beier, 1976f: 211, fig. 11; Harvey, 1985b: 148.

Type locality: Bumbora Reserve, Norfolk Island.
Distribution: Norfolk Island.

Superfamily NEOBISIOIDEA J. C. Chamberlin

Neobisioidea J. C. Chamberlin, 1930: 9; J. C. Chamberlin, 1931a: 215; Hoff, 1949b: 443; Petrunkevitch, 1955: 81; Hoff, 1956b: 2; Morikawa, 1960: 111–112; Hoff, 1964a: 6; Murthy & Ananthakrishnan, 1977: 24; Muchmore, 1982a: 97; Legg & Jones, 1988: 74.
Neobisiides (sic) J. C. Chamberlin: Beier, 1932a: 75; Beier, 1932g: 182; Roewer, 1937: 243.
Neobisiinea J. C. Chamberlin: Beier, 1939f: 165.

Family BOCHICIDAE J. C. Chamberlin

Bochicinae J. C. Chamberlin, 1930: 43; J. C. Chamberlin, 1931a: 220; Beier, 1932a: 168; Roewer, 1937: 255.
Leucohyinae J. C. Chamberlin, 1946: 7.
Bochicidae: Muchmore, 1982a: 98.

Genus **Antillobisium** Dumitresco & Orghidan
Antillobisium Dumitresco & Orghidan, 1977: 101–102.
Type species: *Antillobisium vachoni* Dumitresco & Orghidan, 1977, by original designation.

Antillobisium mitchelli Dumitresco & Orghidan
Antillobisium mitchelli Dumitresco & Orghidan, 1977: 106–107, figs 5b–c, 7a–d.
Antillobisium mitcheli (sic) Dumitresco & Orghidan: Armas & Alayón, 1984: 12.
Type locality: Cueva del Indio, Camagüey, Cuba.
Distribution: Cuba.

Antillobisium vachoni Dumitresco & Orghidan
Antillobisium vachoni Dumitresco & Orghidan, 1977: 102–106, figs 2a–g, 3a–b, 4a–d, 5a, 6, 20; Armas & Alayón, 1984: 12.
Type locality: Cueva del Guano, Oriente, Cuba.
Distribution: Cuba.

Genus **Apohya** Muchmore
Apohya Muchmore, 1973b: 53–54.
Type species: *Apohya campbelli* Muchmore, 1973b, by original designation.

Apohya campbelli Muchmore
Apohya campbelli Muchmore, 1973b: 54, figs 18–23.
Type locality: El Tinieblo, Tamaulipas, Mexico.
Distribution: Mexico.

Genus **Bochica** J. C. Chamberlin
Bochica J. C. Chamberlin, 1930: 43–44; Beier, 1932a: 168; Muchmore, 1984e: 61–62.
Type species: *Ideoroncus withi* J. C. Chamberlin, 1923c, by original designation.

Bochica withi (J. C. Chamberlin)

Ideoroncus mexicanus Banks: With, 1905: 127–131, plate 9 figs 2a–d, plate 10 figs 1a–f (misidentification).
Ideoroncus withi J. C. Chamberlin, 1923c: 359.
Bochica withi (J. C. Chamberlin): J. C. Chamberlin, 1930: 44; J. C. Chamberlin, 1931a: figs 13l, 15h, 28j–k, 36f, 42b; Beier, 1932a: 168–169; Roewer, 1937: 255; Muchmore, 1984e: 62–63, figs 1–2.

Type locality: Chantilly (as Chandilly), Grenada.
Distribution: Grenada, Trinidad.

Genus Leucohya J. C. Chamberlin

Leucohya J. C. Chamberlin, 1946: 7; Muchmore, 1972g: 271; Muchmore, 1973b: 51; Muchmore, 1986b: 26.

Type species: *Leucohya heteropoda* J. C. Chamberlin, 1946, by original designation.

Leucohya heteropoda J. C. Chamberlin

Leucohya heteropoda J. C. Chamberlin, 1946: 8–10, figs 1–14; Muchmore, 1973b: 51–53, figs 15–17.

Type locality: Gruta del Palmito, Bustamante, Nuevo Leon, Mexico.
Distribution: Mexico.

Leucohya magnifica Muchmore

Leucohya magnifica Muchmore, 1972g: 271–272, figs 12–13.

Type locality: Cueva del Carrizal, el Candela Mt, 30 miles N. of Bustamente, Nuevo Leon, Mexico.
Distribution: Mexico.

Leucohya texana Muchmore

Leucohya texana Muchmore, 1986b: 26–27, figs 16–18.

Type locality: Frio Queen Cave, Uvalde County, Texas, U.S.A.
Distribution: U.S.A. (Texas).

Genus Mexobisium Muchmore

Mexobisium Muchmore, 1972g: 272–273; Muchmore, 1973c: 63–65; Muchmore, 1986b: 24.

Type species: *Mexobisium paradoxum* Muchmore, 1972g, by original designation.

Mexobisium armasi Muchmore

Mexobisium armasi Muchmore, 1980d: 123–125, figs 1–6.

Type locality: near Puerto Boniato, Santiago de Cuba, Oriente, Cuba.
Distribution: Cuba.

Mexobisium cubanum Muchmore

Mexobisium cubanum Muchmore, 1973c: 66–67, figs 9–11.

Type locality: Jatibonico, Cuba.
Distribution: Cuba.

Mexobisium goodnighti Muchmore

Mexobisium goodnighti Muchmore, 1973c: 69–71, figs 22–25; Muchmore, 1977: 71.

Type locality: near Augustine, Belize.
Distribution: Belize.

Mexobisium guatemalense Muchmore

Mexobisium guatemalense Muchmore, 1973c: 67–69, figs 18–21; Muchmore, 1977: 71.

Type locality: Cueva Lanquin, Alta Verapaz, Guatemala.
Distribution: Guatemala.

Mexobisium maya Muchmore

Mexobisium maya Muchmore, 1973c: 67, figs 12–17; Muchmore, 1977: 71.

Type locality: Grutas de Coconá, 3 km NE. of Teapa, Tabasco, Mexico.
Distribution: Mexico.

Mexobisium paradoxum Muchmore

Mexobisium paradoxum Muchmore, 1972g: 273–275, figs 14–19; Muchmore, 1973c: 66–67.

Type locality: Cueva del Ojo de Agua de Tlilipan, Tlilipan, Veracruz, Mexico.
Distribution: Mexico.

Mexobisium pecki Muchmore

Mexobisium pecki Muchmore, 1973c: 65–66, figs 1–8.

Type locality: 10 km S. of Valle Nacional, Oaxaca, Mexico.
Distribution: Mexico.

Mexobisium reddelli Muchmore

Mexobisium reddelli Muchmore, 1986b: 24–26, figs 13–15.

Type locality: Agua Fría, 10 km S. of Tamán, San Luis Potosí, Mexico.
Distribution: Mexico.

Mexobisium ruinarum Muchmore

Mexobisium ruinarum Muchmore, 1977: 71–72, figs 18–19; Muchmore, 1986b: 26.

Type locality: Palenque Ruins, Chiapas, Mexico.
Distribution: Mexico.

Mexobisium sierramaestrae Muchmore

Mexobisium sierramaestrae Muchmore, 1980d: 125–126, figs 7–8.

Type locality: Sierra Maestrae near Uvero, El Cobre, Oriente, Cuba.
Distribution: Cuba.

Genus Troglobochica Muchmore

Troglobochica Muchmore, 1984e: 63–64.

Type species: *Troglobochica jamaicensis* Muchmore, 1984e, by original designation.

Troglobochica jamaicensis Muchmore

Troglobochica jamaicensis Muchmore, 1984e: 64–66, figs 3–10.

Type locality: Jackson Bay Great Cave, Jackson Bay, Clarendon Parish, Jamaica.
Distribution: Jamaica.

Troglobochica pecki Muchmore

Troglobochica pecki Muchmore, 1984e: 66–68, figs 11–15.

Type locality: Drip Cave, Stewart Town, Trelawny Parish, Jamaica.
Distribution: Jamaica.

Genus **Troglohya** Beier

Troglohya Beier, 1956f: 83–84; Muchmore, 1973b: 54–55.

Type species: *Troglohya carranzai* Beier, 1956f, by original designation.

Troglohya carranzai Beier

Troglohya carranzai Beier, 1956f: 84–85, figs 2a–e.

Type locality: Cueva de Monteflor, Oaxaca, Mexico.
Distribution: Mexico.

Troglohya mitchelli Muchmore

Troglohya mitchelli Muchmore, 1973b: 55–56, figs 24–31; Muchmore, 1977: 71.

Type locality: Grutas de Zapaluta, 6.5 km SE. of La Trinitaria, Chiapas, Mexico.
Distribution: Mexico.

Family GYMNOBISIIDAE Beier

Gymnobisiidae Beier, 1947b: 287–288; Muchmore, 1972g: 268; Muchmore, 1982a: 98.

Genus **Beierobisium** Vitali-di Castri

Beierobisium Vitali-di Castri, 1970a: 2.

Type species: *Beierobisium oppositum* Vitali-di Castri, 1970a, by original designation.

Beierobisium oppositum Vitali-di Castri

Beierobisium oppositum Vitali-di Castri, 1970a: 2–7, figs 1–11; Vitali-di Castri & Castri, 1970: fig. 6.

Type locality: Port Stanley, Falkland Islands (as Islas Malvinas).
Distribution: Falkland Islands.

Genus **Gymnobisium** Beier

Gymnobisium Beier, 1931c: 304; Beier, 1932a: 161–162; Beier, 1947b: 289–290.

Type species: *Ideobisium quadrispinosum* Tullgren, 1907b, by original designation.

Gymnobisium octoflagellatum Beier

Gymnobisium octoflagellatum Beier, 1947b: 291–292, fig. 4; Vitali-di Castri, 1970c: 126.

Type locality: Viljoenspos (as Viljoens Pass), Cape Province, South Africa.
Distribution: South Africa.

Gymnobisium quadrispinosum (Tullgren)

Ideobisium quadrispinosum Tullgren, 1907b: 230–231, figs 8a–e; Ellingsen, 1912b: 88, 118–119.

Gymnobisium quadrispinosum (Tullgren): Beier, 1931c: 304; Beier, 1932a: 162, fig. 195; Roewer, 1937: 253; Beier, 1947b: 290–291, fig. 3; Beier, 1955l: 278; Beier, 1958a: 162; Beier, 1964k: 35; Beier, 1966k: 456, 469; Vitali-di Castri, 1970c: 125–126; Vitali-di Castri & Castri, 1970: figs 1, 7.

Type locality: Pietermaritzburg (as Maritzburg), Natal, South Africa.
Distribution: Lesotho, South Africa.

Genus **Mirobisium** Beier

Mirobisium Beier, 1931c: 304–305; Beier, 1932a: 169.

Type species: *Ideobisium cavimanum* Beier, 1930a, by original designation.

Mirobisium cavimanum (Beier)

Ideobisium cavimanum Beier, 1930a: 202–204, fig. 1a–b.
Mirobisium cavimanum (Beier): Beier, 1931c: 305; Beier, 1932a: 169–170, figs 199–200; Beier, 1932g: fig. 173; Roewer, 1936: fig. 112f; Roewer, 1937: 256, fig. 213; Vitali-di Castri, 1970c: 127.

Type locality: Bolivia.
Distribution: Bolivia.

Mirobisium chilense Beier

Mirobisium chilense Beier, 1964b: 323–324, fig. 8; Vitali-di Castri, 1970c: 127–128; Vitali-di Castri & Castri, 1970: fig. 2; Cekalovic, 1976: 17–18; Cekalovic, 1984: 12.
Gymnobisium chilense magalhanicum Beier, 1964b: 322 (synonymized by Vitali-di Castri, 1970c: 127).

Type localities: of *Mirobisium chilense*: Rio Caleta, Seno Otway, Magallenes, Chile.
of *Gymnobisium chilense magalhanicum*: Los Robles, Magallanes, Chile.
Distribution: Chile.

Mirobisium dimorphicum Vitali-di Castri

Gymnobisium chilense Beier, 1964b: 320–321 (misidentification, in part).
Mirobisium dimorphicum Vitali-di Castri, 1970c: 129; Cekalovic, 1984: 12.

Type localities: Frutillar, Llanquihue, Chile; and Quilanto, Llanquihue, Chile.
Distribution: Chile.

Mirobisium minore Vitali-di Castri

Gymnobisium chilense Beier: Vitali-di Castri, 1963: 39–49, figs 13–36 (misidentification).
Mirobisium minore Vitali-di Castri, 1970c: 129–131, figs 1–2; Cekalovic, 1984: 12.

Type locality: Parque Nacional de Nahuelbuta, Malleco, Chile.
Distribution: Chile.

Mirobisium patagonicum Beier

Mirobisium patagonicum Beier, 1964j: 489–490, fig. 2; Vitali-di Castri, 1970c: 129–130.
Gymnobisium chilense Beier, 1964b: 320–321, fig. 6 (in part, see *Mirobisium dimorphicum* Vitali-di Castri); Beier, 1964j: 488–489; Vitali-di Castri, 1965b: fig. 2; Mahnert, 1981c: fig. 1 (synonymized by Vitali-di Castri, 1970c: 128).
Not *Gymnobisium chilense* Beier: Vitali-di Castri, 1963: 39–49, figs 13–36 (misidentification; see *Mirobisium minore* Vitali-di Castri).

Type localities: of *Mirobisium patagonicum*: Mt Piltriquitron, El Bolsón, Rio Negro, Argentina.
of *Gymnobisium chilense*: Frutillar, Llanquihue, Chile.
Distribution: Argentina, Chile.

Genus **Vachonobisium** Vitali-di Castri

Vachonobisium Vitali-di Castri, 1963: 31.

Type species: *Vachonobisium troglophilum* Vitali-di Castri, 1963, by original designation.

Vachonobisium heros (Beier)

Gymnobisium heros Beier, 1964b: 322–323, fig. 8.
Vachonobisium heros (Beier): Vitali-di Castri, 1970c: 133; Cekalovic, 1984: 13.

Type locality: Farellones, Chile.
Distribution: Chile.

Vachonobisium intermedium (Vitali-di Castri)

Gymnobisium intermedium Vitali-di Castri, 1963: 49–58, figs 37–54.
Vachonobisium intermedium (Vitali-di Castri): Vitali-di Castri, 1970c: 132–133; Cekalovic, 1984: 13.

Type locality: Polpaico, Santiago, Chile.
Distribution: Chile.

Vachonobisium troglophilum Vitali-di Castri

Vachonobisium troglophilum Vitali-di Castri, 1963: 32–39, figs 1–12; Vitali-di Castri, 1970c: 131–132; Vitali-di Castri & Castri, 1970: figs 11–12; Cekalovic, 1984: 13.
Gymnobisium montanum Vitali-di Castri, 1963: 58–68, figs 55–82; Vitali-di Castri, 1965b: fig. 2 (synonymized by Vitali-di Castri, 1970c: 131).

Type localities: of *Vachonobisium troglophilum*: Cerro El Roble, Cordillera de la Costa, Valparaíso, Chile.
of *Gymnobisium montanum*: Cerro El Roble, Cordillera de la Costa, Valparaíso, Chile.
Distribution: Chile.

Family HYIDAE J. C. Chamberlin

Hyidae J. C. Chamberlin, 1930: 41; J. C. Chamberlin, 1931a: 219–220; J. C. Chamberlin, 1946: 5–6; Murthy & Ananthakrishnan, 1977: 24; Muchmore, 1982a: 98.
Hyinae J. C. Chamberlin: Beier, 1932a: 166; Roewer, 1937: 254–255; J. C. Chamberlin, 1946: 10.

Genus Hya J. C. Chamberlin

Hya J. C. Chamberlin, 1930: 41–42; Beier, 1932a: 166–167; J. C. Chamberlin, 1946: 11; Murthy & Ananthakrishnan, 1977: 24–25.

Type species: *Hya heterodonta* J. C. Chamberlin, 1930, by original designation.

Hya heterodonta J. C. Chamberlin

Hya heterodonta J. C. Chamberlin, 1930: 42, figs 2kk, 3e; J. C. Chamberlin, 1931a: 71, 142, figs 9f, 13q, 15i, 17k, 17p, 19f, 28f, 33h, 33j, 36d–e, 43n, 58; Beier, 1932a: 167–168; Beier, 1932g: fig. 167; Roewer, 1936: figs 30h, 70f; Roewer, 1937: 255; J. C. Chamberlin, 1946: 11–15, figs 15–29; Savory, 1964a: fig. 33a; Savory, 1977: fig. 93a.

Type locality: Mt Makiling, Luzon, Philippines.
Distribution: Philippine Islands.

Hya minuta (Tullgren)

Ideobisium minutum Tullgren, 1905: 44–45, figs 4a–c; Beier, 1930e: 288–289.
Hya minuta (Tullgren): Beier, 1932a: 167, fig. 168; Roewer, 1936: fig. 112c; Roewer, 1937: 255, fig. 212; J. C. Chamberlin, 1946: 16; Weidner, 1959: 114; Beier, 1973b: 42; Beier, 1974e: 1003; Murthy & Ananthakrishnan, 1977: 25.

Type locality: Depok, Java, Indonesia.
Distribution: Indonesia (Java, Sumatra), Sri Lanka.

Genus **Indohya** Beier

Indohya Beier, 1974e: 1003; Murthy & Ananthakrishnan, 1977: 25.

Type species: *Indohya besucheti* Beier, 1974e, by original designation.

Indohya besucheti Beier

Indohya besucheti Beier, 1974e: 1003–1004, fig. 3; Murthy & Ananthakrishnan, 1977: 25.

Type locality: Suruli Falls, Varushanad Hills, Madras, India.
Distribution: India.

Indohya caecata Beier

Indohya caecata Beier, 1974e: 1004–1006, fig. 4; Murthy & Ananthakrishnan, 1977: 25.

Type locality: Kaikatty, Nelliampathi Hills, Kerala, India.
Distribution: India.

Genus **Parahya** Beier

Parahya Beier, 1957d: 15.

Type species: *Parahya pacifica* Beier, 1957d, by original designation.

Parahya pacifica Beier

Parahya pacifica Beier, 1957d: 15–16, figs 4b, 5.

Type locality: Yap Island, Caroline Islands.
Distribution: Caroline Islands.

Parahya submersa (Bristowe), new combination

Obisium submersum Bristowe, 1931a: 465, figs 3–6.

Type locality: Singapore.
Distribution: Singapore.

Genus **Stenohya** Beier

Stenohya Beier, 1967g: 343–344.

Type species: *Stenohya vietnamensis* Beier, 1967g, by original designation.

Stenohya vietnamensis Beier

Stenohya vietnamensis Beier, 1967g: 345, fig. 4; Tenorio & Muchmore, 1982: 383.

Type locality: 6 km S. of Da Lat (as Dalat), Vietnam.
Distribution: Vietnam.

Family IDEORONCIDAE J. C. Chamberlin

Ideoroncidae J. C. Chamberlin, 1930: 42; J. C. Chamberlin, 1931a: 220; Beier, 1932a: 166; Beier, 1932g: 183; Roewer, 1937: 254; Hoff, 1956b: 25; Murthy & Ananthakrishnan, 1977: 25–26; Muchmore, 1982a: 97–98.
Ideoroncinae J. C. Chamberlin, 1930: 44; J. C. Chamberlin, 1931a: 220; Beier, 1932a: 170; Roewer, 1937: 256; Hoff, 1956b: 25.

Family Ideoroncidae

Genus **Afroroncus** Mahnert

Afroroncus Mahnert, 1981a: 539.

Type species: *Afroroncus kikuyu* Mahnert, 1981a, by original designation.

Afroroncus kikuyu Mahnert

Afroroncus kikuyu Mahnert, 1981a: 539–541, figs 1–8, 57a–c.

Type locality: Hunter's Lodge, between Nairobi and Voi, Kenya.
Distribution: Kenya.

Afroroncus sulcatus Mahnert

Afroroncus sulcatus Mahnert, 1981a: 541–543, figs 9–14.

Type locality: Ishiara, Embu, Kenya.
Distribution: Kenya.

Genus **Albiorix** J. C. Chamberlin

Albiorix J. C. Chamberlin, 1930: 44; Beier, 1932a: 172; Hoff, 1945d: 1; Hoff, 1956b: 25; Mahnert, 1984a: 671.
Dinoroncus Beier, 1931c: 305; Beier, 1932a: 171 (synonymized by Mahnert, 1984a: 676).

Type species: of *Albiorix*: *Ideoroncus mexicanus* Banks, 1898, by original designation.
of *Dinoroncus*: *Ideobisium (Ideoroncus) chilense* Ellingsen, 1905e, by original designation.

Albiorix arboricola (Mahnert)

Ideoroncus arboricola Mahnert, 1979d: 753–755, figs 70–74; Adis, Junk & Penny, 1987: 488.
Albiorix arboricola (Mahnert): Mahnert, 1984a: 672–673; Mahnert, 1985b: 78; Mahnert & Adis, 1986: 213; Mahnert, Adis, & Bührnheim, 1987: fig. 10.
Not *Albiorix* aff. *arboricola* (Mahnert): Mahnert, 1984a: 673 (misidentification; see *Albiorix gracilis* Mahnert).

Type locality: Manaus, Amazonas, Brazil.
Distribution: Brazil (Amazonas, Pará).

Albiorix argentiniensis (Hoff)

Dinoroncus chilensis (Ellingsen): Feio, 1945: 4 (misidentification).
Dinoroncus argentiniensis Hoff, 1950b: 229–232, figs 7–9.
Albiorix argentiniensis (Mahnert): Mahnert, 1984a: 675–676, fig. 44.

Type locality: La Sébila, La Rioja, Argentina.
Distribution: Argentina.

Albiorix bolivari Beier

Albiorix bolivari Beier, 1963e: 133–134, fig. 1; Palacios-Vargas, 1981: 64 (not seen); Mahnert, 1984a: 676, fig. 45.

Type locality: Gruta de Acuitlapán, Guerrero, Mexico.
Distribution: Mexico.

Albiorix chilensis (Ellingsen)

Ideobisium (Ideoroncus) chilense Ellingsen, 1905e: 326–327.
Dinoroncus chilense (sic) (Ellingsen): Beier, 1931c: 305.
Dinoroncus chilensis (Ellingsen): Beier, 1932a: 324–325, fig. 202; Roewer, 1937: 257; Beier, 1964b: 324–325; Cekalovic, 1984: 13.

Albiorix chilensis (Ellingsen): Mahnert, 1984a: 676.
Not *Dinoroncus chilensis* (Ellingsen): Feio, 1945: 4 (misidentification; see *Albiorix argentiniensis* (Hoff)).

Type locality: Santiago, Chile.
Distribution: Chile.

Albiorix conodentatus Hoff

Albiorix conodentatus Hoff, 1945d: 8−10, figs 17−20.

Type locality: 5 miles W. of Saltillo, Mexico.
Distribution: Mexico.

Albiorix edentatus J. C. Chamberlin

Albiorix edentatus J. C. Chamberlin, 1930: 46−47; J. C. Chamberlin, 1931a: figs 28l, 33q; Beier, 1932a: 173; Roewer, 1936: fig. 30a; Roewer, 1937: 257; Hoff, 1958: 15.

Type locality: Santa Isabella Creek, Mt Hamilton, California, U.S.A.
Distribution: U.S.A. (California).

Albiorix gracilis Mahnert

Albiorix aff. *arboricola* (Mahnert): Mahnert, 1984a: 673 (misidentification).
Albiorix gracilis Mahnert, 1985c: 223−224, figs 27−28; Mahnert & Adis, 1986: 213.

Type locality: Taruma Mirim, Amazonas, Brazil.
Distribution: Brazil (Amazonas).

Albiorix lamellifer Mahnert

Albiorix lamellifer Mahnert, 1985c: 225−226, figs 29−31; Mahnert & Adis, 1986: 213.

Type locality: 25 km NE. of Manaus, Amazonas, Brazil.
Distribution: Brazil (Amazonas).

Albiorix magnus Hoff

Albiorix magnus Hoff, 1945d: 2−4, figs 1−5.

Type locality: 20 miles E. of San Pedro, Mexico.
Distribution: Mexico.

Albiorix mexicanus (Banks)

Ideoroncus mexicanus Banks, 1898: 289; J. C. Chamberlin, 1923c: 359−360, plate 2 fig. 13, plate 3 figs 14, 34.
Albiorix mexicanus (Banks): J. C. Chamberlin, 1930: 45, figs 2f, 2dd; J. C. Chamberlin, 1931a: figs 9j, 11u, 17q, 19g, 25l, 59; Beier, 1932a: 173; Beier, 1932g: fig. 255; Roewer, 1936: fig. 30c; Roewer, 1937: 257, fig. 215; Vachon, 1949: fig. 203f; Hoff, 1958: 14; Mahnert, 1984a: 673−675, fig 42.
Not *Ideoroncus mexicanus* Banks: With, 1905: 127−131, plate 9 figs 2a−d, plate 10 figs 1a−f (misidentification; see *Bochica withi* (J. C. Chamberlin)).

Type locality: San Miguel de Horcasitas, Baja California, Mexico.
Distribution: Mexico, U.S.A. (California, Utah).

Albiorix mirabilis Muchmore

Albiorix mirabilis Muchmore, 1982f: 75−77, figs 33−36; Mahnert, 1984a: 676, fig. 46.

Type locality: Cueva de las Maravillas, 6 km S. of Acatlán, Oaxaca, Mexico.
Distribution: Mexico.

Albiorix parvidentatus J. C. Chamberlin

Albiorix parvidentatus J. C. Chamberlin, 1930: 45−46; Beier, 1932a: 173; Roewer, 1936: fig. 30b; Roewer, 1937: 257; Hoff, 1958: 15.

Type locality: Palm Canyon, Riverside County, California, U.S.A.
Distribution: U.S.A. (California).

Albiorix reddelli Muchmore

Albiorix reddelli Muchmore, 1982f: 77, figs 37−40.
Albiorix (?) *reddeli* (sic) Muchmore: Mahnert, 1984a: 676−677, fig. 47.

Type locality: Grutas de Monteflor, 6 km N. of Valle Nacional, Oaxaca, Mexico.
Distribution: Mexico.

Albiorix retrodentatus Hoff

Albiorix retrodentatus Hoff, 1945d: 6−8, figs 10−16; Hoff, 1956b: 25−26; Hoff, 1958: 15; Hoff, 1959b: 4, etc.; Rowland & Reddell, 1976: 16.

Type locality: Mexcala, Guerrero, Mexico.
Distribution: Mexico, U.S.A. (New Mexico, Texas).

Albiorix veracruzensis Hoff

Albiorix veracruzensis Hoff, 1945d: 4−7, figs 6−9.

Type locality: La Buena Ventura, Veracruz, Mexico.
Distribution: Mexico.

Genus Dhanus J. C. Chamberlin

Dhanus J. C. Chamberlin, 1930: 47; Beier, 1932a: 173; Murthy & Ananthakrishnan, 1977: 26.

Type species: *Ideoroncus sumatranus* Redikorzev, 1922b, by original designation.

Dhanus afghanicus Beier

Dhanus afghanicus Beier, 1959a: 264, fig. 6; Beier, 1960a: 41; Beier, 1961e: 1; Lindberg, 1961: 31; Beier, 1971a: 360−361.

Type locality: Bagh-Chah Babar, Kabul, Afghanistan.
Distribution: Afghanistan, Iran.

Dhanus doveri Bristowe

Dhanus doveri Bristowe, 1952: 699−700, figs 3−5.

Type locality: Batu Caves, Malaysia.
Distribution: Malaysia.

Dhanus indicus Murthy & Ananthakrishnan

Dhanus indicus Murthy & Ananthakrishnan, 1977: 26−28, figs 7a−b.

Type locality: Poondi, Tamil Nadu, India.
Distribution: India.

Dhanus siamensis (With)

Ideobisium (Ideoroncus) siamensis With, 1906: 81−84, plate 1 figs 4a−i.
Dhanus siamensis (With): J. C. Chamberlin, 1930: 47; Beier, 1932a: 174, fig. 8; Beier, 1932g: figs 200c, 208; Roewer, 1936: figs 37a−b, 83g, 86, 107a−b; Roewer, 1937: 257.

Type localities: Ko Chang (as Koh Chang), Thailand; and Klong Salakpet, Thailand.
Distribution: Thailand.

Dhanus sumatranus (Redikorzev)

Ideoroncus sumatranus Redikorzev, 1922b: 545–550, figs 1–4.
Dhanus sumatranus (Redikorzev): J. C. Chamberlin, 1930: 47; J. C. Chamberlin, 1931a: 128, figs 13n–p, 15j, 25i, 25m–n; Beier, 1932a: 174–175, fig. 203; Roewer, 1936: fig. 30d; Roewer, 1937: 257; Beier, 1963c: 51.

Type locality: Datu Caves, Sumatra, Indonesia (= Batu Caves, Malaysia?).
Distribution: Indonesia (Sumatra), Malaysia.

Genus Ideoroncus Balzan

Ideoroncus Balzan, 1887b: no pagination; Balzan, 1890: 443–444; J. C. Chamberlin, 1930: 44; Beier, 1932a: 171–172; Mahnert, 1984a: 653.
Ideobisium (Ideoroncus) Balzan: Balzan, 1892: 540; With, 1906: 81.

Type species: *Ideoroncus pallidus* Balzan, 1887b, by monotypy.

Ideoroncus anophthalmus Mahnert

Ideoroncus anophthalmus Mahnert, 1984a: 663–665, figs 22–24.

Type locality: Serra da Cantareira, Sao Paulo, Brazil.
Distribution: Brazil (Sao Paulo).

Ideoroncus beieri Mahnert

Ideoroncus lenkoi Beier: Beier, 1974d: 899 (misidentification, in part).
Ideoroncus beieri Mahnert, 1984a: 668–670, figs 34–37.

Type locality: Caioba, Paraná, Brazil.
Distribution: Brazil (Paraná).

Ideoroncus divisus Mahnert

Ideoroncus pallidus Balzan: Beier, 1974d: 899 (misidentification).
Ideoroncus divisus Mahnert, 1984a: 666–668, figs 29–33.

Type locality: Ponte Valle do Diablos, near Santa Maira, Rio Grande do Sul, Brazil.
Distribution: Brazil (Rio Grande do Sul, Santa Catarina).

Ideoroncus lenkoi Beier

Ideoroncus lenkoi Beier, 1970b: 55–56, fig. 3; Beier, 1974d: 899 (in part; see *Ideoroncus beieri* Mahnert); Mahnert, 1984a: 655–656, figs 3–7.
Ideoroncus aff. *lenkoi* Beier: Mahnert, 1984a: 657.
Not *Ideoroncus lenkoi* Beier: Beier, 1974d: 899 (misidentification, in part; see *Ideoroncus beieri* Mahnert and *Ideoroncus paranensis* Mahnert).

Type locality: Barueri, Sao Paulo, Brazil.
Distribution: Brazil (Sao Paulo).

Ideoroncus pallidus Balzan

Ideoroncus pallidus Balzan, 1887b: no pagination, figs; Balzan, 1890: 444–445, figs 23, 23a–b; J. C. Chamberlin, 1930: 44; Beier, 1932a: 171, fig. 201; Roewer, 1937: 257; Mahnert, 1984a: 653–655, figs 1–3.
Ideobisium (Ideoroncus) pallidum (Balzan): Balzan, 1892: 541, 549.
Ideobisium pallidus (Balzan): Ellingsen, 1910a: 395.
Not *Ideoroncus pallidus* Beier: Beier, 1974d: 899 (misidentification; see *Ideoroncus divisus* Mahnert).

Type locality: Rio Apa, Paraguay.
Distribution: Brazil, Paraguay.

Ideoroncus paranensis Mahnert

Ideoroncus lenkoi Beier: Beier, 1974d: 899 (misidentification, in part).
Ideoroncus paranensis Mahnert, 1984a: 657–659, figs 8–14.

Type locality: Caioba, Paraná, Brazil.
Distribution: Brazil (Paraná).

Ideoroncus procerus Beier

Ideoroncus procerus Beier, 1974d: 900–901, fig. 1; Mahnert, 1984a: 665–666, figs 25–28.

Type locality: Nova Teutonia, Santa Catarina, Brazil.
Distribution: Brazil (Santa Catarina).

Ideoroncus setosus Mahnert

Ideoroncus setosus Mahnert, 1984a: 659–663, figs 15–21.

Type locality: Palmeirinha, Sao Paulo, Brazil.
Distribution: Brazil (Sao Paulo).

Genus **Nannoroncus** Beier

Nannoroncus Beier, 1955c: 535–536.

Type species: *Nannoroncus ausculator* Beier, 1955c, by original designation.

Nannoroncus ausculator Beier

Nannoroncus ausculator Beier, 1955c: 536, fig. 6; Mahnert, 1981a: 538–539, figs 56a–c.

Type locality: Suam River, Mt Elgon, Uganda.
Distribution: Kenya, Uganda.

Genus **Negroroncus** Beier

Negroroncus Beier, 1931c: 305; Beier, 1932a: 175; Mahnert, 1984a: 679.

Type species: *Ideoroncus africanus* Redikorzev, 1924a, by original designation.

Negroroncus aelleni Vachon

Negroroncus aelleni Vachon, 1958: 59–62, figs 3–8.

Type locality: Grotte Kila-Tari, near Niari, Congo.
Distribution: Congo.

Negroroncus africanus (Redikorzev)

Ideoroncus africanus Redikorzev, 1924a: 190–192, figs 3–4.
Negroroncus africanus (Redikorzev): Beier, 1931c: 305; Beier, 1932a: 175, fig. 204; Roewer, 1936: fig. 70g: Roewer, 1937: 257; Vachon, 1958: 62–64; Mahnert, 1981a: 543–544, figs 20–23.

Type locality: Taveta, Kenya.
Distribution: Kenya.

Negroroncus azanius Mahnert

Negroroncus azanius Mahnert, 1981a: 553–554, figs 44–48.

Type locality: Jilore, Kilifi, Kenya.
Distribution: Kenya.

Negroroncus densedentatus Mahnert

Negroroncus densedentatus Mahnert, 1981a: 546–547, figs 24–29.

Type locality: Kwale, Shimba Hills, Kenya.
Distribution: Kenya.

Negroroncus gregoryi Mahnert

Negroroncus gregoryi Mahnert, 1981a: 547–549, figs 31–36.

Type locality: N. of Kisumu, Kenya.
Distribution: Kenya.

Negroroncus jeanneli Vachon

Negroroncus jeanneli Vachon, 1958: 57–59, figs 1–2; Mahnert, 1984a: 679.

Type locality: Kulumuzi, Kyomoni, Tanga, Tanzania.
Distribution: Tanzania.

Negroroncus kerenyaga Mahnert

Negroroncus kerenyaga Mahnert, 1981a: 551–553, figs 49–54.

Type locality: Kogari, Embu, Kenya.
Distribution: Kenya.

Negroroncus laevis Beier

Negroroncus laevis Beier, 1972a: 8–9, fig. 4.

Type locality: Parc National Garamba, Zaire.
Distribution: Zaire.

Negroroncus longedigitatus Beier

Negroroncus longedigitatus Beier, 1944: 177–179, fig. 4; Vachon, 1958: 62–64;
    Beier, 1959d: 26–27; Mahnert, 1981a: 547, fig 30.
Negroroncus longidigitatus (sic) Beier: Beier, 1955h: 9.

Type locality: Naramum, Jurkana, Kenya.
Distribution: Kenya, Zaire.

Negroroncus minutus Beier

Negroroncus minutus Beier, 1967d: 73–74, fig. 1.

Type locality: N. of Lake Manyara, Tanzania.
Distribution: Tanzania.

Negroroncus rhodesiacus Beier

Negroroncus rhodesiacus Beier, 1964k: 34–35, fig. 3.

Type locality: Birchenough Bridge, Zimbabwe.
Distribution: Zimbabwe.

Negroroncus silvicola Mahnert

Negroroncus silvicola Mahnert, 1981a: 549–551, figs 37–43, 55b, 55d; Mahnert,
    1981c: fig. 1.

Type locality: Warges, Kenya.
Distribution: Kenya.

Negroroncus tsavoensis Mahnert

Negroroncus tsavoensis Mahnert, 1981a: 544–546, figs 15–19.

Type locality: ca. 20 km W. of Sala Gate, Tsavo National Park, Kenya.
Distribution: Kenya.

Genus **Nhatrangia** Redikorzev

Nhatrangia Redikorzev, 1938: 78–79; Mahnert, 1984a: 680–681.

Type species: *Nhatrangia dawydoffi* Redikorzev, 1938, by monotypy.

Nhatrangia ceylonensis Mahnert

Shravana dawydoffi (Redikorzev): Beier, 1973b: 43 (misidentification).
Nhatrangia ceylonensis Mahnert, 1984a: 681–684, figs 56–61.

Type locality: Medawachchiya, North-central Province, Sri Lanka.
Distribution: Sri Lanka.

Nhatrangia dawydoffi Redikorzev

Nhatrangia dawydoffi Redikorzev, 1938: 79–81, figs 6–9; Roewer, 1940: 345; Beier, 1951a: 67; Mahnert, 1984a: 681, fig. 55.
Shravana dawydoffi (Redikorzev): Beier, 1967g: 343.
Not *Shravana dawydoffi* (Redikorzev): Beier, 1973b: 43 (misidentification; see *Nhatrangia ceylonensis* Mahnert).

Type localities: Nha Trang (as Nhatrang), Vietnam; Pleiku (as Plei-Ku), Vietnam; and Phan Rang (as Phanrang), Vietnam.
Distribution: Cambodia, Vietnam.

Genus Shravana J. C. Chamberlin

Shravana J. C. Chamberlin, 1930: 48; Beier, 1932a: 175–176; Mahnert, 1984a: 679.

Type species: *Ideobisium (Ideoroncus) laminatus* With, 1906, by original designation.

Shravana laminata (With)

Ideobisium (Ideoroncus) laminatus With, 1906: 84–87, plate 1 figs 5a–c.
Shravana laminata (With): J. C. Chamberlin, 1930: 48; J. C. Chamberlin, 1931a: 75, 167, figs 36g–h; Beier, 1932a: 176; Roewer, 1937: 257; Weygoldt, 1969a: fig. 15b; Mahnert, 1984a: 679–680, figs 52–54.
Shrawana (sic) *laminata* (With): Vachon, 1949: fig. 219b.

Type locality: Ko Chang (as Koh Chang), Thailand.
Distribution: Thailand.

Genus Typhloroncus Muchmore

Typhloroncus Muchmore, 1979c: 317–318; Mahnert, 1984a: 677; Muchmore, 1986b: 27–28.

Type species: *Typhloroncus coralensis* Muchmore, 1979c, by original designation.

Typhloroncus attenuatus Muchmore

Typhloroncus attenuatus Muchmore, 1982f: 73–75, figs 30–32; Mahnert, 1984a: 677–678, fig. 50; Muchmore, 1986b: 28.

Type locality: Cueva del Brinco, near Conrado Castillo, about 40 km NW. of Ciudad Victoria, Tamaulipas, Mexico.
Distribution: Mexico.

Typhloroncus coralensis Muchmore

Typhloroncus coralensis Muchmore, 1979c: 318–319, figs 1–5; Muchmore, 1982f: 71; Mahnert, 1984a: 677, fig. 48; Muchmore, 1986b: 28, fig. 19.

Type locality: Coral Bay, St John, U.S. Virgin Islands.
Distribution: U.S. Virgin Islands.

Typhloroncus diabolus Muchmore

Typhloroncus diabolus Muchmore, 1982f: 73, figs 27–29; Mahnert, 1984a: 677, fig. 51; Muchmore, 1986b: 28.

Type locality: Cueva del Diablo, 3 km SSW. of Ciudad Mendoza, Veracruz, Mexico.
Distribution: Mexico.

Typhloroncus troglobius Muchmore

Typhloroncus troglobius Muchmore, 1982f: 71–73, figs 24–26; Mahnert, 1984a: 677, fig. 49; Muchmore, 1986b: 28.

Type locality: Grutas de Atepolihuit, 5 km SW. of Cuetzalan, Puebla, Mexico.
Distribution: Mexico.

Typhloroncus xilitlensis Muchmore

Typhloroncus xilitlensis Muchmore, 1986b: 28–30, figs 20–21.

Type locality: Sótano de Huitzmolotitla, 2 km NNW. of Xilitla, San Luis Potosí, Mexico.
Distribution: Mexico.

Family NEOBISIIDAE J. C. Chamberlin

Obisides (sic) Sundevall, 1833: 33.
Obisia (sic) Sundevall: Schiödte, 1851a: 23.
Obisinae (sic) Sundevall: Stecker, 1874b: 237; Stecker, 1875d: 88.
Obisiinae Sundevall: E. Simon, 1879a: 50; H. J. Hansen, 1884: 533–534; Daday, 1888: 127; H. J. Hansen, 1893: 232; With, 1906: 75–76; Lessert, 1911: 27; Redikorzev, 1924b: 25; Väänänen, 1928a: 16.
Obisiidae Sundevall: Tömösváry, 1882b: 213–214; Daday, 1888: 126–127; Balzan, 1890: 443; H. J. Hansen, 1893: 232; Tullgren, 1899a: 179; Tullgren, 1906a: 204; With, 1906: 74–75; Kew, 1911a: 52; Lessert, 1911: 27; Comstock, 1913: 51; Moles & Moore, 1921: 9; Redikorzev, 1924b: 25; Pratt, 1927: 410; Väänänen, 1928a: 16.
Obisiina (sic) Sundevall: Tömösváry, 1882b: 214.
Microcreagrinae Balzan, 1892: 543.
Neobisiidae J. C. Chamberlin, 1930: 9; J. C. Chamberlin, 1931a: 215–217; Beier, 1932a: 76–77; Beier, 1932g: 182–183; J. C. Chamberlin, 1935b: 479; Roewer, 1937: 244; Beier, 1939f: 167; Hoff, 1949b: 444; G. O. Evans & Browning, 1954: 10; Petrunkevitch, 1955: 81; Hoff, 1956b: 2; Morikawa, 1960: 112; Beier, 1963b: 80; Murthy & Ananthakrishnan, 1977: 28–29; Muchmore, 1979a: 194; Muchmore, 1982a: 97; Legg & Jones, 1988: 74.
Neobisiinae J. C. Chamberlin, 1930: 9; J. C. Chamberlin, 1931a: 217; Beier, 1932a: 77; Roewer, 1937: 244; Hoff, 1949b: 444; Hoff, 1956b: 2; Morikawa, 1960: 112.

Genus **Acanthocreagris** Mahnert

Acanthocreagris Mahnert, 1974d: 848–849; Mahnert, 1976c: 194; Ćurčić, 1985a: 332.

Type species: *Microcreagris gallica* Beier, 1965e, by original designation.

Acanthocreagris aelleni Mahnert

Acanthocreagris aelleni Mahnert, 1978b: 382–384, figs 2–7; Gardini, 1980c: 115, 133; Callaini, 1986b: 351.

Type locality: Sisco Cave, Corsica.
Distribution: Corsica.

Acanthocreagris agazzii (Beier)

Microcreagris agazzii Beier, 1966h: 177–178, fig. 2.
Acanthocreagris (?) *agazzii* (Beier): Mahnert, 1974d: 849.
Acanthocreagris agazzii (Beier): Mahnert, 1976c: 194; Callaini, 1986b: 351, fig. 7.

Type locality: Fondi di Cima Manderiola, Asiago, Vicenza, Veneto, Italy.
Distribution: Italy.

Family Neobisiidae

Acanthocreagris anatolica (Beier)

Microcreagris anatolica Beier, 1963g: 271–272, fig. 3; Beier, 1965f: 90; Beier, 1966e: 162.
Microcreagris cf. *anatolica* Beier: Helversen, 1966b: fig. 4e.
Acanthocreagris anatolica (Beier): Mahnert, 1974d: 878–879; Schawaller, 1985b: 5.

Type locality: Ankara, Turkey.
Distribution: Greece, Turkey.

Acanthocreagris apuanica Callaini

Acanthocreagris apuanica Callaini, 1986b: 359–362, figs 3a–h, 7.

Type locality: Mt Forato, Alpi Apuane, Italy.
Distribution: Italy.

Acanthocreagris apulica Callaini

Acanthocreagris apulica Callaini, 1986b: 356–359, figs 2a–f, 7.

Type locality: Lago di Varano, Cagnano, Gargano, Italy.
Distribution: Italy.

Acanthocreagris balcanica (Hadži)

Microcreagris balcanica Hadži, 1939b: 37–39, figs 10a–h, 11a–d; Beier, 1963b: 207.
Acanthocreagris balcanica (Hadži): Mahnert, 1974d: 849; Schawaller, 1985b: 5.

Type locality: Athos Mountains, Greece.
Distribution: Greece.

Acanthocreagris balearica (Beier)

Microcreagris balearica Beier, 1961b: 27–28, fig. 2; Beier, 1963b: 219–220.
Acanthocreagris balearica (Beier): Mahnert, 1976c: 194; Mahnert, 1989a: 87.

Type locality: Playa Tirant Nou, Menorca, Balearic Islands.
Distribution: Balearic Islands.

Acanthocreagris beieri Mahnert

Acanthocreagris beieri Mahnert, 1974d: 869–871, figs 7a–f; Schawaller, 1985b: 5.

Type locality: Aristi, between Ioannina and Konitza, Epirus, Greece.
Distribution: Greece.

Acanthocreagris callaticola (Dumitresco & Orghidan)

Microcreagris callaticola Dumitresco & Orghidan, 1964: 616–619, figs 28–35, plate 1
   fig. 5.
Acanthocreagris (?) *callaticola* (Dumitresco & Orghidan): Mahnert, 1976c: 194.

Type locality: Grotte de Limanu, Limanu, Romania.
Distribution: Romania.

Acanthocreagris caspica (Beier)

Microcreagris caspica Beier, 1971a: 359–360, fig. 2.
Acanthocreagris (?) *caspica* (Beier): Mahnert, 1974d: 849.
Acanthocreagris caspica (Beier): Mahnert, 1976c: 194.

Type locality: 44 km N. of Chalus, Iran.
Distribution: Iran.

Acanthocreagris corcyraea Mahnert

Acanthocreagris corcyraea Mahnert, 1976c: 196–197, figs 5–8; Schawaller, 1985b: 5.

Type locality: Katsaba Cave, Pantokrator, Kérkira (as Korfu), Greece.
Distribution: Greece.

324

Acanthocreagris corsa Mahnert

Acanthocreagris corsa Mahnert, 1974d: 871–875, figs 8a–k; Mahnert, 1977d: 97; Callaini, 1986b: 351, fig. 7.

Type locality: Dolmen de Cauria, 15 km S. of Sartène, Corsica.
Distribution: Corsica.

Acanthocreagris gallica (Beier)

Microcreagris gallica Beier, 1965e: 301–303, fig. 1.
Acanthocreagris gallica (Beier): Mahnert, 1974d: 852–861, figs 1b, 2b, 3a–d, 4a–d, 5a–e; Heurtault, 1980d: 89; Mahnert, 1981c: fig. 1.

Type locality: Coulounieix near Périgueux, Dordogne, France.
Distribution: France.

Acanthocreagris granulata (Beier)

Roncus (Parablothrus) granulatus granulatus Beier, 1939f: 182–183, fig. 12; Roewer, 1940: 344.
Roncus (Parablothrus) granulatus Beier: Roewer, 1940: 343.
Microcreagris catalonica Beier, 1961b: 28–29; Beier, 1970f: 321 (unnecessary replacement name for *Roncus (Parablothrus) granulatus granulatus* Beier, 1939f; junior secondary homonym of *Microcreagris granulata* Ellingsen, 1912).
Microcreagris catalonica catalonica (Beier): Beier, 1963b: 221.
Acanthocreagris granulata (Beier): Mahnert, 1976c: 194; Mahnert, 1977c: 97.
Acanthocreagris granulata granulata (Beier): Zaragoza, 1986: fig. 1.

Type locality: Cova Vora Major near Terrades, Gerona, Spain.
Distribution: Spain.

Acanthocreagris granulata parva (Beier)

Roncus (Parablothrus) granulatus parvus Beier, 1939f: 183–184, fig. 13; Roewer, 1940: 344.
Microcreagris catalonica parva (Beier): Beier, 1963b: 222.
Acanthocreagris granulata parva (Beier): Mahnert, 1976c: 203; Zaragoza, 1986: fig. 1.

Type locality: Cova Tessana near Cadaqués, Gerona, Spain.
Distribution: Spain.

Acanthocreagris granulata robusta (Beier)

Roncus (Parablothrus) granulatus robustus Beier, 1939f: 184, fig. 14; Roewer, 1940: 344.
Microcreagris catalonica robusta (Beier): Beier, 1963b: 221.
Acanthocreagris granulata robusta (Beier): Mahnert, 1976c: 203; Zaragoza, 1986: fig. 1.

Type locality: Cova dels Hermitons, Cataluña, Spain.
Distribution: Spain.

Acanthocreagris granulata ventalloi (Beier)

Roncus (Parablothrus) granulatus ventalloi Beier, 1939f: 184–185, fig. 15; Roewer, 1940: 344.
Microcreagris catalonica ventalloi (Beier): Beier, 1963b: 222, fig. 229.
Acanthocreagris granulata ventalloi (Beier): Mahnert, 1976c: 203; Mahnert, 1977c: 97–99; Zaragoza, 1986: figs 1–2.

Type locality: Baumes de Brugués near Terrades, Gerona, Spain.
Distribution: Spain.

Acanthocreagris iranica Beier

Acanthocreagris iranica Beier, in Mahnert, 1976c: 199–200, fig. 15; Schawaller, 1983b: 370.

Type locality: 12 km S. of Shah Pasand, Iran.
Distribution: Iran.

Acanthocreagris italica (Beier)

Microcreagris italica Beier, 1958b: 29–30, fig. 2; Beier, 1963b: 207–208, fig. 219; Beier, 1963h: 155; Cerruti, 1968: fig. 7.
Microcreagris italica italica Beier: Lazzeroni, 1969a: 328; Lazzeroni, 1970a: 207.
Acanthocreagris italica italica (Beier): Mahnert, 1974d: 877–878; Mahnert, 1975c: 194; Mahnert, 1976c: 194; Gardini, 1980c: 115, 130; Callaini, 1986b: 351–352, fig. 7.
Acanthocreagris italica (Beier): Beier, 1975a: 56; Callaini, 1982a: 17, 20, fig. 2b.
Acanthocreagris cfr. *italica italica* (Beier); Callaini, 1986e: 253–254, fig. 3e.

Type locality: Foresta Umbra, Gargano, Italy.
Distribution: Italy, Malta, Sicily.

Acanthocreagris italica ruffoi (Lazzeroni)

Microcreagris italica ruffoi Lazzeroni, 1969a: 330–333, figs 2–12; Lazzeroni, 1970a: 207.
Acanthocreagris italica ruffoi (Lazzeroni): Mahnert, 1976c: 203; Callaini, 1983d: 224; Callaini, 1986b: 352–353, fig. 7.

Type locality: Roccamandolfi, Matese, Italy.
Distribution: Italy.

Acanthocreagris lanzai (Beier)

Obisium (Roncus) lucifugum E. Simon: E. Simon, 1900b: 594 (misidentification).
Microcreagris lanzai Beier, 1961f: 125–126, fig. 2; Beier, 1963b: 212–213, fig. 223.
Acanthocreagris (?) *lanzai* (Beier): Mahnert, 1974d: 849.
Acanthocreagris lanzai (Beier): Mahnert, 1976c: 194; Gardini, 1980c: 115, 129; Callaini, 1986b: 353, fig. 7; Callaini, 1986d: 538–541, figs 5a–g.
Acanthocreagris cfr. *lanzai* (Beier): Callaini, 1986b: 353–356, figs 1a–g, 7.

Type locality: Buca del Fumo, Monte Maggio, near Monteriggioni, Italy.
istribution: Italy.

Acanthocreagris leucadia (Mahnert)

Microcreagris leucadia Mahnert, 1972b: 51–54, figs 1–9.
Acanthocreagris leucadia (Mahnert): Mahnert, 1974d: 849; Schawaller, 1985b: 6.

Type locality: between Fryni and Tsoukalades, Levkas, Greece.
Distribution: Greece.

Acanthocreagris leucadia epirensis Mahnert

Acanthocreagris leucadia epirensis Mahnert, 1974d: 875–877, figs 9a–e; Mahnert, 1976c: 194.

Type locality: 2 km E. of Aghios Komasos, Epirus, Greece.
Distribution: Greece.

Acanthocreagris lucifuga (E. Simon)

Obisium (Roncus) lucifugum E. Simon, 1879a: 66–67, plate 18 fig. 25; Caporiacco, 1923: 132; Boldori, 1927: 93.
Ideobisium (Ideoroncus) lucifugum (E. Simon): Balzan, 1892: 541.

Roncus (Roncus) lucifugus (E. Simon): Beier, 1932a: 128−129; Roewer, 1937: 249; Franciscolo, 1955: 126; Beier, 1963b: 186; Costantini, 1976: 124.
Roncus lucifugus (E. Simon): Wolf, 1937: 222; Caporiacco, 1940: 7; Caporiacco, 1936b: 329.
Microcreagris lucifugus (sic) (E. Simon): Heurtault-Rossi, 1966a: 660−664, figs 1−8, 11−12.
Microcregris lucifuga (E. Simon): Lazzeroni, 1969b: 407−408; Lazzeroni, 1969c: 236.
Microcreagris (Roncocreagris) lucifuga (E. Simon): Mahnert, 1974d: 850.
Acanthocreagris lucifuga (E. Simon): Mahnert, 1976c: 195−196; Gardini, 1980c: 115, 120, 122; Bologna & Taglianti, 1985: 228; Heurtault, 1986c: 28; Callaini, 1986b: 350, fig. 7.
Not *Obisium (Roncus) lucifugum* E. Simon: E. Simon, 1900b: 594 (misidentification; see *Acanthocreagris lanzai* (Beier)).
Not *Microcreagris lucifuga* (E. Simon): Lazzeroni, 1969a: 333 (misidentification; see *Acanthocreagris sandaliotica* Callaini).

Type locality: Grotte d'Esparron, Hyères, France.
Distribution: France, Italy.

Acanthocreagris ludiviri Ćurčić

Acanthocreagris ludiviri Ćurčić, 1976a: 159−167, figs 1−14; Ćurčić, 1976b: 178; Ćurčić, 1983b: fig. 2.

Type locality: Pecina Ludi Vir Cave, approximately 6 km SW. of Krivi Vir, Boljevac, Yugoslavia.
Distribution: Yugoslavia.

Acanthocreagris lycaonis Mahnert

Acanthocreagris lycaonis Mahnert, 1978e: 295−296, figs 49−55; Schawaller, 1985b: 6.

Type locality: Spilia tou Garzeniko Cave, Kandila, Arkadhia (as Arkadien), Greece.
Distribution: Greece.

Acanthocreagris mahnerti Dumitresco & Orghidan

Acanthocreagris mahnerti Dumitresco & Orghidan, 1986: 51−57, figs 1a−d, 2a−c, 3a−c.

Type locality: Tismana Forest, Olténie, Romania.
Distribution: Romania.

Acanthocreagris microphthalma Callaini

Acanthocreagris microphthalma Callaini, 1986b: 367−370, figs 5a−h, 7.
Acanthocreagris cfr. *microphthalma* Callaini: Callaini, 1986b: 370, figs 6a−f, 7.

Type locality: Legri, Firenze, Italy.
Distribution: Italy.

Acanthocreagris multispinosa Estany

Acanthocreagris multispinosa Estany, 1978: 35, figs 5−11; Zaragoza, 1986: figs 1, 3−5.

Type locality: Cueva del Forat, Bárig, Valencia, Spain.
Distribution: Spain.

Acanthocreagris obtusa Mahnert

Acanthocreagris obtusa Mahnert, 1976c: 197−199, figs 9−14.

Type locality: Isparta, between Egridir and Candir, Turkey.
Distribution: Turkey.

Family Neobisiidae

Acanthocreagris osellai (Beier)

Microcreagris osellai Beier, 1973c: 233–235, fig. 6.
Acanthocreagris (?) *osellai* (Beier): Mahnert, 1974d: 849.
Acanthocreagris osellai (Beier): Mahnert, 1976c: 194.

Type locality: Amasya, Turkey.
Distribution: Turkey.

Acanthocreagris pyrenaica (Ellingsen)

Ideobisium pyrenaicum Ellingsen, 1909a: 212–214.
Microcreagris pyrenaica (Ellingsen): Beier, 1932a: 156, fig. 188; Roewer, 1937: 251;
    Beier, 1963b: 211–212, fig. 222; Heurtault-Rossi, 1966a: fig. 10.
Acanthocreagris pyrenaica (Ellingsen): Mahnert, 1976c: 194–195, figs 1–4; Heurtault,
    1986c: 28.

Type localities: Grotte des Neufs, Ariège, France; and Massat, Ariège, France.
Distribution: France.

Acanthocreagris redikorzevi Dashdamirov

Acanthocreagris redikorzevi Dashdamirov, 1988: 1414–1416, figs 1–6; Schawaller &
    Dashdamirov, 1988: 32, fig. 54.

Type locality: N. Mardarkert, Azerbaijan, U.S.S.R.
Distribution: U.S.S.R. (Azerbaijan).

Acanthocreagris relicta Mahnert

Acanthocreagris relicta Mahnert, 1977c: 99–101, figs 53–58; Estany, 1978: 34, figs
    3–4; Zaragoza, 1986: fig. 1.

Type locality: Cova Masies d'Abat, Coves de Vinromà, Castellón, Spain.
Distribution: Spain.

Acanthocreagris ressli (Beier)

Microcreagris ressli Beier, 1965f: 90–91, fig. 7; Beier, 1967f: 305.
Acanthocreagris ressli (Beier): Mahnert, 1974d: 849; Mahnert, 1976c: 194.

Type locality: Göksu Nehir (as Göksu-Klamm), Silifke, Turkey.
Distribution: Turkey.

Acanthocreagris ronciformis (Redikorzev)

Microcreagris ronciformis Redikorzev, 1949: 643–644, figs 3, 4a.
Microcreagris ronciformis aucta Redikorzev, 1949: 644, fig. 4b (synonymized by
    Schawaller, 1985a: 4).
Acanthocreagris vachoni Mahnert, 1976c: 200–202, figs 16–19 (synonymized by
    Schawaller, 1985a: 4).
Acanthocreagris aucta (Redikorzev): Ćurčić, 1985a: 332, figs 1–6.
Acanthocreagris abaris Ćurčić, 1985a: 332–333, figs 7–12 (synonymized by Schawaller,
    1985a: 4).
Acanthocreagris ronciformis (Redikorzev): Schawaller, 1985a: 4; Schawaller, 1989:
    10–13, figs 28–33.

Type localities: of *Microcreagris ronciformis*: Tootpi, Uandir' River, Kopet-Dag,
    Turkmenistan, U.S.S.R./Iran.
of *Microcreagris ronciformis aucta*: Agar-cheshma, Serakhs, Tadzhikistan, U.S.S.R.
of *Acanthocreagris vachoni*: 95 km E. of Shah Pasand, Iran.
of *Acanthocreagris abaris*: Kara Kali, Turkmenistan, U.S.S.R.
Distribution: Iran, U.S.S.R. (Turkmenistan).

Acanthocreagris sandaliotica Callaini

Microcreagris lucifuga (E. Simon): Lazzeroni, 1969a: 333 (misidentification).
Acanthocreagris sandaliotica Callaini, 1986b: 362–366, figs 4a–f, 7.

Type locality: Castelsardo, Sardinia.
Distribution: Sardinia.

Acanthocreagris sardoa (Beier)

Microcreagris sardoa Beier, 1959b: 245–246, fig. 1; Beier, 1963b: 220; Cerruti, 1968: 221–222, fig. 7; Lazzeroni, 1969c: 237; Beier, 1970f: 321; Puddu & Pirodda, 1973: 172.
Acanthocreagris sardoa (Beier): Mahnert, 1976c: 194; Gardini, 1980c: 116, 132; Callaini, 1986b: 351, fig. 7.

Type locality: Monte Tuttavista, Galtelli, Sardinia.
Distribution: Sardinia.

Genus Alabamocreagris Ćurčić

Alabamocreagris Ćurčić, 1984b: 160–163; Ćurčić, 1989c: 357.

Type species: *Microcreagris pecki* Muchmore, 1969a, by original designation.

Alabamocreagris mortis (Muchmore)

Microcreagris mortis Muchmore, 1969a: 8–9, figs 6–7.
Alabamocreagris mortis (Muchmore): Ćurčić, 1989c: 357, figs 6, 14.

Type locality: 'The Morgue' Cave, about 1.5 miles ENE. of Paint Rock, Jackson County, Alabama, U.S.A.
Distribution: U.S.A. (Alabama).

Alabamocreagris pecki (Muchmore)

Microcreagris pecki Muchmore, 1969a: 4–5, figs 3–4.
Alabamocreagris pecki (Muchmore): Ćurčić, 1984b: 160–162, figs 12, 35.

Type locality: Beech Spring Cave, 2.5 miles N. of Union Grove, Marshall County, Alabama, U.S.A.
Distribution: U.S.A. (Alabama).

Genus Americocreagris Ćurčić

Americocreagris Ćurčić, 1982d: 48–50; Ćurčić, 1984b: 154.

Type species: *Microcreagris columbiana* J. C. Chamberlin, 1962, by original designation.

Americocreagris columbiana (J. C. Chamberlin)

Microcreagris columbiana J. C. Chamberlin, 1962: 334–336, figs 11a–d; Muchmore, 1969a: 10–11.
Americocreagris columbiana (J. C. Chamberlin): Ćurčić, 1982d: 48–50, figs 1–2; Ćurčić, 1984b: 157.

Type locality: Clatskanie, Oregon, U.S.A.
Distribution: U.S.A. (Oregon, Washington).

Genus Australinocreagris Ćurčić

Australinocreagris Ćurčić, 1984b: 156; Ćurčić, 1989c: 357–359.

Type species: *Microcreagris grahami* Muchmore, 1969a, by original designation.

Australinocreagris grahami (Muchmore)

Microcreagris grahami Muchmore, 1969a: 11–13, figs 8–9.
Australinocreagris grahami (Muchmore): Ćurčić, 1984b: 156, figs 5–6, 30.

Type locality: Pseudoscorpion Cave, 2 miles S. of Vallecito, Calaveras County, California, U.S.A.
Distribution: U.S.A. (California).

Australinocreagris ozarkensis (Hoff)

Microcreagris ozarkensis Hoff, 1945c: 34–37, figs 1–4; Hoff, 1949b: figs 22–23; Hoff, 1958: 12.
Australinocreagris ozarkensis (Hoff): Ćurčić, 1984b: 156, figs 7–8, 28–29.

Type locality: Devil's Den, Washington County, Arkansas, U.S.A.
Distribution: U.S.A. (Arkansas).

Australinocreagris reddelli (Muchmore)

Microcreagris reddelli Muchmore, 1969a: 17–18, fig. 11; Rowland & Reddell, 1976: 14.
Australinocreagris reddelli (Muchmore): Ćurčić, 1989c: 360, figs 7, 16.

Type locality: Schulze Cave, about 2 miles E. of Volente, Travis County, Texas, U.S.A.
Distribution: U.S.A. (Texas).

Australinocreagris texana (Muchmore)

Microcreagris texana Muchmore, 1969a: 18–19, figs 13–14; Rowland & Reddell, 1976: 15.
Australinocreagris texana (Muchmore): Ćurčić, 1989c: 360, figs 8, 15.

Type locality: Tooth Cave, 0.25 mile N. of Hickmuntown, Travis County, Texas, U.S.A.
Distribution: U.S.A. (Texas).

Genus Balkanoroncus Ćurčić

Balkanoroncus Ćurčić, 1975b: 144; Ćurčić, 1978b: 119, 131.

Type species: *Roncus (Parablothrus) bureschi* Hadži, 1939b (now *Balkanoroncus hadzii* Harvey, new name), by original designation.

Balkanoroncus boldorii (Beier)

Neobisium (Blothrus) boldorii Beier, 1931a: 12–13, fig. 4; Beier, 1932a: 110, fig. 129; Roewer, 1937: 248.
Neobisium boldorii Beier: Boldori, 1932: 127.
Roncus (Parablothrus) boldorii (Beier): Beier, 1942: 134; Beier, 1958d: 163; Vachon & Gabbutt, 1964: fig. 14; Beier, 1970f: 320; Gardini, 1980c: 112, 123, 126; Mahnert, 1980g: 98–99, figs 4–7.
Roncus (Parablothrus) ghidinii Beier, 1942: 134–136, fig. 3; Beier, 1972c: 3–5, fig. 1; Costantini, 1976: 123; Gardini, 1980c: 113, 123, 126 (synonymized by Mahnert, 1980h: 98).
Roncus (Parablothrus) boldorii boldorii (Beier): Beier, 1963b: 201, fig. 212; Lazzeroni, 1969b: 407; Costantini, 1976: 123.
Roncus (Parablothrus) boldorii ghidinii Beier: Beier, 1963b: 201.
Balkanoroncus baldensis Gardini, 1982d: 161–169, figs 1–15; Gardini, 1985b: 61 (synonymized by Gardini & Rizzerio, 1986a: 49).
Balkanoroncus boldorii (Beier): Gardini & Rizzerio, 1986a: 49–51.

Type localities: of *Neobisium (Blothrus) boldorii*: Büs Coalghes, Gavardo, Brescia, Italy.

of *Roncus (Parablothrus) ghidinii*: Büs Suradùr (as Bus Suradur), Valle di Lumezzone, Brescia, Italy.
of *Balkanoroncus baldensis*: Grotte dei Cervi, Monte Baldo, Prealpi Venete, Italy.
Distribution: Italy.

Balkanoroncus bureschi (Redikorzev)

Obisium (Blothrus) bureschi Redikorzev, 1928: 129–133, figs 8–9, 10c.
Roncus (Parablothrus) bureschi (Redikorzev): Beier, 1928: 309.
Neobisium (Blothrus) bureschi (Redikorzev): Beier, 1932a: 113, fig. 134; Roewer, 1937: 248; Beier, 1939d: 72.
Neobisium (Heoblothrus) bureschi (Redikorzev): Beier, 1963b: 134–135, fig. 134; Beier, 1970f: 320.
Balkanoroncus praeceps Ćurčić, 1978b: 125–129, figs 7–20 (invalid replacement name for *Obisium (Blothrus) bureschi* Redikorzev).

Type locality: Seewa Dupka Cave, near Malka Bresnitza, S. of Teteven (as Tetewen), Bulgaria.
Distribution: Bulgaria.

Balkanoroncus hadzii, new name

Roncus (Parablothrus) bureschi Hadži, 1939b: 34–37, figs 9a–m (junior secondary homonym of *Balkanoroncus bureschi* (Redikorzev)).
Microcreagris bureschi (Hadži): Beier, 1963b: 219; Beier, 1970f: 321.
Acanthocreagris (?) bureschi (Hadži): Mahnert, 1974d: 849.
Balkanoroncus bureschi (Hadži): Ćurčić, 1975b: 144; Ćurčić, 1978b: 120–125, figs 1–6, 6a–c.

Type locality: Gornjata Peschtera Cave, Bulgaria.
Distribution: Bulgaria.

Genus Bisetocreagris Ćurčić

Bisetocreagris Ćurčić, 1983a: 25; Ćurčić, 1985a: 333–334.

Type species: *Microcreagris annamensis* Beier, 1951a, by original designation.

Bisetocreagris annamensis (Beier)

Microcreagris annamensis Beier, 1951a: 64–65, fig. 11.
Bisetocreagris annamensis (Beier): Ćurčić, 1983a: 25–26, figs 5, 8, 17.

Type locality: Cao Nguyên Lâm Viên (as Plateau von Langbian), Vietnam.
Distribution: Vietnam.

Bisetocreagris furax (Beier)

Microcreagris furax Beier, 1959a: 262–263, fig. 4.
Bisetocreagris furax (Beier): Ćurčić, 1983a: 27–29.

Type locality: Kandahar, Afghanistan.
Distribution: Afghanistan.

Bisetocreagris gracilis (Redikorzev)

Microcreagris gracilis Redikorzev, 1934a: 434–436, figs 13–14; Roewer, 1936: fig. 70e; Roewer, 1937: 251, fig. 210; Redikorzev, 1949: 643.
Bisetocreagris gracilis (Redikorzev): Ćurčić, 1985a: 336–337, figs 19–24; Schawaller, 1985a: 5–6, figs 3–8.

Type locality: Tujuk River, Semiretschje, R.S.F.S.R., U.S.S.R.
Distribution: U.S.S.R. (Kirghizia, R.S.F.S.R.).

Family Neobisiidae

Bisetocreagris kaznakovi (Redikorzev)

Ideobisium (Microcreagris) kaznakovi Redikorzev, 1918: 96–98, figs 5, 6, 6a (in part; see *Bisetocreagris phoebe* Ćurčić).
Microcreagris kaznakovi (Redikorzev): J. C. Chamberlin, 1930: 28; Beier, 1932a: 145; Roewer, 1937: 251.
Bisetocreagris kaznakovi (Redikorzev): Ćurčić, 1985a: 337–338, figs 25–30; Schawaller, 1985a: 209, figs 19–24.
Not *Microcreagris kaznakovi* (Redikorzev): Beier, 1974b: 265; Schawaller, 1983a: 108, figs 4–7 (misidentifications; see *Levigatocreagris gruberi* Ćurčić).

Type locality: Amnenkor Mountains, Tibet, China.
Distribution: China (Tibet), Nepal.

Bisetocreagris kaznakovi lahaulensis (Mani), new combination

Microcreagris kaznakovi lahaulensis Mani, 1959: 15–16; Murthy & Ananthakrishnan, 1977: 29.

Type locality: Pir Panjal Range, Kashmir, India.
Distribution: India.

Bisetocreagris klapperichi (Beier)

Microcreagris klapperichi Beier, 1959a: 262, fig. 3.
Bisetocreagris klapperichi (Beier): Ćurčić, 1983a: 29.

Type locality: Kandahar-Kuna, Afghanistan.
Distribution: Afghanistan.

Bisetocreagris parablothroides (Beier)

Microcreagris parablothroides Beier, 1951a: 65–66, fig. 12.
Bisetocreagris parablothroides (Beier): Ćurčić, 1983a: 26–27, figs 10, 18.

Type locality: Cha Pa (as Chapa), Tonkin, Vietnam.
Distribution: Vietnam.

Bisetocreagris philippinensis (Beier)

Microcreagris philippinensis Beier, 1931c: 300–301; Beier, 1932a: 150–151, fig. 182; Roewer, 1937: 251.
Microcreagris phlippinensis (sic) Beier: Beier, 1966b: 342.
Bisetocreagris philippinensis (Beier): Ćurčić, 1983a: 29, fig. 4.

Type locality: Surigao, Mindanao, Philippines.
Distribution: Philippines.

Bisetocreagris phoebe Ćurčić

Ideobisium (Microcreagris) kaznakovi Redikorzev, 1918: 96–98 (misidentification, in part).
Bisetocreagris phoebe Ćurčić, 1985a: 338–339, figs 31–36.

Type locality: Maychoktschikam, Tschok-tschio River, Tibet, China.
Distribution: China (Tibet).

Bisetocreagris silvicola (Beier)

Microcreagris silvicola Beier, 1979b: 556–557, fig. 3.
Bisetocreagris silvicola (Beier): Ćurčić, 1983a: 27, figs 9, 19; Ćurčić, 1985a: 345; Schawaller, 1989: 13, figs 34–35.

Type locality: Ussuri (as Ussurijsky) Station, Primorskiy Kray, U.S.S.R.
Distribution: U.S.S.R. (R.S.F.S.R.).

Bisetocreagris tenuis (Redikorzev)

Microcreagris tenuis Redikorzev, 1934a: 437–439, figs 15–16 (not 17–18); Redikorzev, 1934b: 152; Roewer, 1937: 251.
Bisetocreagris tenuis (Redikorzev): Ćurčić, 1985a: 339–340, figs 37–42.
Bisetocreagris (?) *tenuis* (Redikorzev): Schawaller, 1985a: 6–8, figs 14–15.

Type locality: Iche-Bogdó, Mongolia.
Distribution: Mongolia, U.S.S.R. (Kirghizia).

Bisetocreagris turkestanica (Beier)

Ideobisium (Microcreagris) turkestanicum Beier, 1929a: 364–365, fig. 15.
Microcreagris turkestanica (Beier): Beier, 1932a: 151, fig. 183; Roewer, 1937: 251; Redikorzev, 1949: 642.
Bisetocreagris egeria Ćurčić, 1985a: 334, figs 13–18 (synonymized by Schawaller, 1985a: 6).
Bisetocreagris turkestanica (Beier): Schawaller, 1985a: 6, figs 9–13.

Type localities: of *Ideobisium (Microcreagris) turkestanicum*: Margelan, Uzbekistan, U.S.S.R.
of *Bisetocreagris egeria*: Kara-Kalinskiy region, Turkmenistan, U.S.S.R.
Distribution: U.S.S.R. (Kirghizia, Turkmenistan, Uzbekistan).

Bisetocreagris ussuriensis (Redikorzev)

Microcreagris ussuriensis Redikorzev, 1934a: 436–437, figs 17–18 (not 15–16); Redikorzev, 1934b: 152; Roewer, 1937: 251.
Bisetocreagris maritima Ćurčić, 1983a: 29–30, figs 24–29 (synonymized by Schawaller, 1989: 13).
Bisetocreagris ussuriensis (Redikorzev): Ćurčić, 1985a: 340–342, figs 43–54; Schawaller, 1989: 13–14, figs 36–37.
Bisetocreagris erytheia Ćurčić, 1985a: 342–343, figs 55–71 (synonymized by Schawaller, 1989: 13).
Bisetocreagris merope Ćurčić, 1985a: 343–344, figs 72–77 (synonymized by Schawaller, 1989: 13.
Bisetocreagris gorgo Ćurčić, 1985a: 344–345, figs 78–82 (synonymized by Schawaller, 1989: 13).

Type localities: of *Microcreagris ussuriensis*: Tigrovaja, Ussur, R.S.F.S.R., U.S.S.R.
of *Bisetocreagris maritima*: Maritime Province, R.S.F.S.R., U.S.S.R.
of *Bisetocreagris erytheia*: Maritime Province, R.S.F.S.R., U.S.S.R.
of *Bisetocreagris merope*: Maritime Province, R.S.F.S.R., U.S.S.R.
of *Bisetocreagris gorgo*: Petrov Island, R.S.F.S.R., U.S.S.R.
Distribution: U.S.S.R. (R.S.F.S.R.).

Genus **Chinacreagris** Ćurčić

Chinacreagris Ćurčić, 1983a: 30–31.

Type species: *Microcreagris chinensis* Beier, 1943, by original designation.

Chinacreagris chinensis (Beier)

Microcreagris chinensis Beier, 1943: 74–75, fig. 1; Beier, 1967g: 342.
Chinacreagris chinensis (Beier): Ćurčić, 1983a: 31, figs 6, 11, 20.

Type locality: Nan-ching (as Nanking), Kiangsu, China.
Distribution: China (Kiangsu).

Chinacreagris kwantungensis (Beier)

Microcreagris kwantungensis Beier, 1967g: 342–343, fig. 2.
Chinacreagris kwantungensis (Beier): Ćurčić, 1983a: 31–32, fig. 22.

Type locality: Tsin Leong San, Kwangtung, China.
Distribution: China (Kwangtung).

Chinacreagris nankingensis Ćurčić

Chinacreagris nankingensis Ćurčić, 1983a: 32–34, figs 7, 12, 21, 30–31.

Type locality: Nan-ching (as Nanking), China.
Distribution: China (Kiangsu).

Genus Cryptocreagris Ćurčić

Cryptocreagris Ćurčić, 1984b: 158–160; Ćurčić, 1989c: 354.

Type species: *Microcreagris laudabilis* Hoff, 1956b, by original designation.

Cryptocreagris laudabilis (Hoff)

Microcreagris laudabilis Hoff, 1956b: 4–9, figs 1–5; Hoff, 1958: 12; Hoff, 1959b: 4, etc.; Zeh, 1987b: 1086.
Cryptocreagris laudabilis (Hoff): Ćurčić, 1984b: 158, figs 14–15, 33; Ćurčić, 1989c: 354–355, figs 3-4, 12–13.

Type locality: Mt Taylor, NE. of Grants, Valencia County, New Mexico, U.S.A.
Distribution: U.S.A. (New Mexico).

Cryptocreagris magna (Banks)

Ideobisium magnum Banks, 1909b: 306–307; Banks, 1911: 639; Moles & Moore, 1921: 8.
Microcreagris magnum (sic) (Banks): J. C. Chamberlin, 1930: 28.
Microcreagris magna (Banks): Beier, 1932a: 147; Roewer, 1937: 252; Hoff, 1958: 11.
Cryptocreagris magna (Banks): Ćurčić, 1984b: 158, fig. 34.

Type locality: Mount Shasta, California, U.S.A.
Distribution: U.S.A. (California).

Genus Electrobisium Cockerell

Electrobisium Cockerell, 1917: 360.

Type species: *Electrobisium acutum* Cockerell, 1917, by original designation.

! Electrobisium acutum Cockerell

Electrobisium acutum Cockerell, 1917: 360, fig. 1; Schawaller, 1978: 3.

Type locality: Burmese Amber.
Distribution: Burmese Amber.

Genus Fissilicreagris Ćurčić

Fissilicreagris Ćurčić, 1984b: 154–156.

Type species: *Microcreagris chamberlini* Beier, 1931c, by original designation.

Fissilicreagris chamberlini (Beier)

Microcreagris macilenta (E. Simon): J. C. Chamberlin, 1930: 31 (misidentification).
Microcreagris chamberlini Beier, 1931c: 301; Beier, 1932a: 153; Roewer, 1937: 252; Hoff, 1958: 12.
Fissilicreagris chamberlini (Beier): Ćurčić, 1984b: 154–156, figs 3–4, 26–27.

Type locality: Claremont, California, U.S.A.
Distribution: U.S.A. (California).

Genus **Globocreagris** Ćurčić

Globocreagris Ćurčić, 1984b: 160.

Type species: *Microcreagris nigrescens* J. C. Chamberlin, 1952, by original designation.

Globocreagris nigrescens (J. C. Chamberlin)

Microcreagris nigrescens J. C. Chamberlin, 1952: 265–268, figs 1a–j; Hoff, 1958: 12; Muchmore, 1971a: 90; Zeh, 1987b: 1086.
Globocreagris nigrescens (J. C. Chamberlin): Ćurčić, 1984b: 160, figs 16, 36.

Type locality: Robertson Creek, Francis Simes Hastings Natural History Reservation, California, U.S.A.
Distribution: U.S.A. (California).

Genus **Halobisium** J. C. Chamberlin

Halobisium J. C. Chamberlin, 1930: 35; Beier, 1932a: 141–142; Morikawa, 1960: 119.

Type species: *Ideobisium orientale* Redikorzev, 1918, by original designation.

Halobisium occidentale Beier

Halobisium orientale (Redikorzev): J. C. Chamberlin, 1930: 35–36, fig. 1p (misidentification, in part); J. C. Chamberlin, 1931a: figs 17h, 25g, 35j–m, 56; Vachon, 1949: fig. 203b (misidentifications).
Halobisium occidentale Beier, 1931c: 299; Beier, 1932a: 142–143, fig. 176; Beier, 1932g: figs 213, 253; Roewer, 1936: fig. 112e; Roewer, 1937: 250; Hoff, 1958: 10; Weygoldt, 1969a: 112; Schulte, 1976: 119–123, figs 1–2.

Type locality: Palo Alto salt marshes, California, U.S.A.
Distribution: Canada (British Columbia), U.S.A. (Alaska, California, Oregon, Washington).

Halobisium orientale (Redikorzev)

Ideobisium orientale Redikorzev, 1918: 98–100, figs 7, 8, 8a.
Halobisium orientale (Redikorzev): J. C. Chamberlin, 1930: 35–36 (in part; see *Halobisium occidentale* Beier); Beier, 1932a: 142, fig. 175; Roewer, 1937: 250; Schawaller, 1989: 10, figs 24–27.
Not *Halobisium orientale* (Redikorzev): J. C. Chamberlin, 1931a: figs 17h, 25g, 35j–m, 56; Vachon, 1949: fig. 203b (misidentifications; see *Halobisium occidentale* Beier).

Type locality: Lake Reineke, R.S.F.S.R., U.S.S.R.
Distribution: U.S.S.R. (R.S.F.S.R.).

Halobisium orientale japonicum Morikawa

Halobisium orientale japonicum Morikawa, 1958: 7–8, figs 1c–d; Morikawa, 1960: 119–120, plate 2 fig. 6, plate 6 fig. 7; Morikawa, 1962: 419.

Type locality: Akkeshi, Hokkaido, Japan.
Distribution: Japan.

Genus **Insulocreagris** Ćurčić

Insulocreagris Ćurčić, 1987: 48–49; Ćurčić, 1988b: 30–31.

Type species: *Insulocreagris regina* Ćurčić, 1987, by original designation.

Insulocreagris regina Ćurčić

Insulocreagris regina Ćurčić, 1987: 49–55, figs 1–8; Ćurčić, 1988b: 31–34, figs 36–43.

Type locality: Kraljicina Spilja Cave, near Okljucna, Vis, Yugoslavia.
Distribution: Yugoslavia.

Insulocreagris troglobia, new name

Roncus (Parablothrus) cavernicola Beier, 1928: 310−312, fig. 9; Beier, 1932a: 134, fig. 165; Roewer, 1937: 249; Beier, 1939d: 81; Beier, 1963b: 202, fig. 214; Beier, 1970f: 320; Ćurčić, 1974a: 24 (junior primary homonym of *Roncus lubricus cavernicola* Tömösváry, 1882b).
Acanthocreagris cavernicola (Beier): Mahnert, 1976c: 195.
Insulocreagris (?) *cavernicola* (Beier): Ćurčić, 1988b: 34−35, figs 44−53.

Type locality: Kallipecina Cave, Herzegovina, Yugoslavia.
Distribution: Yugoslavia.

Genus **Levigatocreagris** Ćurčić

Levigatocreagris Ćurčić, 1983a: 34−35.

Type species: *Levigatocreagris gruberi* Ćurčić, 1983a, by original designation.

Levigatocreagris gruberi Ćurčić

Microcreagris kaznakovi (Redikorzev): Beier, 1974b: 265; Schawaller, 1983a: 108, figs 4−7 (misidentifications).
Levigatocreagris gruberi Ćurčić, 1983a: 35, figs 3, 13, 15, 23, 32−35; Schawaller, 1987a: 205−206, figs 9−11.

Type locality: Khola Valley, Dzunda, Nepal.
Distribution: Nepal.

Levigatocreagris hamatus Leclerc & Mahnert

Levigatocreagris hamatus Leclerc & Mahnert, 1988: 273−277, figs 1−14.

Type locality: Doi Inthanon, Chiang Mai, Thailand.
Distribution: Thailand.

Levigatocreagris heros (Beier)

Microcreagris heros Beier, 1943: 75−76, fig. 2.
Levigatocreagris (?) *heros* (Beier): Leclerc & Mahnert, 1988: 277.

Type locality: Central Asia.
Distribution: U.S.S.R.?

Levigatocreagris kashmirensis Schawaller

Levigatocreagris kashmirensis Schawaller, 1988: 158−160, figs 3−8.

Type locality: Aru, NW. Pahalgam, Kashmir, India.
Distribution: India.

Levigatocreagris lindbergi (Beier)

Microcreagris lindbergi Beier, 1959a: 263, fig. 5.
Levigatocreagris lindbergi (Beier): Ćurčić, 1983a: 35, fig. 14.

Type locality: Djelala, Konar, Afghanistan.
Distribution: Afghanistan.

Levigatocreagris martensi Schawaller

Levigatocreagris martensi Schawaller, 1987a: 207−208, figs 12−18.

Type locality: Tabruk Kharka, NE. Rupina La, Gorkha, Nepal.
Distribution: Nepal.

Genus **Lissocreagris** Ćurčić

Lissocreagris Ćurčić, 1981b: 102; Ćurčić, 1984b: 152−153.
Type species: *Microcreagris pluto* J.C. Chamberlin, 1962, by original designation.

Lissocreagris parva Ćurčić

Lissocreagris parva Ćurčić, 1984b: 154, figs 2, 25, 43–45.

Type locality: Torreya State Park, Florida, U.S.A.
Distribution: U.S.A. (Florida).

Lissocreagris persephone (J. C. Chamberlin)

Microcreagris persephone J. C. Chamberlin, 1962: 348–350, figs 17a–b; Muchmore, 1969a: 7.
Lissocreagris persephone (J. C. Chamberlin): Ćurčić, 1981b: 104, fig. 6; Ćurčić, 1984b: 152.

Type locality: Davidson Cave, Marshall County, Alabama, U.S.A.
Distribution: U.S.A. (Alabama).

Lissocreagris pluto (J. C. Chamberlin)

Microcreagris pluto J. C. Chamberlin, 1962: 345–348, figs 16a–d.
Lissocreagris pluto (J. C. Chamberlin): Ćurčić, 1981b: 102, figs 1, 4, 9; Ćurčić, 1984b: 152.

Type locality: Terrell Cave no. 1, Marshall County, Alabama, U.S.A.
Distribution: U.S.A. (Alabama).

Lissocreagris subatlantica (J. C. Chamberlin)

Microcreagris subatlantica J. C. Chamberlin, 1962: 340–343, figs 14a–d, g–i; Muchmore, 1969a: 2.
Lissocreagris subatlantica (J. C. Chamberlin): Ćurčić, 1981b: 104–105, figs 3, 5, 8; Ćurčić, 1984b: 152–153.

Type locality: Dickey Cave, Colbert County, Alabama, U.S.A.
Distribution: U.S.A. (Alabama, Georgia).

Lissocreagris valentinei (J. C. Chamberlin)

Microcreagris valentinei J. C. Chamberlin, 1962: 350–352, figs 18a–d; Muchmore, 1966a: 97; Holsinger & Culver, 1988: 43, fig. 15.
Lissocreagris valentinei (J. C. Chamberlin): Ćurčić, 1981b: 103–104, figs 2, 7; Ćurčić, 1984b: 152.

Type locality: King Solomon Cave (= Cudjo's Cave), Lee County, Virginia, U.S.A.
Distribution: U.S.A. (Virginia).

Genus **Microbisium** J. C. Chamberlin

Microbisium J. C. Chamberlin, 1930: 20; Beier, 1932a: 136; Beier, 1939f: 186; Hoff, 1949b: 444–445; Hoff, 1956b: 2; Hoff, 1961: 431–432; Beier, 1963b: 202–203; Muchmore, 1979a: 194.
Afrobisium Beier, 1932a: 140 (synonymized by Mahnert, 1981a: 536).

Type species: of *Microbisium*: *Obisium brunneum* Hagen, 1868b, by original designation.
of *Afrobisium*: *Obisium dogieli* Redikorzev, 1924a, by original designation.

Microbisium brevifemoratum (Ellingsen)

Obisium (Obisium) brevifemoratum Ellingsen, 1903: 13–17; Lessert, 1911: 32–33, fig. 25; Kästner, 1928: 11–12, fig. 45.
Obisium brevifemoratum Ellingsen: Schenkel, 1928: 64, figs 20a–b; Schenkel, 1929a: 322–323; Cirdei & Gutu, 1959: 2–3, fig. 2.
Microbisium brevifemoratum (Ellingsen): Beier, 1932a: 138, figs 168, 171; Tumšs, 1934: 13; Roewer, 1936: fig. 74; Roewer, 1937: 249; Lohmander, 1939b: 288–291,

fig. 3; Lohmander, 1945: 8, figs 1, 3−6; Franz & Beier, 1948: 476; Kaisila, 1949a: 74; Kaisila, 1949b: 78−79, fig. 3; Beier, 1952e: 4; Beier & Franz, 1954: 456; Meinertz, 1962a: 392, map 2; Beier, 1963b: 203, fig. 215; Kölzel, 1967: 200; Lehtinen, 1964: 284−285; Ressl, 1965: 289; Rafalski, 1967: 13; Palmgren, 1973: 7; Klausen, 1975: 64; Schawaller, 1985a: 3−4; Jędryczkowski, 1987b: 141, map 5; Schawaller & Dashdamirov, 1988: 25, figs 39, 54; Schawaller, 1989: 10, figs 22−23.

Type locality: Risor, Norway.
Distribution: Austria, Denmark, Finland, Norway, Poland, Sweden, Switzerland, U.S.S.R. (Azerbaijan, Latvia, R.S.F.S.R.), West Germany.

Microbisium brevipalpe (Redikorzev)

Obisium brevipalpe Redikorzev, 1922a: 270−272, figs 12−14.
Microbisium brevipalpe (Redikorzev): Beier, 1932a: 138−139, fig. 172; Roewer, 1937: 249.
Microbisium brevipalpi (sic) (Redikorzev): Mani, 1962: 235.

Type locality: Malyi Javinskij, Kamchatka, R.S.F.S.R., U.S.S.R.
Distribution: U.S.S.R. (R.S.F.S.R.).

Microbisium brunneum (Hagen)

Obisium brunneum Hagen, 1868b: 52; Hagen, 1870: 269; Banks, 1895a: 12; Coolidge, 1908: 113; Ewing, 1911: 76, 79, fig. 13.
Obisium bruneum (sic) Hagen: Banks, 1890: 152.
Obisium parvulum Banks, 1895a: 12 (misidentification, in part).
Microbisium brunneum (Hagen): J. C. Chamberlin, 1930: 20−21; Beier, 1932a: 139− 140, fig. 174; Roewer, 1937: 250; Hoff, 1944b: 125−128, figs 1e−f, 2 (in part; see *Microbisium parvulum* (Banks)); Hoff, 1945a: 109; Hoff, 1945i: 323; Hoff, 1946g: 494−495; Pearse, 1946: 257; Rapp, 1946: 197; Hoff, 1949b: 445, figs 24a−b; Hoff & Bolsterli, 1956: 161; Hoff, 1958: 8−9; Manley, 1969: 6−7, fig. 5; Muchmore, 1971a: 91; Nelson, 1975: 280, figs 18−19; Beier, 1976c: 1; Aitchison, 1979: 85; Nelson, 1984: 345; Sharkey, 1987: 17 (not seen); Koponen & Sharkey, 1989: 388−389.
Microbisium near *brunneum* (Hagen): McClure, 1943: 12.
Not *Microbisium brunneum* (Hagen): Hoff, 1945c: 34 (misidentification; see *Microbisium parvulum* (Banks)).

Type locality: Massachusetts, U.S.A.
Distribution: Canada (Manitoba, Ontario), Costa Rica, U.S.A. (District of Washington, Georgia, Illinois, Iowa, Maine, Massachusetts, Michigan, Missouri, New Jersey, New York, North Carolina, Ohio, South Carolina, Virginia, Utah, Wisconsin).

Microbisium congicum Beier

Microbisium congicum Beier, 1955a: 8−9, fig. 4; Beier, 1959d: 23.

Type locality: Mukana, Upemba, Zaire.
Distribution: Zaire.

Microbisium dogieli (Redikorzev)

Obisium dogieli Redikorzev, 1924a: 192−193, figs 5−6.
Afrobisium dogieli (Redikorzev): Beier, 1932a: 140−141; Roewer, 1937: 250.
Microbisium perpusillum Beier, 1955c: 532−533, fig. 4; Beier, 1959d: 23; Beier, 1962f: 14 (synonymized by Mahnert, 1981a: 536).
Microbisium dogieli (Redikorzev): Mahnert, 1981a: 536−537.

Type localities: of *Obisium dogieli*: Mabira, Tanzania.
of *Microbisium perpusillum*: Mt Elgon, E. side, Kenya.
Distribution: Kenya, Ruanda, Tanzania, Zaire.

Microbisium fagetum Cîrdei, Bulimar & Malcoci

Microbisium fagetum Cîrdei, Bulimar & Malcoci, 1967: 239–240, fig. 6.

Type locality: Masivul Repedea, Iasi, Romania.
Distribution: Romania.

Microbisium lawrencei Beier

Microbisium lawrencei Beier, 1964k: 33–34, fig. 2.

Type locality: Vumba, Zimbabwe.
Distribution: Zimbabwe.

Microbisium manicatum (L. Koch)

Obisium manicatum L. Koch, 1873: 61–62; L. Koch, 1878: 38; E. Simon, 1885a: 349;
Ellingsen, 1909a: 214; Krausse-Heldrungen, 1912: 65; Daday, 1918: 2; Navás,
1925: 124.

Obisium (Obisium) manicatum L. Koch: Daday, 1888: 132, 188, plate 4 fig. 25;
Ellingsen, 1910a: 398–399; Ellingsen, 1912c: 173; Nonidez, 1917: 40; Beier, 1928:
296; Beier, 1929d: 156.

Microbisium manicatum (Ellingsen): Beier, 1932a: 137, fig. 169; Roewer, 1937: 249;
Beier, 1939f: 186; Caporiacco, 1951e: 69; Beier, 1963b: 205, fig. 217; Kobakhidze,
1965b: 541; Kobakhidze, 1966: 704; Lazzeroni, 1969a: 333; Lazzeroni, 1969b: 236;
Cîrdei, Bulimar & Malcoci, 1970: 11–12, fig. 5; Lazzeroni, 1970a: 207.

Type locality: Greece.
Distribution: France, Greece, Hungary, Italy, Romania, Sardinia, Spain, U.S.S.R.
(Georgia), Yugoslavia.

Microbisium parvulum (Banks)

Obisium parvulum Banks, 1895a: 12 (in part; see *Microbisium brunneum* (Hagen));
Banks, 1904a: 141; Coolidge, 1908: 113; Ellingsen, 1909b: 220.

Microbisium parvulum (Banks): J. C. Chamberlin, 1930: 21–22; J. C. Chamberlin,
1931a: figs 28h, 35n; Beier, 1932a: 139, fig. 173; Roewer, 1937: 250; Hoff, 1946g:
495–496; Hoff, 1956b: 3–4; Hoff & Bolsterli, 1956: 161; Hoff, 1958: 9; Hoff,
1959b: 4, etc.; Hoff, 1961: 432, figs 7–9; Rowland & Reddell, 1976: 15; Nelson,
1984: 343–349, maps 1–2.

Microbisium brunneum (Hagen): Hoff, 1944b: 125–128, figs 1e–f, 2 (in part); Hoff,
1945c: 34 (misidentifications).

Microbisium confusum Hoff, 1946g: 496–497; Hoff, 1949b: 446, figs 25a–c; Hoff,
1956b: 3; Hoff & Bolsterli, 1956: 160; Hoff, 1958: 9; Hoff, 1961: 432–433, figs
10–11; Hoff, 1963a: 2–3; Lawson, 1969: 3–4, figs 1–5; Manley, 1969: 6, fig. 4;
Weygoldt, 1969a: fig. 105; Muchmore, 1971a: 91; Knowlton, 1972: 2; Muchmore,
1972d: 110; Nelson, 1973: 232–235, figs a–c; Knowlton, 1974: 3; Nelson, 1975:
280–281, figs 20–24; W. F. Rapp, 1978a: 5–7; Aitchison, 1979: 85; Muchmore,
1979a: 194; Nelson, 1982: 262–274, figs 1–28; W.F. Rapp, 1986: 220–221; Zeh,
1987b: 1086 (synonymized by Nelson, 1984: 349).

Microbisium sp. prob. *confusum* Hoff: Graves & Graves, 1969: 267.

Type localities: of *Obisium parvulum*: North America (Hoff (1958) invalidly declared
Bernalillo County, New Mexico, as type locality).
of *Microbisium confusum* Hoff: Antioche, Illinois, U.S.A.
Distribution: Canada (Manitoba, Ontario, Quebec), Costa Rica, El Salvador, Mexico,
U.S.A. (Arkansas, California, Colorado, Connecticut, Florida, Georgia, Idaho,
Illinois, Indiana, Iowa, Kansas, Kentucky, Maine, Maryland, Massachusetts,
Michigan, Minnesota, Mississippi, Missouri, Nebraska, New Hampshire, New
Jersey, New Mexico, New York, North Carolina, North Dakota, Ohio, Oklahoma,
Pennsylvania, South Carolina, South Dakota, Tennessee, Texas, Utah, Vermont,
Virginia, Wisconsin).

Microbisium pygmaeum (Ellingsen)

Obisium pygmaeum Ellingsen, 1907c: 12−14.
Microbisium pygmaeum (Ellingsen): Beier, 1932a: 140; Roewer, 1937: 250; Morikawa, 1952b: 254−255, fig. 9; Kobari, 1984a: 57−62, figs 1, 2a−f, 3a−e; Sato, 1985: 22, figs 1−2.
Neobisium (Parobisium) pygmaeum (Ellingsen): Morikawa, 1960: 114−115, plate 2 fig. 3; Sato, 1979a: 94; Sato, 1979d: 43; Sato, 1982c: 32; Kobari, 1983: 65−70, fig. 1.
Neobisium pygmaeum (Ellingsen): Sato, 1979b: 13, figs 1−3, 5; Sato, 1983b: 16, figs 2, 4; Sato, 1984c: 14, figs 1, 3; Sato, 1984d: fig. 2; W.K. Lee, 1981: 130, figs 2a−b.

Type localities: Negishi, S. of Yokohama, Japan; Okayama, Bizen, Japan; and Kanagawa, N. of Yokohama, Japan.
Distribution: Japan, Korea.

Microbisium suecicum Lohmander

Microbisium parvulum suecicum Lohmander, 1945: 7−8, figs 2, 7−10.
Microbisium suecicum Lohmander: Beier, 1963b: 204; Rafalski, 1967: 13; Gardini, 1976a: 251−255, figs 1−8; Mahnert, 1983d: 362; Gardini, 1985b: 60; Jędryczkowski, 1987b: 142, map 6.
Microbisium cfr. *suecicum* Lohmander: Callaini, 1988b: 42−46, figs 3c−m.

Type localities: several localities in Sweden.
Distribution: Austria, Hungary, Italy, Morocco, Poland, Sweden, Switzerland.

Microbisium zariquieyi (Navás)

Obisium zariquieyi Navás, 1919: 212−214, figs 9a−c; Navás, 1925: 124, figs 17a−c.
Microbisium (?) *zariquieyi* (Navás): Beier, 1932a: 140; Roewer, 1937: 250.
Microbisium zariquieyi (Navás): Beier, 1939f: 187−188, fig. 17; Beier, 1959f: 122; Beier, 1963b: 204−205, fig. 216.

Type locality: Santa Fe, Barcelona, Spain.
Distribution: Spain.

Genus Microcreagris Balzan

Microcreagris Balzan, 1892: 543−544; Ellingsen, 1907c: 6; J. C. Chamberlin, 1930: 23; Beier, 1932a: 143; Beier, 1939f: 189; Hoff, 1949b: 444; Hoff, 1956b: 4; Morikawa, 1960: 120; Beier 1963b: 206; Muchmore, 1969a: 1−2; Mahnert, 1979a: 339; Muchmore, 1979a: 194−195; Ćurčić, 1983a: 24−25.
Microcregris (Microcreagris) Balzan: Mahnert, 1974d: 850−851.

Type species: *Microcreagris gigas* Balzan, 1892, by monotypy.

Microcreagris abnormis Turk

Microcreagris abnormis Turk, 1946: 67−70, figs 6−10.

Type locality: Almora, Kumaun, Uttar Pradesh, India.
Distribution: India.

Microcreagris afghanica Beier

Microcreagris afghanica Beier, 1959a: 261, fig. 2.

Type locality: Pul-e-Nou, Afghanistan.
Distribution: Afghanistan.

Microcreagris atlantica J. C. Chamberlin

Microcreagris atlantica J. C. Chamberlin, 1930: 29–30; Beier, 1932a: 148; Roewer, 1937: 252; Hoff & Bolsterli, 1956: 163; Hoff, 1958: 11; J. C. Chamberlin, 1962: 338–340, figs 13a–f, 14e–f; Muchmore, 1979a: 195.

Type locality: eastern U.S.A.
Distribution: U.S.A. (Florida, Mississippi, North Carolina).

Microcreagris birmanica Ellingsen

Microcreagris birmanica Ellingsen, 1911b: 142–144; Beier, 1932a: 151; Roewer, 1937: 251.

Type locality: Carin Cheba, Burma.
Distribution: Burma.

Microcreagris brevidigitata J. C. Chamberlin

Microcreagris brevidigitata J. C. Chamberlin, 1930: 26–27; J. C. Chamberlin, 1931a: fig. 28m; Beier, 1932a: 146; Roewer, 1937: 251; Morikawa, 1960: 121.

Type locality: Mt Kirishima, Japan.
Distribution: Japan.

Microcreagris californica (Banks)

Atemnus californicus Banks, 1891: 165–166.
Blothrus californicus (Banks): Banks, 1895a: 13; Banks, 1904b: 364; Coolidge, 1908: 113; Banks, 1911: 640; Moles & Moore, 1921: 9.
Microcreagris (?) *californica* (Banks): Beier, 1932a: 157; Roewer, 1937: 252; Hoff, 1958: 13.
'Blothrus' californicus (Banks): Ćurčić, 1984b: 164, figs 17, 37.

Type locality: California, U.S.A.
Distribution: U.S.A. (California).

Microcreagris cingara J. C. Chamberlin

Microcreagris cingara J.C. Chamberlin, 1930: 29, figs 2gg, 3f, 3r; Beier, 1932a: 147; Roewer, 1936: fig. 30e; Roewer, 1937: 252; Hoff, 1958: 11.

Type locality: St George, Utah, U.S.A.
Distribution: U.S.A. (Oregon, Utah).

Microcreagris eurydice Muchmore

Microcreagris eurydice Muchmore, 1969a: 7–8, fig. 5.
'Microcreagris' eurydice Muchmore: Ćurčić, 1984b: 164, figs 18, 38.

Type locality: Kennamer Cave, 2 miles E. of Paint Rock, Jackson County, Alabama, U.S.A.
Distribution: U.S.A. (Alabama).

Microcreagris ezoensis Morikawa

Microcreagris ezoensis Morikawa, 1972: 34, figs 1a–c.

Type locality: North Cirque, Mt Poroshiri-daké, Japan.
Distribution: Japan.

Microcreagris formosana Ellingsen

Microcreagris granulata formosana Ellingsen, 1912a: 127–128.
Microcreagris formosana Ellingsen: Kishida, 1928: 411–412; J. C. Chamberlin, 1930: 30; Beier, 1932a: 148–149, fig. 180; Roewer, 1937: 251.

Type localities: Takao (now Kao-hsiung), Taiwan; and Gyamma, Taiwan.
Distribution: Philippines, Taiwan.

Microcreagris gigas Balzan

Microcreagris gigas Balzan, 1892: 544—545, figs 34, 34a—b; J. C. Chamberlin, 1930: 28; Beier, 1932a: 145, fig. 178; Roewer, 1937: 251; Mahnert, 1974d: figs 1a, 2a; Mahnert, 1979a: 339—341, figs 1—6.

Type locality: China (possibly Tsin Lin Shan Mountain range).
Distribution: China.

Microcreagris grandis Muchmore

Microcreagris grandis Muchmore, 1962: 1—5, figs 1—2; Muchmore, 1969a: 10.

Type locality: Lehman Caves National Monument, White Pine County, Nevada.
Distribution: U.S.A. (Nevada).

Microcreagris herculea Beier

Microcreagris herculea Beier, 1959a: 260, fig. 1; Lindberg, 1961: 31; Ćurčić, 1983a: 25, figs 1—2, 16.

Type locality: Grotte Laghat, near Ibrahim Khel (Khouguiani), Afghanistan.
Distribution: Afghanistan.

Microcreagris hespera J. C. Chamberlin

Microcreagris hespera J. C. Chamberlin, 1930: 31, fig. 3u; J. C. Chamberlin, 1931a: figs 11q; Beier, 1932a: 153; Roewer, 1937: 252; Hoff, 1958: 11.

Type locality: Pepperwood Creek, near mouth of Gualala River, Mendocino County, California, U.S.A.
Distribution: U.S.A. (California).

Microcreagris imperialis Muchmore

Microcreagris imperialis Muchmore, 1969a: 13—15, fig. 10.
'*Microcreagris*' *imperialis* Muchmore: Ćurčić, 1984b: 164, figs 20, 41.

Type locality: Empire Cave, 1 mile NE. of Santa Cruz, Santa Cruz County, California, U.S.A.
Distribution: U.S.A. (California).

Microcreagris indochinensis Redikorzev

Microcreagris indochinensis Redikorzev, 1938: 76—78, figs 4—5; Roewer, 1940: 344; Beier, 1951a: 63—64.

Type locality: Pleiku, Vietnam.
Distribution: Vietnam.

Microcreagris japonica Ellingsen

Microcreagris gigas japonica Ellingsen, 1907c: 7; Kishida, 1915: 367 (not seen).
Microcreagris japonica Ellingsen: Kishida, 1928: 408; J. C. Chamberlin, 1930: 28, fig. 2bb; J. C. Chamberlin, 1931a: figs 11l, 17i, 25e—f; Beier, 1932a: 148; Roewer, 1937: 251; Morikawa, 1960: 121—123, plate 2 fig. 7, plate 6 fig. 11, plate 8 fig. 7, plate 9 fig. 8, plate 10 fig. 21; Morikawa, 1962: figs 1, 3; Sato, 1978e: 17, figs 3, 7; Sato, 1979a: 97—100, figs 9a—j, 10a—e, plates 9, 14, 16; Sato, 1979b: 13, figs 1—3, 5; Sato, 1979d: 43—44; Sato, 1980a: 58, fig. 1; Sato, 1980c: 10, figs 1—2; Sato, 1980d: fig. 2, plate 4; Sato, 1982c: 32—33; Sato, 1983b: 16, figs 2—4; Sato, 1984c: 14, figs 1, 3; Sato, 1985: 22, figs 1—2; Sato, 1986: 95.
Microcreagris granulata Ellingsen, 1907c: 7—9; Kishida, 1915: 367 (not seen); Kishida, 1928: 411; Beier, 1932a: 149; Roewer, 1937: 251 (synonymized by Morikawa, 1960: 121).

Microcreagris cyclica Kishida, 1928: 409–411, figs 1, 2a–c; Beier, 1932a: 148; Roewer, 1937: 251; Morikawa, 1955a: 216–217; Morikawa, 1957a: 363 (synonymized by Morikawa, 1960: 121).
Microcreagris ishiharanus Morikawa, 1952b: 255–256, figs 10–12 (synonymized with *Microcreagris cyclica* by Morikawa, 1955a: 216).

Type localities: of *Microcreagris gigas japonica*: Yamanaka, Suruga, Japan; Okayama, Bizen, Japan; Takakiyama, near Kanagawa, Japan; Ooyama, Sagami, Japan; and Kuenji, N. of Shizuoka, Koshu, Japan.
of *Microcreagris granulata*: Yamanaka, Suruga, Japan.
of *Microcreagris cyclica*: Shimono, Fukuoka-mura, Enagori, Gifu-ken, Japan.
of *Microcreagris ishiharanus*: Sugitate, near Matsuyama, Shikoku, Japan.
Distribution: China, Japan.

! Microcreagris koellneri Schawaller

Microcreagris koellneri Schawaller, 1978: 13–16, figs 5–10.

Type locality: Baltic Amber.
Distribution: Baltic Amber.

Microcreagris lampra J. C. Chamberlin

Microcreagris lampra J. C. Chamberlin, 1930: 34; Beier, 1932a: 150; Roewer, 1937: 251.

Type locality: Kusang, China?
Distribution: China?

Microcreagris laurae J. C. Chamberlin

Microcreagris laurae J. C. Chamberlin, 1930: 32–33; Beier, 1932a: 153; Roewer, 1937: 252; Hoff, 1958: 12.

Type locality: Berkeley, California, U.S.A.
Distribution: U.S.A. (California).

Microcreagris luzonica Beier

Microcreagris luzonica Beier, 1931c: 300, fig. 1; Beier, 1932a: 149–150, fig. 181; Roewer, 1937: 251.

Type locality: Kalinga Province, Luzon, Philippines.
Distribution: Philippines.

Microcreagris macilenta (E. Simon)

Obisium macilentum E. Simon, 1878b: 157–158; Banks, 1895a: 12; Banks, 1904b: 364; Coolidge, 1908: 113; Banks, 1911: 639, fig. 210g; Moles, 1914c: 195; Moles & Moore, 1921: 9.
Microcreagris duncani J. C. Chamberlin, 1930: 33, fig. 3b; J. C. Chamberlin, 1931a: figs 1a, 1c–d, 1f, 1h–j, 1m–n, 1q, 17j (synonymized by Beier, 1932a: 154).
Microcreagris macilenta (E. Simon): Beier, 1932a: 154, fig. 186; Roewer, 1937: 252; Hoff, 1958: 10.
Not *Microcreagris macilentum* (E. Simon): J. C. Chamberlin, 1930: 31 (misidentification; see *Fissilicreagris chamberlini* (Beier)).

Type localities: of *Obisium macilentum*: Mariposa, California, U.S.A.
of *Microcreagris duncani*: Jasper Ridge, San Mateo County, California, U.S.A.
Distribution: U.S.A. (California).

Microcreagris macropalpus Morikawa

Microcreagris pygmaeum macropalpus Morikawa, 1955a: 217–219, figs 1a, 2a–b, 3b.
Microcreagris macropalpus Morikawa: Morikawa, 1960: 123, plate 4 fig. 9, plate 8

Family Neobisiidae

fig. 8, plate 9 fig. 10, plate 10 fig. 22; Morikawa, 1962: figs 1, 3; Sato, 1979a: plate 1, fig. 4; Sato, 1979b: 13, figs 1–3, 5; Sato, 1979c: 13, fig. 2; Sato, 1980c: 10, figs 1–2; Sato, 1980d: plate 1; Sato, 1982b: 58, figs 1–3; Sato, 1982c: 33; Sato, 1983b: 16, figs 2–4.

Microcreagris macloparpus (sic) Morikawa: Sato, 1979d: 44.

Type locality: Sugitate, near Matsuyama, Shikoku, Japan.
Distribution: Japan.

Microcreagris magna (Ewing)

Blothrus magnus Ewing, 1911: 76–77, 79, fig. 14; Banks, 1911: 640; Moles & Moore, 1921: 8.
Microcreagris (?) *magna* (Ewing): Hoff, 1958: 13.

Type locality: Shasta Springs, California, U.S.A.
Distribution: U.S.A. (California).

Microcreagris microdivergens Morikawa

Microcreagris microdivergens Morikawa, 1955a: 219–220, figs 2c–d, 3c; Morikawa, 1960: 123; Morikawa, 1962: fig. 1; W.K. Lee, 1981: 130–131, figs 3a–b, 4.
Microcreagris microdivergens microdivergens Morikawa: Morikawa, 1960: 124, plate 4 fig. 11, plate 6 fig. 10, plate 8 fig. 9. plate 9 fig. 11, plate 10 fig. 23.

Type locality: Omogo-kei, Shikoku, Japan.
Distribution: Japan, Korea.

Microcreagris nickajackensis Muchmore

Microcreagris nickajackensis Muchmore, 1966a: 98–100, figs 1–2; Muchmore, 1969a: 5–7.
'*Microcreagris*' *nickajackensis* Muchmore: Ćurčić, 1984b: 164, figs 19, 39.

Type locality: Nickajack Cave, 0.6 mile S. of Shellmound Station, Marion County, Tennessee, U.S.A.
Distribution: U.S.A. (Tennessee).

Microcreagris orientalis J. C. Chamberlin

Microcreagris orientalis J. C. Chamberlin, 1930: 34; Beier, 1932a: 150; Roewer, 1937: 251.

Type locality: Lookay, China?
Distribution: China?

Microcreagris pseudoformosa Morikawa

Microcreagris pseudoformosa Morikawa, 1955a: 221, figs 1b, 2i–j, 3e; Morikawa, 1960: 124, plate 4 fig. 10, plate 8 fig. 11, plate 9 fig. 14, plate 10 fig. 26; Sato, 1979a: 95–97.

Type locality: Omogo-kei, Shikoku, Japan.
Distribution: Japan.

Microcreagris pusilla Beier

Ideobisium formosanum Ellingsen, 1912a: 125–127; Beier, 1932a: 160; Roewer, 1937: 252 (junior secondary homonym of *Microcreagris granulata formosana* Ellingsen, 1912a).
Microcreagris pusilla Beier, 1937b: 268–269, fig. 1; Roewer, 1940: 344 (replacement name for *Ideobisium formosanum* Ellingsen).

Type locality: Koroton, Taiwan.
Distribution: Taiwan.

Microcreagris pygmaea Ellingsen

Microcreagris pygmaea Ellingsen, 1907c: 9–10; Kishida, 1915: 367 (not seen); Kishida, 1928: 412, fig. 3; Beier, 1932a: 149; Roewer, 1937: 251; Morikawa, 1960: 123; Sato, 1985: 22, figs 1–2; Sato, 1986: 95.

Type locality: Yamanaka, Suruga, Japan.
Distribution: Japan.

Microcreagris sequoiae J. C. Chamberlin

Microcreagris sequoiae J. C. Chamberlin, 1930: 28–29; J. C. Chamberlin, 1931a: figs 12a–b; Beier, 1932a: 147–148; Roewer, 1937: 252; Vachon, 1949: figs 202a–b; Hoff, 1958: 11.

Type locality: Muir Woods, Marin County, California, U.S.A.
Distribution: U.S.A. (California).

Microcreagris silvestrii J. C. Chamberlin

Microcreagris silvestrii J. C. Chamberlin, 1930: 27; J. C. Chamberlin, 1931a: fig. 40t; Beier, 1932a: 146; Roewer, 1937: 251.

Type locality: Ychyhan, China?
Distribution: China?

Microcreagris tacomensis (Ellingsen)

Ideobisium tacomense Ellingsen, 1909b: 219–220.
Microcreagris tacomensis (Ellingsen): Beier, 1932a: 154, fig. 185; Roewer, 1937: 252; Hoff, 1958: 10.

Type locality: Tacoma, Washington, U.S.A.
Distribution: U.S.A. (Washington).

Microcreagris thermophila J. C. Chamberlin

Microcreagris thermophila J. C. Chamberlin, 1930: 32; Beier, 1932a: 152; Roewer, 1937: 252; Hoff, 1958: 11.

Type locality: Box Springs Grade, near Riverside, California, U.S.A.
Distribution: U.S.A. (California).

Microcreagris theveneti (E. Simon)

Obisium theveneti E. Simon, 1878b: 156–157.
Ideobisium threveneti (sic) (E. Simon): Banks, 1895a: 11; Banks, 1900: 485; Banks, 1901b: 588; Banks, 1904b: 364; Banks, 1911: 639–640, fig. 210a; Moles, 1914c: 196; Moles & Moore, 1921: 8.
Ideobisium threventi (sic) (E. Simon): Coolidge, 1908: 113.
Microcreagris theveneti (E. Simon): Beier, 1932a: 147, fig. 147; Roewer, 1937: 252; Hoff, 1958: 10.

Type locality: Mariposa, California, U.S.A.
Distribution: U.S.A. (Alaska, Arizona, California, Oregon, Washington).

Microcreagris tibialis (Banks)

Ideobisium tibiale Banks, 1909b: 306.
Syarinus (?) *tibialis* (Banks): J. C. Chamberlin, 1930: 40; Beier, 1932a: 164.
Microcreagris (?) *tibialis* (Banks): Hoff, 1956b: 9; Hoff, 1958: 13.

Type locality: Florissant, Colorado, U.S.A.
Distribution: U.S.A. (Colorado).

Family Neobisiidae

Genus **Minicreagris** Ćurčić

Minicreagris Ćurčić, 1989c: 355.

Type species: *Microcreagris pumila* Muchmore, 1969a, by original designation.

Minicreagris pumila (Muchmore)

Microcreagris pumila Muchmore, 1969a: 2–4, figs 1–2.
'*Microcreagris' pumila* Muchmore: Ćurčić, 1984b: 164, figs 22–23, 42.
Minicreagris pumila (Muchmore): Ćurčić, 1989c: 355–357, figs 5, 11.

Type locality: Bryant Cave, 4 miles SE. of Blount Springs, Blount County, Alabama, U.S.A.
Distribution: U.S.A. (Alabama, Georgia, Tennessee).

Genus **Neobisium** J. C. Chamberlin

Obisium Illiger: Walckenaer, 1802: 252–253; Risso, 1826: 158; Théis, 1832: 63; C. L. Koch, 1839: 5; C. L. Koch, 1850: 5; Menge, 1855: 26; McIntire, 1868: 9; Stecker, 1874b: 239; Stecker, 1875d: 89; E. Simon, 1879a: 51–52; Tömösváry, 1882b: 219–220; Canestrini, 1885: no. 4; Daday, 1888: 127; O. P. Cambridge, 1892: 212; Tullgren, 1899a: 180; Tullgren, 1906a: 204; With, 1906: 76; Kew, 1911a: 52; Lessert, 1911: 27; Pratt, 1927: 410.
Blothrus Schiödte, 1847: 80; Schiödte, 1851a: 23; Schiödte, 1851b: 148–149; E. Simon, 1872: 223–224; Tömösváry, 1882b: 233; J. C. Chamberlin, 1930: 9; Harvey, 1985a: 367.
Obisium (Obisium) Illiger: Kew, 1911a: 54; Lessert, 1911: 29.
Obisium (Blothrus): Nonidez, 1925: 45–53.
Neobisium J. C. Chamberlin, 1930: 11–12; Beier, 1932a: 78; Beier, 1939d: 26; J. C. Chamberlin, 1935b: 480; Roewer, 1937: 245; Beier, 1939f: 167; Hoff, 1949b: 444; Petrunkevitch, 1955: 81; Morikawa, 1960: 112; Hoff, 1961: 427; Beier, 1963b: 81; Ćurčić, 1978b: 132; Legg & Jones, 1988: 75.

Type species: of *Blothrus*: *Blothrus spelaeus* Schiödte, 1847, by monotypy.
of *Neobisium*: *Obisium muscorum* Leach, 1817 (junior synonym of *Chelifer carcinoides* Hermann, 1804), by original designation.

Subgenus **Neobisium (Neobisium)** J. C. Chamberlin

Neobisium (Neobisium) J. C. Chamberlin: J. C. Chamberlin, 1930: 13; Beier, 1939d: 26.
Blothrus (Neobisium) J.C. Chamberlin: Harvey, 1985a: 367.

Neobisium (Neobisium) actuarium Ćurčić

Neobisium actuarium Ćurčić, 1984d: 125–127, figs 1–5; Schawaller & Dashdamirov, 1988: 25.

Type locality: Sal'girka, near Simferopl', Krym, Ukraine, U.S.S.R.
Distribution: U.S.S.R. (Ukraine).

Neobisium (Neobisium) aelleni Vachon

Neobisium (Neobisium) aelleni Vachon, 1976: 245–248, figs 1–8.

Type locality: Grotte SO3, Milchlöchii, Hochwald, Soleure, Switzerland.
Distribution: Switzerland.

Neobisium (Neobisium) agnolettii Beier

Neobisium (Neobisium) agnolettii Beier, 1973c: 229–230, fig. 3; Mahnert, 1979b: 260.

Type locality: Grotta di Korükini, Camlik Dalayman, Konya, Turkey.
Distribution: Turkey.

Neobisium (Neobisium) algericum (Ellingsen)

Obisium (Obisium) algericum Ellingsen, 1912c: 170–172.
Neobisium (Neobisium) algericum Ellingsen: Beier, 1932a: 88–89; Roewer, 1937: 246.

Type locality: Ifri Bou Arab, Aït Ali, Dra el Mizan (as Dra-el-Mizan), Algeria.
Distribution: Algeria.

Neobisium (Neobisium) alticola Beier

Neobisium (Neobisium) alticola Beier, 1973c: 226–227, fig. 1.
Neobisium alticola Beier: Schawaller, 1983c: 369; Schawaller & Dashdamirov, 1988: 11–12, figs 28, 33.

Type locality: Büyuk Agri Dagi (as Agri, Mt Ararat), Turkey.
Distribution: Iran, Turkey, U.S.S.R. (Azerbaijan).

Neobisium (Neobisium) anatolicum Beier

Neobisium (Neobisium) anatolicum Beier, 1949: 4–5, fig. 4; Beier, 1965f: 90; Rack, 1971: 114; Beier, 1973c: 225–227.
Neobisium anatolicum Beier: Beier, 1963g: 271; Beier, 1967f: 304; Beier, 1969b: 192; Schawaller & Dashdamirov, 1988: 12–13, figs 29–31, 33.

Type locality: Elazig, Anatolia, Turkey.
Distribution: Turkey, U.S.S.R. (Armenia, Azerbaijan, Georgia, R.S.F.S.R.).

Neobisium (Neobisium) apuanicum Callaini

Neobisium (Neobisium) apuanicum Callaini, 1981: 9–16, figs 1–10.

Type locality: Monte Altissimo, Alpi Apuane, Italy.
Distribution: Italy.

Neobisium (Neobisium) atlasense Leclerc

Neobisium (Neobisium) atlasense Leclerc, 1989: 46–50, figs 1–4.

Type locality: Jbel Ighnayne, Beni Mellal, Atlas Mountains, Morocco.
Distribution: Morocco.

Neobisium (Neobisium) balazuci Heurtault

Neobisium (Neobisium) balazuci Heurtault, 1969b: 955–960, figs 1–15; Leclerc & Heurtault, 1979: 242, map 3; Leclerc, 1982a: plate 2; Heurtault, 1986c: 22–23.
Neobisium balazuci Heurtault: Heurtault & Kovoor, 1980: 325–327.

Type locality: Labeaume, Ardèche, France.
Distribution: France.

Neobisium (Neobisium) beieri Verner

Neobisium (Neobisium) beieri Verner, 1958a: 198–199, fig. 1; Beier, 1963b: 101.
Neobisium beieri Verner: Krumpál, 1980: 23.

Type locality: Lubochna-Tal, Velká Fatra Mountains, Czechoslovakia.
Distribution: Czechoslovakia.

Neobisium (Neobisium) bernardi Vachon

Neobisium bernardi Vachon, 1937b: 40–42, figs 1–2; Vachon, 1938a: fig. 37e; Vachon, 1940g: 142.
Neobisium (Neobisium) bernardi Vachon: Roewer, 1940: 343; Heurtault, 1986c: 23; Mahnert, 1986d: 76–78.
Neobisium (Neobisium) bernardi bernardi Vachon: Beier, 1955n: 91; Beier, 1959f: 116; Beier, 1963b: 88–89.

Type locality: Lac d'Aumar, Hautes-Pyrénées, France.
Distribution: France, Spain.

Neobisium (Neobisium) bernardi franzi Beier

Neobisium (Neobisium) franzi Beier, 1955n: 92–94, fig. 2.
Neobisium (Neobisium) bernardi franzi Beier: Beier, 1959f: 116–118; Beier, 1963b: 89–90, fig. 84.
Neobisium bernardi franzi Beier: Beier, 1961b: 22.

Type locality: Bosque de Muniellos, NE. of Sierra de Ancares, Galicia, Spain.
Distribution: Portugal, Spain.

Neobisium (Neobisium) bernardi gennargentui Callaini

Neobisium (Neobisium) bernardi gennargentui Callaini, 1983c: 282–285, figs 1a–g.

Type locality: Arcu Correboi, Sardinia.
Distribution: Sardinia.

Neobisium (Neobisium) bernardi geronense Beier

Neobisium (Neobisium) bernardi geronense Beier, 1939f: 173–174, fig. 8; Beier, 1955n: 92; Beier, 1959f: 116; Beier, 1963b: 90; Lagar, 1972a: 18.

Type locality: Cova de las Feixassas near Collsacabra, Gerona, Spain.
Distribution: France, Spain.

Neobisium (Neobisium) biharicum Beier

Neobisium (Neobisium) biharicum Beier, 1939c: 10–14, figs 6–8; Beier, 1947a: 180, fig. 5v; Beier, 1963b: 89–90, fig. 123.
Neobisium biharicum Beier: Dumitrescu, 1976: 275.

Type locality: Vascau (as Vascáu), Bihor, Romania.
Distribution: Romania.

Neobisium (Neobisium) blothroides (Tömösváry)

Obisium blothroides Tömösváry, 1882b: 224–225, plate 3 figs 9–10.
Obisium (Obisium) praecipuum E. Simon: Daday, 1888: 130, 186; Ellingsen, 1910a: 400; Daday, 1918: 2; Beier, 1928: 299–300, fig. 5a (misidentifications).
Neobisium (Neobisium) blothroides (Tömösváry): Beier, 1932a: 103–104, fig. 117 (in part; see *Blothrus (Neobisium) gentile* (Beier)); Roewer, 1937: 247; Beier, 1947a: 181, figs 5w, 6; Beier, 1963b: 127, fig. 124; Ćurčić, 1974a: 16.

Type locality: Hungary.
Distribution: Hungary, Romania, Yugoslavia.

Neobisium (Neobisium) bosnicum Beier

Neobisium (Neobisium) bosnicum Beier, 1939d: 27–28, fig. 30.
Neobisium (Neobisium) bosnicum bosnicum Beier: Beier, 1963b: 86, fig. 81; Ćurčić, 1974a: 16.

Type locality: Bjelasnici Planini, Sarajevo, Yugoslavia.
Distribution: Yugoslavia.

Neobisium (Neobisium) bosnicum herzegovinense Beier

Neobisium (Neobisium) herzegovinense Beier, 1939d: 28–29, fig. 31.
Neobisium (Neobisium) bosnicum herzegovinense Beier: Beier, 1963b: 86–87, fig. 82; Ćurčić, 1974a: 16.

Type locality: Volujak, Yugoslavia.
Distribution: Yugoslavia.

Neobisium (Neobisium) bosnicum ondriasi Mahnert

Neobisium (Neobisium) bosnicum ondriasi Mahnert, 1973b: 29–32, figs 12–24.
Neobisium bosnicum ondriasi Mahnert: Schawaller, 1985b: 6.

Type locality: Aenos, Kephallinia, Greece.
Distribution: Greece.

Neobisium (Neobisium) boui Heurtault

Neobisium (Neobisium) boui Heurtault, 1969c: 1171–1173, figs 1–6; Heurtault, 1986c: 23.

Type locality: Mt Marcou, Herault, France.
Distribution: France.

Neobisium (Neobisium) brevidigitatum (Beier)

Obisium (Obisium) doderoi brevidigitatum Beier, 1928: 292.
Neobisium (Neobisium) doderoi brevidigitatum (Beier): Beier, 1932a: 98.
Neobisium (Neobisium) brevidigitatum (Beier): Beier, 1963b: 105.
Neobisium brevidigitatum (Beier): Kobakhidze, 1965b: 541; Kobakhidze, 1966: 703; Rafalski, 1967: 9; Krumpál, 1980: 23; Jędryczkowski, 1987a: 344, fig. 2.

Type locality: Kronstadt (now Brasov), Romania.
Distribution: Czechoslovakia, Poland, Romania, U.S.S.R. (Georgia).

Neobisium (Neobisium) bucegicum Beier

Neobisium (Neobisium) bucegicum Beier, 1964c: 210–211, fig. 1.

Type locality: Bucegi (= Bucecea?) Mountains, Romania.
Distribution: Romania.

Neobisium (Neobisium) caporiaccoi Heurtault-Rossi

Neobisium (Neobisium) caporiaccoi Heurtault-Rossi, 1966b: 606–626, figs 1–26, 29–37; Heurtault, 1969a: 1105–1108, figs 1–7; Heurtault, 1970c: 59; Heurtault & Jézéquel, 1970: figs 1–5; Heurtault, 1971d: 1981–1983, figs 1–5.
Neobisium caporiaccoi Heurtault-Rossi: Heurtault, 1975b: figs 5–13, 15; Heurtault, 1980d: 88.

Type locality: Ru-Fosch, Alleghe, Belluno, Italy.
Distribution: Italy.

Neobisium (Neobisium) carcinoides (Hermann)

Chelifer carcinoides Hermann, 1804: 118, plate 5 figs 6, s; Gervais, 1844: 82.
Obisium muscorum Leach, 1817: 51–52, plate 141 fig. 3; Théis, 1832: 66–67, plate 1 figs 4, 4a; C.L. Koch, 1843: 67–69, fig. 799; Löw, 1871: 842; L. Koch, 1873: 64–65; Ritsema, 1874: xxxiv; Stecker, 1874b: 241; Stecker, 1875d: 89; Canestrini, 1874: 291; L. Koch, 1877: 181; Lebert, 1877: 315 (not seen); L. Koch, 1878: 38; E. Simon, 1878a: 153; Daday, 1880a: 193; Tömösváry, 1882b: 230–232, plate 4 figs 6–7; Becker, 1884: cclxiv; Canestrini, 1884: no. 10, figs 1–5; H. J. Hansen, 1884: 552–553; E. Simon, 1885c: 50; Daday, 1889a: 16; Daday, 1889c: 25; Daday, 1889d: 81; O. P. Cambridge, 1892: 212–214, plate A figs 1a–f, plate B figs 6, 6a–b; H. J. Hansen, 1893: 208–211, plate 4 fig. 15, plate 5 figs 8–11; F. Müller & Schenkel, 1895: 821 (not seen); Krauss, 1896: 628; Cuní y Martornell, 1897: 339; E. Simon, 1899d: 86; Ellingsen, 1897: 16–18; E. Simon, 1898d: 21; Tullgren, 1899a: 180–181, plate 2 figs 4–9; E. Simon, 1900b: 594; Evans, 1901a: 54; Godfrey, 1901a: 118; Godfrey, 1901b: 215–216; Kew, 1901b: 193; Donisthorpe, 1902: 67; Ellingsen, 1903: 13; Kew, 1903: 298, 299; Godfrey, 1904: 195; Tullgren, 1906a: 204–205, fig. 6; With, 1906: 26–27, fig. 4a, plate 1, figs 3a–b; Ellingsen, 1907a: 171;

Godfrey, 1907: 163; E. Simon, 1907: 550; Strand, 1907: 243; Tullgren, 1907e: 246; Whyte & Whyte, 1907: 204; Ellingsen, 1908d: 70; Godfrey, 1908: 158–159; Gozo, 1908: 129 (not seen); Ellingsen, 1909a: 210; Kew, 1909a: 249; Kew, 1909b: 259; Ellingsen, 1910b: 63; Ellingsen, 1910c: 347–348; Godfrey, 1910: 23–28; Kew, 1910a: 64, 66, 68, 73; Kew, 1910b: 109–110; Ellingsen, 1911c: 174; Kew, 1911b: 2; Standen, 1912: 15; Kew, 1914: 107–108; Falconer, 1916: 158; Kew, 1916a: 124; Kew, 1916b: 80; Standen, 1916a: 17; Standen, 1917: 29; Daday, 1918: 2; Standen, 1918: 335; Navás, 1921: 166; Kästner, 1923: 247–252; Redikorzev, 1924b: 26, figs 2, 4; Navás, 1925: 124; Schenkel, 1926: 311, 313; Donisthorpe, 1927: 183; Pratt, 1927: 410; Bacelar, 1928: 190; Redikorzev, 1928: 124; Schenkel, 1928: 67, fig. 26; Väänänen, 1928a: 16–17, figs 2–3, 5(III), 6(1–2, 4); Kew, 1929a: 34; Schenkel, 1929a: 323; Schenkel, 1929b: 14; Beier, 1930e: 288; Kew, 1930: 253–256, fig. 1e; Bartels, 1931: 27; Szalay, 1931b: 371; Berland, 1932: fig. 50; Schenkel, 1933: 29; Bristowe, 1934: 8; Rabeler, 1951: 595; Cirdei & Gutu, 1959: 7, fig. 6; Bapsolle, 1967: 63 (synonymized by Beier, 1963b: 115).

Chelifer muscorum (Leach): Risso, 1826: 157.

Chelifer corticalis C. W. Hahn, 1834: 63, fig. 154; Latreille, 1837: plate 23 fig. 3; Gervais, 1844: 80; Hagen, 1870: 265 (synonymized by E. Simon, 1879a: 54).

Obisium carcinoides (Hermann): C. L. Koch, 1843: 65–67, fig. 798; Löw, 1867: 752; L. Koch, 1873: 64–65; Stecker, 1874b: 240; Stecker, 1875d: 89; L. Koch, 1877: 181; Becker, 1880a: xiv; Tömösváry, 1882b: 232–233, plate 4 fig. 3; Daday, 1889a: 16; Daday, 1889d: 81; Krauss, 1896: 628; E. Simon, 1898d: 21; E. Simon, 1900b: 594; Richters, 1902a: 17 (not seen); Tullgren, 1907e: 246; Daday, 1918: 2; Caporiacco, 1923: 132; Redikorzev, 1928: 124; Schenkel, 1928: 68, fig. 27; Cirdei & Gutu, 1959: 8, fig. 7.

Obisium tenellum C. L. Koch, 1843: 69–70, fig. 800 (synonymized by E. Simon, 1878a: 153).

Obisium gracile C. L. Koch, 1843: 73–75, figs 803–804; Hagen, 1870: 270 (synonymized by Beier, 1932a: 93).

Obisium sylvaticum C. L. Koch: Menge, 1855: 26–28, plate 1 figs 4, 9, 16–17, 20–23, plate 2 fig. 4, plate 3 fig. 3, plate 4 fig. 3 (misidentification).

Obisium (Obisium) muscorum Leach: E. Simon, 1879a: 54–55, plate 19 figs 6, 10, 14; Daday, 1888: 132, 188; Ellingsen, 1910a: 400; Kew, 1911a: 55, fig. 16; Lessert, 1911: 33–34, fig. 26; Nonidez, 1917: 41; Beier, 1928: 288–289, fig. 1c; Kästner, 1928: 11, figs 3, 5, 36, 42–43; Beier, 1929a: 359–360; Beier, 1929d: 155; Beier, 1931d: 92.

Obisium muscorum olivaceum E. Simon, 1879a: 55 (synonymized by Beier, 1932a: 93).

Obisium (Obisium) carcinoides (Hermann): Daday, 1888: 131, 187; Beier, 1928: 294; Kästner, 1928: 11, fig. 44; Beier, 1929c: 219–221, fig. 3.

Neobisium muscorum (Leach): J. C. Chamberlin, 1930: 13–14; J. C. Chamberlin, 1931a: 79, 95, 126, 129, fig. 46b; Kästner, 1931a: fig. 3; Beier, 1932a: fig. 12; Beier, 1932g: figs 172, 178–179, 200a, 209, 212, 229, 230a–b, 237; Fage, 1933: 54; Tumšs, 1934: 13; Caporiacco, 1936b: 328; Ionescu, 1936: 3–4; Roewer, 1936: figs 3–4, 70c, 73, 84, 105, 109–110, 114, 152–155; Vachon, 1936a: 78; Roewer, 1937: figs 156a, 157–158, 160, 184–186; Vachon, 1938a: 60–66, figs 34–36, 37g, 38e; Beier, 1939a: fig. 166b; Lohmander, 1939b: 288; Caporiacco, 1940: 7; Vachon, 1940g: 143; Kaisila, 1947: 86; Beier, 1948b: 456; Franz & Beier, 1948: 446, 447, 448, 471, 474, 495; Beier, 1950: 1002–1004, fig. 1; Gilbert, 1951: 551–552, fig. 3; G. O. Evans & Browning, 1954: 12, fig. 14; Beier, 1956h: 24; George, 1957: 80; Ressl & Beier, 1958: 2; Delany, 1960: 308–309; Strebel, 1961: 107; Meinertz, 1962a: 390–392, map 2; Höregott, 1963: 546, figs 1–2; Gabbutt, 1965: figs 48, 51, 54; Gabbutt & Vachon, 1965: 375–357, figs 1–38; Kobakhidze, 1965b: 541; Pedder, 1965: 108, 109, figs 3–4; Weygoldt, 1965c: 436; Kobakhidze, 1966: 703; H. R. Simon, 1966a: 79–83; Jennings, 1968: figs 1, 3; Legendre, 1968: figs 12–14;

Gabbutt, 1969a: 186, 187, 189, 192, figs 1d–f, 4d–f, 7d–f, 10c–d, 11d–f; Gabbutt, 1969b: 414–427; Gabbutt, 1969c: 230; H.R. Simon, 1969: 149–159, fig. 1; Weygoldt, 1969a: 26, 41, 42, 65, 70–72, 108, 118–120, 122, figs 8, 36a–d, 51, 62a–g, 63–64, 75–78, 83–85, 87, 99, 110; Gabbutt, 1970c: 8–9, figs 1, 3, 4a–b; Goodier, 1970a: 87; Goodier, 1970b: 98; Lazzeroni, 1970a: 207; Legg, 1971a: 476; Howes, 1972a: 108; Salmon, 1972: 66, 67; Legg, 1973b: 430, fig. 2k; Legg, 1975a: 66; Legg, 1975c: 124, figs 1a, 3a–h, 4, 7a, 8a–c, 14a, 15a–c, 16c, 17a–b; Crocker, 1976: 8–9; Ćurčić, 1976b: 176; Goddard, 1976a: 232–234; Goddard, 1976b: 295–298, 300–301, figs 1, 3; Jones, 1978: 93, 95; P. A. Wood & Gabbutt, 1978: 176–182, figs 1–5, 16–20; Bowman, 1979: 11–12; Goddard, 1979: 92–104, figs 1, 3; Jones, 1979a: 201; Judson, 1979a: 59; Judson, 1979b: 7; Rundle, 1979: 47; P. A. Wood & Gabbutt, 1979a: 285–292; P. A. Wood & Gabbutt, 1979b: 329–336; Heurtault, 1980d: 89; Jones, 1980c: map 10; Callaini, 1981d: 217–228, plates 1–5; Beck, 1983: fig. 9; Callaini, 1983f: 375–384, figs 1–15; Callaini, 1983d: 223; Eisenbeis & Wichard, 1987: fig. 46; Krumpál & Cyprich, 1988: 42; Callaini & Dallai, 1989: 89, fig. 7.

Neobisium (Neobisium) germanicum Beier, 1931a: 22; Beier, 1932a: 94, fig. 101; Roewer, 1937: 246; Beier, 1939e: 308; Beier, 1952e: 3; Beier & Franz, 1954: 455; Sanocka-Woloszynowa, 1981: 36 (synonymized by Beier, 1963b: 115).

Neobisium (Neobisium) muscorum (Leach): Beier, 1932a: 93–94, figs 91, 100 (in part; see *Neobisium (Neobisium) chironomus* (L. Koch)); Beier, 1934b: 284–285; Leruth, 1935: 26; Roewer, 1937: 246; Beier, 1939c: 7; Beier, 1939e: 308; Beier, 1939f: 174; Cooreman, 1946b: 8; Leleup, 1948: 9; Kaisila, 1949b: 77, map 2; Beier, 1952e: 3; Beier & Franz, 1954: 455–456; Beier, 1958d: 161; Beier, 1958c: 135, 137; Vachon & Gabbutt, 1964: figs 18–22; Lazzeroni, 1969a: 327; Lazzeroni, 1969b: 399–400; Lazzeroni, 1969c: 229–230; Crocker, 1978: 8; Callaini, 1979c: 349–350; Callaini, 1980a: 230–232, fig. 3; Gardini, 1980c: 106, 121, 123, 127; Mahnert, 1981a: 536; Callaini, 1982a: 17, 18, fig. 1b; Rafalski, 1981: 223; Callaini, 1986e: 249–251; Legg & Jones, 1988: 81–82, figs 5a, 6a–d, 17a, 17b(a–g).

Neobisium (Neobisium) muscorum muscorum (Leach): Beier, 1932a: 94; Hadži, 1939a: 193–196, figs 6a–e; Weidner, 1954b: 108–109; Rafalski, 1967: 9–10; Ćurčić, 1974a: 18; Sanocka-Woloszynowa, 1981: 36.

Neobisium (Neobisium) carcinoides (Hermann): Beier, 1932a: 94–95, fig. 102; Roewer, 1937: 246; Beier, 1939e: 309; Beier & Franz, 1954: 454; Beier, 1963b: 113–115, fig. 109; Beier, 1963h: 153–154; Beier, 1963i: 284; Ressl, 1965: 289; Rafalski, 1967: 10–11; Hammen, 1969: 18–19; Kofler, 1972: 287; Ressl, 1974: 30; Costantini, 1976: 123; Vachon, 1976: 244, 245; Drogla, 1977: 89; Mahnert, 1977d: 97, fig. 1; Leclerc & Heurtault, 1979: 241–242; Sanocka-Woloszynowa, 1981: 36–37; Hammen, 1983: 11; Mahnert, 1983d: 361–362, fig. 1.

Neobisium muscorum muscorum (Leach): Hadži, 1939b: 30–32, figs 7a–e; Caporiacco, 1948c: 240; Conci, 1951: 10, 38; Caporiacco, 1952a: 55.

Neobisium carcinoides (Hermann): Vachon, 1938a: figs 37f, 38b; Caporiacco, 1940: 6; Kölzel, 1963: 175, fig. 1; Lehtinen, 1964: 284; Helversen, 1966a: 137–138; Helversen, 1966b: fig. 3b; Zangheri, 1966: 530; Kölzel, 1967: 200; Turk, 1967: 153; Cîrdei, Bulimar & Malcoci, 1970: 8–10, fig. 3; Ressl, 1970: 250; Helversen & Martens, 1971: 382, fig. 4; Palmgren, 1973: 3–5; Klausen, 1975: 63–64; Schawaller, 1979: 13; Krumpál, 1980: 23–24; Schawaller, 1981b: 50; Almquist, 1982: 104; Hippa, Koponen & Mannila, 1984: 26, 27; Jędryczkowski, 1985: 79; Lippold, 1985: 40; Schawaller, 1985a: 3, figs 1–2; Schawaller, 1985b: 6; Braun & Beck, 1986: 140–148; Sacher & Breinl, 1986: 48; Bliss & Lippold, 1987: 43, 45; Jędryczkowski, 1987b: 140–141, map 4; Mahnert, 1988c: 163–166, figs 1–9, 17–19; Schawaller, 1989: 8.

Neobisium germanicum Beier: Caporiacco, 1948c: 240.

Microbisium dumicola (C. L. Koch): G. O. Evans & Browning, 1954: 10–12; Legg, 1971a: 476 (misidentifications, in part; see *Neobisium (Neobisium) maritimum* (Leach) and *Neobisium (Neobisium) sylvaticum* (C. L. Koch)).

Obisium (Neobisium) muscorum Leach: Andersson, 1961: 140.

Neobisium (Neobisium) muscorum germanicum Beier: Rafalski, 1967: 10.

Neobisium carcionides (sic) (Leach): Frøiland, 1976: 11–12.

Not *Obisium carcinoides* (Hermann): Théis, 1832: 68, plate 2 fig. 1 (misidentification; see *Neobisium (Neobisium) theisianum* (Gervais)); E. Simon, 1879a: 56–57, plate 19 fig. 8; Tullgren, 1900b: 159 (misidentifications; see *Neobisium (Neobisium) sylvaticum* (C. L. Koch)).

Not *Obisium muscorum* Leach: Dale, 1878: 325 (not seen) (misidentification; see *Lamprochernes nodosus* (Schrank)); Börner, 1902: 442, fig. 9 (misidentification; true identification unknown).

Type localities: of *Chelifer carcinoides*: Alsace, 7 km S. of Strasbourg, France.

of *Obisium muscorum*: Scotland, Great Britain.

of *Chelifer corticalis*: Nürnberg, West Germany.

of *Obisium tenellum*: Germany.

of *Obisium gracile*: Oberpflaz, West Germany.

of *Obisium (Obisium) muscorum olivaceum*: not stated, presumably France.

of *Neobisium (Neobisium) germanicum*: Muggendorf, West Germany.

Distribution: Algeria, Austria, Belgium, Bulgaria, Corsica, Czechoslovakia, Denmark, East Germany, Finland, France, Great Britain, Greece, Hungary, Ireland, Italy, Kenya, Morocco, Netherlands, Norway, Poland, Portugal, Romania, Sardinia, Spain, Sweden, Switzerland, Tunisia, U.S.S.R. (Georgia, Latvia, R.S.F.S.R., Ukraine), West Germany, Yugoslavia.

Neobisium (Neobisium) carcinoides balcanicum Hadži

Neobisium (Neobisium) muscorum balcanicum Hadži, 1937: 179–182, figs 14a–e; Roewer, 1940: 343; Ćurčić, 1974a: 18.

Neobisium (Neobisium) carpaticum Beier: Beier, 1963b: 111–112 (in part).

Neobisium muscorum balcanicum Hadži: Ćurčić, 1976b: 176.

Type locality: Topolka, near Veles, Yugoslavia.

Distribution: Yugoslavia.

Neobisium (Neobisium) carinthiacum Beier

Neobisium (Neobisium) carinthiacum Beier, 1939e: 309, fig. 4; Beier, 1952e: 3; Beier, 1956i: 8; Beier, 1963b: 113, fig. 108.

Neobisium carinthiacum Beier: Kölzel, 1967: 200; Palmgren, 1973: 5–6.

Type locality: Hochobir, Austria.

Distribution: Austria.

Neobisium (Neobisium) carpaticum Beier

Neobisium (Neobisium) carpaticum Beier, 1935g: 33–34, fig. 3; Beier, 1939c: 8, fig. 4; Beier, 1963b: 111–112, fig. 106 (in part; see *Neobisium (Neobisium) muscorum balcanicum* Hadži); Beier, 1964c: 210; Rafalski, 1967: 11; Rafalski, 1981: 223.

Neobisium (Neobisium) muscorum carpaticum Beier: Hadži, 1939a: 196–202, figs 7a–g, 8a–e; Roewer, 1940: 343.

Neobisium carpaticum Beier: Ćurčić, 1976b: 174; Dumitrescu, 1976: 275; Ćurčić, 1979c: 226; Ćurčić, 1980a: 10–11, figs 1, 2a–b; Krumpál, 1980: 24; Ćurčić, Krunić & Brajković, 1981: 281–282, figs 2a–c, 3a–b, 4a–c; Krunić & Ćurčić, 1981: 133; Ćurčić, 1982e: 5–6, figs 1–61; Ćurčić & Dimitrijević, 1982: 144–149, figs 1a–e, 2a–j, 3a–d; Ćurčić & Dimitrijević, 1983a: 362–365, figs 1–3; Ćurčić & Dimitrijević, 1983b: 283–284, fig. 1; Ćurčić, Krunić & Brajković, 1983: 245–247, figs 1a–g,

Family Neobisiidae

2a–e, 3, 4a–b; Ćurčić & Dimitrijević, 1984a: 9–10, fig. 1; Ćurčić & Dimitrijević, 1986b: 17–21, figs 1, 2a–f; Jędryczkowski, 1987a: 344, fig. 3; Ćurčić, 1989a: 79–85, figs 1a–h, 2a–f, 3a–e, 4a–f, 5a–f, 6a–d, 7a–f, 8a–f, 9a–f, 10a–c, 11, 12a–d.

Type locality: Muntii Bucegi, Romania.
Distribution: Czechoslovakia, Poland, Romania, Yugoslavia.

Neobisium (Neobisium) carpenteri (Kew)

Obisium carpenteri Kew, 1910a: 70 (nomen nudum).
Obisium carpenteri Kew, 1910b: 110; Kew, 1914: 97; Kew, 1916b: 80.
Obisium (Obisium) carpenteri Kew: Kew, 1911a: 55, fig. 17.
Neobisium (Neobisium) carpenteri Kew: Beier, 1932a: 90; Roewer, 1937: 246; Mackie, 1960: 109; Forcart, 1961: 50; Beier, 1963b: 98–99; Vachon & Gabbutt, 1964: fig. 4; Crocker, 1978: 8; Legg & Jones, 1988: 78–80, figs 16a, 16b(a–g).
Neobisium carpenteri Kew: G. O. Evans & Browning, 1954: 14; Mackie, 1960: 109; Gabbutt, 1965: 360–385, figs 1–5, 11–15, 21–25, 31–38, 47, 50, 53; Gabbutt, 1969a: 186, 187, 189, 192, figs 1g–i, 4g–i, 7g–i, 10a–b, 11g–i; Gabbutt, 1969b: 414–427; Legg, 1971a: 476–477; Legg, 1973b: 430, fig. 2l; Legg, 1975a: 66; Legg, 1975c: 124, figs 1c, 7b, 10a–b, 14b, 15d, 19a–b; Heurtault, 1980d: 88; Jones, 1980c: map 9.

Type locality: Glengariff, Cork, Ireland.
Distribution: Great Britain, Ireland.

Neobisium (Neobisium) carsicum Hadži

Neobisium (Neobisium) carsicum Hadži, 1933b: 146–152, figs 10a–d, 11a–c, 12a–d; Hadži, 1933c: 182–184; Roewer, 1937: 246; Beier, 1939d: 29, fig. 32; Beier, 1947a: 166–168, fig. 1a; Beier, 1963b: 118, fig. 114; Ćurčić, 1974a: 16; Christian & Potocnik, 1985: 15.

Type locality: Omisalj, Yugoslavia.
Distribution: Yugoslavia.

Neobisium (Neobisium) cavernarum (L. Koch)

Obisium cavernarum L. Koch, 1873: 55–56; Gestro, 1887: 497; E. Simon, 1907: 550–551; Ellingsen, 1908a: 418; Ellingsen, 1909a: 214; Navás, 1925: 123.
Obisium (Obisium) cavernarum L. Koch: E. Simon, 1879a: 61–62, 312, plate 19 fig. 7; Ellingsen, 1912c: 172; Nonidez, 1917: 41–42, figs 12, 12a–b.
Obisium (Obisium) myops E. Simon: Ellingsen, 1912c: 173 (misidentification).
Neobisium (Neobisium) cavernarum (L. Koch): Beier, 1932a: 104–105, fig. 119; Roewer, 1937: 247; Beier, 1939f: 174; Beier, 1963b: 75, fig. 126; Lagar, 1972a: 18; Leclerc & Heurtault, 1979: 242; Heurtault, 1986c: 23; Bellés, 1987: fig. 38.
Neobisium cavernarum (C. L. Koch): Brian, 1940: 401; Franciscolo, 1955: 125; Heurtault, 1975b: figs 2–4, 14, 16–19; Juberthie & Heurtault, 1975: 435–438, plates 1–6.

Type locality: Ariège (as Ariége), France.
Distribution: France, Spain.

Neobisium (Neobisium) cephalonicum (Daday)

Obisium (Obisium) cephalonicum Daday, 1888: 130–131, 186–187, plate 4 fig. 22; Beier, 1928: 304, fig. 7a; Beier, 1929a: 361; Beier, 1929e: 451.
Obisium cephalonicum Daday: Daday, 1889d: 81; Ellingsen, 1909a: 216.
Neobisium (Neobisium) cephalonicum (Daday): Beier, 1932a: 87, fig. 93; Roewer, 1937: 246; Hadži, 1939b: 26–27, figs 5a–c; Beier, 1963b: 83–84, fig. 78; Ćurčić, 1974a: 16.

Family Neobisiidae

Neobisium cephalonicum (Daday): Kobakhidze, 1965b: 541; Kobakhidze, 1966: 703;
Beier, 1967f: 304; Ćurčić, 1976b: 174; Ćurčić, 1979c: 226; Ćurčić, 1980a: 12, figs
5a–b, 6; Krunić & Ćurčić, 1981: 133; Ćurčić, 1982e: 18–19, figs 153–208; Ćurčić
& Dimitrijević, 1985: 94, figs 3a–d; Schawaller, 1985b: 6; Ćurčić & Dimitrijević,
1986b: 17–21.
Neobisium cepealonicum (sic) (Daday): Ćurčić, 1988b: 3, figs 16a–b.

Type locality: Kefallinia (as Cephalonia), Greece.
Distribution: Albania, Bulgaria, Greece, Romania, Turkey, U.S.S.R. (Georgia),
Yugoslavia.

Neobisium (Neobisium) chironomum (L. Koch)

Olpium chironomum L. Koch, 1873: 35–36.
Neobisium (Neobisium) carcinoides (Hermann): Beier, 1932a: 94–95 (in part).
Neobisium (Neobisium) chironomum (L. Koch): Mahnert, 1979e: 249–251, figs 1–6;
Callaini, 1986e: 242.
Neobisium (Neobisium) cf. *chironomum* (L. Koch): Mahnert, 1979e: figs 7–8.

Type locality: near Bolzano, Alto Adige, Italy.
Distribution: Italy.

Neobisium (Neobisium) concolor Ćurčić

Neobisium concolor Ćurčić, 1984d: 139–140, figs 54–58; Schawaller & Dashdamirov,
1988: 25.

Type locality: Zurnabad, U.S.S.R.
Distribution: U.S.S.R.

Neobisium (Neobisium) corcyraeum (Beier)

Obisium (Obisium) corcyraeum Beier, 1928: 296–298, fig. 3b.
Obisium sublaeve E. Simon: Beier, 1929e: 451 (misidentification).
Neobisium (Neobisium) corcyraeum (Beier): Beier, 1932a: 103, fig. 116; Roewer, 1937:
247; Beier, 1947a: 175–176, fig. 4p; Beier, 1963b: 115–116, fig. 111.
Neobisium corcyraeum (Beier): Schawaller, 1985b: 6.

Type locality: Kérkira (as Korfu), Greece.
Distribution: Greece.

Neobisium (Neobisium) crassifemoratum (Beier)

Obisium (Obisium) crassifemoratum Beier, 1928: 290–291, fig. 1a.
Obisium crassifemoratum Beier: Beier, 1930e: 288.
Neobisium (Neobisium) crassifemoratum (Beier): Beier, 1932a: 96, fig. 104; Roewer,
1937: 246; Verner, 1958b: 109–110, fig. 1; Beier, 1963b: 101, fig. 94; Beier, 1965f:
89, fig. 6; Rafalski, 1967: 8; Beier, 1973c: 224; Ćurčić, 1974a: 16.
Neobisium crassifemoratum (Beier): Beier, 1963g: 270; Kobakhidze, 1965b: 541;
Kobakhidze, 1966: 703; Beier, 1969b: 192; Ćurčić, 1976b: 174–175; Krumpál,
1980: 24; Schawaller, 1983b: 12–13, figs 29–32; Sacher & Breinl, 1986: 48; Sacher,
1987: 417; Krumpál & Cyprich, 1988: 42; Schawaller & Dashdamirov, 1988: 13–15,
fig. 32; Schawaller, 1989: 8.

Type locality: Retyezàt, Romania.
Distribution: Czechoslovakia, East Germany, Greece, Hungary, Poland, Romania,
Turkey, U.S.S.R. (Georgia, R.S.F.S.R., Ukraine), Yugoslavia.

Neobisium (Neobisium) cristatum Beier

Neobisium (Neobisium) cristatum Beier, 1959f: 119–120, fig. 2; Beier, 1963b: 103.
Type locality: Pantano, E. Orbaiceta, Navarra, Spain.
Distribution: Spain.

Neobisium (Neobisium) delphinaticum Beier

Neobisium (Neobisium) delphinaticum Beier, 1954a: 155−156, fig. 1; Beier, 1963b: 94−95, fig. 89; Vachon & Gabbutt, 1964: fig. 3; Vachon, 1976: fig. 9.

Type locality: Tête de l'Aure, Dauphiné, France.
Distribution: France.

Neobisium (Neobisium) distinctum (Beier)

Obisium (Obisium) distinctum Beier, 1928: 300−301, fig. 5b.
Neobisium (Neobisium) distinctum (Beier): Beier, 1932a: 102, fig. 114; Roewer, 1937: 247; Beier, 1947a: 177, fig. 4r; Beier, 1963b: 119−120, fig. 116; Heurtault-Rossi, 1966b: fig. 28; Lazzeroni, 1969b: 400; Ćurčić, 1974a: 16.
Neobisium distinctum (Beier): Caporiacco, 1940: 6−7; Cîrdei, Bulimar & Malcoci, 1970: 10, fig. 4.

Type locality: Pola, Italy.
Distribution: Italy, Romania?, Yugoslavia.

Neobisium (Neobisium) doderoi (E. Simon)

Obisium doderoi E. Simon, 1896: 373; E. Simon, 1898d: 21; E. Simon, 1900b: 594; Ellingsen, 1909a: 207, 210; Krausse-Heldrungen, 1912: 65; Ellingsen, 1913a: 455; Navás, 1921: 166; Navás, 1925: 122.
Obisium barrosi Navás, 1923: 33−34, fig. 13; Navás, 1925: 122, fig. 15; Bacelar, 1928: 190 (synonymized by Beier, 1932a: 97).
Obisium (Obisium) doderoi E. Simon: Ellingsen, 1910a: 396−397; Nonidez, 1917: 40; Beier, 1928: 291−292; Beier, 1929c: 215−217, fig. 1; Beier, 1929d: 155.
Obisium (Obisium) erythrodactylum mediterraneum Beier, 1929e: 451 (synonymized by Beier, 1963b: 107).
Neobisium (Neobisium) doderoi (E. Simon): Beier, 1932a: 97−98, fig. 106 (in part; see *Neobisium (Neobisium) ischyrus* (Navás); Roewer, 1937: 247; Beier, 1952e: 3; Beier & Franz, 1954: 454; Beier, 1963b: 107, fig. 101; Beier, 1963h: 154; Beier, 1963i: 284; Lazzeroni, 1969a: 326−327; Lazzeroni, 1969c: 229; Ćurčić, 1974a: 16; Callaini, 1979c: 349; Gardini, 1980c: 105, 125; Callaini, 1983c: 288.
Neobisium (Neobisium) doderoi doderoi (E. Simon): Beier, 1932a: 98.
Neobisium (Neobisium) erythrodactylum mediterraneum (Beier): Beier, 1932a: 99; Beier, 1958b: 27; Beier, 1958c: 137.
Neobisium (Neobisium) meridieserbicum Hadži, 1937: 182−187, figs 15a−e, 16a−g, 17a−f; Roewer, 1940: 343; Ćurčić, 1974a: 18 (synonymized by Beier, 1963b: 107).
Neobisium (Neobisium) barrosi (Navás): Roewer, 1940: 343.
Neobisium erythrodactylum mediterraneum (Beier): Caporiacco, 1949b: 122; Caporiacco, 1951e: 69.
Neobisium doderoi (E. Simon): Beier, 1963a: 256; Kobakhidze, 1965b: 541; Kobakhidze, 1966: 703; Zangheri, 1966: 530; Lazzeroni, 1970a: 207; Ćurčić, 1976b: 175; Paoletti, 1978: 91 (not seen).

Type localities: of *Obisium doderoi*: Boccadasse, Italy.
of *Obisium barrosi*: Leça, Portugal.
of *Obisium (Obisium) erythrodactylum mediterraneum*: Potamós, Kérkira (as Corfu), Greece; near Argostólion, Kefallinía, Greece; and Aenos, Greece.
of *Neobisium (Neobisium) meridieserbicum*: Pecina Fusa Spela, Kacanik, Yugoslavia.
Distribution: Albania, Algeria, Austria, France, Greece, Italy, Portugal, Sardinia, Sicily, Spain, Switzerland, U.S.S.R. (Georgia), Yugoslavia.

Neobisium (Neobisium) dolicodactylum (Canestrini)

Obisium dolicodactylum Canestrini, 1874: 229; Canestrini, 1885: no. 5, figs 1−7.
Obisium dolichodactylum (sic) Canestrini: Caporiacco, 1923: 132.

Neobisium (Neobisium) dolicodactylum (Canestrini): Beier, 1932a: 96 (in part; see *Neobisium (Neobisium) sublaeve* (E. Simon)); Roewer, 1937: 246; Beier, 1963b: 108–109, fig. 102; Beier, 1963h: 154; Lazzeroni, 1969b: 399; Lazzeroni, 1969c: 229; Gardini, 1980c: 105, 128; Callaini, 1986e: 242–247, figs 3a–d.
Neobisium dolicodactylum (Canestrini): Caporiacco, 1936b: 328; Vachon, 1940g: 142; Beier, 1948a: 188; Zangheri, 1966: 530–531.
Neobisium (Neobisium) carnicum Beier, 1938b: 79–80, fig. 2; Roewer, 1940: 343; Beier, 1947a: 177, fig. 4q; Beier, 1952e: 3; Beier, 1956i: 8 (synonymized by Beier, 1963b: 109).
Neobisium dolichodactylum (sic) (Canestrini): Caporiacco, 1947c: 252; Caporiacco, 1948c: 240.

Type localities: of *Obisium dolicodactylum*: Trentino, Italy.
of *Neobisium (Neobisium) carnicum*: Wolayer-See, Austria.
Distribution: Austria, Italy, Sardinia.

Neobisium (Neobisium) dolicodactylum latum Cîrdei, Bulimar & Malcoci

Neobisium dolicodactylum latum Cîrdei, Bulimar & Malcoci, 1967: 239–240.

Type locality: Masivul Repedea, Iasi, Romania.
Distribution: Romania.

Neobisium (Neobisium) dolomiticum Beier

Neobisium (Neobisium) dolomiticum Beier, 1952c: 58–59, fig. 2; Beier, 1963b: 93–94, fig. 88; Lazzeroni, 1969b: 398.
Neobisium dolomiticum Beier: Thaler, 1979: 52; Schawaller, 1981c: 159–160, figs 1–3.

Type locality: Pradidali, Italy.
Distribution: Austria, Italy, West Germany.

Neobisium (Neobisium) elegans Beier

Neobisium (Neobisium) elegans Beier, 1939d: 32, fig. 36; Beier, 1947a: 178, fig. 4s; Beier, 1963b: 120, fig. 117; Ćurčić, 1974a: 16.
Neobisium (Neobisium) aff. *elegans* Beier: Ćurčić, 1988b: 40–42, figs 76–80.

Type locality: Rogic Cave, Pazariste, Croatia, Yugoslavia.
Distribution: Yugoslavia.

Neobisium (Neobisium) erythrodactylum (L. Koch)

Obisium erythrodactylum L. Koch, 1873: 63–64; Stecker, 1874b: 240–241; Stecker, 1875d: 89; Tömösváry, 1882b: 229–230, plate 4 fig. 2; Daday, 1889c: 25; Daday, 1889d: 81; Ellingsen, 1909a: 216; Ellingsen, 1910b: 63; Daday, 1918: 2; Schenkel, 1928: 67, fig. 24; Cirdei & Gutu, 1959: 5–7, fig. 5.
Obisium (Obisium) erythrodactylum (L. Koch): Daday, 1888: 132, 188; Ellingsen, 1910a: 397; Beier, 1928: 289, fig. 1b; Kästner, 1928: 13; Beier, 1929e: 450–451.
Neobisium (Neobisium) erythrodactylum (L. Koch): Beier, 1932a: 98–99, fig. 108; Roewer, 1937: 247; Beier, 1952e: 3; Beier, 1963b: 102, fig. 96; Rafalski, 1967: 8–9; Ćurčić, 1974a: 17; Sanocka-Woloszynowa, 1981: 36.
Neobisium (Neobisium) erythrodactylum erythrodactylum (L. Koch): Beier, 1932a: 99.
Neobisium erythrodactylum (L. Koch): Vachon, 1936a: 78; Roewer, 1937: fig. 201c; Kobakhidze, 1965b: 541; Kobakhidze, 1966: 702–703; Lazzeroni, 1970a: 207; Krumpál, 1980: 24; Schawaller, 1980d: 54–56, fig. 1; Callaini, 1983e: 222–223; Schawaller, 1983b: 10–11, figs 24–28; Schawaller, 1983c: 368–369; Jędryczkowski, 1985: 78; Schawaller, 1985b: 6; Jędryczkowski, 1987a: 344; Jędryczkowski, 1987b: 139–140, map 3; Schawaller & Dashdamirov, 1988: 15–18.
Neobisium kelassuriense Kobakhidze, 1960d: 457–459, fig. 1; Kobakhidze, 1965b: 541; Kobakhidze, 1966: 702 (synonymized by Schawaller, 1983b: 10).

Neobisium (Neobisium) erithrodactylum (sic) (L. Koch): Lazzeroni, 1969a: 326.

Type localities: of *Obisium erythrodactylum*: Wroclaw (as Breslau), Poland; and Kraków (as Krakau), Poland.

of *Neobisium kelassuriense*: Kelassuri, Georgia, U.S.S.R.

Distribution: Austria, Czechoslovakia, East Germany, Greece, Hungary, Iran, Italy, Poland, Romania, U.S.S.R. (Armenia, Azerbaijan, Georgia, R.S.F.S.R., Ukraine), West Germany, Yugoslavia.

! Neobisium (Neobisium) exstinctum Beier

Neobisium exstinctum Beier, 1955g: 48–50, fig. 1; Schawaller, 1978: 3.

Type locality: Baltic Amber.
Distribution: Baltic Amber.

Neobisium (Neobisium) fiscelli Callaini

Neobisium fiscelli Callaini, 1983d: 233–238, figs 3a–h.

Type locality: Gran Sasso, Italy.
Distribution: Italy.

Neobisium (Neobisium) fuscimanum (C. L. Koch)

Obisium fuscimanum C. L. Koch, 1843: 63–64, fig. 796; L. Koch, 1873: 60–61; Stecker, 1874b: 240; Stecker, 1875d: 89; Ellingsen, 1905d: 5–6.

Obisium (Obisium) fuscimanum C. L. Koch: Ellingsen, 1910a; 397; Beier, 1928: 295; Beier, 1929c: 217–219, figs 2a–b; Beier, 1929e: 451.

Obisium wächtleri Kästner, 1928: 10, figs 37–39 (synonymized by Beier, 1928: 295).

Obisium waechtleri Kästner: Schenkel, 1928: 69.

Neobisium (Neobisium) fuscimanum (C. L. Koch): Beier, 1932a: 92–93, figs 99a–b; Roewer, 1937: 246; Beier, 1939c: 7–8; Beier, 1939e: 308, fig. 3; Hadži, 1939b: 27–30, figs 6a–g; Beier, 1952e: 3; Beier & Franz, 1954: 454–455; Beier, 1963b: 112–113, figs 107a–b; Beier, 1963h: 154; Ressl, 1965: 289; Kofler, 1972: 287; Ćurčić, 1974a: 17; Drogla, 1977: 89.

Neobisium fuscimanum (C. L. Koch): Ionescu, 1936: 4; Beier, 1948b: 456; Beier, 1950: 1004–1005, fig. 2; Beier, 1956h: 24; Ressl & Beier, 1958: 2; Ressl, 1970: 250; Palmgren, 1973: 6; Ćurčić, 1976b: 175; Dumitrescu, 1976: 275; Ćurčić, 1979c: 226, figs 1a–b; Ćurčić, 1980a: 11–12, figs 3a–b, 4a–d; Krumpál, 1980: 24; Ćurčić, 1982e: 14–18, figs 99–152; Ćurčić, Krunić & Brajković, 1983: 247, figs 2f–h; Schawaller, 1983b: 8–9, figs 18–20; Schawaller, 1983c: 368; Schawaller, 1985b: 6; Ćurčić & Dimitrijević, 1986b: 17–21; Sacher & Breinl, 1986: 48; Schawaller & Dashdamirov, 1988: 18, fig. 33; Ćurčić, 1989b: 3, figs 15a, c.

Neobisium bathumi Kobachidze, 1960c: 465–466, fig. 1; Kobakhidze, 1965b: 541; Kobakhidze, 1966: 702 (synonymized by Schawaller, 1983b: 8).

Neobisium fuscimanum ponticum Beier, 1963g: 269–270, fig. 2; Beier, 1969b: 192 (synonymized by Schawaller, 1983b: 8).

Neobisium (Neobisium) fuscimanum ponticum Beier: Beier, 1965f: 89; Rafalski, 1967: 12.

Neobisium (Neobisium) fuscimanum fuscimanum (C. L. Koch): Rafalski, 1967: 12.

Neobisium (Neobisium) ponticum Beier: Beier, 1973c: 224.

Neobisium fuscimanum fuscimanum (C. L. Koch): Jędryczkowski, 1987a: 345, fig. 5.

Type localities: of *Obisium fuscimanum*: not known.

of *Obisium waechtleri*: Vogtland, East Germany.

of *Neobisium bathumi*: Batumi (as Bathumi), Georgia, U.S.S.R.

of *Neobisium fuscimanum ponticum*: near Bolu, Turkey.

Distribution: Albania, Austria, Bulgaria, Czechoslovakia, East Germany, Greece, Hungary, Iran, Italy, Poland, Romania, Turkey, U.S.S.R. (Georgia), Yugoslavia.

Neobisium (Neobisium) galeatum Beier

Neobisium (Neobisium) galeatum Beier, 1953d: 36–38, fig. 3; Beier, 1956i: 8; Beier, 1963b: 85–86, fig. 80; Vachon & Gabbutt, 1964: figs 7, 10; Lazzeroni, 1969b: 397–398; Callaini, 1980a: 229–230; Gardini, 1980c: 105, 124; Callaini, 1982a: 17, 18.

Type locality: Caverna di Acquaviva N. 83, near Trento, Trentino, Italy.
Distribution: Austria, Italy.

Neobisium (Neobisium) gentile Beier

Obisium praecipuum E. Simon: Beier, 1928: 299 (misidentification, in part).
Neobisium (Neobisium) blothroides (Tömösváry): Beier, 1932a: 103 (misidentification, in part).
Neobisium (Neobisium) blothroides gentile Beier, 1939d: 30, fig. 33.
Neobisium (Neobisium) gentile gentile Beier: Beier, 1947a: 168–169, figs 1b–c, fig. 2; Beier, 1963b: 122–123, fig. 120; Ćurčić, 1974a: 17.

Type locality: Lastve, Rados Planina, Yugoslavia.
Distribution: Yugoslavia.

Neobisium (Neobisium) gentile alternum Beier

Neobisium (Neobisium) flavum alternum Beier, 1939d: 34–35, figs 39–40.
Neobisium (Neobisium) gentile alternum Beier: Beier, 1947a: 171; Beier, 1963b: 125, fig. 122; Ćurčić, 1974a: 17.

Type locality: Torina Cave, near Gradae, Yugoslavia.
Distribution: Yugoslavia.

Neobisium (Neobisium) gentile flavum Beier

Neobisium (Neobisium) flavum flavum Beier, 1939d: 33–34, fig. 38.
Neobisium (Neobisium) gentile flavum Beier: Beier, 1947a: 170, fig. 1f; Beier, 1963b: 124–125, fig. 121; Ćurčić, 1974a: 17.

Type locality: Giuppana Island, near Dubrovnik (as Ragusa), Yugoslavia.
Distribution: Yugoslavia.

Neobisium (Neobisium) gentile giganteum Beier

Neobisium (Neobisium) blothroides giganteum Beier, 1939d: 31, fig. 34.
Neobisium (Neobisium) gentile giganteum Beier: Beier, 1947a: 169–170, fig. 1d; Beier, 1963b: 123; Ćurčić, 1974a: 17.

Type locality: Brasina-Petrace, Dubrovnik (as Ragusa), Yugoslavia.
Distribution: Yugoslavia.

Neobisium (Neobisium) gentile novum Beier

Neobisium (Neobisium) novum Beier, 1939d: 31–32, fig. 35.
Neobisium (Neobisium) gentile novum Beier: Beier, 1947a: 170, fig. 1e; Beier, 1963b: 124; Ćurčić, 1974a: 17.

Type locality: Korito, SW. of Gacko Polje, Yugoslavia.
Distribution: Yugoslavia.

Neobisium (Neobisium) gineti Vachon

Neobisium (Neobisium) gineti Vachon, 1966a: 645–654, figs 1–15; Vachon, 1976: 245; Heurtault, 1986c: 23.

Type locality: St Julien, Labalme-sur-Cerdon, Ain, France.
Distribution: France, Switzerland.

Neobisium (Neobisium) golovatchi Schawaller

Neobisium golovatchi Schawaller, 1983b: 13–14, figs 36–40; Schawaller & Dashdamirov, 1988: 18–19, fig. 33.

Type locality: Saloniki, Lazarevskoye, Sochi, Krasnodar Bezirk, R.S.F.S.R., U.S.S.R.
Distribution: U.S.S.R. (R.S.F.S.R.).

Neobisium (Neobisium) gracilipalpe Beier

Neobisium (Neobisium) gracilipalpe Beier, 1939d: 32–33, fig. 37; Beier, 1947a: 171–172, figs 1g–h, 3; Beier, 1963b: 121–122, fig. 119; Ćurčić, 1974a: 17.

Type locality: Mlacka Jama, Popovo-Polje, Yugoslavia.
Distribution: Yugoslavia.

Neobisium (Neobisium) granulatum Beier

Neobisium (Neobisium) granulatum Beier, 1937a: 108–109, fig. 1; Rafalski, 1949: 92–98, figs 6–9; Beier, 1963b: 96, fig. 91.
Neobisium granulatum Beier: Kobakhidze, 1965b: 541; Kobakhidze, 1966: 702; Schawaller, 1983b: 13, figs 33–35; Ćurčić, 1984d: 136–137, figs 42–47; Schawaller & Dashdamirov, 1988: 19, fig. 33.

Type locality: north-west Caucasus, U.S.S.R.
Distribution: U.S.S.R. (Azerbaijan, Georgia, R.S.F.S.R.).

Neobisium (Neobisium) granulosum Beier

Neobisium (Neobisium) granulatum Beier, 1939c: 9–10, fig. 5 (junior primary homonym of *Neobisium (Neobisium) granulatum* Beier, 1937a); Roewer, 1937: 247.
Neobisium (Neobisium) granulosum Beier, 1963b: 95–96, fig. 90 (replacement name for *Neobisium (Neobisium) granulatum* Beier, 1939c).

Type locality: Sebes, Alba, Romania.
Distribution: Romania.

Neobisium (Neobisium) hellenum (E. Simon)

Obisium hellenum E. Simon, 1885a: 349–350; Ellingsen, 1910a: 400; Redikorzev, 1928: 124–125, fig. 2.
Neobisium (Neobisium) hellenum (E. Simon): Beier, 1932a: 87; Roewer, 1937: 246.
Neobisium hellenum (E. Simon): Caporiacco, 1948b: 34; Caporiacco, 1949b: 122; Schawaller, 1985c: 6.
Neobisium (Neobisium) sylvaticum hellenum (E. Simon): Beier, 1963b: 83.
Neobisium sylvaticum hellenum (E. Simon): Ćurčić, 1976b: 177.
Neobisium (Neobisium) aff. *hellenum* (E. Simon): Mahnert, 1980e: 231.

Type localities: Athens, Greece; and Náxos, Greece.
Distribution: Albania, Bulgaria, Crete, Greece.

Neobisium (Neobisium) helveticum Heurtault

Neobisium (Neobisium) helveticum Heurtault, 1971c: 903–906, figs 1–9; Vachon, 1976: 245.

Type locality: Poteux, Valais, Switzerland.
Distribution: Switzerland.

Neobisium (Neobisium) hermanni Beier

Neobisium (Neobisium) hermanni Beier, 1938b: 78–79, fig. 1; Roewer, 1940: 343; Beier, 1952e: 3; Beier, 1963b: 115, fig. 110; Kreissl, 1969: 43–44; Kofler, 1972: 287.

Type locality: Hermannshöhle, Kirchschlag am Wechsel, Austria.
Distribution: Austria.

Neobisium (Neobisium) hirtum Ćurčić

Neobisium hirtum Ćurčić, 1984d: 134–136, figs 37–41; Schawaller & Dashdamirov, 1988: 25.

Type locality: near Temnolesskaya Station, Stavropol'skaya, R.S.F.S.R., U.S.S.R.
Distribution: U.S.S.R. (R.S.F.S.R.).

Neobisium (Neobisium) improcerum Ćurčić

Neobisium improcerum Ćurčić, 1984d: 137–138, figs 48–53; Schawaller & Dashdamirov, 1988: 25.

Type locality: Sal'girka, near Simferopl', Krym, Ukraine, U.S.S.R.
Distribution: U.S.S.R. (Ukraine).

Neobisium (Neobisium) improvisum Redikorzev

Neobisium improvisum Redikorzev, 1949: 641–642, figs 1–2; Schawaller, 1989: 8, figs 20–21.

Type locality: Tau-chipik, Apmaatinskay, Kazakhstan, U.S.S.R.
Distribution: U.S.S.R. (Kazakhstan).

Neobisium (Neobisium) inaequale J. C. Chamberlin

Neobisium inaequalum (sic) J. C. Chamberlin, 1930: 14–15.
Neobisium (Neobisium) inaequalum (sic) J. C. Chamberlin: Beier, 1932a: 88; Roewer, 1937: 246.
Neobisium (Neobisium) inaequale J. C. Chamberlin: Beier, 1963b: 96.
Neobisium inaequale J. C. Chamberlin: Krumpál & Cyprich, 1988: 42.

Type locality: east Hungary.
Distribution: Czechoslovakia, Hungary.

Neobisium (Neobisium) incertum J. C. Chamberlin

Neobisium incertum J. C. Chamberlin, 1930: 14, figs 1k, 2ee; J. C. Chamberlin, 1931a: figs 13f–g, 17e, 35d, 43g–i; Beier, 1932g: fig. 213; Roewer, 1936: fig. 112a; Vachon, 1940g: 142; Beier, 1948a: 188–189, fig. 1; Vachon, 1949: fig. 203a.
Neobisium (Neobisium) incertum J. C. Chamberlin: Beier, 1932a: 90; Roewer, 1937: 246; Beier, 1963b: 97; Lazzeroni, 1969c: 228–229.

Type locality: Sorgono, Sardinia.
Distribution: Sardinia.

Neobisium (Neobisium) intermedium Mahnert

Neobisium (Neobisium) intermedium Mahnert, 1974b: 205–209, figs 1–6.

Type locality: Höhle Dupkata near Prolaz, Bulgaria.
Distribution: Bulgaria.

Neobisium (Neobisium) intractabile Beier

Neobisium (Neobisium) intractabile Beier, 1973c: 228–229, fig. 2.

Type locality: Tirebolu, Turkey.
Distribution: Turkey.

Neobisium (Neobisium) ischyrum (Navás)

Obisium ischyrum Navás, 1918: 119, 131–132, figs 5a–c; Navás, 1923: 33; Navás, 1925: 123, figs 16a–c.
Neobisium (Neobisium) doderoi (E. Simon): Beier, 1932a: 97 (in part).
Neobisium (Neobisium) ischyrum (Navás): Beier, 1939f: 169–170, fig. 5; Beier, 1953a: 295; Beier, 1955n: 95–96; Beier, 1959f: 120; Mahnert, 1986d: 80–81.

Neobisium (Neobisium) ischyrum ischyrum (Navás): Beier, 1963b: 99.
Neobisium ischyrum (Navás): Vachon, 1940g: 143; Beier, 1961b: 23.

Type locality: Zaragoza, Spain.
Distribution: Portugal, Spain.

Neobisium (Neobisium) ischyrum balearicum Beier

Neobisium (Neobisium) ischyrum balearicum Beier, 1939f: 170–172, fig. 6; Beier, 1963b: 99–100, fig. 93; Estany, 1977b: 30.
Neobisium ischyrum balearicum Beier: Beier, 1961b: 23.

Type locality: Buscatell, Ibiza, Balaeric Islands.
Distribution: Balaeric Islands.

Neobisium (Neobisium) juberthiei Heurtault

Neobisium (Neobisium) juberthiei Heurtault, 1986c: 23–24, fig. 5.

Type locality: Grotte Soubalère no. 1, Pérouse, Hautes-Pyrénées, France.
Distribution: France.

Neobisium (Neobisium) jugorum (L. Koch)

Obisium jugorum L. Koch, 1873: 66–67; Krauss, 1896: 628; E. Simon, 1900b: 594; Schenkel, 1928: 69, figs 29a–b; Schenkel, 1929b: 14; Schenkel, 1933: 17; Cirdei & Gutu, 1959: 8, fig. 8.
Obisium (Obisium) jugorum L. Koch: E. Simon, 1879a: 60–61, plate 19 fig. 4; Lessert, 1911: 35–36, fig. 28; Ellingsen, 1912c: 172–173; Beier, 1928: 294–295; Kästner, 1928: 11, fig. 41.
Obisium jugorum longipalpis Schenkel, 1928: 69, fig. 30 (synonymized by Beier, 1963b: 110).
Neobisium (Neobisium) jugorum (L. Koch): Beier, 1932a: 95, fig. 103; Roewer, 1937: 246; Beier, 1952e: 3; Beier, 1954a: 155; Verner, 1960: 167, fig. 1; Beier, 1963b: 110, fig. 104; Rafalski, 1967: 11; Lazzeroni, 1969c: 244; Callaini, 1986e: 247–252.
Neobisium (Neobisium) longipalpe (Schenkel): Beier, 1932a: 95; Roewer, 1937: 246; Forcart, 1961: 50.
Neobisium jugorum (L. Koch): Vachon, 1936a: 78; Vachon, 1937f: 128; Vachon, 1938a: figs 37b, 38g; Janetschek, 1948: 309–316, figs 1–6; Weidner, 1959: 116; Weygoldt, 1969a: 108; Palmgren, 1973: 6; Thaler, 1980: 391; Thaler, 1981: 101.
Neobisium (Neobisium) muscorum jugorum (L. Koch): Ćurčić, 1974a: 18.

Type localities: of *Obisium jugorum*: La Grance, France; and Stubaier (as Stubayer) Alpen, Tirol, Austria.
of *Obisium jugorum longipalpis*: Albristhorn near Adelboden, Berner Oberland, Switzerland.
Distribution: Austria, France, Italy, Switzerland.

Neobisium (Neobisium) kobachidzei Beier

Neobisium (Neobisium) kobachidzei Beier, 1962b: 146–147, fig. 1; Beier, 1963b: 106–107, fig. 100; Beier, 1973c: 224.
Neobisium kobachidzei Beier: Kobakhidze, 1965b: 541; Kobakhidze, 1966: 701–702; Schawaller, 1983b: 9–10, figs 21–23; Schawaller & Dashdamirov, 1988: 19.

Type locality: near Kurorts Ledarde, Podstilska Grade, Georgia, U.S.S.R.
Distribution: Turkey, U.S.S.R. (Azerbaijan, Georgia, R.S.F.S.R.).

Neobisium (Neobisium) labinskyi Beier

Neobisium (Neobisium) labinskyi Beier, 1937a: 109, fig. 2; Roewer, 1937: 247; Beier, 1963b: 109, fig. 103.

Family Neobisiidae

Neobisium labinskyi Beier: Beier, 1962b: 146; Kobakhidze, 1965b: 541; Kobakhidze, 1966: 702; Beier, 1969b: 192; Schawaller, 1983b: 7, figs 8–11; Schawaller & Dashdamirov, 1988: 20–21, fig. 34.

Type locality: north-west Caucasus, U.S.S.R.
Distribution: Turkey, U.S.S.R. (Azerbaijan, Georgia, R.S.F.S.R.).

Neobisium (Neobisium) latens Ćurčić

Neobisium latens Ćurčić, 1984d: 130–131, figs 17–24; Schawaller & Dashdamirov, 1988: 25.

Type locality: Sal'girka, near Simferopl', Krym, Ukraine, U.S.S.R.
Distribution: U.S.S.R. (Ukraine).

Neobisium (Neobisium) lombardicum Beier

Neobisium (Neobisium) lombardicum Beier, 1934a: 56–58, fig. 5; Roewer, 1937: 246; Beier, 1942: 132; Beier, 1947a: 173, figs 1k–l; Beier, 1953d: 38; Beier, 1958d: 162; Tirini Pavan, 1958: 30 (not seen); Beier, 1966h: 175.
Neobisium (Neobisium) lombardicum lombardicum Beier: Beier, 1963b: 118–119, fig. 115; Heurtault-Rossi, 1966b: fig. 27; Costantini, 1976: 123; Gardini, 1980c: 105–106, 123, 124, 125, 127; Callaini, 1986e: 252.

Type locality: Grotte 125 Lo., Busati, near Brescia, Italy.
Distribution: Italy.

Neobisium (Neobisium) lombardicum martae (Menozzi), new combination

Obisium (Blothrus) martae Menozzi, 1920: 73–74.
Neobisium (Neobisium) lombardicum emiliae Beier, 1963b: 119; Gardini, 1980c: 105, 128. New synonymy.

Type locality: of *Obisium (Blothrus) martae*: Grotta di Santa Maria Maddalena, Emilia-Romagna, Italy.
of *Neobisium (Neobisium) lombardicum emiliae*: Grotta Vei, Monte Rosso, Emilia-Romagna, Italy.
Distribution: Italy.

Neobisium (Neobisium) macrodactylum (Daday)

Obisium (Obisium) macrodactylum Daday, 1888: 132–133, 189, plate 4 fig. 26; Ellingsen, 1910a: 397–398; Beier, 1928: 296, fig. 3a.
Obisium macrodactylum Daday: Daday, 1889a: 16; Ellingsen, 1909a: 216; Daday, 1918: 2; Redikorzev, 1928: 125, fig. 3.
Neobisium (Neobisium) macrodactylum (Daday): Beier, 1932a: 100–101, fig. 111; Roewer, 1937: 247; Beier, 1947a: 175, fig. 4o.
Neobisium (Neobisium) macrodactylum macrodactylum (Daday): Beier, 1963b: 116, fig. 112; Ćurčić, 1974a: 17.
Neobisium macrodactylum (Daday): Ćurčić, 1976b: 175–176; Ćurčić, 1979c: 226; Ćurčić, 1980a: 11; Krumpál, 1980: 24; Krunić & Ćurčić, 1981: 133; Ćurčić, 1982e: 6–14, figs 62–98; Ćurčić & Dimitrijević, 1985: 94–96, figs 4, 5a–b; Ćurčić & Dimitrijević, 1986b: 17–21; Schawaller & Dashdamirov, 1988: 21–22, fig. 34.

Type localities: Mehadia, Romania; and Kérkira (as Corfu), Greece.
Distribution: Bulgaria, Czechoslovakia, Hungary, Romania, U.S.S.R. (Azerbaijan), Yugoslavia.

Neobisium (Neobisium) macrodactylum montenegrense (Ellingsen)

Obisium (Obisium) montenegrense Ellingsen, 1910a: 399–400.
Neobisium (Neobisium) montenegrense (Ellingsen): Beier, 1932a: 101, fig. 112; Roewer, 1937: 247; Beier, 1947a: 175.

Neobisium (Neobisium) macrodactylum montenegrense (Ellingsen): Beier, 1963b: 116–117; Ćurčić, 1974a: 17–18.
Neobisium macrodactylum montenegrense (Ellingsen): Ćurčić, 1976b: 176.

Type locality: Montenegro, Yugoslavia.
Distribution: Yugoslavia.

Neobisium (Neobisium) mahnerti Heurtault

Neobisium (Neobisium) mahnerti Heurtault, 1980d: 95, figs plate 2 figs 1–7.

Type locality: Val d'Escrins, Hautes Alpes, France.
Distribution: France.

Neobisium (Neobisium) mahnerti major Callaini

Neobisium (Neobisium) mahnerti major Callaini, 1982b: 449–452, figs 1a–e, 2a–e.

Type locality: Lago di Mela, Corsica.
Distribution: Corsica.

Neobisium (Neobisium) maritimum (Leach)

Phalangium acaroides Linnaeus: Montagu, 1815: 7–11, plate 2 fig. 4 (misidentification).
Obisium maritimum Leach, 1817: 52; Risso, 1826: 158; Hagen, 1870: 270; Grube, 1872: 119, plate 1 fig. 2 (not seen); Moniez, 1889: 104–105; O. P. Cambridge, 1892: 215–216, plate B figs 8, 8a–c; Imms, 1905: 231–232; Ellingsen, 1907a: 169–171; Godfrey, 1907: 163; Godfrey, 1908: 158; Godfrey, 1909: 161–163; Kew, 1909a: 249–250; Kew, 1910a: 65, 66, 67, 69; Kew, 1910b: 110–111; Kew, 1911b: 1–2; Standen, 1912: 15–16; Kew, 1914: 97; Kew, 1916a: 124; Kew, 1916b: 80–81; Standen, 1917: 29; Barnes, 1925: 746–748; Ewing, 1928: fig. 6b; Schenkel, 1928: 65, fig. 22.
Chelifer maritimus (Leach): Gervais, 1844: 83.
Obisium littorale Moniez, 1889: 104–109, figs 1–4; Ferronière, 1899: 137–138 (synonymized by Ellingsen, 1907a: 169).
Obisium (Obisium) maritimum Leach: Kew, 1911a: 55, fig. 18; Kästner, 1928: 12, fig. 47.
Neobisium (Neobisium) maritimum (Leach): Beier, 1932a: 89, fig. 95; Roewer, 1937: 246; Beier, 1963a: 84–85, fig. 79; Vachon & Gabbutt, 1964: fig. 9; Vachon, 1964a: fig. 10; Crocker, 1978: 8; Legg & Jones, 1988: 75–76, figs 15a, 15b(a–f).
Neobisium maritimum (Leach): Gilbert, 1951: 552–553; G. O. Evans & Browning, 1954: 12–14; Gabbutt, 1962: 87–88, fig. 1; Gabbutt, 1965: 360–385, figs 6–10, 16–20, 26–30, 39–46, 49, 52, 55; Kensler & Crisp, 1965: 509, 512, 515; Pedder, 1965: 108, 109, fig. 2; Gabbutt, 1966a: 338–343; Gabbutt, 1969a: 186, 187, 188, 192, figs 1a–c, 4a–c, 7a–c, 10e–f, 11a–c; Gabbutt, 1969b: 414–427, figs 1–2, 3a–h; Weygoldt, 1969a: 112, 115; Gabbutt, 1970a: 135–140, figs 1a–d, 3a, 4a; Hunt, 1970: 774–776; Legg, 1971a: 476, fig. 2b; Legg, 1975a: 66; Legg, 1975c: 124, figs 1b, 7c, 9a–b, 14c, 15e, 16d, 18a–b; Judson, 1979b: 7; Jones, 1980c: map 8.
Microbisium dumicola (C. L. Koch): G. O. Evans & Browning, 1954: 10–12; Legg, 1971a: 476 (misidentifications, in part; see *Neobisium (Neobisium) carcinoides* (Hermann) and *Neobisium (Neobisium) sylvaticum* (C. L. Koch)).

Type localities: of *Obisium maritimum*: west England, Great Britain.
of *Obisium littorale*: Portel, France.
Distribution: France, Great Britain, Ireland.

Neobisium (Neobisium) maroccanum (Beier)

Obisium (Obisium) maroccanum Beier, 1930f: 70–71, fig. 1.
Neobisium (Neobisium) maroccanum (Beier): Beier, 1932a: 98, fig. 107; Roewer, 1937:
    247; Beier, 1963b: 89–90, fig. 97.
Neobisium maroccanum (Beier): Beier, 1961a: 70; Beier, 1961b: 24; Beier, 1976a: 24;
    Callaini, 1988b: 37–42, figs 2a–i, 3a–b.
Neobisium sp. *maroccanum* (Beier)?: Callaini, 1983: 220.

Type locality: Chorf-el-Akab, Morocco.
Distribution: Algeria, Azores, Morocco.

Neobisium (Neobisium) maxvachoni Heurtault

Neobisium (Neobisium) vachoni Heurtault, 1968c: 315–318, figs 1–8 (junior primary
    homonym of *Neobisium (Blothrus) vachoni* Beier, 1939d).
Neobisium vachoni Heurtault: Heurtault & Kovoor, 1980: 325–327.
Neobisium (Neobisium) maxvachoni Heurtault, 1990a: 128 (replacement name for
    Neobisium (Neobisium) vachoni Heurtault, 1968).

Type locality: Bessèges, Gard, France.
Distribution: France.

Neobisium (Neobisium) medvedevi Ćurčić

Neobisium medvedevi Ćurčić, 1984d: 127–128, figs 6–11; Schawaller & Dashdamirov,
    1988: 25.

Type locality: Azerbaijan, U.S.S.R.
Distribution: U.S.S.R. (Azerbaijan).

Neobisium (Neobisium) minimum (Beier)

Obisium (Obisium) muscorum minimum Beier, 1928: 289.
Neobisium (Neobisium) muscorum minimum (Beier): Beier, 1932a: 94.
Neobisium muscorum minimum (Beier): Vachon, 1935d: 81; Caporiacco, 1951a:
    63.
Neobisium minimum (Beier): Mahnert, 1988c: 167–169, figs 10–16.

Type locality: Wochein (now Bohinska), Yugoslavia.
Distribution: Austria, France, Italy, Romania, Yugoslavia.

Neobisium (Neobisium) moreoticum Beier

Neobisium (Neobisium) moreoticum Beier, 1931a: 22–23; Beier, 1932a: 99–100,
    fig. 109; Roewer, 1937: 247; Rafalski, 1949: 98–100; Beier, 1963b: 106, fig.
    99.
Neobisium aff. *moreoticum* Beier: Schawaller, in Schmalfuss & Schawaller, 1984: 3–4,
    figs 1–3.
Neobisium moreoticum Beier: Schawaller, 1985b: 6.

Type locality: Kumani, Peloponnisos (as Morea), Greece.
Distribution: Greece, U.S.S.R. (Georgia).

Neobisium (Neobisium) nemorale (C. L. Koch)

Obisium nemorale C. L. Koch, 1839: 5; C. L. Koch, 1843: 67; C. L. Koch, 1850:
    5.
Neobisium (Neobisium) (?) *nemorale* (C. L. Koch): Beier, 1932a: 106; Roewer, 1937:
    248.

Type locality: Bayern, West Germany.
Distribution: West Germany.

Neobisium (Neobisium) nivale (Beier)

Obisium (Obisium) nivale Beier, 1929a: 360–361, fig. 12.
Neobisium (Neobisium) nivale (Beier): Beier, 1932a: 104, fig. 118; Roewer, 1937: 247; Beier, 1939f: 174; Beier, 1963b: 128–129, fig. 127.

Type locality: Sierra Nevada, Spain.
Distribution: Spain.

Neobisium (Neobisium) noricum Beier

Neobisium (Neobisium) noricum Beier, 1939e: 310, fig. 5; Beier, 1952e: 3; Beier, 1956i: 8; Beier, 1963b: 110–111, fig. 105.

Type locality: Hohentauern (as Hohe Tauern), Austria.
Distribution: Austria.

Neobisium (Neobisium) osellai Callaini

Neobisium osellai Callaini, 1983d: 228–233, figs 2a–h.

Type locality: Maiella, Italy.
Distribution: Italy.

Neobisium (Neobisium) pallens Ćurčić

Neobisium pallens Ćurčić, 1984d: 133–134, figs 31–36; Schawaller & Dashdamirov, 1988: 25.

Type locality: North Osetia, U.S.S.R.
Distribution: U.S.S.R.

Neobisium (Neobisium) parasimile Heurtault

Neobisium (Neobisium) parasimile Heurtault, 1986c: 24, fig. 6.

Type locality: Grotte d'Antheuil, Côte d'Or, France.
Distribution: France.

Neobisium (Neobisium) pauperculum Beier

Neobisium (Neobisium) pauperculum Beier, 1959f: 118–119, fig. 1; Beier, 1963b: 87.

Type locality: Unquera, Pechon, Spain.
Distribution: Spain.

Neobisium (Neobisium) peloponnesiacum (Beier)

Obisium (Obisium) peloponnesiacum Beier, 1928: 304–305, fig. 7b.
Neobisium (Neobisium) peloponnesiacum (Beier): Beier, 1932a: 87–88, fig. 94; Roewer, 1937: 246; Beier, 1963b: 83.
Neobisium peloponnesiacum (Beier): Schawaller, 1985b: 6.

Type locality: Deomobas, Morea, Greece.
Distribution: Greece.

Neobisium (Neobisium) percelere Ćurčić

Neobisium percelere Ćurčić, 1984d: 132–133, figs 25–30; Schawaller & Dashdamirov, 1988: 25.

Type locality: Kal'girskiy, North Osetia, U.S.S.R.
Distribution: U.S.S.R.

Neobisium (Neobisium) phitosi Mahnert

Neobisium (Neobisium) phitosi Mahnert, 1973b: 27–29, figs 1–11.
Neobisium phitosi Mahnert: Schawaller, 1985b: 6.

Type locality: Gipsberg Skopos, near Zakynthos, Zakynthos, Greece.
Distribution: Greece.

Neobisium (Neobisium) polonicum Rafalski

Neobisium polonicum Rafalski, 1936: 119 (not seen); Rafalski, 1937: 1–14, figs 1–3;
    Cîrdei, Bulimar & Malcoci, 1970: 10–11; Jędryczkowski, 1987a: 344–345, fig. 4.
Neobisium (Neobisium) polonicum Rafalski: Hadži, 1939a: 202–206, figs 10a–d;
    Roewer, 1940: 343; Beier, 1947a: 178, fig. 4t; Beier, 1963b: 125–126; Rafalski,
    1967: 11; Krumpál, 1979a: 429–435, figs 1–8.

Type locality: Ropy po Bukowine, Poland.
Distribution: Czechoslovakia, Poland, Romania.

Neobisium (Neobisium) praecipuum (E. Simon)

Obisium (Obisium) praecipuum E. Simon, 1879a: 59; E. Simon, 1885b: 213; E. Simon,
    1900b: 594; Ellingsen, 1909a: 214.
Obisium praecipuum E. Simon: Beier, 1928: 299 (in part; see Neobisium (Neobisium)
    gentile (Beier)).
Neobisium (Neobisium) praecipuum (E. Simon): Beier, 1932a: 101–102, fig. 113;
    Roewer, 1937: 247; Beier, 1947a: 179; Beier, 1963b: 127, fig. 125; Heurtault, 1968b:
    1077–1081, figs 1–10; Heurtault, 1968c: 318; Lazzeroni, 1969c: 230.
Neobisium praecipuum (E. Simon): Vachon, 1938a: figs 37d, 38c.
Neobisium (Neobisium) cfr. *praecipuum* (E. Simon): Callaini, 1983c: 285–288, figs
    2a–b.
Not *Obisium (Obisium) praecipuum* E. Simon: Daday, 1888: 130, 186; Ellingsen,
    1910a: 400; Daday, 1918: 2; Beier, 1928: 299–300, fig. 5a (misidentifications; see
    Neobisium (Neobisium) blothroides (Tömösváry)).

Type locality: Alpes de Haute Provence (as Basses-Alpes), France.
Distribution: France, Greece, Italy, Sardinia.

Neobisium (Neobisium) pyrenaicum Heurtault

Neobisium (Neobisium) pyrenaicum Heurtault, 1980d: 90–95, plate 1 figs 1–9.

Type locality: Lac Lanoux, Pyrénées Orientales, France.
Distribution: France.

! Neobisium (Neobisium) rathkii (C. L. Koch & Berendt)

Obisium rathkii C. L. Koch & Berendt, 1854: 96–97, fig. 96; Menge, 1855: 28–29,
    plate 4 fig. 4.
Obisium rathki C. L. Koch & Berendt: Menge, in C. L. Koch & Berendt, 1854: 97.
Neobisium rathkii (C. L. Koch & Berendt): Beier, 1937d: 304; Roewer, 1940: 331;
    Petrunkevitch, 1955: fig. 49(1); Schawaller, 1978: 3.

Type locality: Baltic Amber.
Distribution: Baltic Amber.

Neobisium (Neobisium) reductum Mahnert

Neobisium (Neobisium) reductum Mahnert, 1977c: 70–71, figs 10–15.

Type locality: Cova de la Vall. Boixols, Boumort, Lérida, Spain.
Distribution: Spain.

Neobisium (Neobisium) reitteri (Beier)

Obisium (Obisium) reitteri Beier, 1928: 292–294, fig. 2; Beier, 1931d: 92.
Neobisium (Neobisium) reitteri (Beier): Beier, 1932a: 96–97, fig. 105; Roewer, 1937:
    246; Beier, 1963b: 104; Beier, 1966f: 343.

Neobisium reitteri (Beier): Cîrdei, Bulimar & Malcoci, 1967: 238–239; Schawaller, 1985b: 6.

Type locality: Kumani, Morea, Greece.
Distribution: Crete, Greece, Romania.

Neobisium (Neobisium) ressli Beier

Neobisium (Neobisium) ressli Beier, 1965f: 87–88, fig. 5.
Neobisium ressli Beier: Beier, 1967f: 303–304; Schawaller, 1985b: 7.

Type locality: Maras (as Marasch), Turkey.
Distribution: Cyprus, Turkey.

Neobisium (Neobisium) rhodium Beier

Neobisium rhodium Beier, 1962a: 139–141, figs 1–2; Beier, 1966e: 161; Schawaller, 1985b: 7.
Neobisium (Neobisium) rhodium Beier: Beier, 1963b: 91–92, fig. 85.

Type locality: Lindos, Rhodes, Greece.
Distribution: Greece.

Neobisium (Neobisium) ruffoi Beier

Neobisium (Neobisium) ruffoi Beier, 1958c: 135–137, fig. 1; Beier, 1963b: 87–88, fig. 83; Vachon & Gabbutt, 1964: fig. 6; Lazzeroni, 1969a: 325–326.
Neobisium ruffoi Beier: Beier, 1963a: 256; Lazzeroni, 1970a: 207.
Neobisium cfr. *ruffoi* Beier: Callaini, 1983d: 224–228, figs 1a–h.

Type locality: Monte Cervialto, Monte Picentini, Campania, Italy.
Distribution: Italy, Sicily.

Neobisium (Neobisium) seminudum (Daday & Tömösváry)

Obisium seminudum Daday & Tömösváry, in Daday, 1880: 193; Tömösváry, 1882b: 228, plate 4 fig. 1; Daday, 1918: 2.
Neobisium (Neobisium) (?) *seminudum* (Daday & Tömösváry): Beier, 1932a: 106; Roewer, 1937: 248.

Type locality: Hungary.
Distribution: Hungary.

Neobisium (Neobisium) settei Callaini

Neobisium (Neobisium) settei Callaini, 1982b: 452–458, figs 3a–d.

Type locality: Lago di Mela, Corsica.
Distribution: Corsica.

Neobisium (Neobisium) simile (L. Koch)

Obisium simile L. Koch, 1873: 58; Canestrini, 1874: 229; Becker, 1879: clxiii; Tömösváry, 1882b: 225–226, plate 4 fig. 5; Becker, 1884: cclxiv; Daday, 1889d: 81; Lameere, 1895: 474; Becker, 1896: 334 (not seen); Strand, 1907: 243; Ellingsen, 1908d: 70; Ellingsen, 1909a: 214; Ellingsen, 1910c: 348; Ellingsen, 1911c: 174; Caporiacco, 1923: 132; Redikorzev, 1928: 124; Schenkel, 1928: 67, fig. 24; Beier, 1930e: 288; Schenkel, 1933: 29; Savory, 1964a: fig. 32; Savory, 1977: fig. 92.
Obisium (Obisium) simile (L. Koch): E. Simon, 1879a: 58–59, plate 19 figs 11, 15; Daday, 1888: 130, 186; Ellingsen, 1910a: 401; Lessert, 1911: 34–35, fig. 27; Ellingsen, 1912c: 173; Beier, 1928: 303; Kästner, 1928: 12, figs 2b, 4, 35, 40, 50.
Neobisium (Neobisium) simile (L. Koch): Beier, 1932a: 89–90, fig. 96; Beier, 1934b: 284–285; Leruth, 1935: 26; Roewer, 1937: 246; Cooreman, 1946b: 8; Beier, 1952e: 3; Beier, 1959f: 120–121, fig. 4; Beier, 1963b: 101–102, fig. 95; Vachon &

Gabbutt, 1964: fig. 17; Vachon, 1964a: figs 11–12; Rafalski, 1967: 11; Hammen, 1969: 17–18; Lazzeroni, 1969a: 326; Ćurčić, 1974a: 18; Mahnert, 1977d: 97; Vachon, 1976: 244, 245; Leclerc & Heurtault, 1979: 242; Hammen, 1983: 11; Heurtault, 1986c: 24.

Neobisium simile (L. Koch): Fage, 1933: 54; Vachon, 1936a: 78; Vachon, 1937f: 128; Vachon, 1938a: figs 37a, 38d, 57a–b, 58; Caporiacco, 1948c: 240; Caporiacco, 1951a: 63; Vachon, 1964b: 4841; Kobakhidze, 1965b: 541; Vitali-di Castri, 1965b: fig. 2; Helversen, 1966a: 137; Helversen, 1966b: figs 1b, 4f, 5b, 6b; Kobakhidze, 1966: 703; Zangheri, 1966: 530; Lazzeroni, 1970a: 207; Helversen & Martens, 1971: 382, fig. 4; Heurtault, 1975b: figs 20–21.

Type localities: Paris, France; Troyes, France; Sappey, France; and Corsica.

Distribution: Austria, Belgium, Corsica, France, Hungary, Netherlands, Poland, Spain, Switzerland, U.S.S.R. (Georgia), West Germany, Yugoslavia.

Neobisium (Neobisium) simoni (L. Koch)

Obisium simoni L. Koch, 1873: 54–55; Tömösváry, 1882b: 221–222, plate 3 fig. 8; Becker, 1884: cclxiv; Lameere, 1895: 474; Becker, 1896: 333 (not seen); Krauss, 1896: 628; E. Simon, 1898d: 21; E. Simon, 1900b: 593; E. Simon, 1907: 550; Strand, 1907: 243; Ellingsen, 1908a: 416–417; Ellingsen, 1908d: 70; Ellingsen, 1909a: 214; Daday, 1918: 2; Bacelar, 1928: 190; Schenkel, 1928: 64, figs 19a–b.

Obisium (Obisium) simoni L. Koch: E. Simon, 1879a: 53–54, plate 17 figs 14, 14a, plate 19 figs 3, 5; Lessert, 1911: 30–31, fig. 23; Beier, 1928: 295; Kästner, 1928: 12, fig. 46.

Obisium simonii L. Koch: Daday, 1880a: 193.

Neobisium (Neobisium) simoni (L. Koch): Beier, 1932a: 91, fig. 97; Roewer, 1937: 246; Beier, 1939e: 306, figs 1, 2a–b; Cooreman, 1946b: 8; Beier, 1952e: 3; Weidner, 1954b: 109; Beier, 1958c: 137; Beier, 1959f: 121; Leclerc & Heurtault, 1979: 242; Heurtault, 1986c: 24.

Neobisium simoni (L. Koch): Vachon, 1936a: 78; Vachon, 1938a: figs 37c, 38f; Vachon, 1940g: 143; Vachon, 1949: fig. 218a; Weidner, 1959: 116; Beier, 1961b: 23; Helversen, 1966a: 136; Weygoldt, 1969a: fig. 14a; Helversen & Martens, 1971: 382, fig. 4; Eisenbeis & Wichard, 1987: fig. 44a.

Neobisium (Neobisium) simoni simoni (L. Koch): Beier & Franz, 1954: 456; Beier, 1963b: 92–93, fig. 86; Rafalski, 1967: 8; Hammen, 1969: 18; Mahnert, 1986d: 78–80.

Type localities: Paris, France; Basses-Alpes, France; and Troyes, France.

Distribution: Austria, Belgium, France, Hungary, Italy, Netherlands, Poland, Portugal, Spain, Switzerland, West Germany.

Neobisium (Neobisium) simoni petzi Beier

Neobisium (Neobisium) simoni petzi Beier, 1939e: 307, figs 1, 2c–d; Beier, 1952e: 3; Beier & Franz, 1954: 456; Beier, 1956i: 8; Beier, 1963b: 93, fig. 87; Costantini, 1976: 123.

Type locality: Sengfengebirge, Austria.

Distribution: Austria, Italy.

Neobisium (Neobisium) simonioides Beier

Neobisium (Neobisium) simonioides Beier, 1965f: 85–87, fig. 4.

Neobisium simonioides Beier: Beier, 1967f: 303; Beier, 1969b: 192.

Type locality: Göksu Nehir (as Göksu-Klamm), Silifke, Toros Daglari (as Taurus), Turkey.

Distribution: Turkey.

Neobisium (Neobisium) speleophilum Krumpál

Neobisium speleophilum Krumpál, 1986: 163–168, figs 1–7; Schawaller & Dashdamirov, 1988: 25.

Type locality: Perlovodnyi Cave, Krasnodar, R.S.F.S.R., U.S.S.R.
Distribution: U.S.S.R. (R.S.F.S.R.).

Neobisium (Neobisium) speluncarium (Beier)

Obisium (Obisium) speluncarium Beier, 1928: 301–303, fig. 6.
Neobisium (Neobisium) speluncarium (Beier): Beier, 1932a: 105, fig. 120; Roewer, 1937: 247; Beier, 1939d: 29; Beier, 1947a: 172–173, fig. 1i; Beier, 1963b: 117–118, fig. 113; Ćurčić, 1974a: 18.

Type locality: 'Kaludjerovac-See (untere Plitvicer-See)', Yugoslavia.
Distribution: Yugoslavia.

Neobisium (Neobisium) spilianum Schawaller

Neobisium spilianum Schawaller, 1985b: 2–4, 7, figs 1–7.

Type locality: Moni Spilianis Cave, Sámos (as Samos), Greece.
Distribution: Greece.

Neobisium (Neobisium) strausaki Vachon

Neobisium strausaki Vachon, 1966a: 645 (nomen nudum).
Neobisium (Neobisium) strausaki Vachon, 1976: 249–251, figs 10–15; Heurtault, 1986c: 24.

Type locality: Grotte NE10, Grotte de Moron-Ouest, Planchettes, Neuchâtel, Switzerland.
Distribution: Switzerland.

Neobisium (Neobisium) sublaeve (E. Simon)

Obisium (Obisium) sublaeve E. Simon, 1879a: 60; Ellingsen, 1910a: 401; Nonidez, 1917: 40; Beier, 1928: 294; Beier, 1929d: 156.
Obisium sublaeve E. Simon: E. Simon, 1898d: 21; E. Simon, 1900b: 594; Ellingsen, 1905d: 6–8; Ellingsen, 1908a: 417; Ellingsen, 1909a: 210; Navás, 1918: 119; 41; Navás, 1919: 212; Navás, 1921: 166; Navás, 1925: 123; Bacelar, 1928: 190.
Neobisium (Neobisium) dolicodactylum (Canestrini): Beier, 1932a: 96 (in part).
Neobisium (Neobisium) sublaeve (E. Simon): Beier, 1963b: 107–108; Beier, 1974g: 163; Gardini, 1980c: 106, 132; Gardini, 1982c: 89.
Neobisium sublaeve (E. Simon): Puddu & Pirodda, 1973: 171; Schawaller, 1981b: 45, figs 2–4; Schawaller, 1985b: 7.
Not *Obisium sublaeve* E. Simon: Beier, 1929e: 451 (misidentification; see *Neobisium (Neobisium) corcyraeus* (Beier)).

Type locality: Corsica.
Distribution: Corsica, France?, Sardinia.

Neobisium (Neobisium) sylvaticum (C. L. Koch)

Obisium sylvaticum C. L. Koch, 1835: fasc. 132.1, fig.; C. L. Koch, 1843: 61–62, figs 794–795; Hagen, 1870: 270; L. Koch, 1873: 59; Daday, 1880a: 193; Tömösváry, 1882b: 222–224, plate 3 fig. 11; Krauss, 1896: 628; Ellingsen, 1905d: 2–3; Strand, 1907: 243; Ellingsen, 1908d: 70; Daday, 1918: 2; Schenkel, 1928: 65, fig. 21; Beier, 1930e: 288; Cirdei & Gutu, 1959: 3–4, fig. 3.
Obisium dumicola C. L. Koch, 1835: fasc. 132.2, fig.; C. L. Koch, 1843: 64–65, fig. 797; L. Koch, 1873: 62–63; L. Koch, 1877: 181; Daday, 1880a: 193; Tömösváry, 1882b: 226–227, plate 4 fig. 4; Daday, 1889c: 25; Daday, 1889d: 81; Krauss, 1896:

628; Ellingsen, 1905d: 3–4; Strand, 1907: 243; Ellingsen, 1908d: 70; Ellingsen, 1909a: 210, 216; Daday, 1918: 2; Navás, 1925: 124; Bacelar, 1928: 190; Schenkel, 1928: 66, fig. 23; Beier, 1930e: 288; Cirdei & Gutu, 1959: 4–5, fig. 4 (synonymized by Beier, 1963b: 83).

Obisium walckenaerii Théis, 1832: 68–69, plate 2 figs 2, 2a–c; Hagen, 1870: 271 (synonymized by Beier, 1929c: 219).

Obisium elimatum C. L. Koch, 1839: 5; C. L. Koch, 1843: 71–73, figs 801–802; C. L. Koch, 1850: 5 (synonymized by Beier, 1932a: 137).

Obisium dubium C. L. Koch, 1843: 75–76, figs 805 (synonymized by Beier, 1932a: 86).

Obisium sylvaticus (sic) C.L. Koch: Gervais, 1844: 82.

Obisium walknaerii (sic) Théis: Gervais, 1844: 83.

Chelifer dumicolus (C. L. Koch): Gervais, 1844: 83.

Obisium silvaticum (sic) C. L. Koch: Stecker, 1874b: 239–240; Stecker, 1875d: 89; L. Koch, 1877: 181; Ellingsen, 1910c: 348; Navás, 1925: 123; Beier, 1928: 303–304.

Obisium (Obisium) dumicola C. L. Koch: E. Simon, 1879a: 55–56, plate 9; Daday, 1888: 131, 187; Ellingsen, 1910a: 397; Nonidez, 1917: 39–40; Beier, 1928: 296; Kästner, 1928: 12; Beier, 1929a: 360.

Obisium carcinoides Hermann: E. Simon, 1879a: 56–57, plate 19 fig. 8; Tullgren, 1900b: 159 (misidentifications).

Obisium (Obisium) dumicola nitidum Daday, 1888: 131–132, 187–188, plate 4 fig. 24 (synonymized by Beier, 1932a: 137).

Obisium (Obisium) sylvaticum (C. L. Koch): Ellingsen, 1910a: 401; Lessert, 1911: 31–32, fig. 24; Nonidez, 1917: 39; Kästner, 1928: 12, figs 48–49.

Obisium (Obisium) silvaticum (sic) C. L. Koch: Beier, 1929c: 219–221, fig. 4; Beier, 1931d: 92.

Neobisium sylvaticum (C. L. Koch): Roewer, 1931: 42 (not seen); Beier, 1932g: fig. 216; Beier, 1950: 1005–1008, fig. 3; Beier, 1956h: 24; Ressl & Beier, 1958: 2; Strebel, 1961: 107; Beier, 1963g: 269; Kölzel, 1963: 175; Kobakhidze, 1965b: 541; Kobakhidze, 1966: 702; Zangheri, 1966: 530; Beier, 1969b: 191; Weygoldt, 1969a: 103, 109; Lazzeroni, 1970a: 207; Helversen & Martens, 1971: 382; Ćurčić, 1976b: 176–177; Ćurčić, 1979c: 226; Ćurčić, 1980a: 13, figs 7a–b; Ćurčić, Krunić & Brajković, 1981: 281, fig. 1; Ćurčić, 1982e: 23–24, figs 209–257; Schawaller, 1983b: 6, figs 5–7; Ćurčić & Dimitrijević, 1985: 96, figs 6a–b; Schawaller, 1985b: 7; Ćurčić & Dimitrijević, 1986b: 17–21; Sacher & Breinl, 1986: 48; Bliss & Lippold, 1987: 43, 44–45; Schawaller & Dashdamirov, 1988: 22–23, fig. 38; Ćurčić, 1989b: 1–2, figs 13a–d, 14a–b, 15b, d; Schawaller, 1989: 9.

Neobisium (Neobisium) sylvaticum (C. L. Koch): Beier, 1932a: 86–87, fig. 92; Roewer, 1937: 246; Beier, 1939e: 305; Beier, 1939f: 169; Rafalski, 1949: 85–92, figs 3–5; Beier, 1952e: 4; Beier & Franz, 1954: 456; J.C. Chamberlin & Malcolm, 1960: fig. 1c; Beier, 1965f: 85; Helversen, 1966a: 136; Lazzeroni, 1969a: 325; Kofler, 1972: 287; Ćurčić, 1974a: 18; Costantini, 1976: 123; Drogla, 1977: 89; Sanocka-Woloszynowa, 1981: 36.

Microbisium dumicola (Ellingsen): Beier, 1932a: 137–138, fig. 170; Caporiacco, 1936b: 329; Roewer, 1937: 249; Beier, 1939e: 311; Beier, 1939f: 187; Vachon, 1940g: 143; Costantini, 1976: 123.

Microbisium dumicola (C. L. Koch): Boldori, 1934: 59.

Neobisium silvaticum (sic) (C. L. Koch): Caporiacco, 1936b: 328–329.

Neobisium (Neobisium) sylvaticum inaculeatum Hadži, 1939a: 188–192, figs 4a–g, 5a–e; Hadži, 1939b: 24, figs 4a–h (synonymized by Schawaller, 1989: 9).

Neobisium (Neobisium) sylvaticum sylvaticum (C. L. Koch): Beier, 1963b: 82–83, fig. 77; Ressl, 1965: 289; Rafalski, 1967: 7–8; Gardini, 1980c: 106, 122, 123, 127.

Neobisium sylvaticum sylvaticum (C. L. Koch): Palmgren, 1973: 6; Jędryczkowski, 1987a: 343–344; Jędryczkowski, 1987b: 138, map 2.

Not *Obisium sylvaticum* C. L. Koch: Menge, 1855: 26–28 (misidentification; see *Neobisium (Neobisium) carcinoides* (Hermann)).
Not *Obisium sylvaticum* C. L. Koch: O. P. Cambridge, 1892: 214–215 plate B fig. 7 (misidentification; see *Roncus lubricus* L. Koch).
Not *Microbisium dumicola* (C. L. Koch): G. O. Evans & Browning, 1954: 10–12; Legg, 1971a: 476 (misidentifications; see *Neobisium (Neobisium) carcinoides* (Hermann) and *Neobisium (Neobisium) maritimum* (Leach)).

Type localities: of *Obisium sylvaticum*: Regensburg and Frauenholz, West Germany.
of *Obisium dumicola*: Regensburg and Gräfenberg, West Germany.
of *Obisium walckenaerii*: St Gobain, Aisne, France.
of *Obisium elimatum*: Bayern, West Germany.
of *Obisium dubium*: Nürnberg, West Germany.
of *Obisium (Obisium) dumicola nitidum*: Sinnaikö, Hungary.
of *Neobisium (Neobisium) sylvaticum inaculeatum*: near Brno, Czechoslovakia.

Distribution: Albania, Austria, Balearic Islands, Bulgaria, Corsica, Czechoslovakia, France, East Germany, Finland, Greece, Hungary, Italy, Poland, Portugal, Romania, Sardinia, Spain, Switzerland, Turkey, U.S.S.R. (Armenia, Georgia, Moldavia, R.S.F.S.R., Ukraine), West Germany, Yugoslavia.

Neobisium (Neobisium) theisianum (Gervais)

Obisium carcinoides (Hermann): Théis, 1832: 68, plate 2 fig. 1 (misidentification).
Chelifer theisianus Gervais, 1844: 82.
Neobisium (Neobisium) (?) *theisianum* (Gervais): Beier, 1932a: 106; Roewer, 1937: 248.

Type locality: Aisne, France.
Distribution: France.

Neobisium (Neobisium) trentinum Beier

Neobisium (Neobisium) trentinum Beier, 1931a: 23; Beier, 1932a: 102–103, fig. 115; Roewer, 1937: 247; Beier, 1947a: 179–180, fig. 5u.
Neobisium (Neobisium) trentinum trentinum Beier: Beier, 1958d: 161; Beier, 1963b: 120–121, fig. 118; Beier, 1963h: 154; Lazzeroni, 1969b: 400–403, figs 2–8; Costantini, 1976: 123; Callaini, 1980a: 232–233; Gardini, 1980c: 107, 124, 125, 126; Callaini, 1986e: 252.
Neobisium trentinum trentinum Beier: Conci, 1951: 23, 38; Caporiacco, 1952a: 55.

Type locality: Trentino, Italy.
Distribution: Italy.

Neobisium (Neobisium) trentinum ghidinii Beier

Neobisium (Neobisium) trentinum ghidinii Beier, 1942: 131–132, fig. 1; Beier, 1958d: 161–162; Beier, 1963b: 121; Costantini, 1976: 123; Gardini, 1980c: 106–107, 123, 125, 126.
Not *Neobisium trentinum ghidinii* Beier: Caporiacco, 1947d: 138 (misidentification; see *Neobisium (Blothrus) torrei* (E. Simon)).

Type locality: Büs del Trinàl (as Buco del Trinale), near Brescia, Italy.
Distribution: Italy.

Neobisium (Neobisium) schenkeli (Strand), new combination

Obisium simile cavicola Schenkel, 1926: 315–316, fig. 2 (junior primary homonym of *Obisium cavicola* Packard, 1884).
Obisium cavicola Schenkel: Schenkel, 1928: 68–69, figs 28a–b; Kästner, 1928: 13.
Obisium troglodytes Beier, 1928: 303 (replacement name for *Obisium simile cavicola* Schenkel) (junior primary homonym of *Obisium troglodytes* Schmidt, 1848).

Family Neobisiidae

Obisium schenkeli Strand, 1932: 195 (replacement name for *Obisium simile cavicola* Schenkel).

Neobisium (Neobisium) troglodytes (Beier): Beier, 1932a: 95; Roewer, 1937: 246; Aellen, 1952: 146; Forcart, 1961: 50; Beier, 1963b: 109–110; Vachon, 1976: 244, 245; Heurtault, 1986c: 24.

Type locality: near Dornach, Switzerland.
Distribution: Switzerland.

Neobisium (Neobisium) usudi Ćurčić

Neobisium (Neobisium) usudi Ćurčić, 1988b: 42–43, figs 81–85.

Type locality: central Dalmatia, Yugoslavia.
Distribution: Yugoslavia.

Neobisium (Neobisium) validum (L. Koch)

Obisium validum L. Koch, 1873: 56–57; Daday, 1918: 2.
Obisium (Obisium) validum (L. Koch): Daday, 1888: 130, 186.
Obisium (Obisium) caucasicum Beier, 1928: 298–299, fig. 4 (synonymized by Schawaller, 1983b: 7).
Neobisium (Neobisium) validum (L. Koch): Beier, 1932a: 88; Roewer, 1937: 246; Beier, 1955d: 212–213, fig. 1; Beier, 1963f: 185–186; Beier, 1965f: 88; Beier, 1973c: 224–225; Mahnert, 1974c: 378.
Neobisium (Neobisium) caucasicum (Beier): Beier, 1932a: 100, fig. 110; Roewer, 1937: 247; Rafalski, 1949: 100–107, figs 10–13; Beier, 1951d: 96; Beier, 1963b: 104–105, fig. 98.
Neobisium (Neobisium) turcicum Beier, 1949: 2–4, figs 2–3; Beier, 1963b: 96–97; Beier, 1965f: 88; Rack, 1971: 113 (synonymized by Schawaller, 1983b: 7).
Neobisium turcicum Beier: Beier, 1957b: 145; Beier, 1962b: 146; Kobakhidze, 1965b: 541; Kobakhidze, 1966: 702.
Neobisium baniskhevii Kobakhidze, 1960b: 239–240, fig. 1; Kobakhidze, 1965b: 541; Kobakhidze, 1966: 702 (synonymized by Schawaller, 1983b: 7).
Neobisium caucasicum (Beier): Beier, 1962b: 146; Kobakhidze, 1965b: 541; Kobakhidze, 1966: 702; Beier, 1971a: 359.
Neobisium validum (L. Koch): Beier, 1963g: 269; Beier, 1966e: 161–162; Beier, 1967f: 304; Beier, 1969b: 191; Beier, 1971a: 359; Schawaller, 1983b: 7–8, figs 12–17; Schawaller, 1983c: 369; Schawaller, 1985b: 7; Schawaller & Dashdamirov, 1988: 23–24; Schawaller, 1989: 9–10.
Neobisium validum turcicum Beier: Beier, 1967f: 304.
Neobisium (Neobisium) validum caucasicum (Beier): Beier, 1973c: 225.

Type localities: of *Obisium validum*: Syria.
of *Obisium (Obisium) caucasicum*: Ordubad, Armenia, U.S.S.R.
of *Neobisium (Neobisium) turcicum*: Honaz (as Honoz), near Denizli, Anatolia, Turkey.
of *Neobisium baniskhevii*: Baniskhev, Georgia, U.S.S.R.
Distribution: Iran, Israel, Lebanon, Syria, Turkey, U.S.S.R. (Armenia, Arzerbaijan, Georgia, Moldavia, R.S.F.S.R., Turkmenistan, Ukraine).

Neobisium (Neobisium) ventalloi Beier

Neobisium (Neobisium) ventalloi Beier, 1939f: 172–173, fig. 7; Roewer, 1940: 343; Beier, 1955n: 94–95, fig. 3; Beier, 1963b: 97–98, fig. 92; Lagar, 1972a: 18; Lagar, 1972b: 46; Mahnert, 1977c: 70; Bellés, 1987: fig. 37.
Neobisium ventalloi Beier: Beier, 1961b: 24.
Not *Neobisium ventalloi* Beier: Cîrdei, Bulimar & Malcoci, 1970: 8, fig. 2 (misidentification; true identification unknown).

Type locality: Baumas de Aros near Riells, Barcelona, Spain.
Distribution: Spain.

Neobisium (Neobisium) vilcekii Krumpál

Neobisium (Neobisium) vilcekii Krumpál, 1983c: 607–611, figs 1–7.
Neobisium vilcekii Krumpál: Schawaller & Dashdamirov, 1988: 25.

Type locality: Osetinsko, Karmadon, R.S.F.S.R., U.S.S.R.
Distribution: U.S.S.R. (R.S.F.S.R.).

Neobisium (Neobisium) zhiltzovae Ćurčić

Neobisium zhiltzovae Ćurčić, 1984d: 128–130, figs 12–16; Schawaller & Dashdamirov, 1988: 25.

Type locality: near Kara-Kala (as Kara-Kali), mouth of Aydere (as Ay-Dere) River, Turkmenistan, U.S.S.R.
Distribution: U.S.S.R. (Turkmenistan).

Subgenus Neobisium (Blothrus) Schiödte

Obisium (Blothrus) Schiödte: Daday, 1888: 127.
Neobisium (Blothrus) Schiödte: Beier, 1932a: 106; Beier, 1939d: 36–37; Beier, 1939f: 174–175.
Blothrus (Blothrus) Schiödte: Harvey, 1985a: 367.

Neobisium (Blothrus) abeillei (E. Simon)

Blothrus abeillei E. Simon, 1872: 224–226, fig. 10; L. Koch, 1873: 43–44; Berland, 1932: fig. 84.
Obisium (Blothrus) abeillei (E. Simon): E. Simon, 1879a: 68–69, 312, plate 19 fig. 12; Ellingsen, 1912c: 167; Nonidez, 1917: 32.
Obisium abeillei (E. Simon): Ellingsen, 1908a: 416; Navás, 1925: 120.
Blothrus abeilli (E. Simon): J. C. Chamberlin, 1930: 11; J.C. Chamberlin, 1931a: figs 28i, 35a–b; Vachon, 1936a: 78.
Neobisium (Blothrus) abeillei (E. Simon): Beier, 1932a: 107–108, fig. 124; Roewer, 1937: 248; Vachon, 1947c: figs 1, 4, 7, 9; J. C. Chamberlin & Malcolm, 1960: fig. 1d; Beier, 1963b: 138–139, fig. 139; Heurtault, 1986c: 26, fig. 1.
Not *Obisium abeillei* (E. Simon): Lebedinsky, 1904: 80, fig. 24 (misidentification; true identification unknown); Gozo, 1908: 129 (not seen) (misidentification; see *Roncus gestroi* Beier).

Type locality: Grotte d'Estellas, Ariège, France.
Distribution: France, Spain.

Neobisium (Blothrus) absoloni Beier

Neobisium (Blothrus) absoloni Beier, 1938a: 6.
Neobisium (Blothrus) absoloni absoloni Beier: Beier, 1939d: 48–49, fig. 52; Beier, 1963b: 171, fig. 179; Ćurčić, 1974a: 18.

Type locality: Vilina Jama, Lebrsnk Planina, Gacko Polje, Yugoslavia.
Distribution: Yugoslavia.

Neobisium (Blothrus) absoloni grande Beier

Neobisium (Blothrus) absoloni grande Beier, 1939d: 49–50, fig. 55; Beier, 1963b: 172; Ćurčić, 1974a: 18–19.

Type locality: Sertov, Gacko Polje, Yugoslavia.
Distribution: Yugoslavia.

Neobisium (Blothrus) absoloni tacitum Beier

Neobisium (Blothrus) absoloni tacitum Beier, 1939d: 49, fig. 54; Beier, 1963b: 172; Ćurčić, 1974a: 19.

Type locality: Borija Cave, near Krbljina, Yugoslavia.
Distribution: Yugoslavia.

Neobisium (Blothrus) albanicum (G. Müller)

Obisium (Blothrus) albanicum G. Müller, 1931: 126.
Neobisium (Blothrus) albanicum (G. Müller): Beier, 1939d: 68–69, fig. 82; Beier, 1963b: 163, fig. 170; Beier, 1970f: 320.

Type locality: Berat Cave, Paftali, Albania.
Distribution: Albania.

Neobisium (Blothrus) auberti Leclerc

Neobisium (Blothrus) auberti Leclerc, 1982a: 40–43, figs 1–7, plates 1–2; Heurtault, 1986c: 26.

Type locality: Sadoux à Pradelle, Drôme, France.
Distribution: France.

Neobisium (Blothrus) aueri Beier

Neobisium (Blothrus) aueri Beier, 1962i: 2–3, fig. 1; Beier, 1963b: 172–173, fig. 180; Beier, 1970f: 320; Thaler, 1980: 391.

Type locality: Almberg-Eis and Tropfstein Cave near Grundlsee, Austria.
Distribution: Austria.

Neobisium (Blothrus) babusnicae Ćurčić

Neobisium babusnicae Ćurčić, 1980c: 249–254, figs 1–6; Ćurčić, 1983b: fig. 2.

Type locality: Pecina Pripor Cave, Resnik, near Babusnica, Yugoslavia.
Distribution: Yugoslavia.

Neobisium (Blothrus) birsteini (Lapschoff)

Blothrus birsteini Lapschoff, 1940: 62–63, 73, figs 1–2; Kobakhidze, 1965b: 541; Kobakhidze, 1966: 703.
Neobisium (? Blothrus) birsteini (Lapschoff): Mahnert, 1979b: 263–264, figs 7–12.

Type locality: Höhle Tarkyladze, near Gudauten, Sukhumi, Georgia, U.S.S.R.
Distribution: U.S.S.R. (Georgia).

Neobisium (Blothrus) bolivari (Nonidez)

Obisium (Blothrus) bolivari Nonidez, 1917: 32–36, figs 8, 9a–c; Nonidez, 1925: 72–73, figs 11a–b.
Obisium bolivari Nonidez: Navás, 1925: 120.
Neobisium (Blothrus) bolivari (Nonidez): Beier, 1932a: 116, fig. 140; Roewer, 1937: 248; Beier, 1939f: 175; Beier, 1963b: 144–145, fig. 146; Beier, 1970f: 320; Mahnert, 1977c: 81–83, figs 32–33; Estany, 1980b: 526, map 1; Zaragoza, 1986: fig. 1.

Type locality: Cueva de Castromuriel, Castromuriel, near Castrobarco, Sierra de la Magdalena, Burgos, Spain.
Distribution: Spain.

Neobisium (Blothrus) boneti Beier

Neobisium (Blothrus) boneti Beier, 1931a: 11–12, fig. 3; Beier, 1932a: 107, fig. 123; Roewer, 1937: 247; Beier, 1939f: 175; Beier, 1963b: 137, fig. 137; Estany, 1980b: 526, map 1; Zaragoza, 1986: fig. 1; Bellés, 1987: fig. 39.

Type locality: Cueva de Mauloechea, Abaurrea Alta, Navarra, Spain.
Distribution: Spain.

Neobisium (Blothrus) breuili (Bolivar)

Obisium breuili Bolivar, 1924: 103–104, fig. 2; Navás, 1925: 121.
Obisium (Blothrus) breuili Bolivar: Nonidez, 1925: 79–81, figs 17, 18a–b.
Neobisium (Blothrus) breuili (Bolivar): Beier, 1932a: 119–120, fig. 146; Roewer, 1937: 248; Beier, 1939f: 177; Beier, 1963b: 140–141, fig. 142; Beier, 1970f: 320; Zaragoza, 1986: fig. 1; Bellés, 1987: figs 40, 42.
Neobisium (Blothrus) breuli (sic) (Bolivar): Estany, 1980b: 527, map 2.

Type locality: Cuevas de Martinchurito, Lecumberri, Navarra, Spain.
Distribution: Spain.

Neobisium (Blothrus) brevimanum (J. Frivaldsky)

Blothrus brevimanum J. Frivaldsky, 1865b: 80, fig. 3 (not seen).
Obisium (Obisium) brevipes (J. Frivaldsky): Beier, 1928: 306 (in part).
Neobisium (Blothrus) (?) brevimanum (J. Frivaldsky): Beier, 1932a: 124.

Type locality: ?
Distribution: ?

Neobisium (Blothrus) brevipes (J. Frivaldsky)

Blothrus brevipes J. Frivaldsky, 1865a: 38–40; Heyden, 1869: 59; Daday, 1880a: 192; Tömösváry, 1882b: 234–235; Daday, 1918: 2.
Obisium (Blothrus) brevipes (J. Frivaldsky): Daday, 1888: 128, 184; Beier, 1928: 306 (in part; see *Neobisium (Blothrus) brevimanum* J. Frivaldsky and *Neobisium (Blothrus) minutus* Tömösváry); Beier, 1929a: 14a; Beier, 1929e: fig. 5b.
Neobisium (Blothrus) brevipes (J. Frivaldsky): Beier, 1932a: 110–111, fig. 130; Roewer, 1937: 248; Beier, 1939c: 14–15, fig. 9; Beier, 1970f: 320.
Neobisium (Blothrus) brevipes brevipes (J. Frivaldsky): Beier, 1963b: 153–154.

Type localities: Pesterea dela Ferice (as Fericsei barhang), near Ferice, Beius; Pesterea Zemeilor din Onceasa (as Oncsásza barhang), near Budureasa, Beius; and Pesterea dela Fanate (as Fonáczai barhang), Fanate, Orodea, Romania.
Distribution: Hungary, Romania.

Neobisium (Blothrus) brevipes montanum Beier

Neobisium (Neobisium) brevipes montanum Beier, 1939c: 15–16, fig. 10.
Neobisium (Blothrus) brevipes montanum Beier: Beier, 1963b: 154, fig. 159.

Type locality: Pesterea dela Alun, Beius, Bihor, Romania.
Distribution: Romania.

Neobisium (Blothrus) caecum Beier

Neobisium (Neobisium) caecum Beier, 1939d: 35–36, fig. 41; Beier, 1947a: 173–174, fig. 1m.
Neobisium (Blothrus) caecum Beier: Beier, 1963b: 151, fig. 154; Ćurčić, 1974a: 19.

Type locality: Velika Zaba Planina, SW. of Popova-Polje, Yugoslavia.
Distribution: Yugoslavia.

Neobisium (Blothrus) carnae Beier

Neobisium (Blothrus) carnae Beier, 1938a: 7; Beier, 1970f: 320.
Neobisium (Blothrus) carnae carnae Beier: Beier, 1939d: 61–62, fig. 73; Beier, 1963b: 148, fig. 151; Ćurčić, 1974a: 18.

Type locality: Koaca, Rados Planina, N. of Popova-Polje, Yugoslavia.
Distribution: Yugoslavia.

Neobisium (Blothrus) carnae fraternum Beier

Neobisium (Blothrus) carnae fraternum Beier, 1939d: 62, fig. 74; Beier, 1963b: 148–149; Ćurčić, 1974a: 19.

Type locality: Dvogrlama, Rados Planina, Yugoslavia.
Distribution: Yugoslavia.

Neobisium (Blothrus) casalei Gardini

Neobisium (Blothrus) casalei Gardini, 1985a: 57–60, figs 10–16.

Type locality: Drako Trypa Cave, Vitina, Greece.
Distribution: Greece.

Neobisium (Blothrus) cervelloi Mahnert

Neobisium (Blothrus) cervelloi Mahnert, 1977c: 78–80, figs 25–31; Zaragoza, 1986: fig. 1.
Neobisium (Blothrus) aff. *cervalloi* Mahnert: Mahnert, 1977c: 80–81; Zaragoza, 1986: fig. 1.

Type locality: Avenc T-1, Sierra Arañonera, Monte Perdido, Torla, Aragón, Spain.
Distribution: Spain.

Neobisium (Blothrus) closanicum Dumitresco & Orghidan

Neobisium (Blothrus) closanicus Dumitresco & Orghidan, 1970b: 104–111, figs 15–25.

Type locality: Grotte de Closani, Romania.
Distribution: Romania.

Neobisium (Blothrus) coiffaiti Heurtault

Neobisium (Blothrus) coiffaiti Heurtault, 1986c: 26, fig. 4.

Type locality: Grotte de la Palle, Rieulhes, Hautes-Pyrénées, France.
Distribution: France.

Neobisium (Blothrus) creticum (Beier)

Obisium (Blothrus) creticum Beier, 1931d: 92–93, figs 1, 3a.
Neobisium (Blothrus) creticum (Beier): Beier, 1932a: 108–109, fig. 126; Roewer, 1937: 248; Beier, 1939d: 45; Beier, 1963b: 161–162, fig. 168.

Type locality: Katholiko-Höhle near Akrotiri, Crete.
Distribution: Crete.

Neobisium (Blothrus) dalmatinum Beier

Neobisium (Blothrus) dalmatinum Beier, 1938a: 7.
Neobisium (Blothrus) dalmatinum dalmatinum Beier: Beier, 1939d: 51–52, figs 57–58; Beier, 1963b: 149, fig. 152; Beier, 1970f: 320; Ćurčić, 1974a: 19.
Neobisium (Blothrus) dalmatinum aberrans Beier, 1938a: 7; Beier, 1939d: 52, fig. 59; Beier, 1963b: 149–150; Ćurčić, 1974a: 19 (synonymized by Ćurčić, 1988b: 43).
Neobisium dalmatinum Beier: Ćurčić, 1988b: 43–47, figs 86–125.

Type localities: of *Neobisium (Blothrus) dalmatinum dalmatinum*: Kraljeva Pecina, near Spalato, Yugoslavia.
of *Neobisium (Blothrus) dalmatinum aberrans*: Golubinka Jama, Mosor Planina, Yugoslavia.
Distribution: Yugoslavia.

Neobisium (Blothrus) deschmanni (G. Joseph)

Obisium deschmanni G. Joseph, 1882: 22.
Neobisium (Blothrus) (?) *deschmanni* (G. Joseph): Beier, 1932a: 124; Roewer, 1937: 248.

Type locality: Grotte von Luëg, Yugoslavia.
Distribution: Yugoslavia.

Neobisium (Blothrus) dinaricum Hadži

Neobisium (Blothrus) dinaricum Hadži, 1933b: 158–165, figs 18a–c, 19a–c, 20a–d, 21a–b; Hadži, 1933c: 186–189; Roewer, 1937: 248; Beier, 1970f: 320.
Neobisium (Blothrus) dinaricum dinaricum Hadži: Beier, 1939d: 53; Beier, 1963b: 157–158; Ćurčić, 1974a: 19.

Type locality: Crkvice, Yugoslavia.
Distribution: Yugoslavia.

Neobisium (Blothrus) dinaricum caligatum Beier

Neobisium (Blothrus) dinaricum caligatum Beier, 1938a: 7; Beier, 1939d: 54–55, fig. 63; Beier, 1963b: 159, fig. 164; Ćurčić, 1974a: 19.

Type locality: Krk, Yugoslavia.
Distribution: Yugoslavia.

Neobisium (Blothrus) dinaricum tartareum Beier

Neobisium (Blothrus) dinaricum tartareum Beier, 1938a: 7; Beier, 1939d: 53–54, fig. 62; Beier, 1963b: 158–159; Ćurčić, 1974a: 20.

Type locality: Napoda, Krivosije, Yugoslavia.
Distribution: Yugoslavia.

Neobisium (Blothrus) hadzii Beier

Neobisium (Blothrus) hadzii Beier, 1938a: 6; Beier, 1939d: 40, fig. 49; Beier, 1963b: 174–175, fig. 183; Ćurčić, 1974a: 20.

Type locality: Jezero Pecina, Popovo-Polje, Yugoslavia.
Distribution: Yugoslavia.

Neobisium (Blothrus) heros Beier

Neobisium (Blothrus) heros Beier, 1938a: 6; Beier, 1939d: 40–41, fig. 42; Beier, 1963b: 147, fig. 149; Beier, 1970f: 320; Ćurčić, 1974a: 20.

Type locality: Pecina u Kucericama am Orjen, Montenegro, Yugoslavia.
Distribution: Yugoslavia.

Neobisium (Blothrus) hians Mahnert

Neobisium (Blothrus) hians Mahnert, 1979b: 261–263, figs 1–6.

Type locality: Grotte In Dagi, Dösemealth (as Dosemealti), Antalya, Turkey.
Distribution: Turkey.

Neobisium (Blothrus) hiberum Beier

Neobisium (Blothrus) hiberum Beier, 1931a: 13–14, fig. 5; Beier, 1932a: 115, fig. 137; Roewer, 1937: 248; Beier, 1939f: 175; Beier, 1963b: 136, fig. 135; Zaragoza, 1986: fig. 1.

Type locality: Sima de la Raya, Tamajón, Madrid, Spain.
Distribution: Spain.

Family Neobisiidae

Neobisium (Blothrus) hypochthon Beier

Neobisium (Blothrus) hypochthon Beier, 1938a: 6; Beier, 1939d: 50–51, fig. 56; Beier, 1963b: 173, fig. 181; Vachon & Gabbutt, 1964: fig. 15; Ćurčić, 1974a: 20.

Type locality: Babic Pecina, Popovo Polje, Yugoslavia.
Distribution: Yugoslavia.

Neobisium (Blothrus) imbecillum Beier

Neobisium (Blothrus) imbecillum Beier, 1938a: 6; Beier, 1939d: 44, fig. 47; Beier, 1963b: 169, fig. 176; Ćurčić, 1974a: 20.

Type locality: Krbljina, Treskavica Planina, Yugoslavia.
Distribution: Yugoslavia.

Neobisium (Blothrus) infernum Beier

Neobisium (Blothrus) infernum Beier, 1938a: 7; Beier, 1939d: 59–60, fig. 69; Beier, 1963b: 155, fig. 161; Ćurčić, 1974a: 20.

Type locality: Jama kod Komasovice, near Krstaca, Yugoslavia.
Distribution: Yugoslavia.

Neobisium (Blothrus) insulare Beier

Neobisium (Blothrus) insulare Beier, 1938a: 7; Beier, 1939e: 311, fig. 6; Beier, 1963b: 161; Ćurčić, 1974a: 20; Mahnert, 1980g: 97; Christian & Potocnik, 1985: 17.
Neobisium insulare Beier: Ćurčić, 1988b: 49–50, figs 133–137.

Type locality: Insel Deglia, Yugoslavia.
Distribution: Yugoslavia.

Neobisium (Blothrus) jcanneli (Ellingsen)

Obisium (Blothrus) jeanneli Ellingsen, 1912c: 167–169; Nonidez, 1917: 36–37, figs 10a–c; Nonidez, 1925: 82–83, figs 19a–b.
Obisium jeanneli Ellingsen: Navás, 1925: 121.
Neobisium (Blothrus) jeanneli (Ellingsen): Beier, 1932a: 120, fig. 147; Roewer, 1937: 248; Beier, 1939f: 177; Beier, 1963b: 145, fig. 147; Beier, 1970f: 320; Mahnert, 1977c: 83–84, figs 34a–e; Zaragoza, 1986: fig. 1.

Type locality: Cueva del Pindal, Oviedo, Spain.
Distribution: Spain.

Neobisium (Blothrus) karamani (Hadži)

Obisium (Blothrus) karamani Hadži, 1929: 61–71, figs 1–6.
Neobisium (Blothrus) karamani (Hadži): Beier, 1932a: 114; Roewer, 1937: 248; Beier, 1939d: 67; Beier, 1963b: 156, fig. 162; Beier, 1970f: 320; Ćurčić, 1974a: 20; Ćurčić, 1974b: 203–211, figs 13–35.

Type locality: Patiska Pecina, Yugoslavia.
Distribution: Yugoslavia.

Neobisium (Blothrus) korabense Ćurčić

Neobisium korabense Ćurčić, 1982b: 146–149, figs 1a–e; Ćurčić, 1985b: 170–173, figs 1a–f, 2a–e.

Type locality: Pecina Torbeski Most II Cave, near Nicpur, Mavrovi Anovi, Yugoslavia.
Distribution: Yugoslavia.

Neobisium (Blothrus) kosswigi Beier

Neobisium (Blothrus) kosswigi Beier, 1949: 5–6, fig. 5; Beier, 1965f: 90; Rack, 1971: 114; Beier, 1973c: 225; Mahnert, 1979b: 261.

Type locality: Haci Akif Island, Beysehir Gölü, Anatolia, Turkey.
Distribution: Turkey.

Neobisium (Blothrus) kwartirnikovi Mahnert

Neobisium (Blothrus) kwartirnikovi Mahnert, 1972a: 383–388, figs 1–10, plate 1.
Type locality: Duhlata Höhle, Vitosa, S. of Sofia, Bulgaria.
Distribution: Bulgaria.

Neobisium (Blothrus) leruthi Beier

Neobisium (Blothrus) leruthi Beier, 1939c: 16–18, figs 11–12; Beier, 1963b: 154–155, fig. 160; Beier, 1970f: 320.
Type locality: Ghetarul de sub Zguràsti, Câmpeni, Turda, Romania.
Distribution: Romania.

Neobisium (Blothrus) lethaeum Beier

Neobisium (Blothrus) lethaeum Beier, 1938a: 7.
Neobisium (Blothrus) lethaeum lethaeum Beier: Beier, 1939d: 55–56, fig. 64; Beier, 1963b: 165, fig. 171; Ćurčić, 1974a: 20.
Neobisium (Blothrus) lethaeum Beier: Beier, 1970f: 320.
Neobisium lethaeum Beier: Ćurčić, 1988b: 47–48, figs 126–132.

Type locality: Grabovica Pecina, near Grepci, Yugoslavia.
Distribution: Yugoslavia.

Neobisium (Blothrus) lethaeum acherusium Beier

Neobisium (Blothrus) lethaeum acherusium Beier, 1938a: 7; Beier, 1939d: 57, fig. 66; Beier, 1963b: 165; Ćurčić, 1974a: 20.
Type locality: Ilijina Pecina, Trebinje, Yugoslavia.
Distribution: Yugoslavia.

Neobisium (Blothrus) lethaeum parvum Beier

Neobisium (Blothrus) lethaeum parvum Beier, 1938a: 7; Beier, 1939d: 56–57, fig. 65; Beier, 1963b: 164; Ćurčić, 1974a: 21.
Type locality: Mala Zaba Planina, NW. of Popovo-Polje, Yugoslavia.
Distribution: Yugoslavia.

Neobisium (Blothrus) lethaeum superbum Beier

Neobisium (Blothrus) lethaeum superbum Beier, 1938a: 7; Beier, 1939d: 58–59, fig. 68; Beier, 1963b: 164; Ćurčić, 1974a: 21.
Type locality: Sipun Cave, near Dubrovnik (as Ragusa), Yugoslavia.
Distribution: Yugoslavia.

Neobisium (Blothrus) longidigitatum (Ellingsen)

Obisium (Roncus) longidigitatum Ellingsen, 1908a: 416–418; Ellingsen, 1912a: 169.
Roncus (Roncus) longidigitatus (Ellingsen): Beier, 1932a: 129; Roewer, 1937: 249; Beier, 1963b: 186–187.
Neobisium (Blothrus) longidigitatum (Ellingsen): Heurtault, 1974b: 631–635, figs 1–5.
Neobisium (Blothrus) longedigitatum (Ellingsen): Heurtault, 1986c: 26.
Type locality: Grotte d'Istaürdy, Basses-Pyrénées, France.
Distribution: France.

Neobisium (Blothrus) maderi Beier

Neobisium (Blothrus) maderi Beier, 1938a: 8; Beier, 1939d: 66, fig. 79; Beier, 1963b: 162–163, fig. 169; Beier, 1970f: 320; Ćurčić, 1974a: 21.
Neobisium maderi Beier: Ćurčić, 1988b: 50–52, figs 138–163.

Type locality: Ledenica Pecina, Mosor, Yugoslavia.
Distribution: Yugoslavia.

Neobisium (Blothrus) maxbeieri Dumitresco & Orghidan

Neobisium (Blothrus) beieri Dumitresco & Orghidan, 1970b: 101–104, figs 9–14 (junior primary homonym of *Neobisium (Neobisium) beieri* Verner, 1958).
Neobisium (Blothrus) maxbeieri Dumitresco & Orghidan, 1972: 247 (replacement name for *Neobisium (Blothrus) beieri* Dumitresco & Orghidan).

Type locality: Grotte de Topolnita, Romania.
Distribution: Romania.

Neobisium (Blothrus) minutum (Tömösváry)

Blothrus minutus Tömösváry, 1882b: 235–237, plate 4 figs 11–13.
Obisium (Blothrus) brevipes J. Frivaldsky: Beier, 1928: 306 (in part).
Neobisium (Blothrus) minutum (Tömösváry): Beier, 1932a: 113–114, fig. 135; Roewer, 1937: 248; Beier, 1963b: 152–153, fig. 157; Rafalski, 1967: 12.
Neobisium (Blothrus) vulpinum Beier, 1936b: 85–86, fig. 1; Roewer, 1940: 343 (synonymized by Beier, 1963b: 153).

Type localities: of *Blothrus minutus*: Hungary.
of *Neobisium (Blothrus) vulpinum*: 'Schmiedeberg, East Germany', but type material mislabelled (Beier, 1963b: 153).
Distribution: Hungary, Poland.

Neobisium (Blothrus) monasterii Mahnert

Neobisium (Blothrus) monasterii Mahnert, 1977c: 74–78, figs 20–24; Zaragoza, 1986: fig. 1.

Type locality: Cova de Sa Campana, Escorca, Mallorca, Balearic Islands.
Distribution: Balearic Islands.

Neobisium (Blothrus) navaricum (Nonidez)

Obisium (Blothrus) navaricus (sic) Nonidez, 1925: 64–66, figs 5, 6a–b.
Obisium navaricum Nonidez: Navás, 1925: 120.
Neobisium (Blothrus) navaricum (Nonidez): Beier, 1932a: 108, fig. 125; Roewer, 1937: 248; Beier, 1939f: 175; Beier, 1963b: 138, fig. 138; Mahnert, 1977c: 73–74, figs 16–19; Estany, 1980b: 526, map 1; Zaragoza, 1986: fig. 1.
Neobisium (Blothrus) aff. *navaricum* (Nonidez): Zaragoza, 1986: fig. 1.

Type locality: Cueva de Malkorraundi, Gorriti, Navarra, Spain.
Distribution: Spain.

Neobisium (Blothrus) nonidezi (Bolivar)

Obisium nonidezi Bolivar, 1924: 101–103, fig. 1; Navás, 1925: 120.
Obisium (Blothrus) nonidezi Bolivar: Nonidez, 1925: 76–77, figs 14a–b.
Neobisium (Blothrus) nonidezi (Bolivar): Beier, 1932a: 117, fig. 142; Roewer, 1937: 248; Beier, 1939f: 176; Beier, 1963b: 141, fig. 143; Estany, 1980b: 527, map 2; Zaragoza, 1986: fig. 1.

Type locality: Cueva de Akelar, Lecumberri, Navarra, Spain.
Distribution: Spain.

Neobisium (Blothrus) occultum Beier

Neobisium (Blothrus) occultum Beier, 1938a: 7.
Neobisium (Blothrus) occultum occultum Beier: Beier, 1939d: 60–61, figs 70–71;
 Beier, 1963b: 160, fig. 167; Beier, 1970f: 320; Ćurčić, 1974a: 21.

Type locality: Lazaricka Pecina, near Gacko, Yugoslavia.
Distribution: Yugoslavia.

Neobisium (Blothrus) occultum sororium Beier

Neobisium (Blothrus) occultum sororium Beier, 1938a: 7; Beier, 1939d: 61, fig. 72;
 Beier, 1963b: 160–161; Ćurčić, 1974a: 21.

Type locality: Gacko Polje, near Miholjace, Yugoslavia.
Distribution: Yugoslavia.

Neobisium (Blothrus) odysseum (Beier)

Obisium (Blothrus) odysseum Beier, 1929e: 451–453, figs 4, 5a, 6.
Neobisium (Blothrus) odysseum (Beier): Beier, 1932a: 109, fig. 127; Roewer, 1937:
 248; Beier, 1939d: 59; Beier, 1963b: 159, fig. 165; Mahnert, 1975a: 179; Schawaller,
 1985b: 5, 6, figs 8–11.

Type locality: Pantokrator above Barbati, Kérkira (as Korfu), Greece.
Distribution: Greece.

Neobisium (Blothrus) ohridanum Hadži

Neobisium (Blothrus) ohridanum Hadži, 1940: 129–135, figs 1a–c, 2, 3a–b, 4a–c,
 5a–c; Beier, 1963b: 167–168; Ćurčić, 1974a: 21.

Type locality: Meckina, Ohrid, Yugoslavia.
Distribution: Yugoslavia.

Neobisium (Blothrus) peruni Ćurčić

Neobisium peruni Ćurčić, 1988b: 62–64, figs 261–272.

Type locality: Sonjina Jama Pothole, Lovka, Sveti Jure, Mt Biokovo, Yugoslavia.
Distribution: Yugoslavia.

Neobisium (Blothrus) phineum Beier

Neobisium (Blothrus) phineum Beier, 1938a: 6.
Neobisium (Blothrus) phineum phineum Beier: Beier, 1939d: 46–47, fig. 50; Beier,
 1963b: 170–171, fig. 178; Ćurčić, 1974a: 21.

Type locality: Jama Plandiste, Nevesinjsko Polje, Yugoslavia.
Distribution: Yugoslavia.

Neobisium (Blothrus) phineum extensum Beier

Neobisium (Blothrus) phineum extensum Beier, 1939d: 47, fig. 51; Beier, 1963b: 175;
 Ćurčić, 1974a: 21.

Type locality: Nevesinjsko Polje, Yugoslavia.
Distribution: Yugoslavia.

Neobisium (Blothrus) primitivum Beier

Neobisium (Blothrus) primitivum Beier, 1931a: 9–10, fig. 1; Beier, 1932a: 106, fig.
 121; Roewer, 1937: 247; Beier, 1939f: 175; Estany, 1980b: 526, map 1.
Neobisium (Blothrus) primitivum primitivum Beier: Beier, 1963b: 136–137, fig. 136a;
 Zaragoza, 1986: fig. 1.

Type locality: Cueva de Mairuelegorreta, Mt Gorbea, Alava, Spain.
Distribution: Spain.

Neobisium (Blothrus) primitivum primaevum Beier

Neobisium (Blothrus) primaevum Beier, 1931a: 10−11, fig. 2; Beier, 1932a: 107, fig. 122; Roewer, 1937: 247; Beier, 1939f: 175.
Neobisium (Blothrus) primitivum primaevum Beier: Beier, 1963b: 137, fig. 136b; Zaragoza, 1986: fig. 1.

Type locality: Cueva de San Roque, Utzcorta, Bilbao, Vizcaya, Spain.
Distribution: Spain.

Neobisium (Blothrus) princeps Ćurčić

Neobisium (Blothrus) princeps Ćurčić, 1974b: 194−200, figs 1−12.
Neobisium (Blothrus) cf. *princeps* Ćurčić: Gardini, 1985a: 60−64, figs 17−23.

Type locality: Kalina Dupka Cave, 3 km N. of Lazaropolje, Yugoslavia.
Distribution: Greece, Yugoslavia.

Neobisium (Blothrus) pusillum Beier

Neobisium (Blothrus) pusillum Beier, 1939d: 41−42, fig. 43; Beier, 1963b: 167, fig. 173; Vachon & Gabbutt, 1964: figs 1, 12; Ćurčić, 1974a: 21.

Type locality: Cerna Jama, near Planina, Yugoslavia.
Distribution: Yugoslavia.

Neobisium (Blothrus) reimoseri (Beier)

Obisium spelaeum Schiödte: Valle, 1911: 24 (misidentification) (not seen).
Obisium (Blothrus) reimoseri Beier, 1929a: 361−364, figs 13, 14b; Beier, 1929e: fig. 5c.
Neobisium (Blothrus) reimoseri (Beier): Beier, 1932a: 109−110, fig. 128; Roewer, 1937: 248.
Neobisium (Blothrus) reimoseri reimoseri (Beier): Beier, 1939d: 42; Beier, 1963b: 168, fig. 174; Ćurčić, 1974a: 21; Mahnert, 1974b: 210−212, figs 7−10; Gardini, 1980c: 108, 128.
Neobisium (Blothrus) reimoseri histricum Beier, 1939d: 42, fig. 44; Beier, 1963b: 169; Ćurčić, 1974a: 21 (synonymized by Mahnert, 1974b: 210).
Blothrus reimoseri (Beier): Caporiacco, 1949d: 139.

Type locality: of *Obisium (Blothrus) reimoseri*: Draga near Ponikve, Istra (as Istria), Yugoslavia.
of *Neobisium (Blothrus) reimoseri histricum*: Dimnica Jama, near Slivje, Istria, Yugoslavia.
Distribution: Yugoslavia.

Neobisium (Blothrus) reimoseri croaticum Beier

Neobisium (Blothrus) reimoseri croaticum Beier, 1939d: 43, fig. 45; Beier, 1963b: 169, fig. 175; Ćurčić, 1974a: 22.

Type locality: Spilja Pustinja, near Delnice, Croatia, Yugoslavia.
Distribution: Yugoslavia.

Neobisium (Blothrus) remyi Beier

Neobisium (Blothrus) remyi Beier, 1939d: 67−68, fig. 81; Beier, 1963b: 156−157, fig. 163; Beier, 1970f: 320; Ćurčić, 1974a: 22.

Type locality: Zupanska Pecina, near Gorjo-Selo, Yugoslavia.
Distribution: Yugoslavia.

Neobisium (Blothrus) robustum (Nonidez)

Obisium (Blothrus) robustus (sic) Nonidez, 1925: 62−64, figs 3a−b, 4.
Obisium robustum Nonidez: Navás, 1925: 120.

Neobisium (Blothrus) robustum (Nonidez): Beier, 1932a: 115, fig. 138; Roewer, 1937: 248; Beier, 1939f: 176; Beier, 1970f: 320; Estany, 1980b: 526, map 1.
Neobisium (Blothrus) robustum robustum (Nonidez): Beier, 1963b: 139–140, fig. 140; Zaragoza, 1986: fig. 1.
Neobisium (Blothrus) robustus (sic) (Nonidez): Bellés, 1987: fig. 41.

Type locality: Cueva de San Adrin, Cegama, Guipzcoa, Spain.
Distribution: Spain.

Neobisium (Blothrus) robustum escalerai Beier

Neobisium (Blothrus) escalerai Beier, 1931a: 14–15, fig. 6; Beier, 1932a: 115–116, fig. 139; Roewer, 1937: 248; Beier, 1939f: 176.
Neobisium (Blothrus) robustum escalerai Beier: Beier, 1963b: 140, fig. 141; Zaragoza, 1986: fig. 1.

Type locality: Cueva de Aitzquirri, Guipzcoa, Spain.
Distribution: Spain.

Neobisium (Blothrus) sbordonii Beier

Neobisium (Blothrus) sbordonii Beier, 1973c: 230–231, fig. 4.

Type locality: Grotta Guezeu, Afsin, Maras, Turkey.
Distribution: Turkey.

Neobisium (Blothrus) slovacum Gulicka

Neobisium (Blothrus) slovacum Gulicka, 1977: 6–8, figs 1–4.

Type locality: near Ort Brzotn, Slovakia, Czechoslovakia.
Distribution: Czechoslovakia.

Neobisium (Blothrus) spelaeum (Schiödte)

Blothrus spelaeus Schiödte, 1847: 80; Schiödte, 1851a: 23–26, plate 1 figs 2a–f; Schiödte, 1851b: 149, fig. 4; Hagen, 1870: 264; L. Koch, 1873: 41–42; G. Joseph, 1882: 21; J. C. Chamberlin, 1930: 11; J. C. Chamberlin, 1931a: 114, figs 32g–h, 41g–h.
Obisium troglodytes Schmidt, 1848: 10. New synonymy.
Obisium spelaeum (Schiödte): E. Simon, 1881a: fig. 1.
Obisium (Blothrus) spelaeus (sic) (Schiödte): Daday, 1888: 128, 183.
Obisium (Blothrus) spelaeum (Schiödte): Ellingsen, 1910a: 396; Beier, 1928: 306.
Neobisium (Blothrus) spelaeum (Schiödte): Beier, 1932a: 123–124, fig. 151; Roewer, 1937: 248; Beier, 1939d: 75; Beier, 1939e: 311; Beier, 1948b: 456; Beier, 1970f: 320.
Neobisium (Blothrus) spelaeum spelaeum (Schiödte): Beier, 1963b: 177–178, fig. 187; Ćurčić, 1974a: 22.
Neobisium spelaeum (Schiödte): Ćurčić, 1988b: 52–54, figs 164–177.
Not *Blothrus spelaeus* Schiödte: Marchesetti, 1890: 159 (not seen) (misidentification; see *Neobisium (Blothrus) spelaeum istriacum* (G. Müller)).
Not *Obisium spelaeum* (Schiödte): Hamann, 1896: 181 (not seen) (misidentification; see *Neobisium (Blothrus) spelaeum istriacus* (G. Müller)).
Not *Obisium spelaeum* Schiödte: Valle, 1911: 24 (not seen) (misidentification; see *Neobisium (Blothrus) reimoseri* (Beier)).
Not *Neobisium spelaeum* (Schiödte): Roewer, 1931: 42 (not seen) (misidentification; see *Neobisium (Blothrus) spelaeum istriacum* (G. Müller) and *Neobisium (Blothrus) torrei* (E. Simon)).

Type locality: of *Blothrus spelaeus*: Adelsberger Cave, near Postojna, Yugoslavia.
of *Obisium troglodytes*: Adelsberger Cave, near Postojna, Yugoslavia.
Distribution: Yugoslavia.

Family Neobisiidae

Neobisium (Blothrus) spelaeum istriacum (G. Müller)

Blothrus spelaeus Schiödte: Marchesetti, 1890: 159 (not seen) (misidentification).
Obisium spelaeum (Schiödte): Hamann, 1896: 181 (not seen) (misidentification).
Neobisium spelaeum (Schiödte): Roewer, 1931: 42 (not seen) (misidentification, in part).
Obisium (Blothrus) spelaeum istriacum G. Müller, 1931: 126–127.
Neobisium (Blothrus) spelaeum istriacum (G. Müller): Beier, 1939d: 75, fig. 90; Beier, 1963b: 178, fig. 188; Ćurčić, 1974a: 22; Gardini, 1980c: 108–109, 127, 128.

Type locality: Grotta del Fumo, Istria, Yugoslavia.
Distribution: Italy, Yugoslavia.

Neobisium (Blothrus) stankovici Ćurčić

Neobisium (Blothrus) stankovici Ćurčić, 1972e: 86–93, figs 1–19; Ćurčić, 1974a: 22.
Neobisium stankovici Ćurčić: Ćurčić, 1976b: 176, fig. 2; Ćurčić, 1983b: fig. 2.

Type locality: Velika Pecina Cave, Donja Drzina, 7 km SE. of Pirot, Yugoslavia.
Distribution: Yugoslavia.

Neobisium (Blothrus) stygium Beier

Neobisium (Blothrus) stygium Beier, 1931a: 16–17, fig. 8; Beier, 1932a: 122–123, fig. 150; Caporiacco, 1937b: 41; Roewer, 1937: 248; Beier, 1970f: 320; Mahnert, 1974b: 212–216, figs 11–15; Helversen & Martens, 1972: 115–116, fig. 10b.
Neobisium (Blothrus) meuseli Beier, 1931a: 15–16, fig. 7; Beier, 1932a: 122, fig. 149; Roewer, 1937: 248 (synonymized by Mahnert, 1974b: 212).
Obisium (Blothrus) minoum Beier, 1931d: 95–96, figs 3c, 4 (synonymized by Helversen & Martens, 1972: 115).
Neobisium (Blothrus) minoum (Beier): Beier, 1932a: 121–122, fig. 148; Roewer, 1936: fig. 24a; Roewer, 1937: 248; Beier, 1939d: 72–73; Beier, 1963b: 175–176, fig. 184.
Neobisium (Blothrus) stygium stygium Beier: Beier, 1939d: 73; Beier, 1963b: 176, fig. 185; Ćurčić, 1974a: 22.
Neobisium (Blothrus) stygium padewiethi Beier, 1938a: 8; Beier, 1939d: 73, fig. 88; Beier, 1963b: 176–177; Ćurčić, 1974a: 22 (synonymized by Mahnert, 1974b: 212).
Neobisium (Blothrus) stygium meuseli Beier: Beier, 1939d: 74; Beier, 1963b: 177; Ćurčić, 1974a: 22.
Neobisium (Blothrus) stygium csikii Beier, 1938a: 8; Beier, 1939d: 74, fig. 89; Beier, 1963b: 177, fig. 186; Ćurčić, 1974a: 22 (synonymized by Mahnert, 1974b: 212).
Neobisium stygium Beier: Ćurčić, 1988b: 54–57, figs 178–220.

Type locality: of *Neobisium (Blothrus) stygium*: Höhle Duman, Croatia, Yugoslavia.
of *Neobisium (Blothrus) meuseli*: Antrum Kosinski, Croatia, Yugoslavia.
of *Obisium (Blothrus) minoum*: Topolia, Crete, but quite possibly mislabelled (see Helversen & Martens, 1972).
of *Neobisium (Blothrus) stygium padewiethi*: Vlaska Pecina, near Zengg, Croatia, Yugoslavia.
of *Neobisium (Blothrus) stygium csikii*: Risban and Kormas Cave, near Lokve, Croatia, Yugoslavia.
Distribution: Yugoslavia.

Neobisium (Blothrus) svetovidi Ćurčić

Neobisium svetovidi Ćurčić, 1988b: 60–62, figs 241–260.

Type locality: Ivana Jama Pothole, Krasno, Mt Velebit, Yugoslavia.
Distribution: Yugoslavia.

Neobisium (Blothrus) tantaleum Beier

Neobisium (Blothrus) tantaleum Beier, 1938a: 7; Beier, 1970f: 320.
Neobisium (Blothrus) tantaleum tantaleum Beier: Beier, 1939d: 62–63, fig. 75; Beier, 1963b: 166, fig. 172; Ćurčić, 1974a: 22–23.
Neobisium aff. *tantaleum* Beier: Ćurčić, 1988b: 57–60, figs 221–240.

Type locality: Vran Planina, W. of Narenta, Yugoslavia.
Distribution: Yugoslavia.

Neobisium (Blothrus) tantaleum jablanicae Beier

Neobisium (Blothrus) tantaleum jablanicae Beier, 1938a: 8; Beier, 1939d: 64, fig. 77; Beier, 1963b: 166; Ćurčić, 1974a: 23.

Type locality: Jablanica, Yugoslavia.
Distribution: Yugoslavia.

Neobisium (Blothrus) tantaleum rivale Beier

Neobisium (Blothrus) tantaleum rivalis (sic) Beier, 1938a: 7–8; Beier, 1939d: 64, fig. 76.
Neobisium (Blothrus) tantaleum rivale Beier: Beier, 1963b: 165–166; Ćurčić, 1974a: 23.
Neobisium (Blothrus) pluton Beier, 1938a: 8; Beier, 1939d: 65–66, fig. 78 (synonymized by Beier, 1963b: 166).

Type localities: of *Neobisium (Blothrus) tantaleum rivale*: Dragan Selo, near Konjica, Yugoslavia.
of *Neobisium (Blothrus) pluton*: Ostrozac, Narenta, Yugoslavia.
Distribution: Yugoslavia.

Neobisium (Blothrus) tenebrarum Beier

Neobisium (Blothrus) tenebrarum Beier, 1938a: 8; Beier, 1939d: 69, fig. 83; Beier, 1963b: 147, fig. 150; Ćurčić, 1974a: 23.

Type locality: Borija Pecina, near Krbljine, Treskavica Planina, Yugoslavia.
Distribution: Yugoslavia.

Neobisium (Blothrus) tenuipalpe (Nonidez)

Obisium (Blothrus) tenuipalpis (sic) Nonidez, 1925: 73–75, figs 12, 13a–b.
Obisium tenuipalpe Nonidez: Navás, 1925: 121.
Neobisium (Blothrus) tenuipalpe (Nonidez): Beier, 1932a: 116–117, fig. 141; Roewer, 1937: 248; Beier, 1939f: 176; Beier, 1963b: 141–142, fig. 144; Estany, 1980b: 527, map 2; Zaragoza, 1986: fig. 1.

Type locality: Cueva de San Valerio, Mondragn, Guipzcoa, Spain.
Distribution: Spain.

Neobisium (Blothrus) torrei (E. Simon)

Obisium (Blothrus) torrei E. Simon, 1881a: 299–300, fig. 2; Beier, 1929a: fig. 14c; Boldori, 1946: 25 (not seen).
Obisium (Blothrus) roeweri Beier, 1931d: 94–95, figs 2, 3b (synonymized by Helversen & Martens, 1972: 117).
Neobisium spelaeum (Schiödte): Roewer, 1931: 42 (not seen) (misidentification, in part).
Neobisium (Blothrus) roeweri (Beier): Beier, 1932a: 111, fig. 131; Roewer, 1937: 248; Beier, 1939d: 59; Beier, 1963b: 159–160, fig. 166.
Neobisium (Blothrus) torrei (E. Simon): Beier, 1932a: 111–112, fig. 132; Roewer, 1937: 248; Caporiacco, 1948d: 237; Beier, 1958d: 162; Bizzi, 1960: 59 (not seen); Ruffo, 1960: 159 (not seen); Beier, 1970f: 320; Helversen & Martens, 1972: 117–118, fig. 10c.
Blothrus torrei (E. Simon): Caporiacco, 1936c: 90.

Neobisium trentinum ghidinii Beier: Caporiacco, 1947d: 138 (misidentification).
Neobisium (Blothrus) torrei torrei (E. Simon): Beier, 1963b: 145–146, fig. 148; Boscolo,
    1969: 22 (not seen); Lazzeroni, 1969b: 403; Gardini, 1980c: 109, 125, 126, 127, 128;
    Callaini, 1986e: 252.
Neobisium (Blothrus) tonei (sic) (E. Simon): Pellegrini, 1968: 267 (not seen).
Not *Obisium (Blothrus) torrei* E. Simon: Ellingsen, 1905d: 9–11 (misidentification;
    see *Pseudoblothrus ellingseni* (Beier)).

Type localities: of *Obisium (Blothrus) torrei*: Grotta di Oliero, near Bassano,
    Italy.
    of *Obisium (Blothrus) roeweri*: Katavothron, Omalos, Crete, but quite possibly
    mislabelled (see Helversen & Martens, 1972).
Distribution: Italy.

Neobisium (Blothrus) torrei leonidae Beier

Neobisium (Blothrus) leonidae Beier, 1942: 132–133, fig. 2.
Neobisium (Blothrus) torrei leonidae Beier: Beier, 1963b: 146; Gardini, 1980c: 109,
    126, 127.

Type locality: Bocca Lorenza (as Sorenza), near Schio, Italy.
Distribution: Italy.

Neobisium (Blothrus) tuzetae Vachon

Neobisium (Blothrus) tuzeti (sic) Vachon, 1947c: 320–321, figs 2–3, 5–6, 8; Vachon,
    1949: fig. 227; Beier, 1963b: 143–144; Weygoldt, 1969a: fig. 102.
Neobisium (Blothrus) tuzetae Vachon: Heurtault, 1986c: 26, fig. 2.

Type locality: Grotte du Signal de la Montete, Quissac, Gard, France.
Distribution: France.

Neobisium (Blothrus) umbratile Beier

Neobisium (Blothrus) umbratile Beier, 1938a: 7; Beier, 1939d: 52–53, fig. 60; Beier,
    1963b: 152, fig. 156; Ćurčić, 1974a: 23.

Type locality: Lipska Pecina, E. of Cetinje, Montenegro, Yugoslavia.
Distribution: Yugoslavia.

Neobisium (Blothrus) vachoni Beier

Neobisium (Blothrus) vachoni Beier, 1939d: 45, fig. 48; Beier, 1963b: 174, fig. 182;
    Ćurčić, 1974a: 23.

Type locality: Oborova Pecina, Troglav Planina, Yugoslavia.
Distribution: Yugoslavia.

Neobisium (Blothrus) vasconicum (Nonidez)

Obisium (Blothrus) vasconicus (sic) Nonidez, 1925: 67–69, figs 7, 8a–b.
Obisium vasconicum Nonidez: Navás, 1925: 121.
Neobisium (Blothrus) vasconicum (Nonidez): Beier, 1932a: 119, fig. 145; Roewer, 1937:
    248; Beier, 1939f: 176; Beier, 1970f: 320; Mahnert, 1977c: 71–73; Estany, 1980b:
    527, map 2; Zaragoza, 1986: fig. 1.
Neobisium (Neobisium) vasconicum vasconicum Beier: Beier, 1963b: 142, fig. 145a.

Type locality: Cueva de Mendicute, Tolosa, Guipzcoa, Spain.
Distribution: Spain.

Neobisium (Blothrus) vasconicum cantabricum (Nonidez)

Obisium (Blothrus) cantabricus (sic) Nonidez, 1925: 77–79, figs 15, 16a–b.
Obisium cantabricum Nonidez: Navás, 1925: 121.

Neobisium (Blothrus) cantabricum (Nonidez): Beier, 1932a: 118–119, fig. 144; Roewer, 1937: 248; Beier, 1939f: 176.
Neobisium (Blothrus) vasconicum cantabricum (Nonidez): Beier, 1963b: 143, fig. 145b.

Type locality: Cueva de Hernialde, Tolosa, Guipzcoa, Spain.
Distribution: Spain.

Neobisium (Blothrus) vasconicum hypogeum (Nonidez)

Obisium (Blothrus) hypogeus (sic) Nonidez, 1925: 69–72, figs 9, 10a–b.
Obisium hypogaeum (sic) Nonidez: Navás, 1925: 121.
Neobisium (Blothrus) hypogeum (Nonidez): Beier, 1932a: 118, fig. 143; Roewer, 1937: 248; Beier, 1939f: 176.
Neobisium (Blothrus) vasconicum hypogeum (Nonidez): Beier, 1963b: 142–143.

Type locality: Cueva del Chorotte, Tolosa, Guipzcoa, Spain.
Distribution: Spain.

Neobisium (Blothrus) velebiticum Beier

Neobisium (Blothrus) velebiticum Beier, 1938a: 6; Beier, 1939d: 43–44, fig. 46; Beier, 1963b: 169–170, fig. 177; Ćurčić, 1974a: 23.

Type locality: Ostrovica Pecina, Velebit, Croatia, Yugoslavia.
Distribution: Yugoslavia.

Neobisium (Blothrus) verae (Lapschoff)

Blothrus verae Lapschoff, 1940: 63–65, 74, figs 3–4; Kobakhidze, 1965b: 541; Kobakhidze, 1966: 703.
Neobisium (Blothrus) verae (Lapschoff): Mahnert, 1979b: 264–266; Schawaller, 1983a: 15–17, figs 41–46; Schawaller & Dashdamirov, 1988: 24.

Type locality: Höhle Gogoleti, 70 km NE. of Kutais, Georgia, U.S.S.R.
Distribution: U.S.S.R. (Georgia).

Neobisium (Blothrus) vjetrenicae Hadži

Neobisium (Blothrus) vjetrenicae Hadži, 1932: 102–113, figs 1–4; Hadži, 1933a: 49–53, figs 1–2; Beier, 1939d: 70, figs 84–85; Beier, 1963b: 151, fig. 155; Ćurčić, 1974a: 23.

Type locality: Vjetrenica Cave, near Zavale, Yugoslavia.
Distribution: Yugoslavia.

Subgenus Neobisium (Heoblothrus) Beier

Neobisium (Heoblothrus) Beier, 1963b: 132.
Blothrus (Heoblothrus) Beier: Harvey, 1985a: 367.

Type species: *Neobisium (Heoblothrus) beroni* Beier, 1963b, by present designation.

Neobisium (Heoblothrus) beroni Beier

Neobisium (Heoblothrus) beroni Beier, 1963b: 132–133, fig. 131; Beier, 1970f: 320.

Type locality: Höhle Svinskata dupka, Lokatnik, Bulgaria.
Distribution: Bulgaria.

Neobisium (Heoblothrus) bulgaricum (Redikorzev)

Obisium (Blothrus) bulgaricum Redikorzev, 1928: 125–127, figs 4–5, 10b.
Obisium (Blothrus) subterraneum Redikorzev, 1928: 127–129, figs 6–7, 10a (synonymized by Ćurčić, 1978b: 132).
Roncus (Parablothrus) bulgaricus (Redikorzev): Beier, 1928: 309.
Roncus (Parablothrus) subterraneus (Redikorzev): Beier, 1928: 309.

Neobisium (Blothrus) subterraneum (Redikorzev): Beier, 1932a: 112, fig. 133; Roewer, 1937: 248; Beier, 1939d: 72.

Neobisium (Blothrus) bulgaricum (Redikorzev): Beier, 1932a: 114, fig. 136; Roewer, 1936: fig. 66b; Roewer, 1937: 248; Beier, 1939d: 72.

Neobisium (Heoblothrus) subterraneum (Redikorzev): Beier, 1963b: 133–134, fig. 132; Beier, 1970f: 320.

Neobisium (Heoblothrus) bulgaricum (Redikorzev): Beier, 1963b: 134, fig. 133; Beier, 1970f: 320.

Neobisium bulgaricum (Redikorzev): Ćurčić, 1978b: 132–139, figs 21–37.

Type localities: of *Obisium (Blothrus) bulgaricum*: Jalovica (as Jalowitza) Cave, near Golema Zelezna, Bulgaria.
of *Obisium (Blothrus) subterraneum*: Toplja Cave, near Golema Zelezna, Bulgaria.
Distribution: Bulgaria.

Neobisium (Heoblothrus) sakadzhianum Krumpál

Neobisium (Heoblothrus) sakadzhianum Krumpál, 1984a: 637–641, figs 1–10.

Type locality: Sakadzia Cave, near Dorf Kvilishori, Cchaltudo, Georgia, U.S.S.R.
Distribution: U.S.S.R. (Georgia).

Subgenus Neobisium (Ommatoblothrus) Beier

Neobisium (Ommatoblothrus) Beier, 1956d: 131–132.
Blothrus (Ommatoblothrus) Beier: Harvey, 1985a: 367.

Type species: *Neobisium (Ommatoblothrus) sardoum* Beier, 1956d, by original designation.

Neobisium (Ommatoblothrus) battonii Beier

Neobisium (Ommatoblothrus) battonii Beier, 1966l: 35–36, fig. 1; Beier, 1970f: 320; Lazzeroni, 1969c: fig. 2; Gardini, 1980c: 107, 129.

Type locality: Grotte Risorgenze di Stiffe, S. Demetrio, L'Aquila, Abruzzo, Italy.
Distribution: Italy.

Neobisium (Ommatoblothrus) bessoni Heurtault

Neobisium (Ommatoblothrus) bessoni Heurtault, 1979b: 234, plate 1 figs 1–3; Heurtault, 1986c: 27.

Type locality: Liet, Accous, Pyrénées-Atlantiques, France.
Distribution: France.

Neobisium (Ommatoblothrus) cerrutii Beier

Neobisium (Blothrus) cerrutii Beier, 1955j: 26–28, fig. 1; Cerruti, 1959: 50–51.
Neobisium (Ommatoblothrus) cerrutii Beier: Beier, 1956d: 135; Beier, 1963b: 131–132, fig. 130; Cerruti, 1968: fig. 6; Beier, 1970f: 320; Lazzeroni, 1969c: fig. 2; Gardini, 1980c: 107, 129.

Type locality: Grotta di San Luca near Guarcino, Lazio, Italy.
Distribution: Italy.

Neobisium (Ommatoblothrus) gaditanum Mahnert

Neobisium (Ommatoblothrus) gaditanum Mahnert, 1977c: 84–86, figs 35–41; Zaragoza, 1986: fig. 1.

Type locality: Sima del Cacao, Villaluenga del Rosario, E. of Ronda, Cadiz, Spain.
Distribution: Spain.

Neobisium (Ommatoblothrus) gomezi Heurtault

Neobisium (Ommatoblothrus) gomezi Heurtault, 1979b: 233–234, plate 2 figs 1–6; Heurtault, 1986c: 27.

Type locality: Arphidia, Amont, St Engrace, Basses-Pyrénées, France.
Distribution: France.

Neobisium (Ommatoblothrus) gracile Heurtault

Neobisium (Ommatoblothrus) gracilis (sic) Heurtault, 1979b: 232–233, plate 1 figs 4–6; Heurtault, 1986c: 26.

Type locality: Abime du Rabanel, Ganges, Herault, France.
Distribution: France.

Neobisium (Ommatoblothrus) henroti Beier

Neobisium (Blothrus) henroti Beier, 1956b: 55–58, figs 1–2.
Neobisium (Ommatoblothrus) henroti Beier: Beier, 1956d: 135; Beier, 1963b: 129–130, fig. 128; Cerruti, 1968: 220, fig. 6; Lazzeroni, 1969c: 230, fig. 2; Beier, 1970f: 320; Puddu & Pirodda, 1973: 171; Gardini, 1980c: 107, 132.

Type locality: Grotta di Cane Gortoe near Siniscola, Sardinia.
Distribution: Sardinia.

Neobisium (Ommatoblothrus) lulense Gardini

Neobisium (Ommatoblothrus) lulense Gardini, 1982c: 89–94, figs 1–9.

Type locality: Grutta de Nurai, Monte Albo, Lula, Sardinia.
Distribution: Sardinia.

Neobisium (Ommatoblothrus) oenotricum Callaini

Neobisium (Ommatoblothrus) oenotricum Callaini, 1987a: 73–78, figs 2a–g.

Type locality: Grotta di Trecchina, Potenza, Lucania, Italy.
Distribution: Italy.

Neobisium (Ommatoblothrus) pangaeum Gardini

Neobisium (Ommatoblothrus) pangaeum Gardini, 1985a: 53–57, figs 1–9.

Type locality: Mt Pangaeo, NE. Greece.
Distribution: Greece.

Neobisium (Ommatoblothrus) patrizii Beier

Neobisium (Blothrus) patrizii Beier, 1953f: 139–140, fig. 1.
Neobisium (Ommatoblothrus) patrizii Beier: Beier, 1956d: 135; Beier, 1963b: 130, fig. 129; Cerruti, 1968: fig. 6; Beier, 1970f: 320; Lazzeroni, 1969c: fig. 2; Mahnert, 1980d: 23, fig. 6.
Neobisium (Ommatoblothrus) patrizii patrizii Beier: Gardini, 1980c: 107, 129.

Type locality: Grotta degli Ausi, Lazio, Italy.
Distribution: Italy.

Neobisium (Ommatoblothrus) patrizii romanum Mahnert

Neobisium (Ommatoblothrus) patrizii romanum Mahnert, 1980d: 23–25, figs 1–5; Gardini, 1980c: 108, 129.

Type locality: Buco nella Villa, Carpineto Romano, Rome (as Roma), Lazio, Italy.
Distribution: Italy.

Neobisium (Ommatoblothrus) paucedentatum Mahnert

Neobisium (Ommatoblothrus) paucedentatum Mahnert, 1982f: 303−304, figs 17−21; Zaragoza, 1986: fig. 1.

Type locality: Cueva del Agua, Ernalloz, Granada, Spain.
Distribution: Spain.

Neobisium (Ommatoblothrus) phaeacum Mahnert

Neobisium (Ommatoblothrus) phaeacum Mahnert, 1973a: 211−218, figs 7−17, plate 1; Mahnert, 1975a: 179; Gardini, 1989: 61.
Neobisium phaecum (sic) Mahnert: Schawaller, 1985b: 6.

Type locality: Höhle Peristerograva, Lutses, Kérkira (as Korfu), Greece.
Distribution: Greece.

Neobisium (Ommatoblothrus) samniticum Mahnert

Neobisium (Ommatoblothrus) samniticum Mahnert, 1980d: 25−26, figs 7−12; Gardini, 1980c: 108, 129.
Neobisium (Ommatoblothrus) aff. *samniticum* Mahnert: Mahnert, 1980d: 26−28.

Type locality: Grotta della Praie, Lettomanoppello, Pescara, Abruzzo, Italy.
Distribution: Italy.

Neobisium (Ommatoblothrus) sardoum Beier

Neobisium (Ommatoblothrus) sardoum Beier, 1956d: 132−134, figs 1−2; Beier, 1959b: 245; Beier, 1963b: 132; Cerruti, 1968: 220, fig. 6; Lazzeroni, 1969c: 230, fig. 2; Beier, 1970f: 320; Puddu & Pirodda, 1973: 171; Beier, 1974g: 163; Gardini, 1980c: 108, 133.

Type locality: Grotte S'Abba Medica, Oliena, Sardinia.
Distribution: Sardinia.

Neobisium (Ommatoblothrus) staudacheri Hadži

Neobisium (Neobisium) staudacheri Hadži, 1933b: 152−158, figs 13a−b, 14a−b, 15, 16a−e, 17a−c; Hadži, 1933c: 184−186; Roewer, 1937: 246; Beier, 1939d: 26; Beier, 1947a: 174, fig. 1n; Beier, 1970f: 320.
Neobisium (Ommatoblothrus) staudacheri Hadži: Beier, 1963b: 131; Ćurčić, 1974a: 23.
Neobisium (Ommatoblothrus) staudakeri (sic) Hadž: Lazzeroni, 1969c: fig. 2.

Type locality: Biokovo, Makarska, Yugoslavia.
Distribution: Yugoslavia.

Neobisium (Ommatoblothrus) zoiai Gardini & Rizzerio

Neobisium (Ommatoblothrus) zoiai Gardini & Rizzerio, 1986b: 6−9, figs 1−4.

Type locality: Garb del Dighea, Monte Armetta, Alpi Liguri, Italy.
Distribution: Italy.

Subgenus Neobisium (Pennobisium) Ćurčić

Neobisium (Pennobisium) Ćurčić, 1988b: 65−66.

Type species: *Neobisium (Pennobisium) stribogi* Ćurčić, 1988b, by original designation.

Neobisium (Pennobisium) simargli Ćurčić

Neobisium (Pennobisium) simargli Ćurčić, 1988b: 69−71, figs 331−337.

Type locality: Jama II Pothole, Velike Brisnice, Mt Velebit, Yugoslavia.
Distribution: Yugoslavia.

Neobisium (Pennobisium) stribogi Ćurčić

Neobisium (Pennobisium) stribogi Ćurčić, 1988b: 66–69, figs 281–330.

Type locality: Jama Pod Bojinim Kukum Pothole, Bojinac, Mt Velebit, NW. of Starigrad, Yugoslavia.
Distribution: Yugoslavia.

Genus Neoccitanobisium Callaini

Neoccitanobisium Callaini, 1981c: 523.

Type species: *Neoccitanobisium ligusticum* Callaini, 1981c, by original designation.

Neoccitanobisium ligusticum Callaini

Neoccitanobisium ligusticum Callaini, 1981c: 523–533, figs 1a–g, 2a–g, 3a–i, plates 1–6; Gardini, 1985b: 60.

Type locality: Bosco di Rezzo, Italy.
Distribution: Italy.

Genus Nepalobisium Beier

Nepalobisium Beier, 1974b: 263–264.

Type species: *Nepalobisium franzi* Beier, 1974b, by original designation.

Nepalobisium franzi Beier

Nepalobisium franzi Beier, 1974b: 265, fig. 2; Schawaller, 1987a: 204–205, figs 6–8.

Type locality: Taksang, Tukche, Thakkhola, Nepal.
Distribution: Nepal.

Genus Novobisium Muchmore

Novobisium Muchmore, 1967b: 212–214.

Type species: *Obisium carolinense* Banks, 1895a, by original designation.

Novobisium carolinense (Banks)

Obisium carolinensis (sic) Banks, 1895a: 12; Coolidge, 1908: 113; J. C. Chamberlin, 1924c: figs b–c; Berland, 1932: fig. 81.
Neobisium carolinensis (sic) (Banks): J. C. Chamberlin, 1929e: 153; J. C. Chamberlin, 1930: 15, figs 2g, 2m, 2w, 3a, 3d; J. C. Chamberlin, 1931a: figs 4f, 6b, 8c, 9c, 11f, 11x, 12c, 13h, 19c, 25j, 33b–c, 33g, 40o, 45b–d, 50d; J. C. Chamberlin, 1935b: 480.
Neobisium (Neobisium) carolinense (Banks): Beier, 1932a: 91, fig. 9; Roewer, 1937: 246; Hoff, 1958: 7–8; Sweeney, Taylor & Counts, 1977: 55; Zeh, 1987b: 1086.
Neobisium (Neobisium) carolinense carolinense (Banks): Beier, 1932a: 92.
Neobisium carolinense (Banks): Beier, 1932g: figs 174, 228; Roewer, 1936: figs 32, 139; J. C. Chamberlin, 1962: 325–328, figs 8a–e.
Novobisium carolinense (Banks): Muchmore, 1967b: 213–214, fig. 2.

Type locality: Retreat, North Carolina, U.S.A.
Distribution: U.S.A. (Georgia, North Carolina, Pennsylvania, West Virginia).

Novobisium ingratum (J. C. Chamberlin)

Neobisium ingratum J. C. Chamberlin, 1962: 330–333, figs 10a–c.
Novobisium ingratum (J. C. Chamberlin): Muchmore, 1967b: 213.

Type locality: McFarlen Cave, Jackson County, Alabama, U.S.A.
Distribution: U.S.A. (Alabama, Tennessee).

Novobisium tenue (J. C. Chamberlin)

Neobisium carolinensis tenuis (sic) J. C. Chamberlin, 1930: 16–17; J. C. Chamberlin, 1931a: figs 15d, 19d; Hoff, 1949b: 444, figs 20–21.
Neobisium (Neobisium) carolinense tenue J. C. Chamberlin: Beier, 1932a: 92.
Neobisium carolinensis tenue J. C. Chamberlin: Hoff & Bolsterli, 1956: 161–162.
Neobisium carolinense (Banks): Roewer, 1936: fig. 113; J. C. Chamberlin, 1962: 328–330, figs 9a–d.
Novobisium tenue (J. C. Chamberlin): Muchmore, 1967b: 213.

Type locality: Mt Leconte, Tennessee, U.S.A.
Distribution: U.S.A. (Kentucky, North Carolina, Tennessee).

Genus Occitanobisium Heurtault

Occitanobisium Heurtault, 1977: 1122.

Type species: *Occitanobisium coiffaiti* Heurtault, 1977, by original designation.

Occitanobisium coiffaiti Heurtault

Occitanobisium coiffaiti Heurtault, 1977: 1122–1132, figs 1–18; Leclerc & Heurtault, 1979: 245; Heurtault, 1980d: 89.

Type locality: Saint-Guilhem-le-Désert, Herault, France.
Distribution: France.

Genus Orientocreagris Ćurčić

Orientocreagris Ćurčić, 1985a: 345–346.

Type species: *Orientocreagris syrinx* Ćurčić, 1985a, by original designation.

Orientocreagris latona Ćurčić

Orientocreagris (?) *latona* Ćurčić, 1985a: 348–349, figs 89–94.

Type locality: Kondara River, Tadzhikistan, U.S.S.R.
Distribution: U.S.S.R. (Tadzhikistan).

Orientocreagris syrinx Ćurčić

Orientocreagris syrinx Ćurčić, 1985a: 346–348, figs 83–88.

Type locality: Maritime Province, R.S.F.S.R., U.S.S.R.
Distribution: U.S.S.R. (R.S.F.S.R.).

Genus Pedalocreagris Ćurčić

Pedalocreagris Ćurčić, 1985a: 349–350.

Type species: *Pedalocreagris tethys* Ćurčić, 1985a, by original designation.

Pedalocreagris tethys Ćurčić

Pedalocreagris tethys Ćurčić, 1985a: 350, figs 95–99.

Type locality: Ussuri Reservation, R.S.F.S.R., U.S.S.R.
Distribution: U.S.S.R. (R.S.F.S.R.).

Genus Paedobisium Beier

Paedobisium Beier, 1939f: 188; Roewer, 1940: 344; Beier, 1963b: 205–206.

Type species: *Paedobisium minutum* Beier, 1939f, by original designation.

Paedobisium minutum Beier

Paedobisium minutum Beier, 1939f: 188–189, fig. 18; Roewer, 1940: 344; Beier, 1963b: 206, fig. 218; Cîrdei, Bulimar & Malcoci, 1970: 12–13, fig. 6.

Type locality: Cova Maravillas near Llombay, Valencia, Spain.
Distribution: Romania, Spain.

Paedobisium moldavicum Cîrdei, Bulimar & Malcoci

Paedobisium moldavicum Cîrdei, Bulimar & Malcoci, 1967: 240, figs 5a−b.

Type locality: Masivul Repedea, Iasi, Romania.
Distribution: Romania.

Genus Pararoncus J. C. Chamberlin

Pararoncus J. C. Chamberlin, 1938a: 260; Roewer, 1940: 344; Ćurčić, 1979a: 176−177.

Type species: *Pararoncus histrionicus* J. C. Chamberlin, 1938a, by original designation.

Pararoncus chamberlini (Morikawa), new combination

Roncus (Roncus) troglophilus Morikawa, 1957a: 360, figs 2c, 2g (junior primary homonym of *Roncus (Parablothrus) troglophilus* Beier, 1931a).
Roncus (Roncus) chamberlini Morikawa, 1960: 117−118; Morikawa, 1962: 418, fig. 4 (replacement name for *Roncus (Roncus) troglophilus* Morikawa).
Roncus (Roncus) chamberlini chamberlini Morikawa: Morikawa, 1960: 118, plate 6 figs 2−3, plate 9 figs 7, 9; Morikawa, 1962: fig. 4.
Pararoncus troglophilus (Morikawa): Ćurčić, 1979a: 174.

Type locality: Goyomatsu-dani-dô Cave, Dorogawa, Tenkawa-mura, Nara, Japan.
Distribution: Japan.

Pararoncus histrionicus J. C. Chamberlin

Pararoncus histrionicus J. C. Chamberlin, 1938a: 261−263, figs 1a−g; Roewer, 1940: 344, fig. 260; Ćurčić, 1979a: 169.
Roncus (Roncus) japonicus (Ellingsen): Morikawa, 1960: 117−118 (misidentification, in part).

Type locality: Japan.
Distribution: Japan.

Pararoncus japonicus (Ellingsen)

Obisium (Roncus) japonicum Ellingsen, 1907c: 10−12.
Roncus (Roncus) (?) japonicum (Ellingsen): Beier, 1932a: 129; Roewer, 1937: 249.
Roncus (Roncus) japonicus (Ellingsen): Morikawa, 1952b: 253−254, figs 7−8; Morikawa, 1955a: 222; Morikawa, 1960: 117−118, plate 2 fig. 4, plate 7 fig. 11 (in part; see *Pararoncus histrionicus* J. C. Chamberlin); Morikawa, 1962: 418, figs 2, 4; Sato, 1979a: 94−95, figs 7a−k, 8a−e, plates 3, 12.
Pararoncus japonicus (Ellingsen): Ćurčić, 1979a: 172−174; Sato, 1980d: plate 2.
Roncus japonicus (Ellingsen): Sato, 1979d: 43; Sato, 1982b: 58, figs 1−3; Sato, 1982c: 32; Sato, 1983b: 16, fig. 2; Sato, 1984d: fig. 2; Sato, 1986: 95.

Type locality: Yamanaka, Suruga, Japan.
Distribution: Japan.

Pararoncus oinuanensis (Morikawa)

Roncus (Roncus) troglophilus oinuanensis Morikawa, 1957a: 360, figs 2d, 2h.
Roncus (Roncus) chamberlini oinuanensis Morikawa: Morikawa, 1960: 118.
Pararoncus oinuanensis (Morikawa): Ćurčić, 1979a: 174−176.

Type locality: Oinu-ana Cave, Nakakoshizawa, Uéno-mura, Gumma, Japan.
Distribution: Japan.

Pararoncus rakanensis (Morikawa)

Roncus (Roncus) uénoi rakanensis Morikawa, 1957a: 361–363, figs 3b–c, 3e.
Roncus (Parablothrus) rakanensis Morikawa: Morikawa, 1960: 119, plate 2 fig. 5, plate 6 figs 5–6, plate 9 fig. 5.
Pararoncus rakanensis (Morikawa): Ćurčić, 1979a: 170–172, figs 1–7.

Type locality: Rakan-ana Cave, Uchiko-chô, Ehime, Japan.
Distribution: Japan.

Pararoncus uenoi (Morikawa)

Roncus (Roncus) uénoi Morikawa, 1957a: 361, figs 3a, 3d.
Roncus (Parablothrus) uenoi Morikawa: Morikawa, 1960: 119, plate 6 fig. 9, plate 9 fig. 6.
Pararoncus uenoi (Morikawa): Ćurčić, 1979a: 169–170.

Type locality: Yôzawa-dô Cave, Komiya-mura, Tokyo, Japan.
Distribution: Japan.

Pararoncus yosii (Morikawa)

Roncus (Roncus) chamberlini yosii Morikawa, 1960: 118.
Pararoncus yosii (Morikawa): Ćurčić, 1979a: 176.

Type locality: Shirataki-do, Iwato-mura, Miyazaki, Japan.
Distribution: Japan.

Genus **Parobisium** J. C. Chamberlin

Neobisium (Parobisium) J. C. Chamberlin, 1930: 17; Beier, 1932a: 84; Morikawa, 1960: 112–113; Hoff, 1961: 427.
Parobisium J. C. Chamberlin: J. C. Chamberlin & Malcolm, 1960: 112–113; J. C. Chamberlin, 1962: 123.

Type species: *Neobisium (Parobisium) magnum* J. C. Chamberlin, 1930, by original designation.

Parobisium anagamidensis (Morikawa), new combination

Neobisium (Parobisium) anagamidensis Morikawa, 1957a: 357–358, figs 2a, 2e; Morikawa, 1960: 115; Morikawa, 1962: 418, fig. 4; Sato, 1982c: 32.
Neobisium (Parobisium) anagamidensis anagamidensis Morikawa: Morikawa, 1960: 115, plate 6 fig. 1, plate 9 fig. 4.
Neobisium anagamidensis Morikawa: Sato, 1979c: 13, fig. 2; Sato, 1986: 95.

Type locality: Anagami-dô Cave, Nishi-tsugasai, Ogawa-mura, Kochi, Japan.
Distribution: Japan.

Parobisium anagamidensis esakii (Morikawa), new combination

Neobisium (Parobisium) anagamidensis esakii Morikawa, 1960: 116.

Type locality: Omogo-kei, Ehime, Japan.
Distribution: Japan.

Parobisium anagamidensis morikawai, new name

Neobisium (Parobisium) anagamidensis longidigitatus (sic) Morikawa, 1957a: 358; Morikawa, 1960: 115–116; Morikawa, 1962: fig. 4 (junior primary homonym of *Neobisium longidigitatum* (Ellingsen, 1908)).

Type locality: Seiryu-kutsu Cave, Hirao-dai Karst, Hukuoka, Japan.
Distribution: Japan.

Parobisium charlotteae J. C. Chamberlin

Parobisium charlotteae J. C. Chamberlin, 1962: 322–324, figs 7a–h.

Type locality: Redmond Lava Cave, Deschutes County, Oregon, U.S.A.
Distribution: U.S.A. (Oregon).

Parobisium flexifemoratum (J. C. Chamberlin), new combination

Neobisium (Parobisium) flexifemoratum J. C. Chamberlin, 1930: 18; Beier, 1932a:
  85; Roewer, 1937: 245; Morikawa, 1960: 114; Sato, 1979d: 43.
Neobisium flexifemoratum J. C. Chamberlin: J. C. Chamberlin, 1931a: figs 31c–f;
  Roewer, 1936: fig. 116; Vachon, 1949: fig. 219c; Weygoldt, 1969a: fig. 15c; Eisenbeis
  & Wichard, 1987: fig. 44b.

Type locality: Muchogo-o, Japan.
Distribution: Japan.

Parobisium hastatus R. O. Schuster

Parobisium hastatus R. O. Schuster, 1966c: 223–225, figs 1–6.

Type locality: 6 miles S. of El Dorado, El Dorado County, California, U.S.A.
Distribution: U.S.A. (California).

Parobisium hesperum (J. C. Chamberlin)

Neobisium (Parobisium) hesperum J. C. Chamberlin, 1930: 19–20; Beier, 1932a:
  85; Roewer, 1937: 246; Hoff, 1958: 8.
Neobisium hesperum J. C. Chamberlin: J. C. Chamberlin, 1931a: figs 17f, 28g, 40d–l,
  40p, 42c–d; Vachon, 1949: figs 206a–b, 207a.
Parobisium hesperum (J. C. Chamberlin): R. O. Schuster, 1966c: 227, figs 11–13.

Type locality: Cannon Beach, Oregon, U.S.A.
Distribution: U.S.A. (California, Oregon).

Parobisium hesternus R. O. Schuster

Parobisium hesternus R. O. Schuster, 1966c: 225–227, figs 7–10.

Type locality: Riverton, El Dorado County, California, U.S.A.
Distribution: U.S.A. (California).

Parobisium imperfectum (J. C. Chamberlin)

Neobisium (Parobisium) imperfectum J. C. Chamberlin, 1930: 19; Beier, 1932a: 85;
  Roewer, 1937: 246.
Neobisium imperfectum J. C. Chamberlin: J. C. Chamberlin, 1931a: figs 15e, 16e;
  Beier, 1932g: figs 167, 169; Roewer, 1936: figs 22a, 70d; Savory, 1964a: fig. 33c;
  Savory, 1977: fig. 93c.
Neobisium (Parobisium) magnum magnum J. C. Chamberlin: Morikawa, 1960: 113
  (in part).
Parobisium imperfectum (J. C. Chamberlin): Muchmore, 1968g: 533.

Type locality: Muchigo-o, Japan.
Distribution: Japan.

Parobisium magnum (J. C. Chamberlin), new combination

Neobisium (Parobisium) magnum J. C. Chamberlin, 1930: 18; Beier, 1932a: 84–85;
  Roewer, 1937: 245; Morikawa, 1960: 113; Morikawa, 1962: figs 2, 4; Sato, 1979a:
  93–94, plate 15; Sato, 1979d: 43; Sato, 1982c: 32.
Neobisium (Parobisium) magnum magnum J. C. Chamberlin: Morikawa, 1960: 113
  (in part, see *Parobisium imperfectum* (J. C. Chamberlin)).

Neobisium magnum J. C. Chamberlin: J. C. Chamberlin, 1931a: figs 40y–z, 47e, 47h; Sato, 1979b: 13, figs 1–3, 5.

Type locality: Moghi, Kyushu, Japan.
Distribution: Japan.

Parobisium magnum chejuense (Morikawa), new combination

Neobisium (Parobisium) magnum chejuense Morikawa, 1970: 146–147, figs 2f, 4.

Type locality: Mt Halla San (as Halla-san), Cheju do (as Cheju-do), South Korea.
Distribution: South Korea.

Parobisium magnum ohuyeanum (Morikawa), new combination

Neobisium (Parobisium) ohuyeanum Morikawa, 1952b: 252–253, figs 5–6; Morikawa, 1955a: 221–222; Morikawa, 1957a: 358–360, figs 2b, 2f.
Neobisium (Parobisium) magnum ohuyeanum Morikawa: Morikawa, 1960: 113–114, plate 2 fig. 2, plate 6 fig. 4, plate 9 fig. 3; Morikawa, 1962: fig. 3.

Type locality: Sugitate, near Matsuyama, Shikoku, Japan.
Distribution: Japan.

Parobisium utahensis Muchmore

Parobisium utahensis Muchmore, 1968g: 531–533, figs 1–3.

Type locality: Blacksmith Fork Canyon, Cache County, Utah, U.S.A.
Distribution: U.S.A. (Utah).

Parobisium vancleavei (Hoff)

Neobisium (Parobisium) vancleavei Hoff, 1961: 427–431, figs 3–6.
Neobisium vancleavei Hoff: Zeh, 1987b: 1086.
Parobisium vancleavei (Hoff): Muchmore, 1968g: 533.

Type locality: Mesa Verde National Park, Montezuma County, Colorado, U.S.A.
Distribution: U.S.A. (Colorado).

Genus **Protoneobisium** Ćurčić

Protoneobisium Ćurčić, 1988b: 35–38.

Type species: *Obisium (Blothrus) biocovense* G. Müller, 1931, by original designation.

Protoneobisium biocovense (G. Müller)

Obisium (Blothrus) biocovense G. Müller, 1931: 126.
Neobisium (Blothrus) biocovense (G. Müller): Beier, 1939d: 71–72, fig. 87; Beier, 1963b: 150, fig. 153; Beier, 1970f: 320; Ćurčić, 1974a: 19.
Protoneobisium biocovense (G. Müller): Ćurčić, 1988b: 38–40, figs 54–75.

Type locality: Mt Biokova (as Biocovo), Yugoslavia.
Distribution: Yugoslavia.

Genus **Roncobisium** Vachon

Neobisium (Roncobisium) Vachon, 1967a: 363–365.
Roncobisium Vachon: Heurtault, 1974a: 1090.

Type species: *Neobisium (Roncobisium) allodentatum* Vachon, 1967a, by original designation.

Roncobisium allodentatum (Vachon)

Neobisium (Roncobisium) allodentatum Vachon, 1967a: 363–365, figs 1–14; Beier, 1970f: 320.

Roncobisium allodentatum (Vachon): Heurtault, 1974a: 1090; Heurtault, 1986c: 27.

Type locality: Grotte de Blanot, Saône-et-Loire, France.
Distribution: France.

Roncobisium leclerci Heurtault

Roncobisium leclerci Heurtault, 1979a: 225–230, figs 1–7; Leclerc & Heurtault, 1979: 242–244, map 4; Heurtault, 1986c: 27.

Type locality: Bas Chassezac-sous-Planas, Ardèche, France.
Distribution: France.

Genus Roncocreagris Mahnert

Microcreagris (Roncocreagris) Mahnert, 1974d: 849–850.
Roncocreagris Mahnert: Mahnert, 1976c: 209; Legg & Jones, 1988: 87.

Type species: *Roncus cambridgei* L. Koch, 1873, by original designation.

Roncocreagris beieri Mahnert

Roncocreagris beieri Mahnert, 1976c: 209–211, figs 26–30.

Type locality: 'Lens, Wallis, Switzerland', but mislabelled (V. Mahnert, pers. comm.).
Distribution: Portugal?

Roncocreagris blothroides (Beier)

Microcreagris blothroides Beier, 1962c: 25–26, fig. 1; Beier, 1963b: 213–214, fig. 224; Beier, 1970f: 321.
Roncocreagris blothroides (Beier): Mahnert, 1976c: 212; Zaragoza, 1986: fig. 1.

Type locality: Cova da Moura near Arrifana, Coimbra, Portugal.
Distribution: Portugal.

Roncocreagris cambridgei (L. Koch)

Roncus cambridgii L. Koch, 1873: 45–46; O. P. Cambridge, 1892: 217, plate B figs 9, 9a–b; Evans, 1901a: 53; Godfrey, 1904: 195.
Obisium (Roncus) cambridgei (L. Koch): E. Simon, 1879a: 64, plate 18 fig. 26; E. Simon, 1898d: 21.
Ideobisium cambridgii (L. Koch): Ellingsen, 1907a: 167–169; Ellingsen, 1913a: 455.
Ideoroncus cambridgii (L. Koch): Godfrey, 1907: 163; Godfrey, 1908: 158; Godfrey, 1909: 159–161; Donisthorpe, 1910: 84.
Ideoroncus cambridgei (L. Koch): With, 1907: 81–82, fig. 58.
Obisium (Ideoroncus) cambridgii (L. Koch): Kew, 1911a: 53, fig. 14; Kew, 1916b: 79.
Obisium cambridgii (L. Koch): Donisthorpe, 1927: 183.
Obisium (Ideoroncus) cambridgei (L. Koch): Kästner, 1928: 9.
Microcreagris cambridgei (L. Koch): J. C. Chamberlin, 1930: 30; Caporiacco, 1936b: 329; Vachon, 1940g: 143–144; G. O. Evans & Browning, 1954: 14; Beier, 1961b: 26; Heurtault-Rossi, 1966a: fig. 9; Gabbutt & Vachon, 1968: 422–441, figs 1a–d, 2a–c, 3a–d, 4a–e, 5a–h, 66a–h, 7a–b, 8a–c, 9a–d; Gabbutt, 1969a: 186, 187, 190, 195, figs 3a–d, 6a–e, 9a–b, 13a–c; Gabbutt, 1969b: 414–427; Gabbutt, 1969c: 230; Gabbutt, 1970c: 9, fig. 5; Legg, 1971a: 476, fig. 2c; Legg, 1973b: 430, fig. 2m; Legg, 1975a: 66; Legg, 1975c: 125, figs 1e, 5a–b, 7d, 12a–c, 14e, 16e, 21a–b; Crocker, 1978: 8.
Microcreagris cambridgii (L. Koch): Beier, 1932a: 154–155; Roewer, 1937: 251; Beier, 1948b: 457; Beier, 1953a: 296; Beier, 1955n: 97–98, fig. 4; Beier, 1959f: 122; Beier, 1963b: 208–209, fig. 220; Lazzeroni, 1969c: 237; Beier, 1970f: 321.
Roucus (sic) *cambridgei* L. Koch: Caporiacco, 1950: 116.

Microcreagris (Roncocreagris) cambridgei (L. Koch): Mahnert, 1974d: 879–880; Heurtault, 1980d: 89.
Roncocreagris cambridgei (L. Koch): Mahnert, 1976c: 212; Jones, 1980c: map 12; Legg & Jones, 1988: 87–88, figs 19a, 19b(a–h).
Not *Ideoroncus cambridgei* (L. Koch): Beier, 1930f: 71 (misidentification; see *Microcreagrina hispanica* (Ellingsen)).

Type locality: Bloxworth, Dorset, England, Great Britain.
Distribution: Algeria, France, Great Britain, Ireland, Italy, Portugal, Spain.

Roncocreagris cantabrica (Beier)

Microcreagris cantabrica Beier, 1959f: 122–124, fig. 5.
Microcreagris cantabrica cantabrica Beier: Beier, 1963b: 210–211.
Roncocreagris cantabrica cantabrica (Beier): Mahnert, 1976c: 212.

Type locality: Puenteviesgo (as Puente Viesgo), Santander, Spain.
Distribution: Spain.

Roncocreagris cantabrica distinguenda (Beier)

Microcreagris cantabrica distinguenda Beier, 1959f: 124, fig. 6; Beier, 1963b: 211.
Roncocreagris cantabrica distinguenda (Beier): Mahnert, 1976c: 212.

Type locality: Monte de Santoña near Santoña, Spain.
Distribution: Spain.

Roncocreagris cavernicola (Vachon)

Microcreagris cavernicola Vachon, 1946: 333–334, figs 1–6; Beier, 1963b: 213; Beier, 1970f: 321.
Roncocreagris cavernicola (Vachon): Mahnert, 1976c: 212; Zaragoza, 1986: fig. 1.

Type locality: Algar sul das Corujeiras, Abiul, Pombal, Leiria, Portugal.
Distribution: Portugal.

Roncocreagris galeonuda (Beier)

Microcreagris galeonuda Beier, 1955n: 101–104, fig. 6.
Microcreagris galeonuda galeonuda Beier: Beier, 1959f: 125; Beier, 1961b: 27; Beier, 1963b: 217–218.
Roncocreagris galeonuda galeonuda (Beier): Mahnert, 1976c: 212.

Type locality: Agulliero near Pyrnedo, Sierra de Ancares, Galicia, Spain.
Distribution: Spain.

Roncocreagris galeonuda clavata (Beier)

Microcreagris galeonuda clavata Beier, 1955n: 104–106, fig. 7; Beier, 1963b: 218, fig. 227.
Roncocreagris galeonuda clavata (Beier): Mahnert, 1976c: 212.

Type locality: Bosque de Muniellos, NE. Sierra de Ancares, Galicia, Spain.
Distribution: Spain.

Roncocreagris galeonuda nana (Beier)

Microcreagris galeonuda nana Beier, 1959f: 126–127, fig. 8; Beier, 1963b: 218–219.
Roncocreagris galeonuda nana (Beier): Mahnert, 1976c: 212.

Type locality: Pantano, E. of Orbaiceta, Navarra, Spain.
Distribution: Spain.

Roncocreagris galeonuda robustior (Beier)

Microcreagris galeonuda robustior Beier, 1959f: 125–126, fig. 7; Beier, 1963b: 218.
Roncocreagris galeonuda robustior (Beier): Mahnert, 1976c: 212.

Type locality: S. of Villaviciosa near Oviedo, Spain.
Distribution: Spain.

Roncocreagris iberica (Beier)

Microcreagris iberica iberica Beier, 1953a: 296–298, fig. 2; Beier, 1955n: 98; Beier,
1959f: 124; Beier, 1961b: 26–27; Beier, 1963b: 209–210, fig. 221.
Roncocreagris iberica iberica (Beier): Mahnert, 1976c: 212.
Roncocreagris cf. *iberica* (Beier): Mahnert, 1986d: 81.

Type locality: El Pardo near Madrid, Spain.
Distribution: Spain.

Roncocreagris iberica andalusica (Beier)

Microcreagris iberica andalusica Beier, 1953a: 298–299, fig. 3; Beier, 1961b: 27; Beier,
1963b: 210.
Roncocreagris iberica andalusica (Beier): Mahnert, 1976c: 212.

Type locality: Cinca de Pino, Sevilla, Andalucia, Spain.
Distribution: Spain.

Roncocreagris portugalensis (Beier)

Microcreagris portugalensis Beier, 1953a: 299–300, fig. 4; Beier, 1955n: 101; Beier,
1963b: 215.
Roncocreagris portugalensis (Beier): Mahnert, 1976c: 212.

Type locality: Parocco de Peña Maior, Paços de Ferreira, Portugal.
Distribution: Portugal, Spain.

Roncocreagris pycta (Beier)

Microcreagris pycta Beier, 1959f: 127–128, fig. 9; Beier, 1961b: 27; Beier, 1963b: 216.
Roncocreagris pycta (Beier): Mahnert, 1976c: 212.

Type locality: NE. of San Saturnino, La Coruña, Spain.
Distribution: Spain.

Roncocreagris roncoides (Beier)

Microcreagris roncoides Beier, 1955n: 99–101, fig. 5; Beier, 1959f: 125; Beier, 1961b:
27; Beier, 1963b: 216, fig. 226.
Microcreagris (Roncocreagris) roncoides (Beier): Mahnert, 1974d: 850, 861–868, figs
1c, 6a–e.
Roncocreagris roncoides (Beier): Mahnert, 1976c: 212; Mahnert, 1977d: 97.

Type locality: Portomouro, N. Santiago de Compostela, Galicia, Spain.
Distribution: Portugal, Spain.

Genus Roncus L. Koch

Roncus L. Koch, 1873: 44; Stecker, 1875d: 88; Tömösváry, 1882b: 213–214; Canestrini,
1884: no. 2; Balzan, 1887a: no pagination; Balzan, 1890: 445; O.P.–Cambridge,
1892: 216; J.C. Chamberlin, 1930: 12; Beier, 1932a: 124; Beier, 1939f: 177;
Morikawa, 1960: 116; Beier, 1963b: 179; Legg & Jones, 1988: 84.
Obisium (Roncus) L. Koch: Kew, 1911a: 53.
Roncus (Parablothrus) Beier, 1928: 309; Beier, 1932a: 130; Beier, 1939f: 180
(synonymized by Gardini, 1982c: 113).
Roncus (Roncus) L. Koch: Beier, 1932a: 126.

Type species: of *Roncus*: *Roncus lubricus* L. Koch, 1873, by subsequent designation of Beier, 1932a: 124.

of *Roncus (Parablothrus)*: *Obisium (Blothrus) stussineri* E. Simon, 1881a, by present designation.

Roncus abditus (J. C. Chamberlin)

Neobisium (Roncus) abditus J. C. Chamberlin, 1930: 12–13, figs 1h–i, 1w, 1ee, 2ii, 2oo, 3x.

Neobisium abditum J. C. Chamberlin: J. C. Chamberlin, 1931a: figs 16i, 35e, 40q, 50a, 52c.

Roncus (Roncus) abditus (J. C. Chamberlin): Beier, 1932a: 127; Roewer, 1937: 249; Beier, 1963b: 181; Lazzeroni, 1969a: 327–328; Lazzeroni, 1969c: 230–231; Gardini, 1980c: 109, 130.

Roncus abditus (J. C. Chamberlin): Beier, 1948a: 189, fig. 2; Lazzeroni, 1970a: 207; Schawaller, 1981b: 45–46, 50; Callaini, 1983d: 223; Gardini & Rizzerio, 1985: 66–71, figs 58–79.

Roncus arditus (sic) (J. C. Chamberlin): Capolongo, 1969: 202.

Roncus (Roncus) cfr. *abditus* (J. C. Chamberlin): Callaini, 1979c: 346–349, figs 3a–d; Callaini, 1982a: 17, fig. 2a; Callaini, 1983c: 288–289.

Type locality: Sorgono, Sardinia.
Distribution: Corsica, Italy, Sardinia.

Roncus aetnensis Gardini & Rizzerio

Roncus aetnensis Gardini & Rizzerio, 1987b: 74–76, figs 12–17.

Type locality: Grotta Nuovalucello I n. 8 Si/CT, Catania, Sicily.
Distribution: Sicily.

Roncus alpinus L. Koch

Roncus alpinus L. Koch, 1873: 46–47; Canestrini, 1874: 228; Tömösváry, 1882b: 217–218, plate 3 fig. 7; Canestrini, 1884: no. 3, figs 1–3; Krauss, 1896: 628; Daday, 1918: 2; Szalay, 1931b: 371; Vachon, 1936a: 78; Vachon, 1937f: 128; Caporiacco, 1940: 7; Weidner, 1959: 116; Cîrdei, Bulimar & Malcoci, 1970: 11; Palmgren, 1973: 6–7; Paoletti, 1978: 92 (not seen); Gardini, Lattes & Rizzerio, 1981: 59; Gardini & Rizzerio, 1985: 47–53, figs 1–19; Gardini & Rizzerio, 1986a: 46–47.

Obisium (Roncus) alpinum (L. Koch): E. Simon, 1879a: 65, plate 18 fig. 24; Daday, 1888: 129, 185; E. Simon, 1900b: 594; Ellingsen, 1910a: 395; Lessert, 1911: 28–29, fig. 22; Kästner, 1928: 10, fig. 34.

Roncus (Roncus) alpinus (L. Koch): Beier, 1928: 309; Beier, 1932a: 128; Roewer, 1937: 249; Beier, 1952e: 4; Conci & Galvagni, 1956: 21 (not seen); Beier, 1958d: 162; Beier, 1963b: 184–185; Beier, 1963h: 154–155; Lazzeroni, 1969b: 405–406; Lazzeroni, 1969c: 244; Kofler, 1972: 287; Costantini, 1976: 123; Callaini, 1980a: 236–237; Gardini, 1980c: 110, 123, 124, 125, 126, 127; Callaini, 1986e: 253.

Chthonius sp.: Bartholomei, 1957: 60 (not seen) (misidentification).

Roncus (Parablothrus) stussineri dolomiticus Beier, 1952c: 59–60, fig. 3 (synonymized by Gardini & Rizzerio, 1986a: 46).

Roncus (Parablothrus) dolomiticus Beier: Beier, 1963b: 192–193, fig. 201.

Roncus aff. *alpinus* L. Koch: Ćurčić, 1988b: 72–74.

Type localities: of *Roncus alpinus*: Bad Razz, Alto Adige, Italy.
of *Roncus (Parablothrus) stussineri dolomiticus*: Primiero-Mis, Italy.
Distribution: Austria, France, Italy, Romania?, Switzerland.

Roncus andreinii (Caporiacco)

Chelifer andreinii Caporiacco, 1925: 123–124, fig. 1.
Roncus (Roncus) lubricus L. Koch: Beier, 1932a: 126–127 (in part).
Roncus andreinii (Caporiacco): Gardini & Rizzerio, 1985: 63–66, figs 51–57.

Type locality: Calvana, Firenze, Italy.
Distribution: Italy.

Roncus anophthalmus (Ellingsen)

Obisium (Roncus) anophthalmum Ellingsen, 1910a: 395–396.
Roncus (Parablothrus) anophthalmus (Ellingsen): Beier, 1932a: 130, fig. 155; Roewer, 1937: 249; Beier, 1939d: 79–80, fig. 97; Beier, 1963b: 187–188, fig. 195; Vachon & Gabbutt, 1964: fig. 8; Cerruti, 1968: fig. 5; Ćurčić, 1974a: 24.
Roncus (Parablothrus) cyclopius Beier, 1938a: 8 (synonymized by Beier, 1939d: 79).
Roncus anophthalmus (Ellingsen): Schawaller, 1981c: fig. 10.

Type locality: of *Obisium (Roncus) anophthalmum*: south Hercegovina (as Herzegovina), Yugoslavia.
of *Roncus (Parablothrus) cyclopius*: Vucija Pecina, S. of Popovo Polje, Yugoslavia.
Distribution: Yugoslavia.

Roncus antrorum (E. Simon)

Obisium (Blothrus) antrorum E. Simon, 1896: 374–375 (in part; see *Roncus gestroi* Beier).
Blothrus antrorum (E. Simon): Alzona, 1903: 14; Vachon, 1938a: fig. 38h.
Roncus (Parablothrus) antrorus (sic) (E. Simon): Beier, 1930c: 94; Beier, 1932a: 133, fig. 163; Roewer, 1937: 249.
Roncus (Parablothrus) antrorum (E. Simon): Brian, 1940: 401; Beier, 1953e: 108; Franciscolo, 1955: 127; Beier, 1970f: 320.
Roncus (Parablothrus) antrorum antrorum (E. Simon): Beier, 1963b: 199–200, fig. 211; Gardini, 1980c: 111, 122; Bologna & Taglianti, 1985: 227, fig. 9.
Roncus antrorum (E. Simon): Helversen, 1969: fig. 8.
Roncus antrorus (sic) (E. Simon): Gardini & Rizzerio, 1986a: 4–8, figs 5–9.
Not *Obisium (Blothrus) antrorum* E. Simon: E. Simon, 1898d: 22 (misidentification; see *Roncus gestroi* Beier and *Roncus ligusticus* Beier); E. Simon, 1900b: 594 (misidentification; see *Roncus ligusticus* Beier).
Not *Obisium antrorum* (E. Simon): Ellingsen, 1909a: 209 (misidentification; see *Roncus gestroi* Beier).
Not *Roncus (Parablothrus) antrorum* (E. Simon) ssp.: Mahnert, 1974b: 216–217, figs 16–17 (misidentification; see *Roncus paolettii* Mahnert).

Type locality: Grotta della Madonna (now Tana da Roveirola), Bardineto, Liguria, Italy.
Distribution: Italy.

Roncus araxellus Schawaller & Dashdamirov

Roncus araxellus Schawaller & Dashdamirov, 1988: 29–32, figs 42, 46–53.

Type locality: Shikahoh Reserve, Shikahoh, Kafan District, Armenia, U.S.S.R.
Distribution: U.S.S.R. (Armenia, Azerbaijan).

Roncus assimilis Beier

Roncus (Parablothrus) stussineri (E. Simon): Beier, 1928: 310 (misidentification).
Roncus (Parablothrus) assimilis Beier, 1931a: 19–20, fig. 11; Beier, 1932a: 132, fig. 160; Roewer, 1937: 249; Beier, 1939d: 80; Beier, 1958d: 162–163, fig. 1.
Roncus stussineri (E. Simon): Pretner & Strasser, 1931: 88 (not seen) (misidentification).

Roncus (Parablothrus) stussineri assimilis Beier: Beier, 1963b: 195–196, fig. 205; Beier, 1963h: 155; Beier, 1966h: 175; Helversen & Martens, 1972: fig. 9a; Gardini, 1980c: 114, 124, 125, 127, 128; Mahnert, 1980d: 30.

Roncus stussineri assimilis Beier: Paoletti, 1978: 92 (not seen).

Roncus assimilis Beier: Gardini & Rizzerio, 1986a: 22–26, figs 31–37.

Not *Roncus (Parablothrus) stussineri assimilis* Beier: Costantini, 1976: 123 (misidentification; see *Roncus troglophilus* Beier).

Type locality: Cevola della Presa, near Trieste, Italy.
Distribution: Italy.

Roncus baccettii Lazzeroni

Roncus (Parablothrus) baccettii Lazzeroni, 1969c: 232–236, figs 3–5; Puddu & Pirodda, 1973: 172; Beier, 1974g: 163, fig. 3; Mahnert, 1976d: fig. 9; Gardini, 1980c: 112, 131.

Type locality: Grotta San Pietro, Fluminimaggiore, Cagliari, Sardinia.
Distribution: Sardinia.

Roncus barbei Vachon

Roncus (Roncus) barbei Vachon, 1964a: 72–76, figs 1–9, 13; Vachon & Gabbutt, 1964: fig. 16; Heurtault, 1986c: 27.

Type locality: Tournier à Frespach, Lot-et-Garonne, France.
Distribution: France.

Roncus beieri Caporiacco

Roncus (Parablothrus) beieri Caporiacco, 1947c: 256, fig. 1; Beier, 1961f: 125; Beier, 1963b: 199, fig. 210; Gardini, 1980c: 112, 129.

Roncus beieri Caporiacco: Helversen, 1969: fig. 8; Callaini, 1986d: 538; Gardini & Rizzerio, 1986a: 32–34, figs 52–56.

Roncus (Parablothrus) sp. prope *beieri* Caporiacco: Bologna & Taglianti, 1985: 228, fig. 9.

Not *Roncus beieri* Caporiacco: Bologna & Bonzano, 1976: 66 (not seen); Bologna & Taglianti, 1982: 522 (not seen) (misidentifications; see *Roncus ligusticus* Beier).

Type locality: Buca de'Frati, Monteriggioni, Toscana, Italy.
Distribution: Italy.

Roncus bellesi Lagar

Roncus (Parablothrus) bellesi Lagar, 1972a: 18–20, fig. 1; Mahnert, 1977c: 94–96, figs 51–52; Zaragoza, 1986: fig. 1.

Type locality: Forat d'Os, Os de Balaguer, Lérida, Spain.
Distribution: Spain.

Roncus birsteini Krumpál

Roncus birsteini Krumpál, 1986: 168–170, figs 8–15.

Type locality: Snezhnaya, Rayon Sachi, R.S.F.S.R., U.S.S.R.
Distribution: U.S.S.R. (R.S.F.S.R.).

Roncus boneti Beier

Roncus (Parablothrus) boneti Beier, 1931a: 18–19, fig. 10; Beier, 1932a: 131, fig. 131; Beier, 1932g: fig. 170; Roewer, 1937: 249; Beier, 1939f: 180; Beier, 1963b: 193, fig. 202.

Roncus (Parablothrus) boneti boneti Beier: Mahnert, 1977c: 91–92; Zaragoza, 1986: fig. 1.

Type locality: Cueva de las Calaveras, Benidoleig, Alicante, Spain.
Distribution: Spain.

Roncus boneti tarbenae Mahnert

Roncus (Parablothrus) boneti tarbenae Mahnert, 1977c: 92–94, figs 46–50; Estany, 1978: 33–34, figs 1–2; Zaragoza, 1986: fig. 1.

Type locality: Cova del Somo, Tàrbena, Alicante, Spain.
Distribution: Spain.

Roncus caralitanus Gardini

Roncus caralitanus Gardini, 1981b: 129–133, figs 1–7.
Roncus cfr. *caralitanus* Gardini: Callaini, 1989: 144.

Type locality: Is Mortorius, Quartu Sant'Elena, Cagliari, Sardinia.
Distribution: Sardinia, Sicily.

Roncus carinthiacus Beier

Roncus (Parablothrus) stussineri carinthiacus Beier, 1934a: 58–59, fig. 6; Beier, 1952e: 4; Beier, 1956i: 8; Beier, 1963b: 195.
Roncus carinthiacus Beier: Gardini & Rizzerio, 1986b: 26–30, figs 38–45.

Type locality: Eggerloch, near Villach, Carinthia, Austria.
Distribution: Austria.

Roncus carusoi Gardini & Rizzerio

Roncus carusoi Gardini & Rizzerio, 1987b: 69–71, figs 1–7.

Type locality: Grotta Pozzo Baronazzo, Noto, Siracusa, Sicily.
Distribution: Sicily.

Roncus cassolai Beier

Roncus cassolai Beier: Puddu & Pirodda, 1973: 172 (nomen nudum).
Roncus (Parablothrus) cassolai Beier, 1974g: 165–166, fig. 2; Gardini, 1980c: 112, 131.

Type locality: Grotta Cava Romana, Nuxis, Sardinia.
Distribution: Sardinia.

Roncus caucasicus (Beier)

Microcreagris caucasica Beier, 1962b: 148–149, fig. 3; Beier, 1963b: 214, fig. 225; Kobakhidze, 1965b: 541; Kobakhidze, 1966: 704.
Roncus caucasicus (Beier): Mahnert, 1976c: 205.

Type locality: Rayon Mataradsa, Podstoschka, Georgia, U.S.S.R.
Distribution: U.S.S.R. (Georgia).

Roncus cerberus (E. Simon)

Obisium (Blothrus) cerberus E. Simon, 1879a: 67–68, plate 19 fig. 13.
Obisium cerberus E. Simon: Navás, 1925: 120 (possible misidentification).
Roncus (Parablothrus) cerberus (E. Simon): Beier, 1932a: 133, fig. 162; Roewer, 1937: 249; Beier, 1963b: 191–192, fig. 200; Leclerc & Heurtault, 1979: 245; Heurtault, 1986c: 27.

Type localities: Grotte de Vallon, Ardèche, France; Grotte Saint-Isch, Bouches-du-Rhône, France; and Grotte des Demoiselles, Hérault, France.
Distribution: France.

Roncus comasi Mahnert

Roncus (Parablothrus) comasi Mahnert, 1985a: 17–20, figs 1a–g; Gardini & Rizzerio, 1987b: 76.

Type locality: Sidi Bou Zouitine Cave, Djebel Serdj, Tunisia.
Distribution: Tunisia.

Roncus concii (Caporiacco)

Parablothrus concii Caporiacco: Conci, 1949: 121 (not seen); Conci, 1951: 17, 39 (nomen nudum).
Parablothrus concii Caporiacco, 1952a: 55–56, fig. 1.
Roncus (Parablothrus) stussineri concii (Caporiacco): Beier, 1963b: 194–195; Beier, 1963h: 155; Beier, 1966h: 175; Boscolo, 1968: 164 (not seen); Paoletti, 1978: 93 (not seen); Lazzeroni, 1969b: 406–407; Gardini, 1980c: 114–115, 124, 125, 126.
Roncus concii (Caporiacco): Gardini & Rizzerio, 1986a: 35–39, figs 57–64.

Type locality: Bus del Parolet, Monte Zugna, Trentino, Italy.
Distribution: Italy.

Roncus corcyraeus Beier

Roncus (Parablothrus) corcyraeus Beier, 1963b: 189–190, fig. 197; Beier, 1963h: 155; Mahnert, 1973a: 208–211, figs 1–6; Mahnert, 1977d: 97.
Roncus (Roncus) corcyraeus corcyraeus Beier: Mahnert, 1975a: 180.
Roncus corcyraeus Beier: Schawaller, 1981c: fig. 10; Schawaller, 1985b: 7.
Roncus corcyraeus corcyraeus Beier: Gardini, 1989: 61.

Type locality: Höhle Peristerograva, Kérkira (as Korfu), Greece.
Distribution: Greece.

Roncus corcyraeus minor Mahnert

Roncus (Parablothrus) corcyraeus minor Mahnert, 1975a: 180–183, figs 23–28; Schawaller, 1985b: 7.

Type locality: Karoucha Cave, near Sivros, Levkas, Greece.
Distribution: Greece.

Roncus corimanus Beier

Roncus (Roncus) corimanus Beier, 1951d: 96–97, fig. 1a; Beier, 1971a: 360; Mahnert, 1974a: 88.
Roncus (Roncus) glaber Beier, 1962b: 147–148, fig. 2; Beier, 1963b: 185–186, fig. 193 (synonymized by Schawaller, 1983b: 18).
Roncus glabor (sic) Beier: Kobakhidze, 1965b: 541; Kobakhidze, 1966: 703–704.
Roncus corimanus Beier: Schawaller, 1983b: 18–19, figs 52–53; Schawaller, 1983c: 369–370; Schawaller & Dashdamirov, 1988: 28, fig. 42.

Type localities: of *Roncus (Roncus) corimanus*: Lahidschan, Mazandaran, Iran.
of *Roncus (Roncus) glaber*: Rayon Mataradsa, Podstoschka, Georgia, U.S.S.R.
Distribution: Iran, U.S.S.R. (Georgia).

Roncus crassipalpus Rafalski

Roncus (Roncus) crassipalpus Rafalski, 1949: 110–116, figs 16, 17a; Beier, 1963b: 179.

Type locality: Meskish Mountains, Caucasus, U.S.S.R.
Distribution: U.S.S.R. (Caucasus).

Roncus dallaii Callaini

Roncus dallaii Callaini, 1979a: 111–118, figs 1a–f, 2, 3a–e, plates 1–4.

Type locality: Orroli, Sardinia.
Distribution: Sardinia.

Roncus drescoi Heurtault

Roncus drescoi Heurtault, 1986c: 27–28, fig. 3.

Type locality: Grottes Ultrera and Cap Raederis, Pyrénées Orientales, France.
Distribution: France.

Roncus duboscqi Vachon

Roncus duboscqi Vachon, 1937b: 39–40, figs 3–4.
Roncus (Roncus) duboscqi Vachon: Roewer, 1940: 343; Beier, 1955n: 96; Beier, 1959f: 121; Beier, 1961b: 24; Beier, 1963b: 184; Lagar, 1972a: 18; Leclerc & Heurtault, 1979: 245.
Roncus (Roncus) dubosqi (sic) Vachon: Heurtault, 1986c: 27.

Type locality: Mont Canigou, Pyrénées Orientales, France.
Distribution: France, Spain.

Roncus euchirus (E. Simon)

Obisium (Roncus) euchirus (sic) E. Simon, 1879a: 65–66, plate 18 fig 23; Tullgren, 1900: 160; Silvestri, 1922: 19.
Roncus euchirus (E. Simon): Tömösváry, 1882b: 218–219, plate 3 figs 5–6; Daday, 1918: 2; Vachon, 1938a: fig. 38i; Vachon, 1940g: 143; Gardini, 1982b: 151–154, figs 1–8.
Obisium euchirus (sic) E. Simon: Daday, 1889a: 16.
Obisium (Roncus) euchirum E. Simon: Nonidez, 1917: 38–39; Redikorzev, 1928: 123.
Obisium euchirum E. Simon: Ellingsen, 1908a: 416; Navás, 1923: 33; Navás, 1925: 121; Bacelar, 1928: 190.
Roncus (Roncus) euchirus (E. Simon): Beier, 1932a: 127; Roewer, 1937: 249; Franciscolo, 1955: 126; Beier, 1963b: 183; Gardini, 1980c: 110, 122; Bologna & Taglianti, 1985: 226.

Type locality: Martigues, Bouches-du-Rhône, France.
Distribution: Algeria, Balearic Islands, Bulgaria, France, Hungary, Italy, Portugal, Spain.

Roncus gestroi Beier

Obisium (Blothrus) antrorum E. Simon, 1896: 374–375 (misidentification, in part); E. Simon, 1898d: 22 (misidentification, in part; see *Roncus ligusticus* Beier).
Obisium abeillei (E. Simon): Gozo, 1908: 129 (not seen) (misidentification).
Obisium antrorum (E. Simon): Ellingsen, 1909a: 209 (misidentification).
Roncus (Parablothrus) gestroi Beier, 1930c: 95, fig. 2; Beier, 1932a: 135, fig. 166; Roewer, 1937: 249; Beier, 1953e: 108, fig. 3; Beier, 1963b: 201–202, fig. 213; Beier, 1970f: 320; Gardini, 1980c: 112–113, 120, 121.
Parablothrus gestroi (Beier): Caporiacco, 1951b: 103.
Roncus gestroi Beier: Helversen, 1969: fig. 8; Gardini & Rizzerio, 1985: 13–17, figs 16–21.

Type locality: Grotta di Cassana, near Borghetto di Vara, Liguria, Italy.
Distribution: Italy.

Roncus giganteus Mahnert

Roncus (Roncus) giganteus Mahnert, 1973b: 32–33, figs 25–34; Mahnert, 1975a: 180.
Roncus giganteus Mahnert: Schawaller, 1985b: 7.

Type locality: Katastarion, Zakynthos, Greece.
Distribution: Greece.

Roncus grafittii Gardini

Roncus grafittii Gardini, 1982c: 94–97, figs 10–17.

Type locality: Grotta Sa Conca 'e s'Abba, Nughedu S. Nicolò, Sardinia.
Distribution: Sardinia.

Roncus hibericus Beier

Roncus (Parablothrus) hibericus Beier, 1939f: 180–182, fig. 11; Roewer, 1940: 343;
    Beier, 1963b: 197–198, fig. 208; Cerruti, 1968: fig. 5; Mahnert, 1977c: 96; Zaragoza,
    1986: fig. 1.
Roncus (Parablothrus) ibericus (sic) Beier: Lagar, 1972a: 20.

Type locality: Avenc de Can Sadurni, near Begas, Barcelona, Spain.
Distribution: Spain.

Roncus insularis Beier

Roncus (Parablothrus) insularis Beier, 1938a: 8; Beier, 1939d: 80–81, figs 98–99;
    Beier, 1963b: 196–197, fig. 206; Ćurčić, 1974a: 24; Schawaller, 1981c: fig. 10.
Roncus insularis Beier: Ćurčić, 1988b: 82–84, figs 402–417.

Type locality: Zejava Jama, Brac (as Brazza) Island, Yugoslavia.
Distribution: Yugoslavia.

Roncus italicus (E. Simon)

Obisium (Roncus) italicum E. Simon, 1896: 374; E. Simon, 1900b: 594; Ellingsen,
    1910a: 396; Nonidez, 1917: 37.
Obisium italicum E. Simon: E. Simon, 1898d: 22; Navás, 1925: 121.
Roncus italicus (E. Simon): Alzona, 1903: 14; Caporiacco, 1936b: 329; Pavan, 1939:
    23 (not seen); Caporiacco, 1951b: 103; Gardini & Rizzerio, 1985: 53–59, figs 20–39.
Roncus (Roncus) italicus E. Simon: Beier, 1932a: 127; Roewer, 1937: 249; Brian, 1940:
    401; Franciscolo, 1951: 51 (not seen); Beier, 1953e: 107; Franciscolo, 1955: 125–126;
    Beier, 1963b: 180–181, fig. 190; Costantini, 1976: 124; Mahnert, 1976d: 309–311,
    figs 1–3; Gardini, 1980c: 110, 120, 121, 122, 123, 132; Bologna & Taglianti, 1985:
    226, fig. 9.
Roncus aff. *italicus* (E. Simon): Ćurčić, 1988b: 74–76, figs 346–368.

Type locality: Grotta della Madonna (now Tana da Roveirola), Bardineto, Italy.
Distribution: Italy, Sardinia.

Roncus jagababa Ćurčić

Roncus jagababa Ćurčić, 1988b: 76–78, figs 369–379.

Type locality: Tamar, near Planica, Kranjska Gora, Yugoslavia.
Distribution: Yugoslavia.

Roncus jaoreci Ćurčić

Roncus jaoreci Ćurčić, 1984a: 99–102, figs 6–16, map 1.

Type locality: Pestera Jaorec Cave, Velmej, 23 km NNE. of Ohrid, Yugoslavia.
Distribution: Yugoslavia.

Roncus julianus Caporiacco

Roncus julianus Caporiacco, 1949d: 140, fig. 1.
Roncus (Roncus) julianus Caporiacco: Beier, 1963b: 181; Lazzeroni, 1969b: 404; Kofler,
    1972: 287.
Roncus julianus (Caporiacco): Gardini & Rizzerio, 1985: 71–76, figs 80–92.

Type locality: Nevea, Alpi Giulie, Italy.
Distribution: Austria, Italy.

Roncus juvencus Beier

Roncus (Parablothrus) juvencus Beier, 1939f: 185–186, fig. 16; Roewer, 1940: 344;
    Beier, 1963b: 197, fig. 207; Vachon & Gabbutt, 1964: fig. 2; Lagar, 1972a: 20;
    Zaragoza, 1986: fig. 1.

Type locality: Cova Ramé near Capsanes, Tarragona, Cataluna, Spain.
Distribution: Spain.

Roncus lagari Beier

Roncus (Parablothrus) lagari Beier, 1972b: 15–17, fig. 1; Lagar, 1972a: 20.
Roncus (Parablothrus) lagari lagari Beier: Mahnert, 1977c: 96; Zaragoza, 1986: fig. 1.

Type locality: Cova del Cartanyá, La Riba, Tarragona, Spain.
Distribution: Spain.

Roncus lagari sendrai Lagar

Roncus (Parablothrus) lagari sendrai Lagar, 1972b: 47; Zaragoza, 1986: fig. 1.

Type locality: Cova del Codó, Montral, Tarragona, Spain.
Distribution: Spain.

Roncus leonidae Beier

Roncus (Parablothrus) boldorii Beier, 1931a: 20, fig. 12; Beier, 1932a: 132–133, fig.
    161; Roewer, 1937: 249 (junior primary homonym of *Roncus boldorii* (Beier)).
Roncus (Parablothrus) leonidae Beier, 1942: 136; Beier, 1963b: 193–194, fig. 203;
    Gardini, 1980c: 113, 126 (replacement name for *Roncus (Parablothrus) boldorii*
    Beier).
Roncus leonidae Beier: Gardini & Rizzerio, 1986a: 30–32, figs 46–51.

Type locality: Grotta del Covoletto (as Covoletto di Cereda), Valdagno, Veneto, Italy.
Distribution: Italy.

Roncus liebegotti Schawaller

Roncus liebegotti Schawaller, 1981c: 1–7, figs 1–10; Schawaller, 1985b: 7.

Type locality: Guira (= Yioúra), Voríai Sporádhes, Greece.
Distribution: Greece.

Roncus ligusticus Beier

Obisium (Blothrus) antrorum E. Simon: E. Simon, 1898d: 22 (misidentification, in
    part; see *Roncus gestroi* Beier); E. Simon, 1900b: 594 (misidentification).
Roncus (Parablothrus) ligusticus Beier, 1930c: 94–95, fig. 1; Beier, 1932a: 134, fig.
    164; Roewer, 1937: 249; Franciscolo, 1955: 128.
Roncus (Parablothrus) antrorum liguticus Beier: Beier, 1963b: 200; Gardini, 1980c:
    111–112, 122; Bologna & Taglianti, 1985: 227–228.
Roncus beieri Caporiacco: Bologna & Bonzano, 1976: 66 (not seen); Bologna &
    Taglianti, 1982: 522 (not seen) (misidentifications).
Roncus ligusticus Beier: Gardini & Rizzerio, 1986a: 8–13, figs 10–15; Gardini &
    Rizzerio, 1986b: 12–15, figs 5–10.

Type locality: Tana dello Scopeto, near Albenga, Liguria, Italy.
Distribution: Italy.

Roncus lonai Caporiacco

Roncus lonai Caporiacco, 1949b: 122–123, fig. 1.
Roncus (Roncus) lonai Caporiacco: Beier, 1963b: 186.

Type locality: Dukati (as Ducati), Albania.
Distribution: Albania.

Roncus lubricus L. Koch

Roncus lubricus L. Koch, 1873: 44–45; Canestrini, 1874: 227–228; Stecker, 1874b:
    237; E. Simon, 1878a: 153; Daday, 1880a: 192; Tömösváry, 1882b: 215; Canestrini,

1884: no. 4, figs 1–4; O.P.-Cambridge, 1892: 217–218, plate B figs 10, 10a–b; Alzona, 1903: 14; Kew, 1909b: 259; Foster, 1912: 245; Daday, 1918: 2; Schenkel, 1929b: 14; Boldori, 1932: 127; Boldori, 1935: 27, 28; Vachon, 1935j: 188; Caporiacco, 1936b: 329; Wolf, 1937: 222; Brian, 1940: 401; Vachon, 1938a: fig. 37h; Vachon, 1940g: 143; Beier, 1948a: 189; Caporiacco, 1948c: 240; Caporiacco, 1950: 116; Balazuc, Dresco, Henrot & Nègre, 1951: 320; Turk, 1951: 169; G. O. Evans & Browning, 1954: 12, figs 4, 13; Patrizi, 1954: 26 (not seen); Cloudsley-Thompson, 1958: 96; Cirdei & Gutu, 1959: 2, fig. 1; Weidner, 1959: 116; Beier, 1961b: 24; George, 1961: 38; Kölzel, 1963: 175; Muchmore, 1963b: 210; Kobakhidze, 1965b: 541; Kobakhidze, 1966: 704; Zangheri, 1966: 531; Gabbutt & Vachon, 1967: 476–497, figs 1a–d, 2a–c, 3a–d, 4a–e, 5a–h, 6a–h, 7a–b, 8a–c, 9a–d; Turk, 1967: 153; Baccetti & Lazzeroni, 1969: 422, figs 6–7; Gabbutt, 1969a: 186, 187, 190, 194, figs 2a–d, 5a–e, 8a–b, 12a–c; Gabbutt, 1969b: 414–427; Gabbutt, 1969c: 231; Muchmore, 1969d: 66; Weygoldt, 1969a: 122; Gabbutt, 1970c: 9–11, fig. 6; Lazzeroni, 1970a: 207; Legg, 1970a: fig. 2a; Legg, 1970b: fig. 3(6); Legg, 1971a: 476, fig. 2a; Salmon, 1972: 66–67; Delhez, 1973: 220–221 (not seen); Legg, 1973b: 430, fig. 2m; Muchmore, 1973g: 122; Legg, 1975a: 66; Legg, 1975c: 124–125, figs 1d, 7e, 11a–b, 14d, 16a–b, 20a–b; Ćurčić, 1976b: 177–178; Bellés, 1978: 87; Jones, 1978: 93; Paoletti, 1978: 92 (not seen); P. A. Wood & Gabbutt, 1978: 176–182, figs 11–15, 26–30; Ćurčić, 1979c: 226; Judson, 1979a: 59; Rundle, 1979: 47; P. A. Wood & Gabbutt, 1979a: 285–292; P. A. Wood & Gabbutt, 1979b: 329–336; Ćurčić, 1980a: 13, figs 8a–b; Heurtault, 1980d: 89; Jones, 1980c: map 11; Gardini, Lattes & Rizzerio, 1981: 59; Ćurčić, 1982e: 24–28, figs 258–301; Fussey, 1982: 111–112; Gardini, 1982c: 97–99, figs 18–21; Ćurčić & Dimitrijević, 1983b: 283–284, fig. 2; Gardini, 1983a: 78–80, figs 1–14; Ćurčić & Dimitrijević, 1985: 93, figs 1a–b, 2a–b; Schawaller, 1985b: 7; Ćurčić & Dimitrijević, 1986b: 18–21; Judson, 1987: 369; Callaini, 1988b: 42; Callaini & Dallai, 1989: 87, figs 4–6; Ćurčić, 1989b: 3–4, figs 17a–e.

Obisium (Roncus) lubricum (L. Koch): E. Simon, 1879a: 63–64, plate 18 fig. 22; E. Simon, 1880b: xxxv; Daday, 1888: 128–129, 184–185; E. Simon, 1898d: 21; E. Simon, 1899d: 86; E. Simon, 1900b: 594; Gestro, 1904: 14 (not seen); Ellingsen, 1905d: 8–9; Ellingsen, 1910a: 396; Kew, 1911a: 53, fig. 15; Lessert, 1911: 27–28, fig. 21; Ellingsen, 1912c: 169–170; Kew, 1916b: 79–80; Nonidez, 1917: 37–38, figs 11a–c; Navás, 1921: 167; Redikorzev, 1928: 123.

Roncus lubricus cavernicola Tömösváry, 1882b: 216–217, plate 3 figs 1–4 (synonymized by Beier, 1932a: 126).

Obisium lubricum (L. Koch): Daday, 1889c: 25; Daday, 1889d: 81; Fabiani, 1904: 11 (not seen); Ellingsen, 1907a: 169; Ellingsen, 1908a: 416; Ellingsen, 1908c: 670; Gozo, 1908: 130 (not seen); Ellingsen, 1909a: 207, 209–210, 214, 216; Ellingsen, 1910b: 63; Krausse-Heldrungen, 1912: 65; Ellingsen, 1913a: 455; Navás, 1918: 135; Navás, 1919: 212; Navás, 1923: 33; Navás, 1925: 122; Donisthorpe, 1927: 183; Bacelar, 1928: 190.

Obisium sylvaticum C. L. Koch: O.P.-Cambridge, 1892: 214–215, plate B fig. 7 (misidentification).

Obisium (Roncus) tenax Navás, 1918: 115–117, figs 3a–b (synonymized by Beier, 1932a: 126).

Olpium catalaunicum Navás, 1921: 167, figs 4a–c; Navás, 1925: 123 (synonymized by Beier, 1939f: 178).

Obisium tenax Navás: Navás, 1925: 122, figs 13a–b.

Roncus (Roncus) lubricus L. Koch: Beier, 1928: 306–307, fig. 8a; Kästner, 1928: 10, fig. 33; Beier, 1929a: 364; Beier, 1929e: 454; Beier, 1931d: 96; Beier, 1932a: 126–127, fig. 152 (in part; see *Roncus pugnax* (Navás)); Hadži, 1933b: 165–166; Roewer, 1937: 249; Beier, 1939c: 18; Beier, 1939e: 311; Beier, 1939f: 178; Beier, 1942: 134; Beier, 1952e: 4; Beier, 1955n: 96; Franciscolo, 1955: 126; Beier, 1958d: 162; Beier,

1958b: 27; Beier, 1958c: 135, 137; Beier, 1963a: 256; Beier, 1963b: 182, fig. 191; Beier, 1963h: 154; Beier, 1963i: 284; Vachon & Gabbutt, 1964: figs 23–27; Lazzeroni, 1969a: 328; Mahnert, 1975a: 179–180; Mahnert, 1975c: 192–194; Beier, 1977b: 4; Mahnert, 1977c: 88; Mahnert, 1977d: 97, fig. 2; Estany, 1978: 33; Mahnert, 1978e: 295; Crocker, 1978: 8; Callaini, 1982a: 17, 18–19, figs 1c–d; Inzaghi, 1981: 72; Bologna & Taglianti, 1985: 227; Legg & Jones, 1988: 84–86, figs 18a, 18b(a–h).

Neobisium (Roncus) lubricus (L. Koch): J. C. Chamberlin, 1930: 12.

Neobisium lubricum (L. Koch): J. C. Chamberlin, 1931a: fig. 35i.

Neobisium (Neobisium) (?) *catalaunicum* (Navás): Beier, 1932a: 106; Roewer, 1937: 248.

Roncus lubricum (sic) L. Koch: Caporiacco, 1949c: 133; Caporiacco, 1949d: 139–140.

Roncus (Roncus) lubricus lubricus L. Koch: Rafalski, 1949: fig. 7b'; Baccetti & Lazzeroni, 1967: 352–360, plate 29 figs 1–4, plate 30 figs 1–2, plates 31–37, plate 38 figs 1–3, plate 39; Lazzeroni, 1969b: 404–405; Lazzeroni, 1969c: 231–232; Cîrdei, Bulimar & Malcoci, 1970: 11; Lazzeroni, 1970c: 104; Lagar, 1972a: 18; Lagar, 1972b: 47; Puddu & Pirodda, 1973: 171–172; Ćurčić, 1974a: 23–24; Costantini, 1976: 124; Callaini, 1979b: 137; Callaini, 1980a: 233–234; Gardini, 1980c: 110–111, 120, 121, 122, 123, 125, 126, 127, 129; Callaini, 1983a: 150–151; Callaini, 1983c: 289–291.

Roncus cfr. *lubricus lubricus* L. Koch: Callaini, 1980a: 234–236, fig. 4.

Type localities: of *Roncus lubricus*: Bloxworth, Dorset, England, Great Britain.
of *Roncus lubricus cavernicola*: many localities in Hungary.
of *Obisium (Roncus) tenax*: Valmadrid, Spain.
of *Olpium catalaunicum*: Vallvidrera, Spain.
Distribution: Albania, Algeria, Austria, Balearic Islands, Belgium, Bulgaria, Crete, Corsica, Czechoslovakia, France, Great Britain, Greece, Hungary, Ireland, Italy, Malta, Morocco, Portugal, Romania, Sardinia, Sicily, Spain, St Helena, Switzerland, U.S.A. (New York), U.S.S.R. (R.S.F.S.R., Ukraine), Yugoslavia.

Roncus lubricus dalmatinus Hadži

Roncus (Roncus) lubricus dalmatinus Hadži, 1933b: 170–175, figs 25a–b, 26, 27a–b; Hadži, 1933c: 191–193; Beier, 1963b: 181–182; Beier, 1963h: 154; Ćurčić, 1974a: 24; Ćurčić, 1976b: 178.

Type locality: Split, Yugoslavia.
Distribution: Greece, Yugoslavia.

Roncus lubricus tenuis Hadži

Roncus (Roncus) lubricus tenuis Hadži, 1933b: 166–170, figs 22a–c, 23a–d, 24a–d; Hadži, 1933c: 189–191; Hadži, 1938: 13–14, fig. 18; Rafalski, 1949: fig. 7b; Beier, 1963b: 183; Ćurčić, 1974a: 24; Ćurčić, 1976b: 178.

Type locality: Malinska, Krk, Yugoslavia.
Distribution: Yugoslavia.

Roncus lychnidis Ćurčić

Roncus lychnidis Ćurčić, 1984a: 98, figs 1–5, map 1.

Type locality: Pestera Orevce Cave, Pestani, near Ohrid, Yugoslavia.
Distribution: Yugoslavia.

Roncus mahnerti Ćurčić & Beron

Roncus mahnerti Ćurčić & Beron, 1981: 70–72, 83–84, figs 5a–k, 6.

Type locality: Vodnata Dupka Cave, Botunja, Bulgaria.
Distribution: Bulgaria.

Roncus melitensis Gardini & Rizzerio

Roncus melitensis Gardini & Rizzerio, 1987b: 71–73, figs 8–11.

Type locality: Grotta dei Pipistrelli (Tal-Friefet), Birzebbuga, Malta.
Distribution: Malta.

Roncus melloguensis Gardini

Roncus melloguensis Gardini, 1982c: 109–110, figs 44–49.

Type locality: Sa Grutta de S'Ingultidolzu, Santu Giagu, Romana, Sardinia.
Distribution: Sardinia.

Roncus menozzii (Caporiacco)

Obisium (Roncus) menozzii Caporiacco, 1923: 132–134, fig. 1.
Roncus (Roncus) menozzii Caporiacco: Beier, 1932a: 128; Roewer, 1937: 249; Beier, 1963b: 185.
Roncus menozzii Caporiacco: Gardini & Rizzerio, 1985: 59–63, figs 40–50; Gardini & Rizzerio, 1986a: 47–49, figs 71–75.

Type locality: Agro Mutinensi, Castelvetro, Italy.
Distribution: Italy.

Roncus microphthalmus (Daday)

Obisium (Roncus) microphthalmum Daday, 1889a: 21–22, fig. 2.
Roncus (Roncus) microphthalmus (Daday): Beier, 1932a: 128; Roewer, 1937: 249; Rafalski, 1949: 107–110, figs 14–15; Beier, 1963b: 183; Beier, 1973c: 225.
Roncus (Parablothrus) brignolii Beier, 1973c: 232–233, fig. 5 (synonymized by Schawaller, 1983b: 17).
Roncus microphthalmus (Daday): Kobakhidze, 1965b: 541; Kobakhidze, 1966: 703; Schawaller, 1983b: 17–18, figs 47–51; Schawaller, 1983c: 369; Schawaller & Dashdamirov, 1988: 25–28, figs 40–42.

Type localities: of *Obisium (Roncus) microphthalmum*: Lenkoran, Azerbaijan, U.S.S.R.
of *Roncus (Parablothrus) brignolii*: Kalkandere, Rize, Turkey.
Distribution: Iran, Turkey, U.S.S.R. (Azerbaijan, Georgia, R.S.F.S.R.).

Roncus neotropicus Redikorzev

Roncus neotropicus Redikorzev: Bellés, 1987: 68–69; Redikorzev, 1937: 146–147, figs 1–2.
Roncus (Roncus) neotropicus Roewer, 1940: 343.
Roncus (Roncus) balearicus Beier, 1961b: 25–26, fig. 1; Beier, 1963b: 184; Lagar, 1972a: 18; Mahnert, 1977c: 88–89 (synonymized by Bellés, 1987: 68–69).
Rhoncus (sic) *(Roncus) balearicus* Beier: Orghidan, Dumitresco & Georgesco, 1975: 30.

Type locality: of *Roncus neotropicus*: Cueva St. Inez, San Antonio, Ibiza, Balearic Islands.
of *Roncus (Roncus) balearicus*: Sóller, Mallorca, Balearic Islands.
Distribution: Balearic Islands.

Roncus novus Beier

Roncus (Roncus) novus Beier, 1931d: 96–97, fig. 5; Beier, 1932a: 129, fig. 154; Roewer, 1936: fig. 24b; Roewer, 1937: 249; Beier, 1963b: 186, fig. 194; Vachon & Gabbutt, 1964: figs 5, 11.

Type locality: Katholiko near Akrotiri, Crete.
Distribution: Crete.

Roncus numidicus Callaini

Roncus numidicus Callaini, 1983e: 220–224, figs 3a–d.

Type locality: Forêt de Akfadou, Algeria.
Distribution: Algeria.

Roncus paolettii Mahnert

Roncus (Parablothrus) antrorum (E. Simon) ssp.: Mahnert, 1974b: 216–217, figs 16–17 (misidentification).
Roncus (Parablothrus) paolettii Mahnert, 1980d: 29–30; Gardini, 1980c: 113, 125; Mahnert, 1980g: 100.
Roncus paolettii Mahnert: Gardini & Rizzerio, 1986a: 44–46.

Type locality: Al Landre, Revine Lago, Treviso, Italy.
Distribution: Italy.

Roncus parablothroides Hadži

Roncus (Parablothrus) parablothroides Hadži, 1938: 14–20, figs 19a–d, 20a–c, 21a–f; Hadži, 1939b: 32–34, figs 8a–g; Beier, 1949: 6–9, figs 6–7; Beier, 1963b: 190; Ćurčić, 1974a: 24; Mahnert, 1979b: 266.
Roncus parablothroides Hadži: Beier, 1969b: 192; Ćurčić & Beron, 1981: 64–70, 81–83, figs 1a–g, 2a–f, 3a–i, 4a–i, 6; Schawaller, 1981c: fig. 10; Ćurčić, 1982a: 182–184, figs 1a–g, 3; Schawaller, 1985b: 7; Schawaller & Dashdamirov, 1988: 29, figs 42–45.
Not *Roncus parablothroides* Hadži: Schawaller, 1983c: 370 (misidentification; see *Roncus viti* Mahnert).

Type locality: Patiske, Yugoslavia.
Distribution: Bulgaria, Greece, Turkey, U.S.S.R. (Azerbaijan), Yugoslavia.

Roncus peramae Helversen

Roncus (Parablothrus) peramae Helversen, 1969: 225–229, figs 1–6, 7a–c, 8; Mahnert, 1975a: 183.
Roncus peramae Helversen: Schawaller, 1981c: fig. 10; Schawaller, 1985b: 7.

Type locality: near Ioannina (as Joannina), Greece.
Distribution: Greece.

Roncus pljakici Ćurčić

Roncus pljakici Ćurčić, 1973b: 128–130, figs 1a–h, 2a–i; Ćurčić, 1976b: 178, fig. 2; Schawaller, 1981c: fig. 10; Ćurčić, 1983b: fig. 2.
Roncus (Parablothrus) pljakici Ćurčić: Ćurčić, 1974a: 24.
Roncus pljakici pljakici Ćurčić: Ćurčić, 1982a: fig. 3.

Type locality: Pecina u selu Vrelo Cave, Mt Stara Planina, 20 km E. of Pirot, Yugoslavia.
Distribution: Yugoslavia.

Roncus pljakici remesianensis Ćurčić

Roncus pljakici remesianensis Ćurčić, 1981a: 107–109, figs 10–18; Ćurčić, 1982a: fig. 3.

Type locality: Govedja Pecina Cave, Crnokliste, near Bela Palanka, Yugoslavia.
Distribution: Yugoslavia.

Roncus pljakici timacensis Ćurčić

Roncus pljakici timacensis Ćurčić, 1981a: 105–107, figs 1–9, 18.

Type locality: Pecina Bozja Vrata Cave, Beloinje, near Svrljig, Yugoslavia.
Distribution: Yugoslavia.

Roncus podaga Ćurčić

Roncus podaga Ćurčić, 1988b: 79–80, figs 380–395.

Type locality: Pecina Lokvina Cave, Mt Mosec, N. of Split, Yugoslavia.
Distribution: Yugoslavia.

Roncus pripegala Ćurčić

Roncus pripegala Ćurčić, 1988b: 80–81, figs 396–401.

Type locality: Velika Pecina Cave, Neoric, N. of Split, Yugoslavia.
Distribution: Yugoslavia.

Roncus puddui Mahnert

Roncus (Parablothrus) puddui Mahnert, 1976d: 311–314, figs 4–9; Gardini, 1980c: 113–114, 132.

Type locality: Grotte Pirosu, Santadi, Sardinia.
Distribution: Sardinia.

Roncus pugnax (Navás)

Obisium pugnax Navás, 1918: 117–119, figs 4a–c; Navás, 1925: 122, figs 14a–c.
Roncus (Roncus) lubricus L. Koch: Beier, 1932a: 126–127 (misidentification, in part).
Roncus (Roncus) pugnax (Navás): Beier, 1939f: 178–180, figs 9–10; Roewer, 1940: 343; Beier 1955n: 96; Beier, 1959f: 121; Beier, 1963b: 180, fig. 189; Beron, 1972: 13; Lagar, 1972a: 18; Mahnert, 1977c: 88; Estany, 1978: 33; Gardini, 1980c: 111, 133.
Roncus pugnax (Navás): Beier, 1961b: 24; Gardini, 1981b: figs 14–15.
Rhoncus (sic) *(Roncus) pugnas* (sic) (Navás): Orghidan, Dumitresco & Georgesco, 1975: 30.

Type locality: Zaragoza, Spain.
Distribution: Balearic Islands, Corsica, Spain.

Roncus remyi Beier

Roncus (Parablothrus) remyi Beier, 1934a: 59, fig. 7; Roewer, 1937: 249; Beier, 1963b: 190–191, fig. 198; Heurtault, 1986c: 28.

Type locality: Fôret de Sorède, Pyrénées Orientales, France.
Distribution: France.

Roncus sandalioticus Gardini

Roncus sandalioticus Gardini, 1982c: 104–109, figs 37–43.

Type locality: Grotta di Monte Majore, Monte Majore, Thiesi, Sardinia.
Distribution: Sardinia.

Roncus sardous Beier

Roncus (Parablothrus) sardous Beier, 1955k: 44–45, fig. 2; Beier, 1956d: 131; Beier, 1959b: 245; Dell'Oca & Pozzi, 1959: 131 (not seen); Beier, 1963b: 198–199, fig. 209; Vachon & Gabbutt, 1964: fig. 13; Cerruti, 1968: 219, fig. 5; Lazzeroni, 1969c: 232; Beier, 1970f: 320; Puddu & Pirodda, 1973: 172; Gardini, 1980c: 114, 132, 133.

Type locality: Grotte Scavi Taramelli, Dorgali, Sardinia.
Distribution: Sardinia.

Roncus setosus Beier

Roncus (Parablothrus) setosus Zaragoza, 1982: 102–108, figs 1–7; Zaragoza, 1986: fig. 1.

Type locality: Sa Cova des Vells, Tárbena, Alicante, Spain.
Distribution: Spain.

Roncus siculus Beier

Roncus (Parablothrus) siculus Beier, 1963a: 257–258, fig. 2; Beier, 1975a: 56; Gardini, 1980c: 114, 131.
Roncus siculus Beier: Gardini & Rizzerio, 1986a: 39–44, figs 65–70; Gardini & Rizzerio, 1987b: 67–68.

Type locality: Grotta Palombara, Siracusa, Sicily.
Distribution: Sicily.

Roncus sotirovi Ćurčić

Roncus sotirovi Ćurčić, 1982a: 184–187, figs 2a–i, 3.

Type locality: Pecina Djeverica Cave, Vlasi, near Dimitrovgrad, Yugoslavia.
Distribution: Yugolslavia.

Roncus stussineri (E. Simon)

Obisium (Blothrus) stussineri E. Simon, 1881a: 301–302, fig. 3.
Blothrus brevimanus G. Joseph, 1882: 21–22 (synonymized by Beier, 1963b: 194).
Blothrus brachydactylus G. Joseph, 1882: 21 (nomen nudum) (synonymized with *Blothrus brevimanus* by G. Joseph, 1882: 21).
Obisium (Roncus) stussineri E. Simon: E. Simon, 1898d: 21–22; E. Simon, 1900b: 594.
Roncus (Parablothrus) stussineri (E. Simon): Beier, 1929d: 156; Beier, 1932a: 130, fig. 156; Roewer, 1937: 249; Beier, 1939d: 80; Brian, 1940: 401; Franciscolo, 1955: 127; Helversen & Martens, 1972: 114–115, figs 9b, 10a; Mahnert, 1980g: 100.
Roncus (Parablothrus) minoius Beier, 1931d: 97–98, fig. 6; Beier, 1932a: 131–132, fig. 159; Roewer, 1937: 249; Beier, 1939d: 81; Beier, 1963b: 191, fig. 199 (synonymized by Helversen & Martens, 1972: 114).
Roncus (Parablothrus) (?) brevimanus (G. Joseph): Beier, 1932a: 135; Roewer, 1937: 249.
Roncus stussineri (E. Simon): Wolf, 1937: 222; Gardini, Lattes & Rizzerio, 1981: 59; Gardini & Rizzerio, 1986a: 2–4, figs 1–4.
Roncus (Parablothrus) stussineri stussineri (E. Simon): Beier, 1963b: 194, fig. 204; Ćurčić, 1974a: 24.
Roncus aff. stussineri (E. Simon): Ćurčić, 1988b: 71–72, figs 339–345.
Not *Obisium (Roncus) stussineri* E. Simon: E. Simon, 1898d: 21–22; E. Simon, 1900b: 594 (misidentifications; see *Roncus troglophilus* Beier).
Not *Obisium stussineri* E. Simon: Gestro, 1887: 497 (misidentification; see *Roncus troglophilus* Beier).
Not *Roncus (Parablothrus) stussineri* (E. Simon): Beier, 1928: 310 (misidentification; see *Roncus assimilis* Beier).
Not *Roncus stussineri* (E. Simon): Pretner & Strasser, 1931: 88 (not seen) (misidentification; see *Roncus assimilis* Beier)).
Not *Ronchus* (sic) *stussineri* E. Simon: Conci, 1951: 21, 25, 39 (misidentification; see *Roncus troglophilus* Beier).

Type localities: of *Obisium (Blothrus) stussineri*: Jama pod Smarno, Ljubljana, (as Laibach), Yugoslavia.
of *Blothrus brevimanus*: Kevderza jama, Yugoslavia; Ihanska jama, Yugoslavia; and Benkotova jama, Yugoslavia.
of *Roncus (Parablothrus) minoius*: Katholiko Cave near Akrotiri, Crete, but quite possibly mislabelled (see Helversen & Martens, 1972).
Distribution: Italy, Yugoslavia.

! Roncus succineus Beier

Roncus succineus Beier, 1955g: 50–51, fig. 2; Schawaller, 1978: 3.

Type locality: Baltic Amber.
Distribution: Baltic Amber.

Roncus transsilvanicus Beier

Roncus (Roncus) transsilvanicus Beier, 1928: 307–309, fig. 8b; Beier, 1932a: 127–128, fig. 153; Roewer, 1937: 249; Beier, 1939c: 18–19; Hadži, 1939a: 206–208, figs 11a–d; Beier, 1963b: 183–184, fig. 192; Beier, 1964c: 212; Rafalski, 1967: 13.
Roncus transsilvanicus Beier: Ionescu, 1936: 4; Jędryczkowski, 1987a: 345, fig. 6; Schawaller, 1989: 10.
Rhoncus (sic) *(Rhoncus)* (sic) *transsilvanicus* Beier: Dumitrescu, 1976: 274–275.

Type locality: Kronstadt (now Brasov), Romania.
Distribution: Czechoslovakia, Poland, Romania, U.S.S.R. (Ukraine).

Roncus troglophilus Beier

Obisium stussineri E. Simon: Gestro, 1887: 497 (misidentification).
Obisium (Roncus) stussineri E. Simon: E. Simon, 1898d: 21–22; E. Simon, 1900b: 594 (misidentifications).
Roncus (Parablothrus) troglophilus Beier, 1931a: 17–18, fig. 9; Beier, 1932a: 130–131, fig. 137; Roewer, 1937: 249; Beier, 1942: 134; Beier, 1953e: 107; Franciscolo, 1955: 127.
Roncus troglophilus Beier: Boldori, 1932: 127; Gardini & Rizzerio, 1986a: 17–22, figs 22–30.
Ronchus (sic) *stussineri* E. Simon: Conci, 1951: 21, 25, 39 (misidentification).
Roncus (Parablothrus) stussineri troglophilus Beier: Beier, 1963b: 194; Costantini, 1976: 123; Gardini, 1980c: 115, 121, 122.
Roncus (Parablothrus) stussineri assimilis Beier: Costantini, 1976: 123 (misidentification).
Roncus (Parablothrus) strussineri (sic) *troglophilus* (E. Simon): Bologna & Taglianti, 1985: 228.

Type locality: Büs del Füs (as Buco del Fuso), Brione, Lombardia, Italy.
Distribution: Italy.

Roncus trojanicus Ćurčić

Roncus trojanicus Ćurčić, 1988b: 84–86, figs 418–422.

Type locality: Baretina Pecina Cave, Okrug Gornji, near Trogir, Yugoslavia.
Distribution: Yugoslavia.

Roncus turritanus Gardini

Roncus turritanus Gardini, 1982c: 99–103, figs 22–30.
Roncus gr. *turritanus* Gardini: Gardini, 1982c: 103–104, figs 31–36.

Type locality: Grotta di Molafà, Molafà, Sassari, Sardinia.
Distribution: Sardinia.

Roncus vidali Lagar

Roncus (Parablothrus) vidali Lagar, 1972b: 47–49, fig. 1; Mahnert, 1977c: 89–91, figs 42–45; Zaragoza, 1986: fig. 1.
Microcreagris juliae Lagar, 1972b: 49–51, fig. 2 (synonymized by Bellés, 1987: 70).

Type localities: of *Roncus (Parablothrus) vidali*: Cova del Telémetro, Alcudia, Mallorca, Balearic Islands.
of *Microcreagris juliae* : Cinto, Lluc, Mallorca, Balearic Islands.
Distribution: Balearic Islands.

Roncus viti Mahnert

Roncus (Roncus) viti Mahnert, 1974a: 88–90, figs 1–7.
Roncus parablothroides Hadži: Schawaller, 1983c: 370 (misidentification).

Type locality: Guilan, Nav's Valley, Iran.
Distribution: Iran.

Roncus vulcanius Beier

Roncus (Parablothrus) vulcanius Beier, 1938a: 8.
Roncus (Parablothrus) vulcanius vulcanius Beier: Beier, 1939d: 76–78, figs 91–92; Beier, 1963b: 188, fig. 196; Ćurčić, 1974a: 24–25.
Roncus vulcanius vulcanius Beier: Schawaller, 1981c: fig. 10.

Type locality: Poganaca Cave, near Grepci, Yugoslavia.
Distribution: Yugoslavia.

Roncus vulcanius crassimanus Beier

Roncus (Parablothrus) crassimanus Beier, 1938a: 8.
Roncus (Parablothrus) vulcanius crassimanus Beier, 1939d: 78–79, figs 95–96; Beier, 1963b: 188–189; Ćurčić, 1974a: 25.
Roncus vulcanius crassimanus Beier: Schawaller, 1981c: fig. 10.

Type locality: Ostasevica Cave, Mljet (as Meleda) Island, Yugoslavia.
Distribution: Yugoslavia.

Roncus zoiai Gardini & Rizzerio

Roncus zoiai Gardini & Rizzerio, 1987a: 283–289, figs 1–9.

Type locality: Grotta Conca 'e Crapa, Monte Albo, Lula, Sardinia.
Distribution: Sardinia.

Genus Saetigerocreagris Ćurčić

Saetigerocreagris Ćurčić, 1984b: 156–158.

Type species: *Saetigerocreagris setifera* Ćurčić, 1984b, by original designation.

Saetigerocreagris phyllisae (J. C. Chamberlin)

Microcreagris phyllisae J. C. Chamberlin, 1930: 31–32; Beier, 1932a: 153; Roewer, 1937: 252; Hoff, 1958: 11; J. C. Chamberlin, 1962: 336–338, figs 12a–j.
Saetigerocreagris phyllisae (J. C. Chamberlin): Ćurčić, 1984b: 156, figs 11, 32.

Type locality: Coronado, California, U.S.A.
Distribution: U.S.A. (California).

Saetigerocreagris setifera Ćurčić

Saetigerocreagris setifera Ćurčić, 1984b: 158, figs 9–10, 31, 46–49.

Type locality: Altadena, California, U.S.A.
Distribution: U.S.A. (California).

Genus Simonobisium Heurtault

Simonobisium Heurtault, 1974a: 1085.

Type species: *Obisium myops* E. Simon, 1881c, by monotypy.

Simonobisium myops (E. Simon)

Obisium myops E. Simon, 1881c: 91; E. Simon, 1898d: 21; E. Simon, 1900b: 594.
Neobisium (Neobisium) myops (E. Simon): Beier, 1932a: 90; Roewer, 1937: 246; Beier, 1963b: 90–91.
Simonobisium myops (E. Simon): Heurtault, 1974a: 1085–1090, figs 1–4; Callaini, 1986b: 373–375, figs 6g–i.
Not *Obisium (Obisium) myops* E. Simon: Ellingsen, 1912c: 173 (misidentification; see *Neobisium (Blothrus) cavernarum* (L. Koch)).

Type locality: Sospel, Alpes Maritimes, France.
Distribution: France, Italy.

Genus **Tartarocreagris** Ćurčić

Tartarocreagris Ćurčić, 1984b: 163.

Type species: *Microcreagris infernalis* Muchmore, 1969a, by original designation.

Tartarocreagris infernalis (Muchmore)

Microcreagris infernalis Muchmore, 1969a: 15–17, fig. 12; Rowland & Reddell, 1976: 15.
Tartarocreagris infernalis (Muchmore): Ćurčić, 1984b: 163–164, figs 21, 40.

Type locality: Core Hole Cave, Inner Space Caverns, 2 miles S. of Georgetown, Williamson County, Texas, U.S.A.
Distribution: U.S.A. (Texas).

Genus **Trisetobisium** Ćurčić

Trisetobisium Ćurčić, 1982c: 57–59.

Type species: *Microcreagris fallax* J. C. Chamberlin, 1962, by original designation.

Trisetobisium fallax (J. C. Chamberlin)

Microcreagris fallax J. C. Chamberlin, 1962: 343–345, figs 15a–h.
Trisetobisium fallax (J. C. Chamberlin): Ćurčić, 1982c: 60, figs 1–3.

Type locality: Gist Cave, Colbert County, Alabama, U.S.A.
Distribution: U.S.A. (Alabama, North Carolina).

Genus **Tuberocreagris** Ćurčić

Tuberocreagris Ćurčić, 1978a: 112–113; Ćurčić, 1984b: 150–152; Ćurčić, 1989c: 352.

Type species: *Olpium rufulum* Banks, 1891, by original designation.

Tuberocreagris lata (Hoff)

Microcreagris lata Hoff, 1945i: 323–327, figs 10–12; Pearse, 1946: 257; Hoff, 1958: 12; J. C. Chamberlin, 1962: fig. 15f; Zeh, 1987b: 1086.
Tuberocreagris lata (Hoff): Ćurčić, 1989c: 352–354, figs 1–2, 9–10.

Type locality: Duke Forest, Durham, North Carolina, U.S.A.
Distribution: U.S.A. (North Carolina).

Tuberocreagris rufula (Banks)

Olpium rufulum Banks, 1891: 166.
Ideobisium rufulum (Banks): Banks, 1895a: 11; Banks, 1908: 42; Coolidge, 1908: 113; Ewing, 1911: 79.
Microcreagris rufulum (Banks): J. C. Chamberlin, 1930: 30.
Microcreagris rufula (Banks): Beier, 1932a: 152, fig. 184; Roewer, 1937: 252; Beier, 1948b: 457; Hoff & Bolsterli, 1956: 162; Hoff, 1958: 10; Muchmore, 1971a: 87; Rowland & Reddell, 1976: 15.
Tuberocreagris rufula (Banks): Ćurčić, 1978a: 113–117, figs 1–6; Ćurčić, 1984b: 150–152, fig. 1.

Type locality: Washington, District of Columbia, U.S.A.
Distribution: U.S.A. (District of Columbia, Kentucky, Texas, Virginia).

Family SYARINIDAE J. C. Chamberlin

Ideobisini (sic) Banks, 1895a: 2.
Ideobsiidae (sic) Banks: Comstock, 1913: 50; Moles & Moore, 1921: 8.
Ideobisiinae Banks: J. C. Chamberlin, 1930: 22; J. C. Chamberlin, 1931a: 217; Beier,
  1932a: 141; Roewer, 1937: 250; Hoff, 1956b: 4; Morikawa, 1960: 119.
Syarinidae J. C. Chamberlin, 1930: 38; J. C. Chamberlin, 1931a: 218–219; Beier,
  1932a: 162–163; Beier, 1932g: 183; Roewer, 1937: 253; Hoff, 1956b: 9; Hoff,
  1964a: 6; Muchmore, 1979a: 195; Muchmore, 1982a: 97.
Syarininae J. C. Chamberlin, 1930: 39; J. C. Chamberlin, 1931a: 219; Beier, 1932a:
  163; J. C. Chamberlin, 1938b: 109–110; Hoff, 1956b: 10.
Chitrinae J. C. Chamberlin, 1930: 40; J. C. Chamberlin, 1931a: 219.
Chitrellinae Beier, 1932a: 165; J. C. Chamberlin, 1938b: 110; Hoff, 1956b: 20; Hoff,
  1964a: 7.
Microcreagrellinae Beier, 1963b: 222.

Genus **Aglaochitra** J. C. Chamberlin

Aglaochitra J. C. Chamberlin, 1952: 268–269.

Type species: *Aglaochitra rex* J. C. Chamberlin, 1952, by original designation.

Aglaochitra rex J. C. Chamberlin

Aglaochitra rex J. C. Chamberlin, 1952: 269–274, figs 2a–f, 3a–h; Hoff, 1958: 14;
  Muchmore, 1971a: 90.

Type locality: Red Hill, Francis Simes Hastings Natural History Reservation,
  California, U.S.A.
Distribution: U.S.A. (California).

Genus **Alocobisium** Beier

Alocobisium Beier, 1952a: 98–99.

Type species: *Alocobisium malaccense* Beier, 1952a, by original designation.

Alocobisium himalaiense Beier

Alocobisium himalaiense Beier, 1976e: 97–98, fig. 3.

Type locality: Phuntsholing, Bhutan.
Distribution: Bhutan.

Alocobisium malaccense Beier

Alocobisium malaccense Beier, 1952a: 99–100, figs 3a–d.

Type locality: Sungei Buloh Leper Settlement, Selangor, Malaysia.
Distribution: Malaysia.

Alocobisium ocellatum Beier

Alocobisium ocellatum Beier, 1978d: 231–232, fig. 1.

Type locality: Kaziranga, Assam, India.
Distribution: India.

Alocobisium philippinense Beier

Alocobisium philippinense Beier, 1966b: 342; Tenorio & Muchmore, 1982: 383.

Type locality: Tarumpitao Point, Palawan, Philippines.
Distribution: Philippines.

Family Syarinidae

Alocobisium rahmi Beier

Alocobisium rahmi Beier, 1976e: 98–99, fig. 4; Schawaller, 1983a: 108–110, figs 8–11; Schawaller, 1987a: 209.

Type locality: Phuntsholing, Bhutan.
Distribution: Bhutan, Nepal.

Alocobisium solomonense Morikawa

Alocobisium solomonense Morikawa, 1963: 3–4, figs 1d–f; Beier, 1964f: 593–594; Beier, 1965g: 762; Beier, 1966d: 138; Beier, 1970g: 317; Muchmore, 1982c: 218.

Type locality: Honiala, Guadalcanal, Solomon Islands.
Distribution: Solomon Islands.

Genus **Chitrella** Beier

Chitra J. C.Chamberlin, 1930: 40 (preoccupied by *Chitra* Gray, 1844).
Chitrella Beier, 1932a: 165; Hoff, 1956b: 20–21; Muchmore, 1973e: 183 (replacement name for *Chitra* J. C. Chamberlin, 1930).

Type species: *Chitra cala* J. C. Chamberlin, 1930, by original designation.

Chitrella archeri Malcolm & J. C. Chamberlin

Chitrella archeri Malcolm & J. C. Chamberlin, 1960: 3–7, figs 1a–e.

Type locality: Wonder Cave, Monteagle, Tennessee, U.S.A.
Distribution: U.S.A. (Tennessee).

Chitrella cala (J. C. Chamberlin)

Chitra cala J. C. Chamberlin, 1930: 41, figs 2x, 2jj; J. C. Chamberlin, 1931a: 82, figs 13k, 17o, 33p, 36c, 40r, 43l–m, 47z, 57.
Chitrella cala (J. C. Chamberlin): Beier, 1932a: 165–166; Beier, 1932g: fig. 254; Roewer, 1936: fig. 30g; Roewer, 1937: 254; Hoff, 1958: 14; Hoff, 1959b: 4, 26, 34, 60; Malcolm & J. C. Chamberlin, 1960: 11–19, figs 4, 5a–e.

Type locality: San Francisquito Creek, Stanford University, California, U.S.A.
Distribution: U.S.A. (California, Utah).

Chitrella cavicola (Packard)

Obisium caricola (sic) Packard, 1884: 202–203, fig. 1.
Obisium cavicola Packard: Packard, 1886: 42, fig. 11; Banks, 1895a: 11.
Obisium cavicala (sic) Packard: Coolidge, 1908: 113.
Microcreagris (?) *cavicola* (Packard): Beier, 1932a: 157; Roewer, 1937: 252; Hoff, 1958: 12.
Chitrella cavicola (Packard): Muchmore, 1962: 5; Holsinger, 1963: 34; Muchmore, 1963a: 11–14, figs 7–8; Muchmore, 1973e: 184–187, figs 1–4; Holsinger & Culver, 1988: 43.

Type locality: Newmarket Cave, Virginia, U.S.A.
Distribution: U.S.A. (Virginia, West Virginia).

Chitrella muesebecki Malcolm & J. C. Chamberlin

Chitrella muesebecki Malcolm & J. C. Chamberlin, 1960: 9–11, figs 3a–d.

Type locality: Roane County, Tennessee, U.S.A.
Distribution: U.S.A. (Tennessee).

Chitrella regina Malcolm & J. C. Chamberlin

Chitrella regina Malcolm & J. C. Chamberlin, 1960: 7–9, figs 2a–d; Muchmore, 1973e: 187–189, fig. 5.

Type locality: Coffman's Cave, Frankford (as Frankfurt), West Virginia, U.S.A.
Distribution: U.S.A. (West Virginia).

Chitrella superba Muchmore

Chitrella superba Muchmore, 1973e: 189–190, figs 6–8; Holsinger & Culver, 1988: 43, fig. 15.

Type locality: Madden's Cave, Shenandoah County, Virginia, U.S.A.
Distribution: U.S.A. (Virginia).

Chitrella transversa (Banks)

Obisium transversum Banks, 1909b: 307.
Microcreagris transversa (Banks): Beier, 1932a: 157; Roewer, 1937: 252.
Chitrella transversa (Banks): Hoff, 1956b: 21–24, figs 11–12; Hoff, 1958: 14; Hoff, 1959b: 4, etc.; Hoff, 1961: 434–435; Malcolm & J.C. Chamberlin, 1960: 19; Muchmore, 1973e: 190–191, figs 9–10.

Type locality: Pecos, New Mexico, U.S.A.
Distribution: U.S.A. (Colorado, New Mexico).

Genus Hadoblothrus Beier

Hadoblothrus Beier, 1952d: 106; Beier, 1963b: 226–227.

Type species: *Parablothrus gigas* Caporiacco, 1951f, by original designation.

Hadoblothrus aegeus Beron

Hadoblothrus aegeus Beron, 1985: 67, figs 3–7.

Type localities: Zoodochos I and II caves, Santorin, Greece; and Agios Ioannis Cave, Iraklia, Greece.
Distribution: Greece.

Hadoblothrus gigas (Caporiacco)

Parablothrus gigas Caporiacco: Ruffo, 1950: 60 (not seen) (nomen nudum).
Parablothrus gigas Caporiacco, 1951f: 97, fig. 7.
Hadoblothrus gigas (Caporiacco): Beier, 1952d: 106–107, fig. 2; Beier, 1963b: 226, fig. 233; Lazzeroni, 1969a: 333; Beier, 1970f: 322, fig. 2; Lazzeroni, 1970a: 207; Gardini, 1980c: 116, 130; Mahnert, 1980d: 35–37, figs 17–23; Muchmore, 1982c: 218; Inzaghi, 1983: 47–48.

Type locality: Grotta l'Abisso, Castromarina (as Castro), Italy.
Distribution: Italy.

Genus Hyarinus J. C. Chamberlin

Hyarinus J. C. Chamberlin, 1925a: 327–328; Beier, 1932a: 164.

Type species: *Hyarinus hesperus* J. C. Chamberlin, 1925a, by original designation.

Hyarinus hesperus J. C. Chamberlin

Hyarinus hesperus J. C. Chamberlin, 1925a: 329, figs a–d; J. C. Chamberlin, 1931a: fig. 36b; Beier, 1932a: 165, fig. 196; Roewer, 1937: 254; Hoff, 1958: 14.

Type locality: Santa Barbara, California, U.S.A.
Distribution: U.S.A. (California).

Genus Ideobisium Balzan

Ideobisium Balzan, 1892: 539–540; With, 1906: 81; J. C. Chamberlin, 1930: 36; Beier, 1932a: 157; Beier, 1976f: 210; Muchmore, 1982c: 194–195.
Ideobisium (Ideobisium) Balzan: Balzan, 1892: 542.

Family Syarinidae

Type species: *Ideobisium (Ideobisium) crassimanum* Balzan, 1892, by subsequent designation of J. C. Chamberlin, 1930: 36.

Ideobisium antipodum (E. Simon)

Obisium antipodum E. Simon, 1880a: 174–175; Rainbow, 1897: 108.
Ideobisium (Ideobisium) antipodano (sic) (E. Simon): Balzan, 1892: 543.
Ideobisium (?) *antipodum* (E. Simon): Beier, 1932a: 160; J. C. Chamberlin, 1934b: 4; Roewer, 1937: 252.
Ideobisium antipodum (E. Simon): Beier, 1968a: 763–764, fig. 4; Muchmore, 1982c: 206.

Type locality: Nouméa, New Caledonia.
Distribution: New Caledonia, Tuvalu.

Ideobisium balzanii With

Ideobisium balzanii With, 1905: 131–135, plate 10 figs 2a–h; J. C. Chamberlin, 1930: 37; J. C. Chamberlin, 1931a: fig. 35o; Hoff, 1945e: 1; Muchmore, 1982c: 197–198, fig. 10.
Ideobisium balzani With: Beier, 1932a: 158; Roewer, 1937: 252; Heurtault & Rebière, 1983: 598–600, figs 16–21.

Type locality: St Vincent.
Distribution: Dominica, Guadeloupe, St Vincent.

Ideobisium chapmani Muchmore

Ideobisium chapmani Muchmore, 1982c: 198–200, fig. 11.

Type locality: Camburales Cave, 10 km E. of Curimagua, Serrania de San Luis, Falcon, Venezuela.
Distribution: Venezuela.

Ideobisium crassimanum Balzan

Ideobisium (Ideobisium) crassimanum Balzan, 1892: 542–543, figs 33, 33a.
Ideobisium crassimanum Balzan: H. J. Hansen, 1893: figs 5–7; J. C. Chamberlin, 1930: 37; Beier, 1932a: 158, fig. 190; Roewer, 1937: 252; Vachon, 1952a: fig. 16; Beier, 1977c: 100; Beier, 1976d: 46; Muchmore, 1982c: 195–197, figs 2–9.

Type locality: Caracas, Venezuela.
Distribution: Dominican Republic, Ecuador, Venezuela.

Ideobisium ecuadorense Muchmore

Ideobisium ecuadorense Muchmore, 1982c: 201–202, fig. 13.

Type locality: Los Tayos Caves, Cordillera el Condor, Ecuador.
Distribution: Ecuador.

Ideobisium gracile Balzan

Ideobisium (Ideoroncus) gracilis Balzan, 1892: 540–541, figs 31, 31a.
Ideoroncus gracilis (Balzan): J. C. Chamberlin, 1930: 44.
Ideobisium gracile Balzan: Beier, 1932a: 158–159; Roewer, 1937: 252.
Ideobisium (?) *gracile* Balzan: Muchmore, 1982c: 206.

Type locality: San Esteban, Venezuela.
Distribution: Venezuela.

Ideobisium peckorum Muchmore

Ideobisium peckorum Muchmore, 1982c: 200–201, fig. 12; Mahnert, 1985c: 219; Mahnert & Adis, 1986: 214.

Type locality: 7 km N. of Leticia, Amazonas, Colombia.
Distribution: Brazil (Amazonas), Colombia.

Ideobisium peregrinum J. C. Chamberlin

Ideobisium peregrinum J. C. Chamberlin, 1930: 37; J. C. Chamberlin, 1931a: figs 11r, 35f–h, 40s, 43j–k, 50c; Beier, 1932a: 159; J. C. Chamberlin, 1934b: 4; Roewer, 1937: 252; Beier, 1948c: 537; Beier, 1967b: 290; Beier, 1969e: 414; Wise, 1970: 234; Beier, 1976f: 210; Muchmore, 1982c: 205–206.

Type locality: Day's Bay, Wellington, North Island, New Zealand.
Distribution: New Zealand.

Ideobisium puertoricense Muchmore

Ideobisium puertoricense Muchmore, 1982c: 202–203, figs 14–15.

Type locality: Luquillo Mountains, Puerto Rico.
Distribution: Dominican Republic, Puerto Rico.

Ideobisium puertoricense cavicola Muchmore

Ideobisium puertoricense cavicolum (sic) Muchmore, 1982c: 203–204.

Type locality: Aguas Buenas Cave, Aguas Buenas, Puerto Rico.
Distribution: Puerto Rico.

Ideobisium schusteri Mahnert

Ideobisium schusteri Mahnert, 1985c: 220, figs 5–11; Mahnert & Adis, 1986: 214.

Type locality: Taruma Mirim, Amazonas, Brazil.
Distribution: Brazil (Amazonas).

Ideobisium trifidum (Stecker)

Obisium trifidum Stecker, 1875b: 523, plate 4 figs 5–8; With, 1906: 77.
Ideobisium (?) *trifidum* (Stecker): Beier, 1932a: 160; Roewer, 1937: 252.
Genus ? *trifidum* Stecker: Muchmore, 1982c: 207.

Type locality: India.
Distribution: India.

Ideobisium yunquense Muchmore

Ideobisium yunquense Muchmore, 1982c: 204–205, fig. 16.

Type locality: Mt Britton, El Yunque, Puerto Rico.
Distribution: Puerto Rico.

Genus Ideoblothrus Balzan

Ideobisium (Ideoblothrus) Balzan, 1892: 541.
Pachychitra J. C. Chamberlin, 1938b: 111; Hoff, 1945e: 1; Hoff, 1964a: 7–8; Muchmore, 1979a: 195 (synonymized by Muchmore, 1982c: 207).
Ideoblothrus Balzan: Muchmore, 1982c: 207.

Type species: of *Ideobisium (Ideoblothrus)*: *Ideobisium (Ideoblothrus) similis* Balzan, 1892, by subsequent designation of Muchmore, 1982c: 207.
of *Pachychitra*: *Pachychitra maya* J. C. Chamberlin, 1938b, by original designation.

Ideoblothrus amazonicus (Mahnert)

Ideobisium amazonicum Mahnert, 1979d: 743–744, figs 48–52.
Ideoblothrus amazonicus (Mahnert): Muchmore, 1982c: 215; Mahnert & Adis, 1986: 214.

Type locality: Rio Demini (as Demeni), Amazonas, Brazil.
Distribution: Brazil (Amazonas).

Family Syarinidae

Ideoblothrus baloghi (Mahnert)

Ideobisium baloghi Mahnert, 1978f: 90–92, figs 48–50.
Ideoblothrus baloghi (Mahnert): Muchmore, 1982c: 216.

Type locality: Bangu forest, Meya, Kindamba, Congo.
Distribution: Congo.

Ideoblothrus bipectinatus (Daday)

Ideobisium bipectinatum Daday, 1897: 478–479, figs 7, 14, 15; With, 1906: 87–88; Ellingsen, 1910a: 395; Beier, 1932a: 160, fig. 192; J. C. Chamberlin, 1934b: 4; Roewer, 1937: 252; Morikawa, 1963: 4, figs 2a–c; Beier, 1965g: 761–762; Beier, 1967c: 321; Beier, 1982: 43.
Ideoblothrus bipectinatus (Daday): Muchmore, 1982c: 216.

Type locality: Madang (as Friedrich-Wilhelmshafen), Papua New Guinea.
Distribution: Indonesia (Irian Jaya), Papua New Guinea.

Ideoblothrus brasiliensis (Mahnert)

Ideobisium brasiliense Mahnert, 1979d: 747–750, figs 59–64; Adis, Junk & Penny, 1987: 488.
Ideoblothrus brasiliensis (Mahnert): Muchmore, 1982c: 215; Mahnert, 1985b: 76; Mahnert & Adis, 1986: 214; Mahnert, Adis & Bührnheim, 1987: fig. 7b.

Type locality: Belém, Pará, Brazil.
Distribution: Brazil (Amazonas, Pará).

Ideoblothrus caecus (Mahnert)

Ideobisium caecum Mahnert, 1979d: 745–747, figs 53–58.
Ideoblothrus caecus (Mahnert): Muchmore, 1982c: 215; Mahnert & Adis, 1986: 214; Mahnert, Adis & Bührnheim, 1987: fig. 9.

Type locality: Santarém, Pará, Brazil.
Distribution: Brazil (Amazonas, Pará).

Ideoblothrus carinatus (Hoff)

Pachychitra carinata Hoff, 1964a: 13–14, figs 5–6.
Ideoblothrus carinatus (Hoff): Muchmore, 1982c: 215.

Type locality: 5 miles S. of Hardwar Gap, Parish of St Andrew, Jamaica.
Distribution: Jamaica.

Ideoblothrus ceylonicus (Beier), new combination

Ideobisium ceylonicum Beier, 1973b: 42–43, fig. 4.

Type locality: Andapolakanda, 3 miles NE. of Melsiripura, Northwest Province, Sri Lanka.
Distribution: Sri Lanka.

Ideoblothrus colombiae Muchmore

Ideoblothrus colombiae Muchmore, 1982c: 211–212, figs 21–22.

Type locality: between San Pedro and San Javier, Sierra Nevada de Santa Marta, Magdalena, Colombia.
Distribution: Colombia.

Ideoblothrus costaricensis (Beier)

Ideobisium costaricense Beier, 1931c: 302–303, fig. 2; Beier, 1932a: 159, figs 189, 191; Roewer, 1937: 252; Beier, 1977c: 100.

Ideoblothrus costaricensis (Beier): Muchmore, 1982c: 213.

Type locality: Tuis, Costa Rica.
Distribution: Costa Rica, Ecuador.

Ideoblothrus curazavius (Wagenaar-Hummelinck)

Pachychitra curazavia Wagenaar-Hummelinck, 1948: 63−71, figs 22a−e, 23a−g, 24a−f, 25a−f, 26h.
Ideoblothrus curazavius (Wagenaar-Hummelinck): Muchmore, 1982c: 214.

Type locality: Seroe Christoffel, Curaçao.
Distribution: Curaçao.

Ideoblothrus fenestratus (Beier)

Ideobisium fenestratum Beier, 1955m: 3−4, figs 3−5; Weidner, 1959: 115.
Ideoblothrus fenestratus (Beier): Muchmore, 1982c: 214.

Type locality: Sivia, Peru.
Distribution: Peru.

Ideoblothrus floridensis (Muchmore)

Pachychitra floridensis Muchmore, 1979a: 195−197, figs 1−6.
Ideoblothrus floridensis (Muchmore): Muchmore, 1982c: 215.

Type locality: Watson Hammock, Big Pine Key, Monroe County, Florida, U.S.A.
Distribution: U.S.A. (Florida).

Ideoblothrus godfreyi (Ellingsen)

Ideobisium (Ideoblothrus) godfreyi Ellingsen, 1912b: 117−118.
Gymnobisium godfreyi (Ellingsen): Beier, 1932a: 162; Roewer, 1937: 253.
Ideobisium godfreyi (Ellingsen): Beier, 1947b: 290.
Ideoblothrus godfreyi (Ellingsen): Muchmore, 1982c: 215.

Type locality: Frankfort Hill, near King William's Town, Cape Province, South Africa.
Distribution: South Africa.

Ideoblothrus grandis (Muchmore)

Pachychitra grandis Muchmore, 1972g: 266, figs 6−7; Muchmore, 1977: 70.
Ideoblothrus grandis (Muchmore): Muchmore, 1982c: 214.

Type locality: Cueva del Tio Ticho, 1 mile S. of Comitan, Chiapas, Mexico.
Distribution: Mexico.

Ideoblothrus holmi (Beier)

Ideobisium holmi Beier, 1955c: 534, fig. 5; Beier, 1959d: 23; Beier, 1972a: 7.
Ideoblothrus holmi (Beier): Muchmore, 1982c: 216.

Type locality: Bundibugyo, Ruwenzori, Uganda.
Distribution: Uganda, Zaire.

Ideoblothrus insularum (Hoff)

Pachychitra insularum Hoff, 1945e: 1−4, figs 1−5; Wagenaar-Hummelinck, 1948: 73−75, figs 29a−l; Hoff, 1964a: 8−9.
Ideoblothrus insularum (Hoff): Muchmore, 1982c: 214.

Type locality: Desecheo Island, Puerto Rico.
Distribution: Jamaica, Puerto Rico.

Ideoblothrus kochalkai Muchmore

Ideoblothrus kochalkai Muchmore, 1982c: 210, figs 18–20.

Type locality: Casa Antonio, Cuchilla Cebolleta, Sierra Nevada de Santa Marta, Magdalena, Colombia.
Distribution: Colombia.

Ideoblothrus leleupi (Beier)

Ideobisium leleupi Beier, 1959d: 23–25, fig. 10.
Ideoblothrus leleupi (Beier): Muchmore, 1982c: 216.

Type locality: Lubero, Zaire.
Distribution: Zaire.

Ideoblothrus lepesmei (Vachon)

Ideobisium lepesmei Vachon, 1940e: 32; Vachon, 1952a: 24–27, figs 10–11, 13–14.
Ideoblothrus lepesmei (Vachon): Muchmore, 1982c: 216.

Type locality: Sassandra, Ivory Coast.
Distribution: Guinea, Ivory Coast.

Ideoblothrus levipalpus Mahnert

Ideoblothrus levipalpus Mahnert, 1985c: 221, figs 12–13; Mahnert & Adis, 1986: 214.

Type locality: S. of Rio Negro, Amazonas, Brazil.
Distribution: Brazil (Amazonas).

Ideoblothrus maya (J. C. Chamberlin)

Pachychitra maya J. C. Chamberlin, 1938b: 111–113, figs 1a–f; Roewer, 1940: 345, fig. 261; Wagenaar-Hummelinck, 1948: 71–73, figs 26a–g, 27a–c, 28a–f; Muchmore, 1977: 70.
Ideoblothrus maya (J. C. Chamberlin): Muchmore, 1982c: 213.

Type locality: Oxkutzcab, Yucatan, Mexico.
Distribution: Mexico.

Ideoblothrus mexicanus (Muchmore)

Pachychitra mexicana Muchmore, 1972g: 262–264, figs 1–3.
Ideoblothrus mexicanus (Muchmore): Muchmore, 1982c: 214.

Type locality: Rancho del Cielo, 6 miles NW. of Gomez Farais, Tamaulipas, Mexico.
Distribution: Mexico.

Ideoblothrus muchmorei Heurtault

Ideoblothrus muchmorei Heurtault, 1983: 24–26, figs 53–58.

Type locality: Bandama, Ivory Coast.
Distribution: Ivory Coast.

Ideoblothrus occidentalis (Beier)

Ideobisium occidentale Beier, 1959d: 25–26, fig. 11.
Ideoblothrus occidentalis (Beier): Muchmore, 1982c: 216.

Type locality: Inkisi-Kisantu (as d'Inkisi, Kisantu Mission), Zaire.
Distribution: Zaire.

Ideoblothrus palauensis (Beier)

Ideobisium palauense Beier, 1957d: 13–14, fig. 4a.
Ideoblothrus palauensis (Beier): Muchmore, 1982c: 217.

Type locality: East Ngatpang, Babelthuap, Palau Islands, Caroline Islands.
Distribution: Caroline Islands.

Ideoblothrus paraensis Mahnert

Ideoblothrus paraensis Mahnert, 1985b: 76–78, figs 1–6; Mahnert & Adis, 1986: 214.

Type locality: Paricatuba, Benevides, Pará, Brazil.
Distribution: Brazil (Pará).

Ideoblothrus pugil (Beier)

Ideobisium pugil Beier, 1964f: 593, fig. 1; Beier, 1965g: 762; Beier, 1966d: 137; Beier, 1970g: 316.
Ideobisium pugil pugil Beier: Petersen, 1968: 119.
Ideoblothrus pugil pugil (Beier): Muchmore, 1982c: 217.

Type locality: Mt Austen, Guadalcanal, Solomon Islands.
Distribution: Solomon Islands.

Ideoblothrus pugil robustus (Beier)

Ideobisium pugil robustum Beier, 1966d: 137, fig. 3.
Ideoblothrus pugil robustum (Beier): Muchmore, 1982c: 217.

Type locality: Nila, Shortland, Solomon Islands.
Distribution: Solomon Islands.

Ideoblothrus pygmaeus (Hoff)

Pachychitra pygmaea Hoff, 1964a: 9–11, figs 1–2.
Ideoblothrus pygmaeus (Hoff): Muchmore, 1982c: 214; Heurtault & Rebière, 1983: 596–598, figs 11–15.

Type locality: Dolphin Head, 5 miles S. of Lucea, Hanover Parish, Jamaica.
Distribution: Jamaica, Martinique.

Ideoblothrus seychellesensis (J. C. Chamberlin)

Ideobisium seychellesensis J. C.. Chamberlin, 1930: 38, figs 1x, 1dd, 2cc; J. C. Chamberlin, 1931a: figs 13i, 16l, 17g; Beier, 1932a: 160; Roewer, 1937: 252.
Ideoblothrus seychellesensis (J. C. Chamberlin): Muchmore, 1982c: 212–213, figs 23–24.

Type locality: Félicité, Seychelles.
Distribution: Seychelles.

Ideoblothrus similis (Balzan)

Ideobisium (Ideoblothrus) similis Balzan, 1892: 541–542, figs 32, 32a.
Ideobisium simile Balzan: Beier, 1932a: 159; Ellingsen, 1909b: 219; Roewer, 1937: 252; Vachon, 1952a: figs 12, 15; Beier, 1977c: 100–101; Hounsome, 1980: 85.
Ideobisium simile Balzan: Beier, 1974c: 101 (in part; see *Ideoblothrus costaricensis* (Beier)).
Ideoblothrus similis (Balzan): Muchmore, 1982c: 208–210, fig. 17.

Type locality: Petare, Venezuela.
Distribution: Cayman Islands, Galapagos Islands, Mexico, Venezuela.

Ideoblothrus tenuis Mahnert

Ideoblothrus tenuis Mahnert, 1985c: 221–222, figs 14–18; Mahnert & Adis, 1986: 214.

Type locality: Reserva Florestal Ducke, 26 km on Manaus-Itacoatiara Highway, Amazonas, Brazil.
Distribution: Brazil (Amazonas).

Ideoblothrus truncatus (Hoff)

Pachychitra truncata Hoff, 1964a: 11–13, figs 3–4.
Ideoblothrus truncatus (Hoff): Muchmore, 1982c: 215.

Type locality: Maggotty Falls, S. of Maggotty, Parish of St Elizabeth, Jamaica.
Distribution: Jamaica.

Ideoblothrus vampirorum Muchmore

Pachychitra similis Muchmore, 1972g: 264–265, figs 4–5; Reddell & Elliott, 1973b: 183 (junior secondary homonym of *Ideoblothrus similis* (Balzan)).
Ideoblothrus vampirorum Muchmore, 1982c: 214 (replacement name for *Pachychitra similis* Muchmore).

Type locality: Cueva de los Vampiros, 6 miles NNE. of Chamal, Tamaulipas, Mexico.
Distribution: Mexico.

Ideoblothrus zicsii (Mahnert)

Ideobisium zicsii Mahnert, 1978f: 92–93, figs 51–55.
Ideoblothrus zicsii (Mahnert): Muchmore, 1982c: 216.

Type locality: Sibiti, Congo.
Distribution: Congo.

Genus **Microblothrus** Mahnert

Microblothrus Mahnert, 1985c: 222.

Type species: *Microblothrus tridens* Mahnert, 1985c, by original designation.

Microblothrus tridens Mahnert

Microblothrus tridens Mahnert, 1985c: 223–224, figs 19–26; Mahnert & Adis, 1986: 214.

Type locality: Reserva Florestal Ducke, 26 km on Manaus-Itacoatiara Highway, Amazonas, Brazil.
Distribution: Brazil (Amazonas).

Genus **Microcreagrella** Beier

Microcreagrella Beier, 1961a: 70–72; Beier, 1963b: 224.

Type species: *Obisium caecum* E. Simon, 1883, by original designation.

Microcreagrella caeca (E. Simon)

Obisium caecum E. Simon, 1883: 279–280.
Ideobisium (Ideoblothrus) caecum (E. Simon): Balzan, 1892: 542.
Microcreagris caeca (E. Simon): Beier, 1932a: 157; Roewer, 1937: 252; Vachon, 1940b: 108–110, figs 1–6.
Microcreagrella caeca (E. Simon): Beier, 1961a: 72–74, fig. 2; Beier, 1970f: 322, fig. 2; Beier, 1976a: 24; Pieper, 1981: 2–3.
Microcreagris coeca (sic) (E. Simon): Vachon, 1961: 102, 103.
Microcreagrella caeca caeca (E. Simon): Beier, 1963b: 225; Muchmore, 1982c: 218.

Type locality: Ponta Delgada (as Ponta-Delgada), Sao Miguel, Azores.
Distribution: Azores.

Microcreagrella caeca madeirensis Beier

Microcreagrella caeca madeirensis Beier, 1963b: 225, fig. 231; Beier, 1970f: 322, fig. 2; Beier, 1976a: 24.

Type locality: Madeira, Madeira Islands.
Distribution: Madeira Islands.

Genus **Microcreagrina** Beier

Microcreagrina Beier, 1961b: 29–30; Beier, 1963b: 223.

Type species: *Microcreagris maroccana* Beier, 1931c (junior synonym of *Ideobisium hispanicum* Ellingsen, 1910a), by original designation.

Microcreagrina hispanica (Ellingsen)

Ideobisium hispanicum Ellingsen, 1910a: 394; Navás, 1918: 112; Navás, 1925: 114.
Ideobisium (Ideoblothrus) hispanicum Ellingsen: Nonidez, 1917: 32.
Ideoroncus cambridgei (L. Koch): Beier, 1930f: 71 (misidentification).
Microcreagris maroccana Beier, 1931c: 301–302; Beier, 1932a: 155, fig. 187; Roewer, 1937: 251; Vachon, 1940g: 144 (synonymized by Beier, 1970a: 45).
Microcreagris hispanica (Ellingsen): Beier, 1932a: 155–156; Roewer, 1937: 251; Beier, 1939f: 189; Beier, 1963b: 212.
Microcreagris parisi Vachon, 1937e: 107–109, 1–2; Roewer, 1940: 344 (synonymized by Mahnert, 1976d: 206).
Microcreagrina maroccana (Beier): Beier, 1961b: 30; Beier, 1963b: 223–224, fig. 230; Beier, 1965c: 377; Lazzeroni, 1969c: 237–238.
Microcreagrina hispanica (Ellingsen): Beier, 1970a: 45; Beier, 1970f: 322, fig. 2; Beier, 1975a: 56; Beier, 1976a: 24; Orghidan, Dumitresco & Georgesco, 1975: 30; Mahert, 1976c: 206; Mahnert, 1977d: 97; Mahnert, 1980a: 259; Callaini, 1982a: 17, 20, fig. 2c; Mahnert, 1982f: 297; Muchmore, 1982c: 218; Callaini, 1983c: 292; Callaini, 1988b: 46; Mahnert, 1989b: 42.

Type locality: of *Ideobisium hispanicum*: Algeciras, Spain.
of *Microcreagris maroccana*: Korifla, Morocco.
of *Microcreagris parisi*: Skikda (as Philippeville), Algeria.
Distribution: Algeria, Balearic Islands, Canary Islands, Lebanon, Malta, Morocco, Portugal, Sardinia, Sicily, Spain.

Genus **Nannobisium** Beier

Nannobisium Beier, 1931c: 303; Beier, 1932a: 161.
Vescichitra Hoff, 1964a: 15 (synonymized by Mahnert, 1979d: 752).

Type species: of *Nannobisium*: *Nannobisium liberiense* Beier, 1931c, by original designation.
of *Vescichitra*: *Vescichitra mollis* Hoff, 1964a, by original designation.

Nannobisium beieri Mahnert

Nannobisium beieri Mahnert, 1979d: 751–752, figs 65–69; Mahnert, 1985b: 76; Mahnert & Adis, 1986: 214; Mahnert, Adis & Bührnheim, 1987: fig. 8.

Type locality: Santarém, Pará, Brazil.
Distribution: Brazil (Pará).

Nannobisium liberiense Beier

Nannobisium liberiense Beier, 1931c: 303–304, fig. 3; Beier, 1932a: 161, figs 193–194; Roewer, 1936: fig. 75; Roewer, 1937: 252; Vachon, 1941e: 31; Mahnert, 1974d: 851; Muchmore, 1982c: 218.

Type locality: Bolahun, Liberia.
Distribution: Ivory Coast, Liberia, Togo.

Nannobisium mollis (Hoff)

Vescichitra mollis Hoff, 1964a: 15–18, figs 7–8.
Nannobisium mollis (Hoff): Mahnert, 1979d: 752.

Type locality: 1 miles NW. of Ferry, Parish of St Andrew, Jamaica.
Distribution: Jamaica.

Family Syarinidae

Genus **Pseudoblothrus** Beier

Pseudoblothrus Beier, 1931a: 21; Beier, 1932a: 135; Beier, 1963b: 227–228.

Type species: *Ideoblothrus roszkovskii* Redikorzev, 1918, by original designation.

Pseudoblothrus ellingseni (Beier)

Obisium (Blothrus) torrei E. Simon: Ellingsen, 1905d: 9–11 (misidentification).
Blothrus peyerimhoffi E. Simon, 1905: 282–283 (misidentification, in part).
Obisium (Blothrus) ellingseni Beier, 1929a: 363.
Neobisium (Blothrus) ellingseni (Beier): Beier, 1932a: 113; Roewer, 1937: 248.
Pseudoblothrus ellingseni (Beier): Beier, 1963b: 228–229; Gardini, 1980c: 116, 119;
    Mahnert, 1980d: 30–32, figs 13–14; Muchmore, 1982c: 218; Bologna & Taglianti,
    1985: 228–229, fig. 9.
Pseudoblothrus sp. prope *peyerimhoffi* (E. Simon): Vigna Taglianti, 1969: 267
    (misidentification).
Pseudoblothrus ellingseni (Beier) spp.: Mahnert, 1980d: 32–33.

Type locality: Grotta di Bossea, Cuneo, Italy.
Distribution: Italy.

Pseudoblothrus ljovuschkini Krumpál

Pseudoblothrus ljovuschkini Krumpál, 1984a: 642–645, figs 11–16.

Type locality: Peshtshera Egiz, Karabijajra, Krym (as Krim), R.S.F.S.R.,
    U.S.S.R.
Distribution: U.S.S.R. (R.S.F.S.R.).

Pseudoblothrus peyerimhoffi (E. Simon)

Blothrus peyerimhoffi E. Simon, 1905: 282–283 (in part; see *Pseudoblothrus
    ellingseni* (Beier)); Vachon, 1938a: figs 37i, 38a.
Neobisium (Blothrus) peyerimhoffi (E. Simon): Beier, 1932a: 113; Roewer, 1937:
    248.
Pseudoblothrus peyerimhoffi (Beier): Vachon, 1945a: 230–232, figs 1–7; Vachon,
    1952c: 536–537; Balazuc, 1962: 106; Beier, 1963b: 229; Beier, 1970f: 322, fig. 2;
    Mahnert, 1980d: 33–34, figs 15–16; Leclerc, 1983b: 25; Leclerc, 1984a: 56;
    Heurtault, 1986c: 28.
Not *Pseudoblothrus* sp. prope *peyerimhoffi* (E. Simon): Vigna Taglianti, 1969: 267
    (misidentification; see *Pseudoblothrus ellingseni* (Beier)).

Type locality: Grotte de Mélan, Méailles, Basses-Alpes, France.
Distribution: France.

Pseudoblothrus regalini Inzaghi

Pseudoblothrus regalini Inzaghi, 1983: 38–46, figs 1–12.

Type locality: NE. of Monte di Grone, Bergamo, Italy.
Distribution: Italy.

Pseudoblothrus roszkovskii (Redikorzev)

Ideoblothrus roszkovskii Redikorzev, 1918: 94–96, figs 3, 4, 4a.
Pseudoblothrus roszkovskii (Redikorzev): Beier, 1931a: 21–22; Beier, 1932a: 136,
    fig. 167; Roewer, 1937: 249; Roewer, 1940: 345; Beier, 1963b: 228, fig. 234; Beier,
    1970f: 322, fig. 2; Muchmore, 1982c: 218.

Type locality: Suuk-Koba Cave, Krym (as Crimeé), Ukraine, U.S.S.R.
Distribution: U.S.S.R. (Ukraine).

Pseudoblothrus strinatii Vachon

Pseudoblothrus strinatii Vachon, 1954b: 212–217, figs 1–14; Strinati, 1957: 62; Beier, 1963b: 229; Beier, 1970f: 322, fig. 2; Aellen, 1976: 23; Vachon, 1976: 243, 244; Bourne & Cherix, 1981: 24, fig. 1.

Type locality: Grotte de Pertuis, Jura Neuchâtelois, Switzerland.
Distribution: Switzerland.

Pseudoblothrus thiebaudi Vachon

Pseudoblothrus thiebaudi Vachon, 1969: 387–390, figs 1–10; Vachon, 1976: 244, 245; Thaler, 1980: 391.

Type locality: Neuenburgerhöhle (as Neuenbürgenhöhle), LU2, Lucerne, Switzerland.
Distribution: Switzerland.

Genus Syarinus J. C. Chamberlin

Syarinus J. C. Chamberlin, 1925a: 329; J. C. Chamberlin, 1930: 39; Beier, 1932a: 163; Hoff, 1956b: 10.

Type species: *Ideoroncus obscurus* Banks, 1893, by original designation.

Syarinus enhuycki Muchmore

Syarinus enhuycki Muchmore, 1968e: 112–115, figs 1–3; Nelson, 1975: 281–282, figs 25–28.

Type locality: E.N. Huyck Preserve, Rensselaerville, Albany County, New York, U.S.A.
Distribution: U.S.A. (Michigan, New Hampshire, New York, Pennsylvania).

Syarinus granulatus J. C. Chamberlin

Syarinus granulatus J. C. Chamberlin, 1930: 39–40, fig. 2h; J. C. Chamberlin, 1931a: figs 9e, 11v, 13m, 15g, 16j, 17n, 28n, 36a; Beier, 1932a: 164; Roewer, 1937: 254; Hoff & Bolsterli, 1956: 163; Hoff, 1956b: 16–20, fig. 10; Hoff, 1958: 13; Hoff, 1959b: 4, etc.

Type locality: Engleman Canyon, Manitou, Colorado, U.S.A.
Distribution: U.S.A. (Colorado, New Mexico, Wisconsin).

Syarinus honestus Hoff

Syarinus honestus Hoff, 1956b: 14–16, figs 8–9; Hoff, 1958: 14; Hoff, 1959b: 4, etc.

Type locality: Santa Fe ski area, NE. of Santa Fe, Santa Fe County, New Mexico, U.S.A.
Distribution: U.S.A. (New Mexico).

Syarinus obscurus (Banks)

Ideoroncus obscurus Banks, 1893: 66–67; Banks, 1895a: 11; Banks, 1904b: 364; Coolidge, 1908: 113; Ellingsen, 1908e: 163; Banks, 1911: 639; Moles, 1914c: 196–197; Moles & Moore, 1921: 8.
Syarinus obscurus (Banks): J. C. Chamberlin, 1925a: 330; J. C. Chamberlin, 1930: 39; J. C. Chamberlin, 1931a: figs 9d, 15f, 17l–m, 19e, 25k, 43s–t, 47d; Beier, 1932a: 164; Roewer, 1937: 253; Hoff, 1956b: 10–14, figs 6–7; Hoff, 1958: 13; Hoff, 1959b: 4, etc.; Muchmore, 1971a: 82; Knowlton, 1972: 2; Muchmore, 1973g: 122; Knowlton, 1974: 3.

Type locality: Olympia, Washington, U.S.A.
Distribution: Canada (British Columbia, Saskatchewan), U.S.A. (California, Montana, New Mexico, Utah, Washington, Wyoming).

Syarinus palmeni Kaisila

Syarinus palméni Kaisila, 1964: 52—54, figs 1a—e.

Type locality: Holyrood, Newfoundland, Canada.
Distribution: Canada (Newfoundland).

Syarinus strandi (Ellingsen)

Ideobisium (Ideoblothrus) strandi Ellingsen, 1901a: 88—89; Ellingsen, 1903: 12—13.
Ideobisium strandi Ellingsen: Ellingsen, 1910c: 348.
Microcreagris strandi (Ellingsen): Beier, 1932a: 155; Roewer, 1937: 251; Kaisila, 1949b: 79—80, fig. 2, map 4; Vachon, 1954d: 591; Beier, 1963b: 211; Beier, 1970f: 321; Klausen, 1975: 64.
Syarinus strandi (Ellingsen): Mahnert, 1976c: 206—209, figs 20—25; Schawaller, 1987b: 289—292, figs 1—10.

Type locality: Hallingdal, Norway.
Distribution: Austria, Finland, Norway, West Germany.

Genus **Troglobisium** Beier

Troglobisium Beier, 1939f: 189—190; Beier, 1963b: 225—226.

Type species: *Ideobisium racovitzai* Ellingsen, 1912c, by original designation.

Troglobisium racovitzai (Ellingsen)

Ideobisium (Ideoblothrus) racovitzai Ellingsen, 1912c: 164—166; Nonidez, 1917: 32.
Ideobisium racovitzai Ellingsen: Navás, 1925: 114.
Microcreagris racovitzai (Ellingsen): Beier, 1932a: 156; Roewer, 1937: 251.
Troglobisium racovitzai (Ellingsen): Beier, 1939f: 190—191, fig. 19; Beier, 1963b: 226, fig. 232; Beier, 1970f: 322, fig. 2; Lagar, 1972a: 20; Lagar, 1972b: 51; Mahnert, 1977c: 101—102; Muchmore, 1982c: 218; Zaragoza, 1986: fig. 1.
Troglobisium rakovitzai (sic) (Ellingsen): Roewer, 1940: 345.

Type locality: Cova d'en Merla, Tarragona, Spain.
Distribution: Spain.

Family VACHONIIDAE J. C. Chamberlin

Vachoniidae J. C. Chamberlin, 1947a: 3—4; Muchmore, 1972g: 267—268; Muchmore, 1982a: 98.

Genus **Paravachonium** Beier

Paravachonium Beier, 1956f: 81; Muchmore, 1972g: 268—269.

Type species: *Paravachonium bolivari* Beier, 1956f, by original designation.

Paravachonium bolivari Beier

Paravachonium bolivari Beier, 1956f: 82—83, figs 1a—f; Reddell & Mitchell, 1971a: 144; Muchmore, 1972g: 268—269; Muchmore, 1973b: 57, fig. 32; Reddell & Elliott, 1973a: 173.

Type locality: Cueva de Quintero, Tamaulipas, Mexico.
Distribution: Mexico.

Paravachonium delanoi Muchmore

Paravachonium delanoi Muchmore, 1982f: 68–70, figs 12–19.

Type locality: Cueva de Oyamel, SW. of El Barratel, Tamaulipas, Mexico.
Distribution: Mexico.

Paravachonium insolitum Muchmore

Paravachonium insolitum Muchmore, 1982f: 70–71, figs 20–23.

Type locality: Sótano de la Tinaja, 10.5 km NE. of Valles, San Luis Potosí, Mexico.
Distribution: Mexico.

Paravachonium superbum Muchmore

Paravachonium superbum Muchmore, 1972g: 269, figs 8–11; Reddell & Elliott, 1973b: 183.

Type locality: Sotano de Gomez Farais, S. of Gomez Farais, Tamaulipas, Mexico.
Distribution: Mexico.

Genus **Vachonium** J. C. Chamberlin

Vachonium J. C. Chamberlin, 1947a: 4–5.

Type species: *Vachonium boneti* J. C. Chamberlin, 1947a, by original designation.

Vachonium belizense Muchmore

Vachonium belizense Muchmore, 1973b: 59, figs 36–38; Muchmore, 1977: 72.

Type locality: Mountain Cow Cave, Caves Branch, Belize.
Distribution: Belize.

Vachonium boneti J. C. Chamberlin

Vachonium boneti J. C. Chamberlin, 1947a: 6–7, figs 1–23; Muchmore, 1977: 72.

Type locality: Cueva de Sabaca, Yucatan, Mexico.
Distribution: Mexico.

Vachonium chukum Muchmore

Vachonium chukum Muchmore, 1982f: 65–66, figs 3–8.

Type locality: Actún Chukum, 2 km SE. of Maxcanú, Yucatán, Mexico.
Distribution: Mexico.

Vachonium cryptum Muchmore

Vachonium cryptum Muchmore, 1977: 72, figs 20–21.

Type locality: Actún Xkyc, 1 km S. of Calcehtok, Yucatán, Mexico.
Distribution: Mexico.

Vachonium kauae Muchmore

Vachonium kauae Muchmore, 1973b: 57–58, figs 33–35; Muchmore, 1977: 72.

Type locality: Cueva de Kaua, 1 km S. of Kaua, Yucatán, Mexico.
Distribution: Mexico.

Vachonium loltun Muchmore

Vachonium loltun Muchmore, 1982f: 67–68, fig. 11.

Type locality: Actún Lotún, 7 km SSW. of Oxkutzcab, Yucatán, Mexico.
Distribution: Mexico.

Vachonium maya J. C. Chamberlin

Vachonium maya J. C. Chamberlin, 1947a: 8–9, figs 24–36; Muchmore, 1977: 72.

Type locality: Cueva de Balaam Canche, Yucatán, Mexico.
Distribution: Mexico.

Vachonium robustum Muchmore

Vachonium robustum Muchmore, 1982f: 66–67, figs 9–10.

Type locality: Actún Chukum, 2 km SE. of Maxcanú, Yucatán, Mexico.
Distribution: Mexico.

Superfamily CHEIRIDIOIDEA H. J. Hansen

Cheiridioidea H. J. Hansen: J. C. Chamberlin, 1931a: 234–235; Petrunkevitch, 1955: 82; Hoff, 1956c: 3; Morikawa, 1960: 133; Murthy & Ananthakrishnan, 1977: 114; Muchmore, 1982a: 99; Legg & Jones, 1988: 90.
Cheiridiides (sic) H. J. Hansen: Beier, 1932e: 2; Beier, 1932g: 185; Roewer, 1937: 274.

Family CHEIRIDIIDAE H. J. Hansen

Chiridiinae H. J. Hansen, 1893: 232; With, 1906: 116; Lessert, 1911: 924; Väänänen, 1928a: 17.
Cheiridiinae H. J. Hansen: J. C. Chamberlin, 1924a: 33–34; Beier, 1932e: 6–7; Beier, 1932g: 185; Roewer, 1937: 276; Morikawa, 1960: 134; Murthy & Ananthakrishnan, 1977: 115–116.
Cheiridiidae H. J. Hansen: J. C. Chamberlin, 1931a: 236; Beier, 1932e: 3; J. C. Chamberlin, 1935b: 480; Roewer, 1937: 274–275; J. C. Chamberlin, 1938a: 264; Petrunkevitch, 1955: 82; Hoff, 1956c: 3; Morikawa, 1960: 133–134; Vitali-di Castri, 1962: 120; Vitali-di Castri, 1965a: 69; Beier, 1963b: 243; Murthy & Ananthakrishnan, 1977: 115; Muchmore, 1982a: 99–100; Harvey, 1985b: 130; Legg & Jones, 1988: 90.

Genus **Apocheiridium** J. C. Chamberlin

Apocheiridium J. C. Chamberlin, 1924a: 34; J. C. Chamberlin, 1931a: 238; Beier, 1932e: 10; J. C. Chamberlin, 1935b: 480; Hoff, 1952b: 192; Morikawa, 1960: 135; Beier, 1963b: 244–245; Beier, 1976f: 211; Murthy & Ananthakrishnan, 1977: 116; Benedict, 1978b: 232.

Type species: *Apocheiridium ferumoides* J. C. Chamberlin, 1924a, by original designation.

Subgenus **Apocheiridium (Apocheiridium)** J. C. Chamberlin

Apocheiridium (Apocheiridium) asperum Beier
Apocheiridium asperum Beier, 1964k: 68–69, fig. 27.

Type locality: Masite, Lesotho.
Distribution: Lesotho.

Apocheiridium (Apocheiridium) bulbifemorum Benedict
Apocheiridium bulbifemorum Benedict, 1978b: 233–234, fig. 1.

Type locality: 13 km S., 6 km E. of Tiller, Douglas County, Oregon, U.S.A.
Distribution: U.S.A. (Oregon).

Apocheiridium (Apocheiridium) caribicum Beier
Apocheiridium caribicum Beier, 1936a: 446, fig. 3; Roewer, 1940: 346.

Type locality: Bak Ariba, Hato, Curaçao.
Distribution: Curaçao.

Apocheiridium (Apocheiridium) chamberlini Godfrey

Cheiridium ferum E. Simon: Ellingsen, 1912b: 86, 104 (misidentification); Godfrey, 1923: 95–98 (misidentification, in part).
Apocheiridium sp.: J. C. Chamberlin, 1924a: 37.
Apocheiridium chamberlini Godfrey, 1927: 17–18; J. C. Chamberlin, 1932b: 139; Beier, 1932e: 13; Roewer, 1937: 277; Beier, 1964k: 67–68, fig. 26.

Type locality: Lovedale, Cape Province, South Africa.
Distribution: South Africa.

Apocheiridium (Apocheiridium) eruditum J. C. Chamberlin

Apocheiridium eruditum J. C. Chamberlin, 1932b: 139; Beier, 1932e: 13; Roewer, 1937: 277.

Type locality: Los Baños, Luzon, Philippines.
Distribution: Philippines.

Apocheiridium (Apocheiridium) fergusoni Benedict

Apocheiridium fergusoni Benedict, 1978b: 234–235, fig. 2; Zeh, 1987b: 1087.

Type locality: 18 km SE. of Riley, Harney County, Oregon, U.S.A.
Distribution: U.S.A. (Oregon).

Apocheiridium (Apocheiridium) ferum (E. Simon)

Chiridium ferum E. Simon, 1879a: 44–45, plate 18 fig. 21; With, 1907: 80–81, figs 54–57; Lessert, 1911: 25–26, fig. 20; Caporiacco, 1928b: 127.
Cheiridium ferum E. Simon: Ellingsen, 1907a: 166–167; Kew, 1914: 96–97; Godfrey, 1923: 95–98 (in part; see *Apocheiridium (Apocheiridium) chamberlini* Godfrey); Redikorzev, 1924b: 25, fig. 11a; Kästner, 1927a: 15; Kästner, 1928: 4, fig. 11
Apocheiridium ferum (E. Simon): J. C. Chamberlin, 1924a: 37; Beier, 1929a: 358; J. C. Chamberlin, 1931a: fig. 29n; Beier, 1932e: 11–12, fig. 7; J. C. Chamberlin, 1934b: 7; Redikorzev, 1935a: fig. 2; Roewer, 1937: 277, fig. 224; Vachon, 1952c: 537–539; Beier, 1956h: 25; Beier, 1956i: 8; Ressl & Beier, 1958: 2; Beier, 1963b: 245, fig. 249; Dumitresco & Orghidan, 1964: 624; Orghidan & Dumitresco, 1964a: 328; Orghidan & Dumitresco, 1964b: 192; Beier, 1965f: 93; Ressl, 1965: 289; Dumitresco & Orghidan, 1966: 81–82; Helversen, 1966a: 138–139, fig. 2; Zangheri, 1966: 531; Rafalski, 1967: 14–15; Lazzeroni, 1969c: 240; Weygoldt, 1969a: 26, 41, 74, fig. 51; Callaini, 1979c: 350; Thaler, 1979: 52; Callaini, 1982a: 17, 23, fig. 4c; Callaini, 1983a: 152–153; Schawaller & Dashdamirov, 1988: 34, figs 56–57.
Not *Cheiridium ferum* E. Simon: Ellingsen, 1912b: 86, 104 (misidentification; see *Apocheiridium (Apocheiridium) chamberlini* Godfrey).

Type locality: Arcachon, Gironde, France.
Distribution: Austria, France, Italy, Poland, Romania, Samoa, Sardinia, Switzerland, Turkey, West Germany, U.S.S.R. (Azerbaijan, R.S.F.S.R.).

Apocheiridium (Apocheiridium) ferumoides J. C. Chamberlin

Apocheiridium ferumoides J. C. Chamberlin, 1924a: 35–36, figs b–f, h, k–p, r–s, v, x–y, aa (in part, see *Apocheiridium (Apocheiridium) inexpectum* J. C. Chamberlin); Beier, 1930a: 207; J. C. Chamberlin, 1931a: figs 6g, 14g, 15t, 18c–d, 26f, 38e–g, 44h–i, 47u, 51c–e, 52h, 66; J. C. Chamberlin, 1932b: 139; Beier, 1932e: 12, figs 8–9; Beier, 1932g: fig. 261; J. C. Chamberlin, 1935b: 480; Roewer, 1936: figs 20, 43c–d; Roewer, 1937: 277; Gering, 1956: 49 (not seen); Essig, 1958: 11, figs 7a–c; Hoff, 1958: 19; Benedict, 1978b: 235–237; Zeh, 1987b: 1087.
Apocheiridium femuroides (sic) J. C. Chamberlin: Vachon, 1949: figs 200a, 203g.

Type locality: Stanford University, California, U.S.A.
Distribution: U.S.A. (California, Oregon, Utah).

Apocheiridium (Apocheiridium) granochelum Benedict

Apocheiridium granochelum Benedict, 1978b: 237–238, figs 3–5; Zeh, 1987b: 1087.

Type locality: Corvallis, Benton County, Oregon, U.S.A.
Distribution: U.S.A. (Oregon).

Apocheiridium (Apocheiridium) indicum Murthy & Anathakrishnan

Apocheiridium indicum Murthy & Anathakrishnan, 1977: 116–117, fig. 38.

Type locality: Shoranur, Kerala, India.
Distribution: India.

Apocheiridium (Apocheiridium) inexpectum J. C. Chamberlin

Apocheiridium ferumoides J. C. Chamberlin, 1924a: 34 (misidentification, in part).
Apocheiridium inexpectum J. C. Chamberlin, 1932b: 139–140; Beier, 1932e: 13;
   Roewer, 1937: 277; Hoff, 1958: 19; Benedict, 1978b: 238–239; Zeh, 1987b: 1087.

Type locality: Beaumont, Riverside County, California, U.S.A.
Distribution: Mexico, U.S.A. (California).

Apocheiridium (Apocheiridium) minutissimum Beier

Apocheiridium minutissimum Beier, 1964d: 312, fig. 1; Tenorio & Muchmore, 1982:
   378.

Type locality: Subang, Malaysia.
Distribution: Malaysia.

Apocheiridium (Apocheiridium) mormon J. C. Chamberlin

Apocheiridium mormon J. C. Chamberlin, 1924a: 36–37, figs j, jj, p–q, z; Beier,
   1932e: 12–13; Roewer, 1937: 277; Hoff, 1958: 19; Benedict, 1978b: 239–241;
   Zeh, 1987b: 1087.

Type locality: Fish Haven, Bear Lake, Idaho, U.S.A.
Distribution: U.S.A. (Idaho, Oregon, Utah).

Apocheiridium (Apocheiridium) nepalense Ćurčić

Apocheiridium nepalense Ćurčić, 1980b: 80–83, figs 10–14; Schawaller, 1987a: 212.

Type locality: Thini Kola, Nepal.
Distribution: Nepal.

Apocheiridium (Apocheiridium) pallidum Mahnert

Apocheiridium sp.: Mahnert, 1981c: fig. 2.
Apocheiridium pallidum Mahnert, 1982b: 125–127, figs 30–33.

Type locality: Morijo, Loita Hills, Kenya.
Distribution: Kenya.

Apocheiridium (Apocheiridium) pelagicum Redikorzev

Apocheiridium pelagicum Redikorzev, 1938: 87–89, fig. 16; Dawydoff, 1940: 447–449;
   Roewer, 1940: 346; Vachon, 1940f: 3; Beier, 1948b: 461; Beier, 1951a: 70.

Type localities: Poulo Dama, Cambodia; Cu Lao Cham, Nha Trang (as Nhatrang),
   Vietnam; and Poulo Condore, Vietnam.
Distribution: Cambodia, Vietnam.

Apocheiridium (Apocheiridium) pinium Morikawa

Apocheiridium pinium Morikawa 1953a: 347–348, figs 1d–g; Morikawa, 1960: 135,
   plate 3 fig. 5, plate 8 fig. 15, plate 9 fig. 16, plate 10 fig. 20; Morikawa, 1962: 419;

Sato, 1979a: 100–102, figs 11a–o, plates 8, 17; Sato, 1979d: 44; Sato, 1980b: 65–71, figs 1–3; Sato, 1980d: plate 6; Sato, 1982c: 34; Sato, 1984d: fig. 2.

Type locality: Ishite River, Matsuyama, Shikoku, Japan.
Distribution: Japan.

Apocheiridium (Apocheiridium) rossicum Redikorzev

Apocheiridium rossicum Redikorzev, 1935a: 184–185, figs 1–2; Roewer, 1937: 277; Vachon, 1940f: 3; Beier, 1948b: 461; Kaisila, 1949b: 81, map 5.

Type localities: Merreküll, Estonia, U.S.S.R.; Leningrad (as Petersburg), R.S.F.S.R., U.S.S.R.; and Perm, R.S.F.S.R., U.S.S.R.
Distribution: Finland, U.S.S.R. (Estonia, R.S.F.S.R.).

Apocheiridium (Apocheiridium) stannardi Hoff

Apocheiridium stannardi Hoff, 1952b: 193–195, figs 3–4; Hoff & Bolsterli, 1956: 164; Hoff, 1958: 19; Hoff, 1961: 444; Nelson, 1971: 95; Nelson, 1975: 283–284, figs 35–39; Zeh, 1987b: 1087.

Type locality: Herod, Illinois, U.S.A.
Distribution: U.S.A. (Colorado, Illinois, Michigan).

Apocheiridium (Apocheiridium) turcicum Beier

Apocheiridium turcicum Beier, 1967f: 308–310, fig. 3.

Type locality: Namrun, Turkey.
Distribution: Turkey.

Apocheiridium (Apocheiridium) validissimum Beier

Apocheiridium validissimum Beier, 1976f: 212–213, fig. 13.

Type locality: Mangahou Creek, Chatham Island, New Zealand.
Distribution: New Zealand.

Apocheiridium (Apocheiridium) validum Beier

Apocheiridium validum Beier, 1967b: 292–293, fig. 10; Beier, 1976f: 212–213.

Type locality: Totara Reserve, Pohangina Valley, Palmerston North, South Island, New Zealand.
Distribution: New Zealand.

Apocheiridium (Apocheiridium) zealandicum Beier

Apocheiridium zealandicum Beier, 1976f: 212, fig. 12.

Type locality: Stokes Valley, North Island, New Zealand.
Distribution: New Zealand.

Subgenus Apocheiridium (Chiliocheiridium) Vitali-di Castri

Apocheiridium (Chiliocheiridium) Vitali-di Castri, 1969b: 267.

Type species: *Apocheiridium leopoldi* Vitali-di Castri, 1962, by original designation.

Apocheiridium (Chiliocheiridium) chilense Vitali-di Castri

Apocheiridium chilense Vitali-di Castri, 1962: 123–126, fig. 2; Beier, 1964b: 341; Cekalovic, 1984: 17.
Apocheiridium (Chiliocheiridium) chilense Vitali-di Castri: Vitali-di Castri, 1969b: 267.

Type locality: Quebrada El Tigre, Zapallar, Chile.
Distribution: Chile.

Apocheiridium (Chiliocheiridium) leopoldi Vitali-di Castri

Apocheiridium leopoldi Vitali-di Castri, 1962: 121–123, fig. 1; Beier, 1964b: 341; Vitali-di Castri, 1965b: figs 7, 9; Vitali-di Castri, 1966: fig. 3; Cekalovic, 1984: 18.
Apocheiridium (Chiliocheiridium) leopoldi Vitali-di Castri: Vitali-di Castri, 1969b: 267.

Type locality: Cerro El Roble, Chile.
Distribution: Chile.

Apocheiridium (Chiliocheiridium) serenense Vitali-di Castri

Apocheiridium serenense Vitali-di Castri, 1965b: fig. 8 (nomen nudum).
Apocheiridium (Chiliocheiridium) serenense Vitali-di Castri, 1969b: 267–277, figs 1–25; Cekalovic, 1984: 18.

Type locality: Bahía Guanaqueros, La Serena, Coquimbo, Chile.
Distribution: Chile.

Genus Cheiridium Menge

Cheiridium Menge, 1855: 36; Stecker, 1874b: 232; Stecker, 1875d: 87; Tömösváry, 1882b: 209; Balzan, 1887b: no pagination; Balzan, 1890: 409–410; O.P.-Cambridge, 1892: 229–230; Tullgren, 1899a: 177–178; Tullgren, 1906a: 201; Kew, 1911a: 50; J. C. Chamberlin, 1924a: 37–38; Redikorzev, 1924b: 24; Schenkel, 1928: 56; J. C. Chamberlin, 1931a: 237; Beier, 1932e: 7–8; G. O. Evans & Browning, 1954: 14; J. C. Chamberlin, 1938a: 264–265; Hoff, 1952b: 188; Morikawa, 1960: 134; Beier, 1963b: 243–244; Legg & Jones, 1988: 91.
Chiridium Menge: E. Simon, 1879a: 43; H. J. Hansen, 1884: 535; Lessert, 1911: 24.
Cheiridium (Cheiridium) Menge: J. C. Chamberlin, 1938a: 265–266.
Cheiridium (Isocheiridium) J. C. Chamberlin, 1938a: 269 (synonymized by Hoff & Clawson, 1952: 13–14).

Type species: of *Cheiridium*: *Chelifer museorum* Leach, 1817, by subsequent designation of E. Simon, 1879a: 43.
of *Cheiridium (Isocheiridium)*: *Cheiridium (Isocheiridium) minor* J. C. Chamberlin, 1938a, by original designation.

Cheiridium andinum Vitali-di Castri

Cheiridium andinum Vitali-di Castri, 1962: 126–130, figs 3a–b; Beier, 1964b: 341; Vitali-di Castri, 1969b: 266; Cekalovic, 1984: 18.

Type locality: Hacienda San Vicente, Los Andes, Chile.
Distribution: Chile.

Cheiridium angustum Beier

Cheiridium angustum Beier, 1978a: 429–430, fig. 1.

Type locality: Uis Tin mine, Okambahe Reserve, Omaruru, Namibia.
Distribution: Namibia.

Cheiridium aokii Sato

Cheiridium aokii Sato, 1984b: 52–55, figs 14–26.

Type locality: Nakano-daira, Haha-jima, Bonin Islands, Japan.
Distribution: Japan.

Cheiridium capense Beier

Cheiridium capense Beier, 1970c: 60–61, fig. 3.

Type locality: Mafeking, Cape Province, South Africa.
Distribution: South Africa.

Cheiridium chamberlini Dumitresco & Orghidan

Cheiridium chamberlini Dumitresco & Orghidan, 1981: 84–86, figs 1c–e, 2c, 4b, 5b, 6b; Armas & Alayón, 1984: 11.

Type locality: Cueva de Cativar, Oriente, Cuba.
Distribution: Cuba.

Cheiridium congicum Beier

Cheiridium museorum (Leach): Beier, 1955a: 12 (misidentification).
Cheiridium congicum Beier, 1970c: 58–59, fig. 2.

Type locality: Upemba National Park, Zaire.
Distribution: Zaire.

Cheiridium danconai Vitali-di Castri

Cheiridium danconai Vitali-di Castri, 1965a: 70–82, figs 1–28; Vitali-di Castri, 1965b: figs 4, 9; Vitali-di Castri, 1966: figs 2–3; Vitali-di Castri, 1969b: 266; Mahnert, 1981c: fig. 2; Cekalovic, 1984: 18.

Type locality: Monte Amargo, Atacama, Chile.
Distribution: Chile.

Cheiridium fallax Beier

Cheiridium fallax Beier, 1970c: 57–58, fig. 1.

Type locality: Alexandria, Cape Province, South Africa.
Distribution: South Africa.

Cheiridium firmum Hoff

Cheiridium firmum Hoff, 1952b: 188–192, figs 1–2; Park & Auerbach, 1954: 210; Hoff, 1958: 18; Zeh, 1987b: 1087.

Type locality: Pinkstaff, Lawrence County, Illinois, U.S.A.
Distribution: U.S.A. (Illinois).

! Cheiridium hartmanni (Menge)

Chelifer hartmanni Menge, in C. L. Koch & Berendt, 1854: 96; Menge, 1855: 38, plate 5 figs 12–13.
Cheiridium hartmanni (Menge): Hagen, 1870: 264; J. C. Chamberlin, 1931a: 237; Beier, 1932e: 278; Beier, 1937d: 307, fig. 8; Roewer, 1940: 331; Petrunkevitch, 1955: fig. 50(1); Beier, 1955g: 51; Schawaller, 1978: 4.
Cheiridium sp. ? *hartmanni* (Menge): Schawaller, 1978: 18–20, figs 11–14.

Type locality: Baltic Amber.
Distribution: Baltic Amber.

Cheiridium insperatum Hoff & Clawson

Cheiridium insperatum Hoff & Clawson, 1952: 8–14, figs 5–8; Hoff, 1958: 18; Muchmore, 1971a: 90; Zeh, 1987b: 1087.

Type locality: Moab, Grand County, Utah, U.S.A.
Distribution: U.S.A. (Utah).

Cheiridium insulare Vitali-di Castri

Cheiridium insulare Vitali-di Castri, 1984: 1073–1075, figs 38–42.

Type locality: Coreil, Guadeloupe.
Distribution: Guadeloupe.

Family Cheiridiidae

Cheiridium minor J. C. Chamberlin

Cheiridium (Isocheiridium) minor J. C. Chamberlin, 1938a: 270, figs 2a, 2g, 2i, 2m.
Cheiridium minor J. C. Chamberlin: Roewer, 1940: 346; Morikawa, 1953a: 345–347, figs 1a–c; Morikawa, 1960: 134–135, plate 3 fig. 4, plate 10 figs 18, 27; Morikawa, 1962: 419; W. K. Lee, 1981: 129–130, figs 1a–b.

Type locality: China.
Distribution: China, Japan, Korea.

Cheiridium museorum (Leach)

Phalangium cancroides (Linnaeus): Shaw, 1791: no pagination (misidentification).
Chelifer museorum Leach, 1817: 50, plate 142 fig. 4; C.L. Koch, 1843: 43–44, fig. 781; Holmberg, 1876: 28.
Chelifer nepoides Hermann, 1804: 116, plate 5 fig. q; Théis, 1832: 75–77, plate 3 figs 3a–b; Gervais, 1844: 79; Lucas, 1849: 276; Hagen, 1870: 266 (synonymized by Tömösváry, 1882b: 210).
Chélifer musaeòrum Leach: Templeton, 1836: 14.
Chelifer muscorum (sic) (Leach): Gervais, 1844: 78.
Cheiridium museorum (Leach): Menge, 1855: 36–38, plate 1 figs 1, 18, 24, plate 3 figs 4–5, 16–18, plate 4 fig. 11; Hagen, 1870: 264; L. Koch, 1873: 2–3; Stecker, 1874b: 232; Stecker, 1875d: 87–88; Pavesi, 1878: 362 (not seen); E. Simon, 1878a: 152; Tömösváry, 1882b: 210–211, plate 2 figs 16–18; Canestrini, 1884: no. 7; Daday, 1888: 125, 181; Lameere, 1895: 475; Tullgren, 1899a: 178–179, plate 2 figs 1–3; Kew, 1903: 298; Tullgren, 1906a: 201, fig. 3; Ellingsen, 1907a: 167; Whyte & Whyte, 1907: 203; Butterfield, 1908: 112; Godfrey, 1908: 158; Godfrey, 1909: 157–159; Kew, 1909a: 249; Kew, 1909b: 259; Ellingsen, 1911c: 174; Kew, 1911a: 51, fig. 13; Ellingsen, 1912b: 86, 104; Standen, 1912: 14–15; Ellingsen, 1914: 11; Kew, 1914: 96; Falconer, 1916: 191; Kew, 1916a: 124; Kew, 1916b: 79; Standen, 1916b: 124; Standen, 1917: 29; Daday, 1918: 2; J. C. Chamberlin, 1924a: 38, figs a, i, t, u, w, bb; Redikorzev, 1924b: 24–25, figs 10, 11b; Kästner, 1928: 4, figs 7, 9–10; Schenkel, 1928: 56, figs 2a–b; Beier, 1929a: 358; J. C. Chamberlin, 1931a: figs 10b, 15u, 20c, 29k, 33l, 38h; J. C. Chamberlin, 1932b: 137; Beier, 1932e: 8–9, figs 2, 5; Tumšs, 1934: 14, fig. 1; Redikorzev, 1935a: fig. 2; Vachon, 1935a: 330–332; Vachon, 1935j: 187; Nordberg, 1936: 48–49; Roewer, 1936: figs 43a–b, 70r; Vachon, 1936b: 140; Roewer, 1937: 277; Vachon, 1937f: 128–129; J. C. Chamberlin, 1938a: 266–268, figs 2e–f, 2l; Beier, 1939e: 311; Beier, 1939f: 197; Lohmander, 1939b: 292; Vachon, 1940a: 53–54; Cooreman, 1946b: 7–8; Beier, 1947b: 323; Kaisila, 1947: 86; Beier, 1948b: 457, 461; Franz & Beier, 1948: 451; Kaisila, 1949b: 80–81, map 5; Redikorzev, 1949: 645–646; Vachon, 1949: fig. 199f; Beier, 1952e: 4; Beier & Franz, 1954: 457; G.O. Evans & Browning, 1954: 14–15, fig. 5; Weidner, 1954b: 109; Beier, 1956h: 24; George, 1957: 80; Cloudsley-Thompson, 1958: 96; Ressl & Beier, 1958: 2; Cirdei & Gutu, 1959: 1–2; George, 1961: 38; Beier, 1963b: 244, fig. 248; Beier, 1963h: 156; Beier, 1964k: 67; Weygoldt, 1964a: 362–366, figs 6–11; Ressl, 1965: 289; Weygoldt, 1965c: 436; Beier, 1966k: 464; Helversen, 1966a: 138; Weygoldt, 1966b: fig. 9; Rafalski, 1967: 14; Hammen, 1969: 19; Weygoldt, 1969a: 23, 26, 41, 42, 74, 108, 110, figs 23, 38, 51, 89–90, 100; Legg, 1970b: fig. 3(3); Ressl, 1970: 250–251; Helversen & Martens, 1971: 378; Legg, 1971a: 476; Howes, 1972a: 108; Muchmore, 1972d: 109; Lagar, 1972a: 21; Legg, 1973b: 430, fig. 2j; Ćurčić, 1974a: 25; Legg, 1974a: 324–338, figs 1–10; Jones, 1975a: 88; Klausen, 1975: 64; Legg, 1975a: 66; Legg, 1975d: fig. 2; Frøiland, 1976: 12; Crocker, 1978: 8; Jones, 1978: 94; Judson, 1979a: 63; Thaler, 1979: 52; Gardini, 1980c: 116–117, 122, 129; Jones, 1980c: map 13; Jedryczkowski, 1985: 79; Lippold, 1985: 40; Schawaller, 1985a: 10; Judson, 1987: 369; Krumpál & Cyprich, 1988: 42; Legg & Jones, 1988: 91–92, figs 5b, 20a, 20b(a–f).

439

Cheiridium musaeorum (Leach): Ritsema, 1874: xxxiii.
Chiridium museorum (Leach): E. Simon, 1879a: 43–44, plate 18 figs 19–20; Becker, 1884: cclxiv; H. J. Hansen, 1884: 535–536; O.P.-Cambridge, 1892: 230–231, plate C fig. 21; Evans, 1901b: 242; Godfrey, 1901b: 217; Kew, 1901b: 194–195; Lessert, 1911: 24–25; Nonidez, 1917: 26–27; Navás, 1925: 111; Väänänen, 1928a: 17, fig. 6(5).
Cheiridium musearum (sic): Daday, 1880a: 192.
Obisium museorum (Leach): Bernard, 1893b: 428; Bernard, 1894: 410.
Cheiridium muscorum (sic) (Leach): Cerruti, 1959: 51; Meinertz, 1962a: 392, map 3.
Not *Cheiridium museorum* (Leach): Beier, 1955a: 12 (misidentification; see *Cheiridium congicum* Beier).

Type localities: of *Chelifer museorum*: not stated, presumably England, Great Britain.
of *Chelifer nepoides*: not stated, presumably near Strasbourg, France.
Distribution: Algeria, Austria, Belgium, Denmark, Finland, France, Great Britain, Greece, Hungary, India, Ireland, Italy, Mozambique, Netherlands, Norway, Poland, Romania, South Africa, Spain, Sweden, Switzerland, Turkey, West Germany, U.S.A. (Massachusetts), U.S.S.R. (Kirghizia, Latvia, R.S.F.S.R.), Yugoslavia, Zaire.

Cheiridium nepalense Ćurčić

Cheiridium nepalense Ćurčić, 1980b: 83–84, figs 15–20; Schawaller, 1987a: 212.

Type locality: Thini Kola, Nepal.
Distribution: Nepal.

Cheiridium nubicum Beier

Cheiridium nubicum Beier, 1962e: 298–300, fig. 2.

Type locality: Wadi Halfa, Sudan.
Distribution: Sudan.

Cheiridium perreti Mahnert

Cheiridium perreti Mahnert, 1982b: 123–125, figs 25–29.

Type locality: S. of Mpekatoni, Lamu, Kenya.
Distribution: Kenya.

Cheiridium saharicum Beier

Cheiridium saharicum Beier, 1965b: 371–373, fig. 5.

Type locality: Mardengai near Largeau, Chad.
Distribution: Chad.

Cheiridium simulacrum J. C. Chamberlin

Cheiridium (Cheiridium) simulacrum J. C. Chamberlin, 1938a: 268–269, figs 2b–d, 2h, 2j–k.
Cheiridium simulacrum J. C. Chamberlin: Roewer, 1940: 346.

Type locality: Wahiawa, Oahu, Hawaii.
Distribution: Hawaii.

Cheiridium somalicum Mahnert

Cheiridium somalicum Mahnert, 1984b: 58–60, figs 30–37.

Type locality: Sar Uanle, Somalia.
Distribution: Somalia.

Cheiridium tumidum Mahnert

Cheiridium tumidum Mahnert, 1982b: 121–123, figs 17–24.

Type locality: SE. of Kogari, Embu, Kenya.
Distribution: Kenya.

Genus **Cryptocheiridium** J. C. Chamberlin

Cryptocheiridium J. C. Chamberlin, 1931a: 238; Beier, 1932e: 14.

Type species: *Cheiridium subtropicum* Tullgren, 1907b, by original designation.

Subgenus **Cryptocheiridium (Cryptocheiridium)** J. C. Chamberlin

! Cryptocheiridium (Cryptocheiridium) antiquum Schawaller

Cryptocheiridium antiquum Schawaller, 1981d: 3–7, figs 1–12.

Type locality: Dominican Amber.
Distribution: Dominican Amber.

Cryptocheiridium (Cryptocheiridium) australicum Beier

Cryptocheiridium australicum Beier, 1969a: 183–185, fig. 10; Richards, 1971: 19, 24, 25, 30, 32, 43; Harvey, 1981b: 244; Harvey, 1985b: 130; Schawaller, 1981d: figs 11–12.

Type locality: Murra-el-elevyn Cave, Nullarbor Plain, Western Australia, Australia.
Distribution: Australia (Western Australia).

Cryptocheiridium (Cryptocheiridium) elgonense Beier

Cryptocheiridium elgonense Beier, 1955c: 546–548, figs 13–14; Mahnert, 1981c: fig. 2; Mahnert, 1982b: 131–133, figs 34–37; Schawaller, 1981d: figs 11–12.

Type locality: Mt Elgon, E. side, Kenya.
Distribution: Kenya.

Cryptocheiridium (Cryptocheiridium) formosanum (Ellingsen)

Cheiridium formosanum Ellingsen, 1912a: 123–125.
Cryptocheiridium formosanum (Ellingsen): J. C. Chamberlin, 1931a: 238; Beier, 1932e: 15; Roewer, 1937: 277; Schawaller, 1981d: figs 11–12.

Type locality: Takao (now Kao-hsiung), Taiwan.
Distribution: Taiwan.

Cryptocheiridium (Cryptocheiridium) kivuense Beier

Cryptocheiridium kivuense Beier, 1959d: 38–39, fig. 18; Beier, 1972a: 9; Mahnert, 1982b: 133.

Type locality: Itombwe, Mwenga, Zaire.
Distribution: Kenya, Zaire.

Cryptocheiridium (Cryptocheiridium) lucifugum Beier

Cryptocheiridium lucifugum Beier, 1963c: 51–52, fig. 1; Tenorio & Muchmore, 1982: 378.

Type locality: Cavern A, Batu Caves, Malaysia.
Distribution: Malaysia.

Cryptocheiridium (Cryptocheiridium) philippinum Beier

Cryptocheiridium philippinum Beier, 1977a: 187–188, fig. 1; Schawaller, 1981d: figs 11–12.

Family Cheiridiidae

Type locality: Cueva Basilen, Pagbilao, Philippines.
Distribution: Philippines.

Cryptocheiridium (Cryptocheiridium) salomonense Beier

Cryptocheiridium salomonense Beier, 1970g: 319–320, fig. 2; Beier, 1971b: 367.

Type locality: N. of Kuzi, Kolombangara, Solomon Islands.
Distribution: Papua New Guinea, Solomon Islands.

Cryptocheiridium (Cryptocheiridium) somalicum Callaini

Cryptocheiridium somalicum Callaini, 1985: 182–186, figs 1, 2a–g.

Type locality: Mugdile, Somalia.
Distribution: Somalia.

Cryptocheiridium (Cryptocheiridium) subtropicum (Tullgren)

Chiridium subtropicum Tullgren, 1907b: 218–220, figs 5a–b, 5d–f.
Cheiridium subtropicum Tullgren: Ellingsen, 1912b: 86, 104; Godfrey, 1923: 98–99; Godfrey, 1927: 18.
Cryptocheiridium subtropicum (Tullgren): J. C. Chamberlin, 1931a: 238; Beier, 1932e: 14–15, fig. 10; Roewer, 1937: 277; Beier, 1958a: 171; Beier, 1964k: 69; Schawaller, 1981d: figs 11–12.

Type locality: Umfolozi, Natal, South Africa.
Distribution: South Africa.

Subgenus Cryptocheiridium (Cubanocheiridium) Dumitresco & Orghidan

Cryptocheiridium (Cubanocheiridium) Dumitresco & Orghidan, 1981: 77.

Type species: *Cryptocheiridium (Cubanocheiridium) elegans* Dumitresco & Orghidan, 1981, by original designation.

Cryptocheiridium (Cubanocheiridium) elegans Dumitresco & Orghidan

Cryptocheiridium (Cubanocheiridium) elegans Dumitresco & Orghidan, 1981: 77–84, figs 1a–b, 2a–b, 3a–f, 4a, 5a, 6a, 7a–f; Armas & Alayón, 1984: 11.

Type locality: Cueva La Pluma, Cuba.
Distribution: Cuba.

Genus Neocheiridium Beier

Neocheiridium Beier, 1932e: 9.
Neocheiridium (Austrocheiridium) Vitali-di Castri, 1962: 134 (synonymized by Mahnert, 1982b: 130).

Type species: of *Neocheiridium*: *Cheiridium corticum* Balzan, 1887b, by original designation.
of *Neocheiridium (Austrocheiridium)*: *Neocheiridium (Austrocheiridium) chilense* Vitali-di Castri, 1962, by original designation.

Neocheiridium africanum Mahnert

Neochiridium sp.: Mahnert, 1981c: fig. 2.
Neocheiridium africanum Mahnert, 1982b: 127–128, figs 38–45.

Type locality: Warges, Kenya.
Distribution: Kenya.

Neocheiridium beieri Vitali-di Castri

Neocheiridium beieri Vitali-di Castri, 1962: 131–133, figs 5a–b; Beier, 1964b: 341; Vitali-di Castri, 1965b: figs 6, 9; Vitali-di Castri, 1966: fig. 3; Cekalovic, 1984: 19.

Neocheiridium (Neocheiridium) beieri Vitali-di Castri: Vitali-di Castri, 1969b: 266.

Type locality: Hacienda La Rinconada de Maipú, Quebrada La Plata, Chile.
Distribution: Chile.

Neocheiridium chilense Vitali-di Castri

Neocheiridium (Austrocheiridium) chilense Vitali-di Castri, 1962: 134–138, figs 6a–d;
Beier, 1964b: 341; Vitali-di Castri, 1965b: figs 5, 9; Vitali-di Castri, 1966: fig. 3;
Vitali-di Castri, 1969b: 266–267; Cekalovic, 1984: 19.

Type locality: Hacienda La Rinconada de Maipú, Quebrada La Plata, Chile.
Distribution: Chile.

Neocheiridium corticum (Balzan)

Cheiridium corticum Balzan, 1887b: no pagination, figs; Balzan, 1890: 410–411, figs 1,
1a–c; Balzan, 1892: 549.
Apocheiridium (?) *corticum* (Balzan): J. C. Chamberlin, 1931a: 238.
Neocheiridium corticum (Balzan): Beier, 1932e: 9–10, fig. 6; Roewer, 1937: 277;
Mello-Leitão, 1939a: 117; Mello-Leitão, 1939b: 615; Beier, 1977c: 103; Mahnert &
Aguiar, 1986: 499–503, figs 1–8.

Type locality: Rio Apa, Paraguay.
Distribution: Argentina, Galapagos Islands, Paraguay.

Neocheiridium galapagoense Beier

Neocheiridium galapagoense Beier, 1978c: 540–541, fig. 5.

Type locality: Pinzón Island, Galapagos Islands.
Distribution: Galapagos Islands.

Neocheiridium pusillum Mahnert

Neocheiridium pusillum Mahnert, 1982b: 128–130, figs 46–55; Mahnert, Adis &
Bührnheim, 1987: fig. 11b.

Type locality: 10 km S. of Garsen, Tana River, Kenya.
Distribution: Kenya.

Neocheiridium strinatii Mahnert & Aguiar

Neocheiridium strinatii Mahnert & Aguiar, 1986: 503–505, figs 9–12.

Type locality: Grot van San Pedro, Curaçao.
Distribution: Aruba, Curaçao.

Neocheiridium tenuisetosum Beier

Neocheiridium tenuisetosum Beier, 1959e: 202, fig. 14.

Type locality: Lago Trebol, Bariloche, Rio Negro, Argentina.
Distribution: Argentina.

Neocheiridium triangulare Mahnert & Aguiar

Neocheiridium triangulare Mahnert & Aguiar, 1986: 505–508, figs 13–17.

Type locality: Rio Urubu, Amazonas, Brazil.
Distribution: Brazil (Amazonas, Pará).

Genus Nesocheiridium Beier

Nesocheiridium Beier, 1957d: 27.

Type species: *Nesocheiridium stellatum* Beier, 1957d, by original designation.

Nesocheiridium stellatum Beier

Nesocheiridium stellatum Beier, 1957d: 28, figs 13a–c.

Type locality: Mt Magpi, Pidos Kalahe, Saipan, Mariana Islands.
Distribution: Mariana Islands.

Genus **Pycnocheiridium** Beier

Pycnocheiridium Beier, 1964k: 65–66.

Type species: *Pycnocheiridium mirum* Beier, 1964k, by original designation.

Pycnocheiridium mirum Beier

Pycnocheiridium mirum Beier, 1964k: 66–67, fig. 25.

Type locality: Wilton near Alicedale, Cape Province, South Africa.
Distribution: South Africa.

Family PSEUDOCHIRIDIIDAE J. C. Chamberlin

Pseudocheiridiinae J. C. Chamberlin, 1923c: 370; Beier, 1932e: 3–4; Roewer, 1937: 275; Murthy & Ananthakrishnan, 1977: 115.
Pseudocheiridiidae J. C. Chamberlin: J. C. Chamberlin, 1931a: 235–236.
Pseudochiridiidae J. C. Chamberlin: Hoff, 1964c: 89–90; Harvey, 1985b: 149; Muchmore, 1982a: 101.
Pseudochiridiinae J. C. Chamberlin: Vitali-di Castri, 1970b: 1195.

Genus **Paracheiridium** Vachon

Paracheiridium Vachon, 1938d: 237; Roewer, 1940: 346.

Type species: *Paracheiridium decaryi* Vachon, 1938d, by original designation.

Paracheiridium decaryi Vachon

Paracheiridium decaryi Vachon, 1938d: 237–241, figs 1–10; Roewer, 1940: 346; Vitali-di Castri, 1966: fig. 1; Legendre, 1972: 446.

Type locality: Grotte d'Anjohibé, Andranoboka, Madagascar.
Distribution: Madagascar.

Paracheiridium vachoni Vitali-di Castri

Paracheiridium vachoni Vitali-di Castri, 1970b: 1175–1183, figs 1–16; Mahnert, 1981c: fig. 1.

Type locality: Iles Glorieuses, Madagascar.
Distribution: Madagascar.

Genus **Pseudochiridium** With

Pseudochiridium With, 1906: 199–200; Beier, 1932e: 4; Hoff, 1964c: 90.
Pseudocheiridium (sic) With: J. C. Chamberlin, 1931a: 235.
Afrocheiridium Beier, 1932e: 5–6 (synonymized by Beier, 1955l: 306).

Type species: of *Pseudochiridium*: *Pseudochiridium thorelli* With, 1906, by original designation.
of *Afrocheiridium*: *Pseudochiridium traegardhi* Tullgren, 1907b, by original designation.

Pseudochiridium africanum Beier

Pseudochiridium africanum Beier, 1944: 181–183, figs 6–7; Weidner, 1959: 116; Beier, 1965b: 371; Beier, 1972a: 9; Spaull, 1979: 117.

Type locality: Amani, Tanzania.
Distribution: Aldabra Islands, Chad, Tanzania, Zaire.

Pseudochiridium clavigerum (Thorell)

Chelifer claviger Thorell, 1889: 591–594, plate 5 figs 5a–b.
Trachychernes claviger (Thorell): Pocock, 1900: 156.
Pseudochiridium clavigerum (Thorell): With, 1906: 204–206, plate 4 fig. 13a; Beier,
   1932e: 5; Roewer, 1937: 276; Beier, 1955e: 41–43, figs 3–4; Beier, 1965g: 765;
   Beier, 1967c: 318; Beier, 1967g: 352; Beier, 1981: 285; Harvey, 1985b: 149;
   Schawaller, 1987a: 211, figs 25–27.
Pseudocheiridium clavigerum (Thorell): J. C. Chamberlin, 1931a: 236, fig. 29m;
   Chapman, 1984a: 6.
Pseudochiridium sundaicum Beier, 1953g: 83–84, figs 2–3 (synonymized by Beier,
   1965g: 765).

Type localities: of *Chelifer claviger*: Bhamò, Burma.
of *Pseudochiridium sundaicum*: Langgai, Sumba, Indonesia.
Distribution: Andaman Islands, Burma, Christmas Island, India, Indonesia (Irian
   Jaya, Sumba), Malaysia, Nepal, Philippines.

Pseudochiridium heurtaultae Vitali-di Castri

Pseudochiridium heurtaultae Vitali-di Castri, 1970b: 1184–1195, figs 17–42; Mahnert,
   1978f: 94–95.

Type locality: Angola.
Distribution: Angola, Congo.

Pseudochiridium insulae Hoff

Pseudochiridium insulae Hoff, 1964c: 90–92, figs 1–4.

Type locality: Stock Island, Monroe County, Florida, U.S.A.
Distribution: U.S.A. (Florida).

Pseudochiridium kenyense Mahnert

Pseudochiridium sp.: Mahnert, 1981c: fig. 2.
Pseudochiridium kenyense Mahnert, 1982b: 119–120, figs 9–16.

Type locality: Warges, Kenya.
Distribution: Kenya.

Pseudochiridium lawrencei Beier

Pseudochiridium lawrencei Beier, 1964k: 64–65, fig. 24.

Type locality: Pirie Forest, near King William's Town (as Kingwilliamstown), Cape
   Province, South Africa.
Distribution: South Africa.

Pseudochiridium minutissimum Beier

Pseudochiridium minutissimum Beier, 1959d: 36–37, fig. 17.

Type locality: Ile de M'Boko, Lake Tanganyika, Kivu, Zaire.
Distribution: Zaire.

Pseudochiridium thorelli With

Pseudochiridium thorelli With, 1906: 200–204, figs 6a–b, plate 4, figs 12a–g; Beier,
   1932e: 5, fig. 3; Roewer, 1936: figs 22f, 70o; Roewer, 1937: 276, fig. 223; Vachon,
   1949: fig. 199c.

Family Sternophoridae

Pseudocheiridium thorelli With: J. C. Chamberlin, 1931a: 236, figs 10a, 15r, 18a, 20e, 44f–g, 65; Beier, 1932g: fig. 181.
Pseudochiridium thorellii With: Roewer, 1936: fig. 34c, 43e–f, 45e–f.

Type locality: Nancowry (as Nankovry), Nicobar Islands.
Distribution: Nicobar Islands.

Pseudochiridium traegardhi Tullgren

Pseudochiridium trägårdhi Tullgren, 1907b: 225–226, figs 4a–d; Ellingsen, 1912b: 86.
Pseudocheiridium tragardhi Tullgren: J. C. Chamberlin, 1931a: 236.
Afrocheiridium trägardhi (Tullgren): Beier, 1932e: 6; Roewer, 1937: 276.
Pseudochiridium (Afrocheiridium) trägårdhi (Tullgren): Beier, 1955l: 306–307, fig. 25.
Pseudochiridium (Afrocheiridium) tragardhi Tullgren: Beier, 1958a: 171.
Pseudochiridium trägardhi Tullgren: Beier, 1964k: 64.

Type localities: Pietermaritzburg (as Maritzburg), Natal, South Africa; and Van Reenen, Natal, South Africa.
Distribution: South Africa.

Pseudochiridium triquetrum Beier

Pseudochiridium triquetrum Beier, 1965g: 765–767, fig. 8; Tenorio & Muchmore, 1982: 383.

Type locality: Ajappo, south side of Danau Sentani (as Lac Sentani), Irian Jaya, Indonesia.
Distribution: Indonesia (Irian Jaya).

Family STERNOPHORIDAE J. C. Chamberlin

Sternophorinae J. C. Chamberlin, 1923c: 370–371.
Sternophoridae J. C. Chamberlin: J. C. Chamberlin, 1931a: 238; Beier, 1932e: 140; Beier, 1932g: 186; Roewer, 1937: 277–278; Beier, 1954d: 136; Hoff, 1956c: 3–4; Hoff, 1963b: 2; Murthy & Ananthakrishnan, 1977: 117–118; Sivaraman, 1981a: 313; Muchmore, 1982a: 100; Harvey, 1985b: 151; Harvey, 1985c: 144–146.

Genus Afrosternophorus Beier

Sternophorus (Afrosternophorus) Beier, 1967d: 81–82.
Sternophorellus Beier, 1971b: 371–372 (synonymized by Harvey, 1985c: 167).
Indogaryops Sivaraman, 1981a: 322 (synonymized by Harvey, 1985c: 168).
Afrosternophorus Beier: Harvey, 1985c: 167–170.

Type species: of *Afrosternophorus*: *Sternophorus (Afrosternophorus) aethiopicus* Beier, 1967d, by original designation.
of *Sternophorellus*: *Sternophorellus araucariae* Beier, 1971b, by original designation.
of *Indogaryops*: *Indogaryops amrithiensis* Sivaraman, 1981a, by original designation.

Afrosternophorus aethiopicus (Beier)

Sternophorus (Afrosternophorus) aethiopicus Beier, 1967d: 81–82, fig. 6.
Afrosternophorus aethiopicus (Beier): Harvey, 1985c: 173–174, figs 52, 78–79, map 3.

Type locality: Alomata, Ethiopia.
Distribution: Ethiopia.

Afrosternophorus anabates Harvey

Afrosternophorus anabates Harvey, 1985c: 190–193, figs 8, 58, 73, 114–120, 129, map 5; Zeh, 1987b: 1087.

Type locality: 15 km WNW. of Yaapeet, Victoria, Australia.
Distribution: Australia (Victoria).

Afrosternophorus araucariae (Beier)

Sternophorellus araucariae Beier, 1971b: 372–373, fig. 3.
Afrosternophorus araucariae (Beier): Harvey, 1985c: 197–198, figs 61, 130–132, map 6.

Type locality: Mt Dayman, Milne Bay District, Papua New Guinea.
Distribution: Papua New Guinea.

Afrosternophorus cavernae (Beier)

Sternophorellus cavernae Beier, 1982: 44–45, fig. 1.
Afrosternophorus cavernae (Beier): Harvey, 1985c: 199–200, figs 62, 130, 133–134, map 6.

Type locality: Selminum tem Cave, West Sepic District, Papua New Guinea.
Distribution: Papua New Guinea.

Afrosternophorus ceylonicus (Beier)

Sternophorus ceylonicus Beier, 1973b: 47, fig. 11.
Sternophorus indicus Murthy & Ananthakrishnan, 1977: 119–121, figs 39a–c (synonymized by Harvey, 1985c: 174).
Sternophorus (Sternophorus) transiens Murthy & Ananthakrishnan, 1977: 121–123, figs 40a–e (synonymized by Harvey, 1985c: 174).
Sternophorus (Sternophorus) montanus Sivaraman, 1981a: 315–317, figs 1a–b (synonymized by Harvey, 1985c: 174).
Sternophorus (Afrosternophorus) femoratus Sivaraman, 1981a: 317–319, figs 2a–b (synonymized by Harvey, 1985c: 174).
Sternophorus (Afrosternophorus) intermedius Sivaraman, 1981a: 319–321, figs 3a–d (synonymized by Harvey, 1985c: 174).
Indogaryops amrithiensis Sivaraman, 1981a: 322–324, figs 4a–d (synonymized by Harvey, 1985c: 174).
Afrosternophorus ceylonicus (Beier): Harvey, 1985c: 174–177, figs 53, 65–66, 80–85, 90, map 4; Zeh, 1987b: 1087.

Type localities: of *Sternophorus ceylonicus*: Chemiyanpattu, 18 miles SE. of Point Pedro, Northern Province, Sri Lanka.
of *Sternophorus indicus*: Shimoga, Karnataka, India.
of *Sternophorus (Sternophorus) transiens*: Shimoga, Karnataka, India.
of *Sternophorus (Sternophorus) montanus*: Alakarkoil Hill, Madurai, Tamil Nadu, India.
of *Sternophorus (Afrosternophorus) femoratus*: Amrithi forest, North Arcot, Tamil Nadu, India.
of *Sternophorus (Afrosternophorus) intermedius*: Alakarkoil Hill, Madurai, Tamil Nadu, India.
of *Indogaryops amrithiensis*: Amrithi forest, North Arcot, Tamil Nadu, India.
Distribution: India, Sri Lanka.

Afrosternophorus chamberlini (Redikorzev)

Sternophorus chamberlini Redikorzev, 1938: 89–91, figs 17–18; Roewer, 1940: 346; Beier, 1951a: 71–72, fig. 16 (in part; see *Afrosternophorus fallax* Harvey).

Afrosternophorus chamberlini (Redikorzev): Harvey, 1985c: 177–179, figs 54, 67, 86–90, map 4; Zeh, 1987b: 1087.

Type locality: Da Lat (as Dalat), Vietnam.
Distribution: Laos, Vietnam.

Afrosternophorus cylindrimanus (Beier)

Sternophorus cylindrimanus Beier, 1951a: 73–74, fig. 17 (in part).
Afrosternophorus cylindrimanus (Beier): Harvey, 1985c: 182–184, figs 69, 90, 98–100, map 4.

Type locality: Pak Lay (as Paclay), Laos.
Distribution: Laos.

Afrosternophorus dawydoffi (Beier)

Sternophorus dawydoffi Beier, 1951a: 70–71, fig. 15.
Sternophorus cylindrimanus Beier, 1951a: 73–74, fig. 17 (in part).
Afrosternophorus dawydoffi (Beier): Harvey, 1985c: 179–182, figs 55, 68, 90–97, map 4; Zeh, 1987b: 1087.

Type locality: Roessei Chrum (as Rusei), Cambodia.
Distribution: Cambodia, Vietnam.

Afrosternophorus fallax Harvey

Sternophorus chamberlini Redikorzev: Beier, 1951a: 71–72 (misidentification; in part).
Afrosternophorus fallax Harvey, 1985c: 200–201, figs 63, 76, 130, 135–138, map 4; Zeh, 1987b: 1087.

Type locality: Cao Nguyên Lâm Viên, Vietnam.
Distribution: Vietnam.

Afrosternophorus grayi (Beier)

Sternophorus hirsti grayi Beier, 1971b: 370–371, fig. 2.
Afrosternophorus grayi (Beier): Harvey, 1985c: 195–197, figs 60, 75, 125–129, map 6; Zeh, 1987b: 1087.

Type locality: Bulolo, Morobe District, Papua New Guinea.
Distribution: Papua New Guinea.

Afrosternophorus hirsti (J. C. Chamberlin)

Sternophorus hirsti J. C. Chamberlin, 1932b: 143; Beier, 1932e: 18; J. C. Chamberlin, 1934b: 6; Roewer, 1937: 278; Harvey, 1981b: 244.
'*Sternophorus*' *hirsti* J. C. Chamberlin: Harvey, 1982: 192.
Sternophorus (Sternophorus) hirsti hirsti J. C. Chamberlin: Harvey, 1985b: 151.
Afrosternophorus hirsti (J. C. Chamberlin): Harvey, 1985c: 184–187, figs 1–7, 51, 56, 70–71, 101–108, 129, map 5; Zeh, 1987b: 1087.

Type locality: Barringun, New South Wales, Australia.
Distribution: Australia (New South Wales, Queensland).

Afrosternophorus nanus Harvey

Afrosternophorus nanus Harvey, 1985c: 187–190, figs 57, 72, 109–113, 129, map 5; Zeh, 1987b: 1087.

Type locality: Rum Jungle, Northern Territory, Australia.
Distribution: Australia (Northern Territory).

Afrosternophorus papuanus (Beier)

Sternophorus papuanus Beier, 1975b: 211–212, fig. 5.
Afrosternophorus papuanus (Beier): Harvey, 1985c: 193–195, figs 59, 74, 121–124, 129, map 6; Zeh, 1987b: 1087.

Type locality: Americ, Madang, Papua New Guinea.
Distribution: Papua New Guinea.

Afrosternophorus xalyx Harvey

Afrosternophorus xalyx Harvey, 1985c: 201–203, figs 64, 77, 130, 139–143, map 5; Zeh, 1987b: 1087.

Type locality: Townsville, Queensland, Australia.
Distribution: Australia (Queensland).

Genus Garyops Banks

Garyops Banks, 1909b: 305; J. C. Chamberlin, 1931a: 238; Beier, 1932e: 18; Hoff, 1963b: 2–3; Harvey, 1985c: 152–153.
Sternophorus J. C. Chamberlin, 1923c: 371; J. C. Chamberlin, 1931a: 238–239; Beier, 1932e: 16; Murthy & Ananthakrishnan, 1977: 18 (synonymized by Harvey, 1985c: 152).

Type species: of *Garyops*: *Garyops depressus* Banks, 1909b, by monotypy.
of *Sternophorus*: *Sternophorus sini* J. C. Chamberlin, 1923c, by original designation.

Garyops centralis Beier

Garyops centralis Beier, 1953b: 15–16, figs 1–2; Harvey, 1985c: 158–160, figs 22–24, 36, 39, map 1.

Type locality: Cutuco, La Union, El Salvador.
Distribution: El Salvador.

Garyops depressus Banks

Garyops depressa Banks, 1909b: 305–306 (in part); Beier, 1932e: 18; Roewer, 1937: 278; Hoff, 1958: 19; Hoff, 1963b: 4–7, figs 1–4; Beier, 1976d: 46; Brach, 1979: 34.
Garyops depressus Banks: Harvey, 1985c: 153–156, figs 9–16, 28, 34, 39, map 1; Zeh, 1987b: 1087.
Not *Garyops depressa* Banks: Hounsome, 1980: 85 (misidentification; see *Idiogaryops pumilus* (Hoff)).

Type locality: Punta Gorda, Charlotte County, Florida.
Distribution: Dominican Republic, U.S.A. (Florida).

Garyops ferrisi (J. C. Chamberlin)

Sternophorus ferrisi J. C. Chamberlin, 1932b: 143; Beier, 1932e: 18; Roewer, 1937: 278.
Garyops (?) *ferrisi* (J. C. Chamberlin): Harvey, 1985c: 160–161, figs 25–27, 30, 39, map 1.

Type locality: Michoacan, Mexico.
Distribution: Mexico.

Garyops sini (J. C. Chamberlin)

Sternophorus sini J. C. Chamberlin, 1923c: 371–372, plate 1, fig. 6, plate 2, fig. 21, plate 3, figs 6, 15, 22–25; J. C. Chamberlin, 1931a: 192, figs 4d, 10c, 11z, 18e–f, 20f, 52o–q, 67; J. C. Chamberlin, 1932b: 142; Beier, 1932e: 17, figs 11–12; Beier, 1932g: fig. 262; Roewer, 1936: figs 12, 44c, 147; Roewer, 1937: 278; Vachon, 1949: fig. 199b.

Garyops sini (J. C. Chamberlin): Harvey, 1985c: 156–158, figs 17–21, 29, 35, 39, map 1; Zeh, 1987b: 1087.

Type locality: Monument Point, Tiburon Island, Gulf of California, Mexico.
Distribution: Mexico.

Genus **Idiogaryops** Hoff

Idiogaryops Hoff, 1963b: 10–11; Harvey, 1985c: 162–163.

Type species: *Sternophorus paludis* J. C. Chamberlin, 1932b, by original designation.

Idiogaryops paludis (J. C. Chamberlin)

Sternophorus paludis J. C. Chamberlin, 1932b: 142–143; Beier, 1932e: 17–18; Roewer, 1937: 278; Hoff & Bolsterli, 1956: 164–165; Hoff, 1958: 19.
Idiogaryops paludis (J. C. Chamberlin): Hoff, 1963b: 11–13, figs 7–9; Weygoldt, 1969a: 27, fig. 105; Rowland & Reddell, 1976: 19; Brach, 1979: 34; Harvey, 1985c: 164–165, figs 31, 37, 40–45, map 2; Zeh, 1987b: 1087.

Type locality: Alachua County, Florida, U.S.A.
Distribution: U.S.A. (Arkansas, Florida, Georgia, Illinois, Mississippi, Texas).

Idiogaryops pumilus (Hoff)

Garyops depressa Banks, 1909b: 305–306 (misidentification; in part); Hounsome, 1980: 85 (misidentification).
Garyops pumila Hoff, 1963b: 7–10, figs 5–6 (in part).
Idiogaryops pumilus (Hoff): Harvey, 1985c: 165–166, figs 32, 38, 40, 46–50, map 2; Zeh, 1987b: 1087.

Type locality: Parker Islands, near Lake Placid, Highlands County, Florida, U.S.A.
Distribution: Cayman Islands, U.S.A. (Florida).

Superfamily CHELIFEROIDEA Risso

Cheliferoidea Risso: J. C. Chamberlin, 1931a: 239–240; Hoff, 1949b: 449; Petrunkevitch, 1955: 82; Hoff, 1956c: 4; Morikawa, 1960: 135–136; Murthy & Ananthakrishnan, 1977: 123; Muchmore, 1982a: 100; Legg & Jones, 1988: 94, 128.
Cheliferides (sic) Risso: Beier, 1932e: 19; Beier, 1932g: 186; Roewer, 1937: 278.

Family ATEMNIDAE J. C. Chamberlin

Pessigini Navás, 1925: 109.
Atemnidae J. C. Chamberlin, 1931a: 243–244; Beier, 1932b: 548; Beier, 1932e: 20; Beier, 1932g: 186; Roewer, 1937: 279; Hoff, 1956c: 4; Morikawa, 1960: 136; Beier, 1963b: 245; Murthy & Ananthakrishnan, 1977: 124; Muchmore, 1982a: 100; Harvey, 1985b: 128.
Atemninae J. C. Chamberlin: Beier, 1932b: 549; Beier, 1932e: 20–21; Roewer, 1937: 278; Morikawa, 1960: 136–137; Murthy & Ananthakrishnan, 1977: 124.
Miratemninae Beier, 1932b: 608; Beier, 1932e: 77; J. C. Chamberlin, 1933: 262; Roewer, 1937: 285; Murthy & Ananthakrishnan, 1977: 133.
Miratemnidae Beier: Dumitresco & Orghidan, 1970a: 134; Muchmore, 1975c: 231–232; Muchmore, 1982a: 100–101.

Genus **Anatemnus** Beier

Anatemnus Beier, 1932b: 578; Beier, 1932e: 48; Murthy & Ananthakrishnan, 1977: 129.

Type species: *Chelifer javanus* Thorell, 1883, by original designation.

Anatemnus angustus Redikorzev

Anatemnus angustus Redikorzev, 1938: 96–97, figs 24, 25c; Roewer, 1940: 346; Beier, 1951a: 80–82, fig. 21; Beier, 1955e: 43; Beier, 1967g: 353; Beier, 1976e: 100.

Type locality: Djiring, Vietnam.
Distribution: Bhutan, Malaysia, Vietnam.

Anatemnus elongatus (Ellingsen)

Chelifer (Atemnus) elongatus Ellingsen, 1902: 149–151.
Chelifer elongatus Ellingsen: With, 1908a: 324–326, text-fig. 84, figs 37a–g.
Anatemnus elongatus (Ellingsen): Beier, 1932b: 584; Beier, 1932e: 54; Roewer, 1937: 284.

Type locality: Guayaquil, Ecuador.
Distribution: Ecuador, Guyana.

Anatemnus javanus (Thorell)

Chelifer javanus Thorell, 1883: 37–40; With, 1905: 137, plate 8 fig. 1a; With, 1906: 184–185; With, 1907: 59–62, figs 13–14; Ellingsen, 1911b: 141; Ellingsen, 1913b: 127; Ellingsen, 1914: 3.
Chelifer (Lamprochernes) javanus Thorell: E. Simon, 1901: 79.

Anatemnus javanus (Thorell): Beier, 1932b: 579–580; Beier, 1932e: 49, fig. 58; Roewer, 1937: 283, fig. 201b.
Not *Chelifer javanus* Thorell: Pocock, 1900: 156 (misidentification; see *Paratemnoides pococki* (With)).
Not *Chelifer (Atemnus) javanus* Thorell: Ellingsen, 1910a: 358 (misidentification; see *Anatemnus novaguineensis* (With)).

Type locality: Tjibodas, (as Tcibodas), Java, Indonesia.
Distribution: Burma, India, Indonesia (Java), Malaysia, Sri Lanka.

Anatemnus longus Beier

Chelifer (Atemnus) voeltzkowi elongata Ellingsen, 1908b: 488 (junior primary homonym of *Chelifer (Atemnus) elongatus* Ellingsen, 1902).
Anatemnus longus Beier, 1932b: 586; Beier, 1932e: 55; Roewer, 1937: 284; Legendre, 1972: 446.

Type locality: Marovoay, Madagascar.
Distribution: Madagascar.

Anatemnus luzonicus Beier

Anatemnus luzonicus Beier, 1932b: 582, fig. 13; Beier, 1932e: 51–52, fig. 62; Roewer, 1937: 283.

Type locality: Mt Data, Bontoc, Philippines.
Distribution: Philippines.

Anatemnus madecassus Beier

Chelifer rotundus With: Ellingsen, 1910a: 359; Ellingsen, 1912b: 80 (misidentifications).
Anatemnus madecassus Beier, 1932b: 581; Beier, 1932e: 50–51, fig. 60; Roewer, 1937: 283; Legendre, 1972: 446.

Type locality: Andrangoloaba, Madagascar.
Distribution: Madagascar.

Anatemnus megasoma (Daday)

Chelifer megasoma Daday, 1897: 476–477, figs 16–17; With, 1906: 171.
Haplochernes megasoma (Daday): Beier, 1932e: 109; J. C. Chamberlin, 1934b: 8; Roewer, 1937: 295.
Anatemnus megasoma (Thorell): Beier, 1965g: 769–771, fig. 12; Beier, 1967c: 322; Beier, 1982: 43.

Type locality: Madang (as Friedrich-Wilhelmshafen), Papua New Guinea.
Distribution: Indonesia (Irian Jaya), Papua New Guinea.

Anatemnus nilgiricus Beier

Atemnus indicus (With): Beier, 1930e: 297–298 (misidentification).
Anatemnus nilgiricus Beier, 1932b: 582–583; Beier, 1932e: 52, fig. 63; Roewer, 1937: 283; Beier, 1967g: 353; Beier, 1973b: 47; Beier, 1974e: 1009; Murthy & Ananthakrishnan, 1977: 129.

Type locality: Nilgiris, India.
Distribution: India, Sri Lanka.

Anatemnus novaguineensis (With)

Chelifer nova-guineensis With, 1908b: 19–22, figs 5–6, plate 2 figs 9–12.
Chelifer (Atemnus) javanus Thorell: Ellingsen, 1910a: 358 (misidentification).
Paratemnus insularis Beier, 1932b: 568; Beier, 1932e: 38, fig. 41; Roewer, 1937: 283 (junior secondary homonym of *Paratemnoides insularis* (Banks)).

Anatemnus nova-guineensis (With): Beier, 1932b: 584; Beier, 1932e: 53–54; J. C. Chamberlin, 1934b: 8; Roewer, 1937: 283.

Paratemnus histrionicus J. C. Chamberlin, 1934b: 8; Beier, 1964e: 405–406, fig. 2 (replacement name for *Paratemnus insularis* Beier; synonymized by Beier, 1965g: 769).

Anatemnus novaguineensis (With): Beier, 1965g: 768–769, fig. 11; Beier, 1966d: 140; Beier, 1966g: 371; Beier, 1970g: 320; Beier, 1982: 43.

Type localities: of *Chelifer novaguineensis*: Blanche Bay, Papua New Guinea.
of *Paratemnus insularis*: Bismarck Archipelago, Papua New Guinea.
Distribution: Indonesia (Irian Jaya), New Caledonia, Papua New Guinea, Solomon Islands.

Anatemnus orites (Thorell)

Chelifer orites Thorell, 1889: 597–599, plate 5 fig. 7; With, 1906: 188; Tullgren, 1907a: 55–56; Ellingsen, 1911a: 36; Ellingsen, 1911b: 142; Ellingsen, 1914: 4.

Anatemnus orites (Thorell): Beier, 1932b: 580; Beier, 1932e: 50, fig. 59; Roewer, 1937: 283; Beier, 1948b: 461; Weidner, 1959: 114.

Not *Chelifer orites* Thorell: Ellingsen, 1901b: 208–209. (misidentification; see *Anatemnus vermiformis* (With)).

Type locality: Mt Mooleyit, Burma.
Distribution: Burma, India, Indonesia (Sumatra), Philippines, Sri Lanka.

Anatemnus orites major Beier

Anatemnus orites major Beier, 1963d: 510.

Type locality: Rantau Panjang, 5 miles N. of Klang, Selangor, Malaysia.
Distribution: Malaysia.

Anatemnus oswaldi (Tullgren)

Chelifer o'swaldi Tullgren, 1907a: 53–55, figs 14a–c; Ellingsen, 1912b: 80.

Atemnus o'swaldi (Tullgren): J. C. Chamberlin, 1931a: figs 15w, 44j, 44p; Beier, 1932g: fig. 167.

Anatemnus o'swaldi (Tullgren): Beier, 1932b: 583; Beier, 1932e: 52–53, fig. 64; Roewer, 1937: 283; Weidner, 1959: 114.

Atemnus oswaldi (Tullgren): Savory, 1964a: fig. 33d; Savory, 1977: fig. 93d.

Anatemnus oswaldi (Tullgren): Legendre, 1972: 446.

Type locality: Nossi Bé (as Nossibé), Madagascar.
Distribution: Madagascar.

Anatemnus pugilatorius Beier

Anatemnus pugilatorius Beier, 1965g: 768, fig. 10.

Type locality: Finschhafen, Papua New Guinea.
Distribution: Papua New Guinea.

Anatemnus rotundus (With)

Chelifer rotundus With, 1906: 196–199, fig. 24, plate 4 figs 11a–c.

Anatemnus rotundus (With): Beier, 1932b: 580–581; Beier, 1932e: 50; Roewer, 1937: 283.

Not *Chelifer rotundus* With: Ellingsen, 1910a: 359; Ellingsen, 1912b: 80 (misidentifications; see *Anatemnus madecassus* Beier).

Not *Atemnus rotundus* (With): Beier, 1930a: 207–208, fig. 5 (misidentification; see *Paratemnoides magnificus* (Beier)).

Type locality: Little Nicobar (as Nicobar Minor), Nicobar Islands.
Distribution: Nicobar Islands.

Anatemnus seychellesensis Beier

Anatemnus seychellesensis Beier, 1940a: 175–176, fig. 4.

Type locality: Silhouette, Seychelles.
Distribution: Seychelles.

Anatemnus subindicus (Ellingsen)

Chelifer (Atemnus) subindicus Ellingsen, 1910a: 359–360.
Chelifer subindicus Ellingsen: Ellingsen, 1912b: 80.
Atemnus subindicus (Ellingsen): Beier, 1932b: 581; Beier, 1932e: 51, fig. 61; Roewer, 1937: 283; Legendre, 1972: 446.

Type locality: central Madagascar.
Distribution: Madagascar.

Anatemnus subvermiformis Redikorzev

Anatemnus subvermiformis Redikorzev, 1938: 95–96, figs 23, 25b; Roewer, 1940: 346; Beier, 1951a: 78–80, fig. 20.

Type locality: Suoi-Tring, Vietnam.
Distribution: Laos, Vietnam.

Anatemnus tonkinensis Beier

Anatemnus angustus tonkinensis Beier, 1943: 76–77, fig. 3.
Anatemnus tonkinensis Beier: Beier, 1951a: 82–83, fig. 22; Beier, 1967g: 353.

Type locality: Ngai-Tio, Tonkin, Vietnam.
Distribution: Laos, Vietnam.

Anatemnus vermiformis (With)

Chelifer orites Thorell: Ellingsen, 1901b: 208–209 (misidentification).
Chelifer vermiformis With, 1906: 188–191, fig. 21, plate 4 figs 8a–b.
Anatemnus vermiformis (With): Beier, 1932b: 585; Beier, 1932e: 54; Roewer, 1937: 283; Beier, 1967g: 353, fig. 13.

Type locality: Nancowry (as Nankovry), Nicobar Islands.
Distribution: India, Nicobar Islands, Thailand.

Anatemnus voeltzkowi (Ellingsen)

Chelifer (Atemnus) voeltzkowi Ellingsen, 1908b: 487–488; Ellingsen, 1910a: 362–363.
Chelifer voeltzkowi Ellingsen: Ellingsen, 1912b: 81.
Anatemnus voeltzkowi (Ellingsen): Beier, 1932b: 585; Beier, 1932e: 54–55, fig. 66; Roewer, 1937: 284; Legendre, 1972: 446.

Type locality: south-west Madagascar.
Distribution: Madagascar.

Genus Atemnus Canestrini

Acis Canestrini, 1883: no. 9 (junior homonym of *Acis* Billberg, 1820, *Acis* Lesson, 1830, *Acis* Dejean, 1835, and *Acis* Duchassaing & Michelloti, 1860).
Atemnus Canestrini, 1884: no. 1; Beier, 1932b: 574–575; Beier, 1932e: 45; Hoff, 1945c: 1; Beier, 1963b: 246; Murthy & Ananthakrishnan, 1977: 129.
Chelifer (Atemnus) Canestrini: Balzan, 1892: 510; Lessert, 1911: 9–10.
Chelifer (Atemnus): Nonidez, 1917: 9.
Pessigus Navás, 1919: 211 (synonymized by Beier, 1932e: 45).

Type species: of *Acis*: *Acis brevimanus* Canestrini, 1883 (junior synonym of *Atemnus politus* (E. Simon, 1878a), by original designation.

of *Atemnus: Chelifer politus* E. Simon, 1878a, by subsequent designation of Beier, 1930a: 200.

of *Pessigus: Pessigus cabacerolus* Navás, 1919 (junior synonym of *Atemnus politus* (E. Simon, 1878a)), by present designation.

Atemnus letourneuxi (E. Simon)

Chelifer letourneuxi E. Simon, 1881b: 12; Tullgren, 1907c: 3–4, figs 1a–c; Ellingsen, 1912b: 80.

Atemnus letourneuxi (E. Simon): E. Simon, 1885c: 49; Pavesi, 1897: 158; E. Simon, 1897: 158 (not seen); E. Simon, 1890: 121; Beier, 1932b: 577; Beier, 1932e: 47, fig. 57; Roewer, 1937: 283; Beier, 1965b: 373.

Type locality: Mariout, Egypt.
Distribution: Chad, Egypt, Ethiopia, Somalia, South Yemen, Sudan, Tunisia.

Atemnus neotropicus Hoff

Atemnus neotropicus Hoff, 1946c: 1–3, figs 1–2.

Type locality: Mona Island, Puerto Rico.
Distribution: Puerto Rico.

Atemnus politus (E. Simon)

Chelifer politus E. Simon, 1878a: 149–150; E. Simon, 1879a: 35, 312, plate 18 fig. 10; E. Simon, 1898d: 20; Ellingsen, 1909a: 205; Krausse-Heldrungen, 1912: 65; Navás, 1923: 31; Navás, 1925: 106.

Acis brevimanus Canestrini, 1883: no. 10, plate 9 figs 1–4, plate 10 figs 1–4 (synonymized by Ellingsen, 1910a: 358–359).

Atemnus politus (E. Simon): Canestrini, 1884: no. 1, figs 1–3; E. Simon, 1885c: 49; J. C. Chamberlin, 1931a: 244, figs 14j, 38l, 40v, 41l–n, 42o; Beier, 1932b: 576; Beier, 1932e: 46, figs 54–55; Beier, 1932g: fig. 195; Ionescu, 1936: 4; Roewer, 1937: 283, fig. 227; Vachon, 1938a: figs 26b, 53a; Beier, 1939f: 197; Redikorzev, 1949: 646; Caporiacco, 1951e: 69; Beier, 1953a: 301; Beier, 1955d: 216; Beier, 1955n: 107–108; Beier, 1956e: 303; Beier, 1956i: 8; Beier, 1957b: 147; Beier, 1959a: 270; Beier, 1959c: 407; Beier, 1959f: 129; Beier, 1961b: 32–33; Beier, 1963b: 246–247, fig. 250; Beier, 1963g: 272; Beier, 1963i: 285; Dumitresco & Orghidan, 1964: 625–626, figs 47–48; Orghidan & Dumitresco, 1964a: 328; Orghidan & Dumitresco, 1964b: 192; Beier, 1965f: 93; Kobakhidze, 1965b: 541; Beier, 1966f: 345; Kobakhidze, 1966: 704; Zangheri, 1966: 531; Beier, 1967f: 310; Beier, 1967g: 352–353; Beron, 1968: 104; Beier, 1969b: 193; Dumitresco & Orghidan, 1969: fig. 9; Lazzeroni, 1969a: 335; Lazzeroni, 1969c: 240; Weygoldt, 1969a: 44, 51–52, 62, 75, figs 46, 57; Dumitresco & Orghidan, 1970a: plate 1 fig. 2, plate 2 fig. 6; Lazzeroni, 1970a: 208; Beier, 1971a: 363; Beier, 1974e: 1008; Ćurčić, 1974a: 26; Gardini, 1976a: 255–256; Mahnert, 1977b: 67–68; Mahnert, 1977d: 97, fig. 5; Murthy & Ananthakrishnan, 1977: 129; Callaini, 1979b: 139; Estany, 1979a: 222, fig. 1; Leclerc & Heurtault, 1979: 246; Callaini, 1982a: 17, 23, fig. 3d; Callaini, 1983c: 300; Callaini, 1983d: 224; Callaini, 1983e: 224; Schawaller, 1983b: 19; Schawaller, 1983c: 371; Krumpál, 1984b: 63–64; Schawaller, 1985a: 10; Schawaller, 1987a: 212–213; Callaini, 1988b: 55; Schawaller, 1988: 161; Schawaller & Dashdamirov, 1988: 35, figs 56; Callaini, 1989: 145; Schawaller, 1989: 16.

Chelifer (Atemnus) politus (E. Simon): E. Simon, 1900b: 593; Ellingsen, 1910a: 358–359; Lessert, 1911: 10–11, fig. 9; Kästner, 1928: 5, fig. 13; Beier, 1929a: 342; Beier, 1929b: 78; Beier, 1929e: 445.

Chelifer (Atemnus) ariasi Nonidez, 1917: 9–12, figs 1, 2a–c (synonymized by Beier, 1963b: 247).

Chelifer ariasi Nonidez: Navás, 1918: 133–134.

Chelifer (Atemnus) turkestanicus Redikorzev, 1922a: 257–259, figs 1–2 (synonymized by Krumpál, 1984b: 63).
Pessigus ariasi (Nonidez): Navás, 1919: 211; Navás, 1925: 109.
Pessigus cabacerolus Navás, 1919: 211–212, figs 8a–c; Navás, 1925: 109–110 (synonymized by Beier, 1932e: 46).
Atemnus turkestanicus (Redikorzev): Beier, 1932b: 576–577; Beier, 1932e: 47, fig. 56; Roewer, 1937: 283; Redikorzev, 1949: 646; Beier, 1959a: 270; Beier, 1967g: 353; Beier, 1976e: 100.
Atemnus ariasi (Nonidez): Beier, 1932b: 577; Beier, 1932e: 48; Roewer, 1937: 283.
Atemnus balcanicus Hadži, 1938: 20–23, figs 22a–d, 23a–b; Ćurčić, 1974a: 26 (synonymized by Beier, 1963b: 247).
Microcreagris zangherii Caporiacco, 1948c: 241–242, fig. 1; Beier, 1963b: 208; Zangheri, 1966: 531; Cerruti, 1968: fig. 7 (synonymized by Gardini, 1976a: 255).
Atemnus politus turkestanicus (Redikorzev): Beier, 1974b: 269; Beier, 1974h: 167.
Not *Chelifer politus* E. Simon: H. J. Hansen, 1884: 543 (misidentification; see *Oratemnus navigator* (With)).
Type localities: of *Chelifer politus*: Daya, Algeria.
of *Acis brevimanus*: Acireale, Sicily.
of *Chelifer (Atemnus) ariasi*: Algeciras, Cadiz, Spain.
of *Pessigus cabacerolus*: Cabacés, Tarragona, Spain.
of *Chelifer (Atemnus) turkestanicus*: Tashkent, Uzbekistan, U.S.S.R.
of *Atemnus balcanicus*: Klosters Sveti Nikola, Yugoslavia.
of *Microcreagris zangherii*: Pineta de Classe, Italy.
Distribution: Afghanistan, Algeria, Austria, Balearic Islands, Bhutan, Bulgaria, Canary Islands, China (Hupeh), Crete, France, Greece, India, Iran, Italy, Libya, Mongolia, Morocco, Nepal, Romania, Sardinia, Sicily, Spain, Switzerland, Syria, Tunisia, Turkey, U.S.S.R. (Azerbaijan, Georgia, Kazakhstan, Kirghizia, R.S.F.S.R., Turkmenistan, Ukraine, Uzbekistan), Yugoslavia.

Atemnus strinatii Beier

Atemnus strinatii Beier, 1977a: 188–190, fig. 2.

Type locality: Cueva Basilen, Pagbilao, Philippines.
Distribution: Philippines.

Atemnus syriacus (Beier)

Catatemnus syriacus Beier, 1955d: 216, figs 4–5; Beier, 1963g: 272; Beier, 1965f: 93.
Atemnus syriacus (Beier): Beier, 1967f: 310; Mahnert, 1977b: 67.
Atemnus (?) syriacus (Beier): Mahnert, 1974c: 379.

Type locality: Latakia, Syria.
Distribution: Greece, Israel, Syria, Turkey.

Genus **Athleticatemnus** Beier

Athleticatemnus Beier, 1979c: 101–103.

Type species: *Athleticatemnus pugil* Beier, 1979c, by original designation.

Athleticatemnus pugil Beier

Athleticatemnus pugil Beier, 1979c: 103–104, fig. 1.

Type locality: Luki, Boma, Zaire.
Distribution: Zaire.

Genus **Brazilatemnus** Muchmore

Brazilatemnus Muchmore, 1975c: 236.

Type species: *Brazilatemnus browni* Muchmore, 1975c, by original designation.

Brazilatemnus browni Muchmore

Brazilatemnus browni Muchmore, 1975c: 236–239, figs 8–14; Muchmore, 1975d: fig. 14; Mahnert, 1979d: 763–764, figs 91a–c; Adis, 1981: 117, etc.; Adis & Funke, 1983: 358; Mahnert, 1985b: 78; Adis & Mahnert, 1986: 304–305; Mahnert & Adis, 1986: 213; Adis, Junk & Penny, 1987: 488; Mahnert, Adis & Bührnheim, 1987: figs 12a, 13b, 14b; Adis, Mahnert, de Morais & Rodrigues, 1988: 287–290, figs 1–2.

Type locality: 24 km NE. of Manaus, Amazonas, Brazil.
Distribution: Brazil (Amazonas, Pará).

Genus Caecatemnus Mahnert

Caecatemnus Mahnert, 1985c: 226.

Type species: *Caecatemnus setosipygus* Mahnert, 1985c, by original designation.

Caecatemnus setosipygus Mahnert

Caecatemnus setosipygus Mahnert, 1985c: 227, figs 33–36; Mahnert & Adis, 1986: 213.

Type locality: Reserva Florestal Ducke, 26 km from Manaus-Itacoatiara Highway, Amazonas, Brazil.
Distribution: Brazil (Amazonas).

Genus Catatemnus Beier

Catatemnus Beier, 1932b: 593–594; Beier, 1932e: 62–63; Murthy & Ananthakrishnan, 1977: 130.

Type species: *Chelifer birmanicus* Thorell, 1889, by original designation.

Catatemnus birmanicus (Thorell)

Chelifer birmanicus Thorell, 1889: 594–597, plate 5 fig. 6; With, 1906: 176–180, plate 3, fig. 10a, plate 4 figs 4a–f.
Atemnus birmanicus (Thorell): J. C. Chamberlin, 1931a: fig. 18m; Vachon, 1949: fig. 203h.
Catatemnus birmanicus (Thorell): Beier, 1932b: 595; Beier, 1932e: 63–64; Roewer, 1936: figs 22g, 112k, 142; Roewer, 1937: 284; Vachon, 1940f: 3; Weidner, 1959: 113; Murthy & Ananthakrishnan, 1977: 130.
Not *Chelifer birmanicus* Thorell: Tullgren, 1905: 39–40, figs 1a–f (misidentification; see *Catatemnus thorelli* (Balzan)); Tullgren, 1912a: 267; Tullgren, 1912b: 269; Dammerman, 1922: 100; Dammerman, 1929: 114 (not seen); Bristowe, 1931b: 1390; Dammerman, 1948: 496 (misidentifications; see *Paratemnoides assimilis* (Beier)).

Type locality: Bhamó, Burma.
Distribution: Burma.

Catatemnus braunsi (Tullgren)

Chelifer braunsi Tullgren, 1907a: 56–58, fig. 16; Tullgren, 1907b: 224; Ellingsen, 1912b: 78.
Catatemnus braunsi (Tullgren): Beier, 1932b: 599–600; Beier, 1932e: 67–68, fig. 82; Roewer, 1937: 284; Beier, 1958a: 173; Weidner, 1959: 113; Beier, 1964k: 72.

Type locality: Algoa Bay, Port Elizabeth, Cape Province, South Africa.
Distribution: South Africa.

Catatemnus comorensis (Ellingsen)

Chelifer (Lamprochernes) comorensis Ellingsen, 1910a: 366–367.
Chelifer comorensis Ellingsen: Ellingsen, 1912b: 81.
Catatemnus comorensis (Ellingsen): Beier, 1932b: 597–598; Beier, 1932e: 65–66, fig. 78; Roewer, 1937: 284; Mahnert, 1978f: 125–126; Mahnert, 1983b: 362–364, figs 9–15; Mahnert, 1984b: 60–61.

Type locality: Mayotte, Comoro Islands.
Distribution: Comoro Islands, Kenya, Somalia.

Catatemnus concavus (With)

Chelifer concavus With, 1906: 173–176, plate 4 fig. 3a.
Catatemnus concavus (With): Beier, 1932b: 600.
Catatemnus (?) *concavus* (With): Beier, 1932e: 68; Roewer, 1937: 284.

Type locality: Nancowry (as Nankovry), Nicobar Islands.
Distribution: Nicobar Islands.

Catatemnus exiguus Mahnert

Catatemnus exiguus Mahnert, 1978f: 114–119, figs 108–118.

Type locality: Meya, Tal de Niari, Congo.
Distribution: Congo.

Catatemnus fravalae Heurtault

Catatemnus fravalae Heurtault, 1983: 18–21, figs 43–47.

Type locality: Lamto, Ivory Coast.
Distribution: Ivory Coast.

Catatemnus granulatus Mahnert

Catatemnus granulatus Mahnert, 1978f: 112–114, figs 103–107.

Type locality: Sibiti, Congo.
Distribution: Congo.

Catatemnus kittenbergeri Caporiacco

Catatemnus kittenbergeri Caporiacco, 1947a: 98, fig. 1.

Type locality: Moshi, Tanzania.
Distribution: Tanzania.

Catatemnus monitor (With)

Chelifer monitor With, 1906: 180–182, plate 4 figs 5a–b; With, 1908b: 15.
Catatemnus monitor (With): Beier, 1932b: 596; Beier, 1932e: 64; Roewer, 1937: 284.

Type locality: 'Sunda Islands', Indonesia.
Distribution: Indonesia.

Catatemnus nicobarensis (With)

Chelifer nicobarensis With, 1906: 182–184, fig. 17, plate 4 figs 6a–b; Ellingsen, 1911a: 37.
Catatemnus nicobarensis (With): Beier, 1932b: 596; Beier, 1932e: 64; Roewer, 1937: 284.

Type locality: Pulu-Mulu, Nicobar Islands.
Distribution: Indonesia (Sumatra), Nicobar Islands.

Catatemnus schlottkei Vachon

Catatemnus schlottkei Vachon, 1937a: 130–132, figs 3–4; Roewer, 1940: 346.

Type locality: Makapau, Victoria, Zimbabwe.
Distribution: Zimbabwe.

Catatemnus thorelli (Balzan)

Chelifer (Lamprochernes) thorelli Balzan, 1891: 519–520, figs 11, 11a.
Chelifer birmanicus Thorell: Tullgren, 1905: 39–40, figs 1a–f (misidentification).
Chelifer thorellii Balzan: With, 1906: 176.
Chelifer thorelli Balzan: Ellingsen, 1909a: 219.
Atemnus thorelli (Balzan): Beier, 1930a: 209.
Catatemnus thorelli (Balzan): Beier, 1932b: 597; Beier, 1932e: 65, figs 76–77; Roewer, 1937: 284.

Type locality: Sumatra, Indonesia.
Distribution: Indonesia (Borneo, Java, Sumatra), Malaysia (Sarawak).

Catatemnus togoensis (Ellingsen)

Chelifer (Lamprochernes) togoensis Ellingsen, 1910a: 368–369.
Chelifer togoensis Ellingsen: Ellingsen, 1910b: 63; Ellingsen, 1910d: 536; Ellingsen, 1912b: 81.
Catatemnus togoensis (Ellingsen): Beier, 1932b: 598–599; Beier, 1932e: 66–67, fig. 80; Roewer, 1937: 284; Vachon, 1941e: 29; Beier, 1947b: 324–325; Vachon, 1952a: 29, fig. 17; Beier, 1954d: 135; Mahnert, 1978f: 119–126, figs 119–130; Mahnert, 1983b: 361.
Catatemnus congicus Beier, 1932b: 598, fig. 17; Beier, 1932e: 66, fig. 79; Roewer, 1937: 284; Beier, 1947b: 324; Vachon, 1952a: 29; Beier, 1954d: 134; Beier, 1959d: 43 (synonymized by Mahnert, 1978f: 119).
Catatemnus sommerfeldi Beier, 1932b: 599; Beier, 1932e: 67, fig. 81; Roewer, 1937: 284; Vachon, 1941e: 29 (synonymized by Mahnert, 1978f: 119).
Catatemnus similis Vachon, 1938c: 311–313, figs i, m, o–r; Roewer, 1940: 346; Vachon, 1941e: 29 (synonymized by Mahnert, 1978f: 119).
Catatemnus cf. *togoensis* (Ellingsen): Mahnert, 1978f: 126.

Type localities: of *Chelifer (Lamprochernes) togoensis*: Bismarcksburg, Togo.
of *Catatemnus congicus*: Kikiongo Goma, Maduda, Zaire.
of *Catatemnus sommerfeldi*: Yaounde (as Jaunde), Cameroun; and Abong-Mbong (as Abang Mbang), Cameroun.
of *Catatemnus similis*: Bingerville, Ivory Coast.
Distribution: Angola, Cameroun, Congo, Guinea, Ivory Coast, Kenya, Nigeria, Togo, Uganda, Zaire.

Genus Cyclatemnus Beier

Cyclatemnus Beier, 1932b: 560; Beier, 1932e: 30–31.

Type species: *Cyclatemnus centralis* Beier, 1932b, by original designation.

Cyclatemnus affinis Vachon

Cyclatemnus affinis Vachon, 1938c: 307–308, figs d–f, k; Roewer, 1940: 346.

Type locality: Bingerville, Ivory Coast.
Distribution: Ivory Coast.

Cyclatemnus berlandi Vachon

Cyclatemnus berlandi Vachon, 1938c: 309–311, figs b–c, g, j, n; Roewer, 1940: 346.

Type locality: Dakar, Senegal.
Distribution: Senegal.

Cyclatemnus brevidigitatus Mahnert

Cyclatemnus brevidigitatus Mahnert, 1978f: 96, figs 139–144.

Type locality: Lefini reservation, Nanbouli River, Congo.
Distribution: Congo.

Cyclatemnus burgeoni (Beier)

Paratemnus burgeoni Beier, 1932b: 571; Beier, 1932e: 41–42, fig. 47–48; Roewer, 1937: 283; Vachon, 1941e: 30.
Cyclatemnus burgeoni (Beier): Beier, 1944: 184–185, fig. 9; Beier, 1954d: 134; Beier, 1955a: 12; Beier, 1959d: 40; Beier, 1972a: 11.

Type locality: Moto, Uele, Zaire.
Distribution: Cameroun, Uganda, Zaire.

Cyclatemnus centralis Beier

Cyclatemnus centralis Beier, 1932b: 561, fig. 6; Beier, 1932e: 31–32, figs 30–31; Roewer, 1937: 282; Beier, 1967d: 79; Mahnert, 1983b: 375–378, figs 48–57.

Type locality: south-west Rwanda.
Distribution: Kenya, Rwanda.

Cyclatemnus dolosus Beier

Cyclatemnus globosus Beier: Beier, 1958a: 172–173, fig. 7 (misidentification).
Cyclatemnus dolosus Beier, 1964k: 71; Beier, 1966k: 464.

Type locality: Town Bush, near Pietermaritzburg, Natal, South Africa.
Distribution: South Africa, Zimbabwe.

Cyclatemnus fallax Beier

Cyclatemnus fallax Beier, 1955c: 549–551, fig. 15; Beier, 1967d: 80; Mahnert, 1983b: 378–380, figs 58–62.

Type locality: Suam River, Mt Elgon, Uganda.
Distribution: Kenya, Uganda.

Cyclatemnus globosus Beier

Cyclatemnus globosus Beier, 1947b: 323–324, fig. 29.
Cyclotemnus (sic) *globosus* Beier: Beier, 1966k: 470.
Not *Cyclatemnus globosus* Beier: Beier, 1958a: 172–173, fig. 7 (misidentification; see *Cyclatemnus dolosus* Beier).

Type locality: Kakamas, Orange River, Cape Province, South Africa.
Distribution: Namibia, South Africa.

Cyclatemnus globosus parvus Beier

Cyclatemnus globosus parvus Beier, 1964k: 70–71, fig. 28; Beier, 1966k: 464; Lawrence, 1967: 89.

Type locality: Roodeplaat, Pretoria, Transvaal, South Africa.
Distribution: South Africa.

Cyclatemnus granulatus Beier

Cyclatemnus granulatus Beier, 1932b: 561–562; Beier, 1932e: 32, fig. 32; Roewer, 1937: 282; Vachon, 1941e: 29; Vachon, 1952a: 30; Beier, 1954d: 134; Heurtault, 1983: 18.

Type locality: Bosum, Cameroun.
Distribution: Cameroun, Guinea, Ivory Coast, Zaire.

Cyclatemnus minor Beier

Cyclatemnus minor Beier, 1944: 185–187, fig. 10; Beier, 1967d: 80, fig. 5; Mahnert, 1983b: 375.

Type locality: Djem-Djem Wald, Ethiopia.
Distribution: Ethiopia, Kenya.

Cyclatemnus robustus Beier

Cyclatemnus robustus Beier, 1959d: 40–41, fig. 19.

Type locality: Mwenga, Kivu, Zaire.
Distribution: Zaire.

Genus **Diplotemnus** J. C. Chamberlin

Diplotemnus J. C. Chamberlin, 1933: 264–265; Beier, 1963b: 248; Muchmore, 1975c: 232.

Type species: *Diplotemnus insolitus* J. C. Chamberlin, 1933, by original designation.

Diplotemnus afghanicus Beier

Diplotemnus afghanicus Beier, 1959a: 271–272, fig. 14; Beier, 1961e: 3; Beier, 1973c: 226; Beier, 1974h: 167.

Type locality: Khvadjéh Tchicht, Afghanistan.
Distribution: Afghanistan, Mongolia, Turkey.

Diplotemnus ophthalmicus Redikorzev

Atemnus piger (E. Simon): Beier, 1930a: 209–210, fig. 7 (misidentification).
Miratemnus piger (E. Simon): Beier, 1932b: 610; Beier, 1932e: 79–80, fig. 99; Roewer, 1937: 286 (misidentifications).
Diplotemnus piger (Beier): Beier, 1946: 568–569, fig. 2; Beier, 1948b: 442; Beier, 1957b: 147; Verner, 1959: 61–63, figs 1–2; Beier, 1963b: 248, fig. 251; Beier, 1963g: 272; Beier, 1965b: 373; Beier, 1965f: 93; Beron, 1968: 104; Beier, 1971a: 363; Schawaller & Dashdamirov, 1988: 35–36, figs 56, 58; Schawaller, 1989: 16–17, figs 38–42 (misidentifications).
Diplotemnus ophthalmicus Redikorzev, 1949: 646–648, figs 5–7; Schawaller, 1985a: 11, fig. 24.
Diplotemnus pomerantzevi Redikorzev, 1949: 648–649, figs 8–10; Schawaller, 1985a: 11–12, fig. 23 (synonymized by Schawaller, 1989: 16).
Diplotemnus beieri Vachon, 1970: 189, fig. 2; Heurtault, 1970c: 192–194; Mahnert, 1980a: 261–262, figs 7–11. New synonymy.
Withius ophthalmicus (Redikorzev): Krumpál, 1983b: 176.
Withius pomerantzevi (Redikorzev): Krumpál, 1983b: 176.
Diplotemnus milleri Krumpál, 1983b: 173–176, figs 1–7 (synonymized by Schawaller, 1985a: 11).
Diplotemnus turanicus Krumpál, 1983b: 176–178, figs 8–12 (synonymized by Schawaller, 1985a: 11).

Type localities: of *Diplotemnus ophthalmicus*: Duani-tau, Chikmenskei, Kazakhstan, U.S.S.R.; Tuuk River, Ala-tau, Kirghizia, U.S.S.R.
of *Diplotemnus pomerantzevi*: Guzar, Uzbekistan, U.S.S.R.; and Baysun, Uzbekistan, U.S.S.R.
of *Diplotemnus piger*: Guelt-es-stel, E. of Bou Saada, Algeria.
of *Diplotemnus milleri*: Firyuza (as Firjuza), Ashkhabad (as Aschabad), Turkmenistan, U.S.S.R.
of *Diplotemnus turanicus*: Kyzylkum, Uzbekistan, U.S.S.R.

Family Atemnidae

Distribution: Algeria, Bulgaria, Canary Islands, Chad, Czechoslovakia, Iran, Sudan, Turkey, U.S.S.R. (Azerbaijan, Kazakhstan, Kirghizia, Tadzhikistan, Turkmenistan, Uzbekistan).

Diplotemnus ophthalmicus sinensis (Schenkel), new combination

Miratemnus piger sinensis Schenkel, 1953b: 106, figs 47a–d.

Type locality: 'Süden des Ordos', China.
Distribution: China.

Diplotemnus egregius Beier

Diplotemnus egregius Beier, 1959a: 272–273, fig. 15; Lindberg, 1961: 31.

Type locality: Tchachmeh Cher near Pol-Khomri, Afghanistan.
Distribution: Afghanistan.

Diplotemnus garypoides (Ellingsen)

Chelifer (Chelifer) garypoides Ellingsen, 1906: 258–259.
Chelifer garypoides Ellingsen: Ellingsen, 1912b: 83, 92–93.
Miratemnus garypoides (Ellingsen): Beier, 1932e: 79.
Diplotemnus garypoides (Ellingsen): J. C. Chamberlin, 1933: 266; Roewer, 1937: 287.
Not *Chelifer garypoides* Ellingsen: With, 1908b: 15–19, figs 3–4, plate 2 figs 4–8 (misidentification; see *Diplotemnus insularis* J. C. Chamberlin).
Not *Atemnus garypoides* (Ellingsen): J. C. Chamberlin, 1931a: figs 27a (misidentification; see *Diplotemnus insularis* J. C. Chamberlin).

Type locality: Bolama, Guinea-Bissau.
Distribution: Guinea-Bissau, South Africa.

Diplotemnus insolitus J. C. Chamberlin

Diplotemnus insolitus J. C. Chamberlin, 1933: 267, figs b, d–f, h–q, u–w; Roewer, 1937: 287, figs 229–232; Sareen, 1965: 9–17, figs 1–44; Kanwar, 1966: 203–205, figs 1–3; Kanwar, 1968: 369–371, figs 1–4; Kanwar & Kanwar, 1968: 373–375, figs 1–2; Bawa, Sjöstrand, Kanwar & Kanwar, 1971a: 649, figs 1–4.

Type locality: north-west Himalayas.
Distribution: India.

Diplotemnus insularis J. C. Chamberlin

Chelifer garypoides Ellingsen: With, 1908b: 15–19, figs 3–4, plate 2 figs 4–8 (misidentification).
Atemnus garypoides (Ellingsen): J. C. Chamberlin, 1931a: figs 27a (misidentification).
Diplotemnus insularis J. C. Chamberlin, 1933: 266–267, figs c, g, r–t; Roewer, 1937: 287; Beier, 1940a: 177–178, fig. 6; Beier, 1948b: 462.

Type locality: St Paul's Island.
Distribution: St Paul's Island.

Diplotemnus lindbergi Beier

Diplotemnus lindbergi Beier, 1960a: 42–43, fig. 1.

Type locality: Tirgaran between Baharak and Zébak, Afghanistan.
Distribution: Afghanistan.

Diplotemnus namaquensis Beier

Diplotemnus namaquensis Beier, 1947b: 325–326, figs 30–31.

Type locality: Loliefontain, Namaqualand, Cape Province, South Africa.
Distribution: South Africa.

Diplotemnus pieperi Helversen

Diplotemnus pieperi Helversen, 1965: 98–103, figs 2–3; Beier, 1976a: 24; Pieper, 1980: 263.

Type locality: Pitao Grande, Ilhas Selvagens.
Distribution: Ilhas Selvagens.

Diplotemnus pinguis Beier

Chelifer segregatus Ellingsen, 1912b: 84, 100 (misidentification).
Diplotemnus pinguis Beier, 1955l: 308–309, fig. 26; Beier, 1964k: 72–73; Beier, 1966k: 470; Beier, 1973a: 100.

Type locality: Cape Agulhas, Cape Province, South Africa.
Distribution: Namibia, South Africa.

Diplotemnus rothi Muchmore

Diplotemnus rothi Muchmore, 1975c: 232–236, figs 1–7; Muchmore, 1975d: fig. 15.

Type locality: 1 mile SE. of Portal, Cochise County, Arizona, U.S.A.
Distribution: U.S.A. (Arizona).

Diplotemnus rudebecki Beier

Diplotemnus rudebecki Beier, 1955l: 309–310, fig. 27.

Type locality: 7 miles SW. of Bredasdorp, Cape Province, South Africa.
Distribution: South Africa.

Diplotemnus vachoni Dumitresco & Orghidan

Diplotemnus vachoni Dumitresco & Orghidan, 1969: 675–679, figs 1–8; Dumitresco & Orghidan, 1970a: 130–133, plate 1 fig. 1, plate 2 figs 1–4.

Type locality: Retezatului (as Retezat) mountains, Romania.
Distribution: Romania.

Genus Mesatemnus Beier & Turk

Anatemnus (Mesatemnus) Beier & Turk, 1952: 769–770.
Mesatemnus Beier & Turk: Beier, 1963b: 247.

Type species: *Anatemnus (Mesatemnus) cyprianus* Beier & Turk, 1952, by original designation.

Mesatemnus cyprianus (Beier & Turk)

Anatemnus (Mesatemnus) cyprianus Beier & Turk, 1952: 770–771, figs 3–5.
Mesatemnus cyprianus (Beier & Turk): Beier, 1963b: 247.

Type locality: Salamis, Cyprus.
Distribution: Cyprus.

Genus Metatemnus Beier

Metatemnus Beier, 1932b: 606–607; Beier, 1932e: 75.

Type species: *Metatemnus philippinus* Beier, 1932b, by original designation.

Metatemnus heterodentatus Beier

Metatemnus heterodentatus Beier, 1952a: 106–108, figs 9–10.

Type locality: Mehipit, Borneo.
Distribution: Borneo.

Metatemnus philippinus Beier

Metatemnus philippinus Beier, 1932b: 607–608; Beier, 1932e: 75–76, figs 93–95; Roewer, 1937: 285, fig. 228; Beier, 1965g: 772; Beier, 1966b: 343.
Metatemnus major Beier, 1932b: 608; Beier, 1932e: 76, fig. 96; Roewer, 1937: 285 (synonymized by Beier, 1965g: 772).

Type localities: of *Metatemnus philippinus*: Port Banga, Mindanao, Philippines.
of *Metatemnus major*: Sabaan, Philippines.
Distribution: Indonesia (Irian Jaya), Papua New Guinea, Philippines.

Metatemnus superior Muchmore

Metatemnus superior Muchmore, 1972a: 11–14, figs 1–3.

Type locality: Cameron Highlands, Pahang (as Pehang), Malaysia.
Distribution: Malaysia.

Metatemnus unistriatus (Redikorzev)

Anatemnus unistriatus Redikorzev, 1938: 97–99, fig. 26; Roewer, 1940: 346.
Metatemnus unistriatus (Redikorzev): Beier, 1951a: 86.

Type locality: Da Lat (as Dalat), Vietnam.
Distribution: Vietnam.

Genus Micratemnus Beier

Micratemnus Beier, 1932b: 587; Beier, 1932e: 56; Murthy & Ananthakrishnan, 1977: 130.

Type species: *Chelifer (Atemnus) pusillus* Ellingsen, 1906, by original designation.

Micratemnus anderssoni Beier

Micratemnus anderssoni Beier, 1973b: 49, fig. 13.

Type locality: Deerwood Kuruwita, 6 miles NNW. of Ratnapura, Sabaragamuwa Province, Sri Lanka.
Distribution: Sri Lanka.

Micratemnus ceylonicus Beier

Micratemnus ceylonicus Beier, 1973b: 47–49, fig. 12; Beier, 1974e: 1009–1010, fig. 6; Murthy & Ananthakrishnan, 1977: 130.

Type locality: Yalakumbura, 5 miles SSW. of Bibile, Uva Province, Sri Lanka.
Distribution: India, Sri Lanka.

Micratemnus crassipes Mahnert

Micratemnus sulcatus Beier: Beier, 1967d: 81 (misidentification).
Micratemnus crassipes Mahnert, 1983b: 366–368, figs 22–29; Mahnert, 1984b: 60.

Type locality: Lake Baringo Lodge, Lake Baringo, Kenya.
Distribution: Kenya, Somalia.

Micratemnus pusillus (Ellingsen)

Chelifer (Atemnus) pusillus Ellingsen, 1906: 248–250.
Chelifer pusillus Ellingsen: Ellingsen, 1912b: 80; Ellingsen, 1913a: 452.
Micratemnus pusillus (Ellingsen): Beier, 1932b: 587–588; Beier, 1932e: 56–57, fig. 68; Roewer, 1937: 284; Beier, 1948b: 452.

Type locality: Vista Alegre, Sao Tomé.
Distribution: Ghana, Sao Tomé.

Micratemnus sulcatus Beier

Micratemnus sulcatus Beier, 1944: 187–188, fig. 11; Mahnert, 1983b: 364–366, figs 16–21.
Not *Micratemnus sulcatus* Beier: Beier, 1967d: 81 (misidentification; see *Micratemnus crassipes* Mahnert).

Type locality: Tanga, Tanzania.
Distribution: Kenya, Tanzania.

Genus Miratemnus Beier

Miratemnus Beier, 1932b: 609; Beier, 1932e: 77–78.

Type species: *Miratemnus hispidus* Beier, 1932b, by original designation.

Miratemnus hirsutus Beier

Miratemnus hirsutus Beier, 1955l: 310–312, fig. 28; Beier, 1958a: 174; Beier, 1964k: 73.

Type locality: Skoorsteenkop, Houtbaai (as Hout Bay), Cape Peninsula, Cape Province, South Africa.
Distribution: South Africa.

Miratemnus hispidus Beier

Miratemnus hispidus Beier, 1932b: 609–610, fig. 22; Beier, 1932e: 78–79, figs 97–98; Roewer, 1937: 286, fig. 233; Vachon, 1937a: 129, Vachon, 1938a: figs 26a, 53b; Weidner, 1959: 114; Beier, 1964k: 470.

Type locality: Osona (as Osoni), near Okahandja, Namibia.
Distribution: Namibia, Zimbabwe.

Miratemnus kenyaensis Mahnert

Miratemnus kenyaensis Mahnert, 1983b: 358–361, figs 1–8.

Type locality: Lac Elmenteita, Nakura, Kenya.
Distribution: Kenya.

Miratemnus segregatus (Tullgren)

Chelifer segregatus Tullgren, 1908b: 284–285, fig. 2.
Miratemnus segregatus (Tullgren): Beier, 1932e: 79; Roewer, 1937: 286.
Not *Chelifer segregatus* Tullgren: Ellingsen, 1912b: 84, 100 (misidentification; see *Diplotemnus pinguis* Beier).

Type locality: Rooibank, Namibia.
Distribution: Namibia.

Miratemnus zuluanus Lawrence

Miratemnus zuluanus Lawrence, 1937: 269–270, fig. 29; Roewer, 1940: 346; Beier, 1958a: 173–174, fig. 8; Beier, 1964k: 73; Beier, 1966k: 465; Lawrence, 1967: 90.

Type locality: Hluhluwe Game Reserve, Zululand, Cape Province, South Africa.
Distribution: South Africa.

Genus Oratemnus Beier

Oratemnus Beier, 1932b: 588; Beier, 1932e: 57; Morikawa, 1960: 138; Beier, 1976f: 213; Murthy & Ananthakrishnan, 1977: 126; Sivaraman, 1980c: 349.
Steiratemnus Beier, 1948c: 527–529 (synonymized by Beier, 1966a: 296).

Type species: of *Oratemnus*: *Chelifer (Lamprochernes) articulosus* E. Simon, 1899a, by original designation.
of *Steiratemnus*: *Chelifer punctatus* L. Koch, 1885, by original designation.

Oratemnus afghanicus Beier

Oratemnus afghanicus Beier, 1959a: 270–271, fig. 13.

Type locality: Darountah, Jalalabad (as Djelalabad), Afghanistan.
Distribution: Afghanistan.

Oratemnus articulosus (E. Simon)

Chelifer (Lamprochernes) articulosus E. Simon, 1899a: 120–121.
Chelifer articulosus E. Simon: With, 1906: 196; Ellingsen, 1911a: 36.
Oratemnus articulosus (E. Simon): Beier, 1932b: 589; Beier, 1932e: 58, fig. 69;
    Roewer, 1937: 284.

Type locality: Sumatra, Indonesia.
Distribution: Indonesia (Sumatra).

Oratemnus boettcheri Beier

Oratemnus böttcheri Beier, 1932b: 592, fig. 15; Beier, 1932e: 60–61, fig. 73; Roewer,
    1937: 284; Beier, 1966b: 343.
Oratemnus boettcheri Beier: Beier, 1967g: 354.

Type locality: Vinac Island, Philippines.
Distribution: Philippines, Vietnam.

Oratemnus brevidigitatus Beier

Oratemnus brevidigitatus Beier, 1940a: 176–177, fig. 5; Beier, 1948b: 442, 457;
    Mahnert, 1978d: 880–882, figs 24–29.

Type locality: Mahé (as Mahe), Seychelles.
Distribution: Seychelles.

Oratemnus cavernicola Beier

Oratemnus cavernicola Beier, 1976b: 271–272, fig. 1; Harvey, 1981b: 245; Harvey,
    1985b: 128.

Type locality: Jump Up Cave, Gray Range, ca. 48 km N. of Tibooburra, New South
    Wales, Australia.
Distribution: Australia (New South Wales).

Oratemnus confusus Murthy & Ananthakrishnan

Oratemnus confusus Murthy & Ananthakrishnan, 1977: 127–129, figs 41a–b.

Type locality: Coimbatore, Tamil Nadu, India.
Distribution: India.

Oratemnus curtus (Beier)

Steiratemnus curtus Beier, 1954b: 15–16, fig. 6.
Oratemnus curtus (Beier): Beier, 1966a: 278; Harvey, 1981b: 245; Harvey, 1985b:
    128.

Type locality: Pemberton, Western Australia, Australia.
Distribution: Australia (Western Australia).

Oratemnus distinctus (Beier)

Steiratemnus distinctus Beier, 1948c: 531–532, fig. 4.
Oratemnus distinctus (Beier): Beier, 1966a: 278; Beier, 1969a: 187; Richards, 1971:
    18; Harvey, 1981b: 245; Harvey, 1985b: 128.

Type locality: New South Wales, Australia.
Distribution: Australia (New South Wales, Western Australia).

Oratemnus indicus (With)

Chelifer indicus With, 1906: 194–196, fig. 23, plate 4 figs 10a–d; Ellingsen, 1914: 3; Roewer, 1929: 613–614.

Oratemnus indicus (With): Beier, 1932b: 591; Beier, 1932e: 60; Roewer, 1936: figs 44a–b; Roewer, 1937: 284; Beier, 1967g: 353; Beier, 1973b: 49; Murthy & Ananthakrishnan, 1977: 126; Sivaraman, 1981: 22.

Not *Atemnus indicus* (With): Beier, 1930e: 297–298 (misidentification; see *Anatemnus nilgiricus* Beier).

Type locality: Vellore, near Madras, India.
Distribution: India, Sri Lanka.

Oratemnus loyolai Sivaraman

Oratemnus loyolai Sivaraman, 1980c: 349–352, figs 3a–b.

Type locality: Redhills, Madras, Tamil Nadu, India.
Distribution: India.

Oratemnus manilanus Beier

Oratemnus manilanus Beier, 1932b: 590, fig. 14; Beier, 1932e: 59, fig. 70; Roewer, 1937: 284.

Type locality: Manila, Philippines.
Distribution: Philippines.

Oratemnus navigator (With)

Chelifer politus E. Simon: H. J. Hansen, 1884: 543 (misidentification).

Chelifer navigator With, 1906: 191–193, fig. 22, plate 4 fig. 9a; With, 1907: 62–63, figs 15–16; Ellingsen, 1909a: 219; Ellingsen, 1911b: 141; Ellingsen, 1914: 3–4.

Oratemnus navigator (With): Beier, 1932b: 589–590; Beier, 1932e: 58–59; Roewer, 1937: 284; Beier, 1948b: 462; Murthy & Ananthakrishnan, 1977: 126–127.

Not *Chelifer (Atemnus) navigator* With: Ellingsen, 1910a: 358 (misidentification; see *Oratemnus timorensis* Beier).

Type locality: on ship from India.
Distribution: Burma, India, Indonesia (Java), Sri Lanka.

Oratemnus philippinensis Beier

Oratemnus philippinensis Beier, 1932b: 592–593; Beier, 1932e: 61, fig. 74; Roewer, 1937: 284; Beier, 1966b: 342–343.

Type locality: Los Baños, Luzon, Philippines.
Distribution: Philippines.

Oratemnus proximus Beier

Oratemnus proximus Beier, 1932b: 590–591; Beier, 1932e: 59–60, fig. 71; Roewer, 1937: 284; Beier, 1973b: 49–50.

Type locality: Sumatra, Indonesia.
Distribution: Indonesia (Sumatra), Sri Lanka.

Oratemnus punctatus (L. Koch)

Chelifer punctatus L. Koch, 1885: 45–46, plate 4 figs 3, 3a–c.

Chelifer brevidigitatus L. Koch, 1885: 48–49, plate 4 figs 6, 6a–c; With, 1905: 112, 328; Rainbow, 1916: 34 (synonymized by Beier, 1948c: 529).

Chelifer (Atemnus) brevidigitatus L. Koch: Ellingsen, 1910a: 357.

Anatemnus brevidigitatus (L. Koch): Beier, 1932b: 586–587; Beier, 1932e: 56, fig. 67; J. C. Chamberlin, 1934b: 8; Roewer, 1937: 284.

Family Atemnidae

Anatemnus (?) *punctatus* (L. Koch): Beier, 1932e: 56; J. C. Chamberlin, 1934b: 8; Roewer, 1937: 284.

Steiratemnus punctatus (L. Koch): Beier, 1948c: 529–531, figs 2–3.

Oratemnus punctatus (Beier): Beier, 1966a: 296; Beier, 1969a: 187; Beier, 1975b: 203; Beier, 1976f: 213; Harvey, 1981b: 245; Harvey, 1985b: 128–129.

Not *Chelifer brevidigitatus* L. Koch: Ellingsen, 1912a: 121 (misidentification; true identity not known).

Type localities: of *Chelifer punctatus*: Gayndah, Queensland, Australia.

of *Chelifer brevidigitatus*: Rockhampton, Queensland, Australia.

Distribution: Australia (Australian Capital Territory, New South Wales, Queensland, South Australia, Victoria, Western Australia), Lord Howe Island.

Oratemnus saigonensis (Beier)

Chelifer (Atemnus) saigonensis Beier, 1930d: 197–198, figs 1a–b.

Anatemnus saigonensis (Beier): Beier, 1932b: 583; Beier, 1932e: 53, fig. 65; Roewer, 1937: 283.

Oratemnus timorensis Beier: Redikorzev, 1938: 99 (misidentification).

Oratemnus saigonensis (Beier): Beier, 1951a: 84–86, fig. 23; Beier, 1967g: 353–354; Chapman, 1984a: 6.

Type locality: Saigon (now Ho Chi Minh), Vietnam.

Distribution: Cambodia, Laos, Malaysia (Sarawak), Thailand, Vietnam.

Oratemnus samoanus Beier

Oratemnus samoanus Beier, 1932b: 593, fig. 16; Beier, 1932e: 61–62, fig. 75; J. C. Chamberlin, 1934b: 8; Roewer, 1937: 284; J. C. Chamberlin, 1939b: 212–215, figs 3a–m; Beier, 1966d: 140; Beier, 1970g: 320; Beier, 1982: 43.

Oratemnus samoanus samoanus Beier: Beier, 1957d: 30–31.

Type locality: Apia, Samoa.

Distribution: Jamaica, Marquesas Islands, Marshall Islands, Papua New Guinea, Samoa, Solomon Islands, St Kitts.

Oratemnus samoanus whartoni J. C. Chamberlin

Oratemnus whartoni J. C. Chamberlin, 1947b: 308–312, figs 2a–e.

Oratemnus samoanus whartoni J. C. Chamberlin: Beier, 1957d: 30, fig. 15; Morikawa, 1960: 138–139.

Type locality: Ylig Bay, Guam, Mariana Islands.

Distribution: Caroline Islands, Japan, Mariana Islands, Marshall Islands.

Oratemnus semidivisus Redikorzev

Oratemnus semidivisus Redikorzev, 1938: 99–100, figs 27–28; Roewer, 1940: 346; Beier, 1951a: 86.

Type localities: Da Lat (as Dalat), Vietnam; Mt Hon-Ba, Vietnam; and Ream (as Réam), Cambodia.

Distribution: Cambodia, Vietnam.

Oratemnus timorensis Beier

Chelifer (Atemnus) navigator With: Ellingsen, 1910a: 358 (misidentification).

Oratemnus timorensis Beier, 1932b: 591; Beier, 1932e: 60, fig. 72; Roewer, 1937: 284; Petersen, 1968: 120.

Not *Oratemnus timorensis* Beier: Redikorzev, 1938: 99 (misidentification; see *Oratemnus saigonensis* (Beier)).

Type locality: Telang, Timor, Indonesia.

Distribution: Indonesia (Timor), Solomon Islands.

Oratemnus yodai Morikawa

Oratemnus yodai Morikawa, 1968: 262–263, figs 1c, 2a, 2f.

Type locality: Kapure, near Mt Numbur, Nepal.
Distribution: Nepal.

Genus **Paratemnoides**, new name

Paratemnus Beier, 1932b: 562–563; Beier, 1932e: 33; Morikawa, 1960: 137; Murthy
  & Ananthakrishnan, 1977: 125; Sivaraman, 1980c: 345–346 (junior homonym of
  Paratemnus Ameghino, 1904).

Type species: *Chelifer guineensis* Ellingsen, 1906 (junior synonym of *Chelifer pallidus*
  Balzan, 1892), by original designation.

Paratemnoides aequatorialis (Beier), new combination

Chelifer (Atemnus) guineensis Ellingsen, 1906: 246 (misidentification, in part; see
  Paratemnoides guineensis (Ellingsen)).
Paratemnus aequatorialis Beier, 1932b: 565–566; Beier, 1932e: 35, fig. 35; Roewer,
  1937: 282.

Type locality: Guinea-Bissau.
Distribution: Guinea-Bissau.

Paratemnoides assimilis (Beier), new combination

Chelifer birmanicus Thorell: Tullgren, 1912a: 267; Tullgren, 1912b: 269; Dammerman,
  1922: 100; Dammerman, 1929: 114 (not seen); Bristowe, 1931b: 1390; Dammerman,
  1948: 496 (misidentifications).
Paratemnus assimilis Beier, 1932b: 569–570, fig. 9; Beier, 1932e: 40, fig. 45; Roewer,
  1937: 283; Harvey, 1988b: 338–341, figs 86–95.

Type locality: Kolambugan, Mindanao, Philippines.
Distribution: Indonesia (Java, Krakatau Islands), Philippines.

Paratemnoides borneoensis (Beier), new combination

Paratemnus borneoensis Beier, 1932b: 571; Beier, 1932e: 41, fig. 46; Roewer, 1937:
  283; Weidner, 1959: 116.

Type locality: Lebang Hara, Borneo, Indonesia.
Distribution: Indonesia (Borneo).

Paratemnoides ceylonicus (Beier), new combination

Chelifer plebejus With: Tullgren, 1907a: 55, fig. 15 (misidentification); Ellingsen,
  1914: 4 (misidentification, in part).
Atemnus plebejus (With): Beier, 1930a: 208 (misidentification).
Paratemnus ceylonicus Beier, 1932b: 569, fig. 8; Beier, 1932e: 39–40, figs 43–44;
  Roewer, 1937: 283; Beier, 1973b: 47; Murthy & Ananthakrishnan, 1977: 126.

Type locality: Sri Lanka (as Ceylon).
Distribution: India, Sri Lanka.

Paratemnoides curtulus (Redikorzev), new combination

Anatemnus curtulus Redikorzev, 1938: 91–93, figs 19–21; Roewer, 1940: 346.
Paratemnus curtulus (Redikorzev): Beier, 1951a: 78; Beier, 1967g: 352.

Type locality: Arbré-Broyé, Da Lat (as Dalat), Poulo-Condore, Pleiku, Vietnam;
  Angkor, Cambodia.
Distribution: Cambodia, Vietnam.

Family Atemnidae

Paratemnoides ellingseni (Beier), new combination

Chelifer (Atemnus) feae Ellingsen: Ellingsen, 1910a: 357–358 (misidentification, in part).
Chelifer feae Ellingsen: Ellingsen, 1912b: 79, 90–91 (misidentification, in part); Ellingsen, 1913a: 452; Godfrey, 1927: 18 (misidentifications).
Paratemnus ellingseni Beier, 1932b: 572; Beier, 1932e: 42–43, fig. 50; Lawrence, 1937: 268; Roewer, 1937: 283; Caporiacco, 1941: 45; Beier, 1958a: 173; Beier, 1964k: 71–72; Beier, 1966k: 464; Beier, 1967d: 80; Lawrence, 1967: 89; Mahnert, 1983b: 369–371, figs 30–34; Mahnert, 1984b: 64.

Type locality: Delagoa Bay (now Bay de Lourenço Marques), Mozambique.
Distribution: Ghana, Kenya, Mozambique, Somalia, South Africa, Uganda, Zimbabwe.

Paratemnoides elongatus (Banks), new combination

Atemnus elongatus Banks, 1895a: 10; Banks, 1904a: 141; Coolidge, 1908: 112; Banks, 1909c: 174; J. C. Chamberlin, 1931a: fig. 11k.
Atemnus floridanus Tullgren, 1900a: 153–155, plate 1 figs 1–4 (synonymized by Banks, 1904a: 141).
Lustrochernes (?) *floridanus* (Tullgren): Beier, 1932e: 95; Roewer, 1937: 290; Hoff, 1958: 21.
Genus ? *elongatus* Banks: Beier, 1932e: 277; Roewer, 1937: 317.
Paratemnus elongatus (Banks): Hoff, 1946b: 109–113, figs 1–2; Hoff, 1958: 20; Hoff, 1964b: 2–5; Lloyd & Muchmore, 1974: 381; Muchmore, 1974b: 80; Lloyd, Correale & Muchmore, 1975: 241–242, fig. 1; Muchmore, 1975d: fig. 13; Beier, 1976d: 46; Muchmore, 1977: 76; Brach, 1978: 4–11, figs 1–4; Brach, 1979: 34.
Not *Atemnus elongatus* Banks: Berger, 1906: 489–490, fig. 1 (misidentification; see *Paratemnoides nidificator* (Balzan)).

Type localities: of *Atemnus elongatus*: St Lucie River, Sand Point, Enterprise, Punta Gorda, Florida, U.S.A.
of *Atemnus floridanus*: Victoria, Lake County, Florida, U.S.A.; and Apopka, Orange County, Florida, U.S.A.
Distribution: Cuba, Dominican Republic, Guatemala, Mexico, U.S.A. (Alabama, Florida), U.S. Virgin Islands.

Paratemnoides feai (Ellingsen), new combination

Chelifer (Atemnus) feae Ellingsen, 1906: 246–248; Ellingsen, 1910a: 357–358 (in part; see *Paratemnoides ellingseni* (Beier)).
Chelifer feae Ellingsen: Ellingsen, 1912b: 79, 90–91 (in part; see *Paratemnoides ellingseni* (Beier)).
Paratemnus feae (Ellingsen): Beier, 1932b: 571–572; Beier, 1932e: 42, figs 33, 49; Roewer, 1937: 283.
Paratemnus feai (Ellingsen): Vachon, 1956: 2–3.
Not *Chelifer feae* Ellingsen: Ellingsen, 1913a: 452; Godfrey, 1927: 18 (misidentifications; see *Paratemnoides ellingseni* (Beier)).

Type locality: San Thiago, near Praia, Cape Verde Islands.
Distribution: Cameroun, Cape Verde Islands.

Paratemnoides guianensis (Caporiacco), new combination

Paratemnus guianensis Caporiacco, 1947b: 21; Caporiacco, 1948a: 617–618, fig. 11.

Type locality: MacKenzie (now Linden), Guyana.
Distribution: Guyana.

Paratemnoides indicus (Sivaraman), new combination

Paratemnus indicus Sivaraman, 1980c: 346–347, figs 1a–b; Sivaraman, 1982: 187–194, figs 1a–e, 4.

Type locality: Pondicherry, Pondicherry, India.
Distribution: India.

Paratemnoides indivisus (Tullgren), new combination

Chelifer indivisus Tullgren, 1907d: 7–8, figs 5a–c; Ellingsen, 1912b: 79.
Cyclatemnus indivisus (Tullgren): Beier, 1932b: 562.
Cyclatemnus (?) *indivisus* (Tullgren): Beier, 1932e: 32; Roewer, 1937: 282.
Paratemnus indivisus (Tullgren): Mahnert, 1983b: 372–375, figs 40–47.

Type locality: Mombo, Tanzania.
Distribution: Kenya, Tanzania.

Paratemnoides insubidus (Tullgren), new combination

Chelifer insubidus Tullgren, 1907a: 58–59, figs 17a–b; Tullgren, 1908b: 285, fig. 3; Ellingsen, 1912b: 79, 91.
Paratemnus insubidus (Tullgren): Beier, 1932b: 566; Beier, 1932e: 36, fig. 36; Roewer, 1937: 282; Beier, 1947b: 324; Weidner, 1959: 114; Beier, 1964k: 71; Beier, 1966k: 464.

Type locality: Port Elizabeth, Cape Province, South Africa.
Distribution: Namibia, South Africa.

Paratemnoides insularis (Banks), new combination

Atemnus insularis Banks, 1902a: 68–69, plate 2 fig. 11.
Paratemnus insularis (Banks): J.C. Chamberlin, 1934b: 8.

Type locality: Albemarle, Galapagos Islands.
Distribution: Galapagos Islands.

Paratemnoides japonicus (Morikawa), new combination

Paratenmus (sic) *japonicus* Morikawa, 1953a: 348–350, figs 1h–i, 2a–b.
Paratemnus japonicus Morikawa: Morikawa, 1960: 138, plate 3 fig. 6, plate 8 figs 1, 16, plate 9 fig. 17; Morikawa, 1962: 419, fig. 4.

Type locality: Ishite River, Matsuyama, Shikoku, Japan.
Distribution: Japan.

Paratemnoides laosanus (Beier), new combination

Paratemnus laosanus Beier, 1951a: 76–78, fig. 19; Beier, 1967g: 352.

Type locality: Vieng Chang, Mekong, Laos.
Distribution: India, Laos.

Paratemnoides magnificus (Beier), new combination

Atemnus rotundus (With): Beier, 1930a: 207–208, fig. 5 (misidentification).
Paratemnus magnificus Beier, 1932b: 568–569; Beier, 1932e: 39, fig. 42; Roewer, 1937: 283; Beier, 1965g: 767.

Type locality: Hughipagu, Papua New Guinea.
Distribution: Papua New Guinea.

Paratemnoides mahnerti (Sivaraman), new combination

Paratemnus robustus Sivaraman, 1980c: 347–349, figs 2a–b (junior primary homonym of *Paratemnus robustus* Beier, 1932b).

Paratemnus mahnerti Sivaraman, 1981b: 96 (replacement name for *Paratemnus robustus* Sivaraman).

Type locality: Pondicherry, Pondicherry, India.
Distribution: India.

Paratemnoides minor (Balzan), new combination

Chelifer (Atemnus) nidificator minor Balzan, 1892: 510–511, fig. 1; Ellingsen, 1905e: 323–324.
Chelifer nidificator minor Balzan: Ellingsen, 1909b: 216.
Paratemnus minor (Balzan): Beier, 1932b: 568; Beier, 1932e: 38, fig. 40; Roewer, 1937: 283; Feio, 1945: 5; Mello-Leitão & Feio, 1949: 320, plate 4 fig. 1; Beier, 1959e: 203; Mahnert, 1979d: 763; Adis, 1981: 119, etc.; N. S. Hahn & Matthiesen, 1982: 99; Adis & Mahnert, 1986: 311; Mahnert & Adis, 1986: 213; Mahnert, Adis & Bührnheim, 1987: fig. 14a.

Type locality: San Esteban, Venezuela.
Distribution: Argentina, Brazil (Amazonas, Distrito Federal, Pará, Paraná), Colombia, Ecuador, Guatemala, Mexico, Panama, Peru, Venezuela.

Paratemnoides minutissimus (Beier), new combination

Paratemnus minutissimus Beier, 1974d: 901–902, fig. 2.

Type locality: Caioba, Paraná, Brazil.
Distribution: Brazil (Paraná).

Paratemnoides nidificator (Balzan), new combination

Chelifer nidificator Balzan, 1888a: no pagination, figs; Balzan, 1890: 417–418, figs 5, 5a–b; Tullgren, 1907a: 56; With, 1908a: 321–324, text-fig. 83, figs 36a–c.
Chelifer (Atemnus) nidificator Balzan: Balzan, 1892: 547; Ellingsen, 1902: 146–148; Ellingsen, 1905c: 2; Ellingsen, 1910a: 358.
Atemnus elongatus Banks: Berger, 1906: 489–490, fig. 1 (misidentification).
Lamprochernes nidificator (Balzan): Beier, 1930a: 211.
Paratemnus nidificator (Balzan): Beier, 1932b: 567–568; Beier, 1932e: 37–38, fig. 39; Roewer, 1937: 283; Mello-Leitão, 1939a: 118; Mello-Leitão, 1939b: 615; Feio, 1945: 5; Hoff, 1946c: 3; Schenkel, 1953a: 57; Muchmore, 1971a: 82.

Type locality: Asuncion, Argentina; Rio Apa, Paraguay; Mato Grosso (as Mattogrosso), Brazil.
Distribution: Argentina, Brazil (Bahia, Distrito Federal, Mato Grosso, Pará), Costa Rica, Ecuador, Guatemala, Guyana, Haiti, Mexico, Paraguay, St Vincent, Venezuela.

Paratemnoides obscurus (Beier), new combination

Paratemnus obscurus Beier, 1959d: 42–43, fig. 20.

Type locality: Elisabethville (now Lubumbashi), Zaire.
Distribution: Zaire.

Paratemnoides pallidus (Balzan), new combination

Chelifer (Atemnus) pallidus Balzan, 1892: 511–512, figs 2, 2a (incorrectly claimed to be a junior homonym of *Chernes pallidus* Banks, 1890, by Ellingsen, 1912b: 79).
Chelifer (Atemnus) guineensis Ellingsen, 1906: 246 (in part; see *Paratemnoides aequatorialis* (Beier)); Ellingsen, 1910a: 358. New synonymy.
Chelifer guineensis Ellingsen: Ellingsen, 1912b: 79.
Paratemnus guineensis (Ellingsen): Beier, 1932b: 565; Beier, 1932e: 34–35, fig. 34; Roewer, 1937: 282; Vachon, 1952a: 30; Beier, 1972a: 11; Mahnert, 1978f: 97–101, figs 56–68; Heurtault, 1983: 18; Mahnert, 1983b: 369.

Paratemnus congicus Beier, 1932b: 566–567, fig. 7; Beier, 1932e: 36–37, figs 37–38; Roewer, 1936: fig. 83a; Roewer, 1937: 282; Beier, 1948b: 452, 457; Beier, 1954d: 134; Beier, 1959d: 42; Beier, 1967d: 80 (synonymized by Beier, 1972a: 11).

Type localities: of *Chelifer (Atemnus) pallidus*: Sierra Leone.
of *Chelifer (Atemnus) guineensis*: Rio Cassine, Guinea-Bissau; Ribeira Palma, Sao Tomé (as Thomé); Basilé, Fernando Poo (now Macias Nguema), Equatorial Guinea; Punta Frailes, Fernando Poo (now Macias Nguema), Equatorial Guinea; and Fernand-Vaz, Gabon.
of *Paratemnus congicus*: Eala, Zaire.
Distribution: Burundi, Cameroun, Congo, Equatorial Guinea, Gabon, Guinea, Guinea-Bissau, Ivory Coast, Kenya, Sao Tomé, Sierra Leone, Togo, Zaire.

Paratemnoides perpusillus (Beier), new combination

Paratemnus perpusillus Beier, 1935a: 487–488, fig. 3; Roewer, 1937: 283; Beier, 1948b: 457; Muchmore, 1971a: 87.

Type locality: Barbados.
Distribution: Barbados.

Paratemnoides persimilis (Beier), new combination

Paratemnus persimilis Beier, 1932b: 567; Beier, 1932e: 37; Roewer, 1937: 283; Vachon, 1952a: 30, fig. 18; Beier, 1965b: 373.

Type locality: Edea, Cameroun.
Distribution: Cameroun, Chad, Guinea.

Paratemnoides philippinus (Beier), new combination

Paratemnus philippinus Beier, 1932b: 573, fig. 10; Beier, 1932e: 43–44, fig. 52; Roewer, 1937: 283; Beier, 1957d: 28; Morikawa, 1960: 138; Beier, 1966b: 342; Morikawa, 1960: 138.

Type locality: Vinac Island, Philippines.
Distribution: Japan, Philippines.

Paratemnoides plebejus (With), new combination

Chelifer plebejus With, 1906: 185–188, figs 18–20, plate 4 figs 7a–b; Ellingsen, 1910a: 358; Ellingsen, 1911a: 36; Ellingsen, 1911b: 142; Ellingsen, 1914: 4 (in part; see *Paratemnoides ceylonicus* (Beier)).
Paratemnus plebejus (With): Beier, 1932b: 570; Beier, 1932e: 40; Roewer, 1937: 283; Beier, 1952a: 105–106, figs 7–8; Beier, 1955e: 43.
Not *Chelifer plebejus* With: Tullgren, 1907a: 55, fig. 15 (misidentification; see *Paratemnoides ceylonicus* (Beier)).
Not *Atemnus plebejus* (With): Beier, 1930a: 208 (misidentification; see *Paratemnoides ceylonicus* (Beier)).

Type locality: Car Nicobar (as Kar-Nicobar), Nicobar Islands.
Distribution: Burma, Indonesia (Sumatra), Malaysia, Nicobar Islands, Papua New Guinea, Singapore.

Paratemnoides pococki (With), new combination

Chelifer javanus Thorell: Pocock, 1900: 156 (misidentification).
Chelifer pococki With, 1907: 63–65, figs 17–19.
Paratemnus pococki (With): Beier, 1932b: 574; Beier, 1932e: 45; Roewer, 1937: 283; Harvey, 1985b: 129.

Type locality: north coast of Christmas Island.
Distribution: Christmas Island.

Paratemnoides redikorzevi (Beier), new combination

Anatemnus robustus Redikorzev, 1938: 93–95, figs 22, 25a; Roewer, 1940: 346 (junior
  secondary homonym of *Paratemnus robustus* Beier, 1932b).
Paratemnus redikorzevi Beier, 1951a: 74–76, fig. 18; Beier, 1967g: 352 (replacement
  name for *Anatemnus robustus* Redikorzev).

Type locality: Kontum, Vietnam.
Distribution: Thailand, Vietnam.

Paratemnoides robustus (Beier), new combination

Paratemnus robustus Beier, 1932b: 572–573; Beier, 1932e: 43, fig. 51; Roewer, 1937:
  283; Beier, 1966b: 342.

Type locality: Paniol, Manila, Philippines.
Distribution: Philippines.

Paratemnoides salomonis (Beier), new combination

Paratemnus salomonis Beier, 1935e: 639–640, fig. 2; Roewer, 1937: 283; Beier, 1957d:
  28–29, figs 14a–c; Beier, 1964f: 594; Beier, 1965g: 767; Beier, 1966d: 140; Beier,
  1970g: 320; Beier, 1982: 43.
Paratemnus salomonis papuanus Beier, 1948c: 556–558, fig. 19 (synonymized by
  Beier, 1965g: 767).

Type localities: of *Paratemnus salomonis*: Rendova Island, Solomon Islands.
of *Paratemnus salomonis papuanus*: Kokoda, Papua New Guinea.
Distribution: Caroline Islands, Indonesia (Irian Jaya), Mariana Islands, Papua New
  Guinea, Solomon Islands.

Paratemnoides salomonis hebridicus (Beier), new combination

Paratemnus salomonis hebridcus Beier, 1940a: 174–175, fig. 3; Beier, 1966d: 140.

Type locality: Ounua, Malekula, New Hebrides.
Distribution: New Hebrides, Solomon Islands.

Paratemnoides sinensis (Beier), new combination

Paratemnus sinensis Beier, 1932b: 573–574, fig. 11; Beier, 1932e: 44, fig. 53; Roewer,
  1937: 283.

Type locality: near Canton, China.
Distribution: China (Kwangtung).

Paratemnoides singularis (Beier), new combination

Paratemnus singularis Beier, 1965g: 767–768, fig. 9; Tenorio & Muchmore, 1982: 378.

Type locality: Port Moresby, Papua New Guinea.
Distribution: Papua New Guinea.

Paratemnoides sumatranus (Beier), new combination

Paratemnus sumatranus Beier, 1935a: 485–487, fig. 2; Roewer, 1937: 283.

Type locality: Singkarak (as Singkarah), Sumatra, Indonesia.
Distribution: Indonesia (Sumatra).

Genus Progonatemnus Beier

Progonatemnus Beier, 1955g: 51–52.

Type species: *Progonatemnus succineus* Beier, 1955g, by original designation.

! **Progonatemnus succineus** Beier

Progonatemnus succineus Beier, 1955g: 52–53, fig. 3; Schawaller, 1978: 4.

Type locality: Baltic Amber.
Distribution: Baltic Amber.

Genus **Stenatemnus** Beier

Stenatemnus Beier, 1932b: 604; Beier, 1932e: 72; Murthy & Ananthakrishnan, 1977: 130; Sivaraman, 1980c: 352.

Type species: *Chelifer fuchsi* Tullgren, 1907e, by original designation.

Stenatemnus annamensis Beier

Stenatemnus annamensis Beier, 1951a: 86–88, fig. 24.

Type locality: Cao Nguyên Lâm Viên (as Plateau von Langbian), Vietnam.
Distribution: Vietnam.

Stenatemnus asiaticus Sivaraman

Stenatemnus asiaticus Sivaraman, 1980c: 352–354, figs 4a–b.

Type locality: Tambaram, Chengalpet District, Tamil Nadu, India.
Distribution: India.

Stenatemnus boettcheri Beier

Stenatemnus böttcheri Beier, 1932b: 605–606, fig. 19; Beier, 1932e: 74, figs 89, 91; Roewer, 1937: 285; Beier, 1967c: 318.

Type locality: Burauen (as Buranan), Leyte, Philippines.
Distribution: Philippines.

Stenatemnus brincki Beier

Stenatemnus brincki Beier, 1973b: 50, fig. 14; Beier, 1974e: 1010; Murthy & Ananthakrishnan, 1977: 131.

Type locality: Deerwood Kuruwita, 6 miles NNW. of Ratnapura, Sabaragamuwa Province, Sri Lanka.
Distribution: India, Sri Lanka.

Stenatemnus extensus Beier

Stenatemnus extensus Beier, 1951a: 88–89, fig. 25; Beier, 1976e: 100.

Type locality: Cao Nguyên Lâm Viên (as Plateau von Langbian), Vietnam.
Distribution: Bhutan, Vietnam.

Stenatemnus fuchsi (Tullgren)

Chelifer fuchsi Tullgren, 1907e: 247–248, figs a–b; Ellingsen, 1911a: 36.
Stenatemnus fuchsi (Tullgren): Beier, 1932b: 605; Beier, 1932e: 73, fig. 90; Roewer, 1937: 285; Beier, 1964f: 594; Beier, 1965g: 771–772; Beier, 1966d: 140; Beier, 1967c: 322; Beier, 1982: 43.

Type locality: Palembang, Sumatra, Indonesia.
Distribution: Indonesia (Irian Jaya, Sumatra), Papua New Guinea, Solomon Islands.

Stenatemnus indicus Murthy & Ananthakrishnan

Stenatemnus indicus Murthy & Ananthakrishnan, 1977: 131–133, figs 42a–b.

Type locality: Egmore, Madras, Tamil Nadu, India.
Distribution: India.

Stenatemnus kraussi Beier

Stenatemnus kraussi Beier, 1957d: 31–32, figs 16a–c.

Type locality: Ulebsehel (Aurapushekaru), Palau Islands, Caroline Islands.
Distribution: Caroline Islands.

Stenatemnus orientalis Sivaraman

Stenatemnus orientalis Sivaraman, 1980c: 354–356, figs 5a–b.

Type locality: Nungambukkan, Madras, Tamil Nadu, India.
Distribution: India.

Stenatemnus procerus Beier

Stenatemnus procerus Beier, 1957d: 32–33, fig. 17.

Type locality: East Ngatpang, Babelthuap, Palau Islands, Caroline Islands.
Distribution: Caroline Islands.

Stenatemnus sundaicus (Beier)

Atemnus sundaicus Beier, 1930a: 208–209, figs 6a–b.
Stenatemnus sundaicus Beier, 1932b: 606; Beier, 1932e: 74, fig. 92; Roewer, 1937: 285.

Type locality: Borneo.
Distribution: Borneo.

Genus Synatemnus Beier

Synatemnus Beier, 1944: 188.

Type species: *Synatemnus parvulus* Beier, 1944, by original designation.

Synatemnus kilimanjaricus Beier

Synatemnus kilimanjaricus Beier, 1951b: 607–608, fig. 1.

Type locality: Lyamungu, Kilimanjaro, Tanzania.
Distribution: Tanzania.

Synatemnus parvulus Beier

Synatemnus parvulus Beier, 1944: 189, fig. 12; Weidner, 1959: 117.

Type locality: Amani, Tanzania.
Distribution: Tanzania.

Genus Tamenus Beier

Tamenus Beier, 1932b: 600; Beier, 1932e: 68–69; Sivaraman, 1980c: 356.

Type species: *Chelifer camerunensis* Tullgren, 1901, by original designation.

Tamenus aureus Beier

Atemnus camerunensis (Tullgren): Beier, 1930a: 210–211, fig. 8 (misidentification).
Tamenus aureus Beier, 1932b: 601–602, fig. 18; Beier, 1932e: 70, fig. 85; Roewer, 1937: 284; Beier, 1948b: 452.

Type locality: Addah, Ghana.
Distribution: Ghana.

Tamenus camerunensis (Tullgren)

Chelifer camerunensis Tullgren, 1901: 99–100, fig. 4; Ellingsen, 1910b: 63; Ellingsen, 1912b: 81; Ellingsen, 1913a: 452–453.
Chelifer (Lamprochernes) camerunensis Tullgren: Navás, 1921: 26.

Tamenus camerunensis (Tullgren): Beier, 1932b: 601; Beier, 1932e: 69–70, fig. 84; Roewer, 1937: 284.

Not *Chelifer (Lamprochernes) camerunensis* Tullgren: Ellingsen, 1910a: 365–366 (misidentification; see *Tamenus insularis* Beier).

Not *Atemnus camerunensis* (Tullgren): Beier, 1930a: 210–211, fig. 8 (misidentification; see *Tamenus aureus* Beier).

Type locality: Cameroun.
Distribution: Cameroun, Equatorial Guinea, Ghana, Nigeria.

Tamenus equestroides (Ellingsen)

Chelifer (Atemnus) equestroides Ellingsen, 1906: 250–251; Ellingsen, 1910a: 367; Ellingsen, 1912b: 79.
Tamenus (?) *equestroides* (Ellingsen): Beier, 1932b: 603–604; Beier, 1932e: 72; Roewer, 1937: 285.

Type localities: Rio Cassine, Guinea-Bissau; Ribeira Palma, Sao Tomé (as Thomé); Roça, Principe; Bahia do Oeste, Principe; Punta Frailes, Fernando Poo (now Macias Nguema), Equatorial Guinea; and Fernand Vaz, Gabon.
Distribution: Equatorial Guinea, Gabon, Guinea-Bissau, Principe, Sao Tomé.

Tamenus femoratus Beier

Tamenus femoratus Beier, 1932b: 602–603; Beier, 1932e: 71, fig. 87; Roewer, 1937: 284.

Type locality: Tabou (as Tabu), Ivory Coast.
Distribution: Ivory Coast.

Tamenus ferox (Tullgren)

Chelifer ferox Tullgren, 1907a: 49–51, figs 11a–c; Ellingsen, 1912b: 81.
Tamenus ferox (Tullgren): Beier, 1932b: 603; Beier, 1932e: 71–72, fig. 88; Roewer, 1937: 284; Weidner, 1959: 114.

Type localities: Ogowe, Central African Republic; and Gabon.
Distribution: Central African Republic, Gabon.

Tamenus indicus Sivaraman

Tamenus indicus Sivaraman, 1980c: 356–359, figs 6a–b; Sivaraman, 1982: 187–194, figs 2a–e, 5.

Type locality: Chepauk, Madras, Tamil Nadu, India.
Distribution: India.

Tamenus insularis Beier

Chelifer (Lamprochernes) camerunensis Tullgren: Ellingsen, 1910a: 365–366 (misidentification).
Tamenus insularis Beier, 1932b: 602; Beier, 1932e: 70–71, fig. 86; Roewer, 1937: 284.

Type locality: Fernando Poo (now Macias Nguema), Equatorial Guinea.
Distribution: Equatorial Guinea.

Tamenus milloti Vachon

Tamenus milloti Vachon, 1938c: 304–306, figs a, h, l; Roewer, 1940: 346.

Type locality: Bingerville, Ivory Coast.
Distribution: Ivory Coast.

Tamenus schoutedeni Beier

Tamenus schoutedeni Beier, 1954d: 135, fig. 3.

Type locality: Arebi, Bondo-Moto, Zaire.
Distribution: Zaire.

Genus Titanatemnus Beier

Titanatemnus Beier, 1932b: 550–551; Beier, 1932e: 22.

Type species: *Titanatemnus gigas* Beier, 1932b, by original designation.

Titanatemnus alluaudi Vachon

Titanatemnus alluaudi Vachon, 1935b: 513–515, fig. 2; Roewer, 1940: 346; Mahnert, 1983b: 383–386, figs 74–81.
Titanatemnus alluandi (sic) Vachon: Mahnert, 1984b: 64.

Type locality: Voi River, Kenya.
Distribution: Kenya, Somalia.

Titanatemnus chappuisi Beier

Titanatemnus chappuisi Beier, 1935c: 119–120, fig. 3; Roewer, 1940: 346; Beier, 1955c: 549; Beier, 1967d: 79; Mahnert, 1983b: 387–389, figs 82–89.

Type locality: Kitale, Kenya.
Distribution: Kenya, Uganda.

Titanatemnus congicus Beier

Titanatemnus congicus Beier, 1932b: 554–555; Beier, 1932e: 25, fig. 20; Vachon, 1936a: 78; Roewer, 1937: 282; Vachon, 1938a: 43–46, figs 23–25; Vachon, 1951a: 728–730, fig. 1; Beier, 1954d: 134; Beier, 1959d: 39; Beier, 1972a: 9.

Type locality: Kambove, Zaire.
Distribution: Zaire.

Titanatemnus conradti (Tullgren)

Chelifer conradti Tullgren, 1908c: 57–60, figs 1–3; Ellingsen, 1912b: 79.
Titanatemnus conradti (Tullgren): Beier, 1932b: 558; Beier, 1932e: 26; Roewer, 1937: 282.

Type locality: Cameroun.
Distribution: Cameroun.

Titanatemnus coreophilus Beier

Titanatemnus coreophilus Beier, 1948b: 452, 468–469, fig. 1.

Type locality: Cape Coast, Ghana.
Distribution: Ghana.

Titanatemnus equester (With)

Chelifer equester With, 1905: 123–127, plate 8 figs 3a–d, plate 9 figs 1a–f; Ellingsen, 1907b: 28; Ellingsen, 1912b: 79, 90; Ellingsen, 1913a: 452.
Atemnus equester (With): J. C. Chamberlin, 1931a: 134, fig. 32n.
Titanatemnus equester (With): Beier, 1932b: 559; Beier, 1932e: 29–30; Roewer, 1937: 282; Beier, 1948b: 452.
Not *Chelifer equester* With: Tullgren, 1907b: 224 (misidentification; see *Titanatemnus natalensis* Beier); Tullgren, 1907d: 8–10, fig. 6 (misidentification; see *Tamenus orientalis* Beier).
Not *Chelifer (Atemnus) equester* With: Ellingsen, 1910a: 357 (misidentification; see *Titanatemnus natalensis* Beier).

Type locality: Taveta (as Taveita), Kilimanjaro, Kenya.
Distribution: Kenya, Tanzania.

Titanatemnus gigas Beier

Titanatemnus gigas Beier, 1932b: 553–554, fig. 3; Beier, 1932e: 23–24, figs 16–18; Roewer, 1937: 282; Vachon, 1952a: 29; Beier, 1954d: 134; Beier, 1959d: 39; Heurtault, 1983: 21.

Type localities: Moto, Uele, Zaire; and Watsa, Uele, Zaire.
Distribution: Cameroun, Guinea, Ivory Coast, Zaire.

Titanatemnus kibwezianus Beier

Titanatemnus kibwezianus Beier, 1932b: 559–560; Beier, 1932e: 30, fig. 29; Roewer, 1937: 282; Beier, 1948b: 452, 469–470; Mahnert, 1983b: 380–381, figs 66–70.

Type locality: Kibwezi, Kenya.
Distribution: Kenya, Ethiopia.

Titanatemnus monardi Vachon

Titanatemnus monardi Vachon, 1934h: 167–168 (nomen nudum).
Titanatemnus monardi Vachon, 1935b: 510–513, fig. 1; Roewer, 1937: 282; Vachon, 1938a: fig. 53c; Vachon, 1941e: 29; Mahnert, 1978f: 101–104, figs 69–78.

Type locality: Kandingu, Angola.
Distribution: Angola, Congo, Ivory Coast.

Titanatemnus natalensis Beier

Chelifer equester With: Tullgren, 1907b: 224 (misidentification).
Chelifer (Atemnus) equester With: Ellingsen, 1910a: 357 (misidentification).
Titanatemnus natalensis Beier, 1932b: 556; Beier, 1932e: 26–27, fig. 23; Roewer, 1937: 282; Beier, 1948b: 452; Beier, 1955l: 307; Beier, 1958a: 171–172; Beier, 1964k: 69.
Titanotemnus (sic) *natalensis* Beier: Lawrence, 1937: 268.

Type locality: Durban (as Port Natal), Natal, South Africa.
Distribution: South Africa.

Titanatemnus orientalis Beier

Chelifer equester With: Tullgren, 1907d: 8–10, fig. 6 (misidentification).
Titanatemnus orientalis Beier, 1932b: 557–558, fig. 5; Beier, 1932e: 28, figs 26–27; Roewer, 1937: 282; Weidner, 1959: 117; Mahnert, 1983b: 381–383, figs 63–65.

Type locality: Ostufiomi, Kenya.
Distribution: Kenya.

Titanatemnus palmquisti (Tullgren)

Chelifer palmquisti Tullgren, 1907d: 10–12, figs 7a–c; Ellingsen, 1912b: 80.
Chelifer (Atemnus) palmquisti Tullgren: Ellingsen, 1910a: 358.
Anatemnus palmquisti (Tullgren): Beier, 1932b: 586; Beier, 1932e: 55; Roewer, 1937: 284.
Titanatemnus montanus Beier, 1935c: 120–121, fig. 4; Vachon, 1936a: 78; Roewer, 1940: 346; Beier, 1944: 183–184, fig. 8; Caporiacco, 1949a: 318; Beier, 1951b: 606–607; Salt, 1954: 396, 419 (synonymized by Beier, 1962f: 18).
Titanatemnus palmquisti (Tullgren): Beier, 1955c: 548–549; Beier, 1962f: 18–19; Beier, 1967d: 79; Mahnert, 1983b: 389–391, figs 90–97.

Type localities: of *Chelifer palmquisti*: Meru, Tanzania; and Kibo (as Kiboscho), Tanzania.
of *Titanatemnus montanus*: Cherangani, Marakwet, Kenya.
Distribution: Ethiopia, Kenya, Malawi, Tanzania, Uganda.

Titanatemnus regneri Beier

Titanatemnus regneri Beier, 1932b: 557; Beier, 1932e: 27–28, fig. 25; Roewer, 1937: 282; Beier, 1967d: 79; Mahnert, 1983b: 383, figs 71–73.

Type locality: Pangani, N. of Dar es Salam, Tanzania.
Distribution: Kenya, Tanzania.

Titanatemnus saegeri Beier

Titanatemnus saegeri Beier, 1972a: 9–11, fig. 5.

Type locality: Parc National Garamba, Zaire.
Distribution: Zaire.

Titanatemnus serrulatus Beier

Titanatemnus serrulatus Beier, 1932b: 555–556; Beier, 1932e: 26, fig. 22; Roewer, 1937: 282.

Type locality: Moto Uele, Zaire.
Distribution: Zaire.

Titanatemnus similis Beier

Chelifer (Atemnus) sjöstedti Tullgren: Ellingsen, 1910a: 359 (misidentification).
Titanatemnus similis Beier, 1932b: 554; Beier, 1932e: 24–25, figs 14–15, 19; Roewer, 1937: 282; Beier, 1948c: 527; Beier, 1954d: 134; Mahnert, 1978f: 107–112, figs 87–102.

Type locality: Jaunde Station, Cameroun.
Distribution: Cameroun, Congo, Zaire.

Titanatemnus sjoestedti (Tullgren)

Chelifer (Atemnus) sjöstedti Tullgren, 1901: 97–101, figs 1–3; Ellingsen, 1906: 245; Navás, 1921: 26.
Chelifer (Atemnus) sjostedti (sic) Tullgren: E. Simon, 1903a: 124.
Chelifer sjöstedti Tullgren: Ellingsen, 1905a: 2; Ellingsen, 1909a: 218; Ellingsen, 1912b: 80; Ellingsen, 1913a: 452.
Atemnus sjöstedti (Tullgren): J. C. Chamberlin, 1931a: fig. 51k.
Titanatemnus sjöstedti (Tullgren): Beier, 1932b: 556–557; Beier, 1932e: 27, fig. 24; Roewer, 1937: 282, figs 225–226; Beier, 1954d: 134; Beier, 1959d: 39–40.
Titanatemnus sjoestedti (Tullgren): Mahnert, 1978f: 104–107, figs 79–86.
Not *Chelifer (Atemnus) sjöstedti* Tullgren: Ellingsen, 1910a: 359 (misidentification; see *Titanatemnus similis* Beier).

Type locality: Itoki, Cameroun.
Distribution: Cameroun, Congo, Equatorial Guinea, Gabon, Guinea-Bissau, Zaire.

Titanatemnus tanensis Mahnert

Titanatemnus tanensis Mahnert, 1983b: 391–395, figs 98–108; Mahnert, 1984b: 64.

Type locality: S. of Mpekatoni, Lamu, Kenya.
Distribution: Kenya, Somalia.

Titanatemnus tessmanni Beier

Titanatemnus tessmanni Beier, 1932b: 555, fig. 4; Beier, 1932e: 25–26, fig. 21; Roewer, 1937: 282.

Type locality: N. of 'Kolentangan' (= Kolen?), Guinea.
Distribution: Guinea.

Titanatemnus thomeensis (Ellingsen)

Chelifer (Atemnus) sjöstedti thoméensis Ellingsen, 1906: 245–246.
Chelifer sjöstedti thoméensis Ellingsen: Ellingsen, 1912b: 80.
Titanatemnus thoméensis (Ellingsen): Beier, 1932b: 560.
Titanatemnus (?) *thoméensis* (Ellingsen): Beier, 1932e: 30; Roewer, 1937: 282.

Type locality: Agua Izè, Sao Tomé (as Thomé).
Distribution: Sao Tomé.

Titanatemnus ugandanus Beier

Titanatemnus ugandanus Beier, 1932b: 558–559; Beier, 1932e: 29, fig. 28; Roewer, 1937: 282; Beier, 1967d: 79.

Type locality: Bunyali, Bukedi, Uganda.
Distribution: Tanzania, Uganda.

Genus Tullgrenius J. C. Chamberlin

Tullgrenius J. C. Chamberlin, 1933: 263; Murthy & Ananthakrishnan, 1977: 133; Sivaraman, 1980c: 359.

Type species: *Tullgrenius indicus* J. C. Chamberlin, 1933, by original designation.

Tullgrenius afghanicus Beier

Tullgrenius afghanicus Beier, 1959a: 273–274, fig. 16.

Type locality: Bashgultal, Nuristan, Afghanistan.
Distribution: Afghanistan.

Tullgrenius compactus Beier

Tullgrenius compactus Beier, 1951a: 89–91, fig. 26.

Type locality: Phsar Ream, (as Réam), Cambodia.
Distribution: Cambodia.

Tullgrenius indicus J. C. Chamberlin

Atemnus sp.: J. C. Chamberlin, 1931a: figs 27b, 38n.
Tullgrenius indicus J. C. Chamberlin, 1933: 264, fig. a; Roewer, 1937: 287; Murthy & Ananthakrishnan, 1977: 133–134; Sivaraman & Murthy, 1980a: 163–166, figs 1a–b, 2, 5; Sivaraman, 1981: 22; Sivaraman, 1982: 187–194, figs 3a–e, 6.

Type locality: Guindy, Chingleput, India.
Distribution: India.

Tullgrenius orientalis Sivaraman

Tullgrenius orientalis Sivaraman, 1980c: 359–362, figs 7a–b.

Type locality: Tambaram, Chengulpet District, Tamil Nadu, India.
Distribution: India.

Tullgrenius vachoni Murthy

Tullgrenius vachonicus Murthy, 1962: 62–65, figs a–b; Beier, 1974e: 1010; Murthy & Ananthakrishnan, 1977: 134.

Type locality: Krusadi Islands, India.
Distribution: India.

Family CHELIFERIDAE Risso

Cheliferidae Risso, 1826: 157; Hagen, 1879: 400; E. Simon, 1879a: 18; Tömösváry, 1882b: 182–183; H. J. Hansen, 1884: 516–521; Daday, 1888: 113; Balzan, 1890: 409; Balzan, 1892: 509; H. J. Hansen, 1893: 231; Tullgren, 1899a: 164–165; Ellingsen, 1904: 1; Tullgren, 1906a: 201; With, 1906: 115–116; Kew, 1911a: 39; Lessert, 1911: 9; Comstock, 1913: 45; Moles & Moore, 1921: 6; J. C. Chamberlin, 1923c: 369; Redikorzev, 1924b: 21; Pratt, 1927: 409; Schenkel, 1928: 55; Väänänen, 1928a: 17; J. C. Chamberlin, 1931a: 244–246; J. C. Chamberlin, 1931c: 289–290; Beier, 1932e: 191; Beier, 1932g: 188–189; J. C. Chamberlin, 1935b: 481; Roewer, 1937: 304–305; Hoff, 1949b: 485; Petrunkevitch, 1955: 82; Hoff, 1956d: 1–2; Morikawa, 1960: 148; Beier, 1963b: 279–280; Muchmore, 1973b: 58; Murthy & Ananthakrishnan, 1977: 145; Muchmore, 1982a: 101; Harvey, 1985b: 131; Legg & Jones, 1988: 128.

Cheliferinae Risso: Stecker, 1874b: 232; Stecker, 1875d: 87; E. Simon, 1879a: 19; Balzan, 1892: 509–510; H. J. Hansen, 1884: 531–532; Daday, 1888: 113–114; H. J. Hansen, 1893: 232; With, 1906: 116; Lessert, 1911: 9; J. C. Chamberlin, 1923c: 372; Redikorzev, 1924b: 21; Väänänen, 1928a: 18; J. C. Chamberlin, 1931c: 293–294; Beier, 1932e: 226–227; Roewer, 1937: 310; Hoff, 1949b: 485; Hoff, 1956d: 2; Morikawa, 1960: 149–150; Murthy & Ananthakrishnan, 1977: 151.

Cheliferina (sic) Risso: Tömösváry, 1882b: 183.

Cheliferini Risso: Navás, 1925: 101; J. C. Chamberlin, 1932a: 19; Beier, 1932e: 227; Roewer, 1937: 311; Hoff, 1949b: 486; Hoff, 1956d: 2–3; Morikawa, 1960: 150; Murthy & Ananthakrishnan, 1977: 152; Benedict & Malcolm, 1979: 189.

Cheliceridae (sic) Risso: Bacelar, 1928: 190.

Lissocheliferini J. C. Chamberlin, 1932a: 20.

Dactylocheliferini Beier, 1932d: 63; Beier, 1932e: 241; Roewer, 1937: 313; Hoff, 1949b: 491; Hoff, 1956d: 28; Morikawa, 1960: 152; Murthy & Ananthakrishnan, 1977: 157; Benedict & Malcolm, 1979: 197.

Protocheliferini Beier, 1948c: 551–552.

Juxtacheliferini Hoff, 1956d: 22–23.

Genus **Amaurochelifer** Beier
Amaurochelifer Beier, 1951a: 111.

Type species: *Amaurochelifer annamensis* Beier, 1951a, by original designation.

Amaurochelifer annamensis Beier
Amaurochelifer annamensis Beier, 1951a: 112, fig. 40.

Type locality: Pic de Langbian, Vietnam.
Distribution: Vietnam.

Genus **Ancistrochelifer** Beier
Ancistrochelifer Beier, 1951a: 106–107.

Type species: *Ancistrochelifer agniae* Beier, 1951a, by original designation.

Ancistrochelifer agniae Beier
Ancistrochelifer agniae Beier, 1951a: 107–109, figs 37, 38.

Type locality: Phsar Ream (as Réam), Cambodia.
Distribution: Cambodia, Vietnam.

Ancistrochelifer tuberculatus Beier

Ancistrochelifer tuberculatus Beier, 1951a: 109–111, fig. 39.

Type locality: Pak Lay (as Paclay), Laos.
Distribution: Laos.

Genus **Aperittochelifer** Beier

Aperittochelifer Beier, 1955l: 321–322.

Type species: *Chelifer capensis* Hewitt & Godfrey, 1929, by original designation.

Aperittochelifer capensis (Hewitt & Godfrey)

Chelifer capensis Hewitt & Godfrey, 1929: 321–322, fig. 1f, plate 22 fig. 7.
Hansenius (?) *capensis* (Hewitt & Godfrey): Beier, 1932e: 272; Roewer, 1937: 316.
Aperittochelifer capensis (Hewitt & Godfrey): Beier, 1955l: 322–323, fig. 33; Beier,
    1964k: 90; Beier, 1966k: 469.

Type locality: Platteklip, Table Mountain, Cape Town, Cape Province, South Africa.
Distribution: South Africa.

Aperittochelifer minusculus (Ellingsen)

Chelifer minusculus Ellingsen, 1912b: 95–97; Hewitt & Godfrey, 1929: 318–319,
    fig. 1d, plate 21 fig. 5.
Lophochernes minusculus (Ellingsen): Beier, 1932e: 248–249; Roewer, 1937: 315.
Aperittochelifer (?) *minusculus* (Ellingsen): Beier, 1955l: 322.
Aperittochelifer minusculus (Ellingsen): Beier, 1958a: 185–186, fig. 16; Beier, 1964k:
    86–87, fig. 39.

Type locality: Pirie Forest, King William's Town, Cape Province, South Africa.
Distribution: South Africa.

Aperittochelifer protractus (Hewitt & Godfrey)

Chelifer protractus Hewitt & Godfrey, 1929: 319–321, figs 1e, 4a–b, plate 21 fig. 6.
Hansenius (?) *protractus* (Hewitt & Godfrey): Beier, 1932e: 272; Roewer, 1937: 316.
Aperittochelifer protractus (Hewitt & Godfrey): Beier, 1955l: 322.

Type locality: Ngqaba Forest, near Tsolo, Cape Province, South Africa.
Distribution: South Africa.

Aperittochelifer transvaalensis Beier

Aperittochelifer transvaalensis Beier, 1964k: 88–90, fig. 41.

Type locality: Bloemhof, Transvaal, South Africa.
Distribution: South Africa.

Aperittochelifer zumpti Beier

Aperittochelifer zumpti Beier, 1964k: 87–88, fig. 40; Lawrence, 1967: 90.

Type locality: Maseya Spring, Kruger National Park, Transvaal, South Africa.
Distribution: South Africa.

Genus **Aporochelifer** Beier

Aporochelifer Beier, 1953g: 87.

Type species: *Aporochelifer insulanus* Beier, 1953g, by original designation.

Aporochelifer insulanus Beier

Aporochelifer insulanus Beier, 1953g: 87–88, fig. 6.

Type locality: Ende (as Endeh), Flores, Indonesia.
Distribution: Indonesia (Flores).

Genus **Aspurochelifer** Benedict & Malcolm

Aspurochelifer Benedict & Malcolm, 1979: 189.

Type species: *Aspurochelifer littlefieldi* Benedict & Malcolm, 1979, by original designation.

Aspurochelifer littlefieldi Benedict & Malcolm

Aspurochelifer littlefieldi Benedict & Malcolm, 1979: 189–192, figs 1–4.

Type locality: Pinehurst, Jackson County, Oregon, U.S.A.
Distribution: U.S.A. (California, Idaho, Nevada, Oregon, Washington).

Genus **Australochelifer** Beier

Australochelifer Beier, 1975b: 209.

Type species: *Australochelifer pygmaeus* Beier, 1975b, by original designation.

Australochelifer pygmaeus Beier

Australochelifer pygmaeus Beier, 1975b: 209–210, fig. 4; Harvey, 1981b: 244; Harvey, 1985b: 131.

Type locality: near Waste Point, Mt Kosciusko, New South Wales, Australia.
Distribution: Australia (New South Wales).

Genus **Beierius** J. C. Chamberlin

Beierius J. C. Chamberlin, 1932a: 20; Beier, 1932e: 229.

Type species: *Chelifer walliskewi* Ellingsen, 1912b, by original designation.

Beierius aequatorialis Vachon

Beierus (sic) *aequatorialis* Vachon, 1944: 439–441, figs 1–2, 4, 6.

Type locality: Gabon.
Distribution: Gabon.

Beierius semimarginatus Beier

Beierius semimarginatus Beier, 1959d: 64–65, fig. 33; Beier, 1964k: 90; Beier, 1966k: 469.

Type locality: Kundelungu, Zaire.
Distribution: Namibia, Zaire, Zimbabwe.

Beierius simplex Beier

Beierius simplex Beier, 1955l: 323–325, fig. 34.

Type locality: 7 miles SW. of Bredasdorp, Cape Province, South Africa.
Distribution: South Africa.

Beierius walliskewi (Ellingsen)

Chelifer walliskewi Ellingsen, 1912b: 101–103 (in part; see *Chelifer cancroides* (Linnaeus)); Hewitt & Godfrey, 1929: 311–314, figs 1c, 2a–b, plate 21 figs 1–2.
Beierius walliskewi (Ellingsen): J. C. Chamberlin, 1932a: 20; Beier, 1932e: 229; Roewer, 1937: 312; Vachon, 1941a: 80; Beier, 1958a: 186; Beier, 1964k: 90; Newlands, 1978: 700.
Beierus (sic) *walliskewi* (Ellingsen): Vachon, 1944: figs 3–4.
Beierius walliskewi walliskewi (Ellingsen): Beier, 1955l: 325–327, figs 35, 36a.

Type locality: Bushman's Rock, Blythswood, Butterworth, Cape Province, South Africa.
Distribution: South Africa.

Beierius walliskewi legrandi (Vachon)

Dactylochelifer legrandi Vachon, 1939b: 156–159, figs 1–9 (synonymized with *Beierius walliskewi* (Ellingsen) by Vachon, 1941a: 80).
Beierius walliskewi gracilis Beier, 1955l: 328, fig. 36c, 37; Beier, 1964k: 90 (synonymized by Mahnert, 1988a: 77).
Beierus (sic) *walliskewi legrandi* (Vachon): Mahnert, 1988a: 77–78.

Type localities: of *Dactylochelifer legrandi*: Pretoria, Natal, South Africa.
of *Beierius walliskewi gracilis*: Natal, South Africa; and Transvaal, South Africa.
Distribution: Kenya, Lesotho, South Africa.

Beierius walliskewi longipes Beier

Beierius walliskewi longipes Beier, 1955l: 327–328, fig. 36b.

Type localities: Bredasdorp, Cape Province, South Africa; Montagu, Cape Province, South Africa; and Wellington, Cape Province, South Africa.
Distribution: South Africa.

Genus Beierochelifer Mahnert

Beierochelifer Mahnert, 1977b: 73.

Type species: *Rhacochelifer anatolicus* Beier, 1949, by original designation.

Beierochelifer anatolicus (Beier)

Phacochelifer (sic) *anatolicus* Beier, 1949: 15–16, fig. 12; Rack, 1971: 114.
Rhacochelifer anatolicus Beier: Beier, 1957b: 151; Beier, 1965f: 98, fig. 11; Beier, 1967f: 320; Beier, 1969b: 196.
Beierochelifer anatolicus (Beier): Mahnert, 1977b: 73–74, figs 10–11.

Type locality: Dalyan, Köycegiz, Turkey.
Distribution: Greece, Turkey.

Beierochelifer geoffroyi Heurtault

Beierochelifer geoffroyi Heurtault, 1981: 210–213, figs 1–7.

Type locality: Mézel, Alpes de Haute, France.
Distribution: France.

Beierochelifer peloponnesiacus (Beier)

Chelifer (Chelifer) peloponnesiacus Beier, 1929a: 354–356, figs 10a–b, 11b.
Rhacochelifer peloponnesiacus (Beier): Beier, 1932d: 66; Beier, 1932e: 266–267, fig. 288–289; Roewer, 1937: 315; Beier, 1963b: 297–298, fig. 299.
Beierochelifer peloponnesiacus (Beier): Gardini, 1985b: 61.

Type locality: Kumani, Morea, Greece.
Distribution: Greece.

Beierochelifer peloponnesiacus jonicus (Beier)

Rhacochelifer jonicus Beier, 1932d: 65–66; Beier, 1932e: 265–266, figs 286–287; Roewer, 1937: 315.
Rhacochelifer peloponnesiacus jonicus Beier: Beier, 1963b: 298, fig. 300.
Beierochelifer peloponnesiacus jonicus (Beier): Mahnert, 1977b: 74, fig. 12.

Type locality: Aenos, Kefallonía, Greece.
Distribution: Greece.

Genus Canarichelifer Beier

Canarichelifer Beier, 1965c: 379–380.

Type species: *Canarichelifer teneriffae* Beier, 1965c, by original designation.

Canarichelifer teneriffae Beier

Canarichelifer teneriffae Beier, 1965c: 380–381, fig. 3; Beier, 1976a: 25; Mahnert, 1980a: 259; Pieper, 1980: 263.

Type locality: N. of Los Cristianos, Tenerife, Canary Islands.
Distribution: Canary Islands, Ilhas Selvagens.

Genus Centrochelifer Beier

Centrochelifer Beier, 1959a: 276.

Type species: *Centrochelifer afghanicus* Beier, 1959a, by original designation.

Centrochelifer afghanicus Beier

Centrochelifer afghanicus Beier, 1959a: 276–277, fig. 18; Beier, 1974h: 169.

Type locality: Kandahar, Afghanistan.
Distribution: Afghanistan, Mongolia.

Genus Chamberlinarius Heurtault

Chamberlinius Heurtault, 1983: 10 (junior homonym of *Chamberlinius* Wang, 1956).
Chamberlinarius Heurtault, 1990b: 128 (replacement name for *Chamberlinius* Heurtault, 1983).

Type species: *Chamberlinius pujoli* Heurtault, 1983, by original designation.

Chamberlinarius pujoli (Heurtault)

Chamberlinius pujoli Heurtault, 1983: 10–12, figs 24–28.
Chamberlinarius pujoli (Heurtault): Heurtault, 1990b: 128.

Type locality: Lamto, Ivory Coast.
Distribution: Ivory Coast.

Genus Cheirochelifer Beier

Cheirochelifer Beier, 1967f: 321.

Type species: *Cheirochelifer turcicus* Beier, 1967f, by original designation.

Cheirochelifer bigoti Heurtault

Cheirochelifer bigoti Heurtault, 1981: 213–218, figs 8–12.

Type locality: Tour du Valat, Camargue, France.
Distribution: France.

Cheirochelifer heterometrus (L. Koch)

Chelifer heterometrus L. Koch, 1873: 29.
Rhacochelifer heterometrus (L. Koch): Beier, 1932d: 66; Beier, 1932e: 268; Roewer, 1937: 315.
Rhacochelifer (?) *heterometrus* (L. Koch): Beier, 1963b: 298.
Cheirochelifer heterometrus (L. Koch): Mahnert, 1978g: 20–23, figs 16–19; Callaini, 1987b: 291–293.
Not *Rhacochelifer heterometrus* (L. Koch): J. C. Chamberlin, 1949: 23–26, figs 7a–j (true identification unknown).

Type locality: Syros (as Syra), Cyclades, Greece.
Distribution: France, Greece, Turkey.

Cheirochelifer turcicus Beier

Cheirochelifer turcicus Beier, 1967f: 321–323, fig. 8; Beier, 1969b: 196; Mahnert, 1977b: 68.

Type locality: Bulgar-Dagh, Namrun, Turkey.
Distribution: Greece, Turkey.

Genus **Chelifer** Geoffroy

Chelifer Geoffroy, 1762: 617–618; Fourcroy, 1785: 526; Latreille, 1796: 186; Latreille, 1804: 138–141; Latreille, 1810: 118; Latreille, 1817: 108–109; Lamarck, 1818: 79–80; Latreille, 1825a: 131; Risso, 1826: 157; Théis, 1832: 69; Latreille, 1837: 316; C. L. Koch, 1839: 3–4; Gervais, 1849: 11; C. L. Koch, 1850: 3–4; Menge, 1855: 29–30; McIntire, 1868: 8; Stecker, 1874b: 235; Stecker, 1875d: 88; E. Simon, 1879a: 20; Tömösváry, 1882b: 195; Canestrini, 1883: no. 1; H. J. Hansen, 1884: 536–537; Balzan, 1887a: no pagination; Balzan, 1890: 411–412; Banks, 1891: 162; Balzan, 1892: 510; O.P.-Cambridge, 1892: 218–219; Tullgren, 1899a: 165–166; Tullgren, 1906a: 201; With, 1906: 117–130; Kew, 1911a: 39–40; Lessert, 1911: 9; J. C. Chamberlin, 1923c: 372; Pratt, 1927: 409; Schenkel, 1928: 56–57; Väänänen, 1928a: 18; J. C. Chamberlin, 1932a: 19; Beier, 1932e: 235–236; J. C. Chamberlin, 1935b: 481; G. O. Evans & Browning, 1954: 17; Hoff, 1956d: 3; Morikawa, 1960: 150–151; Beier, 1963b: 286–287; Legg & Jones, 1988: 135; International Commission on Zoological Nomenclature, 1989: 143.

Obisium Illiger, 1798: 501.

Chelifer (Chelifer) Geoffroy: E. Simon, 1878a: 146; Balzan, 1892: 528; Kew, 1911a: 47; Lessert, 1911: 18; Nonidez, 1917: 20.

Type species: of *Chelifer*: *Acarus cancroides* Linnaeus, 1758, by subsequent designation of Latreille, 1810: 424.

of *Obisium*: *Acarus cancroides* Linnaeus, 1758, by subsequent designation of J. C. Chamberlin, 1930: 12.

Chelifer cancroides (Linnaeus)

Acarus cancroides Linnaeus, 1758: 616; Linnaeus, 1761: 480; Scopoli, 1763: 390.

Chelifer cancroides (Linnaeus): Geoffroy, 1762: 618; Schrank, 1781: 525; Fourcroy, 1785: 526; Latreille, 1795: 19–20; Latreille, 1796: 109; Donovan, 1797: 83–84, plate 225; Schrank, 1803: 244–245; Hermann, 1804: 114–116, plate 5 figs 6, o, r; Latreille, 1804: 141–142; Latreille, 1810: 424; Leach, 1814: 429; Lamarck, 1818: 80; Latreille, 1825a: 132; Duméril, 1826: 49, plate 56 fig. 4; Brébisson, 1827: 263; Théis, 1832: 69–73, plate 3 figs 1, 1a–c; Anonymous, 1835: 186; Dugès & Edwards, 1836: 84; Latreille, 1837: 316, plate 23 fig. 1; Lamarck, 1838: 108–109; Lamarck, 1839: 301; C. L. Koch, 1843: 41–42, fig. 780; Gervais, 1844: 77; Menge, 1855: 30–31, plate 1 figs 2–3, 5–8, 15, plate 2 figs 1–3, 5–14, plate 3 figs 1–2, 8–12, plate 4 fig. 5; Kolenati, 1857: 431–432; Hagen, 1868b: 51; Packard, 1869: 216, fig. 1; Hagen, 1870: 264–265; L. Koch, 1873: 16–17; Ritsema, 1874: xxxiv; Stecker, 1874b: 236; Stecker, 1875d: 88; E. Simon, 1878a: 146; E. Simon, 1878b: 155; E. Simon, 1879a: 23–25, 312, plate 17 figs 4, 6, plate 18 fig. 2; Daday, 1880a: 192; Tömösváry, 1882b: 206–208, plate 1 figs 20–24; Canestrini, 1883: no. 6, figs 1–3; Packard, 1886: 43; E. Simon, 1887: 36–37; Daday, 1888: 121, 176; Packard, 1888: 43; Daday, 1889c: 25; Daday, 1889d: 80; Banks, 1890: 152; Alfonsus, 1891: 503–506; O.P.-Cambridge, 1892: 220–221, plate B figs 12, 12a–b; Banks, 1895a: 3; Banks, 1895c: 431; Lameere, 1895: 475; E. Simon, 1895: 168; Ellingsen, 1897: 14–16; Kathariner, 1898: 250; E. Simon, 1898d: 20; E. Simon, 1899d: 86; Tullgren, 1899a: 167–169, plate 1 fig. 1; E. Simon, 1900b: 593; Bruntz, 1903: 367–372; Ellingsen, 1903: 11; E. Simon, 1903b: 387; Stschelkanovzeff, 1903a: 318; Banks, 1904b: 364; E. Simon, 1904: 444; Ellingsen, 1905d: 1; Nosek, 1905: 120, 154; Tullgren, 1906a: 203, figs 2c, plate 4 fig. 1; Ellingsen, 1907a: 162; Godfrey, 1907: 162–163; Strand, 1907: 243; Tullgren, 1907e: 246; André, 1908: 289–290; Banks, 1908: 39; Coolidge, 1908: 109; Ellingsen, 1908a: 416; Ellingsen, 1908c: 670; Ellingsen, 1908d: 69–70; Ellingsen, 1908e: 163; Godfrey, 1908: 157; With, 1908a:

221; André, 1909a: 478−479; André, 1909b: 278−279; Banks, 1909c: 173; Ellingsen, 1909a: 209, 216; Godfrey, 1909: 153−154; Kew, 1909a: 249; Ellingsen, 1910b: 62; Banks, 1911: 637; Ellingsen, 1911c: 174; Ewing, 1911: 66−75, 77−78, figs 1−8; Kew, 1911a: 48−49, fig. 11; Ellingsen, 1912b: 83, 92; Ellingsen, 1913a: 454; Kew, 1914: 95; Moles, 1914c: 187−188; Kew, 1916a: 124; Kew, 1916b: 79; Standen, 1916a: 17; Daday, 1918: 2; Navás, 1918: 89; Frickhinger, 1919: 170−171; Navás, 1919: 211; Fage, 1921: 102; Moles & Moore, 1921: 6; Navás, 1921: 165; Filleul, 1922: 140 (not seen); Caporiacco, 1923: 131; Navás, 1925: 107, fig. 4; Pratt, 1927: 409, figs 645a−b; Ewing, 1928: figs 6a; Redikorzev, 1928: 123; Schenkel, 1928: 57, figs 1, 3−6; Väänänen, 1928a: 19, figs 1, 4, 5(I); Väänänen, 1928b: fig. 2; Hewitt & Godfrey, 1929: 309−311; J. C. Chamberlin, 1931a: 168, figs 1b, 1e, 1l, 1r, 14m−n, 15aa, 18s, 20i−j, 45e−f, 45n, 46h, 46k, 47b, 47k, 47n−o, 51n−o, 71; Kästner, 1931a: 73−77, figs 1−2; Beier, 1932e: 236−237, figs 1−2, 244−246; Beier, 1932g: figs 157−158, 200b, 221, 226, 266; Berland, 1932: figs 64, 66−67, 70; J. C. Chamberlin, 1932a: 19; Nester, 1932: 98; Schlottke, 1933: 109−112, figs 1, 2a−c; Vachon, 1932: 21−26, figs 1−7; Vachon, 1933a: 1874−1875; Vachon, 1933b: 59−64, figs 1−8; Tumšs, 1934: 19; Vachon, 1934a: 174−177; Vachon, 1934b: 155−159, figs a−h; Vachon, 1934c: 38−54, figs g, j−p; Vachon, 1934e: 154−155; Vachon, 1934f: 158; Beier, 1935d: 176; J. C. Chamberlin, 1935b: 481; Redikorzev, 1935b: 17; Vachon, 1935c: 21−29, figs a−i; Vachon, 1935j: 187; Nordberg, 1936: 48−49; Roewer, 1936: figs 1−2, 80, 85, 127; Vachon, 1936a: 78; Vachon, 1936b: 140; Vachon, 1936c: 294−298, figs 1−6; Vachon, 1936d: 89−91; Roewer, 1937: 312, figs 179−182, 190−195, 198; Strebel, 1937: 143−155; Vachon, 1937f: 128; Brimley, 1938: 497 (not seen); Hadži, 1938: 31, fig. 28; Vachon, 1938a: 25−35, 90−91, figs 1, 9−14, 15a, 50, 63−73, 75−85; Vachon, 1938b: 200; Beier, 1939a: figs 162−165, 166e, 167−168, 169a−f, 171−180; Beier, 1939e: 312; Beier, 1939f: 199; Hadži, 1939b: 47; Lohmander, 1939b: 319−320; Caporiacco, 1940: 7; Schlottke, 1940: 58−69, figs 1−6; Vachon, 1940f: 2, 3; Vachon, 1943: 299−302, figs 1−8; Beier, 1944: 206; Hoff, 1944b: 123, figs 1a−b; Vachon, 1945a: 85; Cooreman, 1946b: 7; Kaisila, 1947: 86; Vachon, 1947d: 177−178, fig. 1; Beier, 1948a: 191; Beier, 1948b: 445, 459, 462; Caporiacco, 1948c: 242; Levi, 1948: 290−298, figs 1−2, 3a−i; Hoff, 1949b: 486−487, figs 9, 48; Kaisila, 1949b: 86−89, map 13; Redikorzev, 1949: 660−661; Vachon, 1949: figs 199e, 208, 214a−c, 215−216, 221, 222c, 223a, 228−229, 231−233, 235−238; Hoff, 1950a: 1−2; Beier, 1951a: 104; Beier, 1952e: 5; Hoff & Clawson, 1952: 35; Levi, 1953: 55−59, figs 5, 10−11; Woodruffe, 1953: 753; Beier & Franz, 1954: 458; G. O. Evans & Browning, 1954: 17, figs 17, 23; Weidner, 1954b: 110−111; Beier, 1955a: 18; Ketterer, 1955: 93−96; Beier, 1956h: 24; Hoff, 1956d: 3; Hoff & Bolsterli, 1956: 177; Cloudsley-Thompson, 1958: 96; Hoff, 1958: 32; Ressl & Beier, 1958: 2; Beier, 1959a: 276; Beier, 1959d: 63; Hoff, 1959b: 5, etc.; Hoff, 1961: 457; Kobakhidze, 1961b: 471−472; Meinertz, 1962a: 401, map 8; Morikawa, 1962: fig. 3; Beier, 1963b: 287, fig. 288; Beier, 1963f: 197; Beier, 1963g: 275; Beier, 1963h: 156; Höregott, 1963: 546; Hoff, 1964b: 6; Kaisila, 1964: 54; Lehtinen, 1964: 285; Savory, 1964a: figs 30, 38; Vachon, 1964b: 4840, figs 1−4; Beier, 1965f: 97; Kobakhidze, 1965b: 542; Ressl, 1965: 290; Vitali-di Castri, 1965b: fig. 2; Helversen, 1966a: 149; Helversen, 1966b: figs 1c, 4d, 7a−b; Kobakhidze, 1966: 706; Savory, 1966: fig. 1; Zangheri, 1966: 532; Beier, 1967d: 87; Beier, 1967f: 314; Beier, 1967g: 358; Rafalski, 1967: 19−20; Aru, 1968: 607, fig. 1; Legendre, 1968: fig. 15; Levi & Levi, 1968: 121, fig.; Beier, 1969b: 194; Hammen, 1969: 22; Lazzeroni, 1969b: 410; Lazzeroni, 1969c: 243; Manley, 1969: 9, fig. 11; Weygoldt, 1969a: 14, 28, 56−57, 63, 76−79, 100, 105, 106, 108−110, 117, figs 1a−b, 5−7, 26, 50, 56, 68, 69a−b, 70, 95, 97; Legg, 1970a: fig. 2c; Helversen & Martens, 1971: 378; Muchmore, 1971a: 89, 91; Kofler, 1972: 288; Legendre, 1972: 447; Legg, 1972: 576, figs 1(2a, 3b), 3(1); Legg, 1973b: 430, fig. 2c; Ćurčić, 1974a: 27; Beier, 1976a: 25; Jones, 1975a: 88; Klausen,

1975: 64; Legg, 1975a: 66; Legg, 1975d: fig. 2; Nelson, 1975: 293–294, figs 78–81; Beier, 1976f: 244; Cekalovic, 1976: 19; Čurčić, 1976b: 179; Rowland & Reddell, 1976: 18; Drogla, 1977: fig. 1; Estany, 1977b: 32; Mahnert, 1977b: 68; Mahnert, 1977d: 97; Savory, 1977: figs 90, 98; Jones, 1978: 90; Benedict & Malcolm, 1979: 192–193; Judson, 1979a: 61; Jones, 1980c: map 25; Sanocka-Woloszynowa, 1981: 37; Krumpál & Kiefer, 1982: 14; Cekalovic, 1984: 28; Jędryczkowski, 1985: 80; Lippold, 1985: 40; Callaini, 1987b: 286; Schawaller, 1986: 10–11; Eisenbeis & Wichard, 1987: figs 43a–b, 45a–c; Harvey, 1987e: 188–189; Jędryczkowski, 1987b: 146, map 13; Mahnert, Adis & Bührnheim, 1987: fig. 1; Krumpál & Cyprich, 1988: 42; Legg & Jones, 1988: 135–136, figs 5e–f, 35a, 35b(a–j); Mahnert, 1988a: 69; International Commission on Zoological Nomenclature, 1989: 143; Schawaller, 1989: 24.

Phalangium cancroides (Linnaeus): Linnaeus, 1767: 1028; O. F. Müller, 1776: 192; Berendt, 1830: 30.

Chelifer europaeus de Geer, 1778: 355–357, plate 19 figs 14–15 (synonymized by Fabricius, 1793: 436).

Scorpio cancroides (Linnaeus): Fabricius, 1793: 436; Illiger, 1807: 139–140.

Obisium cancroides (Linnaeus): Illiger, 1798: 501; Walckenaer, 1802: 253; Bernard, 1893b: 428.

Chelifer hermanni Leach, 1817: 49, plate 142 fig. 3; Risso, 1826: 157 (synonymized by E. Simon, 1879a: 24).

Chelifer sesamoides Audouin, 1826: 174–175, plate 8 fig. 4; Audouin, 1827: 413–414, plate 8 fig. 4; Dugès & Edwards, 1836: plate 20 bis figs 4, 4a–c; Gervais, 1844: 80, plate 25 fig. 2; Lucas, 1849: 273; Hagen, 1870: 267 (synonymized by Beier, 1932e: 236).

Chélifer cancröìdes (Linnaeus): Anonymous, 1831b: 283; Anonymous, 1832: 754.

Chelifer ixoides C. W. Hahn, 1834: 53, fig. 140 (not seen); C. L. Koch, 1837: fasc. 140.4, fig.; Latreille, 1837: plate 23 fig. 2; C. L. Koch, 1843: 39–40, fig. 779; Gervais, 1844: 79; L. Koch, 1873: 23–24; Stecker, 1874b: 236; L. Koch, 1877: 180 (synonymized by E. Simon, 1878b: 155).

Chelifer granulatus C. L. Koch, 1843: 37, fig. 777; L. Koch, 1873: 21–23; Canestrini, 1874: 225; Ritsema, 1874: xxxiv; Stecker, 1874b: 236; L. Koch, 1877: 180; H. J. Hansen, 1884: 538–540; Croneberg, 1887: 147; Croneberg, 1888: fig. 4; H. J. Hansen, 1893: 205–208, plate 4 figs 10–13; Krauss, 1896: 628; With, 1906: 48 (synonymized by E. Simon, 1878b: 155).

Chelifer grandimanus C. L. Koch, 1843: 38–39, fig. 778 (synonymized by E. Simon, 1879a: 24).

Chelifer cancroïdes (Linnaeus): Lucas, 1849: 273; Becker, 1884: cclxiii; Artault de Vevey, 1901: 105.

Chelifer rhododactylus Menge, 1855: 32, plate 4 fig. 6; Hagen, 1870: 267; Stecker, 1874c: 311–313 (synonymized by Stecker, 1875d: 88).

Chelifer serratus Stecker, 1874a: 235–236; Stecker, 1875d: 88 (synonymized by Beier, 1932e: 236).

Chelifer hermannii Leach: O.P.-Cambridge, 1892: 219–220, plate B figs 11, 11a.

Chelifer ixioides (sic) C. W. Hahn: Krauss, 1896: 628.

Chelifer (Chelifer) cancroides (Linnaeus): Ellingsen, 1905c: 16; Ellingsen, 1910a: 384–385; Lessert, 1911: 19–20, figs 1–2; Standen, 1912: 14; Nonidez, 1917: 20; Redikorzev, 1924b: 24, figs 1, 3; Kästner, 1928: 7, figs 1, 22–23; Beier, 1929a: 347–348; Beier, 1929d: 155.

Chelifer cancroides dentatus Ewing, 1911: 73 (synonymized by Hoff, 1949b: 486–487).

Chelifer walliskewi Ellingsen, 1912b: 101–103 (misidentification, in part).

Chelifer dentatus Ewing: J. C. Chamberlin, 1931a: fig. 10j.

Celifer (sic) *cancroides* (Linnaeus): Crocker, 1978: 9.

Not *Acarus cancroides* Linnaeus: Poda, 1761: 122 (not seen) (misidentification; see *Lamprochernes nodosus* (Schrank)).

Not *Phalangium cancroides* (Linnaeus): Shaw, 1791: no pagination (misidentification; see *Cheiridium museorum* (Leach)).

Not *Chelifer cancroides* (Linnaeus): Anonymous, 1834: 162; Lukis, 1834: 162–163; Moore, 1834: 321–322; Stevens, 1866: xxvii (not seen); Spicer, 1867: 244; Anonymous, 1875: 185; Newman, 1875: 186; Leydig, 1881a: 180 (not seen); Mégnin, 1886: 241 (not seen); Hess, 1894: 120 (misidentifications; see *Lamprochernes nodosus* (Schrank)); Hagen, 1868a: 435 (misidentification; see *Parachelifer muricatus* (Say)); Leidy, 1877: 260 (misidentification; see *Chernes sanborni* Hagen); Fenizia, 1902: 55 (not seen) (misidentification; see *Lamprochernes nodosus* (Schrank)).

Not *Chelifer (Chelifer) cancroides* (Linnaeus): Beier, 1930d: 202 (misidentification; true identification unknown).

Type localities: of *Acarus cancroides*: Europe.
of *Chelifer europaeus*: Europe.
of *Chelifer hermanni*: not stated, presumably England, Great Britain.
of *Chelifer sesamoides*: Egypt.
of *Chelifer ixoides*: Nürnberg, West Germany.
of *Chelifer granulatus*: Gdansk (as Danzig), Poland.
of *Chelifer grandimanus*: Germany.
of *Chelifer rhododactylus*: West Germany.
of *Chelifer serratus*: Prague, Czechoslovakia; and Sobeslav, Czechoslovakia.
of *Chelifer cancroides dentatus*: U.S.A.

Distribution: Afghanistan, Albania, Algeria, Argentina, Austria, Balearic Islands, Belgium, Brazil (Pará, Santa Catarina), Bulgaria, Canada (British Columbia, Newfoundland, Ontario), Canary Islands, Chile, Corsica, Cuba, Czechoslovakia, Denmark, France, East Germany, Egypt, Ethiopia, Finland, France, Ghana, Great Britain, Greece, Hungary, India, Ireland, Israel, Italy, Kenya, Malawi, Mongolia, Netherlands, New Zealand, Norway, Poland, Romania, Sardinia, South Africa, Spain, Sweden, Switzerland, Tanzania, Turkey, U.S.A. (Alaska, California, Colorado, District of Columbia, Florida, Georgia, Idaho, Illinois, Indiana, Kansas, Kentucky, Maine, Maryland, Massachusetts, Michigan, Missouri, Montana, Nebraska, Nevada, New Mexico, New York, North Carolina, North Dakota, Ohio, Oregon, Pennsylvania, Texas, Utah, Virginia, Wisconsin, Wyoming), U.S.S.R. (Georgia, Kazakhstan, Kirghizia, Latvia, R.S.F.S.R., Turkmenistan, Uzbekistan), Vietnam, West Germany, Yugoslavia, Zaire.

Chelifer cancroides orientalis Morikawa

Chelifer cancroides orientalis Morikawa, 1954a: 73–75, figs 1g, 2a–e; Morikawa, 1960: 151, plate 4 fig. 5, plate 7 fig. 20, plate 8 fig. 18, plate 9 fig. 26, plate 10 fig. 23; Morikawa, 1962: 419.

Type locality: Sapporo, Hokkaido, Japan.
Distribution: Japan.

Genus **Cubachelifer** Hoff

Cubachelifer Hoff, 1946c: 21.

Type species: *Cubachelifer strator* Hoff, 1946c, by original designation.

Cubachelifer strator Hoff

Cubachelifer strator Hoff, 1946c: 21–23, figs 27–29; Beier, 1976d: 57.

Type locality: 14 km N. of Vinales, Cuba.
Distribution: Cuba, Dominican Republic.

Genus **Dactylochelifer** Beier

Dactylochelifer Beier, 1932d: 64; Beier, 1932e: 253; Hoff, 1949b: 491; G. O. Evans
   & Browning, 1954: 16; Hoff, 1956d: 28; Beier, 1963b: 288; Legg & Jones, 1988:
   132.

Type species: *Chelifer latreillei* Leach, 1817, by original designation.

Dactylochelifer afghanicus Beier

Dactylochelifer afghanicus Beier, 1959a: 278–279, fig. 20.

Type locality: Tang-Sayed, between Aibak and Tachqourghan, Afghanistan.
Distribution: Afghanistan.

Dactylochelifer amurensis (Tullgren)

Chelifer amurensis Tullgren, 1907a: 38–39, fig. 6.
Dactylochelifer amurensis (Tullgren): Beier, 1932d: 65; Beier, 1932e: 256–257, figs
   267–268; Roewer, 1937: 315; Weidner, 1959: 113; Schawaller, 1989: 26–27, fig. 69.

Type locality: Blagoveshchensk (as Blagowestschensk), Amur, R.S.F.S.R., U.S.S.R.
Distribution: U.S.S.R. (R.S.F.S.R.).

Dactylochelifer anatolicus Beier

Dactylochelifer anatolicus Beier, 1963g: 275–277, fig. 6; Beier, 1965f: 98; Beier,
   1969b: 196; Callaini, 1987b: 282; Schawaller, 1987a: fig. 11.

Type locality: Aksehir (as Akschehir), Sultan Daglari (as Sultan-Dagh), Turkey.
Distribution: Turkey.

Dactylochelifer balearicus Beier

Dactylochelifer balearicus Beier, 1961b: 38–39, figs 6–7; Beier, 1963b: 289–290;
   Mahnert, 1978c: fig. 8; Schawaller, 1987a: fig. 11.

Type locality: Playa Tirant Nou, Balearic Islands.
Distribution: Balearic Islands.

Dactylochelifer beieri Redikorzev

Dactylochelifer beieri Redikorzev, in Beier, 1932e: 257–258, fig. 271; Redikorzev,
   1934b: 152; Roewer, 1937: 315; Redikorzev, 1949: 663, fig. 36.
Chelifer beieri (Redikorzev): Redikorzev, 1934a: 429–430, figs 6–7.

Type locality: Kischlak Baga-Abzal, Bukhara (as Buchara), Uzbekistan, U.S.S.R.
Distribution: U.S.S.R. (Turkmenistan, Uzbekistan).

Dactylochelifer besucheti Mahnert

Dactylochelifer besucheti Mahnert, 1978c: 149–152, figs 1–6; Schawaller, 1987a:
   fig. 11.
Dactylochelifer cf. *besucheti* Mahnert: Mahnert, 1982f: 297.

Type locality: St Avall, Mallorca, Balearic Islands.
Distribution: Balearic Islands, Malta.

Dactylochelifer brachialis Beier

Dactylochelifer brachialis Beier, 1952f: 249–250, fig. 3; Beier, 1959a: 280–281,
   figs 23b, 24; Beier, 1959c: 407; Beier, 1960a: 45; Beier, 1961e: 3–4; Mahnert,
   1978c: figs 12a–b; Schawaller, 1983c: 371; Schawaller, 1986: 14, figs 34, 39;
   Schawaller, 1989: 27.

Type locality: Piustagoli, Koh-i-Baba, Afghanistan.
Distribution: Afghanistan, India, Iran, U.S.S.R. (Tadzhikistan, Turkmenistan).

Dactylochelifer cendsureni Krumpál & Kiefer

Dactylochelifer cendsureni Krumpál & Kiefer, 1982: 8–10, plate 4 figs 1–8.

Type locality: 15 km N. of Dzuun Mod Oase, Bajanchongor Ajmak, Mongolia.
Distribution: Mongolia.

Dactylochelifer changaiensis Krumpál & Kiefer

Dactylochelifer changaiensis Krumpál & Kiefer, 1982: 10–12, plate 5 figs 1–10.

Type locality: 3 km N. of See, Archangaj Ajmak, Mongolia.
Distribution: Mongolia.

Dactylochelifer copiosus Hoff

Dactylochelifer copiosus Hoff, 1945c: 53–57, figs 17–21; Hoff, 1945g: 521; Hoff,
1949b: 491–492, figs 10, 51a–c; Hoff & Bolsterli, 1956: 178; Hoff, 1958: 35;
Hoff, 1964b: 31; Manley, 1969: 9, fig. 12 (in part, see *Parachelifer monroensis*
Nelson); Nelson, 1971: 95; Nelson, 1975: 292–293, figs 76–77; Sweeney, Stryker,
Sweeney, Taylor & Counts, 1977: 98.

Type locality: Farmington, Washington County, Arkansas, U.S.A.
Distribution: U.S.A. (Arkansas, Georgia, Illinois, Kansas, Kentucky, Michigan,
Mississippi, Missouri, Tennessee, West Virginia).

Dactylochelifer dolichodactylus Caporiacco

Dactylochelifer dolichodactylus Caporiacco, 1939a: 116–117; Roewer, 1940:
348.

Type locality: Mogadiscio, Somalia.
Distribution: Somalia.

Dactylochelifer falsus (Beier)

Chelifer (Chelifer) falsus Beier, 1930h: 96–98, figs 2, 3a–b, 4b.
Dactylochelifer falsus (Beier): Beier, 1932d: 65: Beier, 1932e: 258, figs 274–275;
Roewer, 1937: 315; Beier, 1956e: 303; Beier, 1963a: 262; Beier, 1963b: 288; Callaini,
1987b: 282; Schawaller, 1987a: fig. 11; Callaini, 1988b: 58.

Type locality: Tunisia.
Distribution: Morocco, Sardinia, Sicily, Tunisia.

Dactylochelifer gansuensis Redikorzev

Dactylochelifer gansuensis Redikorzev, 1934c: 3–4, figs 1–3; Roewer, 1937: 315;
Roewer, 1940: 348.

Type locality: Tang-ch'ang (as Tan-chang), Kansu (as Kan-su), China.
Distribution: China (Kansu).

Dactylochelifer gobiensis Beier

Dactylochelifer gobiensis Beier, 1969c: 283–285, figs 1–2; Schawaller, 1986: 12–13,
figs 32, 37.

Type locality: 20 km S. of Somon Delgertsogt (as Delgerzogt), Mongolia.
Distribution: Mongolia, U.S.S.R. (Turkmenistan).

Dactylochelifer gobiensis major Beier

Dactylochelifer gobiensis major Beier, 1974h: 169, fig. 2.

Type locality: Derchin-Tsagan-obo, 60 km ENE. of Bajan-Burd, Mongolia.
Distribution: Mongolia.

Dactylochelifer gracilis Beier

Dactylochelifer gracilis Beier, 1951d: 100–101, fig. 1f; Beier, 1957b: 150; Beier, 1965f: 97.

Type locality: Kerman, Iran.
Distribution: Iran, Turkey.

Dactylochelifer gruberi Beier

Dactylochelifer gruberi Beier, 1969b: 194–196, fig. 2; Schawaller, 1983b: 21, figs 57–58; Schawaller, 1987a: fig. 11; Schawaller & Dashdamirov, 1988: 47, figs 85–88, 92.

Type locality: Kizilcahamam, Turkey.
Distribution: Turkey, U.S.S.R. (Armenia, Azerbaijan, Georgia).

Dactylochelifer infuscatus Beier

Dactylochelifer infuscatus Beier, 1967f: 314–316, fig. 5; Mahnert, 1978c: fig. 9; Schawaller, 1987a: fig. 11.

Type locality: Fethiye, Anatolia, Turkey.
Distribution: Turkey.

Dactylochelifer intermedius Redikorzev

Dactylochelifer intermedius Redikorzev, 1949: 664–665, figs 30–32, 36; Schawaller, 1989: 27, figs 63–64, 67–68.

Type locality: Tuuk River, Ala-tau, Kirghizia, U.S.S.R.
Distribution: U.S.S.R. (Kirghizia, Turkmenistan).

Dactylochelifer kaszabi Beier

Dactylochelifer kaszabi Beier, 1970e: 15–17, fig. 1; Beier, 1974h: 169.

Type locality: 54 km W. of Somon Öndörchangaj, Mongolia.
Distribution: Mongolia.

Dactylochelifer kerzhneri Beier

Dactylochelifer kerzhneri Beier, 1974h: 170–171, fig. 3.

Type locality: Chentej aimak, 12 km N. of Somon Galschir, Mongolia.
Distribution: Mongolia.

Dactylochelifer kussariensis (Daday)

Chelifer kussariensis Daday, 1889a: 20–21, fig. 11.
Dactylochelifer (?) *kussariensis* (Daday): Beier, 1932e: 259; Roewer, 1937: 315.
Dactylochelifer kussariensis (Daday): Beier, 1951d: 99–100, figs 1d–e; Beier, 1955d: 219; Beier, 1959a: 277–278, fig. 19; Beier, 1963b: 288–289, fig. 289; Beier, 1963f: 198; Beier, 1971a: 365; Mahnert, 1978b: figs 11a–b; Schawaller & Dashdamirov, 1988: 47, figs 91–92.

Type locality: 'Kussari', Georgia, U.S.S.R.
Distribution: Afghanistan, Iran, Israel, Lebanon, U.S.S.R. (Georgia).

Dactylochelifer kussariensis arenicola Beier

Dactylochelifer kussariensis arenicola Beier, 1967f: 316–318, fig. 6.
Type locality: Finike, Turkey.
Distribution: Turkey, U.S.S.R. (Azerbaijan).

Dactylochelifer ladakhensis Beier

Dactylochelifer ladakhensis Beier, 1978b: 416–417, figs 1–2.
Type locality: Indus Valley near Shey Gompa, Ladakh, India.
Distribution: India.

Dactylochelifer latreillei (Leach)

Chelifer latreillii Leach, 1817: 49–50, plate 142 fig. 5; Gervais, 1844: 78; O.P.-
Cambridge, 1892: 223–224, plate B figs 13, 13a–b; Evans, 1901a: 53; Kew, 1901b: 194; Kew, 1903: 297, 298, 299, fig. 1; Tullgren, 1906a: 204; Tullgren, 1906b: 214; Ellingsen, 1907a: 164–166; Ellingsen, 1908d: 69; Godfrey, 1908: 157; Ellingsen, 1909a: 206, 209, 215; Godfrey, 1909: 154–157; Ellingsen, 1911c: 173; Kew, 1912: 377–381, figs 47, 50a; Krausse-Heldrungen, 1912: 65; Ellingsen, 1913a: 454; Kew, 1914: 96, 107; Kew, 1916a: 124; Väänänen, 1928a: 19, fig. 5(II); Väänänen, 1928b: figs 1, 3, 5a; Kew, 1929a: 34; Syms, 1950: 143–144, fig. 1.
Chelifer degeerii C. L. Koch, 1835: fasc. 132.3, fig.; Gervais, 1844: 80; Crowther, 1882a: 465; Crowther, 1882b: 277; Kew, 1886: 339.
Chelifer fabricii C. L. Koch, 1835: fasc. 132.4, fig.; C. L. Koch, 1843: 50–51, fig. 786; Gervais, 1844: 80 (synonymized by Beier, 1932e: 261).
Chelifer angustus C. L. Koch, 1837: fasc. 140.5, fig.; Gervais, 1844: 79 (synonymized by C. L. Koch, 1843: 53).
Chelifer degeeri C. L. Koch: C. L. Koch, 1843: 53–54, figs 788–789; E. Simon, 1879a: 22–23, plate 18 fig. 4; Tömösváry, 1882b: 204–205, plate 2 figs 10–11; Canestrini, 1883: no. 2, figs 1–2; Becker, 1884: cclxiii; E. Simon, 1885a: 349; E. Simon, 1885c: 48; Lameere, 1895: 475; E. Simon, 1899d: 86; Daday, 1918: 2 (synonymized by Pavesi, 1884: 455).
Chelifer schaefferi C. L. Koch, 1839: 4; C. L. Koch, 1843: 55–56, fig. 790; C. L. Koch, 1850: 4; Hagen, 1870: 267; L. Koch, 1873: 17–19; Ritsema, 1874: xxxiv; Stecker, 1875d: 88; L. Koch, 1877: 180; E. Simon, 1878a: 146; Daday, 1880a: 192; H. J. Hansen, 1884: 541–542; Pavesi, 1884: 455–456; Krauss, 1896: 628; Tullgren, 1899a: 169–170, plate 1 figs 2–5 (synonymized by E. Simon, 1879a: 22).
Chelifer pediculoides Lucas, 1849: 275–276, plate 18 fig. 6; Hagen, 1870: 267; E. Simon, 1878a: 147 (synonymized by E. Simon, 1885c: 48).
Chelifer brevipalpis Canestrini, 1874: 226 (synonymized by E. Simon, 1885c: 48).
Chelifer ninnii Canestrini, 1874: 227 (synonymized by E. Simon, 1885c: 48).
Chelifer schäfferi C. L. Koch: Stecker, 1874a: 237.
Chelifer rutilans Tömösváry, 1882b: 202–203, plate 1 figs 25–26; Daday, 1888: 122, 177; Daday, 1918: 2 (synonymized by Beier, 1932e: 261).
Chelifer latreillei (Leach): E. Simon, 1898d: 20; Ellingsen, 1908c: 669; Falconer, 1916: 191; Navás, 1921: 165; Navás, 1925: 107; Berland, 1932: fig. 79; Nordberg, 1936: 48–49.
Chelifer latraillii (sic) Leach: Evans, 1905: 247.
Chelifer (Chelifer) latreillii Leach: Ellingsen, 1910a: 386; Kew, 1911a: 47–48, fig. 10; Nonidez, 1917: 21.
Chelifer (Chelifer) latreillei Leach: Kästner, 1928: 8, figs 8, 27; Beier, 1929a: 351, fig. 9a (in part; see *Rhacochelifer corcyrensis* (Beier)); Schenkel, 1928: 57, figs 7–8; Beier, 1930h: fig. 4a.
Chelifer cephalonicus Beier, 1929e: 448–450, fig. 3 (synonymized by Mahnert, 1977b: 68–69).
Chelifer (Chelifer) latreilli Leach: Schenkel, 1929a: 321–322.
Ectoceras latreillei (Leach): Beier, 1930a: fig. 17a; Beier, 1930e: 291; Beier, 1931d: 100.
Ectoceras latreilli (Leach): J. C. Chamberlin, 1931a: 79, 178, figs 42r, 48e, 49; J. C. Chamberlin, 1932a: 21.
Ectoceras cephalonicus (Beier): J. C. Chamberlin, 1932a: 21.

Dactylochelifer latreillei (Leach): Beier, 1932e: 254–255, fig. 261–262; Beier, 1932g: fig. 236; Roewer, 1936: figs 98–99; Roewer, 1937: 315; Schenkel, 1938: 37; Vachon, 1938a: 36–40, figs 16a–b, 17–19, 51a; Beier, 1939e: 312; Beier, 1939f: 200; Lohmander, 1939b: 320–322, fig. 13; Cooreman, 1946b: 7; Kaisila, 1949b: 89, map 14; Beier & Franz, 1954: 458; G. O. Evans & Browning, 1954: 17; Kurir, 1954: 137–138, figs 1–2; Beier, 1956h: 24; Ressl & Beier, 1958: 2; Cloudsley-Thompson, 1960: 50, 51, fig. 1; Beier, 1962b: 146; Vachon, 1961: 103; Meinertz, 1962a: 401, map 8; Kobakhidze, 1964b: 445–447; Kobakhidze, 1965b: 542; Pedder, 1965: 108, 109, figs 5–9; Beier, 1966f: 345–346; Helversen, 1966a: 148; Kobakhidze, 1966: 706; Zangheri, 1966: 532; Hammen, 1969: 23; Weygoldt, 1969a: figs 14, 17, 23, 28, 54, 56, 100, 25a–b, 35, 48a–h, 49a–d, 55; Gabbutt, 1970a: 135–140, figs 2a–d, 3b, 4b; Gabbutt, 1970b: 315–334, figs 1a–d, 2a–f, 3a–e, 4a–e, 5a–i, 6a–e; Heurtault, 1971a: fig. 5a; Howes, 1972a: 108–109; Legg, 1972: 576, figs 1(3a), 2(5); Legg, 1973b: 430, fig. 2b; Beier, 1976a: 25; Legg, 1975a: 66; Legg, 1975d: fig. 2; Ćurčić, 1976b: 179; Crocker, 1978: 9; Jones, 1980c: map 26; Callaini, 1982a: 17; Mahnert, 1981c: fig. 1; Pieper, 1981: 4; Schawaller, 1983b: 21, figs 55–56; Schawaller, 1986: 12, figs 31, 36; Krumpál & Cyprich, 1988: 42; Legg & Jones, 1988: 132–134, figs 34a, 34b(a–i); Schawaller & Dashdamirov, 1988: 47, figs 89–90, 92; Schawaller, 1989: 27.
Dactylochelifer latreilli latreilli (Leach): Vachon, 1936a: 78.
Dactylochelifer latreilli (Leach): Vachon, 1949: fig. 230; Gilbert, 1951: 548, fig. 4; Gilbert, 1952a: 32–44, figs 1–7; Gilbert, 1952b: 47–49, figs 1–3; Jennings, 1968: fig. 4.
Dactylochelifer latreillei latreillei (Leach): Beier, 1932e: 255, fig. 263; Caporiacco, 1948c: 242; Franz & Beier, 1948: 446, 447; Caporiacco, 1951a: 63; Beier, 1952e: 6; Weidner, 1954b: 111; Beier, 1963b: 290–291, figs 291–292; Ressl, 1965: 290; Rafalski, 1967: 20; Lazzeroni, 1969c: 243; Heurtault, 1972b: figs 1, 3; Kofler, 1972: 288–289; Lagar, 1972a: 21; Ćurčić, 1974a: 27; Mahnert, 1974a: 90; Mahnert, 1977b: 68–69; Heurtault, 1981: 213; Mahnert, 1983d: 362; Callaini, 1986e: 254; Jędryczkowski, 1985: 81; Lippold, 1985: 40; Callaini, 1987b: 282–286, figs 2a–d, 3a–d.
Dactylochelifer latreillei cephalonicus (Beier): Beier, 1932e: 255–256, figs 265–266; Caporiacco, 1951e: 69; Beier, 1963b: 290, fig. 290; Lazzeroni, 1969a: 338–339; Lazzeroni, 1970a: 208.
Not *Chelifer latreillei* Leach: McIntire, 1869: 246, figs 211–212; McIntire, 1871: 209–210, fig. 3 (misidentifications; see *Dinocheirus panzeri* (C. L. Koch)).

Type localities: of *Chelifer latreillei*: not stated, presumably England, Great Britain.
of *Chelifer fabricii*: Regensburg, West Germany.
of *Chelifer angustus*: Regensburg, West Germany.
of *Chelifer degeeri*: Regensburg, West Germany.
of *Chelifer schaefferi*: Bayern, West Germany.
of *Chelifer pediculoides*: Skikda (as Philippeville), Algeria; and Lacalle, Algeria.
of *Chelifer brevipalpis*: Cervarese, Padova, Italy.
of *Chelifer ninnii*: Valli, Veneto, Italy.
of *Chelifer rutilans*: Hunyadmegye, Hungary.
of *Chelifer cephalonicus*: near Argostólion (as Argostolion), Kefallinia (as Kephalonia), Greece.
Distribution: Albania, Algeria, Austria, Balearic Islands, Belgium, Corsica, Crete, Czechoslovakia, Denmark, East Germany, Finland, France, Great Britain, Greece, Hungary, Iran, Italy, Madeira Islands, Netherlands, Poland, Romania, Sardinia, Sicily, Spain, Sweden, Tunisia, U.S.S.R. (Armenia, Azerbaijan, Georgia, Kazakhstan, R.S.F.S.R., Ukraine), West Germany, Yugoslavia.

Dactylochelifer latreillei septentrionalis Beier

Dactylochelifer latreillei septentrionalis Beier, 1932d: 64; Beier, 1932e: 255, fig. 264; Beier, 1963b: 291, fig. 293; Rafalski, 1967: 20; Palmgren, 1973: 8–9.

Type locality: Friesland, Netherlands.
Distribution: Finland, Netherlands, Poland, Sweden.

Dactylochelifer lindbergi Beier

Dactylochelifer lindbergi Beier, 1959a: 279–280, figs 22, 23a.

Type locality: Douchi, Afghanistan.
Distribution: Afghanistan.

Dactylochelifer lobatschevi Krumpál & Kiefer

Dactylochelifer lobatschevi Krumpál & Kiefer, 1982: 12–14, plate 6 figs 1–5.

Type locality: Oase Echijn gol, Bajanchongor Ajmak, Mongolia.
Distribution: Mongolia.

Dactylochelifer macrotuberculatus Krumpál

Dactylochelifer macrotuberculatus Krumpál, 1987: 221-226, figs 1–15.

Type locality: Ghunsa, Kangchenjunga region, Nepal.
Distribution: Nepal.

Dactylochelifer marlausicola Dumitresco & Orghidan

Dactylochelifer marlausicolus (sic) Dumitresco & Orghidan, 1969: 679–687, figs 10–25; Dumitresco & Orghidan, 1970a: plate 1 fig. 4, plate 2 fig. 5.

Type locality: Suslanesti, near Mt Cicos, Romania.
Distribution: Romania.

Dactylochelifer maroccanus (Beier)

Ectoceras maroccanum Beier, 1930f: 72–75, figs 2a–b, 3.
Ectoceras marocanus (sic) Beier: Beier, 1930e: 291.
Dactylochelifer maroccanus (Beier): Beier, 1932d: 65; Beier, 1932e: 258, figs 272–273; Beier, 1932g: fig. 224; Roewer, 1936: fig. 151; Roewer, 1937: 315; Vachon, 1938e: 205–207, figs a–c; Mahnert, 1978c: fig. 7; Schawaller, 1987a: fig. 11; Callaini, 1988b: 58.

Type locality: Kenitra (as Kénitra), Oued Fouarrat, Morocco.
Distribution: Morocco.

Dactylochelifer mongolicola Beier

Dactylochelifer mongolicola Beier, 1970e: 17–18, fig. 2; Beier, 1974h: 171–172, fig. 4; Krumpál & Kiefer, 1982: 14; Schawaller, 1989: 27, figs 65–66.

Type locality: 17 km SE. of Ulaangom, Mongolia.
Distribution: Mongolia, U.S.S.R. (R.S.F.S.R.).

Dactylochelifer monticola Beier

Dactylochelifer monticola Beier, 1960a: 43–45, fig. 2; Beier, 1961e: 3.

Type locality: Ichkachim, Badakhchan, Afghanistan.
Distribution: Afghanistan.

Dactylochelifer nubicus Beier

Dactylochelifer nubicus Beier, 1962e: 300–301, fig. 3.

Type locality: Wadi Halfa, Sudan.
Distribution: Sudan.

Dactylochelifer pallidus Beier

Dactylochelifer pallidus Beier, 1963f: 198−199, fig. 9; Mahnert, 1974c: 384; Mahnert, 1978c: fig. 10; Schawaller, 1987a: fig. 11.

Type locality: Hula, Israel.
Distribution: Israel.

Dactylochelifer popovi Redikorzev

Dactylochelifer popovi Redikorzev, 1949: 666−667, figs 33−36; Schawaller, 1986: 14, figs 33, 38; Schawaller, 1989: 29.
Dactylochelifer mrciaki Krumpál, 1984b: 64−66, figs 1−4 (synonymized by Schawaller, 1989: 29).

Type localities: of *Dactylochelifer popovi*: Iolotan, Turkmenistan, U.S.S.R.; Kzyl-Orda, Kazakhstan, U.S.S.R.
of *Dactylochelifer mrciaki*: Bukhara (as Buchara), Kyzylkum, Uzbekistan, U.S.S.R.
Distribution: U.S.S.R. (Kazakhstan, Kirghizia, Tadzhikistan, Turkmenistan, Uzbekistan).

Dactylochelifer redikorzevi (Beier)

Chelifer (Chelifer) redikorzevi Beier, 1929a: 351−354, figs 6a−b, 7a−c, 8.
Ectoceras redikorzevi (Beier): J. C. Chamberlin, 1932a: 21; Beier, 1932a: fig. 23.
Dactylochelifer redikorzevi (Beier): Beier, 1932d: 65; Beier, 1932e: 257, figs 269−270; Roewer, 1937: 315; Redikorzev, 1949: 665, fig. 36; Schawaller, 1986: 14, figs 35, 40; Schawaller, 1989: 29.

Type locality: Margelan, Uzbekistan, U.S.S.R.
Distribution: U.S.S.R. (Kazakhstan, Kirghizia, Uzbekistan).

Dactylochelifer ressli Beier

Dactylochelifer ressli Beier, 1967f: 318−319, fig. 7; Schawaller, 1987a: fig. 11.

Type locality: Köycegiz, Turkey.
Distribution: Turkey.

Dactylochelifer saharensis Heurtault

Dactylochelifer saharensis Heurtault, 1971a: 685−688, figs 1−4, 5b, 6; Schawaller, 1987a: fig. 11.

Type locality: Bardai, Chad.
Distribution: Chad.

Dactylochelifer scaurus Mahnert

Dactylochelifer scaurus Mahnert, 1978c: 152−154, figs 13−18; Schawaller, 1987a: fig. 11.

Type locality: Venta del Aire, Jaen, Spain.
Distribution: Spain.

Dactylochelifer scheuerni Schawaller

Dactylochelifer scheuerni Schawaller, 1987a: 277−279, figs 1−11.

Type locality: Delta del Ebro (as Delta-Ebro), Tarragona, Spain.
Distribution: Spain.

Dactylochelifer shinkaii Sato

Dactylochelifer shinkaii Sato, 1982a: 105−109, figs 1−11.

Type locality: Tanzawa, Kanagawa, Japan.
Distribution: Japan.

Family Cheliferidae

Dactylochelifer silvestris Hoff

Dactylochelifer silvestris Hoff, 1956d: 29–33, figs 16–17; Hoff, 1958: 35; Hoff, 1959b: 5, etc.; Hoff, 1961: 460–462; Muchmore, 1971a: 86, 89; Knowlton, 1972: 1–2; Knowlton, 1974: 1–2; Benedict & Malcolm, 1979: 197.

Type locality: near Escabosa, Bernalillo County, New Mexico, U.S.A.
Distribution: U.S.A. (Colorado, Idaho, New Mexico, Oregon, Utah).

Dactylochelifer somalicus Caporiacco

Dactylochelifer somalicus Caporiacco, 1937a: 149, fig. 8; Roewer, 1940: 348.

Type locality: Ola Uager, Somalia.
Distribution: Somalia.

Dactylochelifer spasskyi Redikorzev

Dactylochelifer spasskyi Redikorzev, 1949: 667–668, fig. 37.

Type locality: Kzyl-Dzhar, Amu Dar'ya, Uzbekistan, U.S.S.R.
Distribution: U.S.S.R. (Uzbekistan).

Dactylochelifer syriacus Beier

Dactylochelifer syriacus Beier, 1955d: 219, figs 8–9; Schawaller, 1987a: fig. 11.

Type locality: N. of Himata, Syria.
Distribution: Lebanon, Syria.

Dactylochelifer vtorovi Mahnert

Dactylochelifer vtorovi Mahnert, 1977a: 89–94, figs 1–11.

Type locality: Terskey Alatau (as Terskey-Ala-Too), Tien-Shan, Kirghizia, U.S.S.R.
Distribution: U.S.S.R. (Kirghizia).

Genus Dichela Menge

Dichela Menge, in C. L. Koch & Berendt, 1854: 96.
Oligochelifer Beier, 1937d: 309–310. New synonymy.

Type species: of *Dichela*: *Dichela berendtii* Menge, in C. L. Koch & Berendt, 1854, by monotypy.
of *Oligochelifer*: *Dichela berendtii* Menge, 1854, by original designation.

! Dichela berendtii Menge

Dichela berendtii Menge, in C. L. Koch & Berendt, 1854: 96.
Chelifer berendtii (Menge): Menge, 1855: 32–33, plate 4 fig. 7; Hagen, 1870: 264; Beier, 1932e: 278.
Oligochelifer berendtii (Menge): Beier, 1937d: 310, fig. 12; Roewer, 1940: 331; Vachon, 1940f: 3; Beier, 1948b: 446; Petrunkevitch, 1955: fig. 49(4); Schawaller, 1978: 4.

Type locality: Baltic Amber.
Distribution: Baltic Amber.

! Dichela gracilis (Beier), new combination

Oligochelifer gracilis Beier, 1937d: 310–311, fig. 13; Roewer, 1940: 331; Schawaller, 1978: 4.

Type locality: Baltic Amber.
Distribution: Baltic Amber.

! **Dichela granulatus** (Beier), new combination

Oligochelifer granulatus Beier, 1937d: 311–312, fig. 15–17; Roewer, 1940: 331; Schawaller, 1978: 4.

Type locality: Baltic Amber.
Distribution: Baltic Amber.

! **Dichela serratidentatus** (Beier), new combination

Oligochelifer serratidentatus Beier, 1937d: 311, fig. 14; Roewer, 1940: 331; Schawaller, 1978: 4.

Type locality: Baltic Amber.
Distribution: Baltic Amber.

Genus **Ectoceras** Stecker

Ectoceras Stecker, 1875b: 512–516; J.C. Chamberlin, 1932a: 21; Beier, 1932e: 277.

Type species: *Ectoceras bidens* Stecker, 1875b, by subsequent designation of J. C. Chamberlin, 1932a: 21.

Ectoceras bidens Stecker

Ectoceras bidens Stecker, 1875b: 518–519, plate 1 figs 10–11, plate 2 figs 1–2, 6–7.
Chelifer bidens (Stecker): With, 1906: 164.
Ectoceras (?) *bidens* Stecker; Beier, 1932e: 277; Roewer, 1937: 317.

Type locality: India.
Distribution: India.

Ectoceras helferi Stecker

Ectoceras helferi Stecker, 1875b: 516–517, plate 1 figs 1–5, 7–9.
Chelifer helferi (Stecker): With, 1906: 164.
Ectoceras (?) *helferi* Stecker: Beier, 1932e: 277; Roewer, 1937: 317.

Type locality: India.
Distribution: India.

Genus **Electrochelifer** Beier

Electrochelifer Beier, 1937d: 312.

Type species: *Electrochelifer mengei* Beier, 1937d, by original designation.

! **Electrochelifer bachofeni** Beier

Electrochelifer bachofeni Beier, 1947e: 194–196, figs 3–5; Schawaller, 1978: 4.

Type locality: Baltic Amber.
Distribution: Baltic Amber.

! **Electrochelifer balticus** Beier

Electrochelifer balticus Beier, 1955g: 53–54, fig. 4; Schawaller, 1978: 4.

Type locality: Baltic Amber.
Distribution: Baltic Amber.

! **Electrochelifer mengei** Beier

Electrochelifer mengei Beier, 1937d: 312–313, fig. 18; Roewer, 1940: 331; Schawaller, 1978: 4.

Type locality: Baltic Amber.
Distribution: Baltic Amber.

! **Electrochelifer rapulitarsus** Beier

Electrochelifer rapulitarsus Beier, 1947e: 196—199, figs 6—8; Schawaller, 1978: 4.

Type locality: Baltic Amber.
Distribution: Baltic Amber.

Genus **Ellingsenius** J. C. Chamberlin

Ellingsenius J. C. Chamberlin, 1932a: 35—36; Beier, 1932e: 274; J. C. Chamberlin, 1949: 49—51; Murthy & Ananthakrishnan, 1977: 158.

Type species: *Chelifer sculpturatus* Lewis, 1903, by original designation.

Ellingsenius fulleri (Hewitt & Godfrey)

Chelifer exiguus Tullgren: Ellingsen, 1912b: 83, 92 (misidentification).
Chelifer fulleri Hewitt & Godfrey, 1929: 331—334, figs 1a, 7a—g, plate 22 fig. 11.
Ellingsenius fulleri (Hewitt & Godfrey): J. C. Chamberlin, 1932a: 36; Beier, 1932e: 275; Roewer, 1937: 316; Beier, 1948b: 446, 459; Newlands, 1978: 700.

Type locality: Dunbrody, Cape Province, South Africa.
Distribution: South Africa.

Ellingsenius globosus Beier

Ellingsenius globosus Beier, 1962f: 35—36, fig. 17.

Type locality: Kigali, Rwanda.
Distribution: Rwanda.

Ellingsenius hendrickxi Vachon

Ellingsenius hendrickxi Vachon, 1954c: 284—285, fig. 1; Vachon, 1954d: 591; Weygoldt, 1969a: fig. 107.

Type locality: Kivu, Zaire.
Distribution: Zaire.

Ellingsenius indicus J. C. Chamberlin

Chelifer sp.: J. C. Chamberlin, 1931a: fig. 18t.
Ellingsenius indicus J. C. Chamberlin, 1932a: 36—37; Beier, 1932e: 276—277; Beier, 1937c: 633—634, figs 1—2; Roewer, 1937: 316; Beier, 1948b: 459; J. C. Chamberlin, 1949: 52—57, figs a—k; Murthy & Ananthakrishnan, 1977: 158.

Type locality: Ootacamund, India.
Distribution: India.

Ellingsenius perpustulatus Beier

Ellingsenius perpustulatus Beier, 1962f: 36—37, fig. 18; Mahnert, 1988a: 76—77, figs 76—79.

Type locality: Mugaga, Kenya.
Distribution: Kenya.

Ellingsenius sculpturatus (Lewis)

Chelifer sculpturatus Lewis, 1903: 497—498, figs 1—3; With, 1905: 117—123, 136—137, plate 8 figs 2a—h; Ellingsen, 1912b: 84, 99—100; Godfrey, 1921: 118; Bellhouse Lemare, 1923: 232; Hewitt & Godfrey, 1929: 327—331, figs 6a—e, plate 22 fig. 10; J. C. Chamberlin, 1931a: 68, fig. 30l; Garin, 1937: 33—47, figs 1—9.
Ellingsenius sculpturatus (Lewis): J. C. Chamberlin, 1932a: 36; Beier, 1932e: 275—276, figs 299—300; Beier, 1932g: fig. 267; Roewer, 1936: fig. 28b; Roewer, 1937: 316, fig. 201a; Beier, 1947b: 338; Beier, 1948b: 446, 459; Beier, 1954d: 139; Beier, 1958a:

186; Hoff, 1958: 36; Beier, 1964k: 90; Lawrence, 1967: 90; Muchmore, 1971a: 88; Newlands, 1978: 700.

Type locality: Natal, South Africa.
Distribution: Namibia, South Africa, U.S.A. (California), Zaire, Zimbabwe.

Ellingsenius somalicus Beier

Ellingsenius somalicus Beier, 1932d: 66−67; Beier, 1932e: 275, fig. 298; Roewer, 1937: 316; Beier, 1948b: 459.

Type locality: Afmadù, Somalia.
Distribution: Somalia.

Ellingsenius ugandanus Beier

Ellingsenius ugandanus Beier, 1935a: 488−489, fig. 4; Roewer, 1937: 316; Beier, 1948b: 459.

Type locality: Uganda.
Distribution: Uganda.

Genus Eremochernes Beier

Eremochernes Beier, 1932e: 163; Beier, 1933a: 532.

Type species: *Chelifer (Chelanops) gracilipes* Redikorzev, 1922a, by original designation.

Eremochernes gracilipes (Redikorzev)

Chelifer (Chelanops) gracilipes Redikorzev, 1922a: 265−267, figs 7, 7a, 8.
Eremochernes gracilipes (Redikorzev): Beier, 1932e: 163−164, fig. 176; Roewer, 1937: 298; Beier, 1974h: 172.

Type locality: Sjao-Shao-Bashuj, Gobi Desert, Mongolia.
Distribution: Mongolia.

Eremochernes secundus Beier

Eremochernes secundus Beier, 1937b: 273−274, fig. 4; Roewer, 1940: 347.

Type locality: Taifan, Tien-shan, China.
Distribution: China (Liaoning).

Eremochernes tropicus Beier

Eremochernes tropicus Beier, 1967g: 357−358, fig. 19; Tenorio & Muchmore, 1982: 384.

Type locality: Doi Suthep, Chiang Mai (as Chiangmai), Thailand.
Distribution: Thailand.

Genus Florichelifer Hoff

Florichelifer Hoff, 1964b: 20−21.

Type species: *Florichelifer aureus* Hoff, 1964b, by original designation.

Florichelifer aureus Hoff

Florichelifer aureus Hoff, 1964b: 21−24, figs 7−10.

Type locality: near Archbold Biological Station, Highlands County, Florida, U.S.A.
Distribution: U.S.A. (Florida).

Family Cheliferidae

Genus **Gobichelifer** Krumpál

Gobichelifer Krumpál, 1979b: 667–668.

Type species: *Gobichelifer dashdorzhi* Krumpál, 1979b (junior synonym of *Chelifer chelanops* Redikorzev, 1922a), by original designation.

Gobichelifer chelanops (Redikorzev)

Chelifer chelanops Redikorzev, 1922a: 267–270, figs 9–11; Caporiacco, 1935: 243.
Rhacochelifer chelanops (Redikorzev): Beier, 1932d: 66; Beier, 1932e: 267, fig. 290; Roewer, 1937: 316; Redikorzev, 1949: 668.
Chelifer semenovi Redikorzev, 1934a: 430–432, figs 8–10; Roewer, 1936: fig. 66d (synonymized by Schawaller, 1989: 26).
Rhacochelifer semenovi (Redikorzev): Redikorzev, 1934b: 152; Roewer, 1937: 316; Redikorzev, 1949: 668.
Gobichelifer dashdorzhi Krumpál, 1979b: 668–671, figs 1–12 (synonymized by Schawaller, 1986: 11).
Gobichelifer semenovi (Redikorzev): Schawaller, 1986: 11–12, figs 29–30.
Gobichelifer chelanops (Redikorzev): Schawaller, 1989: 26, figs 57–62.

Type localities: of *Chelifer chelanops*: Sussa-myr River, Alexandrovskij Chain, Pishpek, Semiretshje, U.S.S.R.
of *Chelifer semenovi*: Kischlak Su-Tshar, Turkestan, Kazakhstan, U.S.S.R.
of *Gobichelifer dashdorzhi*: Oesis Echijn gol, Bajanchongor Ajmak, Mongolia.
Distribution: Mongolia, Pakistan, U.S.S.R. (Kirghizia, Kazakhstan, Tadzhikistan, Uzbekistan).

Genus **Hansenius** J. C. Chamberlin

Hansenius J. C. Chamberlin, 1932a: 20–21; Beier, 1932e: 269.

Type species: *Chelifer kewi* Ellingsen, 1907a (junior synonym of *Chelifer torulosus* Tullgren, 1907a), by original designation.

Hansenius basilewskyi Beier

Hansenius basilewskyi Beier, 1962f: 32–33, fig. 15.

Type locality: Kajiado, Kenya.
Distribution: Kenya.

Hansenius fuelleborni (Ellingsen)

Chelifer (Chelifer) kewi fülleborni Ellingsen, 1910a: 385–386.
Chelifer kewi fülleborni Ellingsen: Ellingsen, 1912b: 84.
Hansenius fülleborni (Ellingsen): Beier, 1932e: 270, figs 292–293; Roewer, 1937: 316; Beier, 1955h: 9; Beier, 1959d: 67; Beier, 1962f: 31; Beier, 1967d: 91.
Hansenius cf. *fuelleborni* (Ellingsen): Mahnert, 1984b: 61.
Hansenius fuelleborni (Ellingsen): Mahnert, 1988a: 78–80, figs 80–82.

Type localities: Langenburg, East Africa; and Takaunga (as Takanuga), Kenya.
Distribution: Kenya, Somalia, Tanzania, Zaire.

Hansenius jezequeli Heurtault

Hansenius jezequeli Heurtault, 1983: 2–6, figs 1–11.

Type locality: Lamto, Ivory Coast.
Distribution: Ivory Coast.

Hansenius kilimanjaricus Beier

Hansenius kilimanjaricus Beier, 1962f: 33–34, fig. 16; Beier, 1967d: 91; Mahnert, 1988a: 80–82, figs 83–84.

Type locality: Marangu, Kilimanjaro, Tanzania.
Distribution: Kenya, Tanzania.

Hansenius leleupi Beier

Hansenius leleupi Beier, 1959d: 65–67, fig. 34.

Type locality: Ruzizi, Uvira, Itombwe, Zaire.
Distribution: Zaire.

Hansenius major Beier

Hansenius torulosus major Beier, 1947b: 336–337, fig. 40.
Hansenius major Beier: Beier, 1964k: 84.

Type locality: Doorn River, Montagu, Cape Province, South Africa.
Distribution: South Africa.

Hansenius milloti Vachon

Hansenius milloti Vachon, 1937c: 307–309, figs 1–3; Roewer, 1940: 348; Vachon, 1952a: 41; Heurtault, 1983: 7, figs 12–15.

Type locality: Gabon.
Distribution: Gabon, Guinea, Ivory Coast.

Hansenius mirabilis Beier

Hansenius mirabilis Beier, 1933d: 645–647, figs 2–3; Roewer, 1937: 316.

Type locality: Morogoro, Tanzania.
Distribution: Tanzania.

Hansenius regneri Beier

Hansenius regneri Beier, 1944: 210–211, fig. 23; Beier, 1955c: 558; Beier, 1967d: 91; Mahnert, 1988a: 82–84, figs 87–88.

Type locality: Dar es Salaam (as Daressalam), Tanzania.
Distribution: Kenya, Tanzania.

Hansenius schoutedeni Beier

Hansenius schoutedeni Beier, 1954d: 138–139, fig. 6.

Type locality: Komi, Zaire.
Distribution: Zaire.

Hansenius socotrensis (With)

Chelifer socotrensis With, 1905: 112–116, plate 7 figs 4a–h; With, 1906: 50. 3 Ellingsen, 1912b: 85.
Hansenius socotrensis (With): J. C. Chamberlin, 1932a: 21; Beier, 1932e: 271; Roewer, 1937: 316.

Type locality: Socotra.
Distribution: Socotra.

Hansenius spinosus J. C. Chamberlin

Hansenius spinosus J. C. Chamberlin, 1949: 29–32, figs 9a–k.

Type locality: Zambi, Zaire.
Distribution: Zaire.

Hansenius torulosus (Tullgren)

Chelifer torulosus Tullgren, 1907a: 32–35, figs 4a–f; Tullgren, 1907b: 220; Ellingsen, 1912b: 85; Hewitt & Godfrey, 322–325, fig. 5a, plate 22, fig. 8.

Chelifer kewi Ellingsen, 1907a: 162–164; Ellingsen, 1912b: 83, 93; Godfrey, 1927: 18 (synonymized by Beier, 1955l: 320).
Hansenius kewi (Ellingsen): J. C. Chamberlin, 1932a: 20; Beier, 1932e: 269; Roewer, 1937: 316.
Hansenius torulosus (Tullgren): Beier, 1932e: 271, figs 294–295; Roewer, 1937: 316; Beier, 1947b: 336; Beier, 1955l: 320; Beier, 1958a: 185; Weidner, 1959: 114; Beier, 1964k: 84; Beier, 1966k: 469, 470; Lawrence, 1967: 90.

Type localities: of *Chelifer torulosus*: Port Elizabeth, Cape Province, South Africa. of *Chelifer kewi*: Nieuwveldt, Witte Hardt, Cape Province, South Africa.
Distribution: South Africa.

Hansenius vosseleri Beier

Hansenius vosseleri Beier, 1944: 211–212, fig. 24; Mahnert, 1988a: 82, figs 85–86.
Type locality: Sadani, Tanzania.
Distribution: Kenya, Tanzania.

Genus **Haplochelifer** J. C. Chamberlin

Haplochelifer J. C. Chamberlin, 1932a: 20; Beier, 1932e: 228; J. C. Chamberlin, 1952: 305–306; Hoff, 1956d: 3.
Type species: *Chelifer philipi* J. C. Chamberlin, 1923c, by original designation.

Haplochelifer philipi (J. C. Chamberlin)

Chelifer philipi J. C. Chamberlin, 1923c: 374, plate 2 fig. 8, plate 3 figs 19, 21, 26.
Haplochelifer philipi (J. C. Chamberlin): J. C. Chamberlin, 1932a: 20; Beier, 1932e: 228; Roewer, 1937: 312; Hoff, 1950a: 4; J. C. Chamberlin, 1952: 306–312, figs 15a–b, plates 19–20; Hoff, 1956d: 4; Hoff, 1958: 35; Hoff, 1959b: 5, etc.; Hoff, 1961: 459–460; Knowlton, 1972: 2; Knowlton, 1974: 2; Benedict & Malcolm, 1979: 193–195.

Type locality: Stanford University, California, U.S.A.
Distribution: U.S.A. (Arizona, California, Colorado, Idaho, Nevada, New Mexico, Oregon, Utah).

Genus **Hygrochelifer** Murthy & Ananthakrishnan

Hygrochelifer Murthy & Ananthakrishnan, 1977: 152–153.
Type species: *Hygrochelifer indicus* Murthy & Ananthakrishnan, 1977, by original designation.

Hygrochelifer hoffi Murthy & Ananthakrishnan

Hygrochelifer hoffiae (sic) Murthy & Ananthakrishnan, 1977: 155–157, figs 49a–c.
Type locality: Tambram, Madras, Tamil Nadu, India.
Distribution: India.

Hygrochelifer indicus Murthy & Ananthakrishnan

Hygrochelifer indicus Murthy & Ananthakrishnan, 1977: 153–155, figs 48a–e.
Type locality: Adayar, Tamil Nadu, India.
Distribution: India.

Genus **Hysterochelifer** J. C. Chamberlin

Hysterochelifer J. C. Chamberlin, 1932a: 19; Beier, 1932e: 230; Hoff, 1956d: 9–11; Beier, 1963b: 283.
Karachelifer Hadži, 1938: 31–36 (synonymized by Beier, 1949: 15).

Type species: of *Hysterochelifer*: *Chelifer fuscipes* Banks, 1909b, by original designation.
of *Karachelifer*: *Karachelifer karamani* Hadži, 1938, by original designation.

Hysterochelifer afghanicus Beier
Hysterochelifer afghanicus Beier, 1966i: 259–260, fig. 1; Beier, 1971a: 364.

Type locality: Bala Murghab, Afghanistan.
Distribution: Afghanistan, Iran.

Hysterochelifer allocancroides Redikorzev
Hysterochelifer allocancroides Redikorzev, 1949: 658–660, figs 23–26; Schawaller, 1986: 10, fig. 28.
Centrochelifer allocancroides (Redikorzev): Beier, 1959a: 276.

Type localities: Urgench, Uzbekistan, U.S.S.R.; and Fergana, Uzbekistan, U.S.S.R.
Distribution: U.S.S.R. (Turkmenistan, Uzbekistan).

Hysterochelifer cyprius (Beier)
Chelifer (Chelifer) cyprius Beier, 1929a: 349–351, fig. 5a.
Hysterochelifer cyprius (Beier): Beier, 1932e: 232–233, fig. 239; Roewer, 1937: 312; Beier & Turk, 1952: 771; Beier, 1963b: 284–285, fig. 286; Beier, 1963f: 197; Beier, 1965f: 96–97, fig. 10; Beier, 1967f: 314; Beier, 1969b: 193; Mahnert, 1977b: 68; Halperin & Mahnert, 1987: 128.

Type locality: Prodhromos (as Prodoromus), Cyprus.
Distribution: Cyprus, Greece, Israel, Lebanon, Turkey.

Hysterochelifer distinguendus (Beier)
Chelifer (Chelifer) distinguendus Beier, 1929a: 348–349, fig. 4.
Hysterochelifer distinguendus (Beier): Beier, 1932e: 231–232, fig. 237; Roewer, 1937: 312.

Type locality: Haifa, Israel.
Distribution: Israel.

Hysterochelifer fuscipes (Banks)
Chelifer fuscipes Banks, 1909b: 303; Banks, 1911: 637; Hilton, 1913: 190, figs 3f–g, 4n; Moles, 1914c: 188–192, figs 1–2; Moles & Moore, 1921: 6; J. C. Chamberlin, 1931a: 107, figs 11b, 30o, 40x; Hilton, 1931: figs 105f–g, 106n.
Hysterochelifer fuscipes (Banks): J. C. Chamberlin, 1932a: 19; Beier, 1932e: 230; Roewer, 1937: 312; Hoff, 1958: 34; Benedict & Malcolm, 1979: 195.

Type locality: Claremont, California, U.S.A.
Distribution: U.S.A. (California, Oregon).

Hysterochelifer geronimoensis (J. C. Chamberlin)
Chelifer geronimoensis J. C. Chamberlin, 1923c: 373, plate 2 fig. 6.
Hysterochelifer geronimoensis (J. C. Chamberlin): J. C. Chamberlin, 1932a: 20; Beier, 1932e: 232; Roewer, 1937: 312.

Type locality: San Geronimo Island, Baja California, Mexico.
Distribution: Mexico.

Hysterochelifer gracilimanus Beier
Hysterochelifer gracilimanus Beier, 1949: 13–15, fig. 11; Beier, 1965f: 97; Beier, 1969b: 194; Rack, 1971: 113; Mahnert, 1977b: 68; Halperin & Mahnert, 1987: 128.

Type locality: Dalyanm Köycegiz, Anatolia, Turkey.
Distribution: Greece, Israel, Turkey.

Hysterochelifer meridianus (L. Koch)

Chelifer meridianus L. Koch, 1873: 20–21; Canestrini, 1874: 225; E. Simon, 1878a: 146; E. Simon, 1879a: 25–26, 312, plate 18 fig. 3; E. Simon, 1882b: 365; Canestrini, 1883: no. 4, figs 1–4; Pavesi, 1884: 456; E. Simon, 1885a: 349; E. Simon, 1885c: 47; E. Simon, 1898c: 3, 4; E. Simon, 1900b: 593; Ellingsen, 1905d: 1–2; Ellingsen, 1909a: 206, 209; Krausse-Heldrungen, 1912: 65; Navás, 1918: 88–89; Navás, 1921: 165; Navás, 1923: 31; Caporiacco, 1923: 131; Navás, 1925: 107; Caporiacco, 1928b: 126; Caporiacco, 1929: 240; Beier, 1931d: 100.

Chelifer danaus Tömösváry, 1884: 19–20, fig. 7; Daday, 1888: 123, 179 (synonymized by Beier, 1929a: 347).

Chelifer (Chelifer) meridianus L. Koch: Balzan, 1892: 532, fig. 23; Ellingsen, 1910a: 386; Nonidez, 1917: 21; Beier, 1929a: 348, fig. 5b; Beier, 1929d: 155.

Hysterochelifer meridionalis (sic) (L. Koch): J. C. Chamberlin, 1932a: 20.

Hysterochelifer meridianus (L. Koch): Beier, 1932e: 231, fig. 236; Vachon, 1936a: 78; Roewer, 1937: 312; Vachon, 1938a: fig. 15b; Beier, 1939f: 199; Redikorzev, 1949: 658; Caporiacco, 1950: 116; Beier, 1956e: 303; Beier, 1961b: 37; Beier, 1963b: 283–284, fig. 285; Beier, 1963g: 274; Boissin, 1964: 651–668, figs 1–25; Dumitresco & Orghidan, 1964: 626–629, figs 49–52, plate 1 fig. 7; Orghidan & Dumitresco, 1964a: 328–331, figs 3–4, 6; Orghidan & Dumitresco, 1964b: 192–195, fig. 1; Kobakhidze, 1965b: 541; Beier, 1966f: 345; Boissin & Manier, 1966a: 470–475, plates 1–4; Boissin & Manier, 1966b: 699–705, figs 1–2, plates 1–2; Dumitresco & Orghidan, 1966: 81–83; Kobakhidze, 1966: 705; Tuzet, Manier & Boissin, 1966: 376–378, plate 1 figs 1–2, plate 2 figs 1–5; Boissin, 1967: 480–486, figs 1–8; Boissin & Manier, 1967: 706–710, fig. 1, plates 1–3; Beron, 1968: 105; Beier, 1969b: 193; Boissin & Cazal, 1969: 264–267, plates 1–3; Lazzeroni, 1969a: 338; Lazzeroni, 1969b: 409–410; Lazzeroni, 1969c: 242; Weygoldt, 1969a: 25, 35, 57, 65, 100, 115; Boissin, Bouix & Maurand, 1970: 491–500, fig. 1, plates 1–2; Boissin & Manier, 1970: 49–53, plates 1–2; Dumitresco & Orghidan, 1970a: plate 1 fig. 3; Lazzeroni, 1970a: 208; Boissin, 1971: 114–122, plates 1–2; Schuster, 1972: 239; Anonymous, 1983a: 27; Boissin, 1973: 522–526, fig. 1; Ćurčić, 1974a: 28; Gardini, 1975: 10; Mahnert, 1977b: 68; Heurtault, 1981: 213; Callaini, 1986e: 254; Callaini, 1987b: 275; Callaini, 1988b: 57; Schawaller & Dashdamirov, 1988: 45, figs 80, 82–84.

Karachelifer karamani Hadži, 1938: 31–36, figs 29a–e, 30a–b (synonymized by Beier, 1949: 15).

Hyterochelifer (sic) *meridianus* (L. Koch): Beier, 1955n: 111.

Histerochelifer (sic) *meridianus* (L. Koch): Anonymous, 1983b: 27.

Not *Chelifer meridianus* L. Koch: O.P.-Cambridge, 1892: 221–222 (misidentication; see *Chernes cimicoides* (Fabricius)).

Type localities: of *Chelifer meridianus*: Corsica; Rome, Italy; and Greece.
of *Chelifer danaus*: Kérkira (as Corfu), Greece; and Peloponnisos (as Morea), Greece.
of *Karachelifer karamani*: Skopje (as Skoplje), Yugoslavia.
Distribution: Algeria, Bulgaria, Canary Islands, Crete, France, Greece, Italy, Morocco, Romania, Sardinia, Spain, Tunisia, Turkey, U.S.S.R. (Azerbaijan, Georgia, Kazakhstan, Turkmenistan, Uzbekistan), Yugoslavia.

Hysterochelifer nepalensis Beier

Hysterochelifer nepalensis Beier, 1974b: 278–279, fig. 12; Schawaller, 1983a: 110; Schawaller, 1987a: 217.

Type locality: Gara, Nepal.
Distribution: Nepal.

Hysterochelifer orientalis Beier

Hysterochelifer orientalis Beier, 1967g: 358, fig. 20; Tenorio & Muchmore, 1982: 379.

Type locality: Doi Suthep, Chiang Mai (as Chiangmai), Thailand.
Distribution: Thailand.

Hysterochelifer pauliani Vachon

Hysterochelifer pauliani Vachon, 1938e: 207–209, figs d–h.

Type locality: Sidi Chamarouch, Mt Toubkal, Morocco.
Distribution: Morocco.

Hysterochelifer proprius Hoff

Hysterochelifer proprius Hoff, 1950a: 4–9, figs 1–5; Hoff, 1956d: 11–12; Hoff, 1958: 34; Hoff, 1959b: 5, etc.; Hoff, 1961: 458–459; Benedict & Malcolm, 1979: 195–196.

Type locality: Flagstaff, Arizona, U.S.A.
Distribution: U.S.A. (Arizona, Colorado, New Mexico, Oregon).

Hysterochelifer spinosus (Beier)

Chelifer (Chelifer) spinosus Beier, 1930h: 94–96, fig. 1.
Hysterochelifer spinosus (Beier): Beier, 1932e: 232, fig. 238; Roewer, 1937: 312.

Type locality: Tunisia.
Distribution: Tunisia.

Hysterochelifer tauricus Beier

Hysterochelifer tauricus Beier, 1963g: 274–275, fig. 5; Beier, 1965f: 97; Beier, 1967f: 314; Beier, 1969b: 194.

Type locality: Pozanti-Karakuz, Toros Daglari (as Taurus), Turkey.
Distribution: Turkey.

Hysterochelifer tuberculatus (Lucas)

Chelifer tuberculatus Lucas, 1849: 274–275, plate 18 fig. 5; Hagen, 1870: 267; E. Simon, 1878a: 147; E. Simon, 1885c: 47; Ellingsen, 1908c: 669; Ellingsen, 1909a: 206, 208; Ellingsen, 1910b: 63; Krausse-Heldrungen, 1912: 65; Ellingsen, 1913a: 454; Navás, 1919: 211; Navás, 1923: 31; Navás, 1925: 107; Beier, 1930f: 72; J. C. Chamberlin, 1931a: 126, figs 39k–l.
Chelifer brachydactylus Lucas, 1849: 273–274, plate 18 fig. 4; Hagen, 1870: 264; E. Simon, 1878a: 147 (synonymized by Beier, 1932e: 233).
Chelifer lampropsalis L. Koch, 1873: 19–20; Canestrini, 1874: 224; E. Simon, 1878a: 146; E. Simon, 1879a: 26–27, plate 18 fig. 1; Canestrini, 1883: no. 5, figs 1–2; Daday, 1889d: 80; E. Simon, 1898d: 20; E. Simon, 1899d: 86; With, 1906: 49, figs 4a–b, plate 2 fig. 2a (synonymized by Ellingsen, 1908c: 669).
Chelifer (Chelifer) tuberculatus Lucas: Ellingsen, 1910a: 387; Nonidez, 1917: 21–22; Kästner, 1928: 7, fig. 24; Beier, 1929a: 348; Beier, 1929e: 448.
Chelifer (Chelifer) lampropsalis L. Koch: Lessert, 1911: 21–22, fig. 17.
Hysterochelifer tuberculatus (Lucas): J. C. Chamberlin, 1932a: 20; Beier, 1932e: 233–234, figs 240–241; Caporiacco, 1936b: 330; Roewer, 1936: figs 45a–b, 69, 93; Vachon, 1936a: 78; Roewer, 1937: 312; Vachon, 1937f: 128; Beier, 1939f: 199; Beier, 1948a: 191; Caporiacco, 1948c: 242; Caporiacco, 1951a: 63; Beier, 1953a: 302; Beier, 1955n: 111; Beier, 1959f: 131; Beier, 1961b: 37–38; Beier, 1963a: 262; Beier, 1963f: 197; Beier, 1963g: 274; Beier, 1963h: 156; Beier, 1963i: 286; Beier, 1965f: 96; Zangheri, 1966: 532; Beier, 1969b: 194; Lazzeroni, 1969a: 338; Weygoldt, 1969a: 57, 100, 115; Lazzeroni, 1970a: 208; Ćurčić, 1974a: 28;

Mahnert, 1975c: 196; Estany, 1977b: 31–32; Leclerc & Heurtault, 1979: 246; Callaini, 1982a: 17, 23, fig. 4d; Schawaller, 1981b: 49, 50; Callaini, 1982b: 449; Mahnert, 1982f: 297; Schawaller, in Schmalfuss & Schawaller, 1984: 6–7; Mahnert, 1986d: 82; Callaini, 1987b: 278–282; Callaini, 1988b: 57–58; Callaini, 1989: 145.
Histerochelifer (sic) *tuberculatus* (Lucas): Vachon, 1940g: 145.
Hysterochelifer tuberculatus tuberculatus (Lucas): Beier, 1963b: 285, fig. 287; Lazzeroni, 1969b: 410; Lazzeroni, 1969c: 242–243; Lazzeroni, 1970c: 110–111; Lagar, 1972b: 51; Gardini, 1975: 9–10; Mahnert, 1977b: 68; Callaini, 1979b: 140; Callaini, 1979c: 351; Gardini, 1980c: 119–120, 132; Callaini, 1983a: 156–157; Callaini, 1983c: 310–311; Callaini, 1983d: 224; Callaini, 1983e: 225.

Type localities: of *Chelifer tuberculatus*: Oran, Algeria.
of *Chelifer brachydactylus*: 'Annaba (as Bône), Algeria.
of *Chelifer lampropsalis*: Corsica; Vaucluse, Dryne, France; Rome, Italy; and Merano, Italy.
Distribution: Algeria, Balearic Islands, Corsica, France, Greece, Israel, Italy, Malta, Morocco, Portugal?, Sardinia, Sicily, Spain?, Switzerland, Tunisia, Turkey, U.S.S.R. (R.S.F.S.R.), Yugoslavia.

Hysterochelifer tuberculatus ibericus Beier

Hysterochelifer tuberculatus ibericus Beier, 1955n: 111–112; Beier, 1963b: 285.
Hysterochelifer tuberculatus hibericus (sic) Beier: Orghidan, Dumitresco & Georgesco, 1975: 31.

Type localities: Los Molinos, Spain; Casa del Campo, Spain; Valbonne, Spain; and Costa Brava, 8 km S. of San Feliú, Spain.
Distribution: Balearic Islands, Spain.

Hysterochelifer urbanus Hoff

Hysterochelifer urbanus Hoff, 1956d: 12–18, figs 1–5; Hoff, 1958: 34; Hoff, 1959b: 5, etc.

Type locality: Albequerque, Bernalillo County, New Mexico, U.S.A.
Distribution: U.S.A. (New Mexico).

Genus **Idiochelifer** J. C. Chamberlin

Idiochelifer J.C. Chamberlin, 1932a: 19; Beier, 1932e: 228–229; Hoff, 1946f: 486; Hoff, 1949b: 487.

Type species: *Chelifer cancroides nigripalpus* Ewing, 1911, by original designation.

Idiochelifer nigripalpus (Ewing)

Chelifer cancroides nigripalpus Ewing, 1911: 73.
Chelifer nigripalpus Ewing: J. C. Chamberlin, 1931a: figs 11n–o.
Idiochelifer nigripalpus (Ewing): J. C. Chamberlin, 1932a: 19; Beier, 1932e: 229; Roewer, 1937: 312; Hoff, 1946c: 26–28; Hoff, 1949b: 487–489, figs 49a–c; Hoff, 1950a: 3–4; Hoff & Bolsterli, 1956: 177; Hoff, 1958: 34; Manley, 1969: 9, fig. 13; Nelson, 1975: 297, figs 85–89.
Hysterochelifer longidactylus Hoff, 1945g: 511–515, figs 1–6 (synonymized by Hoff, 1946f: 486).
Hysterochelifer sp. near *longidactylus* Hoff: Théodoridès, 1950: 14.

Type localities: of *Chelifer cancroides nigripalpus*: Ames, Iowa, U.S.A.
of *Hysterochelifer longidactylus*: 1.5 miles NE. of Mahomet, Champaign County, Illinois, U.S.A.
Distribution: U.S.A. (Arkansas, Illinois, Indiana, Iowa, Michigan, Missouri, Wisconsin).

Genus **Juxtachelifer** Hoff

Juxtachelifer Hoff, 1956d: 23.

Type species: *Juxtachelifer fructuosus* Hoff, 1956d, by original designation.

Juxtachelifer fructuosus Hoff

Juxtachelifer fructuosus Hoff, 1956d: 23–28, figs 11–15; Hoff, 1958: 35; Hoff, 1959b: 5, etc.; Muchmore, 1971a: 91.

Type locality: Santa Fe, Santa Fe County, New Mexico, U.S.A.
Distribution: U.S.A. (New Mexico).

Genus **Kashimachelifer** Morikawa

Kashimachelifer Morikawa, 1957b: 399–400, 401; Morikawa, 1960: 151.

Type species: *Kashimachelifer cinnamomeus* Morikawa, 1957b, by original designation.

Kashimachelifer cinnamomeus Morikawa

Kashimachelifer cinnamomeus Morikawa, 1957b: 400, 401–402, figs 1a–h; Morikawa, 1960: 151–152, plate 4 fig. 4, plate 7 fig. 18, plate 6 fig. 13, plate 8 fig. 17, plate 9 fig. 27, plate 10 figs 15, 31; Morikawa, 1962: 419.

Type locality: Kashima Islet, Bay of Tanabe, Wakayama, Honshú, Japan.
Distribution: Japan.

Genus **Levichelifer** Hoff

Levichelifer Hoff, 1946f: 486–487; Hoff, 1956d: 8.
Ocalachelifer J. C. Chamberlin, 1949: 17–19; Hoff, 1964b: 17–18 (synonymized by Muchmore, 1981c: 336).

Type species: of *Levichelifer*: *Idiochelifer fulvopalpus* Hoff, 1946c, by original designation.
of *Ocalachelifer*: *Ocalachelifer cribratus* J. C. Chamberlin, 1949, by original designation.

Levichelifer cribratus (J. C. Chamberlin)

Ocalachelifer cribratus J. C. Chamberlin, 1949: 19–23, fig. 6a–o; Hoff, 1958: 35; Hoff, 1964b: 18–20; Muchmore, 1971a: 86.
Levichelifer cribratus (J. C. Chamberlin): Muchmore, 1981c: 336.

Type locality: Ocala National Forest, Florida, U.S.A.
Distribution: U.S.A. (Alabama, Florida).

Levichelifer fulvopalpus (Hoff)

Idiochelifer fulvopalpus Hoff, 1946c: 23–26, figs 30–32.
Levichelifer fulvopalpus (Hoff): Hoff, 1946f: 487; Hoff, 1950a: 15–17; Hoff, 1956d: 8–9; Hoff, 1958: 34; Hoff, 1959b: 5, etc.; Muchmore, 1971a: 91; Rowland & Reddell, 1976: 18; Muchmore, 1981c: 336.

Type locality: Reynosa, Tamaulipas, Mexico.
Distribution: Mexico, U.S.A. (New Mexico, Texas).

Genus **Lissochelifer** J. C. Chamberlin

Lissochelifer J. C. Chamberlin, 1932a: 20.
Lophochelifer Beier, 1940a: 189–190; Murthy & Ananthakrishnan, 1977: 161. New synonymy.

Type species: of *Lissochelifer*: *Chelifer mortensenii* With, 1906, by original designation.
of *Lophochelifer*: *Lophochelifer insularis* Beier, 1940a, by original designation.

Lissochelifer depressoides (Beier), new combination

Lophochelifer depressoides Beier, 1967g: 362–363, fig. 23; Beier, 1973b: 52–53, fig. 17.

Type locality: 19 km E. of Virajpet, Mysore, India.
Distribution: India, Sri Lanka.

Lissochelifer depressus (C. L. Koch)

Chelifer depressus C. L. Koch, 1843: 57–59, fig. 792; Hagen, 1870: 265; H. J. Hansen,
1884: 540–541; Tullgren, 1899a: 168–169; With, 1906: 47–48, 144–147, plate 2
fig. 10a, plate 3 figs 2a–e; Ellingsen, 1914: 9.
Chelifer (Chelifer) depressus C. L. Koch: Kästner, 1928: 7–8, fig. 25.
Lissochelifer depressus (C. L. Koch): J. C. Chamberlin, 1932a: 20.
Lophochernes depressus (C. L. Koch): Beier, 1932e: 247; Roewer, 1937: 314.
Lophochelifer depressus (C. L. Koch): Beier, 1940a: 190.

Type locality: unsure.
Distribution: India.

Lissochelifer gibbosounguiculatus (Beier), new combination

Lophochelifer gibbosounguiculatus Beier, 1951a: 120–121, fig. 46.

Type locality: Cao Nguyên Läm Viên (as Plateau von Langbian), Vietnam.
Distribution: Vietnam.

Lissochelifer gracilipes (Mahnert), new combination

Lophochelifer gracilipes Mahnert, 1988a: 70–71, figs 60–63.

Type locality: Kimakia Cave, Koboko, Kenya.
Distribution: Kenya.

Lissochelifer hygricus (Murthy & Ananthakrishnan), new combination

Lophochelifer hygricus Murthy & Ananthakrishnan, 1977: 162–163, figs 51a–b.

Type locality: Valparai, Madras, Tamil Nadu, India.
Distribution: India.

Lissochelifer insularis (Beier), new combination

Lophochelifer insularis Beier, 1940a: 191, fig. 16; Beier, 1965g: 790–791; Beier,
1967c: 323–324; Beier, 1971b: 367; Beier, 1982: 44.

Type locality: Tartarii, Espiritu Santo, New Hebrides.
Distribution: New Hebrides, Papua New Guinea.

Lissochelifer mortensenii (With)

Chelifer mortensenii With, 1906: 46–47, 140–143, plate 2 figs 9a–c, plate 3 figs 1a–i.
Chelifer mortenseni With: J. C. Chamberlin, 1931a: fig. 39m; Beier, 1932a: figs 24–25.
Lissochelifer mortenseni (With): J. C. Chamberlin, 1932a: 20.
Lophochernes mortenseni (With): Beier, 1932e: 247–248; Beier, 1932g: figs 201–202.
Lophochernes mortensenii (With): Roewer, 1936: figs 90–92; Roewer, 1937: 314.
Lophochelifer mortenseni (With): Beier, 1940a: 190.

Type locality: Ko Chang (as Koh Chang), Thailand.
Distribution: Thailand.

Lissochelifer nairobiensis (Mahnert), new combination

Lophochelifer nairobiensis Mahnert, 1988a: 71–72, figs 64–68.

Type locality: Rosslyn, Nairobi, Kenya.
Distribution: Kenya.

Lissochelifer novaeguineae (Beier), new combination
Lophochelifer novaeguineae Beier, 1965g: 790, fig. 790; Beier, 1971b: 367.

Type locality: Finschhafen, Papua New Guinea.
Distribution: Papua New Guinea.

Lissochelifer philippinus (Beier), new combination
Lophochernes philippinus Beier, 1937b: 276–277, figs 7–8; Roewer, 1940: 348.
Lophochelifer philippinus (Beier): Beier, 1967c: 319.

Type locality: Limay, Philippines.
Distribution: Philippines.

Lissochelifer strandi (Ellingsen), new combination
Chelifer strandi Ellingsen, 1907b: 28–30; Ellingsen, 1912b: 85.
Lophochernes strandi (Ellingsen): Beier, 1932e: 244; Roewer, 1937: 314.
Lophochelifer strandi (Ellingsen): Beier, 1967d: 87–88, fig. 9.

Type locality: Amani, Tanzania.
Distribution: Tanzania.

Lissochelifer superbus (With)
Chelifer superbus With, 1906: 48, 147–152, plate 3, fig. 3a; Ellingsen, 1913b: 127;
    Ellingsen, 1914: 10.
Lissochelifer superbus (With): J. C. Chamberlin, 1932a: 20.
Lophochernes superbus (With): Beier, 1932e: 247; J. C. Chamberlin, 1934b: 11;
    Roewer, 1937: 314.
Lophochelifer superbus (With): Beier, 1940a: 190; Beier, 1948b: 453.

Type locality: Sulawesi (as Celebes), Indonesia.
Distribution: India, Indonesia (Sulawesi).

Lissochelifer tonkinensis (Beier), new combination
Lophochelifer tonkinensis Beier, 1951a: 122–123, fig. 47.

Type locality: Fan Si Pan (as Fan-Si-Pan), Tonkin, Vietnam.
Distribution: Vietnam.

Genus Litochelifer Beier
Litochelifer Beier, 1948b: 488–489.

Type species: *Litochelifer nidicola* Beier, 1948b, by original designation.

Litochelifer nidicola Beier
Litochelifer nidicola Beier: Beier, 1947b: 336 (nomen nudum).
Litochelifer nidicola Beier, 1948b: 489–491, figs 20–21; Beier, 1955l: 320–321.

Type locality: South Africa.
Distribution: South Africa.

Genus Lophochernes E. Simon
Lophochernes E. Simon, 1878c: 66; Beier, 1932e: 242; Morikawa, 1960: 152; Murthy
    & Ananthakrishnan, 1977: 159.

Type species: *Lophochernes bicarinatus* E. Simon, 1878c, by monotypy.

Lophochernes alter Beier
Lophochernes alter Beier, 1951a: 115–117, fig. 43.

Type locality: Cao Nguyên Läm Viên (as Plateau von Langbian), Vietnam.
Distribution: Vietnam.

Lophochernes balzanii (Thorell)

Chelifer balzanii Thorell, 1890: 352–355; With, 1906: 155.
Lophochernes (?) *balzanii* (Thorell): Beier, 1932e: 250; Roewer, 1937: 315.

Type locality: Pinang, Malaysia.
Distribution: Malaysia.

Lophochernes bicarinatus E. Simon

Lophochernes bicarinatus E. Simon, 1878c: 66–67; Beier, 1932e: 243–244, figs
  250–251; Roewer, 1937: 314; Morikawa, 1954a: 75–77, figs 2f–j; Morikawa,
  1960: 152–153, plate 4 fig. 6, plate 8 figs 4, 19, plate 9 fig. 28, plate 10 figs 14,
  30; Sato, 1979a: plate 2, fig. 7; Sato, 1979d: 44; Sato, 1980d: plate 8.
Chelifer (Chelifer) bicarinatus (E. Simon): Ellingsen, 1907c: 4–5.
Chelifer bicarinatus (E. Simon): Ellingsen, 1912a: 122.

Type locality: Paris, France, from Japan.
Distribution: Japan, Taiwan.

Lophochernes bifissus (E. Simon)

Chelifer bifissus E. Simon, 1899a: 121–122; J. C. Chamberlin, 1931a: figs 10i, 42s,
  46m; With, 1906: 153.
Lissochelifer bifissus (E. Simon): J. C. Chamberlin, 1932a: 20.
Lophochernes bifissus (E. Simon): Beier, 1932e: 246–247, figs 254–255; Roewer,
  1937: 314; J.C. Chamberlin, 1938a: 282–284, figs 6a–f.
Not *Chelifer bifissus* E. Simon: E. Simon, 1900a: 517; With, 1905: 98–101, plate 6
  figs 1a–f (misidentifications; see *Lophochernes cryptus* J. C. Chamberlin).

Type locality: Sumatra, Indonesia.
Distribution: Indonesia (Sumatra).

Lophochernes bisulcus (Thorell)

Chelifer bisulcus Thorell, 1889: 603–606, plate 5 figs 9a–b; With, 1906: 152–153.
Lissochelifer bisulcus (Thorell): J. C. Chamberlin, 1932a: 20.
Lophochernes bisulcus (Thorell): Beier, 1932e: 248; Roewer, 1937: 314; Beier, 1943:
  81, fig. 8; Beier, 1967g: 359–360, fig. 21; Murthy & Ananthakrishnan, 1977: 160.

Type locality: Bhamó, Burma.
Distribution: Bangladesh, Burma, India, Thailand, Vietnam.

Lophochernes brevipes Redikorzev

Lophochernes brevipes Redikorzev, 1938: 113–114, fig. 43; Roewer, 1940: 348; Beier,
  1951a: 115.

Type locality: Phu Tho (as Phu-Tho), Vietnam.
Distribution: Vietnam.

Lophochernes calcaratus Beier

Lophochernes calcaratus Beier, 1967d: 89–90, fig. 11.

Type locality: Maragambo, Uganda.
Distribution: Uganda.

Lophochernes cederholmi Beier

Lophochernes cederholmi Beier, 1973b: 34–35, fig. 19.

Type locality: Maradan Maduwa, 23 miles W. of Anaradhapura, North Central
  Province, Sri Lanka.
Distribution: Sri Lanka.

Lophochernes ceylonicus Beier

Lophochernes ceylonicus Beier, 1973b: 53−54, fig. 18.

Type locality: Deerwood Kuruwita, 6 miles NNW. of Ratnapura, Sabaragamuwa Province, Sri Lanka.
Distribution: Sri Lanka.

Lophochernes cryptus J. C. Chamberlin

Chelifer bifissus E. Simon: E. Simon, 1900a: 517; With, 1905: 98−101, plate 6 figs 1a−f (misidentifications).
Lophochernes cryptus J. C. Chamberlin, 1934b: 11; Roewer, 1937: 315; J. C. Chamberlin, 1938a: 284−285.

Type locality: Olaa, Hawaii.
Distribution: Hawaii.

Lophochernes differens Beier

Lophochernes differens Beier, 1951a: 117−118, fig. 44.

Type locality: Thôn Sông Pha (as Krongpha), Vietnam.
Distribution: Vietnam.

Lophochernes elegantissimus Beier

Lophochernes elegantissimus Beier, 1964k: 83−84, fig. 37; Lawrence, 1967: 90.

Type locality: Skukuza, Kruger National Park, Transvaal, South Africa.
Distribution: South Africa.

Lophochernes flammipes Beier

Lophochernes flammipes Beier, 1951a: 113−115, figs 41, 42.

Type locality: Cha Pa (as Chapa), Tonkin, Vietnam.
Distribution: Vietnam.

Lophochernes frater Beier

Lophochernes frater Beier, 1944: 206−208, fig. 21.

Type locality: Nchanga (as N'Changa), Zambia.
Distribution: Zambia.

Lophochernes gracilis Beier

Lophochernes gracilis Beier, 1943: 78−80, figs 5−6; Weidner, 1959: 115.

Type locality: Futschau, Fukien, China.
Distribution: China (Fukien).

Lophochernes hansenii (Thorell)

Chelifer hansenii Thorell, 1889: 600−603, plate 5 figs 8a−b; With, 1906: 154−155; Ellingsen, 1914: 9−10.
Lophochernes hanseni (Thorell): Beier, 1932e: 246; Roewer, 1937: 314.

Type locality: Bhamó, Burma.
Distribution: Burma, India.

Lophochernes hians (Thorell)

Chelifer hians Thorell, 1890: 355−357; With, 1906: 153−154.
Lophochernes hians (Thorell): Beier, 1932e: 249−250; Roewer, 1937: 314; Beier, 1964f: 598; Beier, 1965g: 789; Beier, 1966d: 155.

Type locality: Pinang, Malaysia.
Distribution: Malaysia, Solomon Islands.

Lophochernes indicus Beier

Lophochernes indicus Beier, 1967g: 361–362, fig. 22; Beier, 1974b: 280; Beier, 1976e: 100; Murthy & Ananthakrishnan, 1977: 160; Schawaller, 1983a: 110; Schawaller, 1987a: 218.

Type locality: 8 km SE. of Ashwaraopet, Andhra Pradesh, India.
Distribution: Bhutan, India, Nepal.

Lophochernes laciniosus (Tullgren)

Chelifer laciniosus Tullgren, 1912a: 263–265, figs 2a–b.
Lophochernes laciniosus (Tullgren); Beier, 1932e: 245–246.
Lophochernes lacinosus (sic) (Tullgren): Roewer, 1937: 314.

Type locality: Nongkodjadjar, Java, Indonesia.
Distribution: Indonesia (Java).

Lophochernes luzonicus Beier

Lophochernes luzonicus Beier, 1937b: 278–279, figs 9–11; Roewer, 1940: 348.

Type locality: Mt Banahao, Luzon, Philippines.
Distribution: Philippines.

Lophochernes mayeti (E. Simon)

Chelifer mayeti E. Simon, 1885c: 48.
Lophochernes mayeti (E. Simon): Beier, 1932e: 245; Roewer, 1937: 315.

Type locality: Gafsa, Tunisia.
Distribution: Tunisia.

Lophochernes melanopygus Redikorzev

Lophochernes melanopygus Redikorzev, 1949: 662–663, figs 27–29.

Type localities: near Shchachrizybza, Uzbekistan, U.S.S.R.; Bukhara, Uzbekistan, U.S.S.R.; Mangit, Uzbekistan, U.S.S.R.; Urgench, Uzbekistan, U.S.S.R.; Chal'sh-tugaj, Uzbekistan, U.S.S.R.; and Kzyl-Orda, Kazakhstan, U.S.S.R.
Distribution: U.S.S.R. (Kazakhstan, Uzbekistan).

Lophochernes mindoroensis Beier

Lophochernes mindoroensis Beier, 1966b: 347–348, fig. 5.

Type locality: San Jose, Mindoro, Philippines.
Distribution: Philippines.

Lophochernes mucronatus (Tullgren)

Chelifer mucronatus Tullgren, 1907a: 30–32, figs 3a–g; Tullgren, 1907b: 220; Ellingsen, 1912b: 84, 97; Hewitt & Godfrey, 1929: 314–316, figs 1b, 3a–c, plate 21 fig. 3.
Chelifer mugronatus (sic) Tullgren: Godfrey, 1927: 18.
Lophochernes mucronatus (Tullgren): Beier, 1932e: 244–245, fig. 252; Roewer, 1937: 314; Lawrence, 1937: 272; Beier, 1958a: 183, fig. 14; Weidner, 1959: 114; Beier, 1964k: 83.

Type locality: Port Elizabeth, Cape Province, South Africa.
Distribution: South Africa.

Lophochernes nilgiricus Murthy & Ananthakrishnan

Lophochernes nilgiricus Murthy & Ananthakrishnan, 1977: 160–161, figs 50a–b.
Type locality: Trichy, Madras, India.
Distribution: India.

Lophochernes obtusecarinatus Beier

Lophochernes obtusecarinatus Beier, 1951a: 118–119, fig. 45; Beier, 1967g: 361.

Type locality: Cao Nguyên Lâm Viên (as Plateau von Langbian), Vietnam.
Distribution: Thailand, Vietnam.

Lophochernes persulcatus (E. Simon)

Chelifer persulcatus E. Simon, 1890: 120; E. Simon, 1900c: 596.
Lophochernes persulcatus (E. Simon): Beier, 1932e: 249, figs 256–257; Roewer, 1937: 315; Beier, 1967d: 88, fig. 10; Mahnert, 1988a: 72.
Lophochernes cf. *persulcatus* (E. Simon): Mahnert, 1984b: 65.

Type locality: Aden, South Yemen.
Distribution: Ethiopia, Kenya, Somalia, South Yemen.

Lophochernes sauteri (Ellingsen)

Chelifer (Chelifer) sauteri Ellingsen, 1907c: 4–5.
Lophochernes sauteri (Ellingsen): Beier, 1932e: 243; Roewer, 1937: 314; Morikawa, 1960: 152; Morikawa, 1962: 419.

Type locality: Okayama, Bizen, Japan.
Distribution: Japan.

Lophochernes semicarinatus Redikorzev

Lophochernes semicarinatus Redikorzev, 1938: 114–116, figs 44–45; Roewer, 1940: 348; Beier, 1951a: 120.

Type localities: Pleiku, Vietnam; Kontum, Vietnam; and Nha Trang (as Nhatrang), Vietnam.
Distribution: Vietnam.

Lophochernes tibetanus Beier

Lophochernes tibetanus Beier, 1943: 80, fig. 7.

Type locality: South-east Tibet, China.
Distribution: China (Tibet).

Genus **Lophodactylus** J. C. Chamberlin

Lophodactylus J. C. Chamberlin, 1932a: 21; Beier, 1932e: 273–274.

Type species: *Chelifer rex* With, 1908a, by original designation.

Lophodactylus rex (With)

Chelifer rex With, 1908a: 225–228, figs 2a–e; J. C. Chamberlin, 1931a: 125, fig. 39n.
Lophodactylus rex (With): J. C. Chamberlin, 1932a: 21; Beier, 1932e: 274; Roewer, 1937: 316.

Type locality: Brazil.
Distribution: Brazil.

Genus **Macrochelifer** Vachon

Macrochelifer Vachon, 1940d: 412–414.

Type species: *Chelifer tibetanus* Redikorzev, 1918, by original designation.

Macrochelifer tibetanus (Redikorzev)

Chelifer tibetanus Redikorzev, 1918: 93–94, fig. 2.
Hysterochelifer tibetanus (Redikorzev): Beier, 1932e: 234, fig. 242; Roewer, 1937: 312.

Macrochelifer tibetanus (Redikorzev); Vachon, 1940d: 412–414; J. C. Chamberlin, 1949: 27–29, figs 8a–c.

Type locality: T'o-so Hu (as Toso-nor), Chinghai, China.
Distribution: China (Chinghai).

Genus **Mesochelifer** Vachon

Mesochelifer Vachon, 1940g: 159; Beier, 1963b: 286.
Hysterochelifer (Mesochelifer) Vachon: Beier, 1956e: 308.

Type species: *Mesochelifer fradei* Vachon, 1940g, by original designation.

Mesochelifer fradei Vachon

Mesochelifer fradei Vachon, 1940g: 155–159, figs 22–27; Beier, 1963b: 286; Vachon, 1966c: fig. 4.
Hysterochelifer (Mesochelifer) fradei (Vachon): Beier, 1956e: 308.

Type locality: Ferreira do Alentejo, Portugal.
Distribution: Portugal.

Mesochelifer insignis Callaini

Mesochelifer insignis Callaini, 1986a: 1–7, figs 1a–g, plates 1–8; Callaini, 1988b: 56.

Type locality: Azeffoun, Algeria.
Distribution: Algeria, Morocco.

Mesochelifer pardoi (Beier)

Hysterochelifer (Mesochelifer) pardoi Beier, 1956e: 307–310, fig. 4.
Mesochelifer pardoi (Beier): Callaini, 1988b: 56.

Type locality: Granja del Muluya, Kebdana, Morocco.
Distribution: Morocco.

Mesochelifer ressli Mahnert

Mesochelifer ressli Mahnert, 1981d: 47–50, figs 1–13; Jędryczkowski, 1985: 80–81; Gardini, 1987: 123 (not seen); Jędryczkowski, 1987b: 146, map 14; Schawaller, 1989: 24.

Type locality: Lunzberg, Lunz, Austria.
Distribution: Austria, Italy, Poland, Switzerland, U.S.S.R. (Kazakhstan).

Mesochelifer thunebergi Kaisila

Mesochelifer thunebergi Kaisila, 1966: 361–363, figs 1–5; Beier, 1976a: 25; Estany, 1979a: 222, fig. 1.

Type locality: Sant André, Anaga mountains, Tenerife, Canary Islands.
Distribution: Canary Islands.

Genus **Metachelifer** Redikorzev

Metachelifer Redikorzev, 1938: 108.

Type species: *Metachelifer duboscqui* Redikorzev, 1938, by monotypy.

Metachelifer duboscqui Redikorzev

Metachelifer duboscqui Redikorzev, 1938: 108–111, figs 39–41; Roewer, 1940: 347; Beier, 1951a: 104–106, fig. 36; Beier, 1966b: 347.

Type localities: Bokor (as Bokkor), Cambodia; Siem Reap (as Siem-Reap), Cambodia; and Da Lat (as Dalat), Vietnam.
Distribution: Cambodia, Laos, Philippines, Vietnam.

Metachelifer hyatti Ćurčić

Metachelifer hyatti Ćurčić, 1980b: 89–90, figs 25–28.

Type locality: Kalapani, Nepal.
Distribution: Nepal.

Genus Mexichelifer Muchmore

Mexichelifer Muchmore, 1973b: 58–60.

Type species: *Mexichelifer reddelli* Muchmore, 1973b, by original designation.

Mexichelifer reddelli Muchmore

Mexichelifer reddelli Muchmore, 1973b: 60–62, figs 39–46; Elliott & Reddell, 1973: 195.

Type locality: Cueva de Carnicerías, Valle de los Fantasmas, 17 km W. of Sta. Catarina, San Luis Potosí, Mexico.
Distribution: Mexico.

Genus Microchelifer Beier

Microchelifer Beier, 1944: 208–209; J. C. Chamberlin, 1952: 32–33; Murthy & Ananthakrishnan, 1977: 158–159.

Type species: *Microchelifer vosseleri* Beier, 1944, by original designation.

Microchelifer acarinatus Beier

Microchelifer acarinatus Beier, 1972a: 17–18, fig. 9.

Type locality: Parc National Garamba, Zaire.
Distribution: Zaire.

Microchelifer dentatus Mahnert

Microchelifer dentatus Mahnert, 1988a: 74–76, figs 72–75.

Type locality: Loita Hills, Morijo, Narok, Kenya.
Distribution: Kenya.

Microchelifer granulatus Beier

Microchelifer granulatus Beier, 1954d: 137–138, fig. 5; Beier, 1959d: 64; Mahnert, 1988a: 72–74, figs 69–71.

Type locality: Mt Wago, Ituri, Zaire.
Distribution: Kenya, Zaire.

Microchelifer lourencoi Heurtault

Microchelifer lourençoi Heurtault, 1983: 7–10, figs 16–23.

Type locality: Lamto, Ivory Coast.
Distribution: Ivory Coast.

Microchelifer minusculoides (Ellingsen)

Chelifer minusculoides Ellingsen, 1912b: 94–95; Hewitt & Godfrey, 1929: 317–318, plate 21 fig. 4.
Lophochernes minusculoides (Ellingsen): Beier, 1932e: 248; Roewer, 1937: 315.
Microchelifer minusculoides (Ellingsen): Beier, 1958a: 184–185, fig. 15; Beier, 1964k: 80.

Type locality: Pirie Forest, King William's Town, Cape Province, South Africa.
Distribution: South Africa.

Microchelifer percarinatus Beier

Microchelifer percarinatus Beier, 1964k: 81–82, fig. 36.

Type locality: Mamathes, Lesotho.
Distribution: Lesotho.

Microchelifer rhodesiacus Beier

Microchelifer rhodesiacus Beier, 1964k: 80–81, fig. 35.

Type locality: Chimanimani, Zimbabwe.
Distribution: Zimbabwe.

Microchelifer sadiya J. C. Chamberlin

Microchelifer sadiya J. C. Chamberlin, 1949: 34–36, figs 10a–f; Murthy & Ananthakrishnan, 1977: 159.

Type locality: Nazira, Atkel, Assam, India.
Distribution: India.

Microchelifer vosseleri Beier

Microchelifer vosseleri Beier, 1944: 209–210, fig. 22.

Type locality: Amani, Tanzania.
Distribution: Tanzania.

Genus Mucrochelifer Beier

Mucrochelifer Beier, 1932d: 63–64; Beier, 1932e: 250.

Type species: *Chelifer borneoensis* Ellingsen, 1901b, by original designation.

Mucrochelifer borneoensis (Ellingsen)

Chelifer borneoensis Ellingsen, 1901b: 206–208; With, 1906: 154; Ellingsen, 1909a: 219; Ellingsen, 1911b: 142; Tullgren, 1912b: 269; Ellingsen, 1914: 6–7.
Chelifer (Chelifer) borneoensis Ellingsen: Ellingsen, 1910a: 383.
Mucrochelifer borneoensis (Ellingsen): Beier, 1932d: 64; Beier, 1932e: 250–251, figs 258–259; Roewer, 1936: fig. 50; Roewer, 1937: 315; Vachon, 1940f: 3; Beier, 1948b: 453.

Type locality: Borneo.
Distribution: Burma, Indonesia (Borneo, Java), Papua New Guinea, Sri Lanka.

Genus Nannochelifer Beier

Nannochelifer Beier, 1967d: 91–92; Harvey, 1984: 291–292.

Type species: *Nannochelifer litoralis* Beier, 1967d, by original designation.

Nannochelifer litoralis Beier

Nannochelifer litoralis Beier, 1967d: 92–93, fig. 12; Harvey, 1984: 293–294, figs 1–5; Mahnert, 1984b: 62; Mahnert, 1988a: 76.

Type locality: Silversands near Malindi, Kenya.
Distribution: Kenya, Somalia.

Nannochelifer paralius Harvey

Nannochelifer paralius Harvey, 1984: 294–296, figs 5–16.

Type locality: Turtle Islet, Coral Sea Islands Territory, Australia.
Distribution: Australia (Coral Sea Islands Territory).

Genus **Nannocheliferoides** Beier

Nannocheliferoides Beier, 1974e: 1014–1015; Murthy & Ananthakrishnan, 1977: 159.

Type species: *Nannocheliferoides mussardi* Beier, 1974e, by original designation.

Nannocheliferoides mussardi Beier

Nannocheliferoides mussardi Beier, 1974e: 1015–1016, fig. 10; Murthy & Ananthakrishnan, 1977: 159.

Type locality: Madurai, India.
Distribution: India.

Genus **Pachychelifer** Beier

Pachychelifer Beier, 1962b: 151–152; Beier, 1963b: 291–292.

Type species: *Pachychelifer caucasicus* Beier, 1962b, by original designation.

Pachychelifer caucasicus Beier

Pachychelifer caucasicus Beier, 1962b: 152–153, figs 8–11; Beier, 1963b: 292, fig. 294; Kobakhidze, 1965b: 541; Kobakhidze, 1966: 706.

Type locality: Kobuleti, Black Sea, Georgia, U.S.S.R.
Distribution: U.S.S.R. (Georgia).

Genus **Paisochelifer** Hoff

Paisochelifer Hoff, 1946f: 487; Hoff, 1949b: 489; Hoff, 1950a: 9–10.

Type species: *Hysterochelifer callus* Hoff, 1945g, by original designation.

Paisochelifer callus (Hoff)

Hysterochelifer callus Hoff, 1945g: 515–521, figs 7–12.
Paisochelifer callus (Hoff): Hoff, 1946f: 487; Hoff, 1949b: 489–491, figs 50a–e; Hoff, 1950a: 10–11; Hoff, 1958: 33; Manley, 1969: 9–10, fig. 14; Nelson, 1975: 298, figs 90–91.

Type locality: Zion, Lake County, Illinois, U.S.A.
Distribution: U.S.A. (Illinois, Maryland, Michigan).

Paisochelifer utahensis Hoff

Paisochelifer utahensis Hoff, 1950a: 11–15, figs 6–9; Hoff, 1958: 33.

Type locality: Utah Lake, Utah, U.S.A.
Distribution: U.S.A. (Utah).

Genus **Papuchelifer** Beier

Papuchelifer Beier, 1965g: 791–793.

Type species: *Papuchelifer nigrimanus* Beier, 1965g, by original designation.

Papuchelifer exiguus Beier

Papuchelifer exiguus Beier, 1965g: 795–796, fig. 32; Tenorio & Muchmore, 1982: 379.

Type locality: near Geelvink Bay, Orensbari, Irian Jaya, Indonesia.
Distribution: Indonesia (Irian Jaya).

Papuchelifer nigrimanus Beier

Papuchelifer nigrimanus Beier, 1965g: 793–794, fig. 30; Beier, 1971b: 367; Tenorio & Muchmore, 1982: 379.

Type locality: Soputa, Papua New Guinea.
Distribution: Papua New Guinea.

Papuchelifer pustulatus Beier

Papuchelifer pustulatus Beier, 1965g: 794–795, fig. 31.

Type locality: Finschhafen, Papua New Guinea.
Distribution: Papua New Guinea.

Genus **Parachelifer** J. C. Chamberlin

Parachelifer J. C. Chamberlin, 1932a: 19; Beier, 1932e: 237–238; J. C. Chamberlin, 1934a: 128–129; J. C. Chamberlin, 1952: 299–300; Hoff, 1956d: 4; Hoff, 1964b: 6–7.

Type species: *Chelifer scabriculus* E. Simon, 1878b, by original designation.

Parachelifer approximatus (Banks)

Chelifer approximatus Banks, 1909a: 146.
Parachelifer approximatus (Banks): Beier, 1932d: 63; Beier, 1932e: 240; Roewer, 1937: 312.

Type localities: Pescadero, Baja California, Mexico; and El Taste, Baja California, Mexico.
Distribution: Mexico.

Parachelifer archboldi Hoff

Parachelifer archboldi Hoff, 1964b: 12–17, figs 1, 3, 6.

Type locality: 8 miles SE. of Lake Pacid, Highlands County, Florida, U.S.A.
Distribution: U.S.A. (Florida).

Parachelifer dominicanus Beier

Parachelifer dominicanus Beier, 1976d: 55–57, fig. 7.

Type locality: Boca Chica, Dominican Republic.
Distribution: Dominican Republic.

Parachelifer ecuadoricus Beier

Parachelifer ecuadoricus Beier, 1959e: 227–228, fig. 34.

Type locality: Ecuador.
Distribution: Ecuador.

Parachelifer hubbardi (Banks)

Chelifer hubbardi Banks, 1901b: 588–589, fig. 9; J. C. Chamberlin, 1923c: 374–375, plate 2 fig. 10, plate 3 fig. 33; Coolidge, 1908: 110.
Chelanops arizonensis Banks, 1901b: 589, fig. 2 (misidentification, in part).
Parachelifer hubbardi (Banks): J. C. Chamberlin, 1932a: 19; Beier, 1932e: 240; Roewer, 1936: fig. 148; Roewer, 1937: 313; Beier, 1948b: 453, 491, fig. 22; Hoff, 1958: 32; Hoff, 1946b: 204–205; Muchmore, 1971a: 84.

Type localities: Catalina Springs, Arizona, U.S.A.; Madera Canyon, Arizona, U.S.A.; and Oracle, Arizona, U.S.A.
Distribution: Mexico, U.S.A. (Arizona).

Parachelifer lativittatus (J. C. Chamberlin)

Chelifer lativittatus J.C. Chamberlin, 1923c: 375, plate 2 fig. 9.
Parachelifer lativittatus (J. C. Chamberlin): J. C. Chamberlin, 1932a: 19; Beier, 1932e: 240; Roewer, 1937: 313.

Type locality: Tapachula, Chiapas, Mexico.
Distribution: Mexico.

Parachelifer longipalpus Hoff

Parachelifer longipalpus Hoff, 1945c: 49–53, figs 13–16; Hoff & Bolsterli, 1956: 177–178; Hoff, 1958: 33; Muchmore, 1971a: 82.

Type locality: Washington County, Arkansas, U.S.A.
Distribution: U.S.A. (Arkansas, Illinois, Nebraska, Tennessee).

Parachelifer mexicanus Beier

Parachelifer mexicanus Beier, 1932d: 63; Beier, 1932e: 239, fig. 248; Roewer, 1937: 312.

Type locality: Teapa, Tabasco, Mexico.
Distribution: Mexico.

Parachelifer monroensis Nelson

Dactylochelifer copiosus Hoff: Manley, 1969: 9, fig. 12 (misidentification, in part).
Parachelifer monroensis Nelson, 1975: 294–297, figs 82–84.

Type locality: Monroe County, Michigan, U.S.A.
Distribution: U.S.A. (Michigan).

Parachelifer montanus J. C. Chamberlin

Parachelifer montanus J. C. Chamberlin, 1934a: 129–131, figs 1a–c, 1h–i; Roewer, 1937: 313, fig. 244; Hoff, 1958: 33.

Type locality: Sula, Montana, U.S.A.
Distribution: U.S.A. (Montana).

Parachelifer muricatus (Say)

Chelifer muricatus Say, 1821: 63–64; Hagen, 1868a: 435; Hagen, 1868b: 51–52; Hagen, 1870: 266; Banks, 1890: 152; Banks, 1895a: 3 (in part; see *Parachelifer superbus* Hoff); Banks, 1904a: 140; Banks, 1904b: 364; Banks, 1908: 39; Coolidge, 1908: 110; Pratt, 1927: 410; Nester, 1932: 98.
Chelifer cancroides (Linnaeus): Hagen, 1868a: 435 (misidentification).
Parachelifer muricatus (Say): J. C. Chamberlin, 1932a: 19; Rowland & Reddell, 1976: 18.
Parachelifer (?) *muricatus* (Say): Beier, 1932e: 240; Roewer, 1937: 313; Hoff, 1958: 33.

Type locality: North America.
Distribution: U.S.A. (Alabama, California, Florida, Illinois, Indiana, Kentucky, Massachusetts, New York, Ohio, Texas, Virginia).

Parachelifer parvus Muchmore

Parachelifer parvus Muchmore, 1981e: 189–191, figs 1–3.

Type locality: Great Lameshur Bay, St John, U.S. Virgin Islands.
Distribution: U.S. Virgin Islands.

Parachelifer persimilis (Banks)

Chelifer persimilis Banks, 1909b: 304.
Parachelifer persimilis (Banks): J. C. Chamberlin, 1932a: 19; Beier, 1932e: 240; J. C. Chamberlin, 1934a: 131; Roewer, 1937: 313; Hoff, 1950a: 2–3; Hoff, 1956d: 4–7; Hoff, 1958: 32; Hoff, 1959b: 5, etc.; Hoff, 1961: 457–458; Hoff, 1963a: 10; Muchmore, 1971a: 85; Benedict & Malcolm, 1979: 196–197.

Type localities: Pecos, New Mexico, U.S.A.; Las Vegas, New Mexico, U.S.A.; Eagle Spring, New Mexico, U.S.A.; and Roswell, New Mexico, U.S.A.
Distribution: Mexico, U.S.A. (Arizona, Colorado, Montana, Nebraska, Nevada, New Mexico, Oregon, South Dakota, Utah).

Parachelifer pugifer Beier

Parachelifer pugifer Beier, 1953b: 26–27, figs 12a–c.

Type locality: Quezaltenango, Guatemala.
Distribution: Guatemala.

Parachelifer scabriculus (E. Simon)

Chelifer scabriculus E. Simon, 1878b: 154–155; Tullgren 1907a: 29–30, fig. 2; Ellingsen, 1909b: 217; J. C. Chamberlin, 1931a: figs 40w, 42t–v.
Chelifer (Chelifer) degeneratus Balzan, 1892: 532–533, fig. 24 (synonymized by Banks, 1901a: 594).
Chelifer scabrisculus (sic) E. Simon: Banks, 1895a: 4; Banks, 1898: 289; Banks, 1904b: 364; Coolidge, 1908: 110; Hilton, 1913: 190, fig. 2; Moles & Moore, 1921: 6.
Chelifer scabrisculis (sic) E. Simon: Banks, 1901a: 594; Banks, 1902b: 220.
Chelifer (Chelifer) scabriculus E. Simon: Ellingsen, 1910a: 386.
Chelifer scabrisulis (sic) E. Simon: Banks, 1911: 637; Moles, 1914c: 192.
Chelifer scabisculus (sic) E. Simon: Hilton, 1931: fig. 104.
Parachelifer scabriculus (E. Simon): J. C. Chamberlin, 1932a: 19; Beier, 1932e: 238–239, fig. 247; Roewer, 1937: 312; Vachon, 1949: fig. 207b; J. C. Chamberlin, 1952: 300–305, figs 13a–c, 14a–j; Hoff, 1956d: 7–8; Hoff, 1958: 32; Hoff, 1959b: 5, etc.; Hoff, 1963a: 10; Benedict & Malcolm, 1979: 196.

Type localities: of *Chelifer scabriculus*: Mariposa, California, U.S.A.
of *Chelifer (Chelifer) degeneratus*: Baja California, Mexico.
Distribution: Mexico, U.S.A. (Arizona, California, New Mexico, Oregon, South Dakota, Utah).

Parachelifer sini (J. C. Chamberlin)

Chelifer sini J. C. Chamberlin, 1923c: 375–376, plate 2 fig. 7, plate 3 figs 18, 20, 32.
Parachelifer sini (J. C. Chamberlin): J. C. Chamberlin, 1931a: 68; J. C. Chamberlin, 1932a: 19; Beier, 1932e: 239; Roewer, 1936: fig. 49; Roewer, 1937: 312.

Type locality: Angeles Bay, Baja California, Mexico.
Distribution: Mexico.

Parachelifer skwarrae Beier

Parachelifer skwarrai (sic) Beier, 1933b: 101, fig. 13; Roewer, 1937: 313.

Type locality: Cordoba, Veracruz, Mexico.
Distribution: Mexico.

Parachelifer superbus Hoff

Chelifer muricatus Say: Banks, 1895a: 3 (misidentification, in part).
Parachelifer superbus Hoff, 1964b: 8–12, figs 2, 4–5; Weygoldt, 1969a: 57, 76, 100; Muchmore, 1971a: 86; Brach, 1979: 34.

Type locality: near Highlands Biological Research Station, Highlands County, Florida, U.S.A.
Distribution: U.S.A. (Alabama, Florida).

Parachelifer tricuspidatus Beier

Parachelifer tricuspidatus Beier, 1953b: 25–26, figs 11a–c.

Type locality: Los Trojados, Guatemala.
Distribution: Guatemala.

Parachelifer viduus Beier

Parachelifer viduus Beier, 1953b: 27–28, fig. 9.

Type locality: Sierra de Chuacus, Guatemala.
Distribution: Guatemala.

Genus **Phorochelifer** Hoff

Phorochelifer Hoff, 1956d: 18.

Type species: *Phorochelifer mundus* Hoff, 1956d, by original designation.

Phorochelifer mundus Hoff

Phorochelifer mundus Hoff, 1956d: 18–22, figs 6–10; Hoff, 1958: 34; Hoff, 1959b: 5, etc.; Muchmore, 1971a: 91.

Type locality: Mt Taylor, near Grants, Valencia County, New Mexico, U.S.A.
Distribution: U.S.A. (New Mexico).

Genus **Pilochelifer** Beier

Pilochelifer Beier, 1935b: 255.

Type species: *Pilochelifer insularis* Beier, 1935b, by original designation.

Pilochelifer insularis Beier

Pilochelifer insularis Beier, 1935b: 256, fig. 3; Roewer, 1937: 316.

Type locality: Mauritius.
Distribution: Mauritius.

Pilochelifer insularis gracilior Mahnert

Pilochelifer insularis gracilior Mahnert, 1975b: 554–558, figs 5a–h.

Type locality: Morne des Patates à Durand, St Denis, Réunion.
Distribution: Réunion.

Genus **Protochelifer** Beier

Protochelifer Beier, 1948c: 552–553; J.C. Chamberlin, 1949: 42–43; Beier, 1976f: 243.

Type species: *Protochelifer novaezealandiae* Beier, 1948c, by original designation.

Protochelifer australis (Tubb)

Ideochelifer (sic) *australis* Tubb, 1937: 414–415, figs 3a–d.
Idiochelifer australis Tubb: Roewer, 1940: 347.
Protochelifer australis (Tubb): Beier, 1948c: 553; J. C. Chamberlin, 1949: 46–49, figs 13a–l; Harvey, 1981b: 244–245; Harvey, 1985b: 131.

Type locality: Lady Julia Percy Island, Victoria, Australia.
Distribution: Australia (Victoria).

Protochelifer brevidigitatus (Tubb)

Ideochelifer (sic) *brevidigitatus* Tubb, 1937: 414, figs 2a–c.
Idiochelifer brevidigitatus Tubb: Roewer, 1940: 347.
Protochelifer brevidigitatus (Tubb): J. C. Chamberlin, 1949: 49; Harvey, 1981b: 245; Harvey, 1985b: 131.

Type locality: Lady Julia Percy Island, Victoria, Australia.
Distribution: Australia (Victoria).

Protochelifer cavernarum Beier

Protochelifer cavernarum Beier, 1967a: 203–205, fig. 4; Beier, 1968a: 764; Beier, 1969a: 187; Richards, 1971: 18, 22, 24, 25, 27, 28, 30, 34, 35, 36, 43, 48, fig. 4a; Beier, 1975b: 203; Harvey, 1981b: 245; Harvey, 1985b: 131–132.
Protochelifer cavernarum aitkeni Beier, 1968a: 764–765, fig. 6; Beier, 1969a: 187 (synonymized by Harvey, 1981b: 245).

Type localities: of *Protochelifer cavernarum*: Murder Cave, Cliefden, New South Wales, Australia; Belfry Cave, Timor, New South Wales, Australia; and Timor Caves, Timor, New South Wales, Australia.
of *Protochelifer cavernarum aitkeni*: Abrakurrie Cave, Nullarbor Plain, Western Australia, Australia.
Distribution: Australia (New South Wales, Victoria, Western Australia).

Protochelifer exiguus Beier

Protochelifer exiguus Beier, 1976f: 243–244, fig. 42.

Type locality: S. end of Lake Wilmot, Fiordland, South Island, New Zealand.
Distribution: New Zealand.

Protochelifer naracoortensis Beier

Protochelifer naracoortensis Beier, 1968a: 763–764, fig. 5; Harvey, 1981b: 245; D. C. Lee, 1983: 183; Harvey, 1985b: 132.

Type locality: Bat Cave, Naracoorte, South Australia, Australia.
Distribution: Australia (South Australia).

Protochelifer novaezealandiae Beier

Protochelifer novae-zealandiae Beier, 1948c: 554–555, figs 16–18.
Protochelifer novaezealandiae Beier: Beier, 1976f: 242–243.
Protochelifer maori J. C. Chamberlin, 1949: 43–46, figs 12a–i (synonymized by Beier, 1976f: 243).

Type localities: of *Protochelifer novaezealandiae*: Kingston, Lake Wakatipu (as Watipu), South Island, New Zealand.
of *Protochelifer maori*: Porirua, North Island, New Zealand.
Distribution: New Zealand.

Protochelifer victorianus Beier

Protochelifer victorianus Beier, 1966a: 301–302, fig. 15; Harvey, 1981b: 245; Harvey, 1985b: 132.

Type locality: Olsens Road, Gippsland, Victoria, Australia.
Distribution: Australia (Victoria).

Genus Pseudorhacochelifer Beier

Pseudorhacochelifer Beier, 1976a: 25–26.

Type species: *Pseudorhacochelifer schurmanni* Beier, 1976a, by original designation.

Pseudorhacochelifer coiffaiti (Vachon)

Rhacochelifer coiffaiti Vachon, 1961: 103–104.
Pseudorhacochelifer coiffaiti (Vachon): Beier, 1976a: 27; Pieper, 1981: 4–5.

Type localities: Queymadas; Rabasal, Madeira Islands; Paol da Serra, Madeira Islands; Deserta Grande, Madeira Islands; Funte das Cruzinhas, Madeira Islands.
Distribution: Madeira Islands.

Pseudorhacochelifer schurmanni Beier
Pseudorhacochelifer schurmanni Beier, 1976a: 26–27, fig. 1.
Type locality: Cañadas, Tenerife, Canary Islands.
Distribution: Canary Islands.

Genus Pugnochelifer Hoff
Pugnochelifer Hoff, 1964b: 31–32.
Type species: *Pugnochelifer amoenus* Hoff, 1964b, by original designation.

Pugnochelifer amoenus Hoff
Pugnochelifer amoenus Hoff, 1964b: 32–35, figs 18–20.
Type locality: Highlands Hammock State Park, Highlands County, Florida, U.S.A.
Distribution: U.S.A. (Florida, Louisiana).

Genus Pycnochelifer Beier
Pycnochelifer Beier, 1937d: 313–314.
Type species: *Chelifer kleemanni* C. L. Koch & Berendt, 1854, by original designation.

! Pycnochelifer kleemanni (C. L. Koch & Berendt)
Chelifer kleemanni C. L. Koch & Berendt, 1854: 95–96, fig. 143; Menge, 1855: 34–35, plate 5 fig. 9; Hagen, 1870: 266; Beier, 1932e: 278.
Pycnochelifer kleemanni (C. L. Koch & Berendt): Beier, 1937d: 314–315, fig. 19; Roewer, 1940: 331; Beier, 1948b: 446; Schawaller, 1978: 4.
Type locality: Baltic Amber.
Distribution: Baltic Amber.

Genus Rhacochelifer Beier
Rhacochelifer Beier, 1932b: 65; Beier, 1932e: 260; Beier, 1963b: 278.
Type species: *Chelifer disjunctus* L. Koch, 1873, by original designation.

Rhacochelifer afghanicus Beier
Rhacochelifer afghanicus Beier, 1959a: 281–282, fig. 25.
Type locality: Kunar-Tal, Asmar, Afghanistan.
Distribution: Afghanistan.

Rhacochelifer andreinii Beier
Rhacochelifer andreinii Beier, 1954c: 329–330, fig. 3.
Type locality: Sliten (now Zlitan), Tripolitania, Libya.
Distribution: Libya.

Rhacochelifer balcanicus (Redikorzev)
Chelifer balcanicus Redikorzev, 1928: 120–123, fig. 1.
Rhacochelifer balcanicus (Redikorzev): Beier, 1932d: 66; Beier, 1932e: 268, fig. 291; Roewer, 1936: fig. 83e; Roewer, 1937: 315.
Rhacochelifer (?) *balcanicus* (Redikorzev): Beier, 1963b: 298.
Type locality: Burgas, Bulgaria.
Distribution: Bulgaria.

Rhacochelifer barkhamae Mahnert
Rhacochelifer barkhamae Mahnert, 1980c: 45–47, figs 36–43.
Type locality: Khureys, Saudi Arabia.
Distribution: Saudi Arabia.

Rhacochelifer brevimanus (Kolenati)

Chelifer brevimanus Kolenati, 1857: 429–431; Hagen, 1870: 264.
Rhacochelifer (?) *brevimanus* (Kolenati): Beier, 1932e: 268; Roewer, 1937: 316.

Type localities: Tbilisi (as Tiflis), Georgia; Sheki (as Scheki), Azerbaijan; 'Schirwan', Caucasus; 'Karabagh', Caucasus; Yerevan (as Eriwan), Armenia, U.S.S.R.
Distribution: U.S.S.R. (Armenia, Azerbaijan, Georgia).

Rhacochelifer caucasicus (Daday)

Chernes (Ectoceras) caucasicus Daday, 1889a: 18–19, fig. 13.
Rhacochelifer (?) *caucasicus* (Daday): Beier, 1932e: 269; Roewer, 1937: 316.

Type locality: 'Kussari', Georgia, U.S.S.R.
Distribution: U.S.S.R. (Georgia).

Rhacochelifer chopardi Vachon

Rhacochelifer chopardi Vachon, 1950: 97–98, figs 5–8.

Type locality: Baguezane (as Baguezans) Mountains, Niger.
Distribution: Niger.

Rhacochelifer corcyrensis (Beier)

Chelifer (Chelifer) latreillei Leach: Beier, 1929a: 351 (misidentification; in part).
Ectoceras corcyrensis Beier, 1930a: 219–221, figs 16a–b, 17b; J. C. Chamberlin, 1932a: 21.
Chelifer (Chelifer) latreillei insularis Hadži, 1930a: 65–72, figs 1–9; Hadži, 1930c: 20–21 (synonymized by Beier, 1932e: 264).
Rhacochelifer corcyrensis (Beier): Beier, 1932d: 66; Beier, 1932e: 264, figs 282–283; Roewer, 1937: 315; Beier & Turk, 1952: 771; Beier, 1962a: 142; Beier, 1963b: 295–296, fig. 298; Beier, 1966f: 346; Beier, 1967f: 319–320; Beier, 1969b: 196–197; Ćurčić, 1974a: 28; Schawaller, 1981c: 9; Callaini, 1987b: 286–287.
Dactylochelifer latreillei insularis (Hadži): Ćurčić, 1974a: 27–28.
Rhacochelifer corcyrensis corcyrensis (Beier): Mahnert, 1977b: 69, figs 2a–b; Halperin & Mahnert, 1987: 127–128.

Type localities: of *Ectoceras corcyrensis*: Kérkira (as Korfu), Greece.
of *Chelifer (Chelifer) latreillei insularis*: Dugi Otok, Yugoslavia.
Distribution: Crete, Cyprus, Greece, Israel, Italy, Turkey, Yugoslavia.

Rhacochelifer corcyrensis bicolor Beier

Rhacochelifer corcyrensis bicolor Beier, 1963f: 202, fig. 12.

Type locality: 'Ein Gedi', Israel?
Distribution: Israel, Jordan.

Rhacochelifer corcyrensis procerus Mahnert

Rhacochelifer corcyrensis procerus Mahnert, 1978g: 17–20, figs 8–15.

Type locality: Djerba Island, Tunisia.
Distribution: Tunisia.

Rhacochelifer disjunctus (L. Koch)

Chelifer disjunctus L. Koch, 1873: 27–28; E. Simon, 1879a: 27–28, plate 18, fig. 5; Daday, 1888: 122, 177; Daday, 1889a: 16; E. Simon, 1898a: 102; E. Simon, 1898d: 20; Daday, 1918: 2; Navás, 1918: 89; Navás, 1919: 211; Navás, 1921: 165; Navás, 1925: 108; Bacelar, 1928: 190.
Chelifer (Chelifer) disjunctus (L. Koch): Nonidez, 1917: 22.
Hansenius disjunctus (L. Koch): J. C. Chamberlin, 1932a: 21.

Rhacochelifer disjunctus (Beier): Beier, 1932d: 65; Beier, 1932e: 260–261, figs 276–277; Roewer, 1937: 315; Beier, 1939f: 200; Vachon, 1940g: 145; Caporiacco, 1950: 116; Beier, 1963b: 295, fig. 296; Beier, 1963i: 286; Lazzeroni, 1969a: 339; Weygoldt, 1969a: 27, 28, 56, 76, 100; Lazzeroni, 1970a: 208; Heurtault, 1980a: 161–172, figs 1, 5, 8; Callaini, 1988b: 58–59.
Rhacochelifer disiunctus (sic) (L. Koch): Caporiacco, 1936b: 330.
Not *Chelifer disjunctus* L. Koch: Tömösváry, 1882b: 199–200, plate 2 figs 14–15 (misidentification; see *Rhacochelifer peculiaris* (L. Koch)).
Not *Chelifer (Chelifer) disjunctus* L. Koch: Ellingsen, 1910a: 385 (misidentification; see *Rhacochelifer maculatus* (L. Koch)).

Type localities: Mont Laberon, Vaucluse, France; and Villafranca (as Villafranka), Spain.
Distribution: France, Italy, Morocco, Portugal, Spain.

Rhacochelifer euboicus Mahnert

Rhacochelifer euboicus Mahnert, 1977b: 72–73, figs 4–9.

Type locality: Dirphys, Évvoia (as Euboea), Greece.
Distribution: Greece.

Rhacochelifer frivaldszkyi (Daday)

Chernes (Ectoceras) frivaldszkyi Daday, 1889a: 19–20, fig. 4.
Rhacochelifer (?) *frivaldszkyi* (Daday): Beier, 1932e: 268; Roewer, 1937: 316.

Type locality: 'Kussari', Georgia, U.S.S.R.
Distribution: U.S.S.R. (Georgia).

Rhacochelifer henschii (Daday)

Chernes (Ectoceras) henschii Daday, 1889d: 81–82.
Rhacochelifer (?) *henschii* (Daday): Beier, 1932e: 268; Roewer, 1937: 316; Beier, 1963b: 298.
Rhacochelifer henschii (Daday): Ćurčić, 1974a: 28.

Type locality: Domanovic, Hercegovina (as Herczegovina), Yugoslavia.
Distribution: Yugoslavia.

Rhacochelifer iranicus Beier

Rhacochelifer iranicus Beier, 1971a: 365–366, fig. 4.

Type locality: 56 km W. of Shiraz, Iran.
Distribution: Iran.

Rhacochelifer lobipes (Beier)

Chelifer (Chelifer) lobipes Beier, 1929a: 356–357, figs 10c, 11a.
Rhacochelifer lobipes (Beier): Beier, 1932d: 66; Beier, 1932e: 265, figs 284–285; Roewer, 1937: 316; Beier, 1963g: 277; Beier, 1965f: 98; Beier, 1967f: 320; Beier, 1969b: 196.

Type locality: Serai-Dagh near Konya (as Konia), Turkey.
Distribution: Turkey.

Rhacochelifer longeunguiculatus Beier

Rhacochelifer longeunguiculatus Beier, 1963f: 199–200, fig. 10.
Rhacochelifer cf. *longeunguiculatus* Beier: Mahnert, 1980c: 47.

Type locality: 'Ein Gedi', Israel?
Distribution: Israel?, Saudi Arabia.

Rhacochelifer maculatus (L. Koch)

Chelifer maculatus L. Koch, 1873: 30–31; E. Simon, 1878a: 147; E. Simon, 1879a: 32, 312, plate 18 fig. 9; E. Simon, 1885c: 49; Daday, 1889d: 80; E. Simon, 1898c: 3, 4; E. Simon, 1898d: 20; E. Simon, 1899d: 86; E. Simon, 1900b: 593; Ellingsen, 1908c: 669; Ellingsen, 1909a: 206, 208, 220; Navás, 1918: 89, 135; Navás, 1919: 211; Navás, 1921: 165; Navás, 1923: 31; Navás, 1925: 107.
Chelifer romanus Canestrini, 1883: no. 3, figs 1–2 (synonymized by Beier, 1932e: 262).
Chernes (Ectoceras) maculatus (L. Koch): Daday, 1888: 120, 175.
Chelifer (Chelifer) disjunctus L. Koch: Ellingsen, 1910a: 385 (misidentification).
Chelifer (Chelifer) maculatus (L. Koch): Nonidez, 1917: 25; Beier, 1929a: 354, fig. 9c; Beier, 1929d: 155; Beier, 1929e: 450.
Ectoceras maculatus (L. Koch): Beier, 1930f: 72; Beier, 1931d: 100.
Hansenius maculatus (L. Koch): J. C. Chamberlin, 1932a: 21.
Rhacochelifer maculatus (Beier): Beier, 1932d: 66; Beier, 1932e: 262–263, figs 280–281; Vachon, 1936a: 78; Roewer, 1937: 315; Vachon, 1937f: 128; Vachon, 1938a: fig. 51b; Beier, 1939f: 200; Beier, 1948a: 191; Vachon, 1953: 573; Beier, 1955n: 112; Beier, 1956e: 303; Beier, 1959f: 131; Weidner, 1959: 114; Beier, 1961b: 39; Beier, 1961d: 95; Beier, 1963b: 293–294, fig. 295; Beier, 1963i: 286; Helversen, 1966b: figs 4c, 5c; Lazzeroni, 1969a: 339; Lazzeroni, 1970a: 208; Heurtault, 1972b: figs 5–8; Lagar, 1972a: 21; Ćurčić, 1974a: 28; Mahnert, 1975c: 196–197; Orghidan, Dumitresco & Georgesco, 1975: 31; Ćurčić, 1976b: 180; Mahnert, 1977b: 72, figs 3a–b; Estany, 1979a: 222, fig. 1; Estany, 1979b: 47–50; Leclerc & Heurtault, 1979: 246; Heurtault, 1980a: 161–172, figs 2, 6, 7; Schawaller, 1981b: 49; Callaini, 1982b: 449; Anonymous, 1983b: 28; Callaini, 1983a: 157; Callaini, 1983d: 224; Schawaller, in Schmalfuss & Schawaller, 1984: 6; Callaini, 1987b: 288–289; Callaini, 1988b: 58; H. Hansen, 1988: 187.
Rhacochelifer maculatus maculatus (L. Koch): Lazzeroni, 1969c: 243.
Rhacochelifer (?) maculatus (L. Koch): Halperin & Mahnert, 1987: 128.

Type localities: of *Chelifer maculatus*: Corsica; and Villafranca, Spain.
of *Chelifer romanus*: Rome, Italy.
Distribution: Algeria, Balearic Islands, Canary Islands, Corsica, France, Greece, Israel, Italy, Malta, Morocco, Nicaragua?, Sardinia, Sicily, Spain, Tunisia, Yugoslavia.

Rhacochelifer maculatus hoggarensis Vachon

Rhacochelifer maculatus hoggarensis Vachon, 1940c: 157–160, figs 1–3.

Type locality: In Ameri, Ahoggar (as Hoggar) Mountains, Algeria.
Distribution: Algeria.

Rhacochelifer massylicus Callaini

Rhacochelifer massylicus Callaini, 1983e: 225–230, figs 4a–f.

Type locality: 5 km N. of Setif, Algeria.
Distribution: Algeria.

Rhacochelifer mateui Heurtault

Rhacochelifer mateui Heurtault, 1971a: 695–697, figs 19–21.

Type locality: Enneri Zouarké, Chad.
Distribution: Chad.

Rhacochelifer mongolicus Beier

Rhacochelifer mongolicus Beier, 1969c: 285–286, figs 3–5; Krumpál, 1984b: 66–69, figs 5–12.

Type locality: Bajanchongor aimak, Oase Echin gol, 90 km NE. of Caganbulag, Mongolia.
Distribution: Mongolia, U.S.S.R. (Tadzhikistan).

Rhacochelifer nubicus Beier

Rhacochelifer nubicus Beier, 1962e: 301−303, fig. 4.

Type locality: Wadi Halfa, Sudan.
Distribution: Sudan.

Rhacochelifer peculiaris (L. Koch)

Chelifer peculiaris L. Koch, 1873: 31−32; E. Simon, 1878a: 147; E. Simon, 1879a: 31, plate 18 fig. 8; Tömösváry, 1882b: 201−202, plate 2 figs 8−9; Pavesi, 1884: 456; E. Simon, 1885c: 48; Daday, 1888: 121−122, 176−177, plate 4 figs 5, 10, 12, 16; E. Simon, 1898d: 20; Ellingsen, 1908c: 669; Ellingsen, 1909a: 206; Daday, 1918: 2.
Chelifer entzii Daday & Tömösváry, in Daday, 1880a: 193; Tömösváry, 1882b: 200−201, plate 1 figs 27−28; Daday, 1918: 2 (synonymized by Beier, 1932e: 261).
Chelifer tegulatus Tömösváry, 1882b: 198−199, plate 2 figs 5−8; Daday, 1888: 122−123, 178, plate 4 figs 11, 13, 15; Daday, 1918: 2 (synonymized by Beier, 1932e: 261).
Chelifer disjunctus L. Koch: Tömösváry, 1882b: 199−200, plate 2 figs 14−15 (misidentification).
Chelifer (Chelifer) peculiaris L. Koch: Lessert, 1911: 20−21, fig. 16; Kästner, 1928: 8, fig. 28; Beier, 1929a: 354, fig. 9b; Beier, 1929e: 450.
Ectoceras peculiaris (L. Koch): Beier, 1930e: 292.
Rhacochelifer peculiaris (L. Koch): Beier, 1932d: 66; Beier, 1932e: 261−262, figs 278−279; Vachon, 1936a: 78; Roewer, 1937: 315; Vachon, 1937f: 128; Beier, 1952e: 6; Beier & Turk, 1952: 771; Beier, 1963b: 295, fig. 297; Beier, 1965d: 633; Beier, 1965f: 98; Beier, 1967f: 320; Beier, 1969b: 196; Ćurčić, 1974a: 28; Heurtault, 1980a: 161−172, figs 3−4, 9.
Rhacochelifer peculiaris peculiaris (L. Koch): Lazzeroni, 1969c: 243−244; Mahnert, 1977b: 69, figs 1a−b.
Not *Chelifer peculiaris* L. Koch: O.P.-Cambridge, 1889: ? (not seen) (misidentification; see *Withius piger* (E. Simon)).

Type localities: of *Chelifer peculiaris*: St Tulle, France.
Chelifer entzii: Hungary.
Chelifer tegulatus: Borbátvíz, Hungary.
Distribution: Algeria, Austria, Cyprus, France, Greece, Hungary, Italy, Sardinia, Sicily, Switzerland, Tunisia, Turkey, Yugoslavia.

Rhacochelifer peculiaris latissimus Beier

Rhacochelifer peculiaris latissimus Beier, 1963f: 201−202, fig. 11; Beier, 1967f: 320; Beier, 1969b: 196; Mahnert, 1977b: 69.

Type locality: Bethlehem (as Betlehem), Israel.
Distribution: Greece, Israel, Turkey.

Rhacochelifer pinicola (Nonidez)

Chelifer (Chelifer) pinicola Nonidez, 1917: 22−25, figs 5, 5a.
Chelifer ibericus Navás, 1918: 89−90, 106, figs 1a−d; Navás, 1925: 108, figs 6a−d (synonymized by Beier, 1932e: 263).
Chelifer pinicola Nonidez: Navás, 1925: 108, fig. 5.
Rhacochelifer pinicola (Beier): Beier, 1932d: 66; Beier, 1932e: 263; Roewer, 1937: 315; Beier, 1939f: 201; Beier, 1963b: 294; Mahnert, 1980a: 262−264, figs 12−16.
Rhacochelifer sp. ? *pinicola* Nonidez: Beier, 1970a: 45; Beier, 1976a: 25.

Type localities: of *Chelifer (Chelifer) pinicola*: Caravaca, Murcia, Spain.
of *Chelifer ibericus*: Zaragoza, Spain.
Distribution: Canary Islands, Spain.

Rhacochelifer quadrimaculatus (Tömösváry)

Chelifer quadrimaculatus Tömösváry, 1882a: 227–228, 296–297, figs 1–6; Tömösváry, 1882b: 196–198, plate 2 figs 1–4; Daday, 1918: 2.
Rhacochelifer quadrimaculatus (Tömösváry): Beier, 1932d: 66; Beier, 1932e: 264–265; Roewer, 1937: 315; Verner, 1960: 168, fig. 2; Beier, 1963b: 296.

Type locality: Homonna, Zemplénmegye, Hungary.
Distribution: Czechoslovakia, Hungary.

Rhacochelifer saharae Beier

Rhacochelifer similis Beier: Vachon, 1950: 98–100, figs 9–12 (misidentification).
Rhacochelifer saharae Beier, 1962e: 303; Beier, 1965b: 374.

Type locality: Baguezane Mountains, Niger.
Distribution: Chad, Niger.

Rhacochelifer samai Callaini

Rhacochelifer samai Callaini, 1987b: 289–291, figs 5a–d.

Type locality: Icef Camliyayla, Turkey.
Distribution: Turkey.

Rhacochelifer similis Beier

Rhacochelifer similis Beier, 1932f: 488–489, figs 1–2; Caporiacco, 1936a: 112, 118; Vachon, 1940: 2; Beier, 1947c: 127–128; Beier, 1948b: 445.
Not *Rhacochelifer similis* Beier: Vachon, 1950: 98–100, figs 9–12 (misidentification; see *Rhacochelifer saharae* Beier).

Type locality: Oasis di Gialo (now Wabat Jalu), Libya.
Distribution: Egypt, Libya.

Rhacochelifer spiniger Mahnert

Rhacochelifer spiniger Mahnert, 1978g: 14–17, figs 1–7.
Rhacochelifer cf. *spiniger* Mahnert: Mahnert, 1980a: 264.

Type locality: Sagres, Algarve, Portugal.
Distribution: Canary Islands, Portugal.

Rhacochelifer subsimilis Vachon

Rhacochelifer subsimilis Vachon, 1940e: 69–70, figs g–j.

Type locality: Adrar des Iforhas, Algeria/Mali.
Distribution: Algeria/Mali.

Rhacochelifer tauricus Beier

Rhacochelifer tauricus Beier, 1969b: 197–198, fig. 3.

Type locality: Namrun, Turkey.
Distribution: Turkey.

Rhacochelifer tenuimanus Heurtault

Rhacochelifer tenuimanus Heurtault, 1971a: 689–692, figs 7–12.

Type locality: Koussi à Koudou, Chad.
Distribution: Chad.

Rhacochelifer tibestiensis Heurtault

Rhacochelifer tibestiensis Heurtault, 1971a: 692–695, figs 13–18.

Type locality: Emi Koussi Gorrom, Chad.
Distribution: Chad.

Rhacochelifer tingitanus (L. Koch)

Chelifer tingitanus L. Koch, 1873: 24–26; E. Simon, 1878a: 147; Navás, 1925: 107.
Chelifer (Chelifer) tingitanus L. Koch: Nonidez, 1917: 20.
Rhacochelifer (?) *tingitanus* (L. Koch): Beier, 1932e: 268; Roewer, 1937: 316.
Chelifer (?) *tingitanus* L. Koch: Callaini, 1988b: 59.

Type locality: Morocco.
Distribution: Morocco, Spain.

Rhacochelifer villiersi Vachon

Rhacochelifer villiersi Vachon, 1938e: 210–212, figs i–l.

Type locality: Demnate (as Demnat), Morocco.
Distribution: Morocco.

Genus **Rhopalochelifer** Beier

Rhopalochelifer Beier, 1964k: 85.

Type species: *Rhopalochelifer lawrencei* Beier, 1964k, by original designation.

Rhopalochelifer lawrencei Beier

Rhopalochelifer lawrencei Beier, 1964k: 85–86, fig. 38; Beier, 1966k: 469.

Type locality: Storms River mouth, Cape Province, South Africa.
Distribution: South Africa.

Genus **Sinochelifer** Beier

Sinochelifer Beier, 1967g: 366–367.

Type species: *Sinochelifer kwantungensis* Beier, 1967g, by original designation.

Sinochelifer kwantungensis Beier

Sinochelifer kwantungensis Beier, 1967g: 367–368, fig. 26.

Type locality: Mei-hsien, Kwangtung, China.
Distribution: China (Kwangtung).

Genus **Stenochelifer** Beier

Stenochelifer Beier, 1967g: 365.

Type species: *Stenochelifer indicus* Beier, 1967g, by original designation.

Stenochelifer indicus Beier

Stenochelifer indicus Beier, 1967g: 365–366, fig. 25.

Type locality: Kankér, Madhya Pradesh, India.
Distribution: India.

Genus **Strobilochelifer** Beier

Strobilochelifer Beier, 1932e: 234–235.

Type species: *Chelifer spinipalpis* Redikorzev, 1918, by original designation.

Strobilochelifer spinipalpis (Redikorzev)

Chelifer spinipalpis Redikorzev, 1918: 91–93, figs 1, 1a.
Strobilochelifer spinipalpis (Redikorzev): Beier, 1932e: 235, fig. 243; Roewer, 1936: fig. 28a; Roewer, 1937: 312; Mahnert, 1980c: 43–45, figs 29–35.
Strobilochelifer grandimanus Beier, 1943: 77–78, fig. 4; Beier, 1951d: 99; Beier, 1971a: 364 (synonymized by Mahnert, 1980c: 43).

Type localities: of *Chelifer spinipalpis*: Bazman, Baluchestan, Iran.
of *Strobilochelifer grandimanus*: Kut al Sayyid, Basrah, Iraq.
Distribution: Iran, Iraq, Saudi Arabia.

Genus Stygiochelifer Beier

Stygiochelifer Beier, 1932d: 66; Beier, 1932e: 272.

Type species: *Chelifer cavernae* Tullgren, 1912a, by original designation.

Stygiochelifer cavernae (Tullgren)

Chelifer cavernae Tullgren, 1912a: 259–263, figs 1a–e.
Stygiochelifer cavernae (Tullgren): Beier, 1932d: 66; Beier, 1932e: 272–273, figs 296–297; Roewer, 1936: fig. 150; Roewer, 1937: 316; Chapman, 1984a: 6.

Type locality: Guwa Lawa, near Babakan, Banjumas, Java, Indonesia.
Distribution: Indonesia (Java), Malaysia (Sarawak).

Genus Telechelifer J. C. Chamberlin

Telechelifer J. C. Chamberlin, 1949: 36–38.

Type species: *Telechelifer lophonotus* J. C. Chamberlin, 1949, by original designation.

Telechelifer lophonotus J. C. Chamberlin

Telechelifer lophonotus J. C. Chamberlin, 1949: 38–41, figs 11a–l.

Type locality: probably Sri Lanka (as Ceylon).
Distribution: Sri Lanka?

Genus Tetrachelifer Beier

Tetrachelifer Beier, 1967g: 363–364.

Type species: *Tetrachelifer vietnamensis* Beier, 1967g, by original designation.

Tetrachelifer pusillus (Redikorzev)

Lophochernes pusillus Redikorzev, 1938: 111–112, fig. 42; Roewer, 1940: 348; Beier, 1951a: 115.
Tetrachelifer pusillus (Redikorzev): Beier, 1967g: 364.

Type locality: Da Lat (as Dalat), Vietnam.
Distribution: Vietnam.

Tetrachelifer vietnamensis Beier

Tetrachelifer vietnamensis Beier, 1967g: 364–365, fig. 24; Tenorio & Muchmore, 1982: 379.

Type locality: 13 km W. of Postal de M'drah, Vietnam.
Distribution: Vietnam.

Genus Trachychelifer Hong

Trachychelifer Hong, 1983: 22, 28.

Type species: *Trachychelifer liaoningense* Hong, 1983, by original designation.

! Trachychelifer liaoningense Hong

Trachychelifer liaoningense Hong, 1983: 24, 28–29, figs 1–3, plate 3 figs 6–7.

Type locality: Chinese Amber.
Distribution: Chinese Amber.

Genus **Tyrannochelifer** J. C. Chamberlin

Tyrannochelifer J. C. Chamberlin, 1932a: 20; Beier, 1932e: 251; Hoff, 1964b: 24–25.

Type species: *Chelifer imperator* With, 1908a, by original designation.

Tyrannochelifer cubanus Hoff

Tyrannochelifer cubanus Hoff, 1964b: 29–30, figs 13–14, 17.

Type locality: near Nuevitas, Camagüey Province, Cuba.
Distribution: Cuba.

Tyrannochelifer floridanus (Banks)

Chelifer floridanus Banks, 1891: 162; Banks, 1895a: 3; Banks, 1904a: 140; Coolidge, 1908: 109; J. C. Chamberlin, 1931a: fig. 39j.
Tyrannochelifer floridanus (Banks): J. C. Chamberlin, 1932a: 20; Beier, 1932e: 253; Roewer, 1937: 315; Hoff, 1958: 35; Hoff, 1964b: 25–29, figs 11–12, 15–16; Beier, 1976d: 58.

Type locality: Biscayne Bay, Dade County, Florida, U.S.A.
Distribution: Dominican Republic, U.S.A. (Florida).

Tyrannochelifer imperator (With)

Chelifer imperator With, 1908a: 221–225, figs 1a–i; J. C. Chamberlin, 1931a: figs 39o–p.
Tyrannochelifer imperator (With): J. C. Chamberlin, 1932a: 20; Beier, 1932e: 251–252; Roewer, 1937: 315; Vachon, 1949: fig. 204a.
Tyrannochelifer imperator (?) (With): Feio, 1945: 8.

Type locality: Brazil.
Distribution: Brazil (Sao Paulo).

Tyrannochelifer macropalpus (Tullgren)

Chelifer macropalpus Tullgren, 1907a: 26–29, fig. 1.
Tyrannochelifer macropalpus (Tullgren): J. C. Chamberlin, 1932a: 20; Beier, 1932e: 252, fig. 260; Roewer, 1936: fig. 149; Roewer, 1937: 315, fig. 245; Weidner, 1959: 114.

Type locality: St Marc, Haiti.
Distribution: Haiti.

Genus **Xenochelifer** J. C. Chamberlin

Xenochelifer J. C. Chamberlin, 1949: 10–12.

Type species: *Xenochelifer davidi* J. C. Chamberlin, 1949, by original designation.

Xenochelifer davidi J. C. Chamberlin

Xenochelifer davidi J. C. Chamberlin, 1949: 12–17, figs 4a–e, 5a–p; Hoff, 1958: 33.

Type locality: Big Rock Creek, Los Angeles County, California, U.S.A.
Distribution: U.S.A. (California).

Family CHERNETIDAE Menge

Chernetidae Menge, 1855: 22; J. C. Chamberlin, 1931a: 241–242; Beier, 1932e: 80–81; Beier, 1932g: 186–187; J. C. Chamberlin, 1935b: 481; Roewer, 1937: 287; Hoff, 1949a: 40–41; Hoff, 1949b: 449; Petrunkevitch, 1955: 82; Hoff, 1956c: 4; Morikawa, 1960: 139; Beier, 1963b: 248; Muchmore, 1974d: 26; Murthy & Ananthakrishnan, 1977: 134; Muchmore, 1982a: 101; Judson, 1985: 321; Harvey, 1985b: 133; Legg, 1987: 180–181; Legg & Jones, 1988: 94.
Chernetinae Menge: Beier, 1932e: 105–106; Beier, 1933a: 509–510; Roewer, 1937: 291; J. C. Chamberlin, 1938b: 118; Hoff, 1949b: 455; Hoff, 1956c: 12–13; Morikawa, 1960: 141–142; Murthy & Ananthakrishnan, 1977: 135; Legg, 1987: 182; Legg & Jones, 1988: 117.
Myrmochernetidae J. C. Chamberlin, 1931a: 240–241; Beier, 1932e: 190; Beier, 1932g: 187; Roewer, 1937: 304; Muchmore, 1982a: 101.
Lamprochernetinae Beier: Beier, 1932e: 81–82; Roewer, 1937: 288; J. C. Chamberlin, 1938b: 114; Hoff, 1949b: 450; Hoff, 1956c: 4–5; Morikawa, 1960: 139–140; Murthy & Ananthakrishnan, 1977: 134–135; Legg, 1987: 181; Legg & Jones, 1988: 94.
Chernetini Menge: Beier, 1932e: 106; Beier, 1933a: 510–511; Roewer, 1937: 292; Morikawa, 1960: 142; Murthy & Ananthakrishnan, 1977: 136.
Hesperochernetini Beier, 1932e: 168–169; Beier, 1933a: 533–534; Roewer, 1937: 299.
Goniochernetinae Beier, 1932e: 187; Beier, 1933a: 547; Roewer, 1937: 303; Vachon, 1939a: 123; Beier, 1954d: 136.
Xenochernetinae Feio, 1945: 36–37.

Genus **Acanthicochernes** Beier

Acanthicochernes Beier, 1964f: 594–595.

Type species: *Acanthicochernes biseriatus* Beier, 1964f, by original designation.

Acanthicochernes biseriatus Beier

Acanthicochernes biseriatus Beier, 1964f: 595–596, fig. 2; Beier, 1965g: 777; Beier, 1966d: 143; Beier, 1967c: 322; Beier, 1970g: 322.

Type locality: Savo Island, Guadalcanal, Solomon Islands.
Distribution: Papua New Guinea, Solomon Islands.

Genus **Acuminochernes** Hoff

Acuminochernes Hoff, 1949b: 476–477; Hoff, 1961: 450.
Phoberocheirus J. C. Chamberlin, 1949: 6–7 (synonymized by Muchmore, 1981c: 335).

Type species: of *Acuminochernes*: *Hesperochernes crassopalpus* Hoff, 1945h, by original designation.
of *Phoberocheirus*: *Phoberocheirus cribellus* J. C. Chamberlin, 1949, by original designation.

Acuminochernes crassopalpus (Hoff)

Hesperochernes crassopalpus Hoff, 1945c: 43–49, figs 7–12.
Acuminochernes crassopalpus (Hoff): Hoff, 1949b: 477–478, figs 44a–d; Park & Auerbach, 1954: 210; Hoff & Bolsterli, 1956: 174; Hoff, 1958: 25; Nelson, 1975: 287–288, figs 54–55; Muchmore, 1981c: 335; Benedict & Malcolm, 1982: 99; Zeh, 1987b: 1084.
Phoberocheirus cribellus J. C. Chamberlin, 1949: 8–10, figs 3a–j; Hoff, 1958: 25; Muchmore, 1971a: 88 (synonymized by Muchmore, 1981c: 335).

Type localities: of *Hesperochernes crassopalpus*: Lake Wedington Wildlife Area, Washington County, Arkansas, U.S.A.
of *Phoberocheirus cribellus*: Vienna, Virginia, U.S.A.
Distribution: U.S.A. (Arkansas, Florida, Georgia, Illinois, Kansas, Michigan, North Carolina, Oregon, Tennessee, Virginia).

Acuminochernes tacitus Hoff

Acuminochernes tacitus Hoff, 1961: 450–455, figs 16–19; Muchmore, 1971a: 89; Nelson, 1975: 287, figs 52–53; Zeh, 1987b: 1084.

Type locality: 2.5 miles E. of Fort Collins, Larimer County, Colorado, U.S.A.
Distribution: U.S.A. (Colorado, Michigan).

Genus Adelphochernes Beier

Adelphochernes Beier, 1937b: 271–272.

Type species: *Adelphochernes mindanensis* Beier, 1937b, by original designation.

Adelphochernes mindanensis Beier

Adelphochernes mindanensis Beier, 1937b: 272–273, fig. 3; Roewer, 1940: 347.

Type locality: Mommangan, Mindanao, Philippines.
Distribution: Philippines.

Adelphochernes mindoroensis Beier

Adelphochernes mindoroensis Beier, 1966b: 346–347, fig. 4.

Type locality: San Jose, Mindoro (as Mindanao), Philippines.
Distribution: Philippines.

Genus Allochernes Beier

Allochernes Beier, 1932e: 145; Beier, 1933a: 524; G.O. Evans & Browning, 1954: 21; Morikawa, 1960: 146; Legg, 1972: 580; Legg, 1987: 181; Legg & Jones, 1988: 111.
Allochernes (Allochernes) Beier: Beier, 1932e: 146; Beier, 1933a: 525; Beier, 1963b: 261–262.

Type species: *Chelifer wideri* C. L. Koch, 1843, by original designation.

Allochernes aetnaeus Beier

Allochernes aetnaeus Beier, 1975a: 56–58, fig. 1; Callaini, 1979c: 148, fig. 3.

Type locality: Mt Minardo, Mt Etna (as Aetna), Sicily.
Distribution: Italy, Sicily.

Allochernes asiaticus (Redikorzev)

Chelifer (Chelanops) asiaticus Redikorzev, 1922a: 262–264, figs 5–6.
Allochernes (Allochernes) asiaticus (Redikorzev): Beier, 1932e: 150–151, fig. 160b.
Allochernes asiaticus (Redikorzev): Roewer, 1937: 298; Schawaller, 1986: 6–7; Schawaller, 1989: 18, figs 43–51.

Type locality: Rchombo-Mtzo Lake, Tibet, China.
Distribution: China (Tibet), U.S.S.R. (Kirghizia, R.S.F.S.R.).

Allochernes asiaticus nepalensis Morikawa

Allochernes (Allochernes) asiaticus nepalensis Morikawa, 1968: 263, figs 1e, 2d–e.

Type locality: Pachkal, near Mt Numbur, Nepal.
Distribution: Nepal.

Allochernes balcanicus Hadži

Allochernes (Allochernes) balcanicus Hadži, 1938: 23–28, figs 24a–f, 25a–b.
Allochernes creticus balcanicus Hadži: Beier, 1963b: 267.
Allochernes balcanicus Hadži: Ćurčić, 1974a: 26; Ćurčić, 1976b: 179.

Type locality: Pecina Rasce, near Skopje (as Skoplje), Yugoslavia.
Distribution: Yugoslavia.

Allochernes brevipilosus Beier

Allochernes brevipilosus Beier, 1967h: 17–18, fig. 1.

Type locality: Tangigaruh, 20 km E. of Kabul, Afghanistan.
Distribution: Afghanistan.

Allochernes bulgaricus Hadži

Allochernes (Allochernes) bulgaricus Hadži, 1939b: 39–43, figs 12a–l.
Allochernes bulgaricus Hadži: Beier, 1963b: 263.

Type locality: Vitosa-Planina, Bulgaria.
Distribution: Bulgaria.

Allochernes elbursensis Beier

Allochernes elbursensis Beier, 1969d: 121–122, fig. 1; Beier, 1971a: 364.

Type locality: Karaj (as Keredj), 40 km W. of Tehran (as Teheran), Iran.
Distribution: Iran.

Allochernes ginkgoanus (Morikawa)

Toxochernes ginkgoanus Morikawa, 1953a: 353–354, figs 3c–f; Morikawa, 1957a: 365.
Allochernes (Toxochernes) ginkgoanus (Morikawa): Morikawa, 1960: 147, plate 4 fig. 2, plate 8 fig. 5, plate 9 fig. 24, plate 10 fig. 17; Morikawa, 1962: 419.
Allochernes ginkgoanus (Morikawa): Mahnert, 1978a: 309, 314.

Type locality: Kami-machi, Setagaya, Tokyo, Japan.
Distribution: Japan.

Allochernes himalayensis Beier

Allochernes himalayensis Beier, 1974b: 272–273, fig. 7.
Allochernes (?) *himalayensis* Beier: Schawaller, 1987a: 215, figs 36–39.

Type locality: Zaral Baira, Dzunda Khola Valley, Nepal.
Distribution: Nepal.

Allochernes japonicus (Morikawa)

Toxochernes japonicus Morikawa, 1953a: 351–353, figs 2g, 3a–b.
Toxochernes dubius japonicus Morikawa: Morikawa, 1955a: 222.
Allochernes (Toxochernes) japonicus (Morikawa): Morikawa, 1960: 147, plate 4 fig. 1, plate 8 fig. 3, plate 9 fig. 21; Morikawa, 1962: fig. 2.
Allochernes japonicus (Morikawa): Mahnert, 1978a: 309; Sato, 1979a: plate 5; Sato, 1982c: 34.

Type locality: Dogo, Matsuyama, Shikoku, Japan.
Distribution: Japan.

Allochernes liwa Harvey

Allochernes liwa Harvey, 1988b: 346–348, figs 113–119.

Type locality: 7 km W. of Liwa, Sumatra, Indonesia.
Distribution: Indonesia (Sumatra).

Allochernes maroccanus Mahnert

Allochernes maroccanus Mahnert, 1976b: 377–381, figs 5–11; Callaini, 1988b: 56.

Type locality: Grotte du Caïd (Ifri el Caïd), Aït Mehammed, Morocco.
Distribution: Morocco.

Allochernes masi (Navás)

Chelifer masi Navás, 1923: 31–32, figs 11a–b; Navás, 1925: 106, fig. 2; Roewer, 1940: 347.
Allochernes (?) *masi* (Navás): Beier, 1932e: 154; Roewer, 1937: 298.
Allochernes masi (Navás): Beier, 1939f: 198–199, fig. 24; Beier, 1962c: 25; Beier, 1963a: 260; Beier, 1963b: 269, fig. 270; Beier, 1963f: 195; Puddu & Pirodda, 1973: 173; Beier, 1974g: 163; Estany, 1977a: 149–152, figs 1–7; Mahnert, 1977c: 102; Gardini, 1980c: 118, 130, 131, 132; Callaini, 1982a: 17, 23, fig. 4c; Callaini, 1986c: 393–394, fig. 4c.
Allochernes (Allochernes) barrosi Vachon, 1940g: 151–153, figs 12–18 (synonymized by Beier, 1962c: 25).
Allochernes wideri phaleratus (E. Simon): Lazzeroni, 1969c: 241 (misidentification).

Type localities: of *Chelifer masi*: Centellas, Barcelona, Spain.
of *Allochernes (Allochernes) barrosi*: Mina dos Mouros, Mexilhoeira Grande, Portugal.
Distribution: Israel, Portugal, Sardinia, Sicily, Spain.

Allochernes microti Beier

Allochernes microti Beier, 1962b: 149–150, figs 4–5; Beier, 1963b: 266, fig. 266; Kobakhidze, 1965b: 541; Kobakhidze, 1966: 704; Beier, 1971a: 364.
Allochernes pauperatus Beier, 1965f: 94, fig. 9 (synonymized by Beier, 1971a: 364).

Type localities: of *Allochernes microti*: Schiraki-Steppe, Georgia, U.S.S.R.
of *Allochernes pauperatus*: Maras (as Marasch), Turkey.
Distribution: Iran, Turkey, U.S.S.R. (Georgia).

Allochernes mongolicus Beier

Allochernes mongolicus Beier, 1966m: 225–227, fig. 1.

Type locality: Cojbalsan aimak, 80 km NW. of Cojbalsan, Mongolia.
Distribution: Mongolia.

Allochernes peregrinus Lohmander

Allochernes peregrinus Lohmander, 1939a: 10–11, figs 5–8; Lohmander, 1939b: 303; Beier, 1948b: 444; Beier, 1963b: 262–263; Rafalski, 1967: 17; Muchmore, 1972d: 109–110; Mahnert, 1983d: 362, figs 2–3; Drogla, 1984: 191.

Type locality: Stockholm, Sweden.
Distribution: Austria, East Germany, Hungary, Poland, Sweden, U.S.A. (New Hampshire).

Allochernes pityusensis Beier

Allochernes pityusensis Beier, 1961b: 35–36, fig. 5; Beier, 1963b: 269–270, fig. 271.

Type locality: Santa Eulalia, Ibiza, Balearic Islands.
Distribution: Balearic Islands.

Allochernes powelli (Kew)

Chelifer (Chernes) wideri C. L. Koch: Standen, 1912: 12–13, fig. 1 (misidentification).
Chelifer (Chernes) powelli Kew, 1916b: 74–75, fig. 2; Standen, 1916b: 124; Standen, 1917: 28.

Chernes creticus Beier, 1931d: 98–99, figs 7, 8a–b, 9a (synonymized by Mahnert, 1980e: 232).
Allochernes (Allochernes) creticus (Beier): Beier, 1932e: 147–148, fig. 157.
Allochernes (Allochernes) powelli (Kew): Beier, 1932e: 149.
Allochernes (Allochernes) italicus Beier, 1932e: 149–150, fig. 159; Beier, 1933a: 526–527, fig. 5 (synonymized by Beier, 1967f: 313).
Allochernes italicus Beier: Vachon, 1934c: 38–54, figs a–f, h–i; Vachon, 1934d: 133; Vachon, 1935j: 187; Vachon, 1936a: 78; Vachon, 1936b: 140; Roewer, 1937: 297; Vachon, 1937e: fig. 7; Vachon, 1937f: 128; Vachon, 1938a: 98–100, figs 33b, 33f, 55a–c, 56j; Vachon, 1957: fig. 7; Beier, 1963b: 268, fig. 269; Beier, 1963h: 156.
Allochernes creticus (Beier): Roewer, 1937: 297.
Allochernes powelli (Kew): Roewer, 1937: 297; Beier, 1953a: 301; G. O. Evans & Browning, 1954: 21; Beier, 1959f: 131; Meinertz, 1962a: 397, map 6; Beier, 1963b: 268–269; Helversen, 1966a: 141–145, figs 3–5; Beier, 1967f: 313; Hammen, 1969: 21; Ressl, 1970: 252–253; Helversen & Martens, 1971: 378; Kofler, 1972: 288; Legg, 1972: 580; Puddu & Pirodda, 1973: 173; Beier, 1974g: 163; Legg, 1975a: 66; Mahnert, 1975a: 184; Crocker, 1976: 9; Crocker, 1978: 9; Mahnert, 1980e: 232, fig. 29; Gardini, 1980c: 118, 132; Jones, 1980c: map 20; Gardini, 1982c: 112; Beron, 1985: 66; Callaini, 1986c: 394, fig. 4d; Judson, 1987: 369; Legg & Jones, 1988: 114–116, figs 28a, 28b(a–i); Mahnert, 1989a: 87.
Allochernes creticus creticus (Beier): Beier, 1963b: 266–267, fig. 267.

Type localities: of *Chelifer (Chernes) powelli*: Surrey, England, Great Britain; Middlesex, England, Great Britain; Essex, England, Great Britain; Lancashire, England, Great Britain; and Montgomeryshire, England, Great Britain.
of *Chernes creticus*: Kumaro Cave, Akrotiri, Crete.
of *Allochernes (Allochernes) italicus*: Genoa (as Genua), Italy.
Distribution: Austria, Balearic Islands, Crete, Denmark, France, Great Britain, Greece, Italy, Netherlands, Portugal, Sardinia, Spain, Turkey, West Germany.

Allochernes rhodius Beier

Allochernes rhodius Beier, 1966e: 164–166, fig. 2.

Type locality: S. of Líndos (as Lindos), Ródhos, Greece.
Distribution: Greece.

Allochernes siciliensis (Beier)

Chernes siciliensis Beier, 1963a: 260–262, fig. 4; Beier, 1963b: 276; Lazzeroni, 1969c: 241–242.
Chernes (?) *siciliensis* Beier: Mahnert, 1975c: 196.
Allochernes siciliensis (Beier): Callaini, 1986c: 389–393, figs 3a–f, 4b.

Type locality: Monte Cronio, Sciacca, Sicily.
Distribution: Italy, Malta, Sardinia, Sicily.

Allochernes solarii (E. Simon)

Chelifer solarii E. Simon, 1898d: 23–24.
Allochernes (Allochernes) solarii (E. Simon): Beier, 1932e: 150, fig. 160a.
Allochernes solarii (E. Simon): Roewer, 1937: 297; Vachon, 1938a: fig. 56g; Beier, 1963b: 267–268, fig. 268; Lazzeroni, 1969b: 409; Lazzeroni, 1969c: 244.

Type locality: Montecapraro, near Tortona, Italy.
Distribution: Italy.

Allochernes tripolitanus Beier

Allochernes tripolitanus Beier, 1954c: 324–326, fig. 1.

Type locality: Misrata (as Misurata), Tripolitania, Libya.
Distribution: Libya.

Family *Chernetidae*

Allochernes tucanus Beier

Allochernes tucanus Beier, 1959c: 407–408, fig. 1.

Type locality: Tasso, Karakorum, Pakistan.
Distribution: Pakistan.

Allochernes turanicus (Redikorzev)

Chelifer (Chelanops) turanicus Redikorzev, 1934a: 425–427, figs 2–3.
Allochernes turanicus (Redikorzev): Redikorzev, 1934b: 152; Roewer, 1937: 298; Redikorzev, 1949: 654; Schawaller, 1989: 21.

Type locality: Iolotan' (as Iolatanj), Turkmenistan, U.S.S.R.
Distribution: U.S.S.R. (Kazakhstan, Turkmenistan).

Allochernes wideri (C. L. Koch)

Chelifer wideri C. L. Koch, 1843: 47–48, fig. 784; Löw, 1867: 746; Stecker, 1875d: 88; H. J. Hansen, 1884: 545–546; Tullgren, 1899a: 174; Ellingsen, 1907a: 162; Ellingsen, 1908c: 669; Donisthorpe, 1926: 54; Donisthorpe, 1927: 183; Kew, 1929c: 21–22.
Chernes oblongus Menge, 1855: 39–40, plate 1 fig. 19, plate 2 fig. 15, plate 3 figs 6, 13–15, plate 5 fig. 14 (junior homonym of *Chelifer oblongus* Say, 1821) (synonymized by Beier, 1932e: 146).
Chernes wideri (C. L. Koch): L. Koch, 1873: 10–11; Canestrini, 1874: 224; Stecker, 1874b: 233; L. Koch, 1877: 180; Kew, 1909b: 259; Daday, 1918: 2; Beier, 1931d: fig. 9b.
Chernes (Trachychernes) wideri (C. L. Koch): Tömösváry, 1882b: 193–194, plate 2 figs 22–23.
Chelifer (Trachychernes) wideri C. L. Koch: Ellingsen, 1910a: 383; Lessert, 1911: 17–18, fig. 15; Beier, 1929a: 343 (in part; see *Chernes similis* (Beier)).
Chelifer (Chernes) wideri (C. L. Koch): Kew, 1911a: 44–45, fig. 6; Kew, 1916b: 74, fig. 1; Kästner, 1928: 6, fig. 16; Schenkel, 1928: 59, figs 12–13.
Allochernes (Allochernes) wideri (C. L. Koch): Beier, 1932e: 146–147, fig. 156a; Beier, 1939e: 312.
Allochernes wideri (C. L. Koch): Tumšs, 1934: 15, fig. 3; Vachon, 1934d: 133; Vachon, 1935j: 187; Roewer, 1937: 297; Vachon, 1937e: fig. 8; Lohmander, 1939b: 300–302, fig. 8; Vachon, 1940f: 2; Cooreman, 1946b: 4; Kaisila, 1947: 86; Beier, 1948b: 457–458, 462; Kaisila, 1949b: 83, map 8; Beier, 1952e: 5; Beier & Franz, 1954: 457; G. O. Evans & Browning, 1954: 21, fig. 15; Weidner, 1954b: 110; Beier, 1956h: 24; Beier, 1957b: 147; Ressl & Beier, 1958: 2; Meinertz, 1962a: 396, map 6; Beier, 1963b: 263–264, fig. 263; Beier, 1963g: 274; Beier, 1965f: 94; Ressl, 1965: 289; Helversen, 1966a: 140–141, figs 3–4; Rafalski, 1967: 16; Weygoldt, 1969a: fig. 96; Ressl, 1970: 252–253; Helversen & Martens, 1971: 379; Gabbutt, 1972a: 37–40, figs 1d, 2d; Gabbutt, 1972b: 2–13; Legg, 1972: 580, figs 2(4c), 4(1); Legg, 1973b: 430, fig. 2d; Ressl, 1974: 27; Klausen, 1975: 64; Legg, 1975a: 66; Legg, 1975d: fig. 2; Crocker, 1976: 9; Drogla, 1977: 89; Klausen & Totland, 1977: 101–108, plates 2–4, figs 1–2; Beier, 1978b: 415; Crocker, 1978: 9; Jones, 1979a: 200; Judson, 1979a: 62; Leclerc & Heurtault, 1979: 246; Jones, 1980c: map 19; Jędryczkowski, 1985: 79; Lippoid, 1985: 40; Callaini, 1986c: 394; Jędryczkowski, 1987b: 143, map 8; Krumpál & Cyprich, 1988: 42; Legg & Jones, 1988: 111–113, figs 27a, 27b(a–k); Schawaller & Dashdamirov, 1988: 36, figs 61, 67; Schawaller, 1989: 21.
Allochernes wideri wideri (C. L. Koch): Mahnert, 1974a: 90; Mahnert, 1977c: 103.
Not *Chelifer wideri* C. L. Koch: Löw, 1866: 944 (misidentification; see *Lamprochernes nodosus* (Schrank)).
Not *Chernes oblongus* Menge: Stecker, 1874c: 313–314 (misidentification; see *Pselaphochernes scorpioides* (Hermann)).

539

Family Chernetidae

Not *Chelifer (Chernes) wideri* C. L. Koch: Standen, 1912: 12–13, fig. 1 (misidentification; see *Allochernes powelli* (Kew)).

Type locality: of *Chelifer wideri*: Erlangen, West Germany.
of *Chernes oblongus*: not stated.

Distribution: Austria, Belgium, Czechoslovakia, Denmark, East Germany, France, Finland, Great Britain, India, Iran, Italy, Norway, Poland, Spain, Sweden, Switzerland, Turkey, U.S.S.R. (Azerbaijan, R.S.F.S.R.), West Germany.

Allochernes wideri phaleratus (E. Simon)

Chelifer phaleratus E. Simon, 1879a: 38, plate 18 fig. 12; Canestrini, 1885: no. 2, figs 1–3; E. Simon, 1898d: 20; Tullgren, 1899a: 173–174, plate 1 fig. 9; Gestro, 1904: 14 (not seen); Tullgren, 1906a: 203; Ellingsen, 1908a: 415; Ellingsen, 1909a: 208, 212; Ellingsen, 1910c: 348; Falcoz, 1912a: 1382; Caporiacco, 1923: 132.
Chernes (Chernes) phaleratus (E. Simon): Daday, 1888: 118, 173.
Chernes phaleratus (E. Simon): Daday, 1889a: 16; Daday, 1889c: 25; Daday, 1889d: 80; Godfrey, 1901b: 216; Daday, 1918: 2.
Chernes (Trachychernes) phaleratus (E. Simon): Ellingsen, 1897: 11–12.
Allochernes (Allochernes) phaleratus (E. Simon): Beier, 1932e: 148–149, fig. 158.
Allochernes phaleratus (E. Simon): Boldori, 1935: 27; Roewer, 1937: 297; Cerruti, 1959: 51; Beier, 1963b: 265, fig. 265; Ćurčić, 1974a: 26.
Allochernes wideri phaleratus (E. Simon): Beier, 1967f: 312; Beier, 1969b: 193; Lazzeroni, 1969a: 337; Lazzeroni, 1970a: 208; Puddu & Pirodda, 1973: 172; Gardini, 1980c: 118, 129, 132, 133; Mahnert, 1983d: 362.
Allochernes sp. ? *phaleratus* (E. Simon): Schawaller, 1981b: 48.
Not *Chelifer phaleratus* E. Simon: O.P.-Cambridge, 1892: 228–229 (misidentification; see *Pselaphochernes scorpioides* (Hermann)).
Not *Chernes phaleratus* (E. Simon): Evans, 1901b: 242 (misidentification; see *Pselaphochernes dubius* (O.P.-Cambridge)); O.P.-Cambridge, 1905: ? (not seen) (misidentification; see *Dinocheirus panzeri* (C. L. Koch)).
Not *Allochernes wideri phaleratus* (E. Simon): Lazzeroni, 1969c: 241 (misidentification; see *Allochernes masi* (Navás)).

Type localities: Fontainebleau, France; and Troyes, France.
Distribution: Algeria, Corsica, France, Hungary, Italy, Norway, Sardinia, Sweden, Turkey, Yugoslavia.

Allochernes wideri transcaucasicus Kobakhidze

Allochernes wideri transcaucasicus Kobakhidze, 1964a: 449–451, fig. 1; Kobakhidze, 1965b: 541; Kobakhidze, 1966: 705.

Type localities: Lagodekhi, Georgia, U.S.S.R.; and Lalbar, Armenia, U.S.S.R.
Distribution: U.S.S.R. (Armenia, Georgia).

Genus Americhernes Muchmore

Americhernes Muchmore, 1976b: 151–152.

Type species: *Chelifer oblongus* Say, 1821, by original designation.

Americhernes andinus (Beier)

Lamprochernes andinus Beier, 1959e: 203–204, fig. 15.
Americhernes andinus (Beier): Mahnert, 1979d: 772.

Type locality: 10 miles N. of Trancos, Tucumán, Argentina.
Distribution: Argentina.

Americhernes bethaniae Mahnert

Americhernes bethaniae Mahnert, 1979d: 769–772, figs 101–107; Adis, 1981: 119, etc.; Mahnert, 1985c: 228; Adis & Mahnert, 1986: 311; Mahnert & Adis, 1986: 213; Adis, Junk & Penny, 1987: 488.

Type locality: Taruma Mirim, Manaus, Amazonas, Brazil.
Distribution: Brazil (Amazonas).

Americhernes chilensis (Beier)

Lamprochernes chilensis Beier, 1964b: 342–343, fig. 19; Cekalovic, 1984: 19.
Americhernes chilensis (Beier): Mahnert, 1979d: 772.

Type locality: Cuya and Rio Camarones, Arcia, Tarapacá, Chile.
Distribution: Chile.

Americhernes eidmanni (Beier), new combination

Pycnochernes eidmanni Beier, 1935f: 45–46, fig. 1; Roewer, 1937: 297; Pereira & Castro, 1944: 240–261, figs 1–29; Beier, 1948b: 457.

Type locality: Mendes, E. of Rio de Janeiro, Rio de Janeiro, Brazil.
Distribution: Brazil (Rio de Janeiro).

Americhernes ellipticus (Hoff)

Lamprochernes ellipticus Hoff, 1944a: 1–3, figs 1–4; Hoff, 1947b: 478; Hoff, 1956c: 5–9; Hoff, 1958: 20; Hoff, 1959b: 4, etc.; Hoff, 1961: 445.
Americhernes ellipticus (Hoff): Muchmore, 1976b: 162, fig. 21; Zeh, 1987b: 1084.

Type locality: Algodones, Baja California, Mexico.
Distribution: Mexico, U.S.A. (Colorado, New Mexico).

Americhernes guarany (Feio), new combination

Pycnochernes guarany Feio, 1946: 168–169, figs 2–14.

Type locality: Isla Valle, Paraguay.
Distribution: Paraguay.

Americhernes incertus Mahnert

Americhernes incertus Mahnert, 1979d: 772–775, figs 108–116; Adis, 1981: 119, etc.; Mahnert, 1985c: 228; Adis & Mahnert, 1986: 311; Mahnert & Adis, 1986: 213; Adis, Junk & Penny, 1987: 488; Mahnert, Adis, & Bührnheim, 1987: fig. 15.

Type locality: Taruma Mirim, Manaus, Amazonas, Brazil.
Distribution: Brazil (Amazonas).

Americhernes kanaka (J. C. Chamberlin), new combination

Lamprochernes kanaka J. C. Chamberlin, 1938a: 279; J. C. Chamberlin, 1939b: 210–212, figs 2a–e; Tenorio & Muchmore, 1982: 380.

Type locality: Tekohepu summit, Uapou, Marquesas Islands.
Distribution: Marquesas Islands.

Americhernes levipalpus (Muchmore)

Lamprochernes levipalpus Muchmore, 1972d: 327–328; Knowlton, 1974: 3.
Lamprochernes lavipalpus (sic) Muchmore: Knowlton, 1972: 2.
Americhernes levipalpus (Muchmore): Muchmore, 1976b: 162–163, fig. 22.

Type locality: Cedar Hills, Curlew Valley, Box Elder County, Utah, U.S.A.
Distribution: U.S.A. (Utah).

Americhernes longimanus Muchmore

Americhernes longimanus Muchmore, 1976b: 156–158, figs 10–13; Brach, 1979: 34; Zeh, 1987b: 1084.

Type locality: Archbold Biological Station, Highlands County, Florida, U.S.A.
Distribution: U.S.A. (Mississippi, Florida).

Americhernes oblongus (Say)

Chelifer oblongus Say, 1821: 64.
Chelifer alius Leidy, 1877: 261 (synonymized by Banks, 1893: 64).
Chernes oblongus (Say): Hagen, 1868b: 51; Banks, 1890: 152; Hagen, 1870: 268.
Chelanops oblongus (Say): Banks, 1895a: 5; Banks, 1904a: 140; Berger, 1905: 407–417, fig. 1, plate 28; Banks, 1908: 39; Coolidge, 1908: 110; Banks, 1909c: 173; Banks, 1911: 638; Ewing, 1911: 79; Moles, 1914c: 192–193; Moles & Moore, 1921: 6; Myers, 1927: 291–292; Pratt, 1927: 410; J. C. Chamberlin, 1931a: 119, fig. 47s; Nester, 1932: 98; Roewer, 1937: fig. 185.
Chelifer (Lamprochernes) oblongus (Say): Ellingsen, 1910a: 368.
Lamprochernes oblongus (Say): Beier, 1932e: 84–85, fig. 103; J. C. Chamberlin, 1935b: 481; Roewer, 1937: 289; Vachon, 1940f: 3; Hoff, 1945c: 37–38; Beier, 1948b: 444, 452; Hoff, 1949b: 450–453, figs 8, 30a–b; Hoff & Bolsterli, 1956: 165–166; Hoff, 1958: 20; Hoff, 1961: 444–445; Manley, 1969: 7, fig. 7; Weygoldt, 1969a: fig. 105; Muchmore, 1971a: 80, 82; Nelson, 1971: 95; Nelson, 1975: 284, figs 40–43; Rowland & Reddell, 1976: 16.
New genus *oblongus* (Say): Muchmore, 1975d: fig. 6.
Americhernes oblongus (Say): Muchmore, 1976b: 153–156, figs 3–9; Dunkle, 1984: 48; Zeh, 1987b: 1084.

Type localities: of *Chelifer oblongus*: Havana, Illinois, U.S.A.
of *Chelifer alius*: Easton, Pennsylvania, U.S.A.
Distribution: Cuba, U.S.A. (Arkansas, California, Colorado, District of Columbia, Florida, Georgia, Illinois, Indiana, Kentucky, Louisiana, Maryland, Massachusetts, Michigan, Mississippi, Missouri, Nebraska, New Jersey, New York, North Carolina, Ohio, Pennsylvania, South Carolina, Tennessee, Texas, Virginia).

Americhernes perproximus (Beier)

Lamprochernes perproximus Beier, 1962h: 133–135, fig. 2.
Americhernes perproximus (Beier): Mahnert, 1979d: 772.

Type locality: Tapia, Tucumán, Argentina.
Distribution: Argentina.

Americhernes plaumanni (Beier)

Lamprochernes plaumanni Beier, 1974d: 902–903, fig. 3.
Americhernes plaumanni (Beier): Mahnert, 1979d: 772.

Type locality: Nova Teutonia, Santa Catarina, Brazil.
Distribution: Brazil (Santa Catarina).

Americhernes puertoricensis Muchmore

Americhernes puertoricensis Muchmore, 1976b: 160–162, figs 18–20.

Type locality: Rio Piedras, Puerto Rico.
Distribution: Puerto Rico.

Americhernes reductus Muchmore

Americhernes reductus Muchmore, 1976b: 158–160, figs 14–17; Zeh, 1987b: 1084.

Type locality: Rattlesnake Lumps, NE. of Sugarloaf Key, Monroe County, Florida, U.S.A.
Distribution: Belize, U.S.A. (Florida).

Americhernes samoanus (J. C. Chamberlin), new combination

Lamprochernes samoanus J. C. Chamberlin, 1938a: 279–282, figs 5a–h; Roewer, 1940: 346.

Type locality: Salailua, Savaii, Samoa.
Distribution: Samoa.

Americhernes suraiurana (Feio)

Lamprochernes suraiurana Feio, 1945: 21–25, figs 13–17.
Americhernes suraiurana (Feio): Mahnert, 1979d: 772.

Type locality: Bom Jesus da Lapa e Fundal, Bahia, Brazil.
Distribution: Brazil (Bahia).

Genus Anaperochernes Beier

Anaperochernes Beier, 1964b: 343–344.

Type species: *Anaperochernes chilensis* Beier, 1964b, by original designation.

Anaperochernes ambrosianus Beier

Anaperochernes ambrosianus Beier, 1964a: 303–305, fig. 1.
Anasperochernes (sic) *ambrosianus* Beier: Cekalovic, 1984: 20.

Type locality: San Ambrosio Island.
Distribution: San Ambrosio Island.

Anaperochernes chilensis Beier

Anaperochernes chilensis Beier, 1964b: 344–345, fig. 20.
Anasperochernes (sic) *chilensis* Beier: Cekalovic, 1984: 20.

Type locality: Zapaller, Aconcagua, Chile.
Distribution: Chile.

Anaperochernes debilis Beier

Anaperochernes debilis Beier, 1964b: 345–347, fig. 21.
Anasperochernes (sic) *debilis* Beier: Cekalovic, 1984: 21.

Type locality: 10 km E. of Zapudo, Aconcagua, Chile.
Distribution: Chile.

Anaperochernes margaritifer Mahnert

Anaperochernes (?) *margaritifer* Mahnert, 1985c: 228–229, figs 37–40; Mahnert & Adis, 1986: 213.

Type locality: 60 km N. of Manaus, Amazonas, Brazil.
Distribution: Brazil (Amazonas).

Genus Ancalochernes Beier

Ancalochernes Beier, 1932e: 180; Beier, 1933a: 540.

Type species: *Ancalochernes mexicanus* Beier, 1932e, by original designation.

Ancalochernes mexicanus Beier

Ancalochernes mexicanus Beier, 1932e: 180, fig. 189; Beier, 1933a: 540–541, fig. 12; Roewer, 1937: 302, fig. 241.

Type locality: Mirador, Mexico.
Distribution: Mexico.

Genus **Anthrenochernes** Lohmander

Anthrenochernes Lohmander, 1939a: 3–4; Beier, 1963b: 252.

Type species: *Anthrenochernes stellae* Lohmander, 1939a, by original designation.

Anthrenochernes stellae Lohmander

Anthrenochernes stellae Lohmander, 1939a: 6–10, figs 1–4; Lohmander, 1939b: 297–298; Meinertz, 1962a: 394, map 3; Beier, 1963b: 252; Rafalski, 1967: 15.

Type locality: Göteborg, Sweden.
Distribution: Denmark, Poland, Sweden.

Genus **Antillochernes** Muchmore

Antillochernes Muchmore, 1984b: 107–109.

Type species: *Antillochernes bahamensis* Muchmore, 1984b, by original designation.

Antillochernes bahamensis Muchmore

Antillochernes bahamensis Muchmore, 1984b: 109–110, figs 1–8.

Type locality: W. end of Grand Bahama Island, Bahamas.
Distribution: Bahamas.

Antillochernes biminiensis Muchmore

Antillochernes biminiensis Muchmore, 1984b: 112–113, figs 12–13.

Type locality: South Bimini Island, Bahamas.
Distribution: Bahamas.

Antillochernes cruzensis Muchmore

Antillochernes cruzensis Muchmore, 1984b: 115–116, figs 17–19.

Type locality: Golden Grove, St Croix, U.S. Virgin Islands.
Distribution: U.S. Virgin Islands.

Antillochernes floridensis Muchmore

Antillochernes floridensis Muchmore, 1984b: 113–115, figs 14–16.

Type locality: St Marks Wildlife Refuge, Leon County, Florida, U.S.A.
Distribution: U.S.A. (Florida).

Antillochernes jamaicensis Muchmore

Antillochernes jamaicensis Muchmore, 1984b: 110–112, figs 9–11.

Type locality: St Ann's Bay, St Ann Parish, Jamaica.
Distribution: Cayman Islands, Jamaica.

Antillochernes muchmorei (Dumitresco & Orghidan)

Parachernes muchmorei Dumitresco & Orghidan, 1977: 118–122, figs 16a–d, 17a–f, 18a–b, 19a–c, 23–24.
Antillochernes muchmorei (Dumitresco & Orghidan): Muchmore, 1984b: 116–117.

Type locality: Río Cacoyugüin, Oriente, Cuba.
Distribution: Cuba.

Genus **Apatochernes** Beier

Apatochernes Beier, 1948c: 540–542; Beier, 1976f: 227.

Type species: *Apatochernes cheliferoides* Beier, 1948c, by original designation.

Apatochernes antarcticus Beier

Apatochernes antarcticus Beier, 1964g: 116–118, fig. 1; Beier, 1964h: 629; Beier, 1969e: 414; Beier, 1976f: 236; Tenorio & Muchmore, 1982: 379.
Apatochernes antarcticus antarcticus Beier: Beier, 1976f: 236–237.

Type locality: North West Bay, Campbell Island, New Zealand.
Distribution: New Zealand.

Apatochernes antarcticus knoxi Beier

Apatochernes antarcticus knoxi Beier, 1976f: 237.

Type locality: Snares Island, New Zealand.
Distribution: New Zealand.

Apatochernes antarcticus pterodromae Beier

Apatochernes pterodromae Beier, 1964h: 628–629, fig. 1.
Apatochernes antarcticus pterodromae Beier: Beier, 1976f: 237.

Type locality: Port Ross, Ocean Island, Auckland Islands, New Zealand.
Distribution: New Zealand.

Apatochernes chathamensis Beier

Apatochernes chathamensis Beier, 1976f: 237–239, fig. 36.

Type locality: Point Munningm Kaingaroa, Chatham Island, New Zealand.
Distribution: New Zealand.

Apatochernes cheliferoides Beier

Apatochernes cheliferoides Beier, 1948c: 542–543, fig. 8; Beier, 1976f: 235.
Not *Apatochernes cheliferoides* Beier: Beier, 1969e: 414 (misidentification; see *Nesochernes gracilis* Beier).

Type locality: Hollyford River, South Island, New Zealand.
Distribution: New Zealand.

Apatochernes cruciatus Beier

Apatochernes cruciatus Beier, 1976f: 230–231, fig. 30.

Type locality: Green Island, North Island, New Zealand.
Distribution: New Zealand.

Apatochernes curtulus Beier

Apatochernes curtulus Beier, 1948c: 545–547, fig. 10; Beier, 1967b: 294, fig. 11; Beier, 1976f: 232–233.

Type locality: Hollyford River, South Island, New Zealand.
Distribution: New Zealand.

Apatochernes gallinaceus Beier

Apatochernes gallinaceus Beier, 1967b: 296–297, fig. 13; Beier, 1976f: 231.

Type locality: East Chicken Island, Hen and Chickens Group, New Zealand.
Distribution: New Zealand.

Apatochernes insolitus Beier

Apatochernes insolitus Beier, 1976f: 235, fig. 34.

Type locality: Totaranui-Anapai Ridge, South Island, New Zealand.
Distribution: New Zealand.

Apatochernes kuscheli Beier

Apatochernes kuscheli Beier, 1976f: 235–236, fig. 35.

Type locality: 22.5 km E. of Lewis Pass, South Island, New Zealand.
Distribution: New Zealand.

Apatochernes maoricus Beier

Apatochernes maoricus Beier, 1966c: 375–376, fig. 10; Beier, 1976f: 233, fig. 32; Tenorio & Muchmore, 1982: 381.

Type locality: Scenic Reserve, Kaituna Valley, Banks Peninsula, South Island, New Zealand.
Distribution: New Zealand.

Apatochernes nestoris Beier

Apatochernes nestoris Beier, 1962g: 400–402, fig. 2; Beier, 1966c: 374; Beier, 1967b: 294; Beier, 1976f: 227, fig. 26.

Type locality: Nelson Creek, Westland, South Island, New Zealand.
Distribution: New Zealand.

Apatochernes obrieni Beier

Apatochernes obrieni Beier, 1966c: 374–375, fig. 9; Beier, 1967b: 295–296, fig. 12; Beier, 1969e: 413; Beier, 1976f: 231; Tenorio & Muchmore, 1982: 384.

Type locality: Ngongotaha, Auckland Province, North Island, New Zealand.
Distribution: New Zealand.

Apatochernes posticus Beier

Apatochernes posticus Beier, 1976f: 227–229, fig. 28; Harvey, 1985b: 133.

Type locality: Mt Pitt, Norfolk Island.
Distribution: Norfolk Island.

Apatochernes proximus Beier

Apatochernes proximus Beier, 1948c: 543–545, fig. 9; Beier, 1976f: 235.

Type locality: Kingston, Lake Wakatipu (as Watipu), South Island, New Zealand.
Distribution: New Zealand.

Apatochernes solitarius Beier

Apatochernes solitarius Beier, 1976f: 231–232, fig. 31.

Type locality: Te Anaputa Point, North Island, New Zealand.
Distribution: New Zealand.

Apatochernes turbotti Beier

Apatochernes turbotti Beier, 1969e: 417–418, fig. 3; Beier, 1976f: 227, fig. 27.

Type locality: South-West Island, Three Kings Islands, North Island, New Zealand.
Distribution: New Zealand.

Apatochernes vastus Beier

Apatochernes vastus Beier, 1976f: 229–230, fig. 29.

Type locality: Kauri Sanctuary, Omahuta Forest, Northland, North Island, New Zealand.
Distribution: New Zealand.

Apatochernes wisei Beier

Apatochernes wisei Beier, 1976f: 233–234, fig. 33.

Type locality: Mangamuka Hills, North Island, New Zealand.
Distribution: New Zealand.

Genus **Asterochernes** Beier

Asterochernes Beier, 1955b: 209–210.

Type species: *Asterochernes vittatus* Beier, 1955b, by original designation.

Asterochernes kuscheli Beier

Asterochernes kuscheli Beier, 1962h: 135 (nomen nudum).
Asterochernes kuscheli Beier, 1964b: 352–354, fig. 25; Cekalovic, 1984: 22.
Asterochernes aff. *kuscheli* Beier: Vitali-di Castri & Castri, 1976: 62, 63.

Type locality: Mirador Alemán, Cerro Caracol, Concepción, Chile.
Distribution: Argentina, Chile.

Asterochernes kuscheli patagonicus Beier

Asterochernes kuscheli patagonicus Beier, 1964j: 493–494, fig. 4.

Type locality: Mt Piltriquitron, El Bolsón, Rio Negro, Argentina.
Distribution: Argentina.

Asterochernes vittatus Beier

Asterochernes vittatus Beier, 1955b: 210–212, fig. 4; Beier, 1957c: 454; Cekalovic, 1984: 22.

Type locality: Miradero, Masatierra, Juan Fernandez Islands.
Distribution: Juan Fernandez Islands.

Genus **Atherochernes** Beier

Atherochernes Beier, 1954f: 140–141.

Type species: *Atherochernes venezuelanus* Beier, 1954f, by original designation.

Atherochernes venezuelanus Beier

Atherochernes venezuelanus Beier, 1954f: 141–142, fig. 6.

Type locality: El Junquito, Venezuela.
Distribution: Venezuela.

Genus **Austrochernes** Beier

Austrochernes Beier, 1932e: 171; Beier, 1933a: 535–536.

Type species: *Chelifer australiensis* With, 1905, by original designation.

Austrochernes australiensis (With)

Chelifer australiensis With, 1905: 101–104, plate 6 figs 2a–g.
Austrochernes australiensis (With): Beier, 1932e: 171; J. C. Chamberlin, 1934b: 10; Roewer, 1937: 301; Harvey, 1981b: 246; Harvey, 1985b: 133.

Not *Chelifer (Trachychernes) australiensis* (With): Ellingsen, 1910a: 373 (misidentification; see *Heterochernes novaezealandiae* (Beier)).

Type locality: Queensland, Australia.
Distribution: Australia (Queensland).

Genus **Bituberochernes** Muchmore

Bituberochernes Muchmore, 1974b: 77–78; Dumitresco & Orghidan, 1977: 108; Muchmore, 1979b: 313.

Type species: *Bituberochernes mumae* Muchmore, 1974b, by original designation.

Bituberochernes jonensis Muchmore

Bituberochernes jonensis Muchmore, 1979b: 314–316, figs 1–4.

Type locality: Reef Bay Trail, Virgin Islands National Park, St John, U.S. Virgin Islands.
Distribution: U.S. Virgin Islands.

Bituberochernes mumae Muchmore

Bituberochernes mumae Muchmore, 1974b: 78–80, figs 1–4; Dumitresco & Orghidan, 1977: 108–113, figs 8a–f, 9a–e, 10a–d, 11a–b, 12a–b, 21–22; Muchmore, 1979b: 313; Armas & Alayón, 1984: 11.

Type locality: Matheson Hammock, Dade County, Florida, U.S.A.
Distribution: Cayman Islands, Cuba, U.S.A. (Florida).

Genus **Byrsochernes** Beier

Byrsochernes Beier, 1959e: 206.

Type species: *Byrsochernes ecuadoricus* Beier, 1959e, by original designation.

Byrsochernes caribicus Beier

Byrsochernes caribicus Beier, 1976d: 50–51, fig. 3; Hounsome, 1980: 85.

Type locality: Boca Chica, Dominican Republic.
Distribution: Cayman Islands, Dominican Republic.

Byrsochernes ecuadoricus Beier

Byrsochernes ecuadoricus Beier, 1959e: 207–208, fig. 17; Beier, 1976c: 1.

Type locality: 27 miles SW. of Quevedo, Los Rios, Ecuador.
Distribution: Ecuador, Mexico.

Genus **Cacoxylus** Beier

Cacoxylus Beier, 1965g: 787–788.

Type species: *Hebridochernes* (?) *echinatus* Beier, 1964f, by original designation.

Cacoxylus echinatus (Beier)

Hebridochernes (?) *echinatus* Beier, 1964f: 598, fig. 4.
Cacoxylus echinatus (Beier): Beier, 1965g: 788, fig. 26; Beier, 1966d: 154; Beier, 1967c: 322–323, fig. 4.

Type locality: Mt Austen, Guadalcanal, Solomon Islands.
Distribution: Papua New Guinea, Solomon Islands.

Genus **Caffrowithius** Beier

Caffrowithius Beier, 1932d: 62; Beier, 1932e: 221.
Adelphochernes 1944: 195–196 (junior homonym of *Adelphochernes* Beier, 1937b).

Plesiochernes Vachon, 1945b: 187 (synonymized by Mahnert, 1982e: 693).
Anepsiochernes Beier, 1947b: 339 (replacement name for *Adelphochernes* Beier, 1944) (synonymized with *Plesiochernes* by Beier, 1955c: 554).

Type species: of *Caffrowithius*: *Chelifer concinnus* Tullgren, 1907a, by original designation.
of *Adelphochernes*: *Adelphochernes aethiopicus* Beier, 1944, by original designation.
of *Plesiochernes*: *Plesiochernes elgonensis* Vachon, 1945b, by original designation.

Caffrowithius aequatorialis Caporiacco

Caffrowithius aequatorialis Caporiacco, 1939b: 312–313, figs 2a–b; Roewer, 1940: 347.

Type localities: Javello, Somalia; and Arero, Somalia.
Distribution: Somalia.

Caffrowithius aethiopicus (Beier)

Adelphochernes aethiopicus Beier, 1944: 196–197, fig. 16.
Anepsiochernes aethiopicus (Beier): Beier, 1947b: 339.
Plesiochernes aethiopicus (Beier): Beier, 1955c: 555.
Caffrowithius aethiopicus (Beier): Mahnert, 1982e: 705.

Type locality: Djem-Djem-Wald, Ethiopia.
Distribution: Ethiopia.

Caffrowithius bergeri (Mahnert)

Plesiochernes bergeri Mahnert, 1978f: 130–131, figs 133–138.
Caffrowithius bergeri (Mahnert): Mahnert, 1982e: 705.

Type locality: Meya, Tal des Niari, Congo.
Distribution: Congo.

Caffrowithius bicolor (Beier)

Plesiochernes bicolor Beier, 1964k: 74–75, fig. 31.
Caffrowithius bicolor (Beier): Mahnert, 1983c: 509.

Type locality: Salisbury (now Hirare), Zimbabwe.
Distribution: Zimbabwe.

Caffrowithius biseriatus Mahnert

Pselaphochernes natalensis Beier: Beier, 1955l: 312 (misidentification, in part).
Caffrowithius biseriatus Mahnert, 1983c: 501–506, figs 12–20.

Type locality: De Hoop Cave, Cape Province, South Africa.
Distribution: South Africa.

Caffrowithius caffer (Beier)

Pselaphochernes caffer Beier, 1947b: 326–328, figs 32–33.
Caffrowithius caffer (Beier): Mahnert, 1983c: 507.

Type locality: Pendeberg, Namaqualand, Cape Province, South Africa.
Distribution: South Africa.

Caffrowithius calvus (Beier)

Plesiochernes calvus Beier, 1959d: 61–62, fig. 32.
Caffrowithius calvus (Beier): Mahnert, 1982e: 695–697, figs 5–11.

Type locality: Kundelungu, Katanga, Zaire.
Distribution: Kenya, Ruanda, Zaire.

Caffrowithius concinnus (Tullgren)

Chelifer concinnus Tullgren, 1907a: 39–41, figs 7a–c; Tullgren, 1907b: 224; Ellingsen, 1912b: 82, 91.
Caffrowithius concinnus (Tullgren): Beier, 1932d: 62; Beier, 1932e: 222, fig. 230; Roewer, 1937: 309; Beier, 1958a: 183; Weidner, 1959: 113; Beier, 1964k: 74 (in part; see *Caffrowithius natalensis* (Beier)); Beier, 1966k: 465; Mahnert, 1983c: 506–507, figs 1–6.

Type locality: Bothaville, Orange Free State, South Africa.
Distribution: South Africa.

Caffrowithius elgonensis (Vachon)

Plesiochernes elgonensis Vachon, 1945b: 188–189, figs 1–6; Beier, 1955c: 551–553, fig. 16.
Caffrowithius elgonensis (Vachon): Mahnert, 1982e: 698–700, figs 14–15.

Type locality: Mt Elgon, Kenya.
Distribution: Kenya, Uganda.

Caffrowithius excellens (Beier)

Plesiochernes excellens Beier, 1958a: 176–178, fig. 10.
Caffrowithius excellens (Beier): Mahnert, 1982e: 701; Mahnert, 1983c: 509.

Type locality: Umhlali, Natal, South Africa.
Distribution: South Africa.

Caffrowithius exiguus (Tullgren)

Chelifer exiguus Tullgren, 1907d: 12–13, figs 8a–b.
Hansenius exiguus (Tullgren): Beier, 1932e: 270; Roewer, 1937: 316.
Plesiochernes exiguus (Tullgren): Beier, 1955c: 555; Beier, 1962f: 22–24, fig. 9.
Caffrowithius exiguus (Tullgren): Mahnert, 1982e: 697–698, figs 12–13.
Not *Chelifer exiguus* Tullgren: Ellingsen, 1912b: 83, 92 (misidentification; see *Ellingsenius fulleri* (Hewitt & Godfrey)).

Type locality: Kibonoto (= Kibongoto?), Tanzania.
Distribution: Kenya, Tanzania.

Caffrowithius facetus (Tullgren)

Chelifer facetus Tullgren, 1907b: 223–224, fig. 3; Ellingsen, 1912b: 83, 92.
Chelifer torulosus facetus Tullgren: Hewitt & Godfrey, 1929: 326–327, fig. 5b, plate 22, fig. 9.
Caffrowithius facetus (Tullgren): Beier, 1932d: 62; Beier, 1932e: 223, fig. 231; Roewer, 1937: 310; Beier, 1958a: 183; Mahnert, 1983c: 509.

Type locality: Stamford Hill, Natal, South Africa.
Distribution: South Africa.

Caffrowithius garambae (Beier), new combination

Plesiochernes garambae Beier, 1972a: 11–13, fig. 6.

Type locality: Parc National Garamba, Zaire.
Distribution: Zaire.

Caffrowithius hanangensis (Beier), new combination

Plesiochernes hanangensis Beier, 1962f: 24–26, fig. 10.

Type locality: Mt Hanang, Tanzania.
Distribution: Tanzania.

Caffrowithius hanangensis curtus (Beier), new combination

Plesiochernes hanangensis curtus Beier, 1962f: 26–27, fig. 11.

Type locality: Ngorongoro, Tanzania.
Distribution: Tanzania.

Caffrowithius lucifugus (Beier), new combination

Plesiochernes lucifugus Beier, 1959d: 58–59, fig. 30.
Allochernes inexpectatus Anciaux de Faveaux, 1980: 345 (nomen nudum).

Type locality: Lubudi, Zaire.
Distribution: Zaire.

Caffrowithius meruensis (Beier), new combination

Plesiochernes meruensis Beier, 1962f: 27–29, fig. 12.

Type locality: Olkokola, Mt Meru, Tanzania.
Distribution: Tanzania.

Caffrowithius natalensis (Beier)

Pselaphochernes natalensis Beier, 1947b: 328–329, fig. 34; Beier, 1955l: 312 (in part; see *Caffrowithius biseriatus* Mahnert); Beier, 1958a: 175–176, fig. 9.
Caffrowithius concinnus (Tullgren): Beier, 1964k: 74 (in part).
Caffrowithius natalensis (Beier): Mahnert, 1983c: 508–509, figs 7–11.

Type locality: Estcourt, Natal, South Africa.
Distribution: South Africa.

Caffrowithius natalicus (Beier)

Plesiochernes natalicus Beier, 1956c: 437–439, fig. 1; Beier, 1958a: 178.
Caffrowithius natalicus (Beier): Mahnert, 1983c: 509.

Type locality: Pietermaritzburg, Natal, South Africa.
Distribution: South Africa.

Caffrowithius planicola Mahnert

Caffrowithius planicola Mahnert, 1982e: 701–705, figs 19–27.

Type locality: near Mundui Estate, Lac Naivasha, Kenya.
Distribution: Kenya.

Caffrowithius procerus Beier

Caffrowithius procerus Beier, 1966k: 465–466, fig. 6.

Type locality: 5 miles from Gravelotte, Transvaal, South Africa.
Distribution: South Africa.

Caffrowithius pusillimus (Beier), new combination

Plesiochernes pusillimus Beier, 1979c: 105–107, fig. 3.

Type locality: Adiopodoumé, Ivory Coast.
Distribution: Ivory Coast.

Caffrowithius rusticus (Beier)

Plesiochernes rusticus Beier, 1955c: 553–554, fig. 17; Beier, 1959d: 60.
Caffrowithius rusticus (Beier): Mahnert, 1982e: 693–695, figs 2–4.

Type locality: Kakamega, Kenya.
Distribution: Kenya, Zaire.

Caffrowithius simplex (Beier), new combination

Plesiochernes simplex Beier, 1955a: 14–16, fig. 7; Beier, 1959d: 60, fig. 31.

Type locality: Mukelengia, Kalumengongo, Zaire.
Distribution: Zaire.

Caffrowithius subfoliosus (Ellingsen)

Chelifer (Trachychernes) subfoliosus Ellingsen, 1910a: 380–382.
Chelifer subfoliosus Ellingsen: Ellingsen, 1912b: 82, 91.
Caffrowithius subfoliosus (Ellingsen): Beier, 1932d: 62; Beier, 1932e: 222–223; Roewer,
    1937: 310; Mahnert, 1983c: 507.

Type locality: Africa.
Distribution: South Africa.

Caffrowithius uncinatus (Beier)

Anepsiochernes uncinatus Beier, 1954e: 84–86, fig. 1.
Plesiochernes uncinatus (Beier): Beier, 1955c: 555; Beier, 1962f: 29; Beier, 1967d: 81.
Caffrowithius uncinatus (Beier): Mahnert, 1982e: 700–701, figs 16–18.

Type locality: Msingi, Tanzania.
Distribution: Kenya, Tanzania.

Genus Calidiochernes Beier

Calidiochernes Beier, 1954f: 136.

Type species: *Calidiochernes musculi* Beier, 1954f, by original designation.

Calidiochernes musculi Beier

Calidiochernes musculi Beier, 1954f: 136–138, fig. 4.

Type locality: Llanos, Venezuela.
Distribution: Venezuela.

Genus Calymmachernes Beier

Calymmachernes Beier, 1954b: 24–25.

Type species: *Calymmachernes angulatus* Beier, 1954b, by original designation.

Calymmachernes angulatus Beier

Calymmachernes angulatus Beier, 1954b: 25–26, fig. 11; Harvey, 1981b: 246; Harvey,
    1985b: 133.

Type locality: Walpole, Western Australia, Australia.
Distribution: Australia (Western Australia).

Genus Caribochernes Beier

Caribochernes Beier, 1976b: 48–49.

Type species: *Caribochernes pumilus* Beier, 1976b, by original designation.

Caribochernes pumilus Beier

Caribochernes pumilus Beier, 1976b: 49–50, fig. 2.

Type locality: Colonia, Dominican Republic.
Distribution: Dominican Republic.

Genus Ceriochernes Beier

Ceriochernes Beier, 1937b: 269–270.

Type species: *Ceriochernes detritus* Beier, 1937b, by original designation.

Ceriochernes amazonicus Mahnert

Ceriochernes (?) *amazonicus* Mahnert, 1985c: 232, figs 48–52.
Ceriochernes amazonicus Mahnert: Adis & Mahnert, 1986: 311; Mahnert & Adis, 1986: 213.

Type locality: Taruma Mirim, Amazonas, Brazil.
Distribution: Brazil (Amazonas).

Ceriochernes besucheti Beier

Ceriochernes besucheti Beier, 1973b: 51, fig. 15.

Type locality: Kegalla, Sabaragamuwa Province, Sri Lanka.
Distribution: Sri Lanka.

Ceriochernes brasiliensis Beier

Ceriochernes brasiliensis Beier, 1974d: 905–906, fig. 5.

Type locality: Nova Teutonia, Santa Catarina, Brazil.
Distribution: Brazil (Santa Catarina).

Ceriochernes detritus Beier

Ceriochernes detritus Beier, 1937b: 270–271, fig. 2; Roewer, 1940: 347.

Type locality: St Cruz, Leyte, Philippines.
Distribution: Philippines.

Ceriochernes foliaceosetosus Beier

Ceriochernes foliaceosetosus Beier, 1974d: 906–907, fig. 6.

Type locality: Nova Teutonia, Santa Catarina, Brazil.
Distribution: Brazil (Paraná, Santa Catarina).

Ceriochernes martensi Beier

Ceriochernes martensi Beier, 1974b: 274–275, fig. 9.
Ceriochernes cf. *martensi* Beier: Ćurčić, 1980b: 84–88, figs 21–24.

Type locality: Jiri, Nepal.
Distribution: Nepal.

Ceriochernes nepalensis Beier

Ceriochernes nepalensis Beier, 1974b: 273–274, fig. 8.
Ceriochernes sp. ? *nepalensis* Beier: Beier, 1978b: 415.

Type locality: Alm Kharana, Nepal.
Distribution: India, Nepal.

Ceriochernes vestitus Beier

Ceriochernes vestitus Beier, 1974b: 275–277, fig. 10.

Type locality: Thaksang, Tukche, Thakkhola, Nepal.
Distribution: Nepal.

Genus Chelanops Gervais

Chelifer (Chelanops) Gervais, 1849: 13.
Chelanops Gervais: Beier, 1932e: 177; Beier, 1933a: 538; Hoff, 1947b: 503; Hoff, 1949b: 460–461.

Type species: *Chelifer (Chelanops) coecus* Gervais, 1849, by monotypy.

Subgenus **Chelanops (Chelanops)** Gervais

Chelanops (Chelanops) affinis Banks

Chelanops affinis Banks, 1894: 314; Banks, 1895a: 8; Banks, 1904a: 141; Coolidge, 1908 112; Hoff, 1947b: 503–506, fig. 15; Hoff, 1958: 24.
Neochernes (?) *affinis* (Banks): Beier, 1932e: 167; Roewer, 1937: 299.

Type locality: Crescent City, Florida, U.S.A.
Distribution: U.S.A. (Florida).

Chelanops (Chelanops) altimanus (Ellingsen)

Chelifer (Trachychernes) altimanus Ellingsen, 1910a: 370–371.
Chelanops altimanus (Ellingsen): Beier, 1932e: 178–179, fig. 187; Roewer, 1937: 302, fig. 240b.

Type locality: St Thomas Island, Virgin Islands.
Distribution: Virgin Islands.

Chelanops (Chelanops) atlanticus Beier

Chelanops atlanticus Beier, 1955f: 7–10, figs 1–2.

Type locality: Tristan da Cunha.
Distribution: Tristan da Cunha.

Chelanops (Chelanops) chilensis Beier

Chelanops chilensis Beier, 1932e: 178, fig. 186; Beier, 1933a: 538–539, fig. 10; Roewer, 1937: 302.
Chelanops (Chelanops) chilensis Beier: Beier, 1964b: 366–367; Cekalovic, 1984: 25.

Type locality: Villa Rica, Panguopullio, Chile.
Distribution: Chile.

Chelanops (Chelanops) coecus (Gervais)

Chelifer (Chelanops) coecus Gervais, 1849: 13, fig. 13.
Chelifer coecus Gervais: With, 1908a: 327.
Chelifer (Trachychernes) rotundimanus Ellingsen, 1910a: 379–380; Beier, 1930d: 199–201, figs 3a–b (synonymized by Beier, 1959e: 215).
Chelanops coecus (Gervais): H. C. Joseph, 1927: 53–56, figs 1–2; Beier, 1959e: 215; Beier, 1962h: 135; Beier, 1964j: 499.
Hesperochernes rotundimanus (Ellingsen): Beier, 1930a: 200.
Chelanops rotundimanus (Ellingsen): Beier, 1932e: 177–178, fig. 185; Roewer, 1937: 302; Mello-Leitão, 1939a: 121.
Chelanops (?) *coecus* (Gervais): Beier, 1932e: 179; Roewer, 1937: 302.
Chelanops (Chelanops) coecus (Gervais): Beier, 1964b: 367–368; Cekalovic, 1976: 18; Cekalovic, 1984: 25.

Type localities: of *Chelifer (Chelanops) coecus*: Chile.
of *Chelifer (Trachychernes) rotundimanus*: Philippi, Chile.
Distribution: Argentina, Chile.

Chelanops (Chelanops) insularis Beier

Chelanops insularis Beier, 1955b: 214–215, fig. 6; Beier, 1957c: 460–463, fig. 4; Cekalovic, 1984: 26.

Type locality: Cumberland Bay, Masatierra, Juan Fernandez Islands.
Distribution: Juan Fernandez Islands.

Chelanops (Chelanops) kuscheli Beier

Chelanops kuscheli Beier, 1955b: 212–213, fig. 5; Beier, 1957c: 460; Cekalovic, 1984: 26.

Type locality: Plateau del Yunque, Masatierra, Juan Fernandez Islands.
Distribution: Juan Fernandez Islands.

Chelanops (Chelanops) nigrimanus Banks

Chelanops nigrimanus Banks, 1902a: 69, plate 2 fig. 6.

Type locality: Albemarle, Galapagos Islands.
Distribution: Galapagos Islands.

Chelanops (Chelanops) occultus Beier

Chelanops (Chelanops) occultus Beier, 1964b: 368–370, fig. 33; Cekalovic, 1984: 26.
Chelanops occultus Beier: Vitali-di Castri, 1965b: fig. 2.

Type locality: Illapal, Chile.
Distribution: Chile.

Chelanops (Chelanops) pugil Beier

Chelanops pugil Beier, 1964a: 305–306, fig. 2; Cekalovic, 1984: 26.

Type locality: San Ambrosio Island.
Distribution: San Ambrosio Island.

Subgenus Chelanops (Neochelanops) Beier

Chelanops (Neochelanops) Beier, 1964b: 370.

Type species: *Chelifer (Chelanops) patagonicus* Tullgren, 1900a, by original designation.

Chelanops (Neochelanops) fraternus Beier

Chelanops (Neochelanops) fraternus Beier, 1964b: 372–373, fig. 35; Beier, 1964j: 500; Cekalovic, 1984: 27.

Type locality: Talinay, Coquimbo, Chile.
Distribution: Argentina, Chile.

Chelanops (Neochelanops) michaelseni (E. Simon)

Chelifer michaelseni E. Simon, 1902: 44–45.
Chelifer (Chernes) michaelseni E. Simon: Ellingsen, 1904: 4–5.
'Chernes' michaelsoni (sic) (E. Simon): J. C. Chamberlin, 1931b: 194.
Parachernes (Parachernes) michaelseni (E. Simon): Roewer, 1937: 295; Mello-Leitão, 1939a: 120; Mello-Leitão, 1939b: 616; Weidner, 1959: 114.
Chelanops (Neochelanops) michaelseni (E. Simon): Beier, 1964b: 373–375, fig. 36; Cekalovic, 1984: 27; Mahnert, 1984c: 28, fig. 30.
Chelanops (Chelanops) michaelseni (E. Simon): Cekalovic, 1976: 18–19.
Not *Chelifer michaelseni* E. Simon: With, 1908a: 282–284, figs 22a–c (misidentification; see *Parachernes (Parachernes) withi* Beier).
Not *Parachernes (Parachernes) michaelseni* (E. Simon): Beier, 1932e: 118–119 (misidentification; see *Parachernes (Parachernes) withi* Beier).

Type localities: Agua Fresca, Punta Arenas, Chile; and Ushuaia (as Uschuaia), Argentina.
Distribution: Argentina, Chile.

Family Chernetidae

Chelanops (Neochelanops) patagonicus (Tullgren)

Chelifer (Chelanops) patagonicus Tullgren, 1900a: 155–157, plate 2 figs 1–5.
Chelifer patagonicus Tullgren: Tullgren, 1908d: 116, fig. 1; With, 1908a: 287.
Parazaona patagonicus (Tullgren): Beier, 1932e: 144; Roewer, 1937: 297; Mello-Leitão, 1939a: 121; Mello-Leitão, 1939b: 617; Feio, 1945: 7; Weidner, 1959: 114.
Chelanops (Neochelanops) exiguus Beier: Beier, 1962h: 135–136 (nomen nudum) (synonymized by Beier, 1964b: 370).
Chelanops (Neochelanops) patagonicus (Tullgren): Beier, 1964b: 370–371, fig. 34; Beier, 1964j: 500; Vitali-di Castri & Castri, 1976: 62, 63; Cekalovic, 1984: 26.

Type localities: Ultima Esperanza, Chile; Mayer, Chile; and between Mayer and Puerto Consuelo, Chile.
Distribution: Argentina, Chile.

Chelanops (Neochelanops) peruanus Mahnert

Chelanops (Neochelanops) peruanus Mahnert, 1984c: 26–28, figs 22–29.

Type locality: Cueva de Catachi, 4 km from Cutervo, Cutervo, Cajamarca, Peru.
Distribution: Peru.

Genus **Chelodamus** R. V. Chamberlin

Chelodamus R. V. Chamberlin, 1925: 236–237; Muchmore, 1984d: 356–357.
Pseudozaona Beier, 1932e: 182; Beier, 1933a: 542–543; Hoff, 1947b: 539; Hoff, 1949b: 471; Hoff & Bolsterli, 1956: 170 (synonymized by Muchmore, 1984d: 356).

Type species: of *Chelodamus*: *Chelodamus atopus* R. V. Chamberlin, 1925, by original designation.
of *Pseudozaona*: *Pseudozaona communis* Beier, 1932e (junior synonym of *Chelodamus mexicolens* R. V. Chamberlin, 1925), by original designation.

Chelodamus atopus R. V. Chamberlin

Chelodamus atopus R. V. Chamberlin, 1925: 237; Muchmore, 1984d: 357–358, figs 5–8.

Type locality: Costa Rica.
Distribution: Costa Rica.

Chelodamus mexicolens R. V. Chamberlin

Chelodamus mexicolens R. V. Chamberlin, 1925: 238; Muchmore, 1984d: 359, figs 9–11.
Pseudozaona communis Beier, 1932e: 182–183, fig. 191; Beier, 1933a: 543–544, figs 13–14; Roewer, 1937: 303, fig. 240c (synonymized by Muchmore, 1984d: 359).

Type localities: of *Chelodamus mexicolens*: Guadalajara, Mexico.
of *Pseudozaona communis*: Cameron, Mexico.
Distribution: Belize, Mexico.

Chelodamus uniformis (Banks)

Chelanops uniformis Banks, 1913: 683–684, figs 16, 18.
Pseudozaona uniformis (Banks): Hoff, 1947b: 540–542, figs 34–35; Beier, 1976c: 2.
Chelodamus uniformis (Banks): Muchmore, 1984d: 360–361, figs 12–13.

Type locality: La Emilia, Juan Viñas, Costa Rica.
Distribution: Costa Rica.

Genus **Chernes** Menge

Chernes Menge, 1855: 39; Stecker, 1874b: 232; Stecker, 1875d: 88; Tömösváry, 1882b: 183–184; Balzan, 1888a: no pagination; Daday, 1888: 115; O.P.-Cambridge,

556

1892: 224; Beier, 1932e: 154; Beier, 1933a: 528; G. O. Evans & Browning, 1954:
20; Beier, 1963b: 273; Legg, 1972: 580; Muchmore, 1974d: 26–27; Legg & Jones,
1988: 122.
Chernes (Trachychernes) Tömösváry, 1882b: 188 (synonymized by Beier, 1932e: 154).
Chernes (Chernes) Menge: Daday, 1888: 115.
Chelifer (Chernes) Menge: Balzan, 1892: 513; Kew, 1911a: 40; Lessert, 1911: 11.
Chelifer (Trachychernes) Tömösváry: Balzan, 1892: 524; Nonidez, 1917: 13.
Reginachernes Hoff, 1949b: 465; Hoff & Clawson, 1956: 176 (synonymized with
Hesperochernes by Hoff, 1963a: 3; synonymized with *Chernes* by Muchmore,
1974d: 31).

Type species: of *Chernes*: *Scorpio cimicoides* Fabricius, 1793, by subsequent designation
of E. Simon, 1879a: 20.
of *Chernes (Trachychernes)*: *Scorpio cimicoides* Fabricius, 1793, by present designation.
of *Reginachernes*: *Reginachernes ewingi* Hoff, 1949b, by original designation.

Chernes amoenus (Hoff)

Hesperochernes amoenus Hoff, 1963a: 3–10, figs 1–2; Nelson, 1975: 292, figs 74–75;
Zeh, 1987b: 1084.
Chernes amoenus (Hoff): Muchmore, 1974d: 31.

Type locality: Horse Thief Lake, Pennington County, South Dakota, U.S.A.
Distribution: U.S.A. (Michigan, South Dakota).

Chernes armenius (Beier)

Chelifer (Trachychernes) armenius Beier, 1929a: 345–346, figs 2a–b, 3.
Chernes armenius (Beier): Beier, 1932e: 158, fig. 169; Roewer, 1937: 298.

Type locality: Ordubad, Araxes, Armenia, U.S.S.R.
Distribution: U.S.S.R. (Armenia).

Chernes beieri, new name

Chernes pallidus Beier, 1936b: 86–87, fig. 2; Roewer, 1940: 347; Beier, 1963b: 277;
Rafalski, 1967: 18 (junior primary homonym of *Chernes pallidus* Banks, 1890).

Type locality: Kaufungen (as Kaufung), West Germany.
Distribution: Poland, West Germany.

Chernes cavicola G. Joseph

Chernes cavicola G. Joseph, 1882: 22; Vachon, 1940f: 3.
Chernes (?) *cavicola* G. Joseph: Beier, 1932e: 159; Roewer, 1937: 298; Beier, 1948b:
445.

Type locality: Grotte von Corgnale, Yugoslavia.
Distribution: Yugoslavia.

Chernes cimicoides (Fabricius)

Scorpio cimicoides Fabricius, 1793: 436.
Obisium cimicoides (Fabricius): Illiger, 1798: 501; Walckenaer, 1802: 253.
Chelifer cimicoides (Fabricius): Latreille, 1804: 142; Latreille, 1806: 133 (not seen);
Leach, 1814: 429; Latreille, 1817: 109; Lamarck, 1818: 80–81; Latreille, 1825a:
133; Duméril, 1826: 49–50; Brébisson, 1827: 263; Anonymous, 1835: 186; Dugès
& Edwards, 1836: 84; Latreille, 1837: 316; Lamarck, 1838: 109; Lamarck, 1839:
301; Gervais, 1844: 78; Hagen, 1870: 268; Stecker, 1875d: 88; E. Simon, 1879a:
39–40, plate 18 fig. 16; Canestrini, 1883: no. 7, figs 1–5; Becker, 1884: cclxiv;
Lameere, 1895: 475; Cuní y Martornell, 1897: 339; E. Simon, 1898d: 20; Tullgren,
1899a: 174–175, plate 1 fig. 10; Ellingsen, 1903: 6; Ellingsen, 1905d: 1; Tullgren,

1906a: 203, plate 4 fig. 3; Tullgren, 1906b: 214; Ellingsen, 1907a: 156; Ellingsen, 1908d: 70; Ellingsen, 1909a: 208, 215; Ellingsen, 1910b: 62; Ellingsen, 1910c: 348; Ellingsen, 1911c: 174; Navás, 1925: 106; Väänänen, 1928a: 18–19; Kew, 1929c: 21–22.

Chelifer fasciatus Leach, 1815: 391; Lamarck, 1818: 80; Lamarck, 1838: 109; Lamarck, 1839: 301; Walckenaer, 1844: 78 (synonymized by E. Simon, 1879a: 39).

Chelifer olfersii Leach, 1817: 50, plate 142 fig. 2; Gervais, 1844: 78 (synonymized with *Chelifer geoffroyi* Leach by C. L. Koch, 1843: 56).

Chelifer geoffroyi Leach, 1817: 50, plate 142 fig. 1; C.L. Koch, 1843: 56–57, fig. 791; Hagen, 1870: 265–266 (synonymized by E. Simon, 1879a: 39).

Chélifer cimicöìdes (Fabricius): Anonymous, 1831b: 284.

Chelifer scorpioides Hermann: Théis, 1832: 73–75, plate 3 figs 2, 2a–b (misidentification).

Chernes cimicoides (Fabricius): Menge, 1855: 40–41, plate 5 fig. 15; Stecker, 1874b: 234; Stecker, 1874c: 306–308; Daday, 1880a: 192; Daday, 1889c: 25; O.P.-Cambridge, 1892: 226–227, plate A fig. 4, plate C fig. 18; H. J. Hansen, 1893: plate 5 fig. 14; Kew, 1903: 299; Stschelkanovzeff, 1902a: 126; Stschelkanovzeff, 1903a: 318; Carr, 1906: 48; Bennett, 1908: 114; Kew, 1909b: 259; Kew, 1914: 96; Daday, 1918: 2; Beier, 1930e: 294; J. C. Chamberlin, 1931a: figs 30g, 47l, 47q; Beier, 1932e: 155–156, fig. 165; Tumšs, 1934: 17–18, fig. 3; Vachon, 1934d: 133; Roewer, 1936: figs 7, 18a–b, 88, 108, 117, 120–123, 124a, 125–126, 135–136, 143; Vachon, 1936a: 78; Roewer, 1937: 298; Vachon, 1937f: 128; Vachon, 1938a: 49–58, 95–97, figs 27–32, 33a, 33e, 54a–c, 56d; Beier, 1939a: fig. 166d; Beier, 1939e: 312; Beier, 1939f: 199; Lohmander, 1939b: 316–318, fig. 12; Caporiacco, 1940: 7; Vachon, 1940f: 2; Cooreman, 1946b: 3; Kaisila, 1947: 86; Beier, 1948b: 445, 458; Caporiacco, 1948c: 242; Kaisila, 1949b: 85–86, fig. 1, map 11; Redikorzev, 1949: 654; Vachon, 1949: figs 220, 223b; Beier, 1952e: 5; Beier & Franz, 1954: 458; G. O. Evans & Browning, 1954: 20; Weidner, 1954b: 110; Beier, 1955n: 109; Beier, 1956h: 24; Pschorn-Walcher & Gunhold, 1957: 346; Vachon, 1957: 389, figs 2, 4, 6; Ressl & Beier, 1958: 2; Beier, 1959f: 131; Beier, 1960b: 100–102, fig. 1; George, 1961: 38; Meinertz, 1962a: 398–399, map 7; Beier, 1963b: 274–275, fig. 276; Lehtinen, 1964: 285; Kobakhidze, 1965b: 541; Ressl, 1965: 289; Kobakhidze, 1966: 705; Zangheri, 1966: 532; Rafalski, 1967: 18; Beron, 1968: 105; Kofler, 1968: 356; Beier, 1969b: 193; Lazzeroni, 1969a: 337; Lazzeroni, 1969b: 409; Lazzeroni, 1969c: 241; M. Smith, 1969: 295–297, figs a–d; Weygoldt, 1969a: 28, 115, figs 16, 43a–c, 53a–c, 82; Lazzeroni, 1970a: 208; Gabbutt, 1972a: 37–40, figs 1f, 2f; Gabbutt, 1972b: 2–13; Howes, 1972a: 109; Kofler, 1972: 288; Lagar, 1972a: 21; Legg, 1972: 580, figs 2(3a), 2(4d), 3(4); Beier, 1973c: 226; Legg, 1973b: 430, fig. 2f; Palmgren, 1973: 9; Ćurčić, 1974a: 26; Muchmore, 1974d: figs 1–4; Beier, 1975a: 58; Jones, 1975a: 88; Klausen, 1975: 64; Legg, 1975a: 66; Legg, 1975d: fig. 2; Muchmore, 1975d: fig. 1; Crocker, 1976: 9; Ćurčić, 1976b: 178–179; Klausen & Totland, 1977: 101–108, plate 8–10, figs 1–2; Crocker, 1978: 9; Jones, 1978: 91, 93, 94, 95; Jones, 1979a: 200; Rundle, 1979: 48; Thaler, 1979: 52; Jones, 1980c: map 22; Jędryczkowski, 1985: 80; Lippold, 1985: 40; Callaini, 1986c: 380–388, figs 1a–f, 4a; Schawaller, 1986: 7, fig. 17; Jędryczkowski, 1987b: 144, map 10; Krumpál & Cyprich, 1988: 42; Legg & Jones, 1988: 122–124, figs 5d, 29b, 31a, 31b(a–j); Schawaller & Dashdamirov, 1988: 42, figs 79; Schawaller, 1989: 21–22.

Chernes mengei L. Koch, 1873: 11–12; Stecker, 1874b: 234; Stecker, 1875d: 88 (synonymized by Beier, 1932e: 155).

Chernes (Trachychernes) cimicoides (Fabricius): Tömösváry, 1882b: 188–190, plate 1 figs 6–12; Ellingsen, 1897: 6–9; Ellingsen, 1910a: 373; Nonidez, 1917: 13; Beier, 1929a: 344.

Chernes (Chernes) cimicoides (Fabricius): Daday, 1888: 117–118, 172; Leleup, 1947: 322.

Chelifer meridianus L. Koch: O.P.-Cambridge, 1892: 221–222, plate 6 fig. 6, plate C fig. 15 (misidentification).
Chelifer (Trachychernes) cimicoides basiléensis Ellingsen, 1906: 251–252 (synonymized by Beier, 1932e: 155).
Chelifer (Chernes) cimicoides (Fabricius): Kew, 1911a: 46–47, fig. 9; Standen, 1912: 13; Kew, 1916a: 124; Kästner, 1928: 6, fig. 18; Schenkel, 1928: 62, figs 17a–b.
Chelifer (Trachychernes) cimicoides (Fabricius): Lessert, 1911: 15–16, figs 7–8, 13; Schenkel, 1929a: 321.
Chelifer cimicoides basiléensis Ellingsen: Ellingsen, 1912b: 82.
Chelifer (Chelanops) cimicoides (Fabricius): Redikorzev, 1924b: 23–24, fig. 9.
Chernes basiléensis (Ellingsen): Beier, 1930a: 212.
Chernes cimicoides caucasicus Kobakhidze, 1965a: 441–443, fig. 1; Kobakhidze, 1965b: 541; Kobakhidze, 1966: 705 (synonymized by Schawaller & Dashdamirov, 1988: 42–43).
Chermes (sic) *cimicoides* (Fabricius): Ressl, 1974: 29.
Not *Chelifer geoffroyi* Leach: Jenyns, 1846: 295 (misidentification; see *Lamprochernes nodosus* (Schrank)).
Not *Chelifer cimicoides* (Fabricius): H. J. Hansen, 1884: 544–545 (misidentification; see *Dinocheirus panzeri* (C. L. Koch)).
Not *Chelifer (Trachychernes) cimicoides* (Fabricius): Beier, 1929e: 445–446 (misidentification; see *Chernes graecus* Beier)).

Type localities: of *Scorpio cimicoides*: Germany.
of *Chelifer fasciatus*: not stated, presumably England, Great Britain.
of *Chelifer olfersii*: not stated, presumably England, Great Britain.
of *Chelifer geoffroyi*: not stated, presumably England, Great Britain.
of *Chernes mengei*: southern Tirol (Italy?).
of *Chelifer (Trachychernes) cimicoides basileensis*: Basilé, Fernando Poo (now Macias Nguema), Equatorial Guinea.
of *Chernes cimicoides caucasicus*: Teberda, R.S.F.S.R., U.S.S.R.; Korul'dashi, Georgia, U.S.S.R.
Distribution: Austria, Belgium, Bulgaria, Cameroun, Czechoslovakia, Denmark, East Germany, Finland, France, Equatorial Guinea, Great Britain, Greece, Hungary, Ireland, Italy, Norway, Poland, Sardinia, Sicily, Spain, Sweden, Turkey, U.S.S.R. (Armenia, Georgia, Kazakhstan, Latvia, R.S.F.S.R.), West Germany, Yugoslavia.

Chernes denisi Vachon

Chernes denisi Vachon, 1937e: 109–111, figs 3–5; Vachon, 1938a: fig. 56b; Roewer, 1940: 347.
Type locality: Skikda (as Philippeville), Algeria.
Distribution: Algeria.

Chernes ewingi (Hoff)

Reginachernes ewingi Hoff, 1949b: 466–467, figs 36a–d; Hoff & Bolsterli, 1956: 176; Hoff, 1958: 25.
Hesperochernes ewingi (Hoff): Hoff, 1963a: 3; Nelson, 1971: 95; Nelson, 1975: 291, figs 70–71.
Chernes ewingi (Hoff): Muchmore, 1974d: 31; Zeh, 1987b: 1084.
Type locality: Muncie, Illinois, U.S.A.
Distribution: U.S.A. (Illinois, Michigan).

Chernes gobiensis Krumpál & Kiefer

Chernes gobiensis Krumpál & Kiefer, 1982: 1–3, plate 1 figs 1–7.
Type locality: Oase Dzuun Mod, Bajanchongor Ajmak, Mongolia.
Distribution: Mongolia.

Family Chernetidae

Chernes graecus Beier

Chelifer (Trachychernes) cimicoides (Fabricius): Beier, 1929e: 445–446 (misidentification).
Chernes graecus Beier, 1932e: 156–157, fig. 166; Beier, 1933a: 529–530, fig. 7; Roewer, 1937: 298; Beier, 1963b: 278, fig. 281.

Type locality: Aenos, Kefallinía (as Kephalonia), Greece.
Distribution: Greece.

Chernes hahnii (C. L. Koch)

Chelifer hahnii C. L. Koch, 1839: 4; C. L. Koch, 1850: 5.
Chelifer hahuii (sic) C. L. Koch: C. L. Koch, 1843: 51–52, fig. 787.
Chernes hahnii (C. L. Koch): L. Koch, 1873: 12–13; Ritsema, 1874: xxxiii; L. Koch, 1877: 180; Croneberg, 1887: 147; Croneberg, 1888: figs 1–3; Krauss, 1896: 628.
Chernes cimicoides hahnii (C. L. Koch): Stecker, 1874c: 308.
Chthonius hahnii (C. L. Koch): Donisthorpe, 1902: 67.
Chernes hahni (C. L. Koch): Beier, 1960b: 100–102, fig. 1; Beier, 1963b: 278, fig. 280; Beier, 1963g: 274; Beier, 1965f: 96; Ressl, 1965: 289; Helversen, 1966a: 148; Helversen, 1966b: figs 3e, 4b, 6d; Beier, 1967f: 313; Rafalski, 1967: 18–19; Beron, 1968: 105; Beier, 1969b: 193; Hammen, 1969: 22; Ressl, 1970: 252; Helversen & Martens, 1971: 379; Mahnert, 1974a: 90; Ressl, 1974: 29; Mahnert, 1977d: 97; Thaler, 1979: 52; Mahnert, 1983d: 362; Schawaller, 1983b: 20–21; Jędryczkowski, 1985: 80; Schawaller, 1986: 7–8, fig. 18; Jędryczkowski, 1987a: 345; Jędryczkowski, 1987b: 145, map 11; Krumpál & Cyprich, 1988: 42; Schawaller & Dashdamirov, 1988: 40–41, figs 72–76, 79; Schawaller, 1989: 22.
Not *Chelifer hahni* C. L. Koch: Schiner, 1872: 75 (not seen) (misidentification; see *Lamprochernes nodosus* (Schrank)).
Not *Chernes hahnii* (C. L. Koch): Wagner, 1892: 434–436 (misidentification; see *Lamprochernes nodosus* (Schrank)).

Type locality: Bayern, West Germany.
Distribution: Austria, Bulgaria, Corsica, Czechoslovakia, France, Great Britain, Hungary, Iran, Netherlands, Poland, Switzerland, Turkey, U.S.S.R. (Armenia, Azerbaijan, Georgia, Kazakhstan, R.S.F.S.R., Ukraine), West Germany.

Chernes hispaniolicus Beier

Chernes hispaniolicus Beier, 1976d: 52–53, fig. 5.

Type locality: San Cristobel, Dominican Republic.
Distribution: Dominican Republic.

Chernes horvathii Daday

Chernes horváthii Daday, 1889a: 17–18, figs 1, 6.
Chernes (?) *horvathii* Daday: Beier, 1932e: 160; Roewer, 1937: 298.
Chernes horvathi Daday: Beier, 1962b: 150–151, figs 6–7; Beier, 1963b: 276–277, fig. 279; Kobakhidze, 1966: 705; Schawaller & Dashdamirov, 1988: 42, figs 77–79.
Chernes chorvathi (sic) Daday: Kobakhidze, 1965b: 541.

Type locality: 'Kussari', Georgia, U.S.S.R.
Distribution: U.S.S.R. (Azerbaijan, Georgia).

Chernes iberus L. Koch

Chernes iberus L. Koch, 1873: 7–8; Beier, 1955n: 110–111, fig. 8; Beier, 1959f: 131; Beier, 1963b: 275–276, fig. 278.
Chelifer (Trachychernes) iberus (L. Koch): Nonidez, 1917: 17–20, figs 4, 4a–b.
Chelifer iberus L. Koch: Navás, 1925: 106, fig. 3.

Allochernes (Allochernes) iberus (L. Koch): Beier, 1932e: 148.
Allochernes iberus (L. Koch): Roewer, 1937: 297; Beier, 1939f: 198.
Chermes (sic) *iberus* (L. Koch): Beier, 1961b: 36.

Type locality: not stated, presumably Spain.
Distribution: Spain.

Chernes lymphatus (Hoff)

Reginachernes lymphatus Hoff, 1949b: 467–471, figs 37a–d; Hoff & Bolsterli, 1956: 176–177; Hoff, 1958: 25.
Hesperochernes lymphatus (Hoff): Nelson, 1971: 95; Nelson, 1975: 291–292, figs 72–73.
Chernes lymphatus (Hoff): Muchmore, 1974d: 31; Zeh, 1987b: 1084.

Type locality: Urbana, Illinois, U.S.A.
Distribution: U.S.A. (Illinois, Michigan, Nebraska).

Chernes mongolicus Beier

Chernes mongolicus Beier, 1974h: 167–169, fig. 1.

Type locality: Hovd (as Kobdo), Mongolia.
Distribution: Mongolia.

Chernes montigenus (E. Simon)

Chelifer montigenus E. Simon, 1879a: 40–41, plate 18 fig. 17; Canestrini, 1885: no. 3, figs 1–3.
Chelifer montigena (sic) E. Simon: E. Simon, 1900b: 593; Navás, 1921: 165; Navás, 1925: 106.
Chernes (Chernes) montigenus (E. Simon): Daday, 1888: 117, 172.
Chelifer (Trachychernes) montigenus E. Simon: Lessert, 1911: 16–17, fig. 14.
Chernes montigenus (E. Simon): Daday, 1918: 2; Schenkel, 1933: 17; Mahnert, 1978a: 313.
Chelifer (Chernes) montigenus E. Simon: Kästner, 1928: 6–7, fig. 19.
Allochernes (Allochernes) montigenus (E. Simon): Beier, 1932e: 153, fig. 164.
Allochernes montigenus (E. Simon): Roewer, 1937: 298; Vachon, 1938a: fig. 56a.
Allochernes (Toxochernes) montigenus (E. Simon): Beier, 1939e: 312.
Allochernes montanus (sic) (E. Simon): Beier, 1939f: 199.
Toxochernes montigenus (E. Simon): Beier, 1954a: 155; Beier, 1963b: 272, fig. 274.
Chernes cfr. *montigenus* (E. Simon): Callaini, 1986c: 388, figs 2a–b.

Type localities: Theodul (as Théodule), Valais, Switzerland; and Mourain, Valais, Switzerland.
Distribution: Austria, France, Italy, Switzerland.

Chernes nigrimanus Ellingsen

Chernes (Trachychernes) montigenus nigrimanus Ellingsen, 1897: 9–11.
Chelifer montigenus nigrimanus (Ellingsen): Tullgren, 1899a: 175.
Allochernes (Toxochernes) nigrimanus (Ellingsen): Beier, 1932e: 153–154.
Allochernes nigrimanus (Ellingsen): Roewer, 1937: 298.
Toxochernes nigrimanus (Ellingsen): Lohmander, 1939b: 314–316, fig. 11; Beier, 1947d: 300–301, fig. 3; Kaisila, 1947: fig. 1; Kaisila, 1949a: 75; Kaisila, 1949b: 85, map 10; Beier, 1952e: 5; Beier & Franz, 1954: 457; Beier, 1956i: 8; Ressl & Beier, 1958: 2; Beier, 1963b: 271–272, fig. 273; Ressl, 1965: 289; Rafalski, 1967: 17; Ressl, 1970: 251–252; Helversen & Martens, 1971: 379, fig. 2; Kofler, 1972: 288; Klausen, 1975: 64.
Chernes nigrimanus (Ellingsen): Mahnert, 1978a: 309–314, figs 1, 9, 12–13; Thaler, 1979: 52.

Family Chernetidae

Type locality: Kragero, Norway.
Distribution: Austria, Finland, Norway, Poland, Sweden.

Chernes rhodinus Beier

Chernes rhodinus Beier, 1966e: 166–167, fig. 3; Beier, 1967f: 313; Mahnert, 1974c: 379–383, figs 1–8.

Type locality: Ródhos (as Rhodes), Greece.
Distribution: Greece, Israel, Turkey.

Chernes sanborni Hagen

Chernes sanborni Hagen, 1868b: 51; Hagen, 1870: 268; Banks, 1890: 152; Muchmore, 1974d: 31; Zeh, 1987b: 1084.
Chelifer cancroides (Linnaeus): Leidy, 1877: 260 (misidentification).
Chernes santorni (sic) Hagen: Hagen, 1879: 400.
Chelanops sanborni (Hagen): Banks, 1895a: 8; Coolidge, 1908: 111; Ewing, 1911: 76, 79, fig. 12; Pratt, 1927: 410.
Neochernes (?) *sanborni* (Hagen): Beier, 1932e: 167; Roewer, 1937: 299; Beier, 1948b: 445.
Neochernes sanborni (Hagen): Vachon, 1940f: 2.
Hesperochernes sanborni (Hagen): Hoff, 1946a: 100–103, figs 1–3; Hoff, 1949b: figs 42–43; Hoff, 1958: 22; Muchmore, 1971a: 81.

Type locality: Andover, Massachusetts, U.S.A.
Distribution: U.S.A. (Massachusetts, New York).

Chernes similis (Beier)

Chelifer (Trachychernes) wideri C. L. Koch: Beier, 1929a: 343 (misidentification, in part).
Allochernes (Toxochernes) similis Beier, 1932e: 152–153, fig. 163; Beier, 1933a: 527–528, fig. 6; Ionescu, 1936: 5.
Allochernes similis Beier: Roewer, 1937: 298.
Allochernes (Toxochernes) karamani Hadži, 1938: 28–31, figs 26a–b, 27a–e; Hadži, 1939b: 43–44, figs 13a–b (synonymized by Beier, 1963b: 273).
Toxochernes karamani (Hadži): Ćurčić, 1974a: 27 (synonymized by Beier, 1963b: 273).
Toxochernes similis (Beier): Beier, 1952e: 5; Beier & Franz, 1954: 457–458; Beier, 1963b: 272–273, fig. 275; Rafalski, 1967: 17; Ressl, 1970: 251; Ćurčić, 1974a: 27; Ressl, 1974: 27; Krumpál, 1980: 24.
Chernes similis (Beier): Mahnert, 1978a: 313, figs 2, 11; Krumpál & Cyprich, 1988: 42.

Type localities: of *Allochernes (Toxochernes) similis*: Kronstadt (now Brasov), Romania.
of *Allochernes (Toxochernes) karamani*: Skopje (as Skoplje), Yugoslavia.
Distribution: Austria, Bulgaria, Czechoslovakia, Poland, Romania, Turkey, Yugoslavia.

Chernes sinensis Beier

Chernes sinensis Beier, 1932e: 157–158, fig. 168; Beier, 1933a: 530–531, fig. 8; Roewer, 1937: 298.

Type locality: Canton, Kwangtung, China.
Distribution: China (Kwangtung).

Chernes vicinus (Beier)

Allochernes (Allochernes) vicinus Beier, 1932e: 147, fig. 156b; Beier, 1933a: 525–526.
Allochernes vicinus Beier: Roewer, 1937: 297; Beier, 1952e: 5; Beier & Franz, 1954: 457; Beier, 1963b: 264–265, fig. 264; Kofler, 1972: 288; Ressl, 1974: 26–27; Krumpál & Cyprich, 1988: 42.

Chernes lasiophilus Cooreman, 1947: 3–6, figs 1–4; Leleup, 1947: 322; Beier, 1963b: 273–274 (synonymized by Helversen, 1966a: 146).
Chernes vicinus (Beier): Helversen, 1966a: 145–147, fig. 6.

Type localities: of *Allochernes (Allochernes) vicinus*: Austria.
of *Chernes lasiophilus*: Houx, Namur, Belgium; and Wavreille, Namur, Belgium.
Distribution: Austria, Belgium, Czechoslovakia, West Germany.

Chernes zavattarii Caporiacco

Chernes zavattarii Caporiacco, 1941: 45–46, fig. 4.

Type locality: El Dire, Ethiopia.
Distribution: Ethiopia.

Genus Chiridiochernes Muchmore

Chiridiochernes Muchmore, 1972e: 427–428.

Type species: *Chiridiochernes platypalpus* Muchmore, 1972e, by original designation.

Chiridiochernes platypalpus Muchmore

Chiridiochernes platypalpus Muchmore, 1972e: 428–432, figs 1–9.

Type locality: Biroro, base of Mt Lampobathang, S. of Melino, Sulawesi, Indonesia.
Distribution: Indonesia (Sulawesi).

Genus Chrysochernes Hoff

Chrysochernes Hoff, 1956c: 25–26.

Type species: *Chrysochernes elatus* Hoff, 1956c, by original designation.

Chrysochernes elatus Hoff

Chrysochernes elatus Hoff, 1956c: 27–31, figs 8–11; Hoff, 1958: 29; Hoff, 1959b: 5, etc.

Type locality: 1 mile W. of Mountainair, Torrance County, New Mexico, U.S.A.
Distribution: U.S.A. (New Mexico).

Genus Cocinachernes Hentschel & Muchmore

Cocinachernes Hentschel & Muchmore, 1989: 345–346.

Type species: *Cocinachernes foiliosus* Hentschel & Muchmore, 1989, by original designation.

Cocinachernes foliosus Hentschel & Muchmore

Cocinachernes foliosus Hentschel & Muchmore, 1989: 346–349, figs 1–7.
Type locality: Islas Cocinas, Estado de Jalisco, Mexico.
Distribution: Mexico.

Genus Conicochernes Beier

Conicochernes Beier, 1948c: 532–533.

Type species: *Chelifer brevispinosus* L. Koch, 1885, by original designation.

Conicochernes brevispinosus (L. Koch)

Chelifer brevispinosus L. Koch, 1885: 46–47, plate 4 figs 4, 4a–b; With, 1905: 110.
Chelifer keyserlingi With, 1907: 53–55, figs 3–5 (synonymized by Beier, 1954b: 24).
Chelifer (Lamprochernes) silvestrii Beier, 1930d: 198–199, figs 2a–b (synonymized by Beier, 1948c: 533).

Haplochernes brevispinosus (L. Koch): Beier, 1932e: 111; J. C. Chamberlin, 1934b: 9; Roewer, 1937: 295.
Haplochernes keyserlingi (With): Beier, 1932e: 112; J. C. Chamberlin, 1934b: 9; Roewer, 1937: 295.
Austrochernes silvestrii (Beier): Beier, 1932e: 171, fig. 179; J. C. Chamberlin, 1934b: 10; Roewer, 1937: 301.
Thalassochernes brevispinosus (L. Koch): Beier, 1940a: 182.
Conicochernes brevispinosus Beier, 1948c: 533–535, fig. 5; Beier, 1966a: 297; Harvey, 1981b: 246; Harvey, 1985b: 134.

Type localities: of *Chelifer brevispinosus*: Gayndah, Queensland, Australia; Rockhampton, Queensland, Australia; and Peak Downs (as Peack Downs), Queensland, Australia.
of *Chelifer keyserlingi*: Rockhampton, Queensland, Australia.
of *Chelifer (Lamprochernes) silvestrii*: Mount Lofty, South Australia, Australia.
Distribution: Australia (Australian Capital Territory, Queensland, South Australia, Victoria).

Conicochernes crassus Beier

Conicochernes crassus Beier, 1954b: 21–24, fig. 10; Harvey, 1981b: 246; Harvey, 1985b: 134.

Type locality: Denmark, near mouth of Denmark River, Western Australia, Australia.
Distribution: Australia (Western Australia).

Conicochernes globosus Beier

Conicochernes globosus Beier, 1954b: 19–21, fig. 9; Harvey, 1981b: 246; Harvey, 1985b: 134.

Type locality: Porongorups, Western Australia, Australia.
Distribution: Australia (Western Australia).

Conicochernes incrassatus (Beier)

Haplochernes incrassatus Beier, 1933d: 644–645, fig. 1; Roewer, 1937: 295.
Conicochernes incrassatus (Beier): Beier, 1966a: 279; Harvey, 1981b: 246; Harvey, 1985b: 134.

Type locality: Upper Ferntree Gully, Victoria, Australia.
Distribution: Australia (Victoria).

Genus Coprochernes Beier

Coprochernes Beier, 1976c: 3.

Type species: *Coprochernes costaricensis* Beier, 1976c, by original designation.

Coprochernes costaricensis Beier

Coprochernes costaricensis Beier, 1976c: 3–5, fig. 2.

Type locality: Reventazon, Costa Rica.
Distribution: Costa Rica.

Genus Cordylochernes Beier

Cordylochernes Beier, 1932c: 265; Beier, 1932e: 99–100.

Type species: *Chelifer macrochelatus* Tömösváry, 1884, by original designation (junior synonym of *Acarus scorpioides* Linnaeus, 1758).

Cordylochernes angustochelatus Hoff

Cordylochernes angustochelatus Hoff, 1944a: 5–6, fig. 8.

Type locality: Boquete, Chiriqui, Panama.
Distribution: Panama.

Cordylochernes costaricensis Beier

Cordylochernes costaricensis Beier, 1932c: 266–267; Beier, 1932e: 103–104, figs 121, 125; Roewer, 1937: 290; Beier, 1939b: 289–290; Roewer, 1940: 346; Beier, 1948b: 453; Muchmore, 1971a: 84; Beier, 1976c: 1.

Type locality: Reventazon, Costa Rica.
Distribution: Bolivia, Costa Rica, Mexico.

Cordylochernes fallax Beier

Cordylochernes fallax Beier, 1933b: 93, fig. 3; Roewer, 1937: 290.

Type locality: Presidio, Veracruz, Mexico.
Distribution: Mexico.

Cordylochernes nigermanus Hoff

Cordylochernes nigermanus Hoff, 1944a: 6–8, figs 9–10; Beier, 1948b: 453; Muchmore, 1971a: 84.

Type locality: Gatun Lock, Panama.
Distribution: Panama.

Cordylochernes octentoctus (Balzan)

Chelifer (Lamprochernes) octentoctus Balzan, 1892: 514–515, fig. 5.
Chelifer octentoctus Balzan: Ellingsen, 1912b: 81.
Cordylochernes octentoctus (Balzan): Beier, 1932e: 104; Roewer, 1937: 290; Vachon, 1942: 181–184, figs 1–4.

Type locality: South Africa, but probably mislabelled (Vachon, 1942).
Distribution: South Africa?

Cordylochernes panamensis Hoff

Cordylochernes panamensis Hoff, 1944a: 8–10, figs 11–15; Beier, 1948b: 453; Muchmore, 1971a: 84.

Type locality: Gatun Lock, Panama.
Distribution: Panama.

Cordylochernes perproximus Beier

Cordylochernes perproximus Beier, 1933b: 93–94, fig. 4; Roewer, 1937: 290.

Type locality: Peñuela, Veracruz, Mexico.
Distribution: Mexico.

Cordylochernes potens Hoff

Cordylochernes potens Hoff, 1947c: 63–64, figs 4–6.

Type locality: Rancho Grande, Venezuela.
Distribution: Venezuela.

Cordylochernes scorpioides (Linnaeus)

Acarus scorpioides Linnaeus, 1758: 616.
Phalangium acaroides Linnaeus, 1767: 1028 (synonymized by Hagen, 1867: 325).

Family Chernetidae

Chelifer americanus de Geer, 1778: 353–355, plate 42 figs 1–5; de Geer, 1783: 137, plate 42 figs 1–5 (not seen); Gervais, 1844: 81; Hagen, 1867: 325; Hagen, 1870: 264; Vachon, 1940f: 3 (synonymized by Hermann, 1804: 117).
Scorpio americanus (de Geer): Fabricius, 1793: 434.
Scorpio acaroides (Linnaeus): Fabricius, 1793: 437.
Chelifer acaroides (Linnaeus): Hermann, 1804: 117; Latreille, 1804: 142.
Chelifer scorpioides (Linnaeus): Hagen, 1867: 325.
Chelifer nodulimanus Tömösváry, 1882b: 244–245; Tömösváry, 1884: 26–27, fig. 14; With, 1906: 171; Tullgren, 1907a: 46–49, figs 10a–f; With, 1908a: 314–316, text-fig. 81, fig. 33a; Ellingsen, 1913a: 453 (synonymized by Beier, 1948b: 470).
Chelifer macrochelatus Tömösváry, 1884a: 20–21, figs 12–13; Ellingsen, 1905b: 1; With, 1908a: 310–314, text-figs 72–73, 78–80, fig. 32a; Ellingsen, 1909a: 219; Ellingsen, 1909b: 216; Ellingsen, 1913a: 453 (synonymized with *Chelifer nodulimanus* Tömösváry by Daday, 1888: 118; synonymized with *Acarus scorpioides* Linnaeus by Beier, 1948b: 470).
Chernes (Ectoceras) nodulimanus (Tömösváry): Daday, 1888: 118–119, 173–174, plate 4 figs 3, 9.
Chelifer (Lamprochernes) nodulimanus Tömösváry: Balzan, 1892: 543–544, figs 4, 4a; Ellingsen, 1905c: 3–6; Ellingsen, 1910a: 368.
Chelifer (Chernes) macrochelatus Tömösváry: Ellingsen, 1902: 152–154, 165.
Chelifer (Lamprochernes) macrochelatus Tömösváry: Ellingsen, 1905c: 6; Ellingsen, 1910a: 368.
Chelanops nodulimanus (Tömösváry): Banks, 1913: 683.
Chelanops (Lamprochernes) macrochelatus (Tömösváry): J. C. Chamberlin, 1923b: 190–192, figs 21a–g.
Chelanops macrochelatus (Tömösváry): J. C. Chamberlin, 1924c: fig. a; Berland, 1932: fig. 80.
Lamprochernes nodulimanus (Tömösváry): Beier, 1930a: 212.
Chernes macrochelatus (Tömösváry): J. C. Chamberlin, 1931a: figs 18u–v, 31g–i.
Cordylochernes macrochelatus (Tömösváry): Beier, 1932c: 265; Beier, 1932e: 101–102, fig. 123; Beier, 1932g: fig. 214; Roewer, 1936: fig. 115; Roewer, 1937: 290; Mello-Leitão, 1939b: 616; Feio, 1945: 6; Vachon, 1949: figs 203d, 219a; Weygoldt, 1969a: fig. 15a.
Cordylochernes peruanus Beier, 1932c: 265; Beier, 1932e: 102, fig. 124; Roewer, 1937: 290; Mello-Leitão & Feio, 1949: 320 (synonymized by Beier, 1948b: 470).
Cordylochernes brasiliensis Beier, 1932c: 265–266; Beier, 1932e: 102–103; Roewer, 1937: 290; Vachon, 1940f: 3; Feio, 1945: 6 (synonymized by Beier, 1948b: 470).
Cordylochernes nodulimanus (Tömösváry): Beier, 1932e: 100–101, fig. 122; Roewer, 1937: 290, fig. 235; Mello-Leitão, 1939a: 120; Mello-Leitão, 1939b: 616.
Cordylochernes scorpioides (Linnaeus): Beier, 1948b: 444, 452–453, 470–472; Beier, 1955m: 5–6; Beier, 1959e: 206; Beier, 1964b: 343; Beck, 1968: 30–32, fig. 8; Weygoldt, 1969a: 117; Muchmore, 1971a: 84; Beier, 1977c: 105; Mahnert, 1979d: 765; Cekalovic, 1984: 20; Mahnert, 1985b: 78; Mahnert & Adis, 1986: 213.
Cordylochernes scorpioides brasiliensis Beier: Beier, 1974d: 899.
Not *Phalangium acaroides* Linnaeus: Montagu, 1815: 7–11, plate 2 fig. 4 (misidentification; see *Neobisium (Blothrus) maritimum* (Leach)).

Type localities: of *Acarus scorpioides*: Surinam.
of *Phalangium acaroides*: tropical America.
of *Chelifer americanus*: America.
of *Chelifer nodulimanus*: Dalmatia, Yugoslavia (presumably incorrect).
of *Chelifer macrochelatus*: Colombia.
of *Cordylochernes peruanus*: Rio Pachitea, Monte Alegre, Peru.
of *Cordylochernes brasiliensis*: Brazil.

Distribution: Argentina, Bolivia, Brazil (Amazonas, Bahia. Espirito Santo, Maranhao, Minas Gerais, Pará, Rio de Janeiro, Santa Catarina, Sao Paulo), Chile, Colombia, Costa Rica, Ecuador, Guatemala, Guyana, Mexico, Panama, Paraguay, Peru, Surinam, Trinidad, Venezuela; records from Indonesia (Sumatra), Ghana and Yugoslavia by Daday (1888) are probably incorrect.

Genus **Corosoma** Karsch

Corosoma Karsch, 1879: 95; Beier, 1932e: 105.

Type species: *Corosoma sellowi* Karsch, 1879, by original designation.

Corosoma sellowi Karsch

Corosoma sellowi Karsch, 1879: 95; Beier, 1932e: 105; Roewer, 1936: fig. 31; Roewer, 1937: 291; Mahnert, 1982a: 12, figs 1–5.
Lamprochernes sellowi (Karsch): Beier, 1930e: 298–300, figs 9–12.

Type locality: Sao Paulo (as St Paul), Brazil.
Distribution: Brazil (Sao Paulo).

Genus **Cyclochernes** Beier

Cyclochernes Beier, 1970g: 322–323.

Type species: *Cyclochernes montanus* Beier, 1970g, by original designation.

Cyclochernes montanus Beier

Cyclochernes montanus Beier, 1970g: 323–324, fig. 5.

Type locality: Mt Popamanusiu, Guadalcanal, Solomon Islands.
Distribution: Solomon Islands.

Genus **Dasychernes** J. C. Chamberlin

Dasychernes J.C. Chamberlin, 1929c: 49–51; Beier, 1932e: 168; Beier, 1933a: 533.

Type species: *Dasychernes inquilinus* J. C. Chamberlin, 1929c, by original designation.

Dasychernes inquilinus J. C. Chamberlin

Dasychernes inquilinus J. C. Chamberlin, 1929c: 51; Salt, 1929: 446–447, plate 25 fig. 1; J. C. Chamberlin, 1931a: 68, 171, 195, figs 6d, 10l, 15y, 18q, 23h, 27c, 33a, 33e–f, 33m, 46e, 47i, 51f; Beier, 1932e: 168; Beier, 1932g: fig. 220; Beier, 1948b: 458; Vachon, 1949: figs 200b, 210; Mahnert, 1982a: 13; Mahnert, 1987: 407.
Dasychernes inuqilinus (sic) J.C. Chamberlin: Roewer, 1937: 299.

Type locality: Rio Frio, Colombia.
Distribution: Colombia.

Dasychernes panamensis Mahnert

Dasychernes (?) *panamensis* Mahnert, 1987: 408–410, figs 16–19.

Type locality: El-Llano-carti Road km 25, Comarca de San Blas, Panama.
Distribution: Panama.

Dasychernes roubiki Mahnert

Dasychernes roubiki Mahnert, 1987: 407–408, figs 10–15.

Type locality: El-Llano-carti Road km 25, Comarca de San Blas, Panama.
Distribution: Panama.

Family Chernetidae

Dasychernes trigonae Mahnert

Dasychernes trigonae Mahnert, 1987: 403–407, figs 1–9.

Type locality: El-Llano-carti Road km 23, Comarca de San Blas, Panama.
Distribution: Panama.

Genus Dendrochernes Beier

Dendrochernes Beier, 1932e: 172–173; Beier, 1933a: 537; Hoff, 1947b: 536; Hoff, 1949b: 464–465; G.O. Evans & Browning, 1954: 18; Hoff, 1956c: 38–39; Beier, 1963b: 278; Legg, 1972: 580; Legg & Jones, 1988: 125.
Pachycheirus J. C. Chamberlin, 1934a: 125–126 (synonymized by Hoff, 1958: 29).

Type species: of *Dendrochernes*: *Chernes cyrneus* L. Koch, 1873, by original designation.

of *Pachycheirus*: *Pachycheirus instabilis* J. C. Chamberlin, 1934a, by original designation.

Dendrochernes crassus Hoff

Dendrochernes crassus Hoff, 1956c: 39–43, figs 16–17; Hoff, 1958: 29; Hoff, 1959b: 5, etc.; Hoff, 1961: 456; Benedict & Malcolm, 1982: 99.

Type locality: Fourth of July Camp area, W. of Tajique, Torrance County, New Mexico, U.S.A.
Distribution: U.S.A. (Colorado, New Mexico, Oregon).

Dendrochernes cyrneus (L. Koch)

Chernes cyrneus L. Koch, 1873: 6–7; Daday, 1889c: 25; Carr, 1906: 48; Kew, 1906: 41–45, plate 5; Bennett, 1908: 114; Daday, 1918: 1; J. C. Chamberlin, 1931a: 79, 178–179, 196, figs 48a–f; Syms, 1950: 144, figs 2a–c.
Chelifer cyrneus (L. Koch): E. Simon, 1878a: 150; E. Simon, 1879a: 36, plate 18 fig. 11; Canestrini, 1885: no. 1, figs 1–3; E. Simon, 1899d: 86; Tullgren, 1899a: 171–172, plate 1 fig. 6; Ellingsen, 1903: 8–11; Tullgren, 1906a: 202–203, plate 4 fig. 4; Ellingsen, 1907a: 156; Ellingsen, 1908d: 70; Ellingsen, 1909a: 215; Kew, 1912: 381–387, figs 48, 49a–c, 50b–c; Kew, 1914: 102–107; Väänänen, 1928a: 18, figs 5(IV), 6(3); Väänänen, 1928b: figs 4a–c, 5a–b; Kew, 1929b: 83–86; Berland, 1932: figs 77–78.
Chernes (Trachychernes) cyrneus L. Koch: Tömösváry, 1882b: 194–195, plate 1 figs 17–19.
Chernes (Chernes) cyrneus hungaricus Daday, 1888: 116–117, 171–172, plate 4 figs 4, 6 (synonymized by Beier, 1929a: 346).
Chernes multidentatus Stschelkanowzeff, 1902b: 350–355, figs 1–2; Stschelkanowzeff, 1903a: 318, fig. 1 (synonymized by Redikorzev, 1928: 120).
Chelifer (Chernes) cyrneus (L. Koch): Kew, 1911a: 46, fig. 8; Kästner, 1928: 7, fig. 20; Schenkel, 1928: 61, figs 16a–c; Kew, 1929a: 34.
Chelifer (Chelanops) cyrneus (L. Koch): Redikorzev, 1924b: 23, fig. 7; Redikorzev, 1928: 119–121.
Chelifer (Trachychernes) cyrneus (L. Koch): Beier, 1929a: 346–347; Beier, 1929e: 446.
Hesperochernes cyrneus (L. Koch): Beier, 1930a: 200.
Dendrochernes cyrneus (L. Koch): Beier, 1932e: 172–173; Beier, 1932g: figs 229, 230c, 234–235; Ionescu, 1936: 5; Vachon, 1936a: 78; Roewer, 1937: 302, figs 156b, 159, 161, 196–197; Vachon, 1937f: 128; Vachon, 1938a: figs 33c, 56c; Lohmander, 1939b: 318–319; Beier, 1948b: 445; Kaisila, 1949b: 86, map 12; Redikorzev, 1949: 658; Beier, 1952e: 5; Beier & Franz, 1954: 458; G. O. Evans & Browning, 1954: 18, fig. 18; Vachon, 1954d: 591; Beier, 1956i: 8; Ressl & Beier, 1958: 2; George, 1961: 38; Kobakhidze, 1961c: 209–211; Beier, 1963b: 278–279, fig. 282; Kobakhidze, 1965b: 542; Ressl, 1965: 290; Helversen, 1966a: 148;

Kobakhidze, 1966: 705; Rafalski, 1967: 19; Beier, 1970a: 45; Kofler, 1972: 288; Legg, 1972: 583, figs 2(4e), 3(3); Beier, 1973c: 226; Beier, 1976a: 24; Legg, 1975a: 66; Legg, 1975d: fig. 2; Crocker, 1978: 9; Jones, 1978: 91; Crocker, 1979: 5–6; Jones, 1979a: 199, 201; Jones, 1980a: 33–36, fig. 1; Jones, 1980c: map 23; Schawaller, 1981b: 47–48; Callaini, 1983e: 225; Cuthbertson, 1984: 3; Jędryczkowski, 1985: 80; Callaini, 1986c: 396, fig. 4i; Schawaller, 1986: 10; Jędryczkowski, 1987b: 145, map 12; Schawaller, 1987a: 216–217, figs 40–42; Krumpál & Cyprich, 1988: 42; Legg & Jones, 1988: 125–126, figs 29c, 32a, 32b(a–j); Schawaller & Dashdamirov, 1988: 43, figs 79; Schawaller, 1989: 22–24.

Type localities: of *Chernes cyrneus*: Corsica.
of *Chernes (Chernes) cyrneus hungaricus*: Szent-Márton, Baranya, Hungary.
of *Chernes multidentatus*: Chernigov (as Tschernigow), Ukraine, U.S.S.R.
Distribution: Albania, Algeria, Austria, Bulgaria, Canary Islands, Corsica, Czechoslovakia, Finland, France, Great Britain, Hungary, Italy, Nepal, Norway, Poland, Romania, Sicily, Sweden, Turkey, U.S.S.R. (Armenia, Azerbaijan, Georgia, Kazakhstan, R.S.F.S.R., Ukraine), West Germany, Yugoslavia.

Dendrochernes cyrneus minor Cîrdei, Bulimar & Malcoci

Dendrochernes cyrneus Cîrdei, Bulimar & Malcoci, 1967: 241–242, fig. 7.

Type locality: Masivul Repedea, Iasi, Romania.
Distribution: Romania.

Dendrochernes instabilis (J. C. Chamberlin)

Pachycheirus instabilis J. C. Chamberlin, 1934a: 126–128, figs 1f, 1j–l, 1n–p, 1r–s, 1v–y; Roewer, 1937: 301, fig. 238; Beier, 1948b: 453.
Dendrochernes instabilis (J. C. Chamberlin): Hoff, 1958: 29; Muchmore, 1971a: 84.

Type locality: Sula, Montana, U.S.A.
Distribution: U.S.A. (Montana).

Dendrochernes morosus (Banks)

Chelanops morosus Banks, 1895a: 7; Coolidge, 1908: 111.
Neochernes morosus (Banks): Beier, 1932e: 165–166; Roewer, 1937: 299.
Dendrochernes morosus (Banks): Hoff, 1947b: 536–539, fig. 33; Hoff, 1958: 29; Manley, 1969: 8–9, fig. 10; Nelson, 1975: 286, figs 50–51; Brach, 1979: 34.
Dendrochernes cf. *morosus* (Banks): Haack & Wilkinson, 1987: 370–372.

Type locality: Isle Royale, Lake Superior, Michigan, U.S.A.
Distribution: U.S.A. (Florida, Michigan).

Genus **Dinocheirus** J. C. Chamberlin

Dinocheirus J. C. Chamberlin, 1929b: 171–172; Beier, 1932e: 137; Beier, 1933a: 522; J. C. Chamberlin, 1934a: 128; Hoff, 1947b: 513–514; Hoff, 1949b: 471–472; Hoff, 1956c: 43–44; Muchmore, 1974d: 31–32; Legg & Jones, 1988: 118.
Allochernes (Toxochernes) Beier, 1932e: 151; Beier, 1933a: 527 (synonymized by Mahnert, 1978a: 314).
Epaphochernes Beier, 1932e: 173; Beier, 1933a: 537 (synonymized by Beier, 1933b: 100).
Toxochernes Beier: G. O. Evans & Browning, 1954: 21; Beier, 1963b: 270.

Type species: of *Dinocheirus*: *Dinocheirus tenoch* J. C. Chamberlin, 1929b, by original designation.
of *Allochernes (Toxochernes)*: *Chelifer panzeri* C. L. Koch, 1837, by subsequent designation of Mahnert, 1978a: 314.
of *Epaphochernes*: *Chelanops arizonensis* Banks, 1901b, by original designation.

Family Chernetidae

Dinocheirus aequalis (Banks)

Chelanops aequalis Banks, 1908: 41.
Epaphochernes aequalis (Banks): Beier, 1932e: 174.
Dinocheirus aequalis (Banks): Roewer, 1937: 302; Hoff, 1946c: 14–18, figs 19–22; Hoff, 1947b: 520–523, figs 24–25; Hoff, 1956c: 61–62; Hoff, 1958: 27; Hoff, 1959b: 5, etc.; Rowland & Reddell, 1976: 17; Zeh, 1987b: 1084.

Type locality: El Paso, Texas, U.S.A.
Distribution: Mexico, U.S.A. (New Mexico, Texas).

Dinocheirus arizonensis (Banks)

Chelanops arizonensis Banks, 1901b: 589, fig. 2 (in part, see *Parachelifer hubbardi* (Banks)); Coolidge, 1908: 112; J. C. Chamberlin, 1923c: 379–380, plate 3 figs 8, 30.
Hesperochernes arizonensis (Banks): Beier, 1930a: 213–214, figs 9a–c.
Epaphochernes arizonensis (Banks): Beier, 1932e: 173–174, figs 181–182; Roewer, 1936: fig. 83c.
Dinocheirus arizonensis (Banks): Beier, 1933b: 100; J. C. Chamberlin, 1934a: figs 1d–e, 1q, 1t, 1u, 1z; Roewer, 1937: 302, fig. 239; Hoff, 1946d: 200–201; Hoff, 1958: 27; Muchmore, 1974d: 34; Zeh, 1987a: 1495–1501, figs 1a–d; Zeh, 1987b: 1077–1084, figs 1a–b, 4.

Type locality: Oracle, Arizona, U.S.A.
Distribution: Mexico, U.S.A. (Arizona).

Dinocheirus astutus Hoff

Dinocheirus astutus Hoff, 1956c: 44–48, figs 18–20; Hoff, 1958: 28; Hoff, 1959b: 5, etc.; Muchmore, 1971a: 90; Muchmore, 1974d: 34; Zeh, 1987b: 1084.

Type locality: near Santa Fe, Santa Fe County, New Mexico, U.S.A.
Distribution: U.S.A. (New Mexico).

Dinocheirus athleticus Hoff

Dinocheirus athleticus Hoff, 1956c: 48–54, figs 21–24; Hoff, 1958: 28; Hoff, 1959b: 5, etc.; Hoff, 1961: 455–456; Zeh, 1987b: 1084.

Type locality: Las Huertas Canyon, Sandia Mountains, Sandoval County, New Mexico, U.S.A.
Distribution: U.S.A. (Colorado, New Mexico).

Dinocheirus bulbipalpis (Redikorzev)

Chernes bulbipalpis Redikorzev, 1949: 656–658, figs 21–22.
Dinocheirus bulbipalpis (Redikorzev): Schawaller, 1986: 4–5, figs 8–12.

Type locality: Tashkent, Uzbekistan, U.S.S.R.
Distribution: U.S.S.R. (Kirghizia, Uzbekistan).

Dinocheirus chilensis Beier

Dinocheirus chilensis Beier, 1964b: 364–366, fig. 31; Cekalovic, 1984: 24.

Type locality: Illapal, Chile.
Distribution: Chile.

Dinocheirus diabolicus Beier

Dinocheirus diabolicus Beier, 1964b: 366, fig. 32; Cekalovic, 1984: 24.

Type locality: Taltal, Antofagasta, Chile.
Distribution: Chile.

Dinocheirus dorsalis (Banks)

Chelanops dorsalis Banks, 1895a: 8; Banks, 1904b: 364; Coolidge, 1908: 112; Banks, 1911: 638; Wheeler, 1911: 167–168; Moles & Moore, 1921: 6.
Neochernes dorsalis (Banks): Beier, 1932e: 166; Roewer, 1937: 299; Beier, 1948b: 458.
Dinocheirus dorsalis (Banks): Hoff, 1947b: 523–526, fig. 26; Hoff, 1958: 27; Muchmore, 1971a: 87; Knowlton, 1972: 2; Knowlton, 1974: 2; Muchmore, 1974d: 34.

Type locality: Lake Tahoe, California, U.S.A.
Distribution: U.S.A. (California, Utah).

Dinocheirus horricus Nelson & Manley

Dinocheirus horricus Nelson & Manley, 1972: 217–221, figs 1–3; Nelson, 1975: 288–289, figs 59–61; Muchmore, 1974d: 34; Zeh, 1987b: 1084.

Type locality: Ingham County, Michigan, U.S.A.
Distribution: U.S.A. (Michigan).

Dinocheirus imperiosus Hoff

Dinocheirus imperiosus Hoff, 1956c: 54–59, figs 25–28; Hoff, 1958: 28; Hoff, 1959b: 5, etc.; Muchmore, 1974d: 34; Zeh, 1987b: 1084.

Type locality: Sandia Mountains, E. of Albequerque, Bernalillo County, U.S.A.
Distribution: U.S.A. (New Mexico).

Dinocheirus magnificus Hoff

Dinocheirus magnificus Hoff, 1946c: 8–12, figs 10–14.

Type locality: Culebra, Virgin Islands.
Distribution: Virgin Islands.

Dinocheirus magnificus superior Hoff

Dinocheirus magnificus superior Hoff, 1946c: 12, fig. 15.

Type locality: Culebra, Virgin Islands.
Distribution: Virgin Islands.

Dinocheirus obesus (Banks)

Chelanops obesus Banks, 1909a: 146–147.
Hesperochernes obesus (Banks): Beier, 1932e: 176; Roewer, 1937: 302.
Dinocheirus obesus (Banks): Hoff, 1947b: 517–520, fig. 23; Hoff, 1958: 27.

Type locality: Tuscon, Arizona, U.S.A.
Distribution: Mexico, U.S.A. (Arizona).

Dinocheirus pallidus (Banks)

Chernes pallidus Banks, 1890: 152.
Chelanops pallidus (Banks): Banks, 1895a: 8; Coolidge, 1908: 112.
Neochernes (?) *pallidus* (Banks): Beier, 1932e: 167; Roewer, 1937: 299.
Hesperochernes paliidus (Banks): Hoff, 1947b: 509–511, fig. 18.
Dinocheirus pallidus (Banks): Hoff, 1949b: 472–474, figs 39a–c; Hoff, 1958: 26; Manley, 1969: 11, fig. 16; Nelson, 1971: 95; Muchmore, 1974d: 34; Nelson, 1975: 289, figs 62–63; Zeh, 1987b: 1084.
Not *Chelanops pallidus* (Banks): Ewing, 1911: 76, 78 (misidentification; see *Parachernes (Parachernes) virginicus* (Banks)).

Type locality: Ithaca, New York, U.S.A.
Distribution: U.S.A. (Arkansas, Illinois, Indiana, Michigan, Pennsylvania, New York).

Family Chernetidae

Dinocheirus panzeri (C. L. Koch)

Chelifer panzeri C. L. Koch, 1837: fasc. 140.6, fig.; C. L. Koch, 1843: 44–46, figs 782–783; Gervais, 1844: 79; Ellingsen, 1907a: 159–162; Standen, 1916b: 124.

Chelifer schrankii C. L. Koch, 1837: fasc. 140.3, fig.; Gervais, 1844: 79; Hagen, 1870: 267 (synonymized by Beier, 1932e: 151).

Chelifer latreillei Leach: McIntire, 1869: 246, figs 211–212; McIntire, 1871: 209–210, fig. 3 (misidentifications).

Chernes cimicoides panzeri (C. L. Koch): Stecker, 1874c: 308.

Chelifer rufeolus E. Simon, 1879a: 41–42, plate 17 fig 1, plate 18 fig. 15 (synonymized by Ellingsen, 1907a: 159).

Chernes (Trachychernes) rufeolus (E. Simon): Tömösváry, 1882b: 190–191, plate 1 figs 15–16.

Chelifer cimicoides (Fabricius): H. J. Hansen, 1884: 544–545 (misidentification).

Chernes rufeolus (E. Simon): O.P.-Cambridge, 1905: ? (not seen); Whyte & Whyte, 1907: 203; Beier, 1932e: 157, fig. 167; Vachon, 1934d: 133; Vachon, 1936a: 78; Roewer, 1937: 298; Vachon, 1937e: fig. 6; Beier, 1958c: 137; Beier, 1963b: 275, fig. 277; Helversen, 1966a: 147–148, fig. 7; Beier, 1967f: 313; Rafalski, 1967: 18; Kofler, 1968: 356; Lazzeroni, 1970a: 208; Ressl, 1970: 252; Beier, 1971a: 364; Kofler, 1972: 288; Mahnert, 1974a: 90; Mahnert, 1977d: 97, fig. 6.

Chernes phaleratus (E. Simon): O.P.-Cambridge, 1905: ? (not seen) (misidentification).

Chernes panzeri (C. L. Koch): Butterfield, 1908: 112; Godfrey, 1908: 157; Godfrey, 1909: 25–26; Kew, 1909b: 259; Falconer, 1916: 192–193; Ressl, 1974: 28–29.

Chelifer (Chernes) panzeri C. L. Koch: Kew, 1911a: 45, fig. 7; Standen, 1912: 13; Standen, 1914: 461–462; Kew, 1916b: 79; Standen, 1917: 28; Standen, 1918: 335; Kästner, 1928: 7, figs 12, 21; Schenkel, 1928: 61, figs 15a–c.

Chelifer (Trachychernes) panzeri C. L. Koch: Beier, 1929a: 344.

Chelifer (Trachychernes) rufeolus E. Simon: Beier, 1929a: 344–345.

Allochernes (Toxochernes) panzeri (C. L. Koch): Beier, 1932e: 151–152, fig. 161; Leleup, 1947: 322.

Allochernes panzeri (C. L. Koch): Tumšs, 1934: 15–17, fig. 2; Roewer, 1937: 298; Vachon, 1940f: 2.

Epaphochernes bouvieri Vachon, 1936b: 141–143, figs 1–2; Vachon, 1938a: figs 33d, 33g, 56m; Roewer, 1940: 347 (synonymized with *Pselaphochernes anachoreta* (E. Simon) by Beier, 1963b: 257; synonymized with *Chelifer panzeri* C. L. Koch by Mahnert, 1978a: 314).

Toxochernes panzeri (C. L. Koch): Lohmander, 1939b: 308–313, fig. 10; Beier, 1948b: 458, 462; Kaisila, 1949a: 75; Kaisila, 1949b: 83, map 9; Beier, 1952e: 5; G. O. Evans & Browning, 1954: 22, fig. 16; Beier, 1955n: 109; Beier, 1956h: 24; Vachon, 1957: 389, figs 1, 9; Ressl & Beier, 1958: 2; George, 1961: 38; Meinertz, 1962a: 398, map 7; Beier, 1963b: 270–271, fig. 272; Ressl, 1965: 289; Rafalski, 1967: 17; Hammen, 1969: 21; Legg, 1970b: fig. 3(5); Gabbutt, 1972a: 37–40, figs 1e, 2e; Gabbutt, 1972b: 2–13; Gabbutt, 1972c: 83–86, figs 3a–b; Howes, 1972a: 109; Kofler, 1972: 288; Legg, 1972: 583, figs 2(1b), 2(4a), 3(5); Legg, 1973b: 430, fig. 2i; Palmgren, 1973: 9; Ćurčić, 1974a: 27; Jones, 1975a: 88; Klausen, 1975: 64; Legg, 1975a: 66; Legg, 1975d: fig. 2; Crocker, 1976: 10; Klausen & Totland, 1977: 101–108, plate 5–7, figs 1–2; Crocker, 1978: 9; Jones, 1978: 94, 95; Jones, 1979a: 200; Schawaller, 1979: 13; Lippold, 1985: 40.

Toxochernes panzeri caucasicus Kobakhidze, 1963: 645–648, fig. 1; Kobakhidze, 1965b: 542; Kobakhidze, 1966: 705 (synonymized by Schawaller & Dashdamirov, 1988: 39).

Chernes rufeulus (sic) (E. Simon): Lazzeroni, 1969a: 337.

Dinocheirus bouvieri (Vachon): Muchmore, 1974d: 34.

572

Dinocheirus panzeri (C. L. Koch): Mahnert, 1978a: 313, figs 3, 4–7, 10; Judson, 1979a: 63; Jones, 1980c: map 21; Jędryczkowski, 1985: 80; Callaini, 1986c: 380; Jędryczkowski, 1987a: 345; Jędryczkowski, 1987b: 143–144, map 9; Judson, 1987: 370; H. Hansen, 1988: 187; Krumpál & Cyprich, 1988: 42; Legg & Jones, 1988: 119–121, figs 29a, 30a, 30b(a–j); Schawaller & Dashdamirov, 1988: 39, figs 67, 69–71.

Type localities: of *Chelifer panzeri*: Regensburg, West Germany.
of *Chelifer schrankii*: not stated, presumably West Germany.
of *Chelifer rufeolus*: Conflans, Savoie, France; Sos, Lot-et-Garonne, France; and Hyères, Var, France.
of *Epaphochernes bouvieri*: Francheville, Côte-d'Or, France.
of *Toxochernes panzeri caucasicus*: Selo Muganlo, ?; Baniskhevi, Georgia, U.S.S.R.; Shostka, Ukraine, U.S.S.R.; Sochi, R.S.F.S.R., U.S.S.R.; and Poti, Georgia, U.S.S.R.
Distribution: Austria, Czechoslovakia, Denmark, East Germany, Finland, France, Great Britain, Iran, Ireland, Italy, Netherlands, Norway, Poland, Sardinia, Spain, Sweden, Switzerland, Turkey, U.S.S.R. (Azerbaijan, Georgia, Latvia, R.S.F.S.R.), West Germany.

Dinocheirus partitus (Banks)

Chelanops partitus Banks, 1909b: 304.
Epaphochernes partitus (Banks): Beier, 1932e: 174.
Dinocheirus partitus (Banks): Roewer, 1937: 302; Hoff, 1947b: 514–517, figs 20–22; Hoff, 1958: 27; Zeh, 1987b: 1084.

Type locality: Fort (as Pt.) Yuma, Arizona, U.S.A.
Distribution: U.S.A. (Arizona).

Dinocheirus proximus Hoff

Dinocheirus proximus Hoff, 1946c: 12–14, figs 16–18.

Type locality: 2 miles E. of Santo Domingo, San Luis Potosí, Mexico.
Distribution: Mexico.

Dinocheirus serratus (Moles)

Chelanops serratus Moles, 1914c: 193–195, fig. 3; Moles & Moore, 1921: 7.
Dinocheirus (?) *serratus* (Moles): Hoff, 1958: 28.

Type locality: Pomona College, California, U.S.A.
Distribution: U.S.A. (California).

Dinocheirus sicarius J. C. Chamberlin

Dinocheirus sicarius J. C. Chamberlin, 1952: 279–292, figs 6a–i, 7a–r, 8a–j, 9a–h; Gering, 1956: 50 (not seen); Knudsen, 1956: 2; Hoff, 1958: 28; Muchmore, 1971a: 81, 90, 92; Muchmore, 1974d: 34; Benedict & Malcolm, 1982: 100; Zeh, 1987b: 1084.

Type locality: Frances Simes Hastings Natural History Reservation, California, U.S.A.
Distribution: U.S.A. (California, Oregon, Utah).

Dinocheirus solus Hoff

Dinocheirus solus Hoff, 1949b: 474–476, figs 40a–b; Hoff & Bolsterli, 1956: 175; Hoff, 1958: 27; Zeh, 1987b: 1084.

Type locality: Rockford, Winnebago County, Illinois, U.S.A.
Distribution: U.S.A. (Illinois).

Dinocheirus subrudis (Balzan)

Chelifer (Trachychernes) subrudis Balzan, 1892: 521–522, figs 13, 13a–b; Ellingsen, 1910a: 382.
Chelifer subrudis (Balzan): With, 1908a: 264–266, figs 14a–f.
Dinocheirus subrudis (Balzan): Beier, 1932e: 139–140, fig. 153; Roewer, 1937: 302.

Type locality: Tovar, Venezuela.
Distribution: Venezuela.

Dinocheirus tenoch J. C. Chamberlin

Dinocheirus tenoch J. C. Chamberlin, 1929b: 172–173; J. C. Chamberlin, 1931a: figs 16h, 16p, 18k–l, 20h, 30h–i; Beier, 1932e: 138; Beier, 1932g: fig. 169; J. C. Chamberlin, 1934a: figs 1g, 1m; Muchmore, 1974d: 34, figs 11–16; Muchmore, 1975d: fig. 3.
Dinocheirus enoch (sic) J. C. Chamberlin: Roewer, 1937: 302.

Type locality: Mexico City, Mexico.
Distribution: Mexico.

Dinocheirus texanus Hoff & Clawson

Dinocheirus texanus Hoff & Clawson, 1952: 27–31, figs 16–19; Hoff, 1958: 28; Muchmore, 1971a: 90; Rowland & Reddell, 1976: 17; Zeh, 1987b: 1084.

Type locality: Laguna Madre, 25 miles SE. of Harlingen, Cameron County, Texas, U.S.A.
Distribution: U.S.A. (Texas).

Dinocheirus topali Beier

Dinocheirus topali Beier, 1964j: 497–498, fig. 6.

Type locality: Mt Piltriquitron, El Bolsón, Rio Negro, Argentina.
Distribution: Argentina.

Dinocheirus transcaspius (Redikorzev)

Chelifer (Chelanops) transcaspius Redikorzev, 1922a: 259–262, figs 3–4.
Chernes transcaspius (Redikorzev): Beier, 1932e: 158–159, fig. 170; Roewer, 1937: 298; Redikorzev, 1949: 654.
Chernes bachardensis Redikorzev, in Beier, 1932e: 159, fig. 171; Roewer, 1937: 298; Vachon, 1938a: fig. 56n; Redikorzev, 1949: 655–656, figs 18–20; Beier, 1959a: 275; Beier, 1960a: 43; Lindberg, 1961: 31 (synonymized by Schawaller, 1986: 5).
Dinocheirus transcaspius (Redikorzev): Schawaller, 1986: 5–6, figs 13–16; Schawaller, 1989: 18.

Type locality: of *Chelifer (Chelanops) transcaspius*: Bacharden, U.S.S.R.
of *Chernes bachardensis*: Bacharden, near Aschabad, U.S.S.R.
Distribution: Afghanistan, U.S.S.R. (Kazakhstan, Kirghizia, Tadzhikistan, Turkmenistan, Uzbekistan).

Dinocheirus uruguayanus Beier

Dinocheirus uruguayanus Beier, 1970d: 53–54, fig. 2.

Type locality: Gruta de Arequita, Lavalleja, Uruguay.
Distribution: Uruguay.

Dinocheirus validus (Banks)

Chelanops validus Banks, 1895a: 7; Banks, 1901a: 594; Banks, 1904b: 364; Coolidge, 1908: 111; Banks, 1911: 638; Moles & Moore, 1921: 6.

Chelifer validus (Banks): Ellingsen, 1909b: 216.
Dinocheirus validus (Banks): Beier, 1932e: 138; Roewer, 1937: 302; Hoff, 1947b: 526–529, figs 27–28; Hoff, 1956c: 59–61; Hoff, 1958: 26; Hoff, 1959b: 5, etc.; Hoff, 1961: 455; Knowlton, 1972: 2; Knowlton, 1974: 2; Muchmore, 1974d: 34; Benedict & Malcolm, 1982: 100–101; Zeh, 1987b: 1084.

Type locality: Lake Tahoe, California, U.S.A.
Distribution: U.S.A. (California, Colorado, New Mexico, Oregon, Utah).

Dinocheirus vastitatis (J. C. Chamberlin)

Chelanops vastitatis J. C. Chamberlin, 1923c: 381, plate 1 fig. 9, plate 2 figs 18–19, plate 3 fig. 29.
Epaphochernes vastitatis (J. C. Chamberlin): Beier, 1932e: 174.
Dinocheirus vastitatis (J. C. Chamberlin): Roewer, 1937: 302.

Type locality: Monument Point, Tiburon Island, Gulf of Distribution: U.S.A. (Kansas, Missouri).
Distribution: Mexico.

Dinocheirus venustus Hoff & Clawson

Dinocheirus venustus Hoff & Clawson, 1952: 31–35, figs 20–23; Hoff & Bolsterli, 1956: 174–175; Hoff, 1958: 28; Muchmore, 1971a: 90; Zeh, 1987b: 1084.

Type locality: Lawrence, Kansas, U.S.A.
Distribution: U.S.A. (Kansas, Missouri).

Genus **Dinochernes** Beier

Dinochernes Beier, 1933b: 99; Muchmore, 1975e: 275–276.

Type species: *Chelanops vanduzeei* J. C. Chamberlin, 1923c, by original designation.

Dinochernes chalumeaui Heurtault & Rebiere

Dinochernes chalumeaui Heurtault & Rebiere, 1983: 600–603, figs 22–28.

Type locality: Duzer, Guadeloupe.
Distribution: Guadeloupe.

Dinochernes vanduzeei (J. C. Chamberlin)

Chelanops vanduzeei J. C. Chamberlin, 1923c: 378–379, plate 2 figs 22–23; J. C. Chamberlin, 1931a: fig. 69; Beier, 1932g: fig. 264.
Dinocheirus vanduzeei (J. C. Chamberlin): Beier, 1932e: 140; Roewer, 1936: fig. 5.
Dinochernes vanduzeei (J. C. Chamberlin): Beier, 1933b: 99–100, fig. 11; Roewer, 1937: 296; Vachon, 1949: fig. 241; Muchmore, 1975e: 276.

Type locality: Coronados Island, Gulf of California, Mexico.
Distribution: Mexico.

Dinochernes wallacei Muchmore

Dinochernes wallacei Muchmore, 1975e: 277–278, figs 1–4.

Type locality: 5 miles W. of Gainesville on Archer Road, Alachua, County, Florida, U.S.A.
Distribution: U.S.A. (Florida).

Genus **Diplothrixochernes** Beier

Diplothrixochernes Beier, 1962h: 136.

Type species: *Diplothrixochernes patagonicus* Beier, 1962h, by original designation.

Diplothrixochernes patagonicus Beier

Diplothrixochernes patagonicus Beier, 1962h: 136–137, figs 3–4.

Type locality: Lanin Reserve, Neuquen, Argentina.
Distribution: Argentina.

Diplothrixochernes simplex Beier

Diplothrixochernes simplex Beier, 1964j: 498–499, fig. 7.

Type locality: Mt Piltriquitron, El Bolsón, Rio Negro, Argentina.
Distribution: Argentina.

Genus Epactiochernes Muchmore

Epactiochernes Muchmore, 1974c: 397–398.

Type species: *Chelanops tumidus* Banks, 1895a, by original designation.

Epactiochernes insularum Muchmore

Epactiochernes insularum Muchmore, 1974c: 405, figs 13–16.

Type locality: Ahogado Key, Puerto Rico.
Distribution: Cuba, Jamaica, Puerto Rico.

Epactiochernes tristis (Banks)

Chelanops tristis Banks, 1891: 163; Banks, 1895a: 7; Coolidge, 1908: 111; Pratt, 1927: 410.
Chelifer scorpioides tristis (Banks): Ellingsen, 1909b: 216–217.
Pselaphochernes (?) *tristis* (Banks): Beier, 1932e: 134; Roewer, 1937: 297.
Dinocheirus tristis (Banks): Hoff, 1947b: 532–536, figs 31–32; Hoff, 1958: 26.
Epactiochernes tristis (Banks): Muchmore, 1974c: 405, fig. 12; Zeh, 1987b: 1084.

Type locality: Long Island, New York, U.S.A.
Distribution: U.S.A. (Connecticut, Massachusetts, New York).

Epactiochernes tumidus (Banks)

Chelanops tumidus Banks, 1895a: 7; Banks, 1904a: 141; Coolidge, 1908: 111.
Dinocheirus tumidus (Banks): Beier, 1932e: 139; Hoff, 1947b: 529–532, figs 29–30; Hoff, 1958: 26; Weygoldt, 1966e: 462–467, figs 1a–f, 2a–d; Weygoldt, 1969a: 30, 49–51, figs 45a–f, 54a–d, 105.
Epactiochernes tumidus (Banks): Muchmore, 1974c: 398–403, figs 1–11; Muchmore, 1975d: fig. 4; Zeh, 1987b: 1084.

Type locality: Indian River inlet, Florida, U.S.A.
Distribution: U.S.A. (Florida, North Carolina).

Genus Epichernes Muchmore

Epichernes Muchmore, in Muchmore & Hentschel, 1982: 41–42.

Type species: *Epichernes aztecus* Hentschel, in Muchmore & Hentschel, 1982, by original designation.

Epichernes aztecus Hentschel

Epichernes aztecus Hentschel, in Muchmore & Hentschel, 1982: 42–45, figs 1–8.

Type locality: El Ajusco, S. of Mexico City, D.F., Mexico.
Distribution: Mexico.

Genus Eumecochernes Beier

Eumecochernes Beier, 1932e: 185; Beier, 1933a: 545.

Type species: *Chelifer hawaiensis* E. Simon, 1900a, by original designation.

Eumecochernes hawaiensis (E. Simon)

Chelifer hawaiensis E. Simon, 1900a: 518.
Chelifer hawaiiensis (sic) E. Simon: With, 1905: 104−107, plate 6 fig. 3a, plate 7 figs 1a−f.
Eumecochernes hawaiensis (E. Simon): Beier, 1932e: 185, fig. 193; Roewer, 1937: 303, fig. 240e.
Eumecochernes hawaiiensis (sic) (E. Simon): J. C. Chamberlin, 1934b: 10; Tenorio & Muchmore, 1982: 380.

Type locality: Kona, Kauai, Hawaii.
Distribution: Hawaii.

Eumecochernes oceanicus Beier

Chelifer hawaiensis E. Simon (variety): With, 1905: 107−108.
Eumecochernes oceanicus Beier, 1932e: 186; Beier, 1933a: 546; J. C. Chamberlin, 1934b: 10; Roewer, 1937: 303.

Type locality: Kauai Island, Hawaii.
Distribution: Hawaii.

Eumecochernes pacificus (With)

Chelifer pacificus With, 1905: 108−110, plate 7 fig. 2a.
Eumecochernes pacificus (With): Beier, 1932e: 186; J. C. Chamberlin, 1934b: 10; Roewer, 1937: 303.

Type locality: Hawaii.
Distribution: Hawaii.

Genus Gelachernes Beier

Gelachernes Beier, 1940a: 180.

Type species: *Gelachernes salomonis* Beier, 1940a, by original designation.

Gelachernes kolombangarensis Beier

Gelachernes kolombangarensis Beier, 1970g: 321−322, fig. 4.

Type locality: N. of Kuzi, Kolombangara, Solomon Islands.
Distribution: Solomon Islands.

Gelachernes novaguineensis Beier

Gelachernes novaguineensis Beier, 1971b: 369−370; Beier, 1982: 43.

Type locality: Lake Kutubu, Bunu, Southern Highlands District, Papua New Guinea.
Distribution: Papua New Guinea.

Gelachernes perspicillatus Beier

Gelachernes perspicillatus Beier, 1966d: 140−142, fig. 5; Beier, 1970g: 320, fig. 3.

Type locality: Mt Popomanaseu (as Mt Popamanusiu), Guadalcanal, Solomon Islands.
Distribution: Solomon Islands.

Gelachernes salomonis Beier

Gelachernes salomonis Beier, 1940a: 180−181, fig. 9; Beier, 1965g: 772; Beier, 1966d: 140; Beier, 1970g: 321.

Type locality: Boneghi, Guadalcanal (as Guadalcanan), Solomon Islands.
Distribution: Solomon Islands.

Family Chernetidae

Genus **Gigantochernes** Beier

Gigantochernes Beier, 1932e: 186; Beier, 1933a: 546−547; Vitali-di Castri, 1972: 24−25.

Type species: *Chelifer rudis* Balzan, 1887b, by original designation.

Gigantochernes franzi Vitali-di Castri

Gigantochernes franzi Vitali-di Castri, 1972: 33−38, figs 5−6, 8, 13, 16, 18, 20, 22−23, 25−26; Cekalovic, 1984: 27.

Type locality: La Plata, Santiago, Chile.
Distribution: Chile.

Gigantochernes hoffi Vitali-di Castri

Gigantochernes hoffi Vitali-di Castri, 1972: 25−33, figs 1−4, 7, 9−12, 14−15, 17, 19, 21, 24, 27.
Giganthochernes (sic) *hoffi* Vitali-di Castri: Cekalovic, 1984: 27−28.

Type locality: La Plata, Santiago, Chile.
Distribution: Chile.

Gigantochernes rudis (Balzan)

Chelifer rudis Balzan, 1887b: no pagination, figs; Balzan, 1890: 423−424, figs 9, 9a−c; With, 1908a: 253−256, figs 10a−f.
Chelifer (Chelifer) rudis Balzan: Balzan, 1892: 548; Ellingsen, 1910a: 386.
Gigantochernes rudis (Balzan): Beier, 1932e: 186−187, fig. 194−195; Roewer, 1936: fig. 83d; Roewer, 1937: 303; Mello-Leitão, 1939b: 617; Feio, 1945: 7.

Type localities: Resistencia, Argentina; and Rio Apa, Paraguay.
Distribution: Argentina, Brazil (Bahia), Paraguay.

Genus **Gobichernes** Krumpál & Kiefer

Gobichernes Krumpál & Kiefer, 1982: 6.

Type species: *Gobichernes changaiensis* Krumpál & Kiefer, 1982, by original designation.

Gobichernes changaiensis Krumpál & Kiefer

Gobichernes changaiensis Krumpál & Kiefer, 1982: 6−8, plate 3 figs 1−12.

Type locality: 10 km S. of Cagaan Chairchan Somon, Dzavchan Ajmak, Mongolia.
Distribution: Mongolia.

Genus **Gomphochernes** Beier

Gomphochernes Beier, 1932c: 262; Beier, 1932e: 95−96.

Type species: *Chelifer depressimanus* With, 1908a, by original designation.

Gomphochernes communis (Balzan)

Chelifer communis Balzan, 1888a: no pagination, figs; Balzan, 1890: 416−417, figs 4, 4a−b; With, 1908a: 299−303, figs 28a−g.
Chelifer (Lamprochernes) communis Balzan: Balzan, 1892: 548; Ellingsen, 1905c: 10; Ellingsen, 1905e: 324; Ellingsen, 1910a: 366.
Chelifer (Chernes) communis (Balzan): Ellingsen, 1902: 167−168.
Lustrochernes communis (Balzan): Beier, 1932e: 90−91, fig. 110; Beier, 1933b: 91; Roewer, 1937: 290; Mello-Leitão, 1939a: 118; Mello-Leitão, 1939b: 616; Hoff, 1944a: 3; Feio, 1945: 6; Hoff, 1946c: 4; Beier, 1959e: 204; Beier, 1970b: 51; Beier, 1970d: 53; Beier, 1974d: 899; Beier, 1976d: 46; Muchmore, 1977: 77.
Gomphochernes communis (Balzan): Mahnert, 1985d: 78.

Not *Lustrochernes communis* (Balzan): Mahnert, 1979d: 765–767, figs 92–95 (misidentification; see *Lustrochernes intermedius* (Balzan)).

Type localities: Resistencia, Argentina; Asuncion, Paraguay; Rio Apa, Paraguay; and Mato Grosso (as Matto-grosso), Brazil.

Distribution: Argentina, Brazil (Bahia, Distrito Federal, Espirito Santo, Mato Grosso, Rio de Janeiro, Santa Catarina, Sao Paulo), Colombia, Dominican Republic, Ecuador, Mexico, Paraguay, Peru, St Vincent, Trinidad, Uruguay, Venezuela.

Gomphochernes depressimanus (With)

Chelifer depressimanus With, 1908a: 319–320, figs 35a–d.
Gomphochernes depressimanus (With): Beier, 1932c: 262; Beier, 1932e: 96, fig. 117; Roewer, 1937: 291; Mello-Leitão, 1939a: 118; Mello-Leitão, 1939b: 616.

Type locality: Uruguay.
Distribution: Argentina, Paraguay, Uruguay.

Gomphochernes perproximus Beier

Gomphochernes perproximus Beier, 1932c: 262–263; Beier, 1932e: 96–97, fig. 118; Roewer, 1937: 291; Feio, 1945: 6.

Type locality: Paraguay.
Distribution: Argentina, Paraguay.

Genus Goniochernes Beier

Goniochernes Beier, 1932e: 188; Beier, 1933a: 547–548.

Type species: *Chelifer goniothorax* Redikorzev, 1924a, by original designation.

Goniochernes beieri Vachon

Goniochernes beieri Vachon, 1952a: 31–36, figs 19–26.

Type locality: Kéoulenta, Nimba Mountains, Guinea.
Distribution: Guinea.

Goniochernes goniothorax (Redikorzev)

Chelifer (Atemnus) goniothorax Redikorzev, 1924a: 188–189, figs 1–2.
Atemnus goniothorax (Redikorzev): J. C. Chamberlin, 1931a: 60.
Goniochernes goniothorax (Redikorzev): Beier, 1932e: 188–189, fig. 196; Roewer, 1937: 304, fig. 242; Beier, 1954d: 135; Beier, 1964k: 73; Beier, 1967d: 81; Heurtault, 1983: 15–17, figs 36–42.

Type locality: Mabira, Tanzania.
Distribution: Ivory Coast, Mozambique, Tanzania, Uganda, Zaire.

Goniochernes lislei Vachon

Goniochernes lislei Vachon, 1941e: 31.

Type locality: Cameroun.
Distribution: Cameroun.

Goniochernes vachoni Heurtault

Goniochernes vachoni Heurtault, 1970c: 194–199, figs 1–10.

Type locality: Emi Koussi, Koudou, Chad.
Distribution: Chad.

Genus **Haplochernes** Beier

Haplochernes Beier, 1932e: 108; Beier, 1933a: 513; J. C. Chamberlin, 1938a: 274–275; Morikawa, 1960: 143; Beier, 1976f: 213.

Type species: *Chelifer boncicus* Karsch, 1881, by original designation.

Haplochernes aterrimus Beier

Haplochernes aterrimus Beier, 1948c: 558–559, fig. 20; Beier, 1965g: 775; Beier, 1971b: 367.

Type locality: Njau Limon, Irian Jaya, Indonesia.
Distribution: Indonesia (Irian Jaya), Papua New Guinea.

Haplochernes atrimanus (Kästner)

Chelifer atrimanus Kästner, 1927a: 16–20, figs 1–5.
Haplochernes atrimanus (Kästner): Beier, 1932e: 112; J. C. Chamberlin, 1934b: 9; Roewer, 1937: 295.

Type locality: Malololelei, Upolu, Samoa.
Distribution: Samoa.

Haplochernes boncicus (Karsch)

Chelifer boncicus Karsch, 1881: 37.
Chelifer (Trachychernes) boncicus Karsch: Ellingsen, 1907c: 2–3; Ellingsen, 1910a: 373 (in part; see *Haplochernes madagascariensis* Beier).
Haplochernes boncicus (Karsch): Beier, 1932e: 109–110, fig. 126; Roewer, 1937: 295; Morikawa, 1960: 143; Morikawa, 1962: 419; Sato, 1978a: 32–38, figs 2–8; Sato, 1979a: 102–104, figs 12a–h, 13a–g, plates 11, 18; Sato, 1979d: 44; Sato, 1980a: 61, fig. 3; Sato, 1980d: fig. 3, plate 7; Sato, 1982c: 34.
Haplochernes boncicus boncicus (Karsch): Morikawa, 1960: 143.
Not *Chelifer boncicus* Karsch: Ellingsen, 1912b: 82 (misidentification; see *Haplochernes madagascariensis* Beier).

Type locality: Japan.
Distribution: Japan.

Haplochernes boncicus hagai Morikawa

Haplochernes hagai Morikawa, 1953a: 350–351, figs 2c–f.
Haplochernes boncicus hagai Morikawa: Morikawa, 1960: 144, plate 3 fig. 8, plate 8 figs 2, 14, plate 9 fig. 22.

Type locality: Kuhon-butsu, Tokyo, Japan.
Distribution: Japan.

Haplochernes boninensis Beier

Haplochernes boninensis Beier, 1957d: 36–37, figs 21a–c; Morikawa, 1960: 144.

Type locality: Chichi Jima, Bonin Islands, Japan.
Distribution: Japan.

Haplochernes buxtoni (Kästner)

Chelifer buxtoni Kästner, 1927a: 16, 20–24, figs 6–11.
Haplochernes buxtoni (Kästner): Beier, 1932e: 110; J. C. Chamberlin, 1934b: 8; Roewer, 1937: 295.

Type locality: Malololelei, Upolu, Samoa.
Distribution: Samoa.

Haplochernes dahli Beier

Chelifer (Lamprochernes) kraepelini (Tullgren): Ellingsen, 1910a: 367 (misidentification, in part).
Haplochernes dahli Beier, 1932e: 114, fig. 131; Beier, 1933a: 515–516, fig. 2; J. C. Chamberlin, 1934b: 9; Roewer, 1937: 295; Beier, 1965g: 774; Beier, 1966d: 143; Beier, 1971b: 367.

Type locality: Ralum Island, Bismarck Archipelago, Papua New Guinea.
Distribution: Indonesia (Irian Jaya), Papua New Guinea, Solomon Islands.

Haplochernes ellenae J. C. Chamberlin

Haplochernes ellenae J. C. Chamberlin, 1938a: 275–278, figs 4a–e; Roewer, 1940: 346.

Type locality: Viti Levu, Colo-i-Suva, Fiji.
Distribution: Fiji.

Haplochernes funafutensis (With)

Chelifer funafutensis With, 1907: 57–59, figs 9–12.
Haplochernes funafutensis (With): Beier, 1932e: 113; J. C. Chamberlin, 1934b: 9; Roewer, 1937: 295; J. C. Chamberlin, 1939a: 203–205, figs a–j; Beier, 1940a: 179–180, fig. 8.

Type locality: Funafuti, Tuvalu.
Distribution: Fiji, Tahiti, Tuvalu.

Haplochernes hebridicus Beier

Haplochernes hebridicus Beier, 1940a: 178–179, fig. 7; Beier, 1966d: 142–143; Beier, 1970g: 320.

Type locality: Ounua, Malekula, New Hebrides.
Distribution: New Hebrides, Solomon Islands.

Haplochernes insulanus Beier

Haplochernes insulanus Beier, 1957d: 34–36, figs 19a–b, 20a–d.

Type locality: Mt Unibot, Ton, Truk Islands, Caroline Islands.
Distribution: Caroline Islands, Mariana Islands, Marshall Islands.

Haplochernes kraepelini (Tullgren)

Chelifer kraepelini Tullgren, 1905: 40–42, figs 2a–d.
Chelifer (Lamprochernes) kraepelini Tullgren: Ellingsen, 1910a: 367 (in part; see *Haplochernes dahli* Beier).
Haplochernes kraepelini (Tullgren): Beier, 1932e: 113, fig. 130; Roewer, 1937: 295; Beier, 1957d: 33, figs 18a–b; Weidner, 1959: 114; Harvey, 1988b: 344–346, figs 105–112.

Type locality: Buitenzorg (now Bogor), Java, Indonesia.
Distribution: Caroline Islands, Indonesia (Java, Krakatau Islands), Papua New Guinea.

Haplochernes madagascariensis Beier

Chelifer (Trachychernes) boncicus (Karsch): Ellingsen, 1910a: 373 (misidentification, in part); Ellingsen, 1912b: 82 (misidentification).
Haplochernes madagascariensis Beier, 1932e: 110, fig. 127; Beier, 1933a: 515, fig. 1; Roewer, 1937: 295; Legendre, 1972: 447.

Type locality: north-west Madagascar.
Distribution: Madagascar.

Haplochernes nanus Mahnert

Haplochernes nanus Mahnert, 1975b: 545–546, figs 3a–f.

Type locality: Forêt de Mare-Longue, St Philippe, Réunion.
Distribution: Réunion.

Haplochernes norfolkensis Beier

Haplochernes norfolkensis Beier, 1976f: 214, fig. 14; Harvey, 1985b: 134.

Type locality: Steels Point, Norfolk Island.
Distribution: Norfolk Island.

Haplochernes ramosus (L. Koch)

Chelifer ramosus L. Koch, 1885: 47–48, plate 4 figs 5, 5a–c.
Austrochernes ramosus (L. Koch): Beier, 1932e: 171; J.C. Chamberlin, 1934b: 10;
  Roewer, 1937: 301.
Haplochernes ramosus (L. Koch): Beier, 1966a: 296–297, fig. 12; Beier, 1969a: 187;
  Harvey, 1981b: 246; Harvey, 1985b: 134.

Type locality: Rockhampton, Queensland, Australia.
Distribution: Australia (New South Wales, Northern Territory, Queensland, Western
  Australia).

Haplochernes warburgi (Tullgren)

Chelifer warburgi Tullgren, 1905: 42–43, figs 3a–b.
Haplochernes warburgi (Tullgren): Beier, 1932e: 112, fig. 129; Roewer, 1937: 295;
  Weidner, 1959: 114; Beier, 1965g: 774–775; Beier, 1971b: 367; Beier, 1973b: 50;
  Harvey, 1988b: 341–344, figs 96–104.

Type locality: Java, Indonesia.
Distribution: Indonesia (Irian Jaya, Java, Krakatau Islands, Sulawesi), Papua New
  Guinea, Sri Lanka.

Genus Hebridochernes Beier

Hebridochernes Beier, 1940a: 183.

Type species: *Hebridochernes paradoxus* Beier, 1940a, by original designation.

Hebridochernes caledonicus Beier

Hebridochernes caledonicus Beier, 1964e: 408–409, fig. 5; Beier, 1979a: 551; Tenorio
  & Muchmore, 1982: 379.

Type locality: Col d'Amieu, New Caledonia.
Distribution: New Caledonia.

Hebridochernes cornutus Beier

Hebridochernes cornutus Beier, 1965g: 786–787, fig. 25; Beier, 1979a: 551; Tenorio
  & Muchmore, 1982: 380.

Type locality: Madang, Papua New Guinea.
Distribution: Papua New Guinea.

Hebridochernes gressitti Beier

Hebridochernes gressitti Beier, 1964e: 409–411, fig. 6; Beier, 1979a: 551; Tenorio
  & Muchmore, 1982: 380.

Type locality: Mt Mou, New Caledonia.
Distribution: New Caledonia.

Hebridochernes maximus Beier

Hebridochernes maximus Beier, 1979a: 549–551, fig. 1.

Type locality: Mt Canala, New Caledonia.
Distribution: New Caledonia.

Hebridochernes monstruosus Beier

Hebridochernes monstruosus Beier, 1966d: 152–154, fig. 11; Beier, 1970g: 328; Beier, 1979a: 551.

Type locality: Soso, Nggela (as Ngela), Solomon Islands.
Distribution: Solomon Islands.

Hebridochernes papuanus Beier

Hebridochernes papuanus Beier, 1965g: 784–786, fig. 24; Beier, 1979a: 551; Tenorio & Muchmore, 1982: 381.

Type locality: Daradae near Javarere, Musgrove River, Papua New Guinea.
Distribution: Indonesia (Irian Jaya), Papua New Guinea.

Hebridochernes paradoxus Beier

Hebridochernes paradoxus Beier, 1940a: 183–185, figs 10–13; Beier, 1979a: 551.

Type locality: New Hebrides.
Distribution: New Hebrides.

Hebridochernes salomonensis Beier

Hebridochernes salomonensis Beier, 1966d: 150–152, fig. 10; Beier, 1970g: 327; Beier, 1979a: 551.

Type locality: Mt Popomanaseu (as Mt Popamanusiu), Guadalcanal, Solomon Islands.
Distribution: Solomon Islands.

Hebridochernes submonstruosus Beier

Hebridochernes submonstruosus Beier, 1970g: 327–328, fig. 7.

Type locality: Astrolabe Bay, St George Island, Santa Isabel (as Ysabel), Solomon Islands.
Distribution: Solomon Islands.

Genus Hesperochernes J. C. Chamberlin

Hesperochernes J. C. Chamberlin, 1924b: 89–90; Beier, 1932e: 174–175; Beier, 1933a: 537–538; Hoff, 1948: 341; Hoff, 1949b: 476; Hoff & Clawson, 1952: 14; Hoff, 1956c: 31; Muchmore, 1974d: 27–28.

Type species: *Hesperochernes laurae* J. C. Chamberlin, 1924b, by original designation.

Hesperochernes canadensis Hoff

Hesperochernes canadensis Hoff, 1945b: 1–4, figs 1–7; Hoff, 1949b: fig. 41; Hoff, 1958: 23; Hoff, 1961: 446–448; Zeh, 1987b: 1084.

Type locality: Medicine Hat, Alberta, Canada.
Distribution: Canada (Alberta), U.S.A. (Colorado).

Hesperochernes globosus (Ellingsen)

Chelifer (Trachychernes) globosus Ellingsen, 1910a: 374–376.
Hesperochernes globosus (Ellingsen): Beier, 1932e: 175, fig. 183; Roewer, 1937: 302.

Type locality: Mexico.
Distribution: Mexico.

Hesperochernes inusitatus Hoff

Hesperochernes inusitatus Hoff, 1946c: 6–8, figs 6–9; Muchmore, 1977: 77.

Type locality: near Catharinas, Chiapas, Mexico.
Distribution: Mexico.

Hesperochernes laurae J. C. Chamberlin

Hesperochernes laurae J. C. Chamberlin, 1924b: 90–92, figs a–l; J. C. Chamberlin, 1931a: figs 12d, 39a, 51g, 52i; Beier, 1932e: 175; Roewer, 1936: fig. 137; Roewer, 1937: 302; Beier, 1948b: 458; Hoff, 1958: 23; Muchmore, 1971a: 87; Muchmore, 1974d: 28–30, figs 5–10; Muchmore, 1975d: fig. 2; Zeh, 1987b: 1084.

Type locality: Stanford University, California, U.S.A.
Distribution: U.S.A. (California).

Hesperochernes mimulus J. C. Chamberlin

Hesperochernes mimulus J. C. Chamberlin, 1952: 292–299, figs 10a–i, 11a–f, 12a–i; Hoff, 1958: 23; Muchmore, 1971a: 90, 92; Muchmore, 1974d: 30; Zeh, 1987b: 1084.

Type locality: Frances Simes Hastings Natural History Reservation, California, U.S.A.
Distribution: U.S.A. (California).

Hesperochernes mirabilis (Banks)

Chelifer mirabilis Banks, 1895a: 4; Coolidge, 1908: 110.
Chelodamus mirabilis (Banks): R. V. Chamberlin, 1925: 237.
Parachelifer (?) *mirabilis* (Banks): Beier, 1932e: 241; Roewer, 1937: 313.
Pseudozaona mirabilis (Banks): Hoff, 1946d: 201–203, figs 6–8; Hoff, 1949b: figs 38a–c; Hoff, 1958: 24.
Hesperochernes mirabilis (Banks): Muchmore, 1974d: 30; Zeh, 1987b: 1084; Holsinger & Culver, 1988: 42.

Type locality: Indian Cave, Barren County, Kentucky, U.S.A.
Distribution: U.S.A. (Kentucky, Virginia).

Hesperochernes molestus Hoff

Hesperochernes molestus Hoff, 1956c: 33–38, figs 12–15; Hoff, 1958: 24; Hoff, 1959b: 5, etc.; Muchmore, 1971a: 90, 92; Muchmore, 1974d: 30; Zeh, 1987b: 1084.

Type locality: near Santa Fe, Santa Fe County, New Mexico, U.S.A.
Distribution: U.S.A. (New Mexico).

Hesperochernes montanus J. C. Chamberlin

Hesperochernes montanus J. C. Chamberlin, 1935a: 37–39, figs a–g; Roewer, 1937: 302; Beier, 1948b: 462; Hoff, 1958: 23; Muchmore, 1971a: 89.

Type locality: Girds Creek, Ravalli County, Montana, U.S.A.
Distribution: U.S.A. (Montana).

Hesperochernes occidentalis (Hoff & Bolsterli)

Pseudozaona occidentalis Hoff & Bolsterli, 1956: 170–174, figs 1–3; Hoff, 1958: 24.
Hesperochernes occidentalis (Hoff & Bolsterli): Muchmore, 1974d: 30–31; Zeh, 1987b: 1084.

Type locality: Fincher Cave, Washington County, Arkansas, U.S.A.
Distribution: U.S.A. (Arkansas).

Hesperochernes pallipes (Banks)

Chelanops pallipes Banks, 1893: 64; Banks, 1895a: 8; Banks, 1895b: 115; Banks, 1895c:
432; Coolidge, 1908: 111; Banks, 1911: 638, fig. 210d; Moles, 1914c: 193; Moles
& Moore, 1921: 6.
Neochernes (?) *pallipes* (Banks): Beier, 1932e: 167; Roewer, 1937: 299.
Hesperochernes pallipes (Banks): Hoff, 1947b: 506–508, figs 16–17; Hoff, 1958: 22;
Muchmore, 1971a: 81.
Neochernes (?) *pallipses* (sic) (Banks): Beier, 1948b: 445.

Type locality: California, U.S.A.
Distribution: U.S.A. (California, Colorado).

Hesperochernes paludis (Moles)

Chelanops paludis Moles, 1914b: 81–83, fig. 1; Moles, 1914c: 193; Moles & Moore,
1921: 7.
Hesperochernes (?) *paludis* (Moles): Hoff, 1958: 24.

Type locality: Chino Swamp, California, U.S.A.
Distribution: U.S.A. (California).

Hesperochernes riograndensis Hoff & Clawson

Hesperochernes riograndensis Hoff & Clawson, 1952: 19–23, figs 11–12; Hoff, 1956c:
31–32; Hoff, 1958: 23; Hoff, 1959b: 5, etc.; Muchmore, 1971a: 91; Muchmore,
1974d: 30; Zeh, 1987b: 1084.

Type locality: Rio Grande near Lajoya, Socorro County, New Mexico, U.S.A.
Distribution: U.S.A. (New Mexico).

Hesperochernes shinjoensis Sato

Hesperochernes shinjoensis Sato, 1983a: 31–33, figs 1–9.

Type locality: Kanezawa, Shinjo-shi, Yamagata, Japan.
Distribution: Japan.

Hesperochernes tamiae Beier

Hesperochernes tamiae Beier, 1930a: 214–216, figs 10a–b, 11, 12a–d, 13a–b; Beier,
1932a: figs 18, 22; Beier, 1932e: 176, figs 1, 184; Beier, 1932g: fig. 177; Vachon,
1936a: 78; Roewer, 1937: 302; Hoff & Clawson, 1952: 15; Hoff, 1958: 23;
Muchmore, 1971a: 82, 90; Nelson, 1971: 95; Muchmore, 1974d: 30; Nelson, 1975:
290, figs 68–69; Zeh, 1987b: 1084.
Hesperochernes taminae (sic) Beier: Roewer, 1936: fig. 47; Roewer, 1937: fig. 240a.

Type locality: Ithaca, New York, U.S.A.
Distribution: U.S.A. (Maine, Michigan, New York, Vermont).

Hesperochernes thomomysi Hoff

Hesperochernes thomomysi Hoff, 1948: 341–345, figs 1–5; Hoff, 1958: 23;
Muchmore, 1971a: 91; Zeh, 1987b: 1084.

Type locality: Huntington Lake, Fresno County, California, U.S.A.
Distribution: U.S.A. (California).

Hesperochernes tumidus Beier

Hesperochernes tumidus Beier, 1933b: 100–101, fig. 12; Roewer, 1937: 302.

Type locality: Coatepec, Veracruz, Mexico.
Distribution: Mexico.

Hesperochernes unicolor (Banks)

Chelanops unicolor Banks, 1908: 39–40.
Neochernes unicolor (Banks): Beier, 1932e: 166–167; Roewer, 1937: 299.
Hesperochernes unicolor (Banks): Hoff, 1947b: 511–513, fig. 19; Beier, 1948b: 458;
 Hoff & Clawson, 1952: 23–24; Hoff, 1958: 22; Muchmore, 1971a: 87, 91; Rowland
 & Reddell, 1976: 16; Zeh, 1987b: 1084.

Type locality: Austin, Texas, U.S.A.
Distribution: U.S.A. (Texas).

Hesperochernes utahensis Hoff & Clawson

Hesperochernes utahensis Hoff & Clawson, 1952: 15–19, figs 9–10; Hoff, 1956c:
 32–33; Hoff, 1958: 23; Hoff, 1959b: 5, etc.; Hoff, 1961: 448–449; Muchmore,
 1971a: 91; Knowlton, 1972: 2; Knowlton, 1974: 2–3; Muchmore, 1974d: 30;
 Benedict & Malcolm, 1982: 101; Zeh, 1987b: 1084.

Type locality: Kanab Sand Dunes, Kanab, Kane County, Utah, U.S.A.
Distribution: U.S.A. (Colorado, Idaho, New Mexico, Oregon, Utah).

Hesperochernes vespertilionis Beier

Hesperochernes vespertilionis Beier, 1976d: 53–55, fig. 6.

Type locality: Sabaneta, Dominican Republic.
Distribution: Dominican Republic.

Genus Heterochernes Beier

Heterochernes Beier, 1966c: 376; Beier, 1976f: 216.

Type species: *Austrochernes novaezealandiae* Beier, 1932e, by original designation.

Heterochernes novaezealandiae (Beier)

Chelifer (Trachychernes) australiensis With: Ellingsen, 1910a: 373 (misidentification).
Austrochernes novae-zealandiae Beier, 1932e: 170–171, fig. 178; Beier, 1933a:
 536–537, fig. 9; J. C. Chamberlin, 1934b: 10; Roewer, 1937: 301.
Heterochernes novaezealandiae (Beier): Beier, 1966c: 376; Beier, 1976f: 216.

Type locality: Stephens Island, South Island, New Zealand.
Distribution: New Zealand.

Genus Hexachernes Beier

Hexachernes Beier, 1953b: 21.

Type species: *Hexachernes pennatus* Beier, 1953b, by original designation.

Hexachernes pennatus Beier

Hexachernes pennatus Beier, 1953b: 21–22, fig. 5.

Type locality: Sahacóc, Alta Vera Paz, Guatemala.
Distribution: Guatemala.

Genus Illinichernes Hoff

Illinichernes Hoff, 1949b: 481; Benedict & Malcolm, 1982: 101–102.

Type species: *Illinichernes distinctus* Hoff, 1949b, by original designation.

Illinichernes distinctus Hoff

Illinichernes distinctus Hoff, 1949b: 481–484, figs 46a–e; Hoff & Bolsterli, 1956: 169;
 Hoff, 1958: 25; Nelson, 1975: 290, figs 64–67; Zeh, 1987b: 1085.

Type locality: Magnolia, Illinois, U.S.A.
Distribution: U.S.A. (Indiana, Illinois, Maryland, Michigan).

Illinichernes stephensi Benedict & Malcolm

Illinichernes stephensi Benedict & Malcolm, 1982: 102–107, figs 1–8; Zeh, 1987b: 1085.

Type locality: 2 miles E. of Canyonville, Douglas County, Oregon, U.S.A.
Distribution: U.S.A. (Oregon).

Genus Incachernes Beier

Incachernes Beier, 1933b: 94–95.

Type species: *Incachernes mexicanus* Beier, 1933b, by original designation.

Incachernes brevipilosus (Ellingsen)

Chelifer (Lamprochernes) brevipilosus Ellingsen, 1910a: 364–365.
Neochernes brevipilosus (Ellingsen): Beier, 1932e: 166, fig. 177; Roewer, 1937: 299.
Incachernes brevipilosus (Ellingsen): Beier, 1933b: 95.

Type locality: central Colombia.
Distribution: Colombia.

Incachernes mexicanus Beier

Incachernes mexicanus Beier, 1933b: 95–96, fig. 5; Roewer, 1937: 289; Beier, 1948b: 445, 458; Muchmore, 1971a: 81, 87.

Type locality: Cuernavaca, Morellos, Mexico.
Distribution: Mexico.

Incachernes salvadoricus Beier

Incachernes salvadoricus Beier, 1955i: 369–370, fig. 1.

Type locality: Finca El Marne, 6 km SW. of Santa Ana, El Salvador.
Distribution: El Salvador.

Genus Indochernes Murthy & Anathakrishnan

Indochernes Murthy & Ananthakrishnan, 1977: 136–137.

Type species: *Indochernes beieri* Murthy & Ananthakrishnan, 1977, by original designation.

Indochernes beieri Murthy & Ananthakrishnan

Indochernes beieri Murthy & Ananthakrishnan, 1977: 137–139, figs 43a–c.

Type locality: Cochin, Kerala, India.
Distribution: India.

Genus Interchernes Muchmore

Interchernes Muchmore, 1980c: 89–90.

Type species: *Interchernes clarkorum* Muchmore, 1980c, by original designation.

Interchernes clarkorum Muchmore

Interchernes clarkorum Muchmore, 1980c: 90–92, figs 1–5.

Type locality: 7 miles S. of Santo Tomas, Baja California, Mexico.
Distribution: Mexico.

Genus Lamprochernes Tömösváry

Chernes (Lamprochernes) Tömösváry, 1882b: 185; Nonidez, 1917: 13.
Lamprochernes Tömösváry: Beier, 1932e: 82; J. C. Chamberlin, 1935b: 481; Hoff, 1949b: 450; G. O. Evans & Browning, 1954: 18; Hoff, 1956c: 5; Beier, 1963b:

Family Chernetidae

249–250; Beier, 1976f: 213; Murthy & Ananthakrishnan, 1977: 135; Harvey, 1987c: 111; Legg, 1987: 181.
Pycnochernes Beier, 1932e: 136; Beier, 1933a: 522; J. C. Chamberlin, 1952: 274–275 (synonymized by Muchmore, 1975d: 19).
Muscichernes Morikawa, 1960: 140 (synonymized by Harvey, 1987c: 111).

Type species: of *Chernes (Lamprochernes)*: *Chelifer nodosus* Schrank, 1803, by subsequent designation of Beier, 1932e: 82.
of *Pycnochernes*: *Chelifer celerrimus* With, 1908a, by original designation.
of *Muscichernes*: *Muscichernes katoi* Morikawa, 1960, by original designation.

Lamprochernes chyzeri (Tömösváry)

Chernes (Lamprochernes) chyzeri Tömösváry, 1882b: 186–187, plate 1 figs 3–5; Ellingsen, 1897: 4–5.
Chelifer nodosus (Schrank): H. J. Hansen, 1884: 548–549 (misidentification; in part).
Chernes (Chernes) chyzeri Tömösváry: Daday, 1888: 115–116, 170, plate 4 fig. 7.
Chernes chyzeri Tömösváry: Daday, 1889c: 25; Daday, 1918: 1.
Chernes insuetus O.P.-Cambridge, 1892: 225–226, plate C fig. 17 (synonymized by Legg & Jones, 1988: 102).
Chelifer chyzeri (Tömösváry): Tullgren, 1899a: 171; Ellingsen, 1907a: 155; Ellingsen, 1909a: 208, 215.
Chelifer mjöbergi Tullgren, 1909a: 92–94, figs 1–3 (synonymized by Beier, 1963b: 251).
Chelifer (Lamprochernes) chyzeri (Tömösváry): Ellingsen, 1910a: 366; Lessert, 1911: 13, fig. 11.
Chelifer (Chernes) chyzeri (Tömösváry): Kew, 1911a: 43, fig. 3; Standen, 1913: 12; Kästner, 1928: 5, figs 14a–b; Schenkel, 1928: 59, fig. 11.
Chelifer (Chelanops) chyzeri (Tömösváry): Redikorzev, 1924b: 22, fig. 6.
Lamprochernes chyzeri (Tömösváry): Beier, 1932e: 84, fig. 102; Tumšs, 1934: 14–15; Roewer, 1937: 289; Lohmander, 1939b: 294–297, fig. 7; Beier, 1948b: 457; Kaisila, 1949a: 75; Kaisila, 1949b: 81–82, map 6; Redikorzev, 1949: 651; G. O. Evans & Browning, 1954: 20, fig. 22; Vachon, 1954d: 591; Weidner, 1954b: 109; Beier, 1958b: 27; Meinertz, 1962a: 393–394, map 4; Beier, 1963b: 251, fig. 252; Beier, 1963i: 285; Kobakhidze, 1965b: 541; Helversen, 1966a: 139; Kobakhidze, 1966: 704; Rafalski, 1967: 15; Lazzeroni, 1969a: 335–336; Lazzeroni, 1970a: 208; Beier, 1973c: 226; Palmgren, 1973: 9; Ćurčić, 1974a: 26; Crocker, 1978: 9; Jones, 1978: 92; Jones, 1980c: map 15; Callaini, 1986c: 398, fig. 4l; Callaini, 1986e: 254; Schawaller, 1986: 2–3; Jędryczkowski, 1987a: 345; Krumpál & Cyprich, 1988: 42; Legg & Jones, 1988: 102–104, figs 21b, 24a, 24b(a–f); Schawaller, 1989: 17.
Lamprochernes mjöbergi (Tullgren): Beier, 1932e: 86; Roewer, 1937: 289.
Allochernes (?) *insuetus* O.P.-Cambridge: Beier, 1932e: 154; Roewer, 1937: 298.
Lamprochernes chryzeri (sic) (Tömösváry): Legg, 1972: 578, fig. 2(2a).
Lamprochernes cfr. *chyzeri* (Tömösváry): Callaini, 1983a: 153–155, fig. 1a–b.

Type localities: of *Chernes (Lamprochernes) chyzeri*: several localities in Hungary.
of *Chelifer mjöbergi*: Gotska Sandön, Gotland, Sweden.
of *Chernes insuetus*: Dover, England, Great Britain.
Distribution: Austria, Czechoslovakia, Denmark, Finland, Great Britain, Hungary, Italy, Norway, Poland, Romania, Sweden, Switzerland, Turkey, U.S.S.R. (Georgia, Kazakhstan, Latvia), West Germany, Yugoslavia.

Lamprochernes foxi (J. C. Chamberlin)

Pycnochernes foxi J. C. Chamberlin, 1952: 278–279, figs 5a–g; Hoff, 1958: 22.
Lamprochernes foxi (J. C. Chamberlin): Muchmore, 1975d: 19.

Type locality: Twin Falls, Idaho, U.S.A.
Distribution: U.S.A. (Idaho).

588

Lamprochernes indicus Sivaraman

Lamprochernes indicus Sivaraman, 1980a: 107–109, figs 1a–b.

Type locality: Bangalore, Karnataka, India.
Distribution: India.

Lamprochernes leptaleus (Navás)

Chelifer leptaleus Navás, 1918: 106–108, figs 2a–d; Navás, 1925: 108, figs 7a–d.
Lamprochernes (?) *leptaleus* (Navás): Beier, 1932e: 87; Roewer, 1937: 289.

Type locality: Valmadrid, Spain.
Distribution: Spain.

Lamprochernes minor Hoff

Lamprochernes minor Hoff, 1949b: 453–454, figs 29a–b; Levi, 1953: 59–61, figs 1–4,
7–9; Hoff & Bolsterli, 1956: 166; Hoff, 1958: 21; Hoff, 1961: 445; Levi & Levi,
1968: 121, fig.; Manley, 1969: 7–8, fig. 8; Nelson, 1975: 284–285, figs 44–45;
Zeh, 1987b: 1085.

Type locality: Urbana, Illinois, U.S.A.
Distribution: U.S.A. (Colorado, Illinois, Michigan, Minnesota, North Dakota,
Wisconsin).

Lamprochernes moreoticus (Beier)

Chelifer (Lamprochernes) nodosus moreoticus Beier, 1929e: 445.
Lamprochernes moreoticus (Beier): Beier, 1932e: 85–86, fig. 104; Roewer, 1937: 289;
Beier, 1963b: 252, fig. 254.

Type locality: Pelopónnisos (as Peloppones), Greece.
Distribution: Greece.

Lamprochernes muscivorus Redikorzev

Lamprochernes muscivorus Redikorzev, 1949: 650–651, figs 11–12.

Type localities: near Shakhrisyabz, Uzbekistan, U.S.S.R.; Kamashi, Uzbekistan,
U.S.S.R.; and Ashkhabad, Turkmenistan, U.S.S.R.
Distribution: U.S.S.R. (Turkmenia, Uzbekistan).

Lamprochernes nodosus (Schrank)

Acarus cancroides Linnaeus: Poda, 1761: 122 (not seen) (misidentification).
Chelifer nodosus Schrank, 1803: 246; Leydig, 1867: 16 (not seen); Macrae, 1869: 283
(not seen); E. Simon, 1879a: 33–34, 312, plate 18 fig. 14; Leydig, 1881a: 180 (not
seen); Canestrini, 1883: no. 8, figs 1–5; Becker, 1884: cclxiv; H. J. Hansen, 1884:
548–549 (in part; see *Lamprochernes chyzeri* (Tömösváry) and *Lamprochernes
savignyi* (E. Simon)); E. Simon, 1885c: 49; Lameere, 1895: 475; Becker, 1896: 329
(not seen); E. Simon, 1898d: 20; Tullgren, 1899a: 171; E. Simon, 1900b: 593;
Pocock, 1905: 604; With, 1906: 29–31, 170, plate 1 fig 8a; Tullgren, 1907e: 246;
Tullgren, 1911: 125; Ellingsen, 1913a: 453; Ellingsen, 1908d: 70; Ellingsen, 1914:
4–5; Navás, 1918: 88; Navás, 1919: 211; Fage, 1921: 102; Godfrey, 1921: 119;
Grimpe, 1921: 628 (not seen); Caporiacco, 1923: 131; Navás, 1923: 31; Navás, 1925:
106, fig. 1; Bacelar, 1928: 190; Conci, 1951: 28, 39.
Chelifer parasita Hermann, 1804: 117, plate 7 fig. 6 (synonymized by E. Simon,
1879a: 33).
Chelifer reussii C. L. Koch, 1843: 48–50, fig. 785 (synonymized by E. Simon, 1879a: 33).
Chelifer corallinus Loew, 1845: 29 (synonymized by Tömösváry, 1882b: 185).
Chelifer inaequalis Curtis, 1849: 112, figs 62, 62c; Hagen, 1870: 266 (synonymized
by Kew, 1916a: 125).

Chelifer cancroides (Linnaeus): Anonymous, 1834: 162; Lukis, 1834: 162–163; Moore, 1835: 321–322; Stevens, 1866: xxvii (not seen); Spicer, 1867: 244; Anonymous, 1875: 185; Newman, 1875: 186; Mégnin, 1886: 241 (not seen); Leydig, 1881a: 180 (not seen); Hess, 1894: 120; Fenizia, 1902: 55 (not seen) (misidentifications).

Chelifer geoffroyi Leach: Jenyns, 1846: 295 (misidentification).

Chelifer wideri C. L. Koch: Löw, 1866: 944 (misidentification).

Chelifer corallifer (sic) Loew: Hagen, 1870: 265.

Chelifer hahni C. L. Koch: Schiner, 1872: 75 (not seen) (misidentification).

Chernes reussii (C. L. Koch): L. Koch, 1873: 5–6; Ritsema, 1874: xxxiii; L. Koch, 1877: 180; Daday, 1880a: 192.

Chernes bohemicus Stecker, 1874b: 232; Stecker, 1875d: 88 (synonymized by Beier, 1932e: 83).

Obisium muscorum Leach: Dale, 1878: 325 (not seen) (misidentification).

Chernes (Lamprochernes) nodosus (Schrank): Tömösváry, 1882b: 185–186, plate 1 figs 1–2.

Chernes nodosus Schrank: O.P.-Cambridge, 1892: 225, plate C fig. 16; Evans, 1901a: 53; Kew, 1901b: 194; Kew, 1903: 298; Kew, 1904: 292; Butterfield, 1908: 112; Godfrey, 1908: 157; Godfrey, 1909: 22–23; Falconer, 1916: 192; Graham-Smith, 1916: fig. 10, plate 30 figs 1–7; Daday, 1918: 1; George, 1955: 123.

Chernes hahnii (C. L. Koch): Wagner, 1892: 434–436 (misidentification).

Chelifer (Lamprochernes) nodosus Schrank: Ellingsen, 1910a: 368; Lessert, 1911: 11–13, fig. 10; Nonidez, 1917: 13; Beier, 1929a: 342; Beier, 1929d: 155.

Chelifer (Chernes) nodosus Schrank: Kew, 1911a: 41–42, fig. 1; Standen, 1912: 11; Kew, 1916b: 78; Standen, 1917: 27; Kästner, 1928: 4, figs 6, 15; Schenkel, 1928: 58, figs 9–10; Berland, 1932: 52.

Chelifer (Chelanops) nodosus (Schrank): Redikorzev, 1924b: 22, fig. 5.

Lamprochernes nodosus (Schrank): Beier, 1932e: 83–84, fig. 101; Vachon, 1934d: 133; Vachon, 1934f: 158; Vachon, 1935j: 187; Roewer, 1937: 289; Vachon, 1938a: fig. 56h; Beier, 1939e: 312; Beier, 1939f: 197; Lohmander, 1939b: 292–294, fig. 5; Vachon, 1940f: 2; Vachon, 1940g: 145; Cooreman, 1946b: 3; Vachon, 1947a: 85; Beier, 1948b: 442–444; Caporiacco, 1948c: 242; Beier, 1949: 10; Kaisila, 1949a: 74–75; Kaisila, 1949b: 81, map 6; Redikorzev, 1949: 649; Sankey, 1949: 247; Beier, 1952e: 4; Vachon, 1953: 572; Beier, 1954d: 135; Beier & Franz, 1954: 457; G. O. Evans & Browning, 1954: 18–20, figs 20–21; Vachon, 1954d: 590; Weidner, 1954b: 109; George, 1957: 80; Vachon, 1957: figs 3, 5; Ressl & Beier, 1958: 2; Beier, 1963b: 251–252, fig. 253; Beier, 1963f: 195; Beier, 1963h: 156; Beier, 1965f: 94; Kobakhidze, 1965b: 541; Pedder, 1965: 108–110, fig. 10; Ressl, 1965: 289; Helversen, 1966a: 139–140; Kobakhidze, 1966: 704; Zangheri, 1966: 531; Rafalski, 1967: 15; Hammen, 1969: 20; Lazzeroni, 1969b: 408; M. Smith, 1969: 297, figs e–h; Jones, 1970b: 118–119; Ressl, 1970: 251; Gabbutt, 1972a: 37–40, figs 1a, 2a; Gabbutt, 1972b: 2–13; Gabbutt, 1972c: 83–86, figs 1a–b; Howes, 1972a: 109; Kofler, 1972: 287; Legg, 1972: 579, figs 1(6b), 2(2b), 4(3); Legg, 1973b: 430, fig. 2g; Mahnert, 1974a: 90; Legg, 1975a: 66; Muchmore, 1975d: fig. 5; Crocker, 1976: 9–10; Klausen, 1977: 83–84; Klausen & Totland, 1977: 101–108, plate 1, fig. 1; Crocker, 1978: 9; Jones, 1978: 90, 92; Jones, 1979a: 201; Judson, 1979a: 62; Leclerc & Heurtault, 1979: 246; Rundle, 1979: 47–48; Gardini, 1980c: 117, 124; Jones, 1980c: map 16; Mahnert, 1982f: 297; Mahnert, 1983d: 362; Lippold, 1985: 40; Andersen, 1987: 23 (not seen); Clements, 1987: 222 (not seen); Judson, 1987: 369; Krumpál & Cyprich, 1988: 42; Legg & Jones, 1988: 99–100, figs 21c, 23a, 23b(a–i); Schawaller & Dashdamirov, 1988: 36, figs 60, 67; Schawaller, 1989: 17.

Genus ? *inaequalis* Curtis: Beier, 1932e: 277; Roewer, 1937: 317.

Chelifer corralifer (sic) Loew: Vachon, 1940f: 2.

Lamprochernes cf. *nodosus* (Schrank): Mahnert, 1978f: 128.

Not *Chelifer parasita* Hermann: Templeton, 1836: 14 (misidentification; see *Pselaphochernes dubius* (O.P.-Cambridge)).
Not *Chelifer nodosus* Schrank: Ellingsen, 1907a: 156 (misidentification; see *Lamprochernes savignyi* (E. Simon)).
Not *Chernes nodosus* (Schrank): Kew, 1909b: 259 (misidentification; see *Lamprochernes savignyi* (E. Simon)).

Type localities: of *Chelifer nodosus*: not stated.
of *Chelifer parasita*: not stated, presumably near Strasbourg, France.
of *Chelifer reussii*: Erlangen, West Germany.
of *Chelifer corallinus*: Ofen, Poland.
of *Chelifer inaequalis*: not stated, presumably England, Great Britain.
of *Chernes bohemicus*: Roztok, near Prague, Czechoslovakia.
Distribution: Austria, Belgium, Congo, Corsica, Czechoslovakia, Denmark, East Germany, Finland, France, Ghana, Great Britain, Greece, Hungary, India, Iran, Ireland, Israel, Italy, Malta, Netherlands, Norway, Poland, Portugal, Spain, Sri Lanka, Sweden, Switzerland, Tunisia, Turkey, U.S.S.R. (Armenia, Azerbaijan, Georgia, Kirghizia, R.S.F.S.R.), West Germany, Zaire.

Lamprochernes nodosus afrikanus (E. Simon)

Chelifer nodosus afrikanus E. Simon, 1885c: 49.
Lamprochernes (?) *afrikanus* (E. Simon): Beier, 1932e: 87; Roewer, 1937: 289.
Lamprochernes nodosus afrikanus (E. Simon): Mahnert, 1982e: 693.

Type locality: Aïn Draham (as Aïn-Draham), Tunisia.   Distribution: Tunisia.

Lamprochernes procer (E. Simon)

Chelifer procer E. Simon, 1878a: 150–151.
Lamprochernes (?) *procer* (E. Simon): Beier, 1932e: 87; Roewer, 1937: 289.

Type locality: Algeria.   Distribution: Algeria.

Lamprochernes savignyi (E. Simon)

Chelifer savignyi E. Simon, 1881b: 12–13.
Chelifer nodosus (Schrank): H.J. Hansen, 1884: 548–549 (misidentification, in part); Ellingsen, 1907a: 156 (misidentification).
Chernes nodosus (Schrank): Kew, 1909b: 259 (misidentification).
Chelifer pygmaeus L. Koch, 1885: 49–50, plate 4 figs 8, 8a–b; With, 1905: 110 (synonymized by Harvey, 1987c: 111).
Chelifer brevifemoratus Balzan, 1887b: no pagination, figs; Balzan, 1890: 420–421, figs 7, 7a–c; With, 1908a: 284–287 (synonymized by Harvey, 1987c: 111).
Chelifer (Trachychernes) brevifemoratus Balzan: Balzan, 1892: 548; Ellingsen, 1905e: 324.
Chelifer (Chernes) brevifemoratus Balzan: Ellingsen, 1902: 156–158.
Chelifer celerrimus With, 1908a: 285–286, figs 23a–e (synonymized by Harvey, 1987c: 111).
Chelifer (Chernes) godfreyi Kew, 1911a: 42, fig. 2; Standen, 1912: 11; Kew, 1916b: 78; Standen, 1917: 27; Kästner, 1928: 5 (synonymized with *Chelifer celerrimus* With by Muchmore, 1975d: 19).
Chernes godfreyi (Kew): Godfrey, 1927: 17.
Lamprochernes savignyi (E. Simon): Beier, 1932e: 86, fig. 105; Roewer, 1937: 289; Beier, 1946: 569; Beier, 1948b: 444; Beier, 1953c: 73; Vachon, 1954d: 590; Beier, 1958a: 175; Beier, 1963f: 195; Beier, 1964k: 75; Beier, 1965b: 374; Beier, 1969e: 413; Beier, 1974e: 1010; Mahnert, 1975b: 545; Beier, 1976f: 213; Murthy & Ananthakrishnan, 1977: 135; Mahnert, 1978d: 884; Harvey, 1987c: 111–115, figs 1–17; Legg & Jones, 1988: 96–98, figs 21a, 22a, 22b(a–f).

Lamprochernes godfreyi (Kew): Beier, 1932e: 86–87; Roewer, 1937: 289; Lohmander, 1939b: fig. 6; Vachon, 1940f: 2; Beier, 1948b: 444; G. O. Evans & Browning, 1954: 18, fig. 19; Hoff & Bolsterli, 1956: 166–167; Hoff, 1958: 20; Meinertz, 1962a: 394, map 4; Beier, 1963b: 250; Muchmore, 1971a: 93; Howes, 1972a: 109; Howes, 1972b: 122; Legg, 1972: 578, fig. 2(6c); Legg, 1975a: 66; Jones, 1978: 90; Crocker, 1978: 9; Judson, 1979a: 62; Jones, 1980c: map 14; Zeh, 1987b: 1085.

Haplochernes pygmaeus (L. Koch): Beier, 1932e: 110–111; J. C. Chamberlin, 1934b: 9; Roewer, 1937: 295; Harvey, 1981b: 246; Harvey, 1985b: 134.

Pycnochernes celerrimus (With): Beier, 1932e: 137; Roewer, 1937: 297.

Pycnochernes brevifemoratus (Balzan): Beier, 1932e: 137; Roewer, 1937: 297; Feio, 1945: 7; Feio, 1946: fig. 1.

Pycnochernes linsdalei J. C. Chamberlin, 1952: 275–277, figs 4a–i; Hoff, 1958: 21; Muchmore, 1975d: fig. 7 (synonymized by Harvey, 1987c: 112).

Muscichernes katoi Morikawa, 1960: 140–141, plate 3 fig. 7, plate 7 fig. 16, plate 9 fig. 19 (synonymized by Harvey, 1987c: 112).

Lamprochernes cf. *savignyi* (E. Simon): Mahnert, 1982e: 692–693, fig. 1.

Not *Chelifer (Trachychernes) brevifemoratus* Balzan: Ellingsen, 1910a: 373 (misidentification; see *Neowithius dubius* Beier).

Type localities: of *Chelifer savignyi*: Ramleh (as Ramlé), Egypt.

of *Chelifer pygmaeus*: Gayndah, Queensland, Australia.

of *Chelifer brevifemoratus*: Resistencia, Argentina; and Asuncion, Paraguay.

of *Chelifer celerrimus*: Balthazar, Grenada.

of *Chelifer godfreyi*: Petersham, England, Great Britain; Newport, Isle of Wight, England, Great Britain; Hatfield, England, Great Britain; South Norwood, England, Great Britain; Oban, Scotland, Great Britain; and Rathmines, County Dublin, Ireland.

of *Pycnochernes linsdalei*: Frances Simes Hastings Natural History Reservation, Monterey, California, U.S.A.

of *Muscichernes katoi*: Shakujii, Tokyo, Japan.

Distribution: Australia (Australian Capital Territory, New South Wales, Queensland, Victoria), Brazil (Mato Grosso), Chad, Denmark, Ecuador, Egypt, Great Britain, Grenada, India, Ireland, Israel, Japan, Kenya, Mauritius, New Zealand, Paraguay, Réunion, Seychelles, South Africa, Sudan, Uruguay, U.S.A. (California, Indiana, Kansas).

Genus **Lasiochernes** Beier

Lasiochernes Beier, 1932e: 134; Beier, 1933a: 521–522; Beier, 1959d: 45; Beier, 1963b: 258.

Type species: *Chelifer (Trachychernes) pilosus* Ellingsen, 1910a, by original designation.

Lasiochernes anatolicus Beier

Lasiochernes anatolicus Beier, 1963g: 272–274, fig. 4; Beier, 1965f: 94.

Type locality: Inhisar near Sögut, Turkey.   Distribution: Turkey.

Lasiochernes congicus Beier

Lasiochernes congicus Beier, 1959d: 45–46, fig. 22.

Type locality: Itombwe, Uvira, Kivu, Zaire.   Distribution: Zaire.

Lasiochernes graecus Beier

Lasiochernes graecus Beier, 1963b: 259–260, fig. 260b; Beier, 1965a: 89–90, fig. 3; Mahnert, 1975a: 183; Mahnert, 1978e: 298.

Type locality: Petalás (as Petalas), Akarnanìa, Greece.
Distribution: Albania, Greece.

Lasiochernes jonicus (Beier)

Chelifer (Trachychernes) jonicus Beier, 1929e: 446–448, figs 1a–b, 2.
Chernes jonicus (Beier): Beier, 1932a: fig. 3.
Lasiochernes jonicus (Beier): Beier, 1932e: 136, fig. 152; Beier, 1932g: fig. 166; Roewer, 1936: fig. 27a; Roewer, 1937: 297; Beier, 1963b: 261, fig. 261; Mahnert, 1978e: 296–298.

Type locality: Hagjos Mathias, Kérkira (as Korfu), Greece.
Distribution: Greece.

Lasiochernes pilosus (Ellingsen)

Chelifer (Trachychernes) pilosus Ellingsen, 1910a: 378–379; Beier, 1929a: 347.
Chelifer falcomontanus Heselhaus, 1914: 77 (synonymized by Beier, 1929a: 347).
Chelifer (Chernes) falcomontanus Heselhaus: Berland, 1925: 213–216, figs 1–4; Berland, 1932: figs 49, 61, 65, 68.
Chelifer (Trachychernes) falcomontanus Heselhaus: Kästner, 1928: 13.
Chernes falcomontanus (Heselhaus): J. C. Chamberlin, 1931a: 60, fig. 10g.
Lasiochernes pilosus (Ellingsen): Beier, 1932e: 134–135, figs 150–151; Vachon, 1936a: 78; Roewer, 1937: 297, fig. 237; Cooreman, 1946b: 4–7, fig. 1; Leleup, 1948: 8–9; Caporiacco, 1949d: 141; Vachon, 1949: fig. 196; Beier, 1952e: 5; Weidner, 1954b: 110; Beier, 1956h: 24; Ressl & Beier, 1958: 2; Beier, 1963b: 258–259, fig. 259; Weygoldt, 1964a: 355–362, figs 1–5; Ressl, 1965: 289; Weygoldt, 1965b: 218, figs 1–3; Weygoldt, 1965c: 436; Hammen, 1969: 20–21; Weygoldt, 1969a: 25–27, 30, 45–47, 110, 117, figs 42a–e, 67, 88, 101; Ressl, 1970: 251; Ćurčić, 1974a: 26; Inzaghi, 1981: 72.

Type localities: of *Chelifer (Trachychernes) pilosus*: Görz, Austria.
of *Chelifer falcomontanus*: near Valkenburg, Netherlands.
Distribution: Austria, Belgium, France, Italy, Luxembourg, Netherlands, West Germany, Yugoslavia.

Lasiochernes punctiger Beier

Lasiochernes punctiger Beier, 1959d: 47–48, fig. 23.
Type locality: SE. of Kahuzi, Katondi, Kivu, Zaire.
Distribution: Zaire.

Lasiochernes siculus Beier

Lasiochernes siculus Beier, 1961d: 92–95, fig. 2; Beier, 1963a: 260; Beier, 1963b: 259, fig. 260a; Beier, 1963h: 156; Beier, 1963i: 286; Lazzeroni, 1969a: 337; Lazzeroni, 1970a: 208; Gardini, 1980c: 117–118, 130, 131.

Type locality: Grotta Chiusazza, Floridia, Siracusa, Sicily.
Distribution: Italy, Sicily.

Lasiochernes turcicus Beier

Lasiochernes turcicus Beier, 1949: 11–13, figs 9–10; Beier, 1957b: 149–150, fig. 3; Beier, 1963f: 195; Beier, 1965f: 94; Rack, 1971: 113; Beier, 1973c: 226; Mahnert, 1979b: 266.

Type locality: Dodurga near Acipayam, Anatolia, Turkey.
Distribution: Israel, Turkey.

Lasiochernes villosus Beier

Lasiochernes villosus Beier, 1957b: 147–149, fig. 2; Beier, 1965f: 94; Mahnert, 1979b: 266.

Type locality: Korha, Turkey.
Distribution: Turkey.

Genus **Lustrochernes** Beier

Pelorus C. L. Koch, 1843: 59 (junior homonym of *Pelorus* Montfort, 1808, and *Pelorus* Fischer-Waldheim, 1821).
Lustrochernes Beier, 1932c: 259–260; Beier, 1932e: 87–88; Hoff, 1956c: 9–10 (synonymized by Beier, 1932c: 267).

Type species: of *Pelorus*: *Pelorus rufimanus* C. L. Koch, 1843, by monotypy.
of *Lustrochernes*: *Chelifer argentinus* Thorell, 1877a, by original designation.

Lustrochernes acuminatus (E. Simon)

Chelifer (Chelanops) acuminatus E. Simon, 1878b: 156.
Chelanops acuminatus (E. Simon): Banks, 1895a: 5; Banks, 1904b: 364; Coolidge, 1908: 111; Banks, 1911: 638, fig. 210e; Moles, 1914c: 193; Moles & Moore, 1921: 6.
Lustrochernes (?) *acuminatus* (E. Simon): Beier, 1932e: 95; Roewer, 1937: 290; Hoff, 1958: 21.

Type locality: Mariposa, California, U.S.A.
Distribution: U.S.A. (California, Washington).

Lustrochernes andinus Beier

Lustrochernes andinus Beier, 1959e: 204–205, fig. 16.

Type locality: 40 miles SW. of Tingo María, E. side of Carpish Mountains, Peru.
Distribution: Peru.

Lustrochernes argentinus (Thorell)

Chelifer argentinus Thorell, 1877a: 216–218; Balzan, 1890: 414–416, figs 3, 3a–b; Ellingsen, 1905b: 1; Tullgren, 1907a: 51–52, figs 12a–f; With, 1908a: 307–310, text-figs 74–77, figs 31a–f (in part; see *Lustrochernes propinquus* Beier); Ellingsen, 1913a: 452.
Chelifer capreolus Balzan, 1888a: no pagination, figs (synonymized by Tullgren, 1907a: 51).
Chelifer (Lamprochernes) argentinus (Thorell): Balzan, 1892: 516–517, fig. 7; Ellingsen, 1905c: 6–8; Ellingsen, 1905e: 324; Ellingsen, 1910a: 363.
Lamprochernes argentinus (Thorell): Beier, 1930a: 211.
Lustrochernes argentinus (Thorell): Beier, 1932c: 259; Beier, 1932e: 91–92, fig. 112; Roewer, 1937: 290; Mello-Leitão, 1939a: 118; Mello-Leitão, 1939b: 616; Vachon, 1940f: 3; Feio, 1945: 6; Hoff, 1946c: 4; Beier, 1948b: 452; Vachon, 1951a: 728; Beier, 1959e: 205; Beier, 1962h: 135; Beier, 1970d: 52; Beier, 1974d: 899; Beier, 1977c: 104; Mahnert, 1984c: 24.

Type localities: of *Chelifer argentinus*: Córdova, Argentina.
of *Chelifer capreolus*: Mato Grosso (as Matto-grosso), Brazil.
Distribution: Argentina, Brazil (Ceará, Espírito Santo, Mato Grosso, Minas Gerais, Paraná, Rio de Janeiro, Rio Grande do Sol, Santa Catarina, Sao Paulo), Colombia, Ecuador, Mexico, Paraguay, Peru, Puerto Rico, Venezuela.

Lustrochernes ariditatis (J. C. Chamberlin)

Chelanops ariditatis J. C. Chamberlin, 1923c: 380, plate 1 fig. 8, plate 3 figs 1–2, 28; J. C. Chamberlin, 1931a: figs 10h, 47r.
Lustrochernes (?) *ariditatis* (J. C. Chamberlin): Beier, 1932e: 95; Roewer, 1937: 290.

Type locality: Las Animas Bay, Baja California, Mexico.
Distribution: Mexico.

Lustrochernes brasiliensis (Daday)

Chernes brasiliensis Daday, 1889b: 23–24, figs 5, 15.
Chelifer brasiliensis (Daday): With, 1908a: 296.
Lustrochernes (?) *brasiliensis* (Daday): Beier, 1932e: 94; Roewer, 1937: 290.

Type locality: Cazaza, Brazil.
Distribution: Brazil.

Lustrochernes caecus Beier

Lustrochernes caecus Beier, 1953b: 16–19, figs 3a–b; Beier, 1955i: 370.

Type locality: Laguna de Jocotal, San Miguel, El Salvador.
Distribution: El Salvador.

Lustrochernes concinnus Hoff

Lustrochernes concinnus Hoff, 1947c: 61–63, figs 1–3; Beier, 1948b: 444.

Type locality: Rancho Grande, Venezuela.
Distribution: Venezuela.

Lustrochernes consocius (R. V. Chamberlin)

Chelanops consocius R. V. Chamberlin, 1925: 238–239.
Lustrochernes consocius (R. V. Chamberlin): Muchmore, 1984d: 355–356, figs 2–4.

Type locality: Barro Colorado Island, Panama.
Distribution: Panama.

Lustrochernes crassimanus Beier

Lustrochernes crassimanus Beier, 1933b: 92, fig. 2; Roewer, 1937: 290; Beier, 1953b: 19; Reyes-Castillo & Hendrichs, 1975: 129.

Type locality: Chilapa, Guerrero, Mexico.
Distribution: Guatemala, Mexico.

Lustrochernes dominicus Hoff

Lustrochernes dominicus Hoff, 1944a: 3–5, figs 5–7.

Type locality: Fore Hunt Flat, Dominican Republic.
Distribution: Dominican Republic.

Lustrochernes gracilis (Banks)

Atemnus gracilis Banks, 1909a: 146.
Lustrochernes (?) *gracilis* (Banks): Beier, 1932e: 95; Roewer, 1937: 290.

Type locality: Sonora, Mexico.
Distribution: Mexico.

Lustrochernes granulosus Beier

Lustrochernes granulosus Beier, 1977c: 104–105, fig. 6.

Type locality: Guayaquil à Cuenca, Cañar, Ecuador.
Distribution: Ecuador.

Lustrochernes grossus (Banks)

Chelanops grossus Banks, 1893: 65; Banks, 1895a: 5; Banks, 1895c: 432; Banks, 1901a: 594; Banks, 1902b: 220; Coolidge, 1908: 110.
Chelanops (?) *grossus* (Banks): Beier, 1932e: 179; Roewer, 1937: 302.
Lamprochernes grossus (Banks): Hoff, 1947b: 475–478, fig. 1.

Family Chernetidae

Lustrochernes grossus (Banks): Hoff, 1956c: 10–12; Hoff, 1958: 21; Hoff, 1959b: 4, etc.; Hoff, 1961: 446; Muchmore, 1971a: 85; Hoff & Jennings, 1974: 21–22; Benedict & Malcolm, 1982: 108.

Type locality: Colorado, U.S.A.
Distribution: U.S.A. (Arizona, Colorado, New Mexico, Oregon).

Lustrochernes intermedius (Balzan)

Chelifer (Lamprochernes) intermedius Balzan, 1892: 515–516, fig. 6, 6a; Ellingsen, 1905c: 8–10; Ellingsen, 1910a: 367.
Chelifer (Atemnus) rotundatus Ellingsen, 1902: 151–152 (synonymized by Beier, 1932e: 88).
Chelifer intermedius (Balzan): Tullgren, 1907a: 52–53, figs 13a–f; With, 1908a: 297–299, figs 27a–d; Ellingsen, 1913a: 453.
Lamprochernes intermedius (Balzan): Beier, 1930a: 211.
Lustrochernes intermedius (Balzan): Beier, 1932e: 88–89, fig. 108; Roewer, 1937: 289; Mello-Leitão, 1939a: 118; Mello-Leitão, 1939b: 616; Feio, 1942: 117; Feio, 1945: 5; Mahnert, 1985b: 78–80, figs 7–13; Mahnert, 1985c: 227; Adis & Mahnert, 1986: 311; Mahnert & Adis, 1986: 213.
Lustrochernes communis (Balzan): Mahnert, 1979d: 765–767, figs 92–95 (misidentification).

Type localities: of *Chelifer (Lamprochernes) intermedius*: Colonia Tovar, Venezuela.
of *Chelifer (Atemnus) rotundatus*: Naranjito, Ecuador.
Distribution: Argentina, Brazil (Amazonas, Distrito Federal, Mato Grosso, Pará, Rio de Janeiro, Santa Catarina, Sao Paulo), Ecuador, Guatemala, Mexico, Paraguay, Venezuela.

Lustrochernes mauriesi Heurtault & Rebière

Lustrochernes mauriesi Heurtault & Rebière, 1983: 604–608, figs 29–34.

Type locality: Basse Terre, Guadeloupe.
Distribution: Guadeloupe.

Lustrochernes minor J. C. Chamberlin

Lustrochernes minor J. C. Chamberlin, 1938b: 114–118, figs 2a–h, 3a–g; Roewer, 1940: 346; Muchmore, 1977: 77.

Type locality: Oxkutzcab, Yucatan, Mexico.
Distribution: Mexico.

Lustrochernes nitidus (Ellingsen)

Chelifer (Chernes) nitidus Ellingsen, 1902: 155–156.
Chelifer (Lamprochernes) nitidus Ellingsen: Ellingsen, 1905e: 324.
Chelifer nitidus Ellingsen: With, 1908a: 303–304, figs 29a–b.
Lustrochernes nitidus (Ellingsen): Beier, 1932e: 93; Roewer, 1937: 290; Beier, 1959e: 206.

Type locality: Naranjito, Ecuador.
Distribution: Ecuador.

Lustrochernes ovatus (Balzan)

Chelifer (Lamprochernes) ovatus Balzan, 1892: 519, fig. 10, 10a; Ellingsen, 1910a: 368 (in part; see *Lustrochernes propinquus* Beier).
Chelifer ovatus Balzan: With, 1908a: 296.
Lustrochernes ovatus (Balzan): Beier, 1932e: 89–90, fig. 109; Roewer, 1937: 289; Feio, 1945: 5; Beier, 1955m: 5; Beier, 1954f: 135.

Type locality: Caraça, Minas Gerais, Brazil.
Distribution: Argentina, Brazil (Bahia, Distrito Federal, Goiás, Minas Gerais, Rio de Janeiro), Peru, Venezuela.

Lustrochernes pennsylvanicus (Ellingsen)

Chelifer communis pennsylvanicus Ellingsen, 1910a: 366.
Lustrochernes (?) *pennsylvanicus* (Ellingsen): Beier, 1932e: 95; Roewer, 1937: 290.
Lustrochernes pennsylvanicus (Ellingsen): Hoff & Bolsterli, 1956: 167–168; Hoff, 1958: 21; Weygoldt, 1969a: fig. 105.

Type locality: Pennsylvania, U.S.A.
Distribution: U.S.A. (Louisiana, Mississippi, North Carolina, Pennsylvania).

Lustrochernes propinquus Beier

Chelifer (Lamprochernes) ovatus Balzan: Ellingsen, 1910a: 368 (misidentification, in part).
Chelifer argentinus Thorell: With, 1908a: 307–310, text-figs 74–77, figs 31a–f (misidentification).
Lustrochernes propinquus Beier, 1932c: 261; Beier, 1932e: 93–94, fig. 115; Roewer, 1937: 290; Beier, 1954f: 135.

Type locality: Caracas, Venezuela.
Distribution: Venezuela.

Lustrochernes reimoseri Beier

Lustrochernes reimoseri Beier, 1932c: 261–262; Beier, 1932e: 94, fig. 116; Beier, 1933b: 92; Roewer, 1937: 290.

Type locality: La Caja, Costa Rica.
Distribution: Costa Rica, Mexico.

Lustrochernes rufimanus (C. L. Koch)

Pelorus rufimanus C. L. Koch, 1843: 59–60, fig. 793; Hagen, 1870: 271.
Lustrochernes (?) *rufimanus* (C. L. Koch): Beier, 1932e: 95; Roewer, 1937: 290.

Type locality: Brazil.
Distribution: Brazil.

Lustrochernes schultzei Beier

Lustrochernes schultzei Beier, 1933b: 91–92, fig. 1; Roewer, 1937: 290.

Type locality: near Chilapa, Guerrero, Mexico.
Distribution: Mexico.

Lustrochernes silvestrii Beier

Lustrochernes silvestrii Beier, 1932c: 260; Beier, 1932e: 91, figs 106–107, 111; Roewer, 1936: fig. 83b; Roewer, 1937: 290, fig. 234; Beier, 1953b: 19.
Lustrochernes silvestri (?) Beier: Feio, 1945: 6.

Type locality: San José, Costa Rica.
Distribution: Argentina, Costa Rica, El Salvador.

Lustrochernes similis (Balzan)

Chelifer (Lamprochernes) similis Balzan, 1892: 517, figs 8, 8a.
Chelifer similis Balzan: With, 1908a: 305–307, figs 30a–f.
Lustrochernes similis (Balzan): Beier, 1932e: 93, fig. 114; Roewer, 1937: 290; Beier, 1954f: 135; Beier, 1959e: 206; Mahnert, 1979d: 767–769, figs 96–100; Mahnert, 1985c: 227–228; Adis & Mahnert, 1986: 311; Mahnert & Adis, 1986: 213; Mahnert, Adis, & Bührnheim, 1987: fig. 12b.

Type locality: Manaus, Amazonas, Brazil.
Distribution: Brazil (Amazonas), Peru, Venezuela.

Lustrochernes subovatus (With)

Chelifer subovatus With, 1908a: 294–296, figs 26a–f.
Lustrochernes subovatus (With): Beier, 1932e: 89; Roewer, 1937: 289; Mello-Leitão, 1939a: 118; Mello-Leitão, 1939b: 616; Feio, 1945: 5.

Type localities: Argentina; Riacho dell'Oro, Brazil; and La Moka, Venezuela.
Distribution: Argentina, Brazil (Distrito Federal, Santa Catarina, Sao Paulo), Venezuela.

Lustrochernes surinamus Beier

Lustrochernes surinamus Beier, 1932c: 261; Beier, 1932e: 92–93, fig. 113; Roewer, 1937: 290.

Type locality: Paramaribo, Surinam.
Distribution: Surinam.

Lustrochernes viniai Dumitresco & Orghidan

Lustrochernes viniai Dumitresco & Orghidan, 1977: 113–118, figs 13a–g, 14a–f, 15a–b; Armas & Alayón, 1984: 12.

Type locality: Cueva de Mexico, Camagüey, Cuba.
Distribution: Cuba.

Genus **Macrochernes** Hoff

Macrochernes Hoff, 1946a: 104; Muchmore, 1969c: 9–11.

Type species: *Chelifer wrightii* Hagen, 1868b, by original designation.

Macrochernes attenuatus Muchmore

Macrochernes attenuatus Muchmore, 1969c: 11–13, figs 1–2.

Type locality: Sobana River, Puerto Rico.
Distribution: Puerto Rico.

Macrochernes wrightii (Hagen)

Chelifer wrightii Hagen, 1868b: 52; Hagen, 1870: 267.
Genus ? *wrightii* Hagen: Beier, 1932e: 277; Roewer, 1937: 317.
Macrochernes wrightii (Hagen): Hoff, 1946a: 105–107, figs 4–6; Muchmore, 1969c: 10.

Type locality: Cuba.
Distribution: Cuba.

Genus **Maorichernes** Beier

Maorichernes Beier, 1932e: 163; Beier, 1933a: 532; Beier, 1976f: 217.

Type species: *Chelifer vigil* With, 1907, by original designation.

Maorichernes vigil (With)

Chelifer vigil With, 1907: 50–53, figs 1–2.
Maorichernes vigil (With): Beier, 1932e: 163; J. C. Chamberlin, 1934b: 9; Roewer, 1937: 298; Beier, 1966c: 372, fig. 7; Beier, 1976f: 216.

Type locality: Taieri, South Island, New Zealand.
Distribution: New Zealand.

Genus **Maxchernes** Feio

Maxchernes Feio, 1960: 74–76.

Type species: *Maxchernes birabeni* Feio, 1960, by original designation.

Maxchernes birabeni Feio

Maxchernes birabeni Feio, 1960: 77–80, figs a–n.

Type locality: Higueritas, Fraile Pintado, Jujuy, Argentina.
Distribution: Argentina.

Maxchernes plaumanni Beier

Maxchernes plaumanni Beier, 1974d: 907–909, fig. 7.

Type locality: Nova Teutonia, Santa Catarina, Brazil.
Distribution: Brazil (Santa Catarina).

Genus **Megachernes** Beier

Megachernes Beier, 1932e: 128; Beier, 1933a: 518; Beier, 1948b: 476; Morikawa, 1960: 144.

Type species: *Chernes grandis* Beier, 1930e, by original designation.

Megachernes afghanicus Beier

Megachernes afghanicus Beier, 1959a: 274–275, fig. 17; Beier, 1960a: 43; Beier, 1961e: 3; Lindberg, 1961: 31.

Type locality: Grotte Boulan, Qalat, Afghanistan.
Distribution: Afghanistan.

Megachernes barbatus Beier

Megachernes barbatus Beier, 1951a: 95–98, figs 30–31; Beier, 1967g: 357, fig. 17–18.

Type locality: Lang Bian Peaks, Vietnam.
Distribution: Vietnam.

Megachernes crinitus Beier

Megachernes crinitus Beier, 1948b: 483–485, figs 12–13.

Type locality: Gedeh, Tjibodas, Java, Indonesia.
Distribution: Indonesia (Java).

Megachernes grandis (Beier)

Chernes grandis Beier, 1930e: 295–297, figs 7a–b, 8; Beier, 1932a: fig. 13.
Megachernes grandis (Beier): Beier, 1932e: 128–129, figs 144a–b; Roewer, 1936: fig. 112i; Roewer, 1937: 296; Beier, 1948b: 485–486, figs 14–15; Beier, 1964d: 313; Beier, 1966b: 343; Durden, 1986: 324, fig. 4.

Type locality: Padang, Sumatra, Indonesia.
Distribution: Indonesia (Java, Sulawesi, Sumatra), Malaysia, Philippines.

Megachernes himalayensis (Ellingsen)

Chelifer himalayensis Ellingsen, 1914: 5–6.
Megachernes himalayensis (Ellingsen): Beier, 1932e: 130; Roewer, 1937: 296; Beier, 1974b: 269–270; Martens, 1975: 84–90, figs 1–2; Beier, 1978b: 415; Schawaller, 1983a: 110; Schawaller, 1987a: 213; Schawaller, 1988: 161.
Megachernes sinensis Beier, 1932e: 129–130, fig. 145; Beier, 1933a: 519–520, fig. 3; Roewer, 1937: 296; Beier, 1948b: 486–487, figs 16–17 (synonymized by Beier, 1974b: 269).

Type localities: of *Chelifer himalayensis*: Mussoorie, India.
of *Megachernes sinensis*: Batang, China.
Distribution: China (Fukien), India, Nepal.

Megachernes limatus Hoff & Parrack

Megachernes limatus Hoff & Parrack, 1958: 4–9, figs 1–2; Beier, 1965g: 772–773, fig. 13.

Type locality: Mt Dayman, Maneau Range, Papua New Guinea.
Distribution: Papua New Guinea.

Megachernes limatus crassus Beier

Megachernes limatus crassus Beier, 1965g: 773–774, fig. 14; Tenorio & Muchmore, 1982: 380.

Type locality: Bokondini, 40 km N. of Baliem Valley, Irian Jaya, Indonesia.
Distribution: Indonesia (Irian Jaya).

Megachernes mongolicus (Redikorzev)

Chelifer (Chelanops) mongolicus Redikorzev, 1934a: 423–425, fig. 1.
Megachernes mongolicus (Redikorzev): Redikorzev, 1934b: 152; Roewer, 1937: 296, fig. 236.

Type locality: Schilbiz-chuduk, Bergkette Churchu, Mongolia.
Distribution: Mongolia.

Megachernes monstrosus Beier

Megachernes monstrosus Beier, 1966b: 344–346, fig. 3; Tenorio & Muchmore, 1982: 381.

Type locality: Mt Mantalingan, Palawan, Philippines.
Distribution: Philippines.

Megachernes ochotonae Krumpál & Kiefer

Megachernes ochotonae Krumpál & Kiefer, 1982: 3–6, plate 2 figs 1–10.

Type locality: 30 km SE. of Bajan Öndör Somon, Övörchangaj Ajmak, Mongolia.
Distribution: Mongolia.

Megachernes papuanus Beier

Megachernes papuanus Beier, 1948b: 481–483, figs 10–11; Hoff & Parrack, 1958: 2–4; Beier, 1965g: 772.

Type locality: Sattelberg, Huon Gulf, Papua New Guinea.
Distribution: Indonesia (Irian Jaya), Papua New Guinea.

Megachernes pavlovskyi Redikorzev

Megachernes pawlovskyi (sic) Redikorzev: Vachon, 1938a: fig. 56k (nomen nudum).
Megachernes pavlovskyi Redikorzev, 1949: 651–652, figs 13–15; Beier, 1959a: 274; Lindberg, 1961: 31; Schawaller, 1986: 3; Schawaller & Dashdamirov, 1988: 43, figs 62, 79.
Megachernes caucasicus Krumpál, 1986: 170–171, figs 16–22 (synonymized by Schawaller & Dashdamirov, 1988: 43).

Type localities: of *Megachernes pavlovskyi*: Gaudan, Turkmenistan, U.S.S.R.; Kadzhi-gal'ton, 25 km from Kulyab, Tadzhikistan, U.S.S.R.; and Bodontu, near Kulyab, Tadzhikistan, U.S.S.R.
of *Megachernes caucasicus*: Azychskaya Cave, Nagorno-Karabakhskaya A.O., Azerbaijan, U.S.S.R.

Distribution: Afghanistan, U.S.S.R. (Azerbaijan, Kirghizia, R.S.F.S.R., Tadzhikistan, Turkmenistan).

Megachernes penicillatus Beier

Megachernes penicillatus Beier, 1948b: 478–479, figs 6–7; Harvey, 1981b: 246; Harvey, 1985b: 135.

Type locality: Dinner Creek, Ravenshoe (as Ravenshor), Queensland, Australia.
Distribution: Australia (Queensland).

Megachernes philippinus Beier

Megachernes philippinus Beier, 1966b: 343–344, fig. 2; Tenorio & Muchmore, 1982: 381.

Type locality: Dapitan Peak, Mindanao, Philippines.
Distribution: Philippines.

Megachernes queenslandicus Beier

Megachernes queenslandicus Beier, 1948b: 480–481, figs 8–9; Harvey, 1981b: 246; Harvey, 1985b: 135.

Type locality: Dinner Creek, Ravenshoe (as Ravenshor), Queensland, Australia.
Distribution: Australia (Queensland).

Megachernes ryugadensis Morikawa

Megachernes ryugadensis Morikawa, 1954b: 84–86, figs 3a–c, 4; Morikawa, 1960: 144–145; Morikawa, 1962: 418; Sato, 1979a: plate 2, fig. 6; Sato, 1979d: 44; Sato, 1982c: 34.
Megachernes ryuaadensis (sic) Morikawa: Morikawa, 1957a: 363–364.
Megachernes ryugadensis ryugadensis Morikawa: Morikawa, 1960: 145, plate 7 fig. 17, plate 10 fig. 28; Morikawa, 1962: fig. 4.

Type locality: Ryuga-do Cave, Shikoku, Japan.
Distribution: Japan.

Megachernes ryugadensis myophilus Morikawa

Megachernes ryugadensis myophilus Morikawa, 1960: 146; Morikawa, 1962: fig. 4.

Type locality: Sasayama-cho, Hyogo, Japan.
Distribution: Japan.

Megachernes ryugadensis naikaiensis Morikawa

Megachernes ryugadensis naikaiensis Morikawa, 1957a: 364; Morikawa, 1960: 145–146, plate 3 fig. 9, plate 9 fig. 23; Morikawa, 1962: fig. 4.

Type locality: Hakuun-dô Cave, Taishaku-kyô, Hiroshima, Japan.
Distribution: Japan.

Megachernes soricicola Beier

Megachernes soricicola Beier, 1974b: 270–272, fig. 6; Martens, 1975: 84–90, fig. 2.

Type locality: Pare, Khumbu, Nepal.
Distribution: Nepal.

Megachernes titanius Beier

Megachernes titanius Beier, 1951a: 93–95, figs 28–29; Beier, 1967g: 355, fig. 15.

Type locality: Lang Bian Peaks, Vietnam.
Distribution: Vietnam.

Megachernes vietnamensis Beier

Megachernes vietnamensis Beier, 1967g: 355–356, fig. 16; Tenorio & Muchmore, 1982: 381.

Type locality: Da Lat (as Dalat), corrected to Thac Da Tan La, Vietnam, by Tenorio & Muchmore, 1982: 384.
Distribution: Vietnam.

Genus Meiochernes Beier

Meiochernes Beier, 1957d: 60–61.

Type species: *Meiochernes dybasi* Beier, 1957d, by original designation.

Meiochernes dybasi Beier

Meiochernes dybasi Beier, 1957d: 61–63, figs 39a–b, 40a–f.

Type locality: Mt Kupuriso, Ponape, Caroline Islands.
Distribution: Caroline Islands.

Genus Mesochernes Beier

Mesochernes Beier, 1932c: 263; Beier, 1932e: 97.

Type species: *Mesochernes gracilis* Beier, 1932c, by original designation.

Mesochernes australis Mello-Leitão

Mesochernes australis Mello-Leitão, 1939a: 118–120, fig. 2.

Type locality: Lago Moreno, Bariloche, Rio Negro, Argentina.
Distribution: Argentina.

Mesochernes costaricensis Beier

Mesochernes costaricensis Beier, 1932c: 263–264; Beier, 1932e: 98, figs 100, 119; Roewer, 1937: 290.

Type locality: Jiminez, Costa Rica.
Distribution: Costa Rica.

Mesochernes elegans (Balzan)

Chelifer (Lamprochernes) elegans Balzan, 1892: 520–521, figs 12, 12a.
Chelifer elegans Balzan: With, 1908a: 293.
Mesochernes elegans (Balzan): Beier, 1932e: 99; Roewer, 1937: 290.

Type locality: Tovar, Venezuela.
Distribution: Venezuela.

Mesochernes gracilis Beier

Mesochernes gracilis Beier, 1932c: 264; Beier, 1932e: 98–99, fig. 120; Roewer, 1937: 290; Beier, 1953b: 20.

Type locality: Moretto, Guatemala.
Distribution: Guatemala.

Mesochernes venezuelanus (Balzan)

Chelifer (Lamprochernes) venezuelanus Balzan, 1892: 518, figs 9, 9a.
Chelifer venezuelanus Balzan: With, 1908a: 293–294.
Mesochernes venezuelanus (Balzan): Beier, 1932e: 99; Roewer, 1937: 290; Vachon, 1938a: fig. 561.

Type locality: San Esteban, Venezuela.
Distribution: Venezuela.

Genus **Metagoniochernes** Vachon

Metagoniochernes Vachon, 1939a: 123; Vachon, 1951c: 168–169.

Type species: *Metagoniochernes picardi* Vachon, 1939a, by original designation.

Metagoniochernes milloti Vachon

Metagoniochernes milloti Vachon, 1951c: 160–168, figs 1–15; Legendre, 1972: 447.

Type locality: Forêt d'Analamazaotra, Madagascar.
Distribution: Madagascar.

Metagoniochernes picardi Vachon

Metagoniochernes picardi Vachon, 1939a: 125–129, figs 1–9; Roewer, 1940: 347; Vachon, 1949: figs 200c, 205.

Type locality: Boda, Central African Republic.
Distribution: Central African Republic.

Genus **Mexachernes** Hoff

Mexachernes Hoff, 1947b: 495–496.

Type species: *Chelanops calidus* Banks, 1909a, by original designation.

Mexachernes calidus (Banks)

Chelanops calidus Banks, 1909a: 147.
Neochernes calidus (Banks): Beier, 1932e: 165; Roewer, 1937: 298.
Mexachernes calidus (Banks): Hoff, 1947b: 496–498, figs 11–12.

Type localities: Sonora, Mexico; and Baja California, Mexico.
Distribution: Mexico.

Mexachernes carminis (J. C. Chamberlin)

Chelanops carminis J. C. Chamberlin, 1923c: 378, plate 1 fig. 10, plate 3 figs 3–5, 27.
Epaphochernes carminis (J. C. Chamberlin): Beier, 1932e: 174.
Dinocheirus carminis (J. C. Chamberlin): Roewer, 1937: 302.
Mexachernes carminis (J. C. Chamberlin): V. F. Lee, 1979a: 21–24, figs 26–27, 31.

Type locality: Puerto Ballandra, Carmen Island, Gulf of California, Mexico.
Distribution: Mexico.

Genus **Mirochernes** Beier

Mirochernes Beier, 1930a: 216; Beier, 1932e: 182; Beier, 1933a: 542; Hoff, 1949b: 478.

Type species: *Chelanops dentatus* Banks, 1895a, by original designation (based upon misidentified type species; an application is currently before the International Commission on Zoological Nomenclature).

Mirochernes dentatus (Banks)

Chelanops dentatus Banks, 1895a: 6; Banks, 1904a: 141; Coolidge, 1908: 111; J. C. Chamberlin, 1931a: 123, figs 30j–k; Nester, 1932: 98; Brimley, 1938: 497 (not seen).
Mirochernes dentatus (Banks): Beier, 1932e: 182; Roewer, 1937: 303; Hoff, 1945c: 49; Hoff, 1947b: 502–503; J. C. Chamberlin, 1949: 8; Hoff, 1949b: 478–481, figs 45a–d; Hoff & Bolsterli, 1956: 176; Hoff, 1958: 25–26; Muchmore, 1971a: 88, 90; Nelson, 1975: 288, figs 56–58.
Microchernes (sic) *dentatus* Banks: Park & Auerbach, 1954: 210; Weygoldt, 1969a: fig. 24; Nelson, 1984: 348–349.

Not *Mirochernes dentatus* (Banks): Beier, 1930a: 217–218, fig. 14 (misidentification; see *Semeiochernes militaris* Beier).

Not *Chernes dentatus* (Banks): Ross, 1944: fig. 56 (misidentification; see *Larca granulata* (Banks)).

Type locality: 'probably' Florida, U.S.A.

Distribution: U.S.A. (Arkansas, Connecticut, Florida, Illinois, Indiana, Michigan, North Carolina, Oklahoma, Virginia).

Genus **Mucrochernes** Muchmore

Mucrochernes Muchmore, 1973a: 43–44.

Type species: *Atemnus hirsutus* Banks, 1914, by original designation.

Mucrochernes hirsutus (Banks)

Atemnus hirsutus Banks, 1914: 203, fig. 1; Moles, 1914c: 195; Moles & Moore, 1921: 7.

Genus ? *hirsutus* Banks: Hoff, 1958: 22.

Mucrochernes hirsutus (Banks): Muchmore, 1973a: 44–47, figs 1–6; Muchmore, 1975d: fig. 11; Muchmore, 1984a: 20–22, figs 1–5.

Type locality: Laguna Beach, California, U.S.A.

Distribution: U.S.A. (California).

Genus **Myrmochernes** Tullgren

Myrmochernes Tullgren, 1907a: 59–60; Beier, 1932e: 190; Judson, 1985: 321–323.

Type species: *Myrmochernes africanus* Tullgren, 1907a, by monotypy.

Myrmochernes africanus Tullgren

Myrmochernes africanus Tullgren, 1907a: 60–61, figs 18a–e; Ellingsen, 1912b: 85; J. C. Chamberlin, 1931a: 241, figs 10k, 14i, 15s, 18g, 20d, 29h, 38o, 68; Beier, 1932e: 190–191, figs 198–199; Beier, 1932g: fig. 265; Roewer, 1936: fig. 23; Roewer, 1937: 304; Beier, 1948b: 458; Weidner, 1959: 115; Beier, 1964k: 75; Newlands, 1978: 700; Judson, 1985: 322–326, figs 1–10.

Type locality: Port Elizabeth, Cape Province, South Africa.

Distribution: South Africa.

Genus **Neoallochernes** Hoff

Neoallochernes Hoff, 1947b: 499.

Type species: *Chelanops garcianus* Banks, 1909a, by original designation.

Neoallochernes garcianus (Banks)

Chelanops garcianus Banks, 1909a: 147–148.

Dinocheirus garcianus (Banks): Beier, 1932e: 139.

Dinocheirus gracianus (sic) (Banks): Roewer, 1937: 302.

Neoallochernes garcianus (Banks): Hoff, 1947b: 500–502, figs 13–14.

Type locality: Havana, Cuba.

Distribution: Cuba.

Genus **Neochernes** Beier

Neochernes Beier, 1932e: 164; Beier, 1933a: 532.

Type species: *Chernes peninsularis* J. C. Chamberlin, 1925a, by original designation.

Neochernes melloleitaoi Feio

Neochernes mello-leitaoi Feio, 1945: 32–35, figs 25–26, 27a–b.

Type locality: Assusques, Santiago, Argentina.
Distribution: Argentina.

Neochernes peninsularis (J. C. Chamberlin)

Chernes peninsularis J. C. Chamberlin, 1925a: 330–331, figs e–g.
Neochernes peninsularis (J. C. Chamberlin): Beier, 1932e: 165; Roewer, 1937: 298.

Type locality: San Lucus, Baja California, Mexico.
Distribution: Mexico.

Genus Nesidiochernes Beier

Nesidiochernes Beier, 1957d: 50; Beier, 1976f: 222.

Type species: *Nesidiochernes maculatus* Beier, 1957d, by original designation.

Nesidiochernes australicus Beier

Nesidiochernes australicus Beier, 1966a: 298–299, fig. 13; Beier, 1975b: 203; Harvey, 1981b: 246; Harvey, 1985b: 135.

Type locality: 3 miles E. of Cobar, New South Wales, Australia.
Distribution: Australia (New South Wales, South Australia, Victoria, Western Australia).

Nesidiochernes caledonicus Beier

Nesidiochernes caledonicus Beier, 1964e: 406–408, figs 3–4; Beier, 1966g: 371; Tenorio & Muchmore, 1982: 380.

Type locality: Mt Koghi, New Caledonia.
Distribution: New Caledonia.

Nesidiochernes carolinensis Beier

Nesidiochernes carolinensis carolinensis Beier, 1957d: 52–53, figs 32a–e.

Type locality: Saliap, Woleai, Caroline Islands.
Distribution: Caroline Islands.

Nesidiochernes carolinensis dybasi Beier

Nesidiochernes carolinensis dybasi Beier, 1957d: 54, figs 33, 34a–b.

Type locality: Kalabera, Saipan, Mariana Islands.
Distribution: Mariana Islands.

Nesidiochernes insociabilis Beier

Nesidiochernes insociabilis Beier, 1957d: 59–60, fig. 38.

Type locality: Moen (Weba), Truk, Caroline Islands.
Distribution: Caroline Islands.

Nesidiochernes kuscheli Beier

Nesidiochernes kuscheli Beier, 1976f: 223–224, fig. 23.

Type locality: Rakeahua Valley, Stewart Island, New Zealand.
Distribution: New Zealand.

Nesidiochernes maculatus Beier

Nesidiochernes maculatus Beier, 1957d: 50–52, figs 31a–d.

Type locality: Lelu, Kusaie, Caroline Islands.
Distribution: Caroline Islands, Marshall Islands.

Nesidiochernes novaeguineae Beier

Nesidiochernes novaeguineae Beier, 1965g: 780–781, fig. 20; Tenorio & Muchmore, 1982: 381.

Type locality: Maprik, Papua New Guinea.
Distribution: Papua New Guinea.

Nesidiochernes palauensis Beier

Nesidiochernes palauensis Beier, 1957d: 55–56, figs 35a–d.

Type locality: Ngergoi (Garakayo), Palau Islands, Caroline Islands.
Distribution: Caroline Islands.

Nesidiochernes plurisetosus Beier

Nesidiochernes plurisetosus Beier, 1965g: 781–782, fig. 21.

Type locality: Maffin Bay, Irian Jaya, Indonesia.
Distribution: Indonesia (Irian Jaya).

Nesidiochernes robustus Beier

Nesidiochernes robustus Beier, 1957d: 57–59, figs 37a–d.

Type locality: Tinian, Mariana Islands.
Distribution: Mariana Islands.

Nesidiochernes scutulatus Beier

Nesidiochernes scutulatus Beier, 1969e: 415–416, fig. 2; Beier, 1976f: 222–223.

Type locality: Hamilton, North Island, New Zealand.
Distribution: New Zealand.

Nesidiochernes slateri Beier

Nesidiochernes slateri Beier, 1975b: 208; Harvey, 1981b: 247; Harvey, 1985b: 135.

Type locality: Peak Head, S. of Albany, Western Australia, Australia.
Distribution: Australia (Western Australia).

Nesidiochernes tumidimanus Beier

Nesidiochernes tumidimanus Beier, 1957d: 56–57, figs 36a–d.

Type locality: Ruul district, Yap, Caroline Islands.
Distribution: Caroline Islands.

Nesidiochernes zealandicus Beier

Nesidiochernes zealandicus Beier, 1966c: 372–373, fig. 8; Beier, 1976f: 222; Tenorio & Muchmore, 1982: 384.

Type locality: Scenic Reserve, Kaituna Valley, Banks Peninsula, South Island, New Zealand.
Distribution: New Zealand.

Genus Nesiotochernes Beier

Nesiotochernes Beier, 1976f: 217–218.

Type species: *Nesiotochernes stewartensis* Beier, 1976f, by original designation.

Nesiotochernes stewartensis Beier

Nesiotochernes stewartensis Beier, 1976f: 218, fig. 17.

Type locality: Bald Cone, Port Pegasus, Stewart Island, New Zealand.
Distribution: New Zealand.

Genus **Nesochernes** Beier

Nesochernes Beier, 1932e: 184; Beier, 1933a: 544; Beier, 1976f: 239.

Type species: *Nesochernes gracilis* Beier, 1932e, by original designation.

Nesochernes gracilis Beier

Nesochernes gracilis Beier, 1932e: 184, fig. 192; Beier, 1933a: 544–545, fig. 15; J. C. Chamberlin, 1934b: 10; Roewer, 1937: 303, fig. 240d.
Apatochernes cheliferoides Beier: Beier, 1969a: 114 (misidentification).
Neoschernes gracilis gracilis Beier: Beier, 1976f: 239.

Type locality: Waitakaruru, near Auckland, North Island, New Zealand.
Distribution: New Zealand.

Nesochernes gracilis norfolkensis Beier

Nesochernes gracilis norfolkensis Beier, 1976f: 239, fig. 37; Harvey, 1985b: 135.

Type locality: Mt Bates, Norfolk Island.
Distribution: Norfolk Island.

Genus **Nudochernes** Beier

Nudochernes Beier, 1935c: 122; Beier, 1959d: 48.

Type species: *Nudochernes montanus* Beier, 1935c, by original designation.

Nudochernes basilewskyi Beier

Nudochernes basilewskyi Beier, 1962f: 21–22, fig. 8.

Type locality: Olkokola, Mt Meru, Tanzania.
Distribution: Tanzania.

Nudochernes crassus Beier

Nudochernes crassus Beier, 1944: 192–193, fig. 14; Beier, 1955c: 551; Mahnert, 1982e: 710.

Type locality: Aberdare, Kenya.
Distribution: Kenya, Uganda.

Nudochernes gracilimanus Mahnert

Nudochernes gracilimanus Mahnert, 1982e: 705–707, figs 28–33.

Type locality: Naromuru Track, Mt Kenya, Kenya.
Distribution: Kenya.

Nudochernes gracilipes Beier

Nudochernes gracilipes Beier, 1959d: 49–50, fig. 24.

Type locality: Isale, Ht. Muhi, Kivu, Zaire.
Distribution: Zaire.

Nudochernes granulatus Beier

Nudochernes granulatus Beier, 1951b: 608–609, fig. 2.

Type locality: Kilimanjaro, Tanzania.
Distribution: Tanzania.

Nudochernes intermedius Beier

Nudochernes intermedius Beier, 1959d: 51–53, fig. 26.

Type locality: Lubero, Zaire.
Distribution: Zaire.

Nudochernes kivuensis Beier

Nudochernes kivuensis Beier, 1959d: 53−55, fig. 27.

Type locality: SE. of Kahuzi, Kabare, Kivu, Zaire.
Distribution: Zaire.

Nudochernes leleupi Beier

Nudochernes leleupi Beier, 1959d: 55−56, fig. 28.

Type locality: Ht. Luvubu, Uvira, Kivu, Zaire.
Distribution: Zaire.

Nudochernes longipes Beier

Nudochernes longipes Beier, 1944: 193−195, fig. 15; Beier, 1959d: 58; Mahnert, 1982e: 707−708, figs 34−35.

Type locality: Aberdare, Kenya.
Distribution: Kenya, Zaire.

Nudochernes lucifugus Beier

Nudochernes lucifugus Beier, 1935c: 125−126, fig. 7; Vachon, 1945b: fig. 16; Beier, 1962f: 19; Mahnert, 1982e: 710.

Type locality: Grotte Shimo Kapseta, Mt Elgon, Uganda.
Distribution: Tanzania, Uganda.

Nudochernes lucifugus meruensis Beier

Nudochernes lucifugus meruensis Beier, 1962f: 19−21, fig. 7.

Type locality: Olkokola, Mt Meru, Tanzania.
Distribution: Tanzania.

Nudochernes montanus Beier

Nudochernes montanus Beier, 1935c: 124−125, fig. 6; Vachon, 1936a: 78; Vachon, 1937f: 128; Vachon, 1938a: fig. 56f; Roewer, 1940: 347; Vachon, 1945b: 191−193, figs 13, 15; Mahnert, 1982e: 709.

Type locality: Mt Elgon, Kenya.
Distribution: Kenya.

Nudochernes nidicola Beier

Nudochernes nidicola Beier, 1935c: 122−123, fig. 5; Vachon, 1936a: 78; Vachon, 1937f: 128; Vachon, 1938a: fig. 56e; Roewer, 1940: 347; Vachon, 1945b: 189−191, figs 7−12, 14; Beier, 1962f: 22; Mahnert, 1982e: 708−709.

Type locality: Mt Elgon, Uganda.
Distribution: Kenya, Uganda.

Nudochernes procerus Beier

Nudochernes procerus Beier, 1959d: 56−58, fig. 29.

Type locality: SE. of Kahuzi, Kabare, Kivu, Zaire.
Distribution: Zaire.

Nudochernes robustus Beier

Nudochernes robustus Beier, 1935c: 126, fig. 8; Roewer, 1940: 347; Vachon, 1945b: fig. 17; Mahnert, 1982e: 710.

Type locality: Mt Elgon, Kenya.
Distribution: Kenya.

Nudochernes setiger Beier

Nudochernes setiger Beier, 1944: 190−191, fig. 13.

Type locality: Kampala, Uganda.
Distribution: Uganda.

Nudochernes spalacis Beier

Nudochernes spalacis Beier, 1955d: 217−218, figs 6−7; Beier, 1963f: 195; Costa & Nevo, 1969: 207, 208, 211.

Type locality: Jerusalem, Israel.
Distribution: Israel.

Nudochernes sudanensis Beier

Nudochernes sudanensis Beier, 1953c: 73−75, fig. 1.

Type locality: Gilo, Sudan.
Distribution: Sudan.

Nudochernes virgineus Beier

Nudochernes virgineus Beier, 1959d: 50−51, fig. 25.

Type locality: Lake Lungwe, Mwenga, Itombwe, Kivu, Zaire.
Distribution: Zaire.

Nudochernes wittei Beier

Nudochernes wittei Beier, 1955a: 12−14, fig. 6.

Type locality: Mubale River, Zaire.
Distribution: Zaire.

Genus Ochrochernes Beier

Ochrochernes Beier, 1932e: 126; Beier, 1933a: 518; Murthy & Ananthakrishnan, 1977: 140.

Type species: *Chelifer (Trachychernes) tenggerianus* Ellingsen, 1910a, by original designation.

Ochrochernes asiaticus Murthy & Ananthakrishnan

Ochrochernes asiaticus Murthy & Ananthakrishnan, 1977: 140−142, figs 44a−b.

Type locality: Shimoga, Karnataka, India.
Distribution: India.

Ochrochernes galatheae (With)

Chelifer galatheae With, 1906: 167−170, plate 4, figs 2a−c; Ellingsen, 1912a: 122; Tullgren, 1912a: 265−267, figs 3a−c.
Ochrochernes galatheae (With): Beier, 1932e: 127−128; Roewer, 1937: 296.

Type locality: Nancowry (as Nankovry), Nicobar Islands.
Distribution: Indonesia (Java), Nicobar Islands, Taiwan.

Ochrochernes granulatus Murthy & Ananthakrishnan

Ochrochernes granulatus Murthy & Ananthakrishnan, 1977: 142−144, figs 45a−d.

Type locality: Gersoppa Falls (as Jog falis), Karnataka, India.
Distribution: India.

Ochrochernes indicus Beier

Ochrochernes indicus Beier, 1974e: 1012–1013, fig. 8; Murthy & Ananthakrishnan, 1977: 140.

Type locality: Coonoor, Nilgiri, Madras, India.
Distribution: India.

Ochrochernes modestus (With)

Chelifer modestus With, 1906: 165–167, plate 4 figs 1a–d.
Chernes modestus (With): Beier, 1930e: 295.
Ochrochernes modestus (With): Beier, 1932e: 127; Roewer, 1937: 296.

Type locality: Teressa Island, Nicobar Islands.
Distribution: Nicobar Islands.

Ochrochernes tenggerianus (Ellingsen)

Chelifer (Trachychernes) tenggerianus Ellingsen, 1910a: 382–383.
Chernes tengerrianus (Ellingsen): Beier, 1930e: 294–295, fig. 6.
Ochrochernes tenggerianus (Ellingsen): Beier, 1932e: 126–127, fig. 143; Roewer, 1937: 296.

Type locality: Tengger Mountains, Java, Indonesia.
Distribution: Indonesia (Java, Sumatra).

Genus Odontochernes Beier

Odontochernes Beier, 1932c: 267; Beier, 1932e: 104.

Type species: *Chelifer cervus* Balzan, 1890, by original designation.

Odontochernes cervus (Balzan)

Chelifer cervus Balzan, 1888a: no pagination, figs; Balzan, 1890: 412–414, figs 2, 2a; With, 1908a: 316–318, text-fig. 82, figs 34a–e.
Chelifer (Lamprochernes) cervus Balzan: Balzan, 1892: 548, fig. 4b.
Chelifer (Chernes) cervus (Balzan): Ellingsen, 1902: 165–167.
Chernes cervus (Balzan): J. C. Chamberlin, 1931a: figs 30d, 42n.
Odontochernes cervus (Balzan): Beier, 1932c: 267; Beier, 1932e: 104–105; Roewer, 1937: 290.

Type locality: Mato Grosso (as Matto-grosso), Brazil.
Distribution: Brazil (Amazonas, Mato Grosso), Surinam.

Genus Oligochernes Beier

Oligochernes Beier, 1937d: 307–308.

Type species: *Oligochernes bachofeni* Beier, 1937d, by original designation.

! Oligochernes bachofeni Beier

Oligochernes bachofeni Beier, 1937d: 308, fig. 10; Roewer, 1940: 331; Beier, 1948b: 446; Petrunkevitch, 1955: fig. 50(2); Schawaller, 1978: 4.

Type locality: Baltic Amber.
Distribution: Baltic Amber.

! Oligochernes wigandi (Menge)

Chelifer wigandi Menge, in C. L. Koch & Berendt, 1854: 96.
Chernes wigandi (Menge): Menge, 1855: 41–42, plate 5 fig. 16; Hagen, 1870: 268; Beier, 1932e: 278.
Oligochernes wigandi (Menge): Beier, 1937d: 308, fig. 9; Roewer, 1940: 331; Schawaller, 1978: 4.

Type locality: Baltic Amber.
Distribution: Baltic Amber.

Genus **Opsochernes** Beier

Opsochernes Beier, 1966c: 369–370; Beier, 1976f: 216.

Type species: *Opsochernes carbophilus* Beier, 1966c, by original designation.

Opsochernes carbophilus Beier

Opsochernes carbophilus Beier, 1966c: 370–371, fig. 6; Beier, 1967b: 294; Beier, 1976f: 216; Tenorio & Muchmore, 1982: 380.

Type locality: Perpendicular Point, 48 km N. of Greymouth, South Island, New Zealand.
Distribution: New Zealand.

Genus **Orochernes** Beier

Orochernes Beier, 1968b: 17; Murthy & Ananthakrishnan, 1977: 145.

Type species: *Orochernes nepalensis* Beier, 1968b, by orignal designation.

Orochernes nepalensis Beier

Orochernes nepalensis Beier, 1968b: 17–18, fig. 1; Murthy & Ananthakrishnan, 1977: 145.

Type locality: Yaral near Pangpoche, Nepal.
Distribution: Nepal.

Orochernes sibiricus Schawaller

Orochernes sibiricus Schawaller, 1986: 8–10, figs 19–27; Schawaller, 1989: 22, figs 52–56.

Type locality: Us River, Krasnoyarsk Bezirk, R.S.F.S.R., U.S.S.R.
Distribution: U.S.S.R. (R.S.F.S.R.).

Genus **Pachychernes** Beier

Pachychernes Beier, 1932e: 114; Beier, 1933a: 516.

Type species: *Chelifer (? Atemnus) subrobustus* Balzan, 1892, by original designation.

Pachychernes baileyi Feio

Pachychernes baileyi Feio, 1945: 26–30, figs 18–24; Mahnert, 1979d: 775–777, figs 117–118; Adis, 1981: 121, etc.; Adis & Mahnert, 1986: 311; Mahnert & Adis, 1986: 213; Adis, Junk & Penny, 1987: 488; Mahnert, Adis, & Bührnheim, 1987: figs 2, 11a.

Type locality: Barreiras, Bahia, Brazil.
Distribution: Brazil (Amazonas, Bahia).

! Pachychernes effossus Schawaller

Pachychernes effossus Schawaller, 1980b: 5–10, figs 1–9.
Pachychernes sp. *effossus* Schawaller?: Schawaller, 1980b: 12–14, figs 10–18.

Type locality: Dominican Amber.
Distribution: Dominican Amber.

Pachychernes gracilis (Ellingsen)

Chelifer (Atemnus) gracilis Ellingsen, 1902: 148–149.
Chelifer gracilis Ellingsen: With, 1908a: 259.
Pachychernes gracilis (Ellingsen): Beier, 1932e: 116; Roewer, 1937: 295.

Type locality: Naranjito, Ecuador.
Distribution: Ecuador.

Family Chernetidae

Pachychernes robustus (Balzan)

Chelifer robustus Balzan, 1888a: no pagination, figs; Balzan, 1890: 418–420, figs 6, 6a–b; With, 1908a: 259.
Chelifer (Atemnus) robustus Balzan: Balzan, 1892: 548.
Pachychernes robustus (Balzan): Beier, 1932e: 116; Roewer, 1936: fig. 70p; Roewer, 1937: 295.

Type locality: Mato Grosso (as Matto-grosso), Brazil.
Distribution: Brazil (Mato Grosso).

Pachychernes shelfordi Hoff

Pachychernes shelfordi Hoff, 1946h: 13–14, figs 1–3: Muchmore, 1975d: fig. 9.

Type locality: Mexico.
Distribution: Mexico.

Pachychernes subgracilis (With)

Chelifer subgracilis With, 1908a: 257–259, figs 11a–b.
Pachychernes subgracilis (With): Beier, 1932e: 116; Roewer, 1937: 295.

Type locality: Brazil.
Distribution: Brazil.

Pachychernes subrobustus (Balzan)

Chelifer (? Atemnus) subrobustus Balzan, 1892: 512–513, figs 3, 3a.
Chelifer subrobustus Balzan: With, 1908a: 260–261, figs 12a–b.
Chelifer (Atemnus) subrobustus Balzan: Ellingsen, 1910a: 360–362.
Pachychernes subrobustus (Balzan): Beier, 1932e: 115–116, figs 132–133; Roewer, 1937: 295; Mello-Leitão, 1939a: 120.

Type locality: Caracas, Venezuela.
Distribution: Argentina, Paraguay, Venezuela.

Genus **Paracanthicochernes** Beier

Paracanthicochernes Beier, 1966d: 144.

Type species: *Paracanthicochernes uniseriatus* Beier, 1966d, by original designation.

Paracanthicochernes uniseriatus Beier

Paracanthicochernes uniseriatus Beier, 1966d: 145–146, fig. 6; Beier, 1970g: 322.

Type locality: Mt Popomanaseu (as Popamanusiu), Solomon Islands.
Distribution: Solomon Islands.

Genus **Parachernes** J. C. Chamberlin

Parachernes J. C. Chamberlin, 1931b: 192–194; Beier, 1932e: 116–117; Beier, 1933a: 516–517; Hoff, 1949b: 456; Hoff, 1956c: 13–14; Muchmore & Alteri, 1974: 477–478; Murthy & Ananthakrishnan, 1977: 139.
Parachernes (Argentochernes) Beier, 1932e: 119; Beier, 1933a: 517 (synonymized by Hoff, 1956c: 14).

Type species: of *Parachernes*: *Parachernes ronnaii* J. C. Chamberlin, 1931b, by original designation.
of *Parachernes (Argentochernes)*: *Chelifer (Trachychernes) albomaculatus* Balzan, 1892, by present designation.

Subgenus **Parachernes (Parachernes)** J. C. Chamberlin

Parachernes Parachernes Beier: Beier, 1932e: 118; Beier, 1933a: 517.

Parachernes (Parachernes) adisi Mahnert

Parachernes adisi Mahnert, 1979d: 786–788, figs 140–145, 187i; Adis, 1981: 121, etc.; Mahnert, 1985c: 230; Mahnert & Adis, 1986: 213; Adis, Junk & Penny, 1987: 488.

Type locality: Manaus, Amazonas, Brazil.
Distribution: Brazil (Amazonas).

Parachernes (Parachernes) albomaculatus (Balzan)

Chelifer (Trachychernes) albomaculatus Balzan, 1892: 526–527, fig. 17; Ellingsen, 1910a: 369–370.
Chelifer albomaculatus Balzan: With, 1908a: 269–272, text-fig. 67, fig. 16a.
Parachernes (Argentochernes) albomaculatus (Balzan): Beier, 1932e: 119, fig. 135; Roewer, 1937: 295; Beier, 1954f: 135.

Type locality: Tovar, Venezuela.
Distribution: Colombia, Venezuela.

Parachernes (Parachernes) arcuodigitus Muchmore & Alteri

Parachernes arcuodigitus Muchmore & Alteri, 1974: 491–492, figs 9, 22–23, 30; Rowland & Reddell, 1976: 18.

Type locality: Victoria, Victoria County, Texas, U.S.A.
Distribution: U.S.A. (Texas).

Parachernes (Parachernes) argentatopunctatus (Ellingsen)

Chelifer (Trachychernes) argentatopunctatus Ellingsen, 1910a: 371–373.
Parachernes (Argentochernes) argentatopunctatus (Ellingsen): Beier, 1932e: 122–123, fig. 138; Roewer, 1937: 296; Beier, 1948b: 444, 473–474, fig. 3.
Parachernes argentatopunctatus (Ellingsen): Mahnert, 1979d: 788.

Type locality: Brazil.
Distribution: Brazil, Guyana.

Parachernes (Parachernes) argentinus Beier

Parachernes argentinus Beier, 1967e: 95–96, fig. 1.

Type locality: St Cruz Alta, La Soledad, Tucumán, Argentina.
Distribution: Argentina.

Parachernes (Parachernes) auster Beier

Parachernes auster Beier, 1964b: 349–350, fig. 23; Cekalovic, 1984: 21.

Type locality: Chepu, Chile.
Distribution: Chile.

Parachernes (Parachernes) bicolor (Balzan)

Chelifer (Trachychernes) bicolor Balzan, 1892: 524–526, figs 16, 16a.
Chelifer bicolor Balzan: With, 1908a: 267–269, figs 15a–e.
Parachernes (Argentochernes) bicolor (Balzan): Beier, 1932e: 120; Roewer, 1937: 295.

Type localities: Caracas, Venezuela; and Petare, Venezuela.
Distribution: Venezuela.

Parachernes (Parachernes) bisetus Muchmore & Alteri

Parachernes bisetus Muchmore & Alteri, 1974: 478–481, figs 4, 10–13; Zeh, 1987b: 1085.

Type locality: islet near Galden Key, Monroe County, Florida, U.S.A.
Distribution: U.S.A. (Florida).

Parachernes (Parachernes) bougainvillensis Beier

Parachernes bougainvillensis Beier, 1965g: 776, fig. 16; Beier, 1966d: 143; Tenorio & Muchmore, 1982: 379.

Type locality: Kokure, Bougainville, Papua New Guinea.
Distribution: Papua New Guinea.

Parachernes (Parachernes) chilensis Beier

Parachernes chilensis Beier, 1964b: 350–352, fig. 24; Cekalovic, 1984: 21.

Type locality: Paposo, Antofagasta, Chile.
Distribution: Chile.

Parachernes (Parachernes) cocophilus (E. Simon)

Chelifer cocophilus E. Simon, 1901: 79–80; E. Simon, 1903a: 124; With, 1906: 171–173, plate 3 figs 9a–b; Ellingsen, 1912b: 81.
Parachernes (Argentochernes) cocophilus (E. Simon): Beier, 1932e: 123–124, fig. 140; Roewer, 1937: 296.
Parachernes cocophilus (E. Simon): Beier, 1967g: 354; Beier, 1973b: 51.

Type locality: Kuala Aring, Kelantan, Malaysia.
Distribution: Equatorial Guinea, Malaysia, Sri Lanka, Vietnam.

Parachernes (Parachernes) confraternus (Banks)

Chelanops confraternus Banks, 1909a: 147.
Parachernes (Argentochernes) (?) confraternus (Banks): Beier, 1932e: 125; Roewer, 1937: 296.
Parachernes (Argentochernes) confraternus (Banks): Hoff, 1947b: 489–492, fig. 7.
Parachernes confraternus (Banks): Mahnert, 1979d: 788, fig. 187k.

Type locality: Poco Grande, Brazil.
Distribution: Brazil.

Parachernes (Parachernes) crassimanus (Balzan)

Chelifer crassimanus Balzan, 1887b: no pagination, figs; Balzan, 1890: 421–424, figs 8, 8a–c; Tullgren, 1907a: 71–72; With, 1908a: 272–274, text-figs 68–69, figs 17a–b.
Chelifer (Trachychernes) crassimanus Balzan: Balzan, 1892: 548; Ellingsen, 1905c: 13–15; Ellingsen, 1910a: 373.
Parachernes (Argentochernes) crassimanus (Balzan): Beier, 1932e: 121–122; Roewer, 1937: 296; Mello-Leitão, 1939a: 120; Mello-Leitão, 1939b: 616; Feio, 1945: 6; Mello-Leitão & Feio, 1949: 321, plate 4 fig. 2; Beier, 1954f: 135.

Type locality: Resistencia, Argentina; Asuncion, Paraguay; Mato Grosso (as Matogrosso), Brazil.
Distribution: Argentina, Brazil (Mato Grosso), Colombia, Ecuador, Paraguay, Peru, Venezuela.

Parachernes (Parachernes) darwiniensis Beier

Parachernes darwiniensis darwiniensis Beier, 1978c: 541–543, fig. 6.

Type locality: Isabela Island, Galapagos Islands.
Distribution: Galapagos Islands.

Parachernes (Parachernes) darwiniensis maculosus Beier

Parachernes darwiniensis maculosus Beier, 1978c: 543–544, fig. 7.

Type locality: Pinzón Island, Galapagos Islands.
Distribution: Galapagos Islands.

Parachernes (Parachernes) dissimilis Muchmore

Parachernes dissimilis Muchmore, 1980f: 227–229, figs 1–3.

Type locality: 4 miles W. of Quezaltepeque, El Salvador.
Distribution: El Salvador.

Parachernes (Parachernes) distinctus Beier

Parachernes (Argentochernes) distinctus Beier, 1933b: 96–97, fig. 7–8; Roewer, 1937: 296.

Type locality: Camarón, Veracruz, Mexico.
Distribution: Mexico.

Parachernes (Parachernes) dominicanus Beier

Parachernes dominicanus Beier, 1976d: 47–48, fig. 1.

Type locality: Boca Chica, Dominican Republic.
Distribution: Dominican Republic.

Parachernes (Parachernes) fallax Beier

Parachernes (Argentochernes) peruanus Beier: Beier, 1955m: 6–7 (misidentification, in part).
Parachernes fallax Beier, 1959e: 212–213, fig. 21; Weidner, 1959: 116.

Type locality: 10 km S. of Chiclayo, Peru.
Distribution: Peru.

Parachernes (Parachernes) floridae (Balzan)

Chelifer (Trachychernes) floridae Balzan, 1892: 524, fig. 15.
Chelanops floridae (Balzan): Banks, 1895a: 5; Banks, 1904a: 140; Ellingsen, 1909a: 219; Coolidge, 1908: 111; Ewing, 1911: 76, 79, fig. 10.
Neochernes floridae (Balzan): Beier, 1932e: 165; Roewer, 1937: 298.
Parachernes (?) *floridae* (Balzan): Hoff, 1958: 301; Muchmore & Alteri, 1974: 499.

Type locality: Florida, U.S.A.
Distribution: Nicaragua, U.S.A. (Florida).

Parachernes (Parachernes) franzi Beier

Parachernes franzi Beier, 1978c: 544–545, fig. 8.

Type locality: Pinta Island, Galapagos Islands.
Distribution: Galapagos Islands.

Parachernes (Parachernes) galapagensis Beier

Parachernes galapagensis Beier, 1977c: 106–108, fig. 7.

Type locality: Turtle Bay, Santa Cruz, Galapagos Islands.
Distribution: Galapagos Islands.

Parachernes (Parachernes) gracilimanus Mahnert

Parachernes gracilimanus Mahnert, 1986b: 813–816, figs 1–6.

Type locality: Cuyabeno, Napo, Ecuador.
Distribution: Ecuador.

Parachernes (Parachernes) indicus Beier

Parachernes indicus Beier, 1967g: 354–355, fig. 14; Beier, 1973b: 50–51; Murthy & Ananthakrishnan, 1977: 139.

Type locality: 6 km SW. of Sidapur, Mysore, India.
Distribution: India, Sri Lanka.

Parachernes (Parachernes) inpai Mahnert

Parachernes inpai Mahnert, 1979d: 781–784, figs 133–137, 187a; Adis, 1981: 121, etc.; Mahnert & Adis, 1986: 213; Adis, Junk & Penny, 1987: 488.

Type locality: Ilha Curari, Rio Solimoes, Manaus, Amazonas, Brazil.
Distribution: Brazil (Amazonas).

Parachernes (Parachernes) insuetus Beier

Parachernes (Argentochernes) insuetus Beier, 1933b: 98–99, fig. 10; Roewer, 1937: 296.

Type locality: Veracruz, Veracruz, Mexico.
Distribution: Mexico.

Parachernes (Parachernes) insularis Beier

Parachernes insularis Beier, 1935e: 640–641, fig. 3; Roewer, 1937: 296; Beier, 1965g: 776; Beier, 1966d: 143.

Type locality: Rendova Island, Solomon Islands.
Distribution: Solomon Islands.

Parachernes (Parachernes) kuscheli Beier

Parachernes (Argentochernes) kuscheli Beier, 1955b: 208–209, fig. 3; Cekalovic, 1984: 21.

Type locality: Plateau del Yunque, Masatierra, Juan Fernandez Islands.
Distribution: Juan Fernandez Islands.

Parachernes (Parachernes) latus (Banks)

Chelanops latus Banks, 1893: 64–65; Banks, 1895a: 5; Banks, 1904a: 140; Coolidge, 1908: 111.
Chelanops latimanus Banks, 1895a: 6; Banks, 1904a: 141; Coolidge, 1908: 111 (synonymized by Muchmore & Alteri, 1974: 492).
Neochernes latus (Banks): Beier, 1932e: 165; Roewer, 1937: 298.
Hesperochernes latimanus (Banks): Beier, 1932e: 176; Roewer, 1937: 302.
Parachernes latus (Banks): Hoff, 1947b: 478–482, figs 2–3.
Parachernes latimanus (Banks): Hoff, 1947b: 482–483, fig. 4.
Parachernes latus (Banks): Hoff, 1958: 29–30; Muchmore & Alteri, 1974: 492–495, figs 5, 14–15, 32; Rowland & Reddell, 1976: 18; Brach, 1979: 34; Zeh, 1987b: 1085.
Parachernes latimanus (Banks): Hoff, 1958: 30; Muchmore, 1971a: 86.

Type localities: of *Chelanops latus*: East Florida, U.S.A.
of *Chelanops latimanus*: Punta Gorda, Florida, U.S.A.
Distribution: U.S.A. (Florida, Texas).

Parachernes (Parachernes) leleupi Beier

Parachernes leleupi Beier, 1977c: 108–110, fig. 8.

Type locality: Baños, Ecuador.
Distribution: Ecuador.

Parachernes (Parachernes) litoralis Muchmore & Alteri

Parachernes litoralis Muchmore & Alteri, 1969: 131–133, figs 1–2, 5–6; Weygoldt, 1969a: 47–49, 115, 121, figs 44, 105; Muchmore & Alteri, 1974: 498.
Parachernes littoralis (sic) Muchmore & Alteri: Rey & McCoy, 1983: 499; Zeh, 1987b: 1085.

Type locality: Beaufort, Carteret County, North Carolina, U.S.A.
Distribution: U.S.A. (Florida, Georgia, North Carolina).

Parachernes (Parachernes) loeffleri Beier

Parachernes loeffleri Beier, 1959e: 209–210, fig. 19.

Type locality: Janganuco Lake, Peru.
Distribution: Peru.

Parachernes (Parachernes) meinertii (With)

Chelifer meinertii With, 1908a: 275–277, figs 18a–g.
Parachernes (Argentochernes) meinertii (With): Beier, 1932e: 122.
Parachernes (Argentochernes) meinerti (With): Roewer, 1937: 296.
Parachernes meinertii (With): Mahnert, 1979d: 784–785, figs 129–130, 187d–e;
    Adis, 1981: 119, etc.; Adis & Mahnert, 1986: 311; Mahnert & Adis, 1986: 213.

Type locality: Caracas, Venezuela.
Distribution: Brazil (Amazonas), Venezuela.

Parachernes (Parachernes) melanopygus Beier

Parachernes melanopygus Beier, 1959e: 213–214, fig. 22; Muchmore, 1977: 77;
    Mahnert, 1979d: 786, figs 131–132, 187f; Adis, 1981: 119, etc.; Adis & Mahnert,
    1986: 311; Mahnert & Adis, 1986: 213.

Type locality: Ecuador.
Distribution: Brazil (Amazonas), Guatemala, Ecuador, Mexico.

Parachernes (Parachernes) nevermanni Beier

Parachernes nevermanni Beier, 1976c: 2–3, fig. 1.

Type locality: Coronado, Costa Rica.
Distribution: Costa Rica.

Parachernes (Parachernes) niger Mahnert

Parachernes niger Mahnert, 1987: 413–415, figs 27–31.

Type locality: Nanchoc Quebrada, Cajamarca, Peru.
Distribution: Peru.

Parachernes (Parachernes) nigrimanus Beier

Parachernes (Argentochernes) nigrimanus Beier, 1948b: 453, 472–473, fig. 2; Weidner,
    1959: 116.
Parachernes nigrimanus Beier: Muchmore, 1971a: 84; Mahnert, 1987: 416.

Type locality: San José, Costa Rica.
Distribution: Costa Rica, Ecuador.

Parachernes (Parachernes) nitidimanus (Ellingsen)

Chelifer (Trachychernes) nitidimanus Ellingsen, 1905c: 11–12.
Chelifer nitidimanus Ellingsen: With, 1908a: 281–282, text-fig. 70, fig. 21a.
Parachernes (Argentochernes) nitidimanus (Ellingsen): Beier, 1932e: 120; Roewer,
    1937: 295.

Type locality: Pará, Brazil.
Distribution: Brazil (Pará), St Vincent.

Parachernes (Parachernes) nubilis Hoff

Parachernes nubilis Hoff, 1956c: 14–21, figs 1–5; Hoff, 1958: 31; Hoff, 1959b: 5,
    etc.; Muchmore & Alteri, 1974: 498, figs 2, 29; Zeh, 1987b: 1085.

Type locality: near Albuquerque, Bernalillo County, New Mexico, U.S.A.
Distribution: U.S.A. (Colorado, New Mexico, Oklahoma).

Parachernes (Parachernes) ovatus Mahnert

Parachernes ovatus Mahnert, 1979d: 791–794, figs 154–158; Adis, 1981: 121, etc.; Mahnert & Adis, 1986: 213; Adis, Junk & Penny, 1987: 488.

Type locality: Rio Solimoes, Manaus, Amazonas, Brazil.
Distribution: Brazil (Amazonas).

Parachernes (Parachernes) pallidus Beier

Parachernes pallidus Beier, 1959e: 211–212, fig. 20.

Type locality: Colombia.
Distribution: Colombia.

Parachernes (Parachernes) peruanus Beier

Parachernes (Argentochernes) peruanus Beier, 1955m: 6–7, figs 6–7 (in part; see *Parachernes (Parachernes) fallax* Beier); Weidner, 1959: 116.

Type locality: Sivia, Peru.
Distribution: Peru.

Parachernes (Parachernes) plumatus Beier

Parachernes (Argentochernes) plumatus Beier, 1933b: 97–98, fig. 9; Roewer, 1937: 296.

Type locality: Palapita, Nayarit, Mexico.
Distribution: Mexico.

Parachernes (Parachernes) plumosus (With)

Chelifer plumosus With, 1908a: 279–280, figs 20a–e.
Parachernes (Argentochernes) plumosus (With): Beier, 1932e: 120–121, fig. 136; Roewer, 1937: 295.
Parachernes plumosus (With): Beier, 1959e: 208; Mahnert, 1979d: 785–786, figs 138–139, 187b–c; Adis, 1981: 119, etc.; Adis & Mahnert, 1986: 311; Mahnert & Adis, 1986: 213.

Type locality: La Moka, Venezuela.
Distribution: Brazil (Amazonas), Ecuador, Venezuela.

Parachernes (Parachernes) pulchellus (Banks)

Chelanops pulchellus Banks, 1908: 41 (in part; see *Parachernes (Parachernes) rasilis* Muchmore & Alteri).
Parachernes (Argentochernes) pulchellus (Banks): Beier, 1932e: 119–120; Beier, 1933b: 96, fig. 6; Roewer, 1937: 295; Hoff, 1947b: 492–493, fig. 8.
Parachernes pulchellus (Banks): Hoff, 1958: 30; Muchmore & Alteri, 1974: 486, fig. 26; Rowland & Reddell, 1976: 17; Zeh, 1987b: 1085.
Parachernes corticis Muchmore & Alteri, 1969: 134–136, figs 3–4, 7–8; Weygoldt, 1969a: 49, 115, 121, fig. 105 (synomymized by Muchmore & Alteri, 1974: 486).

Type locality: of *Chernes pulchellus*: Esperanzo (as Esperanza) Ranch, Brownsville, Texas, U.S.A.
of *Parachernes corticis*: along Highway 58, Carteret County, North Carolina, U.S.A.
Distribution: Mexico, U.S.A. (Louisiana, North Carolina, Texas).

Parachernes (Parachernes) pulcher Mahnert

Parachernes pulcher Mahnert, 1979d: 779–781, figs 124–128; Mahnert & Adis, 1986: 213.

Type locality: Maruim Island, Rio Demini (as Demeni), Amazonas, Brazil.
Distribution: Brazil (Amazonas).

Parachernes (Parachernes) rasilis Muchmore & Alteri

Chernes pulchellus Banks, 1908: 41 (misidentification, in part).
Parachernes rasilis Muchmore & Alteri, 1974: 481–485, figs 1, 7, 18–19, 28; Rowland & Reddell, 1976: 17; Zeh, 1987b: 1085.

Type locality: Brownsville, Cameron County, Texas, U.S.A.
Distribution: U.S.A. (Texas).

Parachernes (Parachernes) robustus Hoff

Parachernes (Argentochernes) robustus Hoff, 1946c: 4–6, figs 3–5.

Type locality: 2 miles E. of Santo Domingo, San Luis Potosí, Mexico.
Distribution: Mexico.

Parachernes (Parachernes) ronnaii J. C. Chamberlin

Parachernes ronnaii J. C. Chamberlin, 1931b: 194–195, figs a–j; Roewer, 1936: fig. 79; Mahnert 1979d: 794.
Parachernes (Parachernes) ronnaii J. C. Chamberlin: Beier, 1932e: 118; Roewer, 1937: 295.

Type locality: Caxias, Rio Grande do Sul, Brazil.
Distribution: Brazil (Rio Grande do Sul).

Parachernes (Parachernes) rubidus (Ellingsen)

Chelifer (Trachychernes) rubidus Ellingsen, 1906: 252–254.
Chelifer rubidus Ellingsen: Ellingsen, 1912b: 82.
Parachernes (Argentochernes) rubidus (Ellingsen): Beier, 1932e: 124–125, fig. 141; Roewer, 1937: 296; Beier, 1954d: 135; Beier, 1972a: 11; Mahnert, 1978f: 129–130, figs 131–132; Heurtault, 1983: 12–14, figs 29–35.

Type localities: Rio Cassine, Guinea-Bissau; and Ribeira Palma, Sao Tomé (as Thomé).
Distribution: Cameroun, Central African Republic, Congo, Ecuador?, Gabon, Guinea-Bissau, Ivory Coast, Sao Tomé, Zaire.

Parachernes (Parachernes) sabulosus (Tullgren)

Chelifer (Parachernes) sabulosus Tullgren, 1909b: 411–412, fig. 1.
Parachernes (Argentochernes) sabulosus (Tullgren): Beier, 1932e: 123, fig. 139; J. C. Chamberlin, 1934b: 9; Roewer, 1937: 296; Weidner, 1959: 114.
Parachernes sabulosus (Tullgren): Beier, 1966a: 279; Harvey, 1981b: 247; Harvey, 1985b: 135.

Type locality: Brown Station, Dirk Hartog Island, Western Australia.
Distribution: Australia (Western Australia).

Parachernes (Parachernes) schlingeri Beier

Parachernes schlingeri Beier, 1959e: 208–209, fig. 18.

Type locality: Tingo María, Monson Valley, Peru.
Distribution: Peru.

Parachernes (Parachernes) semilacteus Beier

Parachernes semilacteus Beier, 1965g: 775–776, fig. 15.

Type locality: Finschhafen, Papua New Guinea.
Distribution: Papua New Guinea.

Parachernes (Parachernes) setiger Mahnert

Parachernes setiger Mahnert, 1979d: 789–791, figs 146–152, 187g–h; Adis, 1981: 119, etc.; Adis & Mahnert, 1986: 311; Mahnert & Adis, 1986: 213; Adis, Junk & Penny, 1987: 488.

Type locality: Taruma Mirim, Manaus, Amazonas, Brazil.
Distribution: Brazil (Amazonas).

Parachernes (Parachernes) setosus Beier

Parachernes (Argentochernes) setosus Beier, 1948b: 453, 475–476, fig. 4; Weidner, 1959: 116.
Parachernes setosus Beier: Muchmore, 1971a: 84.

Type locality: San José, Costa Rica.
Distribution: Costa Rica.

Parachernes (Parachernes) subrotundatus (Balzan)

Chelifer (Trachychernes) subrotundatus Balzan, 1892: 522–523, figs 14, 14a; Ellingsen, 1910a: 382.
Chelifer subrotundatus Balzan: With, 1908a: 277–278, figs 19a–c.
Parachernes (Argentochernes) subrotundatus (Balzan): Beier, 1932e: 121, fig. 137; Roewer, 1937: 296.

Type locality: Tovar, Venezuela.
Distribution: Paraguay, Venezuela.

Parachernes (Parachernes) subtilis Beier

Parachernes subtilis Beier, 1964b: 347–349, fig. 22; Cekalovic, 1984: 21.

Type locality: Batuco, Chile.
Distribution: Chile.

Parachernes (Parachernes) topali Beier

Parachernes topali Beier, 1964j: 491–493, fig. 3.

Type locality: Mt Piltriquitron, El Bolsón, Rio Negro, Argentina.
Distribution: Argentina.

Parachernes (Parachernes) tumimanus (Banks)

Chelanops tumimanus Banks, 1908: 40.
Dinocheirus tumimanus (Banks): Beier, 1932e: 138–139; Roewer, 1937: 302.
Parachernes (Argentochernes) tumimanus (Banks): Hoff, 1947b: 494–495, figs 9–10.
Parachernes tumimanus (Banks): Hoff, 1958: 30; Muchmore & Alteri, 1974: 498–499, figs 6, 24; Rowland & Reddell, 1976: 17.

Type locality: San Antonio, Texas, U.S.A.
Distribution: U.S.A. (Texas).

Parachernes (Parachernes) virginicus (Banks)

Chelanops virginica Banks, 1895a: 6; Coolidge, 1908: 111.
Chelanops pallidus Banks: Ewing, 1911: 76, 78, fig. 11 (misidentification).
Dinocheirus virginicus (Banks): Beier, 1932e: 139; Roewer, 1937: 302.
Parachernes (Argentochernes) virginica (sic) (Banks): Hoff, 1947b: 486–489, fig. 6.
Parachernes squarrosus Hoff, 1949b: 456–460, figs 32a–e; Hoff & Bolsterli, 1956: 169; Hoff, 1958: 30; Muchmore, 1971a: 89; Nelson, 1975: 285, figs 46–47 (synonymized by Muchmore & Alteri, 1974: 486).
Parachernes virginica (sic) (Banks): Hoff, 1958: 30; Muchmore & Alteri, 1974: 486–490; Zeh, 1987b: 1085.

Type localities: of *Chelanops virginica*: Fredericksburg (as Fredricksburg), Virginia, U.S.A.
of *Parachernes squarrosus*: Fowler, Illinois, U.S.A.
Distribution: U.S.A. (Alabama, Arkansas, Delaware, Georgia, Illinois, Indiana, Maryland, Michigan, Missouri, New York, North Carolina, South Carolina, Tennessee, Virginia).

Parachernes (Parachernes) withi Beier

Chelifer michaelseni E. Simon: With, 1908a: 282–284, figs 22a–c (misidentification).
Parachernes (Parachernes) michaelseni (E. Simon): Beier, 1932e: 118–119 (misidentification).
Parachernes withi Beier, 1967e: 96.

Type locality: Rio de Janeiro, Brazil.
Distribution: Brazil (Rio de Janeiro).

Subgenus Parachernes (Scapanochernes) Beier

Scapanochernes Beier, 1932e: 125; Beier, 1933a: 517–518.
Parachernes (Scapanochernes) Beier: Beier, 1976d: 48.

Type species: *Chelifer compressus* Tullgren, 1907a, by original designation.

Parachernes (Scapanochernes) compressus (Tullgren)

Chelifer compressus Tullgren, 1907a: 42–43, figs 8a–b.
Chelanops diversus Banks, 1909b: 304–305 (synonymized by Beier, 1976d: 48).
Scapanochernes compressus (Tullgren): Beier, 1932e: 125–126, fig. 126; Roewer, 1937: 296; Weidner, 1959: 113.
Neochernes diversus (Banks): Beier, 1932e: 167; Roewer, 1937: 299.
Parachernes (Argentochernes) diversus (Banks): Hoff, 1947b: 483–486, fig. 5.
Parachernes diversus (Banks): Hoff, 1958: 30; Muchmore & Alteri, 1974: 495–498, figs 3, 16–17, 25; Zeh, 1987b: 1085.
Parachernes (Scapanochernes) compressus (Tullgren): Beier, 1976d: 48.

Type localities: of *Chelifer compressus*: St Marc, Haiti.
of *Chelanops diversus*: Lake Worth, Florida, U.S.A.
Distribution: Dominican Republic, Haiti, U.S.A. (Florida).

Parachernes (Scapanochernes) cordimanus (Beier), new combination

Scapanochernes cordimanus Beier, 1953b: 20–21, figs 4a–b.

Type locality: Los Cobanos, Sonsonate, El Salvador.
Distribution: El Salvador.

Genus Parapilanus Beier

Parapilanus Beier, 1973b: 51–52.

Type species: *Parapilanus ceylonicus* Beier, 1973b, by original designation.

Parapilanus ceylonicus Beier

Parapilanus ceylonicus Beier, 1973b: 52–53, fig. 16.

Type locality: Kegalla, Sabaragamuwa Province, Sri Lanka.
Distribution: Sri Lanka.

Genus Paraustrochernes Beier

Paraustrochernes Beier, 1966a: 299–300.

Type species: *Paraustrochernes victorianus* Beier, 1966a, by original designation.

Paraustrochernes novaeguineensis Beier

Paraustrochernes novaeguineensis Beier, 1975b: 212–213, fig. 6.

Type locality: Wau, Morobe, Papua New Guinea.
Distribution: Papua New Guinea.

Paraustrochernes victorianus Beier

Paraustrochernes victorianus Beier, 1966a: 300–301, fig. 14; Harvey, 1981b: 247; Harvey, 1985b: 136.

Type locality: Morwell River Road, Victoria, Australia.
Distribution: Australia (Victoria).

Genus Parazaona Beier

Parazaona Beier, 1932e: 142–143; Beier, 1933a: 523.

Type species: *Chelifer bocki* Tullgren, 1907a, by original designation.

Parazaona bocki (Tullgren)

Chelifer bocki Tullgren, 1907a: 44–46, figs 9a–b.
Parazaona bocki (Tullgren): Beier, 1932e: 143–144, fig. 155; Roewer, 1937: 297; Beier, 1939b: 290; Beier, 1955m: 10; Weidner, 1959: 113.

Type locality: Oruro, Bolivia.
Distribution: Bolivia, Peru.

Parazaona bucheri Beier

Parazaona bucheri Beier, 1967e: 96–98, fig. 2.

Type locality: Raco, Tucumán, Argentina.
Distribution: Argentina.

Parazaona cavicola J. C. Chamberlin

Parazaona cavicola J. C. Chamberlin, 1938b: 118–121, figs 4a–j; Roewer, 1940: 347; Muchmore, 1977: 77.

Type locality: San Bulha Cave, Yucatan, Mexico.
Distribution: Mexico.

Parazaona chilensis Beier

Parazaona chilensis Beier, 1964b: 357–359, fig. 28; Cekalovic, 1984: 22.

Type locality: Paposo, Antofagasta, Chile.
Distribution: Chile.

Parazaona ellingsenii (With)

Chelifer ellingsenii With, 1908a: 287–289, text-fig. 71, figs 25a–e.
Parazaona ellingseni (With): Beier, 1932e: 144; Roewer, 1937: 297.

Type localities: Colombia (as New Granada); and Bogota, Colombia.
Distribution: Colombia.

Parazaona klapperichi Beier

Parazaona klapperichi Beier, 1976d: 51–52, fig. 4.

Type locality: Sabaneta, Dominican Republic.
Distribution: Dominican Republic.

Parazaona kuscheli Beier

Parazaona kuscheli Beier, 1964b: 356–357, fig. 27; Cekalovic, 1984: 23.

Type locality: Cuya, Rio Camarones, Arica, Tarapacá, Chile.
Distribution: Chile.

Parazaona morenensis (Tullgren)

Chelifer morenensis Tullgren, 1908c: 60–62, figs 4–6.
Parazaona morenensis (Tullgren): Beier, 1932e: 144; Roewer, 1937: 297; Mello-Leitão,
1939a: 121; Mello-Leitão, 1939b: 617.

Type locality: Puna de Jujuy, Moreno, Argentina.
Distribution: Argentina.

Parazaona nordenskjoeldi (Tullgren)

Chelifer nordenskjöldi Tullgren, 1908c: 63–64, figs 7–8.
Parazaona nordenskjöldi (Tullgren): Beier, 1932e: 145; Roewer, 1937: 297; Mello-
Leitão, 1939b: 617.
Parazaona nordenskioeldi (Tullgren): Mello-Leitão, 1939a: 121.

Type locality: Ultima Esperanza, Chile.
Distribution: Argentina, Chile.

Parazaona pycta Beier

Parazaona pycta Beier, 1964b: 354–356, fig. 26; Cekalovic, 1984: 23.

Type locality: 15 km NW. of Pica, Tarapacá, Chile.
Distribution: Chile.

Genus Petterchernes Heurtault

Petterchernes Heurtault, 1986b: 351–352.

Type species: *Petterchernes brasiliensis* Heurtault, 1986b, by original designation.

Petterchernes brasiliensis Heurtault

Petterchernes brasiliensis Heurtault, 1986b: 352–354, figs 1–9.

Type locality: Exu, Pernambuco, Brazil.
Distribution: Brazil.

Genus Phaulochernes Beier

Phaulochernes Beier, 1976f: 218–219.

Type species: *Phaulochernes maoricus* Beier, 1976f, by original designation.

Phaulochernes howdenensis Beier

Phaulochernes howdenensis Beier, 1976f: 220, fig. 19.

Type locality: Lake Howden, Hollyford Valley, South Island, New Zealand.
Distribution: New Zealand.

Phaulochernes jenkinsi Beier

Phaulochernes jenkinsi Beier, 1976f: 221, fig. 21.

Type locality: Shoe Island, North Island, New Zealand.
Distribution: New Zealand.

Phaulochernes kuscheli Beier

Phaulochernes kuscheli Beier, 1976f: 244.

Type locality: Wairaki Stream, Lynfield, Auckland, North Island, New Zealand.
Distribution: New Zealand.

Phaulochernes maoricus Beier

Phaulochernes maoricus Beier, 1976f: 219–220, fig. 18.

Type locality: Te Anau-Manapouri Road, South Island, New Zealand.
Distribution: New Zealand.

Phaulochernes townsendi Beier

Phaulochernes townsendi Beier, 1976f: 220–221, fig. 20.

Type locality: Mt Stokes, Marlborough, South Island, New Zealand.
Distribution: New Zealand.

Genus **Phymatochernes** Mahnert

Phymatochernes Mahnert, 1979d: 777.

Type species: *Phymatochernes crassimanus* Mahnert, 1979d, by original designation.

Phymatochernes crassimanus Mahnert

Phymatochernes crassimanus Mahnert, 1979d: 777–779, figs 119–123; Mahnert & Adis, 1986: 213.

Type locality: Reserva Ducke, Manaus, Amazonas, Brazil.
Distribution: Brazil (Amazonas).

Genus **Pilanus** Beier

Pilanus Beier, 1930i: 44; Beier, 1932e: 160; Beier, 1933a: 531.

Type species: *Pilanus pilatus* Beier, 1930i, by original designation.

Pilanus pilatus Beier

Pilanus pilatus Beier, 1930i: 44–47, figs 1–4; Beier, 1932e: 160–161, figs 172–173; Roewer, 1936: fig. 67; Roewer, 1937: 298; Beier, 1948b: 458.

Type locality: Dakar, Senegal.
Distribution: Senegal.

Pilanus pilifer Beier

Pilanus pilifer Beier, 1930i: 47, fig. 5; Beier, 1932e: 161, fig. 174; Roewer, 1937: 298; Beier, 1948b: 458.

Type locality: Nefasit, Eritrea, Ethiopia.
Distribution: Ethiopia.

Pilanus proximus Beier

Pilanus proximus Beier, 1955h: 7–9, fig. 1; Mahnert, 1982e: 710.

Type locality: Elmenteita, Kenya.
Distribution: Kenya.

Genus **Pselaphochernes** Beier

Pselaphochernes Beier, 1932e: 130; Beier, 1933a: 520; J. C. Chamberlin, 1935b: 481; Hoff, 1949b: 461; G. O. Evans & Browning, 1954: 20; Beier, 1963b: 253; Legg, 1972: 579–580; Murthy & Ananthakrishnan, 1977: 139; Legg, 1987: 181; Legg & Jones, 1988: 105.

Type species: *Chelifer scorpioides* Hermann, 1804, by original designation.

Pselaphochernes anachoreta (E. Simon)

Chelifer anachoreta E. Simon, 1878a: 151–152; Pavesi, 1884: 456–457; Ellingsen, 1908a: 415; Ellingsen, 1909a: 205–206; Krausse-Heldrungen, 1912: 65.

Chelifer (Trachychernes) anachoreta E. Simon: Ellingsen, 1910a: 371; Beier, 1930h: 95.
Chernes anachoreta (E. Simon): Beier, 1930f: 72.
Pselaphochernes anachoreta (E. Simon): Beier, 1932e: 132−133, fig. 148; Roewer, 1937: 296; Beier, 1948a: 190; Caporiacco, 1951a: 63; Vachon, 1951b: 199; Beier, 1955n: 108; Beier, 1961b: 34; Beier, 1963b: 257, fig. 258; Beier, 1963i: 286; Lazzeroni, 1969a: 336; Lazzeroni, 1969c: 241; Lazzeroni, 1970a: 208; Beier, 1971a: 364; Lagar, 1972b: 51; Gardini, 1980c: 117, 133; Callaini, 1983a: 155−156; Callaini, 1986c: 396, fig. 4f; Callaini, 1988b: 55−56.

Type locality: Daya, Algeria.
Distribution: Algeria, Balearic Islands, France, Iran, Italy, Morocco, Sardinia, Spain, Tunisia.

Pselaphochernes balcanicus Beier

Pselaphochernes balcanicus Beier, 1932e: 132, fig. 147; Beier, 1933a: 521, fig. 4; Roewer, 1937: 296; Beier, 1949: 10; Beier, 1963b: 255, fig. 257.

Type locality: Novoselo (now Mt Botev), Bulgaria.
Distribution: Bulgaria, Turkey.

Pselaphochernes balearicus Beier

Pselaphochernes balearicus Beier, 1961b: 33−34, fig. 4; Beier, 1963b: 257; Orghidan, Dumitresco & Georgesco, 1975: 31.

Type locality: Palma, Mallorca, Balearic Islands.
Distribution: Balearic Islands.

Pselaphochernes becki Hoff & Clawson

Pselaphochernes becki Hoff & Clawson, 1952; 24−27, figs 13−15; Hoff, 1958: 31; Muchmore, 1971a: 90; Zeh, 1987b: 1085.

Type locality: mouth of Provo River, Provo, Utah County, Utah, U.S.A.
Distribution: U.S.A. (Utah).

Pselaphochernes dubius (O.P.-Cambridge)

Chelifer parasita Hermann: Templeton, 1836: 14 (misidentification).
Chelifer dubius O.P.-Cambridge, 1892: 227−228, plate C fig. 19; Ellingsen, 1907a: 156−158; Godfrey, 1907: 162.
Chelifer n. sp.: Tullgren, 1899a: 176−177, plate 1 figs 11−12.
Chelifer tullgreni Strand, 1900: 102; Tullgren, 1906b: 215 (synonymized by Evans, 1903b: 250).
Chernes phaleratus (E. Simon): Evans, 1901b: 242 (misidentification).
Chelifer (Chernes) tullgreni Strand: Evans, 1903a: 120−121.
Chernes dubius (O.P.-Cambridge): Evans, 1903b: 249−250; Whyte & Whyte, 1907: 203−204; Godfrey, 1908: 157; Godfrey, 1909: 23−25; Kew, 1909b: 259; Falconer, 1916: 192.
Chelifer (Chernes) dubius (O.P.-Cambridge): Kew, 1909a: 249; Kew, 1911a: 44, fig. 5; Standen, 1912: 12; Kew, 1916b: 78−79; Kästner, 1928: 6.
Allochernes (Toxochernes) dubius (O.P.-Cambridge): Beier, 1932e: 152, fig. 162.
Allochernes dubius (O.P.-Cambridge): Roewer, 1937: 298; Lohmander, 1939b: 303−308, fig. 9; Vachon, 1947a: 85, 86; Beier, 1948b: 444, 458; Kaisila, 1949b: 83, map 8; G. O. Evans & Browning, 1954: 21; Beier, 1955n: 109; Beier, 1959f: 130; Beier, 1961a: 74; Pax & Paul, 1961: 73; Strebel, 1961: 107; Meinertz, 1962a: 398, map 6; Beier, 1963b: 262, fig. 262; Pedder, 1965: 108, 110; Helversen, 1966a: 140; Rafalski, 1967: 16−17; Gabbutt, 1969c: 231−232; Hammen, 1969: 21; M. Smith, 1969: 297−298, figs i−l; Beier, 1970a: 45; Gabbutt, 1970c: 11−12; Legg, 1970b: fig. 3(1); Gabbutt, 1972a: 37−40, figs 1c, 2c; Gabbutt, 1972b: 2−13; Howes,

1972a: 109; Legg, 1972: 580, fig. 4(2); Salmon, 1972: 66; Legg, 1973b: 430, fig. 2e; Beier, 1976a: 24; Klausen, 1975: 64; Legg, 1975a: 66; Legg, 1975d: fig. 2; Crocker, 1976: 9; Klausen & Totland, 1977: 101–108, plates 11–13, figs 1–2; Crocker, 1978: 9; Jones, 1978: 92; Goddard, 1976b: 296; Jones, 1979a: 200; Judson, 1979a: 62–63; Rundle, 1979: 48; Jones, 1980c: map 18; Pieper, 1981: 4; Cowden, 1983: 5; Mahnert, 1986d: 81–82; Judson, 1987: 369.

Allochernes (Chernes) dubius (O.P.-Cambridge): Cloudsley-Thompson, 1956a: 71.
Pselaphochernes dubius (O.P.-Cambridge): Legg, 1987: 181–182; Legg & Jones, 1988: 105–106, figs 25a, 25b(a–j).

Type localities: of *Chelifer dubius*: Glanvilles' Wooton, England, Great Britain.
of *Chelifer tullgreni*: Öland, Sweden.
Distribution: Balearic Islands, Belgium, Denmark, Finland, France, Great Britain, Madeira Islands, Netherlands, Norway, Poland, Spain, Sweden, West Germany.

Pselaphochernes hadzii Ćurčić

Pselaphochernes hadzii Ćurčić, 1972a: 79–92, figs 1–26; Ćurčić, 1974a: 26–27.

Type locality: Dragos-Sedlo, Mt Maglic, Yugoslavia.
Distribution: Yugoslavia.

Pselaphochernes iberomontanus Beier

Pselaphochernes iberomontanus Beier, 1959f: 129–130, fig. 10; Beier, 1963b: 256.

Type locality: Puerto La d'Home, Serra de Gerez near Lovíos (as Lovios), Spain.
Distribution: Spain.

Pselxphochernes indicus Beier

Pselaphochernes indicus Beier, 1974e: 1011–1012, fig. 7; Murthy & Ananthakrishnan, 1977: 140.

Type locality: 6 km E. of Coonoor, Nilgiri, Madras, India.
Distribution: India.

Pselaphochernes italicus Beier

Pselaphochernes italicus Beier, 1966j: 109–111, fig. 1; Lazzeroni, 1969a: 336; Lazzeroni, 1970a: 208.

Type locality: Scavi di Velia, Salerno, Italy.
Distribution: Italy.

Pselaphochernes lacertosus (L. Koch)

Chernes lacertosus L. Koch, 1873: 9–10; Daday, 1889c: 25; Daday, 1889d: 80; Daday, 1918: 2.
Chelifer lacertosus (L. Koch): E. Simon, 1879a: 37, 312, plate 18 fig. 13; E. Simon, 1898d: 20; E. Simon, 1900b: 593; Ellingsen, 1905d: 1; E. Simon, 1907: 550; Ellingsen, 1909a: 205; Krausse-Heldrungen, 1912: 65; Navás, 1925: 106.
Chernes (Chernes) lacertosus (L. Koch): Daday, 1888: 118, 173.
Chelifer (Trachychernes) lacertosus (L. Koch): Ellingsen, 1910a: 376; Nonidez, 1917: 16; Beier, 1929a: 344.
Pselaphochernes lacertosus (L. Koch): Beier 1932e: 133–134, fig. 149; Roewer, 1937: 297; Beier, 1939f: 198; Beier & Turk, 1952: 771; Beier, 1953a: 301; Beier, 1955n: 108–109; Beier, 1958b: 27; Beier, 1959f: 130; Beier, 1961b: 34; Beier, 1963b: 254, fig. 255; Lazzeroni, 1969a: 336; Lazzeroni, 1969c: 240; Beier, 1970a: 45; Lazzeroni, 1970a: 208; Ćurčić, 1974a: 27; Beier, 1976a: 24; Gardini, 1975: 9; Mahnert, 1975a: 183; Mahnert, 1975c: 195–196; Callaini, 1979b: 139; Callaini, 1982a: 17, 23, fig. 4b; Callaini, 1982b: 449; Callaini, 1983c: 304; Mahnert, 1985a: 17; Callaini, 1986c: 395, fig. 4h; Schawaller & Dashdamirov, 1988: 36, figs 63–67; Callaini, 1989: 145.

Family Chernetidae

Type locality: Corsica.
Distribution: Balearic Islands, Canary Islands, Corsica, Cyprus, France, Greece, Hungary, Italy, Malta, Sardinia, Sicily, Spain, Tunisia, U.S.S.R. (Azerbaijan), Yugoslavia.

Pselaphochernes litoralis Beier

Pselaphochernes litoralis Beier, 1956b: 61–63, fig. 4; R. Schuster, 1956: 245; Beier, 1963b: 253; Weygoldt, 1969a: 114; Lazzeroni, 1970b: 38; Ćurčić, 1974a: 27.
Pselaphochernes cf. *litoralis* Beier: Mahnert, 1985a: 17.

Type locality: Cap l'Abeille, Banyuls sur Mer, France.
Distribution: France, Italy, Tunisia, Yugoslavia.

Pselaphochernes litoralis siculus Beier

Pselaphochernes litoralis siculus Beier, 1963a: 259–260, fig. 3; Beier, 1963b: 245; Gardini, 1980c: 117, 131.
Pselaphochernes litoralis cfr. *siculus* Beier: Callaini, 1983c: 300–304, figs 2c–i.

Type locality: Grotta dell'Acqua, Canicattini, Siracusa, Sicily.
Distribution: Sardinia, Sicily.

Pselaphochernes parvus Hoff

Pselaphochernes parvus Hoff, 1945c: 38–43, figs 5–6; Rapp, 1946: 197; Hoff, 1949b: 461–464, figs 34a–d; Levi, 1953: fig. 6; Hoff & Bolsterli, 1956: 168; Hoff, 1958: 31; Manley, 1959: 8, fig. 9; Muchmore, 1971a: 90; Nelson, 1975: 285–286, figs 2, 48–49; Zeh, 1987b: 1085.

Type locality: Lake Wedington Wildlife Area, Washington County, Arkansas, U.S.A.
Distribution: U.S.A. (Arkansas, Illinois, Michigan, Tennessee, Wisconsin).

Pselaphochernes rybini Schawaller

Pselaphochernes rybini Schawaller, 1986: 3–4, figs 1–7.

Type locality: Gulcha-Koro, Kirghizia, U.S.S.R.
Distribution: U.S.S.R. (Kirghizia).

Pselaphochernes scorpioides (Hermann)

Chelifer scorpioides Hermann, 1804: 116–117, plate 5 figs l–m; Gervais, 1844: 78–79; Lucas, 1849: 276; H. J. Hansen, 1884: 546–548; Tullgren, 1899a: 172–173, plate 1 figs 7–8; Tullgren, 1906a: 203, fig. 2a, plate 4 fig. 5; Tullgren, 1906b: 215; With, 1906: 164–165; Ellingsen, 1907a: 159; Ellingsen, 1909a: 208, 215; Krausse-Heldrungen, 1912: 65; Ellingsen, 1913a: 453; Donisthorpe, 1927: 182–183, fig. 38; Hammen, 1949: 73, 74.
Chernes scorpioides (Hermann): L. Koch, 1873: 8–9; Donisthorpe, 1907: 255–256; Kew, 1909b: 259; Butterfield, 1908: 112; Donisthorpe, 1910: 84; Donisthorpe, 1915: 262; Graham-Smith, 1916: fig. 11; Daday, 1918: 1; Standen, 1922: 23–24; Walsh, 1924: 140; Beier, 1930f: 72; Beier, 1930e: 294; J.C. Chamberlin, 1931a: figs 10d, 52j–k; Nester, 1932: 98.
Chernes oblongus (Say): Stecker, 1874c: 313–314 (misidentification).
Chernes (Trachychernes) scorpioides (Hermann): Tömösváry, 1882b: 192, plate 1 figs 13–14; Beier, 1929a: 342; Beier, 1929d: 155; Beier, 1929e: 445.
Chernes affinis Tömösváry, 1884: 18–19, figs 10–11 (synonymized by Daday, 1888: 116).
Chernes (Chernes) scorpioides Hermann: Daday, 1888: 116, 170–171, plate 4 fig. 8.
Chernes scorpioides bertalanii Daday, 1889c: 26–27, figs 3, 7, 12, 16 (synonymized by Beier, 1929a: 342).

Family Chernetidae

Chelifer phaleratus E. Simon: O.P.-Cambridge, 1892: 228–229, plate C fig. 20 (misidentification).

Chernes (Trachychernes) minutus Ellingsen, 1897: 12–14 (synonymized by Ellingsen, 1907a: 159).

Chelifer minutus (Ellingsen): Tullgren, 1899a: 177.

Chelifer (Chernes) scorpioides Hermann: Ellingsen, 1903: 6–8; Kew, 1911a: 43–44, fig. 4; Standen, 1917: 27; Kästner, 1928: 6, fig. 17; Schenkel, 1928: 60, fig. 14.

Chelifer (Trachychernes) scorpioides Hermann: Lessert, 1911: 14–15, fig. 12.

Chelifer (Chelanops) scorpioides Hermann: Redikorzev, 1924b: 23, fig. 8.

Chelifer (Chelanops) rostombekovi Redikorzev, 1930: 100–103, figs 1–2 (synonymized by Schawaller, 1983b: 19).

Pselaphochernes scorpioides (Hermann): Beier, 1932e: 131, fig. 146; Tumšs, 1934: 15; J. C. Chamberlin, 1935b: 481; Caporiacco, 1936b: 329–330; Roewer, 1937: 296; Vachon, 1938a: fig. 56i; Beier, 1939c: 19; Lohmander, 1939b: 298–300; Beier, 1939e: 312; Vachon, 1940f: 2; Kaisila, 1947: 86; Beier, 1948b: 444, 457, 462; Caporiacco, 1948c: 242; Kaisila, 1949b: 82–83, map 7; Beier, 1952e: 4; Beier & Franz, 1954: 457; G. O. Evans & Browning, 1954: 20, fig. 24; Vachon, 1953: 572; Weidner, 1954b: 110; Beier, 1955d: 217; Beier, 1955n: 108; Beier, 1956h: 24; Hoff & Bolsterli, 1956: 168–169; George, 1957: 80; Vachon, 1957: fig. 8; Hoff, 1958: 31; Ressl & Beier, 1958: 2; Beier, 1959f: 129; Beier, 1961a: 74; Beier, 1961b: 33; Beier, 1961d: 92; Meinertz, 1962a: 394–395, map 5; Beier, 1963b: 255, fig. 256; Beier, 1963h: 156; Weygoldt, 1963: 447–451, figs 1–5; Beier, 1965f: 94; Ressl, 1965: 289; Helversen, 1966a: 140; Beier, 1966e: 164; Beier, 1966f: 345; Zangheri, 1966: 532; Beier, 1967f: 310–311; Rafalski, 1967: 16; Beron, 1968: 105; Beier, 1969b: 193; Hammen, 1969: 20; Lazzeroni, 1969a: 336; Lazzeroni, 1969b: 408; Lazzeroni, 1969c: 240–241; Weygoldt, 1969a: 27, 31, 72–74, 109, 121, figs 12, 20, 27–29, 65a–f, 66, 72, 86a–d, 106, 111; Lazzeroni, 1970a: 208; Ressl, 1970: 251; Muchmore, 1971a: 82, 93–94; Gabbutt, 1972a: 37–40, figs 1b, 2b; Gabbutt, 1972b: 2–13; Gabbutt, 1972c: 83–86, figs 2a–b; Howes, 1972a: 109; Kofler, 1972: 287; Legg, 1972: 580, figs 1(6d), 2(4b), 4(4); Beier, 1973c: 226; Legg, 1973b: 430, fig. 2h; Mahnert, 1974a: 90; Mahnert, 1974c: 384; Ressl, 1974: 29–30; Beier, 1976a: 24; Jones, 1975a: 88; Klausen, 1975: 64; Crocker, 1976: 10; Klausen & Totland, 1977: 101–108, plates 14–16, figs 1–2; Mahnert, 1977d: 97; Crocker, 1978: 9; Jones, 1978: 91, 93, 94, 95; Mahnert, 1978e: 298; Callaini, 1979c: 350; Jones, 1979a: 201; Judson, 1979a: 62; Rundle, 1979: 48; Callaini, 1980c: 117, 131; Jones, 1980c: map 17; Pieper, 1981: 4; Schawaller, 1981b: 48; Cuthbertson, 1982: 4; Callaini, 1983c: 304–305; Mahnert, 1983d: 362; Schawaller, 1983b: 19–20, fig. 54; Jędryczkowski, 1985: 79; Lippold, 1985: 40; Callaini, 1986c: 395–396, fig. 4g; Jędryczkowski, 1987b: 142, map 7; Judson, 1987: 369; Zeh, 1987b: 1085; Callaini, 1988b: 56; Legg & Jones, 1988: 108–110, figs 26a, 26b(a–j); Schawaller & Dashdamirov, 1988: 36, figs 67; Schawaller, 1989: 18.

Pselaphochernes rostombekovi (Redikorzev): Beier, 1932e: 132; Roewer, 1937: 296; Beier, 1963b: 254–255.

Chernes (Pselaphochernes) scorpioides (Hermann): Cooreman, 1946b: 3; Leleup, 1947: 322.

Pselaphochernes macrochaetus Redikorzev, 1949: 653, figs 16–17 (synonymized by Schawaller, 1989: 18).

Not *Chelifer scorpioides* Hermann: Théis, 1832: 73–75, plate 3 figs 2, 2a–b (misidentification; see *Chernes cimicoides* (Fabricius)).

Not *Chernes scorpioides* (Hermann): Daday, 1897: 477, figs 10–13 (possible misidentification; see *Verrucachernes oca* J. C. Chamberlin).

Type localities: of *Chelifer scorpioides*: not stated, presumably near Strasbourg, France. of *Chernes affinis*: Kérkira (as Corfu), Greece.

628

of *Chernes scorpioides bertalanii*: Visz, Hungary.
of *Chernes (Trachychernes) minutus*: Fredrikstad (as Fredriksstad), Norway.
of *Chelifer (Chelanops) rostombekovi*: Kjassim-abadi, Geokchay (as Geoktshaj), Azerbaijan, U.S.S.R.
of *Pselaphochernes macrochaetus*: Urgench, Uzbekistan, U.S.S.R.
Distribution: Algeria, Austria, Azores, Balearic Islands, Belgium, Bulgaria, Corsica, Crete, Denmark, East Germany, Finland, France, Great Britain, Greece, Hungary, Iran, Ireland, Israel, Italy, Lebanon, Morocco, Netherlands, Norway, Poland, Romania, Sardinia, Sicily, Spain, Sweden, Switzerland, Syria, Turkey, U.S.A. (Connecticut, Indiana, Kentucky, Massachusetts, New York), U.S.S.R. (Armenia, Azerbaijan, Latvia, R.S.F.S.R., Ukraine, Uzbekistan), West Germany, Yugoslavia.

Pselaphochernes setiger (L. Koch)

Chelifer setiger L. Koch, 1881: 670–671; Navás, 1918: 88; Navás, 1925: 106.
Chelifer (Trachychernes) setiger (L. Koch): Nonidez, 1917: 13–16, fig. 3.
Pselaphochernes setiger (L. Koch): Beier, 1932e: 133; Roewer, 1937: 297; Beier, 1939f: 197; Beier, 1963b: 257; Estany, 1977b: 31.

Type locality: Marina de Blummajor, Mallorca, Balearic Islands.
Distribution: Balaeric Islands, Spain.

Pselaphochernes turcicus Beier

Pselaphochernes turcicus Beier, 1967f: 311–312, fig. 4.

Type locality: Gaybi near Eregli, Turkey.
Distribution: Turkey.

Genus Pseudopilanus Beier

Pseudopilanus Beier, 1957c: 455.

Type species: *Pseudopilanus fernandezianus* Beier, 1957c, by original designation.

Pseudopilanus chilensis Beier

Pseudopilanus chilensis Beier, 1964b: 362–364, fig. 30; Cekalovic, 1984: 23.

Type locality: Frutillar, Llanquihua, Chile.
Distribution: Chile.

Pseudopilanus crassifemoratus Mahnert

Pseudopilanus crassifemoratus Mahnert, 1985c: 230–231, figs 43–47; Mahnert & Adis, 1986: 213.
Pseudopilanus aff. *crassifemoratus* Mahnert: Mahnert, 1985c: 231; Mahnert & Adis, 1986: 213.

Type locality: Reserva Florestal Ducke, 26 km on Manaus-Itacoatiara Highway, Amazonas, Brazil.
Distribution: Brazil (Amazonas).

Pseudopilanus echinatus (Ellingsen)

Chelifer (Chernes) echinatus Ellingsen, 1904: 2–4.
Chelifer echinatus Ellingsen: With, 1908a: 262.
Rhopalochernes echinatus (Ellingsen): Beier, 1932e: 142; Roewer, 1937: 297; Mello-Leitão, 1939a: 120; Mello-Leitão, 1939b: 617, Feio, 1945: 7.
Pseudopilanus echinatus (Ellingsen): Beier, 1959e: 214–215, fig. 23; Beier, 1962h: 135; Beier, 1964b: 361–362; Cekalovic, 1984: 23.

Type locality: Buenos Aires, Argentina.
Distribution: Argentina, Brazil (Rio de Janeiro), Chile.

Pseudopilanus fernandezianus Beier

Pseudopilanus fernandezianus Beier, 1957c: 455–457, fig. 2; Cekalovic, 1984: 24.

Type locality: Inocentes Bajos, Masafuera, Juan Fernandez Islands.
Distribution: Juan Fernandez Islands.

Pseudopilanus inermis Beier

Pseudopilanus inermis Beier, 1977c: 110–112, fig. 11.

Type locality: San Domingo, Ecuador.
Distribution: Ecuador.

Pseudopilanus kuscheli Beier

Pseudopilanus kuscheli Beier, 1964b: 359–361, fig. 29; Cekalovic, 1984: 24.

Type locality: Concepción, Chile.
Distribution: Chile.

Pseudopilanus topali Beier

Pseudopilanus topali Beier, 1964j: 494–497, fig. 5; Cekalovic, 1984: 23.

Type locality: Mt Piltriquitron, El Bolsón, Rio Negro, Argentina.
Distribution: Argentina.

Genus Reischekia Beier

Reischekia Beier, 1948c: 547–548; Beier, 1976f: 239.

Type species: *Reischekia coracoides* Beier, 1948c, by original designation.

Reischekia coracoides Beier

Reischekia coracoides Beier, 1948c: 549–550, figs 12–14; Beier, 1967b: 297; Beier, 1976f: 239.

Type locality: Hendes Ferrys, central Westland, South Island, New Zealand.
Distribution: New Zealand.

Reischekia exigua Beier

Reischekia exigua exigua Beier, 1976f: 239–240, fig. 39.

Type locality: head of Lake Alabaster, Pyke Valley, Fiordland, South Island, New Zealand.
Distribution: New Zealand.

Reischekia exigua sentiens Beier

Reischekia exigua sentiens Beier, 1976f: 240.

Type locality: Waipoua State Forest, North Island, New Zealand.
Distribution: New Zealand.

Reischekia papuana Beier

Reischekia papuana Beier, 1965g: 789, fig. 27.

Type locality: Maffin Bay, Irian Jaya, Indonesia.
Distribution: Indonesia (Irian Jaya).

Genus Rhinochernes Beier

Rhinochernes Beier, 1955m: 7–8.

Type species: *Rhinochernes granulatus* Beier, 1955m, by original designation.

Rhinochernes ashmolei Muchmore

Rhinochernes ashmolei Muchmore, 1982b: 87–88, figs 1–4.

Type locality: near Los Toyas Caves, Cordillera el Condor, Ecuador.
Distribution: Ecuador.

Rhinochernes granulatus Beier

Rhinochernes granulatus Beier, 1955m: 8–9, figs 8–9; Weidner, 1959: 116.

Type locality: Aina, Peru.
Distribution: Peru.

Genus Rhopalochernes Beier

Rhopalochernes Beier, 1932e: 140; Beier, 1933a: 523.

Type species: *Chelifer ohausi* Tullgren, 1907a, by original designation.

Rhopalochernes antillarum (With)

Chelifer antillarum With, 1908a: 262–264, figs 13a–b.
Rhopalochernes antillarum (With): Beier, 1932e: 142; Roewer, 1937: 297.

Type locality: St Vincent.
Distribution: St Vincent.

Rhopalochernes beckeri Beier

Rhopalochernes beckeri Beier, 1953b: 22–23, fig. 6.

Type locality: Quezaltenango, Guatemala.
Distribution: Guatemala.

Rhopalochernes foliosus (Balzan)

Chelifer foliosus Balzan, 1887b: no pagination, figs; Balzan, 1890: 427–428, figs 12, 12a–c; With, 1908a: 262.
Chelifer (Trachychernes) foliosus Balzan: Balzan, 1892: 548.
Rhopalochernes foliosus (Balzan): Beier, 1932e: 142; Roewer, 1937: 297; Mello-Leitão, 1939a: 120; Mello-Leitão, 1939b: 617; Feio, 1945: 7.

Type localities: Resistencia, Argentina; Asuncion, Paraguay.
Distribution: Argentina, Brazil (Mato Grosso, Rio de Janeiro), Paraguay.

Rhopalochernes germainii (Balzan)

Chelifer germainii Balzan, 1887b: no pagination, figs; Balzan, 1890: 424–426, figs 10, 10a–c; With, 1908a: 262.
Chelifer (Chelifer) germaini Balzan: Balzan, 1892: 548; Ellingsen, 1905e: 324.
Rhopalochernes germaini (Balzan): Beier, 1932e: 141; Roewer, 1937: 297.
Not *Chelifer germainii* Balzan: Ellingsen, 1905b: 1–3 (misidentification; see *Rhopalochernes titschacki* Beier).

Type locality: Mato Grosso (as Matto-grosso), Brazil.
Distribution: Argentina, Brazil (Mato Grosso).

Rhopalochernes insulanus Beier

Rhopalochernes insulanus Beier, 1978c: 545–547, fig. 9.

Type locality: Volcan Sierra Negra, Isabela Island, Galapagos Islands.
Distribution: Galapagos Islands.

Rhopalochernes ohausi (Tullgren)

Chelifer ohausi Tullgren, 1907a: 72–73.
Rhopalochernes ohausi (Tullgren): Beier, 1932e: 141, fig. 154; Roewer, 1937: 297; Weidner, 1959: 114.

Type locality: Petropolis, Rio de Janeiro, Brazil (not Ecuador as stated by Tullgren).
Distribution: Brazil (Rio de Janeiro).

Rhopalochernes titschacki Beier

Chelifer germainii Ellingsen, 1905b: 1–3 (misidentification).
Rhopalochernes titschacki Beier, 1955m: 9–10, fig. 10; Weidner, 1959: 116.

Type locality: Sivia, Peru.
Distribution: Ecuador, Peru.

Genus Semeiochernes Beier

Semeiochernes Beier, 1932e: 180–181; Beier, 1933a: 541.

Type species: *Semeiochernes militaris* Beier, 1932e, by original designation.

Semeiochernes armiger (Balzan)

Chelifer (Trachychernes) armiger Balzan, 1892: 527–528, fig. 18.
Chelifer armiger Balzan: With, 1908a: 261.
Chernes armiger (Balzan): J. C. Chamberlin, 1931a: fig. 30e.
Semeiochernes armiger (Balzan): Beier, 1932e: 181–182; Roewer, 1937: 302; Beier, 1955m: 10.

Type locality: Pebas, Peru.
Distribution: Brazil (Amazonas), Peru.

Semeiochernes extraordinarius Beier

Semeiochernes extraordinarius Beier, 1954f: 138–139, fig. 5.

Type locality: Rancho Grande, Venezuela.
Distribution: Venezuela.

Semeiochernes militaris Beier

Mirochernes dentatus (Banks): Beier, 1930a: 217–218, fig. 14 (misidentification).
Semeiochernes militaris Beier, 1932e: 181, fig. 190; Beier, 1933a: 543; Roewer, 1936: fig. 27b; Roewer, 1937: 302; Vachon, 1949: fig. 204e; Mahnert, 1987: 411–413, figs 20–26.
Chelanops costaricensis Beier, 1932e: 179, fig. 188; Beier, 1933a: 539–540, fig. 11; Roewer, 1937: 302 (synonymized by Beier, 1954f: 139).

Type localities: of *Semeiochernes militaris*: Juan Vinas, Costa Rica.
of *Chelanops costaricensis*: Rio Barbilla, Costa Rica.
Distribution: Brazil (Pará), Costa Rica.

Genus Smeringochernes Beier

Smeringochernes Beier, 1957d: 41; Beier, 1976f: 224.

Type species: *Smeringochernes yapensis* Beier, 1957d, by original designation.

Subgenus Smeringochernes (Smeringochernes)

Smeringochernes (Smeringochernes) Beier, 1957d: 41.

Smeringochernes (Smeringochernes) aequatorialis (Daday)

Chelifer aequatorialis Daday, 1897: 475–476, figs 5–6; With, 1906: 164.

Ochrochernes (?) *aequatorialis* (Daday): Beier, 1932e: 128; J. C. Chamberlin, 1934b: 9; Roewer, 1937: 296.
Smeringochernes aequatorialis (Daday): Beier, 1965g: 783–784, fig. 23; Beier, 1967c: 322.

Type locality: Madang (as Friedrich-Wilhelmshafen), Papua New Guinea.
Distribution: Indonesia (Irian Jaya), Papua New Guinea.

Smeringochernes (Smeringochernes) greensladeae Beier

Smeringochernes greensladeae Beier, 1966d: 146–147, fig. 7; Beier, 1970g: 324.

Type locality: Jonapau, Guadalcanal, Solomon Islands.
Distribution: Solomon Islands.

Smeringochernes (Smeringochernes) guamensis Beier

Smeringochernes (Smeringochernes) guamensis Beier, 1957d: 43–45, figs 26a–c; Beier, 1965g: 784.

Type locality: Oca Point, Guam, Mariana Islands.
Distribution: Caroline Islands, Mariana Islands, Papua New Guinea.

Smeringochernes (Smeringochernes) navigator (J. C. Chamberlin)

Rhopalochernes navigator J. C. Chamberlin, 1938a: 271–274, figs 3a–i; Roewer, 1940: 346; Tenorio & Muchmore, 1982: 381.
Smeringochernes navigator (J. C. Chamberlin): Beier, 1964f: 597.

Type locality: Tutuila, Samoa.
Distribution: Samoa.

Smeringochernes (Smeringochernes) novaeguineae Beier

Smeringochernes (Smeringochernes) novaeguineae Beier, 1965g: 782, fig. 22.

Type locality: E. of Nicakamp, Biak, Irian Jaya, Indonesia.
Distribution: Indonesia (Irian Jaya).

Smeringochernes (Smeringochernes) pauperculus Beier

Smeringochernes pauperculus Beier, 1970g: 325–326, fig. 6.

Type locality: S. of Kuzi, Kolombangara, Solomon Islands.
Distribution: Solomon Islands.

Smeringochernes (Smeringochernes) plurisetosus Beier

Smeringochernes plurisetosus Beier, 1966d: 148–149, fig. 8; Beier, 1970g: 326.

Type locality: Mt Jonapau, Guadalcanal, Solomon Islands.
Distribution: Solomon Islands.

Smeringochernes (Smeringochernes) salomonensis Beier

Smeringochernes (Smeringochernes) salomonensis Beier, 1964f: 596–597, fig. 3.
Smeringochernes salomonensis Beier: Beier, 1965g: 784; Beier, 1966d: 150, fig. 9; Petersen, 1968: 120; Beier, 1970g: 326.

Type locality: Mt Austen, Guadalcanal, Solomon Islands.
Distribution: Solomon Islands.

Smeringochernes (Smeringochernes) yapensis Beier

Smeringochernes (Smeringochernes) yapensis Beier, 1957d: 42–43, figs 24a–d, 25c; Tenorio & Muchmore, 1982: 381.

Type locality: Ruul District, Yap, Caroline Islands.
Distribution: Caroline Islands.

Smeringochernes (Smeringochernes) zealandicus Beier

Smeringochernes zealandicus Beier, 1976f: 224–225.

Type locality: Butterfly Bay, Tauranga Bay, North Island, New Zealand.
Distribution: New Zealand.

Subgenus Smeringochernes (Gressittochernes) Beier

Smeringochernes (Gressittochernes) Beier, 1957d: 45.

Type species: *Smeringochernes (Gressittochernes) carolinensis* Beier, 1957d, by original designation.

Smeringochernes (Gressittochernes) carolinensis Beier

Smeringochernes (Gressittochernes) carolinensis Beier, 1957d: 45–46, figs 17a–b.

Type locality: SE. Nanponmal, Ponape, Caroline Islands.
Distribution: Caroline Islands.

Genus Sphenochernes Turk

Chelifer (Tullgrenia) Mello-Leitão, 1925: 232 (junior homonym of *Tullgrenia* van der Groot, 1912).
Sphenochernes Turk, 1953: 951–952.
Syndeipnochernes Beier, 1970b: 51–52 (synonymized by Mahnert, 1985c: 230).

Type species: of *Chelifer (Tullgrenia)*: *Chelifer bruchi* Mello-Leitão, 1925, by subsequent designation of Mahnert, 1985c.
of *Sphenochernes*: *Sphenochernes schulzi* Turk, 1953, by original designation.
of *Syndeipnochernes*: *Syndeipnochernes camponoti* Beier, 1970b, by original deisgnation.

Sphenochernes bruchi (Mello-Leitão)

Chelifer (Tullgrenia) bruchi Mello-Leitão, 1925: 228–232, figs 1, 1a–b.
Genus ? *bruchi* Mello-Leitão: Beier, 1932e: 277; Roewer, 1937: 317.
Chelifer bruchi Mello-Leitão: Roewer, 1937: fig. 188.
Rhopalochernes bruchi (Mello-Leitão): Mello-Leitão, 1939a: 121; Mello-Leitão, 1939b: 617; Beier, 1948b: 444, 457.
Chelifer bruchi Mello-Leitão: Vachon, 1940f: 2, 3.
Syndeipnochernes bruchi (Mello-Leitão): Beier, 1970b: 53.
Spenochernes bruchi (Mello-Leitão): Mahnert, 1985c: 230.

Type locality: La Plata, Argentina.
Distribution: Argentina.

Sphenochernes camponoti (Beier)

Syndeipnochernes camponoti Beier, 1970b: 52–54, fig. 1.
Sphenochernes camponoti (Beier): Mahnert, 1985c: 230.

Type locality: Barueri, Sao Paulo, Brazil.
Distribution: Brazil (Sao Paulo).

Sphenochernes schulzi Turk

Sphenochernes schulzi Turk, 1953: 952–954, figs 1–8; Weygoldt, 1969a: 118.

Type locality: Castelar, 12 km from Buenos Aires, Argentina.
Distribution: Argentina.

Genus Stigmachernes Beier

Stigmachernes Beier, 1957c: 457.

Type species: *Stigmachernes skottsbergi* Beier, 1957c, by original designation.

Stigmachernes skottsbergi Beier

Stigmachernes skottsbergi Beier, 1957c: 457–460, fig. 3; Cekalovic, 1984: 25.

Type locality: La Correspondencia, Masafuera, Juan Fernandez Islands.
Distribution: Juan Fernandez Islands.

Genus Sundochernes Beier

Sundochernes Beier, 1932e: 162; Beier, 1933a: 531; Beier, 1976f: 225.

Type species: *Chelifer modiglianii* Ellingsen, 1911a, by original designation.

Sundochernes australiensis Beier

Sundochernes australiensis Beier, 1954b: 16–18, fig. 7; Harvey, 1981b: 247; Harvey, 1985b: 135.

Type locality: Denmark, near mouth of Denmark River, Western Australia, Australia.
Distribution: Australia (Western Australia).

Sundochernes brasiliensis Beier

Sundochernes (?) *brasiliensis* Beier, 1974d: 903–904, fig. 4.

Type locality: Nova Teutonia, Santa Catarina, Brazil.
Distribution: Brazil (Santa Catarina).

Sundochernes dewae Beier

Sundochernes dewae Beier, 1967a: 200–202, fig. 1; Harvey, 1981b: 247; Harvey, 1985b: 135.

Type locality: Brewarrina, New South Wales, Australia.
Distribution: Australia (New South Wales).

Sundochernes dubius Beier

Sundochernes dubius Beier, 1954b: 18–19, fig. 8; Harvey, 1981b: 247; Harvey, 1985b: 135.

Type locality: Augusta, Western Australia, Australia.
Distribution: Australia (Western Australia).

Sundochernes grayi Beier

Sundochernes grayi Beier, 1976f: 225, fig. 25; Harvey, 1985b: 135.

Type locality: near Old Settlement, Lord Howe Island.
Distribution: Lord Howe Island.

Sundochernes gressitti Beier

Sundochernes gressitti Beier, 1957d: 46–47, figs 28a–b.

Type locality: Ngaremeskang, Babelthuap, Palau Islands, Caroline Islands.
Distribution: Caroline Islands.

Sundochernes guanophilus Beier

Sundochernes guanophilus Beier, 1967a: 202–203, fig. 3; Harvey, 1981b: 247; Harvey, 1985b: 136.

Type locality: Fig Tree Cave, Wombeyan, New South Wales, Australia.
Distribution: Australia (New South Wales).

Sundochernes malayanus Beier

Sundochernes malayanus Beier, 1963d: 511–512, fig. 2; Tenorio & Muchmore, 1982: 380.

Type locality: Rantau Panjang, 5 miles N. of Klang, Selangor, Malaysia.
Distribution: Malaysia.

Sundochernes modiglianii (Ellingsen)

Chelifer modiglianii Ellingsen, 1911a: 37–40.
Sundochernes modiglianii (Ellingsen): Beier, 1932e: 162, fig. 175; Roewer, 1937: 298; Beier, 1967g: 357.

Type locality: Si-Rambé, Sumatra, Indonesia.
Distribution: Indonesia (Sumatra), Malaysia.

Sundochernes novaeguineae Beier

Sundochernes novaeguineae Beier, 1965g: 779–780, fig. 19; Beier 1982: 44; Tenorio & Muchmore, 1982: 381.

Type locality: Mt Giluwe, Papua New Guinea.
Distribution: Papua New Guinea.

Sundochernes queenslandicus Beier

Sundochernes queenslandicus Beier, 1975b: 207–208, fig. 3; Harvey, 1981b: 247; Harvey, 1985b: 136.

Type locality: Marburg, Queensland, Australia.
Distribution: Australia (Queensland).

Genus Sundowithius Beier

Sundowithius Beier, 1932d: 57; Beier, 1932e: 210; Beier, 1964f: 595.

Type species: *Chelifer sumatranus* Thorell, 1889, by original designation.

Sundowithius sumatranus (Thorell)

Chelifer sumatranus Thorell, 1889: 599–601; With, 1906: 163–164; Ellingsen, 1911b: 142.
Sundowithius sumatranus (Thorell): Beier, 1932d: 57; Beier, 1932e: 210, fig. 216; Beier, 1955e: 46; Beier, 1982: 44.

Type locality: Mt Singalang, Sumatra, Indonesia.
Distribution: Burma, Indonesia (Sumatra), Malaysia, Papua New Guinea.

Genus Systellochernes Beier

Systellochernes Beier, 1964g: 118–119; Beier, 1976f: 217.

Type species: *Systellochernes zonatus* Beier, 1964g, by original designation.

Systellochernes alacki Beier

Systellochernes alacki Beier, 1976f: 217, fig. 16.

Type locality: Devil's Thumb, Wangapeka, South Island, New Zealand.
Distribution: New Zealand.

Systellochernes zonatus Beier

Systellochernes zonatus Beier, 1964g: 119–120, fig. 2; Beier, 1964h: 629; Beier, 1976f: 217; Tenorio & Muchmore, 1982: 382.

Type locality: Beeman Hill, Campbell Island, New Zealand.
Distribution: New Zealand.

Genus **Tejachernes** Hoff

Tejachernes Hoff, 1957: 83–84.

Type species: *Dinocheirus stercoreus* Turk, 1949, by original designation.

Tejachernes stercoreus (Turk)

Dinocheirus stercoreus Turk, 1949: 121–126, figs 1–6; Hoff, 1958: 28.
Tejachernes stercoreus (Turk): Hoff, 1957: 84–88, figs 1–5; Muchmore, 1971a: 89; Muchmore, 1975d: fig. 8; Rowland & Reddell, 1976: 17.

Type locality: Bracken Cave, Comal County, Texas, U.S.A.
Distribution: U.S.A. (Texas).

Genus **Teratochernes** Beier

Teratochernes Beier, 1957d: 37.

Type species: *Teratochernes mirus* Beier, 1957d, by original designation.

Teratochernes mirus Beier

Teratochernes mirus Beier, 1957d: 38–39, figs 22a–b, 25a.

Type locality: Mt Nanalaud, Ponape, Caroline Islands.
Distribution: Caroline Islands.

Genus **Thalassochernes** Beier

Thalassochernes Beier, 1940a: 182; Beier, 1976f: 214.

Type species: *Chelifer pallipes* White, 1849, by original designation (based upon misidentified type species; an application is currently before the International Commission on Zoological Nomenclature).

Thalassochernes kermadecensis Beier

Thalassochernes kermadecensis Beier, 1976f: 215–216, fig. 15.

Type locality: Bell's Flat, Raoul Island, Kermadec Island, New Zealand.
Distribution: New Zealand.

Thalassochernes taierensis (With)

Chelifer taierensis With, 1907: 55–57, figs 6–8.
Haplochernes taierensis (With): Beier, 1932e: 111; J. C. Chamberlin, 1934b: 9; Roewer, 1937: 295; Beier, 1966c: 369; Beier, 1967b: 293.
Thalassochernes taierensis (With): Beier, 1976f: 215.
Chelifer (Trachychernes) pallipes White: Ellingsen, 1910a: 376–377 (misidentification; see *Philomaoria pallipes* (White)).
Haplochernes pallipes (White): Beier, 1932e: 111; J. C. Chamberlin, 1934b: 9; Roewer, 1937: 295 (misidentifications; see *Philomaoria pallipes* (White)).
Thalassochernes pallipes (White): Beier, 1940a: 182; Beier, 1948b: 457; Beier, 1948c: 537–540, figs 6–7; Beier, 1966c: 369; Beier, 1967b: 293; Beier, 1969e: 413 (misidentifications; see *Philomaoria pallipes* (White)).

Type locality: Taieri, South Island, New Zealand.
Distribution: New Zealand.

Genus **Thapsinochernes** Beier

Thapsinochernes Beier, 1957d: 47.

Type species: *Thapsinochernes flavus* Beier, 1957d, by original designation.

Thapsinochernes flavus Beier

Thapsinochernes flavus flavus Beier, 1957d: 48–49, figs 25b, 29a–c.

Type locality: SE. of Asan, Guam, Mariana Islands.
Distribution: Mariana Islands.

Thapsinochernes flavus major Beier

Thapsinochernes flavus major Beier, 1957d: 49, fig. 30.

Type locality: Limestone Ridge, Koror, Palau Islands, Caroline Islands.
Distribution: Caroline Islands.

Genus Troglochernes Beier

Troglochernes Beier, 1969a: 185.

Type species: *Troglochernes imitans* Beier, 1969a, by original designation.

Troglochernes imitans Beier

Troglochernes imitans Beier, 1969a: 185–187, fig. 11; Richards, 1970: 19, 24, 25, 27, 28, 30, 43; Beier, 1975b: 203; Harvey, 1981b: 247; Harvey, 1985b: 136.

Type locality: Dingo Cave, Nullarbor Plain, Western Australia, Australia.
Distribution: Australia (Western Australia).

Genus Tychochernes Hoff

Tychochernes Hoff, 1956c: 21–22.

Type species: *Tychochernes inflatus* Hoff, 1956c, by original designation.

Tychochernes inflatus Hoff

Tychochernes inflatus Hoff, 1956c: 22–25, figs 6–7; Hoff, 1958: 31; Hoff, 1959b: 5, etc.; Muchmore, 1971a: 91.

Type locality: Sandia Mountains, E. of Albequerque, Bernalillo County, New Mexico, U.S.A.
Distribution: U.S.A. (New Mexico).

Genus Verrucachernes J. C. Chamberlin

Verrucachernes J. C. Chamberlin, 1947b: 312–313.
Microchernes Beier, 1951a: 91 (synonymized by Beier, 1957d: 40).

Type species: of *Verrucachernes*: *Verrucachernes oca* J. C. Chamberlin, 1947d, by original designation.
of *Microchernes*: *Microchernes orientalis* Beier, 1951a, by original designation.

Verrucachernes congicus Beier

Verrucachernes congicus Beier, 1959d: 44–45, fig. 21.

Type locality: Kisanga, Elisabethville (now Kubumbashi), Zaire.
Distribution: Zaire.

Verrucachernes montigenus Beier

Verrucachernes montigenus Beier, 1965g: 779, fig. 18; Tenorio & Muchmore, 1982: 381.

Type locality: Kakebe, Mt Otto, Papua New Guinea.
Distribution: Papua New Guinea.

Verrucachernes oca J. C. Chamberlin

Chernes scorpioides (Hermann): Daday, 1897: 477–478, figs 10–13 (possible misidentification).
Verrucachernes oca J. C. Chamberlin, 1947b: 313–316, figs 3a–i; Beier, 1957d: 39–41, figs 23a–d; Beier, 1965g: 777; Beier, 1966d: 147; Beier, 1970g: 324; Harvey, 1988b: 348–351, figs 120–127.
Microchernes orientalis Beier, 1951a: 92–93, fig. 27; Beier, 1973b: 51 (synonymized by Harvey, 1988b: 348).
Microchernes insularis Beier, 1953g: 84–86, fig. 4 (synonymized by Harvey, 1988b: 348).
Verrucachernes orientalis (Beier): Beier, 1975d: 40.
Verrucachernes insularis (Beier): Beier, 1975d: 40.
Micratemnus (sic) *orientalis* Beier: Beier, 1976e: 100.

Type localities: of *Verrucachernes oca*: Oca Point, Guam, Mariana Islands.
of *Microchernes orientalis*: Ha Tien (as Hatien), Vietnam.
of *Microchernes insularis*: Mau Marru, Sumba, Indonesia.
Distribution: Bhutan, Cambodia, Caroline Islands, Indonesia (Krakatau Islands, Sumba), Mariana Islands, Marshall Islands, Papua New Guinea, Solomon Islands, Sri Lanka, Vietnam.

Verrucachernes spinosus Beier

Verrucachernes spinosus Beier, 1979c: 104–105, fig. 2.

Type locality: Adiopodoumé, Ivory Coast.
Distribution: Ivory Coast.

Verrucachernes sublaevis Beier

Verrucachernes sublaevis Beier, 1965g: 777–778, fig. 17.

Type locality: Maffin Bay, Irian Jaya, Indonesia.
Distribution: Indonesia (Irian Jaya).

Genus Wyochernes Hoff

Wyochernes Hoff, 1949a: 41–42.

Type species: *Wyochernes hutsoni* Hoff, 1949b, by original designation.

Wyochernes hutsoni Hoff

Wyochernes hutsoni Hoff, 1949a: 42–48, figs 1–8; Hoff, 1958: 29.

Type locality: Medicine Bow National Forest, Snowy Range, Albany County, Wyoming, U.S.A.
Distribution: U.S.A. (Wyoming).

Genus Xenochernes Feio

Xenochernes Feio, 1945: 37.

Type species: *Xenochernes caxinguba* Feio, 1945, by original designation.

Xenochernes caxinguba Feio

Xenochernes caxinguba Feio, 1945: 37–40, figs 28–32.

Type locality: Pirapora, Minas Gerais, Brazil.
Distribution: Brazil (Minas Gerais).

Genus Zaona J. C. Chamberlin

Zaona J. C. Chamberlin, 1925a: 331; Beier, 1932e: 167; Beier, 1933a: 533.

Type species: *Chelifer biseriatum* Banks, 1895a, by original designation.

Zaona biseriatum (Banks)

Chelifer biseriatum Banks, 1895a: 3; Banks, 1904a: 140; Coolidge, 1908: 109; Pratt, 1927: 409.
Zaona biseriatum (Banks): J. C. Chamberlin, 1925a: 332, figs l–p; J. C. Chamberlin, 1931a: figs 30f, 39b–c; Beier, 1932e: 168; Roewer, 1937: 299; Hoff, 1958: 29; Muchmore, 1975d: fig. 10.

Type locality: Lake Poinsett, Florida, U.S.A.
Distribution: U.S.A. (Florida).

Family WITHIIDAE J. C. Chamberlin

Withiinae J. C. Chamberlin, 1931c: 290; Beier, 1932e: 192; Roewer, 1937: 305; Morikawa, 1960: 148; Murthy & Ananthakrishnan, 1977: 145–146.
Philomaorini (sic) J. C. Chamberlin, 1931c: 291.
Cacodemonini (sic) J. C. Chamberlin, 1931c: 292.
Cacodemoniini J. C. Chamberlin: Beier, 1932e: 192–193; Roewer, 1937: 306.
Withiini J. C. Chamberlin, 1931c: 292; Beier, 1932e: 194–195; Roewer, 1937: 306; Morikawa, 1960: 148–149; Murthy & Ananthakrishnan, 1977: 146.
Philomaoriini J. C. Chamberlin: Beier, 1932e: 225; Roewer, 1937: 310.
Protowithiini Beier, 1955b: 215–216.
Withiidae J. C. Chamberlin: Muchmore, 1982a: 101–102; Harvey, 1985b: 154.

Genus Afrowithius J. C. Chamberlin

Afrowithius J. C. Chamberlin, 1931c: 293; Beier, 1932e: 221.

Type species: *Chelifer paradoxus* Ellingsen, 1912b, by original designation.

Afrowithius paradoxus (Ellingsen)

Chelifer paradoxus Ellingsen, 1912b: 98–99.
Afrowithius paradoxus (Ellingsen): J. C. Chamberlin, 1931c: 293; Beier, 1932e: 221; Roewer, 1937: 309.

Type locality: Ntaba Kandoda, near King William's Town, Cape Province, South Africa.
Distribution: South Africa.

Genus Aisthetowithius Beier

Aisthetowithius Beier, 1967d: 83–84.

Type species: *Aisthetowithius rossi* Beier, 1967d, by original designation.

Aisthetowithius rossi Beier

Aisthetowithius rossi Beier, 1967d: 84, fig. 7; Mahnert, 1988a: 49–50, figs 12–14.

Type locality: Kaimosi Mission, 27 miles NE. of Kisumu, Kenya.
Distribution: Kenya, Tanzania.

Genus Balanowithius Beier

Balanowithius Beier, 1959e: 223.

Type species: *Balanowithius egregius* Beier, 1959e, by original designation.

Balanowithius egregius Beier

Balanowithius egregius Beier, 1959e: 223–224, fig. 31.

Type locality: Pichilingue, Los Rios, Ecuador.
Distribution: Ecuador.

Balanowithius weyrauchi Beier

Balanowithius weyrauchi Beier, 1959e: 224–226, fig. 32; Beier, 1974d: 899; Mahnert, 1975b: fig. 6e.

Type locality: La Balsa, Rio Canchis, Peru.
Distribution: Brazil (Santa Catarina), Peru.

Genus **Beierowithius** Mahnert

Oligowithius Beier, 1937d: 309 (junior homonym of *Oligowithius* Beier, 1936a).
Beierowithius Mahnert, 1979d: 802 (replacement name for *Oligowithius* Beier, 1937d).

Type species: *Chelifer sieboldtii* Menge, 1855, by original designation.

! **Beierowithius sieboldtii** (Menge)

Obisium sieboldtii Menge, in C.L. Koch & Berendt, 1854: 97.
Chelifer sieboldii (sic) Menge, 1855: 35–36, plate 5 fig. 10; Hagen, 1870: 267.
Chelifer sieboldtii (Menge): Beier, 1932e: 278.
Oligowithius sieboldii (sic) (Menge): Beier, 1937d: 309, fig. 11; Beier, 1955g: 53; Schawaller, 1979: 4.
Oligowithius sieboldtii (Menge): Roewer, 1940: 331.
Beierowithius sieboldii (sic) (Menge): Mahnert, 1979e: 802.

Type locality: Baltic Amber.
Distribution: Baltic Amber.

Genus **Cacodemonius** J. C. Chamberlin

Cacodemonius J. C. Chamberlin, 1931c: 292; Beier, 1932e: 193.

Type species: *Withius cactorum* J. C. Chamberlin, 1923c, by original designation.

Cacodemonius cactorum (J. C. Chamberlin)

Withius cactorum J. C. Chamberlin, 1923c: 377, plate 1 fig. 7, plate 2 fig. 14; J. C. Chamberlin, 1931a: fig. 10f.
Cacodemonius cactorum (J. C. Chamberlin): J. C. Chamberlin, 1931c: 292; Beier, 1932e: 193; Roewer, 1937: 306.

Type locality: San Pedro Martir Island, Gulf of California, Mexico.
Distribution: Mexico.

Cacodemonius pusillus Beier

Cacodemonius pusillus Beier, 1953b: 24, figs 7a–b.

Type locality: Laguna de Jocotal, San Miguel, El Salvador.
Distribution: El Salvador.

Cacodemonius quartus Hoff

Cacodemonius quartus Hoff, 1946c: 18–21, figs 23–26; Muchmore, 1977: 77.

Type locality: La Zacualpa, Chiapas, Mexico.
Distribution: Mexico.

Cacodemonius satanas (With)

Chelifer satanas With, 1908a: 245–247, text-fig. 66, figs 8a–c.
Cacodemonius satanas (With): J. C. Chamberlin, 1931c: 292; Beier, 1932e: 194; Roewer, 1937: 306.

Type locality: Los Trinchéras, Venezuela.
Distribution: Venezuela.

Cacodemonius segmentidentatus (Balzan)

Chelifer segmentidentatus Balzan, 1887b: no pagination, figs; Balzan, 1890: 428–430, figs 13, 13a–b; With, 1908a: 242–245, text-fig. 65, figs 7a–j.
Chelifer (Trachychernes) segmentidentatus Balzan: Balzan, 1892: 548; Ellingsen, 1905c: 12–13.
Cacodemonius segmentidentatus (Balzan): J. C. Chamberlin, 1931c: 292; Beier, 1932e: 193–194; Roewer, 1937: 306; Mello-Leitão, 1939a: 121; Mello-Leitão, 1939b: 617; Beier, 1959e: 215, fig. 24.
Cacodemonius serratidentatus (sic) (Balzan): Beier, 1976d: 55.

Type localities: Resistencia, Argentina; Rio Apa, Paraguay; and Mato Grosso, Brazil.
Distribution: Argentina, Brazil (Mato Grosso), Dominican Republic, Ecuador, Paraguay, Venezuela.

Cacodemonius zilchi Beier

Cacodemonius zilchi Beier, 1953b: 23–24, figs 10a–c.

Type locality: Laguna de Zapotitan, La Libertad, El Salvador.
Distribution: El Salvador.

Genus **Cryptowithius** Beier

Cryptowithius Beier, 1967d: 85–86.

Type species: *Cryptowithius inconspicuus* Beier, 1967d, by original designation.

Cryptowithius inconspicuus Beier

Cryptowithius inconspicuus Beier, 1967d: 86–87, fig. 8; Mahnert, 1988a: 45.

Type locality: Simu Beach, Kwale, Kenya.
Distribution: Kenya.

Genus **Cyrtowithius** Beier

Cyrtowithius Beier, 1955l: 314–316.

Type species: *Cyrtowithius capensis* Beier, 1955l, by original designation.

Cyrtowithius capensis Beier

Chelifer tumuliferus Tullgren: Ellingsen, 1912b: 85, 100–101 (misidentification).
Cyrtowithius capensis Beier, 1955l: 316–317, figs 30–31.

Type locality: Yzerfontain, Cape Province, South Africa.
Distribution: South Africa.

Cyrtowithius tumuliferus (Tullgren)

Chelifer tumuliferus Tullgren, 1908b: 283–284, fig. 1.
Caffrowithius tumuliferus (Tullgren): Beier, 1932d: 62; Beier, 1932e: 223; Roewer, 1937: 310.
Cyrtowithius tumuliferus (Tullgren): Beier, 1955l: 316; Beier, 1966k: 467.
Not *Chelifer tumuliferus* Tullgren: Ellingsen, 1912b: 85, 100–101 (misidentification; see *Cyrtowithius capensis* Beier).

Type locality: Port Nolloth, Cape Province, South Africa.
Distribution: Namibia, South Africa.

Genus **Dolichowithius** J. C. Chamberlin

Dolichowithius J. C. Chamberlin, 1931c: 293; Beier, 1932e: 216; Hoff, 1945h: 1.

Type species: *Chelifer longichelifer* Balzan, 1887b, by original designation.

Subgenus **Dolichowithius (Dolichowithius)** J. C. Chamberlin

Dolichowithius (Dolichowithius) argentinus Beier

Dolichowithius argentinus Beier, 1959e: 226–227, fig. 33.

Type locality: Salta, Argentina.
Distribution: Argentina.

Dolichowithius (Dolichowithius) brasiliensis (Beier)

Withius brasiliensis Beier, 1930a: 218–219, figs 15a–b.
Dolichowithius brasiliensis (Beier): Beier, 1932e: 219–220, fig 227; Roewer, 1937: 309.

Type locality: Rio Grande do Sul, Brazil.
Distribution: Brazil (Rio Grande do Sul).

Dolichowithius (Dolichowithius) canestrinii (Balzan)

Chelifer canestrinii Balzan, 1887b: no pagination, figs; Balzan, 1890: 430–431, figs 14, 14a–c; Ellingsen, 1905e: 324; With, 1908a: 236–238, figs 5a–d.
Chelifer (Chelifer) canestrinii Balzan: Balzan, 1892: 532, figs 22, 22a; Ellingsen, 1905c: 16–17; Ellingsen, 1910a: 385 (in part; see *Parawithius iunctus* Beier).
Withius canestrinii (Balzan): J. C. Chamberlin, 1931a: 168.
Dolichowithius canestrinii (Balzan): J. C. Chamberlin, 1931c: 293; Beier, 1932e: 220, fig. 228; Roewer, 1937: 309; Mello-Leitão, 1939a: 122; Mello-Leitão, 1939b: 618; Feio, 1945: 8; Caporiacco, 1948a: 618; Beier, 1977c: 112.

Type locality: Resistencia, Argentina.
Distribution: Argentina, Brazil (Bahia, Pará), Ecuador, Guyana, St Thomas, Venezuela, Virgin Islands.

Dolichowithius (Dolichowithius) centralis Beier

Dolichowithius centralis Beier, 1953b: 25, fig. 8.

Type locality: Laguna de Zapotitan, La Libertad, El Salvador.
Distribution: El Salvador.

Dolichowithius (Dolichowithius) emigrans (Tullgren)

Chelifer emigrans Tullgren, 1907a: 35–36, fig. 5.
Dolichowithius emigrans (Tullgren): Beier, 1932e: 220–221, fig. 229; Roewer, 1937: 309; Weidner, 1959: 114; Mahnert & Adis, 1986: 214.
Dolichowithius (Dolichowithius) emigrans (Tullgren): Mahnert, 1979d: 794–795, figs 168–169, 187i.

Type locality: Manaus (as Manaos), Brazil.
Distribution: Brazil (Amazonas).

Dolichowithius (Dolichowithius) extensus Beier

Chelifer longichelifer Balzan: With, 1908a: 238–242, figs 6a–d (misidentification).
Dolichowithius extensus Beier, 1932d: 59–60; Beier, 1932e: 216–217; Roewer, 1937: 309; Beier, 1959e: 226.

Type locality: Venezuela.
Distribution: Peru, Venezuela.

Dolichowithius (Dolichowithius) granulosus Hoff

Dolichowithius granulosus Hoff, 1945h: 3–5, figs 3–5.

Type locality: Essequibo River, Onoro, Guyana.
Distribution: Guyana.

Dolichowithius (Dolichowithius) intermedius Mahnert

Dolichowithius (Dolichowithius) intermedius Mahnert, 1979d: 795–796, figs 160–162, 187m.
Dolichowithius intermedius Mahnert: Adis, 1981: 119, etc.; Adis & Mahnert, 1986: 312; Mahnert & Adis, 1986: 214; Adis, Junk & Penny, 1987: 489.

Type locality: Taruma Mirim, Manaus, Amazonas, Brazil.
Distribution: Brazil (Amazonas).

Dolichowithius (Dolichowithius) longichelifer (Balzan)

Chelifer longichelifer Balzan, 1887b: no pagination, figs; Balzan, 1890: 433–434, figs 16, 16a–c.
Chelifer (Chelifer) longichelifer Balzan: Balzan, 1892: 534, fig. 26; Ellingsen, 1905e: 324.
Withius longichelifer (Balzan): J. C. Chamberlin, 1931a: 168, 170, fig. 30m.
Dolichowithius longichelifer (Balzan): J. C. Chamberlin, 1931c: 293; Beier, 1932e: 218, fig. 224; Roewer, 1937: 309; Feio, 1945: 7.
Not *Chelifer longichelifer* Balzan: Tullgren, 1907a: 35 (misidentification; see *Dolichowithius (Dolichowithius) vicinus* Beier); With, 1908a: 238–242, figs 6a–d (misidentification; see *Dolichowithius (Dolichowithius) extensus* Beier).
Not *Chelifer (Chelifer) longichelifer* Balzan: Ellingsen, 1910a: 386 (misidentification; see *Dolichowithius (Dolichowithius) modicus* Beier).

Type locality: Resistencia, Argentina; Asuncion and Rio Apa, Paraguay; Mato Grosso (as Matto-grosso), Brazil.
Distribution: Argentina, Brazil (Distrito Federal, Mato Grosso), Ecuador, Paraguay, Venezuela.

Dolichowithius (Dolichowithius) mediofasciatus Mahnert

Dolichowithius (Dolichowithius) mediofasciatus Mahnert, 1979d: 798–801, figs 170–175.
Dolichowithius mediofasciatus Mahnert: Adis, 1981: 119, etc.; Adis & Mahnert, 1986: 312; Mahnert & Adis, 1986: 214; Adis, Junk & Penny, 1987: 489; Mahnert, Adis, & Bührnheim, 1987: figs 11a, 12c.

Type locality: Rio Solimoes, Manaus, Amazonas, Brazil.
Distribution: Brazil (Amazonas).

Dolichowithius (Dolichowithius) minutus Mahnert

Dolichowithius (Dolichowithius) minutus Mahnert, 1979d: 796–798, figs 163–167.
Dolichowithius minutus Mahnert: Adis, 1981: 119, etc.; Adis & Mahnert, 1986: 312; Mahnert & Adis, 1986: 214; Adis, Junk & Penny, 1987: 489.

Type locality: Taruma Mirim, Manaus, Amazonas, Brazil.
Distribution: Brazil (Amazonas).

Dolichowithius (Dolichowithius) modicus Beier

Chelifer (Chelifer) longichelifer Balzan: Ellingsen, 1910a: 386 (misidentification).
Dolichowithius modicus Beier, 1932d: 61; Beier, 1932e: 218–219, fig. 225; Roewer, 1937: 309.

Type locality: Paraguay.
Distribution: Paraguay.

Dolichowithius (Dolichowithius) simplex Beier

Chelifer (Chelifer) simoni Balzan: Ellingsen, 1910a: 387 (misidentification, in part).

Dolichowithius simplex Beier, 1932d: 61; Beier, 1932e: 219, fig. 226; Roewer, 1937: 309; Beier, 1976d: 55.

Type locality: Puerto Rico.
Distribution: Dominican Republic, Puerto Rico.

Dolichowithius (Dolichowithius) solitarius Hoff

Dolichowithius solitarius Hoff, 1945h: 1–3, figs 1–2.

Type locality: Costa Rica.
Distribution: Costa Rica.

Dolichowithius (Dolichowithius) vicinus Beier

Chelifer longichelifer Balzan: Tullgren, 1907a: 35 (misidentification).
Dolichowithius vicinus Beier, 1932d: 60–61; Beier, 1932e: 217, fig. 223; Roewer, 1937: 309.

Type locality: Estancia Postillon, Puerto Max, Paraguay.
Distribution: Paraguay.

Subgenus Dolichowithius (Oligowithius) Beier

Dolichowithius (Oligowithius) Beier, 1936a: 447.

Type species: *Dolichowithius (Oligowithius) abnormis* Beier, 1936a, by original designation.

Dolichowithius (Oligowithius) abnormis Beier

Dolichowithius (Oligowithius) abnormis Beier, 1936a: 446–447, fig. 4.
Dolichowithius abnormis Beier: Roewer, 1940: 347.

Type locality: Estado Trujillo, La Ceiba, Venezuela.
Distribution: Venezuela.

Genus Ectromachernes Beier

Ectromachernes Beier, 1944: 197–198.

Type species: *Ectromachernes mirabilis* Beier, 1944, by original designation.

Ectromachernes elegans Beier

Ectromachernes elegans Beier, 1964k: 77–78, fig. 33.

Type locality: Coldspring near Grahamstown, Cape Province, South Africa.
Distribution: South Africa.

Ectromachernes lamottei Vachon

Ectromachernes lamottei Vachon, 1952a: 38–40, figs 34–41.

Type locality: Mount To, Nimba Mountains, Guinea.
Distribution: Guinea.

Ectromachernes mirabilis Beier

Ectromachernes mirabilis Beier, 1944: 199–200, fig. 17; Mahnert, 1988a: 41–42, figs 1–4.

Type locality: Djem-Djem-Wald, Ethiopia.
Distribution: Ethiopia, Kenya.

Ectromachernes rhodesiacus Beier

Ectromachernes rhodesiacus Beier, 1964k: 76–77, fig. 32.

Type locality: Dombashawa, Zimbabwe.
Distribution: Zimbabwe.

Genus **Hyperwithius** Beier

Hyperwithius Beier, 1951a: 99–100.

Type species: *Sundowithius annamensis* Redikorzev, 1938, by original designation.

Hyperwithius annamensis (Redikorzev)

Sundowithius annamensis Redikorzev, 1938: 101–103, figs 29–31; Roewer, 1940: 347.
Hyperwithius annamensis (Redikorzev): Beier, 1951a: 100.

Type locality: Mt Ba Na (as Bana), Vietnam.
Distribution: Vietnam.

Hyperwithius dawydoffi Beier

Hyperwithius dawydoffi Beier, 1951a: 102–104, figs 34a, 35.

Type locality: Cao Nguyên Lâm Viên (as Plateau von Langbian), Vietnam.
Distribution: Vietnam.

Hyperwithius tonkinensis Beier

Hyperwithius tonkinensis Beier, 1951a: 100–102, figs 33, 34b.

Type locality: Lau Chau, Tonkin, Vietnam.
Distribution: Vietnam.

Genus **Metawithius** J. C. Chamberlin

Metawithius J. C. Chamberlin, 1931c: 293; Beier, 1932e: 200; Murthy & Anantha-krishnan, 1977: 149.

Type species: *Chelifer murrayi* Pocock, 1900, by original designation.

Subgenus **Metawithius (Metawithius)** J. C. Chamberlin

Metawithius (Metawithius) indicus Murthy & Ananthakrishnan

Metawithius (Metawithius) indicus Murthy & Ananthakrishnan, 1977: 149–151, figs 47a–b.

Type locality: Vedanthangal, Tamil Nadu, India.
Distribution: India.

Metawithius (Metawithius) murrayi (Pocock)

Chelifer murrayi Pocock, 1900: 156–157, plate 16 figs 1, 1a; With, 1906: 159–163, plate 3 figs 8a–j; Ellingsen, 1911a: 37; Ellingsen, 1911b: 142.
Chelifer (Trachychernes) murrayi Pocock: Ellingsen, 1910a: 376.
Withius murrayi (Pocock): Beier, 1930e: 292–293.
Metawithius murrayi (Pocock): J. C. Chamberlin, 1931c: 293; Beier, 1932e: 200, fig. 205; Roewer, 1936: figs 63b, 65; Roewer, 1937: 308; Harvey, 1985b: 154.

Type locality: Christmas Island.
Distribution: Burma, Christmas Island, Indonesia (Sumatra, Timor).

Metawithius (Metawithius) parvus (Beier)

Withius parvus Beier, 1930e: 293–294, fig. 5.
Metawithius parvus (Beier): Beier, 1932e: 201, fig. 206; Roewer, 1937: 308.

Type locality: Travancore, India.
Distribution: India.

Metawithius (Metawithius) philippinus Beier

Metawithius philippinus Beier, 1937b: 274–275, fig. 5; Roewer, 1940: 347.

Type locality: Limay, Luzon, Philippines.
Distribution: Philippines.

Metawithius (Metawithius) spiniventer Redikorzev

Metawithius spiniventer Redikorzev, 1938: 103–106, figs 32–34; Roewer, 1940: 347; Beier, 1951a: 98–99, fig. 32; Beier, 1967g: 358.
Metawithius (Metawithius) spiniventer Redikorzev: Beier, 1955e: 43.

Type localities: Mt Hon-Ba, Vietnam; Mt Ba Na (as Bana), Vietnam; and Pursat, Cambodia.
Distribution: Cambodia, Malaysia, Thailand, Vietnam.

Metawithius (Metawithius) spiniventer pauper Beier

Metawithius spiniventer pauper Beier, 1953g: 86, fig. 5.

Type locality: Langgai, Sumba, Indonesia.
Distribution: Indonesia (Sumba).

Subgenus Metawithius (Microwithius) Redikorzev

Microwithius Redikorzev, 1938: 106.
Metawithius (Microwithius) Redikorzev: Beier, 1955e: 45.

Type species: *Microwithius yurii* Redikorzev, 1938, by monotypy.

Metawithius (Microwithius) bulli Sivaraman

Metawithius (Microwithius) bulli Sivaraman, 1980a: 113–115, figs 4a–b.

Type locality: Bangalore, Karnataka, India.
Distribution: India.

Metawithius (Microwithius) chamundiensis Sivaraman

Metawithius (Microwithius) chamundiensis Sivaraman, 1980a: 111–113, figs 3a–b.

Type locality: Chamundi Hills, Mysore, Karnataka, India.
Distribution: India.

Metawithius (Microwithius) tweediei Beier

Metawithius (Microwithius) tweediei Beier, 1955e: 43–45, fig. 5.

Type locality: Telom Valley, near Gunong Siku, Cameron Highlands, Pahang, Malaysia.
Distribution: Malaysia.

Metawithius (Microwithius) yurii (Redikorzev)

Microwithius yurii Redikorzev, 1938: 106–108, figs 35–38; Roewer, 1940: 347; Beier, 1951a: 104.
Metawithius (Microwithius) yurii (Redikorzev): Beier, 1955e: 45.
Metawithius yurii (Redikorzev): Harvey, 1988b: 336–338, figs 79–85.

Type localities: Duong-Dong, Dao Phú Quoc (as Ile Phu-Quoc), Cambodia; and Poulo-Condore, Vietnam.
Distribution: Cambodia, Indonesia (Krakatau Islands), Vietnam.

Genus Nannowithius Beier

Nannowithius Beier, 1932d: 57; Beier, 1932e: 211.
Myrmecowithius Beier, 1963f: 195–196 (synonymized by Mahnert, 1988a: 68).

Type species: of *Nannowithius*: *Chelifer aethiopicus* E. Simon, 1900c, by original designation.
of *Myrmecowithius*: *Myrmecowithius wahrmani* Beier, 1963f, by original designation.

Nannowithius aethiopicus (E. Simon)

Chelifer aethiopicus E. Simon, 1900c: 596.
Nannowithius aethiopicus (E. Simon): Beier, 1932d: 57; Beier, 1932e: 211, fig. 217; Roewer, 1937: 309; Mahnert, 1988a: 68, figs 57−59.

Type locality: Agordat, Ethiopia.
Distribution: Ethiopia.

Nannowithius buettikeri (Mahnert)

Myrmecowithius buettikeri Mahnert, 1980c: 40−42, figs 23−28.
Nannowithius buettikeri (Mahnert): Mahnert, 1988a: 68.

Type locality: Kushm Buwaybiyat, Saudi Arabia.
Distribution: Saudi Arabia.

Nannowithius pakistanicus (Beier)

Myrmecowithius pakistanicus Beier, 1978d: 233−234, fig. 2.
Nannowithius pakestanicus (Beier): Mahnert, 1988a: 68.

Type locality: Kohala, Kashmir, Pakistan.
Distribution: Pakistan.

Nannowithius paradoxus (Mahnert)

Myrmecowithius paradoxus Mahnert, 1980c: 38−40, figs 17−22.
Nannowithius paradoxus (Mahnert): Mahnert, 1988a: 68.

Type locality: Wadi Zabib, Yemen.
Distribution: Yemen.

Nannowithius wahrmani (Beier)

Myrmecowithius wahrmani Beier, 1963f: 196−197, fig. 8; Mahnert, 1974c: 384.
Myrmecowithius wahrmanni (sic) Beier: Mahnert, 1975b: fig. 6d.
Nannowithius wahrmani (Beier): Mahnert, 1988a: 68.

Type locality: Wadi Abyad, Israel.
Distribution: Israel.

Genus Neowithius Beier

Neowithius Beier, 1932d: 62; Beier, 1932e: 223.

Type species: *Chelifer insignis* With, 1908a, by original designation.

Neowithius chilensis (Beier)

Chelifer (Chelanops) chilensis Beier, 1930d: 201, figs 4−6.
Neowithius chilensis (Beier): Beier, 1932d: 62; Beier, 1932e: 224−225, fig. 233; Roewer, 1937: 309.

Type locality: Temuco, Chile.
Distribution: Chile.

Neowithius cubanus (Banks)

Chelanops cabanus (sic) Banks, 1909c: 173, fig. 3.
Neowithius cubanus (Banks): Hoff, 1947b: 545−547, fig. 37.

Type locality: Cayamas, Cuba.
Distribution: Cuba.

Neowithius dubius Beier

Chelifer (Trachychernes) brevifemoratus Balzan: Ellingsen, 1910a: 373 (misidentification).
Neowithius dubius Beier, 1932d: 62; Beier, 1932e: 224, fig. 232; Roewer, 1937: 310; Feio, 1945: 8.

Type locality: Theresopolis, Santa Catarina (as S. Catharina), Brazil.
Distribution: Brazil (Distrito Federal, Santa Catarina).

Neowithius exilimanus (Balzan)

Chelifer exilimanus Balzan, 1887b: no pagination, figs; Balzan, 1890: 426–427, figs 11, 11a–b; With, 1908a: 249.
Chelifer (Chelifer) exilimanus Balzan: Balzan, 1892: 549; Ellingsen, 1910a: 385.
Neowithius exilimanus (Balzan): Beier, 1932d: 62; Beier, 1932e: 225, fig. 234; Roewer, 1937: 309.

Type locality: Rio Apa, Paraguay.
Distribution: Paraguay.

Neowithius insignis (With)

Chelifer insignis With, 1908a: 247–249, figs 9a–e.
Neowithius insignis (With): Beier, 1932d: 62; Beier, 1932e: 224; Roewer, 1937: 310.

Type locality: Contiroguita, ?
Distribution: ? (South America).

Genus Nesowithius Beier

Nesowithius Beier, 1940a: 185–186.

Type species: *Nesowithius seychellesensis* Beier, 1940a, by original designation.

Nesowithius dilatimanus Mahnert

Nesowithius dilatimanus Mahnert, 1988a: 43–45, figs 5–8.

Type locality: Kilifi, Kenya.
Distribution: Kenya.

Nesowithius eburneus Beier

Nesowithius eburneus Beier, 1979c: 112–113, fig. 6.

Type locality: Bingerville, Ivory Coast.
Distribution: Ivory Coast.

Nesowithius seychellesensis Beier

Nesowithius seychellesensis Beier, 1940a: 186–187, fig. 14; Mahnert, 1978d: 883–884, figs 30–32.
Paragoniochernes digitulus Beier, 1966k: 466–467, fig. 7 (synonymized by Mahnert, 1988a: 45).

Type localities: of *Nesowithius seychellesensis*: Praslin, Seychelles.
of *Paragoniochernes digitulus*: Durban, Natal, South Africa, imported from Seychelles.
Distribution: Seychelles, South Africa.

Genus Paragoniochernes Beier

Paragoniochernes Beier, 1932e: 189; Beier, 1933a: 548.

Type species: *Chelifer lamellatus* Tullgren, 1907b, by original designation.

Paragoniochernes lamellatus (Tullgren)

Chelifer lamellatus Tullgren, 1907b: 222−223, figs 2a−b; Ellingsen, 1912b: 81.
Paragoniochernes lamellatus (Tullgren): Beier, 1932e: 189, fig. 197; Roewer, 1937: 304; Beier, 1958a: 178−179, fig. 11; Beier, 1964k: 76.

Type locality: Van Reenen, Natal, South Africa.
Distribution: South Africa.

Paragoniochernes parvulus Beier

Paragoniochernes parvulus Beier, 1955l: 313−314, fig. 29.

Type locality: Storms River mouth, Tzitzikama Forest, Cape Province, South Africa.
Distribution: South Africa.

Genus Parallowithius Beier

Parallowithius Beier, 1955l: 318.

Type species: *Parallowithius pauper* Beier, 1955l, by original designation.

Parallowithius deserticola (Beier)

Allowithius deserticola Beier, 1947b: 334−335; Weidner, 1959: 113.
Parallowithius deserticola (Beier): Beier, 1955l: 318.

Type locality: Okahandja, Namibia.
Distribution: Namibia.

Parallowithius pauper Beier

Parallowithius pauper Beier, 1955l: 318−320, fig. 32.

Type locality: Cape Point, Cape Peninsula, South Africa.
Distribution: South Africa.

Genus Parawithius J. C. Chamberlin

Parawithius J. C. Chamberlin, 1931c: 292; Beier, 1932e: 212; Beier, 1959e: 216.

Type species: *Chelifer nobilis* With, 1908a, by original designation.

Subgenus Parawithius (Parawithius) J. C. Chamberlin

Parawithius (Parawithius) J. C. Chamberlin: Beier, 1959e: 216.

Parawithius (Parawithius) iunctus Beier

Chelifer (Chelifer) canestrinii Balzan: Ellingsen, 1910a: 385 (misidentification, in part).
Parawithius iunctus Beier, 1932d: 57−58; Beier, 1932d: 213, fig. 219; Roewer, 1937: 309.
Parawithius (Parawithius) iunctus Beier: Beier, 1959e: 216.

Type locality: Paraguay.
Distribution: Paraguay.

Parawithius (Parawithius) nobilis (With)

Chelifer nobilis With, 1908a: 234−236, figs 4a−b.
Parawithius nobilis (With): J. C. Chamberlin, 1931c: 292; Beier, 1932e: 213; Roewer, 1937: 309; Beier, 1959e: 216−217, fig. 25.

Type locality: Bogota, Colombia.
Distribution: Colombia.

Parawithius (Parawithius) nobilis ecuadoricus Beier

Parawithius (Parawithius) nobilis ecuadoricus Beier, 1959e: 217–218, fig. 26.

Type locality: Lago Zurucuchu, 11 miles W. of Cuenca, Ecuador.
Distribution: Ecuador, Peru.

Parawithius (Parawithius) pseudorufus Beier

Chelifer (Chelifer) rufus Balzan: Ellingsen, 1910a: 386 (misidentification, in part; see
  Parawithius (Victorwithius) fiebrigi Beier).
Parawithius pseudorufus Beier, 1932d: 58; Beier, 1932e: 214, fig. 220; Roewer, 1937:
  309.
Tropidowithius pseudorufus (Beier): Beier, 1955m: 11.
Parawithius (Parawithius) pseudorufus Beier: Beier, 1959e: 216.

Type locality: Paraguay.
Distribution: Paraguay.

Subgenus Parawithius (Victorwithius) Feio

Victorwithius Feio, 1944: 1–3.
Cacodemoniellus Beier, 1954c: 326–327 (synonymized by Beier, 1959e: 216).
Parawithius (Victorwithius) Feio: Beier, 1959e: 216.

Type species: of *Victorwithius*: *Victorwithius monoplacophorus* Feio, 1944, by original
  designation.
of *Cacodemoniellus*: *Cacodemoniellus mimulus* Beier, 1954c, by original designation.

Parawithius (Victorwithius) coniger Mahnert

Parawithius (Victorwithius) coniger Mahnert, 1979d: 804–806, figs 182–186; Mahnert
  & Adis, 1986: 214; Adis, Junk & Penny, 1987: 489.

Type locality: Rio Solimoes, Manaus, Amazonas, Brazil.
Distribution: Brazil (Amazonas).

Parawithius (Victorwithius) fiebrigi Beier

Chelifer (Chelifer) rufus Balzan: Ellingsen, 1910a: 386 (misidentification, in part; see
  Parawithius (Parawithius) pseudorufus Beier).
Parawithius fiebrigi Beier, 1932d: 59; Beier, 1932e: 215, fig. 222; Roewer, 1937: 309;
  Feio, 1945: 7.
Tropidowithius fiebrigi (Beier): Beier, 1955m: 11.
Parawithius (Victorwithius) fiebrigi Beier: Beier, 1959e: 216.

Type locality: Paraguay.
Distribution: Argentina, Paraguay.

Parawithius (Victorwithius) gracilimanus Mahnert

Parawithius (Victorwithius) gracilimanus Mahnert, 1979d: 802–804, figs 176–181;
  Adis, 1981: 119, etc.; Adis & Mahnert, 1986: 312; Mahnert & Adis, 1986: 214;
  Adis, Junk & Penny, 1987: 489; Mahnert, Adis, & Bührnheim, 1987: fig. 16.

Type locality: Taruma Mirim, Manaus, Amazonas, Brazil.
Distribution: Brazil (Amazonas).

Parawithius (Victorwithius) incognitus Beier

Parawithius (Victorwithius) incognitus Beier, 1959e: 220–221, fig. 29.

Type locality: 18 miles NE. of La Merced, Río Perené, Colonia Perené, Peru.
Distribution: Peru.

Parawithius (Victorwithius) mimulus (Beier)

Cacodemoniellus mimulus Beier, 1954c: 327–329, fig. 2.
Parawithius (Victorwithius) mimulus (Beier): Beier, 1959e: 216.

Type locality: Palmeria, Paraná, Brazil.
Distribution: Brazil (Paraná).

Parawithius (Victorwithius) monoplacophorus (Feio)

Victorwithius monoplacophorus Feio, 1944: 1–3, figs 1–11; Feio, 1945: 8.
Parawithius (Victorwithius) monoplacophorus (Feio): Beier, 1959e: 216.

Type locality: Etienne, Bituruna, Paraná, Brazil.
Distribution: Argentina, Brazil (Paraná), Uruguay.

Parawithius (Victorwithius) proximus (Ellingsen)

Chelifer (Chelifer) proximus Ellingsen, 1905e: 324–326; With, 1908a: 231.
Parawithius proximus (Ellingsen): J. C. Chamberlin, 1931c: 292; Beier, 1932e:
    215; Roewer, 1937: 309; Mello-Leitão, 1939a: 122; Mello-Leitão, 1939b:
    617.
Parawithius (Victorwithius) proximus (Ellingsen): Beier, 1959e: 216.

Type locality: Santa Catalina, Argentina.
Distribution: Argentina.

Parawithius (Victorwithius) rufeolus Beier

Parawithius (Victorwithius) rufeolus Beier, 1959e: 221–222, fig. 30; Muchmore,
    1977: 77.

Type locality: Pichilingue, Los Rios, Ecuador.
Distribution: Ecuador, Mexico.

Parawithius (Victorwithius) rufus (Balzan)

Chelifer rufus Balzan, 1887b: no pagination, figs; Balzan, 1890: 431–432, figs 15,
    15a–b.
Chelifer (Chelifer) rufus Balzan: Balzan, 1892: 533–534, fig. 25.
Parawithius rufus (Balzan): J. C. Chamberlin, 1931c: 292; Beier, 1932e: 214, fig. 221;
    Roewer, 1937: 309; Mello-Leitão, 1939a: 122; Mello-Leitão, 1939b: 617; Feio,
    1945: 7.
Tropidowithius rufus (Balzan): Beier, 1955m: 11.
Parawithius (Victorwithius) rufus (Balzan): Beier, 1959e: 216.
Not *Chelifer (Chelifer) rufus* Balzan: Ellingsen, 1902: 158–159 (misidentification;
    see *Parawithius (Victorwithius) venezuelanus* Beier); Ellingsen, 1910a: 386 (mis-
    identification; see *Parawithius (Victorwithius) fiebrigi* Beier and *Parawithius
    (Parawithius) pseudorufus* Beier).
Not *Chelifer rufus* Balzan: Tullgren, 1907a: 37; With, 1908a: 231–234, text-fig. 64,
    figs 3a–e (misidentification; see *Parawithius (Victorwithius) venezuelanus*
    Beier).

Type localities: Resistencia, Argentina; Asuncion, Paraguay; Rio Apa, Paraguay;
    and Mato Grosso (as Matto-grosso), Brazil.
Distribution: Argentina, Brazil (Mato Grosso), Paraguay, Venezuela.

Parawithius (Victorwithius) schlingeri Beier

Parawithius (Victorwithius) schlingeri Beier, 1959e: 219–220, fig. 28.

Type locality: Tingo María, Monson Valley, Peru.
Distribution: Peru.

Parawithius (Victorwithius) similis Beier

Parawithius (Victorwithius) similis Beier, 1959e: 218, fig. 27.

Type locality: 10 miles N. of Trancas, Tucumán, Argentina.
Distribution: Argentina.

Parawithius (Victorwithius) venezuelanus Beier

Chelifer (Chelifer) rufus Balzan: Ellingsen, 1902: 158–159 (misidentification).
Chelifer rufus Balzan: Tullgren, 1907a: 37; With, 1908a: 231–234, text-fig. 64, figs 3a–e (misidentifications).
Parawithius venezuelanus Beier, 1932d: 58–59; Beier, 1932e: 214–215; Roewer, 1937: 309; Feio, 1945: 7.
Tropidowithius venezuelanus (Beier): Beier, 1954f: 142.
Parawithius (Victorwithius) venezuelanus Beier: Beier, 1959e: 216.

Type locality: Venezuela.
Distribution: Argentina, Brazil (Sao Paulo), Colombia, Ecuador, Venezuela.

Genus Philomaoria J. C. Chamberlin

Philomaoria J. C. Chamberlin, 1931c: 291; Beier, 1932e: 226; Beier, 1976f: 240–241.

Type species: *Philomaoria novazealandica* J. C. Chamberlin, 1931c (junior synonym of *Chelifer pallipes* White, 1849), by original designation.

Philomaoria hispida Beier

Philomaoria hispida Beier, 1976f: 241–242, fig. 41.

Type locality: Pelorus Ridge, Marlborough, South Island, New Zealand.
Distribution: New Zealand.

Philomaoria pallipes (White)

Chelifer pallipes White, 1849: 6; With, 1905: 111–112, plate 7 figs 3a–b.
Philomaoria nova-zealandica J. C. Chamberlin, 1931c: 291–292; Beier, 1932e: 226; J. C. Chamberlin, 1934b: 11; Roewer, 1937: 310; Beier, 1948c: 550–551, fig. 15; Beier, 1962g: 402 (synonymized by Beier, 1976f: 241).
Philomaoria novazealandica J. C. Chamberlin: Beier, 1966c: 376; Beier, 1969e: 414.
Philomaoria pallipes (White): Beier, 1976f: 241–242, fig. 40; Harvey, 1985b: 154–155.
Not *Chelifer (Trachychernes) pallipes* White: Ellingsen, 1910a: 376–377 (misidentification; see *Thalassochernes taierensis* (With)).
Not *Haplochernes pallipes* (White): Beier, 1932e: 111; J. C. Chamberlin, 1934b: 9; Roewer, 1937: 295 (misidentifications; see *Thalassochernes taierensis* (With)).
Not *Thalassochernes pallipes* (White): Beier, 1940a: 182; Beier, 1948c: 537–540, figs 6–7; Beier, 1966c: 369; Beier, 1967b: 293; Beier, 1969e: 413 (misidentifications; see *Thalassochernes taierensis* (With)).

Type localities: of *Chelifer pallipes*: New Zealand.
of *Philomaoria novazealandica*: New Brighton, South Island, New Zealand.
Distribution: Lord Howe Island, New Zealand.

Genus Plesiowithius Vachon

Plesiowithius Vachon, 1954a: 1029.

Type species: *Plesiowithius dekeyseri* Vachon, 1954a, by original designation.

Plesiowithius dekeyseri Vachon

Plesiowithius dekeyseri Vachon, 1954a: 1026–1029, figs 6–11.

Type locality: Atar, Mauritania.
Distribution: Mauritania.

Genus **Pogonowithius** Beier

Pogonowithius Beier, 1979c: 109–111.

Type species: *Pogonowithius donisi* Beier, 1979c, by original designation.

Pogonowithius donisi Beier

Pogonowithius donisi Beier, 1979c: 111–112, figs 5a–b.

Type locality: Yangambi, Zaire.
Distribution: Zaire.

Genus **Protowithius** Beier

Protowithius Beier, 1955b: 216–217.

Type species: *Protowithius fernandezianus* Beier, 1955b, by original designation.

Protowithius fernandezianus Beier

Protowithius fernandezianus Beier, 1955b: 217–218, fig. 7; Cekalovic, 1984: 28.

Type locality: Las Chozas, Masatierra, Juan Fernandez Islands.
Distribution: Juan Fernandez Islands.

Protowithius robustus Beier

Protowithius robustus Beier, 1955b: 218–220, fig. 8; Beier, 1957c: 463; Cekalovic, 1984: 28.

Type locality: Plateau del Yunque, Masatierra, Juan Fernandez Islands.
Distribution: Juan Fernandez Islands.

Genus **Pseudatemnus** Beier

Pseudatemnus Beier, 1947b: 330–331.

Type species: *Pseudatemnus lawrencei* Beier, 1947b, by original designation.

Pseudatemnus lawrencei Beier

Pseudatemnus lawrencei Beier, 1947b: 331–332, figs 35–36.

Type locality: River Jonder End, Cape Province, South Africa.
Distribution: South Africa.

Genus **Pseudochernes** Beier

Pseudochernes Beier, 1954e: 86–87.

Type species: *Pseudochernes crassimanus* Beier, 1954e, by original designation.

Pseudochernes crassimanus Beier

Pseudochernes crassimanus Beier, 1954e: 87–88, figs 2–3.

Type locality: Wahi, Lake Victoria, Tanzania.
Distribution: Tanzania.

Genus **Pycnowithius** Beier

Pycnowithius Beier, 1979c: 107–108.

Type species: *Pycnowithius sambicus* Beier, 1979c, by original designation.

Pycnowithius cavernicola Mahnert

Pycnowithius cavernicola Mahnert, 1988a: 50–52, figs 23–28.

Type locality: Kimakia Cave, Hunter's Lodge, Kiboko, Kenya.
Distribution: Kenya.

Pycnowithius garambicus (Beier)

Allowithius garambicus Beier, 1972a: 13–15, fig. 7.
Pycnowithius garambicus (Beier): Beier, 1979c: 108.

Type locality: Parc National Garamba, Zaire.
Distribution: Zaire.

Pycnowithius sambicus Beier

Pycnowithius sambicus Beier, 1979c: 108–109, figs 4a–b.

Type locality: Hangama, Sambia, Zimbabwe.
Distribution: Zimbabwe.

Genus Scotowithius Beier

Scotowithius Beier, 1977b: 6–7.

Type species: *Scotowithius helenae* Beier, 1977b, by original designation.

Scotowithius helenae Beier

Scotowithius helenae Beier, 1977b: 7–8, fig. 3.

Type locality: Teutonic Hall, St Helena.
Distribution: St Helena.

Genus Sphaerowithius Mahnert

Sphaerowithius Mahnert, 1988a: 68–69.

Type species: *Chelifer (Trachychernes) perpusillus* Ellingsen, 1910a, by original designation.

Sphaerowithius basilewskyi (Beier)

Nannowithius basilewskyi Beier, 1962f: 30–31, fig. 14.
Sphaerowithius basilewskyi (Beier): Mahnert, 1988a: 69.

Type locality: Bunduki, Uluguru Mountains, Tanzania.
Distribution: Tanzania.

Sphaerowithius perpusillus (Ellingsen)

Chelifer (Trachychernes) perpusillus Ellingsen, 1910a: 377–378.
Chelifer perpusillus Ellingsen: Ellingsen, 1912b: 82.
Nannowithius perpusillus (Ellingsen): Beier, 1932d: 57; Beier, 1932e: 211–212, fig. 218; Roewer, 1937: 309.
Sphaerowithius perpusillus (Beier): Mahnert, 1988a: 45–47, figs 9–11.

Type locality: Takaungu (as Takanuga), S. of Malindi, Kenya.
Distribution: Kenya.

Sphaerowithius saegeri (Beier)

Nannowithius saegeri Beier, 1972a: 15–17, fig. 8.
Sphaerowithius saegeri (Beier): Mahnert, 1988a: 69.

Type locality: Parc National Garamba, Zaire.
Distribution: Zaire.

Sphaerowithius salomonensis (Beier)

Nannowithius salomonensis Beier, 1966d: 154–155, fig. 12.
Sphaerowithius salomonensis (Beier): Mahnert, 1988a: 69.

Type locality: Giro (= Gizo?), New Georgia, Solomon Islands.
Distribution: Solomon Islands.

Sphaerowithius vafer (Beier)

Nannowithius vafer Beier, 1966k: 467–469, fig. 8.
Sphaerowithius vafer (Beier): Mahnert, 1988a: 69.

Type locality: 33 miles from Pretoria via Warmbad, Transvaal, South Africa.
Distribution: South Africa.

Genus Sphallowithius Beier

Sphallowithius Beier, 1977b: 8–9.

Type species: *Sphallowithius excelsus* Beier, 1977b, by original designation.

Sphallowithius excelsus Beier

Sphallowithius excelsus Beier, 1977b: 9–10, fig. 4.

Type locality: East Prosperous Bay Plain, St Helena.
Distribution: St Helena.

Sphallowithius inhonestus Beier

Sphallowithius inhonestus Beier, 1977b: 10–11, fig. 5.

Type locality: Entre Lufkins and Peak Gut, St Helena.
Distribution: St Helena.

Genus Stenowithius Beier

Stenowithius Beier, 1932d: 55; Beier, 1932e: 206.

Type species: *Chelifer (Chelifer) buettneri* Ellingsen, 1910a, by original designation.

Stenowithius angulatus (Ellingsen)

Chelifer (Chelifer) angulatus Ellingsen, 1906: 256–258.
Chelifer angulatus Ellingsen: Ellingsen, 1912b: 83.
Stenowithius angulatus (Ellingsen): Beier, 1932d: 57; Beier, 1932e: 209; Roewer, 1937: 309.

Type locality: Roça, Principe.
Distribution: Principe.

Stenowithius bayoni (Ellingsen)

Chelifer bayoni Ellingsen, 1910d: 536–538; Ellingsen, 1912b: 83, 92; Ellingsen, 1913a: 453–454.
Withius bayoni (Ellingsen): Godfrey, 1927: 18.
Stenowithius bayoni (Ellingsen): Beier, 1932d: 57; Beier, 1932e: 207–208; Roewer, 1937: 308; Vachon, 1940f: 3; Beier, 1948b: 462; Beier, 1964k: 79; Beier, 1972a: 15; Mahnert, 1988a: 45.
Stenowithius ugandanus Beier, 1932d: 55–56; Beier, 1932e: 207, fig. 213; Roewer, 1937: 308; Beier, 1948b: 445; Beier, 1955c: 556 (synonymized by Beier, 1964k: 79).
Not *Stenowithius ugandanus* Beier: Caporiacco, 1949a: 318 (misidentification; see *Withius congicus* (Beier)).

Type localities: of *Chelifer bayoni*: Buvama, Uganda; and Bugala, Uganda.
of *Stenowithius ugandanus*: Kampala, Uganda.
Distribution: Kenya, Malawi, Mozambique, South Africa, Uganda, Zaire.

Stenowithius bayoni angustus Beier

Stenowithius bayoni angustus Beier, 1964k: 79–80, fig. 34.

Type locality: Potgietersrus (as Potgietersrust), Transvaal, South Africa.
Distribution: South Africa.

Stenowithius buettneri (Ellingsen)

Chelifer (Chelifer) büttneri Ellingsen, 1910a: 383–384.
Chelifer büttneri Ellingsen: Ellingsen, 1912b: 83.
Stenowithius büttneri (Ellingsen): Beier, 1932d: 55; Beier, 1932e: 209, fig. 215; Roewer, 1937: 309.
Stenowithius buettneri (Ellingsen): Mahnert, 1988a: 66, figs 54–56.

Type locality: Bismarcksburg, Togo.
Distribution: Malawi, Togo.

Stenowithius duffeyi Beier

Stenowithius duffeyi Beier, 1961c: 597–598, fig. 3.

Type locality: Boatswain-bird (as Bos'nbird) Island, Ascension Island.
Distribution: Ascension Island.

Stenowithius parvulus Beier

Stenowithius parvulus Beier, 1954d: 136–137, fig. 4.

Type locality: Flandria, Zaire.
Distribution: Zaire.

Stenowithius persimilis Beier

Stenowithius persimilis Beier, 1932d: 56–57; Beier, 1932e: 208, fig. 214; Roewer, 1937: 309; Beier, 1954d: 136.

Type locality: Rutshum, Buseregenye, Zaire.
Distribution: Zaire.

Stenowithius phagophilus Beier

Stenowithius phagophilus Beier, 1953c: 78, fig. 3.

Type locality: Aminuis Reserve, Namibia.
Distribution: Namibia.

Stenowithius torpidus Beier

Stenowithius torpidus Beier, 1958a: 181–183, fig. 13.

Type locality: Pietermaritzburg, Natal, South Africa.
Distribution: South Africa.

Genus Thaumatowithius Beier

Thaumatowithius Beier, 1940a: 187–188.

Type species: *Thaumatowithius tibialis* Beier, 1940a, by original designation.

Thaumatowithius aberrans Mahnert

Thaumatowithius aberrans Mahnert, 1975b: 549–553, figs 4a–k, 6f.

Type locality: Plaine des Chicots, Réunion.
Distribution: Réunion.

Thaumatowithius tibialis Beier

Thaumatowithius tibialis Beier, 1940a: 188–189, fig, 15.

Type locality: Le Pouce, Mauritius.
Distribution: Mauritius.

Genus **Trichotowithius** Beier

Trichotowithius Beier, 1944: 203–204.

Type species: *Trichotowithius abyssinicus* Beier, 1944, by original designation.

Trichotowithius abyssinicus Beier

Trichotowithius abyssinicus Beier, 1944: 204–206, fig. 20.

Type locality: Djem-Djem Wald, Ethiopia.
Distribution: Ethiopia.

Trichotowithius elgonensis Beier

Trichotowithius elgonensis Beier, 1955c: 556–557, fig. 19; Mahnert, 1988a: 48.

Type locality: Mt Elgon, E. side, Kenya.
Distribution: Kenya.

Genus **Tropidowithius** Beier

Tropidowithius Beier, 1955m: 10–11.

Type species: *Tropidowithius peruanus* Beier, 1955m, by original designation.

Tropidowithius peruanus Beier

Tropidowithius peruanus Beier, 1955m: 11–12, fig. 11; Weidner, 1959: 117.

Type locality: Sivia, Peru.
Distribution: Peru.

Genus **Withius** Kew

Chelifer (Withius) Kew, 1911a: 49.
Withius Kew: J. C. Chamberlin, 1923c: 376–377; J. C. Chamberlin, 1931c: 293; Beier, 1932e: 196; G. O. Evans & Browning, 1954: 16; Morikawa, 1960: 149; Beier, 1963b: 281; Murthy & Ananthakrishnan, 1977: 146; Legg & Jones, 1988: 128.
Allowithius Beier, 1932d: 53; Beier, 1932e: 201 (synonymized by Beier, 1979c: 107).
Xenowithius Beier, 1953c: 75–76 (synonymized by Mahnert, 1988a: 65).

Type species: of *Chelifer (Withius)*: *Chelifer subruber* E. Simon, 1879a (junior synonym of *Chelifer piger* E. Simon, 1878a) by original designation.
of *Allowithius*: *Chelifer (Chelifer) simoni* Balzan, 1892, by original designation.
of *Xenowithius*: *Xenowithius transvaalensis* Beier, 1953c, by original designation.

Withius abyssinicus (Beier)

Allowithius abyssinicus Beier, 1944: 202–203, fig. 19; Beier, 1967d: 83.
Withius abyssinicus (Beier): Mahnert, 1988a: 62–63.

Type locality: Djem-Djem-Wald (as Djem-Dem), Ethiopia.
Distribution: Ethiopia, Kenya.

Withius angolensis (Beier), new combination

Allowithius angolensis Beier: Beier, 1947b: 333. Nomen nudum.
Allowithius angolensis Beier, 1948b: 487–488, figs 18–19; Beier, 1959d: 62–63; Mahnert, 1978f: 132.

Type locality: Solok River, E. of Catengue, Angola.
Distribution: Angola, Congo, Zaire.

Withius angustatus (Tullgren)

Chelifer angustatus Tullgren, 1907d: 13–14, fig. 9; Ellingsen, 1912b: 83.
Allowithius angustatus (Tullgren): Beier, 1932d: 55; Beier, 1932e: 205; Roewer, 1937: 308; Beier, 1964k: 78; Beier, 1967d: 83.
Withius angustatus (Tullgren): Mahnert, 1988a: 56–58, fig. 33.

Type locality: Kibonoto (presumably Kibongoto), Tanzania.
Distribution: Kenya, Mozambique, Tanzania.

Withius arabicus Mahnert

Withius (Allowithius) arabicus Mahnert, 1980c: 36–38, figs 10–16.

Type locality: Dorf Qaraah, Saudi Arabia.
Distribution: Saudi Arabia.

Withius ascensionis (Beier), new combination

Allowithius ascensionis Beier, 1961c: 596–597, fig. 2.

Type locality: Spire Beach, Ascension Island.
Distribution: Ascension Island.

Withius australasiae (Beier), new combination

Chelifer (Chelifer) simoni Balzan: Ellingsen, 1910a: 387 (misidentification, in part).
Allowithius australasiae Beier, 1932d: 54–55; Beier, 1932e: 206, fig. 212; J. C. Chamberlin, 1934b: 11; Roewer, 1937: 308; Beier, 1957d: 63–64; Beier, 1971b: 367.

Type locality: Jaluit, Marshall Islands.
Distribution: Mariana Islands, Marshall Islands, Papua New Guinea.

Withius australasiae formosanus (Beier), new combination

Allowithius australasiae formosanus Beier, 1937b: 275–276, fig. 6; Roewer, 1940: 347.

Type locality: Kao-hsing (as Takao), Taiwan.
Distribution: Taiwan.

Withius brevidigitatus Mahnert

Withius brevidigitatus Mahnert, 1988a: 61, figs 48–50.

Type locality: between Wamunyu and Myondoni, Machakos, Kenya.
Distribution: Kenya.

Withius caecus Beier

Withius caecus Beier, 1929b: 78–79, figs 1a–b; Beier, 1932e: 199–200, fig. 204; Roewer, 1937: 308; Heurtault, 1971a: 706.

Type locality: El Merg, Cyrenaica, Libya.
Distribution: Libya.

Withius capensis (Beier), new combination

Allowithius capensis Beier, 1947b: 335–336, fig. 39.

Type locality: Queenstown, Cape Province, South Africa.
Distribution: South Africa.

Withius ceylanicus (Ellingsen), new combination

Chelifer ceylanicus Ellingsen, 1914: 7–9.
Allowithius ceylanicus (Ellingsen): Beier, 1932d: 55; Beier, 1932e: 205; Roewer, 1937: 308; Beier, 1973b: 53.

Type locality: Peradeniya, Sri Lanka.
Distribution: Sri Lanka.

Withius congicus (Beier)

Allowithius congicus Beier, 1932d: 53; Beier, 1932e: 203–204, fig. 209; Roewer, 1937: 308; Beier, 1954d: 136; Beier, 1959d: 63; Beier, 1967d: 83; Beier, 1972a: 15; Spaull, 1979: 117.
Stenowithius ugandanus Beier: Caporiacco, 1949a: 318 (misidentification).
Withius congicus (Beier): Mahnert, 1988a: 54–55, figs 15–16.

Type locality: Luluabourg (now Kananga), Zaire.
Distribution: Aldabra Islands, Kenya, Tanzania, Zaire.

Withius congicus exiguus (Beier), new combination

Allowithius exiguus Beier, 1955a: 16–18, fig. 8.
Allowithius congicus exiguus Beier: Beier, 1972a: 15.

Type locality: Mabwe, Zaire.
Distribution: Zaire.

Withius crassipes (Lawrence)

Stenowithius crassipes Lawrence, 1937: 270–272, figs 30a–c; Roewer, 1940: 347.
Allowithius crassipes (Lawrence): Beier, 1958a: 180–181, fig. 12; Beier, 1964k: 79; Beier, 1966k: 467, 470.
Withius crassipes (Lawrence): Mahnert, 1988a: 52–54, figs 29–32.

Type locality: Nkandhla Forest, Zululand, Natal, South Africa.
Distribution: Ethiopia, Kenya, South Africa.

Withius despaxi Vachon

Withius despaxi Vachon, 1937b: 42–43, figs 5–7; Roewer, 1940: 347; Beier, 1963b: 281; Heurtault, 1971a: 706.

Type locality: Lectoure, Gers, France.
Distribution: France.

Withius faunus (E. Simon)

Chelifer faunus E. Simon, 1879a: 29–30, plate 18 fig. 18.
Withius faunus (E. Simon): Beier, 1932e: 197, fig. 200; Roewer, 1937: 308; Heurtault, 1971a: 707; Heurtault, 1971b: 1041–1043, 6a–b; Mahnert, 1977b: 68.
Withius hispanus (L. Koch): Beier, 1963b: 281–282 (misidentification; in part).

Type locality: La Teste, Gironde, France.
Distribution: France, Greece.

Withius fuscus Mahnert

Withius fuscus Mahnert, 1988a: 58–59, figs 38–40.

Type locality: between Wamunyu and Myondoni, Machakos, Kenya.
Distribution: Kenya.

Withius glabratus (Ellingsen), new combination

Chelifer (Trachychernes) glabratus Ellingsen, 1910a: 373–374.
Chelifer glabratus Ellingsen: Ellingsen, 1912b: 82.
Allowithius glabratus (Ellingsen): Beier, 1932d: 55; Beier, 1932e: 203, fig. 208; Roewer, 1937: 308; Vachon, 1940e: 71; Beier, 1959d: 63.

Type locality: 'Jos. Albrechtshöhe', Cameroun.
Distribution: Cameroun, Mali, Zaire.

Withius gracilipalpus Mahnert

Withius gracilipalpus Mahnert, 1988a: 56, figs 34–37.

Type locality: near Strassenkreuzung Narok-Nairagi Engare, Narok, Kenya.
Distribution: Kenya.

Withius hispanus (L. Koch)

Chelifer hispanus L. Koch, 1873: 26–27; E. Simon, 1879a: 28–29, plate 18 figs 6,
6a; Daday, 1889a: 16; Daday, 1889c: 25; E. Simon, 1898c: 3; Ellingsen, 1909a:
207; Krausse-Heldrungen, 1912: 65; Navás, 1923: 31; Navás, 1925: 107; Bacelar,
1928: 190.
Chernes (Ectoceras) hispanus (L. Koch): Daday, 1888: 120, 175; Daday, 1889d: 80.
Chelifer (Chelifer) hispanus L. Koch: Ellingsen, 1910a: 385; Lessert, 1911: 22–23,
fig. 18; Nonidez, 1917: 26; Kästner, 1928: 8, fig. 26; Beier, 1929a: 357.
Withius hispanus (L. Koch): J. C. Chamberlin, 1931a: figs 14l, 15z, 18o, 39i, 44m–n,
45l–m, 51i; J. C. Chamberlin, 1931c: 293; Beier, 1932e: 196–197; Vachon, 1936a:
78; Roewer, 1937: 308; Vachon, 1937f: 128; Beier, 1939f: 199; Hadži, 1939b: 44–47,
figs 14a–f, 15a–e; Vachon, 1940g: 145; Beier, 1948a: 190, fig. 3; Beier, 1962b:
146; Beier, 1963b: 281–282, fig. 283 (in part; see *Withius faunus* (E. Simon));
Beier, 1965f: 96; Kobakhidze, 1965b: 541; Kobakhidze, 1966: 705; Zangheri, 1966:
532; Beier, 1967f: 313; Lazzeroni, 1969c: 242; Heurtault, 1971a: 707; Heurtault,
1971b: 1040–1041, 1045, figs 4a–b, 5, 10b, 12c, 13a–c; Heurtault, 1972b: figs 2,
4; Ćurčić, 1974a: 27; Schawaller, 1981b: 48–49, figs 7–8; Callaini, 1986e: 254;
Callaini, 1987b: 275; Callaini, 1988b: 57.

Type locality: Spain.
Distribution: Bulgaria, Corsica, France, Italy, Morocco, Portugal, Sardinia, Spain,
Switzerland, Turkey, U.S.S.R. (Georgia), Yugoslavia.

Withius indicus Murthy & Ananthakrishnan

Withius indicus Murthy & Ananthakrishnan, 1977: 146–149, figs 46a–b.

Type locality: Nungambakkam, Tamil Nadu, India.
Distribution: India.

Withius japonicus Morikawa

Withius japonicus Morikawa, 1954a: 71–73, figs 1a–f; Morikawa, 1960: 149, plate 4
fig. 3, plate 7 fig. 19, plate 8 fig. 13, plate 9 fig. 25, plate 10 figs 16, 29; Morikawa,
1962: 419.

Type locality: Futako-tamagawa, Tokyo, Japan.
Distribution: Japan.

Withius kaestneri (Vachon)

Allowithius kästneri Vachon, 1937a: 132–133, figs 5–6; Vachon, 1938a: fig. 52b;
Roewer, 1940: 347; Beier, 1947b: 333–334, figs 37–38; Beier, 1964k: 78; Lawrence,
1967: 90.
Withius kaestneri (Vachon): Hulley, 1983: 43, 44.

Type locality: Makapau, Victoria, Zimbabwe.
Distribution: South Africa, Zimbabwe.

Withius lagunae (Moles)

Chelanops lagunae Moles, 1914a: 42–44, figs 1–2; Moles, 1914c: 193; Moles &
Moore, 1921: 6.
Withius lagunae (Moles): J. C. Chamberlin, 1923c: 377; J. C. Chamberlin, 1931a:
figs 30n, 45h; J. C. Chamberlin, 1931c: 293; Hoff, 1958: 36.

Family Withiidae

Withius (?) *lagunae* (Moles): Beier, 1932e: 200; Roewer, 1937: 308.

Type locality: Sycamore Canyon, near Laguna Beach, California, U.S.A.
Distribution: U.S.A. (California).

Withius lawrencei (Beier), new combination

Allowithius lawrencei Beier, 1935b: 253−255, figs 1−2; Roewer, 1937: 308; Mahnert, 1975b: 547−549, fig. 6c.

Type locality: Mauritius.
Distribution: Mauritius, Réunion.

Withius laysanensis (E. Simon)

Chelifer laysanensis E. Simon, 1899c: 414−415.
Lophochernes (?) *laysanensis* (E. Simon): Beier, 1932e: 250; J. C. Chamberlin, 1934b: 11.
Withius laysanensis (E. Simon): Roewer, 1937: 308.

Type locality: Laysan Island, Hawaii.
Distribution: Hawaii.

Withius lewisi (Beier)

Allowithius congicus lewisi Beier, 1946: 570−571, fig. 3.
Withius lewisi (Beier): Mahnert, 1988a: 55−56, figs 17−18.

Type locality: Lujulu, Zaire.
Distribution: Kenya, Zaire.

Withius litoreus (Beier)

Allowithius litoreus Beier, 1935c: 128−129, fig. 10; Roewer, 1940: 347.
Withius litoreus (Beier): Mahnert, 1988a: 59.

Type locality: Nanoropus, Lake Rudolf, Kenya.
Distribution: Kenya.

Withius lohmanderi Kobakhidze

Withius lohmanderi Kobakhidze, 1965c: 417−419, fig. 1.

Type locality: Sochi, Krasnodar Province, R.S.F.S.R., U.S.S.R.
Distribution: U.S.S.R. (R.S.F.S.R.).

Withius madagascariensis (Ellingsen), new combination

Chelifer madagascarensis Ellingsen, 1895: 137−138.
Allowithius (?) *madagascariensis* (Ellingsen): Beier, 1932e: 206; Roewer, 1937: 308.
Allowithius madagascariensis (Ellingsen): Legendre, 1972: 447.

Type locality: Annanarivo, Madagascar.
Distribution: Madagascar.

Withius nanus Mahnert

Withius nanus Mahnert, 1988a: 59−61, figs 41−47.

Type locality: Shimba Hills National Park, Coastal Province, Kenya.
Distribution: Kenya.

Withius neglectus (E. Simon)

Chelifer neglectus E. Simon, 1878a: 147−148.
Withius neglectus (E. Simon): Beier, 1932e: 197−198, fig. 201; Roewer, 1937: 308; Heurtault, 1971a: 707; Heurtault, 1971b: 1043, figs 7a−b; Callaini, 1987b: 274.

Type locality: Bou Saâda (as Bou-Saada), Algeria.
Distribution: Algeria, Tunisia.

Withius nepalensis Beier

Withius nepalensis Beier, 1974b: 277–278, fig. 11.
Type locality: Daman, Mahabarat, Nepal.
Distribution: Nepal.

Withius pekinensis (Balzan)

Chelifer (Chelifer) pekinensis Balzan, 1892: 528–529, fig. 19.
Withius pekinensis (Balzan): Beier, 1932e: 199; Roewer, 1937: 308.
Type locality: Pei-ching (as Peking), Hopeh, China.
Distribution: China (Hopeh).

Withius persicus (Redikorzev)

Chelifer persicus Redikorzev, 1934a: 427–428.
Withius persicus (Redikorzev): Redikorzev, 1934b: 152; Roewer, 1937: 308.
Type locality: Kerman, Iran.
Distribution: Iran.

Withius piger (E. Simon)

Chelifer piger E. Simon, 1878a: 148–149.
Chelifer subruber E. Simon, 1879a: 30, plate 18 fig. 7; E. Simon, 1881b: 12; Tömösváry, 1882b: 203–204, plate 2 figs 12–13; H. J. Hansen, 1884: 542–543; E. Simon, 1885c: 47; Daday, 1888: 122, 178; O.P.-Cambridge, 1892: 222–223, plate C figs 14, 14a–b; Kulczynski, 1899: 458; Tullgren, 1899a: 170–171; E. Simon, 1900b: 593; Ellingsen, 1905d: 2; Nosek, 1905: 120, 153; With, 1906: 155–159, figs 5a–b, plate 3, figs 7a–g; Ellingsen, 1907b: 28; Tullgren, 1907a: 37; Ellingsen, 1909a: 209; Ellingsen, 1912a: 122; Ellingsen, 1912b: 85; Ellingsen, 1914: 10; Daday, 1918: 2; Caporiacco, 1923: 131; Schenkel, 1928: 58 (synonymized by Heurtault, 1971b: 1037).
Chelifer peculiaris L. Koch: O.P.-Cambridge, 1889: ? (not seen) (misidentification).
Chelifer (Chelifer) subruber E. Simon: Ellingsen, 1910a: 387; Lessert, 1911: 23–24.
Chelifer (Withius) subruber E. Simon: Kew, 1911a: 49–50, fig. 12; Standen, 1912: 14; Standen, 1916b: 124–125; Standen, 1917: 28–29; Standen, 1918: 335; Kästner, 1928: 8, figs 29–30; Beier, 1929a: 357.
Chelifer (Trachychernes) oculatus Beier, 1929a: 343–344, fig. 1 (synonymized by Heurtault, 1971b: 1037).
Withius subruber (E. Simon): Beier, 1930a: 219; Beier, 1930f: 75; J. C. Chamberlin, 1931a: figs 41i, 52g; Beier, 1932e: 198, fig. 202; Beier, 1932f: 488; Roewer, 1936: figs 45c–d, 63a, 70q, 82, 144; Roewer, 1937: 308; Vachon, 1938a: 40–43, figs 20a–b, 21–22, 52a; Beier, 1952e: 5; G. O. Evans & Browning, 1954: 16; Turk, 1955: 169; Beier, 1956e: 303; Cloudsley-Thompson, 1958: 96; Beier, 1961b: 37; Meinertz, 1962a: 400, map 8; Beier, 1963b: 282–283, fig. 284; Beier, 1963h: 156; Beier, 1963i: 286; Beier, 1964b: 375; Hoff, 1964b: 36–40; Beier, 1966a: 301; Beier, 1966d: 154; Zangheri, 1966: 532; Beier, 1967c: 319; Beier, 1967f: 313; Hammen, 1969: 22; Lazzeroni, 1969a: 337–338; Weygoldt, 1969a: 44, 53, 62, 64, 110, figs 47, 58; Weygoldt, 1969b: 230–234, figs 1–2, 3a–c; Beier, 1970a: 45; Lazzeroni, 1970a: 208; Muchmore, 1971a: 94; Legg, 1972: 576, figs 1(1, 2b), 3(2); Beier, 1976a: 24; Legg, 1975a: 66; Legg, 1975d: fig. 2; Beier, 1976f: 244; Beier, 1977b: 6; Crocker, 1978: 9.
Withius oculatus (Beier): Beier, 1932e: 199, fig. 203; Roewer, 1937: 308.
Withius piger (E. Simon): Vachon, 1970: 186–188, figs 1, 3–6; Heurtault, 1971a: 699–703, 707, figs 22–25, 27; Heurtault, 1971b: 1039–1040, 1045–1046, figs 3a–b,

10a, 11, 12a–b; Mahnert, 1975c: 196; Thaler, 1979: 52; Mahnert, 1980a: 259; Jones, 1980c: map 24; Harvey, 1981b: 245; Pieper, 1981: 4; Callaini, 1983c: 308–310, figs 3d–f; Callaini, 1986e: 254; Harvey, 1985b: 155; Callaini, 1987b: 274; Callaini, 1988b: 56–57; Legg & Jones, 1988: 129–130, figs 33a, 33b(a–i); Schawaller & Dashdamirov, 1988: 44, figs 80–81.

Not *Atemnus piger* (E. Simon): Beier, 1930a: 209–210 (misidentification; see *Diplotemnus ophthalmicus* Redikorzev).

Not *Miratemnus piger* (E. Simon): Beier, 1932b: 610; Beier, 1932e: 79–80, fig. 99; Roewer, 1937: 285 (misidentifications; see *Diplotemnus ophthalmicus* Redikorzev).

Not *Diplotemnus piger* (E. Simon): Beier, 1946: 568–569; Beier, 1957b: 147; Verner, 1959: 61–63, figs 1–2; Beier, 1963b: 248, fig. 251; Beier, 1963g: 272; Beier, 1965b: 373; Beron, 1968: 104; Beier, 1971a: 363; Schawaller & Dashdamirov, 1988: 34, figs 56–57; Schawaller, 1989: 16–17, figs 38–42 (misidentifications; see *Diplotemnus ophthalmicus* Redikorzev).

Type localities: of *Chelifer piger*: Bou Saâda (as Bou-Saada), Algeria.

of *Chelifer subruber*: Hyères, Var, France.

of *Chelifer (Trachychernes) oculatus*: Haifa, Israel.

Distribution: Algeria, Australia (New South Wales), Austria, Balearic Islands, Canary Islands, Chad, Chile, China (Kwangtung), Cuba, Denmark, Egypt, France, Ghana, Great Britain, Hungary, India, Italy, Libya, Malta, Mexico, Morocco, Netherlands, New Zealand, Philippines, Samoa, Sardinia, Senegal, Solomon Islands, Sri Lanka, St Helena, Switzerland, Syria, Taiwan, Tanzania, Tunisia, Turkey, U.S.A. (Arizona, Florida, Georgia), U.S.S.R. (Azerbaijan), West Germany, Yugoslavia.

Withius rebierei Heurtault

Withius rebierei Heurtault, 1971a: 703–706, fig. 26.

Type locality: Goubone, Chad.
Distribution: Chad.

Withius simoni (Balzan)

Chelifer (Chelifer) simoni Balzan, 1892: 529–531, figs 20. 20a; Ellingsen, 1906: 254–255.

Chelifer simoni Balzan: Tullgren, 1901: 100; With, 1907: 65–66; Ellingsen, 1909a: 218; Ellingsen, 1910b: 63; Ellingsen, 1910d: 536; Ellingsen, 1912b: 84, 100; Ellingsen, 1913a: 454.

Withius simoni (Balzan): Godfrey, 1927: 18.

Withius (?) *simoni* (Balzan): J. C. Chamberlin, 1931c: 293.

Allowithius simoni (Balzan): Beier, 1932d: 53; Beier, 1932e: 202–203, fig. 207; Roewer, 1937: 308; Vachon, 1940e: 71; Beier, 1948b: 458; Vachon, 1956: 3; Beier, 1959d: 62; Beier, 1965b: 374; Beier, 1972a: 15.

Not *Chelifer (Chelifer) simoni* Balzan: Ellingsen, 1910a: 387 (misidentification; see *Dolichowithius (Dolichowithius) simplex* Beier and *Withius australasiae* (Beier)).

Type locality: Sierra Leone.
Distribution: Angola, Cameroun, Cape Verde Islands, Chad, Equatorial Guinea, Ethiopia, Guinea-Bissau, Ivory Coast, Malawi, Réunion, Sao Tomé, Senegal, Sierra Leone, South Africa, Togo, Uganda, Zaire.

Withius soederbomi (Schenkel)

Chelifer söderbomi Schenkel, 1937: 311–313, figs 110a–b.

Withius söderbomi Schenkel: Vachon, 1937b: 44; Roewer, 1937: 308; Roewer, 1940: 347.

Type locality: Etsin-gol, Mongolia.
Distribution: Mongolia.

Withius somalicus (Beier)

Allowithius somalicus Beier, 1932d: 54; Beier, 1932e: 204, fig. 210; Roewer, 1937: 308; Caporiacco, 1939b: 311; Beier, 1962f: 29–30, fig. 29; Beier, 1967d: 82.
Allowithius crassus Beier, 1935c: 127–128, fig. 9; Roewer, 1940: 347 (synonymized by Mahnert, 1988a: 63).
Allowithius somalicus major Beier, 1935c: 129; Roewer, 1940: 347; Beier, 1955c: 555, fig. 18; Beier, 1967d: 82; Mahnert, 1975b: fig. 6a (synonymized by Mahnert, 1988a: 63).
Withius somalicus (Beier): Mahnert, 1988a: 63–65, figs 51–53.

Type localities: of *Allowithius somalicus*: Afmadú, Somalia.
of *Allowithius crassus*: Mt Elgon, Uganda.
of *Allowithius somalicus major*: Rift Valley, Kenya.
Distribution: Ethiopia, Kenya, Somalia, Tanzania, Uganda.

Withius suis Sivaraman

Withius suis Sivaraman, 1980a: 109–111, figs 2a–b.

Type locality: Madras, India.
Distribution: India.

Withius tenuimanus (Balzan), new combination

Chelifer (Chelifer) tenuimanus Balzan, 1892: 531–532, fig. 21.
Chelifer tenuimanus Balzan: Ellingsen, 1912b: 85.
Allowithius tenuimanus (Balzan): Beier, 1932d: 55; Beier, 1932e: 205; Roewer, 1937: 308; Legendre, 1972: 447.

Type locality: Nossi Bé, Madagascar.
Distribution: Madagascar.

Withius termitophilus (Tullgren), new combination

Chelifer termitophilus Tullgren, 1907b: 220–221, figs 1a–b; Ellingsen, 1912b: 85.
Allowithius termitophilus (Tullgren): Beier, 1932d: 55; Beier, 1932e: 204–205, fig. 211; Lawrence, 1937: 272; Beier, 1947b: 332; Beier, 1948b: 458–459; Beier, 1955l: 317; Beier, 1958a: 180; Beier, 1959d: 63; Beier, 1964k: 78–79; Beier, 1966k: 470.
Allowithius termitiphilus (sic) (Tullgren): Roewer, 1937: 308.

Type locality: Stamford Hill, Natal, South Africa.
Distribution: South Africa, Zaire.

Withius texanus (Banks)

Chelifer texanus Banks, 1891: 162.
Chelanops texanus (Banks): Banks, 1895a: 5; Banks, 1908: 39; Coolidge, 1908: 110–111.
Parachelifer (?) *texanus* (Banks): Beier, 1932e: 241; Roewer, 1937: 313.
Withius texanus (Banks): Hoff, 1947b: 542–545, fig. 36; Hoff, 1958: 36; Rowland & Reddell, 1976: 18.

Type locality: Brazos County, Texas, U.S.A.
Distribution: U.S.A. (Texas).

Withius transvaalensis (Beier)

Xenowithius transvaalensis Beier, 1953c: 76–77, fig. 2.
Withius transvaalensis (Beier): Mahnert, 1988a: 65–66.
Not *Xenowithius transvaalensis* Beier: Mahnert, 1975b: fig. 6b (misidentification; = *Sphaerowithius* sp.).

Type locality: Johannesburg, Transvaal, South Africa.
Distribution: South Africa.

Withius vachoni (Beier), new combination

Allowithius vachoni Beier, 1944: 200–202, fig. 17; Beier, 1954d: 136; Beier, 1959d: 62.

Type locality: Chiromo, Malawi.
Distribution: Malawi, Zaire.

Withius vagrans J. C. Chamberlin

Withius vagrans J. C. Chamberlin, 1925a: 332, figs h–k; Beier, 1932e: 198–199; J. C. Chamberlin, 1931a: figs 39q, 51l; J. C. Chamberlin, 1931c: 293; Roewer, 1937: 308; Hoff, 1947b: 545; Hoff, 1958: 36; Rowland & Reddell, 1976: 19.

Type locality: Brownsville, Texas, U.S.A.
Distribution: U.S.A. (Texas).

Nomina dubia and nomina nuda

The following species and genera are either so poorly characterized that their true identity is unknown (*nomina dubia*), or they are considered *nomina nuda*. The 'authorship' of *nomina nuda* is attributed to the author who first published the name, even though that author may have been citing an unpublished manuscript name of another author.

Genus-group names

Genus **Megathis** Stecker, nomen dubium

Megathis Stecker, 1875b: 519–521; With, 1906: 74; J. C. Chamberlin, 1925b: 337; J. C. Chamberlin, 1929a: 60; Beier, 1932a: 69.

Type species: *Megathis kochii* Stecker, 1875b, by subsequent designation of J. C. Chamberlin, 1925b: 337.

Genus **Orideobisium** Kishida, nomen nudum

Orideobisium Kishida, 1966: 7.

Species-group names

Neobisium amititae Schawaller, nomen nudum

Neobisium amititae Schawaller, 1983b: 15.

Chernes angustiventris Tömösváry, nomen dubium

Chernes (Trachychernes) cimicoides angustiventris Tömösváry, 1882b: 190.

Type locality: Hungary.
Distribution: Hungary.

Chelifer angustus Gervais, nomen dubium

Chelifer (Obisium) angustus Gervais, 1849: 11–12.
Austrochthonius (?) *angustus* (Gervais): Beier, 1932a: 40; Roewer, 1937: 238.

Type locality: Chile.
Distribution: Chile.

Roncocreagris aurouxi Zaragoza, nomen nudum

Roncocreagris aurouxi Zaragoza, 1986: fig. 1.

Chthonius (Globochthonius) caligatus angustus Beier, nomen nudum

Chthonius (Globochthonius) caligatus angustus Beier, 1939a: 5; Beier, 1939d: 21.

Chernes armatus Tömösváry, nomen dubium

Chernes armatus Tömösváry, 1884: 17–18, figs 8–9; Daday, 1889d: 80.
Chernes (Chernes) armatus (Tömösváry): Daday, 1888: 115, 170.
Chelifer armatus (Tömösváry): Ellingsen, 1912b: 82.
Genus ? *armatus* Tömösváry: Beier, 1932e: 277; Roewer, 1937: 317.

Type locality: Ashanti, Ghana.
Distribution: Ghana, Yugoslavia?.

Apochthonius barbarae Lawson, nomen nudum

Apochthonius barbarae Lawson, 1968: 4351 (not seen).

Chelifer baltistanus Caporiacco, nomen dubium

Chelifer baltistanus Caporiacco, 1935: 243–244, plate 7 figs 9a–b; Roewer, 1940: 347.

Type localities: many localities in Baltistan, Pakistan.
Distribution: Pakistan.

Chelifer bicolor Risso, nomen dubium

Chelifer bicolor Risso, 1826: 158.

Type locality: Nice, France.
Distribution: France.

Olpium bicolor E. Simon, nomen nudum

Olpium bicolor E. Simon, 1899b: 244.

Chelifer cimex Gervais, nomen dubium

Chelifer (Chelifer) cimex Gervais, 1849: 12.
Chelifer cimex Gervais: With, 1908a: 327.
Genus ? *cimex* Gervais: Beier, 1932e: 277; Roewer, 1937: 317.

Type locality: Chile.
Distribution: Chile.

Chelanops corticis Ewing, nomen dubium

Chelanops corticis Ewing, 1911: 75–76, 79, fig. 9; Nester, 1932: 98.
Genus ? *corticis* Ewing: Hoff, 1949b: 484.
Chelanops (?) corticis Ewing: Hoff, 1958: 24.

Type localities: Urbana, Illinois, U.S.A.; and Havana, Illinois, U.S.A.
Distribution: U.S.A. (Illinois, Indiana).

Diplotemnus decebali Dumitresco & Orghidan, nomen nudum

Diplotemnus decebali Dumitresco & Orghidan, 1966: 81.

Megathis desiderata Stecker, nomen dubium

Megathis desiderata Stecker, 1875b: 522–523, plate 4 figs 1–4; Beier, 1932a: 69.
Megathis disiderata (sic) Stecker: With, 1906: 74.
Megatis (sic) *desiderata* Stecker: Roewer, 1937: 240.

Type locality: India.
Distribution: India.

! Chelifer ehrenbergii C. L. Koch & Berendt, nomen dubium

Chelifer ehrenbergii C. L. Koch & Berendt, 1854: 95, fig. 95; Hagen, 1870: 265.
Genus ? *ehrenbergii* C. L. Koch & Berendt: Beier, 1932e: 278.

Type locality: Baltic Amber.
Distribution: Baltic Amber.

Austrochthonius elegans Vitali-di Castri, nomen nudum

Austrochthonius elegans Vitali-di Castri, 1968: 145; Vitali-di Castri: 1975a: 125.

! Chelifer eucarpus Dalman, nomen dubium

Chelifer eucarpus Dalman, 1825: 408–409, fig. 25.

Type locality: Copal.
Distribution: Copal.

Chelifer excentricus Holmberg, nomen dubium

Chelifer excentricus Holmberg, 1874: 299, figs 6, 6a; Holmberg, 1876: 28; With, 1908a: 327.
Genus ? *excentricus* Holmberg: Beier, 1932e: 277; Roewer, 1937: 317.

Type locality: Argentina.
Distribution: Argentina.

Chernes hungaricus Daday, nomen dubium

Chernes (Ectoceras) hungaricus Daday, 1888: 119, 174–175, plate 4 figs 1–2; Daday, 1918: 2. Junior homonym of *Chernes (Chernes) cyrneus hungaricus* Daday, 1888.

Type locality: Paulis, Arad, Romania.
Distribution: Romania.

Chelifer hydaspis Caporiacco, nomen dubium

Chelifer hydaspis Caporiacco: 244–245, plate 7 figs 8a–b; Roewer, 1940: 347.

Type locality: Srinagar, Kashmir, India.
Distribution: India.

Chthonius (Ephippiochthonius) labilus Dumitresco & Orghidan, nomen nudum

Chthonius (Ephippiochthonius) labilus Dumitreco & Orghidan, 1964: figs 21–26, 27a–b, plate 1 fig. 1 (see *Chthonius (Ephippiochthonius) tetrachelatus* (Preyssler)).

Chelifer loewi Hagen, nomen nudum

Chelifer loewi Hagen, 1879: 400; Vachon, 1940f: 2.
Genus ? *loewi* Hagen: Beier, 1932e: 277; Roewer, 1937: 317.
Lamprochernes (?) *loewi* (Hagen): Beier, 1948b: 444.

Obisium longicolle Frauenfeld, nomen dubium

Obisium longicolle Frauenfeld, 1867: 461; With, 1906: 77.
Genus ? *longicolle* Frauenfeld: Beier, 1932e: 277; Roewer, 1937: 317.

Type locality: Nicobar Islands.
Distribution: Nicobar Islands.

Megathis kochii Stecker, nomen dubium

Megathis kochii Stecker, 1875b: 521–522, plate 2 figs 9–10, 12–14, plate 1 figs 1–4, 6; Beier, 1932a: 69.
Megathis kochi Stecker: With, 1906: 74.
Megatis (sic) *kochii* Stecker: Roewer, 1937: 240.

Type locality: India.
Distribution: India.

Chelifer mariannus Gervais, nomen dubium

Chelifer mariannus Gervais, 1844: 80−81.

Type locality: Marianas (as Mariannes) Islands.
Distribution: Marianas Islands.

Chelifer minax Gerstaecker, nomen dubium

Chelifer minax Gerstaecker, 1873: 470−472.
Genus ? *minax* Gerstaecker: Beier, 1932e: 277; Roewer, 1937: 317.

Type locality: Endara, Zanzibar, Tanzania.
Distribution: Tanzania.

Chernes nigricans Tömösváry, nomen dubium

Chernes (Trachychernes) cimicoides nigricans Tömösváry, 1882b: 190.

Type locality: Hungary.
Distribution: Hungary.

Chelifer nipponicus Kishida, nomen dubium

Chelifer nipponicus Kishida, 1927: ? (not seen).
Chelanops nipponicus (Kishida): Kishida, 1947: ? (not seen).

Type locality: ?
Distribution: Japan.

Chelifer parcegranosus Beier, nomen nudum

Chelifer parcegranosus Beier, 1932e: 277.

Obisium pusio Kolenati, nomen dubium

Obisium pusio Kolenati, 1857: 431; Hagen, 1870: 270; With, 1906: 77.
Genus ? *pusio* Kolenati: Beier, 1932e: 277; Roewer, 1937: 317.

Type locality: Calcutta, India.
Distribution: India.

Neobisium redikorzevi Schawaller, nomen nudum

Neobisium redikorzevi Schawaller, 1983b: 15.

Chernes rufescens Tömösváry, nomen dubium

Chernes (Trachychernes) cimicoides rufescens Tömösváry, 1882b: 190.

Type locality: Hungary.
Distribution: Hungary.

Chelifer saltator Brébisson, nomen dubium

Chelifer saltator Brébisson, 1827: 264; Gervais, 1844: 80.
Chthonius (?) *saltator* (Brébisson): Beier, 1932a: 61.

Type locality: Calvados, France.
Distribution: France.

Neobisium saqarthvelosi Kobakhidze, nomen nudum

Neobisium saqarthvelosi Kobakhidze, 1965b: 541.

Chelifer sardous Beier, nomen nudum

Chelifer sardous Beier, 1932e: 277.

Olpium savignyi E. Simon, nomen dubium

Chelifer hermannii Audouin, 1826: 175, plate 8 fig. 5; Gervais, 1844: 82, plate 25 fig. 1 (junior primary homonym of *Chelifer hermannii* Leach).
Chelifer hermanii (sic) Audouin: Audouin, 1827: 414, plate 8 fig. 5.
Olpium hermanni (Audouin): L. Koch, 1873: 37–38.
Olpium savignyi E. Simon, 1879a: 49; Ellingsen, 1910a: 392–393; Beier, 1929a: 359; Beier, 1932a: 183, fig. 209; Beier, 1932f: 487; Roewer, 1937: 261 (replacement name for *Chelifer hermannii* Audoiun, 1826).
Olpium pallipes (Lucas): Tullgren, 1907d: 10, fig. 4 (misidentification).
Minniza savignyi (E. Simon): Harvey & Mahnert, 1985: 86 (designated as nomen dubium).

Type locality: Egypt.
Distribution: Egypt, Libya.

Chelifer setosus Schrank, nomen dubium

Chelifer setosus Schrank, 1803: 245.
Genus ? *setosum* Schrank: Beier, 1932e: 277; Roewer, 1937: 317.

Type locality: Tegernsee, West Germany.
Distribution: West Germany.

Chelifer stellatus Navás, nomen dubium

Chelifer stellatus Navás, 1921: 165–166, figs 3a–b; Roewer, 1937: 317; Roewer, 1940: 347.

Type locality: Macias Nguema Biyogo (as Fernando Póo), Equatorial Guinea.
Distribution: Equatorial Guinea.

Orideobisium takanoanum Kishida, nomen nudum

Orideobisium takanoanum Kishida, 1966: 7.

Allowithius tbilissicus Kobakhidze, nomen nudum

Allowithius tbilissicus Kobakhidze, 1965b: 542.

Obisium stussineri tenuimanus E. Simon, nomen nudum

Obisium (Roncus) stussineri tenuimanus E. Simon, 1898d: 22.

Chelifer timidus Holmberg, nomen dubium

Chelifer timidus Holmberg, 1876: 28; With, 1908a: 327.
Genus ? *timidus* Holmberg: Beier, 1932e: 277; Roewer, 1937: 317.

Type locality: Argentina.
Distribution: Argentina.

Obisium touricum Beier, nomen nudum

Obisium touricum Beier, 1932e: 277 (nomen nudum).
Genus ? *tauricum* (sic) Beier: Roewer, 1937: 317.

Neobisium (Ommatoblothrus) vandeli Beron, nomen nudum

Neobisium (Ommatoblothrus) vandeli Beron, 1972: 13.

Maorigarypus viridis Tubb, nomen dubium

Maorigarypus viridis Tubb, 1937: 412–413, fig. 1; Roewer, 1940: 345; Harvey, 1987a: 53.

Synsphyronus (Maorigarypus) viridis (Tubb): J. C. Chamberlin, 1943: 498–499; Harvey, 1981b: 243; Harvey, 1985b: 145.

Type locality: Lady Julia Percy Island, Victoria, Australia.
Distribution: Australia (Victoria).

Forms and varieties

Under Articles 16 and 45g of the International Code of Zoological Nomenclature (third edition), the following names, which were originally proposed as forms (f.) or varieties (var.), are treated as infrasubspecific. Even though they are not species group names and are not bound by the Code (Articles 1b (5) and 45e), they are listed in the index in italics.

Microcreagris microdivergens form mediocris Morikawa

Microcreagris microdivergens f. *mediocris* Morikawa, 1955a: 220, figs 2e–f; Morikawa, 1960: 124, plate 9 fig. 12, plate 10 fig. 24.

Microcreagris microdivergens variety rectus Morikawa

Microcreagris microdivergens var. *rectus* Morikawa, 1955a: 220–221, figs 1c–d, 2g–h, 3d; Morikawa, 1960: 124, plate 8 fig. 10, plate 9 fig. 13, plate 10 fig. 25.

Ochrochernes indicus form montanus Beier

Ochrochernes indicus f. *montanus* Beier, 1974e: 1014, fig. 9.

Summary of taxonomic changes

Replacement names

Paratemnoides, new name for *Paratemnus* Beier, 1932b (junior homonym of *Paratemnus* Ameghino, 1904 (see p. 469).

Insulocreagris troglobia, new name for *Roncus (Parablothrus) cavernicola* Beier, 1928, junior primary homonym of *Roncus lubricus cavernicola* Tömösváry, 1882b (see p. 336).

Balkanoroncus hadzii, new name for *Roncus (Parablothrus) bureschi* Hadži, 1939b, junior secondary homonym of *Balkanoroncus bureschi* (Redikorzev, 1928) (see p. 331).

Parobisium anaganidensis morikawai, new name for *Neobisium (Parobisium) anagamidensis longidigitatus* Morikawa, 1957a, junior primary homonym of *Neobisium longidigitatum* (Ellingsen, 1908) (see p. 394).

Chernes beieri, new name for *Chernes pallidus* Beier, 1936b junior primary homonym of *Chernes pallidus* Banks, 1890 (see p. 557).

Summary of new type species

Chernes (Trachychernes) Tömösváry, 1882b: type species *Scorpio cimicoides* Fabricius, 1793 (see p. 557).

Chthonius (Globochthonius) Beier, 1931b: type species *Chthonius globifer* E. Simon, 1879a (see p. 175).

Chthonius (Hesperochthonius) Muchmore, 1968b: type species *Chthonius californicus* J. C. Chamberlin, 1929a (see p. 176).

Feaella (Tetrafeaella) Beier, 1955c: type species *Feaella mucronata* Tullgren, 1907b (see p. 231).

Parachernes (Argentochernes) Beier, 1932e: type species *Chelifer (Trachychernes) albomaculatus* Balzan, 1892 (see p. 612).

Roncus (Parablothrus) Beier, 1928: type species *Obisium (Blothrus) stussineri* E. Simon, 1881a (see p. 400).

New synonymies

Neobisium (Neobisium) lombardicum emiliae Beier, 1963b, with *Neobisium (Neobisium) lombardicum martae* (Menozzi, 1920) (see p. 362).

Obisium troglodytes Schmidt, 1848, with *Neobisium (Blothrus) spelaeum* (Schiödte, 1847) (see p. 383).

Diplotemnus beieri Vachon, 1970, with *Diplotemnus ophthalmicus* Redikorzev, 1949 (see p. 461).

Oligochelifer Beier, 1937d, with *Dichela* Menge, in C.L. Koch & Berendt, 1854 (see p. 498).

Lophochelifer Beier, 1940a, with *Lissochelifer* J. C. Chamberlin, 1932a (see p. 509).

New combinations

Chthoniidae

Austrochthonius tullgreni (Beier) (from *Paraustrochthonius*)
Chthonius (Chthonius) paganus (Hoff) (from *Kewochthonius*)
Lagynochthonius arctus (Beier) (from *Tyrannochthonius*)
Lagynochthonius brincki (Beier) (from *Tyrannochthonius*)
Lagynochthonius flavus (Mahnert) (from *Tyrannochthonius*)
Lagynochthonius guasirih (Mahnert) (from *Tyrannochthonius*)
Lagynochthonius himalayensis (Morikawa) (from *Tyrannochthonius*)
Lagynochthonius kenyensis (Mahnert) (from *Tyrannochthonius*)
Lagynochthonius nagaminei (Sato) (from *Tyrannochthonius*)
Lagynochthonius novaeguineae (Beier) (from *Tyrannochthonius*)
Lagynochthonius salomonensis (Beier) (from *Tyrannochthonius*)
Lagynochthonius sinensis (Beier) (from *Tyrannochthonius*)
Pseudotyrannochthonius dentifer (Morikawa) (from *Allochthonius*)
Pseudotyrannochthonius kobayashii akiyoshiensis (Morikawa) (from *Spelaeochthonius*)
Pseudotyrannochthonius kobayashii dorogawaensis (Morikawa) (from *Spelaeochthonius*)
Pseudotyrannochthonius undecimclavatus kishidai (Morikawa) (from *Allochthonius*)

Hyidae

Parahya submersa (Bristowe) (from *Obisium*)

Neobisiidae

Bisetocreagris kaznakovi lahaulensis (Mani) (from *Microcreagris*)
Neobisium (Neobisium) lombardicum martae (Menozzi) (from *Obisium*)
Neobisium (Neobisium) schenkeli (Strand) (from *Obisium*)
Pararoncus chamberlini (Morikawa) (from *Roncus*)
Parobisium anagamidensis anagamidensis (Morikawa) (from *Neobisium*)
Parobisium anagamidensis esakii (Morikawa) (from *Neobisium*)
Parobisium flexifemoratum (J. C. Chamberlin) (from *Neobisium*)
Parobisium magnum magnum (J. C. Chamberlin) (from *Neobisium*)
Parobisium magnum chejuense (Morikawa) (from *Neobisium*)
Parobisium magnum ohuyeanum (Morikawa) (from *Neobisium*)

Syarinidae

Ideoblothrus ceylonicus (Beier) (from *Ideobisium*)

Atemnidae

Diplotemnus ophthalmicus sinensis (Schenkel) (from *Miratemnus*)
Paratemnoides aequatorialis (Beier) (from *Paratemnus*)
Paratemnoides assimilis (Beier) (from *Paratemnus*)
Paratemnoides borneoensis (Beier) (from *Paratemnus*)
Paratemnoides ceylonicus (Beier) (from *Paratemnus*)
Paratemnoides curtulus (Redikorzev) (from *Paratemnus*)
Paratemnoides ellingseni (Beier) (from *Paratemnus*)
Paratemnoides elongatus (Banks) (from *Paratemnus*)
Paratemnoides feai (Ellingsen) (from *Paratemnus*)
Paratemnoides guianensis (Caporiacco) (from *Paratemnus*)
Paratemnoides indicus (Sivaraman) (from *Paratemnus*)
Paratemnoides indivisus (Tullgren) (from *Paratemnus*)
Paratemnoides insubidus (Tullgren) (from *Paratemnus*)

Paratemnoides insularis (Banks) (from *Paratemnus*)
Paratemnoides japonicus (Morikawa) (from *Paratemnus*)
Paratemnoides laosanus (Beier) (from *Paratemnus*)
Paratemnoides magnificus (Beier) (from *Paratemnus*)
Paratemnoides mahnerti (Sivaraman) (from *Paratemnus*)
Paratemnoides minor (Balzan) (from *Paratemnus*)
Paratemnoides minutissimus (Beier) (from *Paratemnus*)
Paratemnoides nidificator (Balzan) (from *Paratemnus*)
Paratemnoides obscurus (Beier) (from *Paratemnus*)
Paratemnoides pallidus (Balzan) (from *Paratemnus*)
Paratemnoides perpusillus (Beier) (from *Paratemnus*)
Paratemnoides persimilis (Beier) (from *Paratemnus*)
Paratemnoides philippinus (Beier) (from *Paratemnus*)
Paratemnoides plebejus (With) (from *Paratemnus*)
Paratemnoides pococki (With) (from *Paratemnus*)
Paratemnoides redikorzevi (Beier) (from *Paratemnus*)
Paratemnoides robustus (Beier) (from *Paratemnus*)
Paratemnoides salomonis salomonis (Beier) (from *Paratemnus*)
Paratemnoides salomonis hebridicus (Beier) (from *Paratemnus*)
Paratemnoides sinensis (Beier) (from *Paratemnus*)
Paratemnoides singularis (Beier) (from *Paratemnus*)
Paratemnoides sumatranus (Beier) (from *Paratemnus*)

Cheliferidae

Dichela gracilis (Beier) (from *Oligochelifer*)
Dichela granulatus (Beier) (from *Oligochelifer*)
Dichela serratidentatus (Beier) (from *Oligochelifer*)
Lissochelifer depressoides (Beier) (from *Lophochelifer*)
Lissochelifer gibbosounguiculatus (Beier) (from *Lophochelifer*)
Lissochelifer gracilipes (Mahnert) (from *Lophochelifer*)
Lissochelifer hygricus (Murthy & Ananthakrishnan) (from *Lophochelifer*)
Lissochelifer insularis (Beier) (from *Lophochelifer*)
Lissochelifer nairobiensis (Mahnert) (from *Lophochelifer*)
Lissochelifer novaeguineae (Beier) (from *Lophochelifer*)
Lissochelifer philippinus (Beier) (from *Lophochelifer*)
Lissochelifer strandi (Ellingsen) (from *Lophochelifer*)
Lissochelifer tonkinensis (Beier) (from *Lophochelifer*)

Chernetidae

Americhernes eidmanni (Beier) (from *Pycnochernes*)
Americhernes guarany (Feio) (from *Pycnochernes*)
Americhernes kanaka (J.C. Chamberlin) (from *Lamprochernes*)
Americhernes samoanus (J.C. Chamberlin) (from *Lamprochernes*)
Caffrowithius garambae (Beier) (from *Plesiochernes*)
Caffrowithius hanangensis hanangensis (Beier) (from *Plesiochernes*)
Caffrowithius hanangensis curtus (Beier) (from *Plesiochernes*)
Caffrowithius lucifugus (Beier) (from *Plesiochernes*)
Caffrowithius meruensis (Beier) (from *Plesiochernes*)
Caffrowithius pusillimus (Beier) (from *Plesiochernes*)
Caffrowithius simplex (Beier) (from *Plesiochernes*)
Parachernes (Scapanochernes) cordimanus (Beier) (from *Scapanochernes*)

Withiidae

Withius angolensis (Beier) (from *Allowithius*)
Withius ascensionis (Beier) (from *Allowithius*)
Withius australasiae australasiae (Beier) (from *Allowithius*)
Withius australasiae formosanus (Beier) (from *Allowithius*)
Withius capensis (Beier) (from *Allowithius*)
Withius ceylanicus (Ellingsen) (from *Allowithius*)
Withius congicus exiguus (Beier) (from *Allowithius*)
Withius glabratus (Ellingsen) (from *Allowithius*)
Withius lawrencei (Beier) (from *Allowithius*)
Withius madagascariensis (Ellingsen) (from *Allowithius*)
Withius tenuimanus (Balzan) (from *Allowithius*)
Withius termitophilus (Tullgren) (from *Allowithius*)
Withius vachoni (Beier) (from *Allowithius*)

Index

Family-group names are capitalized; junior synonyms, junior homonyms, nomina dubia and nomina nuda are italicized. The genus in which a species-group name was originally described is shown in brackets preceded with ' = '.

aalbui, Archeolarca 236
abaris, Acanthocreagris 328
abditus, Roncus 400
abeillei, Neobisium (Blothrus)
  (= Blothrus) 373
aberrans, Neobisium (Blothrus)
  dalmatinum 376
aberrans, Thaumatowithius 657
abnormis, Chthonius (Globochthonius)
  175
abnormis, Dolichowithius
  (Oligowithius) 645
abnormis, Microcreagris 340
absitus, Synsphyronus 244
absoloni, Chthonius (Chthonius) 143
absoloni, Neobisium (Neobisium) 373
aburi, Compsaditha 219
abyssinica, Pycnodithella
  (= Verrucadithella) 224
abyssinicus, Trichotowithius 658
abyssinicus, Withius (= Allowithius)
  658
Acanthicochernes 534
Acanthocreagris 323
acarinatus, Microchelifer 517
acaroides, Phalangium 565
acherusium, Neobisium (Blothrus)
  lethaeum 379
Acis 454
actuarium, Neobisium (Neobisium)
  346

acuminatus, Lustrochernes (= Chelifer)
  594
Acuminochernes 534
acutum, Electrobisium 334
addititius, Tridenchthonius 225
Adelphochernes 535
Adelphochernes 548
adiposum, Pachyolpium 296
adisi, Parachernes (Parachernes)
  613
aegatensis, Chthonius
  (Ephippiochthonius) 163
aegeus, Hadoblothrus 419
aeginense, Cardiolpium
  (= Apolpiolum) 273
aegyptiacum, Halominniza (= Olpium)
  279
aelleni, Acanthocreagris 323
aelleni, Negroroncus 320
aelleni, Neobisium (Neobisium) 346
aequalis, Dinocheirus (= Chelanops)
  570
aequatorialis, Beierius 484
aequatorialis, Caffrowithius 549
aequatorialis, Minniza 284
aequatorialis, Paratemnoides
  (= Paratemnus) 469
aequatorialis, Smeringochernes
  (Smeringochernes) (= Chelifer) 632
aethiopicus, Afrosternophorus
  (= Sternophorus) 446

aethiopicus, Caffrowithius
  (=Adelphochernes) 549
aethiopicus, Nannowithius (=Chelifer)
  648
aetnaeus, Allochernes 535
aetnensis, Roncus 400
affinis, Chelanops (Chelanops) 554
affinis, Chernes 627
affinis, Cyclatemnus 459
affinis, Feaella (Tetrafeaella) 231
affinis, Kleptochthonius
  (Chamberlinochthonius) 179
afghanica, Microcreagris 340
afghanica, Minniza babylonica 284
afghanicum, Indolpium 282
afghanicum, Olpium 290
afghanicus, Centrochelifer 486
afghanicus, Dactylochelifer 491
afghanicus, Dhanus 318
afghanicus, Diplotemnus 461
afghanicus, Garypinus 277
afghanicus, Hysterochelifer 505
afghanicus, Megachernes 599
afghanicus, Oratemnus 466
afghanicus, Rhacochelifer 525
afghanicus, Tullgrenius 481
africanum, Neocheiridium 442
africanum, Pseudochiridium 444
africanus, Calocheiridius 268
africanus, Myrmochernes 604
africanus, Negroroncus (=Ideoroncus)
  320
africanus, Solinus 304
africanus, Tridenchthonius (=Ditha)
  225
afrikanus, Lamprochernes nodosus
  (=Chelifer) 591
Afrobisium 337
Afrocheiridium 444
Afrochthonius 131
Afroditha 225
Afrogarypus 249
Afroroncus 316
Afrosternophorus 446
Afrowithius 640
agassizi, Cerogarypus 233
agazzii, Acanthocreagris
  (=Microcreagris) 323
agazzii, Chthonius (Chthonius) 143
Aglaochitra 417
agniae, Ancistrochelifer 482
agniae, Euryolpium 275
agnolettii, Neobisium (Neobisium) 346
Aisthetowithius 640

aitkeni, Protochelifer cavernarum 524
akiyoshiensis, Pseudotyrannochthonius
  kobayashii (=Spelaeochthonius) 199
Alabamocreagris 329
alacki, Systellochernes 636
albanicum, Neobisium (Blothrus)
  (=Obisium) 374
albidus, Tyrannochthonius
  (=Morikawia) 205
Albiorix 316
albomaculatus, Parachernes
  (Parachernes) (=Chelifer) 613
albus, Geogarypus 252
Aldabrinus 262
aldabrinus, Aldabrinus 262
algerica, Minniza 284
algericum, Neobisium (Neobisium)
  (=Obisium) 347
alienum, Neoamblyolpium 287
alius, Chelifer 542
allocancroides, Hysterochelifer 505
Allochernes 535
Allochthonius 132
Allochthonius, Allochthonius 132
allodentatum, Roncobisium
  (=Neobisium) 396
Allowithius 658
alluaudi, Titanatemnus 478
Alocobisium 417
alpicola, Chthonius (Chthonius) 143
alpinus, Mundochthonius 190
alpinus, Roncus 400
alter, Lophochernes 511
alteriae, Aphrastochthonius 134
alternum, Neobisium (Neobisium)
  gentile 358
alticola, Neobisium (Neobisium) 347
altimanus, Chelanops (Chelanops)
  (=Chelifer) 554
Alura 228
Amaurochelifer 482
amazonicus, Ceriochernes 553
amazonicus, Geogarypus 252
amazonicus, Ideoblothrus
  (=Ideobisium) 421
amazonicus, Tyrannochthonius 205
Amblyolpium 262
amboinense, Euryolpium
  (=Xenolpium) 275
ambrosianus, Anaperochernes 543
americanus, Chelifer 566
Americhernes 540
Americocreagris 329
amititae, Neobisium 667

Ammogarypus 235
amoenus, Chernes (= Hesperochernes) 557
amoenus, Pugnochelifer 525
amplissimus, Synsphyronus 244
amplum, Olpiolum 289
amplus, Neochthonius (= Kewochthonius) 192
amrithiensis, Calocheiridius 269
amrithiensis, Indogaryops 447
amurensis, Dactylochelifer (= Chelifer) 491
anabates, Afrosternophorus 447
anachoreta, Pselaphochernes (= Chelifer) 624
anagamidensis, Parobisium (= Neobisium) 394
Anagarypus 235
Anaperochernes 543
Anatemnus 451
anatolica, Acanthocreagris (= Microcreagris) 324
anatolica, Lechytia 186
anatolicum, Amblyolpium 263
anatolicum, Neobisium (Neobisium) 347
anatolicus, Beierochelifer (= Rhacochelifer) 485
anatolicus, Chthonius (Ephippiochthonius) 163
anatolicus, Dactylochelifer 491
anatolicus, Lasiochernes 592
Anaulacodithella 217
Ancalochernes 543
Ancistrochelifer 482
andalusica, Roncocreagris iberica (= Microcreagris) 399
anderseni, Feaella (Tetrafeaella) 231
anderssoni, Micratemnus 464
andinum, Cheiridium 437
andinum, Teratolpium 306
andinus, Americhernes (= Lamprochernes) 540
andinus, Lustrochernes 594
andreinii, Rhacochelifer 525
andreinii, Roncus (= Chelifer) 401
andrewsi, Hesperolpium 280
Anepsiochernes 549
angolense, Olpium 291
angolensis, Withius (= Allowithius) 658
angulatus, Calymmachernes 552
angulatus, Geogarypus 252
angulatus, Stenowithius (= Chelifer) 656

angustatus, Withius (= Chelifer) 659
angustimana, Anaulacodithella 217
angustiventris, Chernes 667
angustochelatus, Cordylochernes 565
angustula, Compsaditha 219
angustum, Cheiridium 437
angustus, Anatemnus 451
angustus, Chelifer 494
angustus, Chelifer 667
angustus, Chthonius (Globochthonius) caligatus 667
angustus, Stenowithius bayoni 656
Anisoditha 218
annamensis, Amaurochelifer 482
annamensis, Bisetocreagris (= Microcreagris) 331
annamensis, Hyperwithius (= Sundowithius) 646
annamensis, Lagynochthonius (= Tyrannochthonius) 182
annamensis, Stenatemnus 475
anophthalma, Typhloditha 227
anophthalmus, Chthonius (Ephippiochthonius) 163
anophthalmus, Ideoroncus 319
anophthalmus, Kleptochthonius (Chamberlinochthonius) 179
anophthalmus, Roncus (= Obisium) 401
antarcticus, Apatochernes 545
anthracinus, Allochthonius 133
Anthrenochernes 544
antillarum, Rhopalochernes (= Chelifer) 631
Antillobisium 309
Antillochernes 544
Antiolpium 307
antipodum, Ideobisium (= Obisium) 420
antiquum, Cryptocheiridium (Cryptocheiridium) 441
antrorum, Roncus (= Obisium) 401
antushi, Calocheiridius 269
aokii, Cheiridium 437
Apatochernes 545
Aperittochelifer 483
Aphelolpium 264
Aphrastochthonius 134
apimelus, Synsphyronus 244
Apocheiridium 433
Apocheiridium, Apocheiridium 433
Apochthonius 136
Apohya 309
apollinis, Chthonius (Chthonius) 143

Apolpiolum 273
Apolpium 265
Aporochelifer 483
approximatus, Parachelifer
  (=Chelifer) 520
apuanica, Acanthocreagris 324
apuanicum, Neobisium (Neobisium)
  347
apulica, Acanthocreagris 324
apulicus, Chthonius
  (Ephippiochthonius) 163
arabicum, Parolpium (=Olpium) 297
arabicus, Withius 659
araucariae, Afrosternophorus
  (=Sternophorellus) 447
araxellus, Roncus 401
arborea, Lechytia 186
arboreum, Planctolpium 297
arboricola, Albiorix (=Ideoroncus) 316
arboricola, Serianus (=Garypinus) 301
archboldi, Parachelifer 520
Archeolarca 236
archeri, Chitrella 418
arctus, Lagynochthonius
  (=Tyrannochthonius) 183
arcuodigitus, Parachernes
  (Parachernes) 613
arenicola, Dactylochelifer kussariensis
  493
argentatopunctatus, Parachernes
  (Parachernes) 613
argentinae, Austrochthonius 139
argentinae, Serianus 302
argentiniensis, Albiorix (=Dinoroncus)
  316
argentinus, Dolichowithius
  (Dolichowithius) 643
argentinus, Lustrochernes (=Chelifer)
  594
argentinus, Oligomenthus 261
argentinus, Parachernes (Parachernes)
  613
Argentochernes, Parachernes 612
ariasi, Chelifer (Atemnus) 455
ariditatis, Lustrochernes (=Chelanops)
  594
arizonensis, Dinocheirus (=Chelanops)
  570
armasi, Mexobisium 310
armatus, Chernes 668
armeniacus, Garypus 238
armenius, Chernes (=Chelifer) 557
armiger, Semeiochernes (=Chelifer)
  632

articulosus, Oratemnus (=Chelifer)
  466
arubense, Pachyolpium 295
arubensis, Pseudochthonius 195
ascensionis, Withius (=Allowithius)
  659
ashmolei, Rhinochernes 631
asiatica, Lechytia 186
asiaticum, Indolpium 282
asiaticum, Olpium 291
asiaticus, Allochernes (=Chelifer) 535
asiaticus, Geogarypus 252
asiaticus, Ochrochernes 609
asiaticus, Stenatemnus 475
asper, Garypinus 278
asper, Horus 280
asperum, Apocheiridium
  (Apocheiridium) 433
asperum, Stenolpium 305
Aspurochelifer 484
assimilis, Paratemnoides
  (=Paratemnus) 469
assimilis, Roncus 401
astatum, Ectactolpium 274
Asterochernes 547
asturiensis, Chthonius
  (Ephippiochthonius) 163
astutus, Dinocheirus 570
ATEMNIDAE 451
Atemnus 454
aterrimus, Haplochernes 580
Atherochernes 547
Athleticatemnus 456
athleticus, Dinocheirus 570
atlantica, Microcreagris 341
atlanticum, Pachyolpium 295
atlanticus, Chelanops (Chelanops) 554
atlantis, Chthonius
  (Ephippiochthonius) 164
atopos, Calocheirus 273
atopus, Chelodamus 556
atrimanus, Haplochernes (=Chelifer)
  580
attenuatus, Kleptochthonius
  (Chamberlinochthonius) 179
attenuatus, Macrochernes 598
attenuatus, Typhloroncus 322
attiguus, Synsphyronus 244
auberti, Neobisium (Blothrus) 374
aucta, Microcreagris ronciformis
  328
audyi, Geogarypus 255
aueri, Neobisium (Blothrus) 374
aureum, Euryolpium 275

aureum, Olpiolum (= Pachyolpium) 289
aureus, Florichelifer 501
aureus, Tamenus 476
aurouxi, Roncocreagris 667
ausculator, Nannoroncus 320
auster, Parachernes (Parachernes) 613
australasiae, Withius (= Allowithius) 659
australianus, Anagarypus 236
australica, Anaulacodithella 217
australicum, Cryptocheiridium (Cryptocheiridium) 441
australicum, Olpium 291
australicus, Heterolophus 223
australicus, Lagynochthonius (= Tyrannochthonius) 183
australicus, Nesidiochernes 605
australiensis, Austrochernes (= Chelifer) 547
australiensis, Pseudotyrannochthonius 198
australiensis, Solinus 304
australiensis, Sundochernes 635
Australinocreagris 329
australis, Austrochthonius 139
australis, Mesochernes 602
australis, Protochelifer (= Idiochelifer) 523
Australochelifer 484
austriacus, Chthonius (Ephippiochthonius) 166
Austrochernes 547
Austrochthonius 139
Austrohorus 266
azanius, Negroroncus 320
azanius, Paraliochthonius 193
azerbaidzhanus, Chthonius (Chthonius) 144
aztecus, Epichernes 576

babusnicae, Neobisium (Blothrus) 374
babylonica, Minniza 284
baccettii, Chthonius (Chthonius) 144
baccettii, Roncus 402
bachardensis, Chernes 574
bachofeni, Electrochelifer 499
bachofeni, Oligochelifer 610
badonneli, Calocheiridius 269
bagus, Tyrannochthonius 205
bahamensis, Antillochernes 544
bahamensis, Tyrannochthonius 205
baileyi, Pachychernes 611

bakeri, Lagynochthonius (= Tyrannochthonius) 183
Balanowithius 640
balazuci, Chthonius (Chthonius) 144
balazuci, Neobisium (Neobisium) 347
balcanica, Acanthocreagris (= Microcreagris) 324
balcanicum, Neobisium (Neobisium) carcinoides 352
balcanicum, Olpium pallipes 293
balcanicus, Allochernes 536
balcanicus, Atemnus 456
balcanicus, Pselaphochernes 625
balcanicus, Rhacochelifer (= Chelifer) 525
baldensis, Balkanoroncus 330
balearica, Acanthocreagris (= Microcreagris) 324
balearicum, Neobisium (Neobisium) ischyrum 361
balearicus, Chthonius (Ephippiochthonius) 164
balearicus, Dactylochelifer 491
balearicus, Pselaphochernes 625
balearicus, Roncus 410
Balkanoroncus 330
baloghi, Ideoblothrus (= Ideobisium) 422
balticus, Electrochelifer 499
baltistanus, Chelifer 668
balzanii, Ideobisium 420
balzanii, Lophochernes (= Chelifer) 512
baniskhevii, Neobisium 372
banksi, Pseudogarypus 233
Banksolpium 266
barbarae, Apochthonius 668
barbatus, Megachernes 599
barbei, Roncus 402
barkhamae, Rhacochelifer 525
baronii, Garypus 240
barri, Kleptochthonius (Chamberlinochthonius) 179
barrosi, Allochernes (Allochernes) 537
barrosi, Obisium 355
bartolii, Chthonius (Ephippiochthonius) 164
basarukini, Mundochthonius 190
basiléensis, Chelifer (Trachychernes) cimicoides 559
basilewskyi, Afrogarypus (= Geogarypus) 249
basilewskyi, Compsaditha 219
basilewskyi, Hansenius 502

basilewskyi, Nudochernes 607
basilewskyi, Spaerowithius
  (=Nannowithius) 655
bathumi, Neobisium 357
battonii, Neobisium (Ommatoblothrus)
  388
bauneensis, Chthonius
  Ephippiochthonius) 164
bayoni, Stenowithius (=Chelifer) 656
beauvoisii, Garypus (=Chelifer) 238
beckeri, Rhopalochernes 631
becki, Pselaphochernes 625
beieri, Acanthocreagris 324
beieri, Calocheiridius (=Minniza) 269
beieri, Chernes 557
beieri, Chthonius (Ephippiochthonius)
  164
beieri, Dactylochelifer 491
beieri, Diplotemnus 461
beieri, Goniochernes 579
beieri, Ideoroncus 319
beieri, Indochernes 587
beieri, Nannobisium 427
beieri, Neobisium (Blothrus) 380
beieri, Neobisium (Neobisium) 347
beieri, Neocheiridium 442
beieri, Pseudochthonius 195
beieri, Roncocreagris 397
beieri, Roncus 402
beieri, Tridenchthonius 225
beieri, Tyrannochthonius 205
Beierius 484
Beierobisium 312
Beierochelifer 485
Beierolpium 266
Beierowithius 641
belizense, Vachonium 431
bellesi, Chthonius (Ephippiochthonius)
  164
bellesi, Roncus 402
bellum, Amblyolpium 263
benoiti, Beierolpium 266
berendtii, Dichela 498
bergeri, Caffrowithius
  (=Plesiochernes) 549
berlandi, Cyclatemnus 459
bernardi, Neobisium (Neobisium) 347
berninii, Chthonius
  (Ephippiochthonius) 164
beroni, Neobisium (Heoblothrus) 387
bertalanii, Chernes scorpioides 627
bessoni, Neobisium (Ommatoblothrus)
  388
besucheti, Ceriochernes 553

besucheti, Dactylochelifer 491
besucheti, Indohya 315
bethaniae, Americhernes 541
biaroliatum, Amblyolpium (=Olpium)
  263
bicarinatus, Lophochernes 512
bicolor, Caffrowithius
  (=Plesiochernes) 549
bicolor, Chelifer 668
bicolor, Olpium 668
bicolor, Parachernes (Parachernes) 613
bicolor, Rhacochelifer corcyrensis 526
bicornis, Pseudogarypus (=Garypus)
  233
bidens, Ectoceras 499
bidentatus, Chthonius
  (Ephippiochthonius) 164
bifissus, Lophochernes (=Chelifer) 512
bigoti, Cheirochelifer 486
biharicum, Neobisium (Neobisium) 348
biimpressus, Garypinus (=Olpium) 278
billae, Pseudochthonius 195
biminiensis, Antillochernes 544
binoculatus, Kleptochthonius
  (Chamberlinochthonius) 179
biocovense, Protoneobisium
  (=Obisium) 396
biocularis, Allochthonius
  (Urochthonius) 133
bipectinatus, Ideoblothrus
  (=Ideobisium) 422
birabeni, Maxchernes 599
birabeni, Serianus 302
birmanica, Microcreagris 341
birmanicum, Amblyolpium (=Olpium)
  263
birmanicus, Catatemnus (=Chelifer)
  457
birsteini, Neobisium (Blothrus)
  (=Blothrus) 374
birsteini, Roncus 402
biseriatum, Zaona (=Chelifer) 640
biseriatus, Acanthicochernes 534
biseriatus, Caffrowithius 549
Bisetocreagris 331
bisetus, Parachernes (Parachernes)
  613
bispinosus, Tyrannochthonius
  (=Paraliochthonius) 205
bisulcus, Lophochernes (=Chelifer)
  512
Bituberochernes 548
blothroides, Neobisium (Neobisium)
  (=Obisium) 348

blothroides, Roncocreagris
  (=Microcreagris) 397
Blothrus, Neobisium 373
Bochica 309
BOCHICIDAE 309
bocki, Parazaona (=Chelifer) 622
boettcheri, Oratemnus 466
boettcheri, Stenatemnus 475
bogovinae, Chthonius (Chthonius)
  144
bohemicus, Chernes 590
boldorii, Balkanoroncus (=Neobisium)
  330
boldorii, Chthonius
  (Ephippiochthonius) 164
boldorii, Roncus (Parablothrus) 407
bolivari, Albiorix 316
bolivari, Chthonius
  (Ephippiochthonius) 165
bolivari, Neobisium (Blothrus)
  (=Obisium) 374
bolivari, Paravachonium 430
bolivianus, Austrochthonius 140
bolivianus, Serianus (=Paraserianus)
  302
bonairensis, Garypus 238
boncicus, Haplochernes (=Chelifer)
  580
boneti, Neobisium (Blothrus) 374
boneti, Roncus 402
boneti, Vachonium 431
boninensis, Haplochernes 580
borealis, Allochthonius (Allochthonius)
  132
bornemisszai, Beierolpium
  (=Xenolpium) 266
bornemisszai, Pseudotyrannochthonius
  198
borneoensis, Mucrochelifer
  (=Chelifer) 518
borneoensis, Paratemnoides
  (=Paratemnus) 469
bosnicum, Neobisium (Neobisium) 348
*bougainvillense, Xenolpium
  (Xenolpium)* 271
bougainvillensis, Parachernes
  (Parachernes) 614
boui, Neobisium (Neobisium) 349
bounites, Synsphyronus 244
bouvieri, Epaphochernes 572
braccatus, Calocheiridius 269
brachialis, Dactylochelifer 491
brachydactylus, Blothrus 413
brachydactylus, Chelifer 507

brandmayri, Chthonius (Chthonius)
  144
brasiliensis, Ceriochernes 553
brasiliensis, Cordylochernes 566
brasiliensis, Dolichowithius
  (Dolichowithius) (=Withius) 643
brasiliensis, Ideoblothrus
  (=Ideobisium) 422
brasiliensis, Lustrochernes (=Chernes)
  595
brasiliensis, Petterchernes 623
brasiliensis, Pseudochthonius 195
brasiliensis, Sundochernes 635
brasiliensis, Tridenchthonius 225
brasiliensis, Tyrannochthonius 206
braunsi, Catatemnus (=Chelifer)
  457
bravaisii, Chelifer 238
Brazilatemnus 456
breuili, Neobisium (Blothrus)
  (=Obisium) 375
brevidigitata, Microcreagris 341
brevidigitatum, Neobisium (Neobisium)
  (=Obisium) 349
brevidigitatus, Chelifer 467
brevidigitatus, Cyclatemnus 460
brevidigitatus, Oratemnus 466
brevidigitatus, Protochelifer
  (=Idiochelifer) 523
brevidigitatus, Withius 659
brevifemoratum, Ectactolpium 274
brevifemoratum, Microbisium
  (=Obisium) 337
brevifemoratum, Pachyolpium
  (=Olpium) 295
brevifemoratus, Chelifer 591
brevimanum, Neobisium (Blothrus)
  (=Blothrus) 375
brevimanus, Acis 455
brevimanus, Blothrus 413
brevimanus, Rhacochelifer (=Chelifer)
  526
brevimanus, Tyrannochthonius 206
brevipalpe, Microbisium (=Obisium)
  338
brevipalpis, Chelifer 494
brevipes, Horus 281
brevipes, Lophochernes 512
brevipes, Neobisium (Blothrus)
  (=Blothrus) 375
brevipes, Pachylopium (=Olpium) 295
brevipilosus, Allochernes 536
brevipilosus, Incachernes (=Chelifer)
  587

brevis, Chthonius (Chthonius) irregularis 149
brevispinosus, Conicochernes (=Chelifer) 563
brignolii, Roncus (Parablothrus) 410
brincki, Afrochthonius 131
brincki, Lagynochthonius (=Tyrannochthonius) 183
brincki, Stenatemnus 475
browni, Brazilatemnus 457
bruchi, Sphenochernes (=Chelifer) 634
brunneum, Microbisium (=Obisium) 338
buanensis, Allochthonius (Allochthonius) 132
bucculentus, Geogarypus 252
bucegicum, Neobisium (Neobisium) 349
bucheri, Parazaona 622
buchwaldi, Tridenchthonius (=Chthonius) 225
buettikeri, Nannowithius (=Myrmecowithius) 648
buettneri, Stenowithius (=Chelifer) 657
bulbifemorum, Apocheiridium (Apocheiridium) 433
bulbipalpis, Dinocheirus (=Chernes) 570
bulgaricum, Neobisium (Heoblothrus) (=Obisium) 387
bulgaricus, Allochernes 536
bulli, Metawithius (Microwithius) 647
bureschi, Balkanoroncus (=Obisium) 331
bureschi, Roncus (Parablothrus) 331
burgeoni, Cyclatemnus (=Paratemnus) 460
burmiticus, Garypus 239
butleri, Caribchthonius 142
buxtoni, Haplochernes (=Chelifer) 580
Byrsochernes 548

cabacerolus, Pessigus 456
Cacodemoniellus 651
Cacodemonius 641
Cacoxylus 548
cactorum, Cacodemonius (=Withius) 641
caeca, Microcreagrella (=Obisium) 426
caecata, Indohya 315
Caecatemnus 457
caecatus, Tyrannochthonius (=Paraliochthonius) 206

caecum, Neobisium (Blothrus) 375
caecum, Thaumatolpium 306
caecus, Chthonius 146
caecus, Chthonius 202
caecus, Ideoblothrus (=Ideobisium) 422
caecus, Lustrochernes 595
caecus, Pseudohorus 300
caecus, Withius 659
caffer, Caffrowithius (=Pselaphochernes) 549
Caffrowithius 548
cala, Chitrella (=Chitra) 418
calcaratus, Lophochernes 512
caledonicus, Hebridochernes 582
caledonicus, Nesidiochernes 605
Calidiochernes 552
calidus, Mexachernes (=Chelanops) 603
californica, Microcreagris (=Atemnus) 341
californicus, Chthonius (Hesperochthonius) 176
californicus, Garypus 239
californicus, Menthus 261
caligatum, Neobisium (Blothrus) dinaricum 377
caligatus, Chthonius (Globochthonius) 175
callaticola, Acanthocreagris (=Microcreagris) 324
callidus, Tyrannochthonius 206
callus, Paisochelifer (=Hysterochelifer) 519
callus, Synsphyronus 244
Calocheiridius 268
Calocheirus 272
calvus, Caffrowithius (=Plesiochernes) 549
Calymmachernes 552
cambridgei, Roncocreagris (=Roncus) 397
camerunensis, Tamenus (=Chelifer) 476
campbelli, Apohya 309
camponota, Compsaditha 219
camponoti, Sphenochernes (=Syndeipnochernes) 634
canadensis, Hesperochernes 583
Canarichelifer 485
canariense, Olpium 291
canariensis, Calocheirus (=Apolpiolum) 273

canariensis, Chthonius
  (Ephippiochthonius) machadoi 168
canariensis, Geogarypus (= Garypus)
  253
canariensis, Paraliochthonius 193
cancroides, Chelifer (= Acarus) 487
canestrinii, Dolichowithius
  (Dolichowithius) (= Chelifer) 643
cantabrica, Roncocreagris
  (= Microcreagris) 398
cantabricum, Neobisium (Blothrus)
  vasconicum (= Obisium) 386
caoduroi, Chthonius (Chthonius) 144
capense, Cheiridium 437
capensis, Aperittochelifer (= Chelifer)
  483
capensis, Cyrtowithius 642
capensis, Feaella (Tetrafeaella) 231
capensis, Garypinidius (= Garypinus)
  277
capensis, Thaumastogarypus
  (= Garypus) 248
capensis, Withius (= Allowithius) 659
caporiaccoi, Neobisium (Neobisium)
  349
caprai, Chthonius (Chthonius) 144
capreolus, Chelifer 594
caralitanus, Roncus 403
carbophilus, Opsochernes 611
carcinoides, Neobisium (Neobisium)
  (= Chelifer) 349
Cardiolpium 273
Caribchthonius 142
caribicum, Apocheiridium
  (Apocheiridium) 433
caribicus, Byrsochernes 548
Caribochernes 552
carinatus, Ideoblothrus (= Pachychitra)
  422
carinthiacum, Neobisium (Neobisium)
  352
carinthiacus, Chthonius (Chthonius)
  pygmaeus 159
carinthiacus, Roncus 403
carminis, Mexachernes (= Chelanops)
  603
carnae, Neobisium (Blothrus) 375
carnicum, Neobisium (Neobisium) 356
carolinense, Novobisium (= Obisium)
  391
carolinensis, Nesidiochernes 605
carolinensis, Serianus 302
carolinensis, Smeringochernes
  (Gresittochernes) 634

carpaticum, Neobisium (Neobisium)
  352
carpaticus, Mundochthonius 190
carpenteri, Neobisium (Neobisium)
  (= Obisium) 353
carpenteri, Paraliochthonius 193
carranzai, Troglohya 312
carsicum, Neobisium (Neobisium) 353
carusoi, Roncus 403
casalei, Neobisium (Blothrus) 376
caspica, Acanthocreagris
  (= Microcreagris) 324
cassolai, Chthonius
  (Ephippiochthonius) 165
cassolai, Roncus 403
catalaunicum, Olpium 408
catalonica, Microcreagris 325
catalonicus, Chthonius
  (Ephippiochthonius) 165
Catatemnus 457
caucasicum, Obisium (Obisium) 372
caucasicus, Chernes cimicoides 559
caucasicus, Megachernes 600
caucasicus, Pachychelifer 519
caucasicus, Rhacochelifer (= Chelifer)
  526
caucasicus, Roncus (= Microcreagris)
  403
caucasicus, Toxochernes panzeri 572
cavernae, Afrosternophorus
  (= Sternophorellus) 447
cavernae, Stygiochelifer (= Chelifer)
  532
cavernarum, Chthonius (Chthonius)
  145
cavernarum, Neobisium (Neobisium)
  (= Obisium) 353
cavernarum, Protochelifer 524
cavernicola, Chthonius
  (Globochthonius) 175
cavernicola, Mundochthonius 190
cavernicola, Oratemnus 466
cavernicola, Pycnowithius 654
cavernicola, Roncocreagris
  (= Microcreagris) 398
cavernicola, Roncus (Parablothrus) 336
cavernicola, Roncus lubricus 408
cavernicola, Selachochthonius
  (= Chthoniella) 203
cavernicola, Tyrannochthonius
  (= Paraliochthonius) 206
cavicola, Archeolarca 236
cavicola, Austrochthonius 140
cavicola, Chernes 557

cavicola, Chitrella (=Obisium) 418
cavicola, Ideobisium puertoricense 421
cavicola, Lechytia 186
cavicola, Obisium simile 372
cavicola, Parazaona 622
cavicola, Tyrannochthonius
  (=Morikawia) 206
cavimanum, Mirobisium
  (=Ideobisium) 313
cavophilus, Chthonius (Chthonius) 145
caxinguba, Xenochernes 639
cayanum, Aphelolpium 264
cebenicus, Chthonius (Chthonius) 145
Cecoditha 218
cederholmi, Lophochernes 512
celerrimus, Chelifer 591
cendsureni, Dactylochelifer 492
centrale, Indolpium 282
centralis, Calocheiridius (=Minniza)
  269
centralis, Chiliochthonius 142
centralis, Cyclatemnus 460
centralis, Dolichowithius
  (Dolichowithius) 643
centralis, Garyops 449
centralis, Tyrannochthonius 206
Centrochelifer 486
Centrochthonius 142
cephalonicum, Neobisium (Neobisium)
  (=Obisium) 353
cephalonicus, Chelifer 494
cephalotes, Chthonius (Chthonius)
  (=Blothrus) 145
cerberus, Chthonius (Globochthonius)
  175
cerberus, Kleptochthonius
  (Chamberlinochthonius) 179
cerberus, Roncus (=Obisium) 403
Ceriochernes 552
Cerogarypus 232
cerrutii, Neobisium (Ommatoblothrus)
  388
cervelloi, Neobisium (Blothrus) 376
cervus, Odontochernes (=Chelifer) 610
ceylanicus, Withius (=Chelifer) 659
ceylonensis, Nhatrangia 322
ceylonica, Minniza 284
ceylonicum, Olpium 291
ceylonicus, Afrochthonius 131
ceylonicus, Afrosternophorus
  (=Sternophorus) 447
ceylonicus, Geogarypus 253
ceylonicus, Ideoblothrus
  (=Ideobisium) 422

ceylonicus, Lophochernes 513
ceylonicus, Micratemnus 464
ceylonicus, Parapilanus 621
ceylonicus, Paratemnoides
  (=Paratemnus) 469
chalumeaui, Dinochernes 575
chamarro, Tyrannochthonius
  206
Chamberlinarius 486
chamberlini, Afrosternophorus
  (=Sternophorus) 447
chamberlini, Apocheiridium
  (Apocheiridium) 434
chamberlini, Cheiridium 438
chamberlini, Chthonius (Chthonius)
  (=Neochthonius) 145
chamberlini, Fissilicreagris
  (=Microcreagris) 334
chamberlini, Larca 242
chamberlini, Pararoncus (=Roncus)
  393
Chamberlinius 486
Chamberlinochthonius,
  Kleptochthonius 179
chamberlinorum, Haploditha 223
chamundiensis, Metawithius
  (Microwithius) 647
changaiensis, Dactylochelifer 492
changaiensis, Gobichernes 578
chapmani, Ideobisium 420
chappuisi, Titanatemnus 478
charlotteae, Parobisium 395
charon, Kleptochthonius
  (Chamberlinochthonius) 179
chathamensis, Apatochernes 545
CHEIRIDIIDAE 433
CHEIRIDIOIDEA 433
Cheiridium 437
Cheirochelifer 486
chejuense, Parobisium magnum
  (=Neobisium) 396
Chelanops 553
Chelanops, Chelanops 554
chelanops, Gobichelifer (=Chelifer)
  502
chelatus, Tyrannochthonius 207
Chelifer 487
CHELIFERIDAE 482
CHELIFEROIDEA 451
cheliferoides, Apatochernes 545
Chelignathus 218
Chelodamus 556
Chernes 556
CHERNETIDAE 534

chilense, Apocheiridium
(Chiliocheiridium) 436
chilense, Gymnobisium 313
chilense, Mirobisium 313
chilense, Neocheiridium 443
chilensis, Albiorix (=Ideobisium)
316
chilensis, Americhernes
(=Lamprochernes) 541
chilensis, Anaperochernes 543
chilensis, Austrochthonius
(=Chthonius) 140
chilensis, Chelanops (Chelanops) 554
chilensis, Dinocheirus 570
chilensis, Lechytia 186
chilensis, Neowithius (=Chelifer) 648
chilensis, Oligomenthus 261
chilensis, Parachernes (Parachernes)
614
chilensis, Parazaona 622
chilensis, Pseudopilanus 629
Chiliocheiridium, Apocheiridium 436
Chiliochthonius 142
Chinacreagris 333
chinensis, Chinacreagris
(=Microcreagris) 333
Chiridiochernes 563
chironomum, Neobisium (Neobisium)
(=Olpium) 354
Chitra 418
Chitrella 418
chopardi, Rhacochelifer 526
Chrysochernes 563
Chthoniella 203
CHTHONIIDAE 131
chthoniiformis, Lechytia (=Roncus)
186
CHTHONIOIDEA 131
Chthonius 143
Chthonius, Chthonius 143
chukum, Vachonium 431
chyzeri, Lamprochernes (=Chernes)
588
cimex, Chelifer 668
cimicoides, Chernes (=Scorpio) 557
cinereus, Hoffhorus (=Novohorus)
280
cingara, Microcreagris 341
cinnamomeus, Kashimachelifer 509
clarkorum, Interchernes 587
clarum, Beierolpium (=Indolpium) 266
clarus, Pseudochthonius 195
clathratus, Heterolophus (=Chthonius)
224

clavata, Roncocreagris galeonuda
(=Microcreagris) 398
clavigerum, Pseudochiridium
(=Chelifer) 445
closanicum, Neobisium (Blothrus) 376
Cocinachernes 563
cocophilus, Parachernes (Parachernes)
614
coecus, Apochthonius (=Chthonius)
136
coecus, Chelanops (Chelanops)
(=Chelifer) 554
coiffaiti, Occitanobisium 392
coiffaiti, Pseudorhacochelifer
(=Rhacochelifer) 524
coironi, Chthonius (Chthonius)
mazaurici 155
colecampi, Apochthonius 136
colombiae, Ideoblothrus 422
columbiana, Americocreagris
(=Microcreagris) 329
comasi, Roncus 403
communis, Gomphochernes
(=Chelifer) 578
communis, Pseudozaona 556
comorensis, Catatemnus (=Chelifer)
458
comottii, Chthonius (Chthonius) 145
compactus, Tullgrenius 481
compressus, Parachernes
(Scapanochernes) (=Chelifer) 621
Compsaditha 219
concavus, Catatemnus (=Chelifer) 458
concii, Chthonius (Ephippiochthonius)
165
concii, Roncus 404
concinnus, Caffrowithius (=Chelifer)
550
concinnus, Lustrochernes 595
concolor, Neobisium (Neobisium) 354
confraternus, Parachernes
(Parachernes) 614
confundens, Olpiolum
(=Pachyolpium) 289
confusum, Microbisium 339
confusus, Oratemnus 466
confusus, Tyrannochthonius 207
congica, Compsaditha 219
congicum, Cheiridium 438
congicum, Microbisium 338
congicum, Nanolpium 286
congicus, Calocheiridius
(=Pseudohorus) 269
congicus, Catatemnus 459

congicus, Lasiochernes 592
congicus, Paratemnus 473
congicus, Pseudochthonius 196
congicus, Titanatemnus 478
congicus, Verrucachernes 638
congicus, Withius (=Allowithius) 660
Congochthonius 177
Conicochernes 563
coniger, Parawithius (Victorwithius) 651
connatus, Geogarypus 253
conodentatus, Albiorix 317
conradti, Titanatemnus (=Chelifer) 478
consocius, Lustrochernes (=Chelanops) 595
continentalis, Geogarypus (=Garypus) 253
contractus, Tyrannochthonius (=Chthonius) 207
convivus, Tyrannochthonius 207
cooperi, Pseudogarypinus 299
copiosus, Dactylochelifer 492
Coprochernes 564
coracoides, Reischekia 630
coralensis, Typhloroncus 322
corallinus, Chelifer 589
corcyraea, Acanthocreagris 324
corcyraeum, Neobisium (Neobisium) (=Obisium) 354
corcyraeus, Chthonius (Ephippiochthonius) 165
corcyraeus, Roncus 404
corcyrensis, Rhacochelifer (=Ectoceras) 526
cordimanum, Apolpium (=Olpium) 265
cordimanus, Parachernes (Scapanochernes) (=Scapanochernes) 621
Cordylochernes 564
coreanus, Allochthonius (Allochthonius) opticus 133
coreophilus, Titanatemnus 478
corimanus, Roncus 404
cornutus, Hebridochernes 582
Corosoma 567
corsa, Acanthocreagris 325
corsicus, Chthonius (Ephippiochthonius) 165
corticalis, Chelifer 350
corticis, Chelanops 668
corticis, Parachernes 618
corticolus, Solinus (=Garypinus) 304

corticum, Neocheiridium (=Cheiridium) 443
costaricensis, Coprochernes 564
costaricensis, Cordylochernes 565
costaricensis, Ideoblothrus (=Ideobisium) 422
costaricensis, Mesochernes 602
costaricensis, Pseudogarypinus 299
crassichelatum, Pachyolpium (=Olpium) 295
crassidens, Sathrochthonius 201
crassifemoratum, Neobisium (Neobisium) (=Obisium) 354
crassifemoratus, Calocheiridius 269
crassifemoratus, Pseudopilanus 629
crassimanum, Ideobisium 420
crassimanus, Lustrochernes 595
crassimanus, Parachernes (Parachernes) 614
crassimanus, Phymatochernes 624
crassimanus, Pseudochernes 654
crassimanus, Roncus vulcanius 415
crassipalpus, Roncus 404
crassipes, Micratemnus 464
crassipes, Withius (=Stenowithius) 660
crassopalpus, Acuminochernes (=Hesperochernes) 534
crassum, Olpiolum 289
crassus, Allowithius 665
crassus, Conicochernes 564
crassus, Dendrochernes 568
crassus, Megachernes limatus 600
crassus, Nudochernes 607
cretica, Minniza 285
creticum, Neobisium (Blothrus) (=Obisium) 376
creticus, Chernes 538
creticus, Chthonius (Ephippiochthonius) 165
cribellus, Phoberocheirus 534
cribratus, Levichelifer 509
crinitus, Megachernes 599
cristatum, Neobisium (Neobisium) 354
croaticum, Neobisium (Blothrus) reimoseri 382
crosbyi, Kleptochthonius (Kleptochthonius) (=Apochthonius) 178
cruciatus, Apatochernes 545
cruciatus, Elattogarypus 237
cruzensis, Antillochernes 544
crypticum, Olpium 291
Cryptocheiridium 441

Cryptocheiridium, Cryptocheiridium 441
Cryptocreagris 334
Cryptoditha 220
Cryptowithius 642
cryptum, Vachonium 431
cryptus, Chthonius (Chthonius) 146
cryptus, Lophochernes 513
csikii, Neobisium (Blothrus) stygium 384
Cubachelifer 490
Cubanocheiridium, Cryptocheiridium 442
cubanum, Mexobisium 310
cubanus, Aphrastochthonius 135
cubanus, Neowithius (=Chelanops) 648
cubanus, Tridenchthonius (=Ditha) 226
cubanus, Tyrannochelifer 533
curazavius, Ideoblothrus (=Pachychitra) 423
curazavius, Tyrannochthonius 207
curtulus, Apatochernes 545
curtulus, Paratemnoides (=Anatemnus) 469
curtus, Afrogarypus (=Geogarypus) 249
curtus, Caffrowithius hanangensis (=Plesiochernes) 551
curtus, Oratemnus (=Steiratemnus) 466
curvidigitata, Anisoditha (=Chthonius) 218
curvidigitatus, Chthonius 222
cuyabanus, Geogarypus (=Garypus) 253
Cyclatemnus 459
cyclica, Microcreagris 343
Cyclochernes 567
cyclopium, Calocheiridius (=Xenolpium) 270
cyclopius, Roncus (Parablothrus) 401
cylindrimanus, Afrosternophorus (=Sternophorus) 448
cypria, Minniza hirsti 285
cyprianus, Mesatemnus (=Anatemnus) 463
cyprius, Hysterochelifer (=Chelifer) 505
cyrenaicus, Solinus (=Garypinus) 304
cyrneus, Dendrochernes (=Chernes) 568
Cyrtowithius 642

dacnodes, Chthonius (Chthonius) 146
Dactylochelifer 491
daedaleus, Chthonius (Ephippiochthonius) 166
daemonius, Kleptochthonius (Chamberlinochthonius) 180
dahli, Haplochernes 581
dallaii, Roncus 404
dalmatinum, Neobisium (Blothrus) 376
dalmatinus, Chthonius (Chthonius) 146
dalmatinus, Roncus lubricus 409
danaus, Chelifer 506
danconai, Cheiridium 438
darwiniensis, Parachernes (Parachernes) 614
dashdorzhi, Gobichelifer 502
Dasychernes 567
davidi, Xenochelifer 533
dawydoffi, Afrosternophorus (=Sternophorus) 448
dawydoffi, Hyperwithius 646
dawydoffi, Nhatrangia 322
debilis, Anaperochernes 543
decaryi, Paracheiridium 444
decebali, Diplotemnus 668
deciclavatus, Allochthonius (Urochthonius) 134
decolor, Indolpium 282
decoui, Mundochthonius 190
degeerii, Chelifer 494
degeneratus, Chelifer 522
dekeyseri, Plesiowithius 653
delamarei, Lechytia 186
delanoi, Paravachonium 431
delphinaticum, Neobisium (Neobisium) 355
deminuta, Minniza persica 285
Dendrochernes 568
denisi, Chernes 559
densedentatus, Chthonius (Chthonius) 146
densedentatus, Negroroncus 320
densedentatus, Tyrannochthonius (=Morikawia) 207
dentata, Lechytia 186
dentatus, Chelifer cancroides 489
dentatus, Microchelifer 517
dentatus, Mirochernes (=Chelanops) 603
dentifer, Pseudotyrannochthonius (=Allochthonius) 198
depressimanus, Gomphochernes (=Chelifer) 579

depressoides, Lissochelifer
(=Lophochelifer) 510
depressus, Garyops 449
depressus, Lissochelifer (=Chelifer)
510
deschmanni, Neobisium (Blothrus)
(=Obisium) 377
deserticola, Anaulacodithella
(=Verrucadithella) 217
deserticola, Beierolpium
(=Calocheiridius) 267
deserticola, Minniza 284
deserticola, Parallowithius
(=Allowithius) 650
desiderata, Megathis 668
despaxi, Withius 660
detritus, Ceriochernes 553
dewae, Sundochernes 635
dewae, Synsphyronus 245
Dhanus 318
diabolicus, Dinocheirus 570
diabolus, Apochthonius 136
diabolus, Typhloroncus 322
Dichela 498
Difeaella, Feaella 230
differens, Lophochernes 513
difficilis, Horus 281
digitulus, Paragoniochernes 649
digitum, Olpium 291
dilatimana, Verrucadithella
(=Chthonius) 228
dilatimanus, Nesowithius 649
dimidiatus, Garypinus (=Olpium)
278
dimorphicum, Mirobisium 313
dinaricum, Neobisium (Blothrus) 377
Dinocheirus 569
Dinochernes 575
Dinoroncus 316
diophthalmus, Chthonius (Chthonius)
146
Diplotemnus 461
Diplothrixochernes 575
disjunctus, Rhacochelifer (=Chelifer)
526
dissimilis, Parachernes (Parachernes)
615
dissimilis, Protogarypinus 299
distincta, Minniza sola 286
distinctum, Neobisium (Neobisium)
(=Obisium) 355
distinctus, Illinichernes 586
distinctus, Oratemnus (=Steiratemnus)
466

distinctus, Parachernes (Parachernes)
615
distinguenda, Roncocreagris cantabrica
(=Microcreagris) 398
distinguendus, Chthonius
(Ephippiochthonius) 166
distinguendus, Hysterochelifer
(=Chelifer) 505
Ditha 220
Ditha, Ditha 221
Dithella 223
diversus, Chelanops 621
divisa, Neominniza 288
divisus, Ideoroncus 319
doctus, Pseudochthonius 196
doderoi, Chthonius (Chthonius) 147
doderoi, Neobisium (Neobisium)
(=Obisium) 355
dogieli, Microbisium (=Obisium) 338
dogoensis, Tyrannochthonius japonicus
210
dolichodactylus, Dactylochelifer 492
Dolichowithius 642
Dolichowithius, Dolichowithius 643
dolicodactylum, Neobisium
(Neobisium) (=Obisium) 355
dollfusi, Amblyolpium 263
dolomiticum, Neobisium (Neobisium)
356
dolomiticus, Roncus (Parablothrus)
stussineri 400
dolosus, Cyclatemnus 460
dolosus, Serianus 302
dominicanus, Parachelifer 520
dominicanus, Parachernes
(Parachernes) 615
dominicus, Lustrochernes 595
donaldi, Tridenchthonius 226
donisi, Pogonowithius 654
doratodactylus, Troglochthonius 204
dorogawaensis, Pseudo-
tyrannochthonius kobayashii
(=Spelaeochthonius) 200
dorothyae, Synsphyronus 245
dorsalis, Dinocheirus (=Chelanops)
571
doveri, Dhanus 318
Drepanochthonius 177
drescoi, Roncus 404
dubium, Obisium 370
dubius, Garypus 241
dubius, Neowithius 649
dubius, Pselaphochernes (=Chelifer)
625

dubius, Sundochernes 635
duboscqi, Roncus 405
duboscqui, Metachelifer 516
duffeyi, Stenowithius 657
dumicola, Obisium 370
duncani, Microcreagris 343
dybasi, Lagynochthonius
 (=Tyrannochthonius) 183
dybasi, Meiochernes 602
dybasi, Nesidiochernes carolinensis 605

eburneus, Nesowithius 649
echinatus, Cacoxylus 548
echinatus, Pseudopilanus (=Chelifer)
 629
Ectactolpium 274
Ectoceras 499
Ectromachernes 645
ecuadorense, Apolpium 265
ecuadorense, Ideobisium 420
ecuadoricus, Byrsochernes 548
ecuadoricus, Parachelifer 520
ecuadoricus, Parawithius (Parawithius)
 nobilis 651
ecuadoricus, Tyrannochthonius
 (=Morikawia) 207
edentatus, Albiorix 317
effossus, Pachychernes 611
egeria, Bisetocreagris 333
egregius, Balanowithius 640
egregius, Diplotemnus 462
ehrenbergii, Chelifer 668
eidmanni, Americhernes
 (=Pycnochernes) 541
ejuncidus, Synsphyronus 245
Elattogarypus 237
elatus, Chrysochernes 563
elbanus, Chthonius
 (Ephippiochthonius) 166
elbursensis, Allochernes 536
electri, Garypinus 278
Electrobisium 334
Electrochelifer 499
elegans, Austrochthonius 669
elegans, Calocheiridius 270
elegans, Cryptocheiridium
 (Cubanocheiridium) 442
elegans, Cryptoditha
 (=Tridenchthonius) 220
elegans, Ditha (Ditha) 221
elegans, Ectromachernes 645
elegans, Geogarypus (=Garypus) 253
elegans, Mesochernes (=Chelifer) 602
elegans, Neobisium (Neobisium) 356

elegans, Olpiolum (=Olpium) 289
elegans, Synsphyronus 245
elegans, Tyrannochthonius 208
elegantissimus, Lophochernes 513
elegantula, Compsaditha 219
elgonense, Cryptocheiridium
 (Cryptocheiridium) 441
elgonensis, Caffrowithius
 (=Plesiochernes) 550
elgonensis, Trichotowithius 658
elimatum, Obisium 370
ellenae, Haplochernes 581
ellingseni, Chthonius (Chthonius) 147
ellingseni, Paratemnoides
 (=Paratemnus) 470
ellingseni, Pseudoblothrus (=Obisium)
 428
ellingsenii, Parazaona (=Chelifer) 622
Ellingsenius 500
ellipticus, Americhernes
 (=Lamprochernes) 541
elongata, Chelifer (Atemnus)
 voeltzkowi 452
elongatus, Anatemnus (=Chelifer) 451
elongatus, Chthonius (Chthonius) 147
elongatus, Paratemnoides (=Atemnus)
 470
embuensis, Pseudohorus 300
emigrans, Dolichowithius
 (Dolichowithius) (=Chelifer) 643
emiliae, Neobisium (Neobisium)
 lombardicum 362
enhuycki, Syarinus 429
enoshimaensis, Nipponogarypus 288
entzii, Chelifer 529
Epactiochernes 576
Epaphochernes 569
Ephippiochthonius, Chthonius 163
Epichernes 576
epirensis, Acanthocreagris leucadia 326
equester, Titanatemnus (=Chelifer)
 478
equestroides, Tamenus (=Chelifer) 477
erebicus, Kleptochthonius
 (Chamberlinochthonius) 180
Eremochernes 501
Eremogarypus 237
erosidens, Mundochthonius 190
erratum, Pachyolpium 295
eruditum, Apocheiridium
 (Apocheiridium) 434
erytheia, Bisetocreagris 333
erythrodactylum, Neobisium
 (Neobisium) (=Obisium) 356

esakii, Parobisium anagamidensis
(=Neobisium) 394
escalerai, Neobisium (Blothrus)
robustum 383
euboicus, Rhacochelifer 527
eucarpus, Chelifer 669
euchirus, Roncus 405
Eumecochernes 576
europaeus, Chelifer 489
eurydice, Microcreagris 341
Euryolpium 275
ewingi, Chernes (=Reginachernes) 559
exarmatus, Chthonius (Chthonius) 147
excavatus, Pseudohorus 300
excellens, Afrogarypus excelsus
(=Geogarypus) 250
excellens, Caffrowithius
(=Plesiochernes) 550
excelsus, Afrogarypus (=Geogarypus)
250
excelsus, Sphallowithius 656
excentricus, Chelifer 669
exigua, Reischekia 630
exiguus, Caffrowithius (=Chelifer) 550
exiguus, Catatemnus 458
exiguus, Chelanops (Neochelanops) 556
exiguus, Lagynochthonius
(=Tyrannochthonius) 183
exiguus, Papuchelifer 519
exiguus, Protochelifer 524
exiguus, Withius congicus
(=Allowithius) 660
exilimanus, Neowithius (=Chelifer)
649
eximium, Ectactolpium 274
eximius, Eremogarypus 237
exochus, Geogarypus 254
exorbitans, Minniza 284
exstinctum, Neobisium (Neobisium)
357
exsul, Austrohorus 266
extensum, Neobisium (Blothrus)
phineum 381
extensus, Dolichowithius
(Dolichowithius) 643
extensus, Pseudogarypus 233
extensus, Stenatemnus 475
extraordinarius, Semeiochernes 632
ezoensis, Microcreagris 341

fabricii, Chelifer 494
facetus, Caffrowithius (=Chelifer) 550
fagei, Paragarypus 244
fagetum, Microbisium 339

falcatus, Pseudochthonius 196
falcomontanus, Chelifer 593
*fallaciosus, Synsphyronus
(Maorigarypus)* 246
fallax, Afrosternophorus 448
fallax, Cheiridium 438
fallax, Cordylochernes 565
fallax, Cyclatemnus 460
fallax, Parachernes (Parachernes) 615
fallax, Trisetobisium (=Microcreagris)
416
falsum, Nanolpium 286, 287
falsus, Dactylochelifer (=Chelifer) 492
fasciatus, Chelifer 558
fasciculatum, Stenolpium 305
fastuosus, Tyrannochthonius 209
faunus, Withius (=Chelifer) 660
Feaella 230
Feaella, Feaella 230
FEAELLIDAE 230
FEAELLOIDEA 230
feai, Paratemnoides (=Chelifer) 470
*femoratus, Sternophorus
(Afrosternophorus)* 447
femoratus, Tamenus 477
fenestratus, Ideoblothrus
(=Ideobisium) 423
fenestratus, Pseudohorus transvaalensis
301
fergusoni, Apocheiridium
(Apocheiridium) 434
fernandezianus, Protowithius 654
fernandezianus, Pseudopilanus 630
ferox, Tamenus (=Chelifer) 477
ferox, Tyrannochthonius 208
ferrisi, Garyops (=Sternophorus) 449
ferum, Apocheiridium (Apocheiridium)
(=Cheiridium) 434
ferumoides, Apocheiridium
(Apocheiridium) 434
fiebrigi, Compsaditha (=Ditha) 220
fiebrigi, Geogarypus 254
fiebrigi, Parawithius (Victorwithius)
651
firmum, Cheiridium 438
fiscelli, Neobisium (Neobisium) 357
Fissilicreagris 334
flammipes, Lophochernes 513
flavum, Beierolpium 267
flavum, Ectactolpium 274
flavum, Neobisium (Neobisium) gentile
358
flavus, Geogarypus (Geogarypus)
258

flavus, Lagynochthonius
  (= Tyrannochthonius) 183
flavus, Thapsinochernes 638
flexifemoratum, Parobisium
  (= Neobisium) 395
Florichelifer 501
floridae, Parachernes (Parachernes)
  (= Chelifer) 615
floridanus, Aldabrinus 262
floridanus, Atemnus 470
floridanus, Tyrannochelifer
  (= Chelifer) 533
floridensis, Antillochernes 544
floridensis, Garypus 239
floridensis, Ideoblothrus
  (= Pachychitra) 423
floridensis, Tyrannochthonius 208
foliaceosetosus, Ceriochernes 553
foliosus, Cocinachernes 563
foliosus, Rhopalochernes (= Chelifer)
  631
forbesi, Apochthonius 136
formosana, Microcreagris 341
formosanum, Cryptocheiridium
  (Cryptocheiridium) (= Cheiridium)
  441
formosanum, Ideobisium 344
formosanus, Geogarypus 255
formosanus, Withius australasiae
  (= Allowithius) 659
formosus, Geogarypus (= Ideobisium)
  254
foveauxana, Tyrannochthoniella
  zealandica 204
foxi, Lamprochernes (= Pycnochernes)
  588
fradei, Mesochelifer 516
francisi, Verrucadithella 229
Francochthonius 177
franzi, Amblyolpium 263
franzi, Beierolpium soudanense
  (= Xenolpium) 268
franzi, Galapagodinus 277
franzi, Gigantochernes 578
franzi, Neobisium (Neobisium)
  bernardi 348
franzi, Nepalobisium 391
franzi, Parachernes (Parachernes) 615
frater, Lophochernes 513
fraternum, Neobisium (Blothrus)
  carnae 376
fraternus, Chelanops (Neochelanops)
  555
fravalae, Catatemnus 458

frivaldszkyi, Rhacochelifer (= Chelifer)
  527
frontalis, Pseudogarypinus (= Olpium)
  299
fructuosus, Juxtachelifer 509
fuchsi, Stenatemnus (= Chelifer) 475
fuelleborni, Hansenius (= Chelifer)
  502
fuliginosus, Calocheiridius rhodesiacus
  272
fulleri, Ellingsenius (= Chelifer) 500
fulvopalpus, Levichelifer
  (= Idiochelifer) 509
funafutensis, Haplochernes
  (= Chelifer) 581
funebrum, Indolpium (= Xenolpium)
  282
furax, Bisetocreagris (= Microcreagris)
  331
furculiferum, Pachyolpium (= Olpium)
  295
fuscimanum, Neobisium (Neobisium)
  (= Obisium) 357
fuscimanum, Olpium 291
fuscimanus, Chthonius
  (Ephippiochthonius) 166
fuscipalpum, Olpiolum 289
fuscipes, Hysterochelifer (= Chelifer)
  505
fuscus, Withius 660

gabbutti, Calocheiridius 270
gaditanum, Neobisium
  (Ommatoblothrus) 388
galapagensis, Parachernes
  (Parachernes) 615
galapagensis, Pseudochthonius 196
Galapagodinus 277
galapagoense, Neocheiridium 443
galapagoensis, Serianus 302
galatheae, Ochrochernes (= Chelifer)
  609
galeatum, Neobisium (Neobisium) 358
galeonuda, Roncocreagris
  (= Microcreagris) 398
gallica, Acanthocreagris
  (= Microcreagris) 325
gallinaceus, Apatochernes 545
gansuensis, Dactylochelifer 492
garambae, Caffrowithius
  (= Plesiochernes) 550
garambica, Lechytia 187
garambicus, Pycnowithius
  (= Allowithius) 655

garcianus, Neoallochernes
  (= Chelanops) 604
Garyops 449
GARYPIDAE 235
Garypinidius 277
Garypinus 277
GARYPOIDEA 235
garypoides, Diplotemnus (= Chelifer)
  462
garypoides, Ectactolpium 274
Garypus 237
gasparoi, Chthonius
  (Ephippiochthonius) 166
Gelachernes 577
gennargentui, Neobisium (Neobisium)
  bernardi 348
gentile, Neobisium (Neobisium) 358
geoffroyi, Beierochelifer 485
geoffroyi, Chelifer 558
GEOGARYPIDAE 249
Geogarypus 252
geophilus, Kleptochthonius
  (Kleptochthonius) 178
germainii, Rhopalochernes
  (= Chelifer) 631
germanicum, Neobisium (Neobisium)
  351
geronense, Neobisium (Neobisium)
  bernardi 348
geronimoensis, Hysterochelifer
  (= Chelifer) 505
gertschi, Kleptochthonius
  Chamberlinochthonius) 180
gestroi, Chthonius (Ephippiochthonius)
  166
gestroi, Roncus 405
ghidinii, Neobisium (Neobisium)
  trentinum 371
ghidinii, Roncus (Parablothrus) 330
gibbosounguiculatus, Lissochelifer
  (= Lophochelifer) 510
gibbus, Chthonius (Ephippiochthonius)
  167
giganteum, Neobisium (Neobisium)
  gentile 358
giganteus, Garypus 239
giganteus, Protogarypinus 299
giganteus, Pseudogarypinus 300
giganteus, Pseudotyrannochthonius
  198
giganteus, Roncus 405
Gigantochernes 578
gigas, Calocheirus (= Apolpiolum) 273
gigas, Eremogarypus 237

gigas, Hadoblothrus (= Parablothrus)
  419
gigas, Microcreagris 342
gigas, Pseudotyrannochthonius 199
gigas, Synsphyronus 245
gigas, Titanatemnus 479
gigas, Tyrannochthonius 208
gineti, Neobisium (Neobisium) 358
ginkgoanus, Allochernes
  (= Toxochernes) 536
girgentiensis, Chthonius
  (Ephippiochthonius) 167
gisleni, Synsphyronus (Maorigarypus)
  246
giulianii, Neoamblyolpium 287
giustii, Chthonius (Ephippiochthonius)
  167
giustii, Paraliochthonius hoestlandti
  193
glaber, Roncus (Roncus) 404
glabratus, Withius (= Chelifer) 660
gladiatum, Olpium 291
globifer, Chthonius (Globochthonius)
  175
Globochthonius, Chthonius 175
Globocreagris 335
globosus, Conicochernes 564
globosus, Cyclatemnus 460
globosus, Ellingsenius 500
globosus, Hesperochernes (= Chelifer)
  583
globulus, Geogarypus 254
Gobichelifer 502
Gobichernes 578
gobiensis, Chernes 559
gobiensis, Dactylochelifer 492
godfreyi, Afrochthonius (= Chthonius)
  132
godfreyi, Chelifer (Chernes) 591
godfreyi, Ideoblothrus (= Ideobisium)
  423
golovatchi, Neobisium (Neobisium) 359
gomezi, Neobisium (Ommatoblothrus)
  389
Gomphochernes 578
gomyi, Tyrannochthonius 208
Goniochernes 579
goniothorax, Goniochernes
  (= Chelifer) 579
goodnighti, Mexobisium 310
gorgo, Bisetocreagris 333
gracile, Ideobisium 420
gracile, Neobisium (Ommatoblothrus)
  389

gracile, Obisium 350
gracile, Pachyolpium (= Olpium) 297
gracilimanus, Hysterochelifer 505
gracilimanus, Nudochernes 607
gracilimanus, Parachernes (Parachernes) 615
gracilimanus, Parawithius (Victorwithius) 651
gracilior, Pilochelifer insularis 523
gracilipalpe, Neobisium (Neobisium) 359
gracilipalpus, Calocheiridius 270
gracilipalpus, Withius 661
gracilipes, Eremochernes (= Chelifer) 501
gracilipes, Lissochelifer (= Lophochelifer) 510
gracilipes, Nudochernes 607
gracilis, Albiorix 317
gracilis, Beierius walliskewi 485
gracilis, Bisetocreagris (= Microcreagris) 331
gracilis, Chthonius (Chthonius) orthodactylus 146
gracilis, Dactylochelifer 493
gracilis, Dichela (= Oligochelifer) 498
gracilis, Garypus 239
gracilis, Horus 281
gracilis, Lophochernes 513
gracilis, Lustrochernes (= Atemnus) 595
gracilis, Mesochernes 602
gracilis, Nesochernes 607
gracilis, Pachychernes (= Chelifer) 611
gracilis, Pseudohorus 300
gracilis, Pseudotyrannochthonius 199
gracilis, Synsphyronus 245
gracillimus, Stenolpoides 305
graeca, Minniza (= Olpium) 285
graecum, Amblyolpium 263
graecus, Chernes 560
graecus, Chthonius (Chthonius) 148
graecus, Lasiochernes 592
grafittii, Chthonius (Ephippiochthonius) 167
grafittii, Roncus 405
grahami, Australinocreagris (= Microcreagris) 330
graminum, Olpium 291
grande, Neobisium (Blothrus) absoloni 373
grandimanus, Chelifer 489
grandimanus, Strobilochelifer 532

grandis, Ideoblothrus (= Pachychitra) 423
grandis, Megachernes (= Chernes) 599
grandis, Microcreagris 342
grandis, Thaumastogarypus 248
graniferum, Xenolpium 307
granochelum, Apocheiridium (Apocheiridium) 435
granulata, Acanthocreagris (= Roncus) 325
granulata, Larca (= Garypus) 242
granulata, Microcreagris 342
granulatum, Euryolpium 275
granulatum, Neobisium (Neobisium) 359
granulatum, Neobisium (Neobisium) 359
granulatum, Olpium 292
granulatum, Pachyolpium 296
granulatus, Calocheiridius 270
granulatus, Catatemnus 458
granulatus, Chelifer 489
granulatus, Cyclatemnus 460
granulatus, Dichela (= Oligochelifer) 499
granulatus, Geogarypus 254
granulatus, Horus (= Garypinus) 281
granulatus, Microchelifer 517
granulatus, Nudochernes 607
granulatus, Ochrochernes 609
granulatus, Rhinochernes 631
granulatus, Syarinus 429
granulosum, Euryolpium (= Xenolpium) 276
granulosum, Neobisium (Neobisium) 359
granulosus, Dolichowithius (Dolichowithius) 643
granulosus, Lustrochernes 595
gratus, Serianus 302
gratus, Tridenchthonius 226
gravieri, Neogarypus 243
grayi, Afrosternophorus (= Sternophorus) 448
grayi, Sundochernes 635
grayi, Synsphyronus (Maorigarypus) 247
greensladeae, Smeringochernes (Smeringochernes) 633
greensladeae, Synsphyronus 245
gregoryi, Negroroncus 321
gressitti, Compsaditha 220
gressitti, Hebridochernes 582
gressitti, Sundochernes 635

Gressittochernes, Smeringochernes 634
grimmeti, Tyrannochthonius 208
grossus, Lustrochernes (= Chelanops) 595
grubbsi, Aphrastochthonius 135
grubbsi, Apochthonius 137
gruberi, Dactylochelifer 493
gruberi, Levigatocreagris 336
guadalupensis, Archeolarca 236
guadalupensis, Garypus 240
guadeloupensis, Tyrannochthonius 208
guamensis, Smeringochernes (Smeringochernes) 633
guanophilus, Sundochernes 635
guarany, Americhernes (= Pycnochernes) 541
guasirih, Lagynochthonius (= Tyrannochthonius) 183
guatemalense, Mexobisium 311
guglielmii, Chthonius (Chthonius) 148
guianensis, Paratemnoides (= Paratemnus) 470
guineensis, Chelifer (Atemnus) 472
guttiger, Heterolophus 224
GYMNOBISIIDAE 312
Gymnobisium 312

Hadoblothrus 419
hadronennus, Synsphyronus 245
hadzii, Balkanoroncus 331
hadzii, Neobisium (Blothrus) 377
hadzii, Pselaphochernes 626
hagai, Haplochernes boncicus 580
hageni, Kleptochthonius (Chamberlinochthonius) 180
hahnii, Chernes (= Chelifer) 560
halberti, Chthonius (Chthonius) 148
Halobisium 335
Halominniza 279
halophila, Neominniza 288
halophilum, Olpium 292
hamatus, Lagynochthonius 184
hamatus, Levigatocreagris 336
hamiltonsmithi, Pseudo-tyrannochthonius 199
hanangensis, Caffrowithius (= Plesiochernes) 550
hansenii, Lophochernes (= Chelifer) 513
hansenii, Synsphyronus (= Garypus) 246
Hansenius 502
Haplochelifer 504
Haplochernes 580

Haploditha 223
Haplogarypinus 279
hartmanni, Cheiridium (= Chelifer) 438
harveyi, Pycnodithella 224
hastatus, Parobisium 395
hawaiensis, Eumecochernes (= Chelifer) 577
heatwolei, Anagarypus 236
hebridicus, Haplochernes 581
hebridicus, Paratemnoides salomonis (= Paratemnus) 474
Hebridochernes 582
helenae, Hemisolinus 280
helenae, Scotowithius 655
helenae, Tyrannochthonius (= Paraliochthonius) 208
helferi, Ectoceras 499
hellenum, Neobisium (Neobisium) (= Obisium) 359
helveticum, Neobisium (Neobisium) 359
Hemisolinus 279
hemprichii, Pseudogarypus (= Chelifer) 233
hendrickxi, Ellingsenius 500
henroti, Kleptochthonius (Chamberlinochthonius) (= Chamberlinochthonius) 180
henroti, Neobisium (Ommatoblothrus) 389
henschii, Rhacochelifer (= Chelifer) 527
Heoblothrus, Neobisium 387
heptatrichus, Synsphyronus 246
herbarii, Chthonius (Chthonius) 148
herculea, Microcreagris 342
hermanni, Chelifer 489
hermanni, Neobisium (Neobisium) 359
hermannii, Chelifer 671
heros, Levigatocreagris (= Microcreagris) 336
heros, Neobisium (Blothrus) 377
heros, Vachonobisium (= Gymnobisium) 314
herzegovinense, Neobisium (Neobisium) bosnicum 348
hespera, Microcreagris 342
Hesperochernes 583
Hesperochthonius, Chthonius 176
Hesperolpium 280
hesperum, Parobisium (= Neobisium) 395
hesperus, Hyarinus 419
hesperus, Pseudogarypus 233

hesternus, Parobisium 395
Heterochernes 586
Heterochthonius, Apochthonius 177
heterodactylus, Chthonius (Chthonius) 148
heterodentata, Selachochthonius (= Chthoniella) 203
heterodentatus, Geogarypus 254
heterodentatus, Metatemnus 463
heterodentatus, Pseudochthonius 196
heterodentatus, Tyrannochthonius 209
heterodonta, Hya 314
Heterolophus 223
Heterolpium 280
heterometrus, Cheirochelifer (= Chelifer) 486
heteropoda, Leucohya 310
hetricki, Kleptochthonius (Chamberlinochthonius) 180
heurtaultae, Chthonius (Chthonius) 148
heurtaultae, Pseudochiridium 445
heurtaultae, Spelyngochthonius 203
Hexachernes 586
hians, Lophochernes (= Chelifer) 513
hians, Neobisium (Blothrus) 377
hibericus, Roncus 406
hiberum, Neobisium (Blothrus) 377
hiberus, Chthonius (Ephippiochthonius) 167
himalaiense, Alocobisium 417
himalayana, Lechytia 187
himalayensis, Allochernes 536
himalayensis, Lagynochthonius (= Tyrannochthonius) 184
himalayensis, Megachernes (= Chelifer) 599
hirsti, Afrosternophorus (= Sternophorus) 448
hirsti, Minniza 285
hirsuta, Sororoditha (= Chthonius) 225
hirsutus, Francochthonius 177
hirsutus, Miratemnus 465
hirsutus, Mucrochernes (= Atemnus) 604
hirtum, Neobisium (Neobisium) 360
hispanica, Larca 242
hispanica, Microcreagrina (= Ideobisium) 427
hispaniolicus, Chernes 560
hispanus, Chthonius (Ephippiochthonius) 167
hispanus, Solinus 304
hispanus, Withius (= Chelifer) 661

hispida, Philomaoria 653
hispidus, Miratemnus 465
histricum, Neobisium (Blothrus) reimoseri 382
histricus, Chthonius (Globochthonius) spelaeophilus 176
histrionicus, Pararoncus 393
histrionicus, Paratemnus 453
hoestlandti, Paraliochthonius 193
Hoffhorus 280
hoffi, Gigantochernes 578
hoffi, Hygrochelifer 504
hoffi, Lechytia 187
hoggarensis, Rhacochelifer maculatus 528
holmi, Beierolpium 267
holmi, Ideoblothrus (= Ideobisium) 423
holmi, Tyrannochthonius (Tyrannochthonius) 211
holsingeri, Apochthonius 137
holsingeri, Mundochthonius 190
homodentatus, Pseudochthonius 196
honestus, Syarinus 429
horricus, Dinocheirus 571
horridus, Chthonius (Chthonius) doderoi 147
horridus, Drepanochthonius 177
horridus, Tyrannochthonius (= Paraliochthonius) 209
Horus 280
horvathii, Chernes 560
howarthi, Tyrannochthonius 209
howdenensis, Phaulochernes 623
hubbardi, Parachelifer (= Chelifer) 520
hubrichti, Kleptochthonius (Chamberlinochthonius) 180
hungaricus, Chernes (Chernes) cyrneus 568
hungaricus, Chernes 669
hungaricus, Chthonius (Chthonius) 148
hungaricus, Geogarypus (= Garypus) 254
hutsoni, Wyochernes 639
Hya 314
Hyarinus 419
hyatti, Metachelifer 517
hydaspis, Chelifer 669
hygricus, Calocheiridius 270
hygricus, Lagynochthonius 184
hygricus, Lissochelifer (= Lophochelifer) 510
Hygrochelifer 504
HYIDAE 314
Hyperwithius 646

hypochthon, Neobisium (Blothrus) 378
hypogeum, Neobisium (Blothrus)
  vasconicum (= Obisium) 387
hypogeus, Apochthonius 137
hypogeus, Pseudogarypus 234
Hysterochelifer 504

iberica, Roncocreagris
  (= Microcreagris) 399
ibericus, Chelifer 529
ibericus, Hysterochelifer tuberculatus
  508
iberomontanus, Pselaphochernes 626
iberus, Chernes 560
Ideobisium 419
Ideoblothrus 421
IDEORONCIDAE 315
Ideoroncus 319
Idiochelifer 508
Idiogaryops 450
Idiogarypus 244
iguazuensis, Austrochthonius 140
Illinichernes 586
ilvensis, Chthonius (Chthonius) 149
imadatei, Mundochthonius japonicus
  191
imbecillum, Neobisium (Blothrus) 378
imitans, Troglochernes 638
imitatus, Tyrannochthonius 209
imperator, Chthonius (Chthonius) 149
imperator, Tyrannochelifer
  (= Chelifer) 533
imperfectum, Parobisium
  (= Neobisium) 395
imperialis, Microcreagris 342
imperiosus, Dinocheirus 571
impressus, Afrogarypus (= Garypus)
  250
improcerum, Neobisium (Neobisium)
  360
improvisum, Neobisium (Neobisium)
  360
*inaculeatum, Neobisium (Neobisium)
  sylvaticum* 370
inaequale, Neobisium (Neobisium) 360
inaequalis, Afrochthonius 132
inaequalis, Chelifer 589
Incachernes 587
incertum, Neobisium (Neobisium) 360
incertum, Pachyolpium 296
incertus, Americhernes 541
incertus, Geogarypus 254
incognitus, Parawithius (Victorwithius)
  651

incognitus, Pseudotyrannochthonius
  (= Allochthonius) 199
inconspicuus, Cryptowithius 642
incrassatus, Conicochernes
  (= Haplochernes) 564
incrassatus, Pseudohorus 300
incrassatus, Xenolpium
  (= Calocheiridius) 307
indianensis, Apochthonius 137
indica, Compsaditha 220
indica, Feaella (Tetrafeaella) 231
indica, Lechytia 187
indicum, Apocheiridium
  (Apocheiridium) 435
indicum, Euryolpium 276
indicum, Heterolpium 280
indicum, Olpium 292
indicus, Calocheiridius 270
indicus, Dhanus 318
indicus, Ellingsenius 500
indicus, Hygrochelifer 504
indicus, Indogarypus (= Garypus)
  260
indicus, Lagynochthonius 184
indicus, Lamprochernes 589
indicus, Lophochernes 514
indicus, Metawithius (Metawithius) 646
indicus, Ochrochernes 610
indicus, Oratemnus (= Chelifer) 467
indicus, Parachernes (Parachernes) 615
indicus, Paratemnoides (= Paratemnus)
  471
indicus, Pselaphochernes 626
indicus, Stenatemnus 475
indicus, Stenochelifer 531
indicus, Sternophorus 447
indicus, Tamenus 477
indicus, Tullgrenius 481
indicus, Withius 661
indivisus, Paratemnoides (= Chelifer)
  471
Indochernes 587
indochinensis, Microcreagris 342
Indogaryops 446
Indogarypinus 282
Indogarypus 260
Indohya 315
Indolpium 282
inermis, Pseudopilanus 630
inexpectatus, Allochernes 551
inexpectum, Apocheiridium
  (Apocheiridium) 435
infernalis, Kleptochthonius
  (Chamberlinochthonius) 180

infernalis, Tartarocreagris
  (= Microcreagris) 416
infernum, Neobisium (Blothrus) 378
inflatus, Tychochernes 638
infuscatus, Dactylochelifer 493
ingratum, Novobisium (= Neobisium)
  391
inhonestus, Sphallowithius 656
innoxius, Tyrannochthonius 209
inpai, Parachernes (Parachernes) 615
inquilinus, Dasychernes 567
insignis, Mesochelifer 516
insignis, Neowithius (= Chelifer) 649
insociabilis, Nesidiochernes 605
insolitum, Paravachonium 431
insolitus, Apatochernes 546
insolitus, Diplotemnus 462
insperatum, Cheiridium 438
instabilis, Dendrochernes
  (= Pachycheirus) 569
insubidus, Paratemnoides (= Chelifer)
  471
insuetus, Chernes 588
insuetus, Parachernes (Parachernes)
  615
insulae, Paraliochthonius 194
insulae, Pseudochiridium 445
insulae, Tyrannochthonius 209
insulanum, Stenolpium 305
insulanus, Aporochelifer 483
insulanus, Haplochernes 581
insulanus, Rhopalochernes 631
insulanus, Sathrochthonius 202
insulare, Cheiridium 438
insulare, Neobisium (Blothrus) 378
insulare, Xenolpium 307
insularis, Austrochthonius 140
insularis, Chelanops (Chelanops) 554
insularis, Chelifer latreillei 526
insularis, Chthonius
  (Ephippiochthonius) 168
insularis, Diplotemnus 462
insularis, Garypus 240
insularis, Lissochelifer
  (= Lophochelifer) 510
insularis, Microchernes 639
insularis, Parachernes (Parachernes)
  616
insularis, Paratemnoides (= Atemnus)
  471
insularis, Paratemnus 452
insularis, Pilochelifer 523
insularis, Pseudochthonius 196
insularis, Roncus 406

insularis, Tamenus 477
insularum, Epactiochernes 576
insularum, Ideoblothrus
  (= Pachychitra) 423
Insulocreagris 335
Interchernes 587
intermedium, Euryolpium 276
intermedium, Indolpium 283
intermedium, Neobisium (Neobisium)
  360
intermedium, Olpium 292
intermedium, Vachonobisium
  (= Gymnobisium) 314
intermedius, Afrogarypus
  (= Geogarypus) 250
intermedius, Apochthonius 137
intermedius, Calocheiridius 270
intermedius, Dactylochelifer 493
intermedius, Dolichowithius
  (Dolichowithius) 644
intermedius, Lustrochernes
  (= Chelifer) 596
intermedius, Nudochernes 607
intermedius, Sternophorus
  (Afrosternophorus) 447
intermedius, Tyrannochthonius 209
intractabile, Neobisium (Neobisium)
  360
inusitatus, Hesperochernes 584
inversus, Austrochthonius 141
iranica, Acanthocreagris 326
iranicus, Chthonius
  (Ephippiochthonius) 168
iranicus, Rhacochelifer 527
irmgardae, Pachyolpium 296
irmleri, Tyrannochthonius 210
irregularis, Chthonius (Chthonius)
  149
irrugatus, Geogarypus (= Garypus) 255
irusanga, Neoditha 224
irwini, Apochthonius 137
ischnocheles, Chthonius (Chthonius)
  (= Chelifer) 149
ischnocheloides, Chthonius
  (Chthonius) 152
ischyrum, Neobisium (Neobisium)
  (= Obisium) 360
ishiharanus, Microcreagris 343
ishikawai, Allochthonius
  (Urochthonius) 134
Isocheiridium, Cheiridium 437
isolatum, Pachyolpium (= Olpium) 296
istriacum, Neobisium (Blothrus)
  spelaeum (= Blothrus) 384

italica, Acanthocreagris
(= Microcreagris) 326
italica, Larca 243
italicus, Allochernes (Allochernes)
538
italicus, Chthonius (Chthonius) 152
italicus, Pselaphochernes 626
italicus, Roncus (= Obisium) 406
itapemirinensis, Geogarypus 255
iugoslavicus, Chthonius (Chthonius)
152
iunctus, Parawithius (Parawithius)
(= Chelifer) 650
ixoides, Chelifer 489

jablanicae, Neobisium (Blothrus)
tantaleum 385
jacobsoni, Olpium 292
jagababa, Roncus 406
jalzici, Chthonius (Chthonius) 153
jamaicensis, Antillochernes 544
jamaicensis, Troglobochica 311
jaoreci, Roncus 406
japonica, Microcreagris 342
japonicum, Amblyolpium 263
japonicum, Halobisium orientale 335
japonicus, Allochernes
(= Toxochernes) 536
japonicus, Garypus 240
japonicus, Mundochthonius 190
japonicus, Pararoncus (= Obisium) 393
japonicus, Paratemnoides
(= Paratemnus) 471
japonicus, Solinus 304
japonicus, Tyrannochthonius
(= Chthonius) 210
japonicus, Withius 661
javana, Compsaditha 223
javana, Dithella (= Chthonius) 223
javanus, Anatemnus (= Chelifer) 451
javanus, Geogarypus (= Garypus) 255
jeanneli, Negroroncus 321
jeanneli, Neobisium (Blothrus)
(= Obisium) 378
jeanneli, Verrucadithella 229
jenkinsi, Phaulochernes 623
jezequeli, Hansenius 502
johni, Lagynochthonius (= Chthonius)
184
johnstoni, Paraliochthonius
(= Chthonius) 194
jonensis, Bituberochernes 548
jonesi, Pseudotyrannochthonius
(= Tubbichthonius) 199

jonicus, Beierochelifer peloponnesiacus
(= Rhacochelifer) 485
jonicus, Chthonius (Chthonius) 153
jonicus, Lasiochernes (= Chelifer) 593
juberthiei, Neobisium (Neobisium) 361
jugorum, Chthonius (Chthonius) 153
jugorum, Neobisium (Neobisium)
(= Obisium) 361
juliae, Microcreagris 414
julianus, Roncus 406
juvencus, Roncus 406
Juxtachelifer 509
juxtlahuaca, Tridenchthonius 226

kabylicus, Chthonius
(Ephippiochthonius) 168
kaestneri, Withius (= Allowithius) 661
Kafirchthonius 203
kalaharicum, Ectactolpium 274
kalaharicus, Ammogarypus 235
kaltenbachi, Sathrochthonius 202
kanaka, Americhernes
(= Lamprochernes) 541
kapi, Lagynochthonius 184
Karachelifer 504
karamani, Allochernes 562
karamani, Karachelifer 506
karamani, Microchthonius
(= Chthonius) 189
karamani, Neobisium (Blothrus)
(= Obisium) 378
karamanianus, Chthonius (Chthonius)
153
Kashimachelifer 509
kashmirensis, Levigatocreagris 336
kaszabi, Dactylochelifer 493
katoi, Muscichernes 592
kauae, Vachonium 431
kaznakovi, Bisetocreagris
(= Ideobisium) 332
kelassuriense, Neobisium 356
kenyaensis, Miratemnus 465
kenyense, Pseudochiridium 445
kenyensis, Lagynochthonius
(= Tyrannochthonius) 184
kerenyaga, Negroroncus 321
kerioense, Beierolpium 267
kermadecensis, Thalassochernes 637
kermadecensis, Tyrannochthonius
(= Paraliochthonius) 210
kerzhneri, Dactylochelifer 493
kewi, Chelifer 504
kewi, Chthonius (Ephippiochthonius)
168

Kewochthonius 143
keyserlingi, Chelifer 563
kibwezianus, Titanatemnus 479
kikuyu, Afroroncus 316
kilimanjaricus, Hansenius 502
kilimanjaricus, Synatemnus 476
kishidai, Pseudotyrannochthonius
  undecimclavatus (= Allochthonius)
  201
kittenbergeri, Catatemnus 458
kivuense, Cryptocheiridium
  (Cryptocheiridium) 441
kivuensis, Nudochernes 608
klapperichi, Bisetocreagris
  (= Microcreagris) 332
klapperichi, Parazaona 622
kleemanni, Pycnochelifer (= Chelifer)
  525
Kleptochthonius 177
Kleptochthonius, Kleptochthonius 178
knowltoni, Apochthonius 137
knoxi, Apatochernes antarcticus 545
kobachidzei, Neobisium (Neobisium)
  361
kobayashii, Pseudotyrannochthonius
  (= Spelaeochthonius) 199
kochalkai, Ideoblothrus 424
kochi, Olpium 292
kochii, Chelignathus 219
kochii, Megathis 669
koellneri, Microcreagris 343
kolombangarensis, Gelachernes 577
korabense, Neobisium (Blothrus) 378
kosswigi, Neobisium (Blothrus) 378
kozlovi, Centrochthonius
  (= Chthonius) 142
kraepelini, Haplochernes (= Chelifer)
  581
krakatau, Tyrannochthonius 210
kraussi, Stenatemnus 476
krekeleri, Kleptochthonius
  (Chamberlinochthonius) 180
krugeri, Feaella (Difeaella) 230
krusadiensis, Garypus 240
ksenemani, Chthonius (Chthonius)
  147
kubotai, Pseudotyrannochthonius
  (= Spelaeochthonius) 200
kusceri, Garypinus dimidiatus 278
kuscheli, Apatochernes 546
kuscheli, Asterochernes 547
kuscheli, Chelanops (Chelanops) 555
kuscheli, Lechytia 187
kuscheli, Nesidiochernes 605

kuscheli, Parachernes (Parachernes)
  616
kuscheli, Parazaona 623
kuscheli, Phaulochernes 623
kuscheli, Pseudopilanus 630
kuscheli, Thaumatolpium 306
kussariensis, Dactylochelifer
  (= Chelifer) 493
kwantungensis, Chinacreagris
  (= Microcreagris) 334
kwantungensis, Sinochelifer 531
kwartirnikovi, Neobisium (Blothrus)
  379
kyushuensis, Allochthonius
  (Urochthonius) ishikawai 134

labilus, Chthonius (Ephippiochthonius)
  669
labinskyi, Neobisium (Neobisium) 361
lacertosus, Pselaphochernes
  (= Chernes) 626
laceyi, Larca 243
laciniosus, Lophochernes (= Chelifer)
  514
ladakhensis, Dactylochelifer 494
laevis, Negroroncus 321
laevis, Tyrannochthonius 210
lagari, Roncus 407
lagunae, Withius (= Chelanops) 661
Lagynochthonius 182
lahaulensis, Bisetocreagris kaznakovi
  (= Microcreagris) 332
lamellatus, Paragoniochernes
  (= Chelifer) 650
lamellifer, Albiorix 317
laminata, Shravana (= Ideobisium) 322
lamottei, Ectromachernes 645
lampra, Microcreagris 343
Lamprochernes 587
lampropsalis, Chelifer 507
lanzai, Acanthocreagris
  (= Microcreagris) 326
lanzai, Chthonius (Chthonius) 153
laosana, Ditha (Paraditha) 222
laosanus, Paratemnoides
  (= Paratemnus) 471
Larca 242
Lasiochernes 592
lasiophilus, Chernes 563
lata, Larca (= Garypus) 243
lata, Tuberocreagris (= Microcreagris)
  416
latens, Neobisium (Neobisium) 362
lathrius, Synsphyronus 246

latidentatus, Chthonius (Chthonius)
  bogovinae 144
latimana, Ditha (Paraditha)
  (=Paraditha) 222
latimanus, Chelanops 616
latissimus, Rhacochelifer peculiaris 529
lativittatus, Parachelifer (=Chelifer)
  520
latona, Orientocreagris 392
latreillei, Dactylochelifer (=Chelifer)
  494
latum, Neobisium (Neobisium)
  dolicodactylum 356
latum, Xenolpium oceanicum 267
latus, Parachernes (Parachernes)
  (=Chelanops) 616
laudabilis, Cryptocreagris
  (=Microcreagris) 334
laurae, Hesperochernes 584
laurae, Microcreagris 343
lautus, Tyrannochthonius 209
lawrencei, Ammogarypus 235
lawrencei, Beierolpium
  (=Calocheiridius) 267
lawrencei, Microbisium 339
lawrencei, Pseudatemnus 654
lawrencei, Pseudochiridium 445
lawrencei, Rhopalochelifer 531
lawrencei, Withius (=Allowithius) 662
laysanensis, Withius (=Chelifer) 662
Lechytia 186
leclerci, Roncobisium 397
legrandi, Beierius walliskewi
  (=Dactylochelifer) 485
leleupi, Apolpium 265
leleupi, Feaella (Tetrafeaella) 231
leleupi, Hansenius 503
leleupi, Ideoblothrus (=Ideobisium)
  424
leleupi, Lechytia 187
leleupi, Nudochernes 608
leleupi, Parachernes (Parachernes) 616
leleupi, Pseudochthonius 196
lenkoi, Ideoroncus 319
leo, Synsphyronus 246
leoi, Chthonius (Chthonius)
  (=Neochthonius) 154
leonidae, Neobisium (Blothrus) torrei
  386
leonidae, Roncus 407
leopoldi, Apocheiridium
  (Chiliocheiridium) 437
lepesmei, Ideoblothrus (=Ideobisium)
  424

leptaleus, Lamprochernes (=Chelifer)
  589
leruthi, Chthonius (Chthonius) 154
leruthi, Neobisium (Blothrus) 379
lessiniensis, Chthonius (Chthonius) 154
lethaeum, Neobisium (Blothrus) 379
letourneuxi, Atemnus (=Chelifer) 455
leucadia, Acanthocreagris
  (=Microcreagris) 326
Leucohya 310
levantinus, Garypus 240
Levichelifer 509
Levigatocreagris 336
levipalpus, Americhernes
  (=Lamprochernes) 541
levipalpus, Ideoblothrus 424
lewisi, Withius (=Allowithius) 662
liaoningense, Trachychelifer 533
libanoticus, Calocheiridius 270
liberiense, Nannobisium 427
liebegotti, Roncus 407
ligulifera, Tyrannochthoniella 204
ligusticum, Neoccitanobisium 391
ligusticus, Chthonius (Chthonius)
  microphthalmus 156
ligusticus, Roncus 407
limatus, Megachernes 600
lindahli, Menthus (=Minniza) 261
lindbergi, Chthonius (Chthonius) 154
lindbergi, Dactylochelifer 496
lindbergi, Diplotemnus 462
lindbergi, Levigatocreagris
  (=Microcreagris) 336
lindbergi, Minniza babylonica 284
lindbergi, Olpium 292
lineatus, Synsphyronus 246
linsdalei, Pycnochernes 592
lislei, Goniochernes 579
Lissochelifer 509
Lissocreagris 336
Litochelifer 511
litorale, Halominniza aegyptiacum
  (=Olpium) 279
litoralis, Chthonius (Chthonius) 154
litoralis, Garypus 238
litoralis, Nannochelifer 518
litoralis, Parachernes (Parachernes) 616
litoralis, Pselaphochernes 627
litoralis, Serianus (=Garypinus) 302
littlefieldi, Aspurochelifer 484
liwa, Allochernes 536
ljovuschkini, Pseudoblothrus 428
lobatschevi, Dactylochelifer 496
lobipes, Rhacochelifer (=Chelifer) 527

loebli, Calocheiridius 271
loeffleri, Parachernes (Parachernes) 617
loewi, Chelifer 669
lohmanderi, Withius 662
loltun, Vachonium 431
lombardicum, Neobisium (Neobisium) 362
lonai, Roncus 407
longedigitatus, Negroroncus 321
longesetosum, Thaumatolpium 306
longesetosus, Chthonius (Ephippiochthonius) 168
longeunguiculatus, Rhaochchelifer 527
longichelifer, Dolichowithius (Dolichowithius) (= Chelifer) 644
longicolle, Obisium 669
longidactylus, Hysterochelifer 508
longidigitatum, Apolpium (= Olpium) 265
longidigitatum, Neobisium (Blothrus) (= Obisium) 379
longidigitatus, Geogarypus (= Chelifer) 256
longidigitus, Garypus 240
longimanus, Americhernes 542
longimanus, Thaumastogarypus 248
longipalpis, Chthonius 172
longipalpis, Obisium jugorum 361
longipalpus, Parachelifer 521
longipes, Beierius walliskewi 485
longipes, Nudochernes 608
longipes, Progarypus 298
longiventer, Xenolpium (= Olpium) 307
longum, Neopachyolpium 288
longus, Anatemnus 452
Lophochelifer 509
Lophochernes 511
Lophodactylus 515
lophonotus, Telechelifer 532
loricata, Ditha (Ditha) 221
lourencoi, Microchelifer 517
loyolae, Indolpium (= Minniza) 283
loyolai, Oratemnus 467
lubricus, Roncus 407
lucanus, Chthonius (Ephippiochthonius) 168
lucifuga, Acanthocreagris (= Obisium) 326
lucifugum, Cryptocheiridium (Cryptocheiridium) 441

lucifugus, Caffrowithius (= Plesiochernes) 551
lucifugus, Chthonius (Chthonius) 154
lucifugus, Nudochernes 608
ludiviri, Acanthocreagris 327
lulense, Neobisium (Ommatoblothrus) 389
luscus, Pseudohorus 300
lusitanus, Garypus 257
Lustrochernes 594
lutzi, Kleptochthonius (Chamberlinochthonius) 181
luxtoni, Austrochthonius 141
luxtoni, Tyrannochthonius (= Morikawia) 210
luzonica, Microcreagris 343
luzonicus, Anatemnus 452
luzonicus, Lophochernes 514
lycaonis, Acanthocreagris 327
lychnidis, Roncus 409
lymphatus, Chernes (= Reginachernes) 561

macedonicus, Chthonius (Chthonius) 154
machadoi, Chthonius (Ephippiochthonius) 168
machadoi, Olpiolum 290
macilenta, Microcreagris (= Obisium) 343
macrochaetus, Pselaphochernes 628
macrochelatus, Chelifer 566
Macrochelifer 515
Macrochernes 598
macrodactylum, Neobisium (Neobisium) (= Obisium) 362
macrodactylus, Geogarypus 256
macropalpus, Microcreagris 343
macropalpus, Tyrannochelifer (= Chelifer) 533
macrotuberculatus, Dactylochelifer 496
maculatus, Chthonius 172
maculatus, Geogarypus (= Garypus) 256
maculatus, Nesidiochernes 605
maculatus, Rhacochelifer (= Chelifer) 528
maculosus, Parachernes (Parachernes) darwiniensis 614
madagascariense, Xenolpium (= Parolpium) 308
madagascariensis, Haplochernes 581

madagascariensis, Withius (=Chelifer) 662

madecassus, Anatemnus 452

madeirensis, Microcreagrella caeca 426

maderi, Neobisium (Blothrus) 380

madrasensis, Tyrannochthonius (Tyrannochthonius) 209

madrasica, Lechytia 187

magalhanicum, Gymnobisium chilense 313

magalhanicus, Austrochthonius chilensis 140

magna, Cryptocreagris (=Ideobisium) 334

magna, Microcreagris (=Blothrus) 344

magnanimus, Apochthonius 137

magnifica, Leucohya 310

magnificus, Chthonius (Chthonius) 155

magnificus, Dinocheirus 571

magnificus, Paratemnoides (=Paratemnus) 471

magnum, Banksolpium 266

magnum, Parobisium (=Neobisium) 395

magnus, Albiorix 317

magnus, Kleptochthonius (Kleptochthonius) 178

magnus, Mundochthonius 191

magnus, Synsphyronus 247

mahnerti, Acanthocreagris 327

mahnerti, Chthonius (Ephippiochthonius) 169

mahnerti, Neobisium (Neobisium) 363

mahnerti, Paratemnoides (=Paratemnus) 471

mahnerti, Roncus 409

mahunkai, Tyrannochthonius 211

major, Allowithius somalicus 665

major, Anatemnus orites 453

major, Aphrastochthonius 135

major, Dactylochelifer gobiensis 492

major, Geogarypus 256

major, Hansenius 503

major, Metatemnus 464

major, Neobisium (Neobisium) mahnerti 363

major, Thapsinochernes flavus 638

majusculum, Indolpium 283

malaccense, Alocobisium 417

malaccensis, Tyrannochthonius terribilis 215

malatestai, Chthonius (Chthonius) 155

malayanus, Sundochernes 636

malcolmi, Malcolmochthonius 188

Malcolmochthonius 188

maldivensis, Garypus 241

malheuri, Apochthonius 138

maltensis, Chthonius (Ephippiochthonius) 169

mancus, Thaumastogarypus 249

manicatum, Microbisium (=Obisium) 339

manilanus, Oratemnus 467

maori, Protochelifer 524

Maorichernes 598

Maorichthonius 189

maoricus, Apatochernes 546

maoricus, Nelsoninus 287

maoricus, Phaulochernes 624

maoricus, Sathrochthonius 202

Maorigarypus 244

marcusensis, Ditha (Paraditha) (=Verrucaditha) 222

margaritifer, Anaperochernes 543

marginatus, Progarypus 298

marianae, Garypinus 299

mariannus, Chelifer 670

maritima, Bisetocreagris 333

maritimum, Neobisium (Neobisium) (=Obisium) 363

marlausicola, Dactylochelifer 496

marmoratus, Garypus 241

maroccana, Microcreagris 427

maroccanum, Neobisium (Neobisium) (=Obisium) 364

maroccanus, Allochernes 537

maroccanus, Chthonius (Ephippiochthonius) 169

maroccanus, Dactylochelifer (=Ectoceras) 496

maroccanus, Geogarypus 256

marquesianus, Geogarypus 256

martae, Neobisium (Neobisium) lombardicum (=Obisium) 362

martensi, Ceriochernes 553

martensi, Levigatocreagris 336

martini, Paraliochthonius 194

martiniquensis, Lechytia 187

masi, Allochernes (=Chelifer) 537

massylicus, Rhacochelifer 528

mateui, Rhacochelifer 528

mauriesi, Lustrochernes 596

mauritanicus, Chthonius (Chthonius) (=Neochthonius) 155

mavromoustakisi, Calocheiridius 271

maxbeieri, Neobisium (Blothrus) 380

Maxchernes 599

maxima, Lechytia 188

maximus, Apochthonius 138
maximus, Hebridochernes 583
maxvachoni, Neobisium (Neobisium) 364
maya, Ideoblothrus (= Pachychitra) 424
maya, Mexobisium 311
maya, Vachonium 432
mayeti, Lophochernes (= Chelifer) 514
mayi, Chthonius (Chthonius) 155
mazaurici, Chthonius (Chthonius) 155
mediocre, Stenolpium 305
mediocris, Microcreagris microdivergens 672
mediofasciatus, Dolichowithius (Dolichowithius) 644
mediterraneum, Obisium (Obisium) erythrodactylum 355
medium, Olpiolum 290
medium, Pachyolpium 290
medvedevi, Neobisium (Neobisium) 364
Megachernes 599
megaloptera, Verrucaditha 228
meganennus, Synsphyronus 247
megasoma, Anatemnus (= Chelifer) 452
Megathis 667
meinertii, Parachernes (Parachernes) (= Chelifer) 617
Meiochernes 602
Meiogarypus 243
melanochelatus, Synsphyronus (= Maorigarypus) 247
melanopygus, Lophochernes 514
melanopygus, Parachernes (Parachernes) 617
melitensis, Roncus 410
melloguensis, Roncus 410
melloleitaoi, Neochernes 605
meneghettii, Tyrannochthonius (= Parachthonius) 211
mengei, Chernes 558
mengei, Chthonius (Chthonius) 155
mengei, Electrochelifer 499
menozzii, Roncus (= Obisium) 410
MENTHIDAE 261
Menthus 261
meridianus, Hysterochelifer (= Chelifer) 506
meridieserbicum, Neobisium (Neobisium) 355
meridionalis, Garypus 257
merope, Bisetocreagris 333

meruensis, Caffrowithius (= Plesiochernes) 551
meruensis, Nudochernes lucifugus 608
meruensis, Tyrannochthonius 211
Mesatemnus 463
Mesochelifer 516
Mesochernes 602
Metachelifer 516
Metagoniochernes 603
Metatemnus 463
Metawithius 646
Metawithius, Metawithius 646
meuseli, Chthonius (Chthonius) subterraneus 161
meuseli, Neobisium (Blothrus) 384
Mexachernes 603
mexicanus, Albiorix (= Ideoroncus) 317
mexicanus, Ancalochernes 543
mexicanus, Ideoblothrus (= Pachychitra) 424
mexicanus, Incachernes 587
mexicanus, Menthus 261
mexicanus, Mundochthonius 191
mexicanus, Parachelifer 521
mexicanus, Paraliochthonius 194
mexicanus, Tridenchthonius 226
Mexichelifer 517
Mexichthonius 189
mexicolens, Chelodamus 556
Mexobisium 310
michaelseni, Chelanops (Neochelanops) (= Chelifer) 555
michaelseni, Euryolpium (= Olpium) 276
Micratemnus 464
Microbisium 337
Microblothrus 426
Microchelifer 517
Microchernes 638
Microchthonius 189
Microcreagrella 426
Microcreagrina 427
Microcreagris 340
microdivergens, Microcreagris 344
micronesiensis, Geogarypus 256
microphthalma, Acanthocreagris 327
microphthalmus, Chthonius (Chthonius) 155
microphthalmus, Kleptochthonius (Chamberlinochthonius) 181
microphthalmus, Roncus (= Obisium) 410
microstethum, Olpium 293

microti, Allochernes 537
microtuberculatus, Chthonius
   (Ephippiochthonius) 169
Microwithius, Metwithius 647
migrans, Tyrannochthonius 211
milanganum, Nanolpium 286
militaris, Semeiochernes 632
milleri, Diplotemnus 461
milloti, Hansenius 503
milloti, Metagoniochernes 603
milloti, Tamenus 477
mimetus, Synsphyronus 247
mimulus, Hesperochernes 584
mimulus, Parawithius (Victorwithius)
   (= Cacodemoniellus) 652
mimulus, Synsphyronus 247
minax, Chelifer 670
mindanensis, Adelphochernes 535
mindoroensis, Adelphochernes 535
mindoroensis, Lophochernes 514
Minicreagris 346
minima, Typhloditha 228
minimum, Neobisium (Neobisium)
   (= Obisium) 364
minimus, Apochthonius 138
Minniza 283
minnizioides, Olpium 293
minoius, Roncus (Parablothrus) 413
minor, Ammogarypus 235
minor, Apochthonius 138
minor, Cheiridium 439
minor, Cyclatemnus 461
minor, Dendrochernes cyrneus 569
minor, Garypinus afghanicus 277
minor, Geogarypus (= Garypus) 257
minor, Lamprochernes 589
minor, Lustrochernes 596
minor, Paratemnoides (= Chelifer) 472
minor, Parolpium (= Olpium) 297
minor, Pseudogarypus 234
minor, Roncus corcyraeus 404
minor, Tyrannochthonius 211
minore, Mirobisium 313
minoum, Obisium (Blothrus) 384
minous, Chthonius
   (Ephippiochthonius) 169
minusculoides, Microchelifer
   (= Chelifer) 517
minusculus, Aperittochelifer
   (= Chelifer) 483
minuta, Hya (= Ideobisium) 314
minutissimum, Apocheiridium
   (Apocheiridium) 435
minutissimum, Pseudochiridium 445

minutissimus, Paratemnoides
   (= Paratemnus) 472
minutum, Apolpium 265
minutum, Neobisium (Blothrus)
   (= Blothrus) 380
minutum, Paedobisium 392
minutus, Chernes (Trachychernes) 628
minutus, Chthonius
   (Ephippiochthonius) 169
minutus, Dolichowithius
   (Dolichowithius) 644
minutus, Geogarypus (= Garypus) 257
minutus, Indogarypinus 282
minutus, Negroroncus 321
minutus, Serianus (= Olpium) 303
minutus, Serianus 302
mirabilis, Albiorix 317
mirabilis, Ectromachernes 645
mirabilis, Feaella (Feaella) 230
mirabilis, Garypinus 278
mirabilis, Hansenius 503
mirabilis, Hesperochernes (= Chelifer)
   584
mirabilis, Troglochthonius 204
Miratemnus 465
mirei, Geogarypus 257
Mirobisium 313
Mirochernes 603
mirum, Pycnocheiridium 444
mirus, Calocheirus 273
mirus, Meiogarypus 243
mirus, Teratochernes 637
mitchelli, Antillobisium 309
mitchelli, Troglohya 312
mjoebergi, Chelifer 588
moderatus, Calocheiridius
   crassifemoratus 269
modestum, Banksolpium (= Olpium)
   266
modestum, Indolpium 283
modestus, Horus 281
modestus, Ochrochernes (= Chelifer)
   610
modicus, Dolichowithius
   (Dolichowithius) 644
modiglianii, Sundochernes (= Chelifer)
   636
moestus, Apochthonius (= Chthonius)
   138
moldavicum, Paedobisium 393
molestus, Hesperochernes 584
mollis, Garypinidius 277
mollis, Nannobisium (= Vescichitra)
   427

molliventer, Pseudohorus 300
mombasica, Feaella (Feaella) 230
monae, Olpiolum 290
monardi, Titanatemnus 479
monasterii, Neobisium (Blothrus) 380
mongolicola, Dactylochelifer 496
mongolicus, Allochernes 537
mongolicus, Chernes 561
mongolicus, Megachernes (=Chelifer) 600
mongolicus, Rhacochelifer 528
monitor, Catatemnus (=Chelifer) 458
monoplacophorus, Parawithius (Victorwithius) (=Victorwithius) 652
monroensis, Parachelifer 521
monstrosus, Megachernes 600
monstruosus, Hebridochernes 583
montanum, Gymnobisium 314
montanum, Neobisium (Blothrus) brevipes 375
montanus, Chiliochthonius 143
montanus, Cyclochernes 567
montanus, Hepserochernes 584
montanus, Horus 281
montanus, Mundochthonius 191
montanus, Nudochernes 608
montanus, Ochrochernes indicus 672
montanus, Parachelifer 521
montanus, Sternophorus (Sternophorus) 447
montanus, Titanatemnus 479
montenegrense, Neobisium (Neobisium) macrodactylum (=Obisium) 362
monticola, Afrogarypus (=Geogarypus) 250
monticola, Dactylochelifer 496
montigenus, Chernes (=Chelifer) 561
montigenus, Verrucachernes 638
moralesi, Pseudochthonius 196
mordax, Anaulacodithella (=Chthonius) 217
mordax, Austrochthonius 140
mordor, Lagynochthonius 184
morenensis, Parazaona (=Chelifer) 623
moreoticum, Neobisium (Neobisium) 364
moreoticus, Lamprochernes (=Chelifer) 589
morikawai, Parobisium anagamidensis (=Neobisium) 394
Morikawia 193

mormon, Apocheiridium (Apocheiridium) 435
morosus, Dendrochernes (=Chelanops) 569
mortenseni, Maorichthonius 189
mortensenii, Lissochelifer (=Chelifer) 510
mortis, Alabamocreagris (=Microcreagris) 329
motasi, Chthonius (Chthonius) 156
mrciaki, Dactylochelifer 497
muchmorei, Antillochernes (=Parachernes) 544
muchmorei, Ideoblothrus 424
Mucrochelifer 518
Mucrochernes 604
mucronata, Feaella (Tetrafeaella) 232
mucronatus, Lophochernes (=Chelifer) 514
muesebecki, Chitrella 418
multidentatus, Chernes 568
multidentatus, Chthonius (Chthonius) 156
multispinosa, Acanthocreagris 327
multispinosus, Kleptochthonius (Kleptochthonius) (=Heterochthonius) 178
mumae, Bituberochernes 548
mundanus, Pseudochthonius 196
Mundochthonius 190
mundus, Phorochelifer 523
muricatus, Parachelifer (=Chelifer) 521
murrayi, Metawithius (Metawithius) (=Chelifer) 646
murthii, Calocheiridius 271
Muscichernes 588
muscivorus, Lamprochernes 589
muscorum, Obisium 349
musculi, Calidiochernes 552
museorum, Cheiridium (=Chelifer) 439
mussardi, Calocheiridius 271
mussardi, Nannocheliferoides 519
myophilus, Megachernes ryugadensis 601
myopius, Kleptochthonius (Chamberlinochthonius) 181
myops, Simonobisium (=Obisium) 415
Myrmecowithius 647
Myrmochernes 604
mysterius, Apochthonius 138

nagaminei, Lagynochthonius
    (=Tyrannochthonius) 185
naikaiensis, Megachernes
    ryugadensis 601
nairobiensis, Lissochelifer
    (=Lophochelifer) 510
namaquense, Ectactolpium 274
namaquensis, Diplotemnus 462
nana, Feaella (Tetrafeaella) capensis
    231
nana, Roncocreagris galeonuda
    (=Microcreagris) 398
nankingensis, Chinacreagris 334
Nannobisium 427
Nannochelifer 518
Nannocheliferoides 519
Nannoroncus 320
Nannowithius 647
Nanolpium 286
nanus, Afrogarypus intermedius
    (=Geogarypus) 250
nanus, Afrosternophorus 448
nanus, Chthonius (Ephippiochthonius)
    169
nanus, Congochthonius 177
nanus, Haplochernes 582
nanus, Tyrannochthonius
    (=Morikawia) 211
nanus, Withius 662
naracoortensis, Protochelifer 524
naranjitensis, Pseudochthonius
    (=Chthonius) 197
natalensis, Afrochthonius 132
natalensis, Caffrowithius
    (=Pselaphochernes) 551
natalensis, Lechytia (=Chthonius) 188
natalensis, Titanatemnus 479
natalicus, Caffrowithius
    (=Plesiochernes) 551
navaricum, Neobisium (Blothrus)
    (=Obisium) 380
navigator, Oratemnus (=Chelifer) 467
navigator, Smeringochernes
    (Smeringochernes)
    (=Rhopalochernes) 633
neglectus, Withius (=Chelifer) 662
Negroroncus 320
Nelsoninus 287
nemorale, Neobisium (Neobisium)
    (=Obisium) 364
Neoallochernes 604
Neoamblyolpium 287
NEOBISIIDAE 323
NEOBISIOIDEA 309

Neobisium 346
Neobisium, Neobisium 346
Neoccitanobisium 391
Neocheiridium 442
Neochelanops, Chelanops 555
Neochernes 604
Neochthonius 192
Neoditha 224
Neogarypus 243
Neominniza 288
Neopachyolpium 288
Neopseudogarypus 232
neotropicus, Atemnus 455
neotropicus, Roncus 410
Neowithius 648
nepalense, Apocheiridium
    (Apocheiridium) 435
nepalense, Cheiridium 440
nepalensis, Allochernes asiaticus 535
nepalensis, Calocheiridius 271
nepalensis, Ceriochernes 553
nepalensis, Geogarypus 257
nepalensis, Hysterochelifer 506
nepalensis, Orochernes 611
nepalensis, Withius 663
Nepalobisium 391
nepoides, Chelifer 439
Nesidiochernes 605
Nesiotochernes 606
Nesocheiridium 443
Nesochernes 607
Nesowithius 649
nestoris, Apatochernes 546
nevermanni, Parachernes (Parachernes)
    617
newelli, Pseudotyrannochthonius 199
Nhatrangia 321
nickajackensis, Microcreagris 344
nicobarensis, Catatemnus (=Chelifer)
    458
nicobarensis, Garypus 241
nicolaii, Garypinus 278
nidicola, Chthonius
    (Ephippiochthonius) 170
nidicola, Litochelifer 511
nidicola, Nudochernes 608
nidificator, Paratemnoides (=Chelifer)
    472
niger, Parachernes (Parachernes) 617
niger, Synsphyronus 247
nigermanus, Cordylochernes 565
nigrescens, Globocreagris
    (=Microcreagris) 335
nigricans, Chernes 670

nigrimanus, Chelanops (Chelanops) 555
nigrimanus, Chernes 561
nigrimanus, Geogarypus (= Garypus) 258
nigrimanus, Papuchelifer 519
nigrimanus, Parachernes (Parachernes) 617
nigripalpus, Idiochelifer (= Chelifer) 508
nilgiricus, Anatemnus 452
nilgiricus, Lophochernes 514
ninnii, Chelifer 494
nipponicus, Chelifer 670
Nipponogarypus 288
nitens, Heterolophus 224
nitens, Nanolpium (= Olpium) 286
nitidimanus, Parachernes (Parachernes) (= Chelifer) 617
nitidum, Obisium (Obisium) dumicola 370
nitidus, Lustrochernes (= Chelifer) 596
nitrophilum, Stenolpium asperum 305
nivale, Neobisium (Neobisium) (= Obisium) 365
noaensis, Tyrannochthonius 211
nobilis, Garypinus 279
nobilis, Parawithius (Parawithius) (= Chelifer) 650
nodosus, Lamprochernes (= Chelifer) 589
nodulimanus, Chelifer 566
nonidezi, Neobisium (Blothrus) (= Obisium) 380
nordenskjoeldi, Parazaona (= Chelifer) 623
norfolkense, Xenolpium pacificum 308
norfolkensis, Haplochernes 582
norfolkensis, Nesochernes gracilis 607
norfolkensis, Tyrannochthonius (= Paraliochthonius) 212
noricum, Neobisium (Neobisium) 365
notha, Larca 243
novacaledonica, Anaulacodithella 218
novaecaledoniae, Paraldabrinus 296
novaeguineae, Amblyolpium 264
novaeguineae, Ditha (Ditha) 221
novaeguineae, Lagynochthonius (= Tyrannochthonius) 185
novaeguineae, Lissochelifer (= Lophochelifer) 511
novaeguineae, Nesidiochernes 606
novaeguineae, Smeringochernes (Smeringochernes) 633

novaeguineae, Sundochernes 636
novaeguineensis, Paraustrochernes 622
novaezealandiae, Heterochernes (= Austrochernes) 586
novaezealandiae, Protochelifer 524
novaguineense, Calocheiridius (= Xenolpium) 271
novaguineensis, Anatemnus (= Chelifer) 452
novaguineensis, Gelachernes 577
novazealandica, Philomaoria 653
Novobisium 391
Novohorus 288
novum, Neobisium (Neobisium) gentile 358
novus, Progarypus 298
novus, Roncus 410
nubicum, Cheiridium 440
nubicus, Dactylochelifer 496
nubicus, Rhacochelifer 529
nubilis, Parachernes (Parachernes) 617
nudipes, Chthonius (Ephippiochthonius) 170
Nudochernes 607
nullarborensis, Synsphyronus 248
numburensis, Ditha 221
numidicus, Roncus 411
nymphum, Oreolpium 294

obesus, Dinocheirus (= Chelanops) 571
Obisium 487
oblongus, Americhernes (= Chelifer) 542
oblongus, Chernes 539
obrieni, Apatochernes 546
obscurum, Ectactolpium namaquense 274
obscurus, Austrochthonius zealandicus 141
obscurus, Horus (= Garypinus) 281
obscurus, Novohorus (= Olpium) 288
obscurus, Paratemnoides (= Paratemnus) 472
obscurus, Syarinus (= Ideoroncus) 429
obtusa, Acanthocreagris 327
obtusecarinatus, Lophochernes 515
oca, Verrucachernes 639
Ocalachelifer 509
occidentale, Halobisium 335
occidentalis, Apochthonius 139
occidentalis, Hesperochernes (= Pseudozaona) 584
occidentalis, Ideoblothrus (= Ideobisium) 424

occidentalis, Minniza 285
Occitanobisium 392
occultum, Neobisium (Blothrus) 381
occultus, Chelanops (Chelanops) 555
occultus, Chthonius (Chthonius) 156
occultus, Garypus 241
oceanicum, Beierolpium (=Garypinus) 267
oceanicum, Euryolpium 276
oceanicus, Eumecochernes 577
oceanusindicus, Anagarypus 236
ocellatum, Acolobisium 417
ocellatus, Geogarypus 258
ochotonae, Megachernes 600
Ochrochernes 609
octentoctus, Cordylochernes (=Chelifer) 565
octoflagellatum, Gymnobisium 312
octospinosus, Pseudotyrannochthonius 200
oculatus, Chelifer 663
Odontochernes 610
odysseum, Neobisium (Blothrus) (=Obisium) 381
oenotricum, Neobisium (Ommatoblothrus) 389
ogasawarensis, Ditha (Ditha) 221
ohausi, Rhopalochernes (=Chelifer) 632
ohridanum, Neobisium (Blothrus) 381
ohuyeanum, Parobisium magnum (=Neobisium) 396
oinuanensis, Pararoncus (=Roncus) 393
okahandjanus, Thaumastogarypus 249
okinoerabensis, Nipponogarypus enoshimaensis 288
olfersii, Chelifer 558
Oligochelifer 498
Oligochernes 610
Oligomenthus 261
Oligowithius, Dolichowithius 645
Oligowithius 641
olivaceum, Obisium muscorum 350
olivaceus, Geogarypus (=Garypus) 258
olivieri, Calocheiridius (=Olpium) 272
OLPIIDAE 262
Olpiolum 289
Olpium 290
Ommatoblothrus, Neobisium 388
ondriasi, Neobisium (Neobisium) bosnicum 349
ophthalmicus, Diplotemnus 461

oppositum, Beierobisium 312
Opsochernes 611
opticus, Allochthonius (Allochthonius) (=Chthonius) 132
Oratemnus 465
oregonicus, Chthonius (Hesperochthonius) 177
oregonus, Kleptochthonius (Kleptochthonius) 178
oregonus, Malcolmochthonius 188
Oreolpium 294
Orideobisium 667
orientale, Halobisium (=Ideobisium) 335
orientalis, Calocheiridius 272
orientalis, Chelifer cancroides 490
orientalis, Hysterochelifer 507
orientalis, Microchernes 639
orientalis, Microcreagris 344
orientalis, Stenatemnus 476
orientalis, Titanatemnus 479
orientalis, Tullgrenius 481
Orientocreagris 392
orites, Anatemnus (=Chelifer) 453
ornatus, Garypus 241
Orochernes 611
orpheus, Kleptochthonius (Chamberlinochthonius) 181
orpheus, Pseudogarypus 234
orthodactyloides, Chthonius (Chthonius) 157
orthodactylus, Chthonius (Chthonius) (=Obisium) 157
orthodactylus, Pseudochthonius 197
orthodentatus, Caribchthonius 142
ortonedae, Amblyolpium (=Olpium) 264
osellai, Acanthocreagris (=Microcreagris) 328
osellai, Neobisium (Neobisium) 365
oswaldi, Anatemnus (=Chelifer) 453
ovatus, Lustrochernes (=Chelifer) 596
ovatus, Parachernes (Parachernes) 618
ovatus, Tyrannochthonius 212
oxydactylus, Progarypus (=Olpium) 298
ozarkensis, Australinocreagris (=Microcreagris) 330

pacal, Mexichthonius 189
Pachycheirus 568
Pachychelifer 519
Pachychernes 611
Pachychitra 421

Pachyolpium 295
pachysetus, Aphrastochthonius 135
pachythorax, Tyrannochthonius 212
pacifica, Parahya 315
pacificum, Xenolpium (=Olpium) 308
pacificus, Chthonius (Chthonius) 157
pacificus, Eumecochernes (=Chelifer) 577
pacificus, Mundochthonius (=Roncus) 191
packardi, Kleptochthonius (Chamberlinochthonius) (=Blothrus) 181
padewiethi, Neobisium (Blothrus) stygium 384
Paedobisium 392
paganus, Chthonius (Chthonius) (=Kewochthonius) 157
pahangica, Ditha (Paraditha) 222
Paisochelifer 519
pakistanicus, Nannowithius (=Myrmecowithius) 648
palauanus, Geogarypus 258
palauanus, Tyrannochthonius 212
palauense, Xenolpium oceanicum 267
palauensis, Ditha (Ditha) 221
palauensis, Ideoblothrus (=Ideobisium) 424
palauensis, Nesidiochernes 606
pallens, Calocheiridius elegans 270
pallens, Neobisium (Neobisium) 365
pallidum, Apocheiridium (Apocheiridium) 435
pallidum, Parolpium 297
pallidus, Chernes 557
pallidus, Dactylochelifer 497
pallidus, Dinocheirus (=Chernes) 571
pallidus, Garypus 241
pallidus, Ideoroncus 319
pallidus, Parachernes (Parachernes) 618
pallidus, Paratemnoides (=Chelifer) 472
pallidus, Tyrannochthonius 212
pallipes, Hesperochernes (=Chelanops) 585
pallipes, Olpium (=Obisium) 293
pallipes, Philomaoria (=Chelifer) 653
palmeni, Syarinus 430
palmitensis, Aphrastochthonius 135
palmquisti, Titanatemnus (=Chelifer) 479
paludis, Chthonius (Chthonius) (=Neochthonius) 158

paludis, Hesperochernes (=Chelanops) 585
paludis, Idiogaryops (=Sternophorus) 450
panamensis, Cordylochernes 565
panamensis, Dasychernes 567
pancici, Chthonius (Globochthonius) polychaetus 176
pangaeum, Neobisium (Ommatoblothrus) 389
panzeri, Dinocheirus (=Chelifer) 572
paolettii, Chthonius (Chthonius) 158
paolettii, Roncus 411
papuana, Reischekia 630
papuanus, Afrosternophorus (=Sternophorus) 449
papuanus, Hebridochernes 583
papuanus, Megachernes 600
papuanus, Paratemnus salomonis 474
Papuchelifer 519
parablothroides, Bisetocreagris (=Microcreagris) 332
parablothroides, Roncus 411
Parablothrus, Roncus 399
Paracanthicochernes 612
Paracheiridium 444
Parachelifer 520
Parachernes 612
Parachernes, Parachernes 612
Parachthonius 205
Paraditha, Ditha 222
paradoxum, Mexobisium 311
paradoxus, Afrowithius (=Chelifer) 640
paradoxus, Hebridochernes 583
paradoxus, Nannowithius (=Myrmecowithius) 648
paradoxus, Synsphyronus 248
paraensis, Ideoblothrus 425
Paragarypus 243
Paragoniochernes 649
paraguayanus, Geogarypus 258
paraguayensis, Austrochthonius 140
Parahya 315
Paraldabrinus 296
Paraliochthonius 193
paralius, Nannochelifer 518
Parallowithius 650
Paramenthus 262
paranensis, Ideoroncus 320
Parapilanus 621
Pararoncus 393
Paraserianus 301

parasimile, Neobisium (Neobisium) 365
parasita, Chelifer 589
Paratemnoides 469
Paratemnus 469
Paraustrochernes 621
Paraustrochthonius 139
Paravachonium 430
Parawithius 650
Parawithius, Parawithius 650
Parazaona 622
parcegranosus, Chelifer 670
pardoi, Mesochelifer
 (=Hysterochelifer) 516
parentorum, Halominniza 279
parisi, Microcreagris 427
parmensis, Chthonius
 (Ephippiochthonius) 170
Parobisium 394
Parolpium 297
partitus, Dinocheirus (=Chelanops)
 573
parva, Acanthocreagris granulata
 (=Roncus) 325
parva, Cecoditha 218
parva, Compsaditha 220
parva, Feaella (Tetrafeaella) 232
parva, Lissocreagris 337
parvidentatus, Albiorix 318
parvidentatus, Tridenchthonius
 (=Chthonius) 226
parvioculatus, Chthonius (Chthonius)
 158
parvulum, Microbisium (=Obisium)
 339
parvulus, Chthonius (Chthonius) 160
parvulus, Paragoniochernes 650
parvulus, Stenowithius 657
parvulus, Synatemnus 476
parvulus, Tridenchthonius 227
parvum, Apolpium 265
parvum, Neobisium (Blothrus)
 lethaeum 379
parvus, Aphrastochthonius 135
parvus, Cyclatemnus globosus 460
parvus, Metawithius (Metawithius)
 (=Withius) 646
parvus, Parachelifer 521
parvus, Pselaphochernes 627
patagonicum, Mirobisium 313
patagonicus, Asterochernes kuscheli
 547
patagonicus, Chelanops
 (Neochelanops) (=Chelifer) 556
patagonicus, Diplothrixochernes 576

patagonicus, Serianus (=Garypinus)
 303
patei, Aphrastochthonius 135
patrizii, Neobisium (Ommatoblothrus)
 389
paucedentatum, Neobisium
 (Ommatoblothrus) 390
paucedentatus, Lagynochthonius
 (=Tyrannochthonius) 185
paucisetosum, Olpiolum 290
paucispinosus, Apochthonius 139
pauliani, Hysterochelifer 507
pauper, Metawithius (Metawithius)
 spiniventer 647
pauper, Parallowithius 650
pauperatus, Allochernes 537
pauperatus, Haplogarypinus 279
pauperculum, Neobisium (Neobisium)
 365
pauperculus, Smeringochernes
 (Smeringochernes) 633
pavlovskyi, Megachernes 600
pearsei, Chthonius (Chthonius) 158
pecki, Alabamocreagris
 (=Microcreagris) 329
pecki, Aphrastochthonius 135
pecki, Mexobisium 311
pecki, Troglobochica 311
peckorum, Ideobisium 420
peculiaris, Rhacochelifer (=Chelifer)
 529
Pedalocreagris 392
pediculoides, Chelifer 494
pefauri, Sathrochthonius 202
pekinensis, Withius (=Chelifer) 663
pelagicum, Apocheiridium
 (Apocheiridium) 435
peloponnesiacum, Neobisium
 (Neobisium) (=Obisium) 365
peloponnesiacus, Beierochelifer
 (=Chelifer) 485
Pelorus 594
penicillatus, Megachernes 601
peninsulae, Planctolpium 298
peninsularis, Neochernes (=Chernes)
 605
pennatus, Hexachernes 586
Pennobisium, Neobisium 390
pennsylvanicus, Chthonius 149
pennsylvanicus, Lustrochernes
 (=Chelifer) 597
peramae, Chthonius
 (Ephippiochthonius) minous 169
peramae, Roncus 411

percarinatus, Microchelifer 518
percelere, Neobisium (Neobisium)
  365
peregrinum, Apolpiolum 273
peregrinum, Ideobisium 421
peregrinus, Allochernes 537
perfectus, Eremogarypus 237
perplexus, Malcolmochthonius 189
perproximus, Americhernes
  (= Lamprochernes) 542
perproximus, Cordylochernes 565
perproximus, Gomphochernes 579
perpusillum, Microbisium 338
perpusillus, Paratemnoides
  (= Paratemnus) 473
perpusillus, Sphaerowithius
  (= Chelifer) 655
perpusillus, Tyrannochthonius 212
perpustulatus, Ellingsenius 500
perreti, Cheiridium 440
perreti, Feaella (Tetrafeaella) 232
perreti, Pseudochthonius 197
persephone, Lissocreagris
  (= Microcreagris) 337
persica, Minniza 285
persicus, Withius (= Chelifer) 663
persimilis, Austrochthonius 141
persimilis, Chthonius (Chthonius) 158
persimilis, Parachelifer (= Chelifer) 521
persimilis, Paratemnoides
  (= Paratemnus) 473
persimilis, Stenowithius 657
personatus, Garypus 253
personatus, Geogarypus (= Garypus)
  259
perspicillatus, Gelachernes 577
persulcatus, Lophochernes (= Chelifer)
  515
peruanum, Olpiolum 290
peruanum, Stenolpium 305
peruanus, Chelanops (Neochelanops)
  556
peruanus, Cordylochernes 566
peruanus, Parachernes (Parachernes)
  618
peruanus, Progarypus 298
peruanus, Tridenchthonius 227
peruanus, Tropidowithius 658
peruni, Neobisium (Blothrus) 381
Pessigus 454
petrochilosi, Chthonius (Chthonius)
  158
Petterchernes 623
petzi, Neobisium (Neobisium) 368

peyerimhoffi, Pseudoblothrus
  (= Blothrus) 428
phaeacum, Neobisium
  (Ommatoblothrus) 390
phagophilus, Stenowithius 657
phaleratus, Allochernes wideri
  (= Chelifer) 540
Phaulochernes 623
philipi, Haplochelifer (= Chelifer)
  504
philippinense, Alocobisium 417
philippinensis, Bisetocreagris
  (= Microcreagris) 332
philippinensis, Ditha (Ditha) 221
philippinensis, Oratemnus 467
philippinica, Dithella 223
philippinum, Cryptocheiridium
  (Cryptocheiridium) 441
philippinum, Olpium 294
philippinus, Lissochelifer
  (= Lophochernes) 511
philippinus, Megachernes 601
philippinus, Metatemnus 464
philippinus, Metawithius
  (Metawithius) 647
philippinus, Paratemnoides
  (= Paratemnus) 473
philippinus, Tyrannochthonius
  (= Morikawia) 212
Philomaoria 653
phineum, Neobisium (Blothrus) 381
phitosi, Neobisium (Neobisium) 365
Phoberocheirus 534
phoebe, Bisetocreagris 332
Pholeochthonius, Paraliochthonius 205
Phorochelifer 523
phyllisae, Saetigerocreagris
  (= Microcreagris) 415
Phymatochernes 624
picardi, Metagoniochernes 603
pieltaini, Chthonius
  (Ephippiochthonius) 170
pieperi, Diplotemnus 463
piger, Withius (= Chelifer) 663
Pilanus 624
pilatus, Pilanus 624
pilifer, Pilanus 624
Pilochelifer 523
pilosus, Lasiochernes (= Chelifer)
  593
pilosus, Pseudohorus 301
pinai, Chthonius (Ephippiochthonius)
  170
pinguis, Diplotemnus 463

pinicola, Rhacochelifer (= Chelifer)
529
pinium, Apocheiridium
(Apocheiridium) 435
pisinnus, Geogarypus 259
pityusensis, Allochernes 537
Planctolpium 297
planicola, Caffrowithius 551
platakisi, Chthonius
(Ephippiochthonius) 170
platypalpus, Chiridiochernes 563
plaumanni, Americhernes
(= Lamprochernes) 542
plaumanni, Maxchernes 599
plebejus, Paratemnoides (= Chelifer)
473
Plesiochernes 549
Plesiowithius 653
pljakici, Roncus 411
plumatus, Afrogarypus (= Geogarypus)
250
plumatus, Parachernes (Parachernes)
618
plumosus, Parachernes (Parachernes)
(= Chelifer) 618
plurisetosa, Anaulacodithella 218
plurisetosus, Nesidiochernes 606
plurisetosus, Smeringochernes
(Smeringochernes) 633
pluto, Kleptochthonius
(Chamberlinochthonius) 181
pluto, Lissocreagris (= Microcreagris)
337
pluton, Neobisium (Blothrus) 385
pococki, Paratemnoides (= Chelifer)
473
podaga, Roncus 412
poeninus, Chthonius
(Ephippiochthonius) 170
Pogonowithius 654
politum, Indolpium 283
politus, Atemnus (= Chelifer) 455
polonicum, Neobisium (Neobisium)
366
polychaetus, Chthonius
(Globochthonius) 175
pomerantzevi, Diplotemnus 461
ponapensis, Lagynochthonius
(= Tyrannochthonius) 185
ponticoides, Chthonius (Chthonius)
158
ponticum, Neobisium fuscimanum 357
ponticus, Chthonius (Chthonius) 159
popovi, Dactylochelifer 497

portugalensis, Roncocreagris
(= Microcreagris) 399
posticus, Apatochernes 546
potens, Cordylochernes 565
powelli, Allochernes (= Chelifer) 537
praeceps, Balkanoroncus 331
praecipuum, Neobisium (Neobisium)
(= Obisium) 366
primaevum, Neobisium (Blothrus)
primitivum 382
primitivum, Neobisium (Blothrus) 381
princeps, Neobisium (Blothrus) 382
pripegala, Roncus 412
pristinus, Chthonius (Chthonius) 159
procer, Lamprochernes (= Chelifer)
591
procerus, Caffrowithius 551
procerus, Ideoroncus 320
procerus, Nudochernes 608
procerus, Rhacochelifer corcyrensis 526
procerus, Stenatemnus 476
procerus, Tyrannochthonius 212
Progarypus 298
Progonatemnus 474
propinquus, Lustrochernes (= Chelifer)
597
proprius, Hysterochelifer 507
proserpinae, Kleptochthonius
(Chamberlinochthonius) 181
Protochelifer 523
Protogarypinus 299
Protoneobisium 396
Protowithius 654
protractus, Aperittochelifer
(= Chelifer) 483
provincialis, Spelyngochthonius 203
proxima, Ditha (Ditha)
(= Compsaditha) 221
proximosetus, Kleptochthonius
(Chamberlinochthonius) 182
proximus, Apatochernes 546
proximus, Dinocheirus 573
proximus, Oratemnus 467
proximus, Parawithius (Victorwithius)
(= Chelifer) 652
proximus, Pilanus 624
proximus, Tyrannochthonius 212
Pselaphochernes 624
Pseudatemnus 654
Pseudoblothrus 428
Pseudochernes 654
PSEUDOCHIRIDIIDAE 444
Pseudochiridium 444
Pseudochthonius 195

pseudocurtus, Afrogarypus
  (= Geogarypus) 251
pseudoformosa, Microcreagris 344
PSEUDOGARYPIDAE 232
Pseudogarypinus 299
Pseudogarypus 232
Pseudohorus 300
Pseudopilanus 629
Pseudorhacochelifer 524
pseudorufus, Parawithius (Parawithius)
  651
Pseudotyrannochthonius 198
Pseudozaona 556
pterodromae, Apatochernes antarcticus
  545
puddui, Roncus 412
puertoricense, Ideobisium 421
puertoricensis, Americhernes 542
puertoricensis, Olpiolum
  (= Pachyolpium) 290
puertoricensis, Paraliochthonius 194
pugifer, Parachelifer 522
pugil, Athleticatemnus 456
pugil, Chelanops (Chelanops) 555
pugil, Ideoblothrus (= Ideobisium) 425
pugilatorius, Anatemnus 453
pugnax, Roncus (= Obisium) 412
pugnax, Tyrannochthonius 213
Pugnochelifer 525
pujoli, Chamberlinarius
  (= Chamberlinius) 486
pulchellus, Parachernes (Parachernes)
  (= Chelanops) 618
pulchellus, Pseudochthonius
  (= Chthonius) 197
pulcher, Geogarypus 259
pulcher, Parachernes (Parachernes) 618
pumila, Minicreagris (= Microcreagris)
  346
pumilus, Caribochernes 552
pumilus, Idiogaryops (= Garyops) 450
punctatus, Oratemnus (= Chelifer) 467
punctiger, Lasiochernes 593
pupukeanus, Tyrannochthonius 213
purcelli, Geogarypus (= Garypus) 259
pusilla, Microcreagris 344
pusillimus, Caffrowithius
  (= Plesiochernes) 551
pusillimus, Serianus 303
pusillimus, Tyrannochthonius 213
pusillulum, Olpium 294
pusillum, Nanolpium (= Olpium) 287
pusillum, Neobisium (Blothrus) 382
pusillum, Neocheiridium 443

pusillus, Cacodemonius 641
pusillus, Chthonius (Chthonius) 159
pusillus, Micratemnus (= Chelifer) 464
pusillus, Solinus 305
pusillus, Tetrachelifer
  (= Lophochernes) 532
pusillus, Tyrannochthonius 213
pusio, Obisium 670
pustulatus, Geogarypus 259
pustulatus, Papuchelifer 520
Pycnocheiridium 444
Pycnochelifer 525
Pycnochernes 588
Pycnodithella 224
Pycnowithius 654
pycta, Parazaona 623
pycta, Roncocreagris (= Microcreagris)
  399
pygmaea, Compsaditha 220
pygmaea, Microcreagris 345
pygmaeum, Microbisium (= Obisium)
  340
pygmaeus, Australochelifer 484
pygmaeus, Chelifer 591
pygmaeus, Chthonius (Chthonius) 159
pygmaeus, Ideoblothrus
  (= Pachychitra) 425
pyrenaica, Acanthocreagris
  (= Ideobisium) 328
pyrenaicum, Neobisium (Neobisium)
  366
pyrenaicus, Chthonius
  (Ephippiochthonius) 170

quadrimaculatus, Rhacochelifer
  (= Chelifer) 530
quadrispinosum, Gymnobisium
  (= Ideobisium) 312
quartus, Cacodemonius 641
queenslandicus, Megachernes 601
queenslandicus,
  Pseudotyrannochthonius 200
queenslandicus, Sundochernes 636
queenslandicus, Tyrannochthonius
  (= Morikawia) 213

racovitzai, Troglobisium
  (= Ideobisium) 430
radjai, Chthonius (Chthonius) 159
rahmi, Alocobisium 418
rahmi, Tyrannochthonius 213
rakanensis, Pararoncus (= Roncus)
  394
ramicola, Progarypus (= Olpium) 298

ramosus, Haplochernes (=Chelifer) 582
rapax, Austrochthonius 141
rapulitarsus, Electrochelifer 500
raridentatus, Chthonius (Chthonius) 159
rasilis, Parachernes (Parachernes) 619
rathkii, Neobisium (Neobisium) (=Obisium) 366
rayi, Chthonius 149
realini, Garypus bonairensis 238
rebierei, Withius 664
rectus, Microcreagris microdivergens 672
reddelli, Albiorix 318
reddelli, Australinocreagris (=Microcreagris) 330
reddelli, Mexichelifer 517
reddelli, Mexobisium 311
redikorzevi, Acanthocreagris 328
redikorzevi, Chthonius shelkovnikovi 160
redikorzevi, Dactylochelifer (=Chelifer) 497
redikorzevi, Neobisium 670
redikorzevi, Paratemnoides (=Paratemnus) 474
reductum, Neobisium (Neobisium) 366
reductum, Xenolpium oceanicum 267
reductus, Afrochthonius 132
reductus, Americhernes 542
reductus, Chthonius (Chthonius) ischnocheles 151
regalini, Pseudoblothrus 428
regina, Chitrella 418
regina, Insulocreagris 335
Reginachernes 557
regneri, Hansenius 503
regneri, Titanatemnus 480
regulus, Kleptochthonius (Chamberlinochthonius) 182
reimoseri, Lustrochernes 597
reimoseri, Neobisium (Blothrus) (=Obisium) 382
reimoseri, Pachyolpium 296
Reischekia 630
reitteri, Neobisium (Neobisium) (=Obisium) 367
relicta, Acanthocreagris 328
remesianensis, Roncus pljakici 411
remyi, Chthonius (Ephippiochthonius) 171
remyi, Neobisium (Blothrus) 382
remyi, Roncus 412

ressli, Acanthocreagris (=Microcreagris) 328
ressli, Chthonius (Chthonius) 160
ressli, Dactylochelifer 497
ressli, Mesochelifer 516
ressli, Neobisium (Neobisium) 367
reticulata, Anaulacodithella 218
retrodentatus, Albiorix 318
reussii, Chelifer 589
rex, Aglaochitra 417
rex, Kleptochthonius (Chamberlinochthonius) 182
rex, Lophodactylus (=Chelifer) 515
rex, Tyrannochthonius 213
Rhacochelifer 525
rhantus, Geogarypus 259
Rhinochernes 630
rhodesiacum, Nanolpium 287
rhodesiacus, Afrogarypus sulcatus (=Geogarypus) 251
rhodesiacus, Calocheiridius 272
rhodesiacus, Ectromachernes 645
rhodesiacus, Microchelifer 518
rhodesiacus, Negroroncus 321
rhodinus, Chernes 562
rhodium, Apolpiolum 273
rhodium, Neobisium (Neobisium) 367
rhodius, Allochernes 538
rhodius, Solinus 305
rhodochelatus, Chthonius (Chthonius) 151
rhododactylus, Chelifer 489
Rhopalochelifer 531
Rhopalochernes 631
riberai, Tyrannochthonius 213
riograndensis, Hesperochernes 585
rivale, Neobisium (Blothrus) tantaleum 385
robusta, Acanthocreagris granulata (=Roncus) 325
robustior, Roncocreagris galeonuda (=Microcreagris) 399
robustius, Thaumatolpium 306
robustum, Euryolpium 276
robustum, Neobisium (Blothrus) (=Obisium) 382
robustum, Olpium 294
robustum, Stenolpium 306
robustum, Vachonium 432
robustum, Xenolpium 276
robustus, Anatemnus 474
robustus, Cyclatemnus 461
robustus, Geogarypus 259
robustus, Ideoblothrus pugil 425

robustus, Nesidiochernes 606
robustus, Nudochernes 608
robustus, Pachychernes (=Chelifer) 612
robustus, Parachernes (Parachernes) 619
robustus, Paratemnoides (=Paratemnus) 474
robustus, Paratemnus 471
robustus, Protowithius 654
robustus, Thaumastogarypus 249
robustus, Tyrannochthonius 213
roeweri, Lagynochthonius 185
roeweri, Obisium (Blothrus) 385
rogatus, Microchthonius (=Chthonius) 189
rollei, Minniza 285
romanicus, Chthonius (Ephippiochthonius) 171
romanum, Neobisium (Ommatoblothrus) patrizii 389
romanus, Chelifer 528
ronciformis, Acanthocreagris (=Microcreagris) 328
Roncobisium 396
Roncocreagris 397
roncoides, Roncocreagris (=Microcreagris) 399
Roncus 399
ronnaii, Parachernes (Parachernes) 619
rossi, Aisthetowithius 640
rossi, Beierolpium (=Xenolpium) 268
rossi, Menthus (=Minniza) 261
rossi, Mundochthonius 192
rossi, Pseudotyrannochthonius 200
rossi, Stenolpium 306
rossicum, Apocheiridium (Apocheiridium) 436
rostombekovi, Chelifer (Chelanops) 628
roszkovskii, Pseudoblothrus (=Ideoblothrus) 428
rothi, Diplotemnus 463
rotunda, Archeolarca 236
rotundatus, Chelifer (Atemnus) 596
rotundimanus, Chelifer (Trachychernes) 554
rotundimanus, Tyrannochthonius 214
rotundus, Anatemnus (=Chelifer) 453
roubiki, Dasychernes 567
rubida, Minniza (=Olpium) 285
rubidus, Parachernes (Parachernes) (=Chelifer) 619
rudebecki, Diplotemnus 463

rudis, Gigantochernes (=Chelifer) 578
rufeolum, Apolpium (=Olpium) 265
rufeolus, Chelifer 572
rufeolus, Parawithius (Victorwithius) 652
rufescens, Chernes 670
ruffoi, Acanthocreagris italica (=Microcreagris) 326
ruffoi, Chthonius (Chthonius) ischnocheles 152
ruffoi, Neobisium (Neobisium) 367
ruficeps, Amblyolpium 264
rufimanus, Lustrochernes (=Pelorus) 597
rufula, Tuberocreagris (=Olpium) 416
rufus, Parawithius (Victorwithius) (=Chelifer) 652
ruinarum, Mexobisium 311
russelli, Aphrastochthonius 135
russelli, Apochthonius 139
rusticus, Caffrowithius (=Plesiochernes) 551
rutilans, Chelifer 494
rybini, Pselaphochernes 627
ryugadensis, Megachernes 601

sabulosus, Parachernes (Parachernes) (=Chelifer) 619
sacer, Chthonius (Ephippiochthonius) 171
sadiya, Microchelifer 518
saegeri, Sphaerowithius (=Nannowithius) 655
saegeri, Titanatemnus 480
Saetigerocreagris 415
sagittatus, Geogarypus 260
saharae, Rhacochelifer 530
saharensis, Dactylochelifer 497
saharicum, Cheiridium 440
sahariensis, Serianus 303
saigonensis, Oratemnus (=Chelifer) 468
sakadzhianum, Neobisium (Heoblothrus) 388
sakagamii, Lechytia 188
salomonense, Amblyolpium 264
salomonense, Cryptocheiridium (Cryptocheiridium) 442
salomonensis, Hebridochernes 583
salomonensis, Lagynochthonius (=Tyrannochthonius) 185
salomonensis, Serianus 303
salomonensis, Smeringochernes (Smeringochernes) 633

salomonensis, Sphaerowithius
  (=Nannowithius) 655
salomonis, Euryolpium (=Xenolpium)
  276
salomonis, Gelachernes 577
salomonis, Paratemnoides
  (=Paratemnus) 474
saltator, Chelifer 670
salvadoricus, Incachernes 587
salvagensis, Garypus saxicola 241
samai, Rhacochelifer 530
sambicus, Pycnowithius 655
samius, Chthonius (Ephippiochthonius)
  171
samniticum, Neobisium
  (Ommatoblothrus) 390
samoanus, Americhernes
  (=Lamprochernes) 543
samoanus, Oratemnus 468
sanborni, Chernes 562
sandaliotica, Acanthocreagris 329
sandalioticus, Roncus 412
sandersoni, Mundochthonius 192
saqarthvelosi, Neobisium 670
sardoa, Acanthocreagris
  (=Microcreagris) 329
sardoum, Neobisium
  (Ommatoblothrus) 390
sardous, Chelifer 670
sardous, Roncus 412
sardous, Spelyngochthonius 204
satanas, Cacodemonius (=Chelifer)
  641
satapliaensis, Chthonius (Chthonius)
  160
Sathrochthoniella 201
Sathrochthonius 201
sauteri, Lophochernes (=Chelifer)
  515
savignyi, Lamprochernes (=Chelifer)
  591
savignyi, Olpium 671
saxicola, Garypus 241
sbordonii, Neobisium (Blothrus) 383
scabriculus, Parachelifer (=Chelifer)
  522
Scapanochernes, Parachernes 621
scaurus, Dactylochelifer 497
schaefferi, Chelifer 494
schenkeli, Neobisium (Neobisium)
  (=Obisium) 371
scheuerni, Dactylochelifer 497
schlingeri, Parachernes (Parachernes)
  619

schlingeri, Parawithius (Victorwithius)
  652
schlottkei, Catatemnus 459
schnitnikovi, Centrochthonius
  (=Chthonius) 142
schoutedeni, Hansenius 503
schoutedeni, Tamenus 478
schrankii, Chelifer 572
schultzei, Ectactolpium (=Olpium)
  275
schultzei, Lustrochernes 597
schulzi, Sphenochernes 634
schurmanni, Pseudorhacochelifer
  525
schusteri, Ideobisium 421
scitulum, Aphelolpium 264
scolytidis, Mundochthonius japonicus
  191
scorpioides, Cordylochernes (=Acarus)
  565
scorpioides, Pselaphochernes
  (=Chelifer) 627
Scotowithius 655
sculpturatus, Ellingsenius (=Chelifer)
  500
scutellatus, Neopseudogarypus 232
scutulatus, Nesidiochernes 606
secundus, Eremochernes 501
segmentidentatus, Cacodemonius
  (=Chelifer) 642
segregatus, Miratemnus (=Chelifer)
  465
Selachochthonius 203
sellowi, Corosoma 567
Semeiochernes 632
semenovi, Chelifer 502
semicarinatus, Lophochernes 515
semidentatus, Tyrannochthonius
  (=Chthonius) 214
semidivisus, Oratemnus 468
semihorridus, Tyrannochthonius
  (=Morikawia) 214
semilacteus, Parachernes (Parachernes)
  619
semimarginatus, Beierius 484
seminudum, Neobisium (Neobisium)
  (=Obisium) 367
semiserratus, Austrochthonius 141
semivittatum, Olpium 278
sendrai, Chthonius
  (Ephippiochthonius) 171
sendrai, Roncus lagari 407
senegalensis, Afrogarypus (=Garypus)
  251

sentiens, Reischekia exigua 630
septentrionalis, Dactylochelifer
    latreillei 496
sequoiae, Microcreagris 345
serbicus, Chthonius
    (Ephippiochthonius) 171
serenense, Apocheiridium
    (Chiliocheiridium) 437
Serianus 301
serianus, Serianus (= Garypinus) 303
serratidentatus, Dichela
    (= Oligochelifer) 499
serratidentatus, Selachochthonius
    (= Chthonius) 203
serratus, Chelifer 489
serratus, Dinocheirus (= Chelanops)
    573
serrulata, Lechytia 188
serrulatus, Titanatemnus 480
serrulatus, Tridenchthonius
    (= Chthonius) 227
sesamoides, Chelifer 489
sestasi, Chthonius (Chthonius) 160
setifera, Saetigerocreagris 415
setiger, Nudochernes 609
setiger, Parachernes (Parachernes)
    620
setiger, Pselaphochernes (= Chelifer)
    629
setosipygus, Caecatemnus 457
setosus, Chelifer 671
setosus, Ideoroncus 320
setosus, Parachernes (Parachernes)
    620
setosus, Roncus 412
settei, Neobisium (Neobisium) 367
seychellensis, Compsaditha 220
seychellesensis, Afrogarypus
    (= Geogarypus) 251
seychellesensis, Anatemnus 454
seychellesensis, Ideoblothrus
    (= Ideobisium) 425
seychellesensis, Nesowithius 649
shelfordi, Pachychernes 612
shelkovnikovi, Chthonius (Chthonius)
    160
shinjoensis, Hesperochernes 585
shinkaii, Dactylochelifer 497
shintoisticus, Allochthonius
    (Allochthonius) 133
shiragatakiensis, Allochthonius
    (Urochthonius) ishikawai 134
Shravana 322
shulovi, Chthonius (Chthonius) 160

shulovi, Geogarypus 260
shulovi, Paramenthus 262
siamensis, Dhanus (= Ideobisium) 318
sibiricus, Orochernes 611
sicarius, Dinocheirus 573
siciliensis, Allochernes (= Chernes) 538
siculus, Chthonius (Ephippiochthonius)
    171
siculus, Lasiochernes 593
siculus, Pselaphochernes litoralis 627
siculus, Roncus 413
sieboldtii, Beierowithius (= Obisium)
    641
sierramaestrae, Mexobisium 311
silveirai, Synsphyronus 248
silvestrii, Chelifer (Lamprochernes) 563
silvestrii, Lustrochernes 597
silvestrii, Microcreagris 345
silvestrii, Pseudotyrannochthonius
    (= Chthonius) 200
silvestrii, Thaumatolpium
    (= Ideoroncus) 306
silvestris, Dactylochelifer 498
silvicola, Bisetocreagris
    (= Microcreagris) 332
silvicola, Negroroncus 321
simargli, Neobisium (Pennobisium) 390
simberloffi, Solinellus 304
simile, Ectactolpium 275
simile, Neobisium (Neobisium)
    (= Obisium) 367
similidentatus, Tyrannochthonius 214
similis, Afrochthonius 132
similis, Aphrastochthonius 136
similis, Catatemnus 459
similis, Chernes (= Allochernes) 562
similis, Ideoblothrus (= Ideobisium)
    425
similis, Kleptochthonius
    (Chamberlinochthonius) 182
similis, Lustrochernes (= Chelifer) 597
similis, Pachychitra 426
similis, Parawithius (Victorwithius) 653
similis, Rhacochelifer 530
similis, Titanatemnus 480
simillimus, Tyrannochthonius 214
simoni, Amblyolpium 264
simoni, Neobisium (Neobisium)
    (= Obisium) 368
simoni, Pseudochthonius
    (= Chthonius) 197
simoni, Withius (= Chelifer) 664
simonioides, Neobisium (Neobisium)
    368

Simonobisium 415
simplex, Beierius 484
simplex, Caffrowithius
   (= Plesiochernes) 552
simplex, Chthonius (Globochthonius)
   176
simplex, Diplothrixochernes 576
simplex, Dolichowithius
   (Dolichowithius) 644
simulacrum, Cheiridium 440
simulans, Tyrannochthonius 214
sinensis, Chernes 562
sinensis, Diplotemnus ophthalmicus
   (= Miratemnus) 462
sinensis, Lagynochthonius
   (= Tyrannochthonius) 185
sinensis, Megachernes 599
sinensis, Paratemnoides
   (= Paratemnus) 474
singularis, Paraliochthonius
   (= Chthonius) 194
singularis, Paratemnoides
   (= Paratemnus) 474
sini, Garyops (= Sternophorus) 449
sini, Garypus 242
sini, Lechytia 188
sini, Parachelifer (= Chelifer) 522
Sinochelifer 531
sinuata, Ditha (Paraditha)
   (= Chthonius) 222
siscoensis, Chthonius
   (Ephippiochthonius) 172
sjoestedti, Titanatemnus (= Chelifer)
   480
skottsbergi, Stigmachernes 635
skwarrae, Parachelifer 522
slateri, Nesidiochernes 606
slevini, Hesperolpium (= Olpium)
   280
slovacum, Neobisium (Blothrus) 383
Smeringochernes 632
Smeringochernes, Smeringochernes 632
smithersi, Nanolpium 287
socotrensis, Hansenius (= Chelifer) 503
soederbomi, Withius (= Chelifer) 664
sokolovi, Tyrannochthonius
   (= Chthonius) 214
sola, Minniza 286
solarii, Allochernes (= Chelifer) 538
Solinellus 304
Solinus 304
solitarius, Apatochernes 546
solitarius, Dolichowithius
   (Dolichowithius) 645

solitarius, Pseudotyrannochthonius
   (= Tubbichthonius) 200
solomonense, Alocobisium 418
solus, Dinocheirus 573
solus, Serianus (= Garypinus) 303
somalicum, Cheiridium 440
somalicum, Cryptocheiridium
   (Cryptocheiridium) 442
somalicus, Calocheiridius (= Minniza)
   272
somalicus, Dactylochelifer 498
somalicus, Elattogarypus 237
somalicus, Ellingsenius 501
somalicus, Withius (= Allowithius) 665
sommerfeldi, Catatemnus 459
soricicola, Megachernes 601
sororium, Neobisium (Blothrus)
   occultum 381
Sororoditha 225
sotirovi, Roncus 413
soudanense, Beierolpium (= Horus)
   268
spalacis, Nudochernes 609
sparsedentatus, Tyrannochthonius 215
spasskyi, Dactylochelifer 498
spelaea, Larca 242
Spelaeochthonius 198
spelaeophilus, Chthonius
   (Globochthonius) 176
spelaeum, Neobisium (Blothrus)
   (= Blothrus) 383
spelaeus, Pseudogarypus 234
speleophilum, Neobisium (Neobisium)
   369
speluncarium, Neobisium (Neobisium)
   (= Obisium) 369
Spelyngochthonius 203
Sphaerowithius 655
Sphallowithius 656
Sphenochernes 634
spilianum, Neobisium (Neobisium) 369
spingolus, Chthonius
   (Hesperochthonius)
   (= Kewochthonius) 177
spiniger, Rhacochelifer 530
spinipalpis, Strobilochelifer
   (= Chelifer) 532
spiniventer, Metawithius (Metawithius)
   647
spinosa, Verrucaditha (= Chthonius)
   228
spinosus, Hansenius 503
spinosus, Hysterochelifer (= Chelifer)
   507

spinosus, Verrucachernes 639
squalidum, Beierolpium (= Xenolpium) 268
squalidum, Indolpium 283
squamosus, Pseudochthonius 197
squarrosus, Parachernes 620
stammeri, Chthonius (Chthonius) ischnocheles 152
stanfordianus, Neochthonius 192
stankovici, Neobisium (Blothrus) 384
stannardi, Apocheiridium (Apocheiridium) 436
staudacheri, Neobisium (Ommatoblothrus) 390
Steiratemnus 465
stellae, Anthrenochernes 544
stellatum, Nesocheiridium 444
stellatus, Chelifer 671
stellifer, Afrogarypus (= Geogarypus) 251
Stenatemnus 475
Stenochelifer 531
Stenohya 315
Stenolpiodes 305
Stenolpium 305
Stenowithius 656
stephensi, Illinichernes 587
stercoreus, Tejachernes (= Dinocheirus) 637
Sternophorellus 446
STERNOPHORIDAE 446
Sternophorus 449
stevanovici, Chthonius (Chthonius) 161
stewartensis, Nesiotochernes 606
Stigmachernes 634
strandi, Lissochelifer (= Chelifer) 511
strandi, Syarinus (= Ideobisium) 430
strator, Cubachelifer 490
strausaki, Neobisium (Neobisium) 369
striatum, Euryolpium 277
stribogi, Neobisium (Pennobisium) 391
strinatii, Atemnus 456
strinatii, Chthonius (Chthonius) 161
strinatii, Neocheiridium 443
strinatii, Pseudoblothrus 429
strinatii, Pseudochthonius 197
strinatii, Tyrannochthonius (= Paraliochthonius) 215
Strobilochelifer 531
strumosus, Pseudohorus 301
stupidum, Cardiolpium (= Apolpiolum) 273
stussineri, Roncus (= Obisium) 413

Stygiochelifer 532
stygium, Neobisium (Blothrus) 384
stygius, Kleptochthonius (Chamberlinochthonius) 182
styriacus, Mundochthonius 192
subatlantica, Lissocreagris (= Microcreagris) 337
subfoliosus, Caffrowithius (= Chelifer) 552
subgracilis, Pachychernes (= Chelifer) 612
subgrande, Olpium 294
subimpressus, Afrogarypus (= Geogarypus) 251
subindicus, Anatemnus (= Chelifer) 454
sublaeve, Neobisium (Neobisium) (= Obisium) 369
sublaevis, Verrucachernes 639
submersa, Parahya (= Obisium) 315
submonstruosus, Hebridochernes 583
submontanus, Chthonius (Chthonius) 161
subovatus, Lustrochernes (= Chelifer) 598
subrobustus, Pachychernes (= Chelifer) 612
subrotundatus, Parachernes (Parachernes) (= Chelifer) 620
subruber, Chelifer 663
subrudis, Dinocheirus (= Chelifer) 574
subsimilis, Rhacochelifer 530
subterraneum, Obisium (Blothrus) 387
subterraneus, Chthonius (Chthonius) 161
subtilis, Parachernes (Parachernes) 620
subtropicum, Cryptocheiridium Cryptocheiridium) (= Cheiridium) 442
subvermiformis, Anatemnus 454
succineus, Progonatemnus 475
succineus, Roncus 413
sudanensis, Nudochernes 609
suecicum, Microbisium 340
suffuscus, Novohorus 289
suis, Withius 665
sulcatimana, Verrucadithella 229
sulcatus, Afrogarypus (= Geogarypus) 251
sulcatus, Afroroncus 316
sulcatus, Calocheiridius 272
sulcatus, Micratemnus 465
sumatraensis, Ditha (Paraditha) (= Chthonius) 222

sumatranus, Dhanus (= Ideoroncus) 319
sumatranus, Paratemnoides (= Paratemnus) 474
sumatranus, Sundowithius (= Chelifer) 636
sundaicum, Pseudochiridium 445
sundaicus, Stenatemnus (= Atemnus) 476
Sundochernes 635
Sundowithius 636
superba, Chitrella 419
superbum, Neobisium (Blothrus) lethaeum 379
superbum, Paravachonium 431
superbus, Lissochelifer (= Chelifer) 511
superbus, Parachelifer 522
superior, Dinocheirus magnificus 571
superior, Metatemnus 464
superstes, Tyrannochthonius 215
suraiurana, Americhernes (= Lamprochernes) 543
surinamus, Lustrochernes 598
surinamus, Tridenchthonius (= Ditha) 227
suteri, Planctolpium 298
svetovidi, Neobisium (Blothrus) 384
SYARINIDAE 417
Syarinus 429
sylvaticum, Neobisium (Neobisium) (= Obisium) 369
Synatemnus 476
Syndeipnochernes 634
Synsphyronus 244
syriaca, Minniza 286
syriacus, Atemnus (= Catatemnus) 456
syriacus, Dactylochelifer 498
syrinx, Orientocreagris 392
Systellochernes 636

tacitum, Neobisium (Blothrus) absoloni 374
tacitus, Acuminochernes 535
tacomensis, Microcreagris (= Ideobisium) 345
taierensis, Thalassochernes (= Chelifer) 637
takanoanum, Orideobisium 671
takasawadoensis, Allochthonius (Allochthonius) opticus 133
takashimai, Paraliochthonius (= Tyrannochthonius) 194
takensis, Geogarypus (Geogarypus) javanus 255

tamaninii, Chthonius (Chthonius) 161
Tamenus 476
tamiae, Hesperochernes 585
tanense, Beierolpium 268
tanensis, Titanatemnus 480
tantaleum, Neobisium (Blothrus) 385
tantalus, Kleptochthonius (Chamberlinochthonius) 182
tarbenae, Roncus boneti 403
tartareum, Neobisium (Blothrus) dinaricum 377
Tartarocreagris 416
tasmanicus, Pseudotyrannochthonius 201
tauricus, Chthonius (Neochthonius) 160
tauricus, Hysterochelifer 507
tauricus, Rhacochelifer 530
taylori, Geogarypus 260
tbilissicus, Allowithius 671
tegulatus, Chelifer 529
Tejachernes 637
tekauriensis, Tyrannochthonius 215
Telechelifer 532
tenax, Aphrastochthonius 136
tenax, Obisium 408
tenebrarum, Neobisium (Blothrus) 385
tenebrarum, Paraliochthonius 195
tenellum, Obisium 350
teneriffae, Canarichelifer 486
tenggerianus, Ochrochernes (= Chelifer) 610
tenoch, Dinocheirus 574
tenue, Novobisium (= Neobisium) 392
tenue, Olpium 294
tenuichelatus, Chthonius 162
tenuimanus, Obisium stussineri 671
tenuimanus, Rhacochelifer 530
tenuimanus, Withius (= Chelifer) 665
tenuipalpe, Neobisium (Blothrus) (= Obisium) 385
tenuis, Bisetocreagris (= Microcreagris) 333
tenuis, Chthonius (Chthonius) 161
tenuis, Geogarypus 260
tenuis, Ideoblothrus 425
tenuis, Roncus lubricus 409
tenuisetosum, Neocheiridium 443
Teratochernes 637
Teratolpium 306
termitophila, Typhloditha 228
termitophilus, Calocheiridius 272
termitophilus, Withius (= Chelifer) 665

terribilis, Tyrannochthonius
  (= Chthonius) 215
tertiaria, Lechytia 188
tessmanni, Titanatemnus 480
tethys, Pedalocreagris 392
tetrachelatus, Chthonius
  (Ephippiochthonius) (= Scorpio) 172
Tetrachelifer 532
Tetrafeaella, Feaella 231
tetraphthalmum, Cheiridium 257
texana, Australinocreagris
  (= Microcreagris) 330
texana, Leucohya 310
texanus, Dinocheirus 574
texanus, Withius (= Chelifer) 665
Thalassochernes 637
Thapsinochernes 637
Thaumastogarypus 248
Thaumatolpium 306
Thaumatowithius 657
theisianum, Neobisium (Neobisium)
  (= Obisium) 371
thermophila, Microcreagris 345
thessalus, Chthonius (Chthonius) 162
theveneti, Microcreagris (= Obisium)
  345
thevetium, Indolpium 283
thibaudi, Aphelolpium 264
thibaudi, Pseudochthonius 197
thiebaudi, Pseudoblothrus 429
thomeensis, Titanatemnus (= Chelifer)
  481
thomomysi, Hepserochernes 585
thorelli, Catatemnus (= Chelifer) 459
thorelli, Pseudochiridium 445
thorntoni, Lagynochthonius 185
thunebergi, Mesochelifer 516
tibestiensis, Rhacochelifer 531
tibetanus, Lophochernes 515
tibetanus, Macrochelifer (= Chelifer)
  515
tibialis, Microcreagris (= Ideobisium)
  345
tibialis, Thaumatowithius 657
tibium, Olpium 294
timacensis, Roncus pljakici 411
timidus, Chelifer 671
timorensis, Oratemnus 468
tingitanus, Rhacochelifer (= Chelifer)
  531
Titanatemnus 478
titanicus, Apochthonius 139
titanius, Garypus 242
titanius, Megachernes 601

titschacki, Rhopalochernes 632
tlilapanensis, Tyrannochthonius 215
togoensis, Catatemnus (= Chelifer) 459
tokiokai, Xenolpium (Euryolpium) 276
tonkinensis, Anatemnus 454
tonkinensis, Ditha (Paraditha) 223
tonkinensis, Hyperwithius 646
tonkinensis, Lagynochthonius
  (= Tyrannochthonius) 185
tonkinensis, Lissochelifer
  (= Lophochelifer) 511
topali, Dinocheirus 574
topali, Parachernes (Parachernes)
  620
topali, Pseudopilanus 630
torpidus, Stenowithius 657
torrei, Neobisium (Blothrus)
  (= Obisium) 385
torulosus, Hansenius (= Chelifer) 503
touricum, Obisium 671
townsendi, Phaulochernes 624
Toxochernes, Allochernes 569
Trachychelifer 532
Trachychernes, Chernes 557
traegardhi, Pseudochiridium 446
transcaspius, Dinocheirus (= Chelifer)
  574
transcaucasicus, Allochernes wideri 540
transiens, Indolpium 283
transiens, Sternophorus (Sternophorus)
  447
transsilvanicus, Roncus 414
transvaalense, Nanolpium 287
transvaalensis, Aperittochelifer 483
transvaalensis, Horus 281
transvaalensis, Pseudohorus
  (= Minniza) 301
transvaalensis, Thaumastogarypus 249
transvaalensis, Withius
  (= Xenowithius) 665
transversa, Chitrella (= Obisium) 419
transversus, Austrochthonius chilensis
  140
trebinjensis, Chthonius (Chthonius)
  162
trentinum, Neobisium (Neobisium)
  371
triangulare, Neocheiridium 443
triangularis, Geogarypus (= Garypus)
  260
trichoideus, Eremogarypus 237
Trichotowithius 658
tricuspidatus, Parachelifer 522
TRIDENCHTHONIIDAE 217

Tridenchthonius 225
tridens, Microblothrus 426
trifidum, Ideobisium (= Obisium) 421
trigonae, Dasychernes 568
trinidadensis, Tridenchthonius 227
trinitatis, Lechytia 188
tripartitus, Mundochthonius japonicus 191
tripolitanus, Allochernes 538
triquetrum, Pseudochiridium 446
Trisetobisium 416
tristis, Epactiochernes (= Chelanops) 576
troglobia, Insulocreagris 336
Troglobisium 430
troglobius, Chthonius (Chthonius) 162
troglobius, Pseudochthonius 198
troglobius, Typhloroncus 323
troglobius, Tyrannochthonius 215
Troglobochica 311
Troglochernes 638
Troglochthonius 204
troglodites, Chthonius (Chthonius) 163
troglodytes, Neochthonius 193
troglodytes, Obisium 372, 383
troglodytes, Tyrannochthonius 215
Troglohya 312
troglophilum, Vachonobisium 314
troglophilus, Allochthonius (Allochthonius) opticus 133
troglophilus, Chthonius (Ephippiochthonius) 174
troglophilus, Roncus (Roncus) 393
troglophilus, Roncus 414
troglophilus, Tyrannochthonius (= Morikawia) 216
trojanicus, Roncus 414
trombidioides, Chelifer 149
tropicum, Olpium 294
tropicus, Eremochernes 501
Tropidowithius 658
truncatus, Ideoblothrus (= Pachychitra) 426
tsavoensis, Negroroncus 321
Tubbichthonius 198
tuberculatus, Ancistrochelifer 483
tuberculatus, Chthonius (Ephippiochthonius) 174
tuberculatus, Hysterochelifer (= Chelifer) 507
Tuberocreagris 416
tucanus, Allochernes 539
tuena, Sathrochthonius 202

tullgreni, Austrochthonius (= Paraustrochthonius) 141
tullgreni, Chelifer 625
tullgreni, Sathrochthonius 202
Tullgrenia, Chelifer 634
Tullgrenius 481
tumidimanus, Nesidiochernes 606
tumidum, Cheiridium 441
tumidus, Epactiochernes (= Chelanops) 576
tumidus, Hesperochernes 585
tumimanus, Parachernes (Parachernes) (= Chelanops) 620
tumuliferus, Cyrtowithius (= Chelifer) 642
turanicus, Allochernes (= Chelifer) 539
turanicus, Diplotemnus 461
turbotti, Apatochernes 546
turcicum, Apocheiridium (Apocheiridium) 436
turcicum, Neobisium (Neobisium) 372
turcicum, Olpium 292
turcicus, Cheirochelifer 486
turcicus, Lasiochernes 593
turcicus, Pselaphochernes 629
turkestanica, Bisetocreagris (= Ideobisium) 333
turkestanicus, Chelifer (Atemnus) 456
turritanus, Roncus 414
tuxeni, Pseudochthonius 198
tuzetae, Neobisium (Blothrus) 386
tweediei, Metawithius (Microwithius) 647
Tychochernes 638
Typhlochthonius, Chthonius 195
Typhloditha 227
Typhloroncus 322
typhlus, Apochthonius 139
typhlus, Pseudotyrannochthonius 201
Tyrannochelifer 533
Tyrannochthoniella 204
Tyrannochthonius 205
tzanoudakisi, Chthonius (Chthonius) 163

uenoi, Allochthonius (Urochthonius) ishikawai 134
uenoi, Pararoncus (= Roncus) 394
ugandanus, Ellingsenius 501
ugandanus, Stenowithius 656
ugandanus, Titanatemnus 481
umbratile, Neobisium (Blothrus) 386

uncinatus, Caffrowithius
(=Anepsiochernes) 552
undecimclavatus, Pseudo-
tyrannochthonius 201
unicolor, Hesperochernes
(=Chelanops) 586
unicus, Mexichthonius 189
uniformis, Chelodamus (=Chelanops)
556
uniseriatus, Paracanthicochernes 612
unistriatus, Metatemnus
(=Anatemnus) 464
urbanus, Hysterochelifer 508
Urochthonius, Allochthonius 133
uruguayanus, Dinocheirus 574
ussuricus, Mundochthonius 192
ussuriensis, Bisetocreagris
(=Microcreagris) 333
ussuriensis, Centrochthonius 142
usudi, Neobisium (Neobisium) 372
utahensis, Hesperochernes 586
utahensis, Paisochelifer 519
utahensis, Parobisium 396
utahensis, Pseudotyrannochthonius 201
uyamadensis, Allochthonius
(Urochthonius) ishikawai 134

vaccai, Chthonius 167
vachoni, Acanthocreagris 328
vachoni, Antillobisium 309
vachoni, Chthonius
(Ephippiochthonius) 174
vachoni, Diplotemnus 463
vachoni, Garypinus 279
vachoni, Goniochernes 579
vachoni, Neobisium (Blothrus) 386
vachoni, Neobisium (Neobisium) 364
vachoni, Paracheiridium 444
vachoni, Tullgrenius 481
vachoni, Withius (=Allowithius) 666
VACHONIIDAE 430
Vachonium 431
Vachonobisium 313
vafer, Sphaerowithius
(=Nannowithius) 656
vagrans, Withius 666
valentinei, Lissocreagris
(=Microcreagris) 337
validissimum, Apocheiridium
(Apocheiridium) 436
validum, Apocheiridium
(Apocheiridium) 436
validum, Neobisium (Neobisium)
(=Obisium) 372

validus, Dinocheirus (=Chelanops) 574
validus, Garypinus 279
vallei, Chthonius (Chthonius)
microphthalmus 156
vampirorum, Ideoblothrus 426
vampirorum, Tyrannochthonius 216
vancleavei, Parobisium (=Neobisium)
396
vandeli, Chthonius (Globochthonius)
176
vandeli, Neobisium (Ommatoblothrus)
671
vanduzeei, Dinochernes (=Chelanops)
575
vannii, Chthonius (Chthonius) lanzai
154
vasconicum, Neobisium (Blothrus)
(=Obisium) 386
vastitatis, Dinocheirus (=Chelanops)
575
vastum, Apolpium 266
vastus, Apatochernes 547
velebiticum, Neobisium (Blothrus) 387
venezuelanus, Atherochernes 547
venezuelanus, Mesochernes
(=Chelifer) 602
venezuelanus, Parawithius
(Victorwithius) 653
venezuelanus, Sathrochthonius 202
venezuelense, Beierolpium 268
ventalloi, Acanthocreagris granulata
(=Roncus) 325
ventalloi, Chthonius
(Ephippiochthonius) 174
ventalloi, Neobisium (Neobisium) 372
venustus, Dinocheirus 575
veracruzensis, Albiorix 318
verae, Neobisium (Blothrus) 387
verai, Chthonius (Ephippiochthonius)
174
verapazanus, Aphrastochthonius 136
vermiformis, Anatemnus (=Chelifer)
454
vermiformis, Pseudohorus 301
vermis, Minniza 286
vermis, Pseudohorus 301
Verrucachernes 638
Verrucaditha 228
Verrucadithella 228
vespertilionis, Hesperochernes 586
vestitus, Ceriochernes 553
vicinus, Chernes (=Allochernes) 562
vicinus, Dolichowithius
(Dolichowithius) 645

victorianus, Paraustrochernes 622
victorianus, Protochelifer 524
Victorwithius, Parawithius 651
vidali, Roncus 414
viduus, Parachelifer 523
vietnamensis, Megachernes 602
vietnamensis, Stenohya 315
vietnamensis, Tetrachelifer 532
vigil, Maorichernes (=Chelifer) 598
vilcekii, Neobisium (Neobisium) 373
villiersi, Rhacochelifer 531
villosus, Lasiochernes 593
viniai, Lustrochernes 598
virgineus, Nudochernes 609
virginicus, Chthonius
  (Ephippiochthonius) 175
virginicus, Parachernes (Parachernes)
  (=Chelanops) 620
viridans, Progarypus (=Garypus) 299
viridis, Calocheiridius 272
viridis, Maorigarypus 672
viti, Roncus 414
vittatus, Asterochernes 547
vjetrenicae, Neobisium (Blothrus) 387
voeltzkowi, Anatemnus (=Chelifer) 454
volcancillo, Tyrannochthonius 216
volcanus, Tyrannochthonius 216
vosseleri, Hansenius 504
vosseleri, Microchelifer 518
vtorovi, Dactylochelifer 498
vulcanius, Roncus 415
vulpinum, Neobisium (Blothrus) 380

waechtleri, Obisium 357
wahrmani, Nannowithius
  (=Myrmecowithius) 648
walckenaerii, Obisium 370
wallacei, Dinochernes 575
walliskewi, Beierius (=Chelifer) 484
warburgi, Haplochernes (=Chelifer) 582
webbi, Sathrochthonius 202
welbourni, Archeolarca 237
weygoldti, Paraliochthonius 195
weyrauchi, Balanowithius 641
whartoni, Oratemnus samoanus 468
wideri, Allochernes (=Chelifer) 539
wigandi, Oligochernes (=Chelifer) 610
wisei, Apatochernes 547
withi, Bochica (=Ideoroncus) 310
withi, Garypus bonairensis 239
withi, Parachernes (Parachernes) 621
WITHIIDAE 640
Withius 658
wittei, Nudochernes 609

wittei, Tyrannochthonius 216
wlassicsi, Tyrannochthonius
  (=Chthonius) 216
wrightii, Macrochernes (=Chelifer) 598
Wyochernes 639

xalyx, Afrosternophorus 449
Xenochelifer 533
Xenochernes 639
Xenoditha 217
Xenolpium 307
Xenowithius 658
xilitlensis, Typhloroncus 323

yanoi, Allochthonius 133
yapensis, Smeringochernes
  (Smeringochernes) 633
yodai, Oratemnus 469
yosii, Pararoncus (=Roncus) 394
yucatanus, Pseudochthonius 198
yunquense, Ideobisium 421
yurii, Metawithius (Microwithius)
  (=Microwithius) 647

zangherii, Microcreagris 456
Zaona 639
zariquieyi, Microbisium (=Obisium)
  340
zavattarii, Chernes 563
zealandica, Sathrochthoniella 201
zealandica, Tyrannochthoniella 204
zealandicum, Apocheiridium
  (Apocheiridium) 436
zealandicus, Austrochthonius 141
zealandicus, Nesidiochernes 606
zealandicus, Smeringochernes
  (Smeringochernes) 634
zealandiensis, Olpium 308
zhiltzovae, Neobisium (Neobisium) 373
zicsii, Ideoblothrus (=Ideobisium) 426
zicsii, Tyrannochthonius 216
zilchi, Cacodemonius 642
zoiai, Neobisium (Ommatoblothrus) 390
zoiai, Roncus 415
zonatus, Afrogarypus (=Geogarypus)
  252
zonatus, Horus 282
zonatus, Systellochernes 636
zonatus, Tyrannochthonius
  (=Morikawia) 216
zuluanum, Ectactolpium 275
zuluanus, Miratemnus 465
zuluensis, Thaumastogarypus 249
zumpti, Aperittochelifer 483